D0138302

Perspectives on
STRUCTURE AND MECHANISM
in Organic Chemistry

Felix A. Carroll is the Joseph R. Morton Professor and chair of the department of chemistry at Davidson College in Davidson, North Carolina. He earned a B.S. with Highest Honors in chemistry at the University of North Carolina–Chapel Hill, where he worked with David G. Whitten, and a Ph.D. in organic chemistry from California Institute of Technology, where his research director was George S. Hammond. Most of Carroll's research has concerned the kinetics and mechanisms of photochemical reactions, but he has also published in other areas, including a study of synthetic insect pheromones that led to a U.S. patent.

Perspectives on STRUCTURE AND MECHANISM in Organic Chemistry

Felix A. Carroll
Davidson College

Brooks/Cole Publishing Company

I(T)P® An International Thomson Publishing Company

Pacific Grove • Albany • Belmont • Bonn • Boston • Cincinnati • Detroit • Johannesburg • London
Madrid • Melbourne • Mexico City • New York • Paris • Singapore • Tokyo • Toronto • Washington

Sponsoring Editor: *Harvey C. Pantzis*
Marketing Team: *Kathleen Sharp, Margaret Parks*
Editorial Associate: *Beth Wilbur*
Editorial Assistant: *Leigh Hamilton*
Production Coordinator: *John A. Servideo*
Production Editor: *Jamie Sue Brooks*
Permissions Editor: *Lillian Campobasso*
Cover Design: *Roy R. Neuhaus*
Cover Illustration: *Kenneth Eward, BioGrafx*
Typesetting: *Monotype Composition Company, Inc.*
Cover Printing: *Phoenix Color Corp.*
Printing and Binding: *Quebecor/Fairfield*
Illustrations: *Monotype Composition Company, Inc. and Precision Graphics*

For more information, contact:

BROOKS/COLE PUBLISHING COMPANY
511 Forest Lodge Road
Pacific Grove, CA 93950
USA

International Thomson Publishing Europe
Berkshire House 168-173
High Holborn
London WC1V 7AA
England

Thomas Nelson Australia
102 Dodds Street
South Melbourne, 3205
Victoria, Australia

Nelson Canada
1120 Birchmount Road
Scarborough, Ontario
Canada M1K 5G4

International Thomson Editores
Seneca 3
Col. Polanco
11560 México, D. F., México

International Thomson Publishing GmbH
Königswinterer Strasse 418
53227 Bonn
Germany

International Thomson Publishing Asia
221 Henderson Road
#05-10 Henderson Building
Singapore 0315

International Thomson Publishing Japan
Hirakawacho Kyowa Building, 3F
2-2-1 Hirakawacho
Chiyoda-ku, Tokyo 102
Japan

Printed in the United States of America

10 9 8 7 6 5 4 3 2 1

Library of Congress Cataloging-in-Publication Data

Carroll, Felix A.
Perspectives on structure and mechanism in organic chemistry / Felix A. Carroll.
p. cm.
Includes bibliographical references and index.
ISBN 0-534--24948-5
1. Chemistry, Physical organic. I. Title.
QD476.C375 1998
547—dc20 95-37049

THIS BOOK IS PRINTED ON ACID-FREE RECYCLED PAPER

To Carol, Heather, and Brandon

Brief Contents

Contents

Preface

This book is the result of my experience in teaching advanced organic chemistry for the past twenty-four years. During this time I have felt a need for a text that meets two goals: The first is to present those concepts that are central to the understanding and practice of physical organic chemistry. The second is to emphasize the role of complementary models in the formulation of those concepts. I think that students in advanced courses not only need to learn more organic chemistry but also need to learn to think about organic chemistry in new ways.

The organization of the topics is based on the idea that understanding organic reactions is facilitated by an understanding of the structure of organic compounds. The first five chapters discuss complementary views of structure and bonding of stable molecules and of reactive intermediates. There is a chapter on tools used to study the mechanisms of organic reactions, and then the mechanisms of acid-base reactions, substitution reactions, addition reactions, elimination reactions, concerted reactions, and photochemical reactions are considered in detail. The discussion of each of these topics stresses the additional understanding that results from viewing structures and reactions from more than one perspective.

Obviously this book does not contain all known information about each of these topics. An advanced text represents a distillation in which the original ideas presented in research papers are concentrated into a smaller volume. I have had to be selective in the choice of topics and examples in order to avoid making this text an encyclopedia of organic chemistry. My goal has been to write a teaching text that addresses the most essential topics with an emphasis on their foundations and their relationships.

A teaching text may nevertheless serve as a useful reference text if it has complete references. Not every reader will read every footnote, but the more than 3,000 literature citations (ranging from 1851 to 1996) in this text are intended to direct interested readers to further information about all topics discussed. Along with the author index, these literature citations are also intended to acknowledge the researchers whose efforts produced the information summarized here. In addition to the literature references, there are nearly 800 expository footnotes that amplify or extend the discussion in the main body of the text.

Two other features are designed to make this text more useful both as a teaching text and as a reference. Important terms (such as LUMO and carbenium ion) appear in distinctive type when they are first used and defined. A detailed subject index gives page numbers where definitions or introductory discussions of such terms can be found.

A fundamental element of a teaching text is a set of problems of graded difficulty, but the 250 problems here are designed to do more than merely allow students to test their understanding of the facts and concepts presented in a chapter. They are also intended to encourage readers to actively engage the chemical literature and to develop and defend their own ideas. Some problems do represent straightforward applications of the information in the text. However, other problems can best be answered by consulting the literature for background information before attempting a solution. Still other problems (e.g., 1.1–1.3, 1.12–1.22, 4.22–4.23, 5.20, 9.22–9.33, 10.21) are open-ended, with no one "correct" literature answer. I have prepared a solutions manual giving complete, checked solutions for problems in the first two categories, and there are also ideas for solutions to the open-ended problems.

It is important that readers be prepared to develop and defend their own ideas, because a textbook in chemistry cannot provide information with absolute finality. In a marvelous film entitled "Knowledge or Certainty" in his *The Ascent of Man* series, Jacob Bronowski shows many portraits of the same human face. He notes that "We are aware that these pictures do not so much fix the face as explore it . . . and that each line that is added strengthens the picture but never makes it final."[1] So it must be with this book. It is not a photograph but is, instead, a portrait of physical organic chemistry in which I have chosen to emphasize some lines more than others. As is the case with the human face, it is not possible to fix a continually changing science—we can only explore it. It is my hope that those who read this book will be better prepared to continue that exploration.

---

[1] The quotation is taken from the book with the same title as the film series: Bronowski, J. *The Ascent of Man;* Little, Brown and Company: Boston, 1973; p. 353.

Acknowledgments

I am grateful to the following colleagues for giving their time to read and to offer comments on one or more chapters of the manuscript. Not only did these reviewers correct errors, both of commission and of omission, but some contributed apt phrases to clarify ambiguous wording.

Kenneth K. Andersen, University of New Hampshire
John E. Baldwin, Syracuse University
Thomas W. Bell, University of Nevada, Reno
David E. Bergbreiter, Texas A & M University
Claude F. Bernasconi, University of California, Santa Cruz
Silas C. Blackstock, Vanderbilt University
R. Stanley Brown, University of Alberta
Nigel J. Bunce, University of Guelph
Trudy A. Dickneider, University of Scranton
Thomas A. Dix, Medical University of South Carolina
Ralph C. Dougherty, Florida State University
Slayton A. Evans, Jr., University of North Carolina, Chapel Hill
Kendall N. Houk, University of California, Los Angeles
Ronald M. Magid, University of Tennessee
Robert D. Stolow, Tufts University
Nicholas J. Turro, Columbia University
James K. Whitesell, University of Texas, Austin
Howard E. Zimmerman, University of Wisconsin, Madison

In addition to reviewers of complete chapters, several other colleagues read and commented on short excerpts from the manuscript.

Norman L. Allinger, University of Georgia
Eugene C. Ashby, Georgia Institute of Technology
Robert D. Bach, Wayne State University
Richard F. W. Bader, McMaster University
N. C. Baird, University of Western Ontario
Giuseppe Bellucci, Istituto di Chimica Organica della Facoltà di Farmacia, Pisa, Italy

Joseph F. Bunnett, University of California, Santa Cruz
Donald J. Cram, University of California, Los Angeles
Stanley J. Cristol, University of Colorado, Boulder
Jaques-Emile Dubois, Laboratoire de Chimie Organique Physique, Paris, France
William C. Herndon, University of Texas, El Paso
William L. Jorgensen, Yale University
Alexander J. Kresge, University of Toronto
Paul J. Kropp, University of North Carolina, Chapel Hill
G. Marc Loudon, Purdue University
Hubert Maehr, Roche Research Center, Hoffman-LaRoche, Inc.
Oliver R. Martin, State University of New York, Binghamton
Kurt M. Mislow, Princeton University
Robert A. Moss, Rutgers University
P. C. Myhre, Harvey Mudd College
George A. Olah, University of Southern California
William E. Palke, University of California, Santa Barbara
Charles L. Perrin, University of California, San Diego
Ned A. Porter, Duke University
John D. Roberts, California Institute of Technology
William H. Saunders, Jr., University of Rochester
Henry F. Schaefer III, University of Georgia
Thomas T. Tidwell, University of Toronto
Edward A. Walters, University of New Mexico
Edgar W. Warnhoff, University of Western Ontario
Richard G. Weiss, Georgetown University
Frank H. Westheimer, Harvard University
Keith Yates, University of Toronto

Neither those who reviewed entire chapters nor those who commented on brief excerpts are in any way responsible for any errors that remain, however. Most reviewers read only one chapter or a portion of a chapter, often at an early stage in the writing. Sometimes I did not adopt all suggestions of the reviewers because of limitations of time and space, and none of the reviewers saw or approved the final text.

In addition to those who formally reviewed portions of the manuscript, I also want to thank those who helped in other ways. Dr. George D. Purvis of CAChe Scientific provided assistance with some of the computations and the cover design. Dr. O. Bert Ramsay of Chemical Concepts Corporation provided the negative for Figure 3.10. Professor Richard G. Weiss of Georgetown University offered continual encouragement during this project. David L. Dillon of the University of Wyoming, who checked the solutions manual, also made valuable comments about the manuscript of this book. Dr. Mitchell A. Rhea, W. Stephen Aldridge, and Theodore E. Curey of Davidson College gave valuable assistance in the production of the manuscript.

It is a pleasure to acknowledge the staff of Brooks/Cole Publishing Company for their encouragement and support during the writing of this book. Executive editor Harvey Pantzis, chemistry editor Lisa Moller, and editorial associate Beth Wilbur offered invaluable guidance and helped keep the project on track. Elizabeth Rammel was most supportive and helpful in the preparation of the solutions manual. I also appreciate the support of former chemistry editor Sue Ewing, production coordinator John Servideo, senior production editor Jamie Sue Brooks, and permissions coordinator Lillian Campobasso. Senior designer Roy Neuhaus designed and art directed the cover; artist Kenneth Eward of BioGrafx created and revised the artwork to achieve the final handsome effect.

Finally, I thank my wife, Carol, and my children, Heather and Brandon, for their understanding, patience, and assistance during the lengthy process of writing this book.

Introduction

Every organic chemist instantly recognizes the following drawing as benzene, or at least as one of the Kekulé structures of benzene.

However, it is not really benzene. It is only a figure consisting of a regular hexagon enclosing three extra lines. When we look at the drawing, however, we *see* benzene. That is, we visualize a colorless liquid, and we recall a pattern of physical properties and chemical reactivity associated with benzene and with the concept of aromaticity. The drawing is a representation of an entity that we cannot see, so it is only a model of benzene.[1]

It may seem strange to begin a discussion of structure and mechanism in organic chemistry with an exercise in the meaning of the Kekulé representation of benzene. However, a key theme of this text is the development and use of knowledge in organic chemistry. If we can understand how and why we think about chemistry as we do now, it may be easier to think about it in new ways in the future. In particular, we must realize both how extensively we rely on models and how limited those models are.

That all organic chemists instantly recognize a drawing of benzene confirms that they have been initiated into the chemical fraternity. The tie that binds the members of this fraternity is more than a collective interest, however. It is also a common way of viewing problems and their solutions that can lead to considerable conformity of thinking and of behavior. Although such conformity facilitates communication among members of the group, it also tends to limit independent behavior and action.

---

[1]On still another level, the figure is a symbolic representation of the way we think about the substance benzene. For a discussion of "Representation in Chemistry," including the nature of drawings of benzene rings, see Hoffmann, R.; Laszlo, P. *Angew. Chem., Int. Ed. Engl.* **1991,** *30,* 1. For a discussion of the iconic nature of some chemical drawings, see Whitlock, H. W. *J. Org. Chem.* **1991,** *56,* 7297.

This common way of looking at problems was explored by T. S. Kuhn, who noted two related meanings of the term *paradigm*:

> On the one hand, it stands for the entire constellation of beliefs, values, techniques, and so on shared by the members of a given community. On the other it denotes . . . the concrete puzzle solutions which, employed as models or examples, can replace explicit rules as a basis for the solution of the remaining puzzles of normal science.[2, 3]

The role that paradigms play in science deserves emphasis because, once we have become accustomed to thinking about a problem in a certain way, it becomes quite difficult to think about it differently.[4] In some ways, our paradigms are like the operating system of a computer: they dictate the input and output of information and control the operation of logical processes. We must be prepared to ask continually whether an inconsistency between observation and theory is the fault of the observation or theory or whether it is instead the fault of our paradigms.

The history of phlogiston is an example of the way paradigms can dictate chemical thought. Phlogiston was believed to be a real substance that was given off by burning matter and that could be added to certain mineral ores to produce metals.[5] The phlogiston theory was widely accepted and was taught to students as established fact.[6] As is also the case with the ideas we accept, the phlogiston theory could rationalize phenomena for which it was developed (combustion) and could also account for other observations (such as the death of animals confined in air-tight containers).[7] The theory could also be modified when necessary to account for results that did not agree with it.

Phlogiston may seem to belong to the realm of ancient history, but the more recent examples of polywater[8] and distortional isomers[9] remind us

---

[2]Kuhn, T. S. *The Structure of Scientific Revolutions,* 2nd ed.; The University of Chicago Press: Chicago, 1970, pp. 37, 175.

[3]The paradigm that we may think of chemistry only through paradigms may be applicable to Western science only. For an interesting discussion of "Sushi Science and Hamburger Science," see Motokawa, T. *Perspectives in Biology and Medicine* **1989,** *32,* 489.

[4]This difficulty is compounded if we develop the theory ourselves. "The moment one has offered an original explanation for a phenomenon which seems satisfactory, that moment affection for his intellectual child springs into existence. . . . From an unduly favored child, it readily becomes master, and leads its author whithersoever it will." Chamberlin, T. C. *Science* **1965,** *148,* 754, reprinted from *Science* (old series) **1890,** *15,* 92. For further discussion of this view, see Bunnett, J. F. in *Investigation of Rates and Mechanisms of Reactions,* 3rd ed., Part I; Lewis, E. S., ed.; Wiley-Interscience: New York, 1975; pp. 478–479.

[5]*Cf.* White, J. H. *The History of the Phlogiston Theory;* Edward Arnold & Co.: London, 1932.

[6]Conant, J. B. *Science and Common Sense;* Yale University Press: New Haven, 1951; pp. 170–171.

[7]Note the defense of phlogiston by Priestly cited by Pimentel, G. *Chem. Eng. News* **1989** (May 1), p. 53.

[8]Howell, B. F. *J. Chem. Educ.* **1971,** *48,* 663; Christian, P. A.; Berka, L. H. *J. Chem. Educ.* **1971,** *48,* 667. See also Speedy, R. J. *J. Phys. Chem.* **1992,** *96,* 2322.

[9]Amato, I. *Science* **1991,** *254,* 1452; Desrochers, P. J.; Nebesny, K. W.; LaBarre, M. J.; Lincoln, S. E.; Loehr, T. M.; Enemark, J. H. *J. Am. Chem. Soc.* **1991,** *113,* 9193; Mayer, J. M. *Angew. Chem., Int. Ed. Engl.* **1992,** *31,* 286.

that even currently accepted ideas may later be discounted and the experimental data reinterpreted. Who knows whether contemporary ideas such as atomic and molecular orbitals,[10] molecular shape,[11] and quantum mechanics[12] will some day be considered as outmoded as phlogiston?

Recognizing that our contemporary ideas are only models does not mean that we should not use those models. This point was made in 1929 in an address by Irving Langmuir, who was then president of the American Chemical Society.[13]

> Skepticism in regard to an absolute meaning of words, concepts, models or mathematical theories should not prevent us from using all these abstractions in describing natural phenomena. . . . [Once] it was thought that such concepts as energy, entropy, temperature, chemical potential, etc., represented something far more nearly absolute in character than the concept of atoms and molecules, so that nature should preferably be described in terms of the former rather than the latter. We must now recognize, however, that all of these concepts are human inventions and have no absolute independent existence in nature. Our choice, therefore, cannot lie between fact and hypothesis, but only between two concepts (or between two models) which enable us to give a better or worse description of natural phenomena.

Choosing between two concepts (or two models) does not mean that we must thereafter use only the concept that we accept and totally reject the other one. Instead, we must continually make selections among many complementary models—from plastic balls and sticks to wave functions.[14, 15] The choice of models is usually shaped by our need to solve the problems at hand. For example, Lewis electron dot structures and resonance theory provide quite adequate descriptions of the structures and reactions of organic compounds for some purposes, but in other cases we prefer to use molecular orbital theory. Frequently, therefore, we find ourselves intuitively alternating between these models as we consider diverse aspects of organic chemistry.

[10]For a discussion, see Ogilvie, J. F. *J. Chem. Educ.* **1990,** *67,* 280. This paper bears the intriguing subtitle "There are no such *things* as orbitals!" However, also see Simons, J. *J. Chem. Educ.* **1991,** *68,* 131; Pauling, L. *J. Chem. Educ.* **1992,** *69,* 519; Edmiston, C. *J. Chem. Educ.* **1992,** *69,* 600; Scott, J. M. W. *J. Chem. Educ.* **1992,** *60,* 600; Scerri, E. R. *J. Chem. Educ.* **1992,** *69,* 602.

[11]Woolley, R. G. *J. Am. Chem. Soc.* **1978,** *100,* 1073.

[12]See the discussion of the work of one physicist "who has long harbored doubts about quantum mechanics": Freedman, D. H. *Science* **1991,** *253,* 626.

[13]Langmuir, I. *J. Am. Chem. Soc.* **1929,** *51,* 2847.

[14]For an elaboration of the view that wave functions are models, see the discussion in Dewar, M. J. S. *J. Phys. Chem.* **1985,** *89,* 2145.

[15]For other discussions of the role of models in chemistry, see (a) Hammond, G. S.; Osteryoung, J.; Crawford, T. H.; Gray, H. B. *Models in Chemical Science: An Introduction to General Chemistry;* W. A. Benjamin, Inc.: New York, 1971; pp. 2–7; (b) Sunko, D. E. *Pure Appl. Chem.* **1983,** *55,* 375; (c) Bent, H. A. *J. Chem. Educ.* **1984,** *61,* 774; (d) Goodfriend, P. L. *J. Chem. Educ.* **1976,** *53,* 74; (e) Morwick, J. J. *J. Chem. Educ.* **1978,** *55,* 662; (f) Matsen, F. A. *J. Chem. Educ.* **1985,** *62,* 365.

In this text we will explore the use of complementary models to explain and predict the structures and reactions of organic compounds. We will develop alternative explanations for many familiar concepts, such as the sp^3 hybridization of tetrasubstituted carbon, the σ,π description of the carbon-carbon double bond, the electronic nature of substitution reactions, the acid-base properties of organic compounds in solution, and the nature of concerted reactions. Our intent will not be merely to replace conventional ideas and familiar language with different theories and new terminology. Instead, we will seek to underpin our understanding of organic chemistry by gaining additional perspectives on the fundamental concepts that are its foundation.

CHAPTER 1

Concepts and Models in Organic Chemistry

1.1 Models of Atomic and Molecular Structure

Atoms and Molecules

When we think about chemistry, we routinely envision atoms and molecules as basic units of matter. We work with mental pictures of atoms and molecules, just as we twist, rotate, disconnect, and reconnect physical models in our hands.[1,2] But where do these mental images and physical models come from? During the Senate Watergate hearings in 1973, Senator Howard Baker became known for repeatedly asking the question, "What did the president know, and when did he know it?" It is useful to begin thinking about the nature and limits of our knowledge of organic chemistry by asking ourselves a similar question: What do we know about atoms and molecules, and how do we know it? As Kuhn[3] has pointed out

> Though many scientists talk easily and well about the particular individual hypotheses that underlie a concrete piece of current research, they are little better than laymen at characterizing the established bases of their field, its legitimate problems and methods.

For most of us, the majority of what we know in organic chemistry consists of what we have been taught. Behind that teaching, however, what we know results from what we (or others) observe and how we (or they) interpret those observations.

---

[1]For a detailed discussion of physical models in chemistry, see Walton, A. *Molecular and Crystal Structure Models*; Ellis Horwood Limited: Chichester, England, 1978.

[2]For an interesting application of physical models to infer molecular properties, see Teets, D. E.; Andrews, D. H. *J. Chem. Phys.* **1935**, *3*, 175.

[3]Kuhn, T. S. *The Structure of Scientific Revolutions*, 2nd ed.; The University of Chicago Press: Chicago, 1970; p. 47.

The most fundamental observations are those that we can make directly with our senses. We note the physical state of a particular type of matter—solid, liquid, or gas. We note the color of a material. We note whether it dissolves in a given solvent or evaporates if exposed to the atmosphere. We get some sense of its density by seeing it float or sink in an immiscible liquid. These are only qualitative observations, but they provide an important foundation for further experimentation.[4]

It is only a modest extension of direct observation to the use of simple experimental apparatus for quantitative measurements. We use a heat source and a thermometer to determine melting and boiling ranges. We measure surface tensions, indices of refraction, heats of reaction, densities, and solubilities. Through classical elemental analysis of a substance, we can determine what elements are present in a sample and what their mass ratios seem to be. Then we might determine a formula weight (through melting point depression, for example). In all these experiments, *we use some equipment but still make the actual experimental observations by eye*. These limited experimental techniques can provide essential information. For example, if we find that 159.8 grams of bromine will always be decolorized by 82.15 grams of cyclohexene, then we can observe the law of definite proportions. Such data are consistent with a model of matter in which submicroscopic particles combine with each other in definite patterns, just as the macroscopic samples before our eyes do. It is then only a matter of definition to call the submicroscopic particles atoms or molecules and to further study their properties. However, it is essential to remember that our laboratory experiments are conducted with *materials*. While we may talk about the addition of bromine to cyclohexene in terms of individual molecules, we really can only infer that such a process occurs on the basis of experimental data collected with macroscopic samples of the substances involved.[5]

Modern instrumentation has opened the door to a variety of investigations, most unimaginable to early chemists, that expand the range of observations beyond those of the human senses. They extend our eyes from a limited portion of the electromagnetic spectrum through practically the entire spectrum, from X-rays to radio waves, and they let us "see" light in other ways, as in polarimetry. They allow us to use entirely new tools, such as electron or neutron beams, magnetic fields, and electrical potentials or current. They extend the range of conditions for studying matter from near atmospheric pressure to high vacuum or to high pressure. They effectively expand and compress the human time scale, so that we can study events that occur in femtoseconds[6] or detect changes that occurred over eons.

---

[4]Odor and taste could be considered observable properties, but they are not routinely measured.

[5]This statement may be superseded by developments in scanning tunneling microscopy and atomic force microscopy, with which investigators have reported examining the products of reactions carried out on individual molecules. See reference 14.

[6]Rosker, M. J.; Dantus, M.; Zewail, A. H. *Science* **1988**, *241*, 1200 reported that the photodissociation of ICN to I and CN occurs in ca. 100 femtoseconds.

The unifying characteristic of modern instrumental analysis is that we no longer observe the chemical or physical change directly—we observe it only indirectly through the deflection of an indicator, the movement of a pen on a recorder, or the pattern of markings on a photographic film or other recording device. With such instruments, it is more critical than ever that we recognize the limits of our ability to free our observations from constraints imposed by our paradigms. *To a layman*, a UV-vis spectrum may not seem all that different from an upside-down infrared spectrum, and a capillary gas chromatogram of a complex mixture may seem to resemble a mass spectrum. But the chemist sees these traces not as lines on paper but as vibrating or rotating molecules, as electrons moving from one place to another, as substances separated from a mixture, or as fragments from molecular decomposition. Thus, in interpreting instrumental data, it is the paradigm that both makes the observation interpretable and constrains the observation to fit the paradigm.[7]

With that *caveat*, what do we know about molecules and how do we know it? We begin with the idea that organic compounds and all other chemical substances are composed of atoms, the indivisible particles that are the smallest units of that particular kind of matter that still retain all its properties. It is an idea whose ancestry can be traced to ancient Greek philosophers.[8] Moreover, it is convenient to correlate our observation that substances combine only in certain proportions with the notion that these submicroscopic entities called atoms combine with each other only in certain, definite ways.

Much of our fundamental information about molecules has been obtained from spectroscopy.[9] For example, a 4000 V electron beam has a wavelength of 0.06 Å, so it is diffracted by objects larger than that.[10] Interaction of the electron beam with gaseous molecules produces characteristic circular patterns which can be interpreted in terms of molecular dimensions.[11] We can also determine internuclear distance through infrared spectroscopy of diatomic molecules. Similarly, we can use X-ray or neutron scattering to calculate distances of atoms in crystal structures. We can also infer molecular bonding from magnetic resonance data.

"Pictures" of atoms and molecules are recent developments,[12] and re-

---

[7]"Innocent, unbiased observation is a myth."—P. Medawar, quoted in *Science* **1985**, *227*, 1188.

[8]Asimov, I. *A Short History of Chemistry*; Anchor Books: Garden City, New York, 1965; pp. 8–14.

[9]For a review of methods of structure determination, see Gillespie, R. J.; Hargittai, I. *The VSEPR Model of Molecular Geometry*; Allyn and Bacon: Boston, 1991; pp. 25–39.

[10]Moore, W. J. *Physical Chemistry*, 3rd ed.; Prentice-Hall, Inc.: Englewood Cliffs, N.J., 1962; pp. 575 *ff*.

[11]For discussions of molecular structure determination with gas phase electron diffraction, see Karle, J. in *Molecules in Natural Science and Medicine*, Maksić, Z. B.; Eckert-Maksić, M., eds.; Ellis Horwood: Chichester, England, 1991; pp. 17–27; Hedberg, K. *ibid.*; pp. 29–42.

[12]See, for example, Ottensmeyer, F. P.; Schmidt, E. E.; Olbrecht, A. J. *Science*, **1973**, *179*, 175 and references therein; Robinson, A. L. *Science* **1985**, *230*, 304; *Chem. Eng. News*, **1986** (Sep. 1), 4; Hansma, P. K.; Elings, V. B.; Marti, O.; Bracker, C. E. *Science* **1988**, *242*, 209; Parkinson, B. A. *J. Am. Chem. Soc.* **1990**, *112*, 1030; Frommer, J. *Angew. Chem., Int. Ed. Engl.* **1992**, *31*, 1298.

searchers have reported the manipulation of individual molecules and atoms.[13] Moreover, there has been one report in which the scanning tunneling microscope (STM) was used to dissociate an individual molecule and then examine the fragments.[14] Even though seeing is believing, we must recall that in all of these experiments we do not really see molecules, we see only computer graphics. Two examples illustrate this point: STM features that had been associated with DNA molecules were later assigned to the surface used to support the DNA,[15] and an STM image of benzene molecules was reinterpreted later as possibly being groupings of acetylene molecules instead.[16]

Organic chemists also reach conclusions about molecular structure on the basis of reasoning. For example, the fact that one and only one substance has been found with the molecular formula CH_3Cl is consistent with a structure in which three hydrogen atoms and one chlorine atom are attached to a carbon atom in a tetrahedral arrangement. If methane were a trigonal pyramid, then two different compounds with the formula CH_3Cl might be possible. The existence of only one isomer of CH_3Cl does not require a tetrahedral arrangement, however, since we might also expect only one isomer if the four substituents to the carbon atom were arranged in a square pyramid with a carbon atom at the apex, or a square planar structure with a carbon atom at the center. Since we also find one and only one CH_2Cl_2 molecule, we can also rule out the latter two geometries. Therefore we infer that the parent compound, methane, is also tetrahedral, and this view is reinforced by the existence of two different structures (enantiomers) with the formula CHClBrF. Similarly, we infer the flat, aromatic structure for benzene by noting that there are three and only three isomers of dibromobenzene.[17]

Organic chemists do not think of molecules as being made up only of atoms, however. We often envision them as collections of nuclei and electrons, and we consider the electrons to be constrained to certain regions of space (orbitals) around the nuclei. Thus we interpret UV-vis absorption, emission, or scattering spectroscopy in terms of movement of electrons from one of these orbitals to another. The development of the Bohr model of the atom, the Heisenberg uncertainty principle, and the Schrödinger equation laid the foundation for our current paradigm in chemistry. There may be truth in the view that "The why? and how? as related to chemical bonding

---

[13]Weisenhorn, A. L.; MacDougall, J. E.; Gould, S. A. C.; Cox, S. D.; Wise, W. S.; Massie, J.; Maivald, P.; Elings, V. B.; Stucky, G. D.; Hansma, P. K. *Science* **1990**, *247*, 1330; Whitman, L. J.; Stroscio, J. A.; Dragoset, R. A.; Celotta, R. J. *Science* **1991**, *251*, 1206; Leung, O. M.; Goh, M. C. *Science* **1992**, *255*, 64.

[14]Dujardin, G.; Walkup, R. E.; Avouris, P. *Science* **1992**, *255*, 1232.

[15]Clemmer, C. R.; Beebe, Jr., T. P. *Science* **1991**, *251*, 640.

[16]Moler, J. L.; McCoy, J. R. *Chem. Eng. News*, **1988** (Oct. 24), 2.

[17]These examples are discussed in an analysis of "topological thinking" in organic chemistry by Turro, N. J. *Angew. Chem., Int. Ed. Engl.* **1986**, *25*, 882.

were in principle answered in 1927; the details have been worked out since that time."[18] However, we will see that there are still uncharted frontiers of that paradigm to explore.

Molecular Dimensions

As we have noted, data from spectroscopy or from X-ray, electron, or neutron diffraction measurements allow us to determine the distance between atomic centers as well as to measure the angles between sets of atoms in covalently bonded molecules.[19] The most detailed information comes from microwave spectroscopy, although it is most useful for lower molecular weight molecules because the sample must be in the vapor phase.[20] In addition, diffraction methods locate a center of electron density, which is close to the nucleus for atoms that have electrons below the valence shell. For hydrogen, however, the electron density is shifted toward the atom to which it is bonded, and bonds to hydrogen are determined by diffraction methods to be shorter than are bond lengths determined with spectroscopy.[21] With solid state methods, the possible effect of crystal-packing forces must be considered. Therefore, the various techniques give slightly different measures of molecular dimensions.

Table 1.1 shows data for the interatomic distances and angles of the methyl halides.[22] We should emphasize that these distances and angles only provide geometric information about the location of nuclei (or local centers

Table 1.1 Bond lengths and bond angles for methyl halides.*

Molecule	$r_{C—H}$ (Å)	$r_{C—X}$ (Å)	<H—C—H	<H—C—X
CH_3F	1.105	1.385	109°54′	109°2′
CH_3Cl	1.096	1.781	110°52′	108°0′
CH_3Br	1.11	1.939	111°12′	107°14′
CH_3I	1.096	2.139	111°50′	106°58′

*Data from reference 22.

[18] Ballhausen, C. J. *J. Chem. Educ.* **1979**, *56*, 357.

[19] A tabulation of common bond length values has been provided by Allen, F. H.; Kennard, O.; Watson, D. G.; Brammer, L.; Orpen, A. G.; Taylor, R. *J. Chem. Soc., Perkin Trans. 2* **1987**, S1.

[20] Wilson, E. B. *Chem. Soc. Rev.* **1972**, *1*, 293 and references therein; for more recent developments in this field, see Harmony, M. D. *Acc. Chem. Res.* **1992**, *25*, 321.

[21] Clark, T. *A Handbook of Computational Chemistry*; John Wiley and Sons: New York, 1985; chapter 2.

[22] (a) Tabulations of bond length and bond angle measurements for specific molecules are available in *Tables of Interatomic Distances and Configuration in Molecules and Ions*; compiled by Bowen, H. J. M.; Donohue, J.; Jenkin, D. G.; Kennard, O.; Wheatley P. J.; Whiffen, D. H.; Special Publication No. 11, Chemical Society (London): Burlington House, W1, London, 1958. (b) See also the Supplement, 1965.

of electron density) as points in space. We infer that those points are connected by chemical bonds, so that the distance $r_{C—H}$ is the length of a C—H bond and the angle ∠H—C—H is the angle between two C—H bonds. Of course, we don't really think of the ends of the lines corresponding to bonds as nucleus-sized points. We picture the ends of those lines as three-dimensional objects with measurable dimensions.

We may define many different types of atomic dimensions, the most important for us being the ionic radius (r_i), the covalent radius (r_c), and the van der Waals radius (r_{vdW}) of an atom.[23] We visualize the ionic radius as the apparent size of the electron cloud around an ion as deduced from the packing of ions into a crystal lattice.[24] As might be expected, this value varies with the charge on the ion. The ionic radius for a C^{4+} ion is 0.15 Å, while that for a C^{4-} ion is 2.60 Å.[25] The van der Waals radius is the effective size of the atomic cloud around a covalently bonded atom as perceived by another atom to which it is not bonded, and it is also determined from interatomic distances found in crystals. Note that the van der Waals radius is not the distance at which the repulsive interactions of the electrons on the two atoms outweigh the attractive forces between them, as is often assumed. Rather, it is a crystal packing measurement that provides a smaller number.[26,27] The covalent radius of an atom is taken as half the distance between two atoms of the same element that are covalently bonded to each other, and this radius is much less than the van der Waals radius.

Figure 1.1 illustrates these radii for the chlorine atom in different environments. The computer-drawn plots of electron density surfaces represent the following: (a) r_i for chloride ion; (b) r_c and r_{vdW} for chlorine in Cl_2; (c) r_c and r_{vdW} for chlorine in CH_3Cl.[28]

Table 1.2 lists a number of radius values for several atoms. Note that the covalent radius for an atom depends on its bonding. A carbon atom

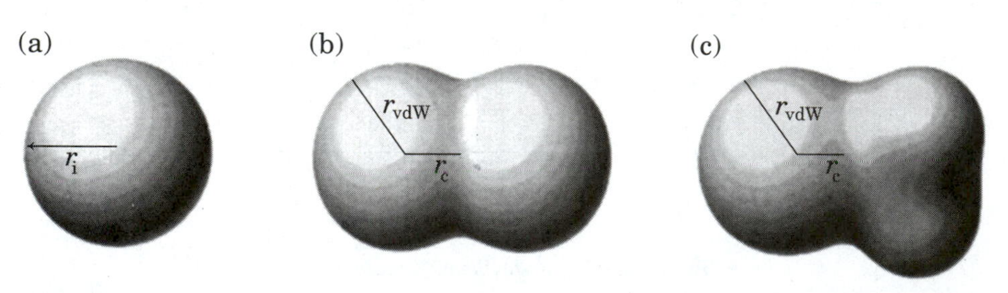

Figure 1.1 Radii values for chlorine.

[23]Pauling, L. *Nature of the Chemical Bond*, 3rd ed.; Cornell University Press: Ithaca, NY, 1960.

[24]For a discussion of estimating sizes of ions, see reference 23, p. 512.

[25]See reference 23, p. 514.

[26]Bondi, A. *J. Phys. Chem.* **1964**, *68*, 441.

[27]The difference is that distances between atoms in a crystal are determined by all of the forces acting on the molecule(s) containing those atoms, not just the forces between those two atoms alone.

[28]The images were produced by the author with a CAChe™ WorkSystem (CAChe Scientific).

bonded to four substituents has a covalent radius of 0.772 Å. The value is 0.67 Å for a carbon atom with one double bond, while the covalent radius for a triply-bonded carbon atom is 0.60 Å. We can also assign an effective van der Waals radius to a group of atoms. The value for a CH_3 or CH_2 group is 2.0 Å, while the van der Waals thickness of half the electron cloud in an aromatic ring is 1.85 Å.[29] Knowledge of van der Waals radii is important in calculations of molecular structure and reactivity.[30]

We may use the atomic dimensions to calculate a value for the volume and for the surface area of the atom. By adding those atomic values, we may approximate values for the volumes and surface areas of molecules. Such calculations have been described by Bondi, and a selected set of atomic volume and surface areas is given in Table 1.3. The numbers there represent

Table 1.2 Comparison of van der Waals, ionic, and covalent radii for selected atoms (Å).

		Ionic Radius[32]		Covalent Radii (r_c)[33]		
Atom	**van der Waals Radius (r_{vdW})[31]**	**Ion**	r_i	**Single bonded**	**Double bonded**	**Triple bonded**
H	1.11 Å	H^-	2.08 Å	0.30 Å[34]		
C	1.68	C^{4-}	2.60	0.772	0.667	0.603
N	1.53	N^{3-}	1.71	0.70		
O	1.50	O^{2-}	1.40	0.66		
F	1.51	F^-	1.36	0.64		
Cl	1.84	Cl^-	1.81	0.99	0.89	
Br	1.96	Br^-	1.95	1.14	1.04	
I	2.13	I^-	2.16	1.33	1.23	
P	1.85	P^{3-}	2.12	1.10	1.00	0.93
S	1.82	S^{2-}	1.84	1.04	0.94	0.87
Si	2.04	Si^{4-}	2.71	1.17	1.07	1.00

[29] L. Pauling (reference 23), pp. 260–261.

[30] For example, see Proserpio, D. M.; Hoffmann, R.; Levine, R. D. *J. Am. Chem. Soc.* **1991**, *113*, 3217.

[31] Many sets of van der Waals radii are available in the literature. The data shown are calculated values reported by Chauvin, R. *J. Phys. Chem.* **1992**, *96*, 9194. These values correlate well with—but are sometimes slightly different from—values given by Pauling (reference 23, p. 260), Bondi (reference 26) and O'Keeffe, M.; Brese, N. E. *J. Am. Chem. Soc.* **1991**, *113*, 3226.

[32] Data from reference 23, p. 514.

[33] Data from reference 23, pp. 224 *ff.*

[34] The covalent radius of hydrogen varies considerably. See the discussion in reference 23, pp. 226–227. The value of r_c for hydrogen is calculated to be 0.30 Å in H_2O and 0.32 Å in CH_4.

Table 1.3 Group contributions to van der Waals atomic volume (V_w) and surface area (A_w).*

Group	V_w, cm^3/mole	A_w, cm^2/mole × 10^9
Alkane, C bonded to four other carbon atoms	3.33	0
Alkane, CH bonded to three other carbon atoms	6.78	0.57
Alkane, CH_2 bonded to two other carbon atoms	10.23	1.35
Alkane, CH_3 bonded to one other carbon atom	13.67	2.12
CH_4	17.12	2.90
F, bonded to a 1° carbon atom	5.72	1.10
F, bonded to a 2° or 3° carbon atom	6.20	1.18
Cl, bonded to a 1° carbon atom	11.62	1.80
Cl, bonded to a 2° or 3° carbon atom	12.24	1.82
Br, bonded to a 1° carbon atom	14.40	2.08
Br, bonded to a 2° or 3° carbon atom	14.60	2.09
I, bonded to a 1° carbon atom	19.18	2.48
I, bonded to a 2° or 3° carbon atom	20.35	2.54

*Data from reference 26.

group increment values. That is, we may estimate the molecular volume of propane by counting 2 × 13.67 cm^3/mol for the two methyl groups and 10.23 cm^3/mol for the methylene, giving a total volume of 37.57 cm^3/mol.

The volume measurements in Table 1.3 are based on empirical X-ray data. Increasingly, however, values for atomic and molecular volume are available from theoretical calculations. The calculated values vary somewhat, depending upon what definition of the interior and exterior of the atom or molecule that one chooses. (Usually the boundary of an atom is a certain minimal value of electron density in units of au; 1.00 au = 6.748 e/Å^3.) Bader[35] has determined that the 0.001 au volumes of methane and ethane are 25.53 and 39.54 cm^3/mol, respectively, while the corresponding 0.002 au volumes are 19.58 and 31.10 cm^3/mol. It thus appears that the 0.002 au values are closer to, but still somewhat larger than, those calculated empirically using the group increment approach of Bondi. The relationships of atomic volume and van der Waals radii are illustrated for cross sections through methane and propane in Figure 1.2. The contour lines represent the electron density contours, and the intersecting arcs represent the van der Waals radii of the atoms.

[35]Bader, R. F. W.; Carroll, M. T.; Cheeseman, J. R.; Chang, C. *J. Am. Chem. Soc.* **1987**, *109*, 7968.

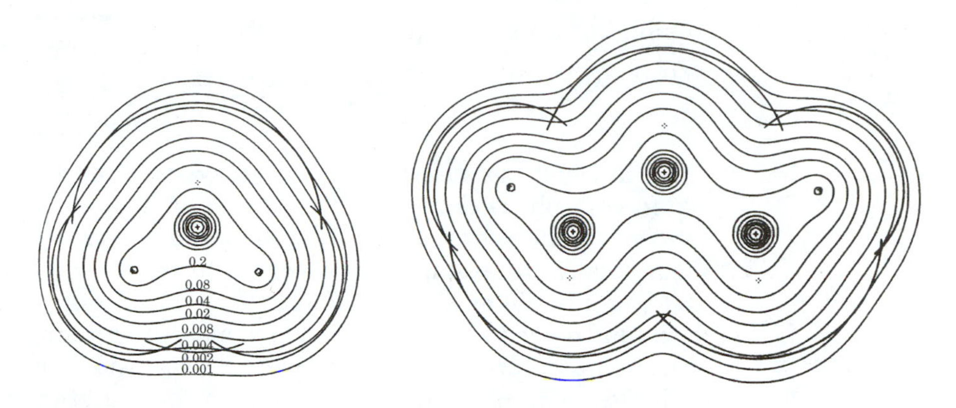

Figure 1.2 Contour maps and van der Waals radii arcs for methane (left) and propane (right). (Reproduced from reference 35.)

Heats of Formation and Reaction

Experimental Determination of Heats of Formation

The **heat of formation** (ΔH_f^0) of an organic compound is defined as the difference in enthalpy between the compound and the starting elements in their standard states.[36] For a hydrocarbon with molecular formula (C_mH_n), we define ΔH_f^0 as the heat of reaction (ΔH_r^0) for the reaction

$$m\ C_{(graphite)} + n/2\ H_{2\ (gas)} \rightarrow C_mH_n \tag{1.1}$$

A fundamental tool in thermochemical measurements is calorimetry, in which the heat of a reaction is determined by accurate measurement of the temperature rise that results from the chemical reaction in a closed system.[37] We determine the heat of formation of an organic compound by determining the heat of reaction of the compound to form other substances for which the heat of formation is known, and the heat of combustion is widely used for this purpose. Consider the combustion of a compound with the formula C_mH_n. The balanced chemical equation is

$$C_mH_n + (m + n/4)\ O_2 \rightarrow m\ CO_2 + n/2\ H_2O \tag{1.2}$$

We know the heats of formation of CO_2 and H_2O:

$$C_{(graphite)} + O_{2\ (gas)} \rightarrow CO_{2\ (gas)} \tag{1.3}$$

$$\Delta H_r^0 = \Delta H_f^0\ (CO_2) \tag{1.4}$$

$$H_{2\ (gas)} + 1/2\ O_{2\ (gas)} \rightarrow H_2O_{\ (liquid)} \tag{1.5}$$

$$\Delta H_r^0 = \Delta H_f^0\ (H_2O) \tag{1.6}$$

[36]For a discussion, see Mortimer, C. T. *Reaction Heats and Bond Strengths*; Pergamon Press: New York, 1962; Clark, T.; McKervey, M. A. in *Comprehensive Organic Chemistry*, Vol. 1; Stoddart, J. F., ed.; Pergamon Press: Oxford, England, 1979; pp. 66 *ff*.

[37]For a discussion of the experimental techniques involved in calorimetry experiments, see (a) Wiberg, K. in *Molecular Structure and Energetics*, Vol. 2; Liebman, J. F.; Greenberg, A., eds.; VCH Publishers, Inc.: New York, 1987; p. 151; (b) Sturtevant, J. M. in *Physical Methods of Chemistry*, Vol. I, Part V; Weissberger, A.; Rossiter, B. W., eds.; Wiley-Interscience: New York, 1971; p. 347.

Combining the above equations, we obtain

$$\Delta H_f^0\,(C_mH_n) = m\,\Delta H_f^0\,(CO_2) + n/2\,\Delta H_f^0\,(H_2O) - \Delta H^0_{(combustion)}\,(C_mH_n) \qquad (1.7)$$

It is sometimes necessary to correct heats of reaction for the heats associated with phase changes in the reactants or products. To convert from a condensed phase to the gas phase (for comparison with values calculated theoretically, for example) the relevant terms are the heat of vaporization (ΔH_v^0) of a liquid or heat of sublimation[38,39] (ΔH_s^0) of a solid. Heat capacity data can be used to correct ΔH values measured at one temperature to another temperature.[40] As an example, let us consider the combustion of 1,3-cyclohexanedione, which was found to have a standard molar heat of combustion of -735.9 kcal/mol.[41,42] Taking -68.32 kcal/mol and -94.05 kcal/mol as the standard heats of formation of H_2O and CO_2, respectively, gives a standard heat of formation for crystalline 1,3-cyclohexanedione of -101.67 kcal/mol. Correcting for the standard heat of sublimation of 1,3-cyclohexanedione of $+21.46$ kcal/mol gives a standard heat of formation in the gas phase of -80.21 kcal/mol.

If we are interested only in the difference between the heats of formation of two compounds, we may be able to measure their relative stability more accurately by measuring the heat of another reaction. That is, we measure very accurately the ΔH of a reaction in which the two different reactants combine with identical reagents to give the same products. Figure 1.3 illustrates how the difference in enthalpy of reactants A and B can be calculated in this manner. If the reaction of A and C to give D has a ΔH_r of $-X$ kcal/mol, and if the reaction of B and C to give D has a ΔH_r of $-Y$ kcal/mol, then the difference in energy between A and B must be $(X - Y)$ kcal/mol. For example, Wiberg and Hao determined that ΔH_r values for the reaction of trifluoroacetic acid with 2-methyl-1-butene and with 2-methyl-2-butene were -10.93 kcal/mol and -9.11 kcal/mol, respectively.[43] Therefore, the internal alkene was judged to be 1.82 kcal/mol lower in energy than the

[38]Determination of heats of sublimation has been discussed by Chickos, J. S. in *Molecular Structure and Energetics*, Vol. 2, Liebman, J. F.; Greenberg, A., eds.; VCH Publishers, Inc.: New York, 1987; p. 67.

[39]The enthalpy associated with transformation of a solid to a liquid is the heat of fusion. For a discussion of calculation of entropies and enthalpies of fusion, see Chickos, J. S.; Braton, C. M.; Hesse, D. G.; Liebman, J. F. *J. Org. Chem.* **1991**, *56*, 927.

[40]Heat capacity is the heat required to raise the temperature of a substance 1°. See Orchin, M.; Kaplan, F.; Macomber, R. S.; Wilson, R. M.; Zimmer, H. *The Vocabulary of Organic Chemistry*; Wiley-Interscience: New York, 1980; pp. 255–256.

[41]Pilcher, G.; Parchment, O. G.; Hillier, I. H.; Heatley, F.; Fletcher, D.; Ribeiro da Silva, M. A. V.; Ferrão, M. L. C. C. H.; Monte, M. J. S.; Jiye, F. *J. Phys. Chem.* **1993**, *97*, 243.

[42]The reported value (converted from kJ/mol) was -735.9 ± 0.2 kcal/mol. Experimental uncertainties will not be carried through this discussion because the emphasis is on the calculation procedure and not the precision of the experimental method.

[43]Wiberg, K. B.; Hao, S. *J. Org. Chem.* **1991**, *56*, 5108.

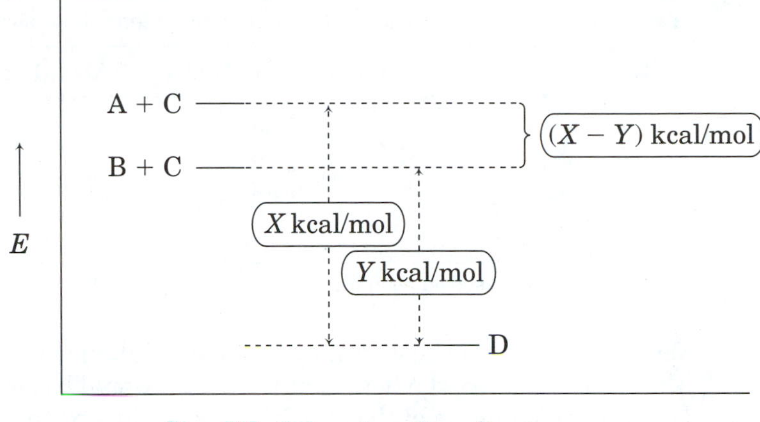

Figure 1.3 Generalized calculation of enthalpy difference of isomers.

external alkene. Heats of hydrogenation are also used to determine the difference in heats of formation of alkenes even though heats of combustion may be measured much more *precisely* than heats of hydrogenation. However, because heats of hydrogenation are smaller in magnitude than are heats of combustion, small differences between isomers may be determined more *accurately* by hydrogenation.[44]

Bond Increment Calculation of Heats of Formation

The estimation of molecular volume by summing the volumes of the individual molecular units (Table 1.3) is one example of the **principle of additivity**, meaning that the properties of a large molecule may be approximated by adding the contributions of its component parts. This concept has been found to be of value for several kinds of molecular properties, particularly for calculation of heats of formation.[45] For example, Table 1.4 shows the experimental heat of formation values for some linear alkanes.[46] There is a general trend in the data, each homologue higher than ethene having a ΔH_f^0 value about 5 kcal/mol more negative than the previous alkane.

The observation that the experimental heat of formation values in Table 1.4 increase regularly from one molecule to the next suggests that it should be possible to approximate the heat of formation of an organic compound by summing the contributions each component makes to that value. The difference between one compound and the next in Table 1.4 is the addition of one carbon atom and two hydrogen atoms. However, we cannot expect heats of

---

[44]Davis, H. E.; Allinger, N. L.; Rogers, D. W. *J. Org. Chem.* **1985**, *50*, 3601.

[45]Benson, S. W. *Thermochemical Kinetics*, 2nd ed.; Wiley-Interscience: New York, 1976; p. 24.

[46]Experimental data for ΔH_f^0 at 298 K are from tabulations in Stull, D. R.; Westrum, Jr., E. F.; Sinke, G. C. *The Thermodynamics of Organic Compounds*; John Wiley & Sons, Inc.: New York, 1969; pp. 243–245.

Table 1.4 Experimental and calculated heats of formation of linear alkanes at 298 K.

Compound	ΔH_f^0 (kcal/mol) obs.	ΔH_f^0 (kcal/mol) calc.*
Methane	−17.89	−15.32
Ethane	−20.24	−20.25
Propane	−24.82	−25.18
Butane	−30.15	−30.11

*Calculations are based on bond increment values in Table 1.5.

formation to be additive with the number of atoms of each element alone. If that were the case, the heat of any reaction would be zero, since the number and kind of atoms on each side of the arrow must be the same. The next simplest approximation is to estimate a heat of formation by counting the number and kind of bonds in a molecule. By this approach, the compounds in Table 1.4 differ by the addition of one C—C single bond and two C—H bonds. Extensive work in this area has been done by Benson, who has published tables of **bond increment contributions** to heats of formation and other thermodynamic properties.[45,47,48,49,50,51] A portion of one such table is reproduced as Table 1.5. Here C_d is a double-bonded carbon atom and (>CO) is a carbonyl carbon atom; thus (>CO)—H represents the C—H bond in an aldehyde.

The heat of formation values for some linear alkanes calculated by the bond increment method are shown in Table 1.4. As an example of such calculations, let us determine the ΔH_f^0 values for methane and ethane. For methane, there are four C—H bonds, each contributing −3.83 kcal/mol, so the ΔH_f^0 value is −15.32 kcal/mol. For ethane, the ΔH_f^0 value is $6(-3.83) + 1(2.73)$ for the six C—H and one C—C bonds, respectively, and the total is −20.25 kcal/mol. As the chain is extended, each additional CH_2 group contributes $2 \times (-3.83) + 1 \times (2.73) = -4.93$ kcal/mol to the ΔH_f^0 value.

There is a problem with the values obtained from very simple bond increment approaches, such as those using the simple bond data in Table 1.5. The five isomers of hexane listed in Table 1.6 all have five C—C bonds and 14 C—H bonds. Using the bond increment values listed in Table 1.5, each would be predicted to have the same heat of formation (−39.97 kcal/mol). As

[47]Benson, S. W.; Buss, J. H. *J. Chem. Phys.* **1959**, *29*, 546.

[48]Benson, S. W.; Cruickshank, F. R.; Golden, D. M.; Haugen, G. R.; O'Neal, H. E.; Rodgers, A. S.; Shaw, R.; Walsh, R. *Chem. Rev.* **1969**, *69*, 279.

[49]For a discussion of the development of bond increment and group increment calculations, see Schleyer, P. v. R.; Williams, J. E.; Blanchard, K. R. *J. Am. Chem. Soc.* **1970**, *92*, 2377.

[50]Calculation of group increments to heats of formation of linear hydrocarbons had been reported by Pitzer, K. S. *J. Chem. Phys.* **1940**, *8*, 711 and to nonlinear hydrocarbons by Franklin, J. L. *Ind. Eng. Chem.* **1949**, *41*, 1070.

[51]Cohen, N.; Benson, S. W. *Chem. Rev.* **1993**, *93*, 2419.

Table 1.5 Bond increment contributions to ΔH_f^0.*

Bond	ΔH_f^0 (kcal/mol)	Bond	ΔH_f^0 (kcal/mol)
C—H	−3.83	O—H	−27.0
C—D	−4.73	O—D	−27.9
C—C	2.73	O—O	21.5
C—F	−52.5	O—Cl	9.1
C—Cl	−7.4	N—H	−2.6
C—Br	2.2	S—H	−0.8
C—I	14.1	S—S	−6.0
C—O	−12.0	C—S	6.7
C_d—C	6.7	(>CO)—H	−13.9
C_d—H	3.2	(>CO)—C	−14.4
C_d—F	−39.0	(>CO)—O	−50.5
C_d—Cl	−5.0	(>CO)—F	−77.0
C_d—Br	9.7	(>CO)—Cl	−27.0
C_d—I	21.7	C_6H_5—H	3.25
C_d—C_d	7.5	C_6H_5—C	7.25

*Data from reference 45, p. 25.

shown in the column labeled ΔH_f^0 (obs.) in Table 1.6, however, the experimental heats of formation become more negative as the branching increases. Specifically, the structure with a quaternary carbon atom is more stable than an isomeric structure with two tertiary carbon atoms, and the structure with two tertiary carbon atoms is more stable than structures with only one tertiary carbon atom, even though all isomers have the same

Table 1.6 Heats of formation (kcal/mol) of isomeric C_6H_{14} structures.

Compound	ΔH_f^0, obs.*	ΔH_f^0, calc.**	ΔH_f^0, corr.***
Hexane	−39.96	−39.96	−39.96
2-Methylpentane	−41.66	−42.04	−41.24
3-Methylpentane	−41.02	−42.04	−41.24
2,2-Dimethylbutane	−44.35	−44.77	−43.16
2,3-Dimethylbutane	−42.49	−44.12	−42.52

*Experimental data for ΔH_f^0 at 298 K are from reference 46, pp. 247–249.
**Calculated from group increments in Table 1.7 without correcting for *gauche* interactions.
***Data from the previous column corrected for *gauche* interactions. See Table 1.7 and Figure 1.4.

number of C—C and C—H bonds. There is a general trend that the more highly branched isomer is the more stable isomer.[52] Thus we must conclude that the heat of formation of a compound depends not only on the number of carbon-carbon bonds, but also on the nature of the carbon-carbon bonds.

The hypothetical conversion of n-hexane to 2,2-dimethylbutane is an example of an **isodesmic reaction**,[53,54] which is a reaction in which both the reactants and the products have the same number of bonds of a given type but there may be changes in the relationship of one bond to another. Both the reactant and the product have five C—C and fourteen C—H bonds. The simple bond increment approach would calculate that the heat of the reaction should be 0, but the data in Table 1.6 indicate that the heat of the reaction should be −4.4 kcal/mol. Therefore, the heat of an isodesmic reaction is an indication of deviation from the additivity of bond energies.[53,55]

Group Increment Calculation of Heats of Formation

The **group increment** approach allows calculation of enthalpy differences that result from the differing arrangement of bonds within molecules. We consider not the bonds holding atoms together but only the groups that are present (on the quite reasonable assumption that they are held in place by some form of bonding). Table 1.7 lists the group increment data for a series of organic functional groups.[48] Using the data in this table, we can closely approximate the heats of formation of the isomeric hexanes. For example, 2-methylpentane is composed of three methyl groups [C—$(H)_3(C)$ in the table], which contribute −10.08 kcal/mol *each* to the heat of formation. Two methylene units [C—$C(H)_2(C)_2$] contribute −4.95 kcal/mol *each*, and one methine unit [C—$(H)(C)_3$] contributes −1.90 kcal/mol. The estimated heat of formation, therefore, is

$$\Delta H_f^0 = 3 \times (-10.08) + 2 \times (-4.95) + 1 \times (-1.90) = -42.04 \text{ kcal/mol} \quad \textbf{(1.8)}$$

which is close to the observed value (−41.66 kcal/mol).[56]

Note that the estimated heats of formation calculated in this way assign the same contribution to each group without regard to its position in the molecule and without regard to strain. In linear alkanes the major form of

[52]For theoretical explanations of this order of stabilities, see Laidig, K. E. *J. Phys. Chem.* **1991**, *95*, 7709; Laurencelle, N.; Pacey, P. D. *J. Am. Chem. Soc.* **1993**, *115*, 625.

[53]Hehre, W. J.; Ditchfield, R.; Radom, L.; Pople, J. A. *J. Am. Chem. Soc.* **1970**, *92*, 4796.

[54]A **homodesmotic reaction** is defined as a reaction in which not only are the number of bonds of each type conserved, but the number of carbon atoms with zero, one, two, or three hydrogen atoms is also conserved. For details, see George, P.; Trachtman, M.; Bock, C. W.; Brett, A. M. *Tetrahedron* **1976**, *32*, 317. Isomers interconverted by homodesmotic reactions are **isologous** (*cf.* Engler, E. M.; Andose, J. D.; Schleyer, P. v. R. *J. Am. Chem. Soc.* **1973**, *95*, 8005).

[55]Isodesmic reactions are widely used in theoretical studies because errors in the energies of reactants and products are more likely to cancel, thereby allowing simple computational approaches to give accurate estimates of heats of reactions. For a discussion, see Hehre, W. J.; Radom, L.; Schleyer, P. v. R.; Pople, J. A. Ab initio *Molecular Orbital Theory*; Wiley-Interscience: New York, 1986.

[56]Reference 46, p. 247.

Table 1.7 Group increment contributions to heats of formation.*

Group	$\Delta H^0_{f\,298}$ (kcal/mol)	Group	$\Delta H^0_{f\,298}$ (kcal/mol)
$C-(H)_3(C)$	−10.08	$C_d-(C_B)(C)$	8.64
$C-(H)_2(C)_2$	−4.95	$C-(C_B)(C)(H)_2$	−4.86
$C-(H)(C)_3$	−1.90	$C-(C_B)(C)_2(H)$	−0.98
$C-(C)_4$	0.50	$C_t-(H)$	26.93
$C_d-(H)_2$	6.26	$C_t-(C)$	27.55
$C_d-(H)(C)$	8.59	$C_t-(C_d)$	29.20
$C_d-(C)_2$	10.34	$C_B-(H)$	3.30
$C_d-(C_d)(H)$	6.78	$C_B-(C)$	5.51
$C_d-(C_d)(C)$	8.88	$C_B-(C_d)$	5.68
$[C_d-(C_B)(H)]$	6.78		

*Reproduced from Table XXXIII, p. 316 of reference 48.

strain to consider is van der Waals strain manifested as butane *gauche* interactions, which may be assigned 0.8 kcal/mol each.[57] Figure 1.4 shows a Newman projection and gives the number of such interactions for each of the isomers of hexane. Correcting the initial ΔH^0_f of 2-methylpentane for one such interaction gives −41.24 kcal/mol, a value close to the experimental value (−41.66 kcal/mol). Angle strain corrections must also be applied for ring compounds. For example, a cyclopropane, cyclobutane, and cyclopentane ring add 27.6, 26.2, and 6.3 kcal/mol, respectively, to the heat of formation calculated from the data in Table 1.7.[48,58]

Figure 1.4 *Gauche* interactions in hexane isomers.

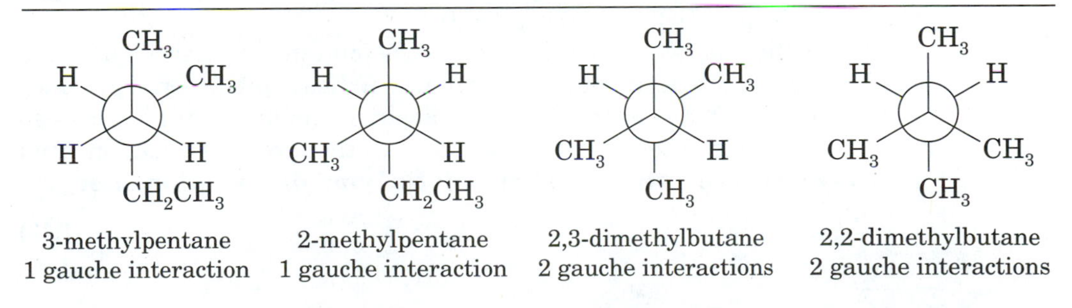

[57]Molecular conformation and van der Waals strain will be discussed in Chapter 3.

[58]These examples only hint at the analysis of heats of formation of organic compounds that is possible. Benson and co-workers have summarized the methods and data for calculations for the major functional groups in organic chemistry.[48,45] In addition, the data allow calculation of heat capacities and entropies of these compounds in the same manner in which heats of formation are determined. Heats of formation are valuable reference points in discussing the stabilities of various isomers or products of reactions, whether they are calculated by bond increments or group increments or are derived as part of a theoretical calculation.

Homolytic and Heterolytic Dissociation Energies

Heats of reaction are important values for processes that involve reactive intermediates. For example, for the gas phase dissociation reaction

$$A-B_{(g)} = A\cdot_{(g)} = B\cdot_{(g)} \quad \textbf{(1.9)}$$

the standard bond dissociation energy,[59,60] $DH^0(A-B)$, for $A-B$ can be calculated from the relationship

$$DH^0(\mathrm{A-B}) = \Delta H^0_r = \Delta H^0_f(\mathrm{A}\cdot) + \Delta H^0_f(\mathrm{B}\cdot) - \Delta H^0_f(\mathrm{A-B}) \quad \textbf{(1.10)}$$

In turn, values of ΔH^0_r for dissociation reactions can be combined to allow prediction of heats of reaction. A familiar example is the calculation of the ΔH^0_f for the reaction of chlorine with methane to produce HCl plus methyl chloride. Using tables of standard bond dissociation energies, we can write the following reactions:

$$CH_3-H \rightarrow CH_3\cdot + H\cdot \qquad \Delta H^0_r = +104 \text{ kcal/mol} \quad \textbf{(1.11)}$$

$$Cl-Cl \rightarrow Cl\cdot + Cl\cdot \qquad \Delta H^0_r = +58 \text{ kcal/mol} \quad \textbf{(1.12)}$$

$$Cl\cdot + CH_3\cdot \rightarrow CH_3Cl \qquad \Delta H^0_r = -84 \text{ kcal/mol} \quad \textbf{(1.13)}$$

$$Cl\cdot + H\cdot \rightarrow H-Cl \qquad \Delta H^0_r = -103 \text{ kcal/mol} \quad \textbf{(1.14)}$$

Summing these four equations and cancelling the radicals that appear on both sides give:

$$CH_3-H + Cl-Cl \rightarrow CH_3Cl + HCl \qquad \Delta H^0_r = -25 \text{ kcal/mol} \quad \textbf{(1.15)}$$

Note that the calculation of ΔH^0_r does *not* presume that the reaction takes place by a radical pathway. Rather, according to Hess' law, the difference in enthalpy between reactants and products is independent of the path of the reaction.[61] The value of ΔH^0_r is quite close to the value of ΔH^0_r value at 300° of −23.7 kcal/mol calculated from the ΔH^0_f values of Cl_2 (0), CH_4 (−17.9), CH_3Cl (−19.6), and HCl (−22.0 kcal/mol).[62]

Technically, the standard bond dissociation energy described above should be called the **standard homolytic bond dissociation energy**, since the molecule A—B dissociates to two radicals. If the dissociation occurs so that one of the species becomes a cation and the other becomes an anion, then the energy is termed a **standard heterolytic bond dissociation energy**:

$$A-B_{(g)} = A^+_{(g)} + B{:}^-_{(g)} \quad \textbf{(1.16)}$$

---

[59] For more details, see Benson, S. W. *J. Chem. Educ.* **1965**, *42*, 502.

[60] A standard bond dissociation energy is a different concept from that of the *average* bond dissociation energy. The latter is just the value obtained by calculating the heat of atomization of a compound (the enthalpy change on converting the molecule to individual atoms) divided by the number of bonds from one atom to another in the molecule. For more details on this distinction, see reference 59.

[61] Reference 46, p. 62.

[62] Reference 45, pp. 290 *ff*.

Therefore

$$\Delta H(\mathrm{A}^+, \mathrm{B}{:}^-) = \Delta H^0_{\mathrm{het}} = \Delta H^0_f\,(\mathrm{A}^+) + \Delta H^0_f\,(\mathrm{B}{:}^-) - \Delta H^0_f\,(\mathrm{A{-}B}) \qquad \textbf{(1.17)}$$

In the gas phase heterolytic bond dissociation energies are much higher than homolytic bond dissociation energies because energy input is needed to separate the two ions, in addition to the energy needed to break the bond. Solvation of the ions can reduce the value of $\Delta H^0_{\mathrm{het}}$ so much, however, that in one case merely adding ether to a pentane solution of a compound was seen to produce heterolytic dissociation.[63]

It is possible to relate homolytic and heterolytic reaction energies by using ionization potential (the energy required to remove an electron from a species) and electron affinity (the energy gained by adding an electron to a species) data.[64] Figure 1.5 shows the approach used by Arnett and co-workers, who have described the use of electrochemical oxidation and reduction potentials to determine ΔH values.[63,65]

Bonding Models

The simple view of chemical bonding developed by G. N. Lewis is still quite useful to organic chemists. Atoms are represented by element symbols with dots around them to indicate the number of electrons in the valence shell of the atom. Covalent bonds are formed by the sharing of one or more pairs of electrons between atoms so that both atoms achieve an electron configuration

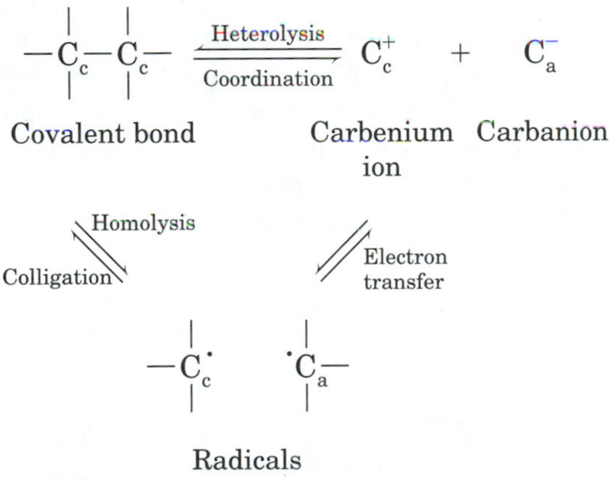

Heterolysis
Coordination
Covalent bond
Carbenium ion
Carbanion
Homolysis
Colligation
Electron transfer
Radicals

Figure 1.5 Enthalpy relationships among homolysis, heterolysis, and electron transfer reactions. (Adapted from reference 63.)

[63]Arnett, E. M.; Amarnath, K.; Harvey, N. G.; Cheng, J. P. *Science* **1990**, *247*, 423.

[64]Unless appropriate corrections for phase are made, the ΔH values are not standard heats of reaction, but they can nevertheless be useful in understanding properties of substances.

[65]Arnett, E. M.; Flowers, II, R. A. *Chem. Soc. Reviews* **1993**, *22*, 9.

Figure 1.6 An electron dot representation of Cl· + Cl· → Cl_2.

$$:\ddot{\underset{..}{Cl}}\cdot + :\ddot{\underset{..}{Cl}}\cdot \longrightarrow :\ddot{\underset{..}{Cl}}-\ddot{\underset{..}{Cl}}:$$

that corresponds to a filled outer shell.[66] Thus, combination of two chlorine atoms can produce a chlorine molecule, as shown in Figure 1.6.

This elementary discussion of bonding has assumed some knowledge of electron shells of the atoms, but it has not presumed any more detailed knowledge of the results of quantum mechanics. We have not specified what atomic orbitals are populated, what the geometric shape(s) of these orbitals might be, or what the distribution of electrons might be in the final molecule of chlorine. We have simply written electron dot structures or Lewis structures to formalize some observations about the patterns in which atoms seem to combine. This approach to describing chemical bonding might be adequate for some purposes, but it leaves many questions unanswered. Moreover, the bonding description has been purely qualitative. We would like a mathematical description of bonding so that we can test quantitative predictions about bonding with the results of experiments.

Before proceeding we need to distinguish two types of information that we wish to acquire about organic molecules. The first type of information is physically observable data that are characteristic of entire molecules or samples of molecules. The dipole moment of a molecule and the heat of formation of a substance belong to this category. The second type includes those nonobservable constituent properties of a molecule that, taken together, give rise to the overall molecular properties. Bond dissociation energies and bond dipole moments belong to this category.

Molecular dipole moments provide an important window into the structures of molecules. A dipole moment is a vector quantity that measures the separation of electrical charge. Dipole moments have units of electrical charge (a full plus or minus charge corresponding to 4.80×10^{-10} esu) times distance, and they are usually expressed in units of Debye, D, with 1 D = 10^{-18} esu cm.[67,68] Thus a system consisting of two atoms, one with a partial

---

[66]Lewis, G. N. *J. Am. Chem. Soc.* **1916**, *38*, 762. It is interesting to note that Lewis proposed in this paper a model for bonding in which electrons were positioned at the corners of a cube, so that an octet meant an electron at every corner. Single bonds were constructed by allowing two cubes to share one edge (and thus one pair of electrons). In the case of a double bond, the two cubes shared a face (and therefore two pairs of electrons). Unfortunately, the cubical model offered no simple representation for triple bonds. For a discussion of the role of G. N. Lewis in the development of structural theory in organic chemistry, see the following papers: Calvin, M. *J. Chem. Educ.* **1984**, *61*, 14; Zandler, M. E.; Talaty, E. R. *J. Chem. Educ.* **1984**, *61*, 124; Saltzman, M. D. *J. Chem. Educ.* **1984**, *61*, 119; Stranges, A. N. *J. Chem. Educ.* **1984**, *61*, 185; Pauling, L. *J. Chem. Educ.* **1984**, *61*, 201.

[67]For background on the theory and measurement of dipole moments, see Minkin, V. I.; Osipov, O. A.; Zhdanov, Y. A. *Dipole Moments in Organic Chemistry*; Hazzard, B. J., trans.; Vaughan, W. E., ed.; Plenum Press: New York, 1970.

[68]Smyth, C. P. in *Physical Methods of Chemistry*, Vol. 1, Part IV; Weissberger, A.; Rossiter, B. W., eds.; Wiley-Interscience: New York, 1972; pp. 397–429.

charge of +0.1 and the other a partial charge of −0.1, located 1.5 Å apart would have a dipole moment of

$$0.1 \times (4.8 \times 10^{-10}\ \text{esu}) \times (1.5 \times 10^{-8}\ \text{cm}) = 0.72 \times 10^{-18}\ \text{esu} \cdot \text{cm}$$
$$= 0.72\ \text{D} \qquad \textbf{(1.18)}$$

Dipole moments can be measured by several techniques, the most important ones being the determination of the dielectric constant of a substance as a gas or in a nonpolar solution and the study of the effect of electrical fields on molecular spectra (Stark effect).

Molecular dipole moments are useful to us primarily as a source of information about molecular structure and bonding. While the center of charge need not coincide with the center of an atom, that is a convenient first approximation. For example, the dipole moment of CH_3F is 1.81 D.[69,70] Since the dipole moment of CH_4 is 0 D, we infer that there is net charge separation in methyl fluoride. We associate the charge separation with the bonding between C and F. Since those atoms are 1.385 Å apart (Table 1.1), the partial charge can be calculated to be +0.27 esu on one of the atoms and −0.27 esu on the other. Thus, CH_3F is presumed to have a bond dipole moment associated with the C — F bond.

If there is more than one dipole in a molecule, then the molecular dipole moment is the vector sum of the individual moments. This idea can be useful in determining the structures and bonding of molecules. For example, Smyth[71] determined that the three isomers of dichlorobenzene have dipole moments of 2.30, 1.55, and 0 D. The dipole moment of chlorobenzene was known to be 1.61 D. Smyth reasoned that in one isomer of dichlorobenzene two chlorobenzene dipole moments added to each other, in the second isomer they cancelled each other partially, and in the third isomer they cancelled each other completely. Using the relationship

$$\mu = 2 \times 1.61 \times 10^{-18} \times \cos(0.5\,A) \qquad \textbf{(1.19)}$$

where A is the angle between the two "bond" dipole moments, Smyth calculated that the three isomers of dichlorobenzene had A values of 89°, 122°, and 180° and that these values corresponded to the *ortho*, *meta*, and *para* isomers of dichlorobenzene, respectively. Not only did this study identify which isomer of dichlorobenzene was which, but it also reinforced the view that benzene is a planar molecule. Alternative structures, such as Baeyer, Körner, or Ladenburg benzene would have given different molecular dipole moments.[72]

---

[69]McClellan, A. L. *Tables of Experimental Dipole Moments*, Vol. 2; Rahara Enterprises: El Cerrito, CA, 1974; p. 167.

[70]A value of 1.857 D is given in reference 22 (b), p. 139. That is a more recent value and may be more accurate than the number used here. The values for the other methyl fluorides there are very similar to those given in reference 69.

[71]Smyth, C. P.; Morgan, S. O. *J. Am. Chem. Soc.* **1927**, *49*, 1030.

[72]The expected angle for *ortho*-dichlorobenzene should be 60°, but Smyth argued that the apparent angle is larger because repulsion of the two adjacent chlorines enlarges the angle between the dipoles but does not appreciably alter the geometry of the benzene ring.

To account for the electric dipole moment associated with a covalent bond, we say that the electrons in the bond are not shared equally between the two atoms. One atom must have more ability to attract the pair of shared electrons so that the bond can be described as having a mixture of both ionic and covalent bonding. It is useful to define a weighting parameter, λ, to indicate how much ionic character is mixed into the covalent bond. Thus we may write

$$\text{Polar bond} = [\text{Covalent bond}] + \lambda\,[\text{Ionic bond}] \tag{1.20}$$

Then the percentage ionic character in the bond is related to λ by equation 1.21[73]

$$\%\ \text{Ionic character} = \frac{\lambda^2}{(1+\lambda^2)} \times 100\% \tag{1.21}$$

Thus, in a HCl molecule with partial charges of $+0.17$ esu on the hydrogen atom and -0.17 esu on the chlorine atom, the value of λ is 0.45.[74]

Electronegativity and Bond Polarity

The extra ionic character in a polar bond gives rise to an increased bond dissociation energy, in comparison with reference nonpolar bonds, since bond dissociation must overcome Coulombic effects as well as the bonding interaction. We rationalize the existence of polar covalent bonds with the concept of **electronegativity**, which Pauling defined as "the power of an atom in a molecule to attract electrons to itself."[75] Pauling obtained electronegativity values (χ_P) by comparing the standard bond dissociation energies between different atoms (A—B) with the average of the standard bond dissociation energies of identical atoms (A—A and B—B).[75] Thus, generally the bond energy of the diatomic molecule A—B will be greater than one half of the sum of the bond strengths of A—A and B—B.[76] For example, the average of the bond strengths of H_2 and Cl_2 is 81.1 kcal/mol, but the bond strength of H—Cl is 103.2 kcal/mol.[77,78] A tabulation of electronegativity differences led to the standard table of Pauling electronegativity values.

---

[73]Coulson, C. A. *Valence*; Clarendon Press: Oxford, England, 1952; p. 128.

[74]The attraction that an atom has for electrons in its covalent bonds is known as electronegativity, a term that will be explored more fully in the following section.

[75]Pauling, L. *J. Am. Chem. Soc.* **1932**, *54*, 3570.

[76]This idea was called the "postulate of the additivity of normal covalent bonds" by Pauling (reference 23, p. 80).

[77]The premise that covalent bonds between atoms with different electronegativity values are stronger than the corresponding bonds between identical atoms is not always found to be true. Benson (reference 59) has pointed out that the reaction of Hg_2 with Cl_2 to produce 2 HgCl, is endothermic by at least 10 kcal/mol.

[78]The relationship between electronegativity values and bond strengths is still studied. See, for example, Reddy, R. R.; Rao, T. V. R.; Viswanath, R. *J. Am. Chem. Soc.* **1989**, *111*, 2914.

On the one hand, the concept of electronegativity is quite simple, and it has been called "perhaps the most popular intuitive concept in chemistry."[79] Yet, on the other hand, it is difficult to determine precise values for electronegativity because "a set of electronegativity values constitutes a chemical pattern recognition scheme which is not amenable to direct physical measurement."[80] Therefore, a great variety of theoretical approaches have been taken in describing and quantifying electronegativity.

The Pauling electronegativity scale is inherently dependent on measurements made on molecules, not on individual atoms.[81] There have been many approaches toward defining electronegativity as an atomic property. Sanderson's definition of electronegativity as "the effectiveness of the nuclear charge as sensed within an outer orbital vacancy" of an atom[82] suggests that atomic properties should be related to electronegativity. Mulliken introduced an electronegativity scale (χ_M) based on the average of the ionization potential (I) and electron affinity (A) of atoms; that is, $\chi = (I + A)/2$.[83] (The greater the electron affinity, the greater the attraction of an atom for an electron from outside the atom; the greater the ionization potential, the greater the affinity of an atom for a nonbonded electron localized on the atom.) Nagle has introduced an electronegativity value based on atomic polarizability.[84] Allen has proposed electronegativity values based on the average ionization potential of all of the p and s electrons on an atom.[85,86] Other treatments have been given by Allred-Rochow and by Gordy.[87] Building on a suggestion of Yuan,[88] Benson proposed another view of electronegativity, V_x, which is calculated by dividing the number of valence electrons about an atom by its covalent radius.[89] Thus, seven electrons in the valence shell of a fluorine atom, divided by 0.706 Å, gives a V_x value of 9.915 for fluorine. Values of V_x correlate well with a number of physical properties.[89]

---

[79]K. D. Sen in the Editor's Note to reference 87.

[80]Allen, L. C.; Egolf, D. A.; Knight, E. T.; Liang, C. *J. Phys. Chem.* **1990**, *94*, 5602 (quotation from page 5605).

[81]Nagle (reference 84) has pointed out that both electronegativity and atomic size are deduced from molecular—not atomic—properties.

[82]Sanderson, R. T. *J. Chem. Educ.* **1988**, *65*, 112, 227. See also Pauling, L. *J. Chem. Educ.* **1988**, *65*, 375.

[83]Mulliken, R. S. *J. Chem. Phys.* **1934**, *2*, 782; **1935**, *3*, 573.

[84]Nagle, J. K. *J. Am. Chem. Soc.* **1990**, *112*, 4741.

[85]Allen, L. C. *J. Am. Chem. Soc.* **1989**, *111*, 9003. Allen has called this definition of electronegativity "the third dimension of the periodic table." For a summary, see Borman, S. A. *Chem. Eng. News* **1990** (January 1), 18.

[86]For other treatments of electronegativity, see Boyd, R. J.; Edgecombe, K. E. *J. Am. Chem. Soc.* **1988**, *110*, 4182; Bratsch, S. G. *J. Chem. Educ.* **1988**, *65*, 223 and references therein.

[87]For a summary of these electronegativity scales, see Mullay, J. in *Electronegativity*, Sen, K. D.; Jorgensen, C. K., eds.; Springer-Verlag: Berlin, 1987.

[88]Yuan, H. C. *Acta Chim. Sin.* **1964**, *30*, 341; *cf.* reference 89; *Chem. Abstr.* **1965**, *62*, 2253h.

[89]Luo, Y.-R.; Benson, S. W. *J. Phys. Chem.* **1988**, *92*, 5255; *J. Am. Chem. Soc.* **1989**, *111*, 2480; *J. Phys. Chem.* **1989**, *93*, 3304. See also Luo, Y.-R.; Pacey, P. D. *J. Am. Chem. Soc.* **1991**, *113*, 1465 and references therein; Luo, Y.-R.; Benson, S. W. *Acc. Chem. Res.* **1992**, *25*, 375.

Table 1.8 compares the electronegativity values reported by Pauling (χ_P), Mulliken (χ_M), Allen (χ_{spec}), Nagle (χ_α) and Benson (V_x).[89,90,91] The Pauling, Allen, and Nagle values are usually quite similar, suggesting that the properties of atoms in molecules can be related to the properties of isolated atoms. However, while the Mulliken values are similar to the other values, there are some differences, particularly for hydrogen. Similarly, the Benson values are larger in magnitude but, except for hydrogen, they generally correlate well with the Pauling values.

Theoretical research on electronegativity has accelerated in recent years. Parr and co-workers[92] defined a quantity, μ, as the "electronic chemi-

Table 1.8 Comparison of electronegativity values.

Atom	χ_P	χ_M	χ_{spec}	χ_α	V_x
H	2.20	3.059	2.300	2.27	2.70
Li	0.91	1.282	0.912	0.94	0.75
Be	1.57	1.987	1.576	1.55	2.08
B	2.04	1.828	2.051	2.02	3.66
C	2.55	2.671	2.544	2.56	5.19
N	3.04	3.083	3.066	3.12	6.67
O	3.44	3.215	3.610	3.62	8.11
F	3.98	4.438	4.193	4.23	9.915
Na	0.93	1.212	0.869	0.95	0.65
Mg	1.31	1.630	1.293	1.32	1.54
Al	1.61	1.373	1.613	1.55	2.40
Si	1.90	2.033	1.916	1.87	3.41
P	2.19	2.394	2.253	2.22	4.55
S	2.58	2.651	2.589	2.49	5.77
Cl	3.16	3.535	2.869	2.82	7.04
K	0.82	1.032	0.734	0.84	0.51
Ca	1.00	1.303	1.034	1.11	1.15
Br	2.96	3.236	2.685	2.56	6.13
I	2.66	2.880	2.359	2.27	5.25

[90]Values for (χ_p), (χ_M), and (χ_{spec}) are taken from the compilation of Allen (reference 85). Values for χ_α are taken from reference 84. Values of V_x are from reference 89.

[91]Table 8 lists χ values for atoms only, but it is also possible to calculate "group electronegativities" to take into account the net effect of a group of atoms. For a tabulation of group electronegativities calculated by a variety of methods, see Bratsch, S. G. *J. Chem. Educ.* **1985**, *62*, 101. As an example, the group electronegativity of the CH_3 group is about 2.3, while that for CF_3 is about 3.5.

[92]Parr, R. G.; Donnelly, R. A.; Levy, M.; Palke, W. E. *J. Chem. Phys.* **1978**, *68*, 3801.

cal potential," which measures "the escaping tendency" of the electrons in the system.[93] The value of μ is approximately the same as $(I + A)/2$, the Mulliken electronegativity, so the value χ_M has been termed "absolute electronegativity."[93] Closely related to the concept of electronegativity is the concept of "chemical potential," which is given the symbol μ and is defined as $\partial E/\partial N$, where E is the energy of the system and N is the number of electrons.[94,95,96] Parr and co-workers have defined

$$\chi = -\mu = -\delta E/\delta\rho \tag{1.22}$$

where the energy is related to a theoretical treatment of electron density.[97,98]

As a result of the many theoretical treatments, chemists now find themselves using one term to mean different things, since "the electronic chemical potential, μ, . . . is an entirely different chemical quantity" from the concept of electronegativity as the origin of bond polarity.[96] As Pearson has noted, "the fact that there are two different measures both called (electronegativity) scales creates considerable opportunity for confusion and misunderstanding. Since the applications are so different, it is not a meaningful question to ask which scale is more correct. Each scale is more correct in its own area of use."[93] Finally, Allen has proposed that the configuration energy, defined as "the average one-electron valence shell energy of a ground-state free atom" should be the third dimension of the periodic table and that this parameter incorporates the information available from electronegativity.[99]

Usually we will use the term *electronegativity* in the sense originally proposed by Pauling, but we must be prepared to recognize the alternative meaning in the literature. Moreover, we see that a simple idea that is intuitively useful in understanding some problems of structure and bonding (dipole moments, heats of formation) may become more difficult to use as we attempt to make it more precise. In the next section we will find the same situation in an introduction to theoretical descriptions of bonding.

---

[93]For a discussion, see Pearson, R. G. *Acc. Chem. Res.* **1990**, *23*, 1.

[94]Pritchard, H. O.; Sumner, F. H. *Proc. Roy. Soc. (London)* **1956**, *A235*, 136.

[95]Iczkowski, R. P.; Margrave, J. L. *J. Am. Chem. Soc.* **1961**, *83*, 3547.

[96]Allen, L. C. *Acc. Chem. Res.* **1990**, *23*, 175.

[97]We will explore some aspects of the relationship of electronegativity, chemical potential and hardness in later chapters. However, the interest in this area in recent years is much too great for it to be covered fully here.

[98]Here ρ is actually a functional, and the approach is known as density-functional theory. For an introduction, see (a) Parr, R. G.; Yang, W. *Density-Functional Theory of Atoms and Molecules*; Oxford University Press: New York, 1989; (b) March, N. H. *Electron Density Theory of Atoms and Molecules*; Academic Press: New York, 1991.

[99]Reference 96; Allen, L. C. *J. Am. Chem. Soc.* **1992**, *114*, 1510. Also see Scerri, E. R. *J. Phys. Chem.* **1993**, *97*, 5786; Allen, L. C. *J. Phys. Chem.* **1993**, *97*, 5787.

1.2 Bonding and Molecular Geometry

Complementary Theoretical Models of Bonding

The Lewis model for forming a chemical bond by sharing an electron pair described on page 17 leads to a theoretical description of bonding known as **valence bond theory (VB theory)**.[100] The key to VB theory is that we consider a structure to be formed by bringing together complete atoms and then allowing them to interact to form bonds. In **molecular orbital theory (MO theory)**, on the other hand, we consider molecules to be constructed by bringing together nuclei (or nuclei and inner filled shells) and then placing the electrons in molecular orbitals.[101] MO theory does not generate discrete chemical bonds; rather, it generates a set of orbitals which allows electrons to roam over many nuclei, perhaps an entire molecule, and does not restrict them to any particular pair of nuclei.[102]

Both VB and MO theories utilize mathematical expressions that can rapidly become complex, even for simple organic molecules. Moreover, VB theory and MO theory are usually described with different symbols, so that it can become difficult to delineate the similarities among and differences between them. Therefore it may be useful to consider a very simple bonding problem, the formation of a hydrogen molecule from two hydrogen atoms. The principles will be the same as for larger molecules, but the comparison between the approaches will be more apparent in the case of H_2.

The discussion that follows has been adapted from several introductory texts on bonding, and these works should be consulted for more details.[103] We begin with two isolated hydrogen atoms, as shown in Figure 1.7. Each atom has one electron in a 1*s* orbital. We can write a wave equation for the 1*s* orbital, since the hydrogen atom can be solved exactly in quantum mechanics. Electron 1 is initially associated with hydrogen **a**, and electron 2 is associated with hydrogen **b**. Bringing the two atoms together allows bonding to occur, as shown in Figure 1.7.

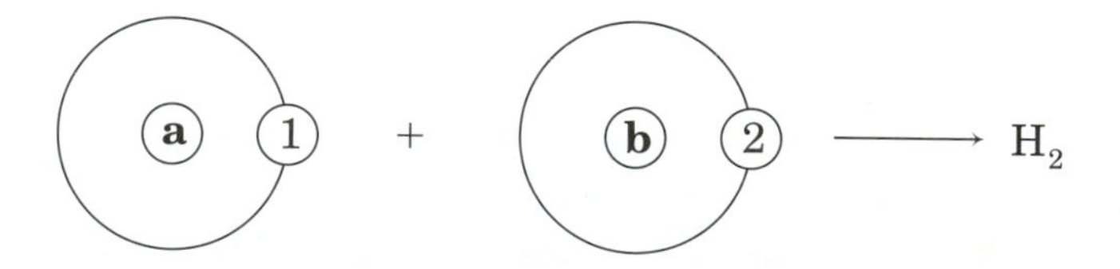

Figure 1.7 Formation of a hydrogen molecule from two hydrogen atoms.

[100]A summary of the development of VB theory and a discussion of the merits of VB and MO theories has been given by Klein, D. J.; Trinajstić, N. *J. Chem. Educ.* **1990**, *67*, 633.

[101]Reference 73, p. 108.

[102]It is possible to combine the ideas of the two theories. See Jug, K.; Epiotis, N. D.; Buss, S. *J. Am. Chem. Soc.* **1986**, *108*, 3640 and references therein, particularly references 4 and 5.

[103]For examples, see (a) reference 10, pp. 517 *ff*; (b) Eyring, H.; Walter, J.; Kimball, G. E. *Quantum Chemistry*; John Wiley & Sons, Inc.: New York, 1944; pp. 212 *ff.* (c) The mathematics is described in some detail in Slater, J. C. *Quantum Theory of Molecules and Solids*, Vol. I. *Electronic Structure of Molecules*; McGraw-Hill: New York, 1963.

Now we want to write a wave equation that will mathematically describe the electron distribution in the hydrogen molecule. The valence bond method initially used by Heitler and London[104] described one possible wave function as being

$$\psi_1 = c\, a(1)b(2) \tag{1.23}$$

in which c is a constant, $a(1)$ is the wave function for electron 1 in a 1s orbital on hydrogen a, and $b(2)$ is the wave function for electron 2 in a 1s orbital on hydrogen b. However, since the electrons are indistinguishable, it should be equally acceptable to write

$$\psi_2 = c\, a(2)b(1) \tag{1.24}$$

Both descriptions are possible and need to be included in the wave function for the molecule. Therefore, Heitler and London wrote that

$$\psi_{VB} = c\, a(1)b(2) + c\, a(2)b(1) \tag{1.25}$$

In this case the constants are chosen so that the overall wave function is properly normalized and made antisymmetric with respect to spin.

The molecular orbital approach[103] to describing hydrogen also starts with two hydrogen nuclei (a and b) and two electrons (1 and 2), but we make no initial presumption about the location of the two electrons. We solve the Hamiltonian for the molecular orbital around the pair of nuclei, and we can then write a wave equation for one electron in the resulting MO.

$$\psi_1 = c_1\, a(1) + c_2\, b(1) \tag{1.26}$$

Note that electron 1 is associated with both nuclei. Similarly,

$$\psi_2 = c_1\, a(2) + c_2\, b(2) \tag{1.27}$$

The combined MO wave function, then, is the product of the two one-electron wave functions:

$$\psi_{MO} = \psi_1\psi_2 = c_1^2\, a(1)a(2) + c_2^2\, b(1)b(2) + c_1c_2\, [a(1)b(2) + a(2)b(1)] \tag{1.28}$$

We see that ψ_{MO} is an apparently more complex wave function than ψ_{VB}. In fact, ψ_{VB} is incorporated into ψ_{MO}; specifically, the third term of ψ_{MO} is the same as ψ_{VB} if the constants are made the same. What is the physical significance of the differences in ψ_{VB} and ψ_{MO}? ψ_{MO} includes two terms that ψ_{VB} does not: $a(1)a(2)$ and $b(1)b(2)$. Each of these terms represents a **configuration** (arrangement of electrons in orbitals) in which *both* electrons are formally localized in what had been a 1s orbital on *one* of the hydrogen atoms, i.e., ionic structures. In other words, $a(1)a(2)$ represents $a{:}^-\ b^+$ and $b(1)b(2)$ represents $a^+\ b{:}^-$. The MO treatment appears to give large weight to terms that represent electronic configurations in which both electrons are on one nucleus, while the VB treatment ignores these terms.

Which approach is correct? Usually our measure of "correctness" of any calculation of a molecular structure is how accurately the calculation repro-

---

[104] Heitler, W.; London, F. *Z. Physik* **1927**, *44*, 455.

duces a known physical property of the molecule. In the case of hydrogen, a relevant property is the bond dissociation energy, which is the energy for the reaction $H_2 \rightarrow H\cdot + H\cdot$. The simple VB calculation described here gives a value of 3.14 eV (72.4 kcal/mol) for H_2 dissociation.[105] The simple MO calculation gives a value of 2.70 eV (62.3 kcal/mol). The experimental value is 4.75 eV (109.5 kcal/mol).[106] Obviously, neither calculation is "correct," unless one takes order of magnitude agreement as satisfactory; in that case, both calculations are correct.

It may appear that the MO method gives a result that underestimates the bond dissociation energy because our wave equation includes patterns of electron density that resemble ionic species such as a^+b^-. But why is the VB result also in error? The answer seems to be that while the MO approach places too much emphasis on these ionic electron distributions, the VB approach underutilizes them. Apparently, a strong bond requires that both electrons spend a lot of time in the area of space between the two protons. Doing so must make it more likely that the two electrons will, at some instant, be on the same atom.[107] Thus we can improve the accuracy of our VB calculation if we add some terms that keep the electrons closer together between the nuclei.

Similarly, we could improve our MO calculation if we added some terms that would keep the electrons out of each other's way. Electrons repel each other, so any excessive amount of ionic character will decrease the calculated stability of the molecule. We separate the electrons by changing the atomic orbitals we use to write ψ_{MO}. If we include a description of the hydrogen atoms in which the electron in each case has some probability of being in an orbital higher than the 1*s* orbital, then in the final wave function the electrons will be calculated to have more "room to maneuver," and excessively repulsive terms will become less important.[108,109] There are other things we can do to improve our calculation, although we will not detail them here. However, Table 1.9 shows how the calculated stability of H_2 varies according to the complexity of the MO calculation.[110,111] Including 13 terms makes a major improvement. From that point on almost any change

[105]Reference 103(b), p. 214 and references therein.

[106]Reference 103(b), p. 212.

[107]Colloquial terminology is used here to be consistent with the level of presentation of the theory.

[108]This procedure is called *configuration interaction*; *cf.* Coffey, P.; Jug, K. *J. Chem. Educ.* **1974**, *51*, 252. See also Coulson, C. A.; Fischer, I. *Phil. Mag.* **1949**, *40*, 386.

[109]This argument is an anthropomorphism, since it ascribes human characteristics to inanimate objects. Such arguments are implicit conceptual models, with the advantages and disadvantages of all other models.

[110](a) Data from McWeeny, R. *Coulson's Valence*, 3rd ed.; Oxford University Press: Oxford, England, 1979; p. 120 and references therein. (b) See also Davis, Jr., J. C. *Advanced Physical Chemistry*; Ronald Press: New York, 1965; p. 426.

[111]For additional references, see King, G. W. *Spectroscopy and Molecular Structure*; Holt, Rinehart and Winston, Inc.: New York, 1964; p. 149.

Table 1.9 Calculated values for H_2 stability.

Calculation Method	DE (calc.)
Simple Valence Bond Theory	3.14 eV
Simple Molecular Orbital Theory	2.70 eV
James-Cooledge (13 parameters)	4.72 eV
Kolos-Wolniewicz (100 terms)	4.7467 eV
Experimental Value	4.7467 eV

decreases the stability of the calculated structure almost as much as it increases it, but small gains can be won. An MO equation with 50 terms does quite well. Similarly, a VB calculation with a large number of terms can also produce an answer that is within experimental error of the measured value. If enough terms are included, the two methods converge and produce equivalent results.[112]

There are several conclusions to be drawn from our analysis:

1. Neither simple VB theory nor simple MO theory produces a value for the dissociation energy that is close to the experimental value.
2. Both VB and MO theories can be modified to produce more accurate results. However, even for a simple molecule such as molecular hydrogen, many terms may be required to produce an acceptable value for a particular molecular property.

More important for our purposes here are the following two conclusions:

3. As both VB and MO theory are modified to become more like reality, they must necessarily produce more nearly equivalent results. In that sense they must become more like each other, and the modifications may make their theoretical bases more nearly equivalent as well.
4. Both VB and MO theories should be regarded only as approaches for the calculation of molecular properties, not as final answers. They represent complementary initial models for computational chemistry.

Why then don't we just talk about high level theoretical calculations and ignore the simple theory? Both elementary theories are useful to us because they provide good conceptual models for the computational process. We can visualize the interactions represented by equation 1.28, as well as the physical situation suggested by the VB equation 1.25. However, it is much more difficult for us to envision the interactions involved in a 50- or 100-term wave function. As the accuracy of the model is increased, its simplicity is decreased. We must choose the model that is sufficiently accurate for our computational purposes, yet still simple enough that we have some

[112]Reference 73, p. 147. See also Kolos, W.; Wolniewicz, L. *J. Chem. Phys.* **1968**, *49*, 404.

understanding of what the model describes. Otherwise, the model is a black box, and we have no understanding of what it does, perhaps even no idea whether the answers it produces are physically reasonable.

Atomic Energy Levels and Bonding Concepts

The description of MO and VB theories presented on page 24 represents an important class of chemistry models, the mathematical model. However, chemists frequently use other models that represent the results of mathematical calculations in schematic or pictorial form. Thus we represent the bonding interactions in molecular hydrogen by diagrams as shown in Figure 1.8. There we represent the combination of two atomic hydrogen 1*s* orbitals to make two new orbitals, called molecular orbitals—regions of space around more than one atom that encompass a certain probability of finding an electron. Just as 1*s* and 2*p* atomic orbitals have three-dimensional spatial characteristics, these molecular orbitals have three-dimensional characteristics as well, and we are familiar with the shapes and shading patterns of the σ and σ* orbitals. The arrow on the left in Figure 1.8 indicates that there is a difference in energy, the σ MO being lower in energy than the original hydrogen 1*s* orbitals, and the σ* orbital being higher in en-

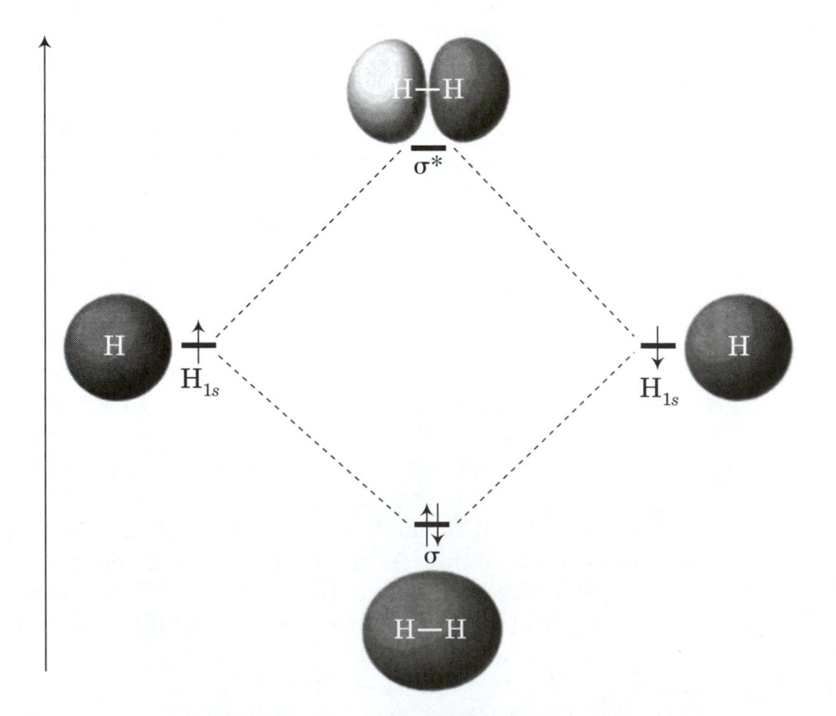

Figure 1.8 Combination of atomic hydrogen orbitals to produce molecular hydrogen.

ergy.[113] The σ orbital is a bonding orbital, and population of the σ MO with two electrons produces a stable H_2 molecule. On the other hand, σ* is an antibonding orbital, and population of this orbital with an electron destabilizes the molecule.

Let us extend this pictorial representation to one of the fundamental questions in organic chemistry—the bonding of methane. We usually begin with one carbon atom and four hydrogen atoms. We represent atomic carbon by an arrangement of horizontal lines, a large vertical arrow, and small arrows that we call an atomic energy level diagram (Figure 1.9). The horizontal lines indicate the energies associated with particular atomic orbitals. We designate the lines with combinations of letters and numbers, such as 1*s* or 2*p*. Each line represents the energy associated with a particular atomic wave function calculated from quantum mechanics. The small arrows represent electrons, the direction of the arrow indicating the spin of each. We know that we may put no more than two arrows on each horizontal line, and then only if the arrows point in opposite directions, meaning two electrons with paired spins in an atomic orbital.

The sp³-Hybridization Model for Methane

If we apply the energy level diagram for carbon (Figure 1.9) with the bonding model shown schematically in Figure 1.8, we can construct a representation of a one-carbon hydrocarbon in which each of the unpaired electrons on the carbon atom is paired with an electron from a hydrogen atom to form a C—H bond (Figure 1.10). Even though there is abundant experimental

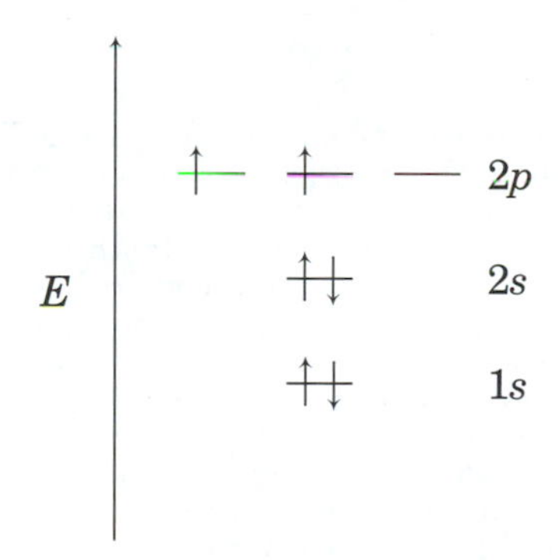

Figure 1.9 Energy levels of carbon atomic orbitals.

[113]This figure is somewhat stylized, since it shows the σ and σ* orbitals symmetrically displaced above and below the energy of the initial hydrogen 1*s* orbitals. In fact, the σ* should be more antibonding than the σ orbital is bonding. For details, see Albright, T. A.; Burdett, J. K.; Whangbo, M.-H. *Orbital Interactions in Chemistry*; Wiley-Interscience: New York, 1985; pp. 12 *ff*. In addition, Willis has calculated that the H-H bond energy is only about 19% of the energy gained by moving two electrons from hydrogen 1*s* orbitals to the H_2 σ bonding orbital. The other 81% of that energy is consumed in offsetting electrical repulsion within the molecule. *Cf.* Willis, C. J. *J. Chem. Educ.* **1988**, *65*, 418.

Figure 1.10 Electron dot structure of "methane" as CH_2.

$$H:\ddot{C}:H$$

data which indicate that methane has the molecular formula CH_4, not CH_2, our simple bonding model does not appear to accommodate such a formula.

Once a model is fixed in our minds, we find it almost impossible to discard or ignore it unless another model is available to take its place. Instead models are almost always modified to fit new data—and that is what we must do here. We use the concept of **hybridization** to change our mental picture of the atomic orbitals of carbon to a more useful one. Figure 1.11 shows how we combine the carbon orbitals to produce four sp^3-hybrid orbitals that are equal in energy.[114,115,116] The energy needed to promote an electron to a higher energy level (about 96 kcal/mol) is more than offset by the energy gained by forming two new C — H bonds (about −200 kcal/mol).[117] The procedure was described by Pauling[118] and has been dis-

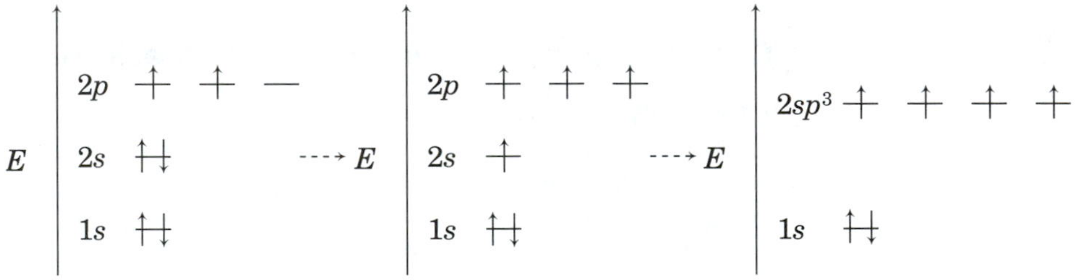

Figure 1.11 Hybridization of carbon orbitals to produce sp^3-hybridized orbitals on carbon.

E	$2p$ ↑ ↑ —	---→ E	$2p$ ↑ ↑ ↑	---→ E	$2sp^3$ ↑ ↑ ↑ ↑
	$2s$ ↑↓		$2s$ ↑		
	$1s$ ↑↓		$1s$ ↑↓		$1s$ ↑↓

[114]The meaning is clearly that **we** hybridize the model to produce a different model. Anyone who believes that atomic orbitals actually hybridize as shown in Figure 1.11 should reread the prologue quite carefully. As Ogilvie (reference 115) has noted "According to *Coulson's Valence*, 'hybridization is not a physical effect but merely a feature of [a] theoretical description' . . . Despite the fact that many authors of textbooks of general chemistry have written that CH_4 has a tetrahedral structure because of sp^3 hybridization, there neither exists now, nor has ever existed, any quantitative experimental or theoretical justification of such a statement." (For a rebuttal of this view, however, see Pauling, L. *J. Chem. Educ.* **1992**, *69*, 519.) Matteson has noted that "Hybridization is not something that atoms do or have done to them. It is purely a mental process gone through by the chemist, who wants to group atomic orbitals according to their symmetry properties so he can talk about one localized bond and ignore the rest. Hybridization does not change the shape of the electron distribution in any atom." Matteson, D. S. *Organometallic Reaction Mechanisms of the Nontransition Elements*; Academic Press: New York, 1974; p. 5.

[115]Ogilvie, J. F. *J. Chem. Educ.* **1990**, *67*, 280.

[116]Note that there are *hybridized orbitals* but not *hybridized atoms*. However, organic chemists frequently use the term "sp^3-hybridized carbon" to refer to a carbon atom with sp^3 hybrid orbitals.

[117]See the discussion in Hameka, H. F. *Quantum Theory of the Chemical Bond*; Hafner Press: New York, 1975; pp. 216 *ff*.

[118](a) Pauling, L. *J. Am. Chem. Soc.* **1931**, *53*, 1367; reference 23, p. 118. (b) See also Slater, J. C. *Phys. Rev.* **1931**, *37*, 481.

cussed by a number of authors.[119,120,121] The wave functions of the hybrid orbitals (using the notation of Bernett) are shown in equations 1.29–1.32.[122]

$$\phi_{sp^3_{(1)}} = \tfrac{1}{2}(C_{2s} + C_{2p_x} + C_{2p_y} + C_{2p_z}) \quad \textbf{(1.29)}$$

$$\phi_{sp^3_{(2)}} = \tfrac{1}{2}(C_{2s} + C_{2p_x} + C_{2p_y} + C_{2p_z}) \quad \textbf{(1.30)}$$

$$\phi_{sp^3_{(3)}} = \tfrac{1}{2}(C_{2s} - C_{2p_x} + C_{2p_y} - C_{2p_z}) \quad \textbf{(1.31)}$$

$$\phi_{sp^3_{(4)}} = \tfrac{1}{2}(C_{2s} - C_{2p_x} + C_{2p_y} + C_{2p_z}) \quad \textbf{(1.32)}$$

Each of the sp^3 orbitals in equations 1.29–1.32 has a large lobe and a small lobe. The two lobes have a different mathematical sign, and the four large lobes point toward the corners of a regular tetrahedron. Figure 1.12 shows the contours of a carbon 2*s* orbital, a carbon 2*p* orbital, and an *sp*, sp^2 and sp^3 hybrid orbital.[123] The shapes of these hybrids are quite different from the representations shown in many chemistry textbooks.[124]

Now we can create methane as CH_4 by combining each carbon sp^3-hybrid orbital with a 1*s* orbital on hydrogen. A molecular orbital diagram for the process is shown in Figure 1.13. The four sp^3-hybrid orbitals point to the corners of a tetrahedron, so each can bond with one hydrogen atom, making

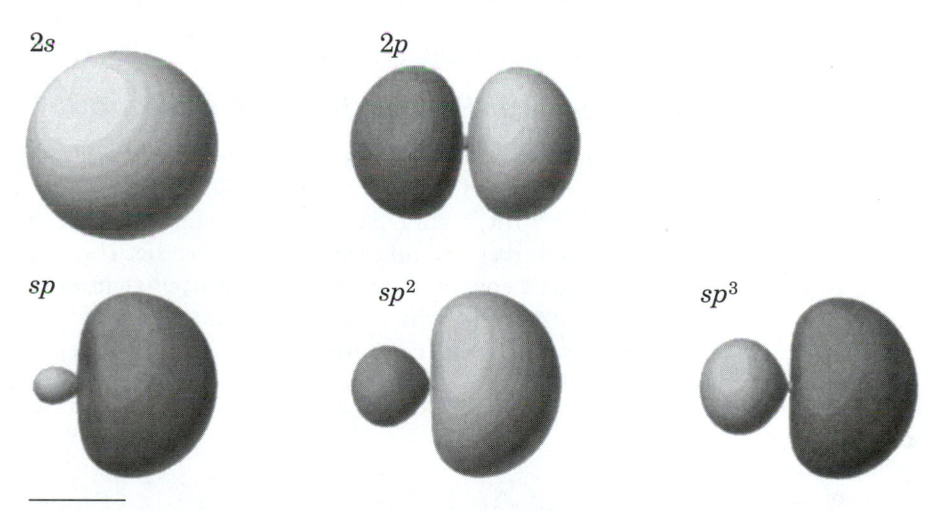

Figure 1.12 Sizes and shapes of carbon atomic and hybrid orbitals.

---

[119] *Cf.* Hsu, C.-Y.; Orchin, M. *J. Chem. Educ.* **1973**, *50*, 114.

[120] Root, D. M.; Landis, C. R.; Cleveland, T. *J. Am. Chem. Soc.* **1993**, *115*, 4201.

[121] For a reconsideration of the concept of hybridization, see Magnusson, E. *J. Am. Chem. Soc.* **1984**, *106*, 1177.

[122] Bernett, W. A. *J. Chem. Educ.* **1969**, *46*, 746. In this notation, C_{2s} represents the wave function for a 2*s* orbital on carbon.

[123] The orbital contours were generated with CAChe™ visualization software.

[124] A discussion of the shapes of hybrid atomic orbitals has been given by Allendoerfer, R. D. *J. Chem. Educ.* **1990**, *67*, 37.

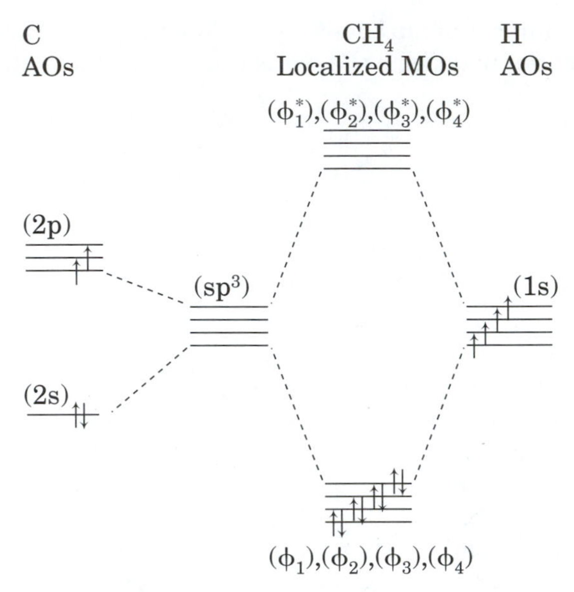

Figure 1.13 Mixing of hydrogen 1*s* and carbon *sp*^{3} MOs to make orbitals of methane. (Adapted from reference 135.)

methane a CH_4 molecule with tetrahedral geometry (bond angles of 109.5°).[125] That prediction is consistent with the experimental data shown in Figure 1.14.[126]

Are There *sp*^{3}-Hybrid Orbitals in Methane?

The hybridization concept is so ingrained in organic chemistry that we often use the concepts of hybridization (*sp*^{3}) and geometry (tetrahedral) interchangeably.[120] Yet, a valid conceptual model must not only explain currently known data, but it should also correctly predict the results of future experiments. We will now discuss one important experimental technique that will cause us to rethink what we have said about methane. The technique is **photoelectron spectroscopy (PES)**, which is used to probe the energy

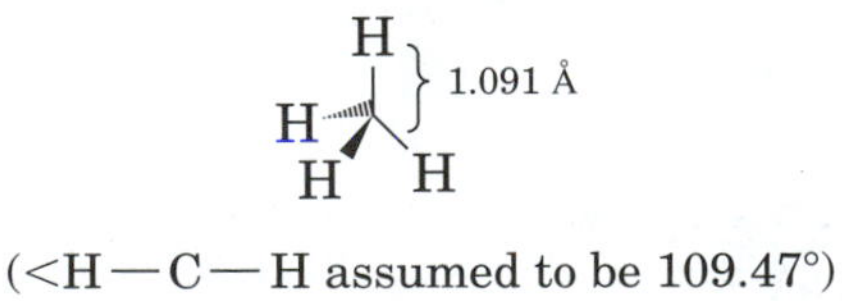

Figure 1.14 Experimental geometry of methane.

1.091 Å

(<H—C—H assumed to be 109.47°)

---

[125]We usually write the bond angle as 109.5°. Mathematically, it is 109°28′, which corresponds to 109.47°. A method for the calculation of *sp*, *sp*^{2}, and *sp*^{3} interorbital angles has been given by Duffey, G. H. *J. Chem. Educ.* **1992**, *69*, 171.

[126]Data from reference 22 (a), p. M113.

levels electrons can occupy within molecules.[127,128] The essence of PES is the measurement of the energies of electrons which have been ejected from molecules (or atoms) by high energy photons (light). As shown in Figure 1.15, the difference between the energies of the electrons (energy out) and the energy of the photons causing the displacement (energy in) is taken to be a measure of the *binding energy* holding the electrons in the molecule or atom. Thus, the higher the energy level from which an electron is removed, the less is its binding energy, and the greater will be its kinetic energy. This relationship is shown in equation 1.33, where $h\nu$ is the energy of the photon, T is the kinetic energy of the electron ejected from the molecule, and E_B is the binding energy of the electron in the molecule. Based on Koopmans' Theorem,[129] we associate the number and position of binding energy levels in a structure with the energies of its atomic or molecular orbitals.

$$h\nu = T + E_B \qquad \textbf{(1.33)}$$

Figure 1.16 shows a PES spectrum of methane.[130] One peak at very high binding energy (>290 eV) is characteristic of molecules with electrons in carbon 1*s* orbitals. However, there are two peaks at lower energy: one at 23.0 eV and one at 12.7 eV. Therefore we are led to the conclusion that the electrons in methane are in three different energy levels, one energy level corresponding to the carbon 1*s* electrons, and two different energy levels corresponding to the other electrons in the molecule. It is somewhat difficult to reconcile this experimental result with an intuitive bonding model in which methane is constructed of four equivalent C(sp^3)-H(1*s*) bonds pro-

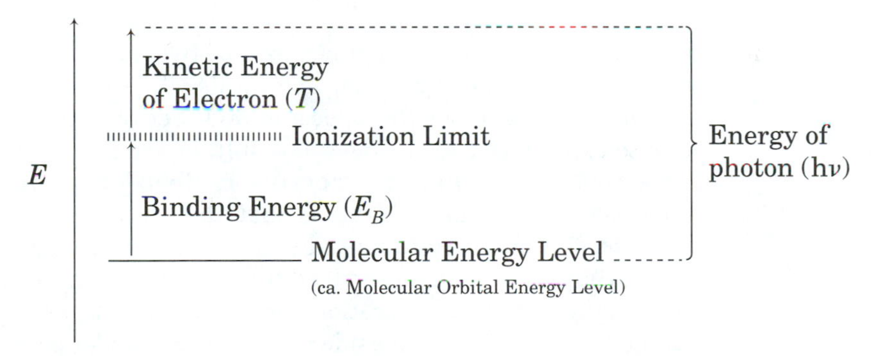

Figure 1.15 Energy relationships in photoelectron spectroscopy.

---

[127]A related procedure is called **ESCA** for **e**lectron **s**pectroscopy for **c**hemical **a**nalysis. For a discussion of PES, ESCA and similar techniques, see Baker, A. D.; Brundle, C. R.; Thompson, M. *Chem. Soc. Rev.* **1972**, *1*, 355; Baker, A. D. *Acc. Chem. Res.* **1970**, *3*, 17. Another technique, Auger spectroscopy, is discussed in reference 129.

[128]Bock, H.; Mollère, P. D. *J. Chem. Educ.* **1974**, *51*, 506; Baker, A. D. *Acc. Chem. Res.* **1970**, *3*, 17; Ballard, R. E. *Photoelectron Spectroscopy and Molecular Orbital Theory*; John Wiley & Sons: New York, 1978.

[129]For a discussion of the application of Koopmans' Theorem in PES, see Albridge, R. G. in *Physical Methods of Chemistry*, Vol. I, Part IIID; Weissberger, A.; Rossiter, B. W., eds.; Wiley-Interscience: New York, 1972; p. 307.

[130]Hamrin, K.; Johansson, G.; Gelius, U.; Fahlman, A.; Nordling, C.; Siegbahn, K. *Chem. Phys. Lett.* **1968**, *1*, 613.

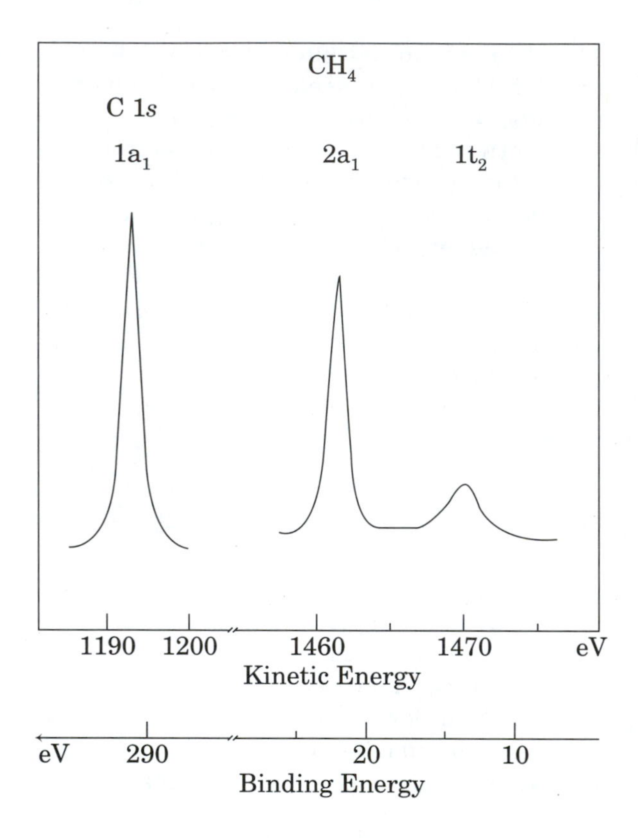

Figure 1.16 PES spectrum of methane. (Adapted from reference 130.)

duced by overlap of four equivalent sp^3-hybrid orbitals on a carbon atom with four equivalent hydrogen 1*s* orbitals.[131]

There is one important aspect of MO theory that not only makes it possible to explain this experimental result, but in fact requires it. That aspect is the concept of **symmetry correct molecular orbitals**. A fundamental property of molecular orbitals is that they have the full symmetry of the basis set of atomic orbitals used to generate the molecular orbitals.[132] This means that the orbitals must be either *symmetric* or *antisymmetric* with respect to the symmetry operations provided for by the symmetry group of the atomic orbitals.[133] If we consider each C—H bond σ bond formed by overlap

[131]Pauling argued that the photoelectron spectrum of methane is consistent with sp^3 hybridization for methane: Pauling, L. *J. Chem. Educ.* **1992**, *69*, 519. See also the discussion by Simons, J. *J. Chem. Educ.* **1992**, *69*, 522.

[132]The term *basis set* refers to the set of atomic orbitals used to construct the molecular orbitals.

[133]The terms *symmetric* and *antisymmetric* mean that the result of any symmetry operation will be an orbital of the same type and in the same location as before the transformation. If the orbital is symmetric with respect to that transformation, then the orbital produced will also have + and − lobes in the same locations as before; if the orbital is antisymmetric, then the resulting orbital will have + lobes where − lobes were, and vice versa. All MOs must be either symmetric or antisymmetric. If a symmetry operation that corresponds to an element of symmetry of the basis set of atomic orbitals transforms a lobe to a position in space in which there was not a *p* orbital lobe, that MO is said to be **asymmetric** (without symmetry) and is not allowable as a symmetry-correct MO for the molecule. This topic is discussed in more detail in Chapter 11.

of an sp^3 orbital on a carbon atom with a 1*s* orbital on a hydrogen atom to be an MO, then clearly each of these MOs lacks the full symmetry of the basis set of *s* and *p* atomic orbitals.

There are two ways to correct our treatment of methane. The first approach is to consider the descriptions of C—H bonds to be **localized molecular orbitals (LMOs)**, that is molecular orbitals localized onto a portion of a molecule. We can then consider these LMOs to be the *basis set* for a new MO calculation to determine the symmetry correct, **delocalized MOs** for the molecule.[134,135,136] The second approach is to calculate delocalized methane molecular orbitals directly from the unhybridized orbitals: a carbon 2*s* orbital, three carbon 2*p* orbitals, and four hydrogen 1*s* orbitals. Both procedures produce four delocalized molecular orbitals, each of which has the full symmetry of the original basis set of tetrahedral methane.[137]

The molecular orbitals for methane are listed in equations 1.34–1.37, where H_1 is the hydrogen 1*s* orbital on hydrogen 1, etc.[122] Figure 1.17 shows the MO diagram and Figure 1.18 shows a calculated 3-D electron contour

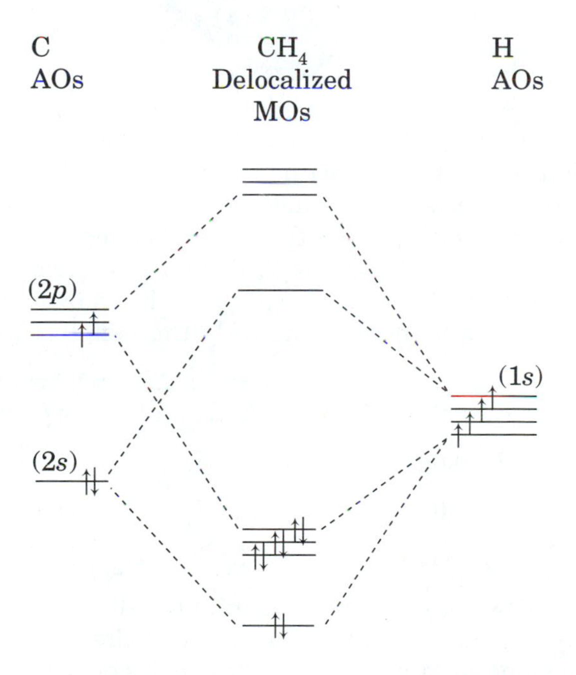

Figure 1.17 Mixing of atomic orbitals on carbon with hydrogen 1*s* orbitals to make orbitals of methane. (Adapted from reference 135.)

[134] Reference 122; Flurry, Jr., R. L. *J. Chem. Educ.* **1976**, *53*, 554; Cohen, I.; Del Bene, J. *J. Chem. Educ.* **1969**, *46*, 487.

[135] Hoffman, D. K.; Ruedenberg, K.; Verkade, J. G. *J. Chem. Educ.* **1977**, *54*, 590.

[136] Dewar, M. J. S.; Dougherty, R. C. *The PMO Theory of Organic Chemistry*; Plenum Press: New York, 1975; pp. 21 *ff.*

[137] Note that the four molecular orbitals are not tetrahedral in shape. Rather, each is either symmetric or antisymmetric with respect to the symmetry operations of the T_d point group.

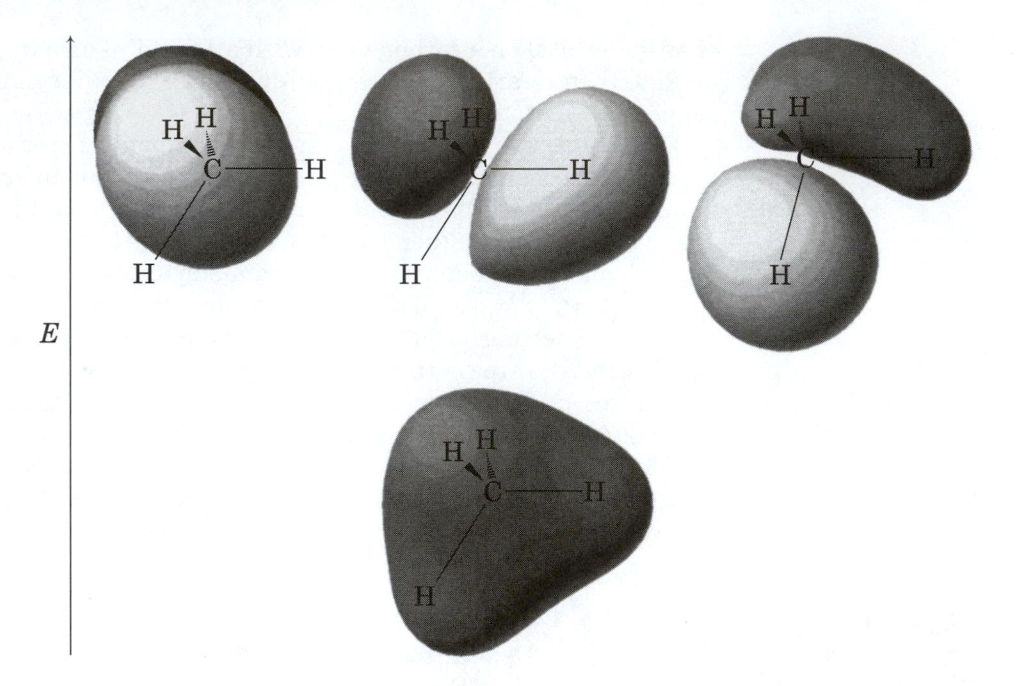

Figure 1.18
Bonding MOs for methane.

plot for the orbitals.[138] We see that one MO (ϕ_1) represents a bonding interaction of the carbon 2*s* orbital with all four of the hydrogen 1*s* orbitals, while the other orbitals have some bonding and some antibonding interactions. Therefore, ϕ_1 is lower in energy than the other three, giving rise to two different energy levels (and thus two PES bands) for the bonding electrons. We conclude, therefore, that the concept of sp^3 hybridization, while useful for predicting geometries, does not provide the simplest explanation for the PES results.

$$\phi_1 = 0.545\mathrm{C}_{2s} + 0.272\,(\mathrm{H}_1 + \mathrm{H}_2 + \mathrm{H}_3 + \mathrm{H}_4) \qquad \textbf{(1.34)}$$

$$\phi_2 = 0.545\mathrm{C}_{2p_x} + 0.272\,(\mathrm{H}_1 + \mathrm{H}_2 - \mathrm{H}_3 - \mathrm{H}_4) \qquad \textbf{(1.35)}$$

$$\phi_3 = 0.545\mathrm{C}_{2p_y} + 0.272\,(\mathrm{H}_1 - \mathrm{H}_2 + \mathrm{H}_3 - \mathrm{H}_4) \qquad \textbf{(1.36)}$$

$$\phi_4 = 0.545\mathrm{C}_{2p_z} + 0.272\,(\mathrm{H}_1 - \mathrm{H}_2 - \mathrm{H}_3 + \mathrm{H}_4) \qquad \textbf{(1.37)}$$

We must now address a fundamental question. Are there C—H bonds in methane? The answer from MO theory is clearly no. Population of the four bonding molecular orbitals with four pairs of electrons leads to a bonding interaction among (not just between) the carbon atom and all of the hydrogen atoms. Thus, we should say that in MO theory, there is *bonding*, but there are not *bonds*—at least not in the sense of discrete electron pairs localized between pairs of atoms.

---

[138]Graphics generated by the author by semi-empirical MO calculations using a CAChe™ WorkSystem. For a more complete discussion of the molecular orbitals of methane, see Jorgensen, W. L.; Salem, L. *The Organic Chemist's Book of Orbitals*; Academic Press: New York, 1973; p. 68.

Valence Shell Electron Pair Repulsion Theory

We could have predicted the tetrahedral shape of methane quite simply by using **valence shell electron pair repulsion (VSEPR) theory**.[139,140,141] The VSEPR method does not require the use of atomic or molecular orbitals; it is simply a mathematical calculation of a points on a sphere problem in which the mutually repulsive points are arranged on the sphere as far apart from each other as possible. (However, the model may be more intuitively useful if we visualize the electron points as three-dimensional objects with shapes similar to those calculated for hybrid atomic orbitals.)[142] In cases such as the ethene molecule, it is useful to think of two of the electron pair points merging to form one point with correspondingly greater repulsion for other electron pairs. The points corresponding to nonbonded electron pairs can be thought of as being more repulsive than bonded electron pairs, while pairs of electrons used for bonding to electronegative groups (and thus pulled away from the central atom) can be considered less repulsive toward other electron pair points.[140] To use the VSEPR theory to predict the geometry of methane, we simply ask the following question: What is the most stable arrangement for four pairs of electrons bonded to a central atom? The tetrahedral arrangement provides for maximum relief of the repulsion of the electrons in one bond for those in another, so we expect it to be the most stable.

The VSEPR approach can easily be extended to ethane. Since each carbon atom has four substituents and since the electronegativity of carbon is close to that of hydrogen, we would predict that the local geometry about each carbon atom would be the same as about methane except that the C—C bond would be longer than the C—H bonds because the covalent radius of a carbon atom is greater than that of a hydrogen atom. Based on our model of the bonding of methane, therefore, we can visualize ethane as also being sp^3-hybridized. The H—C—H bond angles at each carbon atom should remain 109.5°, as should the H—C—C bond angles. Our prediction is quite consistent with the geometry reported for ethane, as shown in Figure 1.19.[143]

Figure 1.19 Experimental geometry of ethane.

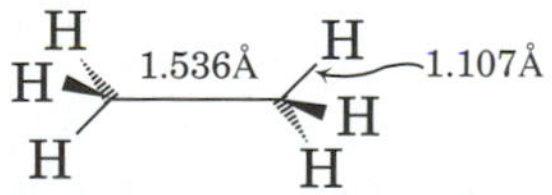

1.536Å 1.107Å

(<H—C—H = 109.3°)

---

[139]For a discussion of VSEPR, see Gillespie, R. J. *J. Chem. Educ.* **1963**, *40*, 295; Bent, H. A. *Chem. Rev.* **1961**, *61*, 275; Burdett, J. K. *Chem. Soc. Rev.* **1978**, *7*, 507. See also the monograph by Gillespie and Hargittai, reference 9.

[140]Gillespie, R. J. *Chem. Soc. Rev.* **1992**, *21*, 59.

[141]Hall, M. B. *J. Am. Chem. Soc.* **1978**, *100*, 6333.

[142]Gillespie, R. J. *J. Chem. Educ.* **1963**, *40*, 295.

[143]Reference 22 (a), p. M135.

Figure 1.20 (a) Qualitative prediction of CH_3Cl geometry. (b) Experimental data.

(a) Predicted

C—Cl 1.76 Å; C—H 1.09 Å

$<$H—C—H = ca. 109.5°

$<$H—C—Cl = ca. 109.5°

(b) Experimental

C—Cl 1.781 Å; C—H 1.096 Å

$<$H—C—H = 110°52'

$<$H—C—Cl = 108°0'

Now let us consider methyl chloride. The covalent radius of chlorine (Table 1.2) is about 0.22 Å greater than that of carbon, so the C—Cl bond distance should be about 1.76 Å. Thus we predict the molecular geometry of methyl chloride to be as shown in Figure 1.20(a). However, spectroscopic data suggest that the experimentally determined structure is like that shown in Figure 1.20(b).[144] While the C—Cl bond length is reasonable, the H—C—H bond angles are greater than 109.5°, and the H—C—Cl angle is smaller than 109.5°. Moreover, the C—H bond distances are shorter than in methane.

We can explain the difference between prediction and experiment with VSEPR theory by noting that the electronegativity of chlorine is much greater than that of carbon (Table 1.8). In a C—H bond in methane, the carbon atom and the hydrogen atom attract the electron pair approximately equally, so that there is no molecular electric dipole. In CH_3Cl, however, the electrons in the C—Cl bond will be *polarized* or pulled toward the chlorine atom and away from the carbon atom. In turn the carbon atom will pull electron density from the hydrogen atoms attached to it. As the electrons in the C—H bonds are pulled closer to the carbon nucleus, so are the hydrogen nuclei, and the C—H bond length gets shorter. Since electron density between the C—H and C—Cl bonds has been increased near the C nucleus, the repulsion between pairs of C—H bonds is now increased, at least by comparison with C—H bond repulsion by the C—Cl bond, so the H—C—H bond angle is expanded in comparison with methane. Therefore the H—C—Cl angle is less than the methane bond angle.

Variable Hybridization and Molecular Geometry

An alternative explanation of the geometry of methyl chloride is based on a modification of the concept of hybridized atomic orbitals. Because methyl chloride is less symmetric than is methane and because chlorine is more electronegative than carbon, we envision the hybrid orbitals on carbon to be

[144]Reference 22 (b), p. M 63s. See also Wiberg, K. B. *J. Am. Chem. Soc.* **1979**, *101*, 1718 and references cited therein.

different in hybridization and, consequently, in energy.[145] Since the electrons in the carbon orbital used to bond to the chlorine will be pulled away from the carbon atom, they will best be described as being in a hybrid orbital having less *s* character and more *p* character than a normal sp^3 hybrid.[146,147,148,149] This hybridization change will produce two important effects. The first is that the C—Cl bond will be longer than a C—H bond because the carbon atom uses an orbital that has more *p* character, that is, character of an orbital further removed from the carbon nucleus. The second effect is that the greater *s* character in the three C—H bonds will change their hybridization to something in which the bond angles are expanded from 109.5° to 110.5°.

The variable hybridization concept provides a quantitative description of bonding that VSEPR theory does not. If a hybrid orbital is described as an sp^n hybrid, where n is the hybridization index, then we may define a hybridization parameter, λ, such that $\lambda^2 = n$.[150,151] Then the fractional *s*-character of the ith orbital is given by the expression $1/(1 + \lambda_i^2)$. For example, for an sp^3-hybrid orbital, the fraction of *s* character is $1/(1 + 3) = 1/4$. Since the total *s*-character of *all* the hybrid orbitals must sum to 1, equation 1.38 holds.

$$\sum_i \frac{1}{1+\lambda_i^2} = 1 \qquad \textbf{(1.38)}$$

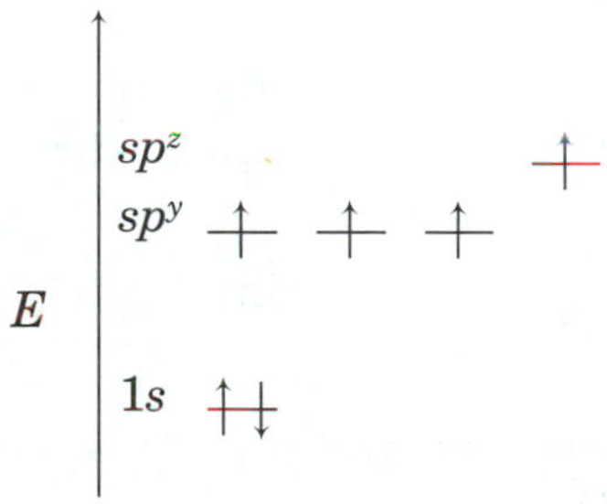

Figure 1.21
Hybridization of carbon proposed for CH_3Cl.

[145]For further reading, see (a) Breslow, R. *Organic Reaction Mechanisms*, 2nd ed.; W. A. Benjamin, Inc.: Menlo Park, CA, 1969; p. 3; (b) Coulson, C. A.; Stewart, E. T. in *The Chemistry of Alkenes*, Vol. 1; Patai, S., ed.; Wiley-Interscience: London, 1964; pp. 98 *ff.*

[146]The fundamental principle is that the more *s* character in a carbon orbital, the more electronegative is that orbital: Walsh, A. D. *Discuss. Faraday Soc.* **1947**, *2*, 18; *J. Chem. Soc.* **1948**, 398.

[147]Bent, H. A. *Chem. Rev.* **1961**, *61*, 275.

[148]Atomic *s* orbitals are lower in energy than are atomic *p* orbitals with the same principal quantum number.

[149]The lower energy of *s* orbitals results because their average distance from the nucleus is less than is the average distance to the nucleus for *p* orbitals of the same principal quantum number. See reference 118(a).

[150]Coulson, C. A. *J. Chem. Soc.* **1955**, 2069.

[151]For further discussion and examples of variable hybridization and orbital angle calculations, see (a) Mislow, K. *Introduction to Stereochemistry*; W. A. Benjamin, Inc.: New York, 1966; pp. 13–23; (b) reference 110 (a), pp. 195 *ff.*

Similarly, the total p character of the hybrid orbitals must sum to the number of p orbitals involved in the hybridization. For an sp^3-hybridized carbon, equation 1.39 applies, where $\lambda_i^2/(1 + \lambda_i^2)$ is the fractional p-character of the ith hybrid orbital.

$$\sum_i \frac{\lambda_i^2}{(1+\lambda_i^2)} = 3 \tag{1.39}$$

This hybridization description can be related to molecular geometry. The *interorbital angle*[152] θ_{ab} between hybrid orbitals from the carbon atom to atom a and to atom b can be determined from equation 1.40. If atoms a and b are identical, then equation 1.41 applies.

$$1 + \lambda_a \lambda_b \cos \theta_{ab} = 0 \tag{1.40}$$

$$1 + \lambda_a^2 \cos \theta_{aa} = 0 \tag{1.41}$$

Note that these equations predict that the interorbital angle increases with greater s-character, which is consistent with the increase in bond angle from 109.5° to 120° to 180° as the hybridization changes from sp^3 to sp^2 to sp, respectively.

Let us consider the case of methyl chloride in some detail. For this monosubstituted methane, CAB_3, there are only two different bond angles, θ_{aa} and θ_{ab}. They are related by the expression[153]

$$3 \sin^2 \theta_{ab} = 2 * (1 - \cos \theta_{aa}) \tag{1.42}$$

so knowledge of only one bond angle is sufficient to calculate the other. If we know only that the H—C—Cl bond angle is 108°, we can calculate the H—C—H bond angle:

$$\cos \theta_{aa} = 1 - \frac{3 \sin^2 \theta_{ab}}{2} \tag{1.43}$$

Substituting and solving for θ_{aa}, we find that $\theta_{aa} = 110.5°$. The value of λ_a can now be calculated from the formula

$$\lambda_a^2 = -\frac{1}{\cos \theta_{aa}} = 2.86 \tag{1.44}$$

So the orbitals used to bond hydrogen to carbon in methyl chloride are $sp^{2.86}$. Since the total s character in all four orbitals to carbon must sum to 1.00, Equation 1.45 holds.

$$3\left[\frac{1}{1+2.86}\right] + \frac{1}{1+\lambda_b^2} = 1 \tag{1.45}$$

---

[152]For strained molecules such as those with small rings, the interorbital bond angle may not be the same as that of the internuclear bond angle. See the discussion of bent bonds on pages 42 and 45.

[153]If there are two sets of identical ligands, CA_2B_2, there will be three bond angles: θ_{aa}, θ_{ab}, and θ_{bb}. They are related by the formula $\cos \theta_{ab} = -\cos \frac{1}{2} \theta_{aa} \cos \frac{1}{2} \theta_{bb}$. For further details and examples, see reference 151(a).

There are three equivalent C — H bonds and one C — Cl bond, so $\lambda_b^2 = 3.50$, and the C — Cl bond uses an $sp^{3.50}$ hybrid orbital. We can rationalize this result by thinking that the more electronegative chlorine atom will pull electrons away from the carbon nucleus, so there is greater stability if the carbon orbital used for bonding to chlorine has more *p* character.[147]

The idea that tetravalent carbon uses sp^3 hybridization is so ingrained in organic chemistry that it may be surprising to find that only for carbon atoms with four identical substituents are the C — C — C bond angles 109.47°. For example, a survey of 3,431 X-ray crystallography measurements by Boese, Schleyer, and their co-workers revealed a range of C — C — C bond angles from 74.88° to 159.66°, with the mean being 113.5° ± 4.5°.[154] (Even in hydrocarbons, the angle is not necessarily 109.5°. The C — C — C angle of propane is 112.4°,[155] and the C — C — C angles in pentane, hexane, and heptane are similar.)[156] These researchers demonstrated that the mean angle C — C — X in a series of ethyl derivatives varies with the electronegativity of the group X, with compounds having a more electronegative X (e.g., F, OH) having smaller C — C — X angles than compounds with less electronegative X groups (e.g., Na, Li).[157]

Now let us consider CH_2Cl_2. Myers and Gwinn determined from the microwave spectra of isotopically substituted methylene chloride that the C — Cl distance was 1.772 Å and the C — H distance was 1.082 Å. The Cl — C — Cl angle was determined to be 111°47′, while the H — C — H angle was found to be 112°0′.[158] Let us first calculate the apparent hybridization of the C — Cl bonds. Using equation 1.44, we calculate λ_{Cl}^2 to be 2.69, so the C — Cl bonding uses a carbon orbital that is an $sp^{2.69}$ hybrid. If we then use an equation analogous to equation 1.45, we calculate that λ_H^2 should be 3.37. That corresponds to an H — C — H angle of 107°. The experimental value is 112°, however, which produces an apparent λ_H^2 of 2.67. Clearly there is an inconsistency between the experimental values and our expectations based on the principles of variable hybridization.

One solution to the problem is to reexamine our view of a covalent bond as a straight line between two atoms. That is, we must consider that the C — Cl or C — H bonds (or both) may actually be curved.[159] Figure 1.22 shows the proposed curved bond structure. We may retain the concept of variable hybridization if we define the **internuclear bond angle** as the

---

[154]Boese, R.; Bläser, D.; Niederprüm, N.; Nüsse, M.; Brett, W. A.; Schleyer, P. von R.; Bühl, M.; Hommes, N. J. R. van E. *Angew. Chem., Int. Ed. Engl.* **1992**, *31*, 314.

[155]Lide, Jr., D. R. *J. Chem. Phys.* **1960**, *3*, 1514.

[156]Bonham, R. A.; Bartell, L. S.; Kohl, D. A. *J. Am. Chem. Soc.* **1959**, *81*, 4765.

[157]Because the C — C — X bond angle also varied with rotation about the C — X bond, electronegativity was judged not to be the only determinant of bond angles. A role was also described for hyperconjugation, a concept that will be explored in later chapters.

[158]Myers, R. J.; Gwinn, W. D. *J. Chem. Phys.* **1952**, *20*, 1420.

[159]An alternative possibility is that the hybrid orbitals in CH_2Cl_2 involve some *d* orbital character, but Myers and Gwinn discounted this possibility because of the much higher energy of *d* orbitals. For further discussion, see reference 158.

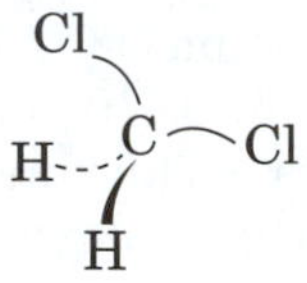

Figure 1.22 Curved bond representation of CH_2Cl_2.

angle measured by the shortest distance between pairs of nuclei and define the **interorbital bond angle** as the angle the hybrid orbitals make as they leave the carbon atom. If there is repulsion between pairs of chlorine atoms or between pairs of hydrogen atoms, or if there is attraction between hydrogens and chlorines, then the H—C—H and Cl—C—Cl internuclear angles could both be greater than 109°, but the interorbital angles could still be consistent with the concept of variable hybridization.

Because of the greater electronegativity of fluorine than chlorine, we might expect the C—F bond to utilize even more *p* character, thereby opening up the H—C—H bond angles and decreasing the H—C—F bond angle even more than the H—C—Cl angle in CH_3CCl. However, the H—C—F bond angle in CH_3F is found to be 108.9°.[160] Wiberg and co-workers have explained this apparent anomaly by suggesting that there is considerable curvature in the C—H bonds, as shown in Figure 1.23.[161] The H—C—F angle made by the C—F bond path and by one of the C—H bond paths *as they leave the carbon atom* is estimated to be 106.7° which is, as expected, smaller than the H—C—Cl bond angle in methyl chloride.[162,163]

The idea of curved bonds in methylene chloride and methyl fluoride may seem unfamiliar, but this explanation has long been invoked to describe the bonding in cyclopropane.[164] The experimental values for the C—C and C—H bond lengths are 1.510 Å and 1.089 Å, respectively, and the H—C—H angle was measured to be 115.1°.[165] The hybridization of the or-

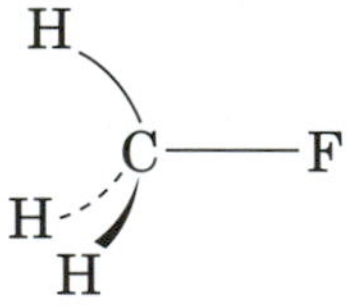

Figure 1.23 Curved bond paths suggested for CH_3F. (Reproduced from reference 162.)

---

[160]Clark, W. W.; De Lucia, F. C. *J. Mol. Struct.* **1976**, *32*, 29.

[161]The curved bond line follows the path of maximum electronic charge density from one atom to another and is known as a **bond path**. See Runtz, G. R.; Bader, R. F. W.; Messer, R. R. *Can. J. Chem.* **1977**, *55*, 3040. For a discussion, see Krug, J. P.; Popelier, P. L. A.; Bader, R. F. W. *J. Phys. Chem.* **1992**, *96*, 7604.

[162]Wiberg, K. B.; Hadad, C. M.; Breneman, C. M.; Laidig, K. E.; Murcko, M. A.; LePage, T. J. *Science* **1992**, *252*, 1266.

[163]For a discussion of theories concerning the stability and geometry of carbon atoms with two, three or four fluorine substituents, see Wiberg, K. B.; Rablen, P. R. *J. Am. Chem. Soc.* **1993**, *115*, 614.

[164]For a leading reference to some theoretical treatments for the bonding in cyclopropane, see Hamilton, J. G.; Palke, W. E. *J. Am. Chem. Soc.* **1993**, *115*, 4159.

[165]Bastiansen, O.; Fritsch, F. N.; Hedberg, K. *Acta Cryst.* **1964**, *17*, 538.

bital on carbon used for C—H bonding is computed to be $sp^{2.36}$, making the hybridization used for C—C bonds $sp^{2.69}$. In turn, that value predicts a C—C—C *interorbital* value of 111.8°.[166] Since this is considerably larger than the 60° internuclear angle required for an equilateral triangle, we must conclude that the orbitals used for C—C bonding point toward each other considerably outside the internuclear bond line (Figure 1.24).[167]

The hybridization parameter is useful as a tool to describe molecular bonding, but it is not a parameter that can be directly measured. However, it can be measured indirectly through the study of physically observable values that correlate with it.[168] Equation 1.46 shows a useful empirical relationship between the nuclear magnetic resonance coupling constant between ^{13}C and H ($J_{^{13}C-H}$) and the hybridization parameter.[169,170,171] In turn,

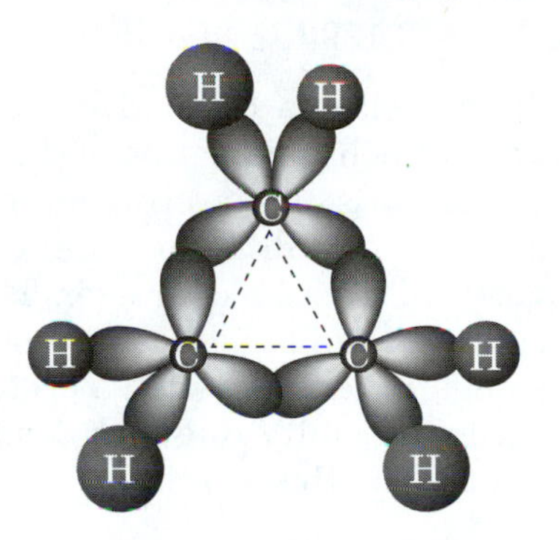

Figure 1.24 Relationship between internuclear and interorbital angles in cyclopropane. (Adapted from reference 145(a).)

[166]The bond path angle for the C—C—C bonds in cyclopropane was determined to be 78° on the basis of the theory of atoms in molecules (reference 35).

[167]The bonding model for cyclopropane assumed here was originated by Coulson, C.A.; Moffitt, W. E. *J. Chem. Phys.* **1947**, *15*, 151; *Phil. Mag.* **1949**, *40*, 1. An alternative description was proposed by Walsh, A. D. *Trans. Faraday Soc.* **1949**, *45*, 179. This mode uses sp^2 hybrid orbitals to produce a central orbital populated by two electrons and exterior orbitals populated by four electrons. For a discussion see Bernett, W. A. *J. Chem. Educ.* **1967**, *44*, 17; Patel, D. J.; Howden, M. E. H.; Roberts, J. D. *J. Am. Chem. Soc.* **1963**, *85*, 3218. (The figures here are reproduced from the last reference.)

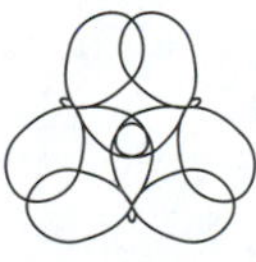

Walsh Coulson and Moffitt

[168]For a discussion and references, see Ferguson, L. N. *Highlights of Alicyclic Chemistry*, Part 1; Franklin Publishing Company, Inc.: Palisade, NJ, 1973; pp. 52 *ff*.

[169]Muller, N.; Pritchard, D. E. *J. Chem. Phys.* **1959**, *31*, 1471.

[170]Reference 151, p. 16.

[171]In addition, Liberles has used hybridization parameters to show that lone pairs in *sp* hybrid orbitals have greater local dipole moments than do lone pairs in any other hybrids: Liberles, A. *J. Chem. Educ.* **1977**, *54*, 479.

Muller and Pritchard determined that the lengths of C—H bonds correlated with values of $J_{^{13}C-H}$ according to the formula in equation 1.47.[172]

$$J_{^{13}C\text{-}H}\ (\text{cps}) = \frac{500}{1+\lambda_a^2} \tag{1.46}$$

$$r_{C-H} = 1.1597 - 4.17 \times 10^{-4} J_{^{13}C\text{-}H} \tag{1.47}$$

Streitwieser and co-workers demonstrated a correlation between $J_{^{13}C-H}$ and the kinetic acidities (the rates of exchange of C—H proton for tritium catalyzed by cesium cyclohexylamide). The data in Table 1.10 show a good correlation of log k_{rel} (rate relative to the rate of cyclohexane exchange) with $J_{^{13}C-H}$ (equation 1.48).[173]

$$\log k_{rel} = 0.129 J_{^{13}C-H} - 15.9 \tag{1.48}$$

The explanation for the correlation of bond length and acidity with $J_{^{13}C-H}$ is that both are related to the hybridization of the carbon orbital used for C—H bonding. Because *s* orbitals are lower in energy than *p* orbitals, a hybrid orbital with more *s* character will be lower in energy and closer to the nucleus than will a hybrid with less *s* character. The lower energy orbital will be more electronegative[147] and will hold a proton tighter and also be better able to stabilize a negative charge when a proton is removed in an acid-base reaction. Differences in acidity due to hybridization differences are more dramatically evident with ethane, ethene, and ethyne, which will be discussed below.

It must be reemphasized that hybridization is only a conceptual and a mathematical model that allows us to calculate (i.e., mathematically visualize) molecular parameters. Changing hybridization is simply modifying that original model to suit a current need, just as the concept of hybridiza-

Table 1.10 Correlation of rates of proton exchange with $J_{^{13}C-H}$.

Compound	Rate (Relative to Cyclohexane)	$J_{^{13}C-H}$
Cyclopropane	7.0×10^4	161
Cyclobutane	28.0	134
Cyclopentane	5.7	128
Cyclohexane	1.00	123, 124
Cycloheptane	0.76	123
Cyclooctane	0.64	122

[172]For a discussion of the relationships among bond lengths, bond angles, and coupling constants for cycloalkanes, see reference 168.

[173]Streitwieser, Jr., A.; Caldwell, R. A.; Young, W. R. *J. Am. Chem. Soc.* **1969**, *91*, 529.

tion represents only a change to the model of atomic energy levels. Variable hybridization is not a more fundamental approach than hybridization itself, so it should not be viewed as being more fundamental than the VSEPR model. The ability to make quantitative predictions about molecular geometry and physical properties makes the variable hybridization model quite useful for some problems. On the other hand, the VSEPR model is also useful because it is an intuitive approach that can readily provide qualitatively correct answers. As is so often the case, we need not decide which of two complementary models to adopt in all cases; we need only to determine which best serves our purposes in each individual situation.

1.3 Complementary Descriptions of the Double Bond

The σ,π Description of Ethene

Let us next consider alternative descriptions of the double bond of ethene. Practically all introductory organic chemistry textbooks[174] describe the double bond in terms of the σ,π formulation. The molecule is said to be held together by two carbon-carbon bonds, one a σ bond made by overlap of an sp^2-hybrid orbital (Figure 1.25) on each of the carbon atoms and the other a π orbital, made by overlap of two parallel $2p$ orbitals on the carbon atoms (Figure 1.26). Overlap of the two sp^2 hybrids produces both a σ bonding and a σ* antibonding orbital, and overlap of the two p orbitals produces both a π bonding and a π* antibonding molecular orbital, with energies qualitatively as shown in Figure 1.26. A pictorial representation of the carbon-carbon σ and π bonds is shown in Figure 1.27.

The σ,π model correctly predicts that the carbon-carbon bond length of ethene should be shorter than the carbon-carbon bond length of ethane. The reason is that the overlap of sp^2 orbitals (which have more s character and

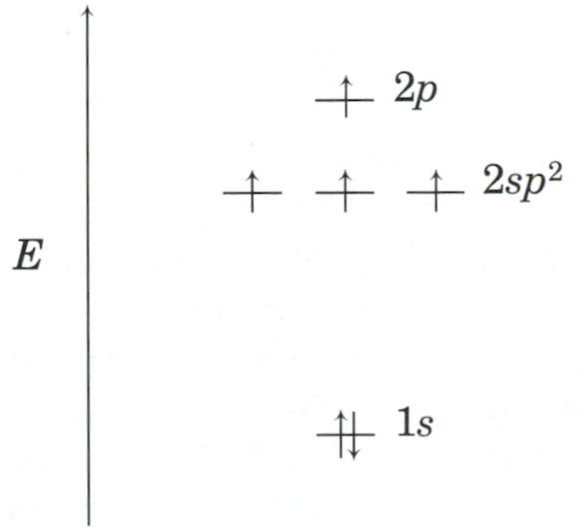

Figure 1.25
*sp*³-Hybrid orbitals on carbon.

$2p$
$2sp^2$
E
$1s$

[174]One text that presented the bent bond formulation also was Roberts, J. D.; Stewart, R.; Caserio, M. C. *Organic Chemistry*; W. A. Benjamin Inc.: Menlo Park, CA, 1971; p. 20.

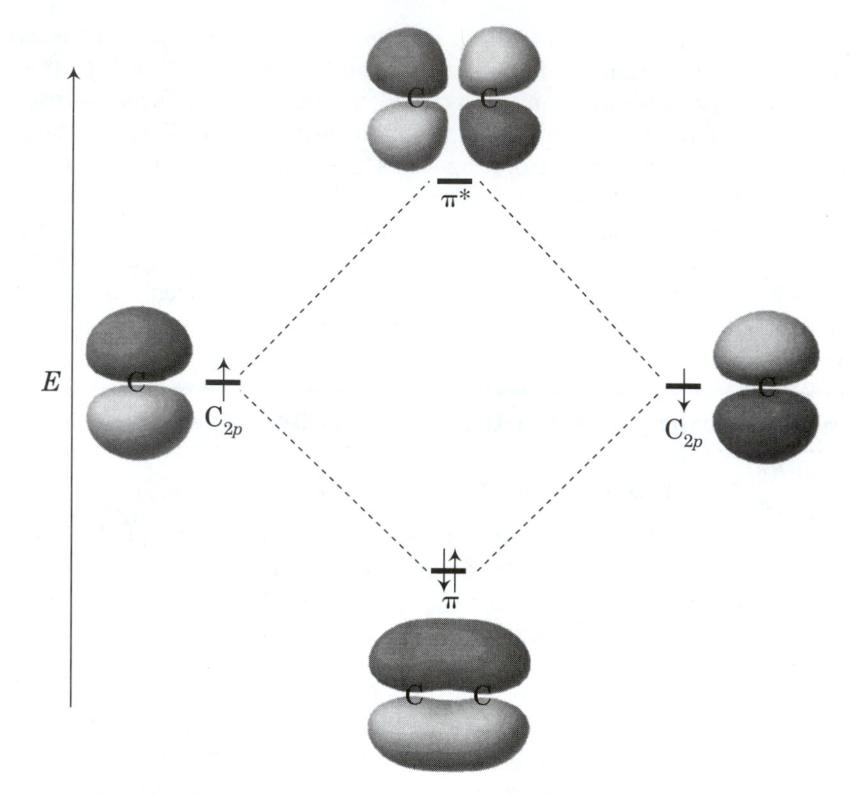

Figure 1.26
Energies of ethene π orbitals.

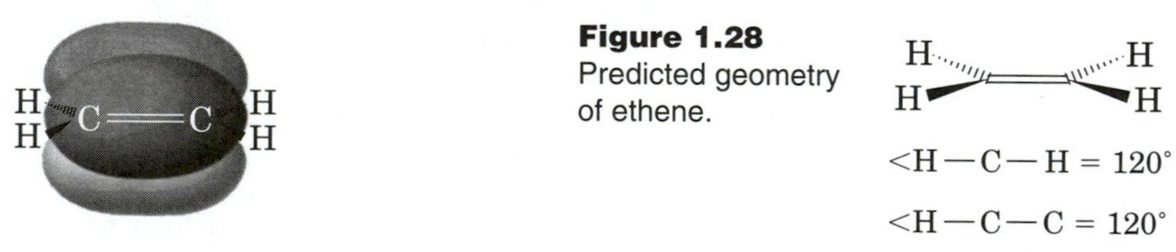

Figure 1.27
The σ,π formulation for ethene.

Figure 1.28
Predicted geometry of ethene.

are therefore closer to the carbon nucleus than are sp^3 hybrids) makes a shorter bond, and the two atoms should be pulled even closer by the second, π bond. Also, since the interorbital angle between two sp^2-hybrid orbitals on the same carbon atom is 120°, the bond angles H—C—H and H—C—C should both be 120°. Thus is derived the predicted geometry in Figure 1.28.

Figure 1.29 shows the geometry of ethene determined from spectroscopic measurements,[175] and the bond angles are 117°, not 120°. However, most organic chemists are bothered very little by the discrepancies between prediction and experiment. We usually argue that the geometry of ethene is an anomaly and that other alkenes would obey our predicted bond angles. We will return to the ethene geometry shortly, but first let us ask whether we can describe ethene by another model.

---

[175]Data from reference 22 (a), p. M78s.

Figure 1.29 Experimental geometry of ethene.

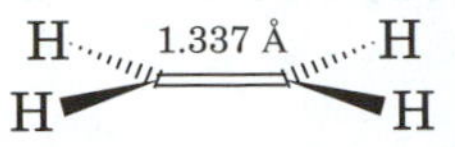

H H (1.337 Å) H H

C—H = 1.086 Å

<H—C—H = 117.3°

<H—C—C = 121.3°
(calculated from <H—C—H)

The Bent Bond Description of Ethene

In fact, there is a much older description of ethene that is known as the bent bond[176] formulation. It has an advantage in that we retain the sp^3-hybrid orbitals of methane. The double bond is viewed as resulting from overlap of two sp^3-hybrid orbitals on each of the two carbon atoms, as shown in Figure 1.30. What does it predict? One prediction is that ethene should be a planar molecule. However, to the first approximation, it predicts H—C—H bond angles of 109.5°, the same as in ethane. That is even further removed from the observed 117° than in our first sp^2 description. Furthermore, except for cyclopropane, we feel uncomfortable about drawing molecular pictures with bonds that curve in space as do those in Figure 1.30. This is an old view of bonding, and it is not taught regularly these days. Does anyone still believe in it?

One advocate of the bent bond description was Linus Pauling, the originator of the concept of atomic orbital hybridization, who wrote that

> There may be chemists who would contend that one innovation of great significance has been made—the introduction of the σ,π description of the double bond and the triple bond and of conjugated systems, in place of the bent bond description. I contend that the σ,π description is less satisfactory than the bent bond description, that this innovation is only ephemeral, and that the use of the σ,π description will die out before long . . .[177]

Figure 1.30 Overlap of sp^3-hybrid orbitals in the bent bond description of ethene.

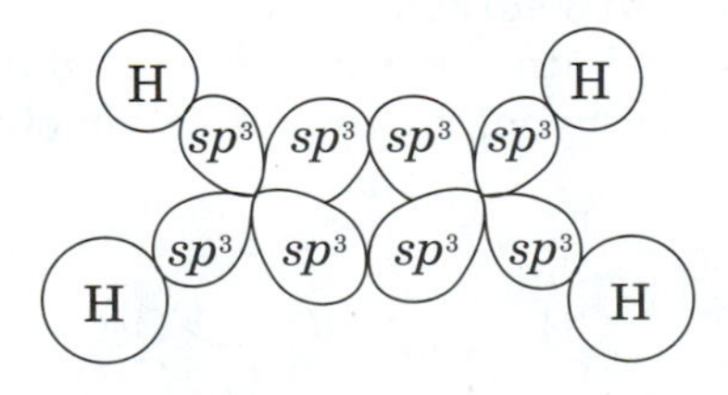

[176] Bent bonds are also known as banana bonds or τ (tau) bonds. For a discussion of the utility of this model in explaining molecular conformation and reactivity, see Wintner, C. E. *J. Chem. Educ.* **1987**, *64*, 587 and references therein.

[177] Pauling, L. in *Theoretical Organic Chemistry, The Kekulé Symposium*; Butterworths Scientific Publications: London, 1959; p. 1.

Predictions of Physical Properties with the Two Models

Geometry of Alkenes

Although Pauling's prediction has not yet come true, there are advantages in using the bent bond formulation. One advantage is conceptual simplicity. If the sp^3-hybridization model can be adapted to give the correct answer in a set of circumstances, why concoct a whole family of explanations (sp^3, sp^2, sp) for these same phenomena? A practical advantage is the ease of construction of physical models. Some molecular model kits designed to be used in introductory organic chemistry courses use bent bonds as the physical model of the double bond. Such models give acceptable structural geometries, require fewer kinds of parts in the model set, and are easier for novices to use than are model kits that attempt to represent double bonds with σ and π bonds. Still another advantage is that the bent bond formulation seems to provide more quantitative answers than does the σ,π formulation. Consider the measurement of the carbon-carbon distances illustrated in Figure 1.31. Taking the C—C bond length of ethane, 1.54 Å, as the length of an arc formed by overlap of sp^3 orbitals outside the internuclear line, Pauling calculated that ethene should have an internuclear distance of 1.32 Å, quite close to the experimental value for ethene.[178] Similarly, three bent bonds arranged as arcs of 1.54 Å in length and directed 109.47° apart produces a model for ethyne in which the C—C internuclear distance is 1.18 Å, which is essentially the same as the experimental value. Robinson and Gillespie have described these calculations as well as the use of molecular models to construct molecules with bent bonds and measure the internuclear distances from the models.[179] However, the σ,π formulation makes no quantitative prediction about the length of the double or triple bond.

Acidities of Hydrocarbons

Let us consider another physical property, acidity. The alkanes, alkenes, and alkynes are not usually considered to be acids, but it is possible to measure rates and equilibria of proton removal in solution and in the gas phase (Chapter 7). Table 1.11 shows some experimental data for the acidities of ethane, ethene, and ethyne in solution.

We explain these results with the σ,π formulation by noting that as we go from ethane to ethene to ethyne, the hybridization of the carbon atom in-

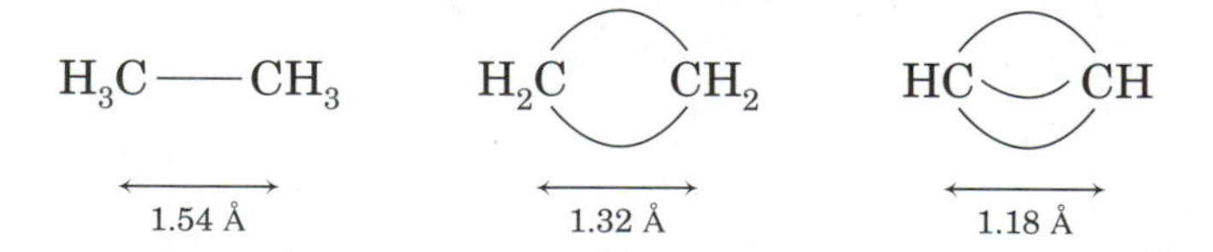

Figure 1.31 Quantitative predictions of the bent bond formulation.

---

[178]Reference 23, p. 138.

[179]Robinson, E. A.; Gillespie, R. J. *J. Chem. Educ.* **1980**, *57*, 329.

Table 1.11 K_a values for C_2 hydrocarbons.*

Compound	K_a
ethane	10^{-42}
ethene	$10^{-36.5}$
ethyne	10^{-25}

*Data from the tabulation by Cram, D. J. *Fundamentals of Carbanion Chemistry*; Academic Press: New York, 1965; p. 19. More recent data suggest that the values for ethane may be too large. The pK_a of methane in dimethyl sulfoxide was reported to be 56 by Bordwell, F. G. *Acc. Chem. Res.* **1988**, *21*, 456. A pK_a value for ethane in water solution was determined theoretically to be 50.6 by Jorgensen, W. L.; Briggs, J. M.; Gao, J. *J. Am. Chem. Soc.* **1987**, *109*, 6857.

volved in the acidic C—H bond changes from sp^3 to sp^2 to sp. Again, the more s character in an atomic orbital, the greater is the electron pulling power of that orbital, and the more stable should be an anion produced by pulling off a proton and leaving behind a pair of electrons.[147] Thus the order of carbanion stability should be $(sp)\text{C:}^- > (sp^2)\text{C:}^- > (sp^3)\text{C:}^-$, so the corresponding acidities should be (sp)C—H > (sp^2)C—H > (sp^3)C—H (see Figure 1.32).

The same phenomena can be rationalized with the bent bond formulation and VSEPR theory by noting the effect of different numbers of C—C bonds on the electron pair repulsions around each carbon atom. As Figure 1.33 shows, formation of the bent bonds pulls the electrons closer to the center of the C—C internuclear line; in turn, this decreases the electron pair–electron pair repulsion between the C—H bond in question and the C—C bond(s). That means that the C—H electrons see more of a bare, unshielded carbon nucleus. They are attracted more strongly to the carbon atom, so an unshared pair of electrons left behind by removal of a proton is much more stable on a triple-bonded carbon atom (Figure 1.33c) than on a

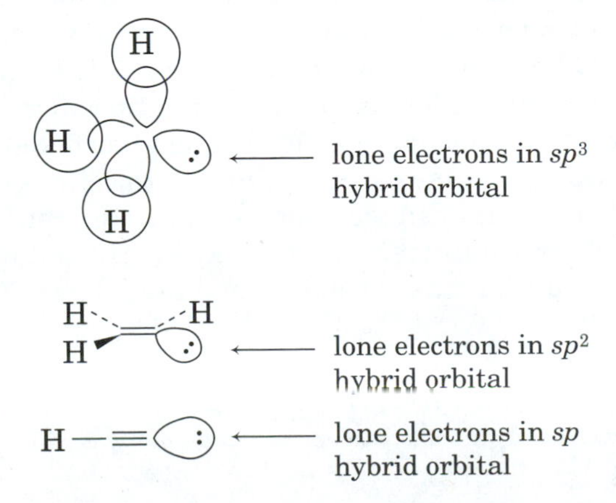

Figure 1.32 σ,π Rationalization of acidities.

Figure 1.33 Bent bond rationalization of acidities.

(a) Ethane

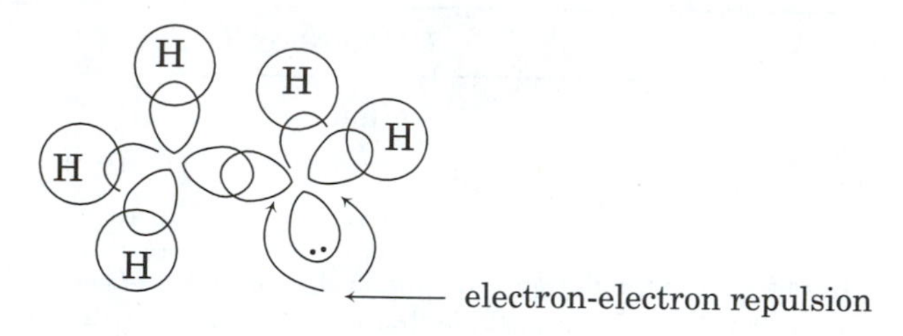

(b) Ethene

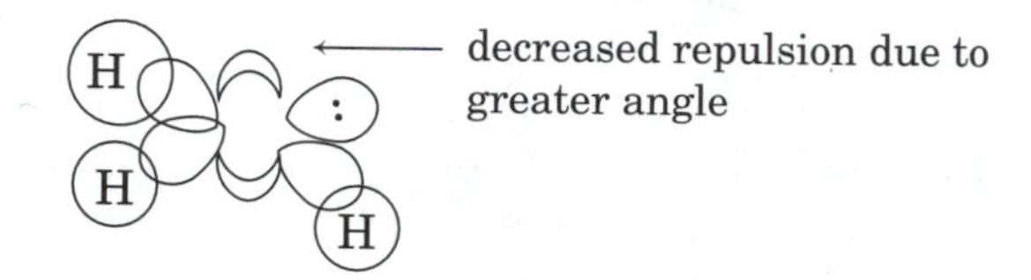

(c) Ethyne

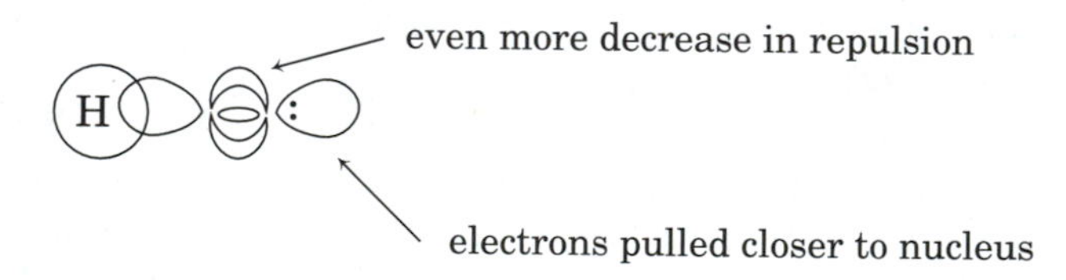

double-bonded carbon atom (Figure 1.33b). Thus the bent bond formulation can at least qualitatively rationalize these experimental observations as well as can the σ,π description.

Conformation of Propene

Is there any situation in which either the bent bond or the σ,π description is more accurate than the other? Let us consider one simple case of a molecular conformation and ask some admittedly naive questions about which conformation should be favored if we view the molecule from the point of view of each of the two models for the double bond. Specifically, what should be the preferred conformation of propene? Walters has noted that the conformers of propene can be visualized as Newman projections observed by sighting down the $CH_3—CH=$ single bond, as shown by conformers I and II in Figure 1.34.[180,181] In conformer I there is a C—H bond eclipsing a carbon-carbon double bond, whereas in conformer II there is a C—H bond eclipsing

[180]Walters, E. A. *J. Chem. Educ.* **1966**, *43*, 134.

[181]Newman projections and other stereochemical representations will be discussed in Chapter 2.

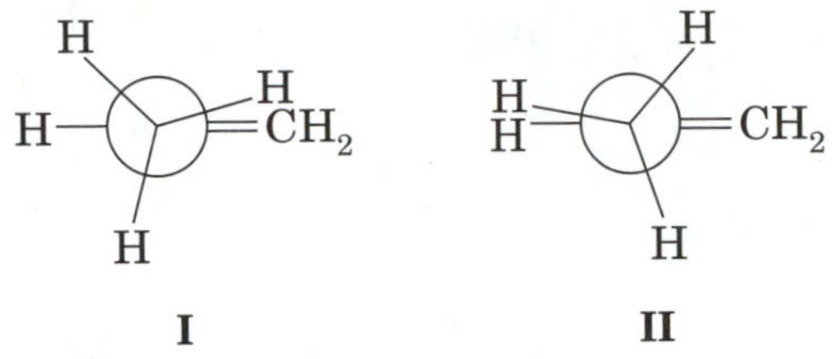

Figure 1.34 Conformations of propene: σ,π description. (Adapted from reference 180.)

another C—H bond. Since there is greater electron density in a double bond than a single bond, we would expect greater repulsion between the C—H bond and the eclipsed bond in conformer I, so we would expect conformation II to be the more stable. However, I is more stable, by about 2 kcal/mol.[182]

If we view the same Newman projections from the bent bond description, as shown in Figure 1.35, we see that conformer II now represents essentially an all-eclipsed conformation. Thus it is easily predicted to be less stable than conformer I, which in fact is an all-staggered arrangement. Thus, if utility is our main criterion for adopting mental/conceptual models, this one experiment would seem to favor the bent bond formulation.[183] Proponents of the σ,π formulation could argue that taking additional factors into account will correct the naive prediction that conformer II is the more stable conformer.[184] However, on the basis of simple prediction of conformational stability, the bent bond model beats the σ,π model in this test.

Pauling's prediction that the use of the σ,π description will wane may yet come true. In recent years some theoreticians have determined that calculations of molecular structure are in better agreement with the bent bond description than with the σ,π description. Figure 1.36 shows calculated contour lines of one orbital of a bonding pair in a plane containing the carbon atoms of cyclopropane (b) and perpendicular to the plane of all the atoms of ethene (a). (The contour lines represent the orbital of only one carbon atom

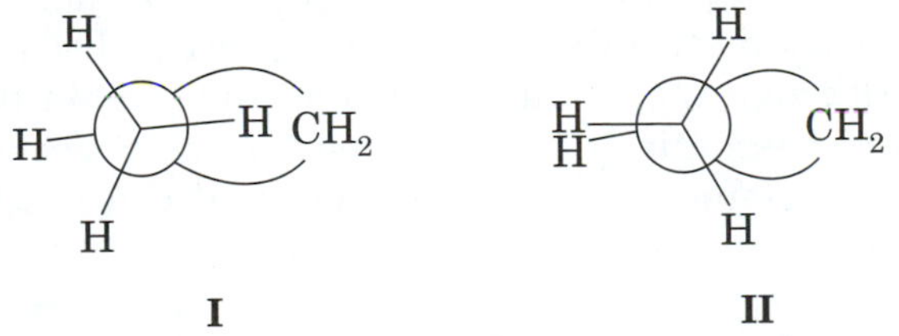

Figure 1.35 Conformations of propene: bent bond formulation. (Adapted from reference 180.)

[182]Herschbach, D. R.; Krisher, L. C. *J. Chem. Phys.* **1958**, *28*, 728.

[183]The emphasis in the present discussion is on the application of two very simple conceptual models to a particular problem. High level calculations provide a much deeper analysis of the conformations of propene and other molecules. For a discussion, see Bond, D.; Schleyer, P. v. R. *J. Org. Chem.* **1990**, *55*, 1003.

[184]When faced with discrepancies between prediction and experiment, the proponents of a particular conceptual model often take the position that consideration of additional factors would favor their model.

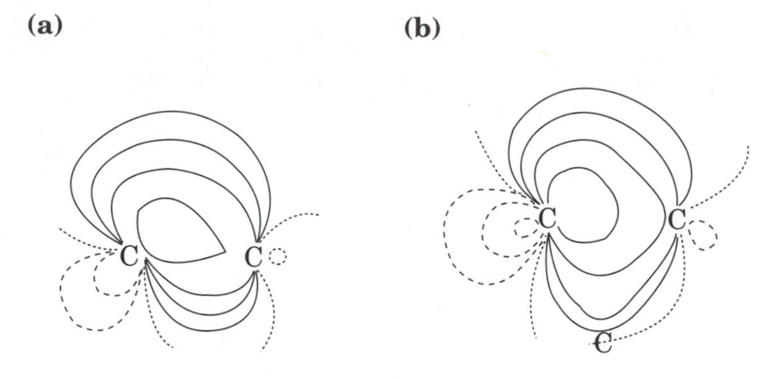

Figure 1.36 Contour lines for a bonding orbital on one carbon atom in (a) ethene and (b) cyclopropane as computed by Hamilton and Palke. (Adapted from reference 164.)

and do not represent the density in a carbon-carbon bond.) There is evidence in both structures that much electron density lies outside the internuclear bond line. It is generally agreed that the bonds in cyclopropane are bent; the picture from this theoretical calculation reinforces the view that they are bent in ethene also. The title of another paper was "Double Bonds are Bent Equivalent Hybrid (Banana) Bonds."[185] Another study concluded that "The GVB description of the double bond in (C_2F_4) is *not* the traditional picture of σ and π bonds but rather a representation in terms of two bent bonds."[186] Still another paper comparing the bent bond and σ,π models concluded that ". . . from an energetical point of view, both constructions provide an equally good starting point for the treatment of correlation effects beyond the one-electron configuration."[187] Another study concluded from a theoretical analysis of ethene and ethyne that "Our results yield bent bonds as the favored bonding description, showing that the σ,π bond descriptions of multiple bonds are artifacts of approximations to the full independent-particle equations."[188] For systems that exhibit resonance, such as benzene and the allyl radical, these authors stated the conclusion even more succinctly: "Bent bonds are better."[189,190]

The distinction between the bent bond and σ,π formulations of the double bond are not as cleanly drawn as the present discussion might suggest, however. Although the results depend on the level of theory used, the two models predict essentially the same result at higher levels of analysis.[191] As Schultz and Messmer put it, "No experiment can possibly distinguish be-

[185]Palke, W. E. *J. Am. Chem. Soc.* **1986**, *108*, 6543.

[186]Schultz, P. A.; Messmer, R. P. *J. Am. Chem. Soc.* **1988**, *110*, 8258.

[187]Karadakov, P. B.; Gerratt, J.; Cooper, D. L.; Raimondi, M. *J. Am. Chem. Soc.* **1993**, *115*, 6863.

[188]Schultz, P. A.; Messmer, R. P. *J. Am. Chem. Soc.* **1993**, *115*, 10925.

[189]Schultz, P. A.; Messmer, R. P. *J. Am. Chem. Soc.* **1993**, *115*, 10943.

[190]For a different view, however, see Carter, E. A.; Goddard III, W. A. *J. Am. Chem. Soc.* **1988**, *110*, 4077.

[191]Gallup, G. A. *J. Chem. Educ.* **1988**, *65*, 671 and references therein.

tween a σ,π double bond and double bent bonds in any system, and therefore neither can be proven to be 'right' in an absolute sense; both are approximate descriptions."[188] The σ,π model for the double bond and the bent bond description should each be taken as viable *starting points* to describe molecular structure but not as complete descriptions.[192] Each approach has its advantages and disadvantages. It is important that we consider both methods and that we know why we choose one over the other when we talk about organic chemistry. In the following chapters we will describe in more detail the types of calculations that organic chemists find useful. We will begin with the simplest form of MO theory used by organic chemists and will proceed to more advanced computational methods. The goal of this discussion is not to become proficient in theoretical chemistry, but to understand the nature and limits of the models used in each case.

Conclusion

Conceptual, mathematical, and physical models are essential tools in organic chemistry because they help correlate and rationalize the results of experiments (observables) with theory (non-observables). Yet the paradox is that these models may be most useful to us when they are oversimplified to the point of being incorrect. Without a straight line or pair of dots to represent a chemical bond, we would find it difficult to describe chemistry in a useful way.[193] Yet in some cases we find it advantageous to draw those lines curved instead of straight, and sp^3-hybrid orbitals cannot be relied on even to predict all of the properties of methane. A more detailed description of that line and of those orbitals can be made only with the help of computers and high level mathematics.

If any one of our models is asked to give a correct answer to all problems, it quickly becomes more and more complex. Electronegativity is useful in a qualitative sense, but attempts to make it more quantitative and therefore more "correct" lead to many different conclusions about what it means and how it should be determined. Elementary VB theory and MO theory are intuitively reasonable, but further development obscures the simple mental pictures. However, we feel a need to retain the simple pictures, even when we know that they cannot be totally accurate.

One solution to the use of oversimplified models in organic chemistry is to mentally hybridize complementary models, just as we hybridize the two Kekulé structures for benzene in our minds in order to understand and describe aromaticity. The σ,π and bent bond descriptions represent a pair of models that serve as useful beginning points or approaches to the description

---

[192] Reference 23, pp. 136 *ff*; England, W. *J. Chem. Educ.* **1975**, *52*, 427; Palke, W. E. *J. Am. Chem. Soc.* **1986**, *108*, 6543.

[193] The future supply of organic chemists would probably be diminished if students were forced to use only quantum mechanics to describe molecules!

of the double bond. Being able to visualize a hybrid of these two mental pictures may be more nearly correct than thinking in terms of either model alone.

Perhaps another metaphor may be useful. If we view complementary models as languages to describe chemistry, we find it better to be multilingual—to be able to converse in many languages, to translate from one language to another, and to think in more than one language. In the end, after all, our models must be described in a language.[194] If we have only one approach, if we can only compute to a certain degree of accuracy by one method, then we are only computers doing what we have been programmed to do. Being able to advance means being able to see new relationships in a new way.[195]

In the following chapters we will consider many diverse aspects of organic chemistry. We will reexamine ideas, such as the mechanism of S_N2 or electrophilic aromatic substitution reactions, that might seem to be well-understood but for which alternative models are useful. In all cases, we will want to remember that our knowledge is limited and that our understanding is still less. We must be prepared to reexamine our preconceived ideas, to defend with fact and with logic any conceptual models that we adopt, and to use complementary models with dexterity.

Problems

1. Kuhn (Reference 3) says that "[scientists] are little better than laymen at characterizing the established bases of their field . . .". Briefly summarize the physical phenomena that support your belief in atomic and molecular theory.
2. Find a popular or scientific article that refers to observations of a single atom or atoms.
   a. What is the nature of the experiment?
   b. What observations are made directly with the human senses?
   c. What paradigms are implicit in the experiment?
   d. What do atoms look like?
3. a. Consider two geometries for methane other than the regular tetrahedron. Show how each alternative geometry is inconsistent with the known number of isomers of some derivative of methane.
   b. Consider at least four alternative geometries for benzene and show how each is inconsistent with the known number of isomers of benzene substitution products with a given molecular formula.

[194] For provocative comments on language and models, see Bent, H. A. *J. Chem. Educ.* **1984**, *61*, 774. In particular, Bent noted that "Indeed, to be useful, *a model must be wrong, in some respects*—else it would be the thing itself. The trick is to see—with the help of a teacher—*where it's right*."

[195] One aspect of this idea is what Turro (reference 17) called "organic thinking," which is one of the ways in which the study of organic chemistry transfers to other areas: "An important value of learning organic chemistry is the mastering of 'organic thinking,' an approach to intellectual processing whereby the 'sameness' of many families of structures and reactions is revealed."

c. In the two previous problems you assumed that the structure of a derivative (e.g., bromobenzene) was essentially the same as that of the parent structure (e.g., benzene). Is that assumption valid? For example, how can we know that methane is not planar, even though chloromethane is roughly tetrahedral?

4. Use the values in Table 1.3 to determine the van der Waals volume and surface area for each of the pentanes. Verify that for *n*-pentane the values are correctly predicted by the formulas $V_w = 6.88 + 10.23\,N_c$ and $A_w = 1.54 + 1.35\,N_c$, where N_c is the number of carbon atoms in the molecule.
5. The heat of formation of corannulene (**1**) in the crystal state is 81.81 kcal/mol. Its heat of sublimation has been calculated to be 29.01 kcal/mol. What is the heat of formation of **1** in the vapor phase?

1

6. A proposed system for the conversion and storage of solar energy was based on the photochemical conversion of norbornadiene (**2**) to quadricyclane (**3**) during sunny periods, with catalytic conversion of **3** to **2** and release of energy at a later time.

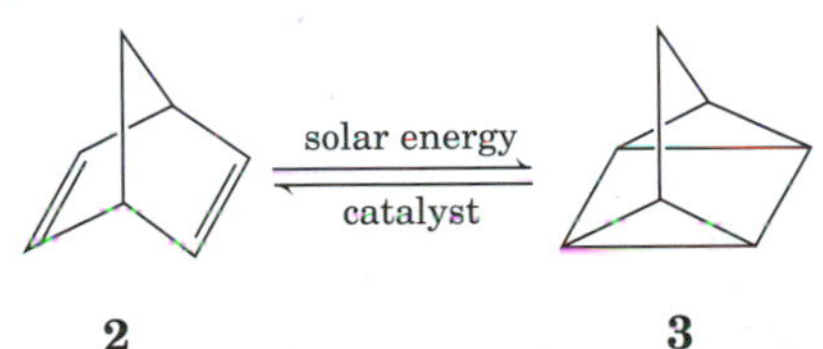

2 **3**

The heats of hydrogenation of **2** and **3** (both in the liquid phase) to norbornane are −68.0 and −92.0 kcal/mol, respectively. What is the potential energy storage density, in kcal/mol, for the photochemical conversion of **2** to **3**?

7. The standard heat of combustion of 4,4-dimethyl-1,3-cyclohexanedione was determined to be −1042.90 kcal/mol. Its standard heat of sublimation was +23.71 kcal/mol. What is the standard heat of formation (gas phase) of this compound?
8. The heat of hydrogenation of phenylethyne was found to be −66.12 kcal/mol, and the heat of formation of phenylethane is 7.15 kcal/mol. What is the heat of formation of phenylethyne?
9. *cis*-3-Methyl-2-pentene has a heat of formation that is 1.65 kcal/mol more negative than that of 2-ethyl-1-butene. The heat of reaction of 2-ethyl-1-butene with trifluoroacetic acid is −10.66 kcal/mol. Predict the heat of reaction of *cis*-3-methyl-2-pentene under the same conditions.
10. The heat of hydrogenation of *cis*-1,3,5-hexatriene is −81 kcal/mol, while that of the trans isomer is −80.0 kcal/mol. Under the same conditions the heat of hydrogenation of 1,5-hexadiene is −60.3 kcal/mol. What would be the heat of hydrogenation for each of the 1,3,5-hexatrienes if one mole of H_2 could be added to the middle double bond of each?
11. The dipole moments of CH_3F, CH_3Cl, CH_3Br, and CH_3I are reported to be 1.81, 1.87, 1.80, and 1.64 D, respectively.[196]

---

[196]Data from McClellan (reference 69).

a. Calculate the partial charge on the halogen in each of the methyl halides (assuming that the halogen bears the partial negative charge and that the partial positive charge is centered on the carbon atom).
b. What trend, if any, do you see in the dipole moments of the methyl halides? How do you rationalize this result?

12. For what category of atoms could Allen or Nagle but not Pauling electronegativity values be determined?
13. Consider the conformations about the O—C(=C) bond that are possible for methyl vinyl ether. (Draw the two lone pairs on oxygen as substituents in appropriate Newman projections.)
    a. What conformation would be preferred according to the σ,π description of bonding?
    b. What conformation would be preferred according to the bent bond description?
    c. After answering parts a and b, use the literature to find a paper on methyl vinyl ether conformations. What do the experimental data tell you?
14. The C—C—C bond angle of propane is reported to be 112.4°. What is the hybridization of the orbitals used for C—C bonding in this compound? What is the hybridization of the orbitals used for C—H bonding in the methylene group?
15. Both experimental and theoretical studies indicate that the C1-C2 bond length (l) and the C1—C2—C3 bond angle (α) in molecules having the following general structure

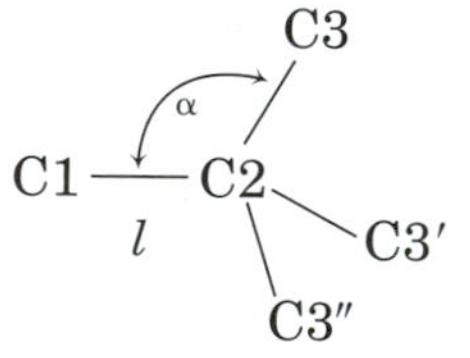

C3
α
C1 — C2
l
C3′
C3″

are strongly correlated. (The C1—C2—C3′ and the C1—C2—C3″ bond angles are also α.) One study found that they could be related by the equation

$$l = 2.0822 - 0.0049\,\alpha \quad \textbf{(1.49)}$$

where distances are in Å and angles are in degrees.
    a. What does this equation indicate about the change in bond length expected when the adjacent angle changes to a value greater than or less than 109.5°?
    b. Rationalize the form of equation 1.49 in terms of the concept of variable hybridization and in terms of VSEPR theory.
16. Show how the VSEPR and variable hybridization descriptions can be used to rationalize the experimentally determined values for the geometry of methyl bromide. Determine the fractional *s* and *p* character of the C—H and C—Br bonds according to the variable hybridization theory.
17. a. Use the bent bond description of double bonds to explain why ethene has H—C—H bond angles of 117° and not 109.5°?
    b. Modify your explanation to predict the H—C—H angle of formaldehyde; should it be larger or smaller than the H—C—H angle of ethene?
18. Consider again ethene and formaldehyde, this time analyzing the double bond in each as a two membered ring. Calculate the fractional *s* character and hybridization parameter for the C—H bond in each. Do the variable hybridization and bent bond descriptions suggest similar properties for the ethene structure?

19. a. Consider the formula $J_{^{13}C-H} = 5.7 \times (\%s) - 18.4$ Hz. Determine whether this formula is equivalent to that in equation 1.46, and explain why or why not.
    b. Use equation 1.47 and equation 1.46 to write an equation in which the length of a C—H bond is related directly to λ^2.
20. Analyze cyclopropane, cyclobutane, and cyclopentane according to both the bent bond (using orbitals that are sp^3 hybrids) and variable hybridization descriptions.
    a. Use the literature value for the H—C—H bond angle to calculate the fractional *s* and *p* character of the carbon orbitals used for C—H bonds, then use the result to calculate the interorbital C—C—C bond angle. How does that differ from the literature value of the internuclear bond angle? What does that tell you about the nature of the strain in the molecule? What does that suggest to you about the chemical properties of cyclopropane (e.g., reaction with electrophiles).
    b. What does each model suggest to you about the acidities of the C—H bonds? How do the literature data for the acidity of cyclopropane agree with the predictions?
21. In bicyclo[1.1.0]butane, the $^{13}C-^{1}H$ coupling constant for the bridgehead C—H groups is 202 Hz. Calculate the percent-*s* character in the bond from carbon to hydrogen at this position. Would you expect the acidity of the bridgehead protons to be greater or less than that of the protons in acetylene?

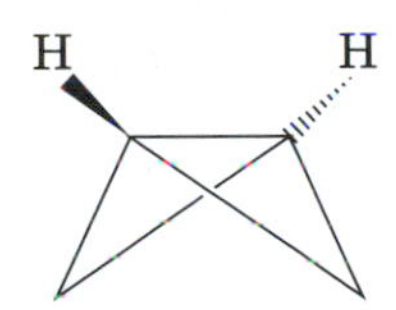

22. On page 34 we saw experimental evidence (PES spectrum) that suggests our views of hybridization are at best oversimplified and at worst wrong. Can you justify the continued use of hybridized orbitals as a conceptual model in view of this result?
23. If λ is a function of molecular geometry and $J_{^{13}C-H}$ is also a function of λ, then is not $J_{^{13}C-H}$ really a function of geometry? Is λ an observable? If not, do we need to define λ at all?
24. Greenberg and Liebman[197] have stated that "... we believe organic chemistry is essentially a pictorial and not a mathematical science ...". Do you agree or disagree that this is the way organic chemistry is now? Do you think organic chemistry should be a mathematical science? Do you believe it will become so in the future?
25. Respond to Coulson's observation about the nature of theory in chemistry: "Sometimes it seems to me that a bond between two atoms has become so real, so tangible, so friendly that I can almost see it. And then I awake with a little shock: for a chemical bond is not a real thing: it does not exist: no one has ever seen it, no-one ever can. It is a figment of our own imagination. . . . Here is a strange situation. The tangible, the real, the solid, is explained by the intangible, the unreal, the purely mental."[150]

---

[197]Greenberg, A.; Liebman, J. F. *Strained Organic Molecules*; Academic Press: New York, 1978; p. 37.

CHAPTER 2

Stereochemistry

2.1 Introduction

The term ***stereochemistry*** refers to the three-dimensional nature of molecules and to their space-filling properties. Our physical models (balls and sticks) and many of our computer models give a perception of three dimensions, but our printed and hand-drawn representations of molecules are two-dimensional images that are meaningful only to those who understand implicit rules for visualizing the third dimension.[1] Moreover, molecular structure drawings have meanings on many levels.

- Structure drawings are not always meant to be taken literally. For example, a Kekulé structure for benzene does not really mean a deformed 1,3,5-cyclohexatriene molecule.
- Our drawings are not designed to show all properties, only the properties that are to be emphasized to the viewer. The Kekulé structure for benzene may not show the hydrogen atom bonded to each carbon atom, since the point of emphasis is the aromatic ring.
- A representation of a carbon-carbon double bond may use the σ,π formulation or the bent bond formulation. An allylic carbocation may be drawn with delocalized charges calculated from molecular orbital theory or may be represented with resonance structures. In such cases, the choice of the representation may imply as much about the conceptual model used to analyze the problem as it states about the chemical substance under consideration.

Let us look more closely at some specific types of structural representations and categorize the information they are intended to convey. Some

[1] Hoffmann and Laszlo have presented an insightful analysis of the way chemists use drawings to convey information: Hoffmann, R.; Laszlo, P. *Angew. Chem. Int. Ed. Engl.* **1991**, *30*, 1.

drawings represent only the **constitution** of a molecule (that is, the order in which the atoms are bonded), without indicating anything about the spatial orientation of the atoms. One way to do that is by representing the molecule as a line of element symbols and number notations, a kind of word structure. For example, pentane could be shown as $CH_3CH_2CH_2CH_2CH_3$. We may also explicitly show bonds connecting the carbon atoms, as in $CH_3—CH_2—CH_2—CH_2—CH_3$, but that is not necessary because bonds between atoms are implicit in structural representations. With the understanding that a methylene group can be considered a unit, we could also write it $CH_3(CH_2)_3CH_3$. Perhaps the most compact representation of all is one designed to represent molecular structures in a format that is easily stored and retrieved electronically. In the Wiswesser Line Notation (WLN) system, for example, pentane is represented as 5H.[2]

Word structures do not work well for larger or more complicated molecules. Instead we use symbols for molecular units and organize these symbols into pictographs (reaction schemes) that must appear to be molecular hieroglyphics to the uninitiated. The common convention in such drawings is that a line represents a chemical bond, that a carbon atom is at each intersection and terminus of a line unless another atom is indicated, and that each carbon atom has sufficient hydrogens as substituents to satisfy normal valency requirements. With this system pentane is the jagged line **1**. That drawing is not much better than $CH_3CH_2CH_2CH_2CH_3$ for pentane, but such representations become more useful as the molecules to be depicted become larger. Consider trying to write a word structure for cholesterol (**2**), for example.

$= CH_3CH_2CH_2CH_2CH_3$

pentane

1

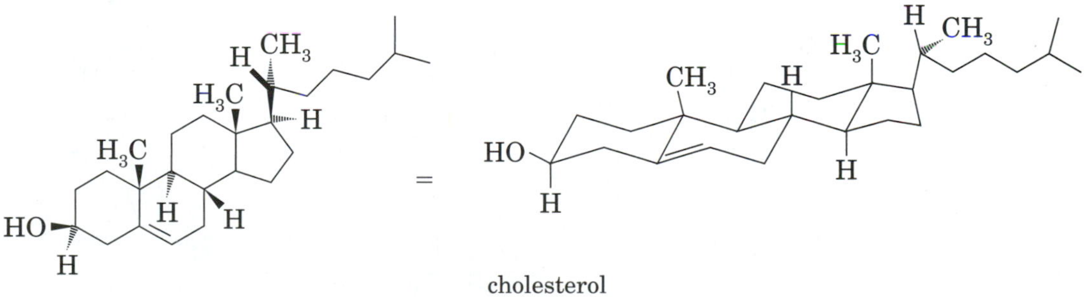

cholesterol

2

[2]*Cf.* Smith, E. G.; Baker, P. A. *The Wiswesser-Line-Formula Chemical Notation (WLN)*, 3rd Ed.; Chemical Information Management Inc.: Cherry Hill, NJ, 1976. See also Vollmer, J. J. *J. Chem. Educ.* **1983**, *60*, 192.

Cholesterol not only has more atoms and more connectivity than does pentane, but it differs in another important way. Physical models tell us that all of the carbon atoms of pentane *could* lie in one plane, but those of cholesterol cannot. Therefore our drawing for cholesterol must have some way to tell not only the connectivity but also the orientation of the various parts of the molecule. We view the drawing as a projection (a shadow, in essence) of a real object, and we often use angles and line thicknesses to convey the three-dimensional information.[3,4] Objects in the distance (behind the plane of the main scene) are reduced in size and/or angled to give an effect of distance, while objects in the front of the main scene are large and bold. Consider the six drawings of 2-chlorobutane **(3)** in Figure 2.1. In representations **3a** and **3b**, a bold line, growing larger as it proceeds, suggests a bond projecting toward the viewer. A hatched line, usually shown growing smaller as it moves from the same atom, suggests a bond projecting away from the viewer. Solid lines that do not change in size suggest bonds in a central plane.

Another way to represent the same molecule is to let all bonds be represented by solid lines, but to represent one of the carbon atoms as a sphere. The sense of perspective is suggested by having bonds penetrate or be eclipsed by the sphere. Thus the ball and stick model for 2-chlorobutane would be as shown in **3c**. A simpler model would be a stick model only, usually referred to as a sawhorse representation because of its similarity to a carpenter's device with the same name (**3d** and **3e**). In some cases we may indicate perspective with dashed/hatched and bold lines or wedges (as in **3d**). In other cases we must use our knowledge that tetrasubstituted carbon

Figure 2.1
Six representations of 2-chlorobutane.

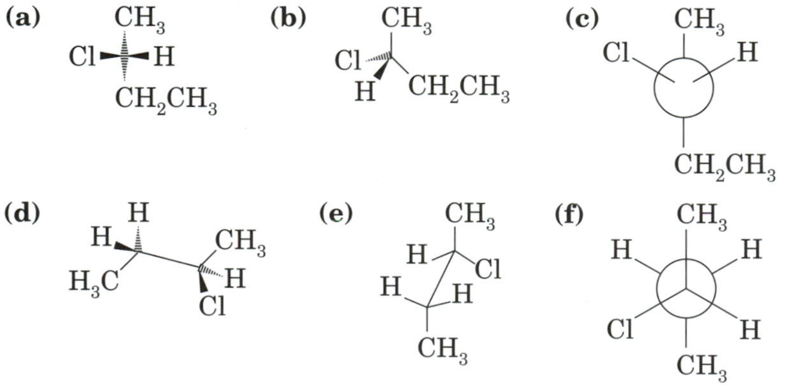

3 a–f

[3]In some cases that representation is clear enough that a computer program can recreate the original object from the projection. Whitlock wrote a program, ZED, to create an "unprojection" from a drawing of a structure: Whitlock, H. W. *J. Org. Chem.* **1991**, *56*, 7297.

[4]In another type of figure, known as the Fischer projection (page 80), the three-dimensional information is conveyed both by the figure itself and by a set of rules for the interpretation of the figure.

atoms are tetrahedral in order to visualize the spatial relationships among the substituents (**3e**).

An important part of stereochemical drawings is the clear representation of **conformations**, which are different spatial relationships that are interconvertible by rotation about single bonds at ambient temperature.[5] In a Newman projection we visualize the view from sighting down a particular bond in such a way that one atom totally eclipses another one. As illustrated in Figure 2.1 for one conformation of 2-chlorobutane (**3f**), we see three substituents on the carbon atom nearer our eyes, and we stylistically show their bonds meeting at the center of a circle. The fourth substituent to the front carbon atom is the other carbon atom. We can see three substituents attached to the second carbon atom, but the lines representing bonds to these substituents are visible only to the edge of the circle representing the front carbon atom. It must be emphasized that a circle and lines are also used in representation **3c**, but the two figures are meant to convey very different kinds of information.

Bold and hashed wedges are commonly used in complex molecules, but an additional convention is required. For example, in the representation of taxol[6] (**4**), it cannot be true that all the bold wedges project toward the viewer to the same plane in space, nor can it be true that all the dashed wedges project to the same plane behind the page. Rather, each wedge is a local descriptor only, and the sense of depth of the viewer is understood to be reset at the origin of each wedge.

taxol

4

2.2 Stereoisomerism

Isomerism

Isomers are different chemical compounds that have the same molecular formula. Whether two compounds are different is not as simple a question

[5]The origin of the term *conformation* is ascribed to Haworth, W. N. *The Constitution of Sugars*; E. Arnold and Co.: London, 1929; p. 90 *ff*. See also Barton, D. H. R.; Cookson, R. C. *Quart. Rev. Chem. Soc.* **1956**, *10*, 44.

[6]Guénard, D.; Guéritte-Voegelein, F.; Potier, P. *Acc. Chem. Res.* **1993**, *26*, 160.

as it might seem. In the first place, we are inherently biased toward human standards of temperature (and pressure), which means that the $\Delta G^{\ddagger}$ for an isomerization reaction must be greater than about 23 kcal/mol in order for us to be able to isolate two species in equilibrium.[7] Thus we do not consider the axial and equatorial[8] conformations of methylcyclohexane to be isomers, since they rapidly interconvert at room temperature.[9,10]

If two molecules are determined to be isomers, we may further describe them as being either (i) **constitutional isomers**,[11] which are pairs of compounds that have the same components connected in different ways or (ii) as **stereoisomers**, which are molecules with the same order of bonding but with different spatial relationships among the atoms (Figure 2.2).[12,13] We

Figure 2.2 Relationships among isomers.

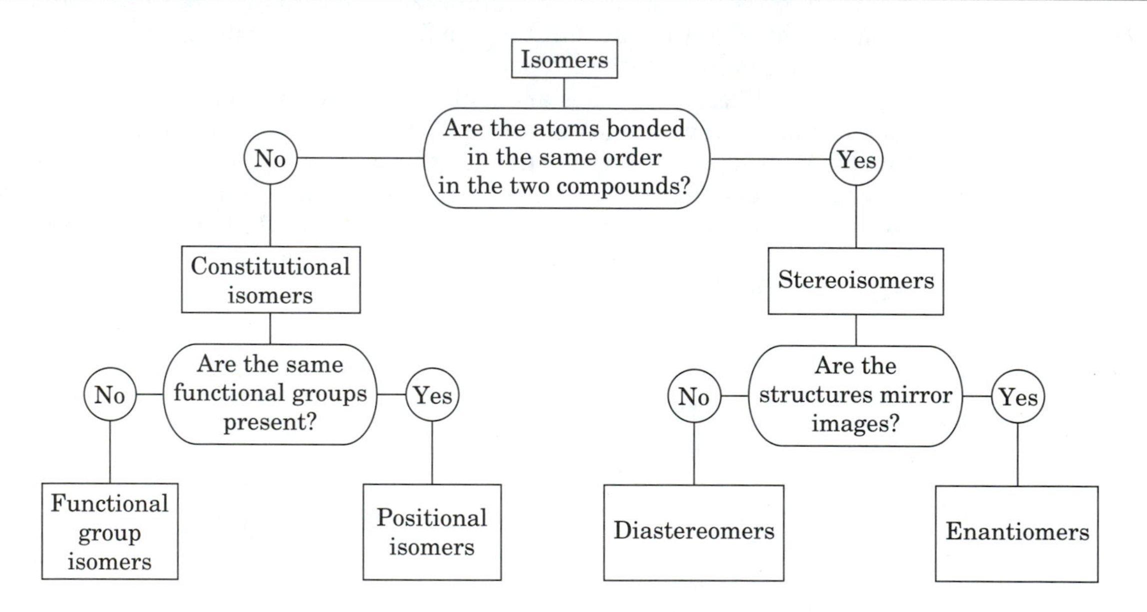

[7]Kalinowski, H.-O.; Kessler, H. *Top. Stereochem.* **1973**, *7*, 295.

[8]Conformational terms (such as *axial* and *equatorial*) are discussed in Chapter 3.

[9]At much lower temperatures, however, the conformational change is slowed. For example, it was reported that pure equatorial chlorocyclohexane can be formed in solution and studied by NMR at −151°. Jensen, F. R.; Bushweller, C. H. *J. Am. Chem. Soc.* **1966**, *88*, 4279.

[10]For a discussion of the concept of isomerism, see Eliel, E. L. *Isr. J. Chem.* **1976/77**, *15*, 7.

[11]Constitutional isomers are sometimes called *structural isomers*, but that term has been criticized because stereoisomerism may be considered to be a component of the structure of a molecule. In the 1974 IUPAC nomenclature rules (reference 75), the term *structural* is "abandoned as insufficiently specific."

[12]For a discussion of stereochemical nomenclature, see Eliel, E. L. *J. Chem. Educ.* **1971**, *48*, 163.

[13]A more detailed flow chart of isomeric relationships was given by Black, K. A. *J. Chem. Educ.* **1990**, *67*, 141.

may further divide constitutional isomers into (a) **functional group isomers**, in which the same atoms are used to form different functional groups, and (b) **positional isomers**, in which the same functional groups are present but are located in different positions in the two structures. Thus ethanol and dimethyl ether are functional group isomers, while 1-chloropropane and 2-chloropropane are positional isomers.

Enantiomers are stereoisomers that are nonsuperimposable mirror images. Such structures are said to be **chiral**, that is "handed," and to possess the property of **chirality**.[14,15] **Diastereomers** are stereoisomers that are not mirror images. These terms are illustrated in Figure 2.3 for a 2,3-disubstituted butane. Structures **5** and **6** are enantiomers, and structures **7** and **8** are enantiomers. Structures **5** and **7** are diastereomers, as are structures **5** and **8**, structures **6** and **7**, and structures **6** and **8**. Note that any one structure can have only one enantiomer, but it may have more than one diastereomer. The term *diastereomers* also includes stereoisomers that are called cis,trans isomers (formerly **geometric isomers**) such as *cis*- and *trans*-2-butene (**9, 10**) and *cis*- and *trans*-1,2-dimethylcyclopropane (**11, 12**). In each case there is no interconversion of the two isomers at room temperature because the C — C double bond or the ring structure prevents rotation about a C — C bond.

Figure 2.3 Stereoisomeric relationships among 2,3-disubstituted butanes.

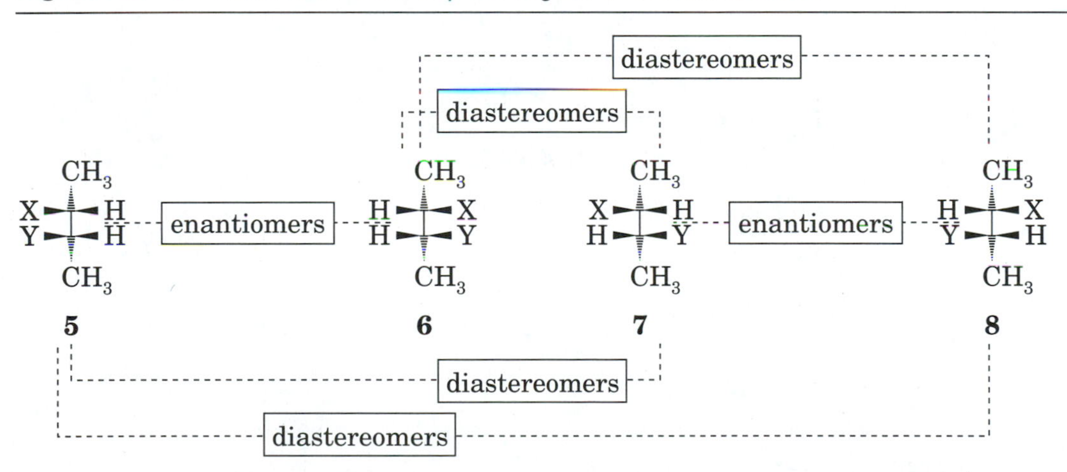

[14]The word *chirality* is attributed to Lord Kelvin (reference 35).

[15]Mezey, P. G., ed. *New Developments in Molecular Chirality*; Kluwer Academic Publishers: Dordrecht, Netherlands, 1991.

The cis and trans isomers of decalin (decahydronaphthalene) could be represented by the line drawings on the left (**13a** and **14a**, respectively) in Figure 2.4 and Figure 2.5. However, we could also show them as **13b** and **14b**, which convey the same information about the ring fusion but do not explicitly show the conformations of each six-membered ring. We could use even simpler representations of the same molecules by adopting the convention that a **solid dot** represents a hydrogen atom coming out of the plane of the molecule toward the viewer, as shown by **13c** and **14c**. The price of simplicity is a greater necessity to understand the implicit meaning of such drawings, or important stereochemical features may be unrecognized.

Symmetric, Asymmetric, Dissymmetric, and Nondissymmetric Molecules

Before continuing our discussion of molecular structures that exhibit enantiomerism, it will be useful to review briefly some designations of molecular

Figure 2.4 Representations of *cis*-decalin.

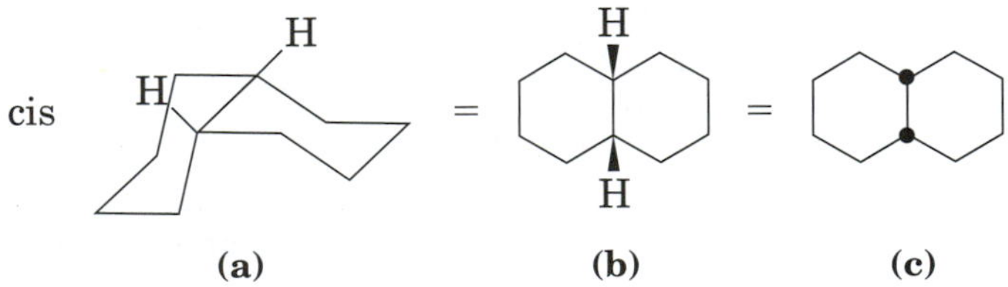

cis (a) = (b) = (c)

13

Figure 2.5 Representations of *trans*-decalin.

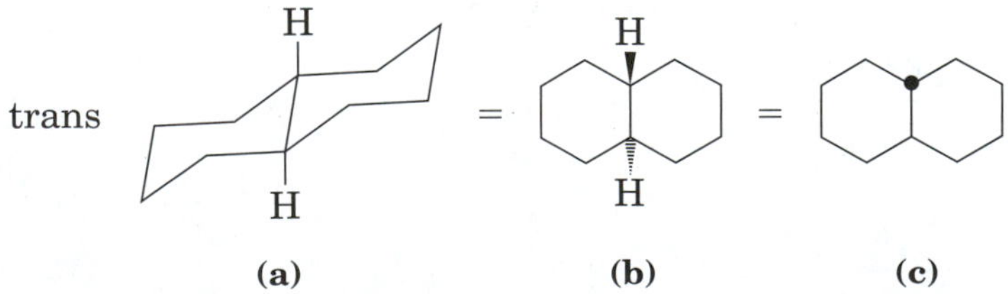

trans (a) = (b) = (c)

14

symmetry.[16,17,18,19] A **symmetry element** corresponds to a plane, point, or line about which (at least in principle) we could carry out a **symmetry operation** on a structure, the result of which would be a structure equivalent to the original one. Four types of symmetry operations (and their associated symmetry elements) need to be considered:[16]

1. reflection in a plane of symmetry
2. inversion of all atoms through a center of symmetry (center of inversion)
3. rotation about a proper axis
4. rotation about an improper axis (corresponding to rotation about an axis, followed by reflection through a plane perpendicular to that axis)

A C_n symmetry element means that there is an axis through the structure such that rotation about 360°/n produces a structure that appears just as it did before the rotation. A C_2 axis means a 180° rotation and is said to be a two-fold rotation axis. *cis*-2-Butene (Figure 2.6) has a C_2 rotation axis. It also has two planes of symmetry: both planes of symmetry are denoted vertical planes of symmetry (σ_v) because they include the C_2 rotation axis. Therefore, one is denoted as σ_v and the other one is designated σ_v'. *trans*-2-Butene (**10**, Figure 2.7) has a C_2 rotation axis and a plane of symmetry that is denoted a horizontal plane of symmetry (σ_h) because it is perpendicular to the C_2 rotation axis.

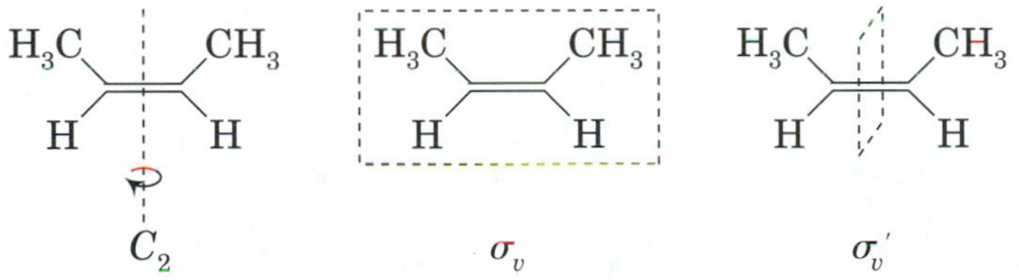

Figure 2.6 C_2 (left) σ_v (center) and σ_v' (right) symmetry elements in *cis*-2-butene.

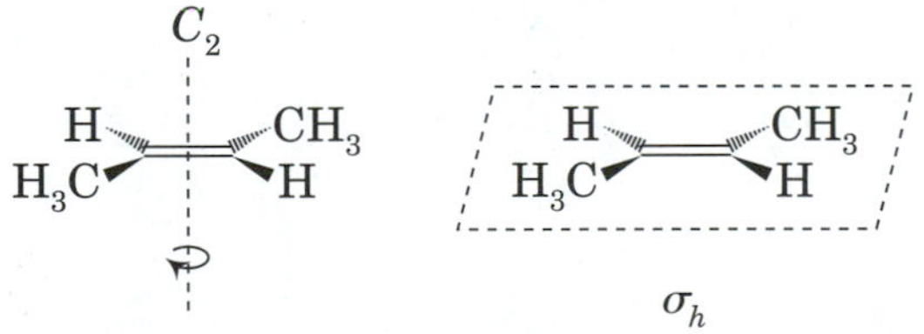

Figure 2.7 C_2 and σ_h symmetry elements in *trans*-2-butene.

[16]Cotton, F. A. *Chemical Applications of Group Theory*, 2nd Ed.; Wiley-Interscience: New York, 1971.

[17]Orchin, M.; Jaffé, H. H. *Symmetry, Orbitals and Spectra (S. O. S.)*; Wiley-Interscience: New York, 1971; pp. 91–136.

[18]Juaristi, E. *Introduction to Stereochemistry and Conformational Analysis*; Wiley-Interscience: New York, 1991.

[19]Heilbronner, E.; Dunitz, J. D. *Reflections on Symmetry*, (Pfalzberger, R., ill.); Verlag Helvetica Chimica Acta: Basel, 1993.

If a center of inversion (i) is present, reflecting each atom through the center of the molecule produces a structure equivalent to the starting structure.[20] This process is illustrated in the top portion of Figure 2.8 for a cyclobutane derivative. It must be emphasized that the atoms are shown with subscripts only to indicate the symmetry operation for the purposes of this discussion. In the absence of these labels, the inversion process would produce a structure that would appear to be identical to the starting structure.

A C_n axis is said to be a proper rotation axis. An improper axis, S_n, exists when equivalency is restored by carrying out two operations consecutively: a rotation about a proper axis, C_n, followed by reflection through a plane perpendicular to that axis. In the cyclobutane shown in Figure 2.8, for example, a C_2 rotation about an axis perpendicular to the plane of the four carbon atoms, followed by a reflection through the plane containing the four carbon atoms, produces a structure identical to the starting structure.[21] Note that neither the C_2 rotation nor the σ_h reflection symmetry elements are *present* in **15**.[22] It is the combination of the C_2 and σ_h operations that

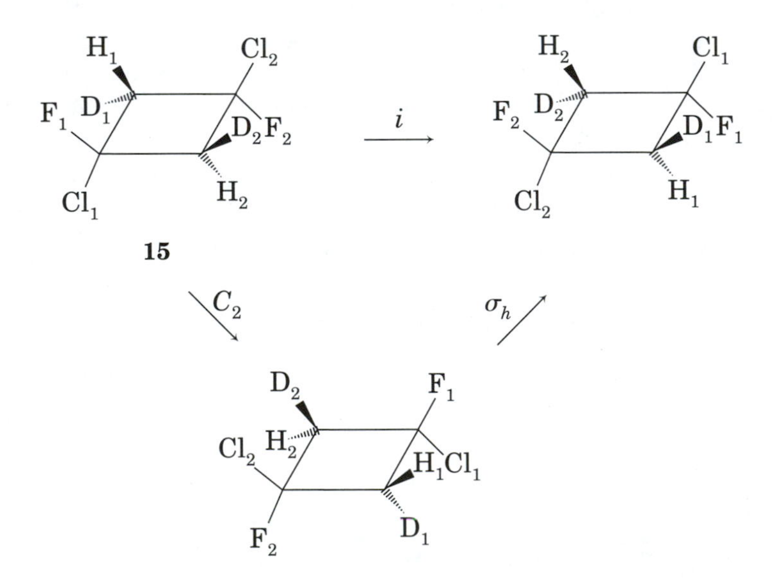

Figure 2.8 Examples of symmetry operations on a cyclobutane derivative.

[20]More precisely, the process of inversion through a point with coordinates of 0,0,0 moves each atom at point (x, y, z) to point $(-x, -y, -z)$. See reference 16, p. 19.

[21]As we will discuss in Chapter 3, cyclobutane itself is not planar but is slightly puckered. However, the planar form is taken to represent the time average of the accessible conformations, so this representation of the structure is used for symmetry analysis.

[22]We may carry out any symmetry *operation* on any structure, whether or not the corresponding symmetry *element* is present in the structure. (If the symmetry element is present, then the operation produces a structure in which every point is oriented the same as an equivalent point in the starting structure. If the symmetry element is not present, then the result of carrying out the symmetry operation is a structure that is not equivalent to the starting structure.)

produces the equivalent structure, so the S_2 symmetry element is present in **15**, but the C_2 and σ_h elements are not.

We commonly describe the symmetry of a structure in terms of a **point group**, which is a symmetry designator determined by the number and kind of symmetry elements present in that structure. Therefore, **15** belongs to the point group S_2. The point group D_n includes structures that contain a C_n plus n C_2 perpendicular to the C_n axis. The point group D_{nh} includes structures that contain a C_n, n C_2 perpendicular to the C_n axis, and a σ_h. For the sake of completeness we define an identity operator, E, that does not move any part of the structure.[23] It is convenient to think of a structure that has only the E element as belonging to point group C_1, since the only operation that restores the original structure is a 360° rotation. Details about the designation of point groups are provided in references 16 and 17, and a classification scheme for point group designations is reproduced in Figure 2.9.[24]

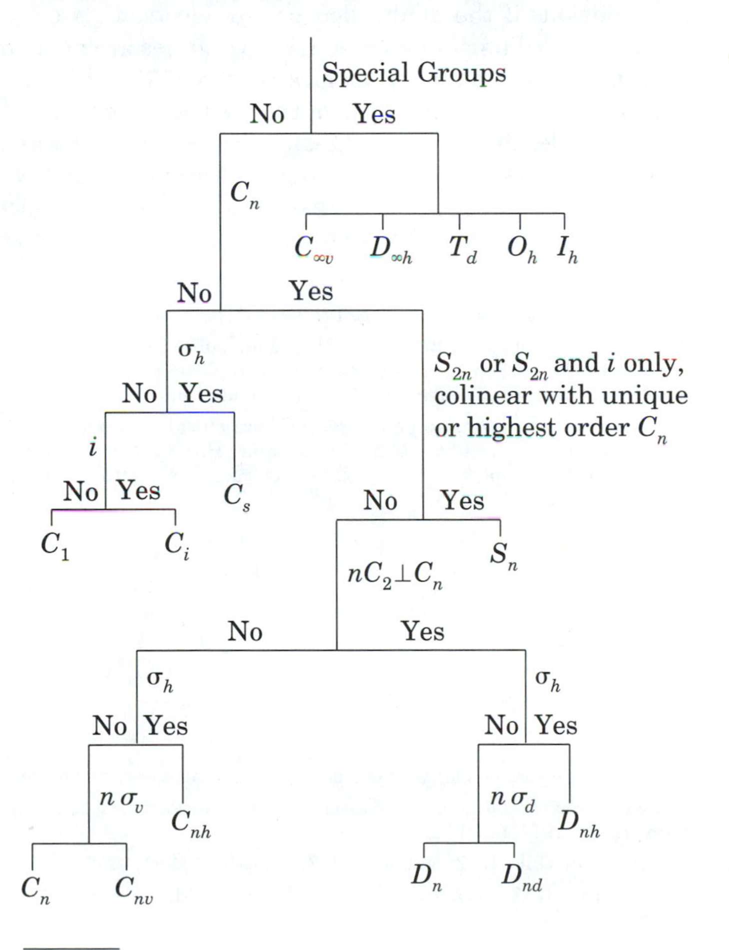

Figure 2.9 Point group classification scheme. (Adapted from reference 24.)

[23]The E operator is also called I (reference 17).

[24]Carter, R. L. *J. Chem. Educ.* **1968**, *45*, 44.

In this scheme, the phrase beside a vertical line is a question. (For example, the notation C_n means "Is there a C_n axis?".) If the answer to the question is yes, the scheme moves along the right line below that question. If the answer is no, the scheme moves along the left line below the question.

Point group designations facilitate discussions of chirality. We generally think first of molecules that show enantiomerism as being asymmetric[25] (without symmetry) due to the presence of a carbon atom with four different substituents, as in the 1-bromo-1-chloro-1-fluoroethane isomer **16**. More generally, asymmetric structures have three or more different substituents bonded to a central atom in such a way that the central atom and the substituents do not lie in one plane.[26] Such structures are said to exhibit enantiomerism due to asymmetry about a point and belong to point group C_1.[27]

Technically, a nitrogen atom with three different substituents meets the requirement for chirality. (The similarity to chirality about a carbon atom is most obvious if the nonbonded pair of electrons is considered to be a "substituent.") Ordinarily, however, trialkylamines are not chiral because rapid inversion produces a mirror image structure (**17**).[28] However, if inversion is slow with respect to the time scale of the measurement, chirality can be observed. For example, the aziridine-2,3-dicarboxylic acid **18** was isolated from a bacterium and was synthesized in high enantiomeric purity.[29] The diaziridine **19** can also be isolated and demonstrated to be optically active.[30] Similarly, the phenyl ethyl sulfoxide **20** can be resolved due to slow inversion at the sulfur.[31]

---

[25]It is a common mistake to misspell this word "assymetric."

[26]The central atom may have more than four substituents if it is further down the periodic table than the first row. Thus some metal complexes may also show enantiomerism due to a pattern of substituents (ligands) about a central atom.

[27]Structures that belong to point group C_1 are chiral, whether or not there are four different groups bonded to a central atom. For example, Hamill, H.; McKervey, M. A. *Chem. Commun.* **1969**, 864 have reported the synthesis of the chiral structure 3-methyl-5-bromoadamantanecarboxylic acid,

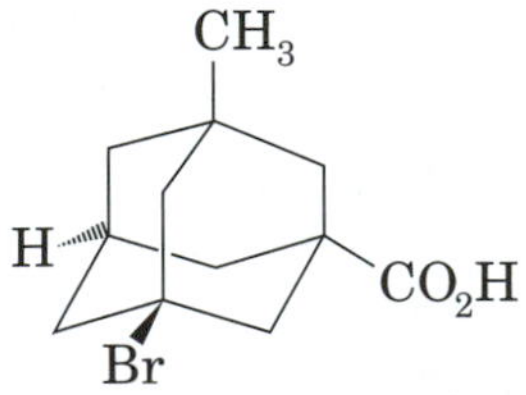

in which the center of chirality is a point in the center of the adamantane skeleton.

[28]Reviews have been given by Binsch, G. *Top. Stereochem.* **1968**, *3*, 97; Lambert, J. B. *Top. Stereochem.* **1971**, *6*, 19.

[29]Legters, J.; Thijs, L.; Zwaneburg, B. *Tetrahedron* **1991**, *47*, 5287.

[30]Dyachenko, O. A.; Atovmyan, L. O.; Aldoshin, S. M.; Polyakov, A. E.; Kostyanovskii, R. G. *J. Chem. Soc., Chem. Commun.* **1976**, 50 and references therein.

[31]Kobayashi, M.; Kamiyama, K.; Minato, H.; Oishi, Y.; Takada, Y.; Hattori, Y. *J. Chem. Soc. D, Chem. Commun.* **1971**, 1577.

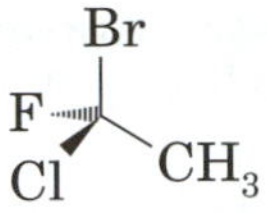

16 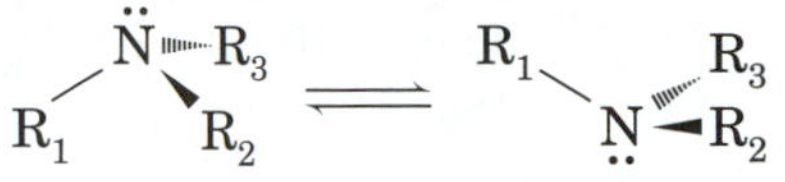 **17**

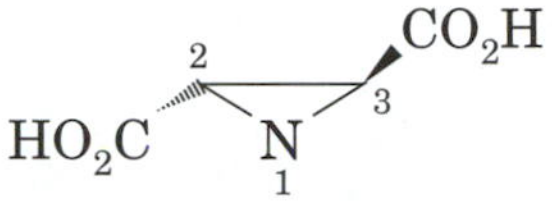

18 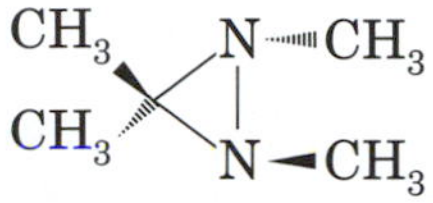 **19** 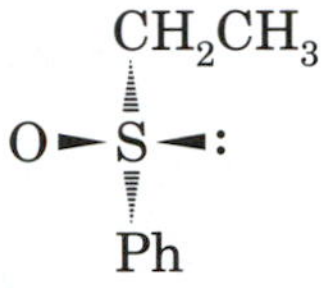 **20**

In addition to molecules belonging to point group C_1, there are structures that exhibit enantiomerism but which nevertheless do have some symmetry.[32] Therefore the more accurate statement of the requirement for enantiomerism is *dissymmetry*, a phrase coined by Pasteur to describe a structure that is not superimposable on its mirror image.[33,34,35] Any structure is chiral if it does not have an improper rotation axis (S_n),[36] that is, if it belongs to point group C_n or D_n. Those that belong to point group C_1 are asymmetric, while those that belong to point group C_2 and higher are dissymmetric. Some examples of chiral C_n structures are illustrated by those

[32]For a compilation of chiral structures, including structures categorized according to point groups, see Klyne, W.; Buckingham, J. *Atlas of Stereochemistry*, 2nd Ed.; Oxford University Press: New York, 1978.

[33]Pasteur, L. *Researches on the Molecular Asymmetry of Natural Organic Products*; Alembic Club Reprint No. 14, William F. Clay: Edinburgh, 1897. See also Shallenberger, R. S.; Wienen, W. J. *J. Chem. Educ.* **1989**, *66*, 67.

[34]For a discussion of the development of the theory of molecular dissymmetry and the introduction of the term *asymmetric* for *dissymmetric*, see O'Loane, J. K. *Chem. Rev.* **1980**, *80*, 41.

[35]Barron, L. D. *J. Am. Chem. Soc.* **1986**, *108*, 5539 noted that there is a difference between the words *dissymmetry* and *chirality*, since dissymmetry is the *absence* of certain symmetry elements, while chirality is the *presence* of the attribute of handedness. However, Mislow defined chirality as the "absence of reflection symmetry" and investigated the quantification of chirality: Buda, A. B.; Auf der Heyde, T.; Mislow, K. *Angew. Chem. Int. Ed. Engl.* **1992**, *31*, 989. Zabrodsky, H.; Peleg, S.; Avnir, D. *J. Am. Chem. Soc.* **1993**, *115*, 8278 have treated symmetry as a continuous and quantifiable property rather than as a "yes or no" condition. Zabrodsky, H.; Avnir, D. *J. Am. Chem. Soc.* **1995**, *117*, 462, have given a quantitative definition of chirality. See also the discussion in reference 15, pp. 241–256.

[36]For a more detailed discussion and a proof of this statement, see reference 16, p. 33.

with C_2 (**21**)[37] and C_6 (**22**)[38] symmetry. Chiral structures with D_n symmetry include those with D_2 (**23**[39] and **24**[40]) and D_3 (**25**)[41] symmetry.[42,43]

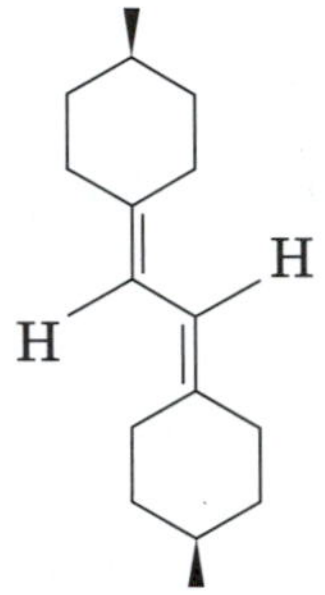

21 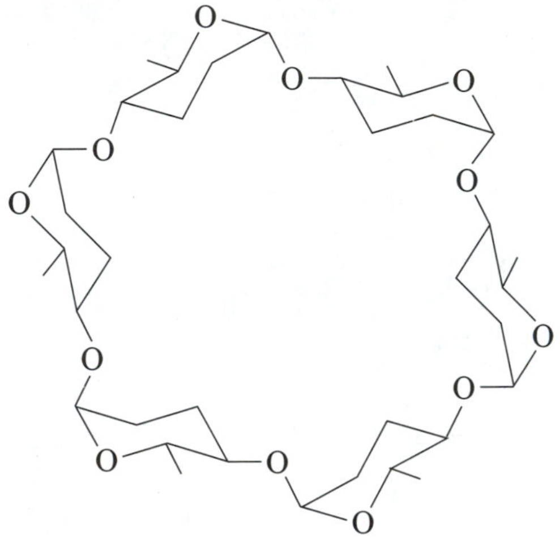 **22**

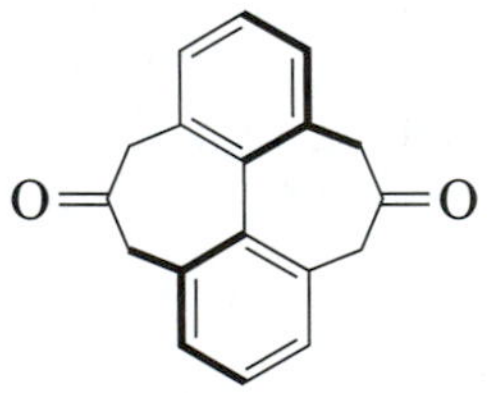

23 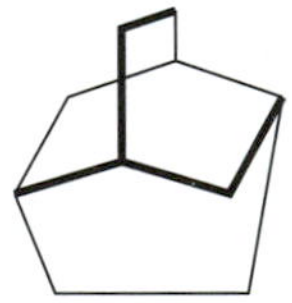 **24** 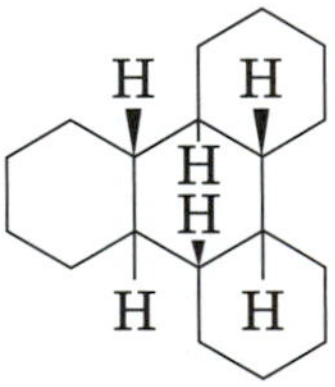 **25**

---

[37]Banks, R. B.; Walborsky, H. M. *J. Am. Chem. Soc.* **1976**, *98*, 3732.

[38]Farina, M.; Morandi, C. *Tetrahedron* **1974**, *30*, 1819. These authors present a general discussion of chiral molecules with high symmetry.

[39]Mislow, K.; Glass, M. A. W.; Hopps, H. B.; Simon, E.; Wahl, Jr., G. H. *J. Am. Chem. Soc.* **1964**, *86*, 1710.

[40]Tichý, M. *Collect. Czech. Chem. Commun.* **1974**, *39*, 2673.

[41]Farina, M.; Audisio, G. *Tetrahedron* **1970**, *26*, 1827, 1839.

[42]Chiral cage-shaped molecules with high symmetry have been reviewed by Naemura, K. in *Carbocyclic Cage Compounds: Chemistry and Applications*, Osawa, E.; Yonemitsu, O., eds.; VCH Publishers, Inc.: New York, 1992; pp. 61–90.

[43]A particularly interesting chiral D_3 structure is that reported for an isomer of the C_{78} fullerene; i.e., a chiral structure composed only of carbon atoms: Diederich, F.; Whetten, R. L.; Thilgen, C.; Ettl, R.; Chao, I.; Alvarez, M. M. *Science* **1991**, *254*, 1768. The C_{76} fullerene was isolated in enantiomeric form by kinetic resolution: Hawkins, J. M.; Meyer, A. *Science* **1993**, *260*, 1918.

A structure with C_i, C_s, C_{nv}, C_{nh}, D_{nd}, D_{nh}, T_d, O_h, or S_n symmetry is achiral and is said to be nondissymmetric.[44] Of particular interest are structures with S_n symmetry. We noted earlier that the S_2 operation is identical to inversion (i). The S_1 operation is identical to σ. It is frequently stated that structures having a σ or i symmetry element are achiral. However, any structure belonging to an S_n point group is nondissymmetric, whether or not it contains a mirror plane or a center of inversion. An example is the achiral isomer of 3,4,3′,4′-tetramethylspiro-(1,1′)bipyrrolidinium ion (**26**) with S_4 symmetry reported by McCasland and Proskow.[45]

26

Some structures are said to have chirality about an axis, which can be considered to be one of the rotational axes in a more symmetric structure that no longer exists because substituents have been moved due to "stretching" of the axis.[46,47] The structure on the left in Figure 2.10 is chiral about a point (X). Stretching the figure as indicated involves motion in a direction along the line XY, which is the axis of chirality. Each of the three elongated structures is now chiral. It should be stressed that the four substituents (a,

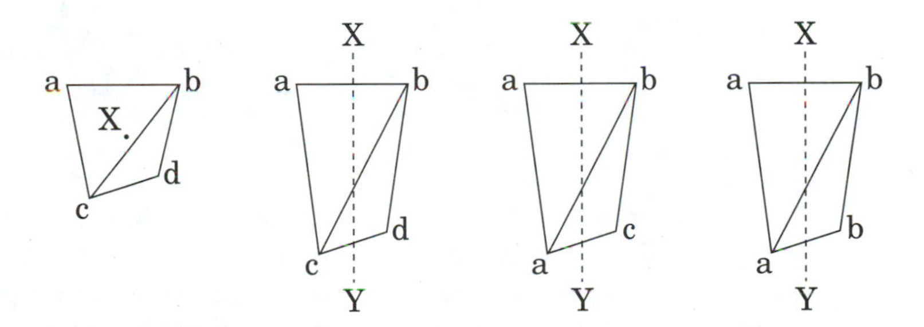

Figure 2.10 "Stretching" a structure with chirality about a point to produce structures with chirality about an axis. (Reproduced from reference 47.)

[44]For further discussions of the relationship of point groups to chirality, see Mislow, K. *Introduction to Stereochemistry*; W. A. Benjamin, Inc.: New York, 1966; pp. 33; reference 34; reference 163; Harris, D. C.; Bertolucci, M. D. *Symmetry and Spectroscopy*; Oxford University Press: New York, 1978; p. 9.

[45]McCasland, G. E.; Proskow, S. *J. Am. Chem. Soc.* **1955**, *77*, 4688; **1956**, *78*, 5646. It is interesting to note that **26** was reported to be "optically inactive within experimental error." The observed specific rotation of **26** was found to be $[\alpha]_D^{20}$ + 0.012° (water, c 5, l 0.5) when prepared from one precursor and $[\alpha]_D^{20}$ −0.027° when prepared from the enantiomeric precursor. The small optical activities apparently resulted from trace impurities in the final products.

[46]For illustrations and examples of both an axis of chirality and a plane of chirality, see Prelog, V. (with Cahn, R. S.) *Chem. in Britain* **1968**, *4*, 382. See also Prelog, V. *Science* **1976**, *193*, 17.

[47]Cahn, R. S. *J. Chem. Educ.* **1964**, *41*, 116.

b, c, d) no longer have to be different to produce chirality. Therefore all three of the elongated structures are chiral.

Chirality about an axis is exemplified by allenes, such as the 2,3-pentadiene enantiomer **27**.[48,49] In **27** the methyl and hydrogen substituents on carbon atom **2** lie in a plane (in the page) that is perpendicular to the plane containing the methyl and hydrogen substituents on carbon atom 4. Here the axis of chirality is coincident with the C2—C3—C4 bond axis. The structure belongs to point group C_2, so it is dissymmetric, not asymmetric. The C_2 rotation axis is perpendicular to the axis of chirality, as illustrated in Figure 2.11.[50] It must be emphasized that not all structures that are chiral about an axis have a C_2 rotation axis. For example, the 2,3-hexadiene enantiomer **28** also has an axis of chirality coincident with the C2—C3—C4 bond axis, but it does not have a C_2 rotation axis. The adamantane derivative **29** and appropriately substituted spiro compounds such as **30**[51] are also chiral about an axis.

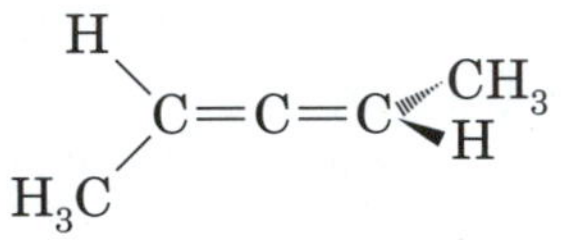

27 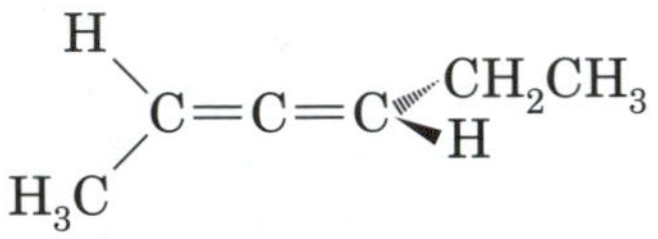 **28**

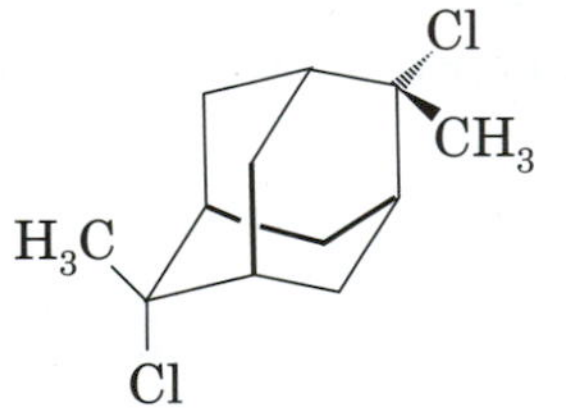

29 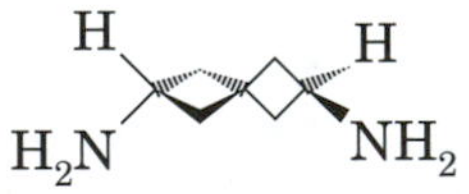 **30**

In examples **27–30**, chirality about an axis was enforced by rings or double bonds. However some molecules may be dissymmetric due to restricted rotation about a carbon-carbon single bond. Compounds such as the biphenyl **31**[52] are called **atropisomers**, since they are resolvable only be-

---

[48]Jones, W. M.; Walbrick, J. M. *Tetrahedron Lett.* **1968**, 5229; Waters, W. L.; Caserio, M. C. *Tetrahedron Lett.* **1968**, 5233.

[49]For a review of the chirality of allenes, see Runge, W. in *The Chemistry of Ketenes, Allenes and Related Compounds*, Part 1; Patai, S., ed.; Wiley-Interscience: Chichester, England, 1980; pp. 99–154.

[50]For a discussion of the C_2 axis and the axis of chirality in such systems, see Bassindale, A. *The Third Dimension in Organic Chemistry*; John Wiley & Sons, Ltd.: Chichester, England, 1984; pp. 34, 104.

[51]Hulshof, L. A.; Wynberg, H.; van Dijk, B.; de Boer, J. L. *J. Am. Chem. Soc.* **1976**, *98*, 2733.

[52]Newman, P.; Rutkin, P.; Mislow, K. *J. Am. Chem. Soc.* **1958**, *80*, 465.

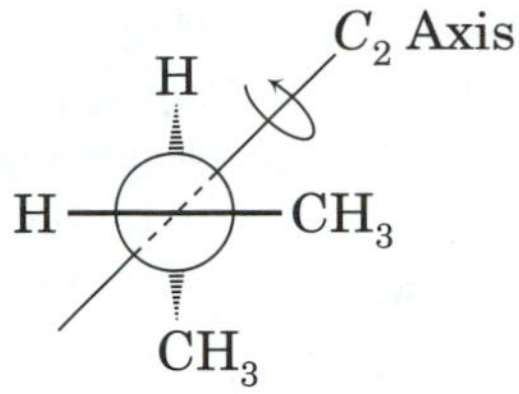

Figure 2.11
C_2 axis in **27**.

cause the activation energy for rotating about the biphenyl single bond is large due to steric hindrance of the pairs of ortho substituents.[53] Although there can be some variation in the angle that the plane of one phenyl group makes with respect to the plane of the other phenyl, any particular conformation of **31** has a C_2 rotation axis and is dissymmetric. As we noted in the case of allenes, however, a C_2 rotation axis may be present in a molecule that is chiral about an axis, but it is not required. Replacing one of the CO_2H groups of **31** with a CH_2OH group would remove the C_2 rotation axis, but the structure would still be chiral about an axis.

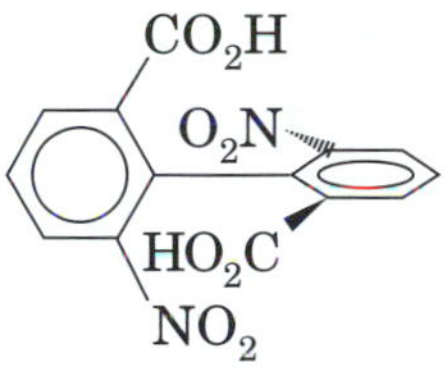

31  **32**

We may consider a plane to be an element of chirality if part of the structure lies in the plane but substituents leave or enter the plane so as to produce chirality. An example is the [6]paracyclophane **33**, in which the plane of chirality can be defined by the three positions marked with a star.[54] In many cases we may consider the chiral element to be either an axis or a plane, and Schlögl has noted the difficulty in defining a plane of chirality.[54] *trans*-Cyclooctene (**32**) is chiral with respect to a plane that includes the two double-bonded carbon atoms and one of the carbon atoms adjacent to the double bond.[55,56] More than one chain can pass through the double bond. For example, one could also consider either a plane of chirality or an axis of chirality associated with the carbon-carbon double bond in the molecule

[53](a) Ōki, M. *Top. Stereochem.* **1983**, *14*, 1 and references therein; (b) see also Ōki, M. *The Chemistry of Rotational Isomers*; Springer-Verlag: Berlin, 1993.

[54]Schlögl, K. *Top. Curr. Chem.* **1984**, *125*, 27. (This example is adapted from the discussion on pp. 30–31.)

[55]The (+) enantiomer had $[\alpha]_D^{25}$ + 414° (c 0.55, methylene chloride). Cope, A. C.; Ganellin, C. R.; Johnson, Jr., H. W.; Van Auken, T. V.; Winkler, H. J. S. *J. Am. Chem. Soc.* **1963**, *85*, 3276.

[56]In an X-ray crystal study, (−)-*trans*-cyclooctene was found to have the (*R*) configuration, and the twist angle of the carbon-carbon double bond was estimated to be 43.5°. Manor, P. C.; Shoemaker, D. P.; Parkes, A. S. *J. Am. Chem. Soc.* **1970**, *92*, 5260.

"[8.10]-betweenanene" (**34**).[57,58] Helical (threaded) structures can also be considered chiral with respect to either an axis or a plane, as in the case of hexahelicene (**35**).[59,60,61,62] Catenanes (structures formed by interlocking rings) can be chiral, as exemplified by the double interlocked **36**.[63] Moreover, the trefoil knot structure **37** is also chiral.[64,65,66] Schill[67] and Sauvage[68] provided a detailed account of the occurrence, synthesis, and chirality of these and other novel molecular structures.

$(CH_2)_2$ CO_2H $(CH_2)_8$ $(CH_2)_{10}$

33 **34** **35**

36 **37**

---

[57]Marshall, J. A. *Acc. Chem. Res.* **1980**, *13*, 213 and references therein.

[58]For a discussion of the stereochemistry of these and other structures with twisted double bonds, see Nakazaki, M.; Yamamoto, K.; Naemura, K. *Top. Curr. Chem.* **1984**, *125*, 1.

[59]Resolution of racemic **35** through formation of diastereotopic charge transfer complexes with optically active 2-(2,4,5,7-tetranitro-9-fluorenylideneaminooxy)propionic acid produced (+)-**35** with $[\alpha]_D^{25}$ + 3707° ($CHCl_3$). Newman, M. S.; Lutz, W. B.; Lednicer, D. *J. Am. Chem. Soc.* **1955**, *77*, 3420.

[60]Newman, M. S.; Lednicer, D. *J. Am. Chem. Soc.* **1956**, *78*, 4765.

[61]For leading references to the racemization of chiral hexahelicene derivatives, see Prinsen, W. J. C.; Hajee, C. A. J.; Laarhoven, W. H. *Polycyclic Aromatic Compounds* **1990**, *1*, 21.

[62]An X-ray crystal structure determination confirmed the helical structure of the molecule: Mackay, I. R.; Robertson, J. M.; Sime, J. G. *J. Chem. Soc. D, Chem. Commun.* **1969**, 1470.

[63]For a discussion and a report of the synthesis of a doubly interlocked catenane, see Nierengarted, J.-F.; Dietrich-Buchecker, C. O.; Sauvage, J.-P. *J. Am. Chem. Soc.* **1994**, *116*, 375. For a discussion of knots in proteins, see Liang, C.; Mislow, K. *J. Am. Chem. Soc.* **1994**, *116*, 11189.

[64]*Cf.* Krow, G. *Top. Stereochem.* **1970**, *5*, 59 *ff.* and references therein.

[65]Boeckmann, J.; Schill, G. *Tetrahedron* **1974**, *30*, 1945.

[66]For assignment of *R* or *S* designations to molecular knots, see Mezey, P. G. *J. Am. Chem. Soc.* **1986**, *108*, 3976.

[67]Schill, G. *Catenanes, Rotaxanes and Knots*; Boeckmann, J., trans.; Academic Press: New York, 1971.

[68]See also Sauvage, J.-P. *Acc. Chem. Res.* **1990**, *23*, 319.

Designation of Molecular Configuration

Chirality about a Point

The goal of organic chemical nomenclature is to provide concise, unambiguous identification of the structures, including stereochemistry, of organic compounds. An important aspect of structure is **configuration**—the three-dimensional arrangement of atoms that characterizes a particular stereoisomer. The most generally used system for specifying configuration is the (*R*) and (*S*) system, which was introduced by Cahn, Ingold, and Prelog.[69,70] For a chiral molecule with four different substituents bonded to a tetrahedral carbon atom, one begins by assigning priorities to the substituents.[71] To assign the (*R*) or (*S*) designation to a particular configuration, one views the structure (rather, a drawing or a model of it) such that the substituent with the lowest priority is held away from one's eye. One then views the other three substituents. If, in looking from the highest priority substituent, to the next highest, to the third highest, one's eye traces a clockwise (to the right) pattern, the configuration is said to be (*R*).[72] If, however, the substituents are arranged such that looking from highest priority to second priority to third priority traces a counterclockwise pattern, the configuration is (*S*).[73]

The priorities of substituents are determined according to a set of criteria, some of which are presented here.[74] The highest priority atom is the one with the highest atomic number. If the chiral center has two identical atoms bonded to it, then one considers the substituents for each of them and assigns priority to the atom that has the higher priority substituent. Thus CH_2Br takes priority over CH_2Cl because the atomic weight of bromine is greater than that of chlorine. Similarly, CH_2CH_2Cl is higher in priority than CH_2CH_2F. However, CH_2CHFCH_3 is higher in priority than $CH_2CH_2CH_2Cl$.

---

[69]Cahn, R. S.; Ingold, C. K.; Prelog, V. *Experientia* **1956**, *12*, 81; *Angew. Chem. Int. Ed. Engl.* **1966**, *5*, 385.

[70]For proposed modifications of the Cahn-Ingold-Prelog system, see Dodziuk, H.; Mirowicz, M. *Tetrahedron: Asymmetry* **1990**, *1*, 171; Dodziuk, H. *Tetrahedron: Asymmetry* **1992**, *3*, 43; Mata, P.; Lobo, A. M.; Marshall, C.; Johnson, A. P. *Tetrahedron: Asymmetry* **1993**, *4*, 657.

[71]Priorities are often indicated numerically. Some authors use the numbers I–IV (or 1–4) with IV (4) being highest priority, but others use the same numbers with I (1) being highest priority. To avoid confusion, we will use the letters A–D, with A being highest priority.

[72]The symbol (*R*) stands for the Latin word *rectus*, which is often said to mean "right." As Todd and Koga have noted, however, *rectus* means "straight" or "proper" and *dexter* means "right." The symbol (*S*) stands for *sinister*, which does mean "left." *Cf.* Todd, D. *J. Chem. Educ.* **1987**, *64*, 732; Koga, G. *Chem. Eng. News* **1988** (April 4,) 3.

[73]It has been noted that the ever-increasing popularity of digital clocks and watches may have subtle, unforeseen effects, and that someday students of organic chemistry may not know the meaning of the terms clockwise and counterclockwise.

[74]Additional criteria apply to situations not considered in the present discussion. Details are provided in reference 69.

Figure 2.12 Priorities for carbon atoms bearing oxygen substituents. (Reproduced from reference 47.)

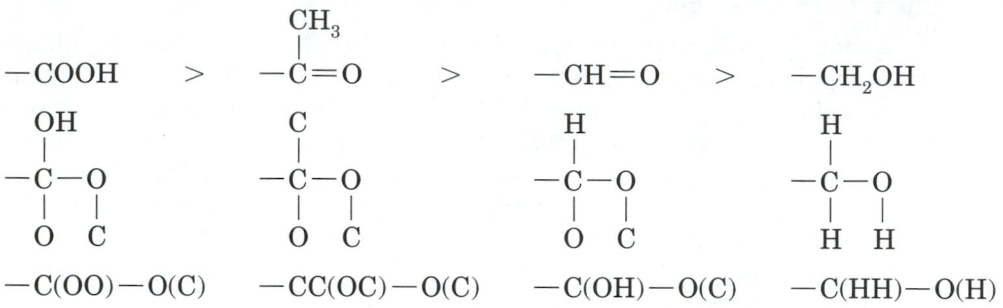

—COOH	>	—C(CH₃)=O	>	—CH=O	>	—CH₂OH
—C(OH)(O)—O—C		—C(C)(O)—O—C		—C(H)(O)—O—C		—C(H)(H)—O—H
—C(OO)—O(C)		—CC(OC)—O(C)		—C(OH)—O(C)		—C(HH)—O(H)

If a chiral center has as substituents CH_2Cl and $CHCl_2$, priority cannot be assigned on the basis of atomic number, since both carbon atoms have a chlorine as a substituent. In that case the group $CHCl_2$ has priority over CH_2Cl, since there are two chlorine atoms. For this purpose, a substituent atom that is part of a double bond is considered to be bonded twice. For example, Figure 2.12 shows that a CO_2H group is considered to be C(OO)—O(C), so it has priority over a $COCH_3$ group, which in turn is higher in priority than a CHO group.

It must be emphasized that the number of atoms of a particular type is only important when comparing substituents on two atoms that have the same atomic number. The $CHCl_2$ substituent on C2 of 1,1,3-trichloro-2-methylpropane (**38**) has greater priority than the CH_2Cl substituent because the atoms attached to the carbon atom are H and Cl in both cases, so the substituent with two chlorine atoms is higher in priority than the substituent with one chlorine atom. However, the CH_2Br substituent in 3-bromo-1,1-dichloro-2-methylpropane (**39**) has higher priority than the $CHCl_2$ group because bromine has a higher atomic number than chlorine. The fact that there are two chlorine atoms does not matter.

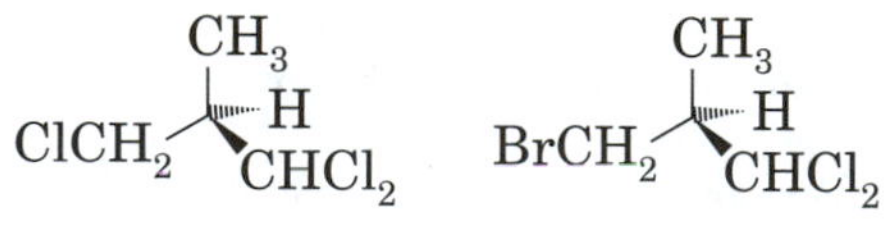

CH_3, H, $ClCH_2$, $CHCl_2$ — (*R*) **38**

CH_3, H, $BrCH_2$, $CHCl_2$ — (*S*) **39**

If two substituents are identical except for isotopic substitution, then the isotope with the greater atomic mass takes priority over the isotope of lower atomic mass. That is, the group CH_2D is higher in priority than the group CH_3, and the group $CH_2—CH_2—{}^{13}CH_3$ is higher in priority than the group $CH_2—CH_2—CH_3$. It must be emphasized, however, that isotopic substitution applies only in cases in which no other criteria apply. The group $CH_2—CH_2—F$ takes priority over the group $CD_2—CH_3$, for example. A more detailed discussion of the sequence rules appears in the IUPAC nomenclature rules,[75] and a listing of groups in order of priority was given by Klyne and Buckingham.[76]

Some examples of the assignment of (*R*) and (*S*) configurations are shown in structures **40–42**.

(*R*) **40** (*R*) **41** (*S*) **42**

The (*R*) and (*S*) nomenclature system is very useful in naming biological products such as α-tocopheryl. Natural vitamin E is (2*R*,4′*R*,8′*R*)-α-tocopherol (**43**), but commercial vitamin capsules contain the acetates of both the natural vitamin and of (2*S*,4′*R*,8′*R*)-α-tocopherol. Both acetates are hydrolyzed to the phenol in the intestine and are absorbed into the body, where they inhibit peroxidation of lipids, but the acetate of natural vitamin E is hydrolyzed more rapidly.[77]

(2*R*, 4′*R*, 8′*R*)-α-tocopherol

43

The compound (*R*)-(−)-muscone (**44**), obtained from the male musk deer, was used in perfumes.[78]

---

[75] IUPAC Organic Chemistry Division Commission on Nomenclature of Organic Chemistry, *Nomenclature of Organic Chemistry*; prepared by Rigaudy, J.; Klesney, S. P.; Pergamon Press: Oxford, England, 1979; pp. 486–490.

[76] Reference 62, p. xiv.

[77] Ingold, K. U. *Aldrichimica Acta* **1989**, *22*, 69.

[78] For a synthesis, see Oppolzer, W.; Radinov, R. N. *J. Am. Chem. Soc.* **1993**, *115*, 1593.

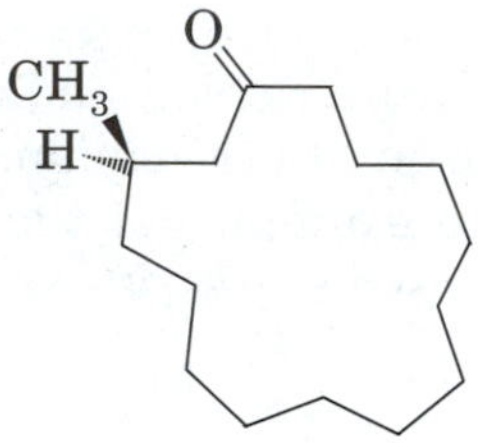

(*R*)-(−)-muscone

44

Chirality about an Axis or a Plane

The (*R*) and (*S*) nomenclature system can also be used for structures with an axis or plane of chirality. The dissymmetry is factorized into stereogenic units with the following order of priority: centers, axes, planes.[64] If an axis or plane of chirality is present, then we apply the additional rule that substituents on the end of an axis or surface of a plane nearer the observer are arbitrarily given higher priority than those further away. This rule takes priority over the rules that atoms with multiple substituents have those substituents multiplied to establish connectivity patterns and that a substituent with a higher atomic number (or mass) is higher in priority than one with a lower atomic number (or mass).

To illustrate the application of the new rule, let us consider first the allenic compound glutinic acid, **45(a)**.[79] Viewing the structure along the C=C=C axis from the right of the drawing (as indicated by the eye symbol) would produce the image **45(b)**. Using the rule that near groups precede far groups in priority, we first assign highest priority (A) to the COOH group and second priority (B) to the hydrogen on the near carbon atom. The COOH group on the far carbon atom is third priority (C), and the hydrogen on that carbon atom is lowest in priority (D). Now we determine the configuration as we would for a chiral carbon having the same substituents with priorities A, B, C and D bonded to it. Thus, the structure is determined to be (*R*).

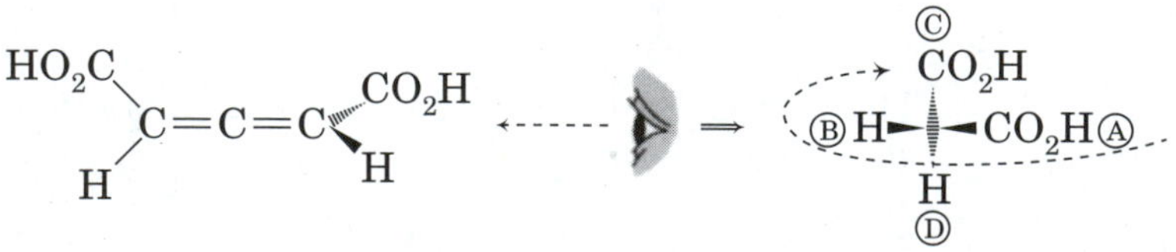

45 (a) **45 (b)**

[79]Agosta, W. C. *J. Am. Chem. Soc.* **1964**, *86*, 2638.

As another example, consider the biphenyl **31(a)** as shown in Figure 2.13.[52] Viewing the three-dimensional representation from the perspective indicated by the top eye symbol would produce the image **31(b)**.[80] Again assigning the two near groups higher priority than the two far ones, the structure is found to be (*S*). It may not be immediately obvious, but the choice of perspective of the original structure does not affect the designation. Viewing **31(a)** from the angle shown by the bottom eye produces **31(c)**, which is also an (*S*). The (*R*) or (*S*) designation of *trans*-cyclooctene (**32**, page 73) can most easily be determined by considering the structure to have chirality about an axis. Viewing **32** as indicated by the eye symbol reveals it to have the (*R*) configuration. Krow detailed the application of the rules for planar chirality to **32** and to other molecules,[64] and Hirschmann and Hanson described approaches to factoring chirality and stereoisomerism.[81]

Frequently designation of helical molecules is done not with (*R*) and (*S*) but with (*P*) (for + or "plus") and (*M*) (for − or "minus") instead.[82] Hexahelicene **35** (page 74) is said to be left-handed since the helix makes a left-handed or counterclockwise turn as it proceeds *away* from the observer. Therefore **35** is a (*M*) helix, which corresponds to the (*R*) designation. (Viewing the helix from the other end would not change this designation.) Similarly, **45** can be considered a (*M*) helical structure in which the CO_2H groups move counterclockwise as they proceed away from the observer. Conversion of (*M*) to (*R*) and (*P*) to (*S*) is easy, since the relative order of the letters *M*

Figure 2.13
Determination of (*R*) or (*S*) designation for an atropisomer.

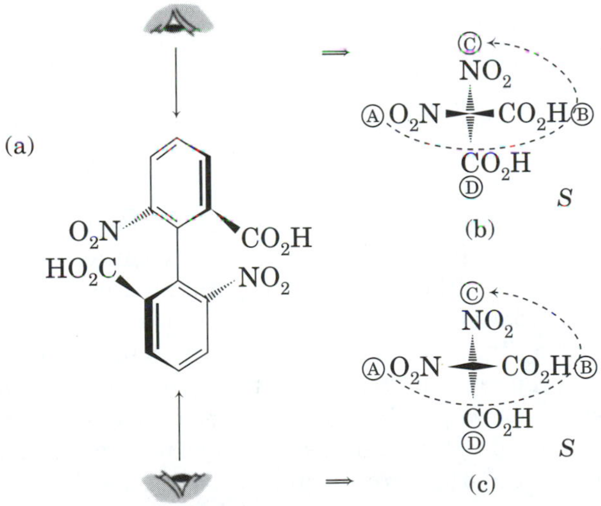

[80]Note that viewing **31(a)** so that the NO_2 and CO_2H groups are horizontal on the near phenyl ring and vertical on the far phenyl ring requires a slight rotation of the entire structure about the bond connecting the two benzene rings.

[81]Hirschmann, H.; Hanson, K. R. *Top. Stereochem.* **1983**, *14*, 183.

[82]Although the symbols (*M*) and (*P*) are not derived from words of foreign origin, they are nonetheless italicized: Dodd, J. S., ed. *The ACS Style Guide*; American Chemical Society: Washington, DC, 1986; p. 75.

and P corresponds to the relative order of the letters R and S in the alphabet.[64]

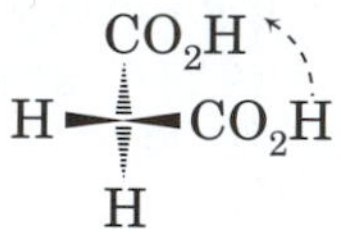

CO_2H
H — — CO_2H
H

45

Fischer Projections

The (R) and (S) system is useful in discussing the Fischer projection, which is one of the most important two-dimensional representations of chiral molecules. This type of drawing was initially developed to display the stereochemical relationships among carbohydrates with several chiral centers,[83] but it is useful in many other applications also. In a Fischer projection, molecules are represented by crossing vertical and horizontal lines, with each intersection representing a carbon atom. Horizontal lines represent bonds that would project forward in space if we drew the molecule using perspective notation for the chemical bonds. Vertical lines represent bonds that would project back from the carbon atom at the point of intersection. For example, the stereoisomer of 2-chlorobutane represented in many different ways in Figure 2.1 (page 60) would be shown as **3(g)**.

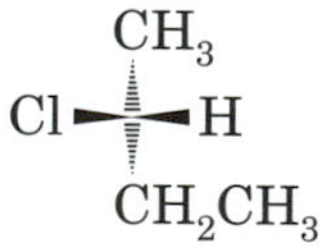

CH_3 / Cl — — H / CH_2CH_3 = CH_3 / Cl — H / CH_2CH_3

3(a) 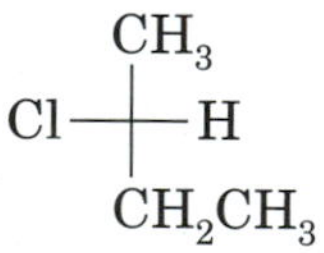 **3(g)**

Figure 2.14 illustrates a Fischer projection for a structure with two chiral centers, and it also shows other representations of the same molecule. One can convert from one representation to another by imagining the view to be had from each perspective. Thus viewing **46(a)** from the perspective indicated by the eye would produce the bold/dashed wedge figure **46(b)**, which is properly oriented (substituents on a horizontal line project *toward* the viewer) to simply draw the Fischer projection, **46(c)**. Clearly a Fischer projection need not, indeed, usually does not, represent the most stable conformation for a structure. Such projections are drawn only to convey information about the three-dimensional arrangement of atoms in structures that have chiral centers.

It must be emphasized that a two-dimensional figure can represent a three-dimensional structure only if we accept implicit rules for drawing and

[83]Lichtenthaler, F. W. *Angew. Chem. Int. Ed. Engl.* **1992**, *31*, 1541. Also see Maehr, H. *Tetrahedron: Asymmetry* **1992**, *3*, 735.

Figure 2.14 Fischer representation and equivalent representations of 2-chloro-3-pentanol.

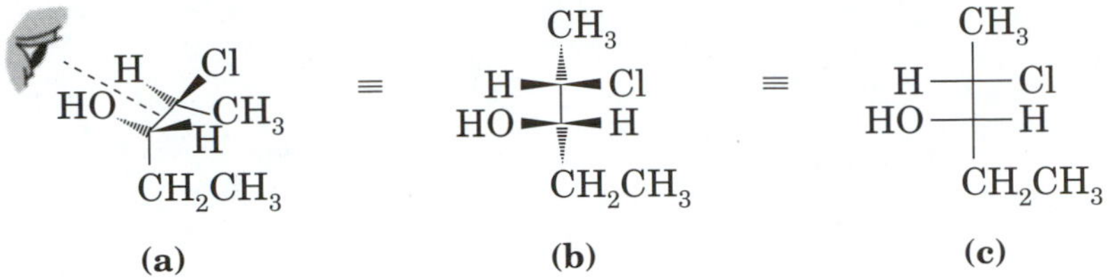

46 (a), (b), (c)

changing it. Rotation of an entire Fischer projection 180° about an axis perpendicular to the page and through the center of the carbon skeleton (for example, conversion of **47(a)** to **47(b)** as shown in Figure 2.15) or the cyclic rotation of any three substituents to a chiral center while holding the fourth substituent fixed (conversion of **47(b)** to **47(c)**, Figure 2.16) preserves the configuration of the structure, as indicated by the (*S*) designation shown for each chiral center. Therefore, these two processes are allowed operations on Fischer projections.

However, a 90° rotation of the figure, exchanging any two groups (**47(a)** to **48**, Figure 2.17) or "pancaking" or flipping the structure (**47(a)** to **49**, Figure 2.18) gives the enantiomer of the original chiral center. Note that **47** and **48** are disastereomers, since only one of the two chiral centers was inverted, while **47** and **49** are enantiomers, because there was inversion at both chiral centers.

Figure 2.15 Effect of rotating a Fischer projection 180° about the midpoint of the C2—C3 bond.

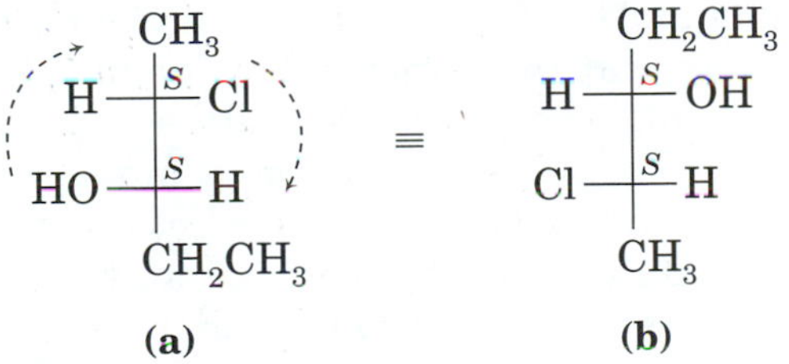

47 (a), (b)

Figure 2.16 Effect of rotating 120° about the C2—C3 bond.

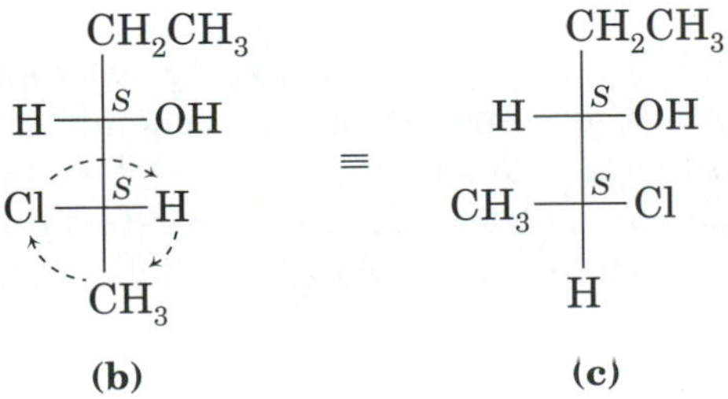

47 (b), (c)

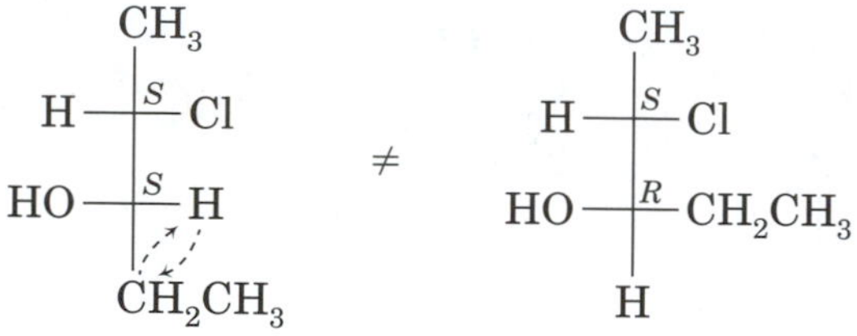

Figure 2.17 Effect of exchanging two groups.

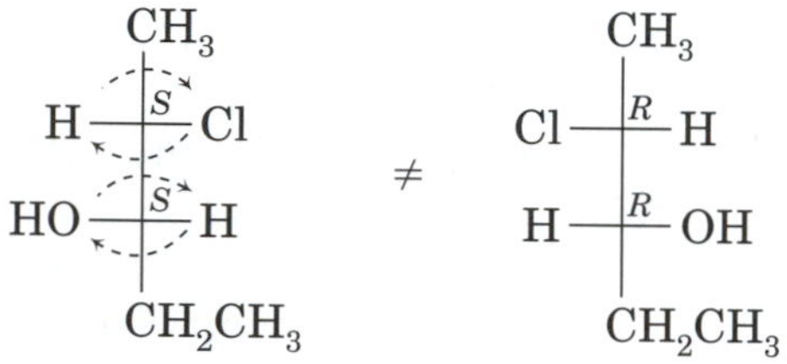

Figure 2.18 Effect of "pancaking" a Fischer projection.

The conventions we have described for representing the stereochemistry of chiral structures are adequate for many representations of molecular structures in organic chemistry. However, a stereochemical drawing, once committed to paper, represents a specific stereoisomer. That is, the image defines the chirality sense of the object it depicts. In the real world, however, chemists must convey information about structures whose absolute configurations may be unknown. For example, suppose we are seeking to identify an unknown substance and have determined from atomic connectivities that it is 2-chloro-3-hydroxypentane. Suppose also that we have determined the relative configurations of the two chiral centers and know that the substance being studied is either the (R,R) or the (S,S) isomer. We might represent the compound with a drawing such as **47(a)**, but we would have to verbally augment the figure by stating that the compound under investigation is either the structure drawn or its mirror image. Similarly, we cannot represent a racemic mixture with just one stereodrawing unless there is an accompanying explanation.

The stereochemical nomenclature is highly developed and can describe, in single terms, all of these aspects of stereochemical information. To provide a graphic counterpart to this uniqueness of verbal stereodescription, Maehr suggested a convention based on targeted redeployment of commonly used stereodescriptors (stereobonds) as follows:[84]

---

[84]Maehr, H. *J. Chem. Educ.* **1985**, *62*, 114.

1. *Solid* and *broken wedges* are descriptors of topography and therefore denote the *absolute configuration* of a chiral structure. Thus (2*S*,3*R*)-2,3-dibromobutanol is represented by Maehr as **50**.

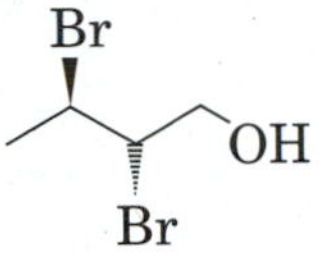

Br

OH

Br

50

2. *Solid* and *broken bold lines* are descriptors of geometry. Employed within a stereostructure, they portray the stereochemical relationship among similarly represented stereocenters within a molecule, but they also imply that a molecule is *racemic*. Therefore, **51(a)** represents both the structure obtained by replacing the bold line by a solid wedge and the dashed line by a broken wedge, (**48**), and the enantiomer of that structure, **51(b)**.

racemic

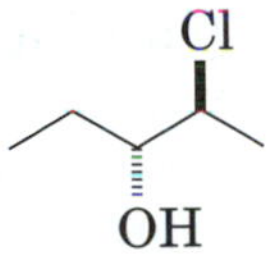

Cl

OH

51(a)

both

Cl

OH

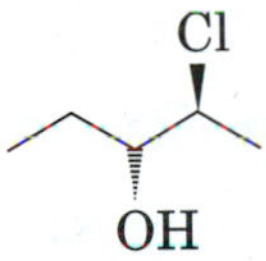

48

enantiopure, absolute configuration known

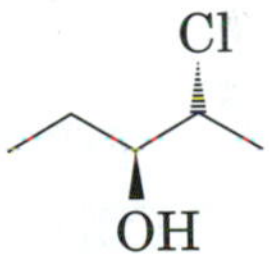

Cl

OH

51(b)

either one or the other, but not both

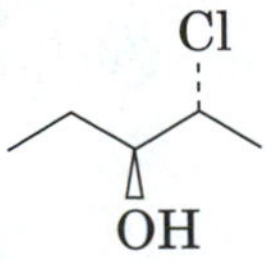

Cl

OH

enantiopure, but absolute configuration unknown

51(c)

3. *Wedge outlines* and *dotted (broken) lines* are also descriptors of geometry and portray the stereochemical relationship among similarly represented stereocenters. They inform the viewer that the compound is enantiopure, but that the chirality sense is unknown. Thus **51(c)** is either

48 or **51(b)**, but not both. Similarly, **52** is either (2*S*,3*R*)- or (2*R*,3*S*)-2,3-dibromobutanol, but not both.

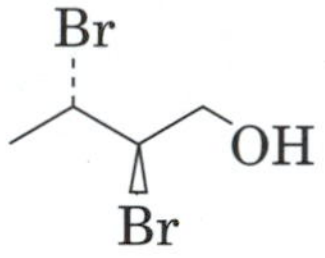

Br
OH
Br

52

The stereochemical information for some molecules with several stereocenters may be limited, as may be the case during the structure elucidation of a new natural product. The absolute configuration may be established for the stereocenters in one part of the molecule, while in another part of the molecule only relative configurations may be known. In such a case, the Maehr convention allows the unambiguous representation of the stereochemical knowledge about the compound with a single diagram incorporating descriptors of type (1) and (3), respectively. A single diagram using type (1) and type (2) descriptors could represent the product obtained by esterifying a racemic carboxylic acid with an optically active alcohol.

Many chemical structure drawing programs include tools for drawing all of the stereobonds needed for the Maehr convention, and the convention has been adopted in some publications.[85] However, alternative verbal stereochemical descriptors are also in use. In particular, Brewster[86] discussed descriptors suggested by Prelog,[87] Carey and Kuehne,[88] and Noyori.[89,90] It remains to be seen which set of descriptors will find widest use among chemists.

Stereochemical Nomenclature

There is an extension of the (*R*) or (*S*) system that is useful for those cases in which we know the relative configuration of the chiral centers in a molecule, but we do not know the actual three-dimensional structure. For example, the drawing of structure **53** indicates that the relationship of the methyl and chloro substituents is cis but that we do not know whether the structure has the configuration shown or its mirror image. In one system of nomenclature,[91] the first-named substituent is arbitrarily assumed to be on

---

[85] *Cf.* Kende, A. S., ed. *Org. Synth.* **1986**, *64*, p. x.

[86] Brewster, J. H. *J. Org. Chem.* **1986**, *51*, 4751.

[87] Prelog, V.; Helmchen, G. *Angew. Chem. Int. Ed. Engl.* **1982**, *21*, 567; Seebach, D.; Prelog, V. *Angew. Chem. Int. Ed. Engl.* **1982**, *21*, 654.

[88] Carey, F. A.; Kuehne, M. E. *J. Org. Chem.* **1982**, *47*, 3811.

[89] Noyori, R.; Nishida, I.; Sakata, J. *J. Am. Chem. Soc.* **1981**, *103*, 2106.

[90] See also the discussion of Tavernier, D. *J. Chem. Educ.* **1986**, *63*, 511.

[91] *Cf.* Fletcher, J. H.; Dermer, O. C.; Fox, R. B. *Nomenclature of Organic Compounds, Adv. Chem. Ser. 126*, American Chemical Society: Washington, DC, 1974; p. 108.

a (R) chiral center, and that center is denoted as (R^*). The configuration of another chiral center is then determined to be either (R^*) or (S^*) according to this arbitrary assumption. Another way to treat the same problem is to use the prefix *rel* (for relative) to indicate that the actual configuration is not known. Thus **53** could be named either as (1R^*,3S^*)-1-chloro-3-methylcyclohexane or as *rel*-(1R,3S)-1-chloro-3-methylcyclohexane.

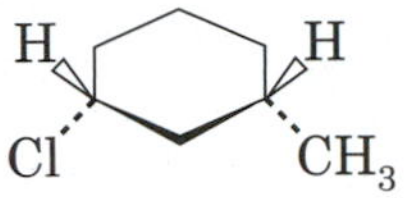

(1R^*,3S^*)-1-Chloro-3-methylcyclohexane
or
rel-(1R,3S)-1-Chloro-3-methylcyclohexane

53

The *r* (for "reference") system has been developed to serve as the reference for the stereochemical placement of substituents and is particularly useful for diastereomers.[92] In the case of **54**, for example, the carboxylic acid function is taken as the reference, and the other substituents are indicated as being either cis (c) or trans (t) to it.

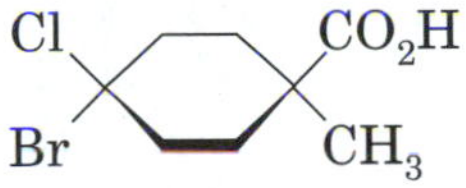

t-4-bromo-*c*-4-chloro-1-methyl-*r*-1-cyclohexanecarboxylic acid

54

The (R) and (S) priority rules have also been adapted to naming geometric isomers for which the designations cis or trans are inconvenient or ambiguous. Assigning priorities to the two substituents on each carbon atom of an alkene results in one of two possible results, as shown at the top of Figure 2.19. We designate as (Z) (an abbreviation for the German word *zusammen*, meaning "together") the isomer in which the two highest priority substituents are cis to each other, and we designate as (E) (German, *entgegen*, "opposite") the isomer in which the two higher priority substituents are trans. (Z) is often equivalent to cis, and (E) is often equivalent to trans, but that is not always the case. In the case of the 2-butenes, the (Z) isomer is the cis isomer, while the (E) isomer is trans. However, in compound **55** the longest continuous chain passes through the double bond in a cis fashion, but the (E) or (Z) designation is (E) because of the priorities of the substituents. There is no longest carbon chain in structure **56**, so applying

---

[92]For further discussion of the *r* notation, see (a) reference 91, pp. 112 *ff*; (b) Orchin, M.; Kaplan, F.; Macomber, R. S.; Wilson, R. M.; Zimmer, H. *The Vocabulary of Organic Chemistry*; John Wiley & Sons, Inc.: New York, 1980; p. 139.

earlier nomenclature rules to determine cis or trans is difficult. However, the (*E*) and (*Z*) system names this compound simply and unambiguously as (*Z*)-1-bromo-1-chloro-2-iodopropene.

Figure 2.19 Examples of (*E*) and (*Z*) nomenclature of alkenes.

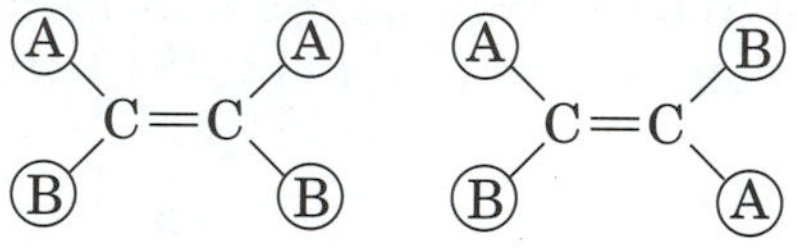

(*Z*) (*E*)

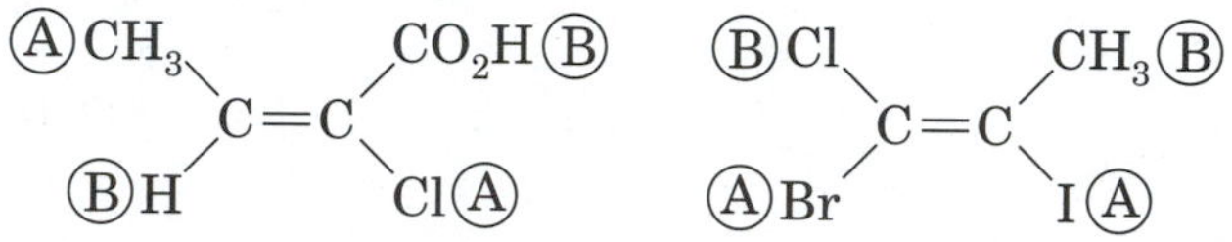

55 **56**

The (*E*) or (*Z*) nomenclature can also replace the older syn, anti terminology. Thus **57** has been called *syn*-propiophenone oxime because the OH and the phenyl are on the same side of the double bond, while **58** has been called the anti isomer because these two larger groups are on opposite sides. With the (*E*) or (*Z*) system, **57** would be the (*Z*) isomer, while **58** is the (*E*) oxime. The distinction between syn and anti depends on the size or complexity of the substituents, and that may not always be unambiguous. Moreover, many authors reserve the terms *syn* and *anti* to describe the stereochemical pathway of a reaction and not the stereochemistry of molecules.

syn-propiophenone oxime or (*Z*)-propiophenone oxime

57

anti-propiophenone oxime or (*E*)-propiophenone oxime

58

The (*R*) or (*S*) and the (*E*) or (*Z*) nomenclature systems can be combined as needed to name specific compounds. For example, the IUPAC name of (−)-matsuone, **59**, the primary sex attractant pheromone of a red pine scale, is (2*E*,4*E*,6*R*,10*R*)-4,6,10,12-tetramethyl-2,4-tridecadien-7-one.[93]

[93]The synthesis of (−)-matsuone was described by Cywin, C. L.; Webster, F. X.; Kallmerten, J. *J. Org. Chem.* **1991**, *56*, 2953.

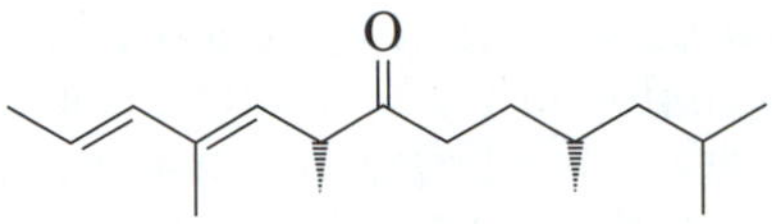

O

Matsuone

59

The terms *endo* and *exo* are useful in naming bicyclic systems. In bicyclic molecules the endo substituent is held toward the inside of a carbon skeleton envelope, while the exo substituent is toward the outside. The nomenclature is exemplified by the 2,4-dimethylbicyclobutanes **60** and **61** and the 2-chloronorbornanes **62** and **63**.[94] A more complicated example is that of 6-*exo*-(acetyloxy)-8-azabicyclo[3.2.1]octan-2-*exo*-ol, **64**, an alkaloid isolated from a Chinese herb that is of medicinal interest because of its use to reduce fever.[95]

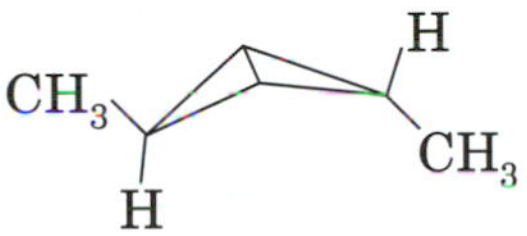

CH_3 H CH_3 H

exo, *endo*-2,4-dimethylbicyclobutane

60

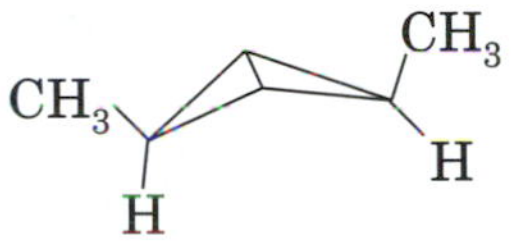

CH_3 CH_3 H H

exo, *exo*-2,4-dimethylbicyclobutane

61

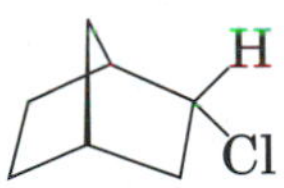

H Cl

endo-2-chloronorbornane

62

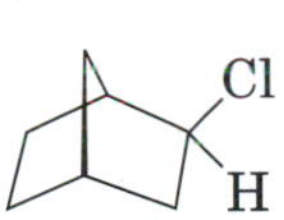

Cl H

exo-2-chloronorbornane

63

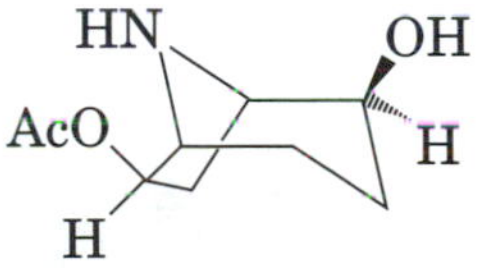

HN OH AcO H H

6-*exo*-(acetyloxy)-8-azabicyclo-[3.2.1]octan-2-*exo*-ol

64

Although the (R) and (S) system is widely accepted for most organic compounds, for historical reasons, we often use the D and L system for carbohydrates and amino acids. By convention, a Fischer projection is drawn vertically with the (IUPAC) atom numbers increasing from top to bottom.[96]

---

[94] In the isomers of **63**, the chlorine sees two envelopes, one due to a six-membered ring and one due to a five-membered ring. The *exo*, *endo* convention refers to the envelope determined by the larger bridge.

[95] Jung, M. E.; Longmei, Z.; Tangsheng, P.; Huiyan, Z.; Yan, L.; Jingyu, S. *J. Org. Chem.* **1992**, *57*, 3528.

[96] An alternative definition specifies that the most highly oxidized carbon atom be at the top of the Fischer projection. For carbohydrates and amino acids, there is ordinarily no conflict between these two definitions. For historical notes and an elaboration of the D and L nomenclature system, see Slocum, D. W.; Sugarman, D.; Tucker, S. P. *J. Chem. Educ.* **1971**, *48*, 597; reference 129 (a), pp. 88–92.

The *highest numbered chiral center* will have two horizontal substituents, normally either H and OH (for carbohydrates) or H and NH_2 (for α-amino acids). If the larger substituent (i.e., not the H) is on the right, the structure is D. If the larger substituent is on the left, the structure is L.

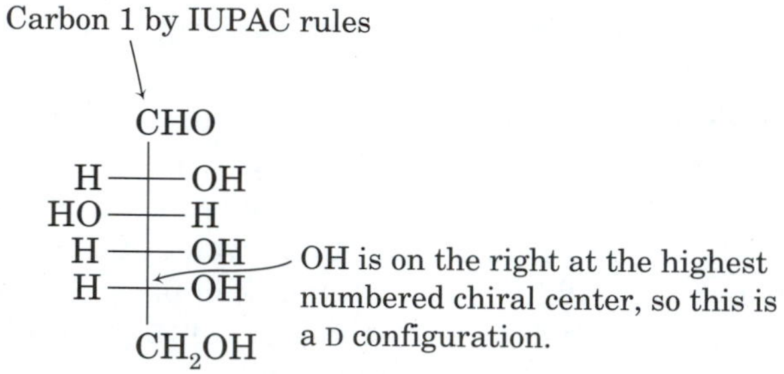

Carbon 1 by IUPAC rules

CHO

H—OH

HO—H

H—OH

H—OH

CH_2OH

OH is on the right at the highest numbered chiral center, so this is a D configuration.

D-(+)-Glucose

71

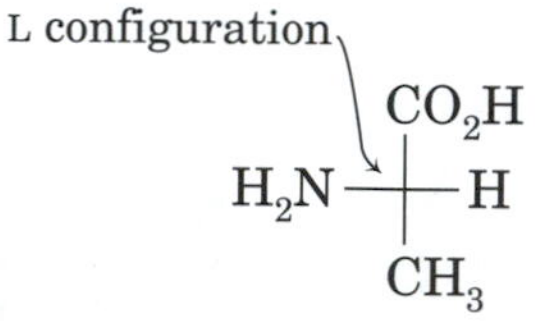

L configuration

CO_2H

H_2N—H

CH_3

L-Alanine

65

The terms α and β are useful in describing the stereochemistry of steroid and carbohydrate systems. For steroid systems drawn as shown in **66**, an α substituent is one that lies beneath the molecular plane.[97,98] On the other hand, a β substituent lies above the plane of the molecule, as shown in 3-β-cholestanol, **67**.[99]

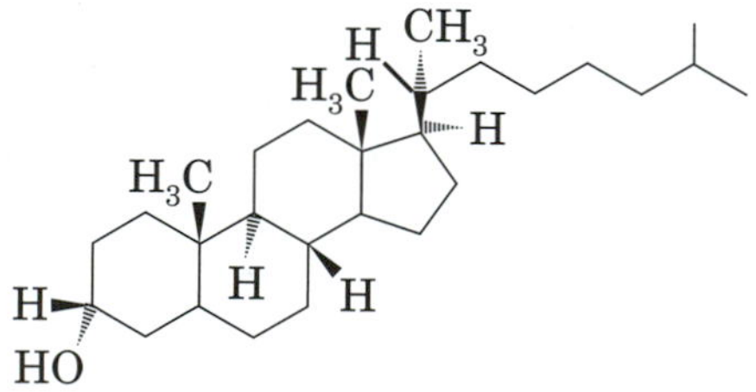

3-α-Cholestanol

66

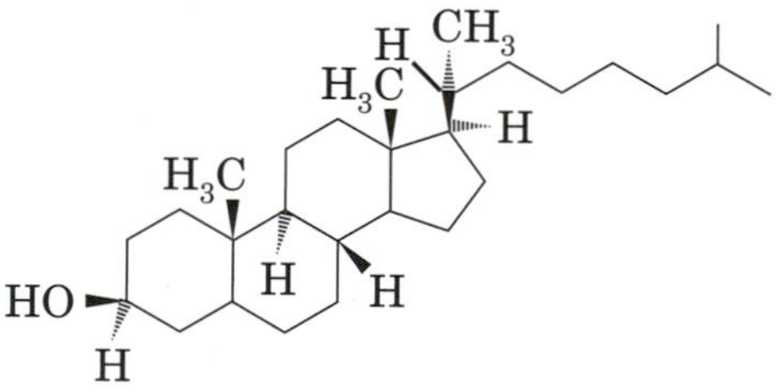

3-β-Cholestanol

67

[97]The rules for conformation nomenclature in cyclic monosaccharides have been summarized by Schwarz, J. C. P. *J. Chem. Soc., Chem. Commun.* **1973**, 505.

[98]For a discussion of the stereochemical designations of steroids, see Fieser, L. F.; Fieser, M. *Steroids*; Reinhold Publishing Corporation: New York, 1959; pp. 2–3 and pp. 330 *ff.*

[99]The designation of substituents as α or β in carbohydrates is not as straightforward as it is with steroids. For D sugars, as proposed by Hudson, the more dextrorotatory of each α,β pair is the α anomer. For L sugars, the more levorotatory of each α,β pair is the α anomer. See Hudson, C. S. *J. Am. Chem. Soc.* **1909**, *31*, 66. For a discussion, see Pigman, W. *The Carbohydrates*; Academic Press: New York, 1953; pp. 42–43.

Similarly, the hemiacetal form of D-glucose, **68**, would be called α-D-glucopyranose, while β-D-glucopyranose is **69**.[100] In these drawings the cyclic hemiacetal form of the carbohydrate is shown as a six-membered ring in the chair conformation. An older representation of carbohydrates is the Haworth drawing **68(b)**.[5] An alternative Haworth representation is shown as **68(c)**. Note that the hydrogens are not shown at the ends of lines to ring carbon atoms in **68(c)**. This representation is an exception to the usual organic convention that a carbon atom terminates a line unless another atom is explicitly shown. Also common are representations similar to **68(c)** in which the lines representing bonds from the carbon atoms to the hydrogen atoms are omitted.

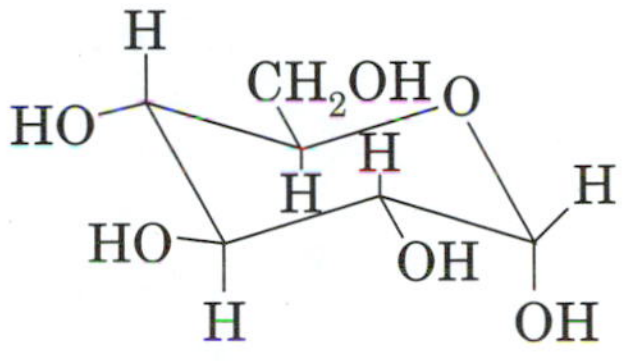

α-D-Glucopyranose

68(a)

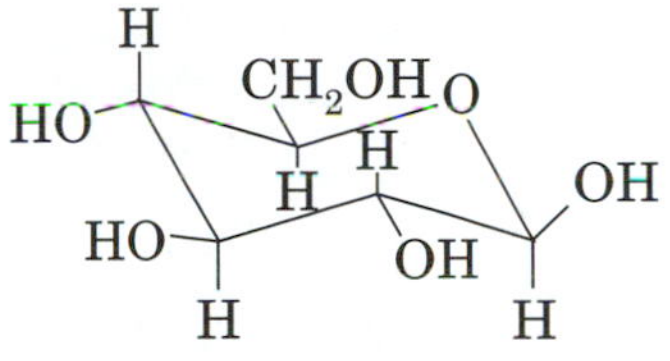

β-D-Glucopyranose

69

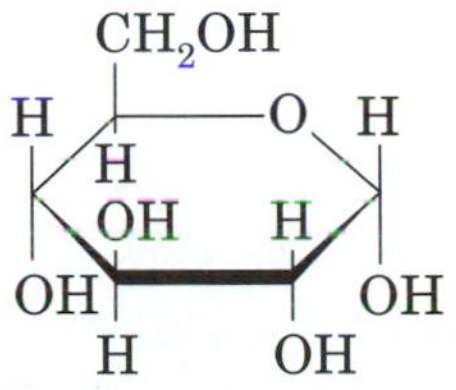

α-D-Glucopyranose

68(b)

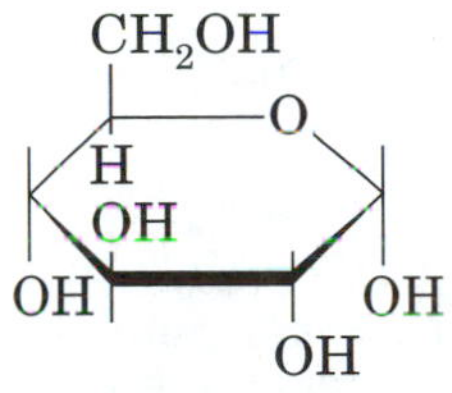

α-D-Glucopyranose

68(c)

When the individual carbohydrate units are joined together to form disaccharides and polysaccharides by converting the hemiacetal groups into acetal linkages, the problem of representing the stereochemistry can become even more complex. For example, sucrose is often represented as **70(a)** (Figure 2.20). However, the same compound is sometimes represented as **70(b)**, which allows the stereochemical aspects of the carbohydrate to be displayed compactly by showing some single bonds with right angles. Thus, there are different conventions for representing molecular structures.

Fischer projections are especially useful for the acyclic forms of monosaccharides (e.g., D-glucose, **71** in Figure 2.21) and can be adapted to display

---

[100]The name *glucopyranose* is derived from the systematic nomenclature of carbohydrates: *gluco* refers to glucose, *pyran* means that the compound is in the hemiacetal form so that there is a six-membered ring containing oxygen, and *ose* is the usual suffix for carbohydrates. A furanose ring would be a five-membered carbohydrate ring containing one oxygen atom.

Figure 2.20 Representations of sucrose.

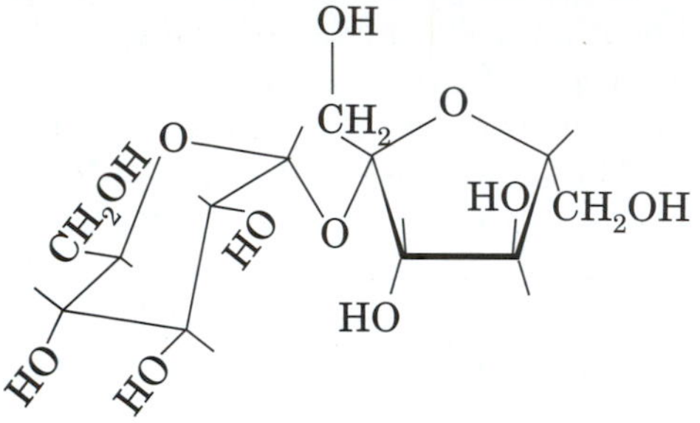

70 (a)

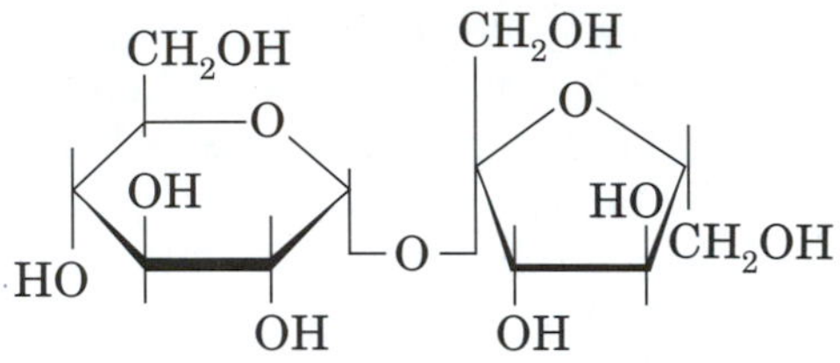

70 (b)

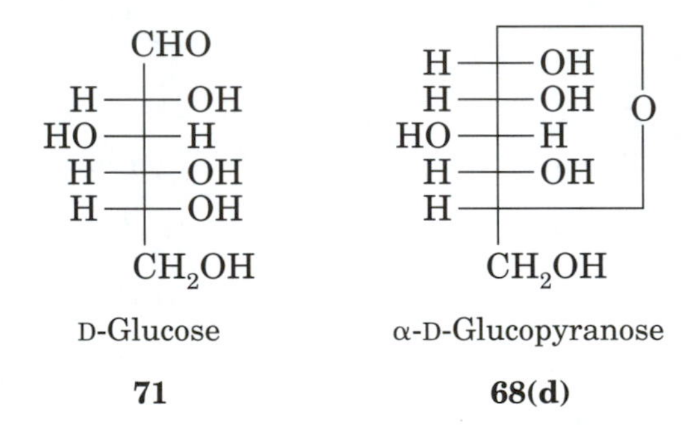

Figure 2.21 Fischer representation of D-glucose (left) and of α-D-glucopyranose (right).

D-Glucose

71

α-D-Glucopyranose

68(d)

their cyclic hemiacetal or hemiketal tautomers, such as **68(d)**. Note, however, that the right angles in **68(d)** do not imply methylene groups. Clearly such a figure is stylistically drawn, which may make it difficult to interpret for one not familiar with the implicit rules.

In some cases, particularly for natural products with many chiral centers, it is convenient to name the mirror image of a substance with a trivial or common name by simply indicating that the structure is the enantiomer (*ent-*) of the natural material. Thus, the enantiomer of cholesterol is *ent*-cholesterol, which was synthesized to study ion transport channels in membranes.[101]

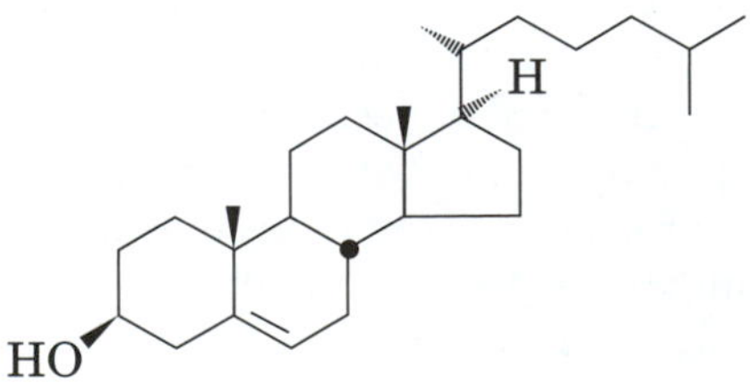

Cholesterol

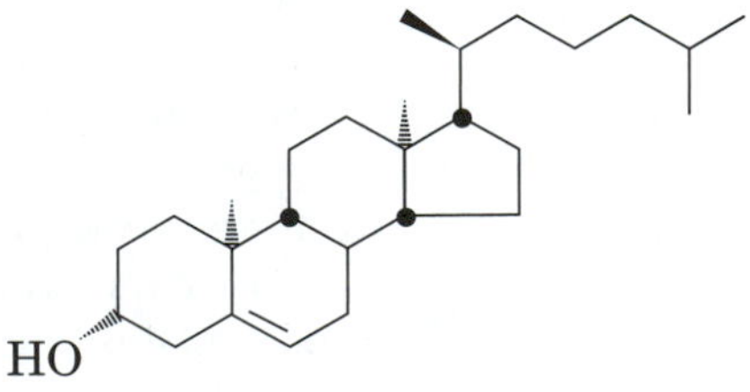

Ent-cholesterol

[101]Rychnovsky, S. D.; Mickus, D. E. *J. Org. Chem.* **1992**, *57*, 2732.

Diastereomers are sometimes identified by special terms. **Epimers** are diastereomers that differ in configuration at only one of several chiral centers. The carbon atom (or other atom) at which the difference in configuration occurs is said to be the **epimeric** center. Structures **72** and **73** are epimeric at C4. **Anomers** are epimers in the carbohydrate series which differ only at the hemiacetal carbon atom. Examples are the α and β forms of D-(+)-gluocopyranose (page 90).

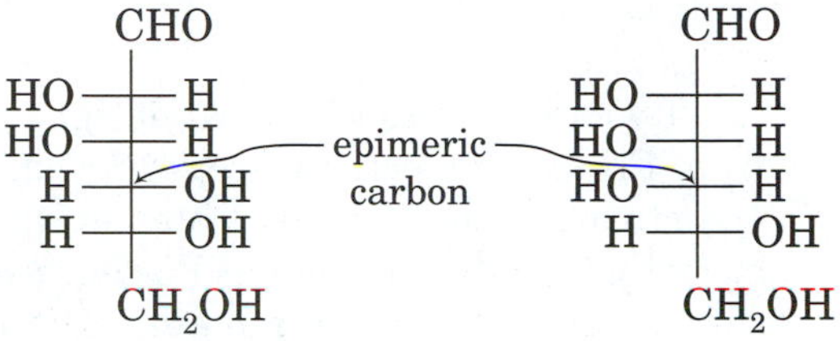

epimeric carbon

D-(+)-Mannose

72

D-(+)-Talose

73

The prefix *epi* is a convenient way to name chiral structures that are epimers of molecules with nonsystematic names. For example, ambrox, **74**, is a substance produced from ambergris, a whale product used in perfumes for hundreds of years.[102] Researchers have found an isomer of ambrox that is epimeric at the 9 position and which offers potential for use in perfume. This compound was named 9-epi-ambrox, **75**, a designation that specifies the structure of the compound, while at the same time relating its properties to those of the parent compound.[103]

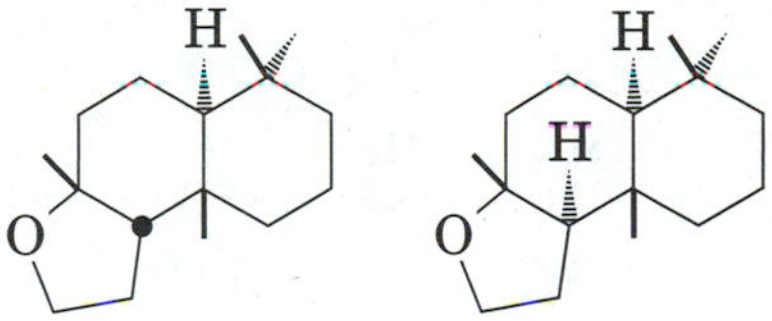

74 **75**

The term *meso* is used to designate a structure that is a diastereomer of enantiomers but which is not chiral. That is, it contains chiral substructures but is not itself optically active. Such structures have a plane of symmetry in at least one conformation as illustrated by **76** and **77**. These structures are said to be internally compensated.[104]

---

[102]Budavari, S., ed. *The Merck Index*, 11th Ed.; Merck & Co., Inc.: Rahway, NJ, 1989; p. 62.

[103]Paquette, L. A.; Maleczka, Jr., R. E. *J. Org. Chem.* **1991**, *56*, 912.

[104]Effectively, a meso compound is a one molecule racemic mixture.

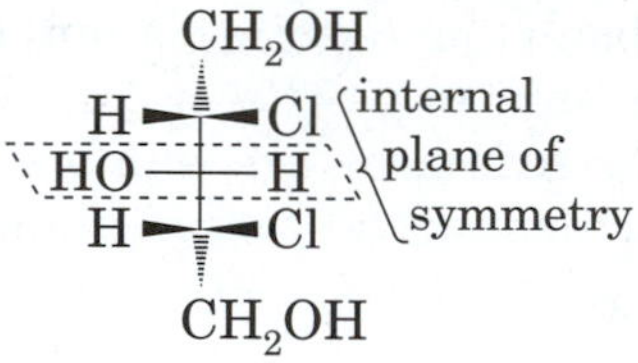

internal plane of symmetry

76

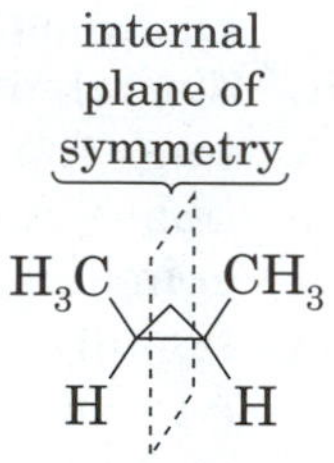

internal plane of symmetry

77

Two other stereochemical designations that are most easily determined from Fischer projections are *erythro* and *threo*. These terms derive from the structures of the carbohydrates erythrose, **78**, and threose, **79**. If the two carbohydrates are drawn as Fischer projections, it is seen that for erythrose the structure would be a *meso* if the end groups (CHO and CH_2OH) were made identical. By analogy, other structures which have a similar stereochemical relationship are called *erythro*. If the two end groups of threose were made identical, the resulting structure would remain chiral. By analogy, structures of similar stereochemical character are termed *threo*.[105] In **78** and **79** both nonhydrogen groups on the chiral carbon atoms are OH groups. In common usage, however, different groups can be considered similar for the purposes of nomenclature. For example, the precursor for the synthesis of the antibiotic thiamphenicol is *threo*-(1*R*,2*R*)-2-amino-1-[(4-methylthio)phenyl]-1,3-propanediol, **80**.[106] Here the *threo* modifier is redundant but is nevertheless useful in visualizing the three-dimensional structure of the compound.

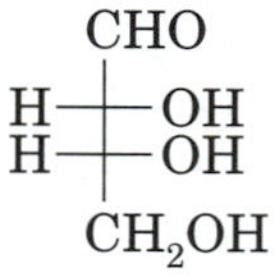

D-Erythrose **78**

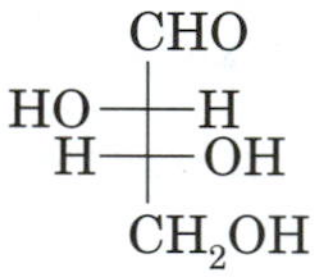

D-Threose **79**

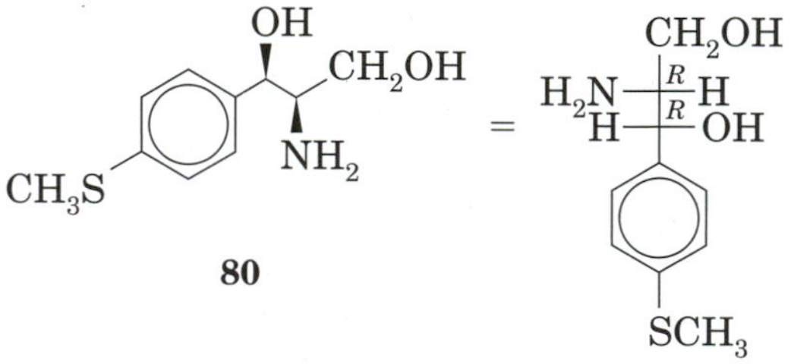

80

One stereochemical term that appears with some regularity in the literature is *homochiral*. Unfortunately, this is a term that has two different

[105]The terms *erythro* and *threo* may be defined in a different fashion using the (*R*) and (*S*) priority designations and Newman projections: reference 84; reference 88. See also Seebach, D.; Prelog, V. *Angew. Chem. Int. Ed. Engl.*, **1982**, *21*, 654; Gielen, M. *J. Chem. Educ.* **1977**, *54*, 673.

[106]The conversion of the (1*S*,2*S*) isomer into the (1*R*,2*R*) enantiomer was described: Giordano, C.; Cavicchioli, S.; Levi, S.; Villa, M. *J. Org. Chem.* **1991**, *56*, 6114.

meanings and has been the subject of some debate. Damewood[107] noted in 1985 that the term as defined by Lord Kelvin[108] meant two species with the same handedness and cited elaboration of this meaning by Anet and Mislow,[109,110] but that the term also had been used to mean a sample in which one enantiomer exists in pure form or is present in greater concentration than its mirror image. A series of letters then appeared in *Chemical & Engineering News* in which several authors proposed and discussed possible alternative terms for enantiomerically pure or enriched substances. Among the suggestions were *enantiopure, scalemic, unichiral, monochiral,* and *aracemic.*[111] These letters were followed by a letter from a member of the IUPAC commission preparing a *Glossary of Stereochemical Terms*, who reported that the committee favored the terms *enantiomerically pure* or *enantiomerically enriched* and specifically discouraged the use of the term *homochiral.*[112] It remains to be seen whether this recommendation will become widely accepted. Even if it is generally accepted, however, readers of the relatively recent literature will need to consider carefully how the word *homochiral* is used in each paper.

Another source of uncertainty in the use of stereochemical terminology concerns the terms *stereospecific* and *stereoselective*, which were proposed by Zimmerman.[113] Adams reviewed the varied definitions of these terms in textbooks and has recommended that these terms should be used as follows:[114]

- **Stereoselective Reaction:** A reaction in which one stereoisomer (or pair of enantiomers) is formed or destroyed at a greater rate or to a greater extent (at equilibrium) than other possible stereoisomers.
- **Stereospecific Reaction:** A reaction in which stereoisomerically different reactants yield stereoisomerically different products.

A familiar example of a stereoselective reaction would be the predominant formation in an E2 reaction of a higher yield of *trans*-2-butene than *cis*-2-butene, whether the starting material is (R)- or (S)-2-bromobutane. However, the addition of bromine to *cis*-2-butene to produce *meso*-2,3-dibromobutane or the addition of bromine to the trans isomer to produce an

---

[107]Damewood, Jr., J. R. *Chem. Eng. News* **1985** (Nov. 4), 5.

[108]*Baltimore Lectures*; Clay: London, 1904; p. 618.

[109]Anet, F. A. L.; Miura, S. S.; Siegel, J.; Mislow, K. *J. Am. Chem. Soc.* **1983**, *105*, 1419.

[110]Mislow, K.; Bickart, P. *Isr. J. Chem.* **1976/77**, *15*, 1.

[111]Eliel, E. L.; Wilen, S. H. *Chem. Eng. News* **1990** (Sept. 10), 2; Heathcock, C. H. *Chem. Eng. News* **1991** (Feb. 4), 3; Gal, J. *Chem. Eng. News* **1991** (May 20), 42; Castrillón, J. *Chem. Eng. News* **1991** (June 24), 94; Eliel, E. L.; Wilen, S. H. *Chem. Eng. News* **1991** (July 22), 3; Brewster, J. H. *Chem. Eng. News* **1992** (May 18), 3.

[112]Halevi, E. A. *Chem. Eng. News* **1992** (Oct. 26), 2.

[113]Zimmerman, H. E.; Singer, L.; Thyagarajan, B. S. *J. Am. Chem. Soc.* **1959**, *81*, 108. The definitions are in footnote 16, p. 110.

[114]Adams, D. L. *J. Chem. Educ.* **1992**, *69*, 451.

equimolar mixture of the two enantiomers of 2,3-dibromobutane is a stereospecific reaction. Note that a reaction that gives only one of a pair of enantiomers is not necessarily stereospecific. Yeast-mediated reduction of 3-chloropropiophenone gives (1*S*)-3-chloro-1-phenylpropan-1-ol, with no evidence for formation of the (1*R*) enantiomer.[115] However, because the reactant cannot exist as stereoisomers, it is not possible for stereoisomerically different reactants to give stereoisomerically different products, and the reaction can only be considered stereoselective, not stereospecific.

Part of the confusion about the terms *stereoselective* and *stereospecific* comes from the use of other terms that sound similar but in which the suffixes *-selective* and *-specific* do not mean the same relationship as they do in stereospecific and stereoselective. The term ***regioselective*** refers to reactions that lead to the formation of one positional isomer in greater yield than another isomer.[116] An example would be the formation of more 2-alkene than the isomeric 1-alkene in the E2 reaction of an alkyl halide. However, the term *regiospecific* means "completely regioselective," so its meaning is not parallel with *stereospecific*. Because of this difficulty, Adams recommended that the term *regiospecific* not be used.[114]

2.3 Manifestations of Stereoisomerism

Optical Activity

Rotation of Plane-Polarized Light

The physical phenomenon that first led to the study of enantiomers is optical activity—the rotation of plane polarized light. We indicate the experimental results by saying that the rotation is clockwise (also denoted as (+), *d*, or *dextrorotatory*) or counterclockwise (also indicated as (−), *l*, or *levorotatory*). Because of the similarity of *d* and *l* with D and L, which have entirely different meanings, the use of (+) and (−) to indicate the direction of rotation of the plane of polarization is now favored.[75]

It must be emphasized that chirality is a *molecular designation*, but optical activity is a *macroscopic phenomenon*. It is possible for a carbon atom to have four different substituents and yet a sample of the substance *not* appear to rotate the plane of polarized light. This can happen most easily if the four substituents are so similar that rotation of light is too small to be observed. For example, optical activity could not be detected in butylethylhexylpropylmethane (5-ethyl-5-propylundecane, **81**).[117] However, substitu-

---

[115] Fronza, G.; Fuganti, C.; Grasselli, P.; Mele, A. *J. Org. Chem.* **1991**, *56*, 6019.

[116] Reference 92 (b), p. 146.

[117] Wynberg, H.; Hekkert, G. L.; Houbiers, J. P. M.; Bosch, H. W. *J. Am. Chem. Soc.* **1965**, *87*, 2635. (Precursors to **81** were chiral structures with significant optical activity, so failure to observe optical activity with **81** was not due to racemization or low optical purity.) See also Wynberg, H.; Hulshof, L. A. *Tetrahedron* **1974**, *30*, 1775.

tion of a deuterium for a hydrogen at a chiral center can produce measurable optical activity, as shown by **82**.[118,119]

n-C_3H_7, n-C_2H_5, n-C_4H_9, n-C_6H_{13}

81

CH_2OH, H, D, C_6H_5

82

Optical activity is reported as **specific rotation**, [α]. Because one needs a longer path length to measure optical rotation precisely than typically is needed for UV-vis spectroscopy, the standard cell length is 1 decimeter (dm), which is 10 cm. In many cases the molecular weight of the sample to be tested is unknown, so the concentration is reported as g/mL (i.e., the density for a pure liquid or the concentration for a solution), as opposed to moles/liter. It is also possible to define a molecular rotation (Φ) when molecular weight is known, however, as shown in equation 2.1.

$$[\Phi] = 0.01 \times [\alpha] \times (\text{molecular weight}) \quad \textbf{(2.1)}$$

Calculating molecular rotation makes it easier to determine the effect of the same chromophore in different molecules.[120] For solutions it is necessary to specify the solvent, since solvent-solute interactions can affect not only the magnitude of optical activity, but perhaps the direction of the rotation as well.[121] It is also important to specify both the temperature at which the measurement is made and the wavelength of light that is used. The temperature is usually room temperature, expressed as °C. The sodium emission D line at 589 nm is a bright, sharp spectral line that can easily be produced, and it is often used for polarimetry measurements. Thus an experimental measurement at 25° in which a pure sample with density 0.8 g/mL gave a rotation of the sodium D line of +36° in a 10 cm cell would give a specific rotation of 45° mL g^{-1} dm^{-1}.[122] Specific rotations are usually reported in an abbreviated format. For example, the specific rotation of cholesterol is given as $[\alpha]^{20}_{D}$ −31.5° (c = 2 in ether); $[\alpha]^{20}_{D}$ −39.5° (c = 2 in chloroform).[123] Note that in this format, the concentration of a solution is usually taken to mean grams per 100 mL.[124]

---

[118]The specific rotation of the (−) enantiomer was determined to be −3.006 ± 0.004° by Streitwieser, Jr., A.; Wolfe, Jr., J. R.; Schaeffer, W. D. *Tetrahedron* **1959**, *6*, 338.

[119]For a review of chirality resulting from isotopic substitution, see Arigoni, D.; Eliel, E. L. *Top. Stereochem.* **1969**, *4*, 127.

[120]P. Crabbé in reference 133, p. 1.

[121]Reference 129(a), p. 8.

[122]While specific rotation is ordinarily reported only as degrees, it is helpful to carry units through calculations to ensure correct path length and concentration units are used.

[123]Reference 102, pp. 341–342.

[124]Reference 102, p. xiii.

Our discussion to this point has assumed that we can obtain pure enantiomers. In many cases we can obtain one pure enantiomer from a natural source, but often we find that enantiomeric species are formed as a **racemic mixture**—an equimolar mixture of the two enantiomers. Racemic mixtures are denoted with the prefixes (±)- or *rac*-.[125] We may try to *resolve* a racemic mixture (separate it into enantiomers) by any of a variety of chemical or biological means.[126] If we are nevertheless unable to obtain one pure enantiomer, we may be forced to use a mixture in which there is more of one enantiomer than its mirror image. In such cases we would like to know the *optical purity* of the sample, which is defined as the percentage of excess of one enantiomer over the other enantiomer. For example, the specific rotation of (+)-glyceraldehyde is +14°. A mixture of 95% (+)-glyceraldehyde and 5% (−)-glyceraldehyde is said to be 90% optically pure because the rotation of the sample is 12.6°, which is 90% of the rotation that would be observed if all of the molecules were (+)-glyceraldehyde. Thus

$$\text{Optical purity} = (\text{fraction major enantiomer} - \text{fraction minor enantiomer}) \times 100\% \quad \textbf{(2.2)}$$

The **enantiomeric excess** is the percent major enantiomer minus the percent minor enantiomer (which is the same percentage as optical purity).[127]

Optical Rotary Dispersion and Circular Dichroism

The sodium D line is chosen for optical activity measurements because it is a bright line that can be easily seen and easily reproduced from instrument to instrument. However, optical activity can be measured at any wavelength (λ). In the technique of **optical rotary dispersion (ORD)**, the value of $[\Phi]$ is determined over the entire near UV-vis spectrum. Optical activity is generally greater at shorter wavelengths than at longer wavelengths, so a (+) compound will generally show a curve that is more dextrorotatory at short wavelengths than at long wavelengths, while a (−) compound will be more levorotatory at shorter wavelengths.

Any optically active compound will have an ORD curve. If the graph of $[\Phi]$ versus λ does not show a maximum or minimum but simply increases in magnitude from longer to shorter wavelengths, the curve is called a **plain curve**. In some cases the ORD curve exhibits a maximum or minimum (or both), so that a Cotton effect is observed. Cotton effects occur when the wavelength is near a UV-vis absorption band of the optically active compound. Figure 2.22 shows the ORD curve for 2α-bromocholestan-3-one re-

---

[125]The term *racemic modification* is still accepted and widely used. However, the word *modification* suggests a chemical or physical change to the structures involved, whereas the word *mixture* more correctly describes the composition of the material.

[126]For a review, see Wilen, S. H. *Top. Stereochem.* **1971**, *6*, 107.

[127]For a review of the determination of enantiomeric purity (a) by NMR, see Parker, D. *Chem. Rev.* **1991**, *91*, 1441; (b) for other methods, see Raban, M.; Mislow, K. *Top. Stereochem.* **1967**, *2*, 1.

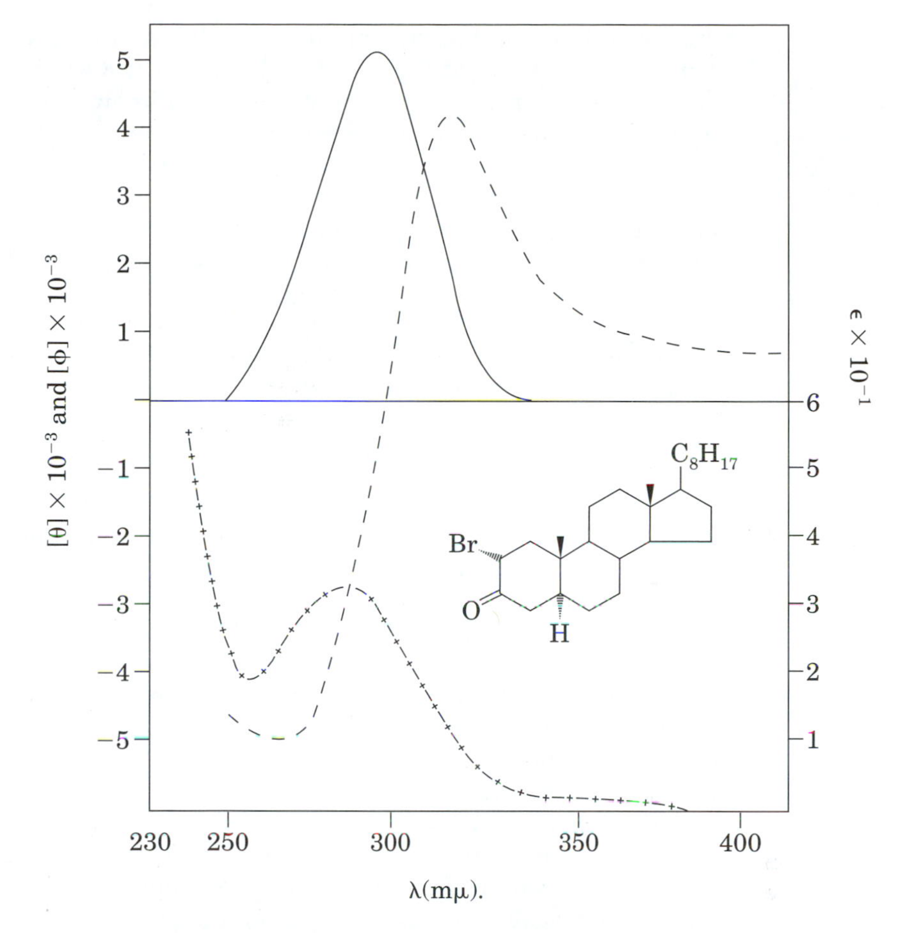

Figure 2.22 ORD (- - -), CD (------), and UV-vis (+-+-) spectra of 2α-bromocholestan-3-one in dioxane solution. (Adapted from reference 128.)

ported by Djerassi.[128] The ORD exhibits a positive Cotton effect, meaning that the magnitude of the observed optical activity increases from 400 nm to about 310 nm, then it decreases and actually becomes negative at wavelengths shorter than 300 nm. The midpoint between the extrema (most positive value and most negative value) of the Cotton effect is 293 nm, which is close to the UV-vis absorption maximum, 286 nm.

Plane polarized light has all of its electric vectors in one plane, but that is physically the same as the result obtained by taking the vector sum of two different circular polarizations of light, one clockwise and one counterclockwise, each of which has an electric vector which rotates about the axis of propagation of the photon.[129] The origin of optical activity is the different in-

[128]Djerassi, C.; Wolf, H.; Bunnenberg, E. *J. Am. Chem. Soc.* **1963**, *85*, 324.

[129]For a discussion of optical activity and circularly polarized light, see (a) Eliel, E. L. *Stereochemistry of Carbon Compounds*; McGraw-Hill: New York, 1962; chapter 14; (b) Eliel, E. L.; Wilen, S. H. *Stereochemistry of Organic Compounds*; Wiley-Interscience: New York, 1994; pp. 992–999; (c) Hill, R. R.; Whatley, B. G. *J. Chem. Educ.* **1980**, *57*, 306; (d) Brewster, J. H. *Top. Stereochem.* **1967**, *2*, 1.

dices of refraction that a medium has for left and right circularly polarized light. This means that the speed of the light through the medium is different for the two polarizations, a phenomenon known as **circular birefringence**. Not only are the two circular polarizations refracted differently, but they are also absorbed differently by an optically active material. The light passing through the sample is said to be elliptically polarized, and the phenomenon of differential absorption is known as **circular dichroism (CD)**. The sample will thus have a different extinction coefficient for left circularly polarized light (ϵ_L) from that for right circularly polarized light (ϵ_R). With instrumentation to measure the absorption spectrum with both circular polarizations, one can determine the differential dichroic absorption, $\Delta\epsilon$, and molecular ellipticity, $[\Theta]$, of the sample:

$$\Delta\epsilon = \epsilon_L - \epsilon_R \tag{2.3}$$

$$[\Theta] = 3300\,\Delta\epsilon \tag{2.4}$$

Figure 2.22 also shows the CD curve for 2α-bromocholestan-3-one. The maximum corresponds closely to the λ_{max} of the first absorption peak in the UV-vis spectrum of the compound.[130] In some cases CD curves can be helpful in locating UV-vis spectra that are too weak to be seen clearly in a normal absorption spectrum.[128]

ORD and CD curves play an important role in determining the structures and stereochemistry of organic compounds. Based on the spectra of compounds of known structure and absolute stereochemistry, empirical rules have been developed to predict ORD and CD spectra for other similar compounds. One of the better known such rules is the octant rule for correlating the sign and magnitude of the Cotton effect to the structure of saturated ketones.[131] The center of the carbon-oxygen double bond is taken as the origin for x, y, and z axes, so that the xy, xz, and yz planes divide the space near the molecule into eight regions or octants. A substituent in a particular octant contributes to a positive or negative Cotton effect according to the product of its x, y, and z coordinates. This system has been particularly useful in determining the structures of steroids and other chiral substances. Additional sources of information about circular dichroism and optical rotary dispersion can be found in papers and monographs by Mason,[132] Snatzke,[133]

[130]Because ORD and CD curves are both determined by the absorption spectrum of the compound, they are related to each other. Tinoco presented an analysis of the relationship of UV, CD and ORD curves for helices, such as those formed by polynucleic acids: Tinoco, Jr., I. *J. Am. Chem. Soc.* **1964**, *86*, 297.

[131]Moffitt, W.; Woodward, R. B.; Moscowitz, A.; Klyne, W.; Djerassi, C. *J. Am. Chem. Soc.* **1961**, *83*, 4013.

[132]Mason, S. F. *Quart. Rev. Chem. Soc.* **1963**, *17*, 20.

[133]Snatzke, G. ed. *Optical Rotary Dispersion and Circular Dichroism in Organic Chemistry*; Heyden and Son: London, 1967.

Crabbé,[134] Djerassi,[135] and Eliel.[129] While the discussion here has concerned only the interaction of UV and visible light with chiral substances, infrared radiation may also be used to study vibrational circular dichroism and Raman optical activity.[136]

Configuration and Optical Activity

Notice that the designators D and L, (*R*) and (*S*), and *erythro* and *threo* are *molecular notations* of the configuration of a molecular model or of a conceptual model. However, the (+) and (−) designators result from *macroscopic observations* made on a bulk sample of material. How can we know, then, what is the absolute configuration of a particular enantiomer that has a particular optical activity? For much of the early period in the development of stereochemical theory, that information was not attainable, and there was no way of knowing whether it would ever be obtained. Nevertheless, there still needed to be some systematic method for discussing the configurations and optical activities of organic compounds, so Fischer arbitrarily proposed that (+)-tartaric acid had the configuration shown in **83**.[137] The usual methods for determining the structure of molecules, such as ordinary X-ray diffraction, do not distinguish between enantiomers. However, Bijvoet and co-workers were able to use anomalous X-ray diffraction, in which the X-rays are not only diffracted but are also absorbed by one or more atoms in the sample, to study sodium rubidium tartrate with X-rays that are absorbed by rubidium.[138] The results allowed Bijvoet to confirm Fischer's assignment.[139] Bijvoet's pioneering work introduced a system for the determination of the absolute configuration of other molecules,[140] and many absolute configurations have been cataloged.[32,141]

---

[134](a) Crabbé, P. *Optical Rotary Dispersion and Circular Dichroism in Organic Chemistry*; Holden-Day: San Francisco, 1965; (b) Crabbé, P. *Top. Stereochem.* **1967**, *1*, 93.

[135]Djerassi, C. *Optical Rotary Dispersion: Applications to Organic Chemistry*; McGraw-Hill: New York, 1960.

[136]Freedman, T. B.; Nafie, L. A. *Top. Stereochem.* **1987**, *17*, 113.

[137]The reference to Fischer is found in the paper by Bijvoet (reference 138).

[138]Bijvoet, J. M.; Peerdeman, A. F.; van Bommel, A. J. *Nature* **1951**, *168*, 271.

[139]It is worthwhile to read Bijvoet's original paper, because it reinforces some ideas that are carried through this text. Bijvoet points out that ". . . Fischer's *convention . . . appears to answer to reality*" (emphasis in the original paper), and he explicitly refers to the assignment of stereochemistry as a model.

[140]For a discussion of the determination of absolute configuration from crystals and for an explanation of Bijvoet's method, see Addadi, L.; Berkovitch-Yellin, Z.; Weissbuch, I.; Lahav, M.; Leiserowitz, L. *Top. Stereochem.* **1986**, *16*, 1.

[141]For further discussion of the correlation of configurations, see Mills, J. A.; Klyne, W. *Prog. Stereochem.* **1954**, *1*, 177.

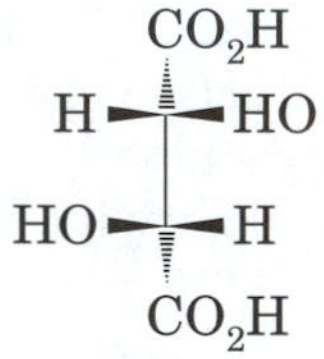

CO_2H
H — — HO
HO — — H
CO_2H

(+)-Tartaric acid

83

The accumulation of extensive experimental data relating absolute configuration to optical activity has made it possible for absolute configuration to be predicted by empirical and semi-empirical correlations. Brewster[142] proposed that the sign of [α] can be predicted by analyzing the polarizabilities of the substituents arranged around a chiral center. If the polarizabilities of the substituents shown in Figure 2.23 are $A > B > C > D$, then the corresponding compound would be expected to be *dextrorotatory*. For the example in which the substituents (and their polarizabilities) are Br (8.7), C_6H_5 (3.4), CH_3 (2.6) and H (1.0), the rotation is +178°. Applications of this model to many kinds of compounds have been reported.

It is also possible to determine the absolute configuration of a molecule indirectly by *relating* the configuration of that structure to another structure for which the absolute configuration has been determined directly. In other words, if we know the *relative configuration* of two compounds and if we also know the absolute configuration of one of them, then we can deduce the absolute configuration of the other.[143] For example, the absolute configuration of (−)-serine, **85**, was determined to be (*S*) by absolute X-ray structure determination.[144] Any reaction carried out on **85** that does not involve breaking any bonds to the chiral center (that is, a reaction that proceeds with *retention of configuration*) must give a product with the same relative configuration as **85**. So the conversion of the OH substituent on C3 to a Cl by a mechanism not involving breaking any bonds to C2 produces (−)-2-amino-3-chloropropanoic acid with the absolute configuration shown in **86**.[145]

Figure 2.23 Chiral structure expected to be dextrorotatory if polarizabilities are $A > B > C > D$.

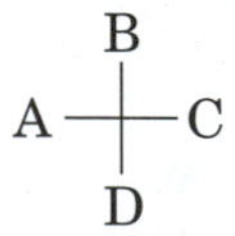

B
A — C
D

[142]Brewster, J. H. *J. Am. Chem. Soc.* **1959**, *81*, 5475 and succeeding papers in that volume.

[143]For a description of the method used to relate absolute configuration to optical activity for tartaric acid, see reference 129(a), pp. 95 *ff*; reference 129(b), pp. 113–114.

[144]Zalkin, A.; Forrester, J. D.; Templeton, D. H. *Science* **1964**, *146*, 261.

[145]Fischer, E.; Raske, K. *Ber.* **1907**, *40*, 3717.

Similarly, hydrolysis of (−)-azaserine, **84**, produces (*S*)-(−)-serine by a mechanism that should involve retention of configuration, so **84** must have the (*S*) configuration.[146]

Note that **86** is an (*R*) configuration, but it was formed by retention of configuration in a reactant that had the (*S*) configuration. This example illustrates that the (*R*) and (*S*) designations are based on a set of arbitrary rules, and the (*R*) or (*S*) designation for a chiral center can change if reactions alter the priorities of substituent atoms attached to the chiral center, even if the configuration of the chiral center is retained.

CO_2H / H_2N—C—H / $CH_2OCOCH_2NH_2$ ⟶ CO_2H / H_2N—C—H / CH_2OH ⟶ CO_2H / H_2N—C—H / CH_2Cl

(*S*)-(−)-Azaserine **84** (*S*)-(−)-Serine **85** (*R*)-(−)-2-Amino-3-chloropropanoic acid **86**

We may also determine the relative configuration of chiral structures even though bonds are broken to the chiral center if we are able to determine whether the reaction proceeds with retention or (more commonly) with inversion of configuration. The S_N2 reaction of (*S*)-(−)-2-chloropropanoic acid, **87**, with hydroxide ion produces (−)-lactic acid, which must have the (*R*) configuration, **88**.[147] (Note that in this case the direction of rotation of plane polarized light stays the same, even though there has been an inversion of configuration at the chiral center.)

CO_2H / Cl—C—H / CH_3 $\xrightarrow{^{-}OH}$ CO_2H / H—C—OH / CH_3

(*S*)-(−)-2-Chloropropanoic acid **87** (*R*)-(−)-Lactic acid **88**

Other Physical Properties of Stereoisomers

Each member of a pair of enantiomers has physical properties that are identical to the other member of the pair for all measurements made with unpolarized light and in the absence of chiral molecules. Thus they have the same melting point, boiling point, index of refraction, etc. Mixtures of enantiomers, however, have different physical properties from those of either enantiomer. In particular, a *racemic mixture* usually has different

---

[146]Fusari, S. A.; Haskell, T. H.; Frohardt, R. P.; Bartz, Q. R. *J. Am. Chem. Soc.* **1954**, *76*, 2881.
[147]Brewster, P.; Hughes, E. D.; Ingold, C. K.; Rao, P. A. D. S. *Nature* **1950**, *166*, 178.

Table 2.1 Physical properties of tartaric acids.

Isomer	Melting Point
+ (or −)	171–174°
(±)	206°
meso	146–148°

properties from those of the pure enantiomers. Table 2.1 shows the physical properties of (+)-, (−)-, (±)-, and *meso* tartaric acid.[148]

Stereochemical relationships play an especially important role in the properties of polymers.[149] Consider the three polypropylene segments illustrated as Fischer projections in Figure 2.24. In the ***isotactic polymer***, all of the methyl-substituted carbon atoms have the same configuration. In the ***syndiotactic polymer***, the methyl-substituted carbon atoms have alternating configurations. In the ***atactic polymer*** there is a random pattern of configurations at the methyl-substituted carbon atoms. As with smaller molecules, the physical properties of stereoisomeric polymers differ. Table 2.2 shows the relationship between tacticity and properties of polypropene.[150,151]

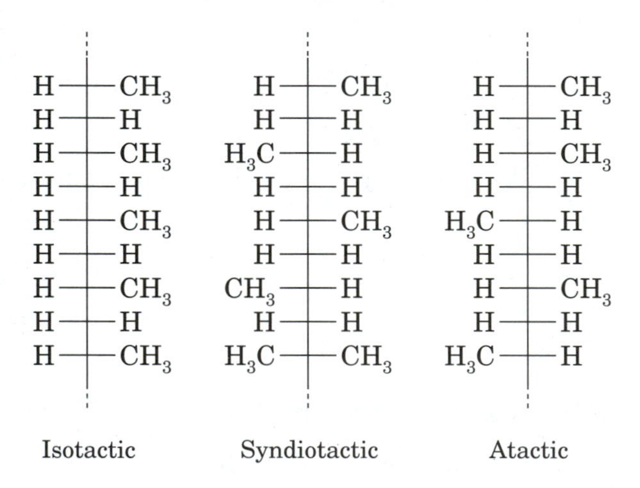

Figure 2.24 Isotactic, syndiotactic, and atactic polypropene.

[148]Data from Lide, D. R., ed. *CRC Handbook of Chemistry and Physics*, 71st Edition; CRC Press: Boca Raton, FL, 1990; p. **3**–475.

[149]For a review, see Goodman, M. *Top. Stereochem.* **1967**, *2*, 73.

[150]These data are from the compilation by Quirk, R. P. *J. Chem. Educ.* **1981**, *58*, 540, which includes references to the literature from which the data shown are taken.

[151]For reviews, see Bawn, C. E. H.; Ledwith, A. *Quart. Rev. Chem. Soc.* **1962**, *16*, 361; Farina, M. *Top. Stereochem.* **1987**, *17*, 1. Also see Coates, G. W.; Waymouth, R. M. *Science* **1995**, *267*, 217.

Table 2.2 Tacticity and physical properties of polypropene.

Tacticity	Density (g/cm^3)	Phase Transition Temperature (°C)	Solubility in *n*-Heptane
Atactic	0.86	−35	soluble
Syndiotactic	0.91	138	soluble
Isotactic	0.94	171	insoluble

(Data from reference 150.)

Stereochemical Relationship of Substituents

Our emphasis to this point has been on the stereochemical relationship of one structure to another. However, the principles developed so far are also relevant to the consideration of symmetry relationships within a single molecule. Many of the labels we use are based on the suffix *-topic*, which is Greek for "place."[152] Terms incorporating this suffix apply both to atoms and to spaces in a molecule, although we usually think of them in terms of atoms. Identical atoms that occupy equivalent environments (both in terms of chemical properties and local or molecular symmetry) are said to be **homotopic**, i.e., to have the same place. Identical atoms in nonequivalent environments are said to be **heterotopic**. Heterotopic substituents can be either **constitutionally heterotopic** or **stereoheterotopic**. Stereoheterotopic substituents can be either **enantiotopic** or **diastereotopic**.

Consider the examples in Figures 2.25–2.28. Any proton on either CH_3 group of propane, **89**, is homotopic with any other methyl proton, since the product of replacing any one proton by another substituent (shown as an X in **90**) is the same as that produced by replacing any other methyl proton. (If

Figure 2.25 Examples of homotopic relationships.

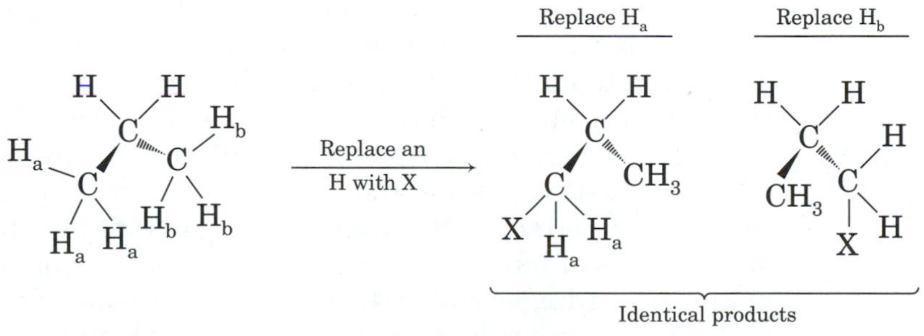

[152]For more details of stereoisomeric relationships of substituents, see Mislow, K.; Raban, M. *Top. Stereochem.* **1967**, *1*, 1.

Figure 2.26 Examples of constitutionally heterotopic relationships.

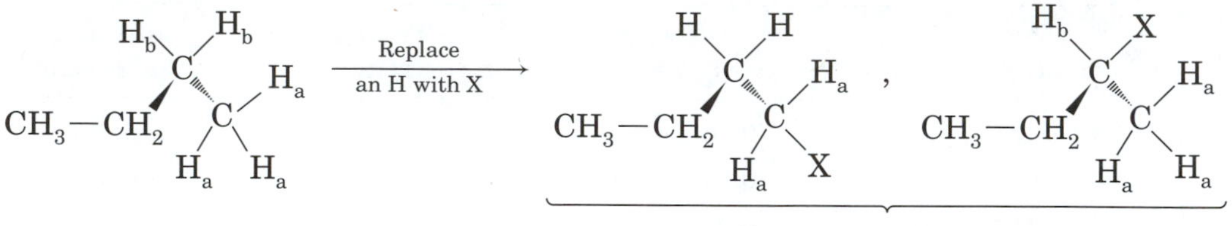

Replace an H with X

Constitutional isomers

91 **92** **93**

Figure 2.27 Example of enantiotopic relationship.

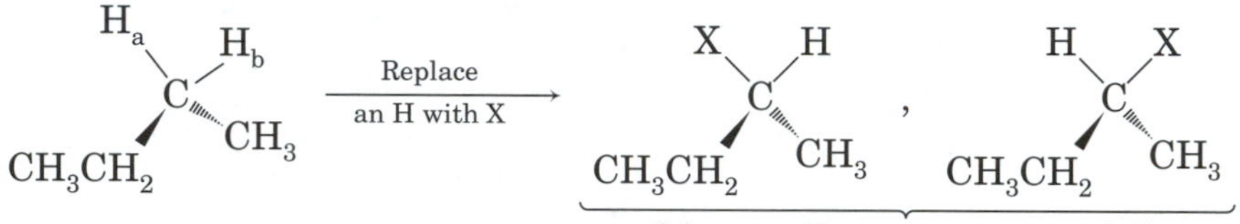

Replace an H with X

94a Enantiomers **94b**

Figure 2.28 Example of diastereotopic relationship.

Replace an H with X

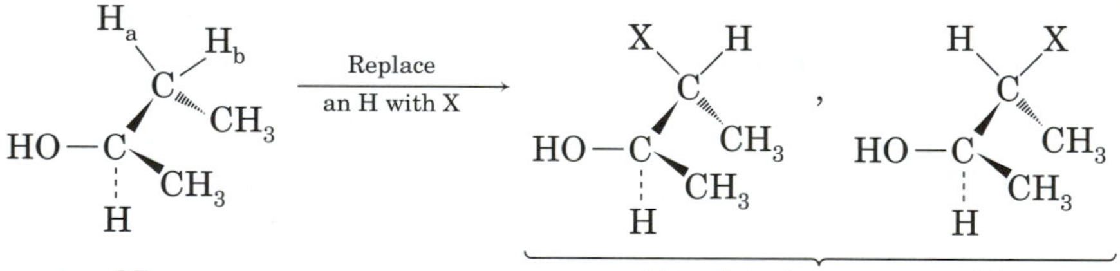

95

96 Diastereomers **97**

necessary, the products can be seen to be identical by rotating the entire molecule, by rotation of one part of the molecule about an internal bond, or both.) The two methylene protons on C2 of propane are also homotopic. On the other hand, a proton on either methyl group in propane is constitutionally heterotopic when compared with a proton on the methylene group, since the products of the two replacements have different connectivities. Similarly, the six methyl protons of butane, **91**, are homotopic, while any one of the methyl protons and any one of the methylene protons are constitutionally heterotopic. However, the two protons on either one of the methylene groups of butane are enantiotopic, since the product of replacing one methylene hydrogen by X, **94(a)**, is the enantiomer of the product formed by replacing the other hydrogen (on the *same* methylene) by X, **94(b)**. Similarly, the two methylene hydrogens on C3 of (*R*)-2-butanol, **95**, are diastereotopic, since replacement of one by X generates a product, **96**, that is the diastereomer of the product, **97**, obtained by replacing the other one.

Frequently we are faced with experimental observations that require understanding enantiotopic and diastereotopic relationships. In particular, we note that enantiotopic protons are equivalent in typical NMR spectroscopy experiments.[153] Diastereotopic protons, on the other hand, are in magnetically different environments.[154] A term used to describe groups with nonequivalent chemical shifts is *anisochronous*; groups with identical chemical shift are said to be *isochronous*.[155] Not only may two diastereotopic protons on the same carbon atom have different chemical shifts, but they will also split each other, resulting in a complex spectrum.[156] The example of the diastereomers of 1,3-diphenyl-1,3-propanediol is informative.[157] In the meso structure, **98**, the two hydrogens on C2 are diastereotopic, so they would have different chemical shifts and would split each other. In the chiral diastereomer, **99**, however, the two central methylene hydrogens are homotopic. They have identical chemical shifts, and they do not split each other.

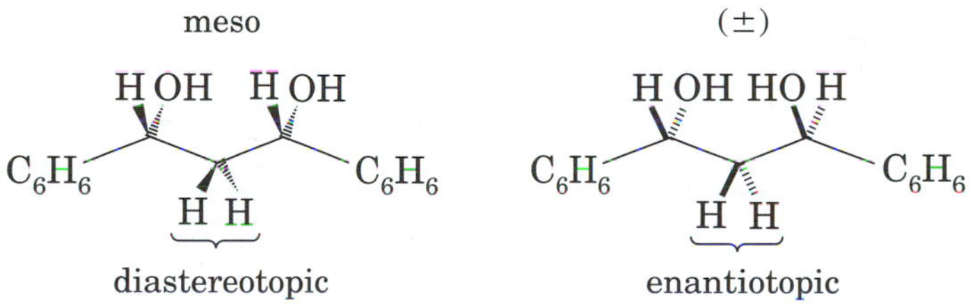

meso

H OH H OH

C_6H_6 C_6H_6

H H

diastereotopic

98

(±)

H OH HO H

C_6H_6 C_6H_6

H H

enantiotopic

99

Two examples discussed by Jennings further illustrate the principles.[155b] In *cis*- and *trans*-1,3-dimethylindane, the two methylene protons are diastereotopic in the meso (cis) structure, **100**, so they have different chemical shifts and split each other. In the chiral trans isomer, **101**, they are isochronous, so they have the same chemical shift and do not split each other. In *N*-benzyl-2,6-dimethylpiperidine, however, the benzyl methylene protons are enantiotopic in the cis isomer, **102**, but are diastereotopic in the trans isomer, **103**.

---

[153] If we use a chiral solvent for NMR spectroscopy, then chemical shifts induced by solute-solvent interactions may be different for enantiotopic protons, and the spectrum may show differences between them. Moreover, chiral shift reagents can also make them distinguishable. See reference 152 for examples.

[154] Ault, A. *J. Chem. Educ.* **1974**, *51*, 729. For the theoretical basis of chemical shift nonequivalence, see Stiles, P. J. *Chem. Phys. Lett.* **1976**, *43*, 23. Note that the term *magnetic nonequivalence* is no longer favored (reference 155(b)).

[155] (a) *Cf.* Binsch, G.; Eliel, E. L.; Kessler, H. *Angew. Chem. Int. Ed. Engl.* **1971**, *10*, 570; (b) Jennings, W. B. *Chem. Rev.* **1975**, *75*, 307 and references therein.

[156] *Cf.* Waugh, J. S.; Cotton, F. A. *J. Phys. Chem.* **1961**, *65*, 562.

[157] For a discussion of the NMR spectra of these compounds, see Deprés, J.-P.; Morat, C. *J. Chem. Educ.* **1992**, *69*, A232.

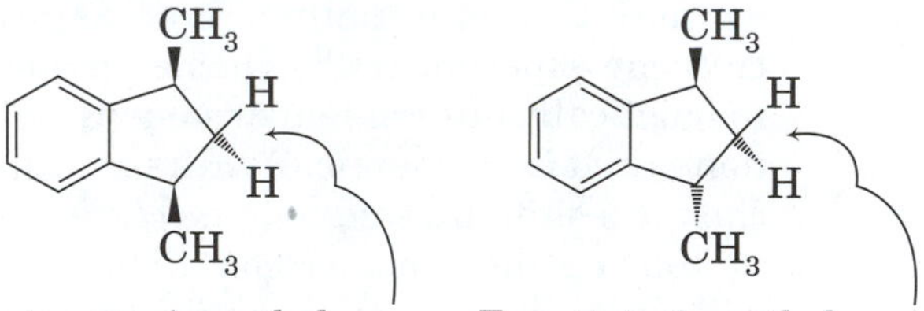

Diastereotopic methylene protons split each other.

100

Enantiotopic methylene protons do not split each other.

101

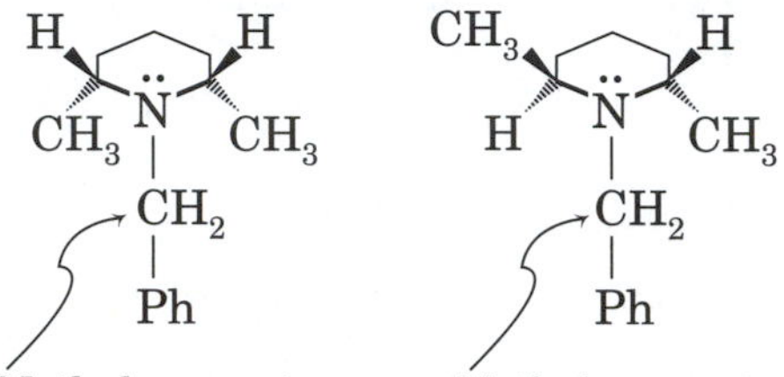

Methylene protons are enantiotopic.

102

Methylene protons are diastereotopic.

103

Structures that are achiral but can be made chiral by substitution are said to be **prochiral**.[158] In the example of butane, we see that replacement of one of the hydrogens on C2 by a deuterium (X = D in Figure 2.29) gives (*R*)-2-deuteriobutane, so that hydrogen is said to be the pro-(*R*) substituent.[159,160] Replacement of the other hydrogen by deuterium gives (*S*)-2-

Figure 2.29 Prochiral relationships.

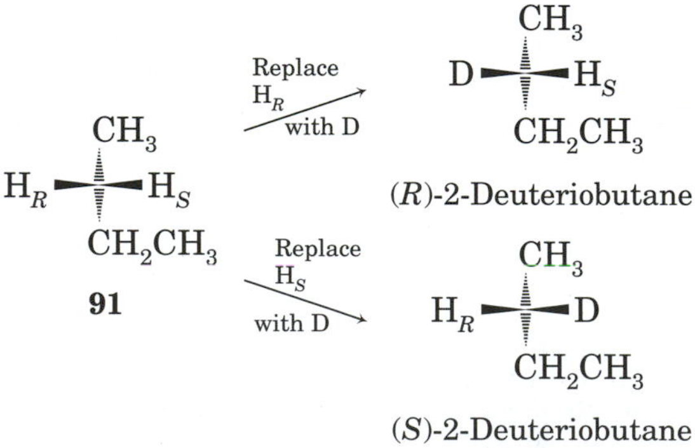

91

Replace H_R with D

(*R*)-2-Deuteriobutane **(*R*)-104**

Replace H_S with D

(*S*)-2-Deuteriobutane **(*S*)-104**

---

[158]Of course, prochiral molecules can be made chiral by other types of reactions also (such as addition reactions, as exemplified by the addition of HCl to 2-butene).

[159]In the example of 2,2-dichlorobutane, replacement of one chlorine with a higher atomic weight isotope would serve to distinguish between the two stereoheterotopic positions. Of course, our analysis is artificial, since Cl atoms occur naturally as different isotopes. Our designation of pro-(*R*) or pro-(*S*) is based on a definition of stereotopicity, not on an isotopic substitution that we are likely to carry out.

[160]The terms pro-(*R*), pro-(*S*), *re* and *si* were defined by Hanson, K. R. *J. Am. Chem. Soc.* **1966**, *88*, 2731.

deuteriobutane, so that hydrogen is the pro-(S) substituent.[161] A subscript R or S can be used to denote the prostereogenic nature of the substituents. Identification of diastereotopic and enantiotopic substituents is important in organic chemistry, but it can be even more important in biochemistry, because enzymes may distinguish between prochiral groups.[162]

Since topicity is a property of *atoms and spaces* in a (prochirotopic) molecule, *spaces around achiral molecules can also be considered prochiral*. For example, reduction of acetophenone, **105**, by lithium aluminum hydride gives either (R)- or (S)-1-phenylethanol, depending on the pathway of the reaction (Figure 2.30). In talking about the pathways of the reaction, it is useful to differentiate the faces of the molecule as the *re* face (from the Latin for *rectus*) or as the *si* face (from the Latin *sinister*),[72] just by applying (R) or (S) sequence rules to the three substituents in the planar molecule as viewed from each face.[160]

Chirotopicity and Stereogenicity

Mislow and Siegel pointed out that the *local geometry* or *symmetry* of a molecule and its stereoisomerism are two distinctly different properties. Furthermore, they distinguished between chirotopicity and stereogenicity.[163]

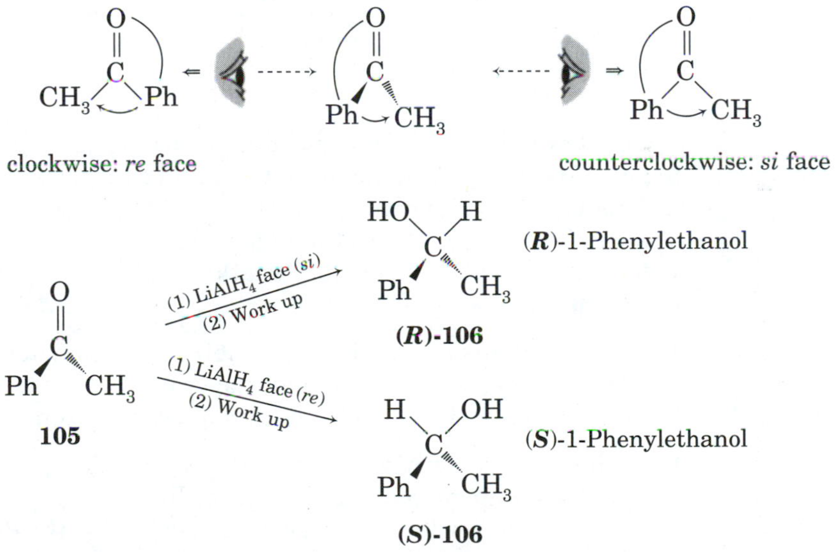

Figure 2.30 Prochiral faces of molecules.

clockwise: *re* face

counterclockwise: *si* face

(1) $LiAlH_4$ face (*si*) (2) Work up → (***R***)-1-Phenylethanol, (***R***)-**106**

(1) $LiAlH_4$ face (*re*) (2) Work up → (***S***)-1-Phenylethanol, (***S***)-**106**

105

[161] Eliel noted some formal statements of pro-(R) and pro-(S) determinations. One easy rule to use is that one arranges the substituents according to normal rules of priority. Then one arbitrarily assigns one of the two identical groups a higher priority than the other one of them. If this arbitrary assignment gives a formal (R) configuration, then the group assigned higher priority is the pro-(R) group. If the arbitrary assignment of higher priority makes the configuration (S), then the group assigned higher priority is the pro-(S) substituent. Eliel, E. L. *J. Chem. Educ.* **1971**, *48*, 163.

[162] See, for example, reference 119.

[163] Mislow, K.; Siegel, J. *J. Am. Chem. Soc.* **1984**, *106*, 3319. For a summary of this paper, see (a) Maugh II, T. H. *Science* **1984**, *225*, 915; (b) Dagani, R. *Chem. Eng. News* **1984** (June 11), 21.

- **Chirotopicity** is a local geometry that produces chirality. Atoms in a chiral environment are said to be *chirotopic*, but so are the spaces around atoms in a chiral environment. Atoms (or spaces) in an achiral environment are said to be *achirotopic*.
- **Stereogenicity** is the property of producing a new stereoisomer resulting from interchange of two bonded atoms in a structure. An atom that displays stereogenicity is said to be *stereogenic*, that is, it is a *stereocenter*. (A space cannot be a stereocenter, since atoms cannot be interchanged about it by breaking bonds to it.)

Mislow and Siegel gave as an example the set of isomers of 2,3,4-trihydroxyglutaric acid, **107–110**. Neither **107** nor **108** is chiral. (Both are meso structures.) However, in both structures C3 has been labeled "undoubtedly an 'asymmetric' carbon atom."[164] According to Mislow and Siegel, however, C3 is stereogenic[165] and achirotopic in **107** and **108**. On the other hand, C3 is nonstereogenic and chirotopic in compounds **109** and **110**.

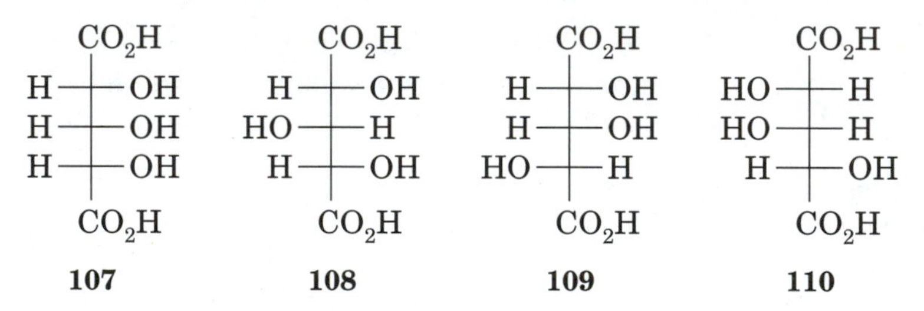

Mislow and Siegel also introduced new definitions to clarify some ambiguous stereochemical nomenclature, and they restated some topicity definitions to make them consistent with the stereochemical terms. The flow chart in Figure 2.31 shows their classification for the kinds of substituents within a molecule.[163] The classification system is based on the answer to three questions:

1. Are the atoms related by a symmetry operation of the molecule?
2. Are they related by a symmetry operation of the first kind?[166]
3. Do they have the same bonding connectivity (constitutions)?

In each case, a yes answer causes branching along the bold line; a no answer causes branching along the lighter line.

In addition, Mislow and Siegel have extended the definition of the term *prochiral* as follows:[167]

---

[164]Jaeger, F. M., cited in reference 163.

[165]It is important to recognize that the definition of stereogenicity calls for interchange of two groups to produce a stereoisomer, not necessarily an enantiomer. In this case the interchange produces a diastereomer.

[166]A symmetry operation of the first kind is a proper rotation (C_n), as opposed to an improper rotation (S_n).

[167]The emphasis in the definition is added.

Figure 2.31 Topic relationships. (Reproduced from reference 163.)

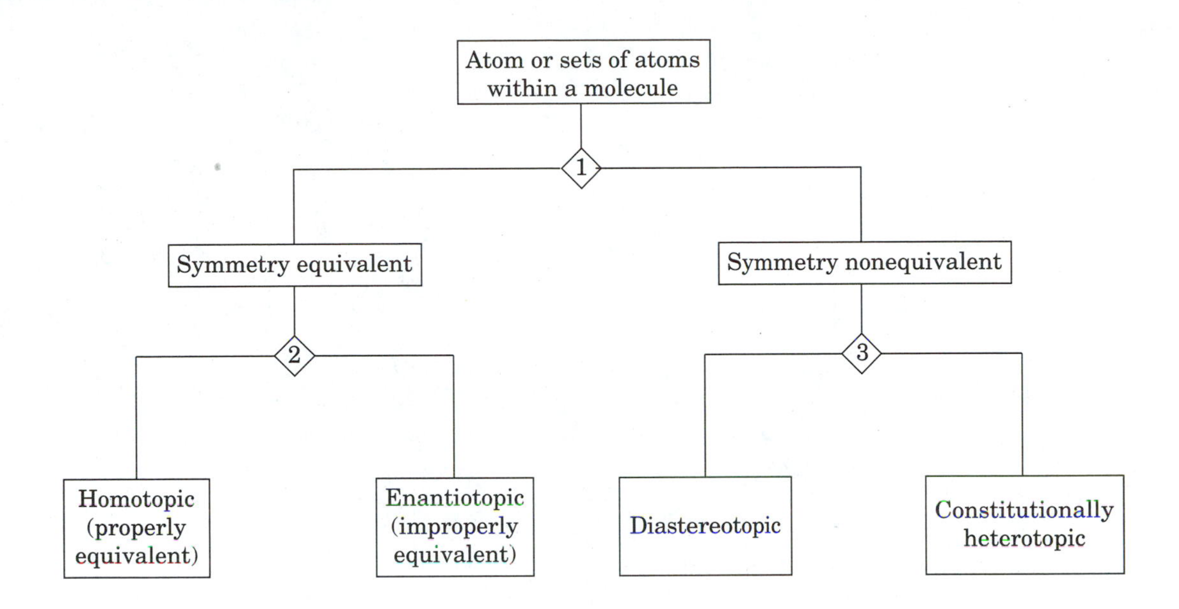

> We define as **(pro)^p-chiral** ($p > 0$) any finite, achiral object that can be desymmetrized into a chiral object by *at most p* stepwise replacements of a point by a differently labeled one[168] and as **(pro)^p-chirality** the corresponding property of an achiral object.

Note that prochirality is now a property of the entire molecule (object), *not* a property of a particular atom within the molecule. Consider the example of triptycene, **111**, in Figure 2.32. The structure can be desymmetrized by several pathways, one of which will produce a chiral structure in only one step. However, the desymmetrization can be accomplished in *at most* three steps (each step involving a different symmetry element), so it is shown to be (pro)^3-chiral. Table 2.3 lists the (pro)^p-chirality of molecules as a function of their symmetry.[169]

It is rare that a chemist wants to accomplish a goal by the maximum number of steps possible. More commonly, we are interested in carrying out a synthesis in the fewest number of steps. Nevertheless, the concepts of prochirality and topicity provide useful tools for understanding molecular structures and properties, as well as the reactions that we will consider in later chapters.

---

[168]Note that each subsequent replacement reduces the symmetry of the previous structure from (pro)$^{(n+1)}$-chiral to (pro)^n-chiral, where n is an integer equal to or greater than 1.

[169]Prochirality was discussed in terms of group theory by Fujita, S. *J. Am. Chem. Soc.* **1990**, *112*, 3390; *Tetrahedron* **1990**, *46*, 5943, **1991**, *47*, 31.

Figure 2.32 Prochirality determination of triptycene. (Reproduced from reference 163.)

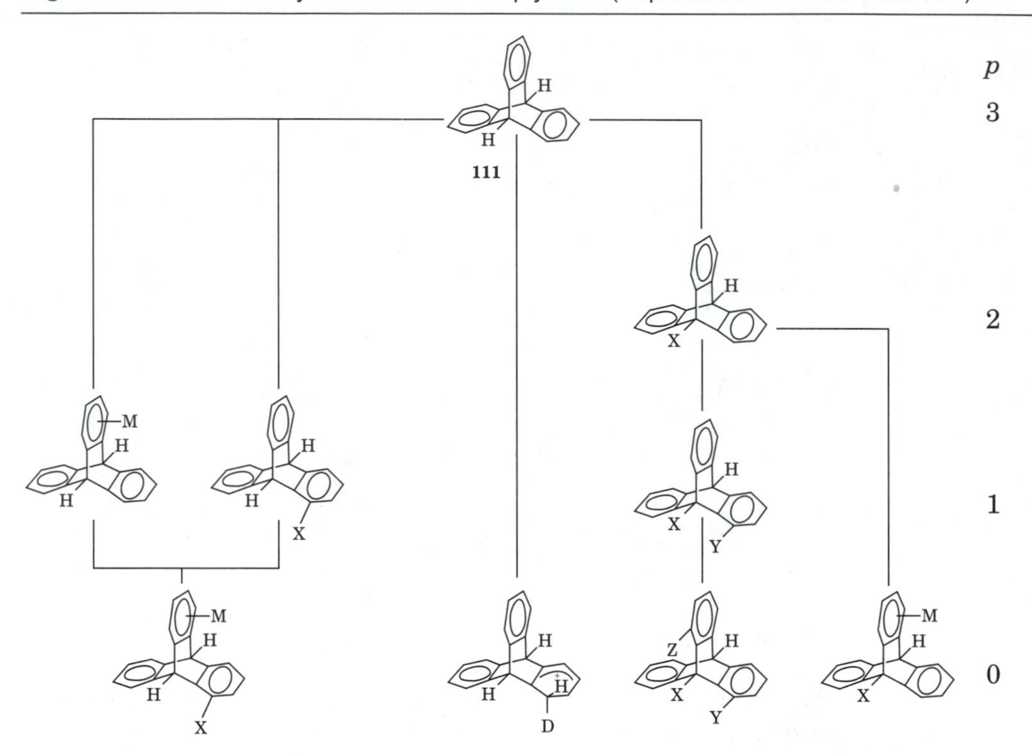

Table 2.3 Prochirality and molecular symmetry (Reproduced from reference 163.) (Pro)^p-chirality and (pro)^p-chirotopicity as attributes of models of molecules and their segments.

Desymmetrization index *p*	0	1	2	3
Description of molecule (segment)	chiral (chirotopic)	(pro)^1-chiral ((pro)^1-chirotopic)	(pro)^2-chiral ((pro)^2-chirotopic)	(pro)^3-chiral ((pro)^3-chirotopic)
Molecular or local symmetry	C_n, D_n, T, O, I	C_s, C_i, S_{2n}	C_{nv}, C_{nh}	D_{nd}, D_{nh}, T_d, T_h, O_h, I_h, K_h
Invariant achirotopic subspace*	none	a plane or the central point	an axis or the central point	the central point

*A set of points that remain stationary under every improper rotation of the point group.

Problems

1. Designate one or more appropriate stereochemical labels, such as *syn*, *anti*, (*E*), (*Z*), *endo*, *exo*, α, β, *r*, *rel*, (*R**), or (*S**) for each of the following structures. Note that parts (f) and (g) use the Maehr notation (page 84).

a. 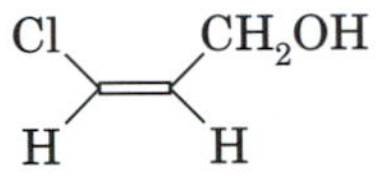

b. 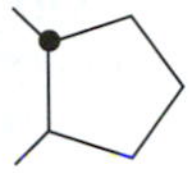

c. 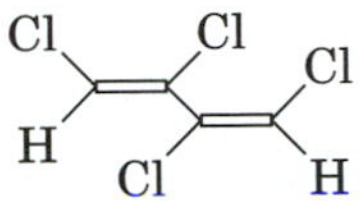

d. 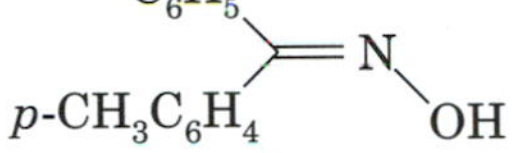

e. 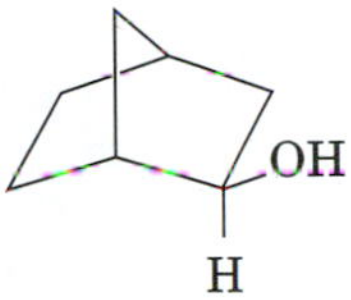

f. 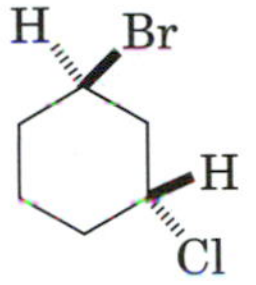

g. 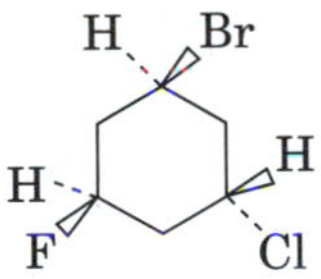

h. 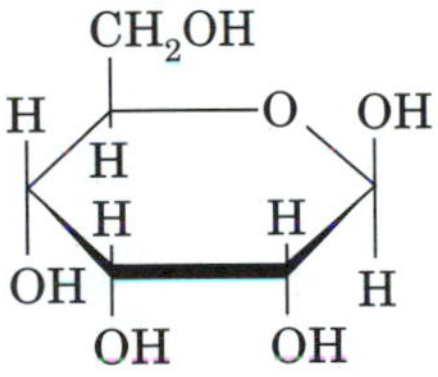

2. Consider the relationship of structure **A** with each of the structures **B**, **C**, **D**, and **E**.

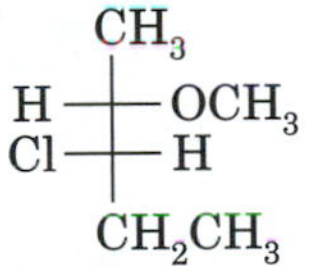

A

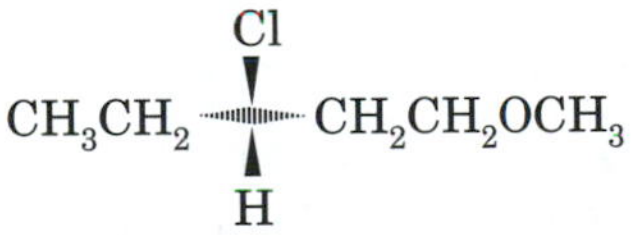

B

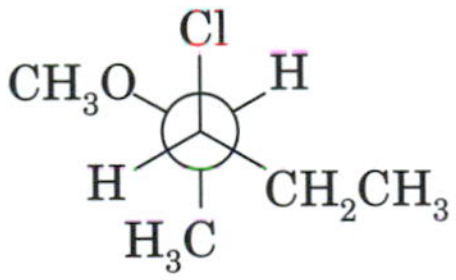

C

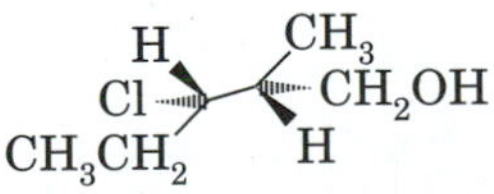

D

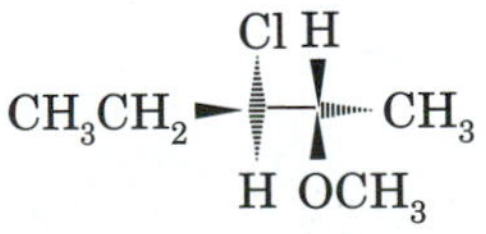

E

In each case, determine whether **A** and any other structure comprise a pair of identical compounds, positional isomers, functional group isomers, enantiomers, diastereomers, or nonisomeric compounds (that is, compounds with different molecular formulas).

3. Name each of the following structures, using the (*R*) or (*S*) or (*M*) or (*P*) designations.

a. 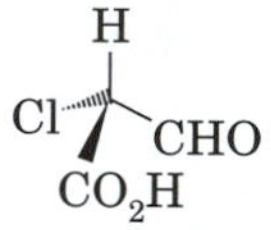 H, Cl, CHO, CO_2H

b. 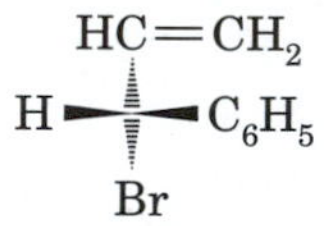 $HC{=}CH_2$, H, C_6H_5, Br

c. H, Br, CH_3, Cl, CH_2CH_3, CH_2CH_3

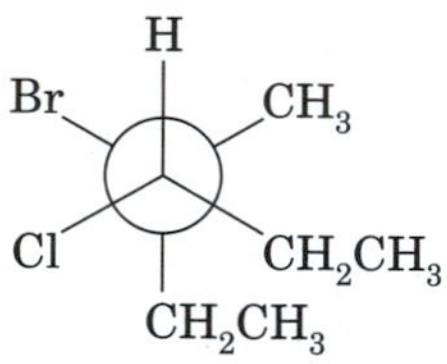

d. 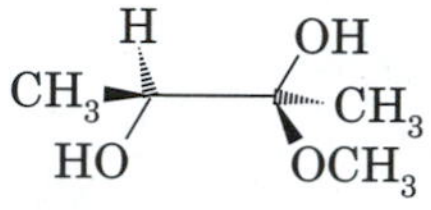 H, OH, CH_3, CH_3, HO, OCH_3

e. 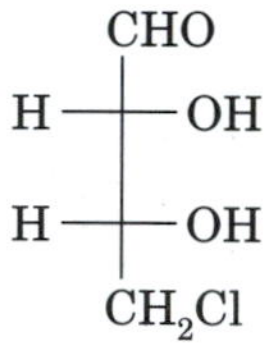 CHO, H, OH, H, OH, CH_2Cl

f. 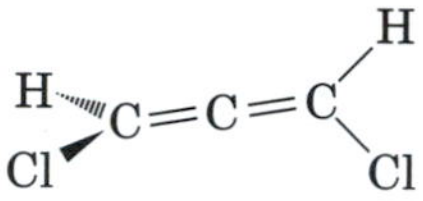 H, H, C=C=C, Cl, Cl

g. 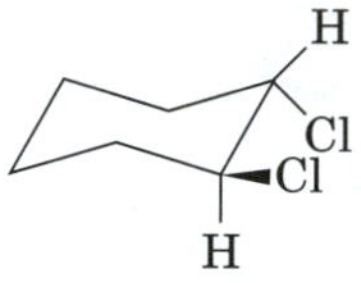 H, Cl, Cl, H

h. Br, H, H, Cl

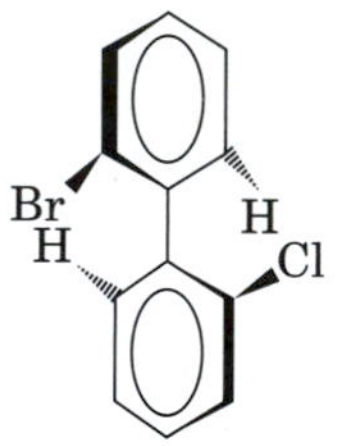

i. 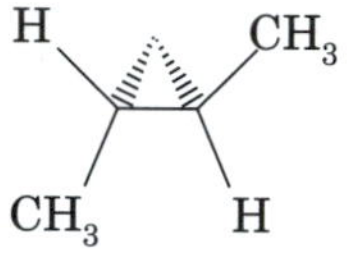 H, CH_3, CH_3, H

j. 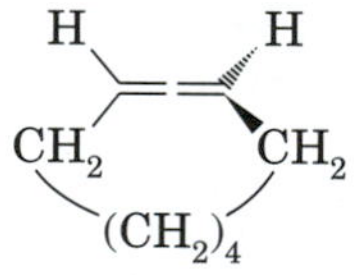 H, H, CH_2, CH_2, $(CH_2)_4$

k. 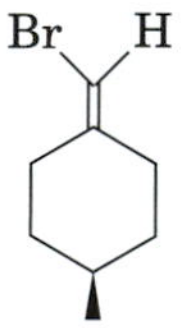 Br, H

l. 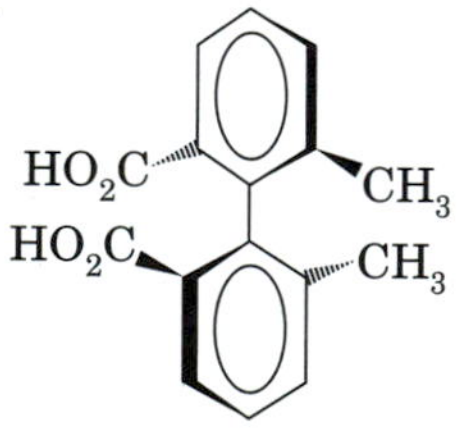 HO_2C, CH_3, HO_2C, CH_3

4. Assign both the (*R*) or (*S*) and the (*M*) or (*P*) designation to the helicene, **112**.

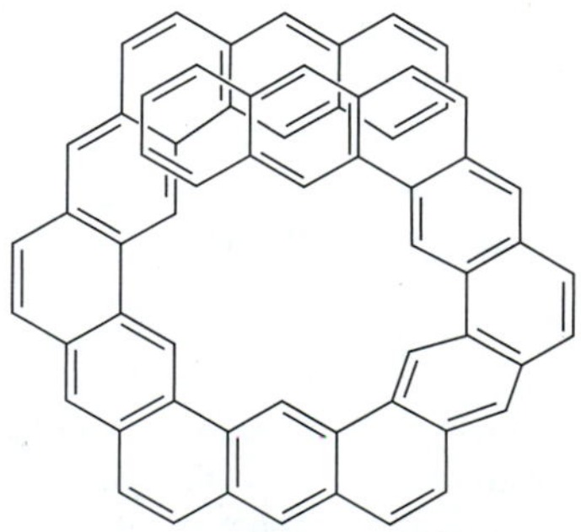

112

5. How many stereoisomers are possible for the dendrimer **113**?

113

6. Indicate the *re* and *si* faces for the following structures. In (c) indicate the prochirality about the atom marked with the asterisk. (CoA is an abbreviation for coenzyme A.)

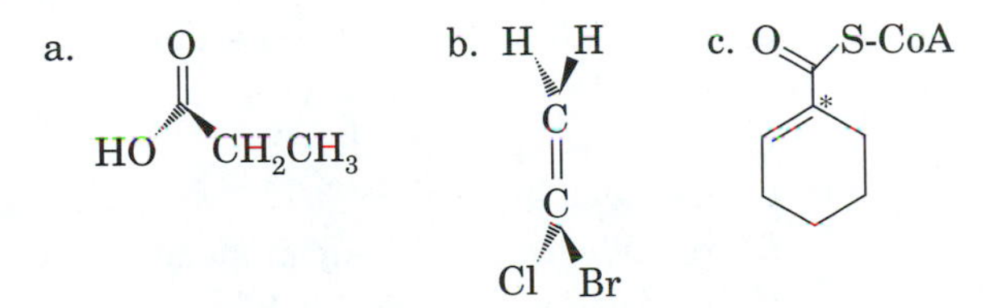

7. In each of the following structures indicate whether the circled groups are
   a. homotopic or heterotopic;
   b. if heterotopic, whether they are structurally (constitutionally) heterotopic or stereoheterotopic;
   c. if stereoheterotopic, whether they are enantiotopic or diastereotopic;
   d. if enantiotopic or diastereotopic, indicate which is pro-(*R*) and which is pro-(*S*).

8. Identify the point group to which each of the following structures belongs, and tell whether each structure is chiral. Note that in part (c) only one of two resonance structures possible for the porphyrin ring is shown.

a.

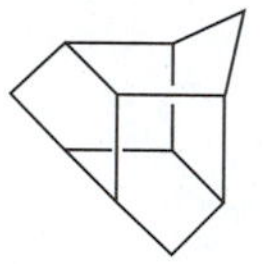

b.

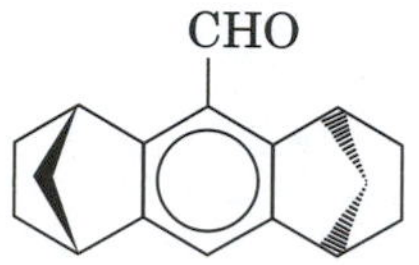

CHO

c.

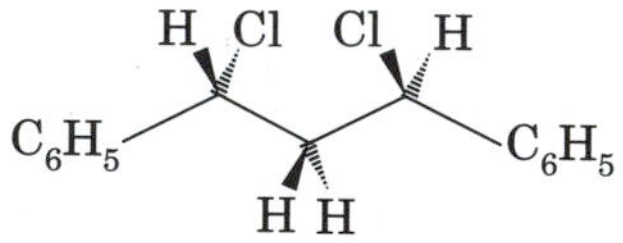

H Cl Cl H

C_6H_5 C_6H_5

H H

9. Draw a clear three-dimensional representation of the following compounds:
   a. (2*E*,4*E*,6*R*,10*R*)-4,6,10,12-tetramethyl-2,4-tridecadien-7-one
   b. (2*S*,4′*R*,8′*R*)-α-tocopheryl acetate
   c. (2*R*,4*R*)-1,2:4,5-diepoxypentane
   d. (*S*,*S*)-cyclopropane-1,2-$^2H_2$
   e. (*S*)-3-butene-1,2-diol
   f. (*R*)-3-hydroxy-3-phenylpropanoic acid
   g. (*Z*,*Z*)-deca-3,7-diene-1,5,9-triyne
   h. (2*R*,6*S*,10*S*)-6,10,14-trimethylpentadecan-2-ol
   i. (*R*)-(−)-(4-methylcyclohexylidene)acetic acid
   j. (*S*)-2,2′-dihydroxy-4,5,6,4′,5′,6′-hexamethoxybiphenyl
   k. (*Z*,*Z*)-2,8-dimethyl-1,7-dioxaspiro[5.5]undecane (found in the rectal gland secretion of the male cucumber fly)
   l. (2*S*,3*S*)-cyclopropane-*1*-^{13}C,^2H-*2,3*-$^2H_2$
   m. (2*R*,3*R*,5*E*)-2-hydroxy-3-methyl-5-heptenal, an intermediate in the synthesis of a possible precursor to the immunosuppressive compound cyclosporine
   n. (*S*)-[2-(phenylmethoxy)ethyl]oxirane
   o. i. (±)-*threo*-1,2-dihydroxybutyric acid
      ii. (±)-*erythro*-1,2-dihydroxybutyric acid
   p. (2*S*,3*S*)-octane-2,3-diol, a pheromone of the grape borer (*Xylotrechus pyrrhoderus*)

q. *r*-1,*c*-3-dimethoxy-*t*-5-methylcyclohexane
r. (1*R*,2*R*,3*R*,4*R*,5*S*)-1-(methylthio)-2,3,4-trihydroxy-5-aminocyclopentane (mannostatin A, a naturally-occurring α-mannosidase inhibitor)
s. (*S*,*S*)-(−)-1,6-bis(*o*-chlorophenyl)-1,6-diphenylhexa-2,4-diyne-1,6-diol
t. (*R*)-dibenzyl ($^{12}CH_2$, $^{13}CH_2$) sulfoxide ($Ph^{12}CH_2SO^{13}CH_2Ph$)

10. Indicate whether each of the following reactions proceeds with inversion of configuration or retention of configuration at each chiral center.
    a. (*R*)-(+)-1,2-epoxybutane → (*R*)-(−)-3-hexanol (upon treatment with ethyllithium)
    b. (*S*)-(+)-2,2-dimethylcyclopropanecarboxamide → (*S*)-(−)-2,2-dimethylcyclopropylamine (Hoffmann conditions)
    c. (*S*)-ethanol-1-*d* → (*S*)-[1-^{2}H,1-^{3}H]ethane (reagents: (i) TsCl, Et_3N; (ii) $LiEt_3B^3H$)
    d. (*S*)-[1-^{2}H,1-^{3}H]ethane → (*R*)-[1-^{2}H,1-^{3}H]ethanol (by action of methane monooxygenase)
    e. (*S*)-(+)-1-bromo-1-methyl-2,2-diphenylcyclopropane → (*R*-(−)-2-methyl-1,1-diphenylcyclopropane
    f. (2*R*,3*S*)-(+)-3-bromobutan-2-ol → (2*S*,3*S*)-(−)-2,3-dibromobutane
    g. D-(+)-1-buten-3-ol → D-(−)-3-methoxy-1-butene

11. How many NMR signals would you expect from the protons on the methylene groups marked with the asterisk in **114**?

F, CH_3: C=C=C: $CH(O\overset{*}{C}H_2CH_3)_2$, H

114

12. Mislow and Siegel noted that C3 in both compounds **109** and **110** is *nonstereogenic and chirotopic*. Use the definitions of these terms and the properties of the molecules to demonstrate that this statement is correct.

13. Use the Mislow and Siegel definition of *(pro)^p-chirality* to determine the order of prochirality of the following structures:
    a. barrelene
    b. cubane
    c. bicyclo[1.1.0]butane
    d. bicyclo[2.2.2]octane

14. A study of the mechanism of the Wurtz reaction required that the configuration of ( ) 3 methylnonane produced in the reaction be known. The researchers treated (*R*)-(−)-2-bromooctane with the product of reaction of sodium and di-

ethyl malonate in ethanol solution. The levorotatory product was then hydrolyzed and heated until 1 mole of CO_2 was lost per mole of product, producing a levorotatory monocarboxylic acid. Treatment of the product with $LiAlH_4$, then PBr_3, and then with $LiAlH_4$ again produced (−)-3-methylnonane. Identify the mechanism of each of the reactions in this sequence and tell whether the product (−)-3-methylnonane has the same configuration as the reactant (*R*)-(−)-2-bromobutane.

15. *Alcaligenes bronchisepticus* KU 1201 decarboxylates α-methyl-α-phenylmalonic acid to form (*R*)-α-phenylpropionic acid.
    a. In the absence of isotopic label, is the parent reaction stereoselective or stereospecific?
    b. In separate experiments, each of the carboxyl carbon atoms was labeled with ^{13}C. When the reactant was (*S*)-(−)-[1-^{13}C]-α-methyl-α-phenylmalonic acid, the product was (*R*)-[1-^{13}C]-α-phenylpropionic acid. When the (*R*)-(+) enantiomer of the starting material was used, the product was (*R*)-(−)-α-phenylpropionic acid (without the ^{13}C label).
       i. Which carboxyl group (pro-*R* or pro-*S*) is removed in the decarboxylation?
       ii. Does the decarboxylation take place with retention or inversion of configuration?

16. *cis*-3,7-Dimethyl-1,5-cyclooctanedione underwent two successive Baeyer-Villager rearrangements to give two products, both with the molecular formula $C_{10}H_{16}O_4$. When one of these products was reduced with $LiAlH_4$, two achiral diols were obtained. When the other was reduced, the single product was a chiral diol (obtained as a racemic mixture). What are the structures of the products obtained in each of the $LiAlH_4$ reductions?

17. a. Assign an appropriate stereochemical designator (*threo, erythro, E, Z,* etc.) to each of the positions marked with an arrow in (+)-uvaricin, **115**.
    b. Draw the structure of (15,16,19,20,23,24)-*hexaepi*-uvaricin and give a stereochemical designator for each of the corresponding positions in this compound.

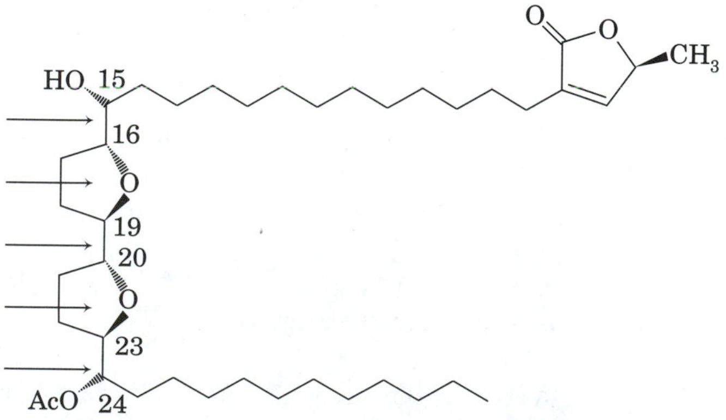

115

18. The specific rotation of (−)-2-bromobutane is reported to be −23.13°. A sample prepared by nucleophilic attack of bromide ion on the tosylate of partially resolved 2-butanol had a measured specific rotation of −16.19°. What is the mole fraction of each enantiomer in the sample of 2-bromobutane?

19. The decahydroquinoline alkaloid **116** is known as 195A. Draw a clear three-dimensional representation of 2-*epi*-195A.

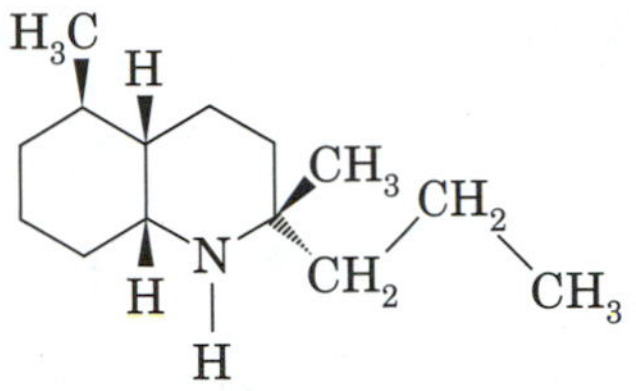

116

20. The ^{1}H NMR spectrum of the methylene group of 2-phenyl-3-methylbutylmagnesium chloride in THF solution indicates that the two protons are magnetically nonequivalent at 66° but become magnetically equivalent at 120°. Propose an explanation for the effect of temperature on the spectrum of this compound.

21. Reaction of acetaldehyde with a chiral, partially deuterated reducing agent produced ethanol-1-*d* with $[\alpha]_D$ −0.123 ± 0.025°. This product was calculated to be 44 ± 9% optically pure. The ethanol-1-*d* was converted to the *p*-nitrobenzenesulfonate, which was then used in an acetoacetic ester synthesis with methyl acetoacetate and sodium methoxide. The product of that reaction was hydrolyzed and decarboxylated to give 2-pentanone-4-*d* having $[\alpha]_D$ + 0.25 ± 0.03°. Clemmensen reduction of that compound produced pentane-2-*d* having $[\alpha]_D$ + 0.19 ± 0.06°. The compound (−)-pentane-2-*d* was reported to have the (*S*) configuration.
   a. If the initially formed ethanol-1-*d* is 44% optically pure, what would be the specific rotation of optically pure ethanol-1-*d*?
   b. What is the absolute configuration of the ethanol-1-*d* formed in the reduction reaction?

22. The biologically important compound (−)-chorismate (**117**) exhibits the circular dichroism spectrum shown in Figure 2.33.
   a. Draw a clear three-dimensional representation of (+)-chorismate.
   b. Sketch the CD spectrum expected for (+)-chorismate.

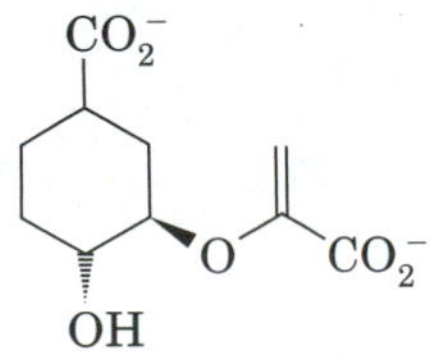

117

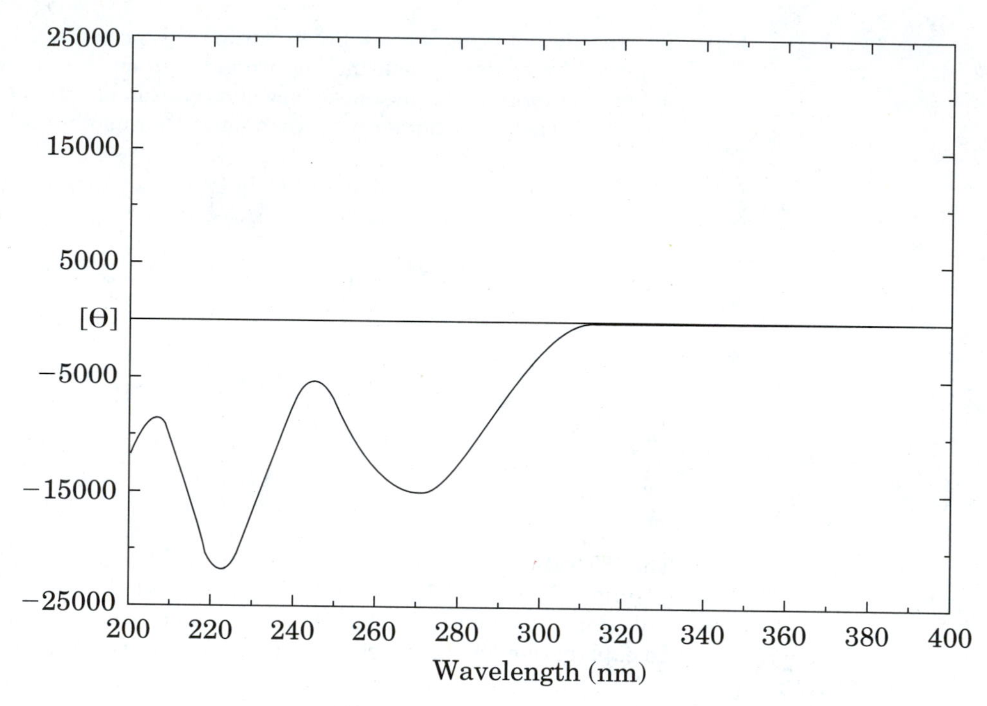

Figure 2.33 Circular dichroism spectrum of (−)-chorismate. (Adapted from Hilvert, D.; Nared, K. D. *J. Am. Chem. Soc.* **1988**, *110*, 5593.)

CHAPTER 3

Conformational Analysis and Molecular Mechanics

3.1 Molecular Conformation

The discussion of stereochemistry in Chapter 2 implicitly assumed that molecules are as rigid as the plastic physical models or two dimensional pictorial representations that we use to study them. However, our current understanding of quantum mechanics, thermodynamics, and spectroscopy tells us that molecules are in constant motion. At room temperature they vibrate and (in the liquid or gas phase) rotate. Not only may the entire molecule rotate, but *internal rotations* within a molecular structure can change the intramolecular spatial relationships depicted in static models.

The term ***conformation*** generally refers to one of several different spatial arrangements that a molecule can achieve by rotations about single bonds between atoms.[1] If a molecule has several bonds about which rotation can occur, then the shape of the molecule can change significantly.[2] Even though they ordinarily cannot be isolated, conformations that correspond to energy minima are known as **conformers**, a contraction of conformational isomers. Occasionally they are called **rotamers**, a shortened form of rotational isomers.

A conformation can be defined by a **dihedral angle**, which is the angle made by two bonds on adjacent atoms when the atoms are eclipsed in a Newman projection.[3] Consider a C — C unit with substituent A on one car-

---

[1]Eliel has noted the difficulties in precisely defining the term *conformation:* (a) Eliel, E. L.; Allinger, N. L.; Angyal, S. J.; Morrison, G. A. *Conformational Analysis;* Wiley-Interscience: New York, 1965; p. 1; (b) Eliel, E. L. *J. Chem. Educ.* **1975**, *52*, 762.

[2]For an example of a molecule in which bond rotation leads to a molecular hinge effect, see Schneider, H.-J.; Werner, F. *J. Chem. Soc., Chem. Commun.* **1992**, 490.

[3]Alternatively, it may be considered the angle made by the intersection of two planes, one defined by the two carbon atoms and atom A, the other defined by the two carbon atoms and atom B.

bon atom and substituent B on the other. The dihedral angle is the angle between the lines A—C and C—B, and the angle is considered positive if the arrow drawn from the near bond line curves clockwise toward the second bond line and negative if it turns counterclockwise. As shown in Figure 3.1, it does not matter which way the Newman projection is viewed.

Several important conformational terms are best described with Newman projections. Consider the set of conformations of the structure A—CH_2—CH_2—B (**1**) in Figure 3.2. In (**a**) the A—C and C—B bonds are eclipsed. If we imagine rotating about the C—C bond so that the substituents on the front carbon atom stay fixed but the substituents on the back carbon atom rotate 60° clockwise, we produce the **gauche** conformation (**b**).[4] The dihedral angle between the lines representing the C—A and C—B bonds in this projection is 60°.

Rotation of the substituents on the back carbon atom another 60° produces another eclipsed conformation (**c**), this time with the C—A bond eclipsing a C—H bond on the back carbon atom, and the C—B bond eclipsing a C—H bond on the front carbon atom. Rotation another 60° produces the **anti** conformation (**d**).[5] Another 60° rotation of the back carbon atom produces another eclipsed conformation, (**e**) and one more 60° rotation leads to another gauche conformation, (**f**).

An alternative notation for conformations, developed by Prelog and Klyne, is illustrated in Figure 3.3.[6,7,8] If the dihedral angles are within 30° of either 0° or 180°, then the substituents A and B are approximately coplanar or periplanar. Therefore if the major substituents of interest on adjacent carbon atoms are in a conformation with dihedral angle of −30° to 0°, then the conformation is *−syn-periplanar* (*−sp*), where the prefix *syn* indi-

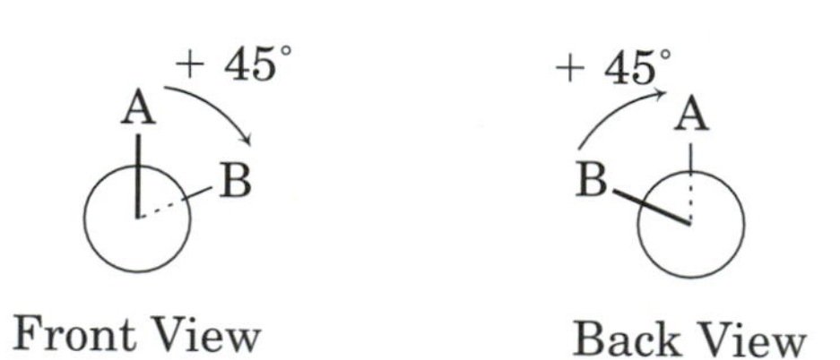

Figure 3.1 Determination of dihedral angle for A—C—C—B bonds.

[4]An alternative term for the *gauche* conformation is *skew*.

[5]In the physical chemistry literature the 180° conformation is often called trans, but organic chemists tend to reserve that term for *cis, trans* isomers. The anti and gauche conformations are occasionally denoted as *staggered*.

[6]Klyne, W.; Prelog, V. *Experientia* **1960**, *16*, 521.

[7]See also the discussion of this system by Hanack, M. *Conformation Theory;* Neumann, H. C., trans.; Academic Press: New York, 1965; pp. 68–69.

[8]IUPAC Organic Chemistry Division Commission on Nomenclature of Organic Chemistry *Nomenclature of Organic Chemistry, Sections A, B, C, D, E, F, and H;* Rigaudy, J.; Klesney, S. P., eds.; Pergamon Press: Oxford, England, 1979; Section E-5.6, p. 484.

Figure 3.2 Conformations of A — CH_2 — CH_2 — B. (Adapted from reference 7, p. 30.)

	Newman Projection	Dihedral Angle	Conformation Description
(a)	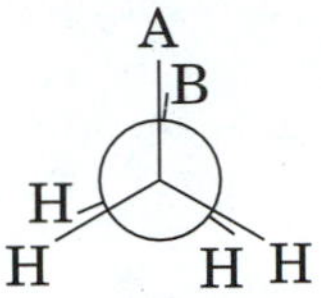	0°	eclipsed
(b)	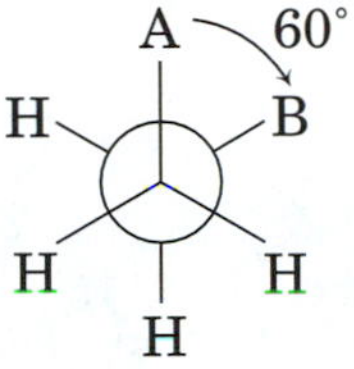	60°	gauche (skew) (syn) (staggered)
(c)	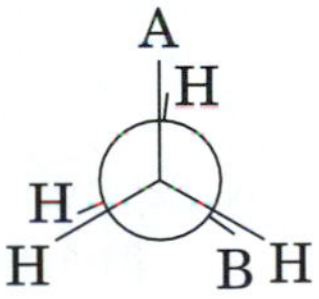	120°	eclipsed
(d)	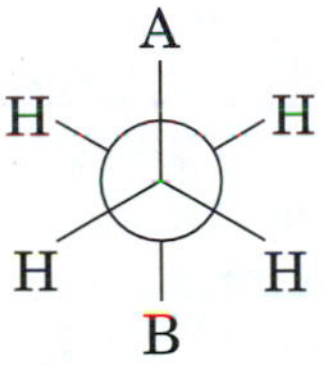	180°	anti (trans) (staggered)
(e)	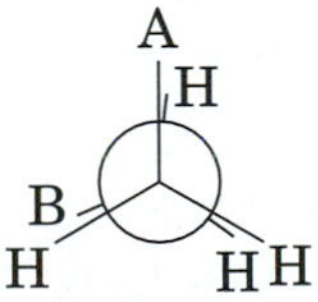	240°	eclipsed
(f)	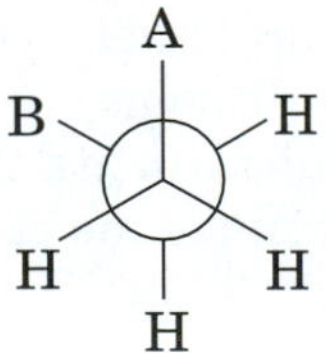	300°	gauche (skew) (syn) (staggered)

1

Figure 3.3 Conformational descriptors.

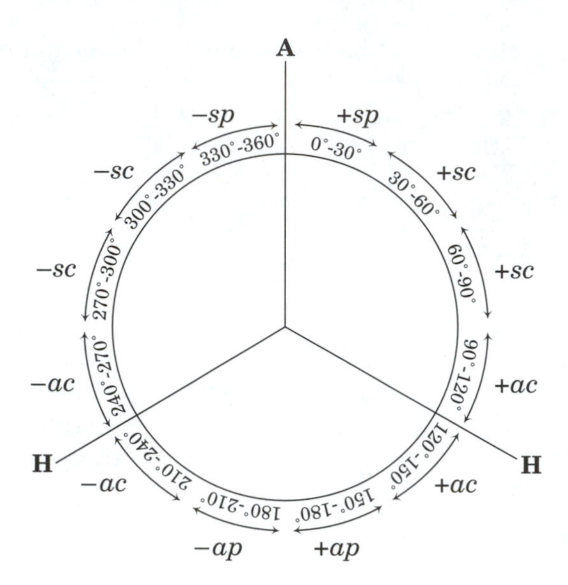

cates that the substituents are located on the same side and the minus sign indicates a negative dihedral. If the dihedral angle is 0° to +30°, then the conformation is +*syn-periplanar* (+*sp*). Similarly, if the dihedral angle is 150° to 180°, the conformation is +*anti-periplanar* (+*ap*), and a conformation with dihedral angle of 180° to 210° is −*anti-periplanar* (−*ap*). Exactly 180° would be *ap* (without a sign), and exactly 0° would be *sp*.

If the dihedral angle A—C—C—B is near 60°, 120°, 240°, or 300°, then one of the substituents is said to be clinal (inclined or slanted) with respect to the other. Thus the notation for conformations with dihedral angles of 30° to 90° is +*syn-clinal* (+*sc*); 90° to 150° is +*anti-clinal* (+*ac*); 210° to 270° is −*anti-clinal* (−*ac*); and 270° to 330° is −*syn-clinal* (−*sc*).

When applying conformational nomenclature, it is essential to identify the substituents on the two carbon atoms that determine the designation of conformation.

1. If there are three different substituents on a carbon atom, the defining group is that which has highest priority according to the sequence rules for (*R*) or (*S*) nomenclature.[9]
2. If the three substituents on a carbon atom are identical, then the one that gives the smallest dihedral angle with the determining substituent on the other carbon atom is chosen.
3. If a carbon atom has two substituents that are identical, then it is the third substituent that determines the classification.

[9]Cahn, R. S.; Ingold, C. K.; Prelog, V. *Experientia* **1956**, *12*, 81; *Angew. Chem. Int. Ed. Engl.* **1966**, *5*, 385.

As an example of the last criterion, consider isopropylcyclohexane (**2**) and ethylcyclohexane (**3**). In **2** each of the carbon atoms has one hydrogen and two other substituents that are identical to each other. Therefore the H—C—C—H dihedral angle determines the conformational classification, and the three conformations shown are the (*ap*), (+*sc*) and (−*sc*) conformations. In **3**, however, the unique atoms are the hydrogen on the cyclohexane ring and the methyl on the substituent group.[10]

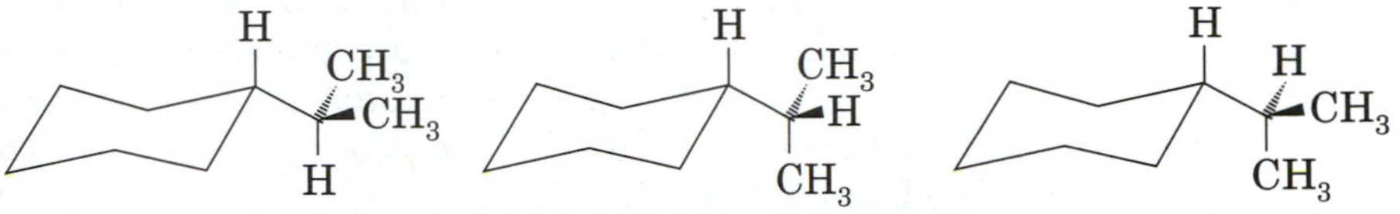

ap +*sc* −*sc*

2

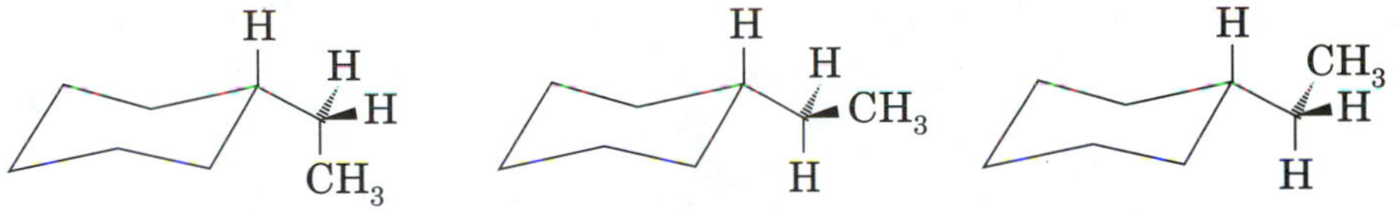

ap +*sc* −*sc*

3

Another notation for conformers has been developed to describe the geometry about the single bond of 1,3-butadiene (**4**). (We will see in Chapter 4 that theory predicts some double bond character for the C2—C3 bond.) There is a barrier to rotation about this bond, and two different conformations are possible. We describe **4**(**a**) as *s-trans*-1,3-butadiene, since the two double bond substituents are **trans** to each other across the formally single bond. Similarly, **4**(**b**) is the *s-cis* conformer.[11] With the advent of the (*E*) and (*Z*) nomenclature system, the notations *s*-(*Z*) and *s*-(*E*) have begun to replace *s-cis* and *s-trans*.

---

[10]These conformations are discussed in Golan, O.; Goren, Z.; Biali, S. E. *J. Am. Chem. Soc.* **1990**, *112*, 9300.

[11]There is some debate about the planarity of the two double bonds in the *s-cis* conformation. Fisher and Michl determined through polarized matrix-isolation IR spectroscopy that the two bonds are nonplanar by no more than 10°–15°, indicating that the term *s-cis* is an appropriate designation for this conformation. (Fisher, J. J.; Michl, J. *J. Am. Chem. Soc.* **1987**, *109*, 1056.) However, both theoretical studies and experimental data suggest that the two double bonds might be nonplanar by about 25° to 35°, making the term *s-gauche* a better descriptor for the less stable conformer for 1,3-butadiene: Wiberg, K. B.; Rosenberg, R. E. *J. Am. Chem. Soc.* **1990**, *112*, 1509.

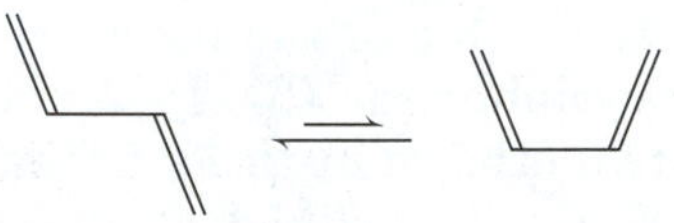

(a) *s–trans* (b) *s–cis*

4

Combination of the (*E*) and (*Z*) nomenclature system with the *s-cis*, *s-trans* system allows us to designate the stereochemistry and conformation of polyenes. For example, the photochemical reaction of 1,2-dihydronaphthalene (**5**) forms first the *cZc* conformer (**6**), which then rotates about the single bond to form the *cZt* isomer (**7**).[12]

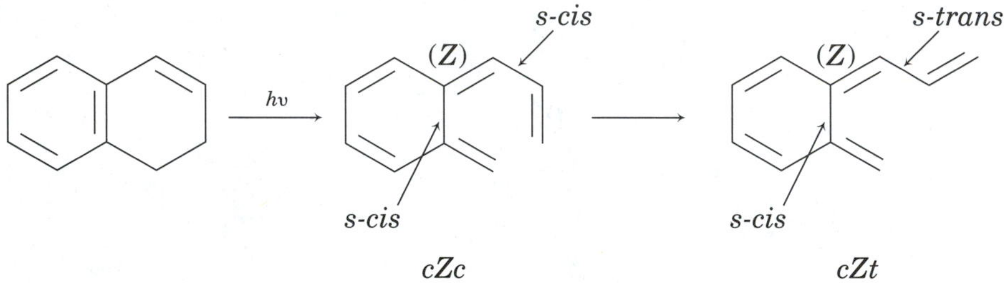

s-cis (Z) *hν* *s-cis* *cZc* *s-trans* (Z) *s-cis* *cZt*

5 **6** **7**

Ordinarily conformational isomers cannot be separated from each other. However, the C(O)—N bond of an amide also shows appreciable double bond character and restricted rotation. Chupp and Olin found a series of 2′,6′-dialkyl-2-halo-*N*-methylacetanilides in which the barrier to rotation was so high conformational isomers could be separated, purified, and characterized. For example, the melting points of **8** and **9** are 64–65° and 105–106°, respectively.[13]

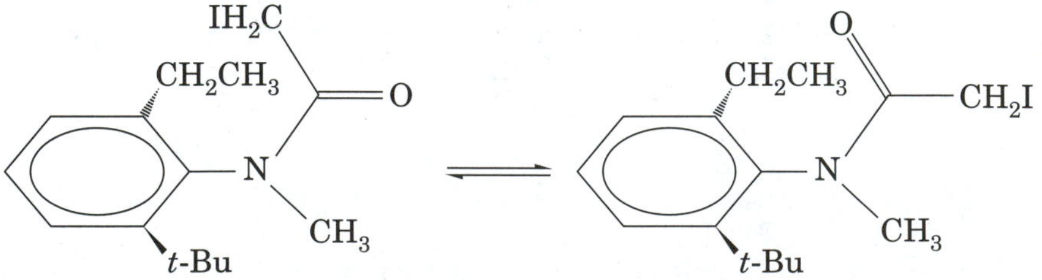

IH_2C CH_2CH_3 O N CH_3 *t*-Bu

O CH_2CH_3 CH_2I N CH_3 *t*-Bu

8 **9**

---

[12]Keijzer, F.; Stolte, S.; Woning, J.; Laarhoven, W. H. *J. Photochem. Photobiol. A: Chem.* **1990**, *50*, 401 and references therein.

[13]Chupp, J. P.; Olin, J. F. *J. Org. Chem.* **1967**, *32*, 2297.

There may be rotational restrictions in other structures as well. Yamamoto has reported that **10** is a conformationally biased molecule due to electrostatic interactions between the carbomethoxy and aldehyde groups.[14,15,16]

10

Molecular conformations are determined by a wide variety of experimental techniques.[17] Conformations of molecules in crystals may be determined by X-ray diffraction,[18] and molecules in the gas phase may be investigated by electron diffraction[19] and microwave spectroscopy.[20] Other techniques used to study molecular conformation include NMR,[21] vibrational spectroscopy,[22] UV spectroscopy,[23] and polarimetry and circular dichroism.[24] Other techniques include calorimetry[17] and the determination

---

[14]Yamamoto, Y.; Taniguchi, K.; Maruyama, K. *J. Chem. Soc., Chem. Commun.* **1985**, 1429; Yamamoto, Y.; Nemoto, H.; Kikuchi, R.; Komatsu, H.; Suzuki, I. *J. Am. Chem. Soc.* **1990**, *112*, 8598.

[15]As noted in Chapter 2, restricted rotation in molecules such as some substituted biphenyls can lead to enantiomerism.

[16]Complexation of a bipyridene side chain with Hg^{2+} has been used to induce a conformational change in a triptycene derivative so that the side chain effectively acts as a "brake" to stop a spinning "molecular wheel." Kelly, T. R.; Bowyer, M. C.; Bhaskar, K. V.; Bebbington, D.; Garcia, A.; Lang, F.; Kim, M. H.; Jette, M. P. *J. Am. Chem. Soc.* **1994**, *116*, 3657.

[17]Reference 1(a), pp. 129–188.

[18]Lipscomb, W. N.; Jacobson, R. A. in *Techniques of Chemistry, Vol. I. Physical Methods of Chemistry. Part IIID;* Weissberger, A.; Rossiter, B. W., eds.; Wiley-Interscience: New York, 1972; pp. 1–123; Cameron, A. F. in *Techniques of Chemistry. Volume IV. Part I*, 2nd ed.; Bentley, K. W.; Kirby, G. W., eds.; Wiley-Interscience: New York, 1972; pp. 481–513.

[19]Bartell, L. S. in *Techniques of Chemistry. Volume I. Physical Methods of Chemistry. Part IIID;* Weissberger, A.; Rossiter, B. W., eds.; Wiley-Interscience: New York, 1972; pp. 125–158.

[20]Flygare, W. H. in *Techniques of Chemistry. Volume I. Physical Methods of Chemistry. Part IIIA;* Weissberger, A.; Rossiter, B. W., eds.; Wiley-Interscience: New York, 1972; pp. 439–497.

[21]McFarlane, W. in *Techniques of Chemistry. Volume IV. Part I*, 2nd ed.; Bentley, K. W.; Kirby, G. W., eds.; Wiley-Interscience: New York, 1972; pp. 225–322; Phillips, L. in *Techniques of Chemistry. Volume IV. Part I*, 2nd ed.; Bentley, K. W.; Kirby, G. W., eds.; Wiley-Interscience: New York, 1972; pp. 323–353.

[22]Devlin, J. P.; Cooney, R. P. J. in *Techniques of Chemistry. Volume IV. Part I*, 2nd ed.; Bentley, K. W.; Kirby, G. W., eds.; Wiley-Interscience: New York, 1972; pp. 121–224.

[23]Timmons, C. J. in *Techniques of Chemistry. Volume IV. Part I*, 2nd ed.; Bentley, K. W.; Kirby, G. W., eds.; Wiley-Interscience: New York, 1972; pp. 58–119.

[24]Barrett, G. C. in *Techniques of Chemistry. Volume IV. Part I*, 2nd ed.; Bentley, K. W.; Kirby, G. W., eds.; Wiley-Interscience: New York, 1972; pp. 515–610.

of physical properties such as pK values[25] and dipole moments.[26] In addition, conformation may be inferred from kinetics of reactions of functional groups that may be in different conformational environments.[17] However, our emphasis in this chapter will not be the methods by which conformations are determined, nor will we consider in detail the effects of conformation on chemical reactivity. Instead, we will emphasize the principles of conformational analysis and molecular mechanics as tools for the exploration of chemical reactivity in later chapters.

3.2 Conformational Analysis

The energy of any molecular conformation is determined by all of the stabilizing and destabilizing forces that act on the atoms in the spatial arrangement at that instant. In a collection of molecules at ambient temperature, we would not expect all molecules to have exactly the same conformation. The relative number of molecules with any conformation can be predicted from knowledge of the energy required for each conformation and the temperature. Molecules may be said to explore all possible arrangements, experiencing all of them at one time or another, but spending most of their time in the more stable arrangements.[27,28] The search for knowledge of this kind comprises an important part of the field of *conformational analysis*.

Usually we are not so much interested in describing all possible spatial arrangements that result from rotation of bonds within a molecule as we are in describing the relative stabilities of the conformations that are energy minima. Often that knowledge will allow us to predict the fraction of the molecules that are in each of the possible conformations at any time. We also want to know what rotations or other changes lead to the interconversion of conformations and what the energetics and kinetics of these changes might be.[29]

Introduction

By using simple graphic or mechanical models of molecular structures, we can identify destabilizing forces that occur due to deviations from ideal

---

[25]Barlin, G. B.; Perrin, D. D. in *Techniques of Chemistry. Volume IV. Part I*, 2nd ed.; Bentley, K. W.; Kirby, G. W., eds.; Wiley-Interscience: New York, 1972; pp. 611–676.

[26]Smyth, C. P. in *Techniques of Chemistry. Volume I. Physical Methods of Chemistry. Part IV;* Weissberger, A.; Rossiter, B. W., eds.; Wiley-Interscience: New York, 1972; pp. 397–429.

[27]This is an anthropomorphic description. While anthropomorphisms can be useful models, we should recognize the danger of imputing motives to molecules.

[28]Except for explicit references to free energy, in this discussion we will follow the convention of using the word *energy* to mean *enthalpy*.

[29]The principles of conformational analysis were established by Barton, D. H. R. *Experientia* **1950**, *6*, 316. For general references, see reference 1(a) and Dauben, W. G.; Pitzer, K. S. in *Steric Effects in Organic Chemistry;* Newman, M. S., ed.; John Wiley & Sons: New York, 1956; pp. 1–60.

bonding parameters. The three types of strain most useful to us in conformational analysis are torsional strain (due to deviation from staggered bonds on adjacent atoms), van der Waals strain (repulsion due to overlap of electron clouds on nonbonded atoms), and angle strain (due to deviation from expected bond angles).[30]

Torsional Strain

Torsional strain results from deviations from staggered conformations. The usual way to analyze torsional energies is to compare energies of various rotations (as represented with Newman projections) on an energy level diagram, as shown in Figure 3.4 for ethane. The eclipsed conformation is about 3 kcal/mol higher in energy than the staggered conformation, and the potential energy varies with the angle of rotation in a sinusoidal fashion.[31] Although it might be considered that torsional strain could arise because of van der Waals repulsion between substituents, theoretical calculations suggest that is not the case.[32,33] Molecular orbital calculations indicate that the

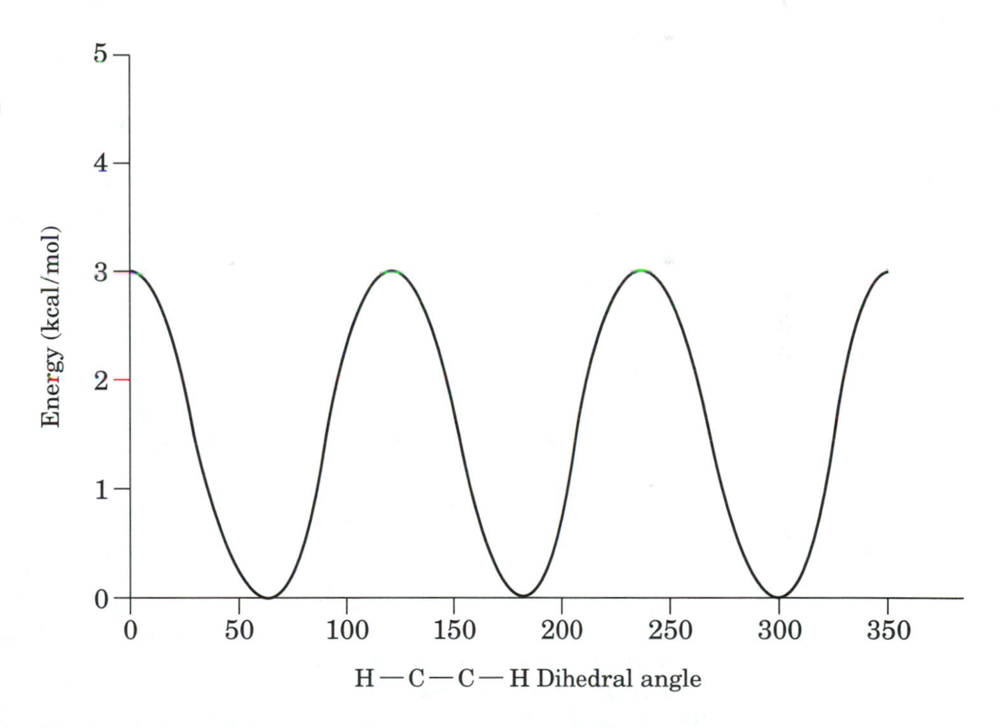

Figure 3.4
Torsional energy due to conformation changes in ethane.

[30]The molecular mechanics method (page 141) also considers the energy associated with stretching or compressing a bond as well as other interactions.

[31]Weiss, S.; Leroi, G. E. *J. Chem. Phys.* **1968**, *48*, 962.

[32]Pitzer, R. M. *Acc. Chem. Res.* **1983**, *16*, 207. Torsional strain is frequently called "Pitzer strain."

[33]Bader, R. F. W.; Cheeseman, J. R.; Laidig, K. E.; Wiberg, K. B.; Breneman, C. *J. Am. Chem. Soc.* **1990**, *112*, 6530.

barrier is due to the overlap (exchange) repulsion between the electron pairs in the coplanar C—H bonds,[34] although alternate explanations have also been advanced.[35,36]

van der Waals Strain

van der Waals strain arises when a conformation forces two atoms that are not bonded to each other to be closer than the sum of their van der Waals radii. Figure 3.5 shows a potential energy curve for rotation about the C2—C3 bond of butane in the gas phase.[37] Although the gauche and anti conformations are both staggered, the anti conformation is more stable by 0.8 kcal/mol.[38,39,40] This energy difference is usually attributed to the van

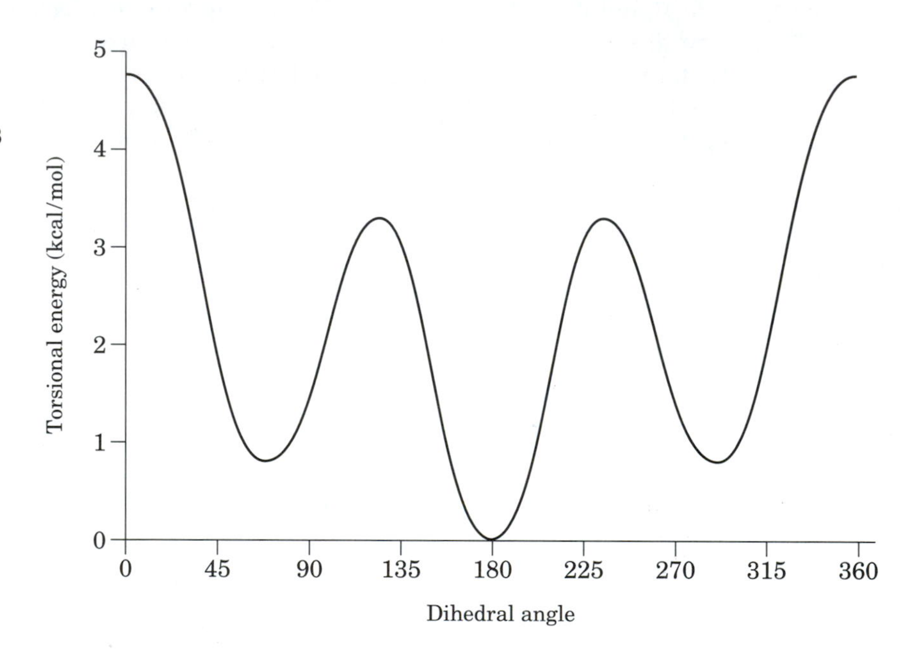

Figure 3.5 Energy changes due to rotation about the C2—C3 bond of butane in the gas phase.

[34]Sovers, O. J.; Kern, C. W.; Pitzer, R. M.; Karplus, M. *J. Chem. Phys.* **1968**, *49*, 2592.

[35]Sacks has presented arguments in favor of a Coulombic explanation for torsional strain: Sacks, L. J. *J. Chem. Educ.* **1986**, *63*, 487.

[36]Lowe, J. P. *J. Am. Chem. Soc.* **1974**, *96*, 3759.

[37]This figure is modified from a figure in reference 42. The curve was calculated from equation (3.7) on page 146 with V_1 = 1.522 kcal/mol, V_2 = −0.315 kcal/mol, and V_3 = 3.207 kcal/mol (as reported in reference 42).

[38]Pitzer, K. S. *J. Chem. Phys.* **1940**, *8*, 711. The exact value depends on the theoretical technique used for measurement and the phase (gas or liquid; potentials are somewhat higher in the gas phase). A value of 0.97 kcal/mol was measured by Verma, A. L.; Murphy, W. F.; Bernstein, H. J. *J. Chem. Phys.* **1974**, *60*, 1540.

[39]Wiberg, K. B.; Murcko, M. A. *J. Am. Chem. Soc.* **1988**, *110*, 8029.

[40]Although it may not be evident from Figure 3.5, the dihedral angle of the gauche conformation is 65°. See the discussion on page 153.

der Waals repulsion of the two methyl groups in the gauche conformation. We also note that there are two different kinds of eclipsed conformations: one has two methyl-hydrogen and one hydrogen-hydrogen eclipsed arrangements, the other has two hydrogen-hydrogen and one methyl-methyl eclipsed arrangements. Both kinds of eclipsed conformation are higher in energy than are the eclipsed conformations in ethane. Apparently the eclipsed conformations in butane include not only torsional strain but also some van der Waals strain. This explanation also rationalizes the higher energy when the two methyl groups are eclipsed, since two methyls should interact more strongly than a methyl and a hydrogen.[41]

It is interesting to ask what these energy differences mean in terms of the distribution of conformations in a sample of butane molecules. The solid line in Figure 3.6 shows the population distributions calculated for butane in the gas phase at 25°.[42] There is a large distribution of conformations about the 180° (*anti*) angle and smaller distributions about the 60° and 300°

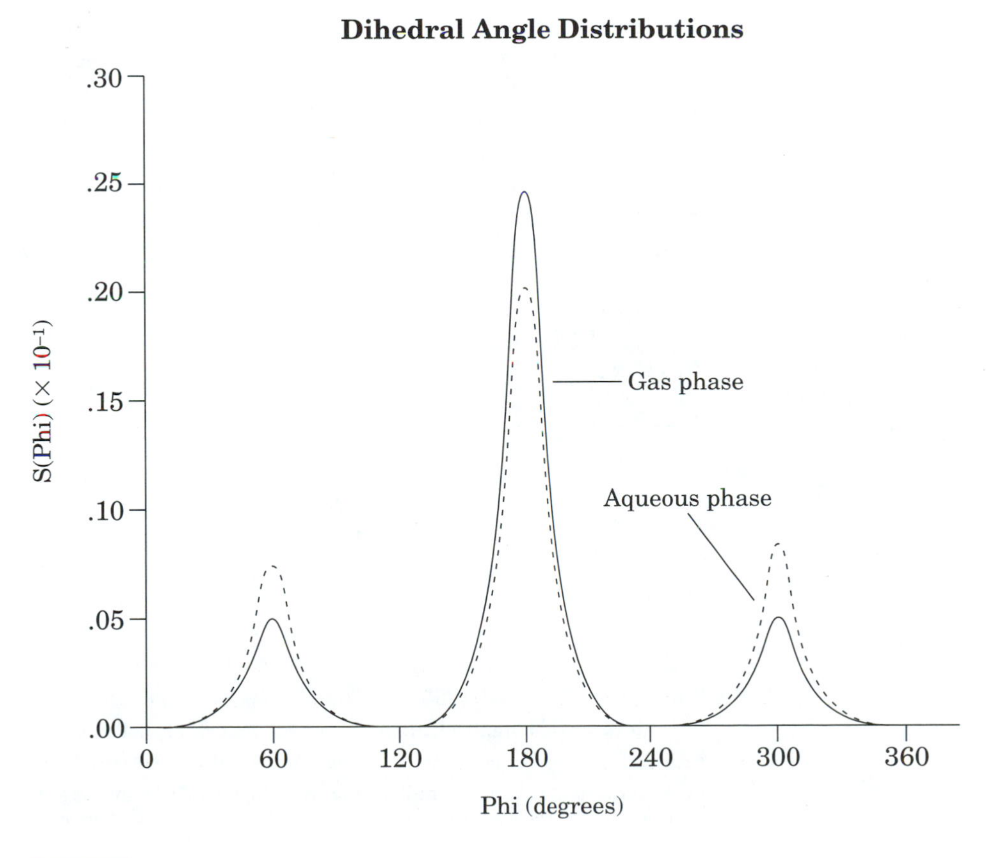

Figure 3.6 Conformational population distributions for butane in the gas phase (solid line) and in water (dashed line) at 25°. Ordinate units are mole fraction per degree. (Adapted from reference 42.)

[41]The *syn* rotational barrier shown is quite close to the value of 4.89 kcal/mol calculated by Allinger, N. L.; Grev, R. S.; Yates B. F.; Schaeffer III, H. F. *J. Am. Chem. Soc.* **1990**, *112*, 114.

[42]Jorgensen, W. L.; Buckner, J. K. *J. Phys. Chem.* **1987**, *91*, 6083.

(gauche) angles. The calculations agree well with theory: about 62% of the molecules were found to be in the most stable (anti) conformation in a gas phase measurement.[43] The potential energy function seems to depend on the environment. Both the energy of the gauche conformation and the barrier to rotation are lower in solution. The dashed line in Figure 3.6 shows the population distribution for butane in water at 25°.

Angle Strain and Baeyer Strain Theory

Strain that results from deviation from standard bond angles is called *angle strain* or *Baeyer strain*. Baeyer proposed that all rings were inherently strained because their bond angles could not be exactly 109.5°.[44] If the rings are required to be planar, then cyclopropane is seen to have 24.75° of angle strain at each carbon atom[45] (see Figure 3.7). For planar cyclobutane, the deviation at each carbon atom is 9.5°; for cyclopentane it is less than 1°; and for cyclohexane, it is −5°. Multiplying the angle strain for each carbon atom by the number of carbon atoms in each molecule, we arrive at the calculated angle strain for each compound, as shown in Table 3.1.

The experimental data in Table 3.1, however, show that these calculations are approximately correct only for cyclopropane, cyclobutane, and cyclopentane.[46,47] Cyclohexane is definitely not the strained compound Baeyer's theory predicts, and the larger ring compounds are also not very strained. Any chemist today can explain the discrepancy between these calculated and experimental values of strain energy: cyclohexane is not planar. In either the chair or boat conformations (Figure 3.8), all bond angles can be approximately 109.5°. Moreover, in the chair conformation of cyclohexane

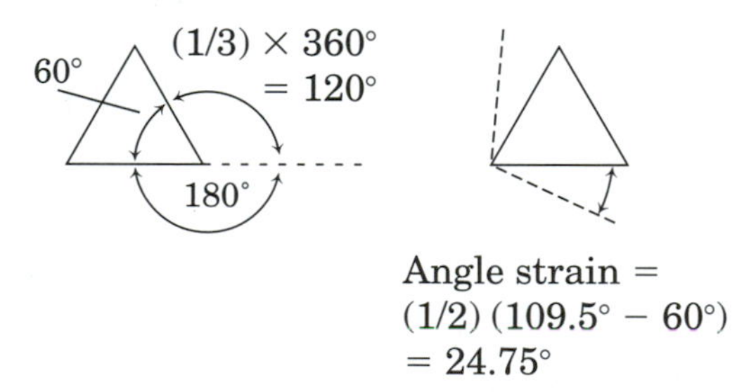

Figure 3.7 Calculation of angle strain for cyclopropane.

---

[43]Murphy, W. F.; Fernández-Sánchez, J. M.; Raghavachari, K. *J. Phys. Chem.* **1991**, *95*, 1124.

[44]Baeyer, A. *Chem. Ber.* **1885**, *18*, 2269; see especially the discussion beginning on p. 2277.

[45]In this discussion angle strain is expressed in degrees, not in kcal/mol.

[46]Eliel, E. L. *Stereochemistry of Carbon Compounds;* McGraw-Hill: New York, 1962; p. 189.

[47]Cremer and Gauss have emphasized that the total "conventional ring strain energies" of cyclopropane and cyclobutane are essentially the same and that calculating strain/CH_2 disguises an important question: is cyclopropane more stable than one would expect from typical conformational analysis considerations or is cyclobutane more strained than we calculate? For a discussion of this point and leading references, see Cremer, D.; Gauss, J. *J. Am. Chem. Soc.* **1986**, *108*, 7467.

Table 3.1 Calculated and experimental angle strain.

Compound	Angle Strain/CH_2	Total Angle Strain	Experimental Strain/CH_2*
Cyclopropane	24°44′	74°12′	9.2
Cyclobutane	9°44′	38°56′	6.55
Cyclopentane	0°44′	3°40′	1.3
Cyclohexane	−5°16′	−31°36′	0.0
Cyclodecane	−17°16′	−172°42′	1.2

*Units are kcal/mol.
(Data from reference 46.)

all bonds are staggered, and there are no apparent van der Waals repulsions in the molecule. Electron diffraction measurements suggest that the preferred conformation of cyclohexane is a somewhat flattened chair, with C—C—C bond angles of 111° and torsional angles of 55.9°.[48]

The boat conformation of cyclohexane is actually an energy maximum; a similar conformation that is an energy minimum is known as the twist boat.[49] Although this conformation increases the angle strain over the boat, lessened torsional strain and van der Waals repulsion of flagpole hydrogens produce an overall lower energy. The twist boat has been spectroscopically observed by rapidly condensing hot cyclohexane vapor into an argon matrix at 20 K. The activation energy required to convert from the twist boat to a chair conformation was determined to be 5.3 kcal/mol, with the twist boat

Figure 3.8 Major conformations of cyclohexane.

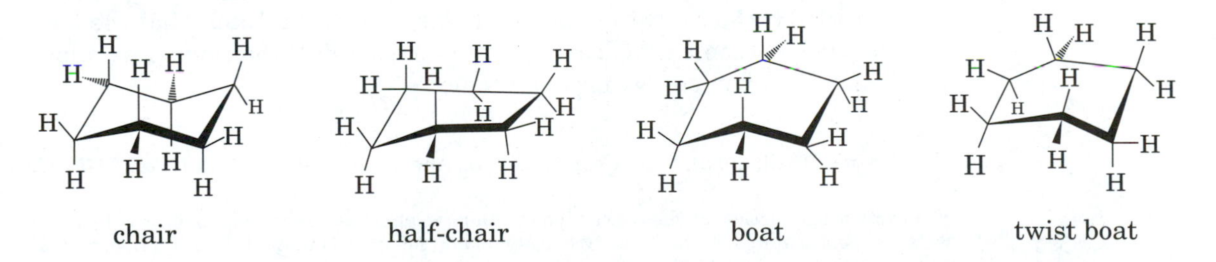

[48]For a review of cyclohexane data, see Ōsawa, E.; Collins, J. B.; Schleyer, P. v. R. *Tetrahedron* **1977**, *33*, 2667. See also the experimental data of Dommen, J.; Brupbacher, T.; Grassi, G.; Bauder, A. *J. Am. Chem. Soc.* **1990**, *112*, 953. For a theoretical analysis, see reference 39.

[49]Johnson, W. S.; Bauer, V. J.; Margrave, J. L.; Frisch, M. A.; Dreger, L. H.; Hubbard, W. N. *J. Am. Chem. Soc.* **1961**, *83*, 606.

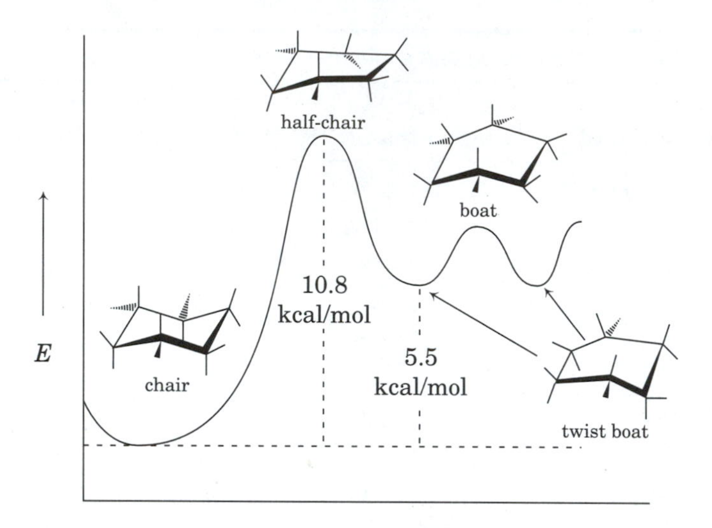

Figure 3.9 Relative energies of cyclohexane conformations.

being 5.5 kcal/mol higher in energy than the chair conformation.[50,51] The energies of the various conformations are shown in Figure 3.9.

Why did Baeyer not consider the possibility that cyclohexane and larger molecules are nonplanar?[52] One explanation reinforces many points we have made about the fallibility of our models. According to Ramsay,[53] Baeyer used a set of Kekulé tetrahedral molecular models that allowed—indeed, required—one to construct a planar representation of a molecule and then to measure angle strain by the deformation of the interatomic bonds from the straight line between them. Thus, cyclohexane would be represented by the model in Figure 3.10. As is so often the case, ideas that become firmly planted in the minds of scientists are difficult to displace. Bassindale has noted that by 1890, only five years after the proposal of strain theory by Baeyer, Sachse pointed out that the angle strain could be relieved by nonplanar conformations of the rings, and the idea was proposed again in 1918 by Mohr. Nevertheless, it was only in the 1950s that the idea became widely accepted.[54] Clearly, we must be careful what model we adopt in calculating strain energy for a molecule.

---

[50]Squillacote, M.; Sheridan, R. S.; Chapman, O. L.; Anet, F. A. L. *J. Am. Chem. Soc.* **1975**, *97*, 3244.

[51]The activation barrier for the conversion of chair cyclohexane to the twist boat was found to be $\Delta G^{\ddagger} = 10.3$ kcal/mol, with $\Delta H^{\ddagger} = 10.8$ kcal/mol and $\Delta S^{\ddagger} = 2.8$ eu: Anet, F. A. L.; Bourn, A. J. R. *J. Am. Chem. Soc.* **1967**, *89*, 760.

[52]Baeyer's theory should not be viewed with disdain. It was experimentally observable that rings larger or smaller than cyclopentane and cyclohexane were difficult to synthesize. Obviously the small rings do have appreciable angle strain. The larger rings can also be difficult to synthesize, however, because of the unfavorable entropy involved in bringing two ends of a long, linear molecule together to make a bond. See the discussion in reference 1(a), pp. 189 *ff.*

[53]Ramsay, O. B. *J. Chem. Educ.* **1977**, *54*, 563.

[54]Bassindale, A. *The Third Dimension in Organic Chemistry;* John Wiley & Sons: Chichester, England, 1984; p. 81.

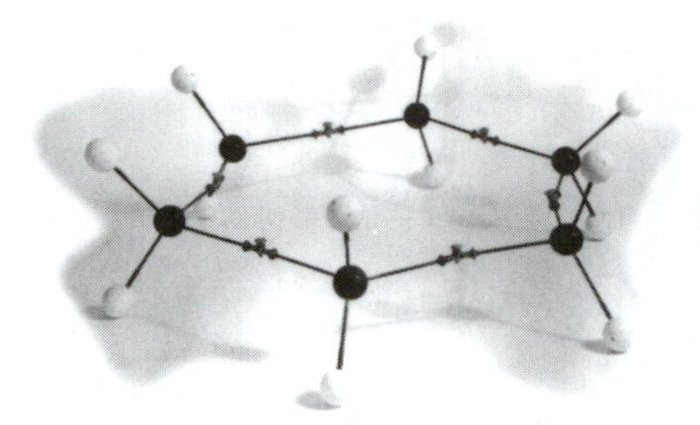

Figure 3.10 Kekulé-Baeyer model of cyclohexane. (Adapted from reference 53.)

Application of Conformational Analysis to Cycloalkanes

Examination of the cyclic alkanes smaller than cyclohexane provides an illustration of the conformational analysis method. Cyclopropane is forced to be planar, since three points (carbon atoms) define a plane.[55] Cyclobutane might be presumed to be planar also, with bond angles of 90°. Any deviation from planarity would make the bond angles even smaller than 90°. Nevertheless, experimental data suggest that cyclobutane is puckered by about 35°, as shown in Figure 3.11. A Newman projection of planar cyclobutane (Figure 3.12) shows that all the C—H bonds are eclipsed. Apparently the torsional strain in the planar conformation is more significant than the increase in angle strain resulting from puckering of the ring.[56]

Figure 3.13 shows the structure of cyclobutane determined through X-ray crystallography of a single crystal at 117°K.[57,58] The X-ray data were

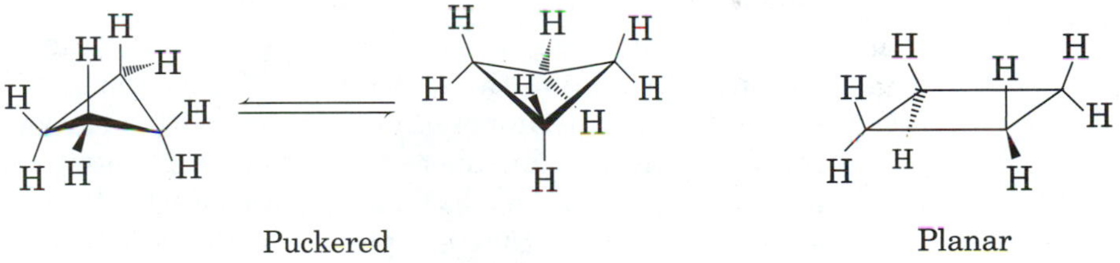

Figure 3.11 Puckered conformation of cyclobutane.

Puckered — Planar

Figure 3.12 Hypothetical planar cyclobutane.

[55]The term *planar* in discussions of cycloalkane conformations refers only to the carbon atoms.

[56]Data from the compilation by Legon, A. C. *Chem. Rev.* **1980**, *80*, 231.

[57]Stein, A.; Lehmann, C. W.; Luger, P. *J. Am. Chem. Soc.* **1992**, *114*, 7684.

[58]The top portion of Figure 3.13 shows an ORTEP representation of a planar conformation of cyclobutane. ORTEP is an acronym for Oak Ridge Thermal Ellipsoid Program, in which atoms are represented by ellipsoids drawn to represent a certain probability of including the nucleus of the atom. For additional reading, see footnote 33 of reference 57.

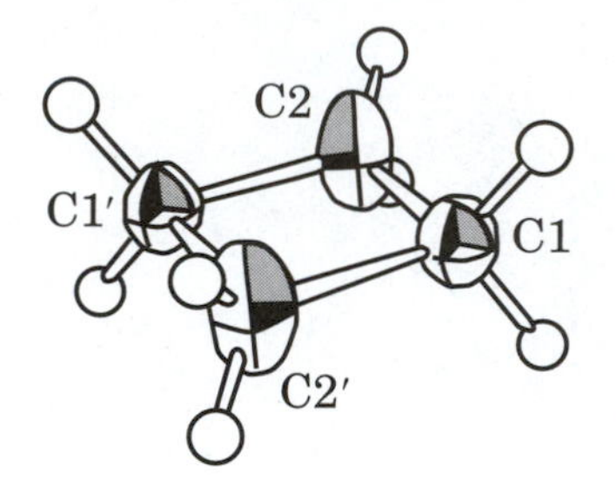

Figure 3.13 Planar (top) and disordered (bottom) views of cyclobutane from X-ray crystallography. (Reproduced from reference 57.)

most consistent with a disordered array (represented by solid and dotted lines) of puckered cyclobutane molecules in the crystal. The calculated value of the dihedral angle for the ring carbon atoms was 31°. Experimental gas phase studies have suggested that the barrier to puckering is less than 1.5 kcal/mol.[59]

One interesting aspect of the X-ray data is the representation of electron density around the cyclobutane ring. Figure 3.14 shows electron density contours superimposed on a square with a carbon atom at each corner, representing the time-average position of the atoms in the ring. Note that the greatest value of electron density is actually outside the lines of the square formed by shortest path internuclear bonds, suggesting that cy-

Figure 3.14 Electron density contours of cyclobutane calculated from X-ray diffraction. (Reproduced from reference 57.)

[59]Cremer, D. *J. Am. Chem. Soc.* **1977**, *99*, 1307.

clobutane has curved or banana bonds. Such bonds had been suggested by Bartell and Andersen, who had also noted that the CH_2 groups should tilt inward toward each other.[60] That tilting interaction relieves additional torsional strain, but does not bring the two axial hydrogens so close that repulsion becomes a problem.[59]

Cyclopentane is also not planar, again apparently for relief of torsional strain (Figure 3.15).[61] Experimental evidence indicates that the molecule exhibits ten different envelope conformations in which one carbon atom at a time is above or below the plane defined by the other four, as well as ten twist or half-chair conformations in which three carbon atoms at a time define a plane, with the fourth carbon atom above the plane and the fifth below it. The conformation is constantly changing, however, so that each carbon atom is 0.458 Å above the molecular plane one-fifth of the time, with the barrier to planarity of 5.16 kcal/mol.[62] The effect of this shifting permutation is the same (unless the atoms were labeled) as rotating the molecule about an axis through its center. For that reason, the process is called **pseudorotation**.[63,64] The energy barrier for pseudorotation is so small that the process is described as essentially barrierless.[65]

Conformational Analysis of Substituted Cyclohexanes

Analysis of the conformations of cycloalkanes becomes even more interesting if we consider a structure with substituents on the ring. As shown in Figure 3.16, there are essentially two kinds of positions for substituents in the chair conformation of cyclohexane: each carbon atom has one sub-

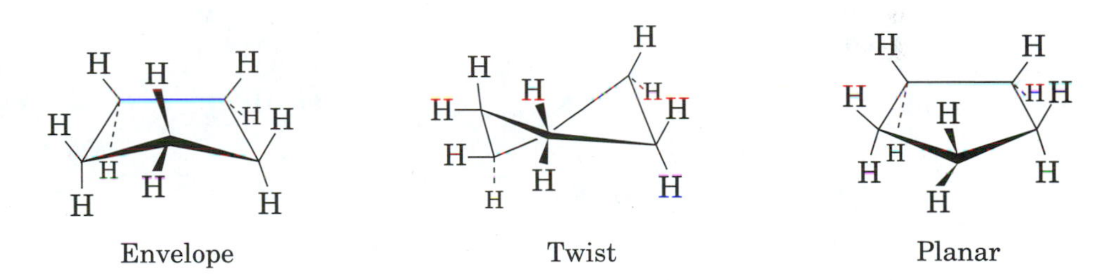

Figure 3.15 Envelope, twist, and (hypothetical) planar conformations of cyclopentane.

[60]Bartell, L. S.; Andersen, B. *J. Chem. Soc. Chem. Commun.* **1973**, 786. It is interesting that the predicted geometry of the molecule correlated well with that predicted by molecular models based on metal connectors for carbon atoms and plastic tubing for bonds.

[61]Kilpatrick, J. E.; Pitzer, K. S.; Spitzer, R. *J. Am. Chem. Soc.* **1947**, *69*, 2483.

[62]Bauman, L. E.; Laane, J. *J. Phys. Chem.* **1988**, *92*, 1040.

[63]This type of pseudorotation is not to be confused with a different use of the term in inorganic chemistry that involves ligand exchange around a central atom; *cf.* Berry, R. S. *J. Chem. Phys.* **1960**, *32*, 933.

[64]A study of the unusual thermodynamic properties of cyclopentane gave rise to the concept of pseudorotation (reference 61).

[65]Variyar, J. E.; MacPhail, R. A. *J. Phys. Chem.* **1992**, *96*, 576; see also MacPhail, R. A.; Variyar, J. E. *Chem. Phys. Lett.* **1989**, *161*, 239.

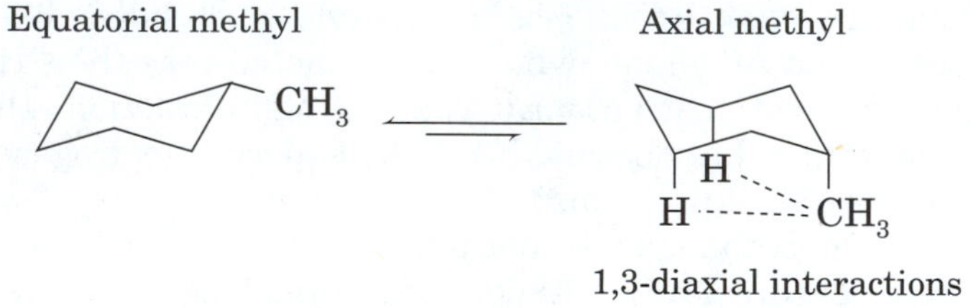

Figure 3.16 Axial and equatorial conformations of a methyl substituent on cyclohexane.

Equatorial methyl CH_3 ⇌ Axial methyl

H, H, CH_3

1,3-diaxial interactions

stituent that is approximately in the plane of the molecule, the equatorial substituent, and another substituent that is perpendicular to the molecular plane, the axial substituent.[66] The axial hydrogen atoms are bonded to one of the three carbon atoms on top of the ring and therefore point up or are bonded to one of the three carbon atoms on the bottom of the ring and therefore point down. Chair-chair conformational change turns axial hydrogens into equatorial hydrogens, and *vice versa*. Thus, at any instant, a substituent such as a methyl group can either be in the axial conformation or in the equatorial conformation.

If we use a double-barreled Newman projection (Figure 3.17) to compare the two conformations of methylcyclohexane, we see that the axial methyl and a C—C bond on an adjacent carbon atom are gauche with respect to each other. Viewing the molecule from another orientation would show that there is another such interaction also. We can describe the interaction of the

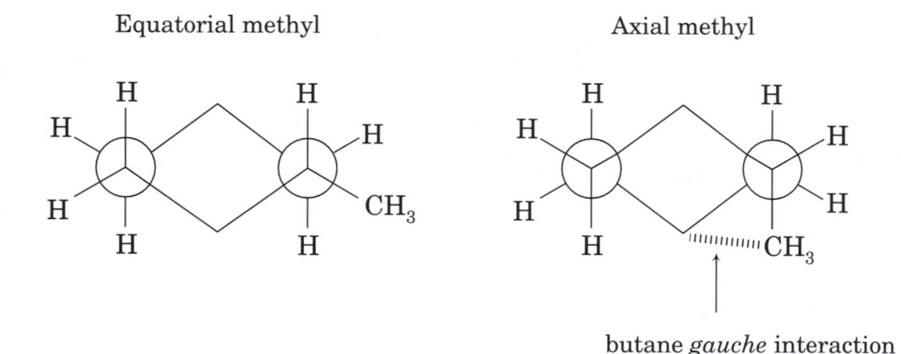

Figure 3.17 Newman projection of monosubstituted cyclohexane conformations.

[66]The classification of hydrogens into what we now call axial and equatorial was by Hassel, O. *Tids. Kjemi, Bergvesen Met.* **1943**, *3*, 32. A translation of this paper by Hedberg, K. appeared in *Top. Stereochem.* **1971**, *6*, p. 11. Note that Hassel termed the hydrogens ϵ (from the Greek εστηκως, "standing") and κ (from the Greek, κειμενος, "reclining") for *axial* and *equatorial*. Pitzer (reference 67) termed the hydrogens *polar* and *equatorial*. Later, Barton, D. H. R.; Hassel, O.; Pitzer, K. S.; Prelog, V. *Science* **1954**, *119*, 49, suggested that the term axial be used instead of polar to distinguish conformations from dipolar properties. The field of conformational analysis was established primarily through the work of Hassel and of Barton, who shared the 1969 Nobel Prize in chemistry. A seminal paper by Barton (reference 29) is also reprinted in *Top. Stereochem.* **1971**, *6*, 1.

axial methyl with the third carbon atom away (from the point of attachment of the methyl) by noting that the methyl and the axial hydrogen on the third carbon atom are closer than the sum of their van der Waals' radii. Thus we term the interaction a *1,3-diaxial interaction*. It is important to recognize that this is the same as a butane gauche interaction, except that it is described in a slightly different way.

Since there are two 1,3-diaxial interactions for each axial methyl, we predict the axial conformation to be less stable than the equatorial conformation by about 1.8 kcal/mol,[67] which is close to the experimental value of 1.74 kcal/mol.[68] Thus, 1.74 kcal/mol is the **equatorial preference** or **A value** for a methyl substituent.[69] If we treat the distribution of a molecule between two conformations as a chemical equilibrium, we can predict there will be 6% axial and 94% equatorial methyl groups, since

$$\Delta G^\circ = -RT \ln K \tag{3.1}$$

For substituents larger than a methyl substituent, we might expect the impact of the 1,3-diaxial or butane gauche interaction to be greater. Table 3.2 provides values of the equatorial preference of other substituent groups as compiled by Hirsch.[70]

Although our discussion has emphasized the role of enthalpic terms, entropy contributions can also play an important role in determining the free energy difference between axial and equatorial conformations.[71] Booth and Everett used variable temperature ^{13}C NMR to determine ΔG°, ΔH°, and ΔS° for the axial to equatorial change of alkylcyclohexanes in $CFCl_3$-$CDCl_3$ solution.[68] The values of $-\Delta H^\circ$ were 1.75, 1.60, and 1.52 kcal/mol for methyl, ethyl, and isopropyl substituents, respectively. That is, the enthalpy of axial to equatorial change was more favorable for methyl than for isopropyl. This result, although perhaps initially surprising, can be rationalized in terms of the number of butane gauche interactions in the likely population of conformers when the alkyl groups are in the axial or equatorial position. However, the ΔS° values for these three substituents had the opposite trend, -0.03, $+0.64$, and $+2.31$ eu, respectively, because of the different number of rotamers possible for each group in the axial and equatorial positions.[68] Therefore the overall ΔG° values at 300 K were 1.74, 1.79, and 2.21 kcal/mol, so that at room temperature the entropy term actually determines the trend in the A values of these three substituents.

---

[67]A butane gauche interaction was calculated to be 0.8 kcal/mol, but a value of 0.9 kcal/mol gave the best fit between calculated and observed values for a series of disubstituted cycloalkanes. Beckett, C. W.; Pitzer, K. S.; Spitzer, R. *J. Am. Chem. Soc.* **1947**, *69*, 2488.

[68]Booth, H.; Everett, J. R. *J. Chem. Soc. Chem. Commun.* **1976**, 278; *J. Chem. Soc., Perkin Trans. 2* **1980**, 255.

[69]Winstein, S.; Holness, N. J. *J. Am. Chem. Soc.* **1955**, *77*, 5562. *A* values are discussed on p. 5574.

[70]Data from Hirsch, J. A. *Top. Stereochem.* **1967**, *1*, 199.

[71]Squillacote, M. E. *J. Chem. Soc., Chem. Commun.* **1986**, 1406. See also Juaristi, E.; Labastida, V.; Antúnez, S. *J. Org. Chem.* **1991**, *56*, 4802.

Table 3.2 Conformational preferences of monosubstituted cyclohexanes.

Substituent	Equatorial Preference (kcal/mol)
-F	0.15
-Cl	0.43
-Br	0.38
-I	0.43
-CN	0.17
$-CH_3$	1.74[68]
$-CH_2CH_3$	1.79[68]
$-CH(CH_3)_2$	2.21[68]
$-C(CH_3)_3$	>5.4[69]
$-C_6H_{11}$	2.15
$-C_6H_5$	3.0
-COOH	1.35
$-COO^-$	1.92
$-CO_2CH_3$	1.27
-OH	0.52 (aprotic solvent)
	0.87 (protic solvent)
$-OCH_3$	0.60
$-NH_2$	1.20 (aprotic solvent)
	1.60 (protic solvent)
$-NH(CH_3)$	1.0 (aprotic solvent)
$-N(CH_3)_2$	2.1 (protic solvent)

(Data from reference 70, except as noted.)

If the two substituents do not directly interact with each other, then the conformational preference of disubstituted cyclohexanes can be computed from the A values of the two substituents individually. For *cis*-1-methyl-4-phenylcyclohexane, ΔG for axial phenyl to equatorial phenyl is -1.13 kcal/mol. Using 1.74 kcal/mol as the A value for methyl and assuming that the effects of the two substituents on the 1 and 4 positions are independent of each other, then the A value for phenyl is 2.87 kcal/mol.[72] With an A value of at least 5.4 kcal/mol, *t*-butylcyclohexane is predicted to exist in solution with only about 0.01% of the *t*-butyl groups in the axial position at 25° or

[72]Eliel, E. L.; Manoharan, M. *J. Org. Chem.* **1991**, *46*, 1959. However, a value of $\Delta G = -0.32$ kcal/mol was determined for axial phenyl to equatorial phenyl in 1-methyl-1-phenylcyclohexane. Here the A values are not additive on the same carbon atom because the presence of the methyl keeps the phenyl form adopting the most stable orientation.

less. This strong preference is the basis for the use of *t*-butyl as a locking group in studies of chemical reactions at axial or equatorial positions. Thus, *trans*-4-*t*-butylcyclohexanol would be expected to allow study of reactions of an equatorial alcohol, while the cis isomer would allow study of the reactions of an axial alcohol group.[69]

For the 1,2-dimethylcyclohexanes, we must not only worry about gauche or 1,3-diaxial interactions of the substituent with the ring, but we must also consider possible interactions of the substituents with each other.[1(a)] Shown in Figure 3.18 are the two chair conformations of *cis*-1,2-dimethylcyclohexane as well as the two corresponding conformations for the trans isomer.[73] We first determine that the e,e conformation of the trans isomer should be about 2.7 kcal/mol more stable than the a,a conformation. The equilibrium mixture of *trans*-1,2-dimethylcyclohexane conformers at room temperature therefore should be composed almost exclusively of molecules in the e,e conformation. Therefore, the energy of the e,e conformation can be used to approximate the energy of the trans isomer. For the cis isomer, both conformations have one axial and one equatorial methyl, so the e,a and a,e conformations are equivalent in energy. Thus we conclude that the cis isomer has two 1,3-diaxial methyl-hydrogen interactions for the axial methyl group, plus one additional gauche interaction between the two

Figure 3.18 Conformations of *trans*- and *cis*-1,2-dimethylcyclohexanes.

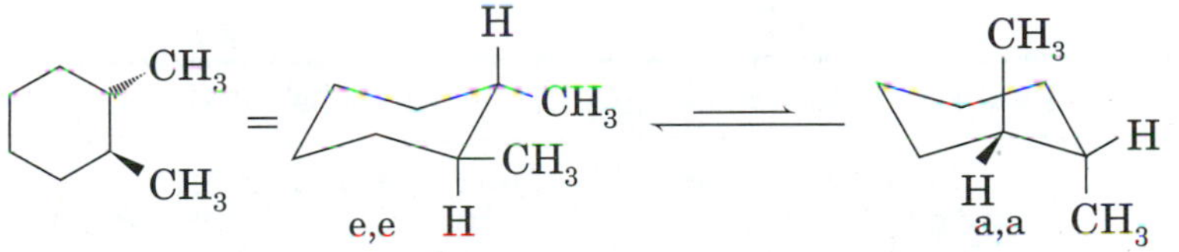

trans-1,2-dimethylcyclohexane

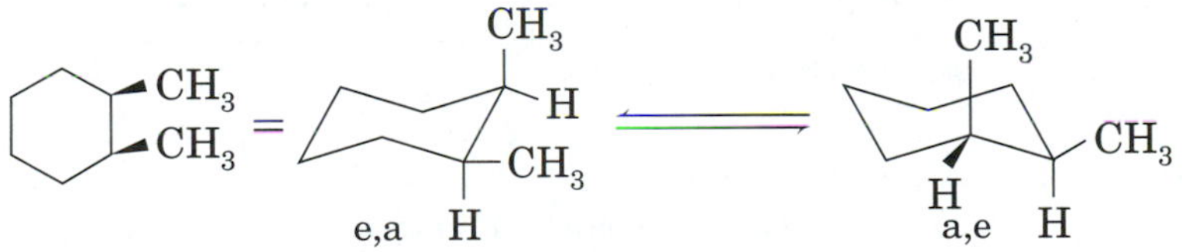

cis-1,2-dimethylcyclohexane

---

[73]Numerous articles in *J. Chem. Educ.* have addressed the problem of recognizing cis and trans isomers of cyclohexanes. (For example, see Richardson, W. S. *J. Chem. Educ.* **1989**, *66*, 478.) A useful technique is to say that each carbon atom has two substituent positions, one more up than down and one more down than up. If two substituents are both more up or are both more down, then they are cis to each other. If one is more up and the other is more down, then they are trans.

methyls.[74] The trans isomer has only the butane gauche interaction between the two methyls. Therefore, we predict that the trans isomer should be more stable than the cis by two methyl 1,3-diaxial interactions or 1.8 kcal/mol. The literature value (Table 3.3) is 1.87 kcal/mol. One can similarly analyze other disubstituted cyclohexanes and compare the results with those shown in Table 3.3. In the 1,3 and 1,4 isomers there are not gauche interactions between the two methyls, so only the number of 1,3-diaxial interactions in each isomer need be considered.

We should not accept without question that all alkyl substituents prefer the equatorial conformation, however. Goren and Biali have reported that the preferred conformation of all-*trans*-1,2,3,4,5,6-hexaisopropylcyclohexane is one in which all six isopropyl groups are axial rather than equatorial![75] They conclude that severe torsional and steric interactions between the equatorial isopropyl groups are responsible for the surprising conformational preference. Moreover, Biali has determined that with an appropriately shaped substituent, it might be possible for a *mono*alkylcyclohexane to be more stable with the substituent in the axial than in the equatorial position.[76]

The relative energies of the dialkylcyclohexanes are not affected significantly by dipole-dipole interactions, but that factor is much more important in the conformational equilibria of molecules that have polar groups.[77] With *trans*-1,2-dichlorocyclohexane, for example, about 25% of the molecules are in the diaxial conformation in benzene solution at room temperature. This

Table 3.3 Conformational preference of dimethylcyclohexanes.

Substitution	More Stable Isomer	ΔH (kcal/mol)
1,2	trans	1.87
1,3	cis	1.96
1,4	trans	1.90

(Data from reference 46, p. 214.)

[74]Although a 1,3-diaxial interaction between a methyl and a hydrogen is often presumed to be exactly equal to a butane gauche interaction, this is not necessarily the case. The gauche butane can adopt a conformation in which the C — C — C — C dihedral angle is more than 60° (see the sample MM2 calculation beginning on page 146) to minimize some van der Waals repulsion, although at the cost of introducing some other strain. In cyclohexane some flattening of the ring can lead to decreased van der Waals repulsion, although again at the cost of other kinds of strain. The data suggest that the A value of methylcyclohexane is more than twice as much as two gauche interactions in butane. For a discussion, see Hendrickson, J. B. *J. Am. Chem. Soc.* **1967**, *89*, 7043.

[75]Goren, A.; Biali, S. E. *J. Am. Chem. Soc.* **1990**, *112*, 893; see also reference 10.

[76]Biali, S. E. *J. Org. Chem.* **1992**, *57*, 2979.

[77]For further discussion, see reference 7, pp. 87–158 and references therein; reference 46, pp. 219 *ff*. and references therein.

arrangement minimizes the unfavorable dipole-dipole interaction present in the diequatorial conformation. With *trans*-1,2-dibromocyclohexane, the population of diaxial conformer is more than 50% under the same conditions. Now there is a van der Waals repulsion as well as a dipole-dipole interaction in the diequatorial conformation.[78] The trans isomer of 1,4-dibromocyclohexane also exhibits a greater population of diaxial than diequatorial conformation in solution, although the diequatorial conformation is favored in the solid state.[79] With polar substituents capable of hydrogen bonding, still another factor comes into play. The diequatorial conformer of *trans*-1,2-cyclohexanediol is favored due to hydrogen bonding, since intramolecular hydrogen bonding is not feasible in the diaxial conformation.[80]

3.3 Molecular Mechanics

One of the limitations of conformational analysis is that we need to have data for rather specific reference compounds for each type of comparison we make. For example, if there is a 1,3-diaxial interaction involving a methyl and a carbomethoxy, we would be hard pressed to estimate the resulting strain from the A values of the two substituents alone. If there is a van der Waals repulsion between those same two groups on a different molecular framework, the A values for cyclohexane would be of limited value to us. Therefore, it would be advantageous to be able to calculate conformational energies from equations in which energy is a function of a structural unit (bond length, bond angle, etc.) and not a particular molecular skeleton. This approach has given rise to a field of computational chemistry that has come to be known as molecular mechanics, and computer programs for such calculations are available for a wide variety of computer platforms.[81]

The term *molecular mechanics* distinguishes this approach from quantum mechanics because the method uses a classical mechanics (mass and spring) approach to ascribe the energy of a particular conformation to specific bonding parameters. Molecular mechanics has also been called the Westheimer method[82] because equations for calculating steric strain were used by Westheimer in a study of racemization of optically active

[78]Bender, P.; Flowers, D. L.; Goering, H. L. *J. Am. Chem. Soc.* **1955**, *77*, 3463.

[79]Kozima, K.; Yoshino, T. *J. Am. Chem. Soc.* **1953**, *75*, 166.

[80]Kuhn, L. P. *J. Am. Chem. Soc.* **1958**, *80*, 5950.

[81]One program for personal computers is MMIIPC: An Interactive Version of MM2 for Microcomputers (Program QCMP010) from the Quantum Chemistry Program Exchange, Bloomington, IN, 1985. The program was written for mainframe computers by N. L. Allinger and Y. H. Yuh and was modified for microcomputers by P. A. Petillo. References to the literature concerning the development of molecular mechanics are provided with the documentation for the program.

[82]For example, see Allinger, N. L.; Hirsch, J. A.; Miller, M. A.; Tyminski, I. J.; Van-Catledge, F. A. *J. Am. Chem. Soc.* **1968**, *90*, 1199.

biphenyls.[83] Other terms include *quantitative conformational analysis*,[84] and (more generally) the *force field method*. A force field is defined as a set of equations and constants that relate energy to internal coordinates, thus allowing calculation of molecular properties such as equilibrium conformation and vibrational modes.[85] Both the equations and the constants are important. A reliable force field must have a correct analytic form (the equations) and a valid parameterization (the specific constants for each type of interaction).[86,87,88] Molecular geometries and energies can also be obtained from molecular orbital calculations, but molecular mechanics calculations are much faster because the time required for a molecular mechanics calculation varies with the *square* of the number of *atoms*, while the computation time for an *ab initio* calculation varies with the *fourth* power of the number of *orbitals*.[89] In addition, molecular mechanics calculations can be quite accurate for structures that have bonds similar to those in the reference compounds used to develop the molecular mechanics program.[90]

Among the more widely used molecular mechanics methods are those developed by Allinger and co-workers. The original formulation was MM1, which was reported in 1973.[89] A modification published in 1977 was named MM2 [frequently cited as MM2(77)].[91] MM2 was widely accepted, and many workers introduced variations suited for particular problems. A newer version by Allinger is known as MM3.[92] Other approaches have been developed

---

[83]See Westheimer, F. H. in *Steric Effects in Organic Chemistry;* Newman, M. S., ed.; John Wiley & Sons, Inc.: New York, 1956; p. 523 *ff.*; Westheimer, F. H.; Mayer, J. E. *J. Chem. Phys.* **1946**, *14*, 733; Westheimer, F. H. *J. Chem. Phys.* **1947**, *15*, 252. Similar ideas were developed independently by Hill, T. L. *J. Chem. Phys.* **1946**, *14*, 465.

[84]Williams, J. E.; Stang, P. J.; Schleyer, P. v. R. *Ann. Rev. Phys. Chem.* **1968**, *19*, 531.

[85]This definition is revised slightly from that given by Hagler, A. T.; Stern, P. S.; Lifson, S.; Ariel, S. *J. Am. Chem. Soc.* **1979**, *101*, 813.

[86]In order to carry out a molecular mechanics calculation, all the force constants for each type of atom under consideration must be known. In the case of MM2, only some atom types were done at the time of development of the procedure. Others were added later, in some cases by other researchers. Lipkowitz has provided a summary of parameter sources: Lipkowitz, K. B. *QCPE Bulletin* **1992**, *12*, 6.

[87]The parameters are based on diverse experimental data and must be optimized so that the force field produces reliable results. For a discussion, see Pearlman, D. A.; Kollman, P. A. *J. Am. Chem. Soc.* **1991**, *113*, 7167.

[88]Parameterization has been called an art and a science. Bowen, J. P.; Allinger, N. L. in *Reviews in Computational Chemistry II;* Lipkowitz, K. B.; Boyd, D. R., eds.; VCH Publishers, Inc.: New York, 1991; pp. 81–97.

[89]Allinger, N. L. *Adv. Phys. Org. Chem.* **1976**, *13*, 1.

[90]For example, the MM3 calculation of the heat of formation of C_{60} was much closer to the experimental value (see reference 171) than were the values calculated by the molecular orbital methods to be discussed in Chapter 4.

[91]Allinger, N. L. *J. Am. Chem. Soc.* **1977**, *99*, 8127.

[92](a) Allinger, N. L.; Yuh, Y. H.; Lii, J.-H. *J. Am. Chem. Soc.* **1989**, *111*, 8551; (b) Lii, J.-H.; Allinger, N. L. *J. Am. Chem. Soc.* **1989**, *111*, 8566; (c) Lii, J.-H.; Allinger, N. L. *J. Am. Chem. Soc.* **1989**, *111*, 8576.

as well, including the method of Schleyer[93] and that of Bartell (MUB-2).[94] A more recent force field method reported by Goddard[95] has been extended by Rappé and co-workers to a force field called UFF (universal force field). UFF incorporates parameters that can be predicted from atomic number, hybridization, and connectivity of the atom.[96] The program is general in that it can be used not only with organic compounds but with main group compounds as well.[97] Some other widely cited force fields are AMBER[98] and CHARMM.[99] For further discussion about molecular mechanics, see the reviews by Schleyer[84,93] or Allinger[89] or the monograph of Burkert and Allinger.[100]

Unlike conformational analysis, in which relative energies are compared by assigning fixed amounts of strain to specific interactions (such as butane gauche interactions), in molecular mechanics the energy of a conformation is determined by computing the value of a mathematical function. An equation such as equation 3.2 can be used to calculate the total **steric energy** of the molecule as the sum of a number of different kinds of interactions.[101,102] Most molecular mechanics methods include as a minimum the components in equation 3.2, and frequently other terms are also included.[103,104,105]

---

[93]Engler, E. M.; Andose, J. D.; Schleyer, P. v. R. *J. Am. Chem. Soc.* **1973**, *95*, 8005.

[94]Fitzwater, S.; Bartell, L. S. *J. Am. Chem. Soc.* **1976**, *98*, 5107.

[95]Mayo, S. L.; Olafson, B. D.; Goddard III, W. A. *J. Phys. Chem.* **1990**, *94*, 8897.

[96]Rappé, A. K.; Casewit, C. J.; Colwell, K. S.; Goddard III, W. A.; Skiff, W. M. *J. Am. Chem. Soc.* **1992**, *114*, 10024.

[97]Casewit, C. J.; Colwell, K. S.; Rappé, A. K. *J. Am. Chem. Soc.* **1992**, *114*, 10035, 10046.

[98]Weiner, S. J.; Kollman, P. A.; Nguyen, D. T.; Case, D. A. *J. Comput. Chem.* **1986**, *7*, 230.

[99]Nilsson, L.; Karplus, M. *J. Comput. Chem.* **1986**, *7*, 591. Also see Smith, J. C.; Karplus, M. *J. Am. Chem. Soc.* **1992**, *114*, 801.

[100]Burkert, U.; Allinger, N. L. *Molecular Mechanics*, ACS Monograph 177; American Chemical Society: Washington, D. C., 1982.

[101]Reference 100, p. 4.

[102]The discussion below is a simplified presentation of the principles of molecular mechanics calculations in MM2. The treatment in MM3 is more complex (reference 92).

[103]There may also be an energy associated with "out of plane deformation" for some molecules. For a discussion of the forces included in a MM calculation, see Susnow, R.; Nachbar, Jr., R. B.; Schutt, C.; Rabitz, H. *J. Phys. Chem.* **1991**, *95*, 8585.

[104]For a discussion of the components of a force field and the use of *ab initio* methods in force field development, see Dinur, U.; Hagler, A. T. in *Reviews in Computational Chemistry II;* Lipkowitz, K. B.; Boyd, D. R., eds.; VCH Publishers, Inc.: New York, 1991; pp. 99–164; Hwang, M. J.; Stockfisch, T. P.; Hagler, A. T. *J. Am. Chem. Soc.* **1994**, *116*, 2515.

[105]In addition to the terms in equation 3.2, other types of interactions may be important. These include dipole-dipole forces (which may be stabilizing or destabilizing), as well as hydrogen bonding, electrostatic effects, donor-acceptor interactions, and solvation effects. Understanding these interactions can be vital to calculating the conformation of molecules more complex than the simple hydrocarbons discussed here. For discussions, see (a) Kingsbury, C. A. *J. Chem. Educ.* **1979**, *56*, 431; (b) Liberles, A.; Greenberg, A.; Eilers, J. E. *J. Chem. Educ.* **1973**, *50*, 676; (c) Juaristi, E. *J. Chem. Educ.* **1979**, *56*, 438; (d) reference 100.

$$E_{\text{steric}} = E(r) + E(\theta) + E(\Phi) + E(d) \quad (3.2)$$

where $E(r)$ is the energy of stretching or compressing an individual bond,
$E(\theta)$ is the energy of distorting a bond angle from the ideal,
$E(\Phi)$ is the torsional strain (due to nonstaggered bonds), and
$E(d)$ is the energy of nonbonded interactions arising from van der Waals forces, which may be stabilizing or destabilizing.

The magnitude of $E(r)$ for each bond can be estimated as a function of the extent of stretching or compression by the formula:

$$E(r) = 0.5\ k_r \times (\Delta r)^2 \times (1 + CS \times \Delta r) \quad (3.3)$$

where k_r is the force constant associated with the deformation, Δr is the deformation of the bond length from the minimum energy length, and CS is a cubic stretching constant. Note that the energy increases with the square of the deformation (although a negative CS allows the bond to weaken somewhat at large deformations). Apparently the resistance to deformation is quite strong; stretching or compressing a carbon-carbon single bond by 0.1 Å is endothermic by about 2.5 kcal/mol. (See Table 3.4.) As a result, carbon-carbon bond lengths usually vary only slightly over a wide variety of strained and unstrained compounds.[106,107] For example, the minimum energy geometry calculated[108] for cyclohexane has a C—C bond length of 1.5356 Å, which is a C—C bond length deformation of 0.0126 Å, and that is calculated to give a stretching energy for each C—C bond of 0.0487 kcal/mol.

The magnitude of $E(\theta)$ is estimated from the formula:

$$E(\theta) = 0.5\ k_\theta \times (\Delta\theta)^2 \times (1 + SF \times \Delta\theta^4) \quad (3.4)$$

Table 3.4 Dependence of carbon-carbon bond deformation energy on Δr.

Δr(Å)	Strain Energy/* (kcal/mol)
0.001	3.2×10^{-4}
0.002	1.3×10^{-3}
0.005	7.8×10^{-3}
0.01	3.1×10^{-2}
0.05	0.71
0.10	2.53

*Calculations based on equations and data from reference 81.

[106]Clark, T.; McKervey, M. A. in *Comprehensive Organic Chemistry*, Vol. 1; Stoddart, J. F., ed.; Pergamon Press: Oxford, Oxford, 1979; p. 37.

[107]Liebman, J. F.; Greenberg, A. *Chem. Rev.* **1976**, *76*, 311.

[108]Calculation done by the author using QCMP010 on a personal computer.

where $\Delta\theta$ is the deviation from ideal bond angles (e.g., 109.5° for sp^3-hybrid orbitals). Note that there is again a dependence of the energy on the square of the magnitude of the deformation, although there is a minor term ($SF \times \Delta\theta^4$) that depends on the sixth power as well.[109] However, in this case the prohibition against deformation (i.e., the force constant, k_θ) is relatively much smaller. Table 3.5 shows the energies calculated for C—C—C angle deviations from 109.5° from equation 3.4. The expected bond angle of 109.5° for tetrasubstituted organic compounds is actually observed for only methane and a few other symmetrically substituted carbon atoms. In other cases the C—C—C bond angles are found to vary from 111° to 113° for acyclic alkanes.

In addition to the pure bending strain given in equation 3.4, it is also useful to calculate a stretch bending strain energy E_{SB}, which incorporates both bond angle and bond length deformations. That is, the energy of an angle deformation will be affected by stretching of the bonds that make that angle.[110] The energy for this term is given by the equation:

$$E_{SB} = k_{SB} \times \Delta\theta_{abc} \times (\Delta r_{a-b} + \Delta r_{b-c}) \tag{3.5}$$

In the example of cyclohexane, the C1—C2—C3 bond angle is calculated to be 110.88°, which differs by 1.415° from the ideal bond angle of 109.47°. The resulting contribution to $E(\theta)$ (for one C—C—C unit) is calculated to be 0.01 kcal/mol. By comparison, the value of E_{SB} for this same C—C—C unit is determined to be 0.02 kcal/mol, based on a C—C bond length deformation of 0.0126 Å.

Table 3.5 Dependence of strain energy on C–C–C angle deviation.

$\Delta\theta$ (deg)	Strain Energy (kcal/mol)
0.01	9.8×10^{-7}
0.1	9.8×10^{-5}
0.2	3.9×10^{-4}
0.5	2.5×10^{-3}
1.0	9.8×10^{-3}
3.0	8.4×10^{-2}
5.0	0.21

[109] In the example on page 152, the value of SF is 0.7×10^{-7}.

[110] Because coordinates for bond lengths, bond angles, and other easily visualized sources of strain are interrelated, the types of strain are not totally separate. (See the discussion in reference 84.) Therefore, it may be necessary to add some cross terms to achieve a better calculation. The choice of such cross terms depends on the model used. The MUB-2 force field included a stretch-torsion term (reference 94).

The torsional strain equation for a pair of tetrasubstituted carbon atoms, such as those in ethane, has the form:

$$E(\Phi) = 0.5\, V_0 \times (1 + \cos 3\,\Phi) \qquad \textbf{(3.6)}$$

where V_0 is the rotational energy barrier and Φ is the torsional (dihedral) angle. The factor of 3 appears because there is a three-fold energy barrier to rotation about C — C single bonds whenever each of the two atoms has three identical substituents (as in ethane, for example). A more general equation[111] would account for different substitutents as follows:

$$E(\Phi) = 0.5\, V_1 \times (1 + \cos \Phi) + 0.5\, V_2 \times (1 - \cos 2\,\Phi) + 0.5\, V_3 \times (1 + \cos 3\,\Phi) \qquad \textbf{(3.7)}$$

where the values of V_1, V_2, and V_3 vary with the particular atom types. In this formulation positive values for V_1 and V_3 destabilize eclipsed forms, while a positive value for V_2 destabilizes 90° arrangements.

The van der Waals strain energy (included in $E(d)$) depends on the extent and pattern of substitution within the structure. As atoms that are not bonded to each other are brought closer together, their electron clouds interact with each other. The interaction is very weak at long distances, so initially there is no appreciable effect. As the two groups come just close enough together that their clouds touch, there is an attractive force. However, if the two groups are pushed too close together, the interaction very quickly becomes quite repulsive. Different force fields use different values for van der Waals radii and different slopes for the increase in energy of nonbonded atoms. The effect of internuclear distance on the van der Waals energy of two nonbonded hydrogen atoms (calculated from the equation shown on page 151) is shown in Figure 3.19.[112]

Let us examine the molecular mechanics approach by carrying out a very simple calculation—the optimization of the gauche conformation of butane. We start with an initial conformation in which the *C — C — C — C dihedral angle is 60°* (atoms 1-2-3-4 in Figure 3.20).[113] The following lines show part of the input file for the gauche conformation of butane in a format similar to that used by many molecular mechanics programs.[114,115] The

[111]Reference 81, p. 54.

[112]For a discussion of different formulations of the van der Waals interaction, see Halgren, T. A. *J. Am. Chem. Soc.* **1992**, *114*, 7827.

[113]The input geometry was created by carrying out a molecular mechanics calculation on a structure in which the C — C — C — C dihedral angle was fixed at 60°.

[114]The initial geometry is specified in an input file in the format required for the particular program that is to be used.

[115]The input file was written with the program PCMODEL (Serena Software, Bloomington, IN), a versatile molecular mechanics program that has a graphical user interface. The force field used in PCMODEL is known as MMX. For a discussion, see Gajewski, J. J.; Gilbert, K. E.; McKelvey, J. in *Advances in Molecular Modeling* Vol. 2; Liotta, D., ed.; JAI Press, Inc.: Greenwich, CT, 1990; pp. 65–92. For clarity, some portions of the input file not discussed here, as well as some portions of the molecular mechanics output (from a calculation carried out with QCMP010), have been omitted.

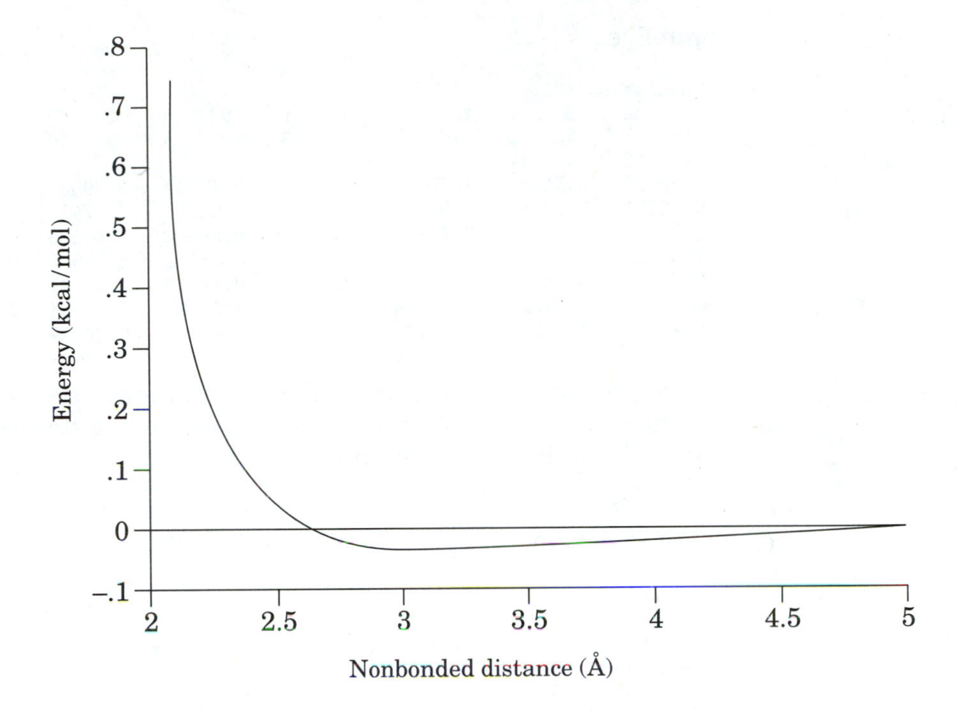

Figure 3.19 van der Waals energy of nonbonded hydrogen atom interactions.

input file includes the name of the compound, any constraints on the program, and information about the connectivity of the atoms. There is a connected atoms list (1 2 3 4 in the third line), and an attached atoms list (1 5 1 6, etc. in the next line, meaning that atom 5 is bonded to atom 1, and so on). Then come the x, y, and z coordinates of each atom and the *atom type*. Type 1 is a tetravalent carbon; type 5 is a hydrogen. We will not discuss the details of the input file here; for a thorough discussion and more extensive examples, see the monograph by Clark.[116]

H13
H6
H12
H7
C4
H14
H5
C1
H11
C3
C2
H9
H10
H8

Figure 3.20 Initial butane gauche conformation with 60° dihedral angle.

[116]Clark, T. *A Handbook of Computational Chemistry;* John Wiley and Sons: New York, 1985.

Input File

```
gauche BUTANE
 1    2    3    4    0    0    0    0    0    0    0    0    0    0    0    0
 1    5    1    6    1    7    2    8    2    9    3   10    3   11    4   12
 4   13    4   14
-0.99450   0.36461   1.25785   1      -0.19237  -0.79026   0.64416   1
 0.16751  -0.57567  -0.83551   1       1.02035   0.67436  -1.08807   1
-1.22001   0.14650   2.32252   5      -0.40988   1.30620   1.19903   5
-1.94901   0.49837   0.70753   5      -0.76482  -1.73514   0.75297   5
 0.73596  -0.94743   1.23223   5       0.70118  -1.46973  -1.22046   5
-0.76285  -0.51485  -1.43806   5       1.24327   0.76931  -2.17124   5
 0.47470   1.58046  -0.75181   5       1.97614   0.60119  -0.52865   5
```

The molecular mechanics program reads the input file and uses the atom coordinates and bonding information to compute the steric energy of the initial conformation of the structure.[117]

Initial Calculation of Coordinates and Energy

```
gauche BUTANE
GEOMETRY AND STERIC ENERGY OF INITIAL CONFORMATION.
CONNECTED ATOMS
     1- 2- 3- 4-
ATTACHED ATOMS
     1- 5, 1- 6, 1- 7, 2- 8, 2- 9, 3-10, 3-11, 4-12,
     4-13, 4-14,
INITIAL ATOMIC COORDINATES
     ATOM          X           Y           Z        TYPE
     C( 1)     -.99450      .36461     1.25785     ( 1)
     C( 2)     -.19237     -.79026      .64416     ( 1)
     C( 3)      .16751     -.57567     -.83551     ( 1)
     C( 4)     1.02035      .67436    -1.08807     ( 1)
     H( 5)    -1.22001      .14650     2.32252     ( 5)
     H( 6)     -.40988     1.30620     1.19903     ( 5)
     H( 7)    -1.94901      .49837      .70753     ( 5)
     H( 8)     -.76482    -1.73514      .75297     ( 5)
     H( 9)      .73596     -.94743     1.23223     ( 5)
     H(10)      .70118    -1.46973    -1.22046     ( 5)
     H(11)     -.76285     -.51485    -1.43806     ( 5)
     H(12)     1.24327      .76931    -2.17124     ( 5)
     H(13)      .47470     1.58046     -.75181     ( 5)
     H(14)     1.97614      .60119     -.52865     ( 5)

INITIAL STERIC ENERGY IS 3.6711 KCAL.
     COMPRESSION         .1756
     BENDING             .6224
     STRETCH-BEND        .0670
     VANDERWAALS
       1,4 ENERGY       2.2947
       OTHER             .1294
     TORSIONAL           .3820
```

[117]For clarity, some parts of the output file not discussed here have been omitted from the listing.

The molecular mechanics calculation thus indicates that the steric energy of the initial 60° conformer is 3.67 kcal/mol, and that this energy is made up mostly of van der Waals repulsion, but with significant contributions from bending and torsion also. The next step is to vary the location of the atoms in a systematic way to achieve a more stable conformation.[118] This process is repeated until the change in steric energy from one such procedure to the next is less than some predetermined value, at which time the energy of the structure is said to be minimized. Initially the hydrogens attached to carbon atoms are moved with the carbon atom, but later the hydrogens are moved independently.

Iterative Minimization of Total Steric Energy

```
    gauche BUTANE
 ENERGY MINIMIZATION
 INITIAL ENERGY

 TOTAL ENERGY IS       3.6711 KCAL.
   COMPRESS     .1756     VANDERWAALS                  TORSION        .3820
   BENDING      .6224        1,4         2.2947
   STR-BEND     .0670        OTHER        .1294      DIPL/CHG        .0000

* * * * * * * * * * * * *   C Y C L E   1  * * * * * * * * * * * * *
                            (CH)-MOVEMENT = 1
       ITERATION    1     AVG. MOVEMENT =   .01320 A
       ITERATION    2     AVG. MOVEMENT =   .00616 A
       ITERATION    3     AVG. MOVEMENT =   .00390 A
       ITERATION    4     AVG. MOVEMENT =   .00237 A
       ITERATION    5     AVG. MOVEMENT =   .00153 A
 TOTAL ENERGY IS       3.0383 KCAL.
   COMPRESS     .1695     VANDERWAALS                  TORSION        .4315
   BENDING      .6027        1,4         2.1050
   STR-BEND     .0752        OTHER       -.3457      DIPL/CHG        .0000

       ITERATION    6     AVG. MOVEMENT =   .00097 A
       ITERATION    7     AVG. MOVEMENT =   .00062 A
       ITERATION    8     AVG. MOVEMENT =   .00040 A
       ITERATION    9     AVG. MOVEMENT =   .00025 A
       ITERATION   10     AVG. MOVEMENT =   .00017 A

 TOTAL ENERGY IS       3.0349 KCAL.
   COMPRESS     .1670     VANDERWAALS                  TORSION        .4490
   BENDING      .5937        1,4         2.1112
   STR-BEND     .0742        OTHER       -.3603      DIPL/CHG        .0000

       ITERATION   11     AVG. MOVEMENT =   .00011 A
       ITERATION   12     AVG. MOVEMENT =   .00008 A
       ITERATION   13     AVG. MOVEMENT =   .00006 A
       ITERATION   14     AVG. MOVEMENT =   .00005 A
       ITERATION   15     AVG. MOVEMENT =   .00004 A

 TOTAL ENERGY IS       3.0348 KCAL.
   COMPRESS     .1667     VANDERWAALS                  TORSION        .4509
   BENDING      .5930        1,4         2.1121
   STR-BEND     .0741        OTHER       -.3620      DIPL/CHG        .0000
```

[118]One procedure for this part of the computation was developed by Wiberg, K. B. *J. Am. Chem. Soc.* **1965**, *87*, 1070.

```
* * * * * * * * * * * * *   C Y C L E   2 * * * * * * * * * * * * *

                         (CH)-MOVEMENT = 0

      ITERATION  16      AVG. MOVEMENT =  .00003 A

 TOTAL ENERGY IS 3.0348 KCAL.
   COMPRESS      .1667     VANDERWAALS                  TORSION        .4510
   BENDING       .5930       1,4          2.1122
   STR-BEND      .0741      OTHER         -.3621      DIPL/CHG        .0000

 * * * * * ENERGY IS MINIMIZED WITHIN .0011 KCAL * * * * *
         * * * * * ENERGY IS 3.0348 KCAL * * * * *
```

Then the program writes out the final coordinates of all atoms and computes the contribution of all interactions to the overall steric energy of the molecule.

Final Coordinates and Energy

```
     gauche BUTANE
GEOMETRY AND STERIC ENERGY OF FINAL CONFORMATION.
CONNECTED ATOMS
     1- 2- 3- 4-

ATTACHED ATOMS
     1- 5, 1- 6, 1- 7, 2- 8, 2- 9, 3-10, 3-11, 4-12,
     4-13, 4-14,

FINAL ATOMIC COORDINATES AND BONDED ATOM TABLE
     ATOM        X          Y          Z       TYPE     BOUND TO ATOMS
     C( 1)   -1.02860     .32977    1.27109   ( 1)     2,   5,   6,   7
     C( 2)    -.17615    -.78428     .64958   ( 1)     1,   3,   8,   9
     C( 3)     .13702    -.55745    -.83879   ( 1)     2,   4,  10,  11
     C( 4)    1.01904     .67110   -1.09710   ( 1)     3,  12,  13,  14
     H( 5)   -1.29907     .08253    2.32393   ( 5)     1,
     H( 6)    -.48525    1.30158    1.29228   ( 5)     1,
     H( 7)   -1.97660     .47462     .70367   ( 5)     1,
     H( 8)    -.72460   -1.75130     .75768   ( 5)     2,
     H( 9)     .77684    -.88816    1.22141   ( 5)     2,
     H(10)     .66177   -1.45957   -1.23684   ( 5)     3,
     H(11)    -.81609    -.45968   -1.41149   ( 5)     3,
     H(12)    1.28620     .74842   -2.17665   ( 5)     4,
     H(13)     .50045    1.61672    -.81973   ( 5)     4,
     H(14)    1.96872     .61329    -.51708   ( 5)     4,

BOND LENGTHS AND STRETCHING ENERGY    (13 BONDS)
     ENERGY = 71.94(KS)(DR)(DR)(1+(CS)(DR))
                        DR = R-RO
                        CS = -2.000
     BOND            LENGTH      R(0)       K(S)      ENERGY
C( 1)- C( 2)        1.5343     1.5230     4.4000      .0395
C( 1)- H( 5)        1.1148     1.1130     4.6000      .0011
C( 1)- H( 6)        1.1136     1.1130     4.6000      .0001
C( 1)- H( 7)        1.1143     1.1130     4.6000      .0006
C( 2)- C( 3)        1.5378     1.5230     4.4000      .0671
C( 2)- H( 8)        1.1170     1.1130     4.6000      .0052
C( 2)- H( 9)        1.1162     1.1130     4.6000      .0034
```

C(3)- C(4)	1.5343	1.5230	4.4000	.0394
C(3)- H(10)	1.1170	1.1130	4.6000	.0052
C(3)- H(11)	1.1162	1.1130	4.6000	.0034
C(4)- H(12)	1.1148	1.1130	4.6000	.0011
C(4)- H(13)	1.1136	1.1130	4.6000	.0001
C(4)- H(14)	1.1143	1.1130	4.6000	.0006

NON-BONDED DISTANCES, VAN DER WAALS ENERGY

```
    54 VDW INTERACTIONS (1,3 EXCLUDED)
      ENERGY = KV*(2.90(10**5)EXP(-12.50/P) - 2.25(P**6))
                       RV = RVDW(I) + RVDW(K)
                       KV = SQRT(EPS(I)*EPS(K))
                        P = (RV/R) OR (RV/R#)
                       (IF P.GT.3.311, ENERGY = KV(336.176)(P**2))
# IN THE VDW CALCULATIONS THE HYDROGEN ATOMS ARE RELOCATED
SO THAT THE ATTACHED HYDROGEN DISTANCE IS REDUCED BY  .915
```

ATOM PAIR	R	R#	RV	KV	ENERGY	(1,4)
C(1), C(3)	2.5685					
C(1), C(4)	**3.1492**		**3.800**	**.0440**	**.0989**	*
C(1), H(8)	2.1649					
C(1), H(9)	2.1784					
C(1), H(10)	3.5141	3.4297	3.340	.0460	-.0527	*
C(1), H(11)	2.8044	2.7677	3.340	.0460	.1037	*
C(1), H(12)	4.1738	4.0849	3.340	.0460	-.0279	
C(1), H(13)	2.8924	2.8985	3.340	.0460	.0172	
C(1), H(14)	3.5017	3.4592	3.340	.0460	-.0520	
C(2), C(4)	2.5686					
C(2), H(5)	2.1945					
C(2), H(6)	2.2044					
C(2), H(7)	2.1976					
C(2), H(10)	2.1718					
C(2), H(11)	2.1824					
C(2), H(12)	3.5320	3.4466	3.340	.0460	-.0524	*
C(2), H(13)	2.8951	2.8519	3.340	.0460	.0418	*
C(2), H(14)	2.8133	2.7760	3.340	.0460	.0964	*
C(3), H(5)	3.5320	3.4465	3.340	.0460	-.0524	*
C(3), H(6)	2.8956	2.8524	3.340	.0460	.0415	*
C(3), H(7)	2.8128	2.7755	3.340	.0460	.0968	*
C(3), H(8)	2.1717					
C(3), H(9)	2.1825					
C(3), H(12)	2.1945					
C(3), H(13)	2.2044					
C(3), H(14)	2.1976					
C(4), H(5)	4.1742	4.0852	3.340	.0460	-.0279	
C(4), H(6)	2.8930	2.8991	3.340	.0460	.0169	
C(4), H(7)	3.5008	3.4583	3.340	.0460	-.0521	
C(4), H(8)	3.5141	3.4297	3.340	.0460	-.0527	*
C(4), H(9)	2.8045	2.7678	3.340	.0460	.1035	*
C(4), H(10)	2.1649					
C(4), H(11)	2.1784					
H(5), H(6)	1.7924					
H(5), H(7)	1.7995					
H(5), H(8)	**2.4791**	**2.3892**	**3.000**	**.0470**	**.2329**	*
H(5), H(9)	**2.5431**	**2.4457**	**3.000**	**.0470**	**.1512**	*
H(5), H(10)	4.3477	4.1964	3.000	.0470	-.0138	
H(5), H(11)	3.8053	3.6823	3.000	.0470	-.0280	
H(5), H(12)	5.2328	5.0511	3.000	.0470	-.0046	
H(5), H(13)	3.9338	3.8523	3.000	.0470	-.0221	
H(5), H(14)	4.3625	4.2569	3.000	.0470	-.0127	
H(6), H(7)	1.8040					

H(6), H(8)	3.1086	2.9421	3.000	.0470	-.0542	*
H(6), H(9)	**2.5284**	**2.4337**	**3.000**	**.0470**	**.1667**	*
H(6), H(10)	3.9161	3.7824	3.000	.0470	-.0244	
H(6), H(11)	3.2437	3.1508	3.000	.0470	-.0517	
H(6), H(12)	3.9341	3.8526	3.000	.0470	-.0221	
H(6), H(13)	**2.3519**	**2.4107**	**3.000**	**.0470**	**.1990**	
H(6), H(14)	3.1256	3.1085	3.000	.0470	-.0531	
H(7), H(8)	2.5544	2.4549	3.000	.0470	.1401	*
H(7), H(9)	3.1156	2.9487	3.000	.0470	-.0544	*
H(7), H(10)	3.8036	3.6800	3.000	.0470	-.0281	
H(7), H(11)	2.5872	2.5856	3.000	.0470	.0276	
H(7), H(12)	4.3609	4.2554	3.000	.0470	-.0127	
H(7), H(13)	3.1243	3.1072	3.000	.0470	-.0532	
H(7), H(14)	4.1322	4.0101	3.000	.0470	-.0178	
H(8), H(9)	1.7929					
H(8), H(10)	**2.4465**	**2.3596**	**3.000**	**.0470**	**.2856**	*
H(8), H(11)	**2.5263**	**2.4297**	**3.000**	**.0470**	**.1720**	*
H(8), H(12)	4.3477	4.1964	3.000	.0470	-.0138	
H(8), H(13)	3.9157	3.7820	3.000	.0470	-.0244	
H(8), H(14)	3.8040	3.6803	3.000	.0470	-.0280	
H(9), H(10)	**2.5264**	**2.4299**	**3.000**	**.0470**	**.1718**	*
H(9), H(11)	3.1070	2.9404	3.000	.0470	-.0542	*
H(9), H(12)	3.8059	3.6828	3.000	.0470	-.0279	
H(9), H(13)	3.2430	3.1501	3.000	.0470	-.0517	
H(9), H(14)	2.5879	2.5862	3.000	.0470	.0272	
H(10), H(11)	1.7929					
H(10), H(12)	**2.4796**	**2.3895**	**3.000**	**.0470**	**.2322**	
H(10), H(13)	3.1086	2.9422	3.000	.0470	-.0542	
H(10), H(14)	**2.5540**	**2.4545**	**3.000**	**.0470**	**.1405**	
H(11), H(12)	**2.5426**	**2.4453**	**3.000**	**.0470**	**.1518**	
H(11), H(13)	**2.5288**	**2.4341**	**3.000**	**.0470**	**.1662**	
H(11), H(14)	3.1155	2.9487	3.000	.0470	-.0544	
H(12), H(13)	1.7924					
H(12), H(14)	1.7995					
H(13), H(14)	1.8040					

BOND ANGLES, BENDING AND STRETCH-BEND ENERGIES (24 ANGLES)

```
EB = 0.021914(KB)(DT)(DT)(1+SF*DT**4)
                DT = THETA-TZERO
                SF =   .00700E-5
ESB(J) = 2.51124(KSB(J))(DT)(DR1+DR2)
                DR(I) = R(I) - RO(I)
```

A T O M S	THETA	TZERO	KB	EB	KSB	ESB
C(2)- C(1)- H(5)	110.875	110.000	.090	.0022	.360	.0060
C(2)- C(1)- H(6)	111.733	110.000	.090	.0044	.360	.0237
C(2)- C(1)- H(7)	111.149	110.000	.090	.0029	.360	.0104
H(5)- C(1)- H(6)	107.095	109.000	.320	.0254		
H(5)- C(1)- H(7)	107.658	109.000	.320	.0126		
H(6)- C(1)- H(7)	108.141	109.000	.320	.0052		
C(1)- C(2)- C(3)	**113.460**	**109.500**	**.120**	**.0311**	**.450**	**.1546**
C(1)- C(2)- H(8)	108.458	109.410	.090	-.0024	.360	.0071
C(1)- C(2)- H(9)	109.536	109.410	.090	.0003	.360	.0001
C(3)- C(2)- H(8)	108.746	109.410	.090	-.0022	.360	.0035
C(3)- C(2)- H(9)	109.615	109.410	.090	.0007	.360	.0003
H(8)- C(2)- H(9)	106.802	109.400	.320	.0473		
C(2)- C(3)- C(4)	**113.461**	**109.500**	**.120**	**.0311**	**.450**	**.1548**
C(2)- C(3)- H(10)	108.751	109.410	.090	-.0022	.360	.0034
C(2)- C(3)- H(11)	109.610	109.410	.090	.0007	.360	.0003
C(4)- C(3)- H(10)	108.459	109.410	.090	-.0024	.360	.0071
C(4)- C(3)- H(11)	109.533	109.410	.090	.0003	.360	.0001

```
H(10)- C( 3)- H(11) 106.801 109.400    .320   .0474
C( 3)- C( 4)- H(12) 110.875 110.000    .090   .0022    .360   .0060
C( 3)- C( 4)- H(13) 111.734 110.000    .090   .0044    .360   .0237
C( 3)- C( 4)- H(14) 111.148 110.000    .090   .0029    .360   .0104
H(12)- C( 4)- H(13) 107.093 109.000    .320   .0255
H(12)- C( 4)- H(14) 107.663 109.000    .320   .0125
H(13)- C( 4)- H(14) 108.137 109.000    .320   .0052

DIHEDRAL ANGLES, TORSIONAL ENERGY (ET)    ( 27 ANGLES)
     ET = (V1/2)(1+COS(W))+(V2/2)(1-COS(2W))+(V3/2)(1+COS(3W))
     SIGN OF ANGLE A-B-C-D 0 WHEN LOOKING THROUGH B TOWARD C,
     IF D IS COUNTERCLOCKWISE FROM A, NEGATIVE.
      A T O M S             OMEGA      V1      V2      V3      ET
C( 1) C( 2) C( 3) C( 4)    65.249    .200    .270    .093    .366
C( 1) C( 2) C( 3) H(10)  -173.963    .000    .000    .267    .007
C( 1) C( 2) C( 3) H(11)   -57.540    .000    .000    .267    .001
C( 2) C( 3) C( 4) H(12)   175.465    .000    .000    .267    .004
C( 2) C( 3) C( 4) H(13)   -65.151    .000    .000    .267    .005
C( 2) C( 3) C( 4) H(14)    55.748    .000    .000    .267    .003
C( 3) C( 2) C( 1) H( 5)   175.388    .000    .000    .267    .004
C( 3) C( 2) C( 1) H( 6)   -65.226    .000    .000    .267    .005
C( 3) C( 2) C( 1) H( 7)    55.677    .000    .000    .267    .003
C( 4) C( 3) C( 2) H( 8)  -173.970    .000    .000    .267    .007
C( 4) C( 3) C( 2) H( 9)   -57.547    .000    .000    .267    .001
H( 5) C( 1) C( 2) H( 8)    54.445    .000    .000    .237    .005
H( 5) C( 1) C( 2) H( 9)   -61.774    .000    .000    .237    .001
H( 6) C( 1) C( 2) H( 8)   173.831    .000    .000    .237    .006
H( 6) C( 1) C( 2) H( 9)    57.613    .000    .000    .237    .001
H( 7) C( 1) C( 2) H( 8)   -65.265    .000    .000    .237    .004
H( 7) C( 1) C( 2) H( 9)   178.516    .000    .000    .237    .000
H( 8) C( 2) C( 3) H(10)   -53.183    .000    .000    .237    .007
H( 8) C( 2) C( 3) H(11)    63.240    .000    .000    .237    .002
H( 9) C( 2) C( 3) H(10)    63.241    .000    .000    .237    .002
H( 9) C( 2) C( 3) H(11)   179.665    .000    .000    .237    .000
H(10) C( 3) C( 4) H(12)    54.513    .000    .000    .237    .005
H(10) C( 3) C( 4) H(13)   173.898    .000    .000    .237    .006
H(10) C( 3) C( 4) H(14)   -65.203    .000    .000    .237    .004
H(11) C( 3) C( 4) H(12)   -61.703    .000    .000    .237    .000
H(11) C( 3) C( 4) H(13)    57.681    .000    .000    .237    .001
H(11) C( 3) C( 4) H(14)   178.580    .000    .000    .237    .000


FINAL STERIC ENERGY IS 3.0348 KCAL.
     COMPRESSION          .1667
     BENDING              .5930
     STRETCH-BEND         .0741
     VANDERWAALS
       1,4 ENERGY        2.1122
       OTHER             -.3621
     TORSIONAL            .4510
```

The steric energy of the minimized gauche conformation of butane is found by this calculation to be more than 0.6 kcal/mol lower than the initial conformation. However, note that the C — C — C — C dihedral angle is no longer 60° but is now 65°. As shown in bold above, the printout indicates that a strain of 0.366 kcal/mol is associated with that dihedral angle, which is slightly larger than the torsional strain of 0.353 kcal/mol calculated for the 60° C — C — C — C dihedral angle.[119] Moreover, other torsional energies are also higher in this conformer, so that there is a total of 0.069 kcal/mol more torsional energy in the 65° conformer. However, the van der Waals strain decreases from 2.42 kcal/mol in the 60° conformer to 1.75 kcal/mol in the 65° conformer,[120] and most of the other strain parameters are also less in the 65° conformer.[121] This example indicates two important points:

1. The minimum energy conformation is that conformation that has the lowest *sum* of the components of steric strain, but some individual components of strain might be higher in a more stable conformation than they are in a less stable conformation.
2. A gauche conformer need not be exactly 60°.

If the initial conformation entered for butane is closer to 180° than to 60°, then the program optimizes the geometry of butane to produce anti butane with a 180° dihedral for the C — C — C — C bonds. That conformation has the following components of steric energy:

```
FINAL STERIC ENERGY IS 2.1714 KCAL.
     COMPRESSION          .1569
     BENDING              .2909
     STRETCH-BEND         .0524
     VANDERWAALS
       1,4 ENERGY        2.0697
       OTHER             -.4058
     TORSIONAL            .0073
```

The difference in steric energy between the 65° gauche conformer and the anti conformer of butane is thus computed to be 0.86 kcal/mol, very close to the value used in our discussion of conformational analysis.

Why did the molecular mechanics program not minimize the initial 60° conformation to the anti conformation? The answer is that the gauche conformation is a local minimum, while the anti conformation represents a global minimum on the butane conformational potential energy surface. (These terms are illustrated for a simplified two-dimensional potential energy surface in Figure 3.21.) A minimization routine that starts with a geometry near a local minimum will generally optimize to that local minimum, even if there is a different geometry that is lower in energy. The problem is simple for butane because we can recognize likely minima and adjust

[119]A C — C — C — C dihedral angle of 60° is not computed to have a torsional energy of 0 kcal/mol in this calculation because the torsional potential energy function is parameterized to incorporate some van der Waals interaction also. For further discussion, see page 158.

[120]There is 2.11 kcal/mol of repulsive 1,4 van der Waals strain, but this is partially compensated for by 0.36 kcal/mol of attractive van der Waals interactions between other atoms.

[121]Some interactions shown in bold above contribute significantly to the total strain energy, even though in many cases we would not have expected energy effects on the basis of a simplified model of conformational analysis.

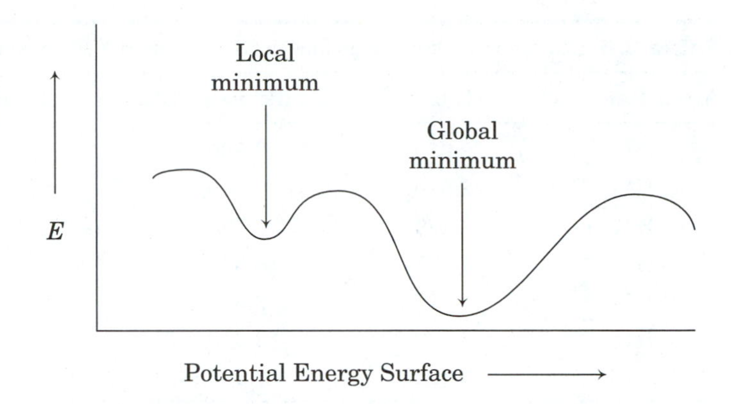

Figure 3.21 Local (left) and global minima on a simplified potential energy surface.

the coordinates of the atoms in the input file so that the global minimum can be found. For large molecules, determining the lowest energy conformation through computation can be difficult, and it is essential to provide the molecular mechanics program with a wide variety of initial conformations so that all possible minima can be located.[122,123]

The output from a molecular mechanics calculation on butane is quite lengthy, but that from a calculation on cyclohexane is even longer. However, several results from the calculation on cyclohexane are worth noting. The minimum energy geometry places atoms C1 and C4 2.959 Å apart, while the sum of their van der Waals radii is 3.800 Å.[124] (See Figure 3.22 for atom numbering.) This gives rise to a van der Waals repulsion of 0.31 kcal/mol. Repulsion between carbon atoms 2 and 5 (as well as between carbon atoms 3 and 6) gives rise to the same interaction energy, so that carbon-carbon van der Waals repulsion accounts for 1.23 kcal/mol of the steric energy calcu-

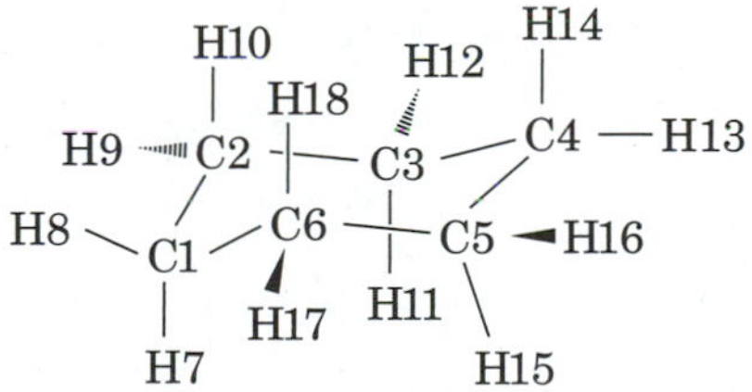

Figure 3.22 Atom numbering for van der Waals forces in cyclohexane.

---

[122]A procedure for finding the global minimum in large molecules has been reported by Saunders, M. *J. Am. Chem. Soc.* **1987**, *109*, 3150; also see Kolossváry, I.; Guida, W. C. *J. Am. Chem. Soc.* **1993**, *115*, 2107.

[123]As one theoretician has put it, when seeking the minimum on a potential energy surface, "You always fall into the nearest hole."

[124]The van der Waals radii used in MM2 calculations differ somewhat from those tabulated in Chapter 1. In general they are 0.2–0.3 Å larger.

Table 3.6 Selected van der Waals interactions from an MM2 calculation.

Atom Pair	R (Å)	Σ vdW radii (Å)	Energy (kcal/mol)
C1 - C4	2.959	3.800	0.312
C1 - H11	2.810	3.340	0.102
C1 - H12	3.499	3.340	−0.053
C1 - H13	3.971	3.340	−0.035
C1 - H14	3.376	3.340	−0.054
C1 - H15	2.769	3.340	0.102
C1 - H16	3.499	3.340	−0.053

lated for this conformation. Similarly, there is a repulsive van der Waals interaction between C1 and H11; they are calculated to be 2.810 Å apart, while the sum of their van der Waals radii is 3.340 Å. There is also 1,4 repulsion between hydrogen atoms on adjacent carbon atoms, such as H9 with H11 (0.204 kcal/mol) and with H12 (0.171 kcal/mol). However, the interaction of a hydrogen atom of the 1,3-diaxial hydrogens such as H10 with H14, is only slightly destabilizing, (0.009 kcal/mol), since they are about the same distance apart as the sum of their van der Waals radii (2.629 Å). Some of the van der Waals interactions are stabilizing, such as the interaction of C1 with H12. A list of some additional van der Waals interactions for C1 and its hydrogens is given in Table 3.6.

The minimum energy conformation of cyclohexane calculated with a molecular mechanics program is shown in Figure 3.24, and the summary of the different contributions to the steric energy are listed in Table 3.7. Note that the C1 — C2 — C3 — C4 dihedral angle is 56.33°, producing a torsional energy, E_T, of 0.343 kcal/mol. This value is quite close to the value of 56.1° measured by Grant and co-workers.[125]

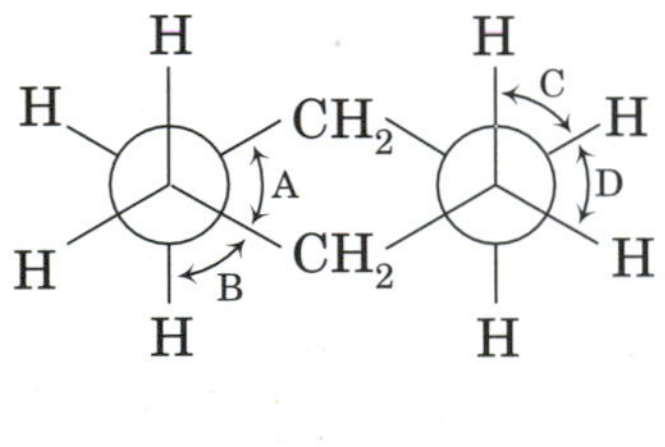

Figure 3.23 Dihedral angles in the chair conformation of cyclohexane from a molecular mechanics calculation.

A = 56.33° B = 64.51°
C = 57.26° D = 60.10°

[125]Curtis, J.; Grant, D. M.; Pugmire, R. J. *J. Am. Chem. Soc.* **1989**, *111*, 7711.

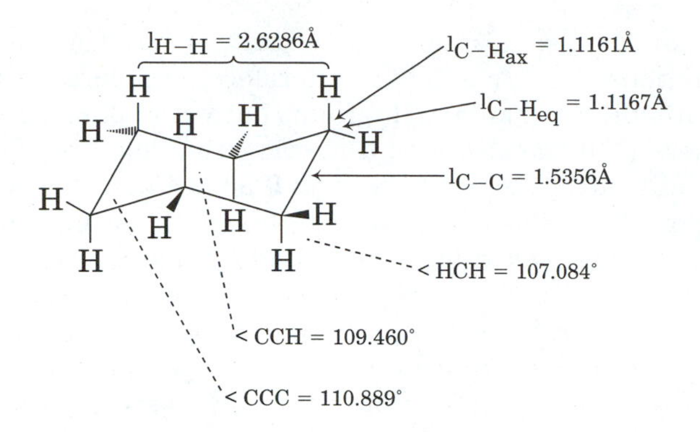

Figure 3.24 Calculated minimum energy conformation of cyclohexane.

In our discussions of conformational analysis, we viewed the chair conformation of cyclohexane as a structure with no strain, and its heat of combustion per methylene unit is the same as that of an acyclic alkane. However, examination of the chair conformation reveals that there are actually six butane gauche interactions, two evident from each conformation seen by a 120° rotation. Moreover, we have noted that there are three repulsive carbon-carbon transannular van der Waals interactions (e.g., the C1 — C4 repulsion). Why, therefore, does cyclohexane not have (6 × 0.8 kcal/mol) + (3 × 0.31 kcal/mol) of strain energy?

Schleyer[126] concluded that the gauche interactions in the cyclohexane skeleton are not comparable to those of gauche butane because the H · · H interactions are minimized in cyclohexane. This question was also addressed by Wiberg,[39] who noted that the heat of combustion comparison may be misleading. Combustion of the alkanes is done not on samples at

Table 3.7 Calculated contributions to steric energy of cyclohexane in the chair conformation.

Parameter	*E* (kcal/mol)
Compression	0.3376
Bending	0.3652
Stretch-bend	0.0826
van der Waals	
1,4-interaction	4.6733
other	−1.0633
Torsional	2.1556
Total	6.5510

[126]Schleyer, P. v. R.; Williams, J. E.; Blanchard, K. R. *J. Am. Chem. Soc.* **1970**, *92*, 2377.

0 K but on samples at room temperature, where the population of gauche conformations is not negligible. Therefore, cyclohexane is not being compared with all-anti alkane CH_2 groups but with alkanes having a mixture of conformers. Wiberg calculated that correcting for this mix of alkane conformations still suggested no more than 0.28 kcal/mol per cyclohexane gauche interaction. A possible explanation based on the electronic properties of the molecular orbitals used for sigma bonding may be responsible,[127] but further investigation may reveal an alternative explanation.

It may be more difficult to rationalize the large van der Waals repulsion due to 1,4 carbon-carbon interactions in cyclohexane. Such interactions are inherent in calculations on all structures incorporating cyclohexane chair conformations.[126] This surprising result should serve to remind us that ball and stick molecular models may severely underestimate the size of the electron clouds we associate with atoms. It may be easy to recognize chair conformations of cyclohexane with plastic models, but we should not be misled by thinking that carbon atoms 1 and 4 are far apart relative to other groups. These models, as are all other models, are useful for only some of our analyses.

We should emphasize that the particular kinds of internal energy determined for molecular structures through molecular mechanics calculation should not be taken too literally. A different molecular mechanics calculation might produce a similar value for the total strain (and thus the resulting ΔH_f^0 of the compounds, *vide infra*), but with different internal contributions to the total strain. For example, the MM1 and the MUB-2 procedures differed significantly in their handling of van der Waals forces, the MM1 procedure using a van der Waals potential for hydrogen that was large and hard but with a potential for carbon that was small. However, the potentials in the MUB-2 force field were smaller and softer for hydrogen but larger for carbon. In the MM2 force field, the butane gauche interaction was modeled with a different torsional potential function, thus making the van der Waals characteristics of the atoms less important.[91] As Allinger has noted,

> If different force-fields give the same results, as far as they can be checked against experimental values, but different results in terms of internal details which are not experimentally accessible, it is clear that one cannot assign physical significance to the different sets of internal details.[89]

In other words, a molecular mechanics calculation, as well as a conformational analysis, is just a particular kind of conceptual model and should not be regarded as a true measure of intramolecular forces.

This conclusion may be reinforced by considering conformations of the class of compounds with molecular formula X — CH_2 — CH_2 — Y. We have

---

[127] Dewar, M. J. S. *J. Am. Chem. Soc.* **1984**, *106*, 669.

seen that when X and Y are both alkyl groups, the anti conformation is favored over the gauche conformation. When X and Y are electronegative groups, we would expect that the gauche conformation would be even more destabilized through a combination of van der Waals forces and the electrostatic repulsion of the two partially negative substituents (Figure 3.25).

It is surprising, therefore, that in the gas phase the gauche conformer of 1,2-difluoroethane is more stable than the anti conformer by about 1 kcal/mol. For the other 1,2-dihaloethanes, the anti conformer is lower in energy, with the anti preference increasing among the series chlorine, bromine and iodine.[128,129] Similar observations are made if X and Y are other electronegative atoms, such as oxygen, and the general trend is that the proportion of gauche conformer increases as X and Y become more electronegative.[130] This phenomenon is known as the **gauche effect**, which is "a tendency to adopt that structure which has the maximum number of gauche interactions between the adjacent electron pairs and/or polar bonds."[131] This idea suggests that there can be a **stereoelectronic effect**, that is, an effect that results from the "stereochemistry of particular electron pairs, bonded or nonbonded."[132]

A similar phenomenon related to the preferred conformations of the hemiacetal forms of carbohydrates is known as the **anomeric effect**. We saw in Chapter 2 that glucose can cyclize to form either the α (axial OH) or β (equatorial OH) hemiacetal. For glucose the β anomer is expected to be the major conformer because all ring substituents are equatorial, and it is present to the extent of 64% at equilibrium, with the α anomer present to the extent of 34%. If the conformational preferences of the tetrahydropyran ring are similar to those of cyclohexane, then an A value of OH of 0.87 in a protic solvent would predict that the equatorial anomer would comprise more than 90% of the mixture. The situation is even more dramatic in the case of D-mannose. Here one OH group is constrained to be axial, but it turns out that the β anomer (**12**) comprises only 32% of the equilibrium

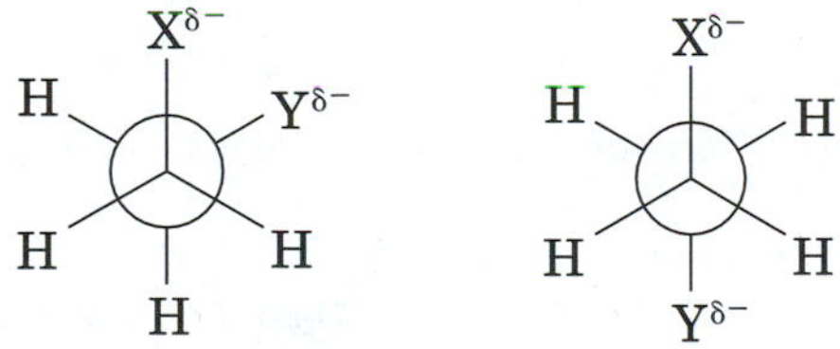

Figure 3.25 Electrostatic effects in gauche (left) and anti (right) conformers of XCH_2CH_2Y.

[128]See, for example, Huang, J.; Hedberg, K. *J. Am. Chem. Soc.* **1990**, *112*, 2070.

[129]For a comparison of theoretical methods used to study the conformations of 1,2-dihaloethanes, see Dixon, D. A.; Matsuzawa, N.; Walker, S. C. *J. Phys. Chem.* **1992**, *96*, 10740.

[130]Phillips, L.; Wray, V. *J. Chem. Soc., Chem. Commun.* **1973**, 90.

[131]Wolfe, S. *Acc. Chem. Res.* **1972**, *5*, 102.

[132]Deslongchamps, P. *Stereoelectronic Effects in Organic Chemistry;* Pergamon Press: Oxford, England, 1983; p. 2.

mixture,[133] although the equatorial anomer would be expected to be more stable in cyclohexane. The experimental and theoretical reasons for the anomeric effect have been extensively discussed.[133,134,135] The most commonly cited explanation is given in terms of an electronic interaction from a nonbonded pair of electrons on oxygen with an antibonding orbital of the bond from the carbon atom to the other electronegative atom.[136,137] However, dipole-dipole interactions have also been proposed as the basis for the anomeric effect.[138]

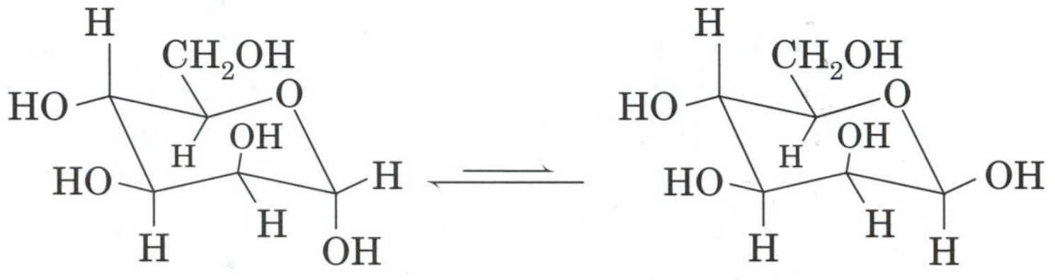

α–D–Mannose

11

β–D–Mannose

12

An interesting alternative explanation has been advanced by Wiberg et al.[139] If the bond paths of methyl fluoride are curved (as was discussed in Chapter 1), then those in 1,2-difluoroethane might be curved also. In Figure 3.26, the angle α_2 designates the difference between the carbon-carbon internuclear bond line and the preferred curved bond path from one carbon atom to another. Wiberg suggested that the curved bond path would lead to poor C — C overlap in the anti conformer, but that bonding would not be adversely affected as much in the case in the gauche conformer. Poorer orbital overlap would be expected to produce a longer carbon-carbon bond distance, and *ab initio* calculations by Durig and co-workers did indicate that the C — C (internuclear) bond distance is 0.010 Å greater in the anti conformer

---

[133]Kirby, A. J. *The Anomeric Effect and Related Stereoelectronic Effects at Oxygen;* Springer-Verlag: Berlin, 1983; p. 7.

[134]Juaristi, E.; Cuevas, G. *Tetrahedron* **1992**, *48*, 5019.

[135]Salzner, U.; Schleyer, P. v. R. *J. Am. Chem. Soc.* **1993**, *115*, 10231 and references therein.

[136]For more details, see chapter 2 of reference 133. For an explanation of the treatment of the gauche effect in molecular mechanics calculations, see reference 100, p. 220; Allinger, N. L.; Rahman, M.; Lii, J.-H. *J. Am. Chem. Soc.* **1990**, *112*, 8293. See also Cramer, C. J. *J. Org. Chem.* **1992**, *57*, 7034.

[137]For an explanation of the anomeric effect in terms of "hardness," see Pearson, R. G. *J. Am. Chem. Soc.* **1988**, *110*, 7684; Hati, S.; Datta, D. *J. Org. Chem.* **1992**, *57*, 6056.

[138]Perrin, C. L.; Armstrong, K. B.; Fabian, M. A. *J. Am. Chem. Soc.* **1994**, *116*, 715 and references therein.

[139]Wiberg, K. B.; Murcko, M. A.; Laidig, K. E.; MacDougall, P. J. *J. Phys. Chem.* **1990**, *94*, 6956; Wiberg, K. B. *Acc. Chem. Res.* **1996**, *29*, 229.

Figure 3.26 Effect of curved bond path on C — C bonding in anti (left) and gauche (right) conformers of 1,2-difluoroethane. (Reproduced from reference 139.)

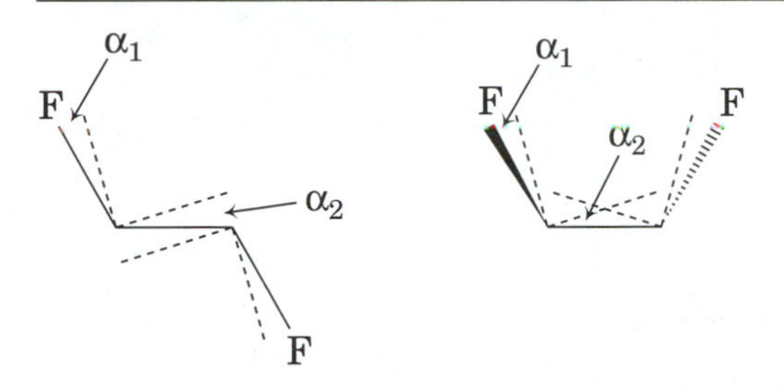

than the gauche conformer.[140] Because bond curvature is expected to increase as the electronegativity of the substituent increases, this explanation provides a simple model to explain a wide array of conformational phenomena. More important for our discussion, though, is that straight bonds are assumed in our treatment of conformational analysis and molecular mechanics. If it develops that curved bonds offer conceptual or computational advantages over straight bonds, then our straight bonds may be a conceptual model as deficient as was the hinged bond model that led Baeyer to propose angle strain for cyclohexane.

3.4 Molecular Strain and Limits to Molecular Stability

Molecular mechanics provides a means to calculate the ΔH_f^0 of a compound (provided the structure is well modeled by the parameters in the force field) that is said to rival experiment for accuracy. Strainless group contributions can then be used to determine a strainless heat of formation for the compound under consideration, and the difference between the two heats of formation can be said to be the strain energy of the structure. The term *strain* is less precise than one might expect, however. Although "Qualitatively, organic chemists usually recognize a strained molecule when they see one,"[107] strain is a parameter that can only be defined with reference to the model structure that is defined to be strain free. For example, one researcher might choose the all-anti (zig-zag) conformation of a linear hydrocarbon as a reference structure, while another might choose the chair conformation of cyclohexane.

The strain energy of a molecule is not the same as the steric energy obtained from a molecular mechanics calculation. Even the minimum energy conformation of a structure has some strain, because it is impossible for every potential energy function to be at its energy minimum in the same geometry.[141] For example, van der Waals interactions between nonbonded

[140]Durig, J. R.; Liu, J.; Little, T. S.; Kalasinsky, V. F. *J. Phys. Chem.* **1992**, *96*, 8224.

[141]Reference 100, p. 185.

Figure 3.27 Qualitative representation of enthalpy relationships used in calculating strain energies.

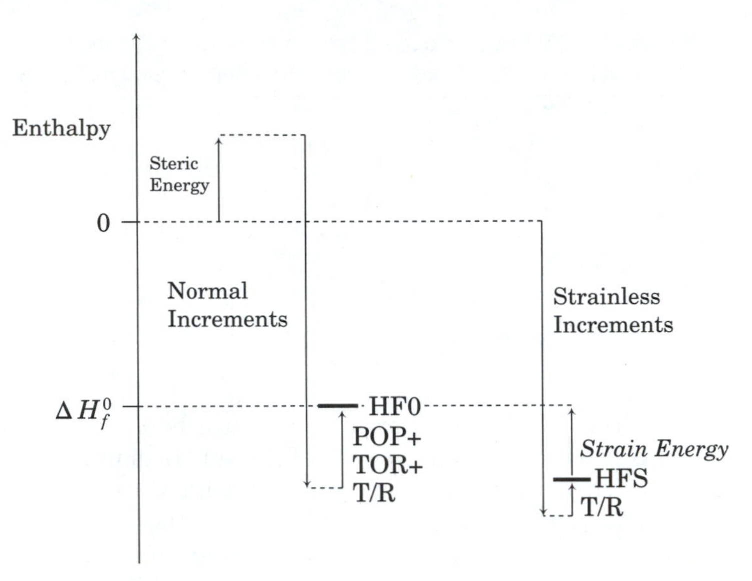

atoms might be optimized at a distance that is greater than the optimum bond distance for atoms to which those groups are bonded. Steric energies do give relative strain energies for conformers because the strainless reference compound is the same in both cases, but the steric energy of butane cannot be compared directly with the steric energy of cyclobutane to determine the strain in cyclobutane. The relationships among strain energies, steric energies, heats of formation, and other energy terms are shown in Figure 3.27.[142]

The relationships of the components used to calculate strain energy from steric energy may be made clearer by considering an example. A final portion of the printout of the molecular mechanics calculation for *anti*-butane is the following:

```
FINAL STERIC ENERGY IS 2.1714 KCAL.
HEAT OF FORMATION AND STRAIN ENERGY CALCULATIONS  (UNITS ARE KCAL.)
BOND ENTHALPY (BE) AND STRAINLESS BOND ENTHALPY (SBE) CONSTANTS AND SUMS
  #   BOND OR STRUCTURE          --- NORMAL ---      --STRAINLESS--
  3   C-C SP3-SP3                 -.004    -.01        .493    1.48
 10   C-H ALIPHATIC              -3.205  -32.05     -3.125  -31.25
  2   C(SP3)-METHYL              -1.510   -3.02     -1.575   -3.15
                                ----------------   ----------------
                                  BE =  -35.08      SBE =  -32.92

PARTITION FUNCTION CONTRIBUTION (PFC)
    CONFORMATIONAL POPULATION INCREMENT (POP)   .00
    TORSIONAL CONTRIBUTION (TOR)                .00
    TRANSLATION/ROTATION TERM (T/R)            2.40
                                         -------------
                                         PFC =  2.40

HEAT OF FORMATION (HFO) = E + BE + PFC          -30.51
STRAINLESS HEAT OF FORMATION (HFS) = SBE + T/R  -30.52
INHERENT STRAIN (SI) = E + BE - SBE                .01
STRAIN ENERGY (S) = POP + TOR + SI                 .01
```

[142]For additional discussion, see reference 116, pp. 20 *ff*.

As indicated in the printout, the heat of formation (HF0) is taken as the sum of three terms. The first is the steric energy (E). The second is the bond enthalpy (BE), which is calculated from normal bond increments.[143,144] The third is a partition function increment (PFC), which is itself the sum of three terms: a conformational population increment (POP), a torsional contribution term (TOR), and a translation/rotation term (T/R).[145] Since the molecular mechanics calculation has considered only one conformation, the POP and TOR terms are defined to be zero for this calculation (but see below). The T/R term is a molecular translational and rotational term that is always taken to be 2.4 kcal/mol at room temperature, since even an *anti*-butane molecule that is not rotating about the C2 — C3 bond will still be undergoing translation and rotation of the entire molecule as a unit. Therefore, HF0 is the total of the bond increment contributions (BE = −35.08) plus the steric energy (E = 2.17) plus the partition function increment (PFC = 2.4). The sum, −30.51 kcal/mol, is slightly lower than the literature value of −30.15 kcal/mol.[146]

We can also calculate a strainless heat of formation of *anti*-butane, which is the heat of formation of an anti structure made up of *hypothetical* strainless bonding interactions. As shown, we add the strainless bond increments to obtain the value of −32.92 kcal/mol for the strainless bond enthalpy (SBE). This value must also be corrected for the T/R term, but the POP and TOR terms are again taken to be zero. Thus the strainless ΔH_f^0 (HFS) is the sum of −32.92 plus 2.4, which is −30.52 kcal/mol. That differs from the heat of formation calculated with normal bond increments (−30.51 kcal/mol) by 0.01 kcal/mol, which is the inherent strain energy of the *anti*-butane.[147]

---

[143] We should note that neither the group increments nor the steric energies from one type of molecular mechanics calculation can be compared to the corresponding terms from calculations involving other force fields. The reason is that one force field might assign a particular interaction to a component of steric energy, but another force field might include some of that same energy in the bond or group increments used for calculation of ΔH_f^0. For example, Schleyer (reference 93) noted that three different force fields calculated values of 49.61, 31.48, and 40.40 kcal/mol as the steric energy of tri-*tert*-butylmethane. However, the group contributions to ΔH_f^0 for these same force fields were −102.69, −88.85, and −96.20 kcal/mol, respectively. Therefore the calculated heats of formation were −53.08, −57.37 and −55.80 kcal/mol. The point is that the calculated heats of formation varied by only about 4 kcal/mol, even though the steric energies varied by more than 18 kcal/mol.

[144] DeTar has developed the concept of Formal Steric Enthalpy as a standardized measure of the steric component of ΔH_f^0. DeTar, D. F. *J. Org. Chem.* **1992**, *57*, 902 and references therein.

[145] For details on the calculation of heats of formation from molecular mechanics, molecular orbital theory, or bond/group increments, see the discussion in Allinger, N. L.; Schmitz, L. R.; Motoc, I.; Bender, C.; Labanowski, J. K. *J. Am. Chem. Soc.* **1992**, *114*, 2880. The strain energy of a conformation can be calculated from values of heats of formation from these sources also. The relationships in Figure 3.27 among heat of formation, T/R, strainless increments, and strain energy hold no matter what the source of the heat of formation value.

[146] Stull, D. R.; Westrum, Jr., E. F.; Sinke, G. C. *The Thermodynamics of Organic Compounds;* John Wiley & Sons, Inc.: New York, 1969; pp. 245.

[147] Burkert and Allinger (reference 100) distinguished between *inherent strain* (shown as SI), which does not include POP and TOR, and *total strain* (shown as S), which does include those terms. By this distinction, one can either consider only the structural components of strain or consider both the structural and conformational components.

There is an important difference between the heat of formation (HF0) determined above and an experimental heat of formation, however. The calculated value is based on only one conformation, while the experimental value is determined from a sample of the *substance*, usually at or above room temperature, in which there is a distribution of conformations. For example, the POP term for butane should include the contribution to the ΔH_f^0 that arises because at 25° some butane molecules will be in the gauche conformation. The excess enthalpy of butane that results from the presence of the higher energy conformers can be calculated from the knowledge of the energies of both the gauche and anti conformers and the Boltzmann distribution, and it is found to be 0.3 kcal/mol.[148] We must also use a nonzero value for TOR to account for the change in enthalpy with temperature that is associated with the presence of nonstaggered conformations, and that term is calculated to be 0.36 kcal/mol.[149] The T/R term remains 2.4 kcal/mol, since it does not vary with conformation. Therefore the value of PFC becomes (0.3 + 0.36 + 2.4), which is 3.06 kcal/mol. Adding that value to the sum of the bond increments for butane (−35.08 kcal/mol) and the steric energy calculated for the anti conformer (2.17 kcal/mol) gives a ΔH_f^0 of butane of −29.85 kcal/mol, which is closer to the literature value.

The strain energy at 25° of the substance butane, then, should be taken to be the difference between the strainless heat of formation and the heat of formation value that includes estimates of POP and TOR, and that difference is 0.67 kcal/mol. Notice that this value is not shown on the printout. Again, it must be emphasized that *the strain energy shown on the molecular mechanics printout above reflects only the inherent strain of the particular* ***conformer*** *calculated, not the strain in the* ***substance*** *represented by the computer model.* A value of zero for POP and TOR on the printout is an indication that appropriate values must be entered if the calculated value of ΔH_f^0 is expected to reproduce an experimental value and if a more accurate value for the strain energy is desired.

For molecules with many bonds about which rotation can occur and many conformations that might be populated at room temperature, the POP and TOR corrections can be significant. For relatively rigid molecules, however, these terms are less important, and the strain energies calculated from molecular mechanics can more closely approximate those that would be determined with consideration of POP and TOR. For example, consider the results of a molecular mechanics calculation for puckered cyclobutane, the final portion of which is reproduced here.

---

[148]Reference 100, p. 187.

[149]Reference 100, p. 176. (Rotation about the C — CH_3 bond is included in the bond increments.)

```
FINAL STERIC ENERGY IS 29.2202 KCAL.
     COMPRESSION         .8020
     BENDING           16.2111
     STRETCH-BEND       -.9344
     VANDERWAALS
       1,4 ENERGY       2.3242
       OTHER            -.2521
     TORSIONAL         11.0693

HEAT OF FORMATION AND STRAIN ENERGY CALCULATIONS (UNITS ARE KCAL.)
BOND ENTHALPY (BE) AND STRAINLESS BOND ENTHALPY (SBE) CONSTANTS AND SUMS
   #   BOND OR STRUCTURE          --- NORMAL ---       --STRAINLESS--
   4   C-C SP3-SP3                -.004    -.02          .493    1.97
   8   C-H ALIPHATIC            -3.205   -25.64       -3.125  -25.00
                                ---------------      ---------------
                                  BE =  -25.66         SBE =  -23.03

PARTITION FUNCTION CONTRIBUTION (PFC)
    CONFORMATIONAL POPULATION INCREMENT (POP)   .00
    TORSIONAL CONTRIBUTION (TOR)                .00
    TRANSLATION/ROTATION TERM (T/R)            2.40
                                        ------------
                                         PFC =  2.40

HEAT OF FORMATION (HFO) = E + BE + PFC            5.96
STRAINLESS HEAT OF FORMATION (HFS) = SBE + T/R  -20.63
INHERENT STRAIN (SI) = E + BE - SBE              26.59
STRAIN ENERGY (S) = POP + TOR + SI               26.59
```

The steric energy for cyclobutane is calculated to be 29.22 kcal/mol. Because the molecule is cyclic, there should not be free rotation about any bonds. However, the conformational dynamics associated with inversion of puckered geometries (for example, see Figure 3.13) means that some internal energy must be included in the calculation of the heat of formation, and the total correction for POP and TOR amounts to 0.36 kcal/mol.[150] Thus, the strain energy of cyclobutane should be greater than the amount shown by 0.36 kcal/mol, or a total of 26.95 kcal/mol.

Cyclobutane shows a significant amount of strain, but not as much as many compounds. One of the continuing challenges in organic chemistry is the design, synthesis, and characterization of novel molecular structures. Molecular mechanics calculations (as well as molecular orbital calculations) are particularly helpful in these studies because they allow calculation of the heats of formation of highly strained molecules and the assignment of the individual components of the strain energy of the structures.[151] This information can be a very useful guide to chemists trying to synthesize ever-

[150]Reference 100, p. 188.

[151]As we have seen, however, we should not put too much emphasis on the specific components of strain calculated for a given molecule because the assignment of steric energies is force-field dependent.

more-strained organic compounds. Some of these structures are of interest primarily for their aesthetic appeal and intellectual challenge.[152] Others allow chemists to put theories about molecular structure and bonding to extreme tests.

Some molecules are designed to test the limits on torsional strain. Anderson has reported that the preferred conformation of 1,1,2-tri-*t*-butylethane is nearly eclipsed because of the steric requirements of the *t*-butyl groups.[153] Many interesting molecules that exhibit torsional strain are polycyclic compounds. For example, the compound iceane incorporates three cyclohexane boat conformations.[154]

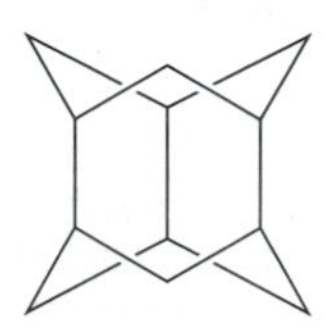

Iceane

13

The limits on the lengths of carbon-carbon bonds can be estimated by experimental data on molecules with very long or very short bond distances. There are reports of structures having carbon-carbon bond lengths around 1.8 Å or more in highly strained molecules.[155] On the other hand, the carbon-carbon nonbonded distance in derivatives of bicyclo[1.1.1]pentane is quite short. Adcock and co-workers have identified one derivative in which two carbon atoms are 1.80 Å apart.[156] Thus it appears that two carbon atoms that are formally bonded to each other in one molecule may be further apart than two carbon atoms in another molecule that are not bonded to each other![157]

Chemists have also sought to synthesize compounds with unusual bond angles to carbon atoms. One goal has been to prepare compounds with **pyramidal carbon atoms**, meaning atoms in which the four valences are

---

[152] For an introduction to the chemistry of strained organic molecules, see (a) reference 107; (b) Greenberg, A.; Liebman, J. F. *Strained Organic Molecules;* Academic Press: New York, 1978.

[153] Anderson, J. E. *J. Chem. Soc., Perkin Trans. 2* **1991**, 299.

[154] Cupas, C. A.; Hodakowski, L. *J. Am. Chem. Soc.* **1974**, *96*, 4668. See also Fărcasiu, D.; Wiskott, E.; Ōsawa, E.; Thielecke, W.; Engler, E. M.; Slutsky, J.; Schleyer, P. v. R. *J. Am. Chem. Soc.* **1974**, *96*, 4669.

[155] Bianchi, R.; Mugnoli, A.; Simonetta, M. *J. Chem. Soc., Chem. Commun.* **1972**, 1073; Bianchi, R.; Morosi, G.; Mugnoli, A.; Simonetta, M. *Acta Crystallogr., Sect. B*, **1973**, *29*, 1196; Zhou, X.; Liu, R.; Allinger, N. L. *J. Am. Chem. Soc.* **1993**, *115*, 7525.

[156] Adcock, J. L.; Gakh, A. A.; Pollitte, J. L.; Woods, C. *J. Am. Chem. Soc.* **1992**, *114*, 3980.

[157] For a discussion of long bonds in strained molecules, see Ōsawa, E.; Kanematsu, K. in *Molecular Structure and Energetics*, Vol. 3; Liebman, J. F.; Greenberg, A., eds.; VCH Publishers, Inc.: Deerfield Beach, FL, 1986; pp. 329 *ff.*

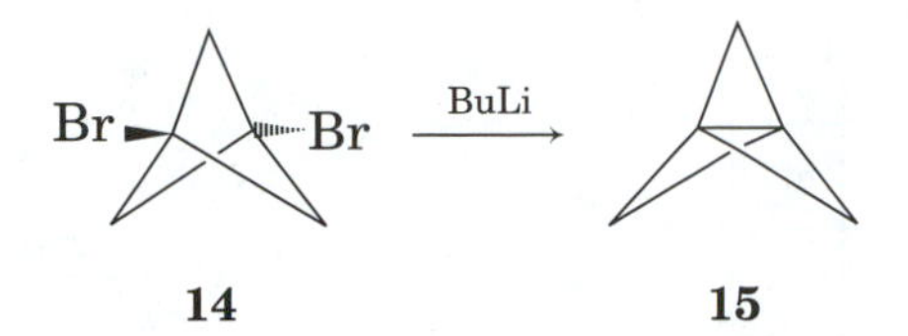

Figure 3.28 Synthesis of [1.1.1]propellane.

all directed toward one side of a plane passing through the carbon atom.[158] Interesting examples of compounds with this structural feature are the *propellanes*,[159] such as the known [1.1.1]propellane (**15**).[160] The compound has surprising stability, with a half-life for thermal rearrangement of 5 minutes at 114°C. This compound illustrates that we should not necessarily equate high strain with high chemical reactivity. A strained molecule might be relatively stable because of a high activation barrier for chemical reaction.

One might expect that if a structure with a pyramidal carbon atom is capable of existence, then one with a flattened carbon atom should be more stable and therefore easier to synthesize. That has turned out not to be the case. A great deal of interest has been expressed in fenestranes, named for the Latin word for "window."[161,162] Structure **16** would be [4.4.4.4]fenestrane, since each "pane" is composed of four carbon atoms. The figure suggests that the central carbon atom should be planar, but more detailed consideration suggests that nonplanar geometries are more stable.[163] Another approach to planar carbon is the proposed class of compounds named alkaplanes, such as hexaplane (**17**), in which the central carbon atom is held with the right symmetry for the carbon atom to be planar.[164]

Many compounds are interesting because they represent novel geometric shapes that have not previously been identified in naturally occurring products. Tetrahedrane[165] (**18**), not yet synthesized, is one of the platonic solids. (However, tetra-*t*-butyltetrahedrane has been reported.[166]) Others are cubane (**19**), which was synthesized by Eaton in 1964,[167] and dodecahedrane

---

[158]For reviews, see Wiberg, K. B. (a) *Acc. Chem. Res.* **1984**, *17*, 379; (b) *Chem. Rev.* **1989**, *89*, 975.

[159]The term *propellane* was coined by D. Ginsburg to denote a structure "having three nonzero bridges and one zero bridge between a pair of bridgehead carbons." For leading references, see reference 158(b).

[160]Wiberg, K. B.; Walker, F. H. *J. Am. Chem. Soc.* **1982**, *104*, 5239.

[161]Georgian, V.; Saltzman, M. *Tetrahedron Lett.* **1972**, 4315. See also the discussion in Nickon, A.; Silversmith, E. F. *The Name Game. Modern Coined Terms and Their Origins;* Pergamon Press: New York, 1987; pp. 55–56.

[162]For a review, see Venepalli, B. R.; Agosta, W. C. *Chem. Rev.* **1987**, *87*, 399.

[163]Liebman and Greenberg (reference 107, p. 350) have argued that a nonplanar conformation would be more stable.

[164]McGrath, M. P.; Radom, L. *J. Am. Chem. Soc.* **1993**, *115*, 3320.

[165]Scott, L. T.; Jones, Jr., M. *Chem. Rev.* **1972**, *72*, 181.

[166]Maier, G. *Angew. Chem., Int. Ed. Engl.* **1988**, *27*, 309.

[167]Eaton, P. E.; Cole, Jr., T. W. *J. Am. Chem. Soc.* **1964**, *86*, 3157.

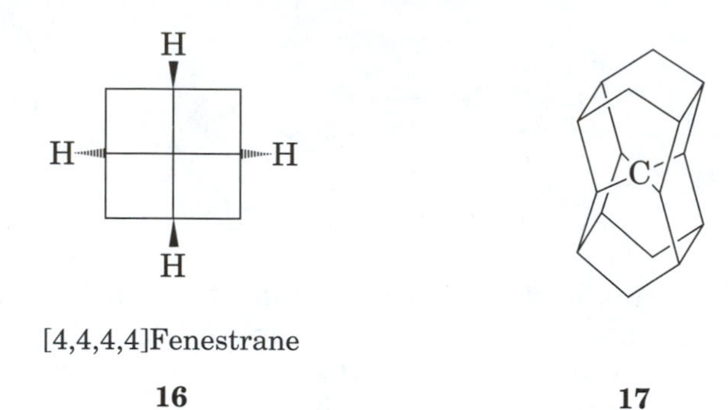

[4,4,4,4]Fenestrane

16

17

(**20**), synthesized by Paquette.[168] A highly symmetric compound that has attracted great interest in recent years is a compound known by a variety of names, including C_{60}, buckyball, Buckminsterfullerene, and soccerballene[169] (**21**).[170] Although it appears not to be highly strained, the π system cannot achieve maximum overlap due to nonplanarity. The ΔH_f^0 of C_{60} on a *per carbon* basis was determined to be 9.08 kcal/mol.[171] Other structures with interesting symmetries, such as those shown in Figure 3.29, are novel compounds that have provided challenges to synthetic chemists. Some, but not all, of the compounds in Figure 3.29 have been synthesized.[172]

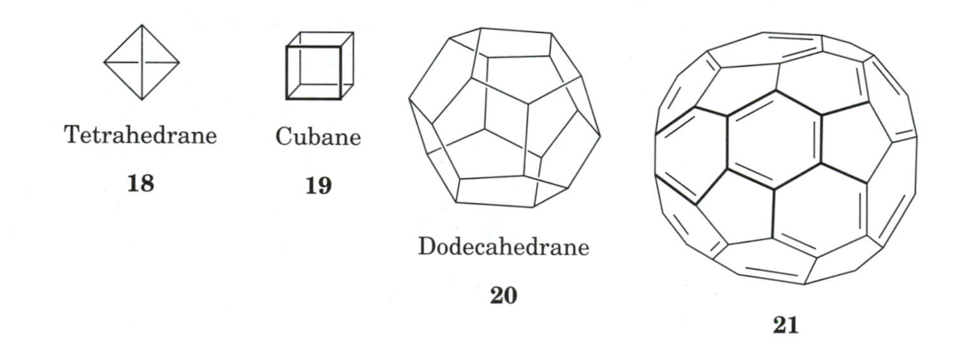

Tetrahedrane **18** Cubane **19** Dodecahedrane **20** **21**

---

[168]Ternansky, R. J.; Balogh, D. W.; Paquette, L. A. *J. Am. Chem. Soc.* **1982**, *104*, 4503.

[169]Haymet, A. D. J. *J. Am. Chem. Soc.* **1986**, *108*, 319.

[170]There is no shortage of references on this subject. Three among many are the following: (a) Kroto, H. W.; Heath, J. R.; O'Brien, S. C.; Curl, R. F.; Smalley, R. E. *Nature* **1985**, *318*, 162; (b) Kroto, H. W.; Allaf, A. W.; Balm, S. P. *Chem. Rev.* **1991**, *91*, 1213; (c) Boo, W. O. J. *J. Chem. Educ.* **1992**, *69*, 605.

[171]The corresponding values are 0.4 kcal/mol for diamond and 0.0 kcal/mol (by definition) for graphite. Beckhaus, H.-D.; Rüchartd, C.; Kao, M.; Diederich, F.; Foote, C. S. *Angew. Chem., Int. Ed. Engl.* **1992**, *31*, 63.

[172]For a discussion of structures with multiple carbon rings, see Ōsawa, E.; Yonemitsu, O., eds., *Carbocyclic Cage Compounds: Chemistry and Applications;* VCH Publishers: New York, 1992.

Figure 3.29 Challenging molecular structures.

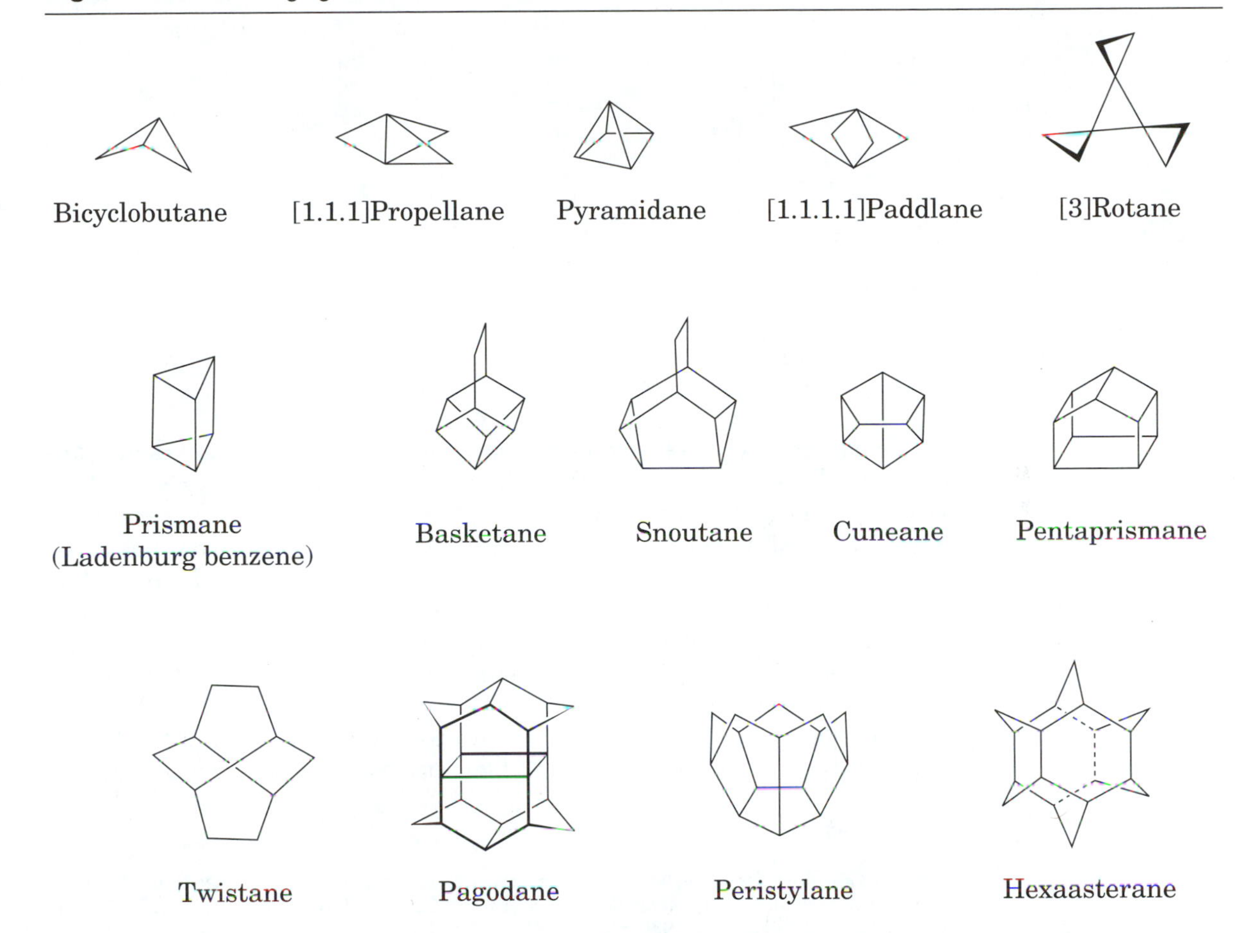

The compounds in Figure 3.29 formally have only single carbon-carbon bonds. Even more strain can be introduced with structures that incorporate double or triple bonds if the alkene unit is nonplanar or the alkyne function is nonlinear. Gano, Lenoir and co-workers have studied the isomers **22**(**a**) and **22**(**b**). Both are distorted because of steric strain associated with the bulky alkyl substituents about the carbon-carbon double bond. Compound **22**(**a**) is stable, and an X-ray analysis indicated that the double bond is twisted from planarity by 36.7°. However, **22**(**b**) is even more strained and is unstable, isomerizing with an activation energy of only about 21 kcal/mol. Molecular mechanics calculations suggest that the substituents on **22**(**b**) are twisted 73°.[173]

[173]Gano, J. E.; Park, B. S.; Pinkerton, A. A.; Lenoir, D. *J. Org. Chem.* **1990**, *55*, 2688.

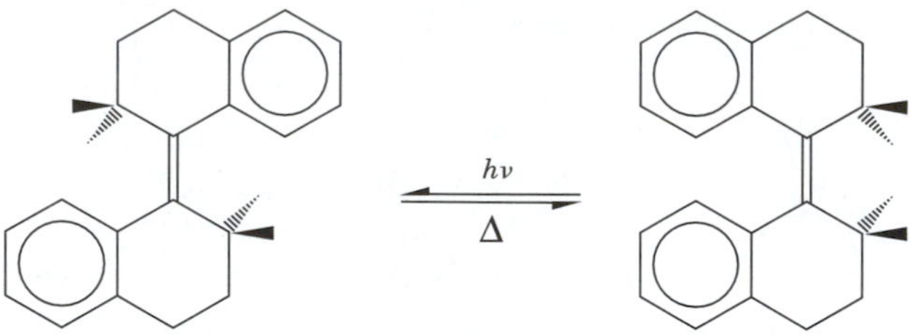

hν

Δ

(a) (b)

22

Cycloalkenes with a trans double bond must show significant deviations from normal bond angles when the rings become small.[174] *trans*-Cyclooctene has been isolated,[175] and *trans*-cycloheptene and *trans*-cyclohexene have been implicated as reaction intermediates.[176,177,178,179] Cycloalkynes must also be highly strained if the ring does not allow the acetylenic carbon atoms to have 180° bond angles. A cycloalkyne with eight or more carbon atoms in the ring is generally large enough to be stable.[180] The smallest stable cycloalkyne reported is cyclooctyne,[181] but smaller ring cycloalkynes have been proposed as unstable species and as reactive intermediates in chemical reactions.[182,183,184] Rings made up only of alternating triple and single bonds have been considered.[185] Isomeric with cycloalkynes are 1,2-cy-

---

[174]For a review of strained cycloalkenes, cycloalkynes and cyclocumulenes, see Johnson, R. P. in *Molecular Structure and Energetics*, Vol. 3; Liebman, J. F.; Greenberg, A., eds.; VCH Publishers: Deerfield Beach, FL, 1986; pp. 85–140.

[175]Cope, A. C.; Pike, R. A.; Spencer, C. F. *J. Am. Chem. Soc.* **1953**, *75*, 3212; Turner, R. B.; Meador, W. R. *J. Am. Chem. Soc.* **1957**, *79*, 4133.

[176]Corey, E. J.; Carey, F. A.; Winter, R. A. E. *J. Am. Chem. Soc.* **1965**, *87*, 934.

[177]*trans*-Cycloheptene was prepared photochemically at −78°: Inoue, Y.; Ueoka, T.; Kuroda, T.; Hakushi, T. *J. Chem. Soc., Perkin Trans. 2* **1983**, 983.

[178]Kropp, P. J. *Mol. Photochem.* **1978–79**, *9*, 39.

[179]For a theoretical study of *trans*-cyclohexene, see Verbeek, J.; van Lenthe, J. H.; Timmermans, P. J. J. A.; Mackor, A.; Budzelaar, P. H. M. *J. Org. Chem.* **1987**, *52*, 2955.

[180](a) Brandsma, L.; Verkruijsse, H. D. *Synthesis of Acetylenes, Allenes and Cumulenes: a Laboratory Manual;* Elsevier: Amsterdam, 1981; (b) For a discussion of triple bonds in small rings, see Sander, W. *Angew. Chem., Int. Ed. Engl.* **1994**, *33*, 1455.

[181]Blomquist, A. T.; Liu, L. H. *J. Am. Chem. Soc.* **1953**, *75*, 2153.

[182]Erickson, K. L.; Wolinsky, J. *J. Am. Chem. Soc.* **1965**, *87*, 1142.

[183]Tseng, J.; McKee, M. L.; Shevlin, P. B. *J. Am. Chem. Soc.* **1987**, *109*, 5474.

[184]Krebs, A.; Kimling, H. *Angew. Chem.* **1971**, *83*, 540, Sander, W. *Angew. Chem., Int. Ed. Engl.* **1994**, *33*, 1455.

[185]For an overview, see Diederich, F.; Rubin, Y. *Angew. Chem., Int. Ed. Engl.* **1992**, *31*, 1101.

cloalkadienes or cyclocumulenes. Cyclocumulenes have also been proposed, and 1,2,5-cyclononatriene has been reported.[186]

Even structures with only one double bond can be highly strained if the double bond occurs at a bridgehead carbon atom.[187] This structural feature was originally noted by Bredt,[188] and the idea that such compounds are incapable of existence has become known as Bredt's rule. A formal statement of the rule has been given by Fawcett:[189]

> In polycyclic systems having atomic bridges, the existence of a compound having a carbon-carbon or carbon-nitrogen double bond at a bridgehead position is not possible, except when the rings are large, because of the strain which would be introduced in its formation by the distortion of bond angles and/or distances. As a corollary, reactions which should lead to such compounds will be hindered or will give products having other structures.

Violations of the rule are possible,[190] and structures such as **23** are known as anti-Bredt compounds. Warner has reviewed the wide variety of structures with strained bridgehead double bonds that have been studied.[191] An intriguing example is the highly strained 1,2-dehydrocubane (**24**).[192]

Bridgehead CO_2R O

23 **24**

Alkenes with unusual arrangements of double bonds offer possibilities to test theories of molecular orbital effects on molecular energy.[193] Among those that have been studied are [4]radialene (**25**),[194] barrelene (**26**),[195] and

---

[186]Baird, M. S.; Reese, C. B. *Tetrahedron* **1976**, *32*, 2153.

[187]A bridgehead carbon atom is a carbon atom that is part of two different ring systems.

[188]Bredt, J.; Thouet, H.; Schmitz, J. *Liebigs Ann. Chem.* **1924**, *437*, 1.

[189]Fawcett, F. S. *Chem. Rev.* **1950**, *47*, 219.

[190]Prelog, V. *J. Chem. Soc.* **1950**, 420.

[191]Warner, P. M. *Chem. Rev.* **1989**, *89*, 1067.

[192]Eaton, P. E.; Maggini, M. *J. Am. Chem. Soc.* **1988**, *110*, 7230. Also see Hrovat, D. A.; Borden, W. T. *J. Am. Chem. Soc.* **1988**, *110*, 4710.

[193]For a review of pyramidalized alkenes, see Borden, W. T. *Chem. Rev.* **1989**, *89*, 1095.

[194]Griffin, G. W.; Peterson, L. I. *J. Am. Chem. Soc.* **1963**, *85*, 2268. See also Hopf, H.; Maas, G. *Angew. Chem., Int. Ed. Engl.* **1992**, *31*, 931.

[195]Zimmerman, H. E.; Paufler, R. M. *J. Am. Chem. Soc.* **1960**, *82*, 1514.

Dewar benzene (**27**).[196] Dewar benzene and benzvalene (**28**)[197] are valence isomers of benzene because they have the formula $(CH)_6$. Another isomer is fulvene (**29**), which is formed photochemically from benzene.[198]

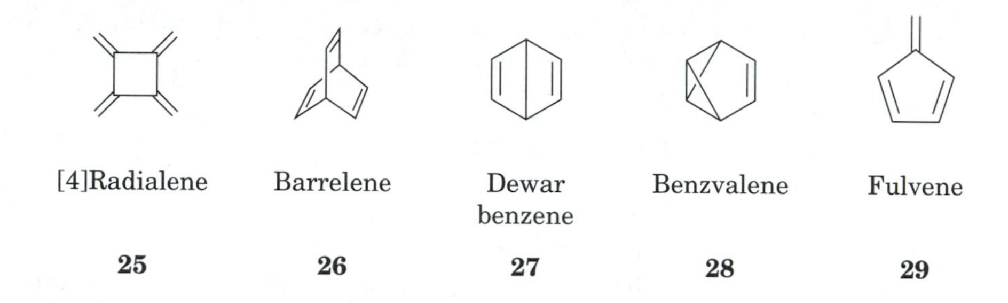

[4]Radialene **25** — Barrelene **26** — Dewar benzene **27** — Benzvalene **28** — Fulvene **29**

A related family of compounds are the isoaromatics, such as 1,2,4-cyclohexatriene, an isobenzene (**30**).[199]

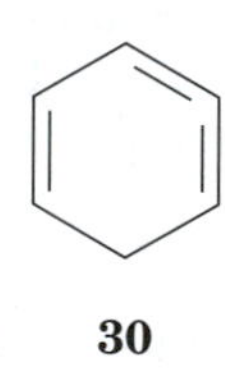

30

Arynes are structures having an additional bond in an aromatic ring. The classic aryne is benzyne (**31**),[200] which is thought to be an intermediate in some substitution reactions.[201] Pyridyne (**32**) has also been detected.[202]

31 **32**

[196]van Tamelen, E. E.; Pappas, S. P. *J. Am. Chem. Soc.* **1963**, *85*, 3297.

[197]Wilzbach, K. E.; Ritscher, J. S.; Kaplan, L. *J. Am. Chem. Soc.* **1967**, *89*, 1031.

[198]The heats of formation of the valence isomers of benzene were reported by Schulman, J. M.; Disch, R. L. *J. Am. Chem. Soc.* **1985**, *107*, 5059. The values for Dewar benzene, benzvalene, prismane and 3,3′-bicyclopropenyl were 94.0, 90.2, 136.4 and 137.6 kcal/mol, respectively.

[199]Christl, M.; Braun, M.; Müller, G. *Angew. Chem., Int. Ed. Engl.* **1992**, *31*, 473.

[200]The structure shown is that of 1,2-dehydrobenzene, more commonly known as *o*-benzyne. The 1,3-dehydro- and 1,4-dehydrobenzenes are known as *m*-benzyne and *p*-benzyne, respectively. The heats of formation of *o*-, *m*-, and *p*-benzyne have been determined to be 106 $\pm$ 3, 116 $\pm$ 3, and 128 $\pm$ 3 kcal/mol, respectively: Wenthold, P. G.; Paulino, J. A.; Squires, R. R. *J. Am. Chem. Soc.* **1991**, *113*, 7414.

[201]Benzyne chemistry will be discussed in Chapter 8.

[202]Nam, H.-H.; Leroi, G. E. *J. Am. Chem. Soc.* **1988**, *110*, 4096.

Cyclophanes are aromatic rings joined at two positions to form a cyclic structure.[203] The smallest known is [6]paracyclophane (**33**), which is stable at room temperature. The carbon atoms attached to the alkyl chain are bent about 20° out of the plane defined by the other four carbon atoms.[204] Another way to deform aromatic rings is to tie them together in rings. For example, cycloanthracene (**34**) has been proposed.[205]

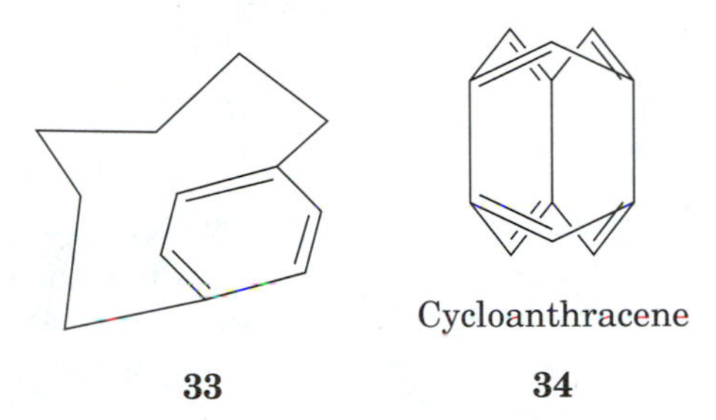

33 Cycloanthracene **34**

Just as analytical chemists seem able to detect ever smaller quantities, so organic chemists seem able to synthesize—or at least to calculate the hypothetical properties of—ever more strained species. Michl noted that

> The concept of strain has fascinated organic chemists for about a century, and the interest shows no sign of abatement; if anything it is growing. Over this period of time, our attitudes have changed dramatically. Today, we accept the remarkable stability of [1.1.1]propellane and tetra-*tert*-butyltetrahedrane casually. Two decades ago, a student who would draw such "impossible" structures during an examination would have surely failed. We wonder what other marvels lie in store.[206]

Problems

1. Use the principles of conformational analysis to quantitatively predict the difference in heats of combustion of *cis*- and *trans*-1,3-dimethylcyclohexane. How does your calculated value compare with the data in Table 3.3?
2. Rationalize the observation that only the more stable of the two isomers of 1,2-dimethylcyclohexane is capable of optical activity, but only the less stable isomer of 1,3-dimethylcyclohexane is capable of optical activity.

---

[203]This definition only hints at the structures of cyclophanes that are possible. For further reading, see Keehn, P. M.; Rosenfeld, S. M., eds., *Cyclophanes*, Vols. I and II; Academic Press: New York, 1983. For a discussion of "super" phanes, see Gleiter, R.; Kratz, D. *Acc. Chem. Res.* **1993**, *26*, 311.

[204]Tobe, Y.; Takemura, A.; Jimbo, M.; Takahashi, T.; Kobiro, K.; Kakiuchi, K. *J. Am. Chem. Soc.* **1992**, *114*, 3479. (See reference 1 in that paper.)

[205]For a theoretical study and leading references, see Haase, M. A.; Zoellner, R. W. *J. Org. Chem.* **1992**, *57*, 1031.

[206]Michl, J. *Chem. Rev.* **1989**, *89*, 973.

3. We expect larger substituents on cyclohexane rings to have greater equatorial preferences than those of smaller substituents. However, the increase in equatorial preference is not linear with increasing size of alkyl substituents. Rationalize the following data.

Substituent	**Preference (kcal/mol)**
Methyl	1.74
Ethyl	1.79
Isopropyl	2.21
t-Butyl	>5.4

4. At low temperature (202 K), the conformational equilibrium of *cis*-1-benzyl-4-methylcyclohexane favors the (chair) conformation with the benzyl group axial by 0.08 kcal/mol. However, at room temperature or higher, the equilibrium favors the (chair) conformation with benzyl group equatorial by 0.04 kcal/mol. Explain this result.
5. Suggest explanations for the following observations concerning equatorial preferences (Table 3.2):
   a. The *A* values of the OH and NH_2 groups are different in protic and aprotic solvents.
   b. The *A* value of the CO_2^- group is greater than that of the CO_2H group.
   c. The *A* values of the halogens do not increase in the order $F < Cl < Br < I$.
6. The interconversion of the rotamers of the 2′,6′-dialkyl-2-halo-*N*-methylacetanilides **8** and **9** (page 124) is slower in polar and hydrogen-bonding solvents than in nonpolar solvents. Rationalize this result.
7. The gas phase molecular structure of 2-fluoroethanol has been studied by electron diffraction and microwave spectroscopy at temperatures of 20°, 156°, and 240°. At the two lower temperatures only the gauche conformation was seen; at 240°, the anti conformer was found to comprise 9.8% of the sample, while the gauche conformer comprised 90.2% of the sample.
   a. Why is the anti conformer not the predominant conformer of 2-fluoroethanol?
   b. Calculate the difference in free energy (ΔG) between the gauche and anti conformers at 240°.
   c. The calculated vibrational and rotational contributions to the entropies of each conformer were found to be very nearly equal. However, there are two gauche conformers (in each of which the OH group does not rotate because of hydrogen bonding), while there is one anti conformer (in which there are three potential energy minima for rotation of the OH group about the carbon-oxygen bond). Therefore, the ΔS^0 for conversion of the anti to the gauche conformer was determined to be $R(\ln 3 - \ln 2) = 0.81$ cal K^{-1} mol^{-1}. What is the ΔH^0 for conversion of the gauche conformer to the anti conformer?
8. Using the data in Table 1.7 (page 15), calculate ΔH_f^0 for the following compounds with molecular formula C_7H_{16}: *n*-heptane, 2-methylhexane, 3-methylhexane, 2,2-dimethylpentane, 2,3-dimethylpentane, and 3,3-dimethylpentane. Remember to include butane gauche interactions in your calculations. Arrange the isomers in order of decreasing calculated ΔH_f^0. Can you rationalize any apparent trends in this order? How do your values compare with literature values?

9. Use *Chemical Abstracts* to find one or more recent articles about at least one of the structures listed below. (Some of them have been synthesized; some are still unknown.) Briefly summarize what the article says about the compound or its attempted synthesis. What types of strain would you expect to see in each?

fenestrane	twistane	cubane
basketane	tetrahedrane	tricyclo[2.1.0.0^{1,3}]pentane

10. Tetrahedrane has not been synthesized, but the compound tetra-*t*-butyltetrahedrane has been prepared. Suggest reasons why the substituted compound has been made but the parent compound has not.

11. The heats of hydrogenation of 2,4,4-trimethyl-1-pentene and of 2,4,4-trimethyl-2-pentene are −25.5 kcal/mol and −26.8 kcal/mol, respectively. Which isomer is more stable? Is the result consistent with the generalization that the more stable alkene isomer is the one with the greater number of alkyl substituents on the carbon-carbon double bond? If not, explain why the generalization does not offer the correct prediction in the case of these two compounds.

12. Classify each of the following conformations according to the nomenclature shown in Figure 3.3.

a. CH_3, H, H, C, H

b. CH_3, H, CH_3, C, H

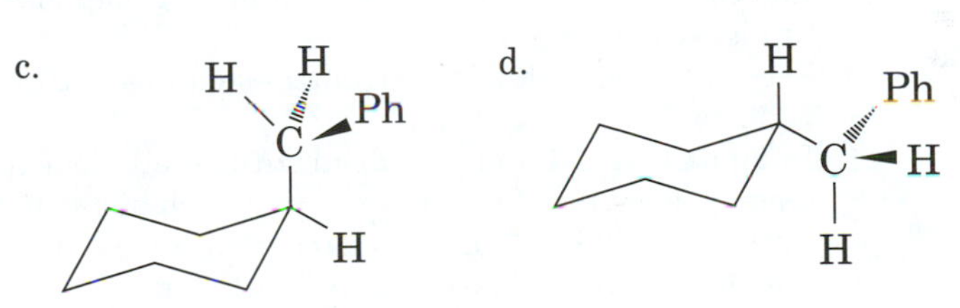

13. Predict whether *cis*- (**35**) or *trans*-decalin (**36**) should be more stable, and estimate the energy difference between them.

cis-decalin

35

trans-decalin

36

14. Use the principles of conformational analysis to predict the relative stability of *trans-anti-trans*-perhydrophenanthrene (**37**) and *trans-syn-trans*-perhydrophenanthrene (**38**).

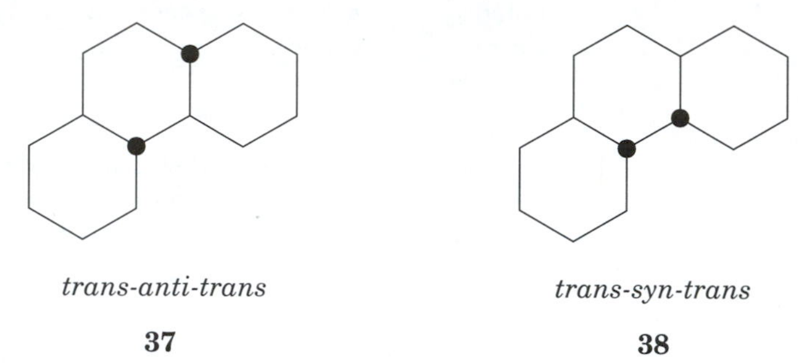

trans-anti-trans

37

trans-syn-trans

38

15. The A value for the acetoxyl ($—O_2CCH_3$) group at −88° has been determined to be 0.71 kcal/mol. Calculate the fraction of equatorial and axial conformers of cyclohexyl acetate at that temperature.
16. At −80° the ratio of equatorial to axial cyclohexyl isocyanate was determined by NMR to be 3.74 : 1.0. Determine the equatorial preference (A value) for the NCO group.
17. At 40 K, the equatorial preference for alkyl substituents on cyclohexane is methyl > ethyl > isopropyl, which is opposite the trend observed at room temperature. Explain this result.
18. For the conformational change of the vinyl group from the axial position to the equatorial position in *cis*-1-methyl-4-vinylcyclohexane, ΔG was +0.06 kcal/mol. What is the A value for the vinyl group?
19. Identify the local maxima, local minima, global maximum and global minimum in Figure 3.9, page 132.
20. Predict the major kinds of strain present in any three of the structures in Figure 3.29, page 169.
21. In dilute CCl_4 solution, 1,2-disubstituted ethylene glycols (RCHOH — CHOHR) usually show two different O — H stretch peaks in the infrared spectrum. One peak near 3635 cm^{-1} is associated with the free OH stretch, while a peak in the region of 3585 cm^{-1} is associated with hydrogen-bonded OH stretch. The intensity of the OH stretch varies with the stereochemistry of the diol and the nature of the R groups, however. Hydrogen bonded OH stretch is seen for the *racemic* diols when R = methyl, isopropyl, or *t*-butyl. For the corresponding *meso* diols, however, the signal is strong for R = methyl, weak for R = isopropyl, and undetectable for R = *t*-butyl. Explain this result.
22. Table 3.8 shows the results of two different molecular mechanics calculations on cyclobutane.[207] In one calculation, the C1 — C2 — C3 — C4 dihedral angle was fixed at 0°, but all other parameters were optimized to obtain the lowest steric energy for the planar conformation. In the second calculation, that dihedral angle was also allowed to vary in the geometry optimization, and the final geometry of the cyclobutane ring was puckered. The results of one calculation are shown below the heading Conformation A, and the results of the other calculation are shown below the heading Conformation B.

[207]The calculations were carried out with the program PCMODEL (Serena Software).

Table 3.8 Results of molecular mechanics calculations on planar and puckered conformations of cyclobutane.

Steric Energy Component	Cyclobutane Conformation A	Cyclobutane Conformation B
Stretch	0.68 kcal/mol	0.80 kcal/mol
Bend	13.47	16.24
Stretch-bend	−0.78	−0.94
Torsional	14.82	11.03
van der Waals	1.95	2.08
Total steric energy	30.14	29.22

a. Which conformer is puckered cyclobutane and which is planar cyclobutane?
b. Rationalize the relative magnitudes of stretch, bend, torsional, and van der Waals energies for the two conformations.

(Note: it may be helpful to use molecular models.)

23. A molecular mechanics calculation was used to identify two conformers of bicyclo[3.2.1]-octane.[207] The components of steric energy for each are listed in Table 3.9.
    a. Which conformer is lower in steric energy?
    b. Using a set of molecular models and the energy parameters given in Table 3.9, suggest the most likely geometry for each of the two conformers.
    c. Can you relate the overall energy difference between the two conformers to the energy difference between conformers of a model monocyclic compound?

Table 3.9 Steric energy parameters for two conformers of bicyclo[3.2.1]octane.

Steric Energy Component	Bicyclo[3.2.1]octane Conformation A	Bicyclo[3.2.1]octane Conformation B
Stretch	0.76 kcal/mol	0.86 kcal/mol
Bend	3.62	4.38
Stretch-bend	−0.05	−0.02
Torsional	9.70	12.55
van der Waals	5.24	6.97

CHAPTER 4

Applications of Molecular Orbital Theory and Valence Bond Theory

4.1 Introduction to Molecular Orbital Theory

Hückel Molecular Orbital Theory

The discussions of stereochemistry and molecular mechanics in Chapters 2 and 3 were implicitly based on a localized bond model of chemical structures. In this chapter we will discuss applications of molecular orbital theory in organic chemistry. The approach we will consider in greatest detail is Hückel MO (HMO) theory, a relatively simple approach that yields useful insights into structures and properties of organic compounds.[1,2] Later we will discuss some more advanced molecular orbital calculations.

The fundamental assumption of HMO theory is that we may calculate molecular orbitals through a process known as **LCAO-MO**: **M**olecular **O**rbitals produced by the **L**inear **C**ombination of **A**tomic **O**rbitals.[3,4,5] That is, we use some combination of the wave functions of the atomic p orbitals to produce a set of molecular π orbitals. For a set of n parallel p orbitals, for example, the molecular orbitals have the form shown in equation 4.1. In this equation ϕ_n is the wave function for a p orbital on the nth carbon atom and c_n is a constant.

---

[1] Hückel, E. *Grundzüge der Theorie ungesättigter und aromatischer Verbindungen*; Verlag Chemie, G.m.b.H.: Berlin, 1938, and references therein.

[2] Sources for further reading include references cited below as well as (a) Flurry, Jr., R. L. *Molecular Orbital Theories of Bonding in Organic Molecules*; Marcel Dekker, Inc.: New York, 1968; (b) Salem, L. *Electrons in Chemical Reactions: First Principles*; John Wiley & Sons: New York, 1982; (c) Coulson, C. A.; Streitwieser, Jr., A. *Dictionary of π-Electron Calculations*; W. H. Freeman and Co.: San Francisco, 1965; see especially reference 9; (d) Wiberg, K. B. *Physical Organic Chemistry*; John Wiley & Sons, Inc.: New York, 1964; (e) Zimmerman, H. E. *Quantum Mechanics for Organic Chemists*; Academic Press: New York, 1975.

[3] Coulson, C. A. *Quart. Rev. Chem. Soc.* **1947**, *1*, 144.

[4] Coulson, C. A. *J. Chem. Soc.* **1955**, 2069.

[5] Pople, J. A. *Acc. Chem. Res.* **1970**, *3*, 217.

$$\psi_i = c_1\phi_1 + c_2\phi_2 + c_3\phi_3 + \ldots + c_n\phi_n = \sum_{\mu=1}^{n} c_{i\mu}\,\phi_\mu \tag{4.1}$$

In order to illustrate the nature and some of the limits of the HMO procedure, we will begin with the simplest possible π molecular system, that of ethene. As shown in Figure 4.1, the HMO model uses the σ,π formulation for carbon-carbon double bonds.[6] We assume that two carbon atoms are close enough to each other that a carbon-carbon σ bond can be formed by overlap of an sp^2 hybrid orbital of each. The other two sp^2 orbitals on each carbon atom are used to form C—H bonds.[7] Each carbon atom has a remaining p orbital that is perpendicular to the plane defined by the sp^2 orbitals. It is the interaction of the two p orbitals that will produce the π molecular orbitals. Rewriting equation 4.1 for the specific case of two p orbitals produces equation 4.2.

$$\psi_\pi = c_1\phi_1 + c_2\phi_2 \tag{4.2}$$

Now we solve for ψ_π by solving the Schrödinger equation, $\mathcal{H}\psi = E\psi$, and calculate the energies of the HMOs produced.[8] Details of the procedure have been given by Roberts,[9] so only an outline is provided here. Multiplying both sides of $\mathcal{H}\psi = E\psi$ by ψ (or ψ^* as is appropriate), dividing by ψ^2, and then integrating over all space in both the numerator and denominator gives equation 4.3.

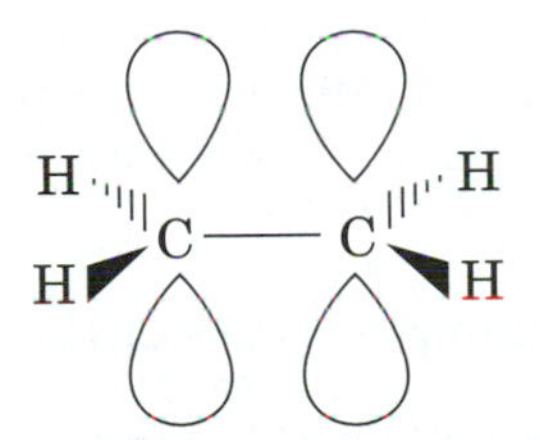

Figure 4.1 Relationship of *p* orbitals assumed in an HMO calculation for ethene.

[6]The popularity of HMO theory has helped make the σ,π formulation (rather than the bent bond formulation) become the standard pictorial representation of double bonds. The applicability of the bent bond model to benzene has been discussed by Schultz, P. A.; Messmer, R. P. *J. Am. Chem. Soc.* **1993**, *115*, 10943.

[7]Our model is one in which the set of π molecular orbitals is superimposed on a skeleton composed of σ valence bonds. In HMO theory we assume that the σ and π systems may be treated separately and that the σ bonds and the π bonds do not interact. It has been said that this is "an assumption that is often no more than a convenient fiction . . .": Coulson, C. A.; Stewart, E. T. in *The Chemistry of the Alkenes*, Vol. I.; Patai, S., ed.; Wiley-Interscience: London, 1964; p. 106. For example, the epr spectrum of the benzene radical anion shows splitting from all six of the protons: Tuttle, Jr., T. R.; Weissman, S. I. *J. Am. Chem. Soc.* **1958**, *80*, 5342. Nonetheless, the approximation made here is sufficiently accurate for our purposes.

[8]An alternative method equation is based on density-functional theory, an approach that is finding increasing use in theoretical chemistry. We will not discuss this method, but an introduction may be found in Parr, R. G.; Yang, W. *Density-Functional Theory of Atoms and Molecules*, Oxford University Press: New York, 1989.

[9]Roberts, J. D. *Molecular Orbital Calculations*; W. A. Benjamin, Inc.: New York, 1961.

$$E = \frac{\int \psi^* \mathcal{H} \psi \, d\tau}{\int \psi^2 \, d\tau} \tag{4.3}$$

We can use equation 4.3 to solve for ψ with a technique known as the **variation principle**. In essence, we assume that any wave function that we write down will *underestimate* the stability of a molecule. That was what we saw in the case of the MO and VB calculations of molecular hydrogen in Chapter 1, but must it always be so? A number of texts offer proofs of the variation principle,[10,11] but an intuitive approach will be sufficient here. A stable molecule exists in the energetically most favorable arrangement of electrons and nuclei possible. The best we can hope to do is describe that arrangement. If we err, it will be on the side of a higher energy.[12,13] A mathematical statement of the variation principle is shown in equation 4.4.

$$E_{\text{calc}\ (\psi\ \text{proposed})} \geq E_{\text{calc}\ (\psi\ \text{correct})} \tag{4.4}$$

In HMO theory we assume that we know the $\mathcal{H}$ for the system and that we know all but the coefficients c_1 and c_2 of ψ. Therefore if we minimize the energy of the wave function with respect to c_1 and c_2 (i.e., take the partial derivatives of E with respect to c_1 and with respect to c_2), we will have obtained the best estimate of the energy of the system that is possible with our theoretical model.

Expanding equation 4.3 by substituting $c_1\phi_1 + c_2\phi_2$ for ψ_π and then multiplying all the terms produces a complex equation that can be simplified by making the following substitutions: First we note that the Hamiltonians are Hermitian (equation 4.5).[14] Then we make the definitions shown in equations 4.6 through 4.9.

$\int \phi_1 \mathcal{H} \phi_2 \, d\tau = \int \phi_2 \mathcal{H} \phi_1 \, d\tau$	Hermitian properties of Hamiltonian	(4.5)
$\int \phi_1 \mathcal{H} \phi_1 \, d\tau \equiv H_{11}$	Coulomb integral (giving the energy of an electron in an isolated p orbital)	(4.6)
$\int \phi_1 \mathcal{H} \phi_2 \, d\tau \equiv H_{12}$	Resonance integral	(4.7)

[10]For example, see reference 2(d), pp. 42 *ff.*

[11]Eyring, H.; Walter, J.; Kimball, G. E. *Quantum Chemistry*; John Wiley & Sons: New York, 1944; p. 99.

[12]Reed and Murphy have pointed out that applying the variation principle to obtain the best energy for a structure is often assumed to also produce the best geometry and other observable properties, but that this is not necessarily the case. For an elaboration of this point, see Reed, L. H.; Murphy, A. R. *J. Chem. Educ.* **1986**, *63*, 757 and references therein.

[13]This argument covers only electron distribution or slight bond length or angle changes within the same structure, not isomerization to produce a different molecular structure.

[14]Pilar, F. L. *Elementary Quantum Chemistry*; McGraw-Hill Book Company: New York, 1968; p. 71.

$\int \phi_1\phi_1 \, d\tau \equiv S_{11}$ Normalization integral for identical atoms[15] **(4.8)**

$\int \phi_1\phi_2 \, d\tau \equiv S_{12}$ Overlap integral for adjacent atoms **(4.9)**

Now we take $\partial E/\partial c_1 \equiv 0$ to determine the lowest energy that can be obtained by variation of the coefficients c_1 and c_2. Making further substitutions and rearranging the terms leads to equation 4.10.

$$c_1(H_{11} - ES_{11}) + c_2(H_{12} - ES_{12}) = 0 \tag{4.10}$$

Similarly, taking $\partial E/\partial c_2 \equiv 0$ produces equation 4.11.

$$c_1(H_{12} - ES_{12}) + c_2(H_{22} - ES_{22}) = 0 \tag{4.11}$$

In these equations it is the coefficients c_1 and c_2 that we do not know—we presume that we know H_{11}, H_{12}, S_{11}, and S_{12}. Thus we have two equations (equation 4.10 and equation 4.11) in two unknowns (c_1 and c_2). In order for there to be a solution for the two unknowns (other than the trivial one, $c_1 = c_2 = 0$), it must be true that the secular determinant of the coefficients *of the unknowns* equals 0. The coefficients of the unknowns are the terms in parentheses in equation 4.10 and equation 4.11. Thus we must have equation 4.12.

$$\begin{vmatrix} H_{11} - ES_{11} & H_{12} - ES_{12} \\ H_{21} - ES_{21} & H_{22} - ES_{22} \end{vmatrix} = 0 \tag{4.12}$$

For the more general HMO calculation (equation 4.2), the corresponding determinant can be shown to be equation 4.13, where k is the number of p orbitals combined.

$$\begin{vmatrix} H_{11} - ES_{11} & \ldots & H_{1k} - ES_{1k} \\ \ldots & \ldots & \ldots \\ H_{k1} - ES_{k1} & \ldots & H_{kk} - ES_{kk} \end{vmatrix} = 0 \tag{4.13}$$

Before trying to solve the determinant in equation 4.12, we need to make some simplifying assumptions and approximations. The **overlap integral**, S, is a measure of the degree to which two orbitals include the same volume of space. It is not too difficult to see the rationale for assuming that $S_{ii} \equiv 1$, since it is a property of *normalized* atomic orbitals that

$$\int \phi_i \phi_i \, d\tau = 1 \tag{4.14}$$

However, what is the value of $\int \phi_i \phi_j \, d\tau$ if $\phi_i \neq \phi_j$? The overlap between two atomic orbitals on different atoms varies according to the distance between the two atomic nuclei and (for all except spherically symmetric s orbitals) with the orientation of the two orbitals in space. A calculation by Mulliken determined that the value of S_{ij} for two adjacent parallel p orbitals in

[15] The value of this integral is unity if the wave functions are normalized.

ethene (1.34 Å apart) is 0.27, while that expected for two adjacent p orbitals in benzene (1.39 Å) is about 0.21.[16] In general, we will not know the distance between p orbitals before starting a calculation. More important, trying to assign a value of S_{12} would make the computation much more complex. Therefore, we make the obviously incorrect but very convenient assumption that $S_{ij} \equiv 0$ unless $\phi_i = \phi_j$.

H_{11} represents the energy of an electron in an isolated p orbital and is given the symbol α. Our reference system is an empty p orbital and an electron at infinite separation. If we define the energy of that reference system as 0, then as the electron and carbon atoms approach each other, the energy of the system is reduced. Therefore, α is a negative number, but we will find it convenient to draw energy level diagrams in which the value of α is taken as the reference point.

A more difficult problem is that of H_{ij}, where $i \neq j$. There are two situations to consider. If atoms i and j are σ-bonded to each other, H_{ij} represents the extra stability of an electron brought in from infinity and placed in the field of both nuclei. An intuitive model is that it is the "strength of the π-electron bond between these atoms."[17] Therefore H_{ij} is also a negative number. Since we do not have a numerical value for it now, we use the symbol β for the value of H_{ij} when i and j are σ-bonded to each other.

Determining the value of H_{ij} when atoms i and j are not σ-bonded to each other is more difficult. In general atoms that are not σ-bonded are further apart than those that are bonded, and we would expect H_{ij} to be in the range $0 - \beta$, depending on the distance separating the two atoms and on the spatial relationship of the p orbitals involved. Here we run into the same problem we faced in evaluating the value of the S_{ij}. Not only is it difficult to determine what value we should use, but also it will complicate our analysis if we use anything other than 0. Hence, we define $H_{ij} \equiv 0$ for $i \neq j$.[18] Gutman has noted that "It would be completely outdated to search for some physical justification of the above approximations," which were introduced before computers became available.[19] Nevertheless, the conceptual simplicity of the HMO method allows us to visualize some of the significant results of HMO calculations.

Having made these assumptions, we can now rewrite the determinant in equation 4.12 as equation 4.15.

$$\begin{vmatrix} \alpha - E & \beta \\ \beta & \alpha - E \end{vmatrix} = 0 \tag{4.15}$$

[16]Mulliken, R. S. *J. Am. Chem. Soc.* **1950**, *72*, 4493.

[17]Brown, R. D. *Quart. Rev. Chem. Soc.* **1952**, *6*, 63.

[18]Reference 9, p. 33.

[19]Gutman, I. *Top. Curr. Chem.* **1992**, *162*, 29.

Dividing through by β and representing $(\alpha - E)/\beta$ as X produces the determinant in equation 4.16.

$$\begin{vmatrix} X & 1 \\ 1 & X \end{vmatrix} = 0 \tag{4.16}$$

Solving the determinant by cross multiplication[20] gives equation 4.17.

$$X^2 - 1 = 0 \tag{4.17}$$

The solutions to this equation are $X = 1$ and $X = -1$. Therefore, $E = \alpha + \beta$ and $E = \alpha - \beta$ are the two solutions to our problem of two equations in two unknowns, equation 4.10 and equation 4.11.[21] However, in those equations the unknowns were the coefficients, c_1 and c_2. What we have determined are the energy levels of the two molecular orbitals created by taking the linear combination of two atomic orbitals on the two carbon atoms. Usually it is the energies of the molecular orbitals that we want to know, even though the procedure is formally designed to find the coefficients that determine the molecular orbitals. Furthermore, we know the energies of these two molecular orbitals. One has energy $\alpha + \beta$. Since α and β are both negative numbers, this orbital represents a very stable orbital. The other orbital has energy $\alpha - \beta$. This orbital is less stable, and in fact an electron in this orbital is less stable than an electron in an isolated p orbital (with energy

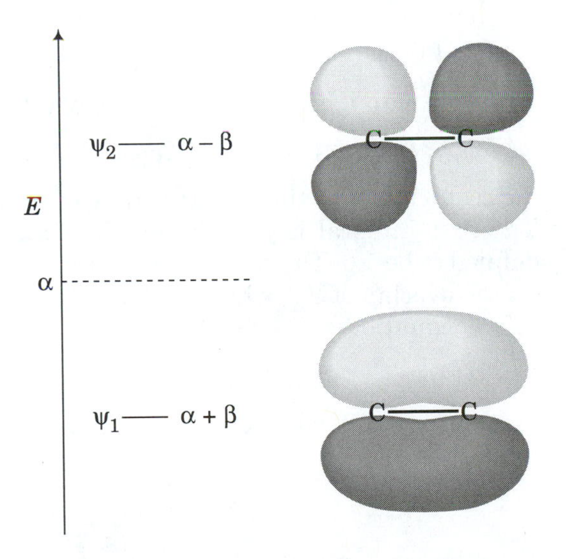

Figure 4.2 HMO terms for ethene.

[20]For this and other ways of solving determinants, see (a) reference 2(e), Chs. 1–3; (b) reference 2(d), pp. 442–458.

[21]As in the case of molecular hydrogen, combining two atomic orbitals produces two molecular orbitals, one bonding and one antibonding. In general, combining n atomic orbitals will always produce n molecular orbitals.

equal to α). A schematic representation of the orbital energy levels is shown in Figure 4.2.

To calculate the coefficients c_1 and c_2, we choose one of the energy levels and substitute that value for E into both equation 4.10 and equation 4.11. Let us choose first the lower energy molecular orbital, $E = \alpha + \beta$. As shown in equation 4.18 and equation 4.19, we now generate two new equations in two unknowns, and this time the unknowns really are c_1 and c_2. Solving the two equations for the case $E = \alpha + \beta$ reveals that $c_1 = c_2 = c$.

$$c_1(H_{11} - ES_{11}) + c_2(H_{12} - ES_{12}) = c_1(\alpha - (\alpha + \beta)) + c_2\beta$$

$$= -c_1\beta + c_2\beta = 0 \tag{4.18}$$

$$c_1(H_{12} - ES_{12}) + c_2(H_{22} - ES_{22}) = c_1\beta + c_2(\alpha - (\alpha + \beta))$$

$$= c_1\beta - c_2\beta = 0 \tag{4.19}$$

$$c_1 = c_2 \equiv c \tag{4.20}$$

We know that molecular orbital wave functions must be normalized, just as atomic orbital wave functions must be.[22] Therefore,

$$\int \psi^*\psi \, d\tau = 1 \tag{4.21}$$

which means that the probability of finding one electron in the molecular orbital should be calculated as being exactly 1.0. Substituting for ψ produces equation 4.22

$$\int (c_1\phi_1 + c_2\phi_2)(c_1\phi_1 + c_2\phi_2) \, d\tau \equiv 1 \tag{4.22}$$

which we can multiply through to produce equation 4.23.

$$c_1^2 \int \phi_1\phi_1 \, d\tau + c_2^2 \int \phi_2\phi_2 \, d\tau + 2c_1c_2 \int \phi_1\phi_2 \, d\tau = 1 \tag{4.23}$$

Ignoring the coefficients for the moment, we can evaluate the integrals individually. The first integral is just the overlap integral S_{11}, which we have already defined to be 1.0. The second integral is S_{22} and is also 1.0. The third integral is the overlap of S_{12}, which we have defined as being equal to 0 earlier. Therefore, equation 4.23 becomes

$$c_1^2 + c_2^2 = 1 \tag{4.24}$$

By symmetry, c_1 must equal c_2, so let us define $c_1 = c_2 = c$. Therefore, we are left with the result that

$$2\,c^2 = 1 \tag{4.25}$$

so

$$c = \frac{1}{\sqrt{2}} \tag{4.26}$$

[22] *Cf.* equation 4.14.

(An equally acceptable result is $c = -1/\sqrt{2}$. In that case all coefficients in the following discussion would be the negative of the ones shown, but bonding or antibonding interactions would be unchanged, and the physical picture of the orbital interactions, Figure 4.2, would also not be changed.) Thus, for $E = \alpha + \beta$,

$$\psi_1 = (1/\sqrt{2})(\phi_1 + \phi_2) \tag{4.27}$$

Similarly, solving for the coefficients for $E = \alpha - \beta$ produces

$$\psi_2 = (1/\sqrt{2})(\phi_1 - \phi_2) \tag{4.28}$$

The subscripts on the ψ terms follow the convention that the lowest energy orbital is numbered ψ_1, with higher energy orbitals having higher numbered subscripts.

The MO terms ψ_1 and ψ_2 are the familiar π and π^* orbitals, so ψ_1 may be called π, and ψ_2 may be called π^*. The physical meaning of the wave function we calculated in equation 4.27 is that the *p* orbitals on carbon atoms 1 and 2 combine to produce a molecular orbital with electron density over both carbon atoms. The shading of this wave function shown in Figure 4.2 indicates that it has a positive mathematical sign above the plane of the molecule and a negative mathematical sign of the wave function below the plane of the molecule. The bonding combination is thus formed by taking the combination of two *p* orbitals with the positive sign of the wave function in the same region of space (as indicated by the fact that both c_1 and c_2 are positive numbers in equation 4.27). Hence, ψ_π is a *bonding* orbital.

In equation 4.28, on the other hand, c_1 has a positive sign and c_2 has a negative sign. This means that the mathematical sign is positive for the top lobe of the *p* orbital ϕ_1 but is negative for the top lobe of the *p* orbital ϕ_2. The physical meaning of such an interaction is interference of one *p* orbital with another. Effectively, one orbital cancels the other in the region of space between the two carbon atoms. It also means that in much of the space between the two carbon atoms, the interference produces zero electron density above and below the plane of the molecule, as well as within the molecular plane. In other words, for ψ_2 there is a node[23] perpendicular to the plane of the molecule. Since now the electrons are not likely to be found between the two carbon nuclei, they cannot provide a force of attraction to hold the molecule together. Therefore ψ_2 is an *antibonding orbital*, since electrons in this orbital produce molecular instability.

Now let us consider combining three *p* orbitals to determine the HMOs of an allyl system. It does not matter whether we are interested in the allyl cation, radical, or anion. In HMO theory the molecular orbitals are calculated on the basis of the number of *p* orbitals in the π system, not on how many electrons are in the π system. We presume that we can determine the

---

[23]If one is needed, a useful mnemonic is that node stands for *no de*nsity. In both Ψ_π and $\Psi_\pi{}^*$ there is also a node in the plane of the molecule that arises from node at the carbon atom in a 2*p* orbital.

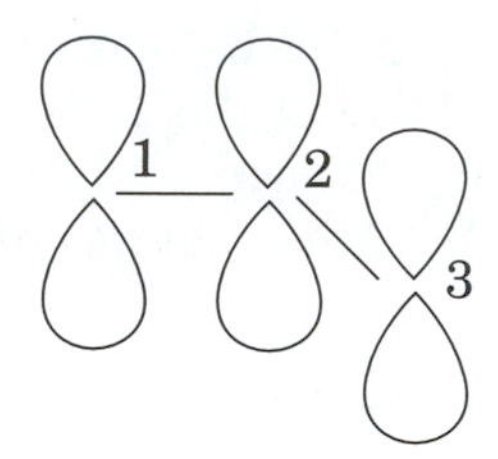

Figure 4.3
Basis set atomic orbitals for the allyl system.

HMOs for the assembled nuclei and then add the electrons later. As shown in Figure 4.3, there are three adjacent parallel *p* orbitals in the allyl system. Combining three *p* orbitals should produce three molecular orbitals, and the general format of each HMO should be

$$\psi = c_1\phi_1 + c_2\phi_2 + c_3\phi_3 \tag{4.29}$$

Fortunately, we don't have to go back to equation 4.1, make substitutions, take partial derivatives, etc. Instead, we proceed directly to the secular determinant, equation 4.13. In this case the determinant is as shown in equation 4.30:

$$\begin{vmatrix} X & 1 & 0 \\ 1 & X & 1 \\ 0 & 1 & X \end{vmatrix} = 0 \tag{4.30}$$

It may be helpful to point out some features of this determinant that will show up in other cases. The *ii*th element of the determinant (the term on the *i*th column and *i*th row, i.e., on the diagonal from top left to bottom right) is X **if** C*i* has a *p* orbital participating in the π bonding. The *ij*th element of the determinant is 1 if C*i* and C*j* are σ-bonded to each other and 0 if they are not. That is a direct result of the assumptions about the overlap integrals and resonance integrals that we made on page 182.

We can now solve the determinant to produce the equation

$$X^3 - 2X = 0 \tag{4.31}$$

Rearranging the terms on the left side of the equation produces

$$X(X - \sqrt{2})(X + \sqrt{2}) = 0 \tag{4.32}$$

So the three solutions are

$$X = \pm\sqrt{2}, 0 \tag{4.33}$$

The three molecular orbitals, then, include a bonding orbital (ψ_1) with energy of $\alpha + 1.414\ \beta$ and an antibonding orbital (ψ_3) with energy $\alpha - 1.414\ \beta$, as shown in Figure 4.4. The solution $X = 0$ produces a molecular orbital (ψ_2) with $E = \alpha$. Therefore, the energy change associated with bringing an electron from a great distance and placing it into ψ_2 is the same as the energy change in bringing an electron from a great distance and placing it in an isolated *p* orbital. Adding an electron to that orbital thus does not decrease the total bonding energy of the π system, nor does it increase it. Hence, ψ_2 is a *nonbonding* molecular orbital.

We can determine the actual coefficients for each of the three molecular orbitals as before. The results are given in equations 4.34–4.36 and are shown schematically in Figure 4.4.[24] We see that the orbitals we produce do have electron density over the entire π system and thus are true molecular orbitals, not orbitals localized between a pair of atoms. Also, it is easy to see why ψ_2 is nonbonding. There is a node *through* C2, so electron density in ψ_2 cannot contribute to bonding between C1 and C2 or between C2 and C3, nor can an electron in that orbital contribute to any antibonding relationships. On the other hand, ψ_1 allows an electron to roam over all three carbon atoms in a bonding relationship. ψ_3 also allows electron movement over the entire molecule, but all interactions are antibonding.

$$\psi_3 = \tfrac{1}{2}\phi_1 - (\sqrt{2}/2)\phi_2 + \tfrac{1}{2}\phi_3 \qquad \textbf{(4.34)}$$

$$\psi_2 = (\sqrt{2}/2)\phi_1 - (\sqrt{2}/2)\phi_3 \qquad \textbf{(4.35)}$$

$$\psi_1 = \tfrac{1}{2}\phi_1 + (\sqrt{2}/2)\phi_2 + \tfrac{1}{2}\phi_3 \qquad \textbf{(4.36)}$$

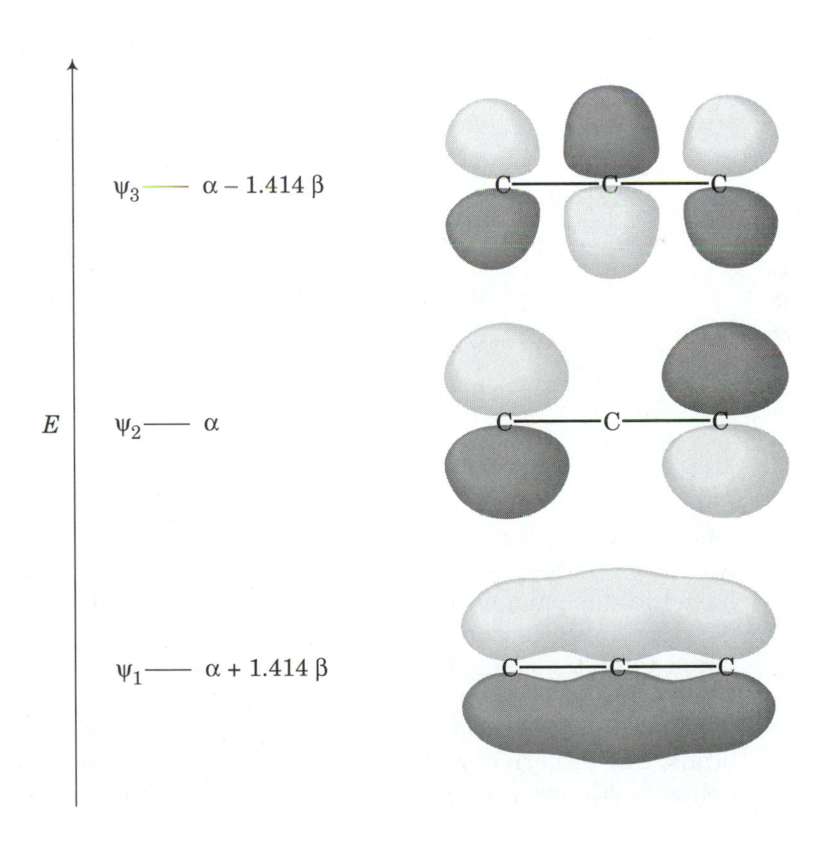

Figure 4.4
Energy levels for allyl molecular orbitals.

---

[24]Notice that there are no nodes in ψ_1, one node in ψ_2, and two nodes in ψ_3.

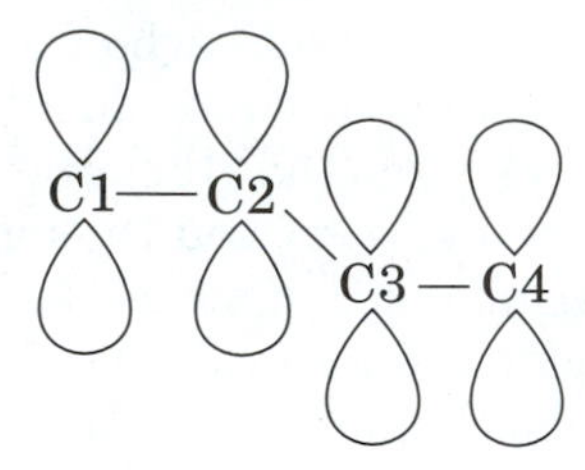

Figure 4.5
The butadiene system.

We will use the HMOs of the allyl system later, but let us first generate the HMOs of some larger systems. 1,3-Butadiene has one more sp^2-hybridized carbon atom than does allyl. If we number the carbon atoms as shown in Figure 4.5, then the secular determinant can be written by inspection as shown in equation 4.37.

$$\begin{vmatrix} X & 1 & 0 & 0 \\ 1 & X & 1 & 0 \\ 0 & 1 & X & 1 \\ 0 & 0 & 1 & X \end{vmatrix} = 0 \qquad \textbf{(4.37)}$$

The solution to the determinant is

$$X^4 - 3X^2 + 1 = 0 \qquad \textbf{(4.38)}$$

from which can be extracted the solutions for X and the molecular orbital energy levels shown in Figure 4.6 and equations 4.39–4.42.

$$\psi_4 = 0.372\phi_1 - 0.602\phi_2 + 0.602\phi_3 - 0.372\phi_4 \qquad \textbf{(4.39)}$$

$$\psi_3 = 0.602\phi_1 - 0.372\phi_2 - 0.372\phi_3 + 0.602\phi_4 \qquad \textbf{(4.40)}$$

$$\psi_2 = 0.602\phi_1 + 0.372\phi_2 - 0.372\phi_3 - 0.602\phi_4 \qquad \textbf{(4.41)}$$

$$\psi_1 = 0.372\phi_1 + 0.602\phi_2 + 0.602\phi_3 + 0.372\phi_4 \qquad \textbf{(4.42)}$$

Note that the magnitude of the coefficients for ϕ_1 and ϕ_4 are the same in all the orbitals, as are the magnitudes of the coefficients for ϕ_2 and ϕ_3 in all four cases. This result must occur by symmetry: It should not matter whether we call the carbon atom on the left end of butadiene C1 or whether we start numbering from the right. As long as the two end carbon atoms have the same magnitude for the coefficient of a particular orbital, then the bonding energies and other parameters we will calculate below will be found to be the same, no matter which end of the molecule we start from when writing the determinant.

Figure 4.6 illustrates several other principles of Hückel molecular orbitals. The pattern of nodes is the same as in the case of the allyl system: 0 nodes in ψ_1, one node in ψ_2, two nodes in ψ_3, and continuing the pattern there are three nodes in ψ_4. This leads to the generalization that *for linear π systems, we will always see $n - 1$ nodes in ψ_n*. Notice also that the bonding and antibonding orbitals are symmetrically placed above and below the

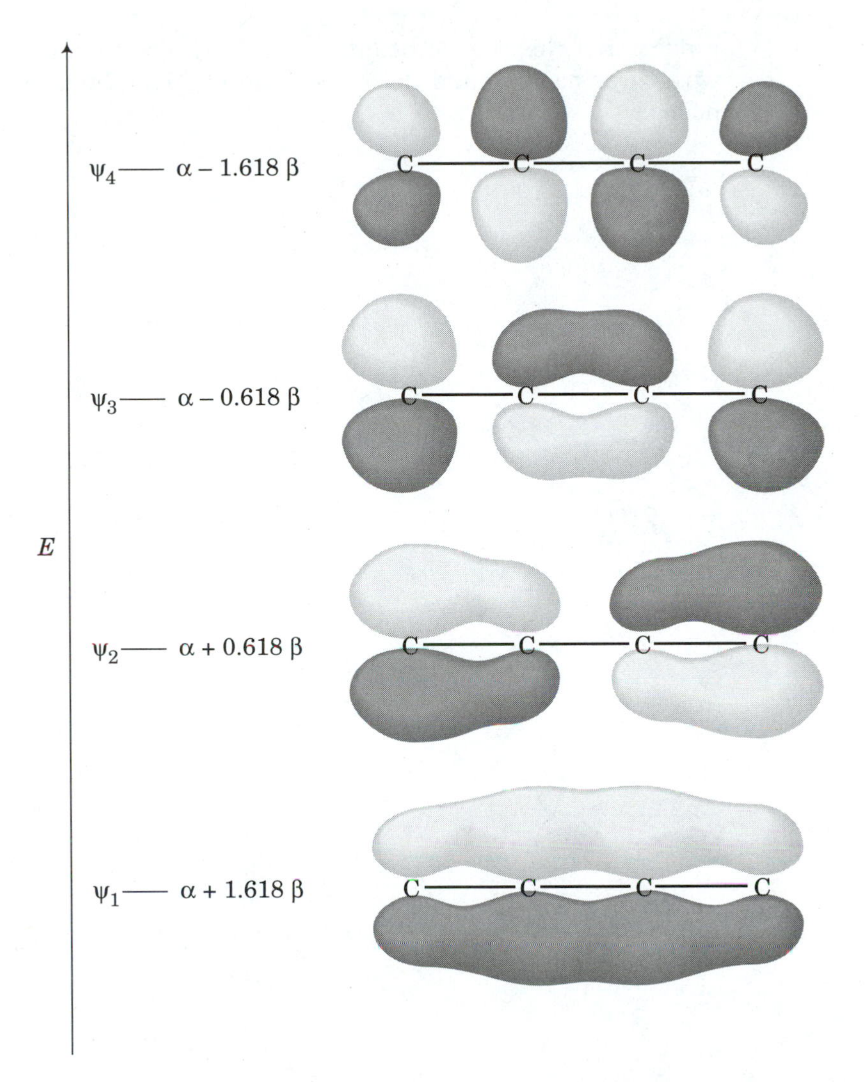

Figure 4.6
Butadiene orbitals and energy levels.

nonbonding ($E = \alpha$) level. That is a general result for linear π systems with an even number of carbon atoms. For a linear π system with an odd number of p orbitals, one orbital will have $E = \alpha$, and the rest of the orbitals will be displaced symmetrically above and below this orbital on an energy diagram.

It is the regularity of these features of molecular orbitals that makes them so useful to us in predicting molecular properties. Without actually doing the calculations, we can predict that the general shape of the MOs for the pentadienyl system will be as shown in Figure 4.7(a), with two bonding MOs, one nonbonding MO, and two antibonding MOs.[25] Similarly, we can

[25]The general shape of the molecular orbitals of linear systems can also be predicted by the free electron model, which essentially treats the electron as a particle in a molecule-sized box. For a leading reference and application to pentadienyl, see Jaffé, H. H. *J. Chem. Phys.* **1952**, *20*, 1646.

predict that the MOs of hexatriene will appear as shown in Figure 4.7(b), with three bonding and three antibonding MOs. In each case ψ_n has $n - 1$ nodes.

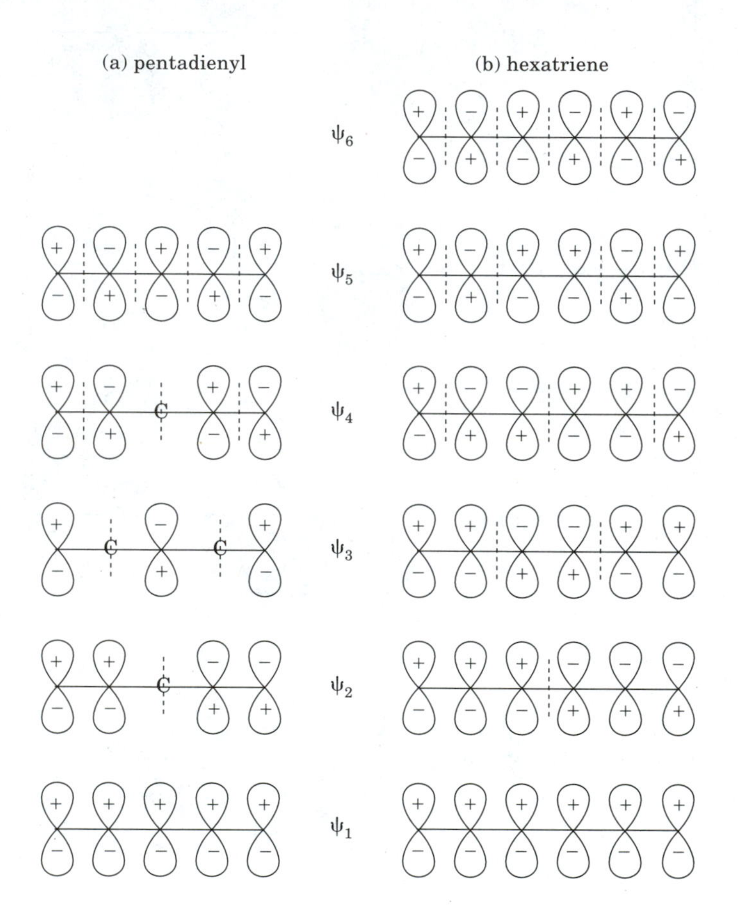

Figure 4.7 Qualitative representation of the molecular orbitals for linear π systems: (a) pentadienyl; (b) hexatriene. (Representation adapted from reference 9, p. 51.)

The HMO method produces different patterns of energy levels for cyclic π systems. For example, writing the 6×6 determinant for benzene gives equation 4.43.

$$\begin{vmatrix} X & 1 & 0 & 0 & 0 & 1 \\ 1 & X & 1 & 0 & 0 & 0 \\ 0 & 1 & X & 1 & 0 & 0 \\ 0 & 0 & 1 & X & 1 & 0 \\ 0 & 0 & 0 & 1 & X & 1 \\ 1 & 0 & 0 & 0 & 1 & X \end{vmatrix} = 0 \qquad \textbf{(4.43)}$$

Solving the determinant for X and finding the energy levels for the six molecular orbitals gives the energy level diagram in Figure 4.8. Now we find a bonding orbital at $\alpha + 2\beta$, two bonding orbitals at $\alpha + \beta$,[26] two antibonding orbitals at $\alpha - \beta$, and an antibonding orbital at $\alpha - 2\beta$. The shapes of the MOs are shown in Figure 4.9.

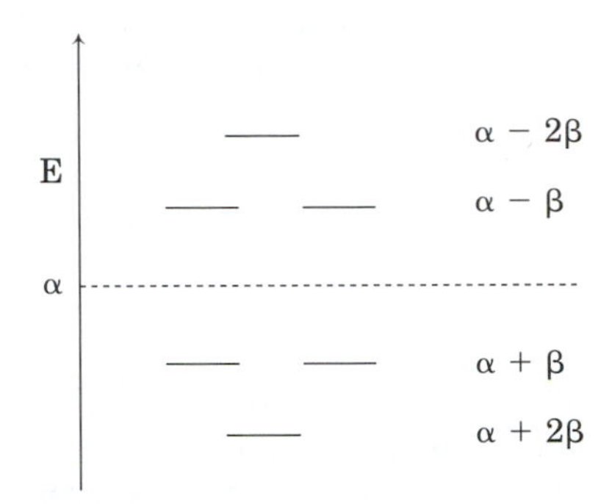

Figure 4.8
Benzene HMO energy levels.

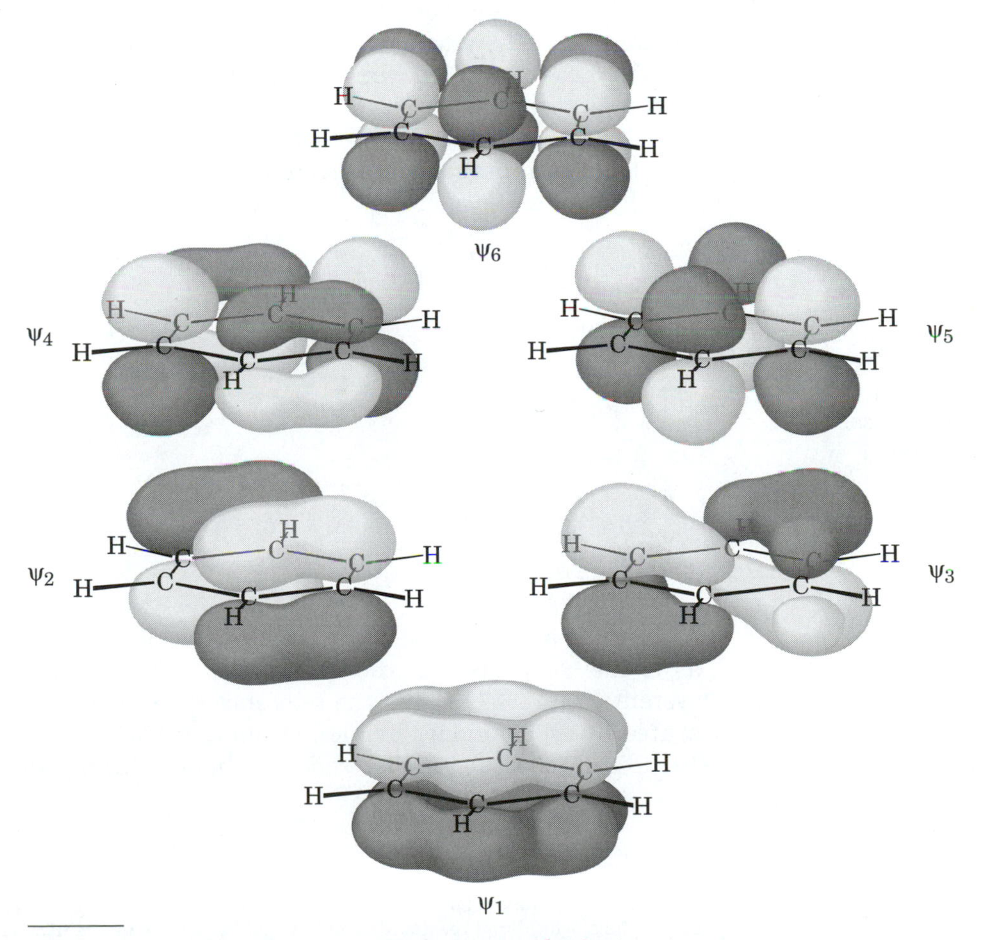

Figure 4.9
Schematic representation of Hückel molecular orbitals of benzene.

[26]Molecular orbitals at the same energy level are said to be degenerate.

Correlation of Physical Properties with Results of HMO Calculations

Molecular orbital energy level diagrams are useful in understanding the energy associated with π electrons. If electrons are placed into the molecular orbitals according to the aufbau principle, the first two electrons go into the lowest energy orbital, the next two electrons go into the next lowest energy orbital, and so forth. In the case of ethene, there are two electrons in the π system (because we assumed that each neutral carbon atom had one electron in its *p* orbital at the start of the calculation). Both electrons can go into the orbital with energy $E = \alpha + \beta$. Therefore, the energy of the π system is $2(\alpha + \beta) = 2\alpha + 2\beta$.

The number of electrons in the allyl system depends upon the charge. The allyl cation can be thought of as being formed from an empty *p* orbital and a double bond, so there are a total of two electrons in the π HMOs of allyl cation. The allyl radical can be thought of as a double bond and a radical (having one electron in a *p* orbital), so there are three electrons in the π system. The allyl anion has four electrons (two from the double bond and two from the carbanion carbon atom) which are placed into the π HMOs as shown in Figure 4.10. The π energy of the allyl cation is $2\ (\alpha + 1.414\ \beta) = 2\alpha + 2.828\ \beta$. For the allyl radical it is $2\alpha + 2.828\ \beta$ (from the two electrons in ψ_1) plus α from the one electron in ψ_2, so $E_\pi = 3\alpha + 2.828\ \beta$. Similarly, for the allyl anion the energy is $4\alpha + 2.828\ \beta$.

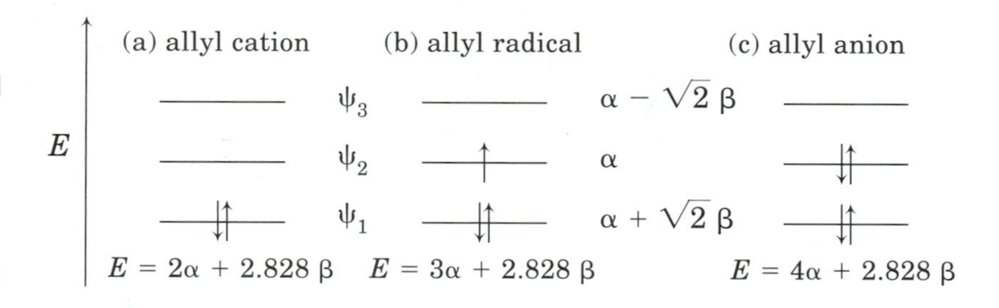

Figure 4.10 Electron population of HMOs in the allyl system: (a) cation, (b) radical, (c) anion.

The E_π of a particular system is often not so important to us, however, as is the comparison of the delocalized system with a reference localized system. Figure 4.11 shows the reference system for allyl: a double bond separated by an imaginary barrier[27] from a *p* orbital that may have 0 (cation), 1 (radical), or 2 (anion) electrons. The molecular orbital description of that

[27]A chemical structure in which one *p* orbital is perpendicular (and therefore orthogonal to) the *p* orbitals in the double bond, or one in which there is an insulating tetravalent carbon atom between the *p* orbital and the double bond, would have the same HMOs as does the reference system here.

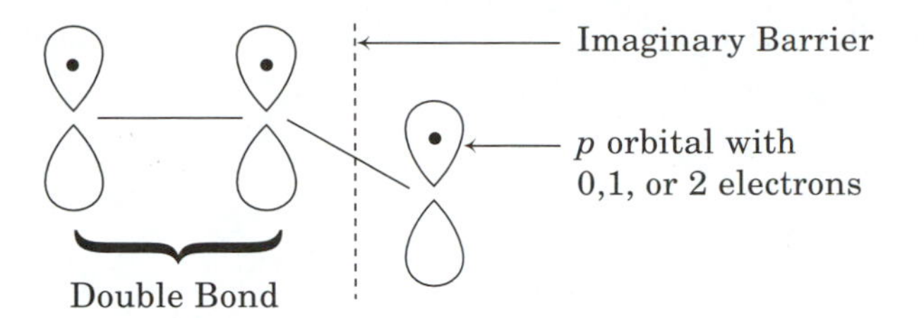

Figure 4.11 Reference system for allylic delocalization. (Adapted from reference 2e, p. 18.)

system, then, is simply a sum of the HMOs of the double bond and of the *p* orbital. Again, it does not matter whether we are talking about the cation, radical, or anion in Figure 4.11. The HMOs of the reference system are simply those of ethene ($E = \alpha + \beta$, $E = \alpha - \beta$) superimposed on the one HMO for an isolated *p* orbital, $E = \alpha$.

Putting electrons into the orbitals of the reference system according to the aufbau principle, we calculate that the E_π value of the cation, radical, and anion are $2\alpha + 2\beta$, $3\alpha + 2\beta$, and $4\alpha + 2\beta$, respectively (Figure 4.12). The difference in E_π between the delocalized system and the reference localized system is the *delocalization energy* of the structure. For the allyl cation, radical, or anion, that difference is 0.828 β. Therefore all three species are predicted to be more stable than a reference system with the same charge.[28]

An alternative description of allyl species is provided by the concept of *resonance*, in which the best description of a structure (the resonance hybrid) is considered to be the result of a superposition of two or more classical valence bond structures.[29,30] For example, in Figure 4.13 the resonance hybrids of the allyl cation, radical, and anion are each shown as the average of

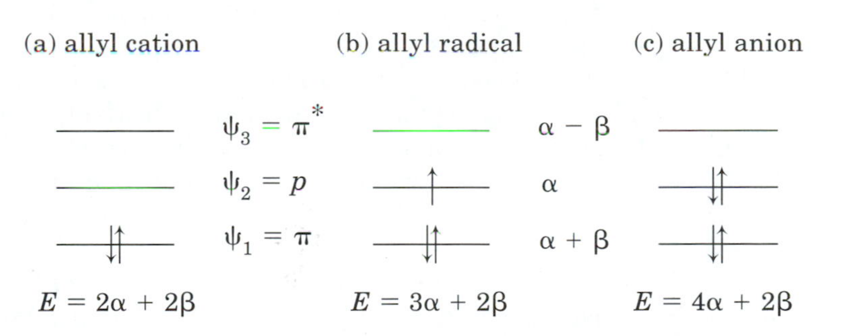

Figure 4.12 Molecular Orbitals for the allyl reference system shown in Figure 4.11.

---

[28]Although the one electron in ψ_2 in the radical does not contribute to bonding, the two electrons in ψ_1 produce bonding that causes the radical to be planar. Electron diffraction measurements on the radical indicated a planar structure with C-C bond length of 1.428 Å ± 0.013 Å and a C-C-C bond angle of 124.6° ± 3.4°. Vajda, E.; Tremmel, J.; Rozsondai, B.; Hargittai, I.; Maltsev, A. K.; Kagramanov, N. D.; Nefedov, O. M. *J. Am. Chem. Soc.* **1986**, *108*, 4352. For a theoretical study of the allyl cation, radical, and anion, see Gobbi, A.; Frenking, F. *J. Am. Chem. Soc.* **1994**, *116*, 9275 and references therein.

[29]Wheland, G. W. *Resonance in Organic Chemistry*; John Wiley & Sons, Inc.: New York, 1955.

[30]For a discussion, see Coulson, C. A. *Endeavour* **1947**, *6*, 42.

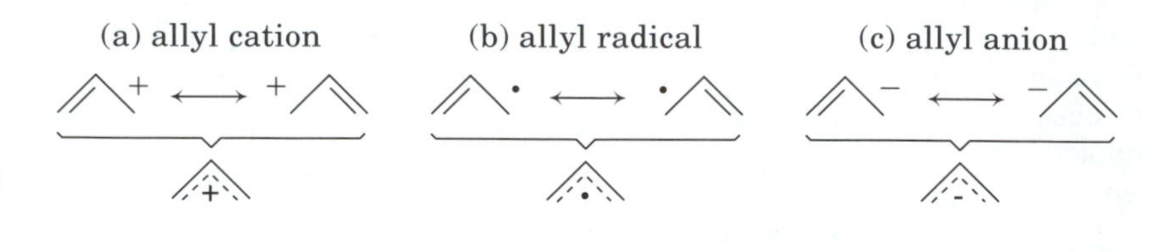

Figure 4.13 Resonance structures (top) and resonance hybrid (bottom) for (a) allyl cation, (b) allyl radical, and (c) allyl anion.

a pair of contributing resonance structures. In each case the resonance hybrid is lower in energy than any one of the contributing resonance structures by an amount known as the *resonance energy*. Although resonance energy is derived from valence bond theory,[31] while delocalization energy is derived from molecular orbital theory, it is easy to relate the two concepts to the same phenomena—in this case, the extra stabilization of allyl systems.

The value of β that appears most frequently in the literature is based on thermochemical data for benzene, which is considered to be the prototype "aromatic" organic compound. With the six π electrons in the three bonding MOs shown in Figure 4.8, the value of E_π is $6\alpha + 8\beta$. A system of three noninteracting double bonds would have an energy of $3\ (2\alpha + 2\beta)$ which equals $6\alpha + 6\beta$. Therefore, the delocalization energy of benzene is 2 β. If we equate this value to the commonly cited figure of 36 kcal/mol as the resonance energy for benzene,[32] then β is 18 kcal/mol.

Conjugated dienes such as 1,3-butadiene exist in two main conformations about the C2—C3 bond, the *s-cis* conformation and the *s-trans* conformation. Both experimental data and theoretical calculations suggest that the barrier to rotation is around 4 kcal/mol.[33] Interconversion of the two conformations by rotating about the C2—C3 bond as shown in Figure 4.14, involves a transition structure in which one π system is perpendicular to the other. At that conformation the two π systems are orthogonal and therefore noninteracting. Thus the HMOs are effectively those of two independent ethene systems, and the energy of the structure is $2\ (2\alpha + 2\beta)$. The conformational change would require an increase in the energy of the system of 0.47 β, which could account, at least qualitatively, for the barrier to rotation.

The barrier to rotation about the C2—C3 bond of 1,3-butadiene can also be predicted quantitatively. Using 18 kcal/mol as the value of β means that the minimum activation energy for the rotation should be about 8 kcal/mol,

[31]Epiotis, N. D. with Larson, J. R.; Eaton, H. L. *Unified Valence Bond Theory of Electronic Structure*; Springer-Verlag: Berlin, 1982; pp. 63 *ff.*

[32]This value of 36 kcal/mol is not universally accepted; see the discussion on page 212.

[33]Wiberg, K. B.; Rosenberg, R. E.; Rablen, P. R. *J. Am. Chem. Soc.* **1991**, *113*, 2890 and references therein.

which is considerably larger than the experimental value of 4 kcal/mol. One explanation for the discrepancy may be that our original formulation for butadiene assumed a set of four *p* orbitals, each of which was capable of interacting equally with one or two *p* orbitals beside it. In fact, the C1—C2 bond length of butadiene is shorter than the C2—C3 bond length, and some authors contend that the C2—C3 bond length is as short as it is primarily because it happens to be formed from sp^2-sp^2 hybrid orbitals, not because of appreciable conjugation between C2 and C3.[34] Thus the interaction we calculate by simple HMO theory significantly overestimates the barrier to rotation.

Figure 4.14 HMOs for butadiene conformational change.

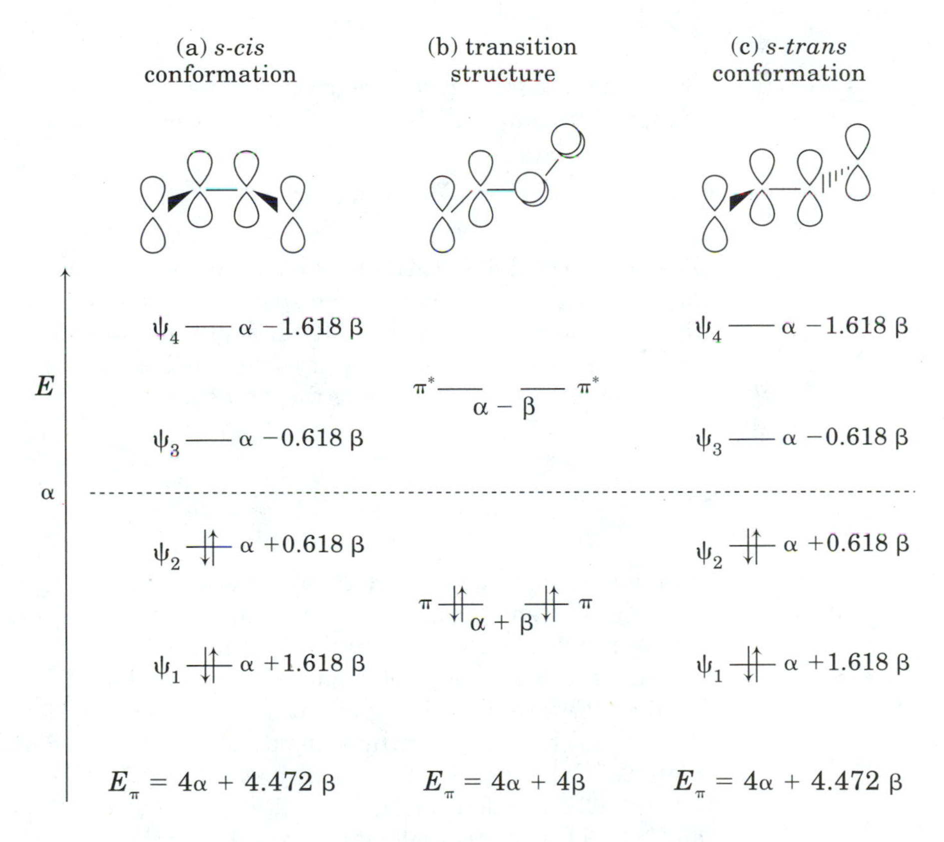

[34]Skaarup, S.; Boggs, J. E.; Skancke, P. N. *Tetrahedron* **1976**, *32*, 1179.

Other Parameters Generated through HMO Theory

In addition to energy comparisons, there are several other results of HMO theory that are useful to the organic chemist. Among them are electron density, charge density, bond order, and free valence. We calculate the **electron density** at each atom by summing the electron density at that atom for each occupied molecular orbital. We have defined ψ in equation 4.1:

$$\psi = c_1\phi_1 + c_2\phi_2 + \cdots + c_k\phi_k$$

The wave functions must be normalized. Solving equation 4.21 and taking the cross terms[35] equal to 0, gives:

$$c_1^2[\int \phi_1\phi_1\, d\tau] + c_2^2[\int \phi_2\phi_2\, d\tau] + \cdots + c_k^2[\int \phi_k\phi_k\, d\tau] = 1 \qquad \textbf{(4.44)}$$

Since the integrals in square brackets are unity (see equation 4.8), then the sum of the squares of the coefficients for each MO must sum to 1; i.e.,

$$c_1^2 + c_2^2 + \cdots + c_k^2 = 1 \qquad \textbf{(4.45)}$$

The square of the coefficient for a carbon atom in each ψ is thus considered to be the fraction of electron density to be found at that carbon atom when there is *one* electron in that MO. If there are two electrons in a particular MO, then the electron density at the ith carbon atom is $2\,c_i^2$. The general expression for *electron density* at the ith position is then

$$\rho_i = \sum_{\text{occupied }\psi} n\, c_i^2 \qquad \textbf{(4.46)}$$

where n is the number of electrons in each ψ, so ρ is the sum of the squares of the coefficients at that carbon atom for each occupied orbital multiplied by the number of electrons in that orbital.

As an example of electron density calculations, consider again the allyl system (Figure 4.4). For the cation, the electron density for C1 is $\rho_1 = 2(\tfrac{1}{2})^2 = \tfrac{1}{2}$. The density is the same at C3, as we would expect from the symmetry of the molecule. The density at C2 is $2(1/\sqrt{2})^2 = 1$. For the allyl radical, the electron density on C1 is: $2(\tfrac{1}{2})^2 + 1(1/\sqrt{2})^2 = 1$. The electron density on C2 and C3 are also calculated to be 1. For the anion the electron densities on the three carbon atoms are $1\tfrac{1}{2}$, 1, and $1\tfrac{1}{2}$, respectively.

---

[35]A cross term is the integral $\int \phi_j\phi_k\, d\tau$ for $j \neq k$.

Electron densities are primarily useful because they allow us to calculate charge densities. The **charge density** on the ith atom (denoted as q_i) is defined[36] as

$$q_i = 1 - \rho_i \tag{4.47}$$

An example will rationalize this relationship. In ethene, there are two electrons in the π bond, and by symmetry they must be distributed equally between the two carbon atoms. Since the molecule is electrically neutral, each carbon atom must also be electrically neutral,[37] so $q_i = 0$. Since $\rho_i = 1$ and $q_i = 0$, equation 4.47 holds.

Equation 4.47 can be used to calculate the charge density for each position of the allyl cation, radical, and anion, and the results are shown in Table 4.1. These values are comforting, because they agree with our chemical experience with allylic systems.[38] The resonance description of the allyl cation and radical (Figure 4.13) suggests that exactly half of the charge or unpaired electron density is associated with each of the terminal carbon atoms, and the HMO result is the same.

Bond order is an estimate of the amount of bonding between two atoms.[39] Since there is usually a σ bond between the same pair of atoms, we are most often interested in the **π bond order**, sometimes called mobile bond order. As in earlier examples (e.g., allyl, butadiene), a pair of atoms may have the same sign for their atomic orbital coefficients (therefore a bonding relationship) for one molecular orbital but may have a different sign for their coefficients (thus an antibonding relationship) for a different molecular orbital. Bond order calculations simply sum the contributions to bonding between each pair of atoms over all the occupied molecular orbitals,

Table 4.1 HMO charge densities for allyl cation, radical, and anion.

Carbon Atom	Cation	Radical	Anion
1	$+\frac{1}{2}$	0	$-\frac{1}{2}$
2	0	0	0
3	$+\frac{1}{2}$	0	$-\frac{1}{2}$

[36]Wheland, G. W.; Pauling, L. *J. Am. Chem. Soc.* **1935**, *57*, 2086.

[37]This statement is rigorously true only for HMO theory, which does not consider the possibility of charge on the carbon atoms arising due to opposite charge on the hydrogen atoms, because the hydrogen atoms are considered not to be involved in the π system bonding.

[38]We should not take great comfort from the correspondence of the simple HMO result with that from resonance (VB) theory. Other calculations give very different charges; *cf.* Slee, T. S. *J. Am. Chem. Soc.* **1986**, *108*, 7541.

[39]The concept of bond order was introduced by Coulson, C. A. *Proc. Roy. Soc. (London)* **1939**, *A169*, 413. Carbon-carbon bond lengths correlate with bond orders; see Jug, K. *J. Am. Chem. Soc.* **1977**, *99*, 7800. For further discussion of bond order and related parameters, see Sannigrahi, A. B.; Kar, T. *J. Chem. Educ.* **1988**, *64*, 674.

$$P_{ij} = \sum_{\text{occupied } \psi} nc_i c_j \qquad \textbf{(4.48)}$$

where P_{ij} is the π bond order between atoms i and j, n is the number of electrons in a particular MO, and c_i and c_j are the coefficients for the ith and jth carbon atoms, respectively, for that MO. If we consider butadiene (Figure 4.6) we may calculate

$$P_{12} = \underset{(\text{from } \psi_1)}{2(0.37)(0.60)} + \underset{(\text{from } \psi_2)}{2(0.60)(0.37)} = 0.895 \qquad \textbf{(4.49)}$$

$$P_{23} = \underset{(\text{from } \psi_1)}{2(0.6)(0.6)} + \underset{(\text{from } \psi_2)}{2(0.37)(-0.37)} = 0.444 \qquad \textbf{(4.50)}$$

These numbers tell us that there is about 90% of a π bond between carbon atoms 1 and 2 and about 44% of a π bond between carbon atoms 2 and 3. By symmetry we also know $P_{34} = P_{12}$. These values are for the π bond order. If we add the σ bond between each of the carbon atoms, then the **total bond order** is the number indicated above the dotted line in Figure 4.15.

Calculating bond order allows us to determine one more property of π systems that will be useful to us in correlating theory with chemical experience. We would like to have a measure of the reactivity of each carbon atom in a π system. We define the parameter **free valence,** ($\mathcal{F}_i$) as the difference between the maximum possible bond order of an atom and the actual total bond order.[40] What is the maximum possible bonding power of an atom? It would be difficult to imagine a π carbon atom being more strongly bonded than C1 in trimethylenemethane (Figure 4.16). Calculation of the HMOs of that system shows that the π bond order for each of the C1—C2, C1—C3 and C1—C4 bonds is $(\sqrt{3}/3)$, so the total π bond order to C1 is $3(\sqrt{3}/3) = 1.732$. Since C1 also has three σ bonds, its total bond order is 4.732. Therefore we define

$$\mathcal{F}_i = 4.732 - \sum_{\psi \text{ occupied}, j} P_{ij} \qquad \textbf{(4.51)}$$

where P_{ij} includes sigma bond order (usually 3) as well as π bond order. For example, P_{12} for butadiene is 0.895 (equation 4.49). The σ bond order is three (two σ bonds to hydrogen atoms and one σ bond to a carbon atom), so the total bond order is 3.895. Subtracting 3.895 from 4.732 gives an $\mathcal{F}_1$ for butadiene of 0.837. We represent $\mathcal{F}_i$ as a number written at the end of an arrow drawn from a particular carbon atom. Figure 4.17 shows $\mathcal{F}_i$ values for

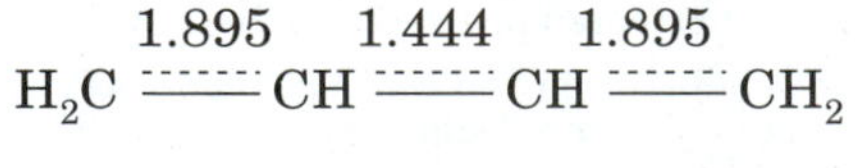

Figure 4.15 Total bond order for butadiene bonds.

$$\mathrm{H_2C} \overset{1.895}{=\!=\!=} \mathrm{CH} \overset{1.444}{=\!=\!=} \mathrm{CH} \overset{1.895}{=\!=\!=} \mathrm{CH_2}$$

[40]Coulson, C. A. *Trans. Faraday Soc.* **1946**, *42*, 265; Coulson, C. A. *Faraday Soc. Discuss.* **1947**, *2*, 9; Coulson, C. A. *J. chim. phys.* **1948**, *45*, 243; Burkitt, F. H.; Coulson, C. A.; Longuet-Higgins, H. C. *Trans. Faraday Soc.* **1951**, *47*, 553.

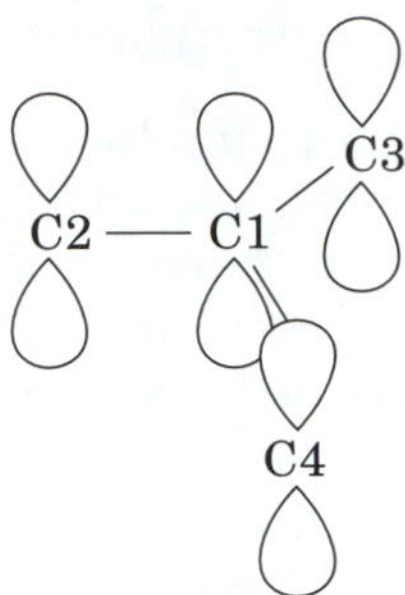

Figure 4.16 The trimethylene-methyl radical.

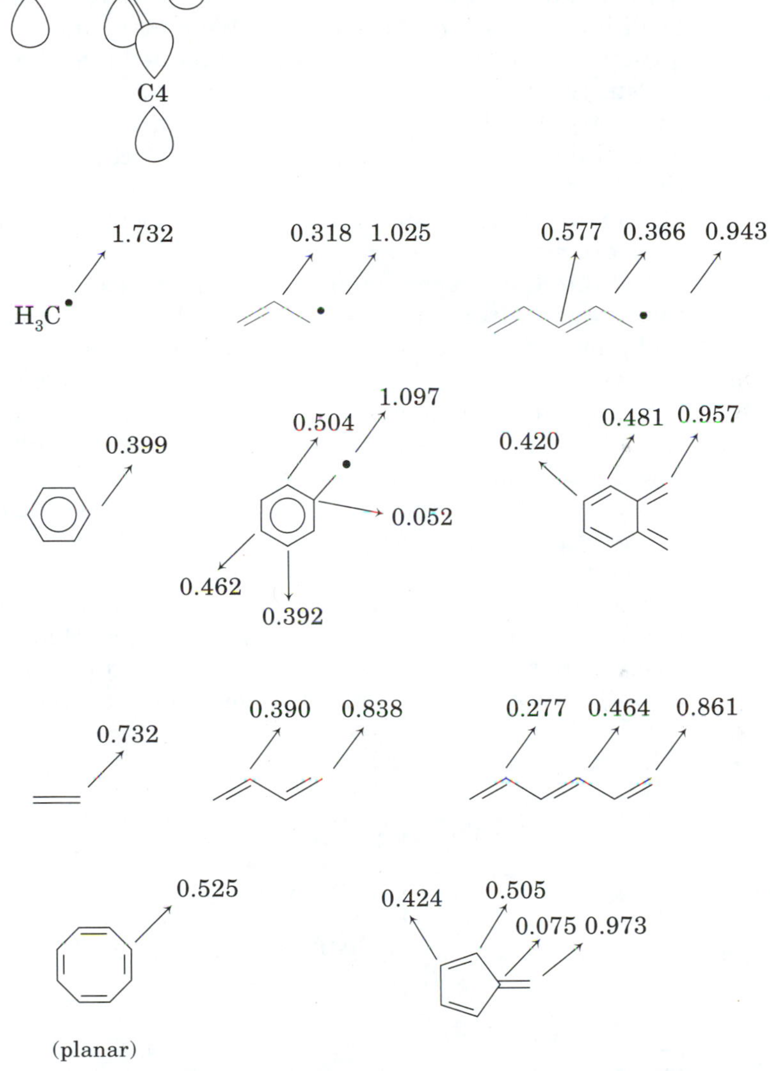

Figure 4.17 $\mathcal{F}_i$ values for unique positions in selected structures.

the species we have considered above and for several other interesting structures. A free valence index of about 1 is usually associated with high chemical reactivity, particularly with radicals.[41]

---

[41]Reactivity indices that give better predictions for nucleophilic or electrophilic reactions have been developed by (a) Fukui, K.; Yonezawa, T.; Shingu, H. *J. Chem. Phys.* **1952**, *20*, 722; (b) Jug, K.; Köster, A. M. *J. Phys. Org. Chem.* **1990**, *3*, 599.

Properties of Odd Alternant Hydrocarbons

For very large molecules, it is not trivial to produce the molecular orbital wave functions needed to calculate electron and charge densities, bond orders, and free valence indices. Fortunately, there are shortcuts that greatly facilitate our use of HMO theory.[42] We may define two types of molecular structures. A molecule is said to be **alternant** if alternating π centers[43] can be "starred" with no stars adjacent to each other.[44,45,46] It is **nonalternant** if two starred or two nonstarred positions are adjacent. Figure 4.18 shows several molecules with stars. All linear structures are alternant, whether they have an even or odd number of π centers. All even-membered rings are alternant, but all odd-membered rings are nonalternant. Some bicyclic systems are alternant and some are not.

Within the category of alternant structures, we further distinguish between **even alternant** systems, which have the same number of starred and nonstarred π centers, and **odd alternant** systems which do not. (We always label π centers so as to produce the greater number of starred positions, so odd alternant structures have more starred than nonstarred π centers.) In Figure 4.18 structures (a) and (b) are even alternant, structure (c) is odd alternant, and structure (d) is nonalternant.

Alternant hydrocarbons are of interest because the carbon atoms can be divided into two sets (starred and nonstarred) so that no carbon atoms in the same set have overlapping p orbitals.[45] This property means that some of the fundamental assumptions of HMO theory are correct for these structures. For example, in HMO theory we assume that all carbon atoms are electrically neutral. If that is true, then all values of H_{ii} are the same (α) and all values of H_{ij} for bonded atoms i and j are the same (β). However, if

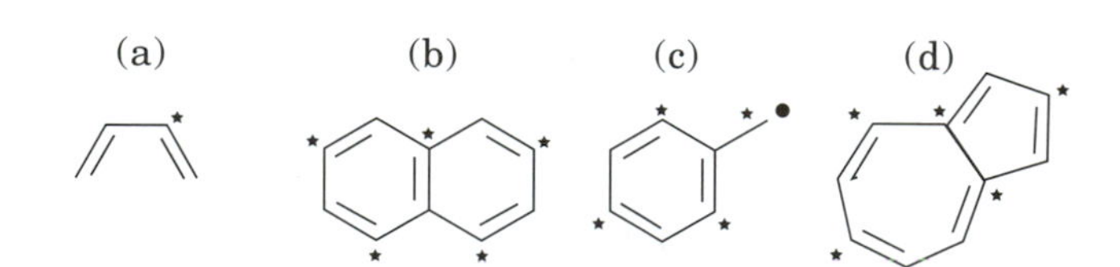

Figure 4.18 Alternant (a–c) and nonalternant (d) systems.

[42]Our goal is to obtain the *results* of HMO calculations in order to solve chemical problems, not just to solve mathematical problems for the sake of solving them. In that sense we as organic chemists are best served by any shortcuts that produce useful results without our having to do more math than necessary. It is more than just an anecdote that back of the envelope or napkin calculations done during lunch may be the most stimulating to some chemists.

[43]π centers are those atoms that contribute a p orbital to the basis set of orbitals used to calculate the π molecular orbitals.

[44]Coulson, C. A.; Longuet-Higgins, H. C. *Proc. Roy. Soc. (London)* **1947**, *A192*, 16.

[45]Longuet-Higgins, H. C. *J. Chem. Phys.* **1950**, *18*, 265.

[46]Often the symbol is an asterisk, but the terminology is *starred* nonetheless.

there are charges on any of the atoms, then the values of H_{ii} and H_{ij} are altered, and our conclusions are incorrect. A specific case is the nonalternant compound azulene (Figure 4.18(d)), which has a dipole moment and for which the charge densities calculated with HMO theory are not 0. Therefore there is a fundamental inconsistency between the assumptions of an HMO calculation on a *nonalternant* structure and the calculated results. For alternant structures, however, the charge densities calculated from the HMOs are found to be 0.

An important property of alternant systems is that[47]

1. every even alternant conjugated π system has the bonding and antibonding MOs arranged symmetrically above and below the energy $E = \alpha$, and
2. every odd alternant conjugated π system has a nonbonding molecular orbital, NBMO, characterized by having a molecular orbital with $E = \alpha$. Moreover, the bonding and antibonding MOs are arranged symmetrically above and below the NBMO.

With this knowledge we could have predicted the qualitative ordering of the energy levels of the allyl system (Figure 4.4) without doing the HMO calculation.

Another important property of odd alternant systems will allow us to calculate directly the wave function of the NBMO without having to calculate the entire set of HMOs.

> If we star an odd alternant system (so as to have more starred than nonstarred positions) the NBMO will have nonzero coefficients only at the starred positions.

This was, in fact, the result obtained for the allyl system, page 187. The NBMO (ψ_2) has nonzero coefficients only on C1 and C3, which would be the starred positions if allyl were designated as an odd alternant system, and ψ_1 and ψ_3 are arranged symmetrically above and below the NBMO.

Since the benzyl system (Figure 4.19) is also an odd alternant system, we can immediately determine that the NBMO of benzyl will have nonzero coefficients only at carbon atoms 1, 3, 5, and 7. Furthermore, we can determine the actual coefficients of the NBMO by using another result of HMO theory:

> The sum of the coefficients of the atomic orbitals of the starred atoms directly linked to a given nonstarred atom is zero.[48]

---

[47]For mathematical proofs of the properties of alternant systems, see Dewar, M. J. S. *The Molecular Orbital Theory of Organic Chemistry*; McGraw-Hill: New York, 1969; pp. 199 *ff*.

[48]See reference 2(e), pp. 148–149.

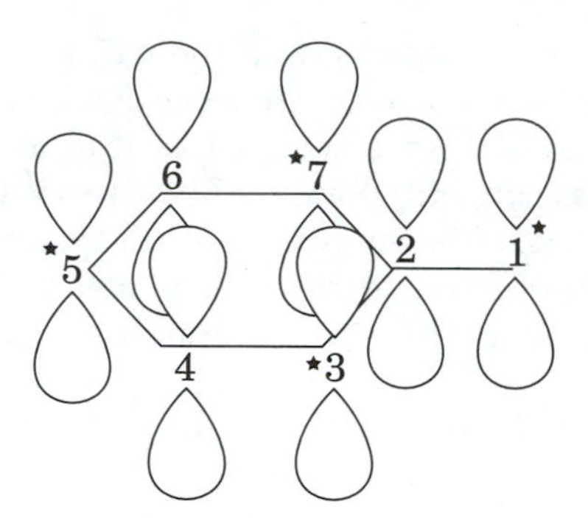

Figure 4.19 Determination of the NBMO of benzyl.

Therefore, we can write the following equations:

$$c_3 + c_5 = 0 \tag{4.52}$$

$$c_5 + c_7 = 0 \tag{4.53}$$

$$c_1 + c_3 + c_7 = 0 \tag{4.54}$$

If we let $c_5 = c$, then we have

$$c_3 = -c \tag{4.55}$$

$$c_7 = -c \tag{4.56}$$

and

$$c_1 = 2c \tag{4.57}$$

Since the wave function for the NBMO must be normalized, we know that

$$c_1^2 + c_3^2 + c_5^2 + c_7^2 = 1 \tag{4.58}$$

Substituting for each unknown c with the values in equations 4.55–4.57 gives

$$4_c^2 + c^2 + c^2 + c^2 = 1 \tag{4.59}$$

so

$$c = \pm\frac{1}{\sqrt{7}} \tag{4.60}$$

Thus

$$\psi_{\text{NBMO}} = \frac{2}{\sqrt{7}}\,\phi_1 - \frac{1}{\sqrt{7}}\,\phi_3 + \frac{1}{\sqrt{7}}\,\phi_5 - \frac{1}{\sqrt{7}}\,\phi_7 \tag{4.61}$$

At first glance it may not seem important to be able to calculate NBMOs for odd alternant systems. First, most stable molecules have an even number of π centers. Second, the method presented here allows us to determine only one of several molecular orbitals. However, the method is very valuable because odd alternant systems may be important intermediates in chemical reactions. For example, the benzyl radical is an intermediate in the free radical α-halogenation of toluene, and benzyl carbocations may be intermediate in S_N1 substitutions with benzyl halides. Therefore, calculating properties of these systems can give useful information about reacting molecules.

The square of the coefficient of each ϕ in the NBMO indicates the relative electron density on each atom due to one electron in that orbital. That

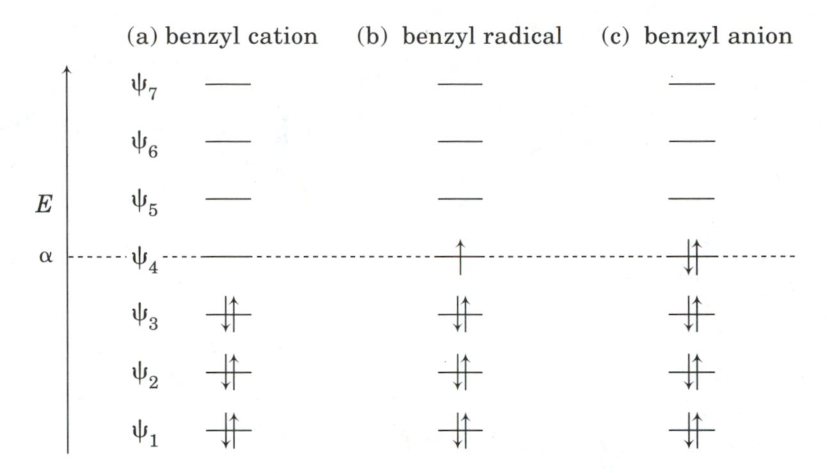

Figure 4.20
Population of HMOs in benzyl systems.

is, in the benzyl radical the unpaired electron density on C1 is 4/7, while the unpaired electron density is 1/7 on C3, C5, and C7. In addition, the electron densities in the NBMO determine the locations of charge density in an anion or cation produced by adding an electron to or removing an electron from the radical, as shown in Figure 4.20.[47] For example, conversion of the benzyl radical to the benzyl cation removes electron density only from ψ_4. Each carbon atom in the radical was electrically neutral, so removing 1/7 of the electron density from carbon atom 3 produces a positive charge of +1/7 at that position. Similarly, reduction of the radical to the anion adds electron density only to ψ_4. Again, adding 1/7 of the charge of an electron to carbon atom 3 gives a net charge of −1/7 there. The unpaired electron densities and charge densities predicted for the benzyl radical, cation, and anion are shown in Figure 4.21.

It is interesting at this point to compare the results of our HMO analysis with the results of analysis by resonance theory. The resonance structures for the benzyl systems are shown in Figure 4.22. The two resonance structures of benzyl in which a benzene ring is maintained should be much more stable than the other three. Therefore, they should make a greater contribution to the resonance hybrid, and the charge on C1 should probably be greater than that at C3, C5, and C7. However, we do not know how to quantify the difference in energy of the resonance structures, so we must treat them equally. There are five resonance structures in each case, so we conclude that C1 has a charge of +2/5 in the benzyl cation.

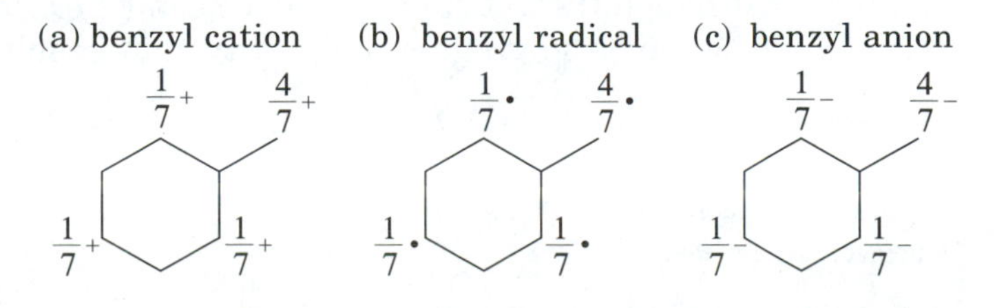

Figure 4.21
Charge and unpaired electron densities from HMO theory.

Figure 4.22 Resonance theory treatment of benzyl systems.

(a) benzyl radical

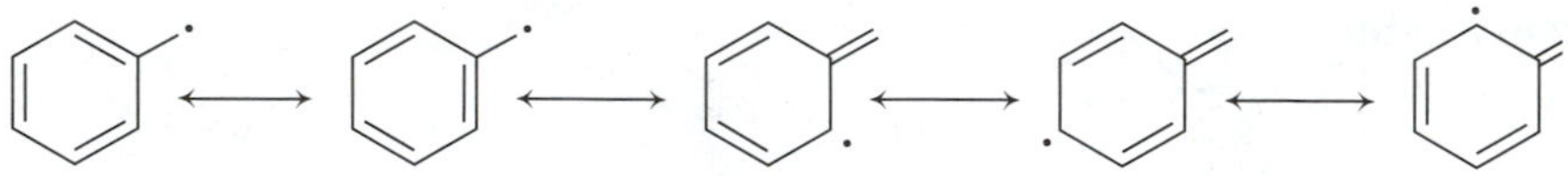

(a) benzyl cation

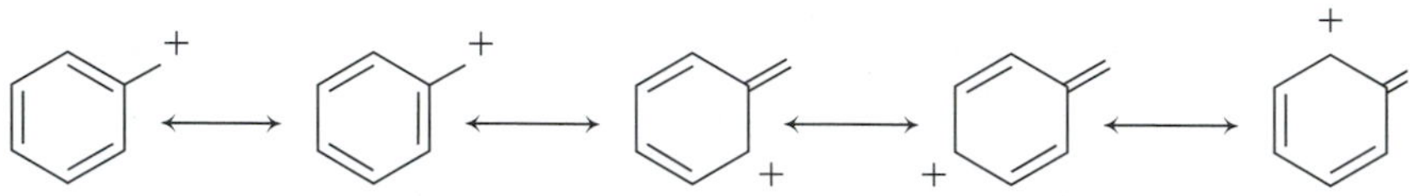

(a) benzyl anion

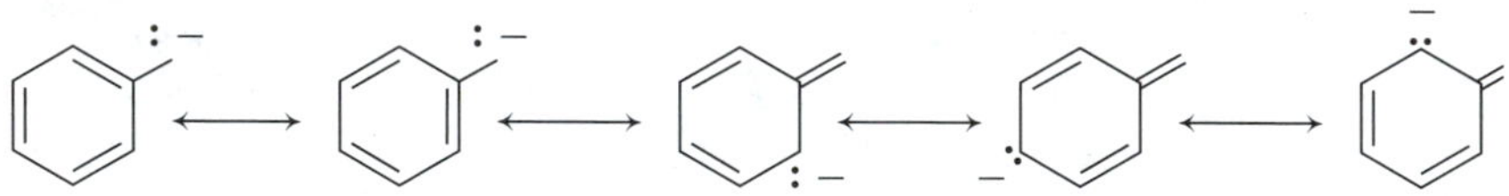

Here we have the first occasion in which the predictions of HMO theory and resonance theory are different. Is one theory correct and the other incorrect? The electron paramagnetic resonance spectrum of the benzyl radical has been interpreted to mean that about 50% of the unpaired electron density is on the benzylic carbon atom, while 15.8% is on each of the two ortho carbon atoms and 18.6% is on the para carbon atom.[49] These results are intermediate between the predictions of simple resonance theory and simple Hückel theory. By this measure, the two procedures are nearly equally "correct" (or equally "wrong," if one prefers). It is easy to account for differences between experiment and either theory. Because of the many assumptions and simplifications in the HMO analysis, it would be surprising if predictions of electron distribution by this method are quantitatively correct. Moreover, we have not really tried to quantify our resonance method (by giving greater weight to some resonance structures than to others, for example).[50] Although theoretical approaches can be extended to yield more accurate predictions, the greater complexity that results may reduce the

[49]See the discussion by Fleming, I. *Frontier Orbitals and Organic Chemical Reactions*; Wiley-Interscience: London, 1976; p. 60.

[50]However, see pp. 68 *ff*.

utility that comes from working with simple conceptual models. Therefore, we will continue to use both simple resonance theory and Hückel molecular orbital theory, but with the understanding that they are only beginning points for the calculation of molecular structure and properties.

The Circle Method

There is another shortcut to the results of HMO theory, and it is one that organic chemists find to be particularly useful. As was noted by Frost and Musulin,[51] the energy levels of a monocyclic π system can be determined directly from the number of carbon atoms in the ring. The procedure can be

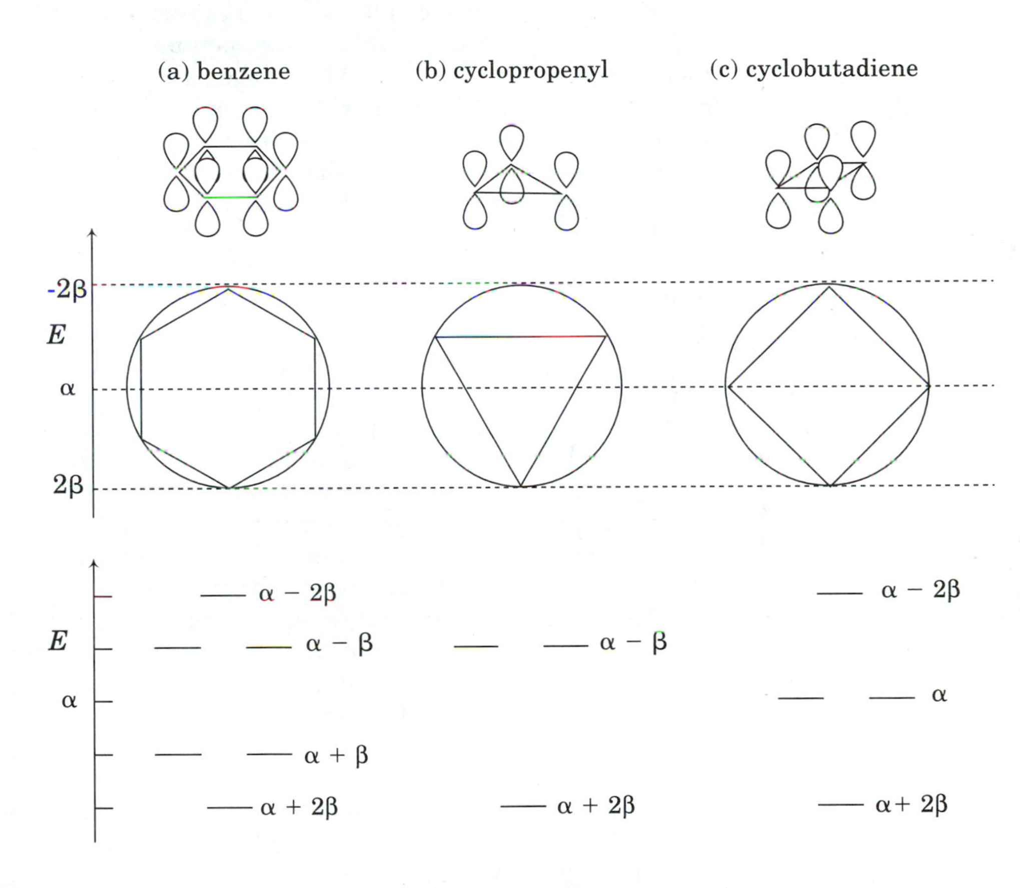

Figure 4.23 Application of the circle method to benzene, cyclopropenyl, and cyclobutadiene.

[51]Frost, A. A.; Musulin, B. *J. Chem. Phys.* **1953**, *21*, 572.

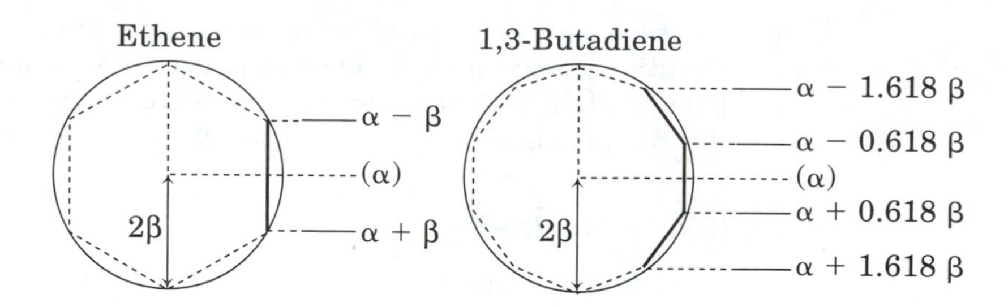

Figure 4.24 Application of the circle method to linear polyenes. (Adapted from reference 2e.)

carried out either mathematically or graphically. The graphical procedure involves inscribing the polygon corresponding to the shape of the planar, monocyclic π system in a circle (with center at $E = \alpha$ and radius 2 β) *with one corner of the polygon at the bottom of the circle*. The height of each corner (as measured on a scale drawn to one side of the circle) then gives the energy level of a molecular orbital for the π system. This procedure is illustrated in Figure 4.23 for benzene, cyclopropenyl, and cyclobutadiene. It may be seen that it easily predicts the energy of benzene to be the values we saw earlier (Figure 4.8). For cyclopropenyl and for cyclobutadiene the energy levels are the same as those calculated by the HMO methods we have just described.

The circle method has also been adapted for use with linear systems. A linear polyene with m p orbitals is transformed into a monocyclic polyene with $2\ m + 2$ sides, and this polyene is inscribed in a circle of radius 2 β, just as we did with the monocyclic systems above. In this case, however, the energy levels correspond *only to the heights of corners that correspond to carbon atoms in the original linear structure*. Figure 4.24 illustrates this technique for ethene and butadiene. It may be seen that the energy levels calculated are the same as those we derived earlier.

We should note two points about the circle method. First, we are not really *calculating* molecular orbital energy levels, we are simply using a graphical method to reproduce the results of HMO calculations. Second, the graphing method works best for small linear or monocyclic polyenes—the very molecules for which we may most easily calculate the HMO energy levels by solving the secular determinant. With larger rings or longer polyenes, the levels pack together more tightly, making the graphical approach tedious. Thus the circle method should be considered a mnemonic device to remind us of the pattern of the energy levels, not to derive them.

4.2 Aromaticity

The circle mnemonic leads directly to a familiar result of HMO theory, as illustrated in Figure 4.25. The molecular orbitals of monocyclic conjugated π systems have energy levels that follow a recurring pattern. There is always

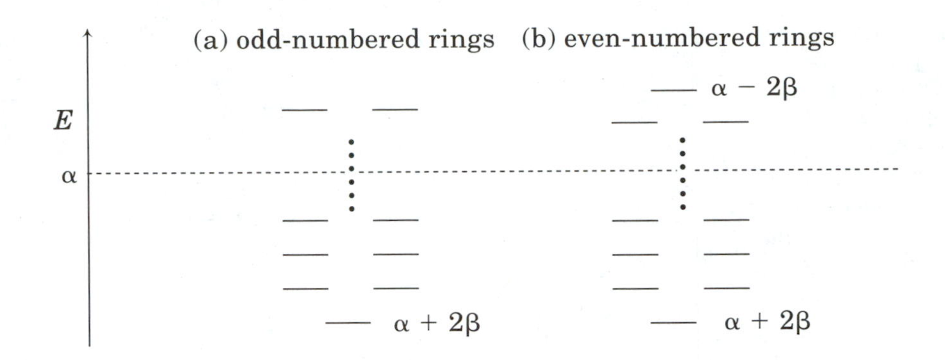

Figure 4.25 Energy levels of monocyclic systems.

one energy level at $E = \alpha + 2\,\beta$, and then there are pairs of degenerate energy levels at successively higher energy. The highest energy level may be a single orbital at $E = \alpha - 2\,\beta$ (if the ring has an even number of π centers) or there may be a pair of energy levels slightly lower in energy if the ring has an odd number of π centers.

Figure 4.26 shows the HMOs for cyclopropenyl carbocation, *square planar* cyclobutadiene, cyclopentadienyl anion, benzene, cycloheptatrienyl carbocation, and *planar* (D_{8h}) cyclooctatetraene.[52,53] If we place electrons into the molecular orbitals of each species according to the aufbau principle, we notice an important relationship between the stability of the systems and the *number of π electrons* (not necessarily with the number of carbon atoms). Systems with 2 or 6 electrons in the π system have the highest energy populated orbitals fully occupied. Therefore, these systems exhibit what is known as a **closed shell configuration**, meaning that there is not a partially filled HMO energy level, nor are there any unpaired electrons. However, systems with 4 or 8 π electrons have two unpaired electrons, each in a different molecular orbital, and are said to exhibit an **open shell configuration**.[54] Because there are both unpaired electrons and high energy orbital vacancies in these orbitals, structures with 4 or 8 electrons exhibit high chemical reactivity. Moreover, the E_π energies of the cyclic π systems with 2 or 6 electrons are much lower than those of reference (localized or acyclic) systems with the same number of π electrons. However, cyclic systems with 4 or 8 electrons do not have large π delocalization energies.

The results observed in Figure 4.26 for cyclic systems with 2 or 6 π electrons can be shown to be true also for those with 10, 14, 18, . . ., π electrons. Similarly, the trends observed for *planar cyclic systems* with 4 or 8 π electrons can be shown to be true for those with 12, 16, 20, . . ., π electrons.

[52]Note the emphasis on *planar* cyclooctatetraene. We will consider the geometry of cyclooctatetraene later.

[53]For a tabular summary, see Yates, K. *Hückel Molecular Orbital Theory*; Academic Press: New York, 1978; p. 143.

[54]By Hund's rule, two electrons in degenerate orbitals are more stable if they are in separate orbitals.

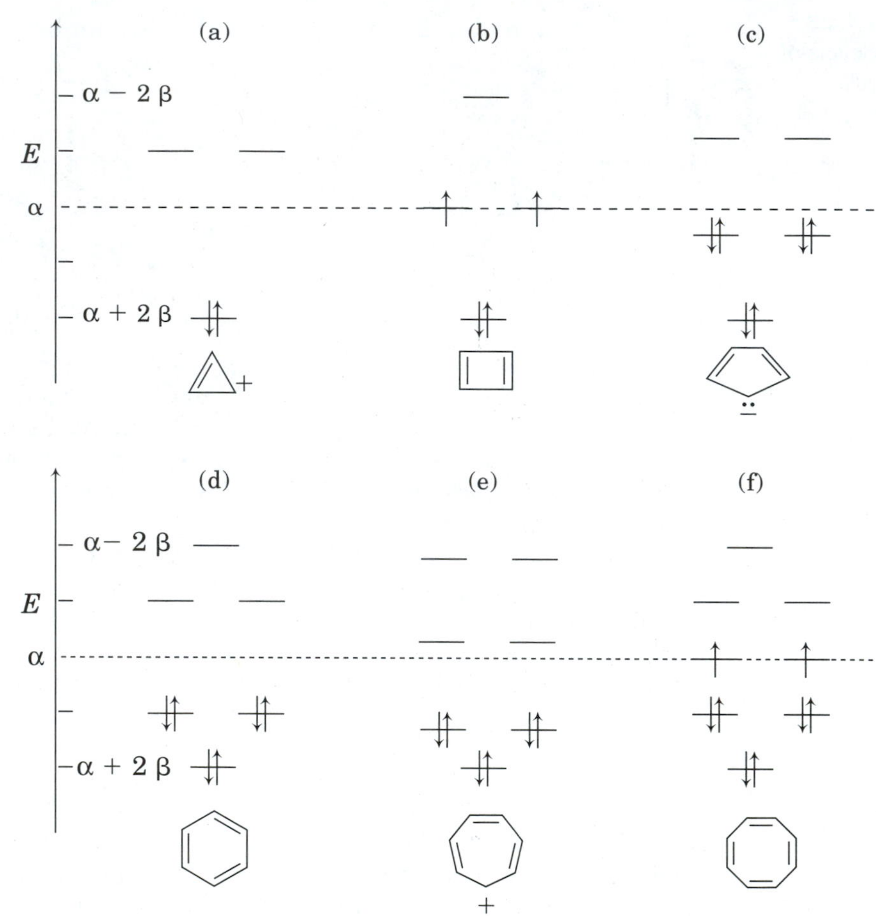

Figure 4.26 Energy levels of HMOs of cyclic π systems. Note: All structures are assumed to be planar, with all carbon-carbon bond lengths the same.

Rewriting these series as arithmetical progressions leads to the Hückel $4n + 2$ rule: planar cyclic systems with $(4n + 2)$ π electrons are said to be especially stable in comparison with acyclic analogues, while those with $4n$ π electrons are unstable in comparison with acyclic analogues. The name that we apply to monocyclic systems with $4n + 2$ electrons is *aromatic*, and we associate **aromaticity** with the properties of benzene. Compounds with $4n$ electrons in a cyclic π system are unstable in comparison with acyclic analogues and are said to be **antiaromatic**.[55] Let us consider what aromaticity means in the context of HMO theory as well as chemical experience.

Benzene

In the early days of chemical science, molecular structures were not known for many familiar substances, including benzaldehyde (from almonds) and

[55]Breslow, R. *Chem. Eng. News* **1965** (June 28), *43*, 90.

methyl salicylate (from oil of wintergreen). These compounds have distinctive aromas, so they came to be called aromatic. As chemical science developed, it was recognized that many of these aromatic compounds share the benzene ring as a structural feature. In time the term *aromatic* came to mean a certain kind of structure and not a certain kind of smell.

Benzene itself is the simplest such molecule, of course. It was isolated by Faraday in 1825, and the accepted structure for benzene is usually attributed to a proposal by Kekulé in 1865.[56,57] Most organic chemists are familiar with the (perhaps apocryphal) "snake tale" of Kekulé's vision of cyclic benzene molecules.[58] In the late 1800s there was great debate about the molecular structure of benzene, and a number of possible structures were proposed for it. The electron diffraction results, shown in Figure 4.27, indicate a planar structure with hexagonal symmetry.[59] With other physical measurements, benzene has been determined to have the structure shown in Figure 4.28.[60]

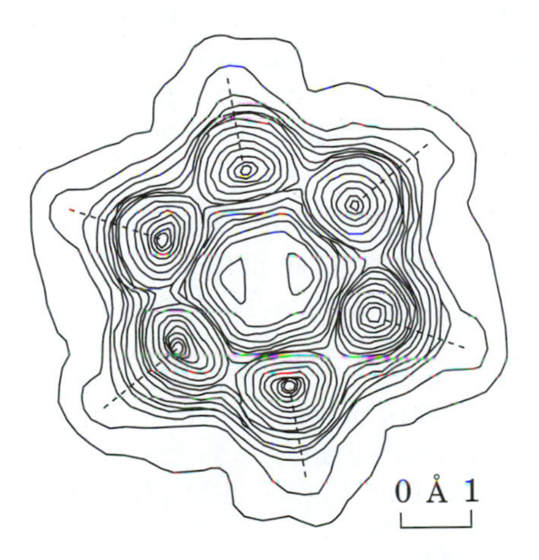

Figure 4.27
Electron density data for benzene. (Reproduced from reference 59.)

[56]For an interesting account of the discovery of benzene and of the many names applied to it, see Badger, G. M. *Aromatic Character and Aromaticity*; Cambridge University Press: Cambridge, 1969; pp. 1 *ff*. Visitors to London may visit the Faraday Museum in the Royal Institution, which has had on display what was reported to be the very sample of benzene isolated by Faraday in 1825.

[57]Wiswesser has concluded that credit for the structure of benzene should be given to Johann Josef Loschmidt: Wiswesser, W. J. *Aldrichimica Acta* **1989**, *22*, 17.

[58]In most versions of the story, Kekulé was sitting before a fire in his room when he had a daydream of writhing snakes, one of which formed a circle by biting its tail. Some recent papers by historians of chemistry have cast doubt on the validity of this part of chemical "mythology." For examples, see (a) Wotiz, J. H.; Rudofsky, S. *Chem. Br.* **1984**, 720; *J. Chem. Educ.* **1982**, *59*, 23; (b) Wotiz, J. H., ed. *The Kekulé Riddle*; Cache River Press: Vienna, IL, 1993; (c) Borman, S. *Chem. Eng. News*, **1993** (Aug. 23), 20; (d) Heilbronner, E.; Dunitz, J. D. *Reflections on Symmetry*; Verlag Helvetica Chemica Acta: Basel, 1993; pp. 47 *ff*. See also Seltzer, R. J. *Chem. Eng. News* **1985** (Nov. 4), 22. For a totally different interpretation of the snake legend, see Reese, K. M. *Chem. Eng. News*, **1984** (Aug. 13), 80

[59]Cox, E. G.; Cruickshank, D. W. J.; Smith, J. A. S. *Proc. Roy. Soc. A* **1958**, *247*, 1. See also reference 56, pp. 7 *ff*.

[60]Schomaker, V.; Pauling, L. *J. Am. Chem. Soc.* **1939**, *61*, 1769.

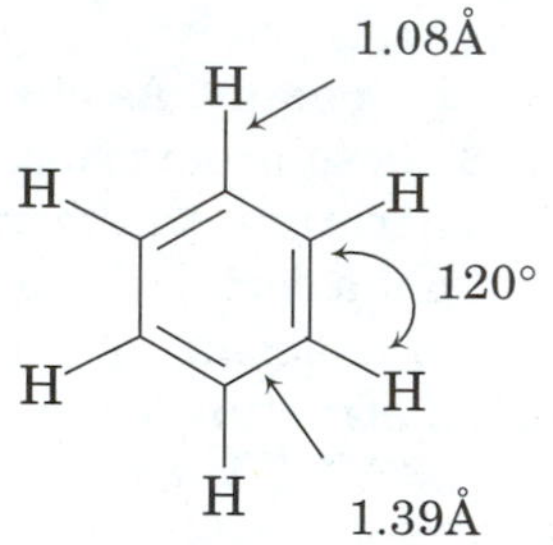

Figure 4.28
Experimental bond lengths and angles in benzene.

The structure of benzene might not have been seen as such an important problem if it were not clear to early researchers that there is something special about benzene and its related compounds. As Badger has noted, benzene derivatives were seen to have four unusual traits:[56]

1. thermal stability, including ease of formation by pyrolytic methods;
2. a pattern of substitution, not addition, with electrophilic reagents;
3. an unusual resistance to oxidation; and
4. different physical properties from analogous aliphatic compounds. For example, aniline is less basic than cyclohexylamine, and phenol is more acidic than cyclohexanol.

We have come to call this set of molecular properties aromatic, and we associate them with benzene and similar structures.[61] However, we must answer an important question: *Is benzene so stable because it is aromatic, or is it aromatic because it is so stable*?

First let us examine the question of stability and attempt to quantify it. The traditional method used to determine the resonance energy of benzene is to go through an analysis based on heats of hydrogenation of cyclohexene, 1,3-cyclohexadiene and benzene. Figure 4.29 shows the ΔH values for these three reactions.[62]

We may conclude from this analysis that a conjugated diene is 1.8 kcal/mol more stable than two nonconjugated double bonds and that benzene is 36 kcal/mol more stable than three nonconjugated double bonds.[63] However, it seems just as reasonable to expect the ΔH value to be $3\,(-55.4/2)$ or -83.1 kcal/mol, since the double bonds in benzene really ought to be compared to linearly conjugated double bonds, not to isolated double bonds. By that reasoning, the resonance energy of benzene is only 33.3 kcal/mol.

---

[61]Aromaticity has more recently been characterized by conformational—Podlogar, B. L.; Glauser, W. A.; Rodriguez, W. R.; Raber, D. J. *J. Org. Chem.* **1988**, *53*, 2129—and kinetic—Bofill, J. M.; Castells, J.; Olivella, S.; Solè, A. *J. Org. Chem.* **1988**, *53*, 5148—parameters.

[62]Data from Kistiakowsky, G. B.; Ruhoff, J. R.; Smith, H. A.; Vaughan, W. E. *J. Am. Chem. Soc.* **1932**, *58*, 137, 146.

[63]A value of 36 kcal/mol can also be calculated by comparing the observed heat of formation of benzene with that expected on the basis of group additivity values: Franklin, J. L. *J. Am. Chem. Soc.* **1950**, *72*, 4278.

Figure 4.29 Heat of hydrogenation data.

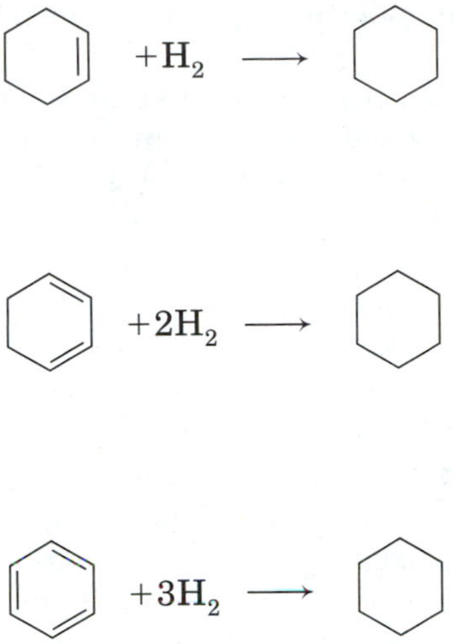

$+H_2 \longrightarrow$ $\Delta H = -28.6$ kcal/mol

$+2H_2 \longrightarrow$ $\Delta H = -55.4$ kcal/mol (expected: 2 * −28.6 or −57.2 kcal/mol; $\Delta\Delta H = 1.8$ kcal/mol)

$+3H_2 \longrightarrow$ $\Delta H = -49.8$ kcal/mol (expected: 3 * −28.6 or −85.8 kcal/mol; $\Delta\Delta H = 36$ kcal/mol)

The preceding analysis has omitted another important factor. Bond lengths and hybridizations change when a double bond is converted to a single bond, and there are energies associated with those changes. What we would like to calculate is the difference in energy between benzene and a cyclohexatriene *with the same geometry as benzene but with imaginary barriers* between the double bonds. Let us begin by presenting the data in Figure 4.29 in a slightly different fashion in Figure 4.30. Now it is clear that the ΔH value of 36 kcal/mol is the enthalpy for the sum of two different processes: One is the removal of the imaginary barrier to resonance, and the

Figure 4.30 Enthalpy of conversion of 1,3,5-cyclohexatriene to benzene.

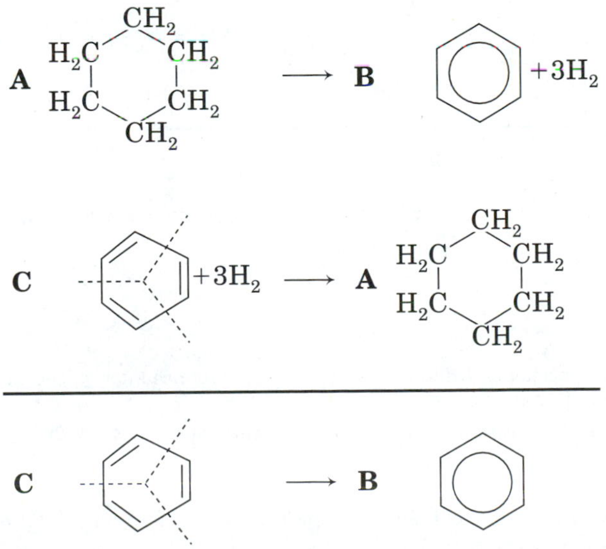

A $\longrightarrow$ B $+3H_2$ $\Delta H = +49.8$ kcal/mol

C $+3H_2 \longrightarrow$ A $\Delta H = -85.8$ kcal/mol

C $\longrightarrow$ B $\Delta H = -36.0$ kcal/mol

other is the distortion of the molecule so that all carbon-carbon bonds are the same length.

If we define resonance energy as the extra stabilization that a structure gains by delocalization of electrons without movement of nuclei, then we should separate the two factors that determine the net ΔH in Figure 4.30.[64] Coulson determined that the distortion of a 1,3,5-cyclohexatriene molecule with short double bonds and long single bonds into a 1,3,5-cyclohexatriene structure in which all the carbon-carbon bonds are the same length requires a compression energy of +27 kcal/mol. Writing such a process in the reverse, as shown in Figure 4.31, allows us to determine that the resonance energy of benzene should be 63 kcal/mol.[65,66,67] Still other analyses are possible, and various authors have determined values for the resonance energy of benzene that have ranged from 13 to 112 kcal/mol![68] Clearly, however, it would appear that simply measuring heats of hydrogenation is far from adequate to determine the resonance energy of any structure.

If it is surprising that there is considerable uncertainty about the exact resonance energy of benzene, it will be even more surprising to learn that some theoreticians have reported results suggesting that the special properties of benzene, while associated with the delocalized $4n + 2$ electrons, do not result from the propensity of these electrons to be delocalized. In partic-

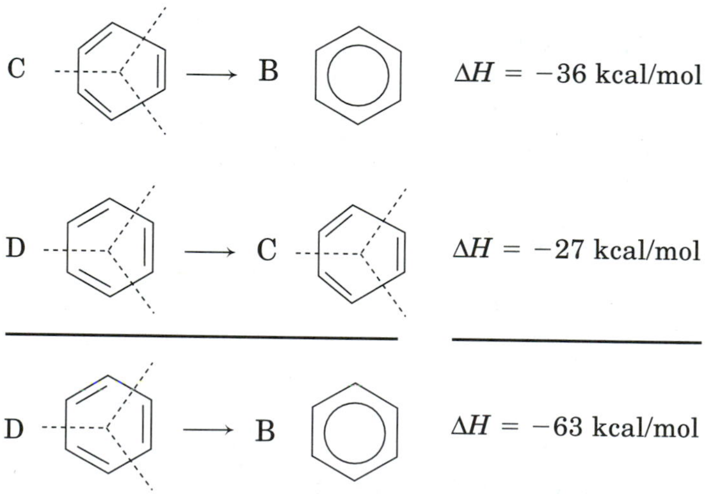

C $\longrightarrow$ B $\quad \Delta H = -36$ kcal/mol

D $\longrightarrow$ C $\quad \Delta H = -27$ kcal/mol

D $\longrightarrow$ B $\quad \Delta H = -63$ kcal/mol

Figure 4.31 Determination of the resonance energy of 1,3,5-cyclohexatriene with all carbon-carbon bonds the same length.

---

[64]For a discussion of different definitions of resonance energy, see the discussion by Dewar in reference 110(a).

[65]Coulson, C. A.; Altmann, S. L. *Trans. Faraday Soc.* **1952**, *48*, 293.

[66]For a detailed discussion, see Streitwieser, Jr., A. *Molecular Orbital Theory for Organic Chemists*; John Wiley and Sons, Inc.: New York, 1961; p. 245.

[67]See also the analysis of George, P.; Bock, C. W.; Trachtman, M. *J. Chem. Educ.* **1984**, *61*, 225.

[68]Reference 66, p. 247.

ular, Shaik and Hiberty have argued that "electronic delocalization in . . . C_6H_6 turns out to be a byproduct of the σ-imposed geometric symmetry and not a driving force by itself."[69,70] Later calculations indicated that the method of analysis can affect the degree to which delocalization is responsible for the hexagonal symmetry of benzene.[71] Nevertheless, it is somewhat unsettling to find called into question a concept that has been called "the most important general concept for the understanding of organic chemistry in general. . . ."[72]

If the analysis of the stability of benzene can be so complicated, what about the analysis of benzene derivatives or of polynuclear aromatic compounds such as naphthalene, anthracene, and phenanthrene? Their experimental resonance energies are found to be 61, 83.5, and 91.3 kcal/mol, respectively.[73] Naphthalene is like benzene in many ways, giving electrophilic aromatic substitution when treated with bromine, for example. Phenanthrene, however, undergoes electrophilic addition with bromine,[74] which suggests that a greater resonance energy does not necessarily result in more aromatic behavior. On the other hand, aromatic character is also associated with thiophene (resonance energy 28 kcal/mol) and furan (resonance energy 16 kcal/mol).[73] At what point in a series of increasing experimental resonance energy values does aromaticity begin?

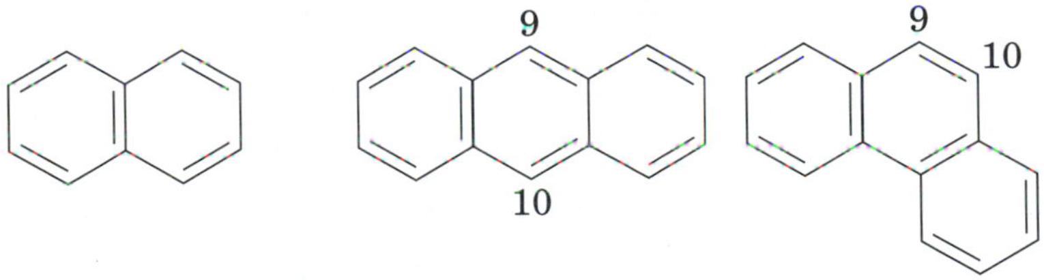

Figure 4.32 Polynuclear aromatic compounds.

9
10
9
10

Naphthalene Anthracene Phenanthrene

[69]Shaik, S. S.; Hiberty, P. C.; Lefour, J.-M.; Ohanessian, G. *J. Am. Chem. Soc.* **1987**, *109*, 363; Shaik, S. S.; Hiberty, P. C.; Ohanessian, G.; Lefour, J.-M. *J. Phys. Chem.* **1988**, *92*, 5086. This conclusion was supported by other researchers: Stanger, A.; Volhardt, K. P. C. *J. Org. Chem.* **1988**, *53*, 4889. A related paper reached the same conclusion about naphthalene from the VB approach: Sironi, M.; Cooper, D. L.; Gerratt, J.; Raimondi, M. *J. Chem. Soc., Chem. Commun.* **1989**, 675. See also Jug, K.; Köster, A. M. *J. Am. Chem. Soc.* **1990**, *112*, 6772.

[70]It has also been suggested that aromaticity in benzene results from highly correlated Cooper pairs of electrons: Squire, R. H. *J. Phys. Chem.* **1989**, *91*, 5149. See also Van Hooydonk, G. *J. Phys. Chem.* **1988**, *92*, 1700; Squire, R. H. *J. Phys. Chem.* **1988**, *92*, 1701.

[71]Glendening, E. D.; Faust, R.; Streitwieser, A.; Volhardt, K. P. C.; Weinhold, F. *J. Am. Chem. Soc.* **1993**, *115*, 10952.

[72]Katritzky, A. R.; Barczynski, P.; Musumarra, G.; Pisano, D.; Szafran, M. *J. Am. Chem. Soc.* **1989**, *111*, 7.

[73]Reference 29, pp. 98 *ff*.

[74]Reference 56, p. 22.

Aromaticity in Small Ring Systems

Whatever may be the uncertainties about the aromaticity of polynuclear aromatics or heteroaromatics, aromaticity does seem to be a useful concept in discussing small, monocyclic hydrocarbons. Let us summarize some of the results for a few important systems. The cyclopropenyl carbocation has only 2 π electrons. Both may be placed in the lowest energy MO, so the molecular energy is calculated to be $E_\pi = 2\alpha + 4\beta$. This is 2 β lower in energy than the energy of a reference system composed of a double bond and a nondelocalized carbocation. Therefore the cyclopropenyl carbocation is predicted to be much more stable than other carbocations, including allyl. In fact, its delocalization energy is calculated to be the same as that of benzene. The cyclopropenyl carbocation **2** is relatively stable. It has been made by the reaction in Figure 4.33 and has been isolated.[75] The triphenyl derivative has been isolated in crystal form, and its structure has been determined by X-ray crystallography. The C-C bonds in the three membered ring are found to be 1.40 Å, which is very similar to the 1.39 Å bond lengths in benzene. The dipropylcyclopropenium ion is soluble and stable in 1 N aqueous HCl.[76] Clark has calculated that the unsubstituted cation is most stable as a planar structure with maximum *p*-overlap, the aromaticity being calculated at 49 kcal/mol.[77]

The cyclopropenyl anion is unstable.[78,79] Calculations by Clark suggested that the anion is antiaromatic by 143 kcal/mol if all three carbon atoms lie in the same plane, but that deformation of the molecule can lower the energy by decreasing the overlap of the *p* orbitals.[77] Calculations also indicated that the vinyl hydrogen atoms of cyclopropene are more acidic than are the methylene hydrogen atoms.[80] That conclusion is supported by ex-

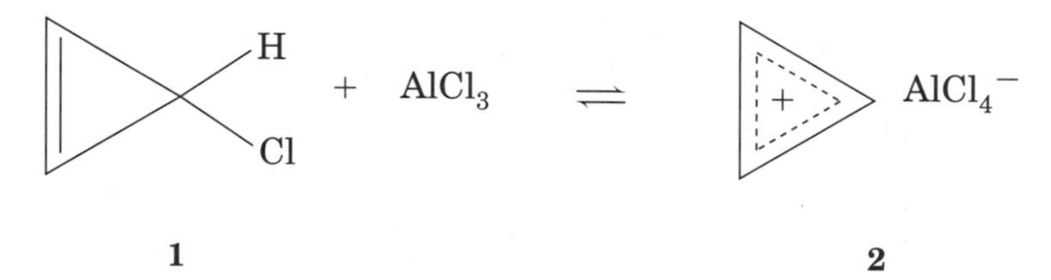

Figure 4.33 Synthesis of cyclopropenyl carbocation.

[75]Breslow, R.; Groves, J. T. *J. Am. Chem. Soc.* **1970**, *92*, 984.

[76]Breslow, R.; Höver, H. *J. Am. Chem. Soc.* **1960**, *82*, 2644.

[77]Clark, D. T. *J. Chem. Soc. D, Chem. Commun.* **1969**, 637. For *ab initio* studies on the aromaticity of three-membered rings, see Byun, Y.-G.; Saebo, S.; Pittman, Jr., C. U. *J. Am. Chem. Soc.* **1991**, *113*, 3689.

[78]The 3-carbomethoxycyclopropen-3-yl anion has been studied in the gas phase: Sachs, R. K.; Kass, S. R. *J. Am. Chem. Soc.* **1994**, *116*, 783.

[79]The cyclopropenyl radical does not exhibit aromatic character either. Both theoretical and experimental studies indicated that the cyclopropenyl radical distorts from a structure with D_{3h} symmetry to form a structure with C_{2v} or C_s symmetry. *Cf.* Davidson, E. R.; Borden, W. T. *J. Chem. Phys.* **1977**, *67*, 2191 and references therein; Closs, G. L.; Evanochko, W. T.; Norris, J. R. *J. Am. Chem. Soc.* **1982**, *104*, 350; Chipman, D. M.; Miller, K. E. *J. Am. Chem. Soc.* **1984**; *106*, 6236.

[80]Dorko, E. A.; Mitchell, R. W. *Tetrahedron Lett.* **1968**, 341.

Figure 4.34

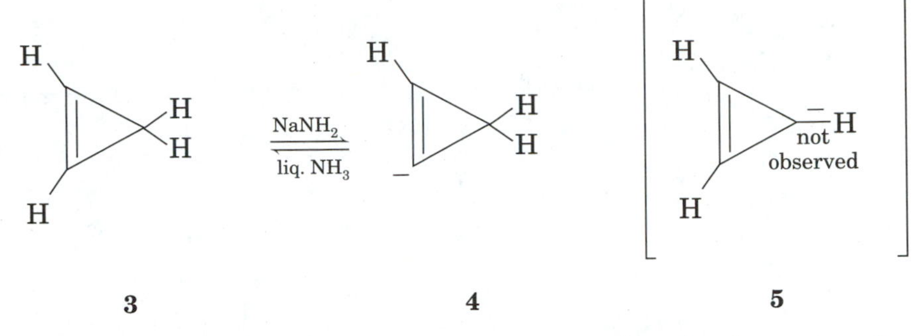

perimental data. Breslow found that 1,2,3-triphenylcyclopropene does not undergo hydrogen exchange in ammonia,[81] so the cyclopropenyl anion is apparently much harder to form than the anion of, for example, cyclopentadiene. Cyclopropene itself does undergo proton loss in liquid ammonia, and in fact the resulting carbanion is sufficiently concentrated in solution that an NMR spectrum can be obtained.[82] Surprisingly, however, the spectrum indicates not the presence of the expected resonance stabilized carbanion **5**, but rather the vinyl carbanion **4**. Thus the carbanion **5** seems from these experiments to show antiaromatic character; at least, there is no evidence for any resonance stabilization.

Cyclobutadiene also has two electrons in an orbital of energy $\alpha + 2\beta$, but it has two other electrons which must go into orbitals having energy α. The π energy of the cyclobutadiene molecule is calculated to be $4\alpha + 4\beta$. This is the same as the energy of two nondelocalized double bonds, so its delocalization energy is 0. Therefore there appears to be no appreciable stabilization to be gained from delocalizing electrons into molecular orbitals in this system.

Cyclobutadiene has been the subject of much experimental effort. A number of reactions that were expected to yield cyclobutadiene were found to produce other products instead.[83] The observation that an apparent dimer of cyclobutadiene (**9**) was found in some cases suggested that the elusive compound might be formed as a transient species but was so reactive that it could not be detected in fluid media. Therefore, some studies of cyclobutadiene utilized an elegant technique known as matrix isolation spectroscopy.[84] In this method, a photochemically reactive compound is deposited in a matrix of an unreactive substance such as argon on optical windows inside a spectrophotometer. The apparatus is designed to maintain a temperature low enough to keep the matrix solid, yet to allow irradiation

---

[81]Breslow, R.; Dowd, P. *J. Am. Chem. Soc.* **1963**, *85*, 2729.

[82]Schipperijn, A. J. *Recl. Trav. Chim. Pays-Bas* **1971**, *90*, 1110.

[83]However, researchers have synthesized substituted cyclobutadienes in which substituents either retard reaction through steric effects or stabilize the ring through push-pull electronic effects. For a discussion see Hess, Jr., B. A.; Schaad, L. J. *J. Org. Chem.* **1976**, *41*, 3058 and references therein.

[84]Barnes, A. J.; Hallam, H. E. *Quart. Rev. Chem. Soc.* **1969**, *23*, 392.

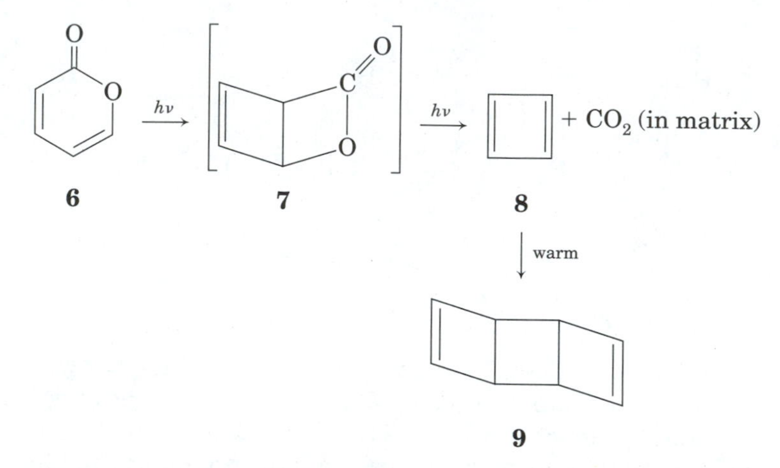

Figure 4.35 Matrix synthesis of cyclobutadiene.

of the sample by light for photochemical reaction and then for spectroscopic analysis of the products.

Chapman, Krantz, and others used bicyclopyranone, a compound believed to generate cyclobutadiene by photochemical elimination of CO_2. Analysis of the infrared spectrum of the photoproduct suggested that cyclobutadiene existed as a square planar molecule.[85] This result caused considerable discussion among theoreticians, since some calculations indicated that cyclobutadiene should exist as a rectangular molecule, with mostly single and mostly double bonds[86] as shown in Figure 4.36. This deformation would be expected to reduce, to some extent at least, the anti-aromatic character of cyclobutadiene itself.

Later the experimental evidence for square cyclobutadiene was called into question. Krantz reported the photolysis of bicyclopyranone in which the carbon atom eliminated with CO_2 was labeled with ^{13}C. One important infrared band which had been assigned to a vibration of square planar cyclobutadiene in earlier studies was found to be altered by the isotopic change, suggesting that this band was due to CO_2 trapped with the cyclobutadiene in the rigid rare gas matrix.[87] Thus the exerimental data did not really answer the question of the structure of cyclobutadiene. Later work on the theoretical determination of the infrared spectrum of cyclobutadiene[88]

Figure 4.36 Distortion of cyclobutadiene to a rectangular structure.

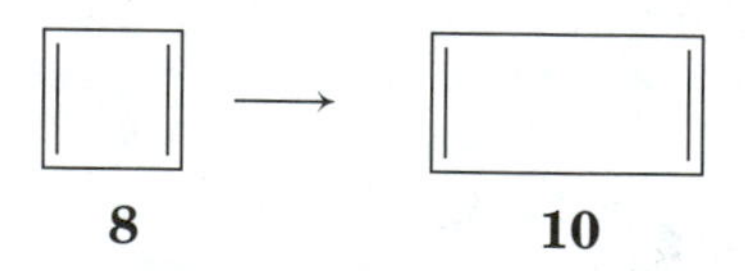

[85]Chapman, O. L.; McIntosh, C. L.; Pacansky, J. *J. Am. Chem. Soc.* **1973**, *95*, 614.

[86]Kollmar, H.; Staemmler, V. *J. Am. Chem. Soc.* **1977**, *99*, 3583.

[87]Pong, R. G. S.; Huang, B.-S.; Laureni, J.; Krantz, A. *J. Am. Chem. Soc.* **1977**, *99*, 4153.

[88]Kollmar, H.; Staemmler, V. *J. Am. Chem. Soc.* **1978**, *100*, 4304.

Figure 4.37
Reaction of cyclopentadiene as an acid.

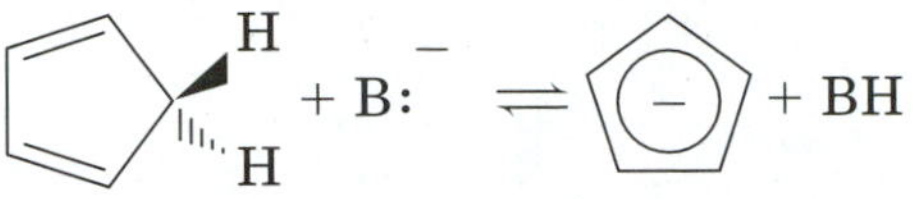

11 + B:⁻ ⇌ 12 + BH

and further matrix isolation spectroscopy experiments,[89] including the use of polarized IR spectroscopy and ^{13}C NMR of labeled cyclobutadiene,[90] appear to have settled the question in favor of a rectangular structure for cyclobutadiene.[91] Interestingly, Carpenter has suggested that different rectangular forms of cyclobutadiene can interconvert by tunneling.[92,93,94,95]

The cyclopentadienyl anion has six electrons in the π system, giving the molecule a π energy of $6\,\alpha + 6.47\,\beta$ and a delocalization energy of $2.47\,\beta$. The relatively high acidity of cyclopentadiene suggests that the cyclopentadienyl anion is more stable than other anions. For example, the pK_a of cyclopentadiene (**11**, Figure 4.37) is 16, which is nearly the same as that of water.[96] In contrast, the very unstable unsubstituted cyclopentadienyl cation was found to be a ground state triplet (that is, a species with two unpaired electrons) at 78 K.[97]

The cycloheptatrienyl carbocation is also a six electron system, and its six electrons also go into purely bonding orbitals. Its delocalization energy is calculated to be $2.99\,\beta$. The cycloheptatrienyl cation is an especially stable carbocation,[98] although it is still a cation and is certainly not as stable as benzene.

[89]Masamune, S.; Souto-Bachiller, F. A.; Machiguchi, T.; Bertie, J. E. *J. Am. Chem. Soc.* **1978**, *100*, 4889.

[90]Orendt, A. M.; Arnold, B. R.; Radziszewski, J. G.; Facelli, J. C.; Malsch, K. D.; Strub, D. H.; Grant, M.; Michl, J. *J. Am. Chem. Soc.* **1988**, *110*, 2648.

[91]Dewar, M. J. S.; Merz, Jr., K. M.,; Stewart, J. J. P. *J. Am. Chem. Soc.* **1984**, *106*, 4040.

[92]Carpenter, B. K. *J. Am. Chem. Soc.* **1983**, *105*, 1700. See also Huang, M.-J.; Wolfsberg, M. *J. Am. Chem. Soc.* **1984**, *106*, 4039.

[93]See also Anold, B. R.; Radziszewski, J. G.; Campion, A.; Perry, S. S.; Michl, J. *J. Am. Chem. Soc.* **1991**, *113*, 692; Lefebvre, R.; Moiseyev, N. *J. Am. Chem. Soc.* **1990**, *112*, 5052.

[94]The cyclobutadiene dication should have two electrons, both in ψ_1, so it is predicted to have DE = $2\,\beta$, the same as cyclopropenyl cation. There is a report that the dication has been observed: Olah, G. A.; Mateescu, G. D. *J. Am. Chem. Soc.* **1970**, *92*, 1430.

[95]The cyclobutadiene monocation should have a delocalization energy of $1\,\beta$, and the electron paramagnetic resonance (epr) spectra of substituted monocations have been reported: Courtneidge, J. L.; Davies, A. G.; Lusztyk, E.; Lusztyk, J. *J. Chem. Soc., Perkin Trans. 2* **1984**, 155.

[96]Streitwieser, Jr., A.; Nebenzahl, L. L. *J. Am. Chem. Soc.* **1976**, *98*, 2188. See also Stewart, R. *The Proton: Applications to Organic Chemistry*; Academic Press: New York, 1985; pp. 72–74.

[97]Saunders, M.; Berger, R.; Jaffe, A.; McBride, J. M.; O'Neill, J.; Breslow, R.; Hoffman, J. M.; Perchonock, C.; Wasserman, E.; Hutton, R. S.; Kuck, V. J. *J. Am. Chem. Soc.* **1973**, *95*, 3017.

[98]Doering, W. v. E.; Knox, L. H. *J. Am. Chem. Soc.* **1954**, *76*, 3203.

13

Planar cyclooctatetraene (COT) is predicted to have a π energy of 8 α + 9.657 β, giving it a delocalization energy of 1.66 β. However, the HMO energy level diagram (Figure 4.26f) also predicts it to be an open shell molecule, and thus unstable. Although the original synthesis by Willstatter was a 13-step process,[99] it was subsequently found that nickel-catalyzed tetramerization of acetylene gives COT in high yields. Today, COT is a commercially available compound regarded as a stable substance. Studies of aromaticity indicate that it is nonaromatic, that is, neither aromatic nor antiaromatic.[100]

The stability of COT is easily rationalized if we recall one of the fundamental assumptions of HMO theory—that the structure is planar with the *p* orbitals all parallel to each other. However, because a planar COT molecule would have ring bond angles of 135°, there would be considerable angle strain in addition to the electronic destabilization. COT can adopt the conformation shown in Figure 4.38 to reduce both angle strain and electronic strain, and experimental evidence indicates that it is predominantly tub shaped, as shown.[101] Thus the double bonds in the molecule are nearly perpendicular to each other, which makes them orthogonal (noninteracting). Therefore, the actual HMOs of COT are more like those of a system of four noninteracting double bonds, not an aromatic or antiaromatic system.[102]

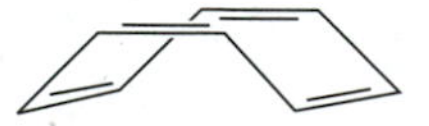

Figure 4.38
Tub conformation of COT.

Adding two more electrons to the COT system would fill the vacancies in the partially filled HMOs, making the molecule an aromatic, closed shell configuration structure. In fact, COT dianion is well known in organic chemistry.[103] Similarly, removing two electrons from COT should generate a *relatively* stable dication, and Olah and co-workers have studied the properties of substituted COT dications.[104]

---

[99]Willstätter, R.; Heidelberger, M. *Chem. Ber.* **1913**, *46*, 517.

[100]For a review, see Craig, L. E. *Chem. Rev.* **1951**, *49*, 103.

[101]Karle, I. L. *J. Chem. Phys.* **1952**, *20*, 65. Dewar has reported calculations showing that the chair conformation of cyclooctatetraene is unstable with respect to the tub conformations shown here: Dewar, M. J. S.; Merz, Jr., K. M. *J. Chem. Soc., Chem. Commun.* **1985**, 343.

[102]Theoretical calculations indicate that a D_{8h} (planar, antiaromatic) structure is a transition state in the double bond shifting reaction of cyclooctatetraene: Hrovat, D. A.; Borden, W. T. *J. Am. Chem. Soc.* **1992**, *114*, 5879.

[103]Katz, T. J. *J. Am. Chem. Soc.* **1960**, *82*, 3784, 3785.

[104]Olah, G. A.; Staral, J. S.; Liang, G.; Paquette, L. A.; Melega, W. P.; Carmody, M. J. *J. Am. Chem. Soc.* **1977**, *99*, 3349.

Larger Annulenes

The term *annulene* refers to cyclic compounds composed of alternating single and double bonds. We denote the annulenes by the terminology [*n*]annulene, where *n* is the number of π carbon atoms in the ring. By this definition cyclooctatetraene is [8]annulene. The next annulene is [10]annulene or cyclodecapentaene, which is predicted by HMO theory to be aromatic. As was evident for cyclooctatetraene, however, a planar $C_{10}H_{10}$ structure with all-cis double bonds may result in a strained σ skeleton. For example, the C—C—C bond angle for D_{10h} cyclodecapentaene (**14**) would be 144°, making the molecule highly strained. Masamune and co-workers reported the isolation of two different $C_{10}H_{10}$ isomers, neither of which was the aromatic **14**, leading to the conclusion that $C_{10}H_{10}$ is not aromatic.[105] A theoretical study by Schaefer and co-workers led to the conclusion that **14** is more stable than the planar (D_{5h}) species with alternating single and double bonds (**15**) by about 8 kcal/mol, suggesting that electron delocalization is stabilizing in this system. However, the study also led to the conclusion that a pair of C_2 structures are 14 and 13 kcal/mol, respectively, lower in energy than **14**, and that several other nonplanar structures are also lower in energy than **14**.[106]

Annulenes with some trans double bonds can have lower angle strain than the corresponding all-cis structures, and van der Waals strain from substituents on the internal double bonds can be minimized if the ring size is large enough. For example, [18]annulene (**17**) is planar because it has six trans and three cis double bonds, an arrangement that puts six vinyl protons on the inside of the ring and twelve on the outside. The interior protons show resonance at −1.0 δ, while the twelve exterior protons are at 8.8 δ.[107]

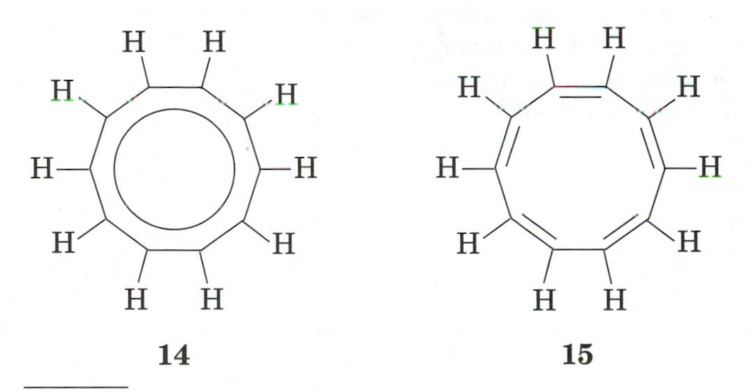

Figure 4.39
D_{10h} (left) and D_{5h} (right) structures for [10]annulene.

---

[105] Masamune, S.; Seidner, T. *J. Chem. Soc., Chem. Commun.* **1969**, 542; Masamune, S.; Hojo, K.; Bigam, G.; Rabenstein, D. L. *J. Am. Chem. Soc.* **1971**, *93*, 4966; Masamune, S.; Darby, N. *Acc. Chem. Res.* **1972**, *5*, 272.

[106] Xie, Y.; Schaefer, III, H. F.; Liang, G.; Bowen, J. P. *J. Am. Chem. Soc.* **1994**, *116*, 1442. However, an aromatic [10]annulene was identified in high level calculations: Sulzbach, H. M.; Schleyer, P. v. R.; Jiao, H.; Xie, Y.; Schaefer, III, H. F. *J. Am. Chem. Soc.* **1995**, *117*, 1369.

[107] The PMR spectrum is temperature-dependent. At 100° all protons are in a single band at δ 5.5. At 20° there are two bands, one at δ −1.0 and one at δ 8.8. At −60° the spectrum consists of a quintet at δ −4.2 and a quartet at δ 9.25. Calder, I. C.; Garratt, P. J.; Sondheimer, F. *Chem. Commun.* **1967**, 41.

Figure 4.40
PMR data for benzene (left) and [18]annulene (right).

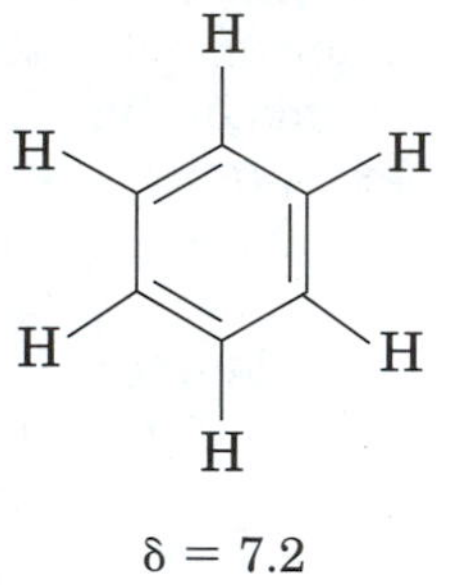

$\delta = 7.2$

16

$\delta = -1.0$

$\delta = 8.8$

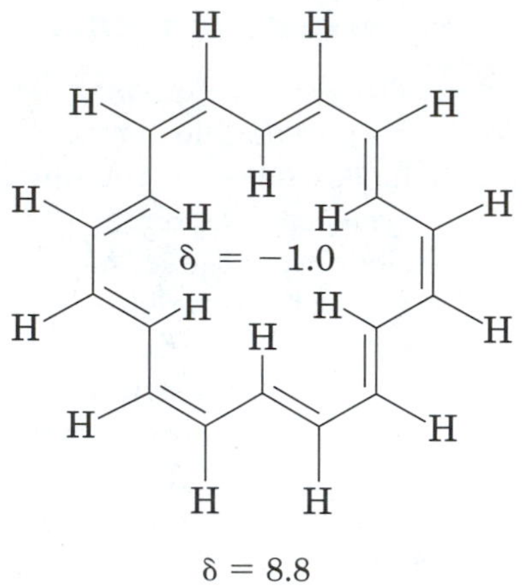

17

These shifts are evidence for a very strong ring current effect, which is one of the experimental criteria for aromaticity.[108]

Chemical reactivity data suggest that larger rings may be less aromatic than are the smaller ones. As shown in Figure 4.25, the energy levels of monocyclic conjugated π systems are constrained to lie between $\alpha + 2\,\beta$ and $\alpha - 2\,\beta$. As the ring includes more and more π centers, there is an ever increasing number of HMO energy levels between those two limits, and the difference in delocalization energy between a [4*n*]annulene and a [4*n* + 2]-annulene becomes ever smaller. Moreover, the HMO method has ignored overlap and the problem of electron-electron interactions. Figure 4.41 shows the resonance energy for a series of annulenes calculated by Dewar and Gleicher by a more advanced method.[109] If we take a large delocalization energy as a measure of aromaticity, then the distinction between what is aromatic and what is antiaromatic becomes smaller and smaller with increasing ring size, and for a ring of around 26 π centers, even a member of the $4n + 2$ series no longer is calculated to be more stable than its acyclic analogue. Now we are faced with a quantitative question: At what point in the series of annulenes does aromatic character end?[110] It seems that the concept of aromaticity can be seen only as a convenient paradigm. It is useful for categorizing the properties and reactions of many compounds and for com-

[108][18]Annulene exhibits *diamagnetic anisotropy*, meaning that the effect of the right current is not the same in every direction from a given point. Thus the lines of force reinforce the applied field at the 12 exterior protons, but it is opposite the applied field in the vicinity of the six interior protons.

[109]Dewar, M. J. S.; Gleicher, G. J. *J. Am. Chem. Soc.* **1965**, *87*, 685.

[110]For further reading, see (a) Chung, A. L. H.; Dewar, M. J. S. *J. Chem. Phys.* **1965**, *42*, 756; (b) Schaad, L. J.; Hess, Jr., B. A. *J. Chem. Educ.* **1974**, *51*, 640.

Figure 4.41 Change in calculated resonance energies for annulenes. (Figure based on data in reference 109.)

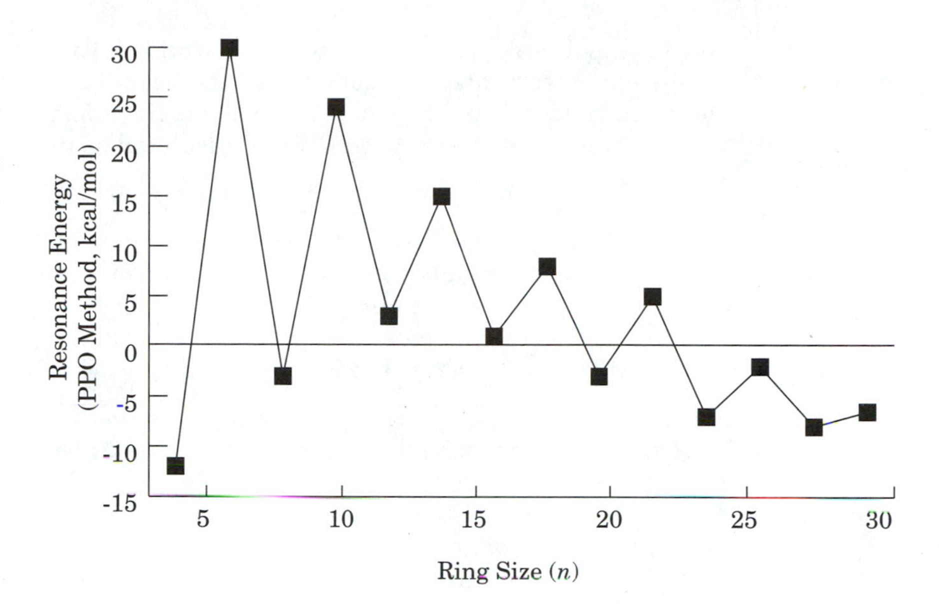

paring some compounds with others, but a simple definition of aromaticity is elusive.[111]

Dewar Resonance Energy and Absolute Hardness

There are two other approaches to defining aromaticity that we will consider. The first is based on a comparison of the heat of formation of an organic compound with the heat of formation calculated for a nondelocalized reference system. The difference between those two values would then be the resonance energy for this system.

$$\text{Resonance Energy}_{(\text{calc})} = \Delta H_f^0 - \Delta H_{f\,\text{ref}}^0 \quad \textbf{(4.62)}$$

At first glance this approach would seem to be dependent on uncertain reference systems, but it has been given firm footing by the work of Dewar,[112] and Baird has elaborated Dewar's approach.[113] Now, calculated *values of resonance energy can be based only on knowledge of the molecular structure and experimental data for a few reference compounds*. The resonance energy

[111]If *aromaticity* is defined in other terms, then quantitative definitions are possible. Dauben developed the concept of exaltation of magnetic susceptibility (the difference in weight of a substance in and out of a magnetic field) as a criterion for aromaticity: Dauben, Jr., H. J.; Wilson, J. D.; Laity, J. L. *J. Am. Chem. Soc.* **1968**, *90*, 811. Schleyer, P. v. R.; Jiao, H.; Herges, R. *Abstracts of the 207th National Meeting of the American Chemical Society*, San Diego, CA, March 13–17, 1994; Abstract ORGN 247, have reported that magnetic susceptibility exaltation can be measured or computed and can be used to define aromaticity, both for ground state structures and for transition structures.

[112]Reference 47, Chapter 5.

[113]Baird, N. C. *J. Chem. Educ.* **1971**, *48*, 509.

so calculated is denoted the **Dewar Resonance Energy** or **DRE**. For a completely unsaturated compound with the formula $C_nH_mO_p\ddot{O}_q$, where p is the number of carbonyl-type oxygens and q is the number of ether-type oxygens, Baird has shown that the DRE is calculated as in equation 4.63:

$$\text{DRE} = 7.435n - 0.605m - 32.175p - 29.38q - \Delta H_f^0\,(C_nH_mO_p\ddot{O}_q) \qquad \textbf{(4.63)}$$

where m is the number of C—H bonds present
p is the number of C=O bonds present
$(n - p)/2$ is the number of C=C bonds present
$2q$ is the number of C—O bonds present and
$\{n - q - m/2\}$ is the number of $C(sp^2)$—$C(sp^2)$ single bonds present.

Note that this approach allows one to calculate indirectly the heat of formation of $C(sp^2)$—$C(sp^2)$ single bonds, which was difficult to determine directly. With this approach, the **definition of aromaticity** is that an *aromatic compound has DRE > 0, an antiaromatic compound has a DRE < 0*, and a compound that is neither aromatic nor antiaromatic has a DRE about 0. Results of DRE calculations by Baird are listed in Table 4.2.[114] They confirm our experience regarding the $4n + 2$ rule for [n]annulenes, and they also provide new ideas about polynuclear aromatics.

The Dewar-Baird approach offers a number of insights into molecular structure and reactivity. For example, the DREs given in Table 4.3 show that for many benzene compounds conjugated with other groups the DRE depends more on the number of benzene rings than on the degree of conju-

Table 4.2 DRE values for selected hydrocarbons.

Compound	Experimental RE[a]	Calculated DRE[b]
Cyclobutadiene[c]	—	− 17 kcal/mol
Benzene	+ 21 kcal/mol	+ 21
Cyclooctatetraene[d]	—	− 10
Cyclodecapentaene	+ 10	+ 6
Naphthalene	+ 33	+ 33
Anthracene	+ 43	+ 42
Phenanthrene	+ 49	+ 49

[a]Calculated from the experimental heat of formation given in reference 113 and equation 4.63.
[b]Values taken from Tables 3 and 4 of reference 113.
[c]Experimental heats of formation are not available.
[d]Assumed to be planar.

[114]Reference 113, pp. 512–513.

Table 4.3 DRE values (kcal/mol) for conjugated benzenes.

Compound	DRE	Benzene Rings	DRE/Benzene Ring
Benzene	21.2	1	21.2
Styrene	21.3	1	21.3
Biphenyl	43.6	2	21.8
Stilbene	42.1	2	21.0
Benzophenone	43.8	2	21.9

Data from reference 113, p. 511.

gation of double bonds with those rings. DREs are also useful in correlating chemical reactivity. As shown in Figure 4.42, addition by the top pathway produces a structure with a naphthalene moiety, so its DRE is 33 kcal/mol. Reaction by the bottom pathway produces a product with two benzene rings, so its DRE is 2 (21) = 42 kcal/mol. Since the DRE of the reactant is 42 kcal/mol, the ΔDRE is +10 kcal/mol for the top pathway, but ≈ 0 for the lower pathway. Therefore the lower pathway should be favored.

Another definition of aromaticity that does not depend explicitly on either experimental results or on comparison with reference compounds is the concept of **absolute hardness**, η, which is defined as one-half the energy difference between the highest occupied molecular orbital (HOMO) and the lowest unoccupied molecular orbital (LUMO) in a system (equation 4.64).[115] According to Koopmans' Theorem, E_{HOMO} is related to the ionization potential of the species, while E_{LUMO} is related to its electron affinity. Thus a large gap between these two orbitals would imply resistance to both oxidation and reduction, which would suggest low chemical reactivity.

$$\eta = \frac{E_{\text{LUMO}} - E_{\text{HOMO}}}{2} \qquad \textbf{(4.64)}$$

Zhou, Parr, and Garst showed that absolute hardness correlates well with theoretical measures of aromaticity,[116] but η alone does not allow the

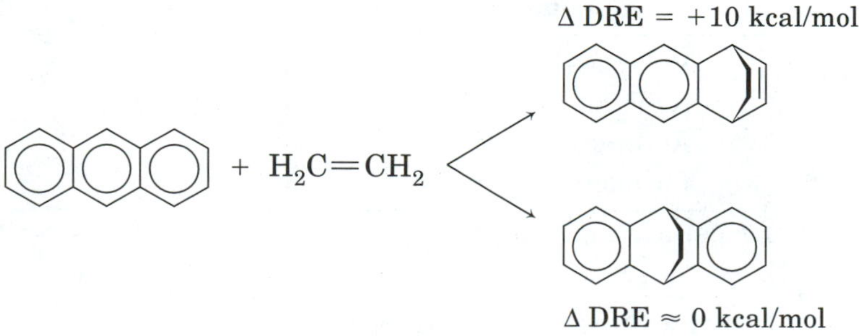

Δ DRE = +10 kcal/mol

+ $H_2C{=}CH_2$

Δ DRE ≈ 0 kcal/mol

Figure 4.42 DRE differences for addition reactions to anthracene. (Adapted from reference 113, p. 512.)

[115] Pearson, R. G. *Acc. Chem. Res.* **1993**, *26*, 250 and references therein.

[116] Zhou, Z.; Parr, R. G.; Garst, J. F. *Tetrahedron Lett.* **1988**, 4843.

categorization of aromaticity. Zhou and Parr later defined the *relative hardness* of a species as the difference between its hardness and the hardness of an acyclic reference compound.[117] Based upon these correlations, these authors proposed that compounds are aromatic if their Hückel absolute hardness values (that is, hardness determined from HMO HOMO-LUMO gaps) are less than $-0.2\ \beta$, antiaromatic if the value of η is greater than $-0.15\ \beta$, and nonaromatic if η is between those two values. The corresponding division based on relative hardness is 0; that is, a cyclic molecule that is harder than an acyclic analogue is aromatic, while one that is not as hard as an acyclic analogue is antiaromatic.

4.3 Quantitative Methods Using Valence Bond Theory

In discussing molecular hydrogen in Chapter 1, we noted that both the MO and VB descriptions should be considered only as starting points for calculating molecular properties and not as ultimate solutions. In this chapter we have used HMO theory to compare theoretical predictions with chemical experience, but chemists have long used resonance theory, which is an approximate form of valence bond theory, to predict properties of molecules. For example, the partial charges in the resonance hybrid of the allyl cation (Figure 4.13) are the same as those determined in the HMO calculation. Furthermore, chemical experience indicates a close relationship between the number of resonance structures and the resonance energy of the system, as shown for several polynuclear aromatics in Table 4.4.[118]

The calculation of resonance energies from valence bond descriptions of contributing resonance structures was discussed in a comprehensive treatise by Wheland.[119] We will not discuss the details of the method because we will not use it again, but it is instructive to consider the example of the cal-

Table 4.4 Resonance structures and resonance energies.

Compound	Number of Kekulé Structures	Resonance Energy (kcal/mol)
Benzene	2	36.0
Naphthalene	3	61.0
Anthracene	4	83.5
Phenanthrene	5	91.3

Data from reference 29, p. 98.

[117]Zhou, Z.; Parr, R. G. *J. Am. Chem. Soc.* **1989**, *111*, 7371.

[118]Kekulé resonance structures are those structures with only single and double bonds, no extended or formal bonds or diradicals.

[119]Reference 29. See especially Chapter 9.

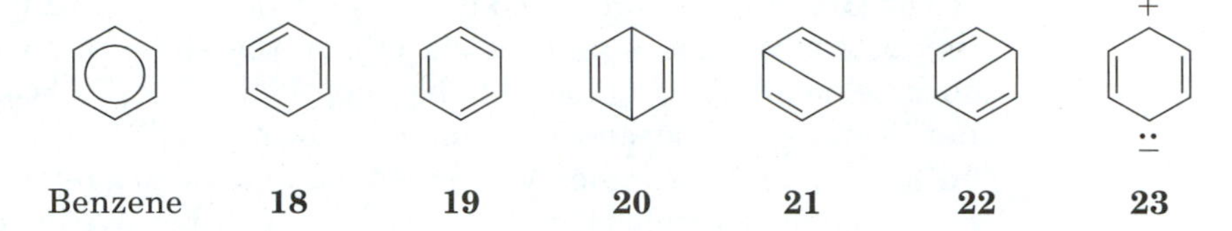

Figure 4.43 Benzene resonance structures.

culation of the resonance energy of benzene.[120,121] One begins by writing valence bond wave functions for each resonance structure that contributes to the resonance hybrid (Figure 4.43).[122] Structures **18** and **19** are the classical Kekulé structures. However, the VB method is not restricted to Kekulé structures, and other structures may also be included, such as those with elongated or formal bonds (**20**, **21**, and **22**) and ionic structures (e.g., **23**). Because the latter structures seem high in energy, we suspect that they will make only a minor contribution to the overall properties of the resonance hybrid. However, some contribution of high energy structures may be needed to accurately calculate the properties of the molecule.

The VB wave function for benzene is a linear combination of the VB wave functions for each of the contributing resonance structures in the general format shown in equation 4.65

$$\Phi_{\text{Benzene}} = k_{18}\theta_{18} + k_{19}\theta_{19} + \cdots + = \textstyle\sum_G k_G\theta_G \quad \textbf{(4.65)}$$

where θ_G is the VB wave function of the Gth contributing resonance structure. We determine the coefficients (k) in the wave function as well as the resonance energy for benzene by use of the variation method, as in HMO theory. That is, we find the mix of contributing resonance structures that minimizes the calculated energy of benzene.[29] In essence one sets the partial derivative of the energy of benzene with respect to each of the ks equal to zero and then finds the solution to the secular determinant that results. Consideration of five contributing structures (**18–22**, Figure 4.43) leads to a 5×5 determinant. There are assumptions, approximations, and substitutions that facilitate the calculation, just as there are in the HMO method. Solving the determinant leads to a set of energy levels, and determination of the coefficients for the lowest energy level leads to the VB function for benzene reported by Wheland (equation 4.66).

$$\Phi_{\text{Benzene}} = 0.37(\theta_{18} + \theta_{19}) + 0.16\,(\theta_{20} + \theta_{21} + \theta_{22}) \quad \textbf{(4.66)}$$

The resonance energy determined in this calculation is the difference in energy between benzene and the most stable contributing resonance structure

[120]Reference 29, pp. 629–641.

[121]See also the discussion of the valence bond calculation for benzene in reference 6.

[122]As was the case with HMO theory, the valence bond description of these resonance structures is limited to wave functions involving the resonance electrons (i.e., those which correspond to the π electrons in the MO method). The σ bonds are not explicitly considered.

(**18** or **19**). The value obtained is $-1.11J$, where J is the adjacent exchange integral, a term that is reminiscent of β in the HMO method. In a number of such calculations, J is found to be approximately -30 kcal/mol, so the resonance energy of benzene is calculated to be about 33 kcal/mol.[123] That value is quite close to the commonly accepted value of 36 kcal/mol, although we have noted the questionable significance of that number. We have not carried out a detailed analysis, but it appears from this example that resonance theory provides an alternative model for calculating properties of benzene.[124]

The method just described is more difficult to apply to larger molecules. For naphthalene the HMO method leads to a 10×10 determinant, because the size of the determinant is fixed by the number of p orbitals being mixed to make the MO. For the resonance method, however, the size of the determinant is limited only by our willingness to include resonance structures. In general there are many more resonance structures that should be considered than there are p orbitals in the MO method, and a reasonable resonance treatment of naphthalene would require a 42×42 determinant.[125,126] For anthracene and phenanthrene, the HMO determinant would be 14×14, but the resonance method would require solution of a 429×429 determinant in each case. Use of symmetry could reduce the problem so that anthracene would require a determinant no larger than 126×126, and phenanthrene would require a determinant no larger than 232×232.[127] Still, these determinants are much more difficult to solve than are the smaller determinants obtained from the HMO method. Therefore, what makes HMO theory the more widely accepted approach to calculating molecular structure in organic chemistry is not that it is inherently superior to the resonance model, but that the calculations are easier to carry out.[128,129]

---

[123]See also Pauling, L.; Wheland, G. W. *J. Chem. Phys.* **1933**, *1*, 362.

[124]Cooper, D. L.; Gerratt, J.; Raimondi, M. *Nature* **1986**, *323*, 699 and references therein have reported a spin-coupled valence bond method for calculation of molecular electronic structure. They report that "our results suggest that the Kekulé description of benzene, as expressed in the classic VB form, is in fact much closer in reality than is a description in terms of delocalized molecular orbitals."

[125]This determinant size is a strict minimum. Even for benzene, determinants can be quite large. For example, Norbeck and Gallup used 175 resonance structures in an *ab initio* valence bond calculation for benzene and found that the ionic structures are major contributors to the calculated structure: Norbeck, J. M.; Gallup, G. A. *J. Am. Chem. Soc.* **1973**, *95*, 4460.

[126]For a valence bond calculation of naphthalene, representations of the valence bond structures used in the calculation, and the relative contribution of these structures to the resonance hybrid, see Sherman, J. *J. Chem. Phys.* **1934**, *2*, 488.

[127]Reference 29, p. 638.

[128]Wheland, G. W. *J. Chem. Phys.* **1934,** *2*, 474, has argued that the valence bond method as described by Heitler, London, Slater, and Pauling gives results closer to experimentally observed values than does the molecular orbital method attributed to Hund, Mulliken, and Hückel. The latter method, however, can more easily treat a wide variety of problems. As in other cases, we choose the model that is more useful, not necessarily the one that is more correct.

[129]A summary of the development of VB theory and a discussion of the merits of VB and MO theories has been given by Klein, D. J.; Trinajstić, N. *J. Chem. Educ.* **1990**, *67*, 633. The authors note that the criticism that the valence bond method requires the inclusion of ionic structures does not hold if antiorthogonalized atomic orbitals are used in the calculation. Moreover, they note that the computational effort required for a VB calculation should be considered in view of the computational effort required for MO calculations incorporating configuration interaction.

There is one other aspect of the comparison of HMO and resonance methods that we have not discussed in sufficient detail. In previous discussions of HMO theory and its applications to organic chemistry, we occasionally referred to comparisons of the predictions of HMO theory with those of the resonance treatment. In general, the two approaches were found to give at least qualitative agreement (e.g., benzyl cation charge densities), and in some cases quantitative agreement as well (e.g., allyl cation charge densities). However, that is not always the case. As shown in Figure 4.44, both benzene and cyclobutadiene have two Kekulé resonance structures. Why, therefore, is cyclobutadiene an elusive, exceedingly unstable compound, while benzene is the paradigm of aromatic stability? The answer is that the intuitive form of resonance theory we often use in organic chemistry is an oversimplified version of valence bond theory, and that stability cannot be predicted just by counting the number of resonance structures possible for a structure.[130] A more complete form of valence bond theory correctly predicts the stabilities of aromatic systems and the instabilities of antiaromatic systems.[31,130,131]

Valence bond theory is the subject of continuing investigation in theoretical organic chemistry.[129,132,133] Goddard has developed the generalized valence bond (GVB) theory.[134] Several workers have described valence bond calculations for conjugated and aromatic π systems,[135] and Hiberty has used valence bond methods to calculate the effect of strained annelated rings on aromatic systems.[136] Relationships between molecular orbital wave functions and valence bond wave functions have been developed,[137] and Fox

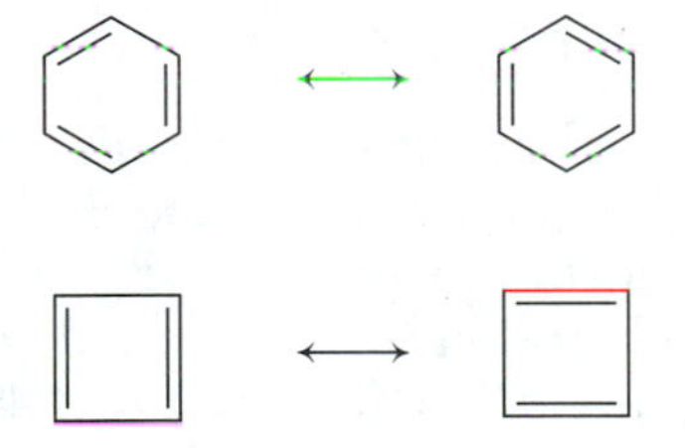

Figure 4.44 Kekulé resonance structures for benzene and cyclobutadiene.

[130]Fischer, H.; Murrell, J. N. *Theoret. chim. Acta (Berl.)* **1963**, *1*, 463.

[131]Klein and Trinajstić (reference 129) have noted that the prediction of simple HMO theory overestimates the difference in energy between rings with $4n$ electrons and those with $4n + 2$ electrons, so that the correct result is in between the MO and VB predictions.

[132]Mizoguchi, N. *J. Am. Chem. Soc.* **1985**, *107*, 4419; Shaik, S. S.; Hiberty, P. C. *J. Am. Chem. Soc.* **1985**, *107*, 3089.

[133]Klein, D. J.; Trinajstić, N., eds. *Valence Bond Theory and Chemical Structure*; Elsevier: Amsterdam, 1990.

[134]For application of a form of GVB theory and leading references, see Voter, A. F.; Goddard, III, W. A. *J. Am. Chem. Soc.* **1986**, *108*, 2830.

[135]Kuwajima, S. *J. Am. Chem. Soc.* **1984**, *106*, 6496; Klein, D. J. *Pure Appl. Chem.* **1983**, *55*, 299.

[136]Hiberty, P. C.; Ohanessian, G.; Delbecq, F. *J. Am. Chem. Soc.* **1985**, *107*, 3095.

[137]Živković, T. P. *Theoret. chim. Acta (Berl).* **1983**, *62*, 335; Hiberty, P. C.; Leforestier, C. *J. Am. Chem. Soc.* **1978**, *100*, 2012; Chen, C. *J. Chinese Chem. Soc.* **1973**, *20*, 1; Imkampe, K. *J. Chem. Educ.* **1975**, *52*, 429; Sardella, D. J. *J. Chem. Educ.* **1977**, *54*, 217.

and Matsen have reported a description of π-electron systems that employs features of both molecular orbital and valence bond theory.[138]

It is not necessary to carry out a complete valence bond calculation to obtain useful quantitative predictions of resonance energies, however. Herndon developed a structure-resonance theory (SRT) method that enables one to calculate resonance energies *using only Kekulé structures*.[139] The methods described in the references present

1. some relatively easy ways to determine the number of Kekulé structures for a molecule (without having to draw all the imaginable structures and then determine which ones are redundant);
2. the definitions of some permutations of these structures; and
3. some formulas for calculation of resonance energies and other parameters.

The results of these procedures are even easier to derive than are the results of simple HMO calculations, but they are very close to the results obtained with the more advanced MO methods we will describe shortly.[140]

The first step in the SRT method is the determination of the number of Kekulé structures for a given molecular structure, which is called the structure count (SC). One procedure for obtaining the SC is described in reference 139(a):[141] A polycyclic molecule is constructed by drawing a single chain or ring containing all atoms of the molecule, followed by insertion of lines one at a time until the molecule is complete. With each line insertion the SC is obtained from equation 4.67:

$$SC_a = SC_b + (SC_c \times SC_d) \quad \textbf{(4.67)}$$

where SC_a is the SC of the molecule or some fragment thereof, SC_b is the SC of the previous fragment, and SC_c and SC_d are the SCs of the fragments of the molecule without the line just inserted or its vertices.

This procedure may be illustrated by the example of anthracene. The outline of the molecule shown on the left of Figure 4.45 represents the σ

Figure 4.45 Kekulé structures for the open chain graph of anthracene.

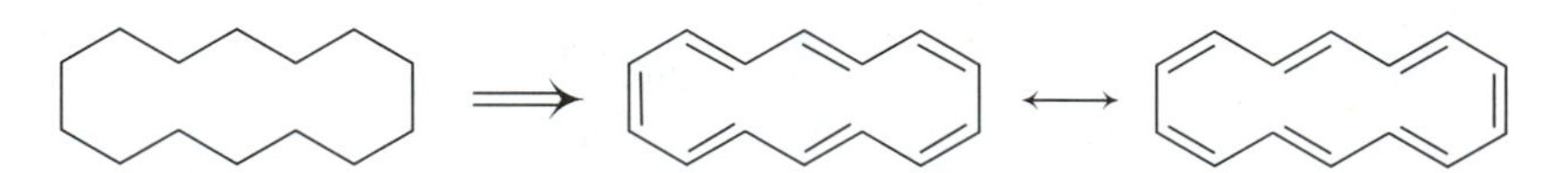

[138]Fox, M. A.; Matsen, F. A. *J. Chem. Educ.* **1985**, *62*, 367, 477, 551.

[139](a) Herndon, W. C. *J. Chem. Educ.* **1974**, *51*, 10; (b) Herndon, W. C. *J. Am. Chem. Soc.* **1973**, *95*, 2404; (c) Herndon, W. C.; Ellzey, Jr., M. L. *J. Am. Chem. Soc.* **1974**, *96*, 6631.

[140]Although we will not explicitly treat antiaromatic systems here, the methods described do incorporate such structures.

[141]The method was developed by Wheland, G. W. *J. Chem. Phys.* **1935**, *3*, 356.

Figure 4.46
Kekulé structures after inserting one σ bond.

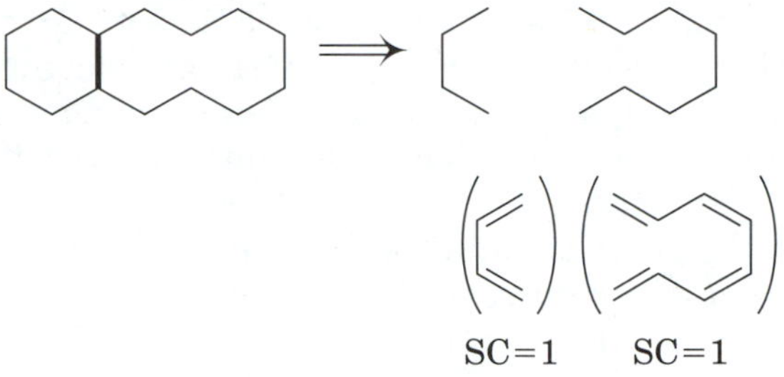

SC=1 SC=1

skeleton of a cyclic heptaene. Two resonance structures can be written for this structure, so the SC for this fragment is 2.[142]

Next we insert a line representing the sigma bond from C11—C12 in anthracene (shown in bold on the top left of Figure 4.46), then we delete that line *and its vertices* as indicated on the top right of Figure 4.46. The SC for each of the fragments so produced is 1, since each fragment would have only one Kekulé structure.[118] Therefore the SC for the initial fragment on the top left of Figure 4.46 (*with the line in place*) is $2 + 1 \times 1 = 3$.

Now we insert the line corresponding to the C13—C14 σ bond of anthracene, then delete this line and its vertices to produce the two fragments shown on the top right in Figure 4.47. Both of these fragments have only one Kekulé structure, so the SC for the anthracene molecule is $3 + 1 \times 1 = 4$. That result agrees with the number of Kekulé resonance forms for anthracene determined by inspection (Table 4.4).

This method of determining the SC is somewhat tedious for large molecules, so Herndon has also described another approach that is easier to use.[143] To determine the SC by this method, one deletes a vertex from the graph of the molecule and then writes the nonarbitrary vertex coefficients (with no coefficient smaller than 1) that sum to zero around every vertex in

Figure 4.47
Kekulé structures after inserting a second σ bond.

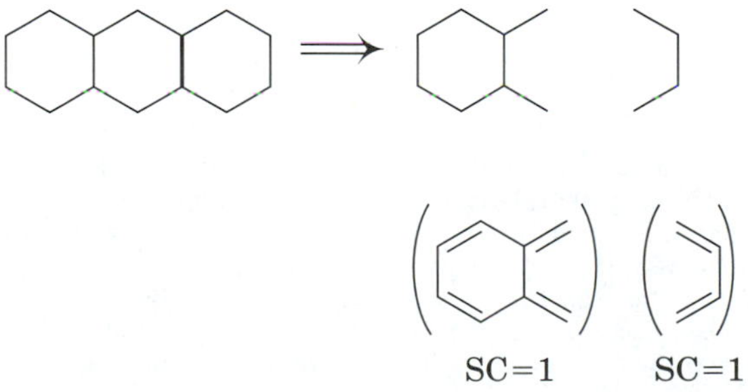

SC=1 SC=1

[142]Because anthracene is planar, the cyclic polyene shown in Figure 4.45 is assumed to be planar for the purposes of this calculation.

[143]In addition to reference 139(a), see also Herndon, W. C. *Tetrahedron* **1973**, *29*, 3.

the residual graph.[144] Note that these coefficients are the *unnormalized* coefficients of the NBMO for the odd alternant structure left by removing a vertex from the graph. The SC is then determined by taking the sum of the absolute values of the coefficients adjacent to the deleted vertex.

Again, it is instructive to consider the example of anthracene. Deleting a vertex from the graph of anthracene as shown on the left in Figure 4.48 generates the *odd alternant structure* on the right. We proceed as though we were determining the MO wave function for the NBMO of this fragment. Only the starred positions have nonzero coefficients, and the method requires that the smallest coefficient be unity. (If a trial run produces a coefficient smaller than 1 for a given carbon atom, we simply repeat the process starting at that position and assigning its coefficient to be 1.) The resulting coefficients are shown in Figure 4.48. Because the coefficients adjacent to the deleted vertex are -3 and -1, the SC of anthracene is $|-3| + |-1| = 4$.

Next we must define different permutations of electrons. A permutation of three pairs of electrons in a single ring is denoted by Γ_1, and the permutation of five pairs of electrons in two rings is denoted by Γ_2 (Figure 4.49).

The number of Γ_1 permutations for each ring in the molecule is the SC for the *residual molecule with that particular ring excised from the structure*. The number of Γ_2s is found by deleting *two adjacent rings* and summing the SCs for the residual system. Let us illustrate by continuing with the example of anthracene. If we identify the rings as A, B, and C, then we first delete the A ring and determine that the SC for the residual fragment is 1 (Figure 4.50). By symmetry, deleting the C ring gives an SC of 1 as well. Deleting the B ring also gives an SC of 1 (Figure 4.51). Therefore $n_1 = 3$.

Figure 4.48 Determination of structure count by vertex deletion method.

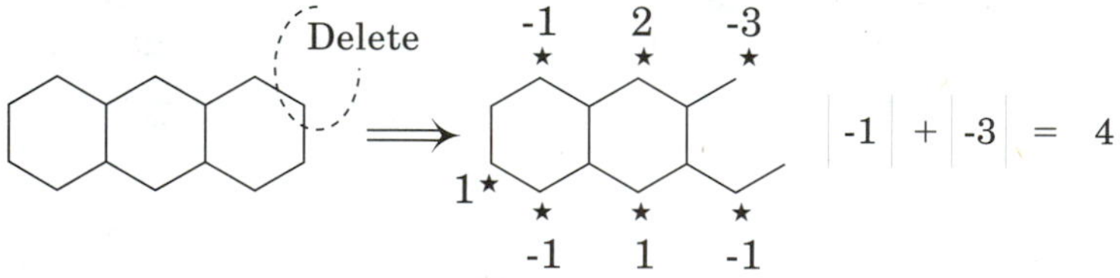

[144]The graph of the molecule is the outline of the molecule showing all σ bonds that comprise the skeleton of the π system. Much of the recent development in this area is inherently related to a branch of mathematics known as graph theory. See, for example, Hansen, P. J.; Jurs, P. C. *J. Chem. Educ.* **1988**, *65*, 574, 661; Trinajstić, N.; Nikolić, S.; Knop, J. V.; Müller, W. R.; Szymanski, K. *Computational Chemical Graph Theory*; Ellis Horwood: London, 1991; Trinajstić, N. *Chemical Graph Theory*, 2nd Ed.; CRC Press: Boca Raton, FL, 1992. For application of graph theory and the concept of conjugated circuits in determining aromaticity, see Randić, M.; Trinajstić, N. *J. Am. Chem. Soc.* **1987**, *109*, 6923; Klein, D. J. *J. Chem. Educ.* **1992**, *69*, 691. Graph theory has been used to extend the utility of valence bond calculations of aromatic systems: Alexander, S. A.; Schmalz, T. G. *J. Am. Chem. Soc.* **1987**, *109*, 6933. Graph theoretical methods may also be used in MO theory. For example, see Dias, J. R. *J. Chem. Educ.* **1989**, *66*, 1012; Mizoguchi, N. *J. Phys. Chem.* **1988**, *92*, 2754 and references therein; Dias, J. R. *J. Chem. Educ.* **1992**, *69*, 695. Graph methods can also be applied to structure-property relationships; Mihalić, Z.; Trinajstić, N. *J. Chem. Educ.* **1992**, *69*, 701.

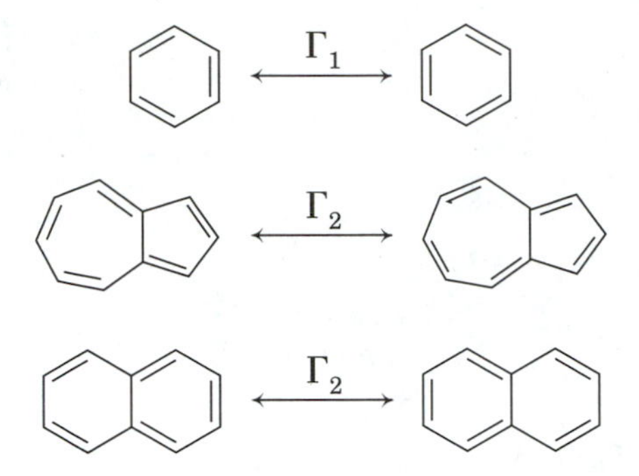

Figure 4.49 Γ_1 and Γ_2 permutations.

To compute the Γ_2s, we delete the A and B rings together (Figure 4.52). The SC of the residual fragment is 1. Then we delete the B and C rings together (Figure 4.53). The SC of the residual system is also 1, as it must be by symmetry. Therefore $n_2 = 2$.

The resonance energy associated with a particular structure is determined from the formula

$$\text{RE} = 2(n_1\Gamma_1 + n_2\Gamma_2)/K_{\text{SC}} \tag{4.68}$$

in which K_{SC} is the number of Kekulé structures for the molecule. The ratio Γ_2/Γ_1 is taken to be 0.40, which is an empirical value determined from the UV-vis absorption spectra of benzene and azulene,[145] and Γ_1 is assigned a value of 0.838 eV (19.3 kcal/mol). For anthracene, therefore,

$$\text{RE} = 2(3\Gamma_1 + 2\Gamma_2)/4 \tag{4.69}$$

Substituting 0.4 Γ_1 for Γ_2 produces

$$\text{RE} = 0.5[3\Gamma_1 + 2(0.4\Gamma_1)] \tag{4.70}$$

so

$$\text{RE} = 1.9\Gamma_1 = 1.59 \text{ eV} \tag{4.71}$$

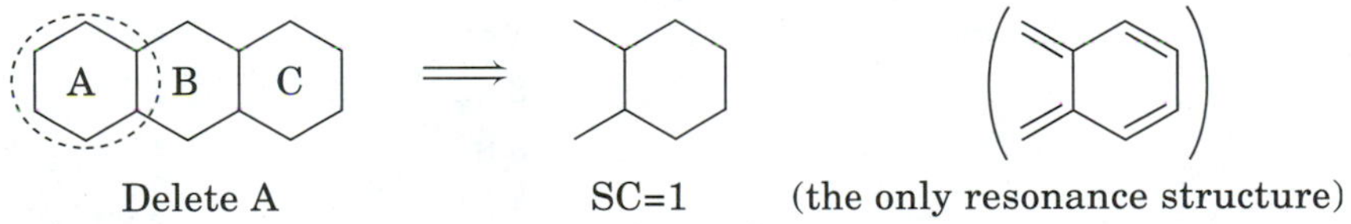

Figure 4.50 Determination of n_1 for anthracene; step 1, deleting ring A.

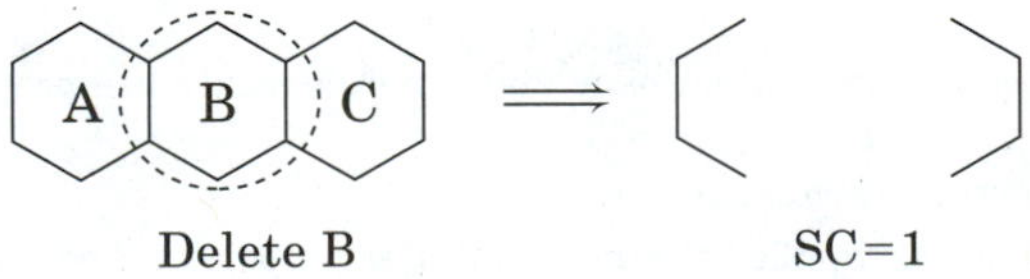

A B C ⟹

Delete A SC=1 (the only resonance structure)

Figure 4.51 Determination of n_1 for anthracene; step 2, deleting ring B.

A B C ⟹

Delete B SC=1

[145]Initially a ratio of 0.37 was obtained theoretically. See reference 139(c), p. 6632.

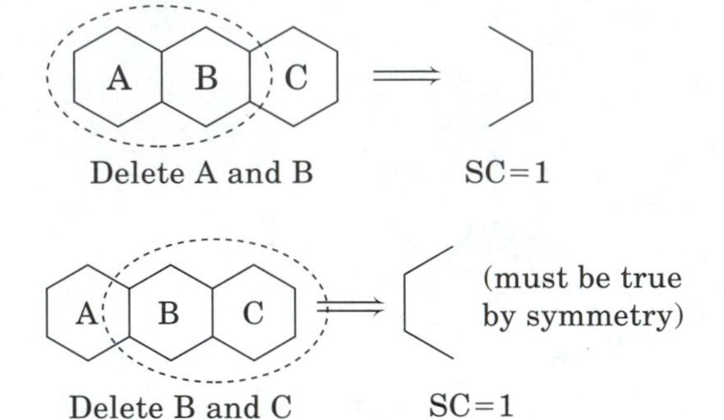

Figure 4.52 Determination of n_2 for anthracene: deletion of A and B rings.

Figure 4.53 Determination of n_2 for anthracene: deletion of B and C rings.

This result can be compared with the HMO calculation of delocalization energy, which is 5.38 β. Taking β to be 18 kcal/mol means the DE is 4.2 ev. By contrast, more advanced molecular orbital calculations (SCF MO) give a delocalization energy of 1.60 eV.[146] Clearly the structure-resonance method gives results that are closer to the more advanced molecular orbital calculations than does simple Hückel theory. Table 4.5 gives some results for other hydrocarbons, and it may be seen that the SRT method generally gives results that are very close to those obtained by SCF-MO calculation methods.[147,148] Even though we will not use the SRT procedure again, this discussion serves as a useful introduction to some applications of graph theory in chemistry. In addition, it provides a reminder of the utility of complementary theoretical methods in organic chemistry.

Table 4.5 Comparison of measures of delocalization and resonance energy.

Compound	DE_{HMO} (eV)*	$DE_{SCF\ mo}$ (eV)	RE_{SRT} (eV)
Benzene	1.56	0.87	0.84
Naphthalene	2.87	1.32	1.35
Phenanthrene	4.25	1.93	1.95
Pyrene	5.08	2.10	2.13
Styrene	1.89	0.86	0.84
Biphenyl	3.42	1.70	1.68
Stilbene	3.81	1.71	1.68
Ovalene	11.31	4.54	4.44

*Published results in units of β are converted to eV: 1 β ≡ 0.78 eV.

[146]SCF MO theory is discussed on page 239. Energy values are presented here in eV (1 eV = 23.06 kcal/mol).

[147]Data from reference 139(c), p. 6631.

[148]This method has been applied to other molecular properties, and a number of papers have examined its theoretical basis. See, for example, Herndon, W. C.; Párkányi, C. *J. Chem. Educ.* **1976**, *53*, 689; Herndon, W. C. *Tetrahedron* **1973**, *29*, 3; Gutman, I. *Chem. Phys. Lett.* **1984**, *103*, 475; Gutman, I.; Trinajstić, N.; Wilcox, Jr., C. F. *Tetrahedron* **1975**, *31*, 143, 147; Cvetković, D.; Gutman, I.; Trinajstić, N. *J. Chem. Phys.* **1974**, *61*, 2700; Wilcox, Jr., C. F.; Gutman, I.; Trinajstić, N. *Tetrahedron* **1975**, *31*, 147; Gutman, I.; Herndon, W. C. *Chem. Phys. Lett.* **1984**, *105*, 281.

4.4 Additional Molecular Orbital Methods

In this chapter we have seen that both HMO theory and resonance theory are useful models for understanding the structure and reactivity of organic compounds. In the following chapters we will also need to use the results of other models to understand reactive intermediates and reaction mechanisms. In this section we will introduce some qualitative aspects of perturbational MO theory, and we will survey some types of computer calculations.

Perturbational Molecular Orbital Theory

Perturbational MO (PMO) theory allows us to estimate the change in electronic energy levels and molecular orbitals resulting from the formation of a new structure by perturbing a structure whose molecular orbitals and energies we already know.[149] In a sense we are making molecular orbitals by taking a combination of molecular orbitals, which is only a variation of the process of taking a combination of atomic orbitals (LCAO-MO) that we carried out in the HMO method. However, we are not so much seeking complete MO descriptions of the new structure as we are seeking the difference in energy between the new structure and the starting structure.

Let us consider the case of a perturbation involving the union of two systems with nondegenerate orbitals.[150] The Hamiltonian of the initial system is $\mathcal{H}^0$. The set of wave functions describing the system is ψ_n^0 and their energy levels are E_n^0. Now we introduce some perturbation, and we write the new Hamiltonian for the system as $(\mathcal{H}^0 + \mathcal{H}')$. E_i, the energy of the ith level after the perturbation is given by equation 4.72 for $i \neq j$.

$$E_i = E_i^0 + \sum_{i,j} \frac{|H'_{ij}|^2}{E_i^0 - E_j^0} \tag{4.72}$$

where E_i^0 is the energy of the ith wave function before the perturbation and

$$H'_{ij} = \int \psi_i^0 \mathcal{H}' \psi_j^0 \, d\tau \tag{4.73}$$

---

[149]For a general introduction to PMO theory and its applications, see Dewar, M. J. S.; Dougherty, R. C. *The PMO Theory of Organic Chemistry*; Plenum Press: New York, 1975. See also Freeman, F. *J. Chem. Educ.* **1978**, *55*, 26; Cooper, C. F. *J. Chem. Educ.* **1979**, *56*, 568; Smith, W. B. *J. Chem. Educ.* **1971**, *48*, 749; Whangbo, M.-H. in *Computational Theoretical Organic Chemistry*; Csizmadia, I. G.; Daudel, R., eds.; D. Reidel Publishing Company: Boston, 1981; pp. 233–252. Herndon has applied PMO theory to saturated systems: Herndon, W. C. *J. Chem. Educ.* **1979**, *56*, 448. Durkin and Langler have offered modifications of the Dewar-Dougherty PMO method discussed here: Durkin, K. A.; Langler, R. F. *J. Phys. Chem.* **1987**, *91*, 2422.

[150]For further discussion, see Hoffmann, R. *Acc. Chem. Res.* **1971**, *4*, 1.

Equation 4.72 tells us that the changes in energy levels and wave functions can be predicted by determining the effect on E_i of the interaction of ψ_i with each of the other wave functions. In other words, the perturbations are pairwise additive.

Let us consider a simple case involving the joining of two components, each of which has one molecular orbital that will perturb (and be perturbed by) one molecular orbital on the other system. We also assume that the two energy levels are not degenerate. The energy of the *i*th MO (E_i) after the perturbation will be

$$E_i' = E_i^0 + \frac{|H_{ij}'|^2}{E_i^0 - E_j^0} \tag{4.74}$$

Since $E_i^0 < E_j^0$, the second term on the right in 4.74 is a negative quantity. Therefore, $E_i' < E_i^0$, meaning that the perturbation has lowered the energy of the *i*th level. We also have

$$E_j' = E_j^0 + \frac{|H_{ij}'|^2}{E_j^0 - E_i^0} \tag{4.75}$$

Since $E_i^0 < E_j^0$, the difference $E_j^0 - E_i^0$ is a positive quantity. Therefore, $E_j' > E_j^0$, meaning that the perturbation has raised the energy of the *j*th level. Thus the interaction raises the higher energy level and lowers the lower one (Figure 4.54).

The calculation to this point has considered only the effect of a perturbation on the energies of molecular orbitals. The effect of a perturbation on the energy and electron distribution of a molecular structure will depend on the population of electrons in the energy levels that are perturbed. Figure 4.55 shows two situations of interest. In Figure 4.55(a), there are two electrons in ψ_i and two electrons in ψ_j. To the first approximation, the perturbation lowers the energy of ψ_i as much as it raises the energy of ψ_j. Since there are two electrons in ψ_i and two in ψ_j, the perturbation causes no net change in the energy of the system. However, in Figure 4.55(b) there are two electrons in ψ_i and no electrons in ψ_j. Therefore, there is a net lowering of the energy of the system, since this perturbation allows those two electrons to be in a more stable molecular orbital.

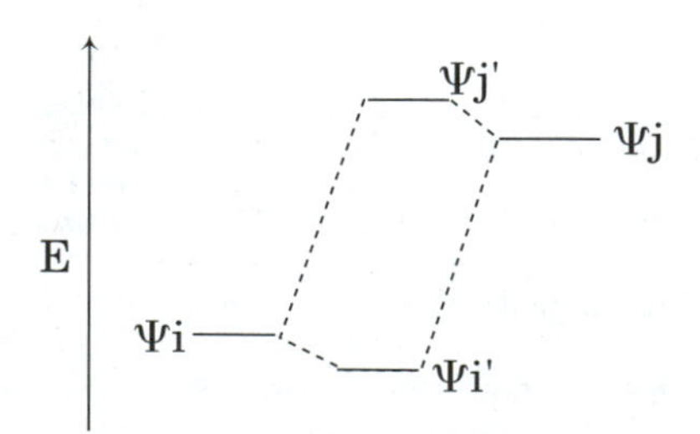

Figure 4.54
Effect of interaction on two nondegenerate orbitals.

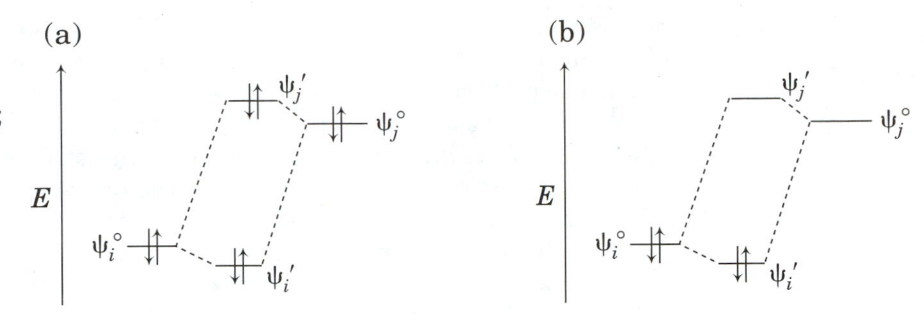

Figure 4.55
PMO interactions: (a) both ψ_j^0 and ψ_i^0 are doubly occupied; (b) only ψ_i^0 is doubly occupied.

The simple molecular orbital energy level diagrams in Figure 4.55 show only one orbital on each of the molecular fragments. In general there will be many orbitals, some populated with electrons and some vacant. The **h**ighest **o**ccupied **m**olecular **o**rbital in a π system is called the HOMO, while the **l**owest **u**noccupied **m**olecular **o**rbital is called the LUMO.[151] If there are noninteracting MOs lower in energy than ψ_i and ψ_j in Figure 4.55, then it is the interaction between the HOMO of fragment j and the LUMO of fragment i that determines the energy change of the union.[41a] The HOMO and LUMO of a molecular system are sometimes called the **frontier molecular orbitals (FMOs)**, since they are the orbitals from which it is easiest to remove an electron (HOMO) or add an electron (LUMO), and thus they are often involved in important chemical processes.[152]

One process that can be described by FMO theory is the formation of a complex between the two molecular systems. Although we might imagine the extent of orbital interaction (and hence perturbation) to be much less than would be the case if a new σ bond were formed between the π systems, the principles are the same. The complex formation would involve transfer of electron density from an orbital localized on fragment j to an orbital that has electron density both on fragment i and fragment j (Figure 4.55(b)). Therefore this type of interaction is called **charge transfer complex formation.**[153] An alternative representation of the process is depicted in equation 4.76.

$$I + J \rightleftharpoons (I^{\delta-} \cdots J^{\delta+}) \tag{4.76}$$

A Survey of Theoretical Calculations

Our approach to theoretical chemistry so far has been to see what chemical insight we can gain from application of very simple back of the envelope HMO or resonance methods in organic chemistry. We recognize that our

---

[151]For radicals, a **s**ingly **o**ccupied **m**olecular **o**rbital is called a **SOMO**.

[152]For a discussion, see reference 49.

[153]Murrell, J. N. *Quart. Rev. Chem. Soc.* **1961**, *15*, 191; Bender, C. J. *Chem. Soc. Rev.* **1986**, *15*, 475.

models are limited, but the results are useful to us nonetheless. However many chemists find that these simple computational procedures are not adequate to calculate molecular properties and structures accurately and that precise computations are necessary.[154] Before leaving this chapter we should touch on the increasingly important subject of computational chemistry. There are many types of advanced calculations, and each computational approach is useful to different chemists. We cannot undertake a detailed analysis of these methods here, but we can introduce some terms that appear with increasing frequency in the current chemical literature.[155]

The simple Hückel calculations described in this chapter assumed that the σ and π systems could be treated independently, but this approximation is not made in more advanced methods. Instead, *all valence electron calculations* compute molecular orbitals by taking linear combinations of all the valence shell orbitals, not just the carbon $2p$ orbitals. For example, the **Extended Hückel Theory (EHT)** developed by Hoffmann[156] computes molecular orbitals that include the $2s$ and all three $2p$ atomic orbitals of carbon. The lowest energy MO (ψ_1) of ethene was determined by an EHT calculation[157] to be

$$\begin{aligned}\psi_1 = &-0.419\ 2s_{\text{C1}} + 0.042\ 2p_{x\text{C1}} + 0.0\ 2p_{y\text{C1}} + 0.0\ 2p_{z\text{C1}} \\ &- 0.419\ 2s_{\text{C2}} - 0.042\ 2p_{x\text{C2}} + 0.0\ 2p_{y\text{C2}} + 0.0\ 2p_{z\text{C2}} \\ &- 0.153\ 1s_{\text{H3}} - 0.153\ 1s_{\text{H4}} - 0.153\ 1s_{\text{H5}} - 0.153\ 1s_{\text{H6}}\end{aligned} \tag{4.77}$$

Analysis of the coefficients of each atomic orbital in this wave function indicates that there are bonding interactions between the $2s$ orbitals on the carbon atoms with each other and with the $1s$ orbitals on the hydrogen atoms to which they are bonded. As shown in Figure 4.56, this combination of atomic orbitals produces a molecular orbital which encompasses all of the σ bonds in the molecule.

Figure 4.56 also shows the wave function for ψ_3 of ethene from the same EHT calculation. This is a much more complicated wave function, but there are bonding interactions between the $2p_x$ orbitals on the carbon atoms with each other and with the $1s$ orbitals on the hydrogen atoms to which each of the carbon atoms is bonded.

$$\begin{aligned}\psi_3 = &+ 0.0\ 2s_{\text{C1}} + 0.0\ 2p_{x\text{C1}} + 0.0\ 2p_{y\text{C1}} - 0.355\ 2p_{z\text{C1}} \\ &+ 0.0\ 2s_{\text{C2}} + 0.0\ 2p_{x\text{C2}} - 0.0\ 2p_{y\text{C2}} - 0.355\ 2p_{z\text{C2}} \\ &+ 0.267\ 1s_{\text{H3}} - 0.267\ 1s_{\text{H4}} + 0.267\ 1s_{\text{H5}} - 0.267\ 1s_{\text{H6}}\end{aligned} \tag{4.78}$$

---

[154]Indeed, for some theoretical chemists, the goal is to reduce chemistry entirely to computation. For one perspective, see the discussion by Dewar, M. J. S. *J. Phys. Chem.* **1985**, *89*, 2145.

[155]For further reading, see reference 2(e), pp. 161–196.

[156]Hoffmann, R. *J. Chem. Phys.* **1963**, *39*, 1397.

[157]Calculation by the author using a CAChe WorkSystem.

Figure 4.56
Molecular orbitals of ethene calculated by extended Hückel theory.

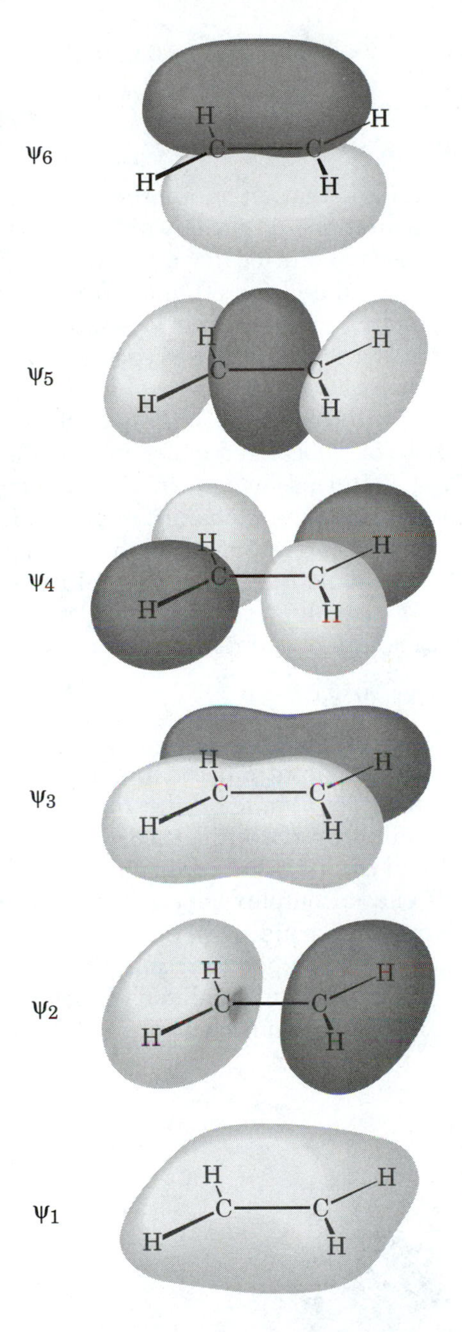

ψ_6 of ethene is the HOMO. It is reassuring that this orbital includes contributions only from the $2p_y$ orbitals (perpendicular to the molecular plane in this calculation) and not from the other orbitals. The shape of this MO (Figure 4.56) is very similar to the shape of the π bonding orbital of ethene that was determined by the simple HMO calculation done earlier.

$$\begin{aligned}\psi_6 = &+ 0.0\ 2s_{C1} + 0.0\ 2p_{xC1} - 0.612\ 2p_{yC1} + 0.0\ 2p_{zC1} \\ &+ 0.0\ 2s_{C2} + 0.0\ 2p_{xC2} - 0.612\ 2p_{yC2} + 0.0\ 2p_{zC2} \\ &+ 0.0\ 1s_{H3} + 0.0\ 1s_{H4} + 0.0\ 1s_{H5} + 0.0\ 1s_{H6}\end{aligned} \tag{4.79}$$

The result of an EHT calculation on another simple structure suggests that more advanced calculations do not always yield the results we might have predicted on the basis of our localized bond view of molecules. Figure 4.57 shows the HOMO calculated for ethanol. We might have expected to find an orbital corresponding to a lone pair localized on oxygen, but instead we see an orbital that has density on the oxygen atom and on the adjacent carbon atom and the two hydrogen atoms to which it is bonded. As we move into the study of reactive intermediates in Chapter 5, we will see further examples of the importance of delocalized molecular orbitals.

In equations 4.77–4.79, terms such as $2p_{zC1}$, $1s_{H3}$, etc., represent mathematical equations that describe the probability amplitude of finding an electron around an atom. The entire set of such equations for the atomic orbitals in a molecule is called a ***basis set***. The solution to the Schrödinger equation for one-electron atomic systems gives rise to the hydrogenic orbitals that have both an angular and radial components

$$\chi(r,\theta,\phi) = R_{nl}(r)Y_{lm}(\theta,\phi) \tag{4.80}$$

where θ and ϕ are polar coordinates, r is the distance from the nucleus, and n, l, and m are the principal, azimuthal, and magnetic quantum numbers.[158] The angular component of the atomic orbital function gives rise to the familiar shapes of the atomic orbitals, while the radial component determines the size of the orbital and can be quite complex. The radial component of a hydrogenic orbital often has a complex dependence on the distance from the nucleus, so evaluation of integrals involving such terms is difficult and time-consuming. A type of basis set function proposed by Slater uses a somewhat simpler radial component, and such functions are called **Slater Type Orbitals (STO)**. Another approach is to use **Gaussian Type Or-**

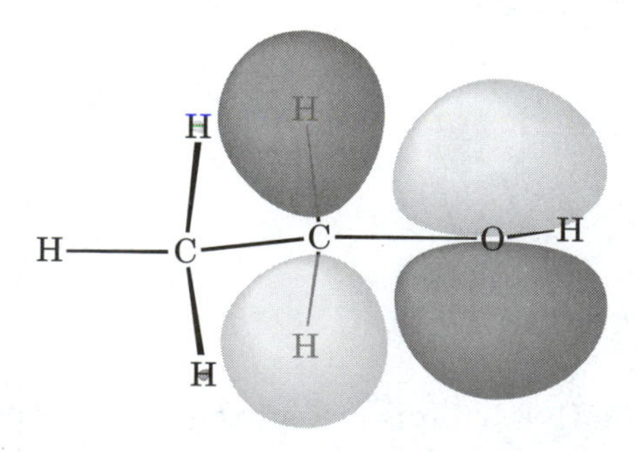

Figure 4.57
HOMO of ethanol.

[158]For a review of the wave functions for atomic orbitals and an introduction to semiempirical MO theory, see Pople, J. A.; Beveridge, D. L. *Approximate Molecular Orbital Theory*; McGraw-Hill Book Company: New York, 1970.

bitals (GTO) which have a radial dependence that is even more approximate but which is still easier to solve analytically. It is possible to write expressions for orbitals that are close in properties to Slater type orbitals by taking combinations of Gaussian type orbitals. Thus a calculation may be done using the STO-3G (**S**later **t**ype **o**rbitals composed of **3** **G**aussian functions) basis set, for example.

Many chemical calculations are done with what is called *ab initio* (from the Latin for "from the beginning") theory since all of the complex integrals that arise during the calculation are computed.[159,160,161] Molecular orbitals are constructed as linear combinations of atomic orbitals, then the molecular orbitals are optimized for energy by an SCF (**s**elf-**c**onsistent **f**ield) procedure.[162,163] In this approach, the initial set of molecular orbitals is used to calculate a new set of molecular orbitals, and the process is repeated until the molecular orbitals no longer change. At that point the molecular orbitals have converged and the calculation is said to be self-consistent.[164] There is an important trade-off between time and accuracy in *ab initio* calculations. More accurate results require extended basis sets, in which each atomic orbital is represented by two basis functions, or diffuse basis sets that include many additional terms.[165] However, larger basis sets are only practicable for smaller structures. Larger molecules can be studied only with minimal basis sets that use only one basis function for each orbital in an occupied shell. In such cases, the accuracy of the calculation is limited by the basis set of atomic orbitals used.

It is possible to obtain useful theoretical results on larger molecules if some of the complex integrals that arise during the calculation of molecular orbitals are replaced by experimental data.[166] (This process is called *para-*

---

[159]For an introduction, see Richards, W. G.; Cooper, D. L.; Ab Initio *Molecular Orbital Calculations for Chemists*, 2nd Ed.; Clarendon Press: Oxford, England, 1983.

[160]For a review, see Whitten, J. L. *Acc. Chem. Res.* **1973**, *6*, 238.

[161]For a commentary, see Davidson, E. R. in *Reviews in Computational Chemistry*; Lipkowitz, K. B.; Boyd, D. B., eds.; VCH Publishers, Inc.: New York, 1990; p. 373.

[162]Roothaan, C. C. J. *Rev. Mod. Phys.* **1951**, *23*, 69 and later papers.

[163]Davidson and Feller have reviewed basis set selection for molecular calculations: Davidson, E. R.; Feller, D. *Chem. Rev.* **1986**, *86*, 681; Feller, D.; Davidson, E. R. in *Reviews in Computational Chemistry*; Lipkowitz, K. B.; Boyd, D. B., eds.; VCH Publishers, Inc.: New York, 1990; p. 1. For "an experimental chemist's guide to *ab initio* quantum chemistry," see Simons, J. *J. Phys. Chem.* **1991**, *95*, 1017.

[164]Reference 11, p. 165.

[165]For a concise summary, especially with regard to MO calculations for carbanions, see Nobes, R. H.; Poppinger, D.; Li, W.-K.; Radom, L. in *Comprehensive Carbanion Chemistry. Part C. Ground and Excited State Reactivity*; Buncel, E.; Durst, T., eds.; Elsevier: Amsterdam, 1987; pp. 1–14.

[166]As a comparison, Zerner noted that the technology of the 1990s allowed molecular mechanics calculations to be carried out on molecules with thousands of atoms, semi-empirical calculations to be done on molecules with hundreds of atoms, and *ab initio* calculations to be done on molecules containing tens of atoms. Zerner, M. C. in *Reviews in Computational Chemistry II*; Lipkowitz, K. B.; Boyd, D. R., eds.; VCH Publishers, Inc.: New York, 1991; pp. 313–365.

meterization.) For example, in the EHT method, values of H_{ii} for each type of orbital are evaluated from experimental ionization potential data.[156] Methods that incorporate experimental data are termed *semi-empirical MO calculations*. To increase the speed of the calculation or the size of the structures that can be studied, these methods also make some approximations in—or even neglect entirely—terms involving the overlap of atomic orbitals.[158,167] Some of the types of semi-empirical calculations that have been developed include the following:

CNDO	**C**omplete **N**eglect of **D**ifferential **O**verlap
NDDO	**N**eglect of **D**iatomic **D**ifferential **O**verlap
PNDO	**P**artial **N**eglect of **D**ifferential **O**verlap
INDO	**I**ntermediate **N**eglect of **D**ifferential **O**verlap
MINDO	**M**odified **INDO** (variations of this have evolved into **MINDO/3**, **MNDO**, and **AM1**).

It might appear that *ab initio* calculations would be preferred over the semi-empirical methods, but *ab initio* calculations require much greater computation time and are limited to much smaller structures than is the case with the semi-empirical methods. Moreover, each of the semi-empirical methods has its own strengths. Some are parameterized for properties such as geometries, energies, or spectroscopic predictions and can yield accurate results for those properties if the structure being calculated is similar to the structures used to parameterize the computational procedure.

Chemists interested in carrying out advanced calculations can obtain tested computer programs from research consortia and corporations. The easy availability of these programs however, makes it easy to lose sight of the many approximations and assumptions upon which they are based. If computer programs are treated as black boxes without an understanding of how they work, it may be possible to read too much into them. Also, different theoretical models can give quite different answers to the same question, and one may choose an approach that is not suited for a particular type of chemical problem. For example, different programs produce very different answers for the problem of the most stable structure for the 7-norbornyl carbocation.[168]

The results of any advanced calculation must always be examined to see whether it is reasonable in view of chemical intuition, which will most likely be expressed in terms of a simpler HMO or resonance model, and whether it reproduces experimental results for test structures. We must not forget that computer calculations, even those done with very large programs on very powerful computers, are only a special type of conceptual model. Computer

---

[167]For a discussion of the approximations and applicability of semi-empirical methods, see (a) Zerner, M. C. reference 166; (b) Stewart, J. J. P. in *Reviews in Computational Chemistry*; Lipkowitz, K. B.; Boyd, D. B., eds.; VCH Publishers, Inc.: New York, 1990; p. 45.

[168]Sunko, D. E. *Pure Appl. Chem.* **1983**, *55*, 375.

calculations are subject to the same limitations as other models, and we should not put more faith into them than is warranted.[169]

Problems

1. Demonstrate that using the relationship $E = \alpha - \beta$ in equations 4.10 and 4.11 produces the wave function for ψ^* shown in equation 4.28.
2. Use the wave functions shown in equations 4.34 through 4.36 to calculate the π bond orders and the free valence indices for the allyl carbocation, radical and carbanion.
3. Write the secular determinant for each of the following structures. Use the atom numbers indicated.

a.

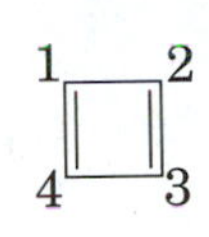

b.

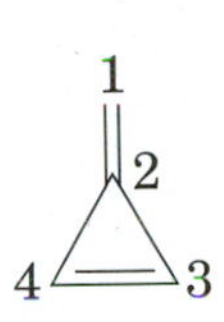

c.

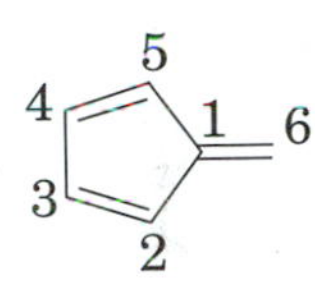

d.

6 7
5 8
1
4 2
3

e.

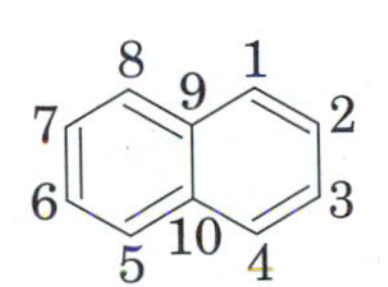

f.

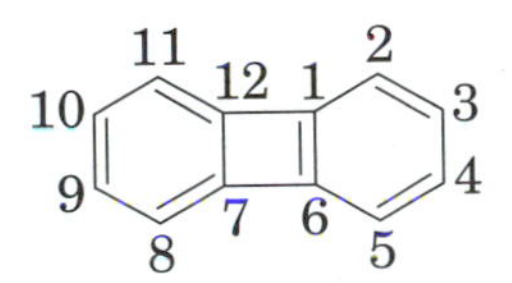

4. Using an HMO program on a personal or time-shared computer, obtain the energy levels of the structures in problem 3.[170] Draw an energy level diagram and calculate the delocalization energy of each system.
5. Using the results of your HMO calculations from problem 4, calculate $\mathcal{F}_i$, ρ_i, and q_i for each carbon atom and determine π bond order and total bond order for each bond. (Take advantage of molecular symmetry to minimize the number of calculations you must make.)

---

[169]This point has been emphasized by Ross, J. *Science* **1992**, *257*, 860: "Computers do calculations. The equation solved may be that of a theory, such as Schrödinger's equation, or that of a model, such as a reaction-diffusion equation for a particular reaction mechanism. Solutions of such equations, whether obtained by analysis or numerical methods, are *predictions*, not results of experiments. The validation of a prediction is confirmation by experiment. No one doubts the ability of computers to solve the equations of models; it is the models that require validation by experiments . . . All this may seem to be only a matter of proper usage of words; it is not. It is a matter of recognizing what science is about."

[170]If you cannot calculate HMOs directly, obtain energy levels and MOs from a reference work.

6. Use the symmetry properties of the Hückel molecular orbitals of linear, alternant hydrocarbons to sketch a qualitative representation of each of the MOs of heptatrienyl, octatetraene, nonatetraenyl, and decapentaene. (You should not carry out a calculation or consult the literature.) You will not be able to give a quantitative estimate of the coefficient of each atomic orbital, but your representation should indicate for each MO which coefficients are positive, which are negative, and which (at a node) are zero. For each structure rationalize the energy of each MO in terms of the bonding and antibonding interactions between atomic orbitals.
7. Identify each of the radicals and diradicals below as an even alternant, odd alternant, or nonalternant hydrocarbon structure. Use stars to illustrate your assignment.

a. b. c. d.

8. Use the properties of odd alternant hydrocarbons to determine the unpaired electron density in the cyclobutadiene-substituted methyl radical **24**. Is this result surprising to you? Show how resonance theory can easily rationalize the result you obtain.

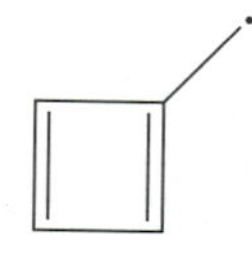

24

9. Methylenecyclopropene (structure b in problem 3) is a nonalternant hydrocarbon. Naphthalene is an alternant hydrocarbon. What did the results of the MO calculations you did in problem 4 tell you about the charge densities on each atom? Do those results agree with the discussion on page 201 about the nature of alternant and nonalternant systems? How could you have predicted the polarity of methylenecyclopropene on the basis of resonance theory?
10. Cyclopropenone has a higher dipole moment than would be expected on the basis of a carbonyl group alone. Can you use the results of your MO analysis of methylenecyclopropene to rationalize the dipole moment? How does resonance theory explain the polarity?
11. Phenanthrene has more Kekulé resonance structures than does anthracene, hence it should be more aromatic, and the data in Table 4.4 indicate that it does have a larger resonance energy. However, while anthracene does not undergo electrophilic addition, phenanthrene readily undergoes electrophilic addition across the 9,10 bond.

a. Either by drawing all possible resonance structures or by applying the procedure described by Herndon, determine that the number of resonance structures indicated in Table 4.4 is correct.
b. Use the nature of the resonance structures in each case to rationalize why phenanthrene adds electrophiles but anthracene tends to undergo substitution instead. Can MO theory predict the same result?

12. Calculate the activation energy for a rotation of an allyl radical about C2—C3 as shown in Figure 4.58. How does your result compare with an experimental value? Do the same for the allyl cation and anion. Also calculate barriers for rotation about the Ar—CH_2 bond in the benzyl cation, radical, and anion.

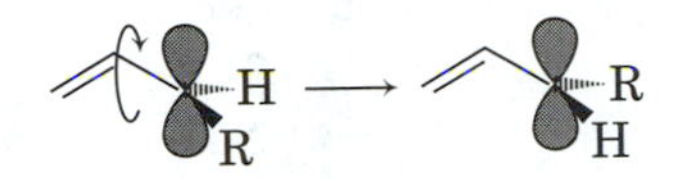

Figure 4.58 Rotation of an allyl radical.

13. Consider the $\mathcal{F}_i$ values shown in Figure 4.17. The value of $\mathcal{F}_i$ for the terminal carbon atom varies from 1.732 to 1.025 to 0.943 as the structure changes from methyl radical to allyl radical to pentadienyl radical. However, the $\mathcal{F}_i$ value for the terminal carbon atom varies from 0.732 to 0.838 to 0.861 as the structure varies from ethene to butadiene to hexatriene.
   a. Predict the value of $\mathcal{F}_i$ for the next member of each of these two series.
   b. Why do the $\mathcal{F}_i$ values increase in one case but decrease in the other?
   c. Why should the $\mathcal{F}_i$ value of a methylene carbon atom on 1,2-dimethylene-3,5-cyclohexadiene be nearly the same as for the terminal carbon atom of an allyl or pentadienyl radical?
   d. The value of 0.525 for $\mathcal{F}_i$ for a carbon atom in cyclooctatetraene is determined from a calculation that assumes the molecule is planar. What would you expect the $\mathcal{F}_i$ value to be if the molecule is assumed to be tub shaped?
14. Work through the SRT method for phenanthrene and reproduce the resonance energy cited in Table 4.5. Do the same for styrene and for pyrene. Consult the literature to find other applications of the SRT method (ionization potential, for example) and work through the analysis of some of the structures.
15. Consult the literature to find discussions of the aromaticity or antiaromaticity of the following structures: tropylium ion, cyclopentadienyl cation, COT dianion. What do these articles indicate about our generalizations of aromatic character?
16. Calculate how distortion of square cyclobutadiene to a rectangular structure would affect the π energy of the molecule. To do this, solve the determinant for a system such as the one in Figure 4.59, where H_{12} is β and H_{23} varies from 1.0 β to 0.7 β to 0.5 β to 0.3 β to 0.0 β. What is the effect of the distortion on the π energy of the system? What factors have we ignored in this analysis?

4 1 ⇒ 4 1 ⇒ 4 1

3 2 ⇒ 3 2 ⇒ 3 2

Figure 4.59 Distortion of cyclobutadiene.

17. Equation 4.63 is valid because the expression $7.435n - 0.605m - 32.175p - 29.38q$ represents the calculated heat of formation for a structure that does not exhibit resonance. Verify that this is so by using the expression to calculate ΔH_f^0 for glyoxal (O=CH—CH=O) and comparing your answer to the accepted value of −50.7 kcal/mol.[171]
18. a. Calculate ΔDRE for addition of ethylene across the 9,10 positions in anthracene and also for addition across the 1,2 positions. Which addition pathway should be favored?
    b. Also calculate ΔDRE for addition of ethylene across the 9,10 position and across the 1,2 position in phenanthrene. Which reaction should be favored in this case?
19. Explain why tris(9-fluorenylidene)cyclopropane (**25**) undergoes two electron reduction at a "remarkably positive" reduction potential.

25

20. Ordinarily the barrier to rotation about a carbon-carbon double bond is quite high, but compound **26** was observed by NMR to have a rotational barrier of only about 20 kcal/mol. Explain this result.

$CH_2CH_2CH_3$

CHO $CH_2CH_2CH_3$

26

[171]Reference 113, p. 510.

21. Predict the electronic and physical properties of the benzene dianion, $C_6H_6^{2-}$.
22. The Dewar resonance energy of triphenylene (**27**) is reported to be 65 kcal/mol. On the basis of the DRE value, is the central ring aromatic?

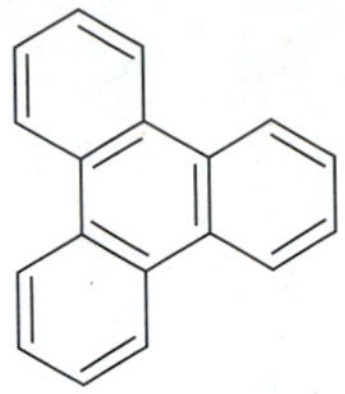

27

23. What is your response to the question posed on page 210: Is benzene aromatic because it is especially stable, or is it unusually stable because it is aromatic?
24. Consider the following statement made by E. Heilbronner after the opening paper of the Jerusalem symposium on Aromaticity, Pseudoaromaticity and Antiaromaticity.[172]

   . . . I think we should all realize that we are united here in a symposium on a non-existent subject. It must also be stated quite clearly at the beginning that aromaticity is not an observable property, i.e., it is not a quantity that can be measured and is not even a concept which, in my experience, has proved very useful. . . .

   Do you agree with Heilbronner, or do you think that aromaticity is an existent subject? Discuss the role of aromaticity as a model of chemical structure and reactivity.

[172]Bergmann, E. D.; Pullman, B., eds.; Academic Press: New York, 1971.

CHAPTER 5

Reactive Intermediates

5.1 Potential Energy Surfaces, Reaction Coordinate Diagrams, and Reactive Intermediates

In previous chapters we have considered in some detail the models we use to describe chemical structures. Our emphasis has been on structures corresponding to stable molecules, that is, molecules that persist for a *long* period of time and under *moderate* conditions as defined by the human perspective. In this chapter we discuss some relatively less stable structures that extend and deepen understandings of both the structural diversity of organic chemistry and the indirect pathways organic reactions sometimes follow. Some of these less stable structures appear to meet normal valency requirements (such as having all atoms with filled outer shells) but are characterized by unusual geometries and bonding arrangements (e.g., benzyne). Other species, among them radicals and ions, have charges or unfilled outer shells (or both) and are formed as short-lived reactive intermediates in chemical reaction sequences.

The term *reactive intermediate* implies a substance that cannot be isolated and characterized under ordinary conditions, so the methods we use to study reactive intermediates are frequently even more indirect than those we use to study stable structures. Moreover, we often study reactive intermediates under conditions that are very different from the conditions in which these species are presumed to be formed during chemical reactions. Therefore we must be especially cautious about drawing conclusions about chemical reactions from these kinds of studies.[1]

[1]For examples of reactions which proceed differently under typical reaction conditions and under the conditions used to study the reactive intermediates, see Bethell, D.; Whitaker, D. in *Reactive Intermediates*, Vol. 2; Jones, Jr., M.; Moss, R. A., eds.; Wiley-Interscience: New York, 1981; pp. 236 *ff*. and references therein.

One tool that is helpful in relating reactive intermediates to chemical reactions is a graphical model known as a **reaction coordinate diagram**—a two-dimensional sketch that represents the potential energy of a chemical system during a reaction. In a typical diagram for a one-step process, the vertical axis is free energy or enthalpy and the horizontal axis is a reaction coordinate sometimes labeled *progress of reaction*. Usually the reactant is on the left of the figure and the product is on the right. For a simple reaction such as dissociation of Cl_2 to two chlorine atoms, Figure 5.1(a), the reaction coordinate corresponds to the internuclear separation of the atoms. The curve in Figure 5.1(a) is only a portion, indicated with dotted lines, of a larger curve representing the ground electronic state of the chlorine molecule, Figure 5.1(b). In both cases the *x*-axis is the distance between the two chlorine atoms. Here the nature of the *x*-axis is clear, but for more complex processes (such as the hydrolysis of a protein by α-chymotrypsin) the meaning of the reaction coordinate is more complex.

Let us illustrate a simple reaction coordinate diagram by considering the conformational energy of butane. Figure 5.2 shows an arbitrary initial conformation for butane. If we represent the angle of rotation about the C1—C2 bond as θ_1, then the potential energy of the molecule varies with rotation as shown in Figure 5.3. Similarly, if the angle of rotation about the C2—C3 bond is represented as θ_2, then the potential energy varies as shown in Figure 5.4.

Although we have calculated the energy of rotation about each of these carbon-carbon bonds individually, we recognize that a molecule can exhibit different conformations about both of them. Thus it is instructive to consider Figure 5.5, which shows the potential energy calculated for rotation about both the C1—C2 and C2—C3 bonds in butane.[2] The calculations generate a three-dimensional *potential energy surface* made up of points characterized by three coordinates: θ_1, θ_2, and energy (vertical scale). We can

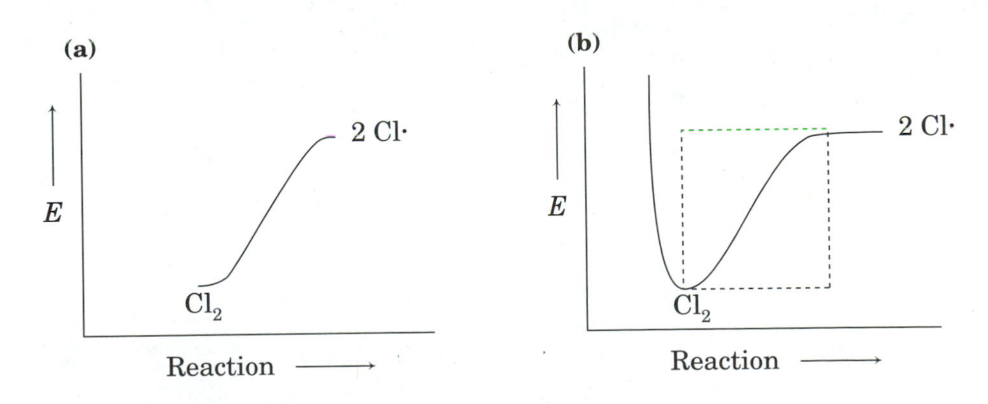

Figure 5.1 (a) Reaction coordinate diagram for dissociation of Cl_2; (b) electronic energy curve for Cl—Cl σ bond.

[2]Figure 5.5 was produced with a CAChe molecular mechanics calculation. For an MO study of rotation about all three torsional angles of butane, see Peterson, M. R.; Csizmadia, I. G. *J. Am. Chem. Soc.* **1978**, *100*, 6911.

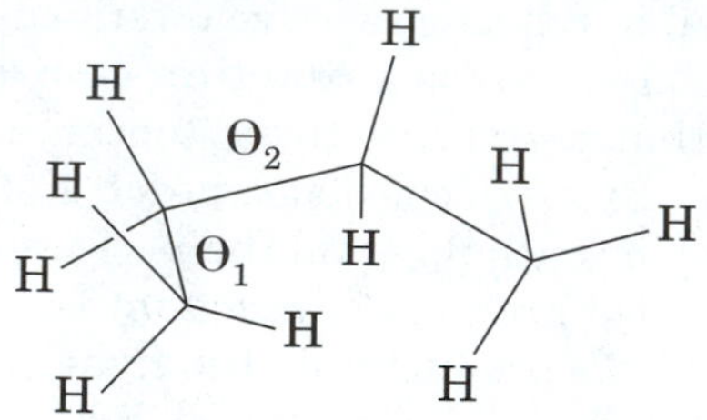

Figure 5.2 Initial eclipsed conformation of butane.

now see that the axes of Figure 5.5 represent a two-dimensional, cross-section (slice) of the three-dimensional potential energy surface. Thus, Figure 5.3 is equivalent to the angle 1 edge of Figure 5.5, while Figure 5.4 is equivalent to the angle 2 edge. Figure 5.3 and Figure 5.4 represent the energy of the molecule when only one molecular parameter is allowed to vary. Of course, we should consider the potential energy surface possible for rotation about all three of the C—C bonds, but that would be a four-dimensional surface, which we cannot envision. Thus even Figure 5.5 is a simplified model of a more complex surface.

Now let us consider a case more closely related to a chemical reaction. Dauben and Funhoff[3] used molecular mechanics to study the compound pre$_3$ (structure **1** in Figure 5.6, where R is CH_3), an analogue of previtamin D_3 (structure **1** in Figure 5.6, where R is C_8H_{17}). Photochemical reaction of previtamin D_3 produces lumisterol D_3 (**2**), provitamin D_3 (**4**), and tachysterol D_3 (**5**), while thermal reaction leads to vitamin D_3 (**3**). These products can be rationalized as reactions proceeding through one of two conformations of previtamin D_3, the *s-cis,(Z),s-cis* conformer and the *s-trans,(Z),s-cis* conformer.[4]

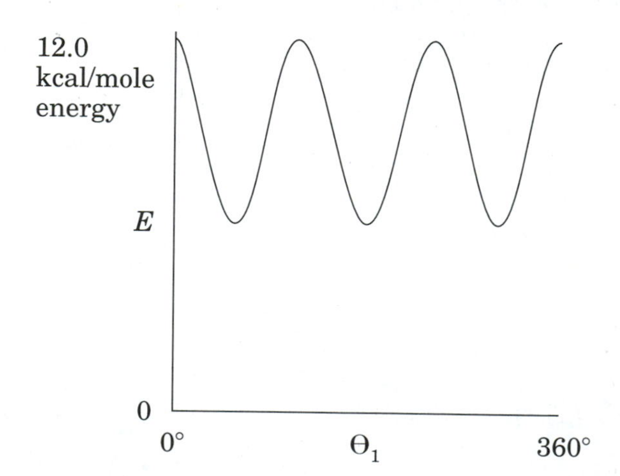

Figure 5.3 Potential energy diagram for rotation about C1—C2 bond of butane.

[3]Dauben, W. G.; Funhoff, D. J. H. *J. Org. Chem.* **1988**, *53*, 5070.

[4]The mechanism of these reactions will be discussed in Chapter 11.

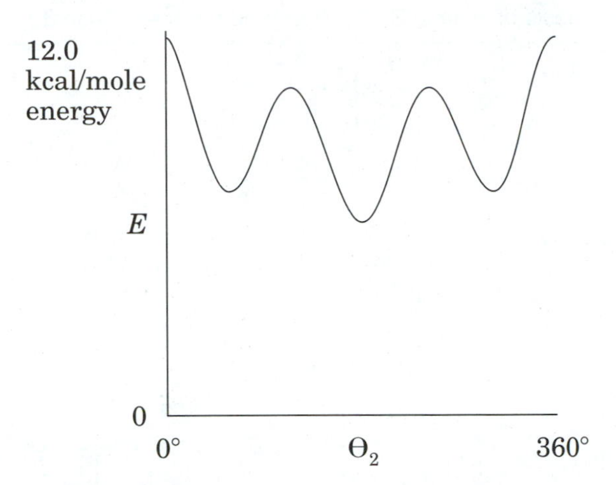

Figure 5.4
Potential energy diagram for rotation about C2—C3 bond of butane.

Dauben and Funhoff generated a three-dimensional potential energy surface analogous to Figure 5.5 for pre_3 by considering rotations about the Cl0—C5—C6—C7 and C5—C6—C7—C8 dihedral angles.[5] Four local minima were identified and are represented as **A**, **B**, **C**, and **D** in Figure 5.7. **A** and **D** are *cZc* conformations, while **B** and **C** are *tZc* conformations. Figure 5.8 (page 251) shows the location of **A–D** on the potential energy surface, and the same surface is represented as a contour plot in Figure 5.9 (page 252).

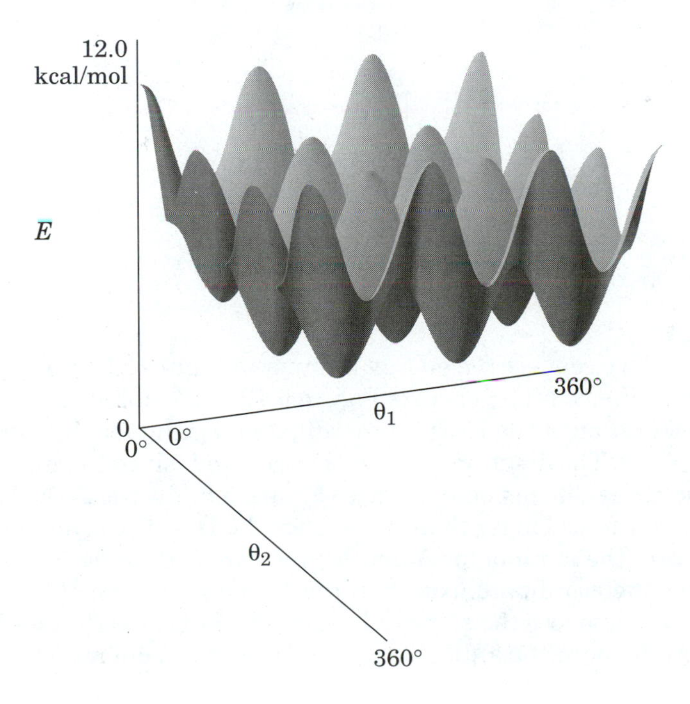

Figure 5.5
Three-dimensional potential energy surface for butane conformations.

[5]For conformations considered here the OH group was pseudoequatorial. Similar results were obtained when the OH group was held pseudoaxial, but the pseudoequatorial conformations produced the global minimum.

Figure 5.6 Reactions in the synthesis of vitamin D_3 (**3**). (Adapted from reference 3.)

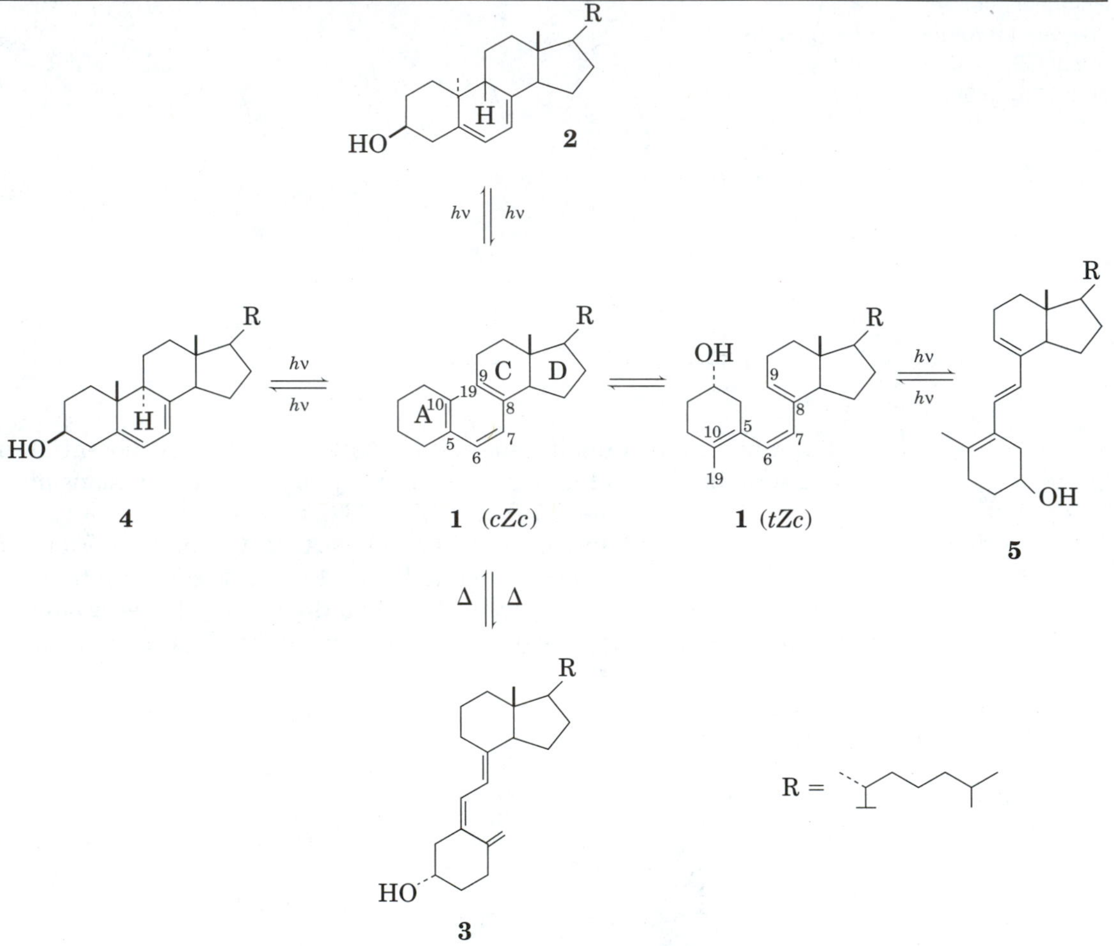

If we connect the energy minima in Figure 5.9 (page 252) with a continuous line, we can generate Figure 5.10 (page 253) which is a schematic representation of the reaction coordinate diagram for the interconversion of **A**, **B**, **C**, **D**. The diagram is somewhat stylized, since the line we would draw on the three-dimensional surface (Figure 5.8) for the **A** → **B** conversion would appear to be longer than the one for the **B** → **C** conversion. Thus, we would expect the minima for **A** and **B** in Figure 5.10 to lie farther apart along the reaction coordinate axis than the **B** and **C** minima. The fact that we draw these distances the same in Figure 5.10 indicates that we frequently use reaction coordinate diagrams that are only schematic representations of chemical transformations.

Reaction coordinate diagrams provide a graphical distinction between a reaction in which an intermediate occurs and one in which there is not an

Figure 5.7 Conformational minima **A, B, C,** and **D** of Pre_3 (R = CH_3).

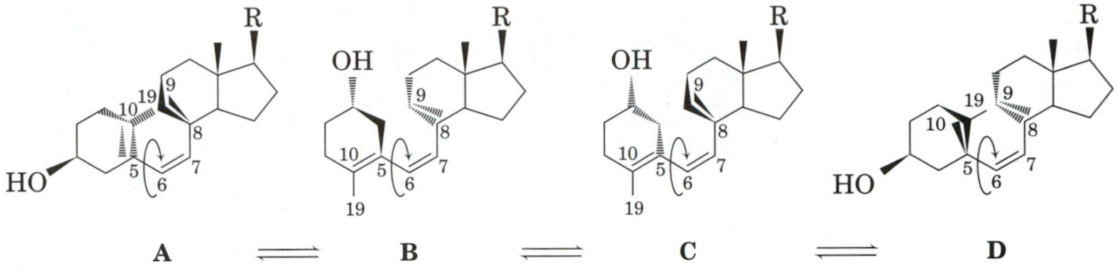

	A	B	C	D
Dihedral C10-C5-C6-C7	34°	168°	−159°	−37°
Dihedral C5-C6-C7-C8	9°	−17°	10°	−9°

intermediate. Figure 5.11 (page 253)[6] shows two possibilities proposed for the ring inversion of a disubstituted bicyclo[2.1.0]pentane. If the reaction follows the dashed line, the diradical is a local minimum on the potential energy surface and is an intermediate. That is, it is stable with respect to small deformations, but it is not sufficiently stable to be isolated for structural analysis. We quantify the requirement for the depth of the energy well by noting that to qualify as a chemical substance, the structure must have a lifetime long enough for at least one molecular vibration, or about 10^{-13} seconds.[7] However, if the solid line describes the reaction, the diradical does not occupy a local energy minimum on the potential energy surface, so it is

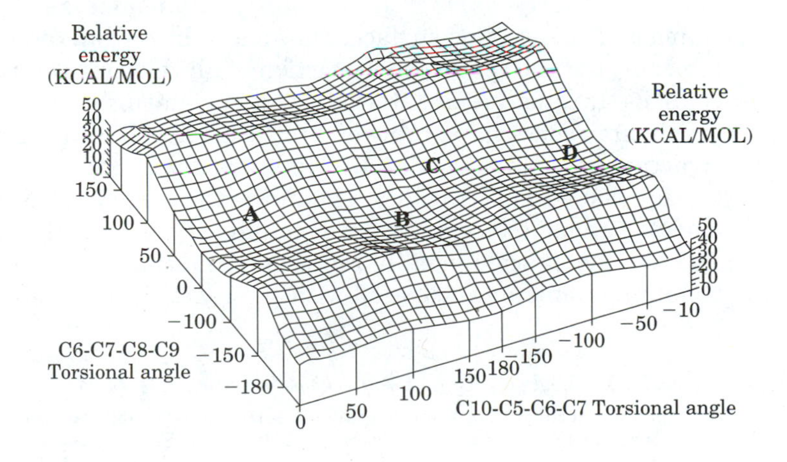

Figure 5.8 Three-dimensional plot of potential energy surface for vitamin D_3 intermediates. (Adapted from reference 3.)

[6]Reproduced from reference 8.

[7]Wentrup, C. *Reactive Molecules*; John Wiley & Sons: New York, 1984; p. 2.

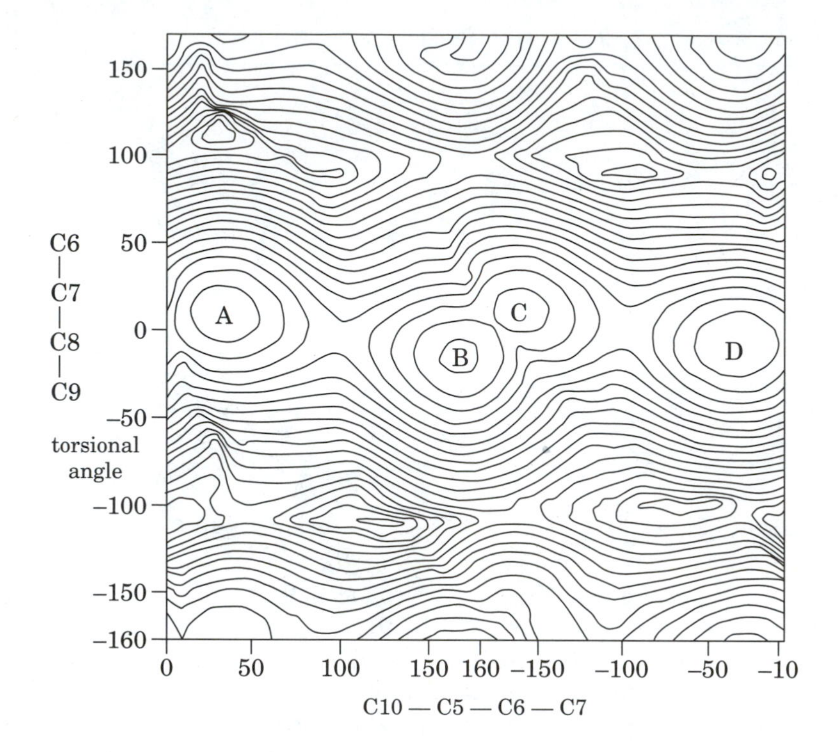

Figure 5.9 Contour plot for conformations of vitamin D_3 precursors. (Adapted from reference 3.)

a transition structure.[8] Both experimental and theoretical evidence suggest that the singlet diradical (in which the two unpaired electrons have opposite spins) is a shallow minimum.[9]

To represent a more complex chemical reaction, we should construct a potential energy surface in as many dimensions as there are bonding changes. The resulting surface, known as a **hypersurface**,[10] would indicate the energy of the species as a function of all relevant bond angle and length changes. It is not possible to represent such a surface as a two-dimensional graph on paper. A large table of numerical data indicating the energy of the structure as a function of each variable might be contemplated, but its complexity would limit its utility. Therefore we do what we usually do when our models become too complex—we simplify them until they are more useful. The horizontal scale usually represents the change in the most significant bonding parameter associated with the reaction. If no single bonding para-

[8]Coms, F. D.; Dougherty, D. A. *J. Am. Chem. Soc.* **1989**, *111*, 6894.

[9]For theoretical calculations and leading references to previous work, see Sherrill, C. D.; Seidl, E. T.; Schaefer III, H. F. *J. Phys. Chem.* **1992**, *96*, 3712. For 4-methylene-1,3-cyclopentanediyl, the energy well was determined to be 2.7 kcal/mol: Roth, W. R.; Bauer, F.; Breuckmann, R. *Chem. Ber.* **1991**, *124*, 2041.

[10]For a discussion, see Mezey, P. G. *Potential Energy Hypersurfaces*; Elsevier: Amsterdam, 1987.

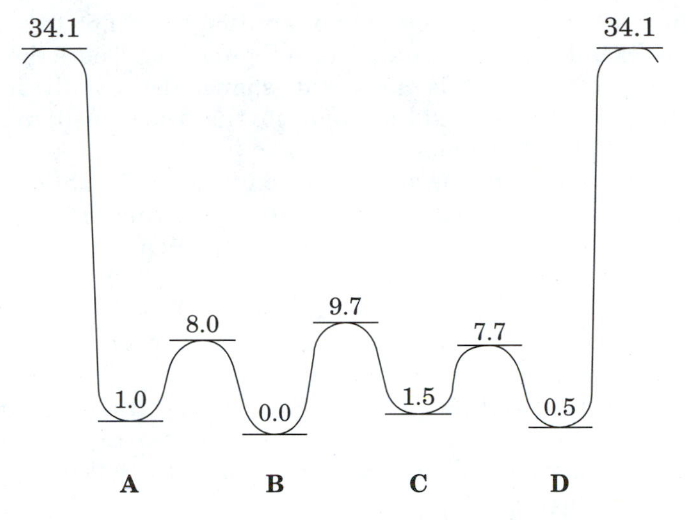

Figure 5.10 Schematic representation of energies and rotational barriers (kcal/mol) for **A**, **B**, **C**, and **D**. (Adapted from reference 3.)

meter is physically meaningful for the changes we intend to represent, then we may draw a figure in which the horizontal scale is only a graphic convenience. Often in such cases the x-axis is labeled as *extent of reaction*.

Cruickshank et al. have criticized the practice of using the terms *reaction coordinate* and *extent of reaction* interchangeably.[11] They prefer *extent of reaction* to mean a macroscopic quantity—the composition of a reaction mixture as a function of time. They prefer *reaction coordinate* to be a molecular quantity—the progress toward product formation of an individual molecule (in a unimolecular process) or set of molecules (in a polymolecular process). With terms defined this way, they note that it is inappropriate to label the vertical axis of a reaction coordinate diagram as *free energy*, since a reaction coordinate does not actually define a thermodynamic state. These

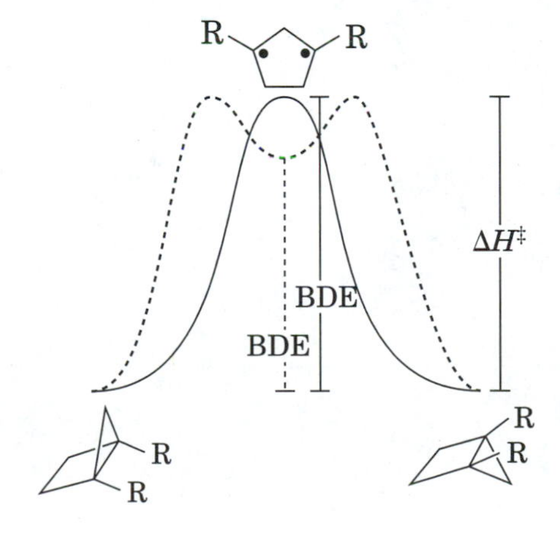

Figure 5.11 Reaction coordinate diagram for 1,3-cyclopentanediyl as an intermediate (dashed line) or transition structure (solid line) in the ring inversion of bicyclo[2.1.0]pentane.

[11]Cruickshank, F. R.; Hyde, A. J.; Pugh, D. *J. Chem. Educ.* **1977**, *54*, 288.

authors prefer reaction coordinate diagrams not to be shown as smooth curves but to be represented as in Figure 5.12. There the vertical scale is G^{θ}, which refers to a mole of the state shown, not to individual molecular units. The horizontal broken lines indicate that the transition states do not represent stable compounds.

In spite of the merits of restricting the information implied in reaction coordinate diagrams, the practice of drawing smooth curves is common practice in organic chemistry. Moreover, such figures also incorporate additional implicit information in the shapes of the curves that connect the species of different energy. Consider the three possible diagrams shown for an (a) endothermic, (b) thermoneutral, and (c) exothermic reaction in Figure 5.13. In each case the drawing in the middle is the one we expect to see, not the drawing to either side of it. Why are the other drawings not used?

Consideration of the interpretation of reactions based on reaction coordinate diagrams led Hammond to propose the following postulate.[12,13]

> If two states, as for example, a transition state and an unstable intermediate, occur consecutively during a reaction process and have nearly the same energy content, their interconversion will involve only a small reorganization of the molecular structures.

Hammond added that "in highly exothermic steps it will be expected that the transition states will resemble reactants closely and in endothermic steps the products will provide the best models for the transition states." In other words, the postulate tells us that reaction coordinate dia-

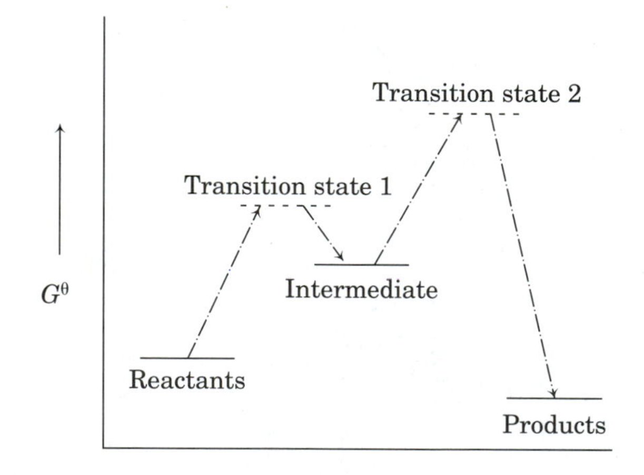

Figure 5.12 Alternative representation for structure-energy diagram. (Reproduced from reference 11.)

---

[12] Hammond, G. S. *J. Am. Chem. Soc.* **1955**, *77*, 334.

[13] Credit is often given to Leffler for an earlier, independent analysis along the same lines as the Hammond postulate: Leffler, J. E. *Science* **1953**, *117*, 340.

grams should look like the middle drawing in each of the examples in Figure 5.13.[14]

The Hammond postulate represents ideas that are so familiar to organic chemists that the postulate seems almost intuitive. For example, suppose we are considering a series of reactions that are believed to proceed through carbocation intermediates. We know that the stability of carbocations decreases in the order 3° > 2° > 1° > methyl. Suppose that the rates of reaction of a series of reactants (R-X) is a function of the R group and follows the same pattern. If the rate-limiting step in the chemical reaction is the endothermic formation of an intermediate carbocation, then the transition state leading to the intermediate must be closer in energy to the intermediate than to the reactant, and the Hammond postulate tells us that it must be closer in structure also.[15] That is, the energy maximum lies to the right on the reaction coordinate diagram (Figure 5.14). Since the transition state and the intermediate are similar in structure, they have similar bonding forces, and whatever stabilizes a 3° carbocation will, to some extent at least, stabilize the transition state leading to it. We conclude, therefore, that the transition state leading to a 3° carbocation should be lower in energy than one leading to a 2° carbocation, which should in turn be lower in

Figure 5.13 Hypothetical shapes for reaction coordinate diagrams.

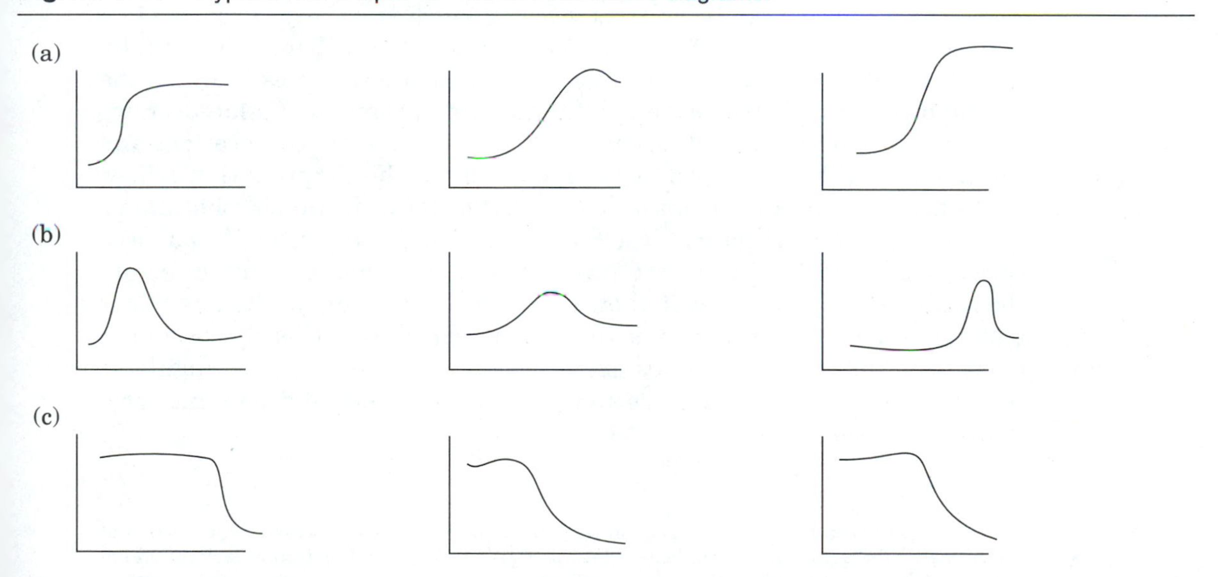

[14]For examples of reaction coordinate diagrams that do not follow the predictions of the Hammond postulate, see Petersson, G. A.; Tensfeldt, T. G.; Montgomery, Jr., J. A. *J. Am. Chem. Soc.* **1992**, *114*, 6133; Figure 5 (page 268) of reference 176.

[15]In much of the literature of organic chemistry, the term *rate-determining step* has been used, but the term *rate-limiting step* is becoming more common.

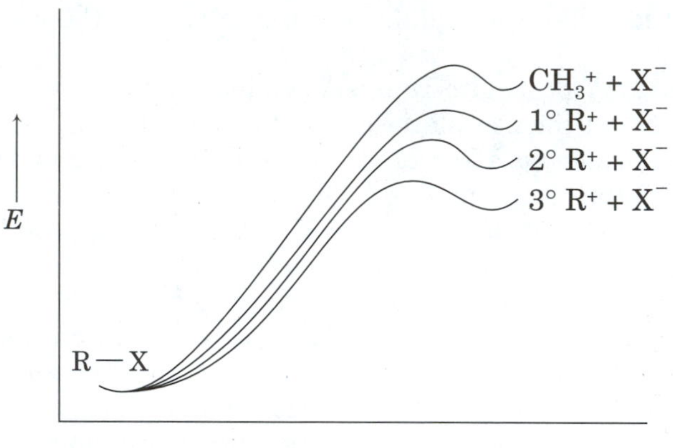

Figure 5.14 Schematic representation of structure-energy relationships in S_N1 reactions.

energy than 1° and methyl carbocations. Lower transition state energies mean lower activation energies and therefore faster rates of reaction.[16,17,18]

Before we develop the relationship between reaction rates and reactive intermediates in more detail in the following chapters, we will discuss in some detail the nature of carbocations and other reactive intermediates. However, there is a certain provincialism in the discussion that follows because the emphasis will be on structures in which reactivity is centered on carbon atoms. That is, we will discuss ions in which a carbon atom formally bears the positive or negative charge or radicals and carbenes in which the nonbonded electrons are formally localized on a carbon atom. Moreover, we will emphasize radicals and carbenes in this chapter, since carbocations and carbanions will be developed more extensively in the chapters that follow. Certainly there are important reactive intermediates in which other atoms exhibit similar characteristics. Focusing on carbon-centered radicals, cations and anions will provide a framework for discussing those factors that affect reactivity and will allow us to introduce many of the important techniques used to study organic reactive intermediates. Clearly we cannot treat even these intermediates exhaustively. Our purpose is to highlight some features of their structures and bonding as a basis for discussing reaction mechanisms.

---

[16]Again, we are assuming that the reaction mechanism is the same in all cases. Thus we expect the Arrhenius A factor (or $\Delta S^\ddagger$) to be constant so that our arguments based on activation energy or enthalpy are valid. We generally also assume that the barrier to reaction arises primarily from enthalpy terms, not from entropy.

[17]The Hammond postulate is fundamentally related to some alternative descriptions of the relationship between transition state structures and chemical reactivity. Jencks has proposed the acronym *Bema Hapothle* to incorporate references to the work of Bell, Marcus, Hammond, Polanyi, Thornton, and Leffler: Jencks, W. P. *Chem. Rev.* **1985**, *85*, 511.

[18]For a mechanical representation of the Hammond postulate, see Nyquist, H. L. *J. Chem. Educ.* **1991**, *68*, 731.

5.2 Reactive Intermediates with Neutral, Electron Deficient Carbon Atoms

Radicals

A carbon-centered radical is a structure having an unpaired electron on a carbon atom and a formal charge of 0.[19] Examples include the methyl radical (**6**), the vinyl radical (**7**), the phenyl radical (**8**), and the triphenylmethyl radical (**9**). Radicals are often called ***free radicals***, a term that arose from early nomenclature systems in which a "radical" was a substituent group that was preserved as a unit through a chemical transformation.[20,21] Thus the CH_3 group as a substituent was known as the methyl radical, so a neutral $\cdot CH_3$ group became a free radical. The terms *radical* and *free radical* are now used interchangeably.[22,23]

$\dot{C}H_3$ **6** $H_2C{=}\dot{C}H$ **7** **8** $Ph_3C\cdot$ **9**

Early Evidence for the Existence of Radicals

Around 1900 chemists had come to believe that carbon was tetrasubstituted in virtually all organic compounds. However, Gomberg's attempt to prepare hexaphenylethane from the reaction of triphenylmethyl chloride with zinc

---

[19]The definition of radical given here is sufficient for our discussions, but more precise definitions are needed for theoretical chemistry. One definition is that a radical is "a species with one or more (spin-free) natural orbitals whose occupation numbers are near 1." Klein, D. J.; Alexander, S. A. in *Graph Theory and Topology in Chemistry*; King, R. B.; Rouvray, D. H., eds.; Elsevier: Amsterdam, 1987; p. 404.

[20]The origin of the concept of radicals was attributed to Lavoisier; see Gomberg, M. *Chem. Rev.* **1924**, *1*, 91.

[21]Walling, C. *Free Radicals in Solution*; John Wiley & Sons, Inc.: New York, 1957; p. 2.

[22]Forrester, A. R.; Hay, J. M.; Thomson, R. H. *Organic Chemistry of Stable Free Radicals*; Academic Press: New York, 1968; p. 1.

[23]For an overview of the chemistry of radicals, see Kochi, J. K., ed. *Free Radicals*, Vols. I and II; John Wiley & Sons: New York, 1973; Leffler, J. E. *An Introduction to Free Radicals*; Wiley-Interscience: New York, 1993.

dust (equation 5.1) gave colored solutions that reacted with reagents such as iodine (equation 5.2). The results were interpreted in terms of an equilibrium between hexaphenylethane (**10**) and triphenylmethyl radicals (**11**).[24,25]

$$Ph_3CCl + Zn \rightarrow Ph_3C{-}CPh_3\ (\mathbf{10}) \rightleftharpoons 2Ph_3C\cdot\ (\mathbf{11}) \qquad (5.1)$$

$$2\ Ph_3C\cdot + I_2 \longrightarrow 2\ Ph_3CI \qquad (5.2)$$

Hexaphenylethane was accepted as the structure for the dimer of triphenylmethyl for over sixty years. However, studies of dimers of diarylalkylmethyl radicals led Lankamp, Nauta, and MacLean to examine the PMR spectrum of the supposed "hexaphenylethane," which suggested that the dimer is most likely **12** (equation 5.3).[26] The structural assignment was strengthened by the finding of Guthrie and Weisman that treating the dimer with *t*-BuOK produces **13**.[27,28]

$$2\ (C_6H_5)_3C\cdot \ (\mathbf{11}) \rightleftharpoons (C_6H_5)_2C{=}C_6H_5(H)C(C_6H_5)_3\ (\mathbf{12}) \xrightarrow{t\text{-BuOK}} (C_6H_5)_2CH{-}C_6H_4{-}C(C_6H_5)_3\ (\mathbf{13}) \qquad (5.3)$$

Even though the proposal of **10** as the structure of the dimer was in error, the demonstration of the existence of radicals as transient species led to a great effort to study other radicals. Radicals reported in the literature range from some that are extremely unstable, short-lived species to others that can be isolated as pure substances.[29] For example, Koelsch reported that the radical, α,γ-bisdiphenylene-β-phenylallyl (**14**) could be recovered in part after being subjected to molecular oxygen in boiling benzene for six hours.[30] A well-known example of a nitrogen-centered radical is 2,2-diphenyl-1-picrylhydrazyl, **15**, which is commercially available.

---

[24]Gomberg, M. *Chem. Ber.* **1900**, *33*, 3150; *J. Am. Chem. Soc.* **1900**, *22*, 757.

[25]Flürscheim, B. *J. Prakt. Chem.* **1905**, *71*, 497; see the discussion in Skinner, K. J.; Hochster, H. S.; McBride, J. M. *J. Am. Chem. Soc.* **1974**, *96*, 4301.

[26]Lankamp, H.; Nauta, W. T.; MacLean, C. *Tetrahedron Lett.* **1968**, 249.

[27]Guthrie, R. D.; Weisman, G. R. *J. Chem. Soc. D.* **1969**, 1316.

[28]McBride, J. M. *Tetrahedron* **1974**, **30**, 2009, noted that **12** had been proposed as the structure for the dimer in 1904 and was supported by several lines of evidence in the following two years, and he discussed the process by which the incorrect structure became accepted.

[29]Longer-lived radicals are termed *kinetically stable* or *persistent*. Extremely persistent radicals have been termed *inert*; Juliá, L.; Ballester, M.; Riera, J.; Castañer, J.; Ortin, J. L.; Onrubia, C. *J. Org. Chem.* **1988**, *53*, 1267 and references therein.

[30]Koelsch, C. F. *J. Am. Chem. Soc.* **1957**, *79*, 4439.

14 **15**

Detection and Characterization of Radicals

If conditions can be found in which transient species can be produced in higher concentration and with longer lifetimes than is the case under typical reaction conditions, then experimental structure elucidation techniques can be brought to bear. The relative stability of the triphenylmethyl radical allowed it to be studied by magnetic susceptibility measurement, which involves weighing it both inside and outside a magnetic field.[31] The unpaired electron makes the radical paramagnetic, so the sample is drawn into the magnetic field. By this technique the dissociation of hexaphenylethane to the triphenylmethyl radical was determined to occur to the extent of 2% in a 0.1 M sample.[32]

A technique that has been of more value to organic chemists is that of electron paramagnetic resonance (EPR) spectroscopy.[33,34] There is associated with the electron a spin quantum number, m_s, which may have the value $+\frac{1}{2}$ or $-\frac{1}{2}$. If electrons are paired in an orbital, then one has $m_s = +\frac{1}{2}$ and the other has $m_s = -\frac{1}{2}$, so there is no net spin magnetic moment. If an electron is unpaired, however, there will be a net magnetic moment of 9.284×10^{-19} erg·gauss^{-1}. The direction of the magnetic moment depends on the spin quantum number, so there will be two spin states. The two states are equal in energy in the absence of an external magnetic field. In the presence of an external magnetic field (with strength H_0) the two spin states have

[31]Selwood, P. W. in *Techniques of Organic Chemistry*, Vol. I, Part IV; Weissberger, A., ed.; John Wiley and Sons: New York, 1960; pp. 2873 *ff.*

[32]Roy, M. F.; Marvel, C. S. *J. Am. Chem. Soc.* **1937**, *59*, 2622.

[33]The technique is also known as electron spin resonance spectroscopy, or ESR. For a brief summary of the procedure, see Rieger, P. H. in *Physical Methods of Organic Chemistry*, Vol. I, Part IIIA; Weissberger, A.: Rossiter, B. W., eds.; Wiley-Interscience: New York, 1972; pp. 499–598.

[34]Ayscough, P. B. *Electron Spin Resonance in Chemistry*; Methuen & Co. Ltd.: London, 1967.

different energies, with the electron having $m_s = -\frac{1}{2}$ being lower in energy. The energy difference between the two states, ΔE, is

$$\Delta E = h\nu = g\beta_e H_0 \tag{5.4}$$

In this equation, g is the spectroscopic splitting factor, which is the ratio of the magnetic moment, μ, to the angular momentum of the electron. This value is 2.002319 for a free electron, and does not vary greatly for unpaired electrons in organic radicals.[35] The term β_e is the Bohr magneton, a constant with the value 9.2732×10^{-21} erg/gauss. Thus at a field of 3400 gauss, the energy difference is

$$\Delta E = (2.0023) \times (9.2732 \times 10^{-21}\ \text{erg/gauss}) \times (3400\ \text{gauss})$$
$$= 0.63 \times 10^{-16}\ \text{erg} \tag{5.5}$$

which corresponds to 0.91 cal/mol and to a frequency of 9500 MHz.[36,37]

There are some similarities between proton magnetic resonance (PMR) and EPR spectroscopy. In both, the sample is placed in a magnetic field and is irradiated with electromagnetic energy of an appropriate frequency. The magnetic field is swept or varied linearly, and absorption of energy by the sample is recorded as a decrease in intensity of the radio frequency energy received by a detector. Thus we could observe an absorption or *resonance* when the changing magnetic field strength makes the energy of the electromagnetic signal exactly equal to the energy difference between the two spin states, Figure 5.15(a). For experimental reasons, however, the EPR signal is

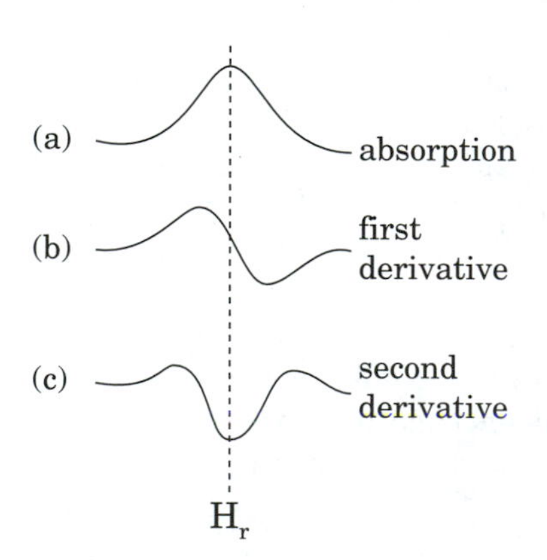

Figure 5.15 Representations of EPR spectra: (a) resonance signal; (b) first derivative; (c) second derivative. (Reproduced from reference 38.)

---

[35]Gordon, A. J.; Ford, R. A. *The Chemist's Companion*; John Wiley & Sons, Inc.: New York, 1972; p. 336.

[36]Relationships between frequency and energy are discussed in Chapter 12.

[37]Applying the Boltzmann distribution to this energy difference gives a relative population of the two spin states of 1,000,000 to 1,001,500. Thus at the same magnetic field strength, the difference between the two spin populations is greater than the corresponding difference in spin populations in proton magnetic resonance. This is fortunate because EPR measurements are usually made on materials in which only a fraction of the substance is in the free radical form.

more often recorded as the first derivative of absorption, Figure 5.15(b), or as the second derivative of the absorption, Figure 5.15(c).[38]

The g value in equation 5.4 has been compared to the chemical shift (δ) in PMR spectroscopy. However, the very slight variation in g values from radical to radical means that g values seldom give chemically useful information for most organic compounds. What is of more value is the splitting of the EPR signal due to interaction of the electron spin with the spins of nearby magnetic nuclei. This effect has been explained by McConnell et al. in terms of spin polarization involving the unpaired electron on carbon with the electrons in the C—H bond, Figure 5.16.[39] Thus, the EPR signal due to the methyl radical (shown as a second derivative spectrum in Figure 5.17)[40] is split into a four-line pattern with relative intensities of 1 : 3 : 3 : 1 because of coupling with the three methyl protons. Similarly, the unpaired electron in the benzene radical anion is split by six equivalent protons, so the signal is a septet with relative intensities 1 : 6 : 15 : 20 : 15 : 6 : 1.[41,42,43]

Figure 5.16
Mechanism of spin polarization for planar methyl radical. (Reproduced from reference 39.)

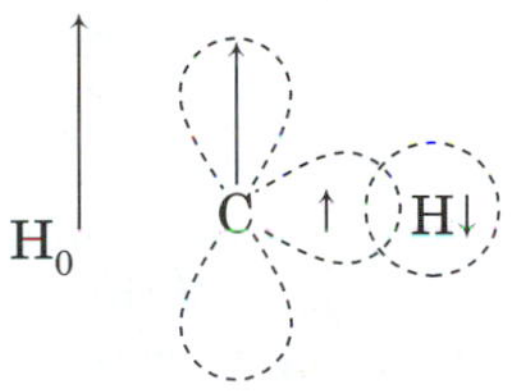

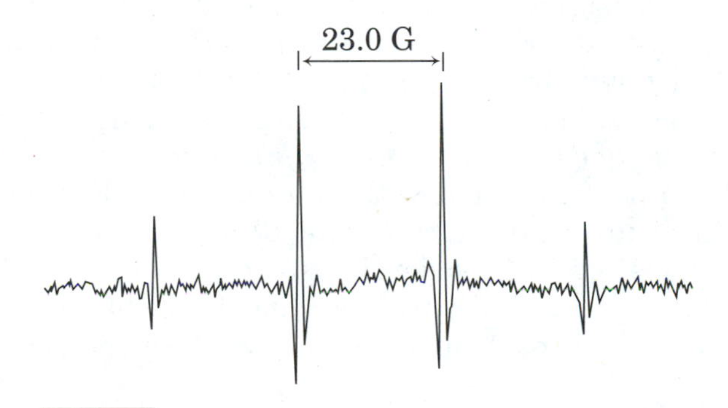

Figure 5.17
EPR spectrum for methyl radical. (Reproduced from reference 40.)

[38]The figures are adapted from those presented by Bunce, N. J. *J. Chem. Educ.* **1987**, *64*, 907.

[39]McConnell, H. M.; Heller, C.; Cole, T.; Fessenden, R. W. *J. Am. Chem. Soc.* **1960**, *82*, 766.

[40]Fessenden, R. W.; Schuler, R. H. *J. Chem. Phys.* **1963**, *39*, 2147. Figure 5.17 is shown as the negative of the second derivative spectrum.

[41]Tuttle, Jr., T. R.; Weissman, S. I. *J. Am. Chem. Soc.* **1958**, *80*, 5342.

[42]Carrington, A. *Quart. Rev. Chem. Soc.* **1963**, *17*, 67.

[43]The spectrum of the *t*-butyl radical shows a ten-line pattern with relative intensities 1 : 9 : 36 : 81 : 126 : 126 : 81 : 36 : 9 : 1 due to splitting by the nine equivalent protons. See reference 34, p. 67 and references therein.

Figure 5.18
EPR spectrum for ethyl radical. (Reproduced from reference 40.)

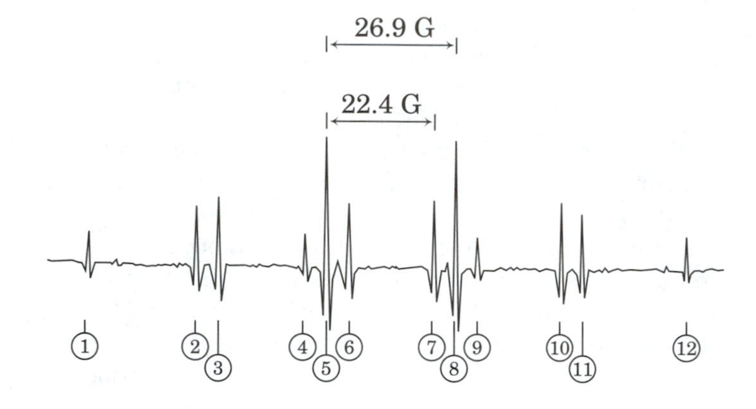

Figure 5.18 shows the EPR spectrum of the ethyl radical (also as a second derivative).[40] Splitting of the signal by the two protons on the α carbon atom (a_α = 22.4 G) gives rise to a three line pattern, but each of those three lines is further split into four lines by the β protons (a_β = 26.9 G). The combination of the two sets of splittings gives rise to a twelve-line pattern. Also of interest here are the magnitudes of the two a values. The magnitude of the hyperfine coupling depends on the density of the unpaired electron in an orbital with s character about a magnetic nucleus.[44] Therefore it might seem surprising that the values of a_α and a_β for the ethyl radical are approximately equal, because such a result is not consistent with a simple model for the ethyl radical (Figure 5.19) in which all of the unpaired electron density is on the α carbon atom. Nevertheless, it is common for values of a_α and a_β to be similar.[45] However, values of a_γ (for example, the hyperfine coupling constant for splitting the signal of a 1-propyl radical by the terminal methyl group) are much smaller, typically less than 1 G.[38,46]

Figure 5.19
Localized model for ethyl radical.

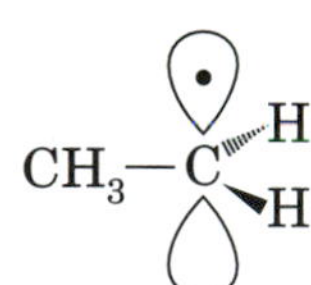

The valence bond explanation for the similarity of a_α and a_β values is based on the concept of ***hyperconjugation***, which holds that a C-H or C-C σ bond can be conjugated with an adjacent radical or cation center. As shown in Figure 5.20, we can write **17** as a resonance structure for **16**.[47] Now the unpaired electron density is localized on one of the hydrogen atoms of the methyl group, and by rotation about the C-C bond all three of the hydrogen atoms can participate equally in the hyperconjugation. The view

[44]This mechanism for hyperfine coupling is known as the *isotropic mechanism*, since it is independent of the orientation of the structure. Another mechanism known as the *anisotropic mechanism* becomes important in solids or other samples in which rotation of the radical is slow. For more details on the coupling mechanism, see reference 22, pp. 18 *ff*.

[45]Symons, M. C. R. *J. Chem. Soc.* **1959**, 277 and references therein.

[46]In special cases longer range coupling can be seen. 3-[*n*]staffyl radicals (based on repeating units of [1.1.1]propellane) have been reported to show coupling with ϵ, ζ, and even ι protons. McKinley, A. J.; Ibrahim, P. N.; Balaji, V.; Michl, J. *J. Am. Chem. Soc.* **1992**, *114*, 10631.

[47]Wheland, G. W. *J. Chem. Phys.* **1934**, *2*, 474.

Figure 5.20
Hyperconjugation model for ethyl radical.

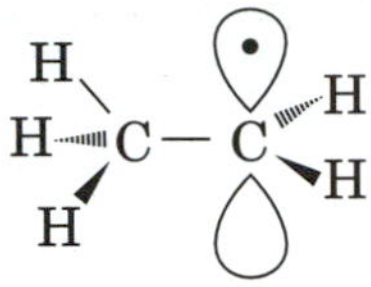

16

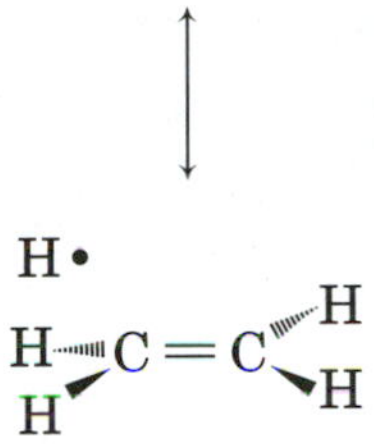

17

that hyperconjugation can be rationalized meaningfully through the use of resonance structures such as **17** has been a source of considerable controversy.[48] There is something unsettling about having each hydrogen in turn (as rotation occurs) appear to become unbonded. Of course, we must remember that drawing structures such as **17** does not mean that the radical is in equilibrium with structures without C_β—H bonds, only that such resonance structures contribute to the properties of the actual resonance hybrid.

The PMO description of radical stabilization is shown in Figure 5.21. The singly occupied *p* orbital is the SOMO, which is higher in energy than a π-like orbital on the methyl fragment. The HOMO is a C—H bonding orbital localized on the methyl group and parallel with the *p* orbital. The interaction of the SOMO and the HOMO produces two new orbitals, one lower in energy than the HOMO and one higher in energy than the SOMO. However, since two electrons go into the lower energy orbital while only one goes into the higher energy orbital, the effect is overall stabilizing.

Additional insight into the structure of the ethyl radical comes from an extended Hückel calculation. Figure 5.22 (page 264) shows the orbital contours for the SOMO, ψ_7, calculated for a conformation in which none of the C_α—H bonds eclipse either of the C_β bonds. We note first that this orbital does have π character, and the portion of the orbital near the CH_2 group appears to be much like a *p* orbital. However, there is also density associated with the CH_3 end of the structure. Specifically, there is significant density on the hydrogen with the C_β—H bond that is parallel to the *p* orbital on the adjacent carbon atom, and there is some density on the other two hydrogen atoms also.[49] Although an analysis of all the populated MOs would be necessary to draw firm conclusions, the orbital contour of the ethyl SOMO is consistent with the view from hyperconjugation and from PMO theory that electron density from the methyl group is delocalized into the half-empty *p* orbital, thus producing stabilization.[50]

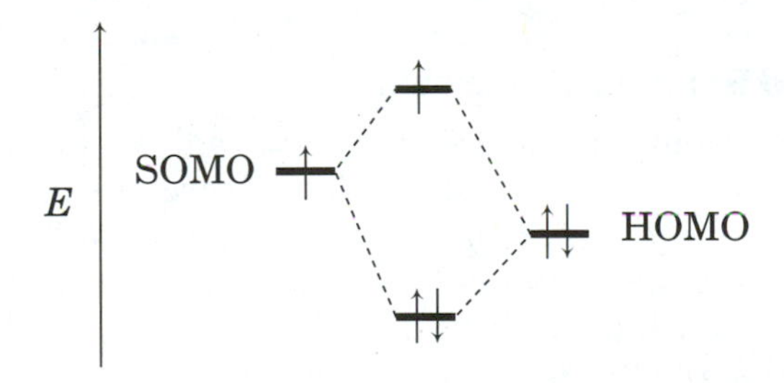

Figure 5.21
PMO description of radical stabilization: interaction of SOMO with donor HOMO.

[48]Dewar has discussed the origin of the theory of hyperconjugation and the debate over its importance: Dewar, M. J. S. *Hyperconjugation*; Ronald Press Company: New York, 1962. See also Shiner, V. J.; Campaigne, E., eds. *Conference on Hyperconjugation*, Proceedings of a Conference Held at Indiana University 2–4 June, 1958; Pergamon Press: New York, 1959.

[49]We should not suggest that one of the CH_3 hydrogen atoms on the ethyl radical is different from the other two. Rotation about the C-C bond will introduce the same amount of unpaired electron density onto all of the hydrogen atoms.

[50]For a summary of the shapes of all the orbitals of the $CH_3CH_2\cdot$ radical, see Jorgensen, W. L.; Salem, L. *The Organic Chemist's Book of Orbitals*; Academic Press: New York, 1973; p. 96.

Figure 5.22 (a) side and (b) end views of SOMO of ethyl radical in bisected conformation. (c) Side and (d) end views of SOMO of ethyl radical in eclipsed conformation.

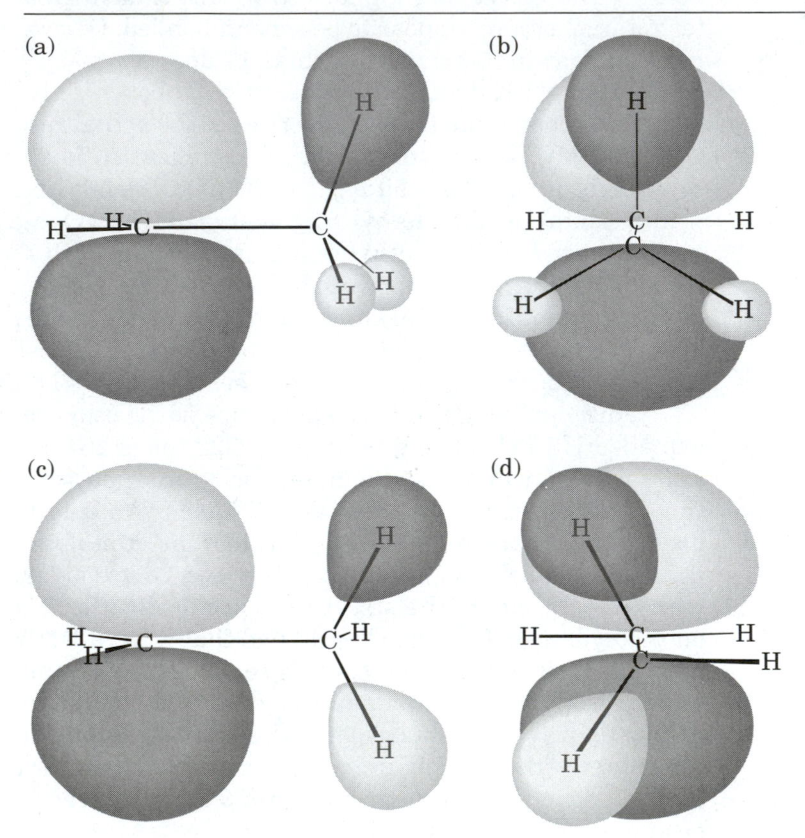

Structure and Bonding of Radicals

There is experimental evidence that alkyl radicals are planar or are pyramidal structures that can become planar through inversion (like ammonia).[51] For example, chlorination of (+)-1-chloro-2-methylbutane (**18**) produced *racemic* 1,2-dichloro-2-methylbutane (**20**). The results are most consistent with an intermediate radical, **19**, which is planar or which has only a small barrier to inversion, so that reaction of **19** can occur equally well from either face of the radical center.[52]

Spectroscopic data can be used to distinguish between planar and nonplanar, rapidly inverting radical centers. The hyperfine coupling constant

[51]For a discussion of the stereochemistry of radicals, see Eliel, E. L. *Stereochemistry of Carbon Compounds*; McGraw-Hill: New York, 1962; pp. 380–384. For a discussion of the stereochemistry of vinylic and cyclohexyl radicals, see Simamura, O. *Top. Stereochem.* **1969**, *4*, 1.

[52]Brown, H. M. C. ; Kharasch, M. S.; Chao, T. H. *J. Am. Chem. Soc.* **1940**, *62*, 3435.

Figure 5.23 Racemization accompanying chlorination of (+)-1-chloro-2-methylbutane.

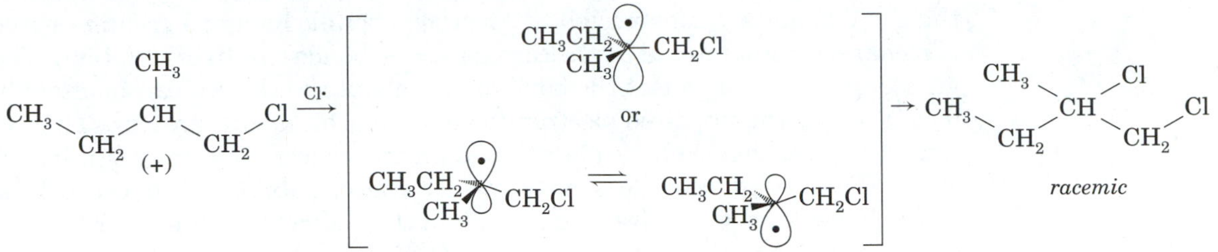

a_α in the methyl radical is 23.0 G, which is a typical value for the splitting of an EPR signal by protons attached to a radical center.[53] Theoretical analysis of the spectrum suggested that the methyl radical is probably flat, although a deviation from planarity of 10°–15° could not be ruled out.[54] There is also spectroscopic evidence that the methyl radical in the gas phase is essentially planar.[55] Thus the methyl radical is conveniently described by sp^2 hybridization, with the unpaired electron located primarily in the p orbital.

Estimates of the geometry of substituted radicals can be obtained from analysis of species labeled with ^{13}C. Fessenden and Schuler determined that the angle F—C—F in the trifluoromethyl radical is 111.1° (very nearly sp^3 hybridization) and that increasing fluorine substitution causes the radicals to go from planar, sp^2 hybridization for $\cdot CH_3$ to nearly tetrahedral, sp^3 hybridization for $\cdot CF_3$.[40] This conclusion is consistent with results of INDO calculations, and the geometries of fluoromethyl radicals calculated by this method are shown in Figure 5.24.[56,57]

Figure 5.24 Calculated geometries of fluoro-substituted methyl radicals. (Reproduced from reference 56.)

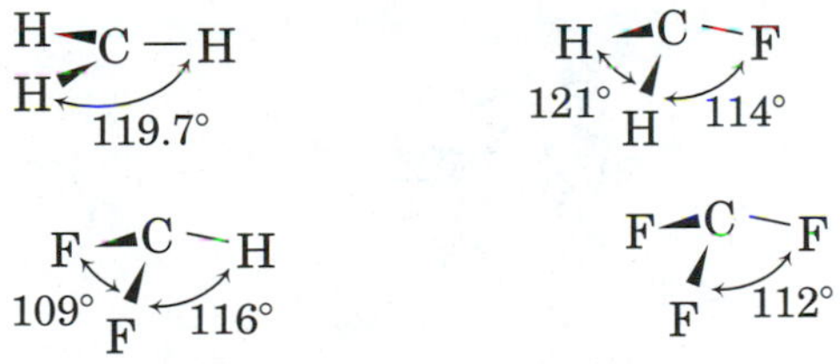

[53]More correctly, the *magnitude* of a_α is 23.0 G. In this experiment, only the magnitude of a_α is evident. There is evidence that the value is actually −23.0 G.

[54]Karplus, M. *J. Chem. Phys.* **1959**, *30*, 15.

[55]Ellison, G. B.; Engelking, P. C.; Lineberger, W. C. *J. Am. Chem. Soc.* **1978**, *100*, 2556 and references therein.

[56]Beveridge, D. L.; Dobosh, P. A.; Pople, J. A. *J. Chem. Phys.* **1968**, *48*, 4802.

[57]The fluoro-substituted radicals have been studied by theoretical methods. For example, see Bernardi, F.; Cherry, W.; Shaik, S.; Epiotis, N. D. *J. Am. Chem. Soc.* **1978**, *100*, 1352 and references therein.

A qualitative explanation of the geometry of fluoro-substituted radicals was earlier given by Pauling:[58] Because of the electronegativity of fluorine, the C—F bonds are constructed with carbon orbitals having a greater degree of *p* character than would be the case for C—H bonds (sp^2 hybridization). The greater *p* character causes the bond angles about the central carbon atom to decrease and the unpaired electron to be in an orbital with greater *s* character. Of course, this explanation is based on the concept of straight (internuclear) C—F bonds and does not consider the possibility of curved bonds, which have since been advanced to explain the geometry of fluorocarbons.

Not only are fluoro-substituted radicals nonplanar, but alkyl-substituted radicals are found to be nonplanar also. An explanation for the nonplanar nature of hydrocarbon radicals was provided by Paddon-Row and Houk, who ascribed the pyramidalization of the radical center to two effects: (1) increased staggering of bonds to the radical center with bonds on adjacent atoms, and (2) increased hyperconjugation of the *p* orbital with one of the adjacent σ bonds.[59] As shown in Figure 5.25, pyramidalization of the ethyl radical makes the C—H bonds on the CH_2 more nearly staggered with respect to two of the C—H bonds on the CH_3 group. At the same time, the *p* orbital on the CH_2 becomes more parallel with the orbitals comprising the third C—H bond, thus stabilizing the unfilled orbital system of the radical.

An especially interesting case is the *t*-butyl radical. We might at first expect the radical to be planar (**21**) because that geometry would minimize the steric repulsion of methyl groups. However, the a_c value was found experimentally to be 46.2 G, which was interpreted in terms of a C—C—C bond angle of 117.3° (**22**),[60] suggesting that the electronic stabilization resulting from pyramidalization (Figure 5.25) outweighs the increase in energy due to steric effects.[61]

Rates of radical reactions at bridgehead carbon atoms do not indicate significant strain due to pyramidalization, at least not in comparison with

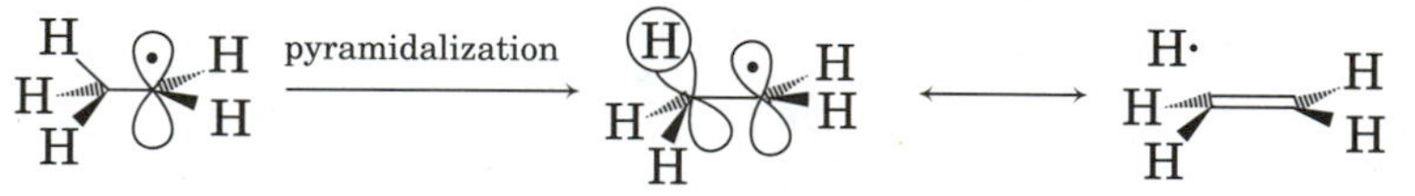

Figure 5.25 Stabilization of ethyl radical through pyramidalization. (Adapted from reference 59.)

[58] Pauling, L. *J. Chem. Phys.* **1969**, *51*, 2767.

[59] (a) Paddon-Row, M. N.; Houk, K. N.; *J. Am. Chem. Soc.* **1981**, *103*, 5046; (b) *J. Phys. Chem.* **1985**, *89*, 3771. To rephrase the explanation in terms used in these papers, pyramidalization will minimize four-electron closed-shell electronic repulsions (torsional strain) and maximize the stabilizing two- or three-electron interactions that can be predicted through PMO theory.

[60] Wood, D. E.; Williams, L. F.; Sprecher, R. F.; Lathan, W. A. *J. Am. Chem. Soc.* **1972**, *94*, 6241.

[61] However, this conclusion has been questioned: Griller, D.; Ingold, K. U. *J. Am. Chem. Soc.* **1973**, *95*, 6459.

Figure 5.26 Planar (left) and nonplanar (right) geometries for *t*-butyl radical.

CH_3 CH_3 CH_3 **21** CH_3 CH_3 CH_3 **22**

the strain due to bridgehead carbenium carbon atoms.[62] Radicals such as 1-adamantyl (**23**) and 7-norbornyl (**24**) can be formed, although their stabilities and rates of formation decrease with increasing deviation from the geometry observed for acyclic analogues.[63,64] The general view that emerges from these studies is that only the methyl radical prefers to be planar, and even it has a low barrier to deformation.

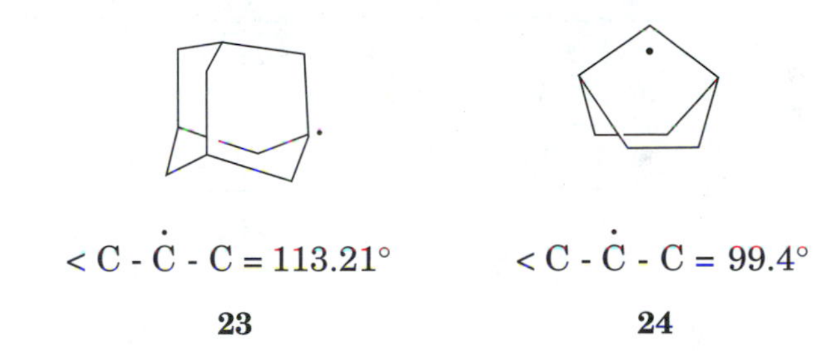

Thermochemical Data on Radicals

The conclusion that nonplanar radicals can be formed without unusual difficulty is obviously a qualitative statement. We need to apply to radicals and other reactive intermediates the same criteria we use for other structures: that is, to determine their heats of formation and/or strain energies and correlate those parameters with their structures. In principle the heat of formation of a radical (R·) formed by homolytic dissociation of R—X (equation 5.6) can be calculated according to equation 5.7, where all data refer to 298 K.[65]

$$\mathrm{RX} \rightleftharpoons \mathrm{R\cdot} + \mathrm{X\cdot} \tag{5.6}$$

$$\Delta H_f^0(\mathrm{R\cdot}) = \Delta H_{rxn}^0 - \Delta H_f^0(\mathrm{X\cdot}) + \Delta H_f^0(\mathrm{RX}) \tag{5.7}$$

[62] *Cf.* Applequist, D. E.; Roberts, J. D. *Chem. Rev.* **1954**, *54*, 1065 and references therein.
[63] Danen, W. C.; Tipton, T. J.; Saunders, D. G. *J. Am. Chem. Soc.* **1971**, *93*, 5186.
[64] For a review of bridgehead radicals, see Walton, J. C. *Chem. Soc. Rev.* **1992**, 105.
[65] Egger, K. W.; Cocks, A. T. *Helv. Chim. Acta* **1973**, *56*, 1516.

The heat of reaction can be estimated from the activation energy for the dissociation if we assume that there is no activation energy for the recombination process.[66]

Values for the heats of formation of radicals and for the heats of dissociation reactions have been compiled and are available in the literature.[65] However, many values for the heat of formation of alkyl radicals have been subsequently revised upward.[67,68,69,70] Table 5.1 lists later data for several small alkyl radicals and gives the bond dissociation energies for the corresponding alkanes. The regular decrease in the C-H bond dissociation energies to form methyl, ethyl, *i*-propyl or *t*-butyl radicals is the basis for the familiar generalization that the stability of radicals is methyl $< 1° < 2° < 3°$. As was the case for the ethyl radical, we can ascribe this increasing stability with increasing alkyl substitution at the radical center either to hy-

Table 5.1 ΔH_f^0 values for selected radicals and bond dissociation energies for selected alkanes.

Radical (R$^\cdot$)	ΔH_f^0 (kcal/mol)	BDE (R-H) (kcal/mol)
H	52.1[a]	104.2[a]
Methyl	34.9 ± 0.2[b]	104.9[e]
Ethyl	28.3 ± 0.4[b]	100.6[e]
n-Propyl	24.0 ± 0.5[c]	100.9[e]
i-Propyl	22.3 ± 0.6[c]	99.3[c]
	21.3 ± 0.7[b]	
sec-Butyl	16.7 ± 0.5[c]	99.0[c]
t-Butyl	11.7 ± 0.4[d]	95.9 ± 0.4[d]
	12.4 ± 0.5[c]	

[a]Data from reference 65. [b] Data from reference 67. [c]Data from reference 68. [d]Data from reference 69. [e]Calculated from ΔH_f^0 values of the radicals and the alkane.

---

[66]There clearly can be activation energies for radical recombination processes. The recombination of two triethylcarbinyl radicals to give hexaethylethane has been reported to have an activation energy of about 20 kcal/mol (Beckhaus, H.-D.; Rüchardt, C. *Tetrahedron Lett.* **1973**, 1971). Such activation energy is usually thought to arise primarily from steric hindrance in the dimer: Wentrup C. (reference 7), p. 27. In addition, if one or both of the radicals are delocalized, then part of the activation energy for combination is a "localization energy" resulting from loss of radical delocalization energy. For a discussion, see Dannenberg, J. J.; Tanaka, K. *J. Am. Chem. Soc.* **1985**, *107*, 671.

[67]Seetula, J. A.; Russell, J. J.; Gutman, D. *J. Am. Chem. Soc.* **1990**, *112*, 1347.

[68]Tsang, W. *J. Am. Chem. Soc.* **1985**, *107*, 2872.

[69]Russell, J. J.; Seetula, J. A.; Timonen, R. S.; Gutman, D.; Nava, D. F. *J. Am. Chem. Soc.* **1988**, *110*, 3084.

[70]See also Seakins, P. W.; Pilling, M. J.; Niiranen, J. T.; Gutman, D.; Krasnoperov, L. N. *J. Phys. Chem.* **1992**, *96*, 9847.

perconjugation (Figure 5.20) or delocalization (Figure 5.21). We note also that the values of bond dissociation energies are essentially the same for ethyl and n-propyl, in agreement with our expectation that the heat of reaction to form a primary radical is independent of the particular molecule involved.

Generation of Radicals

Radicals may be generated directly by thermal or photochemical processes that accomplish homolytic dissociation of a 2-electron bond. Organic peroxides (equation 5.8) and azo compounds (equation 5.9) have weak bonds that undergo dissociation to radicals.[71]

$$C_6H_5C(=O)O{-}OC(=O)C_6H_5 \xrightarrow{\Delta} C_6H_5C(=O)O\cdot + \cdot OC(=O)C_6H_5 \quad \textbf{(5.8)}$$

$$H_3C{-}N{=}N{-}CH_3 \xrightarrow{h\nu} N{\equiv}N + H_3C\cdot + \cdot CH_3 \quad \textbf{(5.9)}$$

Radicals may also result from chemical or electrochemical oxidation or reduction of stable molecules. Single electron transfer processes initially generate radical cations (for oxidation) or radical anions (for reduction), which may then fragment to radicals and ions.[72] For example, Sargent and co-workers determined that in 1,2-dimethoxyethane solution the radical anion of naphthalene (sodium naphthalenide, $Na^+ Ar^{\cdot -}$) transferred an electron to propyl iodide. Subsequent loss of iodide ion from the propyl iodide radical anion produced the propyl radical (equation 5.10).[73]

$$CH_3CH_2CH_2I \xrightarrow{Na^+\ Ar^{\cdot -}} [CH_3CH_2CH_2I]^{\cdot -} \xrightarrow{-I^-} CH_3CH_2CH_2\cdot \quad \textbf{(5.10)}$$

---

[71]Radicals can also be formed in photochemical reactions (see Chapter 12).

[72]Radical ions are subjects of continually increasing interest. For an introduction, see Evans, J. E.; Emes, P. J. in *Free Radical Reactions*; Waters, W. A., ed.; Butterworths: London, 1973; pp. 293–319. For a review of the structure and reactivity of organic radical cations, see Roth, H. D. *Top. Curr. Chem.* **1992**, *163*, 131.

[73]Sargent, G. D.; Cron, J. N.; Bank, S. *J. Am. Chem. Soc.* **1966**, *88*, 5363.

Reactions of Radicals

Radical reactions are frequently found to occur as chain reactions composed of three types of processes:

1. an initiation step (for example, one of the generation reactions discussed in the previous section),
2. a series of propagation steps, and
3. one or more termination steps that stop the chain reaction.

Each propagation step in a radical chain involves the reaction of a species with one unpaired electron to produce another species having an unpaired electron. Moreover, the reactant in one step is a product in a subsequent step. One example of a propagation step in a radical chain reaction is the abstraction of a hydrogen atom by a halogen in free radical halogenation (equation 5.11).

$$Cl\cdot + CH_4 \longrightarrow HCl + \cdot CH_3 \qquad \textbf{(5.11)}$$

A carbon-centered radical can also abstract an atom from another molecule (or from another atom in the same molecule) to fill its outer shell if the free energy change for the abstraction is favorable. One example of this process is *radical trapping*, in which a radical abstracts a hydrogen atom from a facile hydrogen atom donor such as tri-*n*-butyltin hydride (Bu_3SnH).[74,75] The process is illustrated with methyl radical in equation 5.12. This reaction can serve as a means of detecting radical intermediates, because the appearance of the species R—H upon addition of Bu_3SnH to a reaction suggests the intermediacy of R· in the reaction.

$$H_3C\cdot + HSn(Bu)_3 \longrightarrow CH_4 + \cdot Sn(Bu)_3 \qquad \textbf{(5.12)}$$

Trialkyltin hydrides are effective hydrogen atom transfer agents, and trialkyltin radicals will abstract chlorine, bromine, or iodine atoms from alkyl halides. Together with a radical initiator such as azobisisobutyronitrile (AIBN), trialkyltin hydrides and alkyl halides can give reduction or other radical-derived products.

$$\text{AIBN} \longrightarrow N_2 + 2\,R\cdot \qquad \textbf{(5.13)}$$

$$R\cdot + Bu_3SnH \longrightarrow R{-}H + Bu_3Sn\cdot \qquad \textbf{(5.14)}$$

$$Bu_3Sn\cdot + R_1{-}X \longrightarrow Bu_3SnX + R_1\cdot \qquad \textbf{(5.15)}$$

$$R_1\cdot \longrightarrow \text{further reactions} \qquad \textbf{(5.16)}$$

Another common propagation step is addition of a radical to a double or triple bond as in the anti-Markovnikov addition of HBr to an alkene (equation 5.17).

---

[74]Menapace, L. W.; Kuivila, H. G. *J. Am. Chem. Soc.* **1964**, *86*, 3047 and references therein.

[75]Walling, C.; Cooley, J. H.; Ponaras, A. A.; Racah, E. J. *J. Am. Chem. Soc.* **1966**, *88*, 5361.

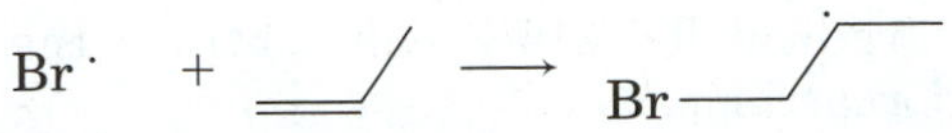

$$\text{Br}^{\cdot} + \text{CH}_2{=}\text{CH}{-}\text{CH}_3 \longrightarrow \text{BrCH}_2{-}\dot{\text{C}}\text{H}{-}\text{CH}_3 \quad \textbf{(5.17)}$$

Radical addition to a multiple bond is the key step in radical polymerization. For example, in the polymerization of styrene, the initiation step is homolysis of an initiator, which produces a radical that adds to a styrene molecule to begin the polymer chain. The propagation step illustrated in equation 5.18 is then repeated hundreds or thousands of times. Two carbon-carbon single bonds are more stable than one carbon-carbon double bond by about 20 kcal/mol, so the difference in the strengths of carbon-carbon single and double bonds provides the driving force for the reaction.[76] Increasingly, radical addition reactions are finding application in organic synthesis. Because charged species are not involved in the reaction, subtle effects due to orbital interactions and steric interactions can provide opportunities for stereoselective syntheses.[77]

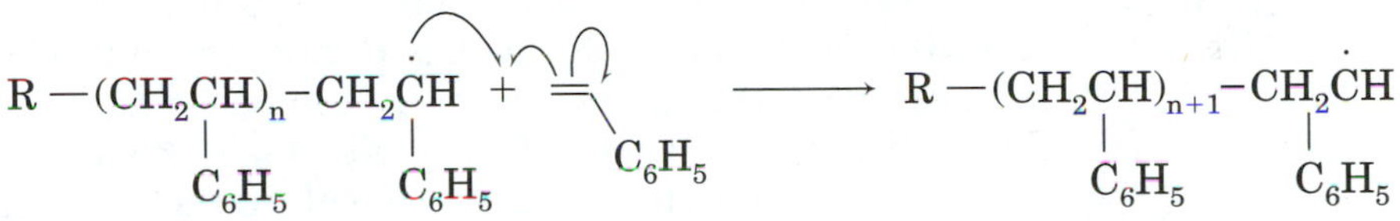

$$\text{R}{-}(\text{CH}_2\text{CH}(\text{C}_6\text{H}_5))_n{-}\text{CH}_2\dot{\text{C}}\text{H}(\text{C}_6\text{H}_5) + \text{CH}_2{=}\text{CH}(\text{C}_6\text{H}_5) \longrightarrow \text{R}{-}(\text{CH}_2\text{CH}(\text{C}_6\text{H}_5))_{n+1}{-}\text{CH}_2\dot{\text{C}}\text{H}(\text{C}_6\text{H}_5) \quad \textbf{(5.18)}$$

Another useful chain reaction involves the PTOC (**p**yridine-2-**t**hione-N-**o**xy**c**arbonyl) esters developed by Barton (Figure 5.27).[78] Reaction of a carboxylic acid chloride (RCOCl) with the sodium salt of N-hydroxypyridine-2-thione produces an ester designated as R-PTOC. Addition of radical Y· (formed by an earlier initiation step) to the R-PTOC leads to the carboxy radical $RCO_2\cdot$. The carboxy radical then decarboxylates to produce the

Figure 5.27 Generation of a radical through a PTOC ester.

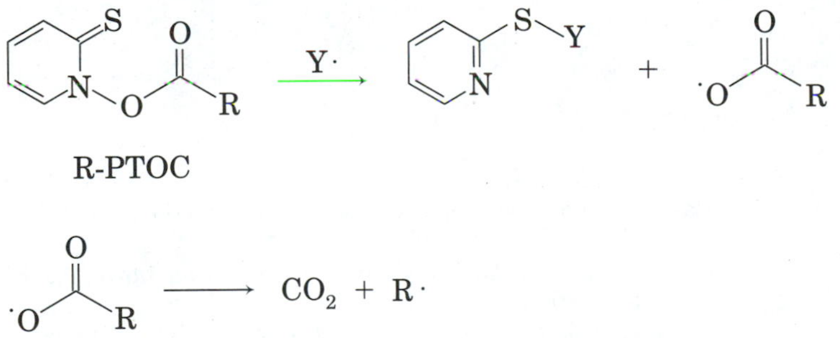

$$\text{R-PTOC} \xrightarrow{\text{Y}\cdot} \text{2-(YS)-pyridine} + \text{R}{-}\text{C}(=\text{O}){-}\text{O}\cdot$$

R-PTOC

$$\text{R}{-}\text{C}(=\text{O}){-}\text{O}\cdot \longrightarrow \text{CO}_2 + \text{R}\cdot$$

[76]For an introduction, see Bevington, J. C. *Radical Polymerization*; Academic Press: London, 1961.

[77]Giese, B. *Radicals in Organic Synthesis: Formation of Carbon-Carbon Bonds*; Pergamon Press: Oxford, England, 1986; Jasperse, D. P.; Curran, D. P.; Fevig, T. L. *Chem. Rev.* **1991**, *91*, 1237; Porter, N. A.; Giese, B.; Curran, D. P. *Acc. Chem. Res.* **1991**, *24*, 296.

[78]Barton, D. H. R.; Crich, D.; Motherwell, W. B. *Tetrahedron* **1985**, *41*, 3901.

radical R·, which can continue the chain reaction or can undergo other reactions.[79]

Another reaction of radicals is rearrangement. Radicals are generally less susceptible to rearrangement than are carbocations,[80] and the 1,2 hydrogen or carbon shifts seen with carbocations are not observed with radicals. However, apparent phenyl migration has been observed. Treatment of neophyl chloride with phenylmagnesium bromide and cobaltous chloride produced isobutylbenzene (15%), 2-methyl-3-phenyl-1-propene (9%), and β,β-dimethylstyrene (4%).[81] The results suggest formation of **27** by rearrangement of neophyl radical (β,β-dimethylphenylethyl, **25**).[82] The 1,1-dimethylspiro[2.5]octadienyl radical **26** has been proposed as an intermediate or transition structure in the rearrangement.[83]

Intramolecular radical rearrangements may be classified as analogues of bimolecular addition or abstraction reactions. In some cases these rearrangements are considered to be diagnostic for the presence of radical intermediates in reactions under study.[84] As illustrated for the 5-hexenyl radical (**28**) in equation 5.19, radicals easily cyclize to give a 5-membered ring (**29**). Although the radical center is 1° in both reactant and product, the conversion of a double bond to two single bonds provides the driving force for the reaction. Formation of a 2° radical and a six-membered ring (**30**) would be favored thermodynamically, but the stereoelectronic requirement of the addition of the radical center to the alkene moiety favors the formation of the five-membered ring by a factor of 50 : 1.[85,86] Such a process pro-

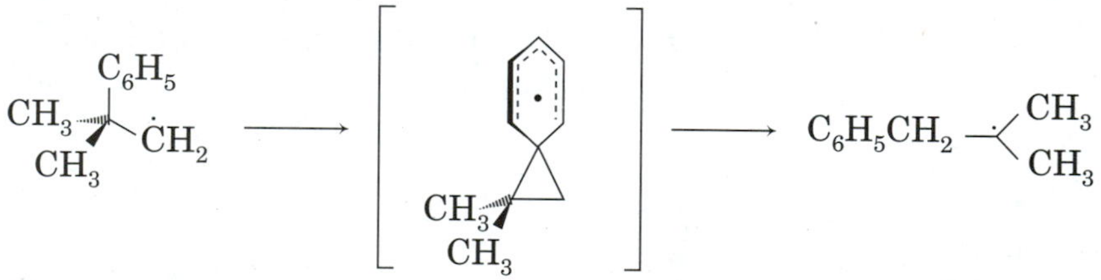

Figure 5.28
Rearrangement of neophyl radical.

C_6H_5 CH_3 CH_3 $\dot{C}H_2$ → [CH_3 CH_3] → $C_6H_5CH_2$ CH_3 CH_3

25 **26** **27**

[79]PTOC esters are also useful in radical clock reactions: Newcomb, M.; Park, S. U. *J. Am. Chem. Soc.* **1986**, *108*, 4132; reference 86.

[80]Walling, C. in *Molecular Rearrangements*, Vol. 1; de Mayo, P., ed.; Wiley-Interscience: New York, 1963; pp. 407–455.

[81]Urry, W. H.; Kharasch, M. S. *J. Am. Chem. Soc.* **1944**, *66*, 1438.

[82]Structure **25** cannot disproportionate because there is no hydrogen bonded to the β carbon atom. Moreover, steric hindrance may reduce the rate of its dimerization.

[83]For studies of the parent spiro[2.5]octadienyl radical, see Effio, A.; Griller, D.; Ingold, K. U.; Scaiano, J. C.; Sheng, S. J. *J. Am. Chem. Soc.* **1980**, *102*, 6063.

[84]For example, see the discussion of single electron transfer involvement in the S_N2 reaction in Chapter 8.

[85]Beckwith, A. L. J. *Chem. Soc. Rev.* **1993**, *22*, 143.

[86]Newcomb, M. *Tetrahedron* **1993**, *49*, 1151.

vides a diagnostic tool for radicals, since cations tend to cyclize to give 6-membered rings, while anions do not readily cyclize.[87] Radical rearrangements can also provide kinetic information by serving as a free radical clock.[88] If the radical partitions between one reaction with a known rate constant and another reaction with an unknown rate constant, the ratio of products can provide an estimate of the unknown rate constant.[89]

(5.19)

Relative yield 50:1

28 **29** **30**

The characteristic of radical termination steps is that two radical centers react to produce a product or products, each of which has an even number of electrons. The two most important processes are dimerization and disproportionation. Dimerization (equation 5.20) is the reverse of the thermal dissociation of a σ bond to produce two radicals. This process is usually quite exothermic and has a small or negligible activation energy. However, significant steric hindrance can lower the dissociation energy and introduce a barrier to recombination. For example, the dissociation energy of 1-(2,6-dimethylphenyl)-2-(2,6-dimethylphenyl)ethane is only 22 kcal/mol.[90]

$$H_3C\cdot + \cdot CH_3 \longrightarrow H_3C{-}CH_3 \qquad (5.20)$$

Disproportionation can be considered to be hydrogen abstraction from another radical.[91] The overall result is the destruction of two radicals, one radical center being converted to a carbon-hydrogen bond, the other being converted into a double bond (equation 5.21). Both dimerization and disproportionation represent bimolecular processes of radicals. Dimerization is more exothermic than is disproportionation, but the ΔS for dimerization is

---

[87]Ingold, K. U. in *Organic Free Radicals*; Pryor, W. A., ed.; *ACS Symposium Series* **69**, American Chemical Society: Washington, DC, 1978; pp. 187 *ff*.

[88]Griller, D.; Ingold, K. U. *Acc. Chem. Res.* **1980**, *13*, 317.

[89]For an overview of this area, see reference 86; Newcomb, M. in *Advances in Detailed Reaction Mechanisms*, Vol. 1; Coxon, J. M., ed.; JAI Press, Inc.: Greenwich, CT, 1991; pp. 1–33. Also see (a) Newcomb, M.; Glenn, A. G.; Williams, W. G. *J. Org. Chem.* **1989**, *54*, 2675; (b) Fossey, J.; Lefor, D.; Sorba, J. *J. Org. Chem.* **1986**, *51*, 3584; (c) Newcomb, M.; Glenn, A. G. *J. Am. Chem. Soc.* **1989**, *111*, 275; (d) Bowry, V. W.; Lusztyk, J.; Ingold, K. U. *J. Am. Chem. Soc.* **1989**, *111*, 1927; (e) Hollis, R.; Hughes, L.; Bowry, V. W.; Ingold, K. U. *J. Org. Chem.* **1992**, *57*, 4284; (f) Newcomb, M.; Johnson, C. C.; Manek, M. B.; Varick, T. R. *J. Am. Chem. Soc.* **1992**, *114*, 10915; (g) Branchaud, B. P.; Glenn, A. G.; Stiasny, H. C. *J. Org. Chem.* **1991**, *56*, 6656; (h) Nonhebel, D. C. *Chem. Soc. Rev.* **1993**, *22*, 347.

[90]Fleurke, K. H.; De Jong, J.; Nauta, W. T. *Rec. Trav. Chim. Pays Bas* **1965**, *84*, 1380.

[91]For a review, see Gibian, M. J.; Corley, R. C. *Chem. Rev.* **1973**, *73*, 441.

much more negative than is ΔS for disproportionation. Therefore, the reaction of two radicals depends strongly on temperature, higher temperature favoring disproportionation, lower temperature favoring dimerization.[92]

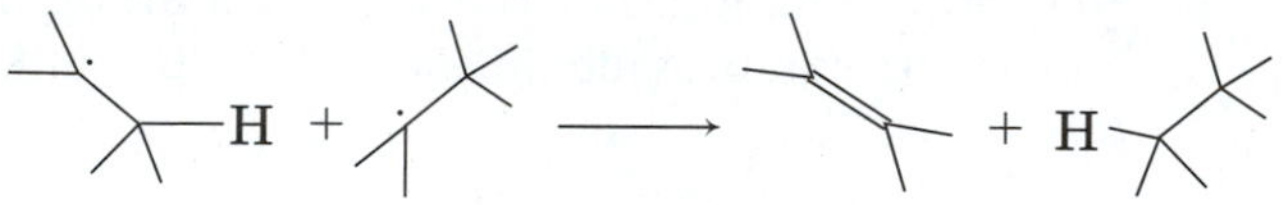

(5.21)

There is one important exception to the statement that radical termination steps produce products with an even number of electrons. A radical addition step may produce a radical product that is much less reactive than the reacting precursor, so further addition may be precluded. This process, which is known as *spin trapping*, is primarily useful as a means of studying radicals that cannot be studied directly by EPR. Adding a nitroso compound or a nitrone to a reaction mixture involving short-lived radicals can produce a *spin adduct*, a longer-lived species that can be studied directly by EPR spectrometry (Figure 5.29). The spectrum of the product is often diagnostic of its radical precursor.[93,94]

Figure 5.29 Reaction of a radical with a nitrone or a nitroso compound in a spin trapping reaction.

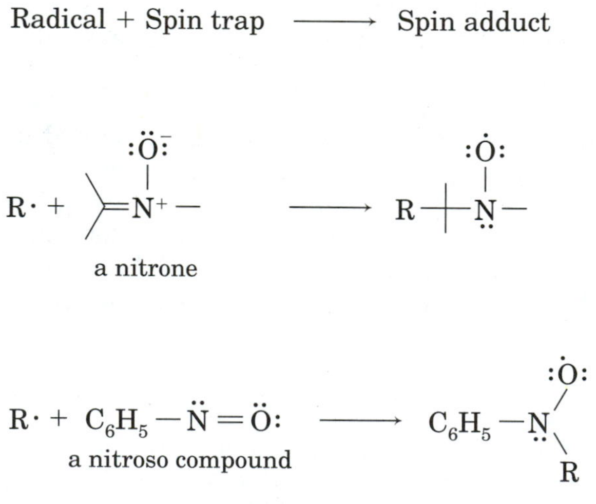

Radical + Spin trap ⟶ Spin adduct

a nitrone

$R\cdot + C_6H_5-\ddot{N}=\ddot{O}:$ ⟶ $C_6H_5-N(\dot{O})R$

a nitroso compound

[92]Bevington, J. C. *Trans. Faraday Soc.* **1952**, *48*, 1045.

[93]For an introduction to spin trapping, see Janzen, E. G. *Acc. Chem. Res.* **1971**, *4*, 31; Perkins, M. J. *Adv. Phys. Org. Chem.* **1980**, *17*, 1.

[94]Janzen, E. G.; Evans, C. A.; Davis, E. R. in *Organic Free Radicals, ACS Symposium Series*, **69**; American Chemical Society: Washington, D. C., 1978; pp. 433 *ff.*

Carbenes

Structure and Geometry of Carbenes

A ***carbene*** is a structure containing a neutral carbon atom that is bonded to only two other atoms, which has a formal charge of 0, and which has two nonbonded electrons localized on that carbon atom.[95] The simplest chemical example of such a structure is CH_2, commonly called *methylene*. Figure 5.30 shows some representations for three different *electronic states* that could be proposed for methylene.[96] In **31(a)** the two electrons are both in the same orbital and must be paired. In **31(b)** and **31(c)** they are in separate orbitals, so these two electrons may or may not have paired spins. Structures **31(a)** and **31(b)** are called *singlet states*, since the nonbonded electrons have their spins paired. In **31(c)** the spins of the nonbonded electrons are not paired, so this is a *triplet state*.[97] In theory all of the states in Figure 5.30 are possible, although they may have different energies, reactivities, and lifetimes under the reaction conditions.[98]

Structure **31(a)** has two electrons in one orbital and no electrons in another orbital. We would expect it to be sp^2 hybridized with the nonbonded electrons in an sp^2 orbital (having 1/3 *s* character) as the *p* orbital is empty. Structures **31(b)** and **31(c)** should have the same hybridization for the two

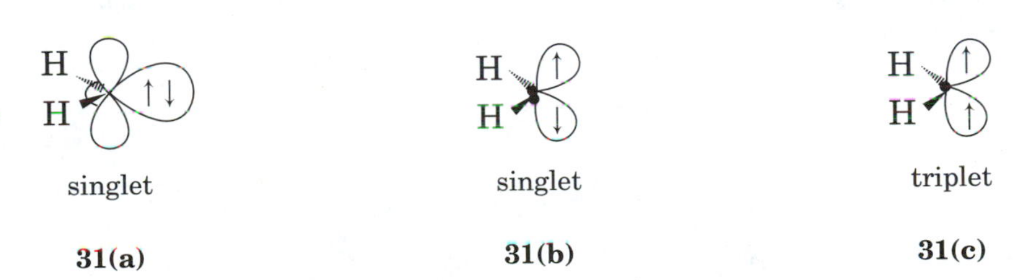

Figure 5.30 Singlet (a and b) and triplet (c) states of methylene.

---

[95]For an introduction to the structures and reactions of carbenes, see (a) Closs, G. L. *Top. Stereochem.* **1968**, *3*, 193. (b) Liebman, J. F.; Simons, J. in *Molecular Structure and Energetics, Volume 1: Chemical Bonding Models*; Liebman, J. F.; Greenberg, A., eds.; VCH Publishers: Deerfield Beach, Florida, 1986; p. 51; (c) Moss, R. A.; Jones, Jr., M. in *Reactive Intermediates*, Vol. 2; Jones, Jr., M.; Moss, R. A., eds.; John Wiley & Sons: New York, 1981; p. 59; (d) Kirmse, W. *Carbene Chemistry*; Academic Press: New York, 1964; (e) Gilchrist, T. L.; Rees, C. W. *Carbenes, Nitrenes and Arynes*; Appleton-Century-Crofts: New York, 1969; (f) Hine, J. *Divalent Carbon*; Ronald Press: New York, 1964; (g) reference 7, pp. 162 *ff*.

[96]For further discussion, see (a) Eisenthal, K. B.; Moss, R. A.; Turro, N. J. *Science* **1984**, *225*, 1439; Gaspar, P. P.; Hammond, G. S. in reference 95 (d), pp. 235–274.

[97]The terms *singlet* and *triplet* arose from experiments in which singlet states gave only one emission line (wavelength) in certain spectroscopic experiments, while triplet states gave three lines (that is, emission at three wavelengths). The distinction between singlet and triplet states will be especially important in the discussion of photochemistry (Chapter 12).

[98]The **lifetime**, τ, of a transient species is the inverse of the sum of the rate constants for its disappearance (i.e., $\tau = 1/\Sigma k_i$).

singly occupied orbitals. If there is greater repulsion between pairs of electrons in the C—H bonds with each other than there is between the single electrons on carbon, then VSEPR theory would suggest the H—C—H bond angle to be greater than 109.5°. The extreme angle of 180°, however, would be *sp* hybridization, and that would require that the two unpaired electrons be in orbitals with no *s* character. There is no obvious basis for predicting the hybridization expected for **31(b)** and **31(c)**.

The geometry of triplet methylene has been the subject of intensive theoretical and experimental work. Indeed, methylene has been called the paradigm of theoretical chemistry because the success of computation in describing methylene, even in the face of initially contradictory experimental evidence, was part of the establishment of credibility for theory in chemical calculations.[99] Both theoretical calculations[100] and experimental work (based on EPR measurements[101,102]) now suggest that triplet methylene is slightly bent, with an H—C—H angle of 136°. As shown in Figure 5.31, however, the molecule is thought to have only a small barrier to inversion in a process reminiscent of the inversion of nitrogen in amines.

Both theoretical and experimental determinations indicate that the lowest energy electronic state of methylene is the triplet, **31(c)**, with the singlet **31(a)** being a higher energy *excited* state, and the singlet **31(b)** an

Figure 5.31 Calculated potential energy surface for the lowest triplet state of methylene. (Adapted from reference 99.)

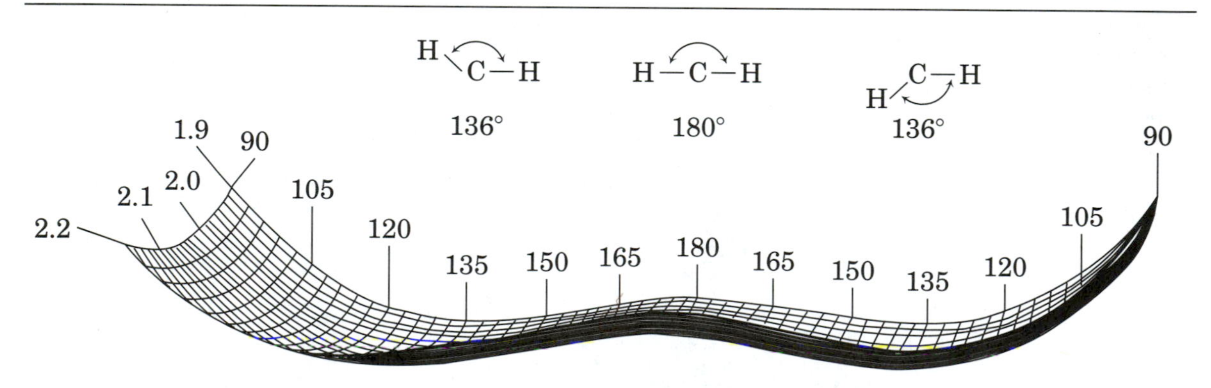

[99]Schaefer III, H. F. *Science* **1986**, *231*, 1100. See also Wasserman, E. *Science* **1986**, *232*, 1319; Schaefer III, H. F. *Science* **1986**, *232*, 1319; Wasserman, E.; Schaefer III, H. F. *Science* **1986**, *233*, 829.

[100]Bender, C. F.; Schaefer III, H. F. *J. Am. Chem. Soc.* **1970**, *92*, 4984.

[101]Wasserman, E.; Yager, W. A.; Kuck, V. J. *Chem. Phys. Lett.* **1970**, *7*, 409.

[102]Bunker, P. R.; Sears, T. J.; McKellar, A. R. W.; Evenson, K. M.; Lovas, F. J. *J. Chem. Phys.* **1983**, *79*, 1211; McKellar, A. R. W.; Yamada, C.; Hirota, E. *J. Chem. Phys.* **1983**, *79*, 1220; Bunker, P. R.; Jensen, P. *J. Chem. Phys.* **1983**, *79*, 1224.

even higher energy state.[103] However, the difference in energy between **31(c)** and **31(a)** was the subject of some debate. Most theoretical calculations indicated that the energy separation was about 10 kcal/mol, while the experimental value was reported to be about 20 kcal/mol. The issue was resolved when further experimental work gave a revised experimental value of 9 kcal/mol, which is very close to the calculated value.[104]

The discussion to this point has concerned only the parent carbene, methylene. The same structural principles generally apply to other carbenes, but there are some additional considerations. For example, diphenylcarbene is more nearly linear due to the steric repulsion of the two phenyl groups. In a study of diphenylcarbene generated in diphenylethene crystals, the $C_{Ph}—C—C_{Ph}$ bond angle was found to be 148°, and the phenyl rings were twisted 36° out of the plane defined by the carbene carbon atom and the two-ring carbon atoms bonded to it.[105]

Generation of Carbenes

Carbenes may be synthesized by a variety of pathways, but all of them accomplish the elimination of two bonds from a tetravalent carbon atom. Consider the following reactions.

1. Thermolysis or photolysis of diazoalkanes and dialkyldiazirines

$$R_2C{=}\overset{+}{N}{=}\overset{-}{\ddot{N}}{:} \longleftrightarrow R_2\overset{-}{\ddot{C}}{-}\overset{+}{N}{\equiv}N{:} \xrightarrow[\text{or } h\nu]{\Delta} R_2C{:} + N_2 \qquad (5.22)$$

$$(CH_3)_2C\langle N{=}N \rangle \xrightarrow{-N_2} (CH_3)_2C{:} \qquad (5.23)$$

2. Reactions of *N*-nitrosoureas with base to generate diazoalkanes, which can then eliminate nitrogen[106]

$$R_2HC{-}N(NO){-}C({=}O){-}NH_2 \xrightarrow{\text{base}} R_2CN_2 \xrightarrow{\Delta} R_2C{:} + N_2 \qquad (5.24)$$

---

[103] Based on Hund's rule, we expect the triplet to be lower in energy than the singlet, so the electronic arrangement in **31(b)** is an excited state of the triplet state in **31(c)**.

[104] Leopold, D. G.; Murray, K. K.; Lineberger, W. C. *J. Chem. Phys.* **1984**, *81*, 1048. See also *Chem. Eng. News* **1984** (Nov. 26), 30.

[105] Doetschman, D. C.; Hutchison, Jr., C. A. *J. Chem. Phys.* **1972**, *56*, 3964.

[106] Jones, W. M.; Grasley, M. H.; Brey, Jr., W. S. *J. Am. Chem. Soc.* **1963**, *85*, 2754.

3. Reactions of tosylhydrazones with base[107]

$$H_3C-C_6H_4-S(=O)_2-\ddot{N}H-\ddot{N}=CR_2 \xrightarrow{n\text{-BuLi}} H_3C-C_6H_4-S(=O)_2-\ddot{N}^{-}-\ddot{N}=CR_2 \longrightarrow R_2C: + N_2 + H_3C-C_6H_4-SO_2^- \quad (5.25)$$

4. α-Elimination reactions[108]

$$HCCl_3 + Base^- \rightleftharpoons Base\text{-}H + :CCl_3^- \rightleftharpoons Cl^- + :CCl_2 \quad (5.26)$$

5. Generation of carbenoids (Simmons-Smith reaction)[109]

$$CH_2I_2 \xrightarrow{ZnCu} ICH_2ZnI \quad (5.27)$$

Reactions 1–4 produce carbenes that are initially in the singlet state, but they can relax to the triplet ground state if reaction does not occur first. The Simmons-Smith reaction produces a *carbenoid*, in which a carbene is stabilized by association with a metal, which reacts as a singlet carbene. It is possible to produce a triplet carbene directly through a process known as ***sensitization***, in which a photoexcited triplet molecule S transfers energy to a carbene precursor and returns to its ground (singlet) electronic state.[110] Conservation of electron spin requires that the carbene be produced in its triplet state.

$$S^{\uparrow\uparrow} + CH_2=\overset{(+)}{N}=\overset{(-)}{N} \longrightarrow S^{\uparrow\downarrow} + CH_2^{\uparrow\uparrow} + N_2 \quad (5.28)$$

Reactions of Carbenes

Singlet and triplet carbenes exhibit similar reaction types, but there are some important differences between them.[95] Because it has both an empty *p* orbital (like a carbocation) and a nonbonded pair of electrons (like a carbanion), the singlet carbene exhibits both carbocation and carbanion character. However, the triplet carbene behaves more as a diradical. These character-

[107]For a mechanism, see Figure 5.33.

[108]The base-promoted α-cleavage of chloroform to dichlorocarbene was reported by Hine, J.; Dowell, Jr., A. M. *J. Am. Chem. Soc.* **1954**, *76*, 2688.

[109]Simmons, H. E.; Smith, R. D. *J. Am. Chem. Soc.* **1958**, *80*, 5323.

[110]A sensitizer is an electronically excited species that transfers energy to a ground state molecule, leading to the ground electronic state of the sensitizer and an electronically excited state of the energy acceptor. The symbol $S^{\uparrow\uparrow}$ means that the sensitizer is a triplet (i.e., the two electrons in singly-occupied orbitals have the same spin). These terms will be defined more extensively in Chapter 12.

istics influence the types and stereochemistries of carbene reactions. One of the major reactions of carbenes is cycloaddition with double bonds:

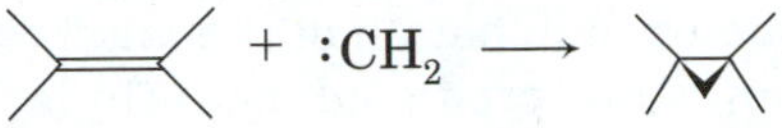

$$+ :CH_2 \longrightarrow \quad (5.29)$$

A singlet carbene inserts stereospecifically, meaning that the cis alkene gives only *cis*-cyclopropane, and the trans alkene gives *trans*-cyclopropane. However, the reaction of the triplet carbene is not stereospecific, the product being a mixture of isomers. The difference in stereochemistry arises because the singlet carbene can add in one step (equation 5.30)

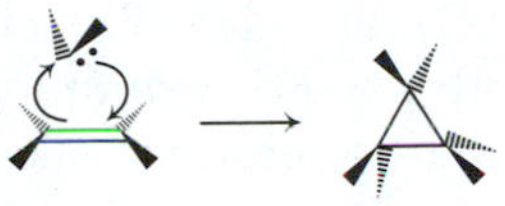

$$\longrightarrow \quad (5.30)$$

while the triplet cannot. (If it were to add in one step, there would be two unpaired electrons in one σ bond; i.e., there would be an excited state of a σ bond, a very high energy species.) The triplet carbene can add first to one end of the double bond to produce a diradical, which can only close to the cyclopropane after a time sufficient for one electron to flip its spin. As shown in Figure 5.32, this time may allow rotation to occur around the C—C single bond, so that a mixture of isomers results. Although addition of singlet carbenes to conjugated dienes most often results in 1,3-addition, 1,4-addition has been observed.[111,112]

Insertion into carbon-hydrogen single bonds may be considered analogous to cycloaddition to alkenes:

$$R{-}H + :CH_2 \longrightarrow R{-}CH_2{-}H \quad (5.31)$$

Insertion into C-H bonds is more probable than insertion into C—C bonds. For example, photolysis of diazomethane in cyclopentane at −75° produced

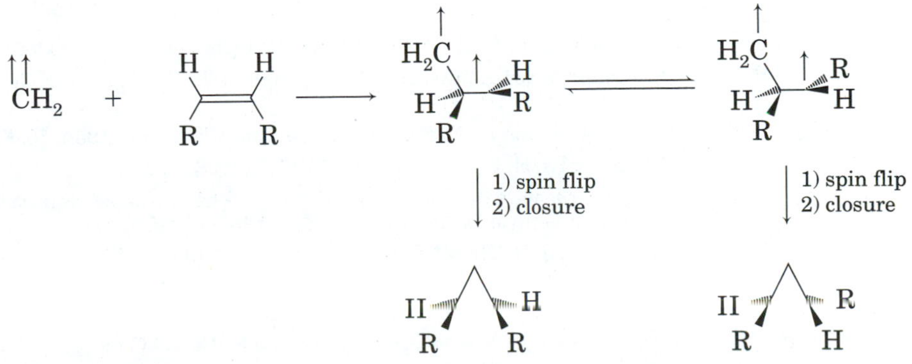

Figure 5.32 Nonstereospecific addition of triplet methylene to a *cis*-alkene.

only methylcyclopentane, with cyclohexane not being observed.[113] Singlet carbenes are thought to add to C—H bonds by a concerted process, while triplet carbenes can produce net addition through hydrogen abstraction and then recombination of the alkyl radicals. Irradiation of diazomethane in cyclohexene produced 1-methylcyclohexene (10%), 3-methylcyclohexene (25%), 4-methylcyclohexene (25%), and norcarane (40%). Thus, singlet methylene appeared to be *indiscriminate* with regard to allylic, 2°, or vinylic C—H bonds.[113] Different carbenes exhibit different selectivities toward insertion and cycloaddition reactions, however. Reaction of cyclohexene with dichlorocarbene resulted in a 60% isolated yield of the dichloro derivative of norcarane, 7,7-dichlorobicyclo-[4.1.0]heptane.[114]

Singlet carbenes will rearrange to isoelectronic structures in which all atoms have an octet of electrons if hydrogen atoms are located on adjacent carbon atoms. For example, alkyl carbenes readily rearrange by a 1,2-hydrogen shift to produce alkenes.[115,116] Evanseck and Houk determined from *ab initio* calculations that the activation energy for the rearrangement of methylcarbene to ethene was 0.6 kcal/mol.[117]

H C → H (5.32)

This rearrangement is the basis of the Bamford-Stevens reaction, in which the tosylhydrazone of an aliphatic ketone is converted to an alkene by the action of strong base, such as the sodium salt of ethylene glycol in ethylene glycol as solvent (Figure 5.33).[118]

The extent to which any particular carbene exhibits these reactions depends on its structure and electronic state.[119] Alkyl and dialkyl carbenes undergo such rapid intramolecular reactions that intermolecular reactions are

---

[111]For a theoretical study of the 1,2- and 1,4-addition pathways and for references to literature reports of 1,4-addition, see Evanseck, J. D.; Mareda, J.; Houk, K. N. *J. Am. Chem. Soc.* **1990**, *112*, 73.

[112]Turkenburg, L. A. M.; de Wolf, W. H.; Bickelhaupt, F. *Tetrahedron Lett.* **1982**, *23*, 769.

[113]Doering, W. von E.; Buttery, R. G.; Laughlin, R. G.; Chaudhuri, N. *J. Am. Chem. Soc.* **1956**, *78*, 3224.

[114]Doering, W. von E.; Hoffmann, A. K. *J. Am. Chem. Soc.* **1954**, *76*, 6162.

[115]Schaefer III, H. F. *Acc. Chem. Res.* **1979**, *12*, 288.

[116]Some carbenes with α-hydrogen atoms have been trapped and studied in matrices: Sander, W.; Bucher, G.; Wierlacher, S. *Chem. Rev.* **1993**, *93*, 1583.

[117]Evanseck, J. D.; Houk, K. N. *J. Phys. Chem.* **1990**, *94*, 5518.

[118]Bamford, W. R.; Stevens, T. S. *J. Chem. Soc.* **1952**, 4735. For a mechanistic discussion and review, see Shapiro, R. H. *Org. React.* **1976**, *23*, 405.

[119]For a review, see Nickon, A. *Acc. Chem. Res.* **1993**, *26*, 84.

not competitive. For example, products from the decomposition of diazocyclooctane are shown in equation 5.33.[120]

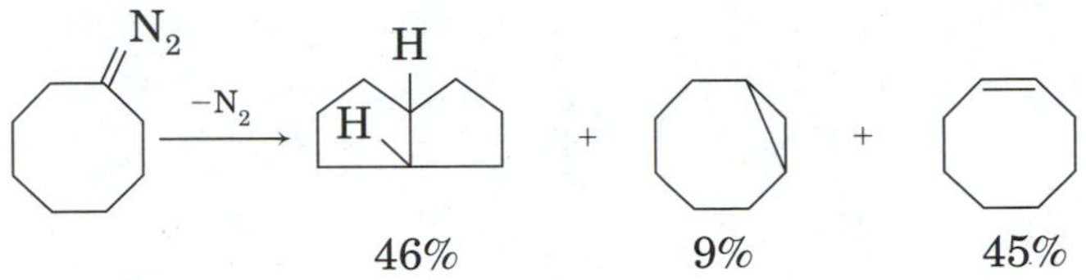

$$\text{diazocyclooctane} \xrightarrow{-N_2} \text{bicyclo[3.3.0]octane (46\%)} + \text{bicyclo[5.1.0]octane (9\%)} + \text{cyclooctene (45\%)} \quad (5.33)$$

If intramolecular reactions cannot occur, the carbanion-carbocation character of a singlet carbene can lead to reactions that occur through ionic species. Singlet carbenes react with methanol by nucleophilic abstraction of a proton, resulting in a carbocation which subsequently reacts with the alcohol to produce an ether.[121] For example, the diphenyl carbene singlet (**32**), which cannot undergo rearrangement to form an alkene, abstracts a proton from methanol to form the diphenylmethyl carbocation (**33**). Methanol then adds as a nucleophile to **33** to produce the ether **34**.[122]

The propensity for intramolecular rearrangement means that the lifetime and bimolecular reactivity of carbenes depend strongly on structure. Dialkylcarbenes that do not have an α-C—H bond have longer lifetimes since a hydrogen shift is not possible. Two have been studied spectroscopically:

Figure 5.33 Conversion of ketone to alkene through the Bamford-Stevens reaction.

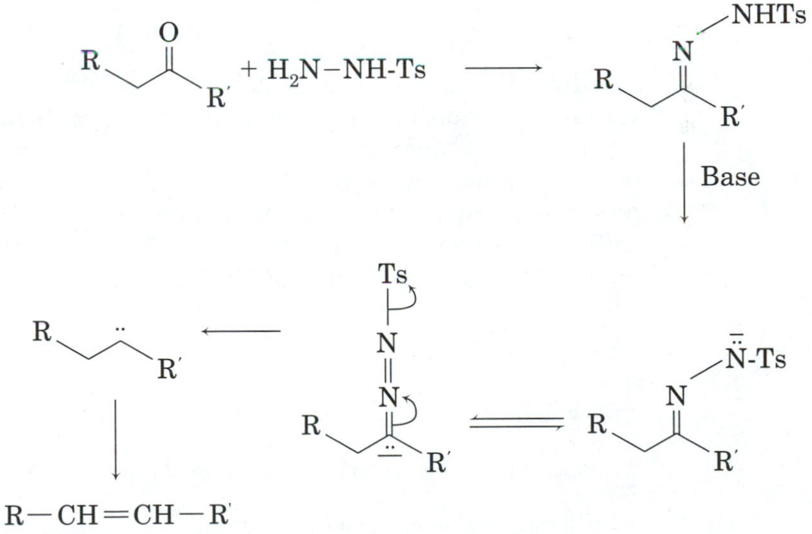

$$\text{RCH}_2\text{C(=O)R}' + \text{H}_2\text{N–NH-Ts} \longrightarrow \text{RCH}_2\text{C(=N–NHTs)R}' \xrightarrow{\text{Base}} \text{RCH}_2\text{C(=N–}\ddot{\text{N}}^{-}\text{-Ts)R}' \rightleftharpoons \text{R}\bar{\text{C}}\text{H–C(=N=N–Ts)R}' \longrightarrow \text{RCH}_2\ddot{\text{C}}\text{R}' \longrightarrow \text{R–CH=CH–R}'$$

[120]Friedman, L.; Shechter, H. *J. Am. Chem. Soc.* **1961**, *83*, 3159.

[121]For a discussion of the mechanism of conversion of diphenyldiazomethane to methyl benzhydryl ether, see reference 95(d).

[122]Reference 95(d), pp. 90–91 and references therein.

Figure 5.34 Reaction of diphenylcarbene with methanol to form benzhydryl methyl ether.

$$C_6H_5\ddot{C}C_6H_5 \xrightarrow{CH_3OH} C_6H_5CH^{+}C_6H_5 \xrightarrow{CH_3OH} C_6H_5CH(OCH_3)C_6H_5$$

32 **33** **34**

di-*t*-butylcarbene[123] and diadamantylcarbene (**35**).[124] Spectroscopic studies indicate that both are ground state triplets, but the reactions of diadamantylcarbene in solution suggested reactions involving both the singlet and triplet states. Spin-equilibrated **35** reacted with methanol ($k = 2 \times 10^7$ L mol^{-1} sec^{-1}) to produce methyl diadamantylmethyl ether.[125] In addition reaction of the triplet state of **35** with oxygen (a triplet ground state molecule) gave a carbonyl oxide, which could be reduced to diadamantyl ketone, confirming triplet reactivity.[126]

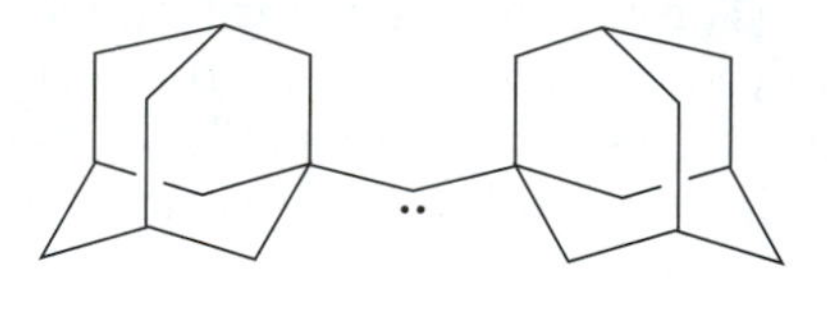

35

Moss and Mamantov found that a halogen bonded to the carbene center stabilizes the singlet state of the carbene.[127,128] Because of the higher activation energy for rearrangement, alkylchlorocarbenes can take part in both intramolecular and intermolecular reactions and the rate constants for both processes can be measured. For benzylchlorocarbene, the rate constant for rearrangement to the isomeric β-chlorostyrenes through a 1,2-hydrogen shift was estimated to be between 4.9 and 6.7×10^7 sec^{-1}, while the rate constant for addition of the carbene to *trans*-3-hexene was 6.8×10^7 L mol^{-1}

---

[123] Gano, J. E.; Wettach, R. H.; Platz, M. S.; Senthilnathan, V. P. *J. Am. Chem. Soc.* **1982**, *104*, 2326.

[124] Myers, D. R.; Senthilnathan, V. P.; Platz, M. S.; Jones, Jr., M. *J. Am. Chem. Soc.* **1986**, *108*, 4232.

[125] The term *spin-equilibrated* means that there is an equilibrium population of the singlet and triplet states.

[126] Morgan, S.; Platz, M. S.; Jones, Jr., M.; Myers, D. R. *J. Org. Chem.* **1991**, *56*, 1351.

[127] Moss, R. A.; Mamantov, A. *J. Am. Chem. Soc.* **1970**, *92*, 6951.

[128] Rate constants for rearrangements of alkylfluorocarbenes are about an order of magnitude slower than those of the alkylchlorocarbenes: Moss, R. A.; Ho, G.-J.; Liu, W. *J. Am. Chem. Soc.* **1992**, *114*, 959.

sec^{-1}.[129,130] For cyclopropylchlorocarbene, the rate constant for rearrangement was determined to be $(8.5 \pm 0.5) \times 10^5$ sec^{-1}, with a rate constant of 1.2×10^7 M^{-1} sec^{-1} for addition to 2,3-dimethylbutene. The activation energy for the rearrangement of cyclopropylchlorocarbene was found to be 3.0 $\pm$ 0.4 kcal/mol, with $\Delta G^\ddagger$ of about 9 kcal/mol, $\Delta H^\ddagger$ of about 2.5 kcal/mol, and $\Delta S^\ddagger$ of about -20 e.u.[131] α-Methyl substitution on the alkyl portion of an alkylchlorocarbene increases the rate constant for rearrangement by a factor of 30 or more.[130(a),132]

5.3 Intermediates with Charged Carbon Atoms

Carbocations

Carbonium Ions and Carbenium Ions

Carbocations are reactive intermediates having a formal charge of +1 on a carbon atom. During much of the recent history of organic chemistry, a structure with a positively charged carbon atom was called a *carbonium ion*, a term reminiscent of other positively charged species, such as ammonium, phosphonium, sulfonium, etc.[133] However, these latter terms all refer to species formed by adding a positively charged atom such as a proton to an atom with a nonbonded pair of electrons to form the positively charged ion. To keep the nomenclature of organic chemistry consistent, it was proposed that a species such as CH_3^+ should be thought of as being the addition product of methylene and a proton (equation 5.34), so it should more properly be termed a ***carbenium ion***, and that is the term now in general use for species in which a trivalent carbon atom bears a positive charge. Often, however, the more general term *carbocation* is used instead.[134]

$$:CH_2 + H^+ \longrightarrow CH_3^+ \qquad \textbf{(5.34)}$$

To continue the analogy of adding the suffix *-ium* to the term for a neutral species, Olah proposed that the term *carbonium ion* refer to a species that could be considered to be formed by adding a positive charge to a neu-

---

[129]Jackson, J. E.; Soundararajan, N.; White, W.; Liu, M. T. H.; Bonneau, R.; Platz, M. S. *J. Am. Chem. Soc.* **1989**, *111*, 6874.

[130]The rate constant for the rearrangement of chloromethylcarbene to vinyl chloride was determined to be 3×10^6 sec^{-1}: (a) Bonneau, R.; Liu, M. T. H.; Rayez, T. *J. Am. Chem. Soc.* **1989**, *111*, 5973; (b) Liu, M. T. H.; Bonneau, R. *J. Am. Chem. Soc.* **1989**, *111*, 6873.

[131]Ho, G.-J.; Krogh-Jespersen, K.; Moss, R. A.; Shen, S.; Sheridan, R. S.; Subramanian, R. *J. Am. Chem. Soc.* **1989**, *111*, 6875; Moss, R. A.; Ho, G.-J.; Shen, S.; Krogh-Jespersen, K. *J. Am. Chem. Soc.* **1990**, *112*, 1638.

[132]LaVilla, J. A.; Goodman, J. L. *J. Am. Chem. Soc.* **1989**, *111*, 6877.

[133]The "waxing and waning" of the term *carbonium ion* has been described by Traynham, J. G. *J. Chem. Educ.* **1986**, *63*, 930.

[134]Olah, G. A. *J. Am. Chem. Soc.* **1972**, *94*, 808, suggested the term *carbocations* as a generic term for all cations of carbon compounds.

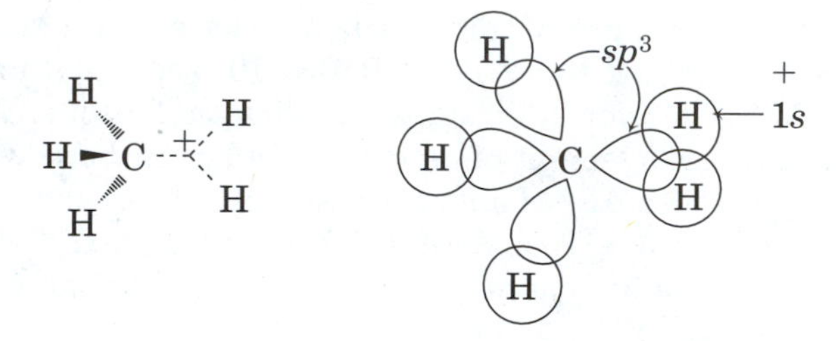

Figure 5.35 Representations of the structure of CH_5^+. (Reproduced from reference 135.)

tral, tetravalent carbon atom, as in equation 5.35. Such a species in which a carbon atom appears to be bonded to more than 4 atoms at once is known as a ***hypercoordinate carbon compound***.[135,136,137] Figure 5.35 shows the bonding in a CH_5^+ ion as suggested by Olah. The relationship between the two types of carbocations is summarized in Figure 5.36.[138]

$$CH_4 + H^+ \longrightarrow CH_5^+ \qquad \textbf{(5.35)}$$

In spite of the conceptual model that gave rise to the name, carbenium ions are usually formed by *heterolytic dissociation* of a bond to carbon. In particular, around 1900 it was observed that triarylmethyl halides could be dissolved in SO_2 to yield electrically conducting solutions, which suggested that carbon atoms could be positively charged.[139,140] As charged intermediates, they require more energy for formation in the gas phase than would the corresponding trivalent carbon radicals because energy is required both

Figure 5.36 Designations of carbocations. (Redrawn from reference 135.)

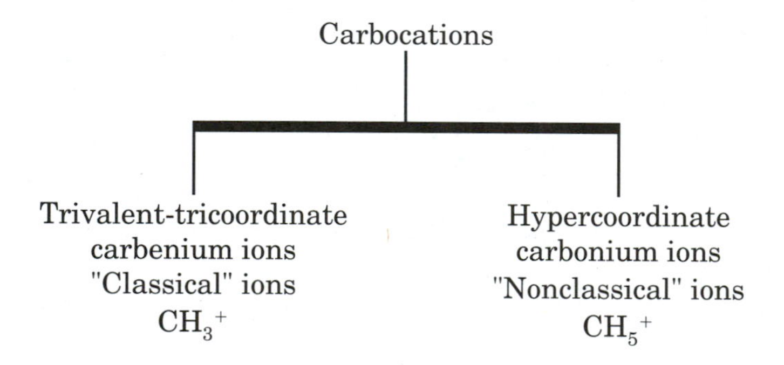

---

[135] Olah, G. A.; Prakash, G. K. S.; Williams, R. E.; Field, L. D.; Wade, K. *Hypercarbon Chemistry*; John Wiley & Sons, Inc.: New York, 1987.

[136] For a discussion of the terms *hypercoordinate* and *hypervalent*, see (a) Schleyer, P. v. R. *Chem. Eng. News* **1984** (May 28), 4; (b) Martin, J. C. *Chem. Eng. News* **1984** (May 28), 4.

[137] CH_6^{2+}, a carbodication, is predicted to be an energy minimum also. For a discussion, see (a) Lammertsma, K.; Olah, G. A.; Barzaghi, M.; Simonetta, M. *J. Am. Chem. Soc.* **1982**, *104*, 6851; (b) Lammertsma, K. *J. Am. Chem. Soc.* **1984**, *106*, 4619 and references therein.

[138] Figure adapted from Reference 135, Page 141.

[139] Bethell, D.; Gold, V. *Carbonium Ions: An Introduction*; Academic Press: New York, 1967.

[140] Carbocations were established by the work of Whitmore, F. C. *J. Am. Chem. Soc.* **1932**, *54*, 3274. For a perspective on the development of the concept, see Traynham, J. G. *J. Chem. Educ.* **1989**, *66*, 451.

to break the bond to carbon and to separate the charged ions. The additional energy may be small in solution, especially with very polar solvent molecules to solvate and stabilize the ions, but it can be considerable in the gas phase. However, we must remember that the stabilization afforded by polar solvents is accompanied by perturbation of the environment of the cation, so any investigation of the cation is necessarily an investigation of both the ion and its environment.

Structure and Geometry of Carbocations

Figure 5.37
Simple representation of a methyl carbocation.

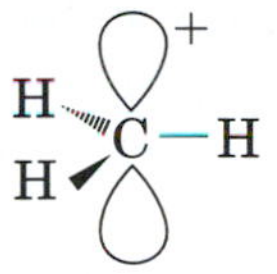

A common model for carbocation structure is a planar species exhibiting sp^2 hybridization, as shown in Figure 5.37. A structure with an empty *p* orbital should be more stable than a structure in which an orbital with *s* character is empty. Moreover, sp^2 hybridization allows the three substituents bonded to the carbon atom to be as far apart from each other as possible. The sp^2-hybridized model is consistent with the observed geometry for the *t*-butyl carbocation, which has been determined through both NMR[141] and X-ray crystallographic[142] studies to be planar, with 120° bond angles about the central carbon atom. Other evidence supporting a planar structure for carbocations comes from the racemization of chiral alkyl halides under solvolysis conditions (discussed in Chapter 8). Moreover, correlation of decreased solvolysis rate constants with increasing angle strain at a cationic carbon atom[143] is also consistent with the view that carbenium ions prefer a planar geometry (sp^2 hybridization).[144,145]

One familiar aspect of carbocation chemistry is the large dependence of the energy of the cation on the substituents attached to the positively charged carbon atom. Table 5.2 shows thermodynamic data for selected cations.[146] The hydride ion affinity, HIA(R^+), is defined as the negative of the ΔH for the attachment of a hydride ion to the cation in the gas phase. That is, the greater the HIA(R^+), the more endothermic is the removal of a hydride ion from an alkane. We expect the trends to be the same for heterolytic dissociation of alkyl halides or other species that produce carbocations.[147] The BDE values are much smaller *in the gas phase* than are the hydride ion affinities because

---

[141]Yannoni, C. S.; Kendrick, R. D.; Myhre, P. C.; Bebout, D. C.; Petersen, B. L. *J. Am. Chem. Soc.* **1989**, *111*, 6440.

[142]An X-ray crystal structure of the *t*-butyl cation as an $Sb_2F_{11}^-$ salt indicated that the structure is planar: Hollenstein, S.; Laube, T. *J. Am. Chem. Soc.* **1993**, *115*, 7240.

[143]*Cf.* Gleicher, G. J.; Schleyer, P. v. R. *J. Am. Chem. Soc.* **1967**, *89*, 582 and references therein. The authors conclude that the increase in angle strain with solvolysis to a carbenium ion is the most important effect, but by no means the only effect, in correlating molecular structure with reaction rate.

[144]Bartlett, P. D.; Knox, L. H. *J. Am. Chem. Soc.* **1939**, *61*, 3184. For a review, see Applequist, D. E.; Roberts, J. D. *Chem. Rev.* **1954**, *54*, 1065.

[145]For a discussion of the electronic structure and stereochemistry of simple carbocations, see Buss, V.; Schleyer, P. v. R.; Allen, L. C. *Top. Stereochem.* **1973**, *7*, 253.

[146]Data in Table 5.2 are reproduced in part from the compilation by Aue, D. H.; Bowers, M. T. in *Gas Phase Ion Chemistry*, Vol. 2; Bowers, M. T., ed.; Academic Press: New York, 1979; p. 1.

of the additional energy required for charge separation. Although dissociation to ions should be considerably facilitated in polar solvents, the much greater difference in energy of cations than radicals is responsible for the much greater tendency of cations to undergo rearrangement.

The data in Table 5.2 indicate that the ΔH for heterolytic dissociation of alkanes in the gas phase varies with the alkyl group as follows: methyl > ethyl > isopropyl > *t*-butyl,[148] which is consistent with the generalization that the ease of formation of carbocations is 3° > 2° > 1°. What is the source of this increase in stability? We might be tempted to conclude that alkyl groups are electron donating relative to hydrogen (Figure 5.38). Yet, how can a carbon atom be electron donating relative to hydrogen when both have essentially the same electronegativity?[149] We can explain some, but not all, of the results by saying that an sp^3-hybrid orbital on carbon has a Pauling electronegativity of 2.5, while an sp^2-hybrid orbital on carbon is about 0.25 units more electronegative.[150] This expectation that the energy of a system is lowered due to electron polarization through a σ bond system is known as *induction*.

We may also explain the electron donating ability of a methyl or other alkyl group in terms of hyperconjugation, a lowering of the energy of a system by delocalization of electrons through π bonds involving sp^3-hybridized

Table 5.2 Thermodynamic data for selected alkyl cations. (Reproduced from the compilation in reference 146.)

Cation	$\Delta H_f^0(R^+)$ (kcal/mol)	HIA (R^+) (kcal/mol)
Methyl	261	312
Ethyl	219	273
Isopropyl	192	247.4
t-Butyl	164	229.6
Allyl	226	256
Vinyl	266	287
Benzyl	213	234

[147] For comparison, Table 5.1 shows the bond dissociation energies of many of the alkanes corresponding to the carbocations listed.

[148] Data of Stevenson, D. P.; cited by Streitwieser, Jr., A. *Chem. Rev.* **1956**, *56*, 571. The ΔH values for allyl and benzyl chlorides were reported to be 158 and 152 kcal/mol, respectively.

[149] In some semiempirical MO calculations of alkanes, the C-H bond is found to be slightly polarized toward carbon; in others, it is found to be slightly polarized toward hydrogen. See the discussion in Dewar, M. J. S.; Thiel, W. *J. Am. Chem. Soc.* **1977**, *99*, 4899 and references therein.

[150] See the discussion in Huheey, J. C. *Inorganic Chemistry*, 3rd Ed.; Harper and Row Publishers: New York, 1983; p. 153.

Figure 5.38 Inductive model for stabilization of a carbocation by a methyl group.

$CH_3 \longrightarrow \overset{+}{C}H_2$

carbon atoms adjacent to the carbocation center, just as we did earlier for radicals.[151] This model for chemical behavior has been the subject of considerable discussion in organic chemistry.[48] In one formulation of the interaction, a methyl group can conjugate with an adjacent positively charged carbon through the resonance structure shown on the right in Figure 5.39.

Figure 5.39 Stabilization of a carbocation by hyperconjugation with an adjacent methyl group.

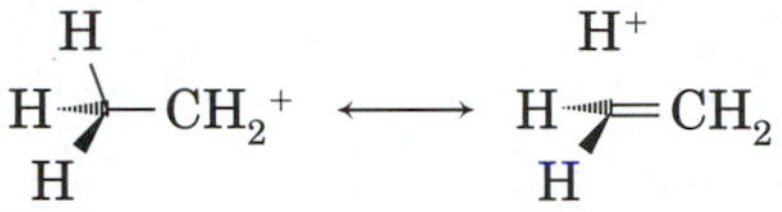

$H_3C{-}CH_2^+ \longleftrightarrow H^+ \; H_2C{=}CH_2$

A somewhat similar result can be obtained by applying PMO theory to the problem.[152] Figure 5.40 shows the PMO description for the interaction of an empty *p* orbital with a $\pi(CH_3)$ localized methyl group orbital. As was discussed in Chapter 4, the net effect is to distribute the electron density from the methyl portion of the molecule into a new orbital that has density on both the methyl group and the adjacent CH_2 group, thus delocalizing the positive charge and stabilizing the carbocation. Figure 5.41 shows four perspectives of the LUMO determined in extended Hückel calculations for the ethyl cation. Clearly, neither the valence bond notion of hyperconjugation nor the MO description give us justification for declaring the methyl to be electron donating by induction.

The preceding discussion was based on the implicit assumption that the ethyl cation could be described as a methyl-substituted methyl cation. This assumption may be unwarranted. *Ab initio* calculations indicate that a

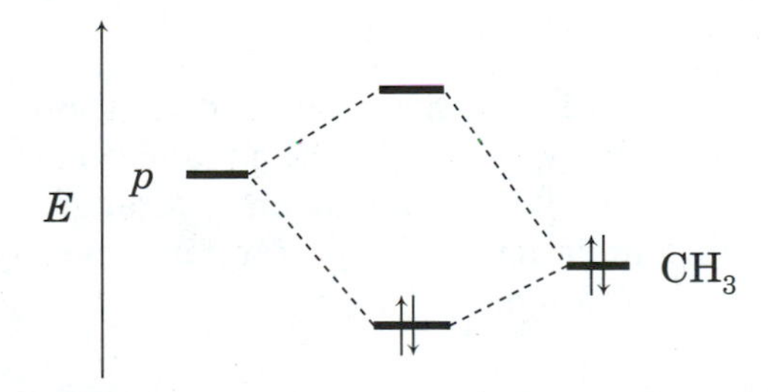

Figure 5.40 PMO description of stabilization of carbocation by methyl group.

[151]Hyperconjugation has also been described as a through-space delocalization of electrons. For a comparison of induction and hyperconjugation, see White, J. C.; Cave, R. J.; Davidson, E. R. *J. Am. Chem. Soc.* **1988**, *110*, 6308 and references therein.

[152]Hoffmann, R.; Radom, L.; Pople, J. A.; Schleyer, P. v. R.; Hehre, W. J.; Salem, L. *J. Am. Chem. Soc.* **1972**, *94*, 6221 have discussed the PMO description for methyl-substituted cations and anions.

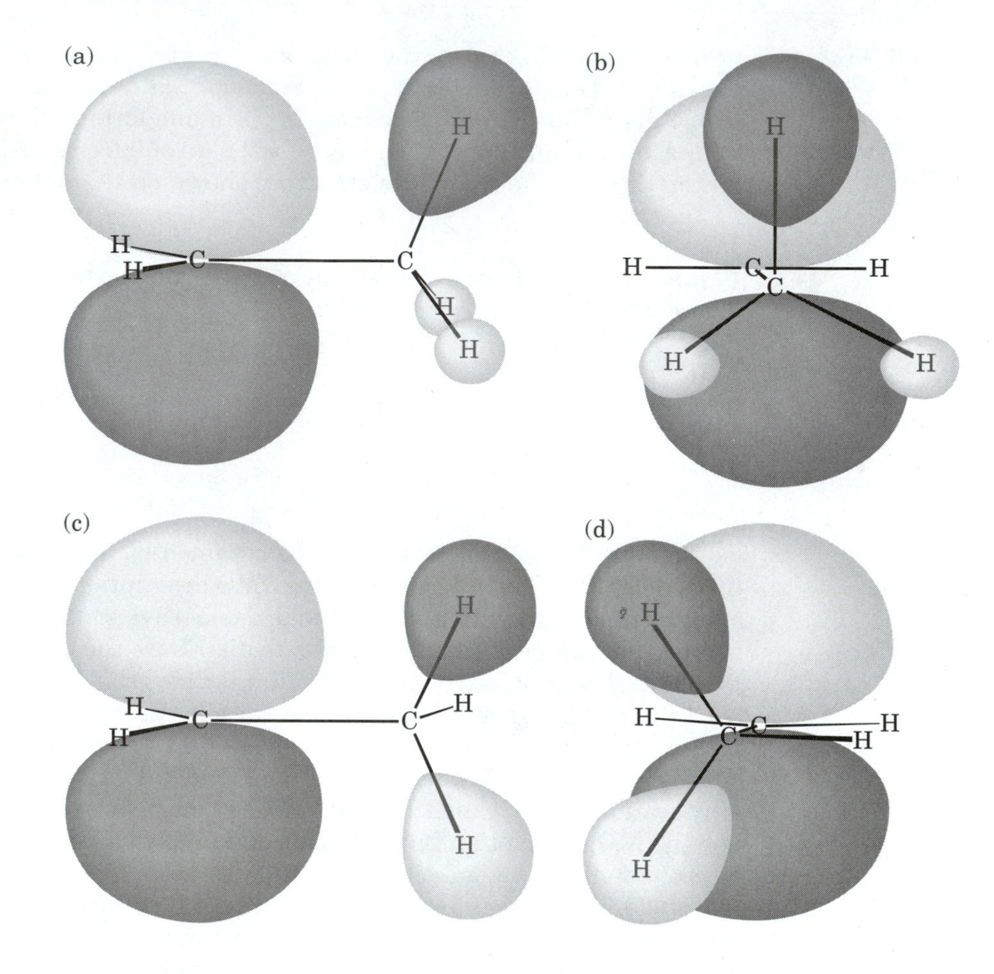

Figure 5.41 (a) Side and (b) end views of LUMO of ethyl cation in bisected conformation; (c) side and (d) end views of LUMO of ethyl cation in eclipsed conformation.

bridged C_{2v} structure (**36**) is more stable than the classical structure. In fact, the latter geometry was found to be a transition state for proton scrambling in the $C_2H_5^+$ ion.[153] This conclusion is strengthened by experimental data showing randomization of the protons in the ethyl cation both in the gas phase[154] and in solution.[155,156]

---

[153]Raghavachari, K.; Whiteside, R. A.; Pople, J. A.; Schleyer, P. v. R. *J. Am. Chem. Soc.* **1981**, *103*, 5649.

[154]Ausloos, P.; Rebbert, R. E.; Sieck, L. W.; Tiernan, T. O. *J. Am. Chem. Soc.* **1972**, *94*, 8939.

[155]Vorachek, J. H.; Meisels, G. G.; Geanangel, R. A.; Emmel, R. H. *J. Am. Chem. Soc.* **1973**, *95*, 4078 and references therein.

[156]Other carbocations are also calculated to have nonclassical structures: Hehre, W. J.; Radom, L.; Schleyer, P. v. R.; Pople, J. A.; Ab Initio *Molecular Orbital Theory*; John Wiley & Sons: New York, 1986; pp. 379 *ff*; see also Klopper, W.; Kutzelnigg, W. *J. Phys. Chem.* **1990**, *94*, 5625.

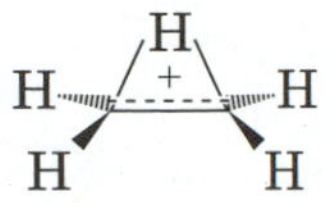

36

Detection of Carbocations

The study of carbocations illustrates many of the ways in which reactive intermediates are studied. Novel, creative techniques have been brought to bear and have led to new insights.[157] In particular, both proton NMR and ^{13}C NMR have been among the primary instrumental methods applied to the structure and properties of carbocations.[158,159] A pioneer in the field was Olah, who found that under appropriate conditions organic precursors dissolved in superacid media (such as $SO_2ClF\text{-}SbF_5$ solution, often at low temperature), gave solutions with spectra consistent with the presence of relatively long-lived carbocations.[160]

Generally, ^{13}C NMR chemical shifts for *carbenium ions* are observed at very low field. For example, the chemical shift for the 3° carbon atom in isobutane is 25.2 ppm, whereas the chemical shift for the corresponding carbon atom in $(CH_3)_3C^+$ is 330.0 ppm (in $SO_2ClF\text{-}SbF_5$ solution).[161] The large shift appears to result from decreased shielding due to the decreased electron density at the carbenium center. Studies of substituent effects on carbocation shifts reinforce this view. For example, the ^{13}C NMR shifts of the benzylic carbon atoms in the series of carbocations **37** (Figure 5.42) range from 219 ppm for *p*-methoxy to 269 ppm for *p*-CF_3.[162]

Intuitively, we might expect *electron-donating alkyl substituents* on a carbenium center to increase the local electron density and thus the shielding at that site, leading to a smaller downfield shift in the carbon resonance.[163] Surprisingly, however, there is evidence to suggest that alkyl substitution may lower the energy of a carbocation without decreasing the

---

[157]In addition to the techniques summarized here, carbocations have also been studied by many other techniques. For leading references, see (a) reference 139; (b) reference 158; (c) Olah, G. A.; Schleyer, P. v. R., eds.; *Carbonium Ions*, Vols. I–V; Wiley-Interscience: New York, 1968–1976.

[158]Olah, G. A. *Carbocations and Electrophilic Reactions*; John Wiley & Sons, Inc.: New York, 1974.

[159]Saunders, M.; Jiménez-Vázquez, H. A. *Chem. Rev.* **1991**, *91*, 375.

[160]Olah, G. A.; Prakash, G. K. S.; Sommer, J. *Superacids*; John Wiley & Sons, Inc.: New York, 1985.

[161]Reference 163(a), pages 306–307 and references therein.

[162]Data from reference 163(a), page 309, and references therein.

Figure 5.42
^{13}C NMR chemical shifts of substituted carbocations.

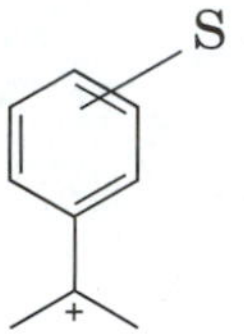

37

S	$\delta^{13}C$
4-OCH_3	219
4-CH_3	243
4-H	255
4-CF_3	269

deshielding of the carbenium carbon atom. The ^{13}C NMR chemical shift for C2 of the isopropyl carbocation is -125.0 ppm from the signal for CS_2, while that of the 3° carbon atom of the *t*-butyl carbocation is -135.4 ppm from the signal for CS_2. Thus, substitution of a hydrogen on C2 by a methyl group appears to deshield the carbocation,[164] suggesting electron withdrawal, not donation. In fact, extended Hückel calculations indicate that the charge on C2 of isopropyl is $+0.611$, while that of the central carbon atom of *t*-butyl is $+0.692$.[165] Thus, in contrast to our expectation that the methyl group is *electron donating* relative to hydrogen, both theory and experiment indicate that the carbenium ion center has a greater positive charge with the methyl substituent present.

Rearrangements of Carbocations

Equation 5.36 shows the familiar rearrangement of *n*-butyl to *sec*-butyl carbenium ion. This kind of rearrangement is a key step in common chemical reactions, such as the formation of the 2-butenes by acid-catalyzed dehydration of 1-butanol, or the formation of *sec*-butylbenzene from the reaction of benzene with 1-chlorobutane in the presence of $AlCl_3$.

[163](a) Absolute chemical shifts do not necessarily correlate with charge densities. The shielding constant for a carbon nucleus in ^{13}C NMR is thought to be the sum of several terms, and an intuitive belief that chemical shifts can be correlated with charge density on carbon nuclei is not entirely correct. For example, see Nelson, G. L.; Williams, E. A. *Prog. Phys. Org. Chem.* **1976**, *12*, 229. (b) However, we would expect such a correlation within a family of closely related compounds. A relationship between ^{13}C shifts and π-electron densities for aromatic compounds was developed by Spiesecke, H.; Schneider, W. G. *Tetrahedron Lett.* **1961**, 468. The correlation was extended by Olah, G. A.; Mateescu, G. D. *J. Am. Chem. Soc.* **1970**, *92*, 1430.

[164]Olah, G. A.; White, A. M. *J. Am. Chem. Soc.* **1969**, *91*, 3954, 3958, 5801.

[165]Hoffmann, R. *J. Chem. Phys.* **1964**, *40*, 2480.

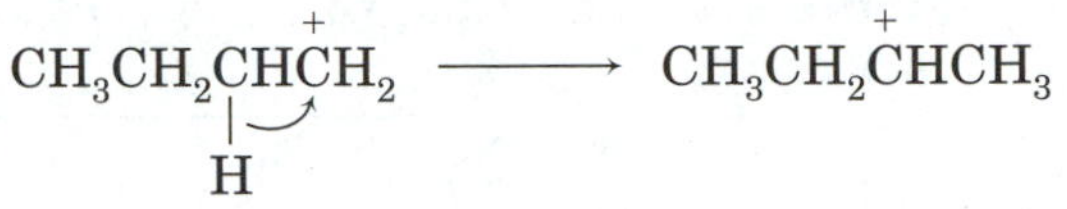

$$CH_3CH_2\underset{H}{\overset{}{C}}H\overset{+}{C}H_2 \longrightarrow CH_3CH_2\overset{+}{C}HCH_3 \qquad (5.36)$$

We might not expect to see rearrangement from a 3° carbocation, since it is not evident how a more stable carbocation could be formed. However, there is evidence that carbocations have a surprising facility for rearrangement, even if the rearrangement is not evident from the chemical reaction. For example, the proton NMR spectrum of the cyclopentyl carbocation (**38**, Figure 5.43) at −70° is a singlet (indicating the equivalence of all the protons), not the multiplet expected for a system with three sets of magnetically nonequivalent protons.[166] Similarly, the ^{13}C NMR spectrum indicates one carbon signal coupled with *nine* equivalent protons. Apparently a rearrangement takes place rapidly on the NMR time scale so that all 9 protons in the molecule are equivalent.[167] A study of the line broadening of the ^{13}C NMR spectrum of **38** by Saunders revealed an isomerization rate of 3.1×10^7 sec^{-1} at −139°, with a $\Delta G^{\ddagger}$ of 3.1 kcal/mol.[168] The ESCA analysis of **38**, however, suggested the presence in the ion of four uncharged carbon atoms and one positively charged carbon atom.[169] The difference between the NMR and ESCA results is that NMR is a "slow camera" that sees only an average of the environments a nucleus experiences during its ca. 10^{-7} second "shutter speed."[170] On the other hand, ESCA is a "fast camera" with a "shutter speed" of 10^{-16} seconds, so it is able to detect the discrete cyclopentyl ions.[171]

---

[166]Reference 158, page 78, and references therein.

[167]A theoretical study has suggested that the carbocation is a twisted structure with partial bridging of protons on C2 and C5 with the carbenium center on C1, as shown in the figure: Schleyer, P. v. R.; Carneiro, J. W. de M.; Koch, W.; Raghavachari, K. *J. Am. Chem. Soc.* **1989**, *111*, 5475.

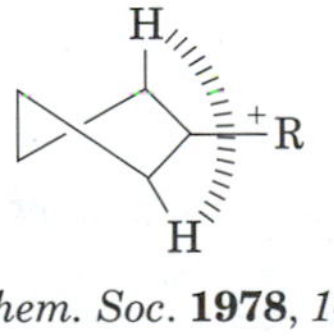

H H +R

[168]Saunders, M.; Kates, M. R. *J. Am. Chem. Soc.* **1978**, *100*, 7082.

[169]For a discussion of ESCA, see Chapter 1.

[170]This is a very rough estimate based on rate constants of processes that lead to coalescence of peaks in variable temperature NMR; see reference 135, page 144. The general rule is that the mean lifetime of a species to be detected by any spectroscopic means must be greater than the inverse of $2\pi\Delta\nu$, where $\Delta\nu$ is the difference in frequencies of the two species in the spectroscopic technique being used. In NMR spectroscopy, exchange rates over 100 sec^{-1} can average peaks. For a discussion, see (a) Ōki, M. *Top. Stereochem.* **1983**, *14*, 1; (b) Bushweller, C. H. in *Stereodynamics of Molecular Systems*, Sarma, R. H., ed.; Pergamon Press: New York, 1979; pp 39–51.

[171]Olah, G. A.; Mateescu, G. D.; Riemenschneider, J. L. *J. Am. Chem. Soc.* **1972**, *94*, 2529.

Figure 5.43 Equilibration of protons due to rapid rearrangement in cyclopentyl carbocation.

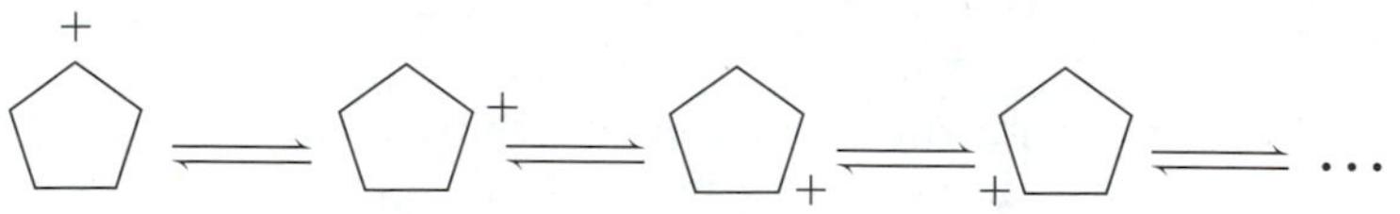

38

We might expect that the isopropyl cation would be immune to the 1,2-hydride shift exhibited by the cyclopentyl cation, since a simple hydride shift would convert a 2° carbocation to a 1° carbocation in an endothermic process. Nevertheless, scrambling of hydrogen atoms in the isopropyl cation was inferred by Saunders from proton NMR spectra,and the E_a for the reaction was determined to be 16.4 kcal/mol.[172] More surprising than the proton rearrangement is the rearrangement of the carbon skeleton. Olah observed that a sample of isopropyl cation labeled at the 2-position with ^{13}C underwent carbon skeletal rearrangement with a half-life of one hour at −78°, and after several hours the label was evenly distributed along the carbon chain.[164] The mechanism shown in Figure 5.44 accounts for both proton and carbon scrambling.

The situation for the *sec*-butyl cation is even more interesting. We might expect it to behave as does the cyclopentyl cation, that is, to undergo degenerate rearrangement among the two 2° carbocations at a rapid rate. At −110° the proton NMR spectrum of this species showed two sets of protons, consistent with expectation.[164,173] However, warming the sample to −40° led to coalescence of the two peaks, indicating scrambling of all nine protons in the ion. (Heating the sample above −40° led to isomerization to the *t*-butyl cation.) Saunders suggested that the proton scrambling reaction arises through isomerization of the *sec*-butyl cation to a protonated methylcyclopropane, followed by opening to an isomerized *sec*-butyl cation.[173]

There is also ^{13}C NMR evidence for carbon skeletal rearrangement for the *sec*-butyl cation. Myhre and Yannoni obtained a solid state ^{13}C NMR spectrum of the *sec*-butyl cation prepared from ^{13}C-labeled 2-chlorobutane

Figure 5.44 Proposed mechanism for rearrangement of isopropyl cation. (Adapted from reference 164.)

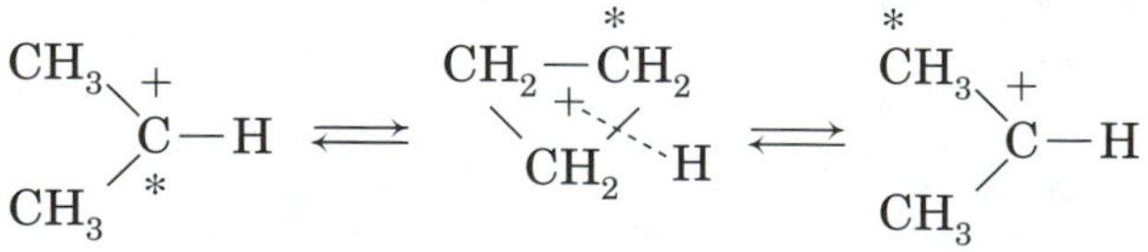

[172]Saunders, M.; Hagen, E. L. *J. Am. Chem. Soc.* **1968**, *90*, 6881.

[173]Saunders, M.; Hagen, E. L.; Rosenfeld, J. *J. Am. Chem. Soc.* **1968**, *90*, 6882.

(with 75% of the label at the 3 position and 25% of the label at the 2 position). The spectrum of the carbocation at −85° indicated that the label was distributed evenly over the four carbon atoms in the chain.[174] The results are consistent with the mechanism proposed for the isomerization by Saunders (Figure 5.45).[175,176]

The solid state ^{13}C NMR spectrum of the *sec*-butyl cation at −85° showed two peaks, one assigned to the C2 and C3 carbon atoms at 22 ppm and another assigned to the C1 and C4 carbon atoms at 178 ppm relative to TMS. There was some broadening of the solid state spectrum as the sample temperature was lowered to −190°, but there was no indication of the decoalescence of bands expected if site exchange becomes slower at lower temperature. Thus the rate constant of the hydride shift that makes C1 and C2 equivalent was determined to be at least 4×10^9 sec^{-1} at that temperature, and the process was calculated to have a $\Delta G^{\ddagger}$ of no more than 2.4 kcal/mol,[168] perhaps even lower.[174] In other words, the activation energy for the hydride shift must be less than that for rotation about the single bond of ethane! Moreover, Olah and Donovan concluded from an analysis of ^{13}C NMR chemical shifts that the *sec*-butyl ion exists as an equilibrium mixture of the two 1,2-hydride shift species and a third, bridged ion intermediate, in which one hydrogen is bonded to both C2 and C3 (Figure 5.46).[177]

Other researchers have studied the *sec*-butyl cation experimentally and theoretically and have proposed at least four other possible structures for it.[178] More recent theoretical calculations[179] and experimental ESCA

Figure 5.45 Mechanism proposed for scrambling of the carbon atoms in the *sec*-butyl carbocation. (Adapted from reference 174.)

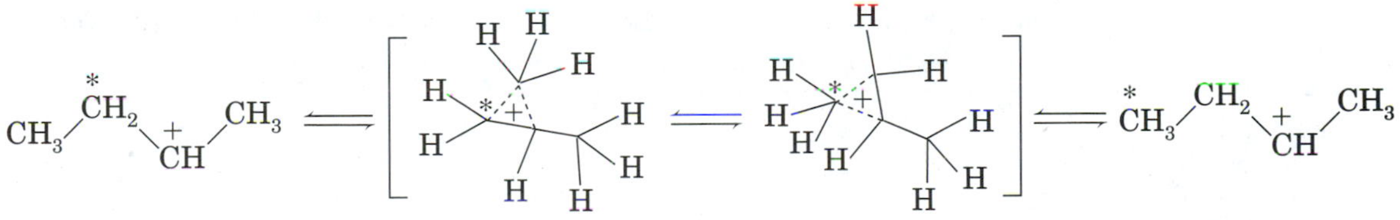

[174] Myhre, P. C.; Yannoni, C. S. *J. Am. Chem. Soc.* **1981**, *103*, 230.

[175] Saunders has presented evidence for protonated cyclopropane intermediates in a variety of carbocation isomerization processes: Saunders, M.; Vogel, P.; Hagen, E. L.; Rosenfeld, J. *Acc. Chem. Res.* **1973**, *6*, 53.

[176] For references to carbocation rearrangements and a discussion of the $C_4H_9^+$ potential energy surface, see Sieber, S.; Buzek, P.; Schleyer, P. v. R.; Koch, W.; Carneiro, J. W. de M. *J. Am. Chem. Soc.* **1993**, *115*, 259.

[177] Olah, G. A.; Donovan, D. J. *J. Am. Chem. Soc.* **1977**, *99*, 5026.

[178] Johnson, S. A.; Clark, D. T. *J. Am. Chem. Soc.* **1988**, *110*, 4112 and references therein.

[179] Clark, D. T.; Harrison, A. *Chem. Phys. Lett.* **1981**, *82*, 143.

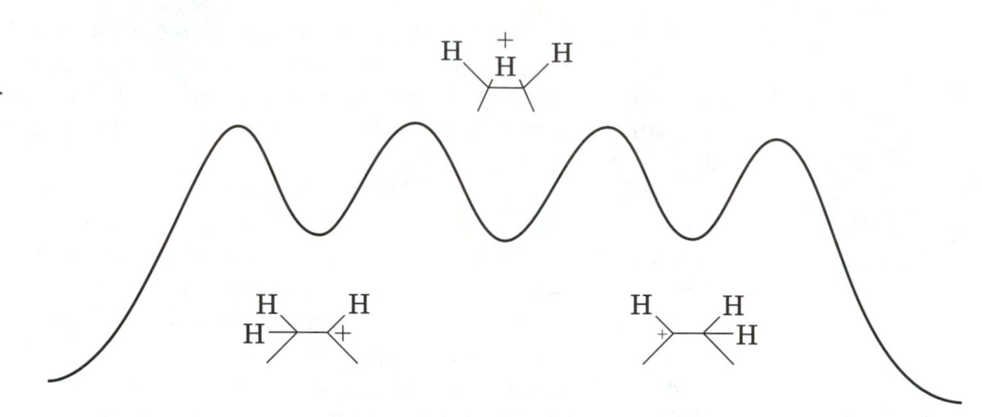

Figure 5.46
A potential energy diagram for the 1,2-hydride shift in the 2-butyl cation. (Reproduced from reference 177.)

results[178] have suggested that a *partially bridged* structure such as that shown in Figure 5.47 may be the best description for this cation. However, high level calculations by Schleyer and co-workers determined that the hydrogen-bridged structure was more stable by 0.4 kcal/mol than the methyl-bridged structure. It was suggested that the ESCA results might have arisen through surface effects.[180] In any case, both theoretical calculations and experimental results lead to the conclusion that alkyl carbocations are incompletely described by the simple drawings we often use to depict them.

Figure 5.47
Partially bridged model for *sec*-butyl cation. (Adapted from reference 179.)

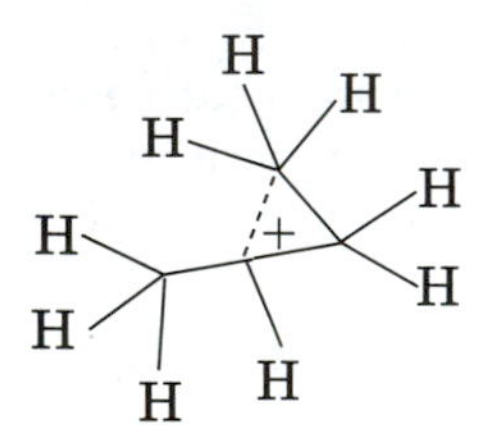

Nonclassical Ions

There has been considerable debate among organic chemists concerning the role of nonclassical structures as intermediates in reactions under normal conditions in solution. Perhaps the "classic" example of such a nonclassical ion is the 2-norbornyl cation. In Chapter 8 we will review the experimental evidence, largely based on solvolysis reactions, that led to the proposal of the nonclassical or hypercoordinate carbonium ion structure shown in Figure 5.48.

Figure 5.48
Nonclassical carbonium ion model for 2-norbornyl cation.

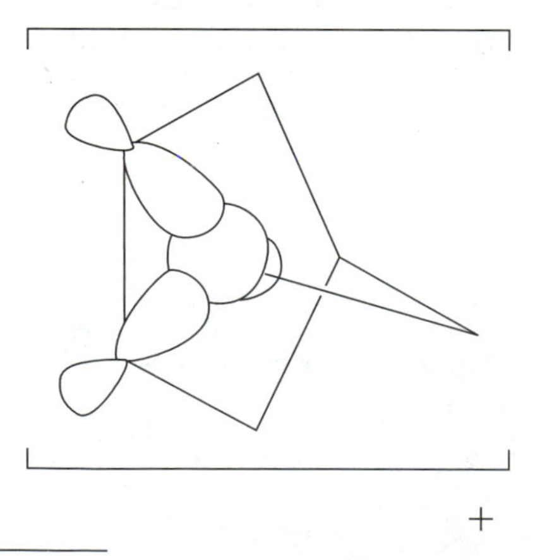

[180]Carneiro, J. W. de M.; Schleyer, P. v. R.; Koch, W.; Raghavachari, K. *J. Am. Chem. Soc.* **1990**, *112*, 4064.

However, this description was not accepted by all researchers, and the 2-norbornyl ion was also described as a pair of rapidly equilibrating classical (carbenium) ions, as shown in Figure 5.49.
Many papers relating to the development of contrasting ideas in this area were published in a reprint and commentary volume by Bartlett.[181,182] We will summarize some more recent data here and will also explore the role of nonclassical ions in reactions to be discussed in more detail in later chapters.

As with other carbocations, NMR spectroscopy has been utilized to determine the structure of the 2-norbornyl cation. The ^{13}C NMR spectrum at −70° showed three peaks: one peak at +101.8 ppm (J = 53.3 Hz) was assigned to carbon atoms 1, 2, and 6; another peak at +162.5 ppm (J = 140 Hz) was assigned to carbon atoms 3, 5, and 7; a third peak at +156.1 ppm (J = 153 Hz) was assigned to carbon atom 4.[164] We expect carbon atoms 1 and 2 to be equivalent, whether we think the carbocation is classical or nonclassical, but their equivalence with carbon atom 6 is a surprise. Apparently there is a rapid hydride shift that interconverts these positions (Figure 5.50). Below 150 K the hydride shift could be frozen out and a spectrum ascribed to a single ion observed. The lowest temperatures used to record a spectrum for this system are 5 K (nonspinning) and 6 K (spinning) by Yannoni, Myhre, and co-workers.[183] Between 150 K (where the hydride shift is

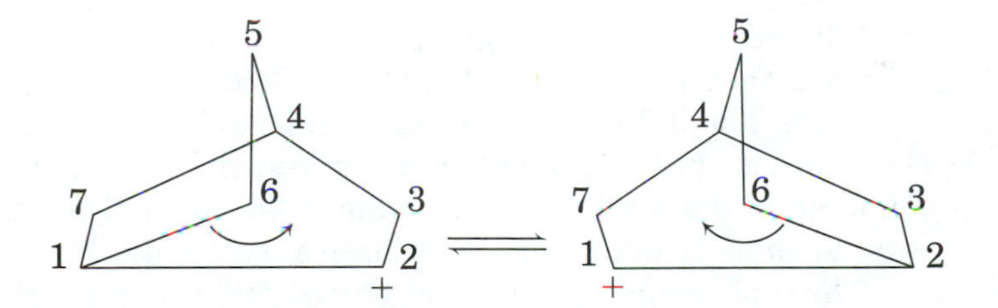

Figure 5.49 Rapidly equilibrating classical carbocation (carbenium ion) model for 2-norbornyl cation.

[181]Bartlett, P. D. *Nonclassical Ions: Reprints and Commentary*; W. A. Benjamin, Inc.: New York, 1965.

[182]Sargent, G. D. *Quart. Rev. Chem. Soc.* **1966**, *20*, 301, has suggested that the “classical” period in carbocation history lasted less than ten years, from the time of Whitmore's publication (Reference 140) until the first report of a bridged ion (Nevell, T. P.; de Salas, E.; Wilson, C. L. *J. Chem. Soc.* **1939**, 1188; see also Saltzman, M. D.; Wilson, C. L. *J. Chem. Educ.* **1980**, *57*, 289).

[183](a) Yannoni, C. S.; Macho, V.; Myhre, P. C. *J. Am. Chem. Soc.* **1982**, *104*, 7380; (b) Myhre, P. C.; Webb, G. G.; Yannoni, C. S. *J. Am. Chem. Soc.* **1990**, *112*, 8991; (c) Similar experiments on the C_4H_7+ ion were consistent with *ab initio* calculations suggesting that both a partially delocalized bisected cyclopropylcarbinyl cation and a nonclassical symmetrical bicyclobutonium ion (puckered cyclobutyl cation) are local energy minima on a very flat potential energy surface. Myhre, P. C.; Webb, G. C.; Yannoni, C. S. *J. Am. Chem. Soc.* **1990**, *112*, 8992 and references therein.

Figure 5.50 1,2,6-Hydride shift in the 2-norbornyl cation.

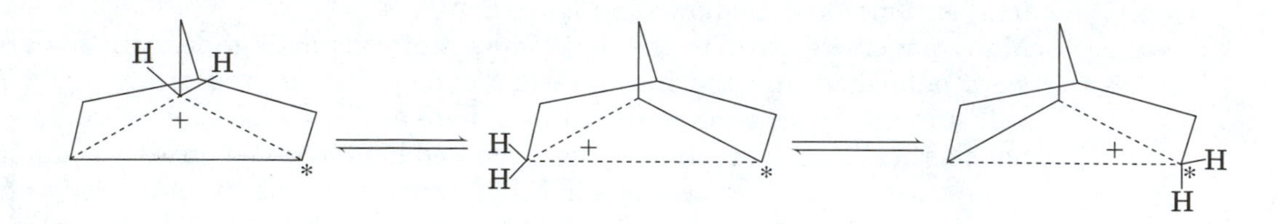

effectively frozen out) and 5 K, the peak observed for C1 and C2 was at 125 ppm and did not change with temperature, leading to the conclusion that the 2-norbornyl cation was indeed a nonclassical carbonium ion (Figure 5.48). The authors concluded that if it were a pair of rapidly equilibrating classical ions (Figure 5.49), the activation energy for the equilibrium could be no more than 0.2 kcal/mol![184] ESCA analysis[185,186] and studies of carbocation stability in the gas phase[187] also were interpreted as favoring the nonclassical structure.

Further support for the nonclassical structure of the 2-norbornyl cation came from an application of ^{13}C NMR spectroscopy that is based on the difference of the total chemical shift of a carbocation and the corresponding alkane. Differences in total chemical shift of 350 ppm or more suggest classical carbocations, while differences of less than 220 ppm indicate nonclassical, bridged carbonium ions. For example, the sum of the total ^{13}C NMR chemical shift of propane is 47 ppm, while the sum for the 2-propyl cation is 423 ppm. The difference, 376 ppm, indicates that the 2-propyl cation is a classical ion. For the 2-norbornyl system the total of the ^{13}C shifts is 408 ppm, while the total for norbornane is 233 ppm. The difference, 175 ppm, was taken as evidence for a nonclassical structure.[188]

The conclusion that the 2-norbornyl cation is a nonclassical carbocation has been strengthened by the experimental determination of its infrared spectrum when the cation is generated in a cryogenic SbF_5 matrix. The experimental IR spectra agree with those calculated for a nonclassical struc-

[184]This conclusion was based on the premise that tunneling of the carbon atoms was not important in the equilibration mechanism. For a discussion of this point and for experimental evidence suggesting that tunneling is not a problem in the data analysis, see (a) Myhre, P. C.; McLaren, K. L.; Yannoni, C. S. *J. Am. Chem. Soc.* **1985**, *107*, 5294 and references therein; (b) reference 183(b).

[185]Olah, G. A.; Mateescu, G. D.; Riemenschneider, J. L. *J. Am. Chem. Soc.* **1972**, *94*, 2529.

[186]Johnson, S. A.; Clark, D. T. *J. Am. Chem. Soc.* **1988**, *110*, 4112.

[187]Kaplan, F.; Cross, P.; Prinstein, R. *J. Am. Chem. Soc.* **1970**, *92*, 1445.

[188]Schleyer, P. v. R.; Lenoir, D.; Mison, P.; Liang, G.; Prakash, G. K. S.; Olah, G. A. *J. Am. Chem. Soc.* **1980**, *102*, 683.

ture.[189] Additional experimental and theoretical evidence also lead to the conclusion that the 2-norbornyl cation is a nonclassical ion.[190,191]

Radical Cations

Each of the carbocations discussed to this point has been a species in which all of the electrons were spin-paired. Another type of positively charged reactive intermediate is the radical cation—a species that both has an unpaired electron and a positive charge. Such species can be produced by the one-electron oxidation of a neutral species having no unpaired electrons.[192,193] Radical cations play important roles in many radiochemical and photochemical reactions, and they may also be important in biological processes,[194] including photosynthesis[193b] and the biosynthesis of natural products.[195]

Organic radical cations can be generated from neutral organic compounds through:[193a]

1. chemical oxidation by a wide variety of oxidizing agents, including Brønsted acids, Lewis acids, metal ions and oxides, nitrosonium ions, other organic radical cations, semiconductor materials, and some zeolites;[196]

---

[189]Koch, W.; Liu, B.; DeFrees, D. J.; Sunko, D. E.; Vančik, H. *Angew. Chem., Int. Ed. Engl.*, **1990**, *29*, 183.

[190]Olah, G. A.; Prakash, G. K. S.; Saunders, M. *Acc. Chem. Res.* **1983**, *16*, 440. Also see Walling, C. *Acc. Chem. Res.* **1983**, *16*, 448.

[191]Lenoir, D.; Apeloig, Y.; Arad, D.; Schleyer, P. v. R. *J. Org. Chem.* **1988**, *53*, 661 and references therein; Schleyer, P. v. R.; Sieber, S. *Angew. Chem., Int. Ed. Engl.* **1993**, *32*, 1606.

[192]Such species are also termed *cation radicals*, but the name *radical cation* is consistent with IUPAC terminology. (Commission on Physical Organic Chemistry, IUPAC, *Pure Appl. Chem.* **1994**, *66*, 1077.) In the recommended nomenclature, the radical cation of molecule **A** is denoted $\mathbf{A}^{\cdot +}$; that is, the symbols for cation and radical are written in the same order as indicated by the term *radical cation*.

[193]For a discussion of radical ions in organic chemistry, see (a)Roth, H. D. *Top. Curr. Chem.* **1992**, *163*, 131; (b) Kaiser, E. T.; Kevan, L., eds. *Radical Ions*; Wiley-Interscience: New York, 1968; (c) Lund, A.; Shiotani, M., eds. *Radical Ionic Systems: Properties in Condensed Phases*. (*Topics in Molecular Organization and Engineering, Vol. 6*; Kluwer Academic Publishers; Dordrecht, 1991; (d) Bauld, N. L. *Advances in Electron Transfer Chemistry*, Vol. 2; Mariano, P. S., ed., JAS Press: Greenwich, 1992; pp. 1–66; (e) Chanon, M.; Rajzmann, M.; Chanon, F. *Tetrahedron* **1990**, *46* 6193; (f) Hammerich, O.; Parker, V. D. *Adv. Phys. Org. Chem.* **1984**, *20*, 55; (g) Shida, T.; Haselbach, E.; Bally, T. *Acc. Chem. Res.* **1984**, *17*, 180; (h) Nelsen, S. F. *Acc. Chem. Res.* **1987**, *20*, 269; (i) Roth, H. D. *Acc. Chem. Res.* **1987**, *20*, 343; (j) Bauld, N. L.; Bellville, D. J. *Acc. Chem. Res.* **1987**, *20*, 371.

[194]For an experimental study related to the possible role of radical cations in the cytochrome P-450 oxidative dealkylation of amines, see Dinnocenzo, J. P.; Karki, S. B.; Jones, J. P. *J. Am. Chem. Soc.* **1993**, *115*, 7111.

[195]Hoffmann, U.; Gao, Y.; Pandey, B.; Klinge, S.; Warzecha, K.-D.; Krüger, C.; Roth, H. D.; Demuth, M. *J. Am. Chem. Soc.* **1993**, *115*, 10358.

[196]Thermally activated Na-ZSM-5 zeolite was found to have a redox potential of 1.65 ± 0.1 V vs. SCE in a study of the conversion of α,ω-diphenylpolyenes to their radical cations. Ramamurthy, V.; Caspar, J. V.; Corbin, D. R. *J. Am. Chem. Soc.* **1991**, *113*, 594.

2. electrochemical oxidation;
3. radiolysis with ionizing radiation (X-rays and γ-rays);
4. photoinduced electron transfer (PET) resulting from bimolecular reaction of a photoexcited molecule with a ground state molecule;[197]
5. electron impact ionization (commonly used for the production of radical cations in mass spectrometry).

Among the major analytical tools for detecting and studying radical cations are mass spectrometry, electron spin resonance spectroscopy, nuclear magnetic resonance spectrometry, and particular variations of these methods.

Organic radical cations can be classified according to the type of orbital from which an electron is removed from a neutral parent compound. Compounds with nonbonded electrons, such as amines, ethers, and ketones, can be oxidized to produce *n*-radical cations. For example, the one-electron oxidation of methanol is conveniently viewed as the removal of an electron from a nonbonding orbital associated with oxygen.[198]

$$H_3C-\ddot{\underset{\cdot\cdot}{O}}-H \xrightarrow{-e^-} H_3C-\overset{\cdot+}{\underset{\cdot\cdot}{O}}-H \qquad (5.37)$$

Figure 5.51 Calculated geometry of methane radical cation. (Adapted from reference 199.)

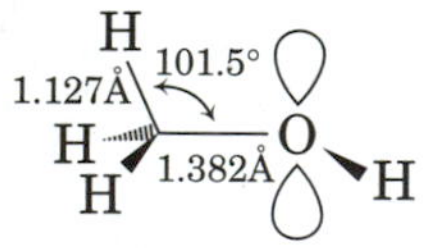

The formulation of the radical cation of methanol shown in equation 5.37 might suggest a species in which both the positive charge and the unpaired electron density are localized on oxygen. As indicated by the HOMO of ethanol in Figure 4.57 (page 238), however, the HOMO of an alcohol is delocalized onto other atoms as well. Consistent with that model, the calculated geometry of the methanol radical cation (Figure 5.51) indicates that one C—H bond is aligned with the *p* orbital on oxygen, this C—H distance is lengthened, and the O—C distance is shortened in comparison with neutral methanol.[199] The implication of this geometry is that the delocalization results in shifting of electron density from C—H bonding to C—O bonding.

One-electron oxidation of alkenes, alkynes, and arenes produces species known as π-radical cations, formed by removal of an electron from a π molecular orbital, as suggested by the generalized equation 5.38.

---

[197]Generation of radical cations by photoinduced electron transfer reactions of neutral molecules generates radical anion/radical cation pairs, which can undergo back electron transfer to regenerate neutral molecules. Polar solvents such as CH_3CN can make diffusive separation of the radical ions more probable, but the use of polar solvents increases the opportunity for reaction of solvent with the radical ions. The use of cationic acceptors for PET from photoexcited organic compounds forms a neutral radical/radical cation pair. With no Coulombic barrier to separation, yields of separated radical cations are increased. For a discussion, see Todd, W. P.; Dinnocenzo, J. P.; Farid, S.; Goodman, J. L.; Gould, I. R. *J. Am. Chem. Soc.* **1991**, *113*, 3601.

[198]Budzikiewicz, H.; Djerassi, C.; Williams, D. H.; *Mass Spectrometry of Organic Compounds*; Holden-Day, Inc.: San Francisco, 1967; p. 94.

[199]Ma, N. L.; Smith, B. J.; Pople, J. A.; Radom, L. *J. Am. Chem. Soc.* **1991**, *113*, 7903.

$$R_2R_1C{=}CR_4R_3 \xrightarrow{-e^-} \left[R_2R_1C\cdots CR_4R_3\right]^{\cdot +} \quad (5.38)$$

From an analysis of the photoelectron spectrum of ethene, the ethene radical cation was found to have a torsion angle of 25°. The C—C bond length is 1.405 Å, and the C—H bond length is 1.091 Å. The H—C—H bond angle is 117°51′.[200] The geometry of the ethene radical cation (Figure 5.52) has been explained on the basis of a compromise between some remaining π bonding in the SOMO (optimized at a torsional angle of 0°) and hyperconjugative interaction of the *p* orbital on each carbon with the C—H bonding orbitals on the adjacent methylene group (optimized at a torsional angle of 90°).[201]

For larger alkenes, hyperconjugation with an alkyl group α to an olefinic carbon atom eliminates the need for rotation, so the radical cations of almost all alkenes other than ethene are planar.[202] The constitution of the alkene affects the geometry of the resulting cation. Eriksson and co-workers found that the radical cation formed by radiolysis of 1-pentene in a fluorochlorocarbon matrix at 77 K exhibits an EPR spectrum that suggests some delocalization of the SOMO over C3. In the case of the isomeric 2-pentenes, the results suggested that the SOMO is localized on the carbon atoms of the former double bond, C2 and C3. The difference in bonding was attributed to a lower ionization energy for the longer alkyl group attached to the double bond in 1-pentene in comparison with the higher ionization energy of the shorter alkyl group attached to the double bond in the 2-pentenes.[203]

One-electron oxidation of alkanes leads to σ-radical cations.[204]

$$R_1R_2R_3C{-}CR_4R_5R_6 \xrightarrow{-e^-} \left[R_1R_2R_3C{-}CR_4R_5R_6\right]^{\cdot +} \quad (5.39)$$

Such ionization removes an electron from an orbital associated with σ bonding among carbon atoms. The radical cation of butane, for example, shows

[200]Köppel, H.; Domcke, W.; Cederbaum, L. S.; von Niessen, W. *J. Chem. Phys.* **1978**, *69*, 4252.

[201]Mulliken, R. S.; Roothaan, C. C. J. *Chem. Rev.* **1947**, *41*, 219.

[202]a) Bellville, D. J.; Bauld, N. L. *J. Am. Chem. Soc.* **1982**, *104*, 294; b) Clark, T.; Nelsen, S. F. *J. Am. Chem. Soc.* **1988**, *110*, 868.

[203]Eriksson, L. A.; Sjöqvist, L.; Lunell, S.; Shiotani, M.; Usui, M.; Lund, A. *J. Am. Chem. Soc.* **1993**, *115*, 3244.

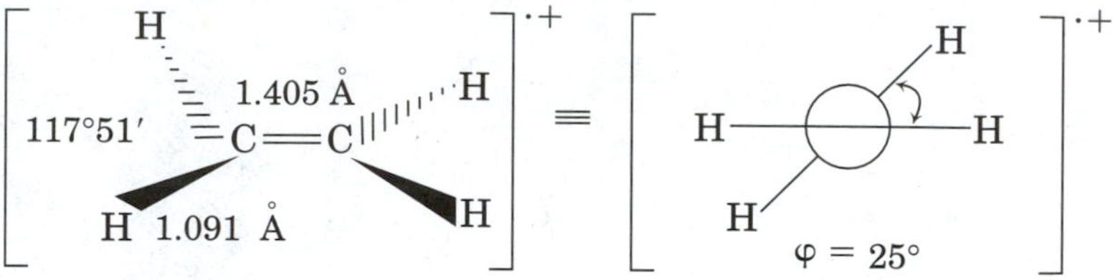

Figure 5.52 Geometry of radical cation of ethene.

elongation of the C2—C3 bond to a distance of about 2.0 Å and a much lower difference in energies of the anti and gauche conformers than is the case with the parent hydrocarbon.[204] Ionization of methane produces a distorted structure having two longer C—H bond distances, with an H—C—H angle of about 60° for these atoms, and two shorter C-H′ bond distances associated with an H′—C—H′ angle near 125°.[205]

Alkyl groups may also delocalize the unpaired electron and charge density in radical cations formed from strained alkanes. For example, radical cations of bicyclo[1.1.0]butane (**39**) and 1,3-dimethylbicyclo[1.1.0]butane (**40**) were detected in Freon matrix following γ-irradiation of the parent hydrocarbon.

Bicyclo[1.1.0]butane

39

1,3-Dimethylbicyclo[1.1.0]butane

40

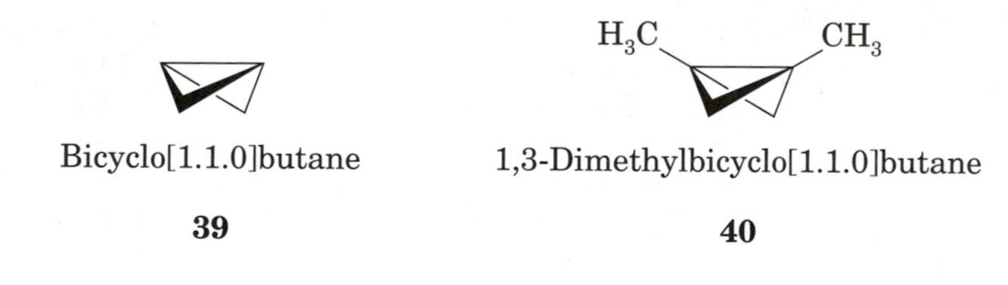

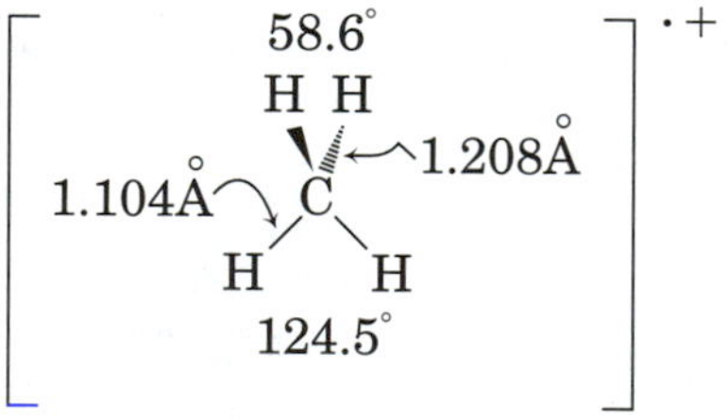

Figure 5.53 Geometry of methane radical cation. (Adapted from reference 204.)

[204]For a theoretical study of the radical cations of alkanes and references to experimental data for these species, see Eriksson, L. A.; Lunell, S.; Boyd, R. J. *J. Am. Chem. Soc.* **1993**, *115*, 6896.

[205]The distortion from the tetrahedral geometry of methane is a Jahn-Teller effect: Jahn, H. A.; Teller, E. *Phys. Rev.* **1936**, *49*, 874; *Proc. R. Soc. London, Ser. A* **1937**, *161*, 220. Experimental evidence for a C_{2v} structure for the methane radical cation (studied in a neon matrix at 4 K) was reported by Knight, Jr., L. B.; Steadman, J.; Feller, D.; Davidson, E. R. *J. Am. Chem. Soc.* **1984**, *106*, 3700. The EPR spectrum of $CH_4^{\cdot+}$ at 4 K suggests four magnetically equivalent protons, but the EPR spectrum of $CH_2D_2^{\cdot+}$ indicates the C_{2v} geometry shown. The apparent equivalence of the four protons in $CH_4^{\cdot+}$ is attributed to dynamic Jahn-Teller distortion so that all protons are equivalent on the PMR time scale. For a discussion and theoretical study, see Paddon-Row, M. N.; Fox, D. J.; Pople, J. A.; Houk, K. N.; Pratt, D. W. *J. Am. Chem. Soc.* **1985**, *107*, 7696.

The data suggest that much of the unpaired electron density in **39**$^{\cdot+}$ is associated with the two bridgehead carbon atoms, but that in **40**$^{\cdot+}$ about 15% of the unpaired electron density is associated with the methyl substituents at these positions.[206]

In each of the radical cation structures discussed so far, the unpaired electron and the positive charge have been closely associated. However, there are also radical cations in which the radical center and the cation center are separate from each other. For example, electron ionization of 1,4-dioxane (**41**) in a mass spectrometer produces the radical cation of **41**, which eliminates formaldehyde to form **42**.[207] Species such as **42** were termed *distonic* by Radom to emphasize the distance of the charge and radical sites, and they can be significantly lower in energy than radical cations in which the charge and unpaired electron density are coincident.[208]

$$\underset{\mathbf{41}}{\text{1,4-dioxane}} \xrightarrow[\text{ionization}]{\text{70 eV electron}} \underset{\mathbf{41}^{\cdot+}}{\text{1,4-dioxane}^{\cdot+}} \longrightarrow \underset{\mathbf{42}}{H_2\dot{C}{-}CH_2{-}\overset{+}{O}{=}CH_2} + \underset{\mathbf{43}}{O{=}CH_2} \qquad (5.40)$$

Radical cations exhibit a wide variety of reactions,[193] including unimolecular reactions such as rearrangement,[209] fragmentation,[210] and intramolecular bond formation as well as bimolecular reactions with ionic, radical, or ground state species. Notable processes include reaction with nucleophiles to produce radicals, reaction with radicals to produce cations, reaction with electron donors to produce biradicals, and reaction with ground state molecules to give addition products.[211] Often the products of reactions of radical cations with neutral species are different from those observed by reaction of the corresponding carbocation with the same reactant.[212]

---

[206] Arnold, A.; Burger, U.; Gerson, F.; Kloster-Jensen, E.; Schmidlin, S. P. *J. Am. Chem. Soc.* **1993**, *115*, 4271.

[207] A radical cation can be considered to be a cationized diradical in the sense that it is the hypothetical product of one-electron oxidation of a diradical. Conversely, neutralization of **41**$^{\cdot+}$ produces the 1,4-biradical $\cdot CH_2CH_2CH_2\dot{C}H_2$, which offers a convenient method for studying its reactions in the gas phase.: Polce, M. J.; Wesdemiotis, C. *J. Am. Chem. Soc.*, **1993**, *115*, 10849.

[208] Yates, B. F.; Bouma, W. J.; Radom, L. *J. Am. Chem. Soc.*, **1984**, *106*, 5805.

[209] The radical cation of benzvalene was found to isomerize to the radical cation of benzene at a temperature of 135 K: Arnold, A.; Gerson, F.; Burger, U. *J. Am. Chem. Soc.* **1991**, *113*, 4359.

[210] For a discussion of bond cleavage reactions of radical cations, see (a) Popielarz, R.; Arnold, D. R. *J. Am. Chem. Soc.* **1990**, *112*, 3068; (b) Camaioni, D. M. *J. Am. Chem. Soc.* **1990**, *112*, 9475.

[211] For a theoretical study of the reaction of σ-radical cations with nucleophiles, see Shaik, S.; Reddy, A. C.; Ioffe, A.; Dinnocenzo, J. P.; Danovich, D.; Cho, J. K. *J. Am. Chem. Soc.* **1995**, *117*, 3205.

[212] See, for example, Gassman, P. G.; Singleton, D. A. *J. Am. Chem. Soc.* **1984**, *106*, 7993.

Radical cations of weak acids may react either by heterolytic cleavage (loss of a proton to produce a radical) or homolytic cleavage (loss of a hydrogen atom to form a carbocation).[213] In a polar solvent heterolytic cleavage is usually favored because of the favorable solvation energy of the proton, and radical cations are ordinarily much more acidic than the corresponding neutral compounds. For example, the pK_{HA} value of toluene in DMSO is 43, while the $pK_{HA^{\cdot+}}$ value for the radical cation of toluene is -20. Therefore, the difference in pK_a values of toluene and its radical cation is 63![214]

In the gas phase, heterolytic cleavage is favored because the positive charge can be stabilized by charge delocalization in a larger ion. Zhang and Bordwell determined that N-H and O-H bond dissociation energies ($BDE_{HA^{\cdot+}}$) of radical cations are only slightly lower than those of their parent nitrogen or oxygen acid compounds. However, when the proton being detached is bonded to a carbon atom, the $BDE_{HA^{\cdot+}}$ value is typically 30–50 kcal/mol lower than the BDE of the corresponding neutral compound.[214(b)]

Reaction of a radical cation with a nucleophile can occur either by electron transfer to give a diradical or by attachment to give a radical. For example, the reaction of azide ion (N_3^-) with the radical cation of 4-methoxystyrene and its β-methyl and β,β-dimethyl derivatives was found to occur by electron transfer in acetonitrile (CH_3CN) solution but by nucleophilic attachment in 2,2,2-trifluoroethanol (TFE) solution. The change in mechanism was ascribed to a change in the oxidation potential of azide ion with solvent (the value being 0.5 V more positive in TFE than in CH_3CN), suggesting that redox properties of nucleophiles and cation radicals under reaction conditions can determine reaction pathways.[215]

Carbanions

Structure and Geometry of Carbanions

Carbanions are anions that contain an even number of electrons and that have an unshared pair of electrons on a tervalent carbon atom.[216,217,218] In contrast to the case with carbenes, carbanions would appear to offer little choice in the way of structure.[219] For the simplest carbanion, $H_3C{:}^-$, we expect four pairs of electrons to be arranged around the carbon atom, with the nonbonded pair of electrons in an orbital that is approximately an sp^3 hy-

---

[213]Alkane radical cations in liquid hydrocarbon solution undergo ion-molecule reactions, such as proton transfer or hydrogen atom transfer on a sub-millisecond scale. Werst, D. W.; Bakker, M. G.; Trifunac, A. D. *J. Am. Chem. Soc.* **1990**, *112*, 40.

[214](a) Bordwell, F. G.; Cheng, J.-P. *J. Am. Chem. Soc.* **1989**, *111*, 1792; (b) Zhang, X.-M.; Bordwell, F. G. *J. Am. Chem. Soc.* **1994**, *116*, 4251.

[215]Workentin, M. S.; Schepp, N. P.; Johnston, L. J.; Wayner, D. D. M. *J. Am. Chem. Soc.* **1994**, *116*, 1141.

[216]Reference 192, p. 1092.

[217]The term *carbanion* was proposed by Wallis, E. S.; Adams, F. H. *J. Am. Chem. Soc.* **1933**, *55*, 3838.

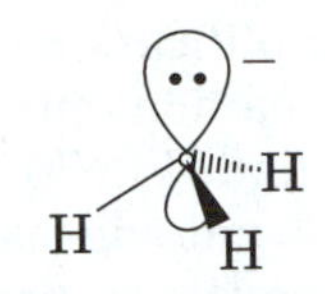

Figure 5.54 sp^3-Hybridized model for methyl anion.

brid (Figure 5.54).[220] Thus, a trigonal pyramid structure, similar to the structure of ammonia, is predicted. This expectation is in accordance with experimental evidence indicating that the methyl anion in the gas phase is tetrahedral (counting the nonbonded pair of electrons as a substituent).[221] NMR studies of alkyl Grignard reagents in solution also show evidence of a tetrahedral carbanion that can undergo inversion of the lone pair similar to ammonia.[222,223] By VSEPR theory we expect the nonbonded electron pair to repel the remaining bonded electrons more than the bonded electrons repel each other, so the bond angles involving the carbon atom and any two of its three substituents may be less than 109.5°. Because there is not a preference for a planar geometry, we would expect bridgehead carbanions to form without difficulty, and except for steric inhibition, that is found to be true.[224]

By similar arguments, we would predict vinyl carbanions to exhibit sp^2 hybridization, and acetylenic carbanions to show *sp* hybridization. Based on the idea that the greater the *s* character in an orbital, the more easily it can accommodate a negative charge, we would expect carbanions in which the negative charge is associated with a pair of nonbonded electrons in an *sp*-hybrid orbital to be more easily formed than those in which the electrons are in sp^2-hybrid or sp^3-hybrid orbitals.[225] Similarly, electron withdrawal by induction can stabilize a carbanion, as can conjugation/resonance interactions.

---

[218]In addition to carbanions in which all electrons are spin-paired, there are also radical carbanions, which have both an unpaired electron and a negative charge. Radical anions can be made by one-electron reduction with metals such as potassium or by ionizing radiation. For an example, see Stevenson, G. R.; Burton, R. D.; Reiter, R. C. *J. Am. Chem. Soc.* **1992**, *114*, 4514. Many radical anions are difficult to study because their large, negative electron affinities (on the order of −2.3 eV) make them very unstable, and they are prone to fragmentation to an anion and a radical. Such processes are especially probable if a portion of the molecule can be detached as a relatively stable anion, such as a halide ion. For a discussion and leading references, see Maslak, P.; Narvaez, J. N.; Kula, J.; Malinski, D. S. *J. Org. Chem.* **1990**, *55*, 4550. However, the ESR spectrum of the *trans*-3-hexene radical *anion* in *trans*-3-hexene/*n*-hexane-d_{14} mixed crystals at 4.2 K has been studied. The results suggested pyramidalization of the olefinic carbon atoms, with half of the unpaired electron density delocalized onto the carbon atoms α to the olefinic carbon atoms: Muto, H.; Nunome, K.; Matsuura, K. *J. Am. Chem. Soc.* **1991**, *113*, 1840.

[219]The application of molecular orbital theory to carbanions has been discussed by Nobes, R. H.; Poppinger, D.; Li, W.-K.; Radom, L. in *Comprehensive Carbanion Chemistry. Part C. Ground and Excited State Reactivity*; Buncel, E.; Durst, T., eds.; Elsevier: Amsterdam, 1987; pp. 1–92.

[220]Often carbanions are shown without explicit depiction of the nonbonded pair of electrons. Thus, the methyl anion is also indicated by H_3C^-.

[221]Ellison, G. B.; Engelking, P. C.; Lineberger, W. C. *J. Am. Chem. Soc.* **1978**, *100*, 2556.

[222]Whitesides, G. M.; Witanowski, M.; Roberts, J. D. *J. Am. Chem. Soc.* **1965**, *87*, 2854.

[223]In general, we expect inversion to be facilitated when conditions allow a relatively free carbanion and to be inhibited when conditions cause the counterion to be held closely to the carbanion.

[224]*Cf.* Applequist, D. E.; Roberts, J. D. *Chem. Rev.* **1954**, *54*, 1065.

[225]Experimental results are consistent with this view. For details, see Cram, D. J. *Fundamentals of Carbanion Chemistry*; Academic Press: New York, 1965; pp. 49 *ff*.

A negatively charged carbon atom is only a convenient simplification of carbanion structure. In some cases the species we use as carbanions may have much less than a full negative charge. For example, ^{13}C and ^{7}Li NMR studies of *n*-butyllithium and *t*-butyllithium in hydrocarbon solvent suggest a polar covalent C-Li bond with only a small negative charge on the carbon atom.[226,227] Moreover, the species present appears to be an associated, not a monomeric species. *n*-Butyllithium is reported to be hexameric in hydrocarbon solutions but to exist as an equilibrium mixture of tetramers and dimers in THF solution.[228,229]

Carbanions in solution can be considered as resonance hybrids of two contributing structures, one purely covalent and one purely ionic (equation 5.41). The degree of ionic character depends on the nature of the metal, the medium, and the substituents on the carbanionic carbon atom. For simple alkyls (those in which resonance stabilization of the negative charge is not expected), the nature of the metal is especially important. The percent ionic character (alternatively, the percent contribution of the ionic resonance structure to the hybrid) increases with increasing difference in electronegativity of the two atoms.[230,231]

$$C^{\delta-}M^{\delta+} = C{-}M \leftrightarrow C\!:^{-} + M^{+} \tag{5.41}$$

The covalent character present in many carbanion carbon-metal bonds means that we must use caution in discussing the properties of carbanions based on reactions of organometallics. One way to study the structures of

---

[226] McKeever, L. D.; Waack, R. *J. Chem. Soc. D, Chem. Commun.* **1969**, 750.

[227] The investigation of carbanions by nuclear magnetic resonance has been reviewed by O'Brien, D. H. in *Comprehensive Carbanion Chemistry. Part A. Structure and Reactivity*; Buncel, E.; Durst, T., eds.; Elsevier: Amsterdam, 1980; pp. 271–322.

[228] For leading references, see Nichols, M. A.; Williard, P. G. *J. Am. Chem. Soc.* **1993**, *115*, 1568.

[229] In addition to nuclear magnetic resonance, carbanions may also be investigated by UV-vis spectrophotometry (Buncel, E.; Menon, B. in *Comprehensive Carbanion Chemistry. Part A. Structure and Reactivity*; Buncel, E.; Durst, T., eds.; Elsevier: Amsterdam, 1980; pp. 97–124) and by infrared and Raman spectroscopy (Corset, J. in *Comprehensive Carbanion Chemistry. Part A. Structure and Reactivity*; Buncel, E.; Durst, T., eds.; Elsevier: Amsterdam, 1980; pp. 125–195).

[230] The fraction of ionic character can be estimated from an empirical relationship. Pauling suggested the equation

$$\text{Fraction ionic character} = 1-\exp(-0.25 \times (\chi_A-\chi_B)^2)$$

where $\chi_A-\chi_B$ is the electronegativity difference between the two atoms involved. Pauling, L. *The Nature of the Chemical Bond*, 3rd Ed.; Cornell University Press: Ithaca, New York, 1960; p. 98. An alternative equation

$$\text{Fraction ionic character} = 0.16\,(\chi_A-\chi_B) + 0.035\,(\chi_A-\chi_B)^2$$

was suggested by Hannay, N. B.; Smyth, C. P. *J. Am. Chem. Soc.* **1946**, *68*, 171. Both equations predict comparable fractional ionic character for most atom pairs, although the latter equation correlates better with the experimental result that the H-F bond is 43% ionic and may serve better for atoms with larger differences in electronegativity.

[231] For a tabulation of the partial ionic character of carbon-metal bonds, see Haiduc, I.; Zuckerman, J. J. *Basic Organometallic Chemistry*; Walter de Gruyter: Berlin, 1985; pp. 9 *ff.*

carbanions is to determine whether chiral carbanions undergo racemization. Studies of noncyclic carbanions indicate that the retention of configuration at a chiral carbanionic center depends on solvent and temperature,[232,233] with solvents such as ether decreasing the covalent character of the carbon-metal interaction, thus facilitating epimerization at the chiral center.[234]

Racemization at a chiral center can be retarded if the rate of inversion can be slowed. Although cyclopropyl radicals racemize,[235] cyclopropyl carbanions retain their configuration.[236] For example, lithiation of the optically active bromocyclopropane derivative **44** and subsequent addition of CO_2 was reported to produce the cyclopropanecarboxylic acid **46** with 100% retention of configuration, which suggests that the three-membered ring inhibits the inversion of the carbanion **45** (Figure 5.55).[234,237]

In contrast, however, substituents that accept electron density by resonance can stabilize carbanionic centers and lessen the interaction with metal ions, leading to racemization. For example, the 1-lithio derivative of (−)-(*R*)-1-cyano-2,2-diphenylcyclopropane is alkylated with methyl iodide to yield *racemic*-1-methyl-1-cyano-2,2-diphenylcyclopropane, indicating racemization at the carbanionic center.[238] This result could be ascribed either to a planar carbanion with appreciable $C{=}N{=}N^-$ character or to a rapidly inverting tetrahedral carbanion. The X-ray crystal structure of 1-cyano-2,2-dimethylcyclopropyllithium indicated a tetrahedral carbanion carbon atom, supporting the latter explanation.[239,240]

Figure 5.55 Retention of configuration by a chiral cyclopropyl carbanion. (Adapted from reference 234.)

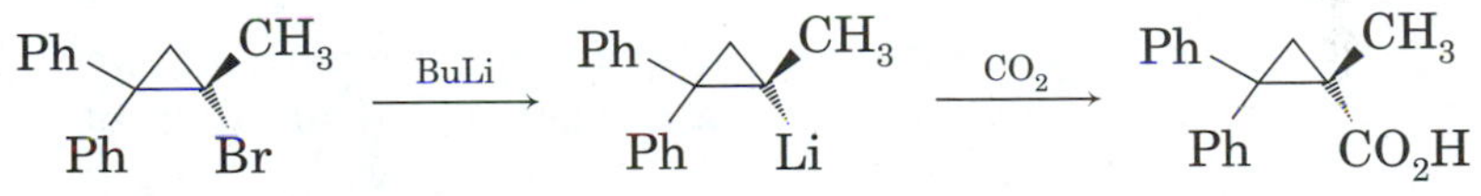

Ph, Ph, CH_3, Br (**44**) $\xrightarrow{\text{BuLi}}$ Ph, Ph, CH_3, Li (**45**) $\xrightarrow{CO_2}$ Ph, Ph, CH_3, CO_2H (**46**)

[232]Letsinger, R. L. *J. Am. Chem. Soc.* **1950**, *72*, 4842.

[233]Curtin, D. Y.; Koehl, Jr., W. J. *J. Am. Chem. Soc.* **1962**, *84*, 1967.

[234]Walborsky, H. M.; Impastato, F. J.; Young, A. E. *J. Am. Chem. Soc.* **1964**, *86*, 3283.

[235]Applequist, D. E.; Peterson, A. H. *J. Am. Chem. Soc.* **1960**, *82*, 2372.

[236]This statement is a generalization. Whether one observes racemization with either radicals or carbanions depends on the reaction conditions and the rates of inversion and of bimolecular reaction.

[237]Applequist, D. E.; Peterson, A. H. *J. Am. Chem. Soc.* **1961**, *83*, 862.

[238]Walborsky, H. M.; Hornyak, F. M. *J. Am. Chem. Soc.* **1955**, *77*, 6026.

[239]Boche, G.; Harms, K.; Marsch, M. *J. Am. Chem. Soc.* **1988**, *110*, 6925.

[240]See also the theoretical calculations reported by Kaneti, J.; Schleyer, P. v. R.; Clark, T.; Kos, A. J.; Spitznagel, G. W.; Andrade, J. G.; Moffat, J. B. *J. Am. Chem. Soc.* **1986**, *108*, 1481.

Generation of Carbanions

A common procedure for the synthesis of organometallic compounds is the reduction of a carbon-halogen bond with a metal (M), as illustrated in equation 5.42.[241] This simple equation ignores the role of solvent molecules and aggregated species in some of these reactions.

$$R_3C{-}X + 2\,M \rightleftharpoons R_3C{:}^- + M^+ + MX \qquad \textbf{(5.42)}$$

Reagents such as *n*-butyllithium, methyllithium, and phenyllithium manufactured by this process are commercially available.[242] Carbanions can also be formed by an acid-base reaction involving heterolytic dissociation of a carbon-hydrogen bond, as illustrated by the hypothetical example in equation 5.43:

$$R_3C{-}W + \text{Base}{:}^- \rightleftharpoons R_3C{:}^- + \text{Base}{-}W \qquad \textbf{(5.43)}$$

Because carbon-hydrogen bonds exhibit very low acidity (see Chapter 7), very strong bases are required for such reactions. However, C—H bonds adjacent to substituents such as carbonyl or cyano groups are stabilized by resonance and induction and are more acidic. Nitrogen bases have been used effectively in these reactions to minimize the nucleophilic addition that can compete with proton removal when an organometallic compound such as *n*-butyllithium is used as the base. For example, methyl ketones react with lithium diisopropylamide (LDA) to form the enolate ion (equation 5.44),[243,244] and even more sterically hindered amides have been used.[245]

$$R{-}C({=}O){-}CH_3 \xrightarrow[\text{pentane-hexane-ether}]{\text{LDA, -60°}} R{-}C({=}O){-}CH_2^- \longleftrightarrow R{-}C(O^-){=}CH_2 \qquad \textbf{(5.44)}$$

In some cases metals will react directly with alkenes (equation 5.45),[246] and alkenes sometimes form carbanions by addition of nucleophiles (equation 5.46).[247]

---

[241]For an introduction to the formation and reaction of carbanions, see Bates, R. B.; Ogle, C. A. *Carbanion Chemistry*; Springer-Verlag: Berlin, 1983; reference 225; Stowell, J. C. *Carbanions in Organic Synthesis*; John Wiley & Sons: New York, 1979; Ayres, D. C. *Carbanions in Synthesis*; Oldbourne Book Co., Ltd.: London, 1966; Terrier, F. *Chem. Rev.* **1982**, *82*, 77.

[242]For a discussion of experimental techniques in carbanion chemistry, see Durst, T. in *Comprehensive Carbanion Chemistry. Part B. Selectivity in Carbon-Carbon Bond Forming Reaction*; Buncel, E.; Durst, T., eds.; Elsevier: Amsterdam, 1984; pp. 239–291.

[243]House, H. O.; Phillips, W. V.; Sayer, T. S. B.; Yau, C.-C. *J. Org. Chem.* **1978**, *43*, 700.

[244]An NMR spectrum of the lithium enolate of *t*-butyl acetate indicated that the carbon-carbon double-bonded species was dominant. Rathke, M. W.; Sullivan, D. F. *J. Am. Chem. Soc.* **1973**, *95*, 3050.

[245]Olofson, R. A.; Dougherty, C. M. *J. Am. Chem. Soc.* **1973**, *95*, 582.

[246]Yasuda, H.; Ohnuma, Y.; Yamauchi, M.; Tani, H.; Nakamura, A. *Bull. Chem. Soc. Jpn.* **1979**, *52*, 2036.

[247]Fraenkel, G.; Estes, D.; Geckle, M. J. *J. Organometal. Chem.* **1980**, *185*, 147.

M^+

M

THF, Et_3N

(5.45)

(M = Li, Na, K, Rb, Cs)

R

(5.46)

Ph R-Li Ph + Li^+

Addition of nucleophiles to aromatic compounds with many nitro or cyano groups leads to the formation of resonance-stabilized carbanions known as Jackson-Meisenheimer complexes. For example, addition of sulfite ion to 1,3,5-trinitrobenzene leads to the complex **47**.[241]

H $S\bar{O}_3$

O_2N NO_2

NO_2

47

Stability of Carbanions

We frequently measure anion stability in terms of the acidity of the corresponding protonated species. Thus, the acidity of carbon acids, in which the acidic proton is removed from a carbon atom, provides one measure of carbanion stability. There has been a rapid change in our models of carbanion stability in recent years. In 1979 a leading researcher summarized what was at that time the prevailing model of carbanion stabilities with a statement to the effect that alkyl substitution at the carbanionic site results in an intensification of the carbanionic character because of the electron-donating character of the alkyl groups. As we will see in more detail in Chapter 7, this view is no longer supported by theoretical or experimental evidence. Figure 5.56 shows the HOMO of ethyl anion as calculated with Extended Hückel theory. It shows evidence of mixing of the carbanion orbital with much *p* character with the orbitals of the methyl substituent, resulting in delocalization of negative charge and stabilization of the anionic

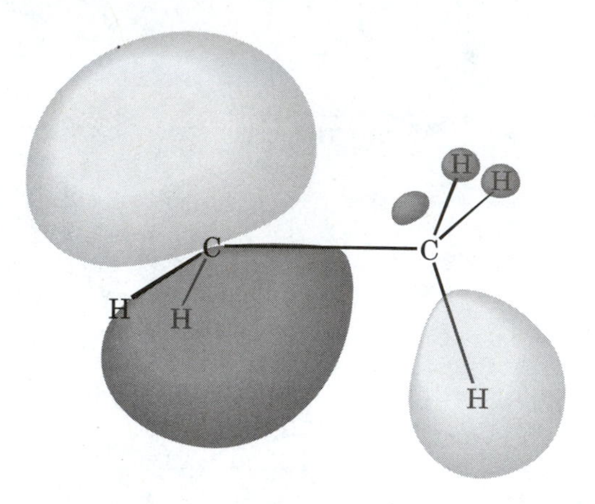

Figure 5.56
HOMO of ethyl anion.

species.[248] This result makes untenable an explanation for decreasing acidity with increasing alkyl substitution that is based on electron-donating alkyl groups.

Major new insights into the electronic effects of substituents have resulted from measurements conducted in the gas phase, where the effects of solvent are eliminated.[249] DePuy and co-workers determined that the order of acidity was found to be ethane < propane (2° hydrogen atoms) < methane < isobutane (3° hydrogen atom) in the gas phase.[250] This rather surprising result that methane is more acidic than ethane has some support in theoretical calculations.[251] Therefore the effect of alkyl substitution on a carbanion in solution may be better described as a result of the interference of solvation by bulky substituents. As le Noble put it in 1978[252]

> It becomes clear that much of the 'truth' about carbanions to be found in current textbooks, however logical and convenient from a pedagogical point of view, is incorrect.

As we discuss reactions involving carbanions in later chapters, especially acid-base reactions, substitution reactions, and elimination reactions, we will see how new views of carbanion structure have emerged in recent years.

---

[248]Theoretical calculations are in accord with stabilization of both cationic and anionic centers by methyl groups, with the affected C-H bond lengthened by the interaction. See the discussion by Forsyth, D. A.; Yang, J.-R. *J. Am. Chem. Soc.* **1986**, *108*, 2157 and references therein.

[249]Three main types of experimental techniques have been used: high pressure mass spectrometry (HPMS), flowing afterglow (FA or FA-SIFT) studies, and pulsed ion cyclotron resonance (ICR) spectrometry. See Chapter 7 for further discussion and leading references.

[250]DePuy, C. H.; Bierbaum, V. M.; Damrauer, R. *J. Am. Chem. Soc.* **1984**, *106*, 4051; DePuy, C. H.; Gronert, S.; Barlow, S. E.; Bierbaum, V. M.; Damrauer, R. *J. Am. Chem. Soc.* **1989**, *111*, 1968.

[251]Kollmar, H. *J. Am. Chem. Soc.* **1978**, *100*, 2660.

[252]le Noble, W. J. in *Reactive Intermediates*, Vol. 1; Jones, Jr., M.; Moss, R. A., eds.; John Wiley & Sons, Inc.: New York, 1978; pp. 27–67.

Reactions of Carbanions

Because of the localization of electron density and negative charge, carbanions are both bases and nucleophiles. As a base, a carbanion can abstract a proton from any substance with a pK_a smaller than that of the protonated carbanion. The use of isotopically labeled proton donors affords a useful synthesis of labeled compounds. For example, abstraction of a proton from the methyl group of *exo*-3-acetyl-*endo*-tricyclo[3.2.1.0^{2,4}]octane (**48**) gave an enolate ion that could abstract a deuterium ion from solvent to produce the monodeuterated compound. Repeated exchange of the methyl protons led to a nearly quantitative yield of trideutero product **49**.[253]

$$\mathbf{48} \xrightarrow[\substack{CH_3OD \\ D_2O}]{NaOCH_3} \mathbf{49} \qquad (5.47)$$

48 **49**

As a nucleophile, a carbanion can react readily with a carbon atom bearing a good leaving group, which is a useful method for forming new carbon-carbon bonds. The aldol reaction and the Claisen condensation are familiar examples of carbanions (as enolates) undergoing nucleophilic addition to carbon-oxygen double bonds. Carbanions may also act as nucleophiles in S_N2 reactions. For example, the synthesis of 2,3-diphenylpropionic acid (**52**, equation 5.48) takes advantage of the reaction of a carbanion with benzyl chloride.[254]

$$\underset{\mathbf{50}}{C_6H_5\text{-}CH_2-CO_2H} \xrightarrow[NH_3\,(liq)]{2\ NaNH_2} \underset{\mathbf{51}}{C_6H_5\text{-}\underset{\displaystyle Na}{\underset{|}{CH}}-CO_2Na}$$

$$\xrightarrow[EtOEt]{C_6H_5CH_2Cl} \xrightarrow{H^+} \underset{\mathbf{52}}{C_6H_5\text{-}\underset{\displaystyle CH_2C_6H_5}{\underset{|}{CH}}-CO_2H} \qquad (5.48)$$

---

[253]Creary, X. *J. Org. Chem.* **1976**, *41*, 3740.

[254]Hauser, C. R.; Dunnavant, W. R. *Org. Synth. Collect. Vol. V* **1973**, 526.

In the absence of substrates for nucleophilic or basic attack, carbanions can also undergo intramolecular reaction, including rearrangement, although less readily than do carbocations. One example is epimerization to a more stable structure. The reaction shown in equation 5.49 was used in the synthesis of (±)-*trans*-chrysanthemic acid by Welch and Valdes.[255]

O H (CH$_3$O)$_2$HC H **53** —*t*-BuOK / *t*-BuOK→ O H (CH$_3$O)$_2$HC H **54** (5.49)

The 5-hexenyl carbanions, where M is sodium, potassium, rubidium, and cesium undergo rearrangement from a primary carbanion to the isomeric allylic carbanions in THF at −50°. However, 5-hexenyllithium rearranges to a cyclopentyl structure.[256]

M, M = Na, K, Rb, Cs —THF / −50° 1 hr→ M$^+$ (5.50)

Li —THF / −50° 1 hr→ Li (5.51)

Another example is the rearrangement of the carbanion formed by deprotonation of the spirodiene **55** by *n*-BuLi in THF. The product is the 2-phenylethyllithium species **56**.[257]

55 —*n*-BuLi / THF, hexane 25°→ CH$_2$ $\bar{C}H_2$ Li$^+$ **56** (5.52)

Carbanions can take part in elimination reactions, as will be discussed in the context of the E1cb reaction (Chapter 8). The elimination of carbanions

[255]Welch, S. C.; Valdes, T. A. *J. Org. Chem.* **1977**, *42*, 2108.

[256]Punzalan, E. R.; Bailey, W. F. *Abstracts of the 207th National Meeting of the American Chemical Society*; San Diego, CA, March 13–17, 1994; Abstract ORGN 384.

[257]Staley, S. W.; Cramer, G. M.; Kingsley, W. G. *J. Am. Chem. Soc.* **1973**, *95*, 5052.

derived from THF and its derivatives (equation 5.53) were studied by Bates et al.[258]

$$\text{THF} \xrightarrow{\text{BuLi}} \text{THF}^{:-} \longrightarrow CH_2{=}CH_2 + CH_2{=}CH{-}O^- \longleftrightarrow \overset{-}{C}H_2{-}CH{=}O \qquad (5.53)$$

Carbanions are also susceptible to oxidation.[259] For example, ketone enolates can undergo oxidative coupling in presence of $CuCl_2$.[260]

$$c\text{-}C_3H_5COCH_3 \xrightarrow[\text{THF, } -78°]{\text{LDA}} c\text{-}C_3H_5COCH_2^- \xrightarrow[\text{DMF}]{CuCl_2} c\text{-}C_3H_5COCH_2CH_2COC_3H_5\text{-}c \qquad (5.54)$$

Oxidation of carbanions by molecular oxygen is also an important reaction. Equation 5.55 shows the oxidation of the anion of triphenylmethane by molecular oxygen in a solution composed of 80% DMF and 20% *t*-butyl alcohol. Addition of water to the reaction mixture allowed isolation of triphenylmethyl hydroperoxide in high yields (e.g., 87%). When the reaction was carried out in 80% DMSO–20% *t*-butyl alcohol mixtures, the product isolated was the triphenylmethanol, apparently due to reaction of the hydroperoxide with the solvent.[261]

$$(C_6H_5)_3C{:}^- + O_2 \longrightarrow (C_6H_5)_3COO{:}^- \qquad (5.55)$$

Conclusion

In this chapter we have developed models of reactive intermediates that are more complex than the models that organic chemists often use. Just as our representation of cyclohexane in a chair conformation ignores rapid conformational changes, a picture of a cyclopentyl or isopropyl carbocation as a static carbenium ion is not consistent with the rapid internal bonding changes that can occur, even at low temperature. Moreover, there is experimental evidence that the 2-norbornyl cation is a nonclassical, hypercoordinate carbocation and that other ions exhibit varying degrees of bridging or nonclassical behavior.[188] The simple picture of a carbanion ignores the

---

[258] Bates, R. B.; Kroposki, L. M.; Potter, D. E. *J. Org. Chem.* **1972**, *37*, 560.

[259] The electrochemistry of carbanions has been discussed by Fox, M. A. in *Comprehensive Carbanion Chemistry. Part C. Ground and Excited State Reactivity*; Buncel, E.; Durst, T., eds.; Elsevier: Amsterdam, 1987; pp. 93–174.

[260] Ito, Y.; Konoike, T.; Harada, T.; Saegusa, T. *J. Am. Chem. Soc.* **1977**, *99*, 1487.

[261] Russell, G. A.; Bemis, A. G. *J. Am. Chem. Soc.* **1966**, *88*, 5491.

critical role of metal atoms, solvent molecules, and the model of a monomeric species ignores what may be very important effects of oligomerization.

In later chapters we will consider in more detail the energies and structures of reactants, transition states, intermediates, and products. It will be important to remember that the environment and lifetime of a reactive intermediate in a spectroscopic study, such as a carbocation in a frozen matrix or in a solution of magic acid at low temperature, are quite different from those of the species presumed to be involved in chemical reactions under much milder conditions or at room temperature. In all cases, it will be important to recognize that the structures and diagrams that we draw are at best simplified models of complex molecular systems in which certain components of a larger picture are emphasized and others ignored. Because the reaction coordinate diagram model will be drawn from the properties of unstable intermediates and transition states, as well as from the properties of ground state molecules, it is even further removed from the kinds of direct and indirect knowledge that we have about pure substances. Thus, we will be studying models based on models derived from models, so any conclusions that we draw must carefully consider the limitations inherent in our analysis.

Problems

1. Use VSEPR/variable hybridization theory to rationalize the fact that cyclopropyl radicals are nonplanar at the radical center.
2. Heats of formation of alkyl radicals in a given series follow the trend of alkanes, that is, each additional $—CH_2—$ group adds about -4.95 kcal/mol to the heat of formation. However, for the linear carbocations, the heats of formation do not follow the same trend. The reported values for the primary carbocations are: ethyl (219), propyl (208), butyl (218), pentyl (171), and hexyl (170 kcal/mol). Explain this difference between radicals and carbocations.
3. The two diradicals **57** and **58** show rather different properties. Structure **57** shows predominantly singlet character at low temperature, although triplet character becomes evident at higher temperature. On the other hand, **58** displays predominantly triplet character under all conditions. Rationalize these results on the basis of the structures of the two diradicals.

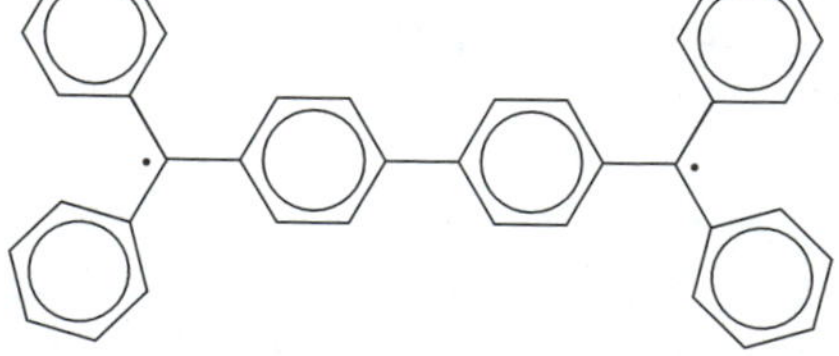

57 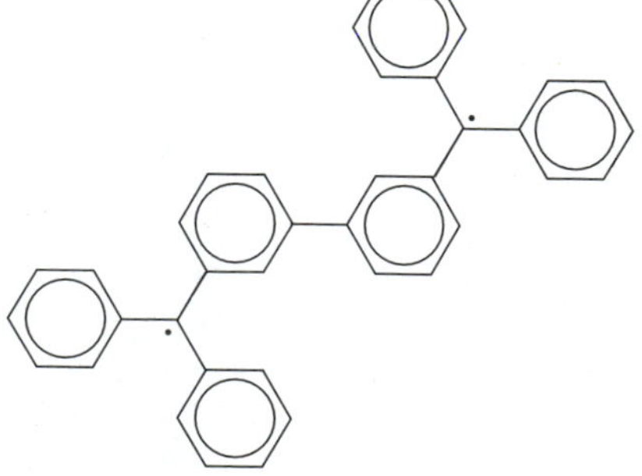 **58**

4. The EPR spectrum of the benzyl radical in solution shows the following values for a for the α, ortho, meta, and para protons: 16.4 G, 5.17 G, 1.77 G, and 6.19 G, respectively. Interpret these results in terms of the relative unpaired electron density at each of these four positions. How do the results compare with the predictions of HMO theory or simple resonance theory from Chapter 4?
5. Predict the relative yield of products expected when 6-bromo-1-hexene reacts with Bu_3SnH (AIBN initiator) in benzene at 100°.
6. In the ^{13}C NMR spectrum of the isopropyl cation, the coupling constant between C2 and the hydrogen bonded to it is 169 Hz. Based on the relationship between ^{13}C NMR coupling constants and hybridization (see Chapter 1), what is the apparent hybridization of C2 in this ion?
7. When formic acid is dissolved in superacid media, the proton NMR spectrum indicates the presence of two isomeric species in a ratio of 2 : 1. Suggest a structure for each of these two species.
8. Although the ESCA spectrum of the *t*-butyl carbocation shows C 1*s* levels for both the carbenium carbon atom and the three methyl carbon atoms, the ESCA spectrum of the trityl cation (Ph_3C^+) suggests that all the C 1*s* electrons have the same binding energy. Explain this result.
9. Rationalize the observation that the NMR spectrum of the 2,6-dimethyl-2-heptyl cation at $-100°$ shows only one peak for the four methyl groups at the ends of the chain.
10. When the *t*-butyl carbocation labeled with ^{13}C at the 3° position is heated in the superacid HSO_3F:SbF_5:SO_2ClF at 70° for 20 hours, complete scrambling of the label is observed. The activation energy for the process appears to be 30 kcal/mol or more. Based on rearrangement processes of carbocations, suggest a mechanism by which the scrambling could occur.
11. The 1-lithio derivative of (−)-(*R*)-1-cyano-2,2-diphenylcyclopropane is alkylated with methyl iodide to yield *racemic*-1-methyl-1-cyano-2,2-diphenylcyclopropane.[262] However, the 1-lithio derivative of the isomeric (+)-(*S*)-2,2-diphenylcyclopropyl isocyanide reacts with methyl iodide to yield (+)-(*S*)-1-methyl-2,2-diphenylcyclopropyl isocyanide. Explain the racemization in the former compound and the retention of configuration in the latter.
12. In a study of 2,2-diphenylpropyllithium, a product isolated after addition of CO_2 and water was *o*-(1-phenyl-1-methylethyl)benzoic acid. Propose a mechanism to account for its formation.
13. Reaction of the diene **59** with *n*-butyllithium in THF gives the anion **60**. Why should the reaction occur with a proton allylic to the double bond external to the ring and not the endocyclic double bond?

n-BuLi

THF

59 **60**

[262]Walborsky, H. M.; Hornyak, F. M. *J. Am. Chem. Soc.* **1955**, 77, 6026.

14. Propose a mechanism for the following acid-catalyzed rearrangement.

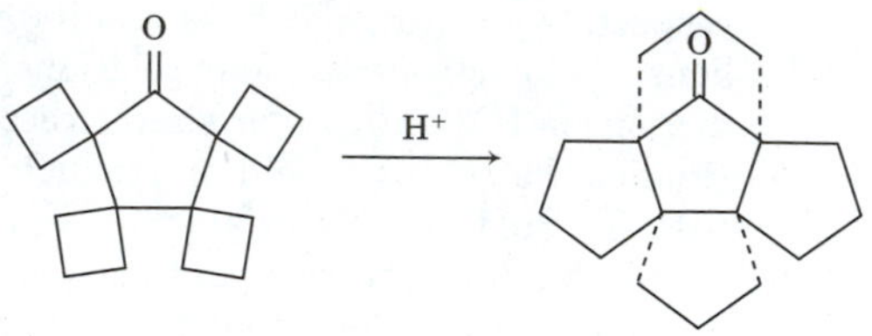

O

H^+

O

15. Propose a mechanism for the following reaction.

CCl_3

Bz_2O_2

CCl_4

Cl

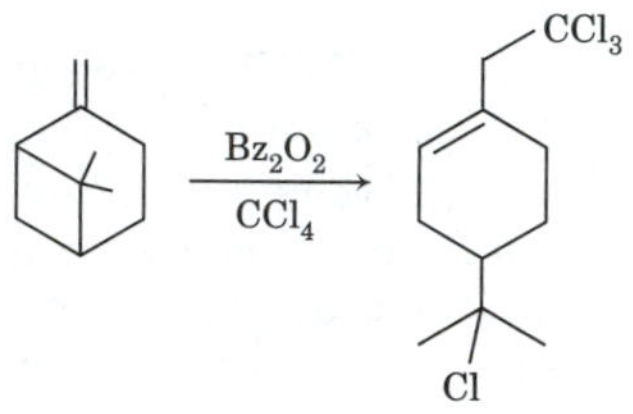

16. In contrast to the reactivity of most simple carbenes, dimethoxycarbene does not readily undergo addition to alkenes unless the alkene is substituted with electron-withdrawing substituents. Propose an explanation for this behavior.
17. Propose a detailed mechanism for the reaction of a radical with α-phenyl *N*-*t*-butyl nitrone, which is shown below, and explain how such a reaction could be useful in studying the structures and properties of free radicals.

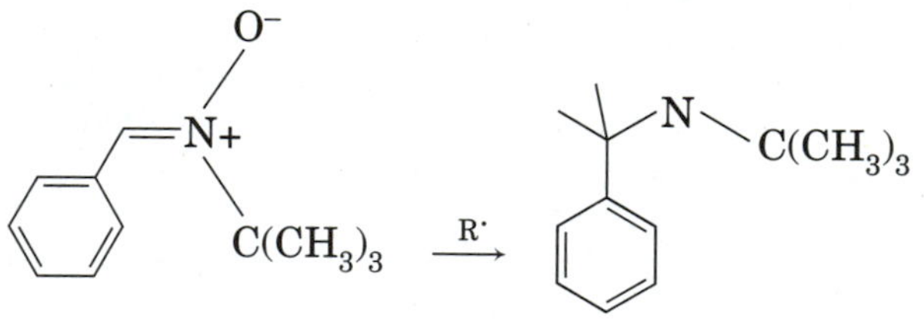

O^-

$=N+$

$C(CH_3)_3$

$R^{\cdot}$

N

$C(CH_3)_3$

18. When cyclobutanone is subjected to the conditions for the Bamford-Stevens reaction, the major products isolated are cyclobutene and methylenecyclopropane. Propose a mechanism that accounts for their formation.
19. Propose a mechanism to account for the formation of products formed by decomposition of diazocyclodecane in the following reaction.

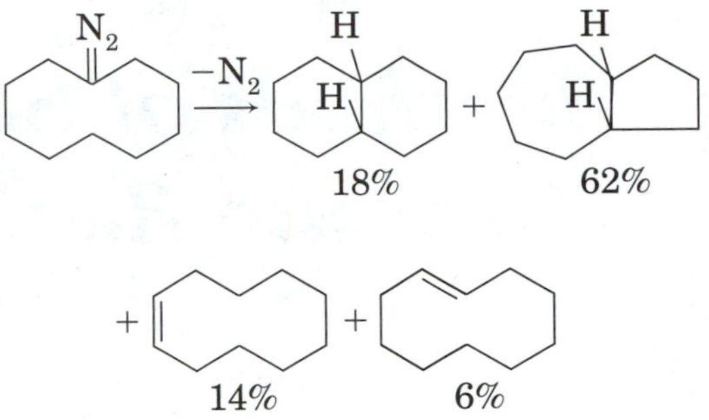

N_2 → $-N_2$ → H, H (18%) + H, H (62%) + (14%) + (6%)

20. Deuterium-labeled *cis*-1,5-hexadiyn-3-ene (**61**) was found to undergo thermal rearrangement to **61′** (equation 5.56), suggesting that the reaction involves cyclization to an intermediate or transition structure having C_2 or higher symmetry. Based on this result and other data, the investigators proposed that 1,4-dehydrobenzene (benzene-1,4-diyl), **62**, is an intermediate in the reaction.

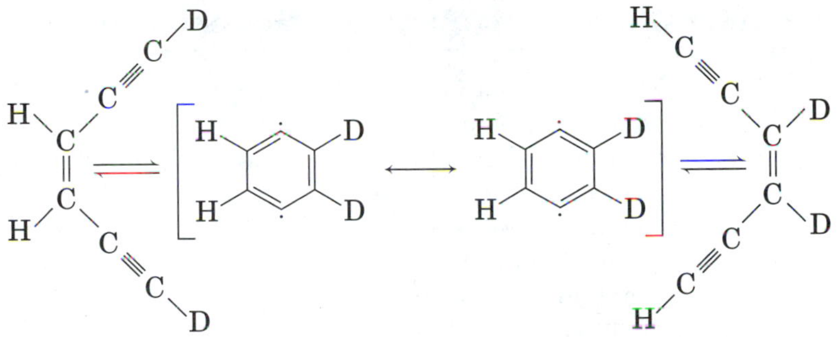

$$\textbf{61} \rightleftharpoons [\textbf{62}] \rightleftharpoons \textbf{61'} \qquad (5.56)$$

a. Propose other possible C_6H_4 structures that have C_2 or higher symmetry and which might also be considered as possible intermediates in this reaction. Consider particularly candidate structures having (i) both carbocation and carbanion centers, (ii) two carbene centers, (iii) a highly strained bicyclic triene structure, (iv) two allenic units.
b. When the reaction was carried out in hydrocarbon solvent, benzene was formed as a byproduct of the reaction. When the reaction was carried out in methanol, benzene and some benzyl alcohol were observed, but anisole was not detected. Reaction in toluene produced diphenylmethane, and reaction in CCl_4 produced 1,4-dichlorobenzene. (In all cases, deuterium-labeled product was observed from the deuterium-labeled reactant.) How do these results reinforce the conclusion that the reactive intermediate is **62** and not one of the alternative structures?

21. McBride has written about the hexaphenylethane story that "This glaring, if relatively harmless, error dramatizes the importance of reviewing the basis in fact on which the classic generalizations of chemistry, however reasonable, are founded."[28] Do you agree with this view? Could there be other errors that are not so harmless that may persist only because no one has repeated crucial early experiments with more modern techniques?

CHAPTER 6

Methods of Studying Organic Reactions

6.1 Molecular Change and Reaction Mechanisms

The sheer volume of material that chemists must assimilate requires that we categorize information, so the concept of the *functional group* is crucial to organic chemists. For example, we expect that a double bond will undergo characteristic reactions—such as addition of bromine—whether it is in ethene or cholesterol.[1] This way of thinking is reinforced by introductory organic chemistry textbooks that are organized as studies of first one functional group and then another. Similarly, we categorize the transformations of functional groups by naming types of reactions (addition, elimination, substitution, . . .) that have similar features. We also must categorize the pathways (S_N1, S_N2, . . .) by which each functional group undergoes each type of reaction. In other words, we want to be able to describe the *mechanism* of each process and to categorize the types of mechanisms observed for particular functional groups and reaction types.

Mechanism is one of those terms used frequently by chemists that sometimes have different meanings to the speaker and hearer. One definition is that it is a step-by-step description of a chemical transformation at the molecular level, giving information about the location of all nuclei and electrons, including those of solvent and other species present, as well as the total energy of the system. Gould called a mechanism a motion picture of the chemical transformation—one that we can stop and analyze frame by

[1]Generalizations about functional groups are useful, but we must recognize that they are only generalizations and that exceptions may be found. For example, addition of bromine to adamantylideneadamantane yields not the dibromide but, rather, the bromonium ion–tribromide salt: Strating, J.; Wieringa, J. H.; Wynberg, H. *J. Chem. Soc. D, Chem. Commun.* **1969**, 907. See also Slebocka-Tilk, H.; Ball, R. G.; Brown, R. S. *J. Am. Chem. Soc.* **1985**, *107*, 4504; Bennet, A. J.; Brown, R. S.; McClung, R. E. D.; Klobukowski, M.; Aarts, G. H. M.; Santarsiero, B. D.; Bellucci, G.; Bianchini, R. *J. Am. Chem. Soc.* **1991**, *113*, 8532.

frame.[2] We should understand, however, that such a motion picture should be viewed as animation or simulation. That is, we cannot see the molecular events; we can only depict what we infer them to be.

We must also remember that the methods we use to study mechanisms are never conclusive but are only indicative. That is, a mechanism can be disproved, but it can never be proved.[3] A mechanism may become *established*, which merely means that only one of several proposed mechanisms is able to predict the results of experiments designed to test possible mechanisms. Scientists are always open to the possibility of new information making their ideas untenable, and they treat mechanisms as more or less thoroughly tested models and not as timeless realities.

6.2 Tools to Determine Reaction Mechanisms

Identification of Reaction Products

The most fundamental step in any investigation of a chemical reaction is the determination of the products of the reaction.[4] This might appear to be a trivial statement, but it is not. Consider the study of the reaction of benzyl chloride (**1**) with hydroxide to give benzyl alcohol (**2**). One way to study the mechanism of the reaction is to determine what effect substituents on the benzene ring have on the rate constant for the reaction. Therefore one might first determine the kinetics of the reaction of benzyl chloride with hydroxide ion, then determine the kinetics of the reactions of hydroxide with a series of substituted benzyl chlorides, including *p*-nitrobenzyl chloride (**3**). However, the product of the reaction of **3** is 4,4′-dinitrostilbene (**4**), not *p*-nitrobenzyl alcohol.[5] Obviously kinetic studies or other mechanistic investigation of all of the reactions could not be interpreted validly in terms of one uniform mechanistic pattern.[6,7]

---

[2]Gould, E. S. *Mechanism and Structure in Organic Chemistry*; Holt, Rinehart and Winston: New York, 1959.

[3]The idea that scientific evidence can never prove a hypothesis but can only refute it is closely associated with the work of Karl Popper. See, for example, Popper, K. R. *Objective Knowledge*; Clarendon Press: Oxford, 1972.

[4]This statement assumes that the reactants are well characterized. Obviously, a mechanistic study is also futile if the identity and purity of the reactants are not known.

[5]For leading references, see Tewfik, R.; Fouad, F. M.; Farrell, P. G. *J. Chem. Soc., Perkin Trans. 2* **1974**, 31.

[6]A similar example was presented in reference 2, p. 131.

[7]However, if the products are identified from each reaction, kinetic studies and other mechanistic investigations might provide useful insights into both the hydrolytic reactions of some substituted benzyl halides and the reactions of other benzyl halides that led to substituted stilbenes.

Cl $\xrightarrow[aq.\ \text{Dioxane, air}]{^{-}OH}$ OH

1 2 **(6.1)**

O_2N Cl $\xrightarrow[aq.\ \text{Dioxane, air}]{^{-}OH}$ NO_2 O_2N

3 4 **(6.2)**

Determination of Intermediates

The reactants and products fix only the end points of a reaction mechanism; to be complete, it is then necessary to fill in details of all that is between them. One important concept in mechanistic studies is the determination of the number of steps in the reaction. A reaction that involves only one step (reactants to transition structure to products) is called an ***elementary reaction***. If there is more than one elementary step, then at least one intermediate must be involved in the reaction. As we discussed in Chapter 5, we may consider two hypothetical reaction coordinate diagrams (Figure 6.1) for a reaction in which at least one bond is broken and at least one bond is formed. Figure 6.1(a) represents a one-step reaction that is a ***concerted process***, in which bonds are formed and broken simultaneously, while the presence of an intermediate as in Figure 6.1(b) or Figure 6.1(c) requires the mechanism to be ***step-wise***.[8] This distinction between step-wise and concerted processes requires that there be a minimum in the curve in Figure

Figure 6.1 (a) Elementary reaction; (b) reaction with an unstable intermediate; (c) reaction with a more stable intermediate.

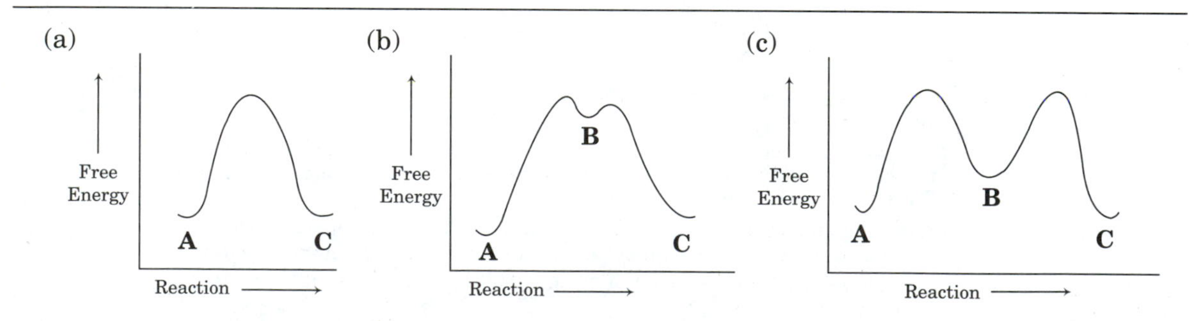

[8]Lowe noted that the term *concerted* must be defined with care: for a reaction in which several bonds are broken and several bonds are formed, bond breaking and bond formation might be synchronous for some processes but not for others. Lowe, J. P. *J. Chem. Educ.* **1974**, *51*, 785. See also the discussion by Bernasconi, C. F. *Acc. Chem. Res.* **1992**, *25*, 9.

6.1(b) that is *at least* as deep as the energy available to the system at the reaction temperature.[9,10] The deeper the dip in the potential energy surface, as in Figure 6.1(c), the more stable is the intermediate and the more likely we are to be able to trap it or perhaps even to isolate it.

Intermediates can be demonstrated by several techniques. If a substance isolated from a reaction mixture is subjected to the reaction conditions and is found to proceed to the same products as the reactants, there is strong evidence that the isolated compound is indeed an intermediate (and not just a byproduct) in the reaction. For example, Winstein and Lucas proposed that 3-acetoxy-2-butanol, 2-acetoxy-3-bromobutane, and 3-bromo-2-butanol were intermediates in the conversion of 2,3-diacetoxybutane to 2,3-dibromobutane by fuming aqueous HBr. Each of the proposed intermediates was isolated from reaction mixtures before the reaction had gone to completion, and subjecting each proposed intermediate to the reaction conditions produced 2,3-dibromobutane. The possibility that 2,3-butanediol might also be an intermediate, even though it was not isolated from reaction mixtures, was eliminated by showing that it did not produce 2,3-dibromobutane under the reaction conditions.[11,12,13]

If the intermediate is not sufficiently stable to be isolated, it may be formed in sufficiently large concentration to be detected spectroscopically. This may mean UV-vis spectroscopy (e.g., stopped-flow kinetics experiments) for relatively stable intermediates or IR spectroscopy in frozen argon

---

[9]As we noted in Chapter 5, the intermediate must be able to survive for at least one molecular vibration before the reaction proceeds (i.e., the energy well must be at least as deep as the 0th vibrational level), otherwise it is simply a transition structure and not an actual intermediate.

[10]We usually assume that the intermediate is a minimum on an *enthalpy* surface, but that need not be the case. Tetramethylene has been reported to be an intermediate without an enthalpy minimum because it occurs in an "entropy dominated free energy minimum": Doubleday, Jr., C.; Camp, R. N.; King, H. F.; McIver, J. W.; Mullally, D.; Page, M. *J. Am. Chem. Soc.* **1984**, *106*, 447; Doubleday, Jr., C. *J. Am. Chem. Soc.* **1993**, *115*, 11968.

[11]Winstein, S.; Lucas, H. J. *J. Am. Chem. Soc.* **1939**, *61*, 1581. The stereochemistry of these interconversions will be considered in Chapter 8.

[12]For further discussion and another example, see Kubiak, G.; Cook, J. M.; Weiss, U. *J. Org. Chem.* **1984**, *49*, 561.

[13]Finding that a substance proposed as an intermediate does produce the expected products when it is subjected to the conditions of a reaction under investigation suggests, but does not confirm, that the substance actually is an intermediate. It may be possible for the product to be formed by two different pathways, one from the intermediate, and a totally different pathway from the reactant. In such a case, the role of a substance as an intermediate can be supported further if it can be shown that the proposed intermediate produces product at the rate expected under the reaction conditions. Conversely, a substance can be eliminated as an intermediate in a reaction if it can be shown that the actual concentration of the substance in the reaction mixture must be less than would be required to produce the product at the observed rate. One way to do this utilizes the technique of isotopic dilution. For a discussion, see Swain, C. G.; Powell, A. L.; Sheppard, W. A.; Morgan, C. R. *J. Am. Chem. Soc.* **1979**, *101*, 3576. A general discussion of the use of isotopes to study organic reaction mechanisms was given by Collins, C. J. *Adv. Phys. Org. Chem.* **1964**, *2*, 1.

in matrix isolation spectroscopy.[14,15] For photochemical reactions we can detect transient spectra of intermediates in the millisecond to microsecond ("conventional" flash spectroscopy) or nanosecond to picosecond[16] (laser flash spectroscopy) time scale. In all cases we must be certain that the spectra observed are indeed indicative of the presence of the proposed intermediate and only the proposed intermediate. Theoretical calculations have been useful in determining whether a proposed intermediate is likely to be sufficiently stable for detection, its spectroscopic properties, and the type of experiment most likely to detect it.[17] In addition, kinetic studies may suggest optimum conditions for spectroscopic detection of an intermediate.[18]

Even if the intermediate cannot be observed by spectroscopy, it might be trapped by added reagents. For example, we discussed in Chapter 5 the use of spin trap reagents to trap transient radicals for analysis by EPR spectrometry. Another example can be seen in the studies that led to the bromonium ion mechanism for the addition of bromine to alkenes, which will be discussed in more detail in Chapter 9. Several mechanisms can be proposed for the reaction of the 2-butenes with bromine in CCl_4 solution to give 2,3-dibromobutane. However, the observation that adding a nucleophile such as methanol to the reaction mixture leads to products incorporating the nucleophile, as shown in Figure 6.2, suggests the possibility that an intermediate such as a carbocation formed by initial electrophilic addition to the alkene may be intermediate in the reaction.[19] Additional evidence is necessary to determine the structure of the intermediate (i.e., bromonium ion or bromine-substituted carbocation), but that is a matter of detail once the existence of an intermediate of some kind is established.[20]

Trapping reactions can also be useful to demonstrate that a compound is *not* likely to be an intermediate in a reaction. This approach requires showing that a proposed intermediate cannot be trapped in a reaction but that it should be trapped if it actually were present. Porter and Zuraw used this approach in studying the rearrangement of allylic hydroperoxides,

---

[14] *Cf.* Gebicki, J.; Krantz, A. *J. Chem. Soc., Perkin Trans. 2* **1984**, 1623; Kesselmayer, M. A.; Sheridan, R. S. *J. Am. Chem. Soc.* **1986**, *108*, 99.

[15] For a review of techniques to measure extremely fast reactions, see Zuman, P.; Patel, R. C. *Techniques in Organic Reaction Kinetics*; John Wiley & Sons: New York, 1984; pp. 247–327; Krüger, H. *Chem. Soc. Rev.* **1982**, *11*, 227. Bell has summarized the use of a variety of experimental methods for studying fast reactions with time scales ranging from 10^{-3} to 10^{-10} sec: Bell, R. P. *The Proton in Chemistry*, 2nd ed.; Cornell University Press: Ithaca, New York, 1973; pp. 111 *ff.*

[16] Simon, J. D.; Peters, K. S. *Acc. Chem. Res.* **1984**, *17*, 277.

[17] For an example, see Hu, C.-H.; Schaefer III, H. F. *J. Phys. Chem.* **1993**, *97*, 10681.

[18] Bernasconi, C. F.; Fassberg, J.; Killion, Jr., R. B.; Rappoport, Z. *J. Am. Chem. Soc.* **1990**, *112*, 3169.

[19] For an analysis of products formed by trapping the intermediate with nucleophiles, see Nagorski, R. W.; Brown, R. S. *J. Am. Chem. Soc.* **1992**, *114*, 7773.

[20] The use of competition methods to detect and characterize intermediates has been reviewed by Huisgen, R. *Angew. Chem., Int. Ed. Engl.* **1970**, *9*, 751.

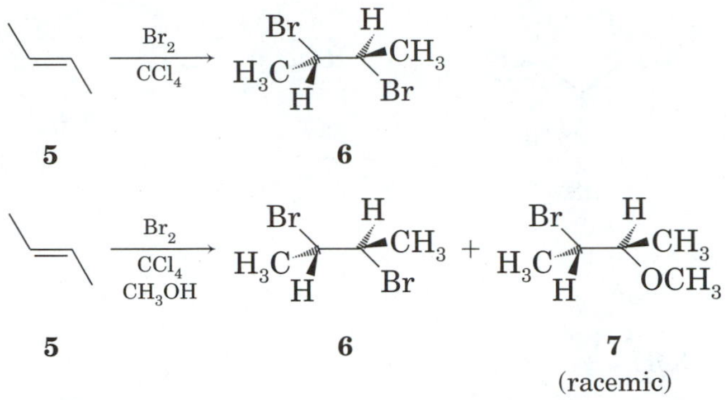

Figure 6.2
Reaction of bromine with *trans*-2-butene in CCl_4 solution (top) and with added CH_3OH (bottom).

5 $\xrightarrow[CCl_4]{Br_2}$ **6**

5 $\xrightarrow[CCl_4,\ CH_3OH]{Br_2}$ **6** + **7** (racemic)

shown in Figure 6.3. The mechanism that had been proposed involved intramolecular reaction of a peroxy radical (**9**) with the double bond to produce intermediate **10**. The proposed intermediate **10** had not been trapped in earlier studies of the allylic hydroperoxide rearrangement, even when the reaction was conducted under 500 lb/in^2 of oxygen.[21] Porter and Zuraw synthesized **10** by an unambiguous pathway (Figure 6.4) and established that it could be trapped with oxygen at atmospheric pressure to yield **12–15**. The results indicated that **10** could not be a "competent intermediate" in the allylic hydroperoxide reaction and that another mechanism must be followed.[22]

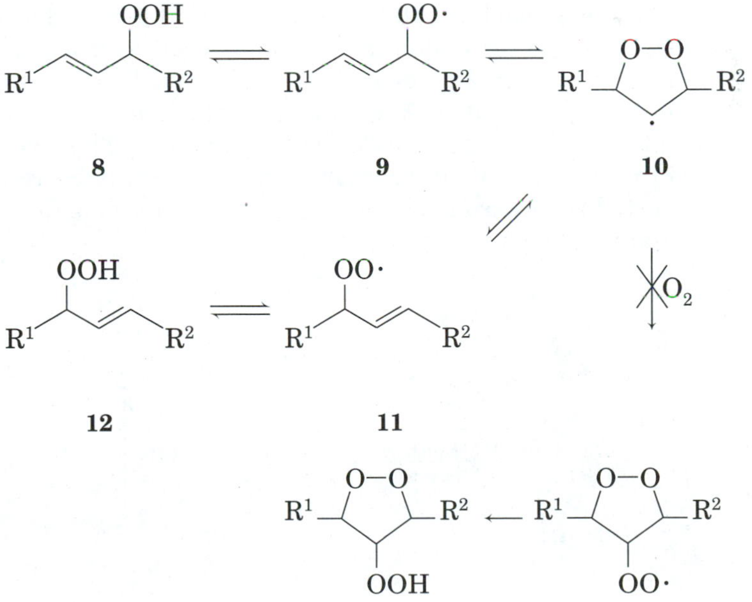

Figure 6.3
Failure to observe **10** in the rearrangement of an allylic hydroperoxide. (Reproduced from reference 22.)

8 ⇌ **9** ⇌ **10** ⇌ **11** ⇌ **12**; **10** $\xrightarrow{O_2}$ (not observed)

[21]Brill, W. F. *J. Chem. Soc., Perkin Trans. 2* **1984**, 621.

[22]Porter, N.; Zuraw, P. *J. Chem. Soc., Chem. Commun.* **1985**, 1472.

Figure 6.4 Demonstration that **10** can be trapped by oxygen. (Reproduced from reference 22.)

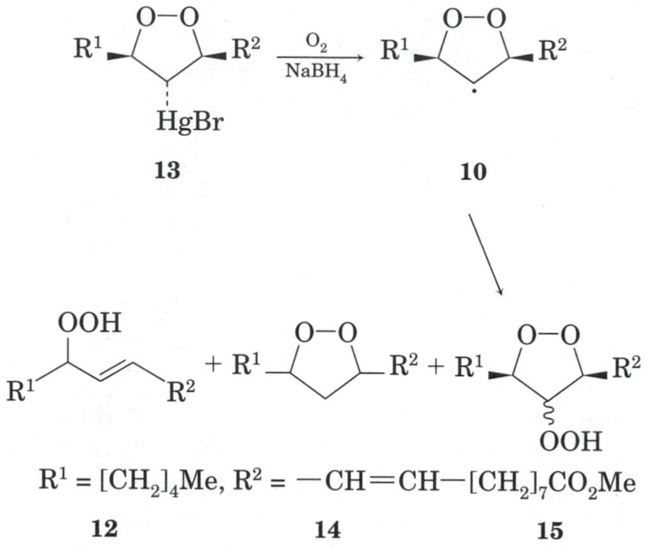

$R^1 = [CH_2]_4Me$, $R^2 = —CH{=}CH—[CH_2]_7CO_2Me$

12 **14** **15**

The nature of an intermediate can also be inferred from experiments in which changing reaction conditions provide a new reaction pathway for the intermediate. For example, a carbanion had been proposed as an intermediate in the interconversion of the enantiomers of the mandelate ion by the enzyme mandelate racemase, as shown in Figure 6.5. The most likely reaction for the carbanion intermediate derived from mandelate ion itself is reprotonation. However, Lin et al. found that treatment of *p*-(bromomethyl)mandelate (**17**) with mandelate racemase formed *p*-methylbenzoylformate (**18**), as shown in Figure 6.6. That result suggested the intermediacy of a carbanion that can eliminate bromide ion, which supported the intermediacy of a carbanion intermediate in the racemization of mandelate ion.[23]

Figure 6.5 Racemization of mandelate via carbanion intermediate.

(*R*)-**16** (*S*)-**16**

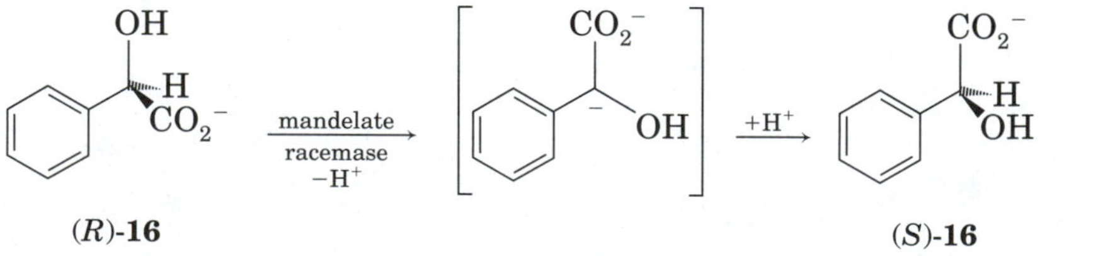

[23](a) The elimination is essentially an E1cb type process. See the discussion of such reactions in Chapter 10; (b) Lin, D. T.; Powers, V. M.; Reynolds, L. J.; Whitman, C. P.; Kozarich, J. W.; Kenyon, G. L. *J. Am. Chem. Soc.* **1988**, *110*, 323; (c) For a review of mandelate racemase, see Kenyon, G. L.; Gerlt, J. A.; Petsko, G. A.; Kozarich, J. W. *Acc. Chem. Res.* **1995**, *28*, 178.

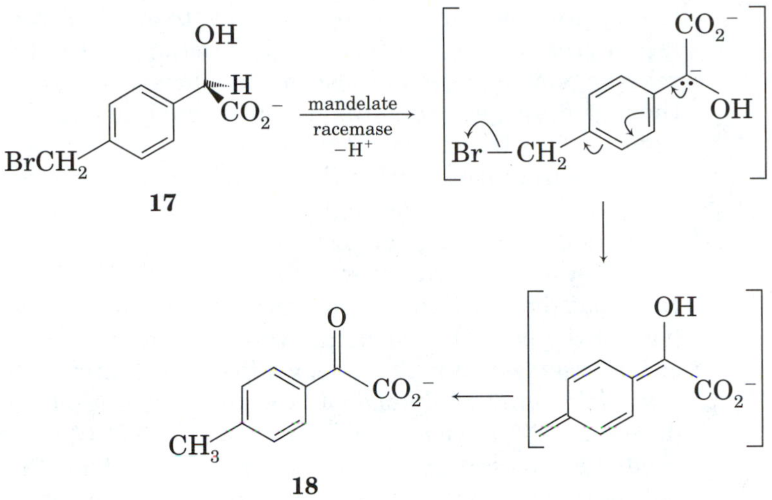

Figure 6.6 Elimination of bromide from carbanion intermediate.

Crossover Experiments

A method of studying reaction mechanisms that is particularly relevant for many molecular rearrangements is the technique of *crossover experiments*. We prepare a mixture of two reactants such that both the migrating group and the rest of the molecule are labeled (for example, with methyl substituents) in one of the reactants. After the reaction, we analyze the products to see whether a portion of the labeled reactant has become bonded to a portion of the unlabeled reactant. As an example of a crossover experiment, consider the Claisen reaction, Figure 6.7(a), in which an allyl group appears

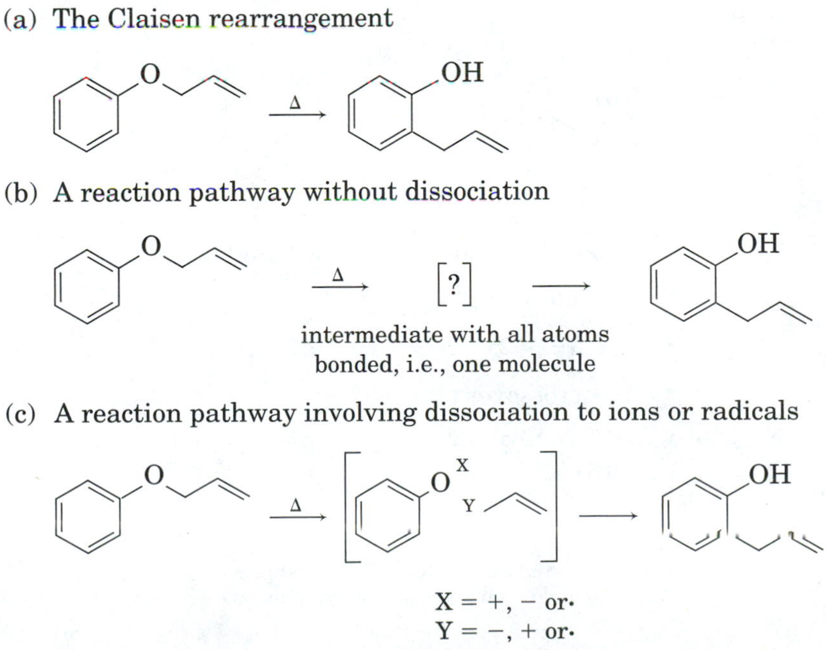

Figure 6.7 (a) The Claisen rearrangement; (b) a reaction pathway without dissociation; (c) a reaction pathway involving dissociation.

to migrate from one place in a molecule to another. We may propose two different *types* of mechanisms for the reaction. In the first, Figure 6.7(b), the rearrangement takes place such that there is continuous bonding between the migrating group and the rest of the molecule at all times. In the second, Figure 6.7(c), the migrating group separates from the rest of the molecule to form two fragments, which then recombine. (The notations X and Y in that figure indicate that the fragments may both be radicals, or one may be an anion and the other a cation.)

To apply the crossover technique to the Claisen rearrangement, we carry out the reaction on a mixture in which one substance bears a label on both of the possible fragments, while the other substance is not labeled on either fragment. For example, we first demonstrate that phenyl allyl ether (**19**) yields only *o*-allylphenol (**20**) and that the labeled analogue (**21**) produces only **22**. If we then find that a mixture of **19** and **21** produces only the products expected from each of the reactants alone (**20** and **22**), then the result of the crossover experiment is negative, suggesting that each reacting molecule does not separate into fragments during the reaction. If we observe the new products **23** and **24**, however, then we conclude that there was recombination of fragments produced during the reaction.[24]

Figure 6.8 (a) Reactions of labeled reactants carried out on individual samples of each, (b) possible outcomes of crossover experiments.

(a) Reactions of labeled reactants carried out on individual samples of each

(i) **19** $\xrightarrow{\Delta}$ **20**

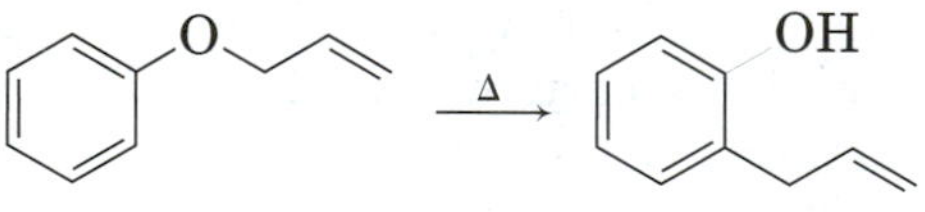

(ii) **21** $\xrightarrow{\Delta}$ **22**

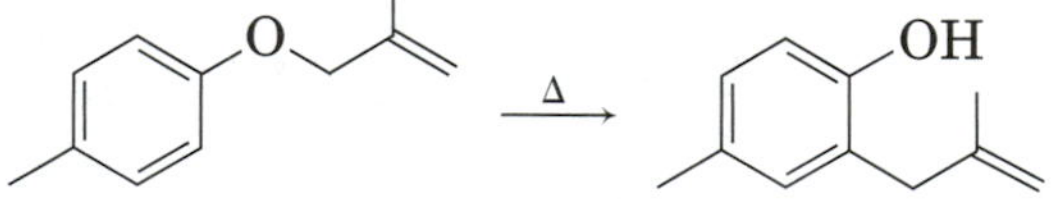

(b) Possible outcomes of crossover experiments

(i) no crossover

19 + **21** ⟶ **20** + **22** only

(ii) crossover observed

19 + **21** ⟶ **20** + **22** + **23** + **24**

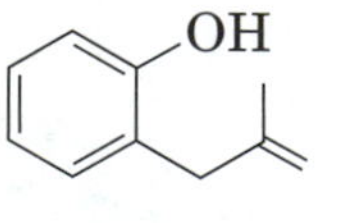

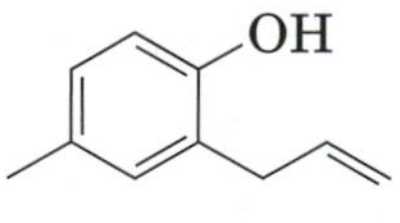

[24]Crossover studies of the Claisen rearrangement have not supported dissociative pathways for the reaction. For a discussion and references, see Tarbell, D. S. *Org. React.* **1944**, *2*, 1.

We must be particularly careful in our interpretation of crossover experiments. If we find evidence of mixed products, we can safely conclude that the reaction mechanism involves fragmentation of the reactants and recombination of the fragments.[25] However, if we do not see crossover products, we are more limited in our conclusions. It may be that dissociation occurs, perhaps in a solvent stage, but that recombination to give product is faster than diffusion out of the solvent stage. Thus observation of the outcome predicted on the basis of a proposed mechanism can be taken as support for that mechanism, but failure to observe the outcome predicted on the basis of a mechanism does not necessarily rule out that mechanism.

Isotopic Labeling

Crossover experiments often use methyl or other alkyl groups as labels. There is always the possibility, however, that a label may alter the course of a reaction. Therefore, the minimum label that will allow us to distinguish the reaction pathway is always best, and an atomic isotope is the smallest perturbation to the molecular structure we can envision. In the case of the Claisen reaction, labeling the allyl functionality with an isotope of carbon as shown in Figure 6.9 leads only to **26**, not to **27**.[26] This information reconfirms our view that the mechanism does not involve molecular fragmentation, since a reaction involving an allyl radical would be expected to give both products.[27,28]

In contrast to the thermal Claisen rearrangement, studies of isotopic labeling in the photochemical Claisen rearrangement support a dissociative mechanism. Figure 6.10 shows the results obtained by Schmid and Schmid

Figure 6.9
Isotopic label study of the Claisen rearrangement.

O Δ OH OH

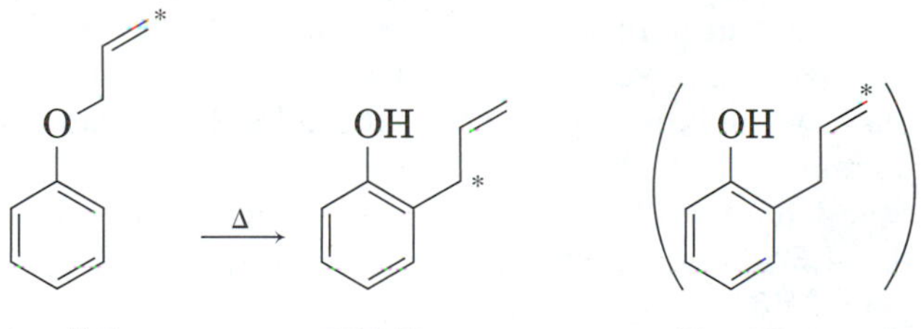

$* = {}^{14}C$ ONLY Not Observed

25 **26** **27**

[25]Actually, we cannot conclude from this information alone that **all** product is formed by a dissociative pathway, only that the crossover products are.

[26]Ryan, J. P.; O'Connor, P. R. *J. Am. Chem. Soc.* **1952**, *74*, 5866; Schmid, H.; Schmid, K. *Helv. Chim. Acta* **1952**, *35*, 1879; **1953**, *36*, 489.

[27]The mechanism of the thermal Claisen rearrangement will be discussed in Chapter 11. Photodissociative reactions are discussed in Chapter 12.

[28]With isotopic labels, it is essential to verify that the label does not exchange with isotopes of the same atom in the solvent or other species present in the reaction mixture. For example, see Hengge, A. C. *J. Am. Chem. Soc.* **1992**, *114*, 2747.

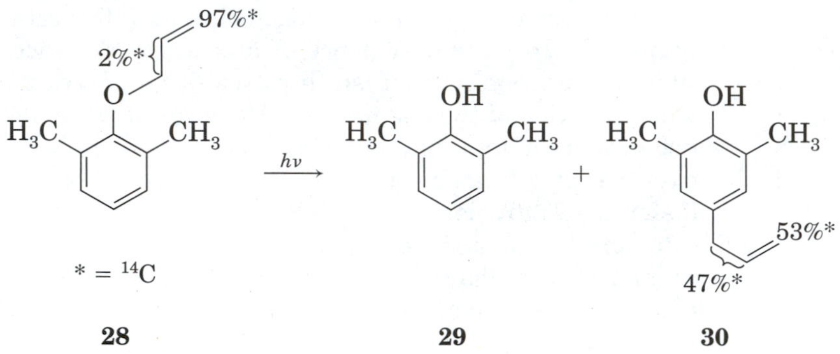

Figure 6.10 Scrambling of isotopic label in photochemical Claisen reaction.

for irradiation of 3-^{14}C-allyl 2,6-dimethylphenyl ether (**28**). The 4-allyl-2,6-dimethylphenol (**30**) produced in the reaction has the ^{14}C label nearly equally distributed between the α and γ positions, suggesting that it was formed by recombination of 2,6-dimethylphenoxy and allyl intermediates and that rotation of the allyl fragment allows the label to become effectively scrambled between the two end carbon atoms of the allyl group by the time of recombination. With other information, these results suggest that the photochemical reaction involves the intermediacy of free radicals.[29]

In the previous example, fragmentation to allyl and phenoxy groups appeared to be the most likely dissociative mechanism for the Claisen rearrangement. In other cases, however, there may be many possible pathways. Often the nature of intermediates in these reactions can be inferred from the results of labeling experiments. For example, two mechanisms have been considered for the thermal isomerization of benzocyclobutene to styrene:

1. Homolysis to a diradical, followed by a 1,3-hydrogen atom transfer. If this mechanism is operative, then the ^{13}C-labeled benzocyclobutene shown as the starting material in Figure 6.11 would produce styrene having the label equally distributed in both atoms of the olefinic unit.
2. Isomerization through a series of electrocyclic valence isomerizations (see Chapter 11), hydrogen migrations, and phenylcarbene–cyclohepta-1,2,4,6-tetraene interconversions (Figure 6.12). According to this more

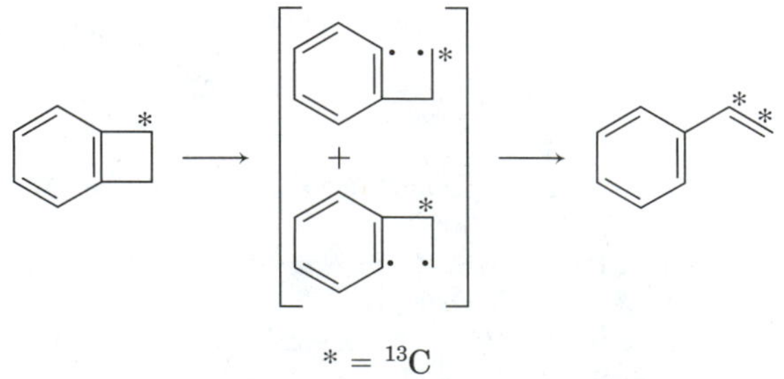

Figure 6.11 Possible mechanism for conversion of benzocyclobutene to styrene. Asterisks indicate possible isotope locations. (Reproduced from reference 30.)

[29]Schmid, K.; Schmid, H. *Helv. Chim. Acta* **1953**, *36*, 687.

Figure 6.12 Alternative mechanism for formation of styrene from thermolysis of benzocyclobutene. Asterisks indicate possible isotope locations. (Adapted from reference 30.)

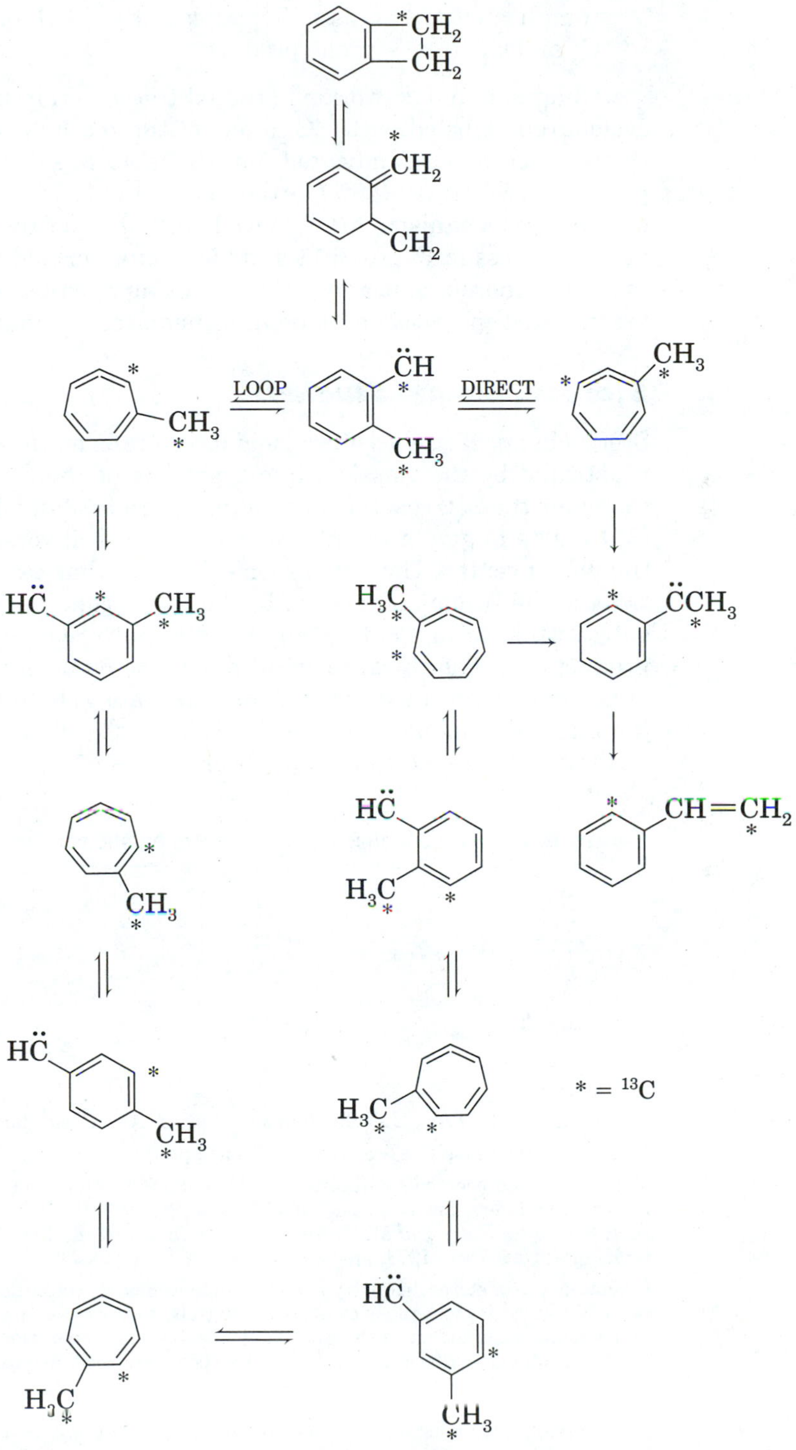

* = ^{13}C

complicated mechanistic hypothesis, the ^{13}C labels would be located at C_β and C_2 of the styrene product.

Chapman and co-workers studied this reaction by thermolyzing benzocyclobutene labeled with ^{13}C in one of the methylene groups.[30] Analysis of the product mixture indicated that the label was distributed on the styrene product as follows: β (48%), ortho (30%), α (14%).[31] These results suggested that both mechanisms are followed, with 75% of the product being formed by the process in Figure 6.11 and 25% being formed by the pathway in Figure 6.12. Therefore, it appears that in this reaction both radical and equilibrating carbenecyclocumulene intermediates are involved.

Stereochemical Studies

Stereochemical studies often yield useful information about mechanisms, as illustrated by the classical investigations of the S_N1 and S_N2 reactions.[32] Consider the S_N2 reaction shown in Figure 6.13 in which a chiral substrate (**31**) reacts to give a chiral product (**32**) with inversion of configuration at the chiral center. The simplest mechanism that can explain both observations is the Walden inversion. In the S_N1 second reaction (Figure 6.14), the substitution occurs so that an optically pure reactant (**33**) gives a racemic product (**34**). The chiral reactant must lose its asymmetry due to formation of an achiral intermediate at some point along the reaction coordinate. This familiar mechanism involving formation of a planar carbocation fits comfortably with the experimental observations.[33]

Figure 6.13 Stereochemical labels in mechanistic studies: S_N2 reaction.

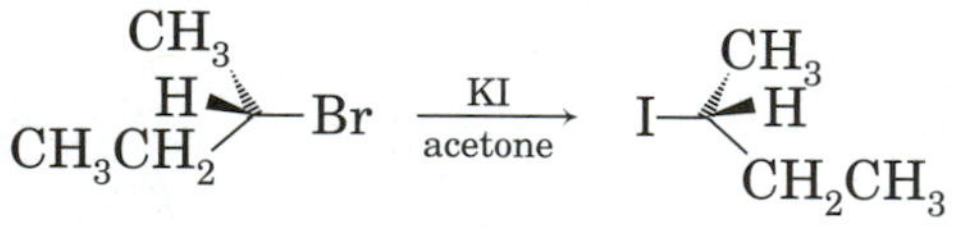

CH_3, H, CH_3CH_2, Br — KI / acetone → I, CH_3, H, CH_2CH_3

(*S*)-**31** (*R*)-**32**

---

[30]Chapman, O. L.; Tsou, U.-P. E.; Johnson, J. W. *J. Am. Chem. Soc.* **1987**, *109*, 553.

[31]Other pathways led to 4% meta and 4% para product.

[32]The use of stereochemistry to investigate reaction mechanisms will be developed further in following chapters. For additional reading, see Stevens, R. V.; Billups, W. E.; Jacobson, B. in *Investigation of Rates and Mechanisms of Reactions*, 3rd ed., Part I; Lewis, E. S., ed.; Wiley-Interscience: New York, 1974; pp. 285–366.

[33]Another use of stereochemistry is a technique known as the *endocyclic restriction test*. In this procedure a cyclic molecule is constructed so that two substituents might react with each other in a mechanism analogous to a bimolecular reaction. However, because the transition structure for the reaction must be cyclic, one of the two possible mechanisms is sterically feasible, but the other is not. Determining whether the intramolecular reaction can occur for transition structures with various ring sizes provides some insight into the probable geometry of the reactants in the bimolecular reaction. For an example, see Li, J.; Beak, P. *J. Am. Chem. Soc.* **1992**, *114*, 9206. See also Beak, P. *Acc. Chem. Res.* **1992**, *25*, 215; *Pure Appl. Chem.* **1993**, *65*, 611.

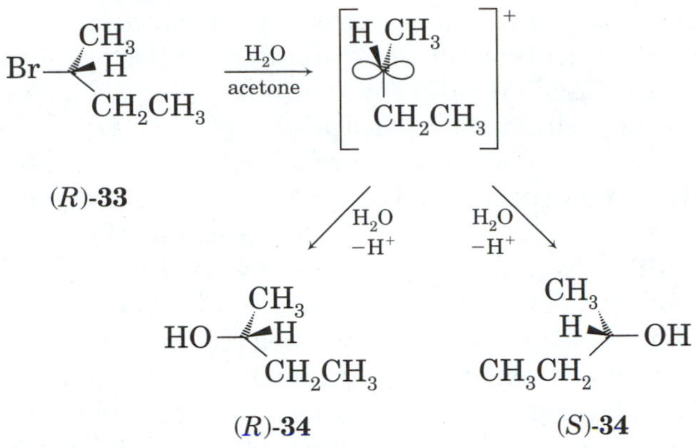

Figure 6.14 Stereochemical labels in mechanistic studies: S_N1 reaction.

Solvent Effects

Determining the effect of solvent on the rate or course of a reaction can often provide insight into the reaction mechanism. If we find, for example, that the rate of a reaction increases with increasing solvent polarity, as is the case with S_N1 reactions, then we may conclude that the transition structure has more ionic character than do the reactants. Conversely, if we find that increasing solvent polarity decreases the rate of the reaction, we deduce that there is less ionic character in the transition structure than in the reactants. Such a trend is observed in S_N2 reactions with anionic nucleophiles, for example, because a more polar solvent stabilizes the nucleophile more than it stabilizes the transition structure.

There is no single measure of solvent polarity, however. Polarity is sometimes characterized by molecular dipole moment (μ), but that is a property of an individual molecule. Polarity is also characterized by the dielectric constant (ϵ) of a material, a property measured by observing the effect of a substance on the electric field between two parallel, oppositely charged plates. Both polarity and polarizability influence dielectric constant, but the orientation of solvent molecules around a charged or polar solute would be very different from the parallel orientation of solvent molecules between two charged metal plates.[34,35]

Many attempts have been made to develop empirical parameters of solvent polarity which reflect the interaction of polar molecules with solutes and which correlate better with rates of chemical reactions.[36] While some

[34]For definitions and compilations of solvent parameters, see Riddick, J. A.; Bunger, W. B.; Sakano, T. K. *Organic Solvents: Physical Properties and Methods of Purification*, 4th ed. (*Techniques of Chemistry*, Vol. II); Wiley-Interscience: New York, 1986.

[35]See also Kosower, E. M. *An Introduction to Physical Organic Chemistry*; John Wiley & Sons, Inc.: New York, 1968; pp. 259–382.

[36]In addition to those considered here, Reichardt has discussed 35 additional parameters of solvent polarity (reference 39).

scales are based on the effect of solvent on a model reaction,[37] others are based on measurements of physical properties of solutes in a solvent. Most scales are based on **solvatochromism**, which is a change in the shape, intensity, or position of an absorption band in a UV-vis spectrum. Among the more widely cited scales are the Kosower Z scale, based on the charge-transfer absorption spectra of 1-ethyl-4-carbomethoxypyridinium iodide (**35**)[38,35] and the $E_T(30)$ scale, based on the spectrum of a pyridinium *N*-phenol betaine (**36**).[39] Kamlet, Taft, and co-workers proposed a general dipolarity/polarizability index π^* to measure the ability of a solvent to stabilize an ionic or polar species by means of its dielectric effect. Values of π^* are based on solvochromatic parameters of many dyes, not just one.[40] Buncel and co-workers developed the π^*_{azo} scale based on the spectral properties of azo merocyanine dyes.[41] A compilation of ϵ, μ, Z, $E_T(30)$, and π^* values for selected solvents is given in Table 6.1.

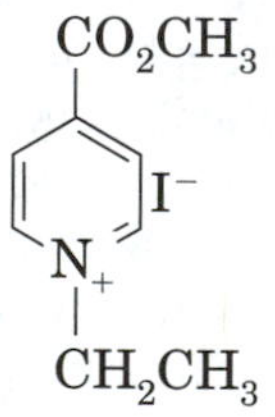

CO_2CH_3 I^- N^+ CH_2CH_3

35

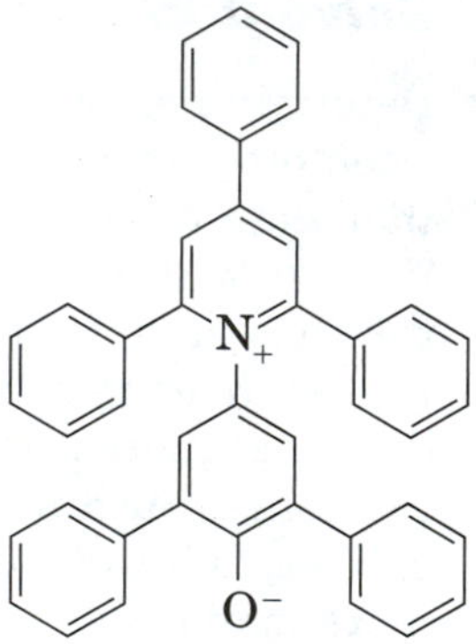

N^+ O^-

36

We should not overemphasize polarity as the determinant of solvent influence on a chemical reaction. Solvent molecules may also play important roles as hydrogen bond donors or acceptors. Kamlet, Taft, and co-workers also developed the parameter α as a measure of the ability of solvent to act as a proton acceptor in a solvent-solute hydrogen bond and the parameter β to describe the ability of solvent to act as a proton donor in a solvent-solute hydrogen bond.[40,42] Values of α and β for selected solvents are also shown in Table 6.1.

---

[37]In Chapter 8 we will examine some empirical measures of solvent polarity that correlate with rates of substitution reactions.

[38]Kosower, E. M. *J. Am. Chem. Soc.* **1958**, *80*, 3253.

[39]Reichardt, C. *Solvents and Solvent Effects in Organic Chemistry*, 2nd ed.; VCH: Weinheim, 1988.

[40]Kamlet, M. J.; Abboud, J.-L. M.; Abraham, M. H.; Taft, R. W. *J. Org. Chem.* **1983**, *48*, 2877.

[41]Buncel, E.; Rajagopal, S. *Acc. Chem. Res.* **1990**, *23*, 226 and references therein.

[42]For a review of these solvation parameters and their relationships with other parameters, see Marcus, Y. *Chem. Soc. Rev.* **1993**, *22*, 409.

Table 6.1 Solvent parameters.

Solvent	ε[a]	μ(D)[a]	Z[b]	$E_T(30)$[c]	π*	β	α[f]
Formamide	111.0	3.37	83.3	56.6	0.97	0.48[g]	0.71
Water	78.4	1.8	94.6	63.1	1.09	0.47[g]	1.17
Formic acid	58.5	1.82		54.3	0.65[g]	0.38[g]	1.23[g]
Dimethyl sulfoxide	46.5	4.06	71.1	45.1	1.00	0.76	0.00
N,N-Dimethylformamide	36.7	3.24	68.5	43.8	1.00[g]	0.76[g]	0.00
Nitromethane	35.9	3.56		46.3	0.85	0.06[g]	0.22
Acetonitrile	35.9	3.53	71.3	45.6	0.75	0.40[g]	0.19
Methanol	32.7	2.87	83.6	55.4	0.60	0.66[g]	0.93
Hexamethylphosphoramide	29.3	4.31	62.8[d]	40.9	0.87[g]	1.05[g]	0.00
Ethanol	24.5	1.66	79.6	51.9	0.54	0.75[g]	0.83
1-Propanol	20.4	3.09	78.3	50.7	0.52	0.90[g]	0.84[g]
1-Butanol	17.5	1.75	77.7	50.2	0.47	0.84[g]	0.84[g]
Acetone	20.6	2.69	65.7	42.2	0.71	0.43[g]	0.08
2-Propanol	19.9	1.66	76.3	48.4	0.48	0.84[g]	0.76
Pyridine	12.9	2.37	64.0	40.5	0.87	0.64	0.00
t-Butyl alcohol	12.5	1.66	71.3	43.3	0.41	0.93[g]	0.42[g]
Methylene chloride	8.9	1.14	64.2	40.7	0.82	0.10[g]	0.13[g]
Tetrahydrofuran	7.58	1.75		37.4	0.58	0.55	0.00
1,2-Dimethoxyethane	7.20	1.71	62.1[d]	38.2	0.53	0.41	0.00
Acetic acid	6.17	1.68	79.2	51.7	0.64	0.45[g]	1.12
Ethyl acetate	6.02	1.82		38.1	0.55	0.45	0.00
Chloroform	4.80	1.15	63.2[e]	39.1	0.58	0.10[g]	0.20[g]
Diethyl ether	4.2	1.15		34.5	0.27	0.47	0.00
Benzene	2.27	0	54.0[d]	34.3	0.59	0.10	0.00
Carbon tetrachloride	2.23	0		32.4	0.28	0.10[g]	0.00
n-Hexane	1.89	0.085		31.0	−0.04[g]	0.00	0.00

[a]Data from the compilation in reference 34. [b]Data from reference 38. [c]Data from the compilation in reference 39, pp. 365–371. [d]Data from the compilation in reference 35, p. 301. [e](0.13 M EtOH). [f]Values for π*, β, and α from reference 40. [g]Data from the compilation in reference 42.

Solvent molecules may also act as electron pair donors or acceptors, as evidenced by the formation of charge transfer complexes or by the participation of solvent molecules as nucleophiles or electrophiles (Lewis bases or acids) in reactions. Solvent molecules may also behave as acids or bases in the Brønsted-Lowry sense.[39] Moreover, the medium in which a reaction occurs may consist of molecules of more than one substance. Often reactions are carried out in a solvent mixture consisting of several nonionic (polar and nonpolar) and ionic species. The properties of polar solvents can be altered by the addition of ionic species, and the resulting *salt effects* can increase the rates of organic reactions that occur faster in more polar solvents.[43] One other solvent property that may be important in extremely fast reactions is viscosity. If a reaction is *diffusion limited*, then chemical reaction occurs upon every collision of the reacting species. In that case, changing from a low viscosity solvent to a higher viscosity solvent with similar polarity (e.g., from hexane to tetradecane) can decrease the rate of the reaction.

6.3 Applications of Kinetics in Studying Reaction Mechanisms

A kinetic study is often the first tool chemists consider when planning a mechanistic investigation, but we should emphasize a point made earlier with regard to methods of studying reactions in general. Kinetic studies do not "prove" any reaction mechanism. Rather, kinetics can rule out some proposed mechanisms, leaving one or more mechanisms that are consistent with the kinetic observations.[44]

The rate of a chemical reaction can be expressed as the time dependence of the appearance of a product or, alternatively, as the time dependence of the disappearance of a reactant.[45] If the stoichiometry of a reaction is such that n_A molecules of A combine with n_B molecules of B and n_C molecules of C (and so on) to produce n_P molecules of P, which we express as

$$n_A A + n_B B + n_C C \rightarrow n_P P \text{ (product)} \tag{6.3}$$

we may find that the rate of the reaction depends on the concentration of some or all of them according to the expression in equation 6.4,

$$\text{Rate} = \frac{1}{n_P}\frac{d[P]}{dt} = -\frac{1}{n_A}\frac{d[A]}{dt} = -\frac{1}{n_B}\frac{d[B]}{dt} = -\frac{1}{n_C}\frac{d[C]}{dt} = k_r[A]^a[B]^b[C]^c \tag{6.4}$$

where k_r is the rate constant for the reaction and a,b, and c do not necessarily equal n_A, n_B, and n_C, respectively. In this expression, the **overall order** of the reaction is $a + b + c$, and the **order with respect to A** is a, the order with respect to B is b, the order with respect to C is c (and so forth). The

[43]However, addition of ions identical to those that are produced by the reaction (such as halide ions produced in elimination or substitution reactions) can decrease the rate of a reaction through a "common ion effect." As discussed in Chapter 8, "special salt effects" can have dramatic effects on the rates of some reactions. For a general discussion, see Loupy, A.; Tchoubar, B. *Salt Effects in Organic and Organometallic Chemistry*; VCH: Weinheim, 1992.

[44]In a sense, a kinetic study is like a patent. It has been said that a patent does not entitle the inventor to do anything—it only allows the patent holder to prevent others from doing something. Similarly, a kinetic study does not enable us to conclude that a particular mechanism operates in a reaction we are studying. It might, however, allow us to state that some other reaction mechanism does not operate, since that mechanism would not give the observed kinetics.

[45]The rate is sometimes called the *velocity* of the reaction, although that term occurs more frequently in biochemistry than in organic chemistry.

order of a reaction with respect to a certain reagent is frequently a whole number, but fractional order is possible, and the order may even be 0 in some cases.

Reactions composed of two or more elementary reactions are ***complex reactions***.[46,47] For an ***elementary reaction*** the overall order of the reaction is the same as the ***molecularity*** (the number of reacting molecules). However, a complex reaction may yield a complicated rate expression that includes concentration terms in the denominator or sums of terms that are similar to the right hand side of equation 6.4, in which case the term *overall order of the reaction* is not defined.[48,49] For example, we will see in Chapter 9 that the kinetic expression for the addition of bromine to an alkene in CCl_4 solution may take the form

$$-d[Br_2]/dt = [\text{alkene}](k_2[Br_2] + k_3[Br_2]^2 + k_{Br_3^-}[Br_3^-]) \tag{6.5}$$

The presence in equation 6.5 of a term that is second order with respect to bromine does not require that there be a step in which an alkene collides simultaneously with two bromine molecules. The more likely possibility, which is consistent with other experimental evidence, is that an alkene and one bromine molecule first react to produce an intermediate that subsequently reacts with another bromine molecule.

Let us consider in more detail a reaction with simpler stoichiometry in which reactant A goes to product P. If the rate constant for that step is k_1, then the rate expression is

$$d[P]/dt = -d[A]/dt = k_1[A] \tag{6.6}$$

The reaction is first order with respect to A and is overall first order. That order is consistent with a mechanism in which the rate-limiting step is unimolecular, meaning that one molecule of A reacts to give P. We can exclude mechanisms such as one in which two molecules of A collide to give a molecule of P and one of A, since that process would give a different rate expression:

$$d[P]/dt = -d[A]/dt = k_2[A]^2 \tag{6.7}$$

Note that the term $d[P]/dt$ has units of concentration/time. The right hand side of a rate equation must also have units of concentration/time, so the units of k_1 must be 1/time, while those of k_2, must be 1/(time-concentration).

A ***differential rate equation*** such as equation 6.7 is generally not as useful to us as is the ***integrated rate equation***, which allows us to

---

[46]For a discussion see Connors, K. A. *Chemical Kinetics*; VCH Publishers, Inc.: New York, 1990; pp. 3–4.

[47]The kinetics of complex reactions have been discussed by Noyes, R. M. in *Investigation of Rates and Mechanisms of Reactions*, 3rd. ed., Part I, Lewis, E. S., ed.: Wiley-Interscience: New York, 1974; pp. 489–538.

[48]Laidler, K. J. *Chemical Kinetics*, 2nd ed.; McGraw-Hill Book Company: New York, 1965; p. 4.

[49]For a commentary on the use of the term *overall reaction order*, see Reeve, J. C. *J. Chem. Educ.* **1991**, *68*, 728.

compare experimental concentration data with that predicted by the rate expression. For a first order reaction the integrated rate expression is

$$\ln [A] = \ln [A]_0 - k_1 t \tag{6.8}$$

where t is the time and $[A]_0$ is the value of [A] at $t = 0$. In this case plotting the natural logarithm of the concentration of A versus time should produce a linear correlation with slope $-k_1$. For a second order elementary reaction with stoichiometry $A + A \rightarrow P$, the integrated expression is

$$1/[A] - 1/[A]_0 = kt \tag{6.9}$$

Now plotting the reciprocal of the concentration of A versus time should produce a linear correlation, with slope $-k_1$. This same expression can be derived for the situation described in equation 6.11 if $[A]_0 = [B]_0$.[50,51]

A reaction of the type $A + B \rightarrow P$ may be said to be ***pseudo–first order*** if the rate expression is of the form[52]

$$d[P]/dt = k_{observed}[A] \tag{6.10}$$

if the concentration of B is held essentially constant during the experiment. This situation can arise if B is an acid or base in a buffered solution, if B is a catalyst that is not consumed in the reaction, or if the ratio [B] to [A] is very large. The last situation is typical in solvolysis reactions, in which the concentration of solvent is large and essentially invariant.[53]

If the conditions for pseudo-first order kinetics are not met, then the rate expression for an elementary reaction with stoichiometry $A + B \rightarrow P$ would be

$$d[P]/dt = k_2[A][B] \tag{6.11}$$

Now the integrated expression becomes

$$\ln\frac{[A]}{[B]} + \ln\frac{[B]_0}{[A]_0} = ([A]_0 - [B]_0)kt \tag{6.12}$$

---

[50]In general, for a reaction that is n^{th} order (for $n > 1$) in A, the integrated rate expression is

$$\frac{1}{(n-1)}\left(\frac{1}{[A]^{n-1}} - \frac{1}{[A]_0^{n-1}}\right) = kt$$

For a discussion, see reference 54, pp. 142. For a detailed discussion of the kinetics of many kinds of reactions, see Capellos, C.; Bielski, B. H. J. *Kinetic Systems: Mathematical Description of Chemical Kinetics in Solution*; Wiley-Interscience: New York, 1972.

[51]We may develop similar expressions for second and third order reactions (in which two or three molecules of A collide in the rate-limiting step). See, for example, Gordon, A. J.; Ford, R. A. *The Chemist's Companion*; John Wiley & Sons, Inc.: New York, 1972; p. 135, for a listing of the differential and integrated forms of zero, first, and second order rate equations.

[52]The symbol $k_{observed}$ is usually abbreviated k_{obs}. The symbols k_ψ and k_{app} ($k_{apparent}$) have also been used for the rate constant in a pseudo-order reaction. See the discussion in reference 46, p. 23 *ff*.

[53]Reference 46, pp. 23–24.

so plotting the natural logarithm of the term [A]/[B] versus time should yield a linear graph with slope $k([A]_o - [B]_o)$.[54]

Now let us consider a consecutive reaction

$$A \xrightarrow{k_1} B \xrightarrow{k_2} C \qquad (6.13)$$

In this discussion the use of forward arrows (as opposed to equilibrium arrows) in the stoichiometric expressions is meant to imply that the two elementary steps are both ***irreversible reactions***, meaning that all molecules of B go on to C and do not revert to A. Similarly, no C molecules revert to B. In theory, however, all reactions are reversible, and the reverse operation of the reaction mechanism we write should convert product to reactant. If the potential energy surfaces that connect reactants to products are not discontinuous, then the energy and geometry of the transition structure will be the same in both cases.[55] If a particular step in a reaction is highly exothermic, the reverse reaction will be highly endothermic and may not be observed under our reaction conditions. By the term *irreversible*, therefore, we really mean that the rate of the reverse reaction of each step is so slow as to be negligible under the given reaction conditions.

We can write an expression for the rate of product formation in equation 6.13 as follows:

$$\text{Rate} = d[C]/dt = k_2[B] \qquad (6.14)$$

The concentration of [B] varies with time, and at the beginning of the reaction [B] is 0. The rate expression for B is

$$d[B]/dt = k_1[A] - k_2[B] \qquad (6.15)$$

The exact solution to this problem can be obtained, and equations for the time-dependence of the concentrations [A], [B], and $[C]_{exact}$ are available.[56] The relative concentrations of these species as a function of time are calculated for relative rate constants $k_1/k_2 = 3.0$ (Figure 6.15), $k_1/k_2 = 0.33$ (Figure 6.16), and $k_1/k_2 = 0.033$ (Figure 6.17).

In Figure 6.16 and Figure 6.17 we notice that [B] does not change significantly once the reaction is under way. This observation is the basis for a method that is commonly used to approximate the kinetics of systems

---

[54]Bunnett, J. F. in *Investigation of Rates and Mechanisms of Reactions*, 3rd ed., Part I, Lewis, E. S., ed.; Wiley-Interscience: New York, 1974; pp. 140–141.

[55]A formal statement of this theory is the **principle of microscopic reversibility**. For leading references, see (a) Morrissey, B. W. *J. Chem. Educ.* **1975**, *52*, 296; (b) Mahan, B. H. *J. Chem. Educ.* **1975**, *52*, 299.

[56]*Cf.* Frost, A. A.; Pearson, R. G. *Kinetics and Mechanism*, 2nd ed.; John Wiley & Sons, Inc.: New York, 1961; pp. 14 *ff.*, p. 166.

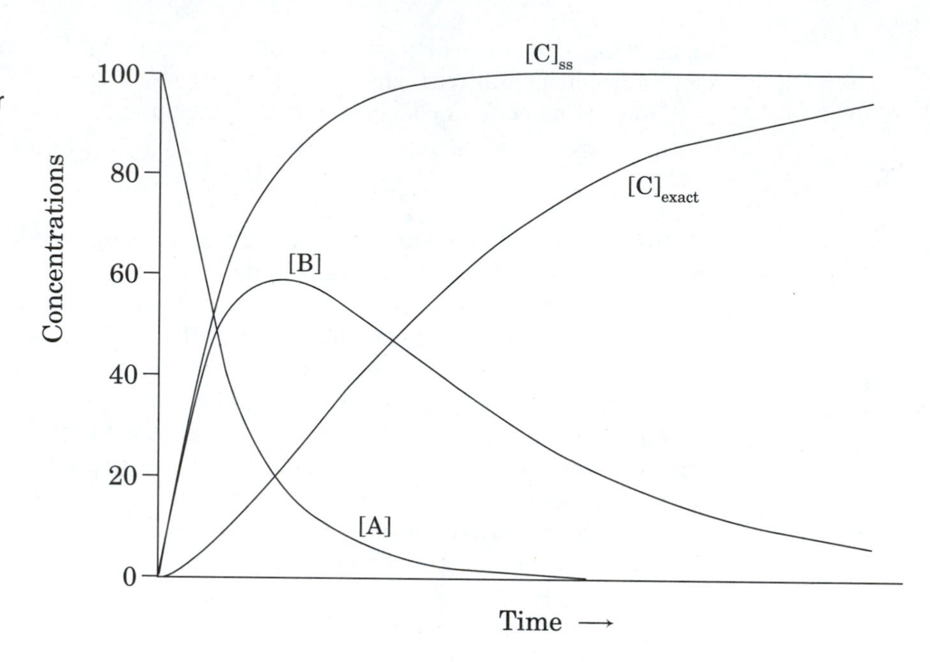

Figure 6.15 Variation of [A], [B], and [C] with time for $k_1/k_2 = 3.0$.

involving intermediates, the ***steady state approximation***.[57,58,59,60] Thus, we assume that

$$d[\mathrm{B}]/dt = k_1[\mathrm{A}] - k_2[\mathrm{B}] \approx 0 \tag{6.16}$$

so

$$k_2[\mathrm{B}] \approx k_1[\mathrm{A}]. \tag{6.17}$$

Then

$$\text{Rate} = d[\mathrm{C}]/dt = k_2[\mathrm{B}] \approx k_1[\mathrm{A}] \tag{6.18}$$

---

[57](a) The steady state approximation is commonly attributed to Bodenstein (Bodenstein, M.; Dux, W. *Z. Phys. Chem.* **1913**, *85*, 297; Bodenstein, M. *Z. phys. Chem.* **1913**, *85*, 329). (b) However, Laidler, K. J. *Acc. Chem. Res.* **1995**, *28*, 187 has noted that the method was earlier used by Chapman, D. L.; Underhill, L. K. *J. Chem. Soc.* **1913**, *103*, 496.

[58]See also the discussion by Bunnett, J. F. in *Investigation of Rates and Mechanisms of Reactions*, 4th ed., Volume VI, Part I; Bernasconi, C. F., ed.; John Wiley & Sons: New York, 1986; p. 251.

[59]Gilbert, H. F. *J. Chem. Educ.* **1977**, *54*, 492, presented a useful rule of thumb for writing steady state equations in the general case. See also the discussion of the steady state approximation and free energy profiles by Raines, R. T.; Hansen, D. E. *J. Chem. Educ.* **1988**, *65*, 757.

[60]For a discussion of the validity of the steady state approximation, see Viossat, V., Ben-Aim, R. I. *J. Chem. Educ.* **1993**, *70*, 732.

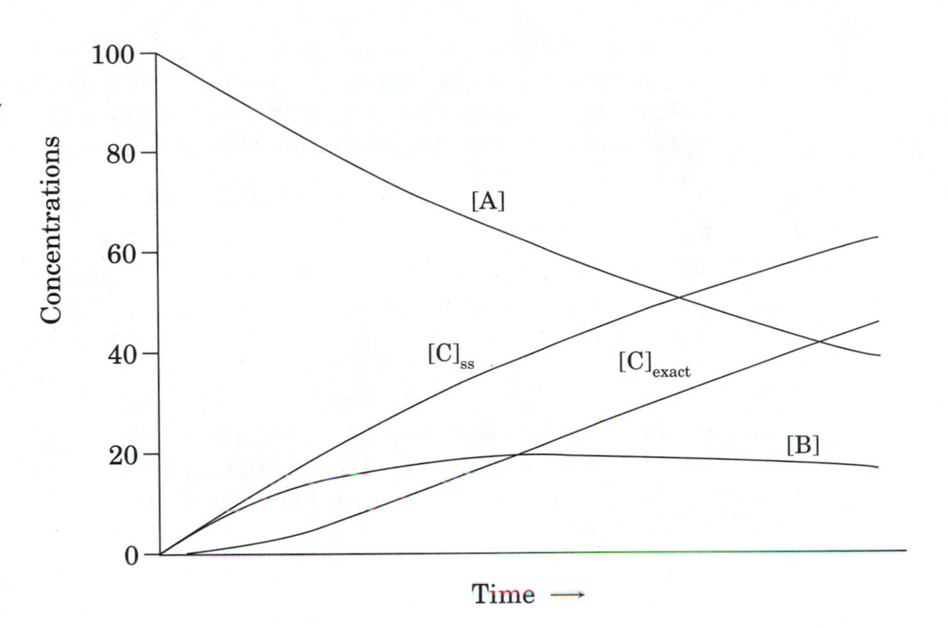

Figure 6.16
Variation of [A], [B], and [C] with time for $k_1/k_2 = 0.33$.

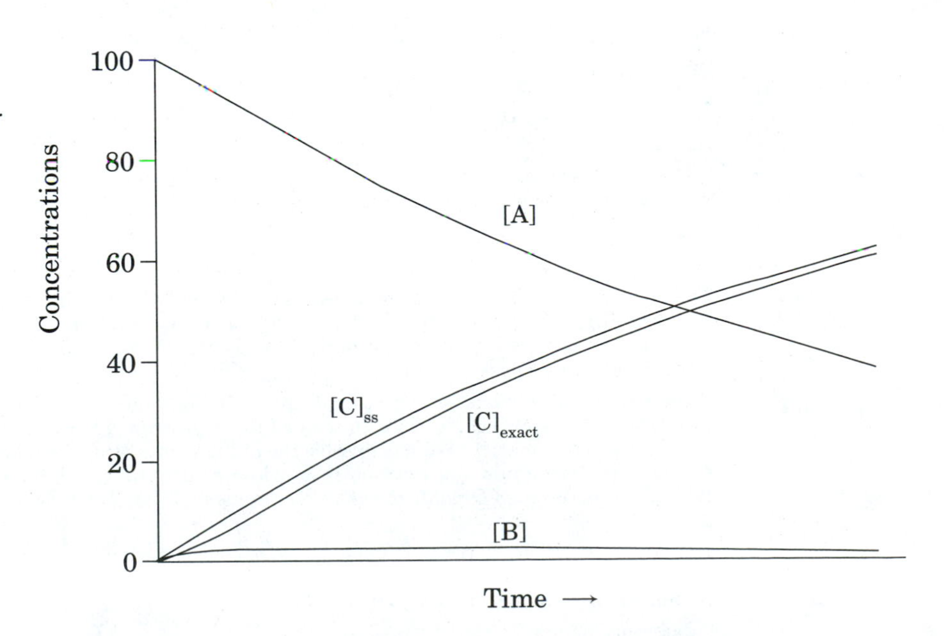

Figure 6.17
Variation of [A], [B], and [C] with time for $k_1/k_2 = 0.033$.

In other words, the overall reaction rate appears to be equal to the rate of the slowest step in the reaction sequence. If we make this assumption, the integrated expression for $[C]_{ss}$ (the concentration of C calculated on the basis of steady state approximation) becomes

$$[C]_{ss} = [A]_0 - [A] \tag{6.19}$$

Figures 6.15, 6.16, and 6.17 illustrate the value of $[C]_{ss}$ as well as the value of [C] calculated without using the steady state approximation, $[C]_{exact}$, for the rate constants shown. The figures suggest that the validity of the steady state approximation depends on the relative magnitudes of the rate constants, k_1 and k_2, and on the time scale of the measurement of [C]. Essentially, $[C]_{ss} = [C]_{exact} + [B]$. The larger the value of [B], the greater will be the error in the value of $[C]_{ss}$. As the ratio k_2/k_1 becomes larger, both $[B]_{ss}$ and the difference between $[C]_{ss}$ and $[C]_{exact}$ become smaller.[61,62]

Now let us consider a more complicated reaction involving at least one *bimolecular step*.

$$A + B \underset{k_{-1}}{\overset{k_1}{\rightleftharpoons}} C \xrightarrow{k_2} D \tag{6.20}$$

Then

$$\text{Rate} = d[D]/dt = k_2[C] \tag{6.21}$$

The exact solution to this problem is not known, but it is possible to approximate a solution using numerical integration methods such as the Runge-Kutta method.[63,64] We may also apply the steady state approximation to C:

$$d[C]/dt = k_1[A][B] - k_2[C] - k_{-1}[C] = 0 \tag{6.22}$$

so

$$[C] \equiv \{k_1[A][B]\}/\{k_2 + k_{-1}\} \tag{6.23}$$

---

[61]The set of two consecutive irreversible reactions just considered is actually a subset of the more general case in which both reactions are reversible:

$$A \underset{k_{-1}}{\overset{k_1}{\rightleftharpoons}} B \underset{k_{-2}}{\overset{k_2}{\rightleftharpoons}} C$$

The equations for this problem have been solved exactly for two limiting cases: (i) $[A]_0 = 1.00$ and both $[B]_0$ and $[C]_0 = 0$, and (ii) A is present as a saturated solution so that the concentration does not change with time. For details, see Lowry, T. M.; John, W. T. *J. Chem. Soc.* **1910**, *97*, 2634 and references therein.

[62]The kinetic expressions for more complicated reactions can become much more complex. Fortunately, there are exact solutions to some of these problems. For example, Carpenter has described the use of matrix methods to solve the problem of a "unimolecular array" in which a set of isomeric compounds and reaction intermediates are directly interconvertible. Carpenter, B. K. *Determination of Organic Reaction Mechanisms*; Wiley-Interscience: New York, 1984; pp. 52 *ff*.

[63]The exact solution for the time dependence of concentration for A, B, C, and D for the set of *consecutive irreversible* reactions $A + B \rightarrow C \rightarrow D$ has been reported: Anderson, R. L.; Nohr, R. S.; Spreer, L. O. *J. Chem. Educ.* **1975**, *52*, 437.

[64]Reference 62, pp. 77 *ff*.

Thus we say that

$$d[\mathrm{D}]/dt = \{k_2k_1/(k_2 + k_{-1})\}[\mathrm{A}][\mathrm{B}] \tag{6.24}$$

If we can write that $k_2 \ll k_{-1}$, the term $k_2 + k_{-1} \approx k_{-1}$, and equation 6.24 becomes

$$d[\mathrm{D}]/dt = (k_2k_1/k_{-1})[\mathrm{A}][\mathrm{B}] = k_2(k_1/k_{-1})[\mathrm{A}][\mathrm{B}] \tag{6.25}$$

where k_1/k_{-1} represents the equilibrium constant of the preequilibrium for formation of the species involved in the rate-limiting step. As was the case for the irreversible reaction, the concentrations calculated from the steady state approximation will be in error, although the magnitude of the error depends on the relative rate constants. If, however, $k_2 \gg k_1$, then equation 6.24 simplifies to

$$d[\mathrm{D}]/dt = k_1[\mathrm{A}][\mathrm{B}] \tag{6.26}$$

If neither simplification is appropriate, then the reaction described by equation 6.20 may be studied by carrying out the reaction with a large initial concentration of A or B so that the first elementary reaction is carried out under pseudo-first order conditions.

Another application of kinetics involves the comparison of rate data with predictions made on the basis of mechanistic schemes that involve one or more intermediates. In such cases the sensitivity of the reaction rate to catalysts, solvent effects, and varying structural parameters can be analyzed for consistency with the proposed mechanisms.[65] Each of these approaches will be considered in more detail in the chapters that follow as we explore additional applications of kinetic relationships. However, we must not lose sight of the approximations and assumptions that pervade many of these treatments. A kinetic expression becomes a model for chemical behavior, just as a picture of a double bond or a wave function is a model. Every chemical system is different, and we must be wary of force fitting experimental data to the kinetics provided by a simple model system.

6.4 Arrhenius Theory and Transition-State Theory

One of the most important methods of investigating a reaction mechanism is the determination of the rate of the reaction as a function of temperature. The treatment of temperature effects that may be most familiar to organic chemists is the Arrhenius equation, which relates the rate constant of a reaction to the temperature and to the *activation energy*, E_a.[66,67] As shown in

[65]For an example of the use of kinetics to interpret a reaction mechanism in terms of a proposed intermediate, see Bunnett, J. F.; Garst, R. H. *J. Am. Chem. Soc.* **1965**, *87*, 3875, 3879.

[66]See "The Development of the Arrhenius Equation," by Laidler, K. J. *J. Chem. Educ.* **1984**, *61*, 494; also see the comments regarding van't Hoff's pioneering work noted by Laidler in reference 57(b).

[67]E_a is also written as E_{act}.

Figure 6.18, the activation energy is defined as the difference between the energy of the starting material(s) and the energy of the transition structure, which is the arrangement of electrons and nuclei at the high point on the reaction coordinate diagram for a one step reaction.[68,69,70]

The expression of the Arrhenius equation is

$$k = A \exp(-E_a/RT) \tag{6.27}$$

in which the preexponential term A is the probability factor for the reaction.[71] Taking the natural logarithm of both sides gives

$$\ln k = -E_a/RT + \ln A \tag{6.28}$$

so plotting $\ln k$ versus $1/T$ (in K) would be expected to give a straight line with slope $-E_a/R$. Because of experimental uncertainties, a plot covering a larger temperature range will produce a greater precision in the value determined for E_a. The intercept of the plot corresponds to a value of $1/T$ of 0, and is equal to $\ln A$. A long extrapolation of a plot of experimental data may produce an inaccurate value for A, however. An alternative approach is to calculate $\ln A$ from the E_a value and data for the reaction rate constant

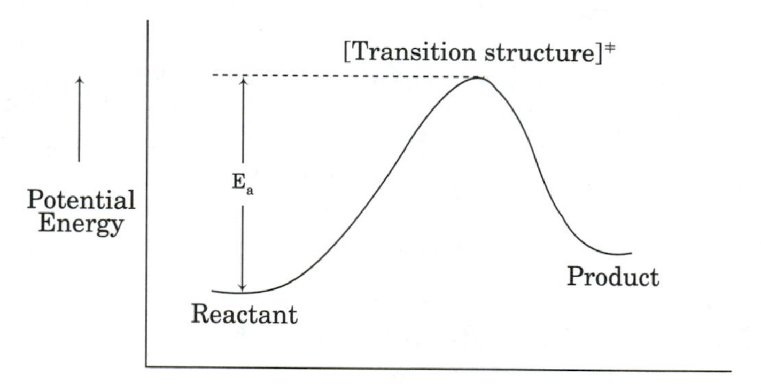

Figure 6.18
Reaction coordinate diagram for a one-step reaction.

[68] Laidler, K. J. *J. Chem. Educ.* **1988**, *65*, 540.

[69] For a discussion of the nature and depiction of transition states, see the anonymous note in *J. Chem. Educ.* **1987**, *64*, 208.

[70] For a discussion of the terms *transition state* and *transition structure*, see Williams, I. H. *Chem. Soc. Rev.* **1993**, *22*, 277.

[71] It may be that a slow reaction is slow at high temperature as well as low temperature because of a small preexponential factor; i.e., the reaction is improbable under all temperature conditions. The meaning of A in several theoretical treatments has been discussed by Gowenlock, B. G. *Quart. Rev. Chem. Soc.* **1960**, *14*, 133.

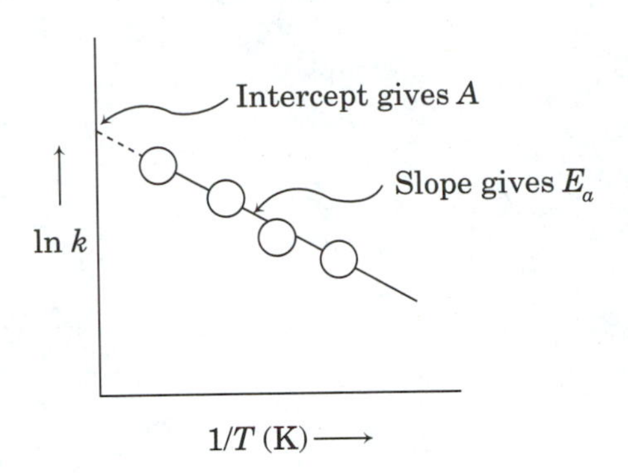

Figure 6.19 Determination of Arrhenius E_a and A from kinetic data.

at some temperature in the middle of the range of temperature-rate data.[72,73]

An alternative to Arrhenius theory is ***transition-state theory***, which was put forward by Eyring and others in the 1930s.[74,75] In this approach the *transition state*, also termed the *activated complex*, is considered a chemical species and treated from the point of view of thermodynamics. Thus we can define the activation free energy ($\Delta G^‡$), activation enthalpy ($\Delta H^‡$), and activation entropy ($\Delta S^‡$) for the transformation from reactants to the activated complex, as shown in Figure 6.20.[76] The units of free energy and enthalpy are given in energy/mole, e.g., kcal/mol. The units of entropy are often given as cal degree^{-1} mole^{-1}, which is often written as eu or entropy units.

According to transition-state theory, the rate constant k_r is defined as

$$k_r = (\kappa \mathbf{k} T/h) \times \exp(-\Delta G^‡/RT) \tag{6.29}$$

where h is Planck's constant (6.626×10^{-27} erg-sec); T is the absolute temperature (0 K = $-273.16°$); R is the gas constant, 1.987 cal deg^{-1} mol^{-1}; κ is the transmission coefficient, a factor included because not all activated

---

[72]Kalantar has discussed the precision of E_a and A measurements: Kalantar, A. H. *J. Phys. Chem.* **1986**, *90*, 6301. For a discussion of nonlinear Arrhenius plots, see Hulett, J. R. *Quart. Rev. Chem. Soc.* **1964**, *18*, 227.

[73]In Figure 6.19 the slope of the line is negative, so E_a is a positive number. This is usually the situation that is observed. However, if there is a preequilibrium between monomeric and complexed reactants prior to the rate-limiting step, the slope of the Arrhenius plot may be positive, and the calculated E_a may be negative. Note that this does *not* mean that there is an elementary reaction step that takes place with a negative activation energy. For example, see Albrecht-Gary, A.-M.; Dietrich-Buchecker, C.; Saad, Z.; Sauvage, J.-P. *J. Chem. Soc., Chem. Commun.* **1992**, 280.

[74]See "The Development of Transition-State Theory" by Laidler, K. J.; King, M. C. *J. Phys. Chem.* **1983**, *87*, 2657.

[75]For a review of developments in this area, see "Current Status of Transition-State Theory," Truhlar, D. G.; Hase, W. L.; Hynes, J. T. *J. Phys. Chem.* **1983**, *87*, 2664; "Transition-State Theory Revisited," Albery, W. J. *Adv. Phys. Org. Chem.* **1993**, *28*, 139.

[76]Another mechanistic tool is the activation volume, $\Delta V^‡$, which is determined by plotting ln $k_{observed}$ versus pressure. For a review, see Van Eldik, R.; Asano, T.; le Noble, W. J. *Chem. Rev.* **1989**, *89*, 549; Whalley, E. *Adv. Phys. Org. Chem.* **1964**, *2*, 93.

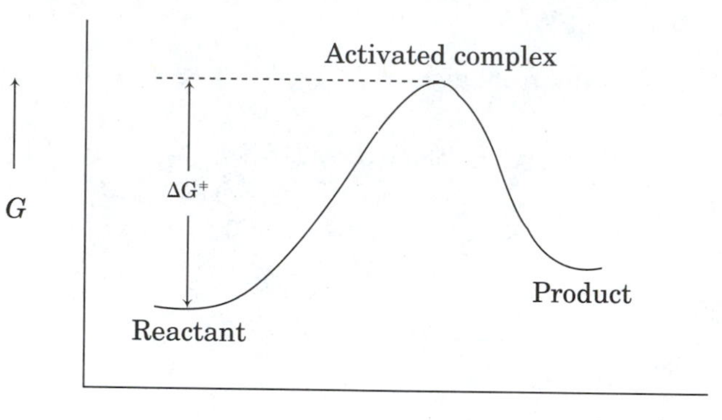

Figure 6.20 Identification of activated complex and $\Delta G^{\ddagger}$ in an elementary reaction.

complexes go on to reaction products;[77] and **k** is the Boltzmann constant (1.380×10^{-16} erg/deg). Substituting the relationship $\Delta G^{\ddagger} = \Delta H^{\ddagger} - T\Delta S^{\ddagger}$ gives

$$k_r = (\kappa \mathbf{k} T/h) \times \exp(\Delta S^{\ddagger}/R) \times \exp(-\Delta H^{\ddagger}/RT) \qquad \textbf{(6.30)}$$

Dividing by T and then taking the natural logarithm of both terms in equation 6.30 gives

$$\ln(k_r/T) = \ln(\kappa \mathbf{k}/h) + \Delta S^{\ddagger}/R - \Delta H^{\ddagger}/RT \qquad \textbf{(6.31)}$$

Taking $\log_{10}$ instead of ln gives

$$\log(k_r/T) = \log(\kappa \mathbf{k}/h) + [\Delta S^{\ddagger}/R - \Delta H^{\ddagger}/RT]/2.303 \qquad \textbf{(6.32)}$$

Replacing R with 1.987 cal deg^{-1} mol^{-1} produces the equivalent equation

$$\log(k_r/T) = \log(\kappa \mathbf{k}/h) + \Delta S^{\ddagger}/4.576 - \Delta H^{\ddagger}/4.576T \qquad \textbf{(6.33)}$$

Therefore $\Delta H^{\ddagger}$ can be calculated from the slope of a plot of $\log(k_r/T)$ versus $1/T$ by the relationship

$$\Delta H^{\ddagger} = -4.576 \times \text{slope} \qquad \textbf{(6.34)}$$

Note that the units of $\Delta H^{\ddagger}$ calculated from equation 6.34 are **cal**/mol, not kcal/mol. In solution, $\Delta H^{\ddagger}$ is usually about 0.6 kcal/mol less than the Arrhenius E_a value, since[78]

$$E_a = \Delta H^{\ddagger} + RT \qquad \textbf{(6.35)}$$

The value of $\Delta S^{\ddagger}$ can be calculated by rewriting equation 6.33 as

$$\Delta S^{\ddagger} = R \times 2.303 \times \log(k_r/T) + (\Delta H^{\ddagger}/T) - R \times 2.303 \times \log(\kappa \mathbf{k}/h) \qquad \textbf{(6.36)}$$

[77]The transmission coefficient is assumed to be a constant for all the variants of one reaction type.

[78]Leffler, J. E.; Grunwald, E. *Rates and Equilibria of Organic Reactions*; John Wiley and Sons, Inc.: New York, 1963; p. 71.

and using the calculated value $\Delta H^{\ddagger}$ with values of k_r and T for one experiment.[79] There is a close relationship between $\Delta S^{\ddagger}$ and the Arrhenius A value, as shown in equation 6.37 for A values expressed in units of sec^{-1} and temperatures around 300 K[78]:

$$\Delta S^{\ddagger} = 4.575 \log A - 60.53 \tag{6.37}$$

The value of $\Delta S^{\ddagger}$ for a reaction provides an estimate of the change in the order of the system on going from reactants to activated complex.[80] Consider the reactions shown in Figures 6.21, 6.22, and 6.23. The dissociation of di-*t*-butyl peroxide (**37**, Figure 6.21) has been found to have an activation entropy of +13.8 eu in chlorobenzene solution,[81] suggesting that the activated complex is less ordered than the reactant. This result is compatible with a mechanism involving rate-limiting breaking of the O—O bond.[82]

On the other hand, the Cope rearrangement of *cis*-1,2-divinylcyclobutane to *cis,cis*-1,5-cyclooctadiene (Figure 6.22) was found to have an activation entropy of −11.7 eu,[83] meaning that the transition structure is *more* ordered than the reactant. This result suggests that the reaction does not proceed by a rate-limiting step involving breaking of the bond between the two carbon atoms with vinyl substituents. It is more consistent with a mechanism in which the two vinyl groups, which can rotate freely in the reactant, are aligned in the transition structure.[84]

The dimerization of cyclopentadiene has a $\Delta S^{\ddagger}$ of −26 eu.[85] This large, negative activation entropy suggests a very highly ordered transition structure in which free rotations have been lost as two different molecules come together in a very precise orientation. This result is consistent with the mechanism suggested in Figure 6.23. While this discussion has emphasized the activation entropy associated with the reacting molecules, values of $\Delta H^{\ddagger}$ and $\Delta S^{\ddagger}$ for reactions in solution also include contributions from solvent molecules.

Figure 6.21 Cleavage of di-*t*-butyl peroxide.

$$(CH_3)_3CO{-}OC(CH_3)_3 \xrightarrow{\Delta} \left[(CH_3)_3CO\cdots OC(CH_3)_3\right]^{\ddagger} \longrightarrow 2\,(CH_3)_3CO\cdot$$

$$\Delta S^{\ddagger} = +13.8 \text{ eu}$$

37

---

[79]Several simplified relationships are given in reference 51, p. 136.

[80]For a discussion, see Schaleger, L. L.; Long, F. A. *Adv. Phys. Org. Chem.* **1963**, *1*, 1.

[81]Bartlett, P. D.; Hiatt, R. R. *J. Am. Chem. Soc.* **1958**, *80*, 1398.

[82]Raley, J. H.; Rust, F. F.; Vaughan, W. E. *J. Am. Chem. Soc.* **1948**, *70*, 1336.

[83]Hammond, G. S.; DeBoer, C. D. *J. Am. Chem. Soc.* **1964**, *86*, 899.

[84]In addition, the puckered cyclobutane ring may become more planar in the transition structure. The mechanism of this and similar reactions will be considered in Chapter 11.

[85]Reference 56, p. 104.

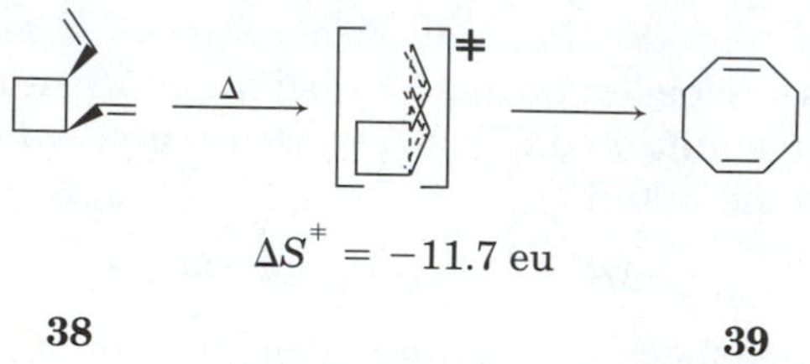

Figure 6.22 Rearrangement of *cis*-1,2-divinylcyclobutane to *cis,cis*-1,5-cyclooctadiene.

$\Delta S^{\ddagger} = -11.7$ eu

38 **39**

While $\Delta H^{\ddagger}$ and $\Delta S^{\ddagger}$ provide important clues to reaction mechanisms, we must not lose sight of the fact that it is $\Delta G^{\ddagger}$ that determines the rate of a reaction. Knowledge of activation free energies is particularly useful in understanding parallel organic reactions, since relatively small differences in activation free energies can produce significant differences in product distributions. Consider the reaction of compound **A** to give **B** or **C** in irreversible, forward reactions with rate constants k_B and k_C, respectively. If the reactions are truly irreversible, then the product distribution ratio ([**C**]/[**B**]) at any time will reflect the ratio of rate constants, k_C/k_B. Then

$$k_B = (\kappa \mathbf{k} T/h) \times \exp(-\Delta G^{\ddagger}_B/RT) \tag{6.38}$$

and

$$k_C = (\kappa \mathbf{k} T/h) \times \exp(-\Delta G^{\ddagger}_C/RT) \tag{6.39}$$

so

$$\ln(k_C/k_B) = \ln([\mathbf{C}]/[\mathbf{B}]) = \Delta\Delta G^{\ddagger}/RT \tag{6.40}$$

That is, the natural log of the product ratio should tell us the difference in activation free energies for the two pathways (Figure 6.25). Product ratios expected for different values of $\Delta\Delta G^{\ddagger}/RT$ are listed in Table 6.2. As Saunders noted, a change in the product ratio of **B** to **C** from 1 : 2 to 2 : 1 corresponds to a $\Delta\Delta G^{\ddagger}$ of only about 0.8 kcal/mol at 25°, so small changes in relative activation free energies can have pronounced changes in product distributions.[86]

We have to be very careful in using product distributions to estimate differences in activation free energies unless we are certain that the competing reactions are irreversible, that is, that the product distribution actually represents the relative magnitude of the rate constants for product formation. For example, Conant and Bartlett studied the formation of semi-

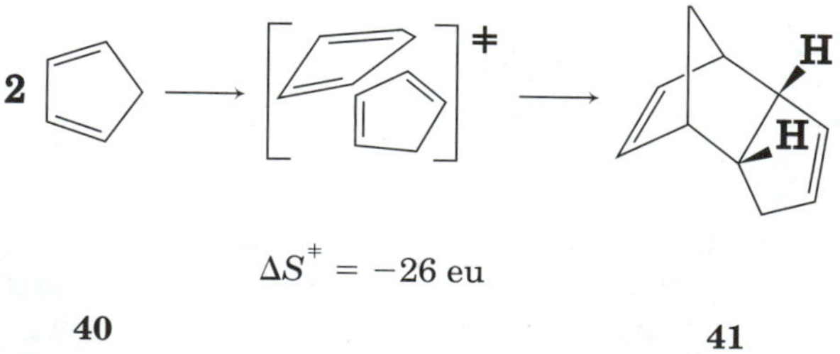

Figure 6.23 Dimerization of cyclopentadiene.

2

H

H

$\Delta S^{\ddagger} = -26$ eu

40 **41**

[86]Saunders, Jr., W. H. in *The Chemistry of Alkenes*; Patai, S., ed.; Wiley-Interscience: London, 1964; p. 184.

carbazones from a solution composed initially of equal concentrations of semicarbazide hydrochloride, furfural, and cyclohexanone (in a solution of KOAc in alcohol). The product isolated after a reaction period of twenty seconds was cyclohexanone semicarbazone. However, when the reaction was allowed to proceed for 2.5 hours, the product isolated was furfural semicarbazone.[87] Similarly, reaction of chlorobenzene with isopropyl chloride under Friedel-Crafts conditions produced primarily ortho and para alkylation product after a short reaction period. Allowing the reaction mixture to stand for a week before workup, however, produced primarily meta product.[88] In both of these cases, the product isolated after a short time is said to be the product of ***kinetic control*** of the product distribution, since the product isolated is the one that is formed faster. The product isolated after a longer time is said to be the product of ***thermodynamic control*** of product distribution, since it is the thermodynamically more stable product.

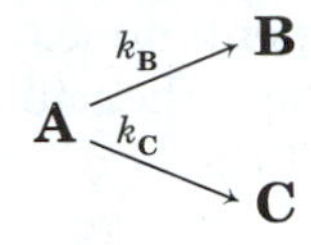

Figure 6.24 Competitive irreversible reactions of **A** to produce **B** and **C**.

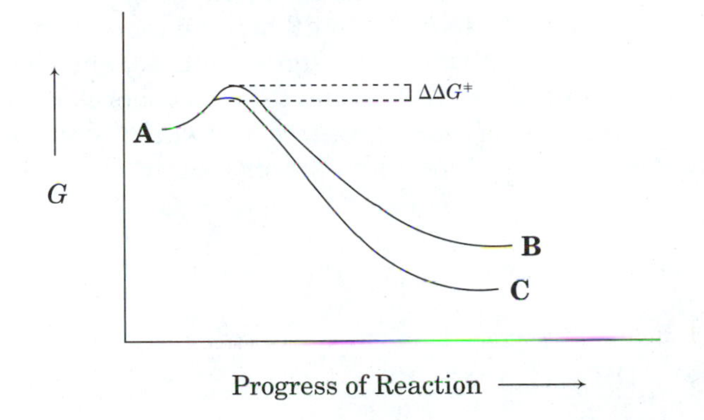

Figure 6.25 Illustration of the difference in free energies of activation for two competitive reactions.

Table 6.2 Product ratios and differences in activation free energy for competitive, irreversible reactions.

[C]/[B]	$\Delta\Delta G^{\ddagger}$ (25°, kcal/mol)
1	0.0
2	0.41
5	0.96
10	1.36
100	2.73
1,000	4.09
10,000	5.45

[87]Conant, J. B.; Bartlett, P. D. *J. Am. Chem. Soc.* **1932**, *54*, 2881.

[88]Kolb, K. E.; Standard, J. M.; Field, K. W. *J. Chem. Educ.* **1988**, *65*, 367.

Questions about kinetic and thermodynamic control of product distribution often arise in connection with reactions in which different products seem to form under different temperature conditions. Consider the example cited by Youssef and Ogliaruso (Figure 6.26), in which an anion (**44**) derived from base catalyzed rearrangement of 1,2,3,4,5-pentaphenyl-2,4-cyclopentadien-1-ol (**42**) was protonated to give the pentaphenylcyclopentenones **45** and **46**.[89] In a control experiment, it was found that subjecting either **45** or **46** to the reaction conditions produced the same 99 : 1 ratio of the two compounds.

These results can be rationalized by the reaction coordinate diagram shown in Figure 6.27. Because $\Delta G^{\ddagger}$ for conversion of **44** to **45** is smaller than the value of $\Delta G^{\ddagger}$ for conversion of **44** to **46**, formation of **45** is faster, so more of **45** is formed during the early stages of the reaction. Therefore the distribution of products in the early stage of the reaction is said to reflect kinetic control. As the reaction proceeds, the concentration of **44** is reduced, while the concentrations of **45** or **46** increase, and the much slower processes that convert **45** to **44** to **46** become more significant. Eventually, an equilibrium is established between **45** and **46**, and the product distribution at the end of the reaction is said to reflect thermodynamic control.

The terms *thermodynamic control* and *kinetic control* have been criticized by Brown et al., since both "thermodynamics *and* kinetics control *all*

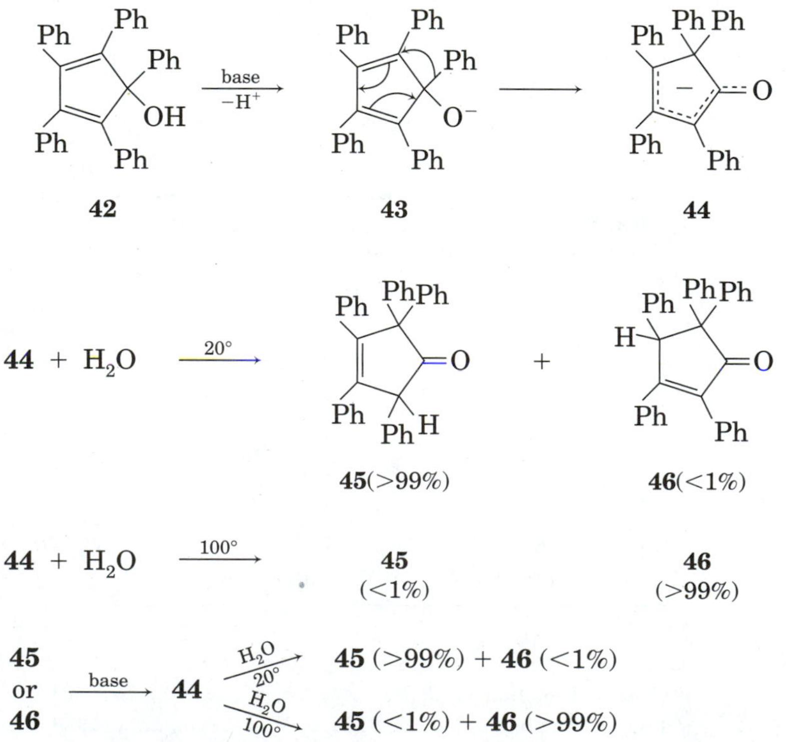

Figure 6.26
Temperature effects on the reaction of **44** with water.

42 —(base, $-H^+$)→ **43** → **44**

44 + H_2O —(20°)→ **45**(>99%) + **46**(<1%)

44 + H_2O —(100°)→ **45** (<1%) + **46** (>99%)

45 or **46** —(base)→ **44** —(H_2O, 20°)→ **45** (>99%) + **46** (<1%)

44 —(H_2O, 100°)→ **45** (<1%) + **46** (>99%)

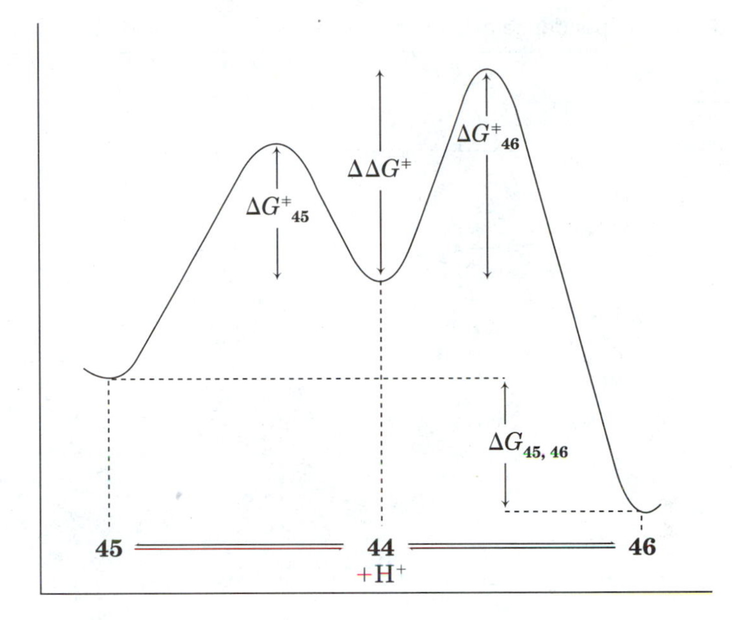

Figure 6.27
Reaction coordinate diagram for formation of kinetic and thermodynamic products. (Adapted from reference 91.)

products," and the terms *kinetic* and *thermodynamic control* sometimes are thought to imply that temperature controls the reaction.[90] Snadden has emphasized that the distinction between thermodynamic and kinetic control of product distributions is really one of *time* and not of *temperature* control of the reaction.[91] The time dependence of reactant and product concentrations for a reaction such as that in Figure 6.26 would be qualitatively similar to that shown in Figure 6.28. (Note that the x-axis is logarithmic.) Because the equilibrium between the two products will be established more rapidly at a higher temperature, the effect of increasing the temperature of the reaction would be to compress the curves along the x-axis, so that the thermodynamic product would be isolated in a shorter period. Keeping the temperature low would expand the curves along the x-axis, ensuring that the kinetic product would be isolated during the same time period. In other words, quenching of **44** at 20° would also give a 99 : 1 ratio of **46** to **45** if we were to wait long enough before working up the reaction mixture.[92]

---

[89]Youssef, A. K.; Ogliaruso, M. A. *J. Chem. Educ.* **1975**, *52*, 473.

[90]Brown, M. E.; Buchanan, K. J.; Goosen, A. *J. Chem. Educ.* **1985**, *62*, 575. Alternative terms are *rate control* and *equilibrium control*, which more precisely describe the factors that determine product distribution. For an example of this usage, see Solomons, T. W. G. *Organic Chemistry*, 4th ed.; John Wiley & Sons, Inc.: New York, 1988; p. 470.

[91]Snadden, R. B. *J. Chem. Educ.* **1985**, *62*, 653.

[92]This discussion ignores a small temperature effect on the equilibrium between the two products.

Figure 6.28 Time dependence of the concentrations of **44**, **45**, and **46**. (Adapted from reference 91.)

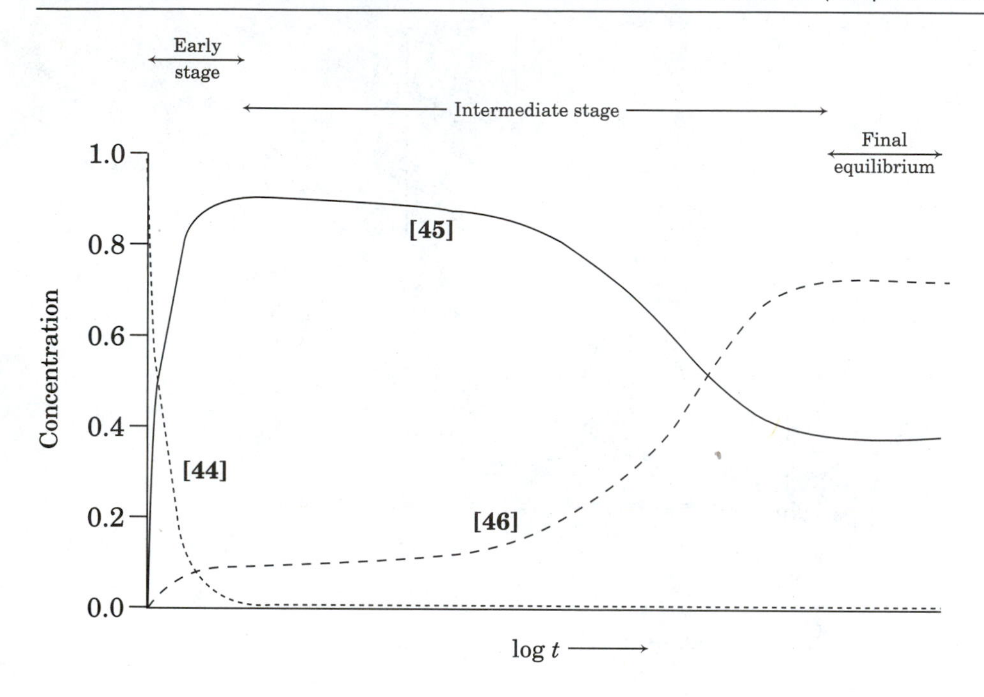

Another aspect of kinetics that is important in organic chemistry is called the **Curtin-Hammett principle.**[93,94] The chemical reaction that led to the statement of the principle is the E2 elimination of 1-chloro-1,2-diphenylethane (**47**) to give *cis*-stilbene (**48**) and *trans*-stilbene (**49**). If we assume that an *anti-periplanar* relationship is required for the reaction, then the two conformations of the reactant should give two stereochemically different products.

The *anti* conformer **47(a)** should be more stable than the gauche conformer **47(g)**. If we observe that the yield of **49** is greater than that of **48**, then it is tempting to conclude that the product ratios provide a measure of equilibrium between **47(a)** and **47(g)** in the reactant. In other words, we might assume that

$$[\mathbf{47(a)}]/[\mathbf{47(g)}] \equiv [\mathbf{49}]/[\mathbf{48}] \tag{6.41}$$

However, this conclusion would be correct *only* if k_3 and k_4 are both much greater than k_1 and k_2. In general, however, we would expect the rate of rotation about a C—C single bond to be much faster than the rate of a bimol-

[93]Curtin, D. Y. *Record Chem. Prog.* **1954**, *15*, 111. In the second edition of *Physical Organic Chemistry* (reference 95, p. 119), Hammett refers to the "Curtin Principle." Then, in a footnote, he says that "because Curtin is very generous in attributing credit, this is sometimes referred to as the Curtin-Hammett principle."

[94]For a memorial lecture summarizing the contributions of L. P. Hammett, see Shorter, J. *Prog. Phys. Org. Chem.* **1990**, *17*, 1.

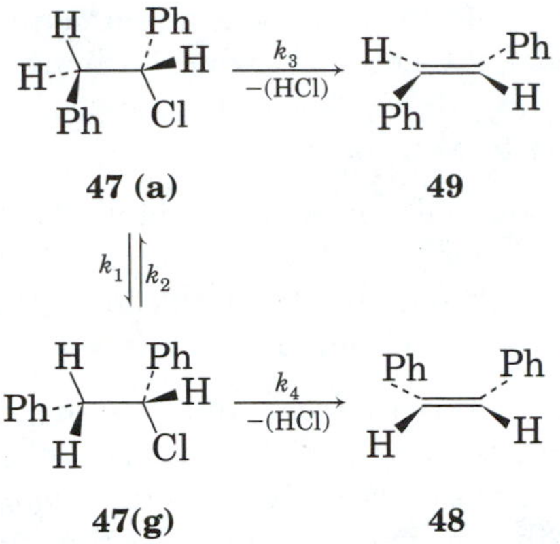

Figure 6.29
Illustration of the Curtin-Hammett principle.

47 (a) $\xrightarrow[-(\mathrm{HCl})]{k_3}$ 49

k_1 ⇅ k_2

47(g) $\xrightarrow[-(\mathrm{HCl})]{k_4}$ 48

ecular elimination reaction. As illustrated in Figure 6.30, the rates of formation of the products *cis*- and *trans*-stilbene are determined only by the difference in activation free energies ($\Delta\Delta G^{\ddagger}$) for the two elimination steps, provided that the conformations of the precursor interconvert faster than the elimination steps.[95,96,97]

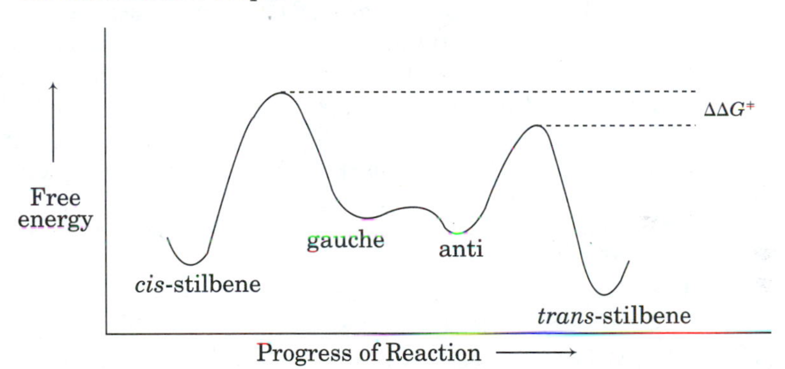

Figure 6.30
Graphical representation of the Curtin-Hammett principle. (Adapted from reference 96.)

6.5 Kinetic Isotope Effects

In the examples discussed so far, we have considered reactions that were either one-step reactions or were multi-step reactions in which the rate-limiting step is known. In many cases reactions involve more than one step.

[95]Hammett, L. P. *Physical Organic Chemistry*, 2nd ed.; McGraw-Hill: New York, 1970; p. 120.

[96]See also the discussion by Eliel, E. L. *Stereochemistry of Carbon Compounds*; McGraw-Hill Book Company: New York, 1962; pp. 151–152, 237–239; Eliel, E. L.; Wilen, S. H. *Stereochemistry of Organic Compounds*; Wiley-Interscience: New York, 1994: pp. 647–656.

[97]Reaction rate constants and conformational equilibria are related through the Winstein-Holness equation. See Winstein, S.; Holness, N. J. *J. Am. Chem. Soc.* **1955**, *77*, 5562. See also Eliel, E. L. *J. Chem. Educ.* **1960**, *37*, 126 and references therein. For a discussion of these relationships, see reference 96; Seeman, J. I.; Farone, W. A. *J. Org. Chem.* **1978**, *43*, 1854; Seeman, J. I. *J. Chem. Educ.* **1986**, *63*, 42; Perrin, C. L.; Seeman, J. I. *J. Org. Chem.* **1984**, *49*, 2887. See also DeTar, D. F. *J. Org. Chem.* **1986**, *51*, 3749.

Even if we have enough information about a reaction that we can propose a general mechanism, we may not know what is the rate-limiting step. Consider the oxidation of isopropyl alcohol by aqueous chromic acid (Figure 6.31). Two steps appear to be involved:[98,99]

1. formation of a chromate ester from isopropyl alcohol and chromic acid, and
2. decomposition of the chromate ester with concurrent oxidation of the alcohol and reduction of the chromium.

A central question about this reaction concerns the H—C bond shown in bold in Figure 6.31. Does the rate-limiting step occur before this bond is broken (perhaps during formation of the chromate ester), does it occur during the breaking of this bond, or does it occur after this bond is broken (for example, through the decomposition of a carbanionic intermediate in which the oxygen-chromium bond is still intact)?

There is a kinetic method that can in principle tell us about bonding changes in the rate-limiting step of a reaction and which was used to answer the questions just posed. The method is the study of ***kinetic isotope effects***, in which the rate of a reaction may vary when an atom is replaced by a different (usually heavier) isotope. In one approach we replace with a heavier isotope an atom to which a bond is broken, and the technique is called the ***primary (1°) kinetic isotope effect (PKIE)***. In other cases we place isotopes on sites near the center of bonding changes, and that technique is the study of ***secondary (2°) kinetic isotope effects***.[100]

Figure 6.31
Two steps in the oxidation of isopropyl alcohol by chromic acid.

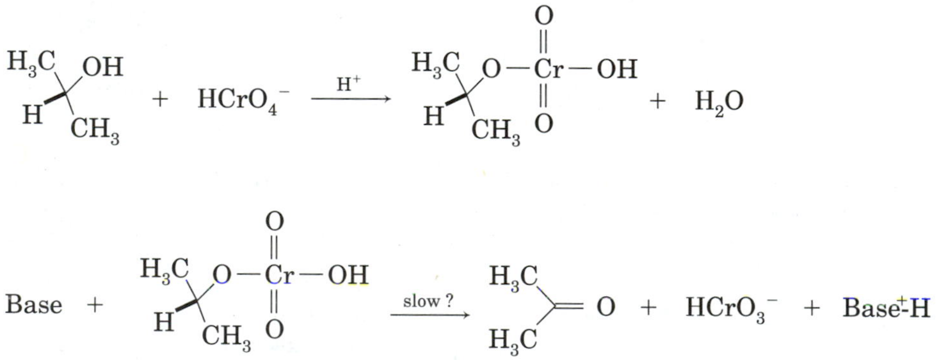

$$(CH_3)_2CH\text{–}OH + HCrO_4^- \xrightarrow{H^+} (CH_3)_2CH\text{–}O\text{–}CrO_2\text{–}OH + H_2O$$

$$\text{Base} + (CH_3)_2CH\text{–}O\text{–}CrO_2\text{–}OH \xrightarrow{\text{slow ?}} (CH_3)_2C{=}O + HCrO_3^- + \text{Base}^+\text{–}H$$

[98]Holloway, F.; Cohen, M.; Westheimer, F. H. *J. Am. Chem. Soc.* **1951**, *73*, 65; Roček, J.; Westheimer, F. H.; Eschenmoser, A.; Moldoványi, L.; Schreiber, J. *Helv. Chim. Acta* **1962**, *45*, 2554.

[99]The overall mechanism is quite complex, and additional processes are involved in the reaction. For further discussion, see (a) Westheimer, F. H. *Chem. Rev.* **1949**, *45*, 419; (b) Watanabe, W.; Westheimer, F. H. *J. Chem. Phys.* **1949**, *17*, 61; (c) Wiberg, K. B.; Schäfer, H. *J. Am. Chem. Soc.* **1969**, *91*, 927, 933.

[100]For further reading, see (a) Melander, L.; Saunders, Jr., W. H. *Reaction Rates of Isotopic Molecules*; John Wiley & Sons: New York, 1980; (b) Saunders, Jr., W. H. in *Investigation of Rates and Mechanisms of Reactions*, 4th ed., Vol. VI, Part I, Bernasconi, C. F., ed.; John Wiley & Sons: New York, 1986, p. 565; (c) Buncel, E.; Lee, C. C., eds., *Isotopes in Organic Chemistry*, Vol. 7; Elsevier: Amsterdam, 1987.

Primary Kinetic Isotope Effects

We will approach the theory of kinetic isotope effects with a somewhat intuitive conceptual model, leaving the detailed mathematical treatment to other references.[101] Consider first a unimolecular thermal dissociation of a C—H bond. The electronic energy of the ground state of the molecule varies as a function of the internuclear distance and can be described by the curve in Figure 6.32. Superimposed on the electronic energy curve of the molecule is a set of vibrational energy levels ($v = 0, 1, 2, \ldots$) with widths that represent the vibrational amplitudes of the bond. The vibrational energy levels become compressed near the dissociation limit because of the anharmonicity of the bond vibration.[102] The energy of each vibrational level is given (to a first approximation) by the formula[103]

$$E = h\nu(v + \tfrac{1}{2}) \qquad \textbf{(6.42)}$$

where

$$\nu = (2\pi)^{-1}(k/\mu)^{1/2} \qquad \textbf{(6.43)}$$

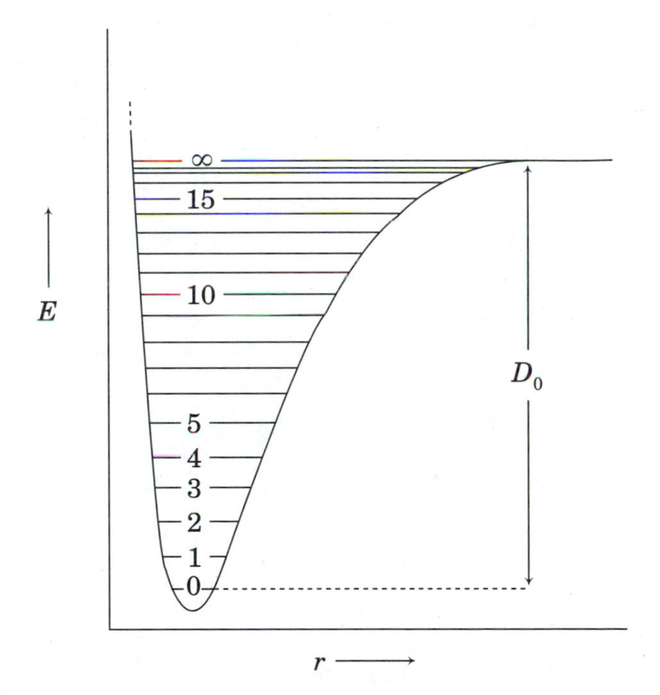

Figure 6.32 Vibrational energy levels and bond dissociation energy. (Adapted from reference 102.)

[101]See, for example, (a) Kresge, A. J. *J. Am. Chem. Soc.* **1980**, *102*, 7797; (b) Wiberg, K. B. *Chem. Rev.* **1955**, *55*, 713; (c) Westheimer, F. H. *Chem. Rev.* **1961**, *61*, 265; (d) Collins, C. J.; Bowman, N. S., eds., *Isotope Effects in Chemical Reactions*; Van Nostrand Reinhold Company: New York, 1970.

[102]For a discussion, see King, G. W. *Spectroscopy and Molecular Structure*; Holt, Rinehart and Winston, Inc.: New York, 1964; pp. 160 *ff*.

[103]This equation implicitly models the bond vibration after a harmonic oscillator. Bonds are anharmonic oscillators, especially at higher vibrational levels. However, at very low vibrational levels the approximation is not too bad.

and where k is the force constant for the molecular vibration and μ is the reduced mass, which is defined as

$$\mu = m_1 m_2/(m_1 + m_2) \tag{6.44}$$

for masses m_1 and m_2. At room temperature 99% of the molecules are in their 0th vibrational level,[104] so the vibrational energy of the molecule is greater than the hypothetical energy at the electronic energy minimum by an amount equal to ½ $h\nu$. This energy is called the ***zero point energy*** or ***ZPE***. The energy associated with the bond breaking (the **dissociation energy**, denoted E_D or D_0), is the difference in energy between the 0th vibrational energy level and the dissociation limit of the bond.[105]

There is a difference between the dissociation energy for a C—H bond in a CH_4 molecule and that of a C—D bond in a CH_3D molecule. If the CH_4 and CH_3D molecules are in their ground electronic states, they should have a common potential energy surface, and their energies should differ only in the vibrational and rotational energy levels associated with the C—H and C—D bonds.[106,107] Due to the greater mass of deuterium, E_0 is lower for a C—D bond than for a C—H bond (Figure 6.33).

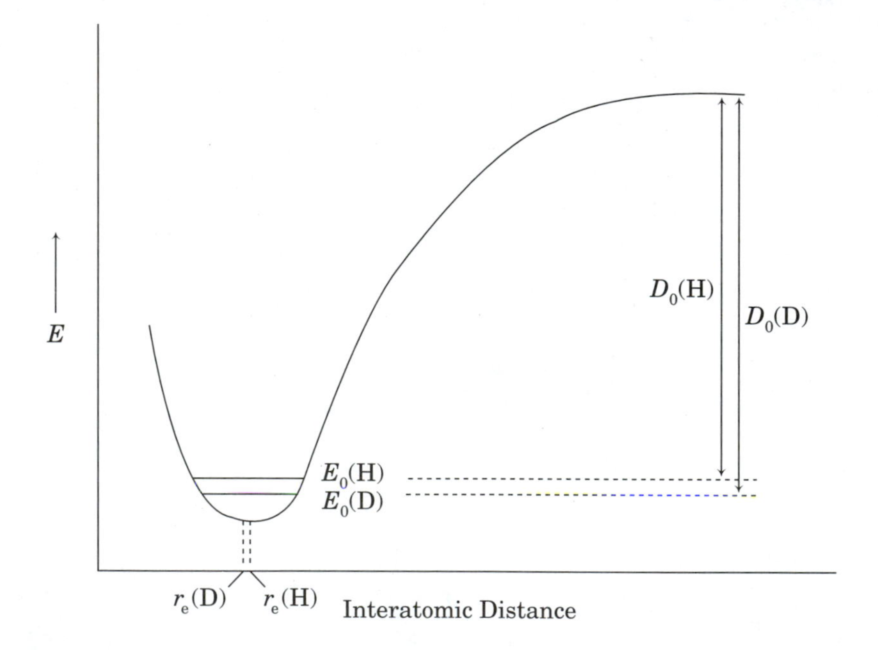

Figure 6.33 Difference in zero point energies for C—H and C—D bonds. (Adapted from reference 101(b), p. 715.)

---

[104] Reference 101(b), p. 715.

[105] Figure 6.32 is adapted from the electronic curve and vibrational levels calculated for the HF molecule in reference 102.

[106] In this discussion we are presuming that we can describe the bonding in terms of valence bond theory or in terms of "bonds" composed of localized molecular orbitals. An analysis using delocalized molecular orbitals would be equivalent, although less intuitive.

[107] Huskey, W. P. *J. Phys. Chem.* **1992**, *96*, 1263.

We could calculate the difference in E_0 values for C—H and C—D bonds from the reduced masses and an estimate of the stretching force constant. Alternatively, we can determine the energy difference empirically through infrared spectroscopy. As noted on page 352, the zero point energy for a bond stretching vibration is $\frac{1}{2}h\nu$. The zero point energy for a C-H bond is then determined from the wavelength of the infrared radiation absorbed in a C-H stretching vibration, which is about 3000 cm^{-1}.

$$E_0(\mathrm{H}) = \tfrac{1}{2}h\nu = (\tfrac{1}{2})\,3000\ \mathrm{cm}^{-1} = 1500\ \mathrm{cm}^{-1} \qquad (6.45)$$

For a C—D bond the corresponding zero point energy is[108]

$$E_0(\mathrm{D}) = \tfrac{1}{2}h\nu = (\tfrac{1}{2})\,2200\ \mathrm{cm}^{-1} = 1100\ \mathrm{cm}^{-1} \qquad (6.46)$$

Then

$$\Delta E_0 = 1500\ \mathrm{cm}^{-1} - 1100\ \mathrm{cm}^{-1} = 400\ \mathrm{cm}^{-1} \approx 1.15\ \mathrm{kcal/mol} \qquad (6.47)$$

This dissociation energy difference means that the activation energy for the thermal dissociation reaction should be 1.15 kcal/mol greater for the C—D dissociation than for the C—H dissociation.[101c] The rate constants for the C—H dissociation (k_{H}) and for the C—D dissociation (k_{D}) are given by the equations

$$k_{\mathrm{H}} = A_{\mathrm{H}} \exp(-E_a(\mathrm{H})/RT) \qquad (6.48)$$

$$k_{\mathrm{D}} = A_{\mathrm{D}} \exp(-E_a(\mathrm{D})/RT) \qquad (6.49)$$

The terms $E_a(\mathrm{H})$ and $E_a(\mathrm{D})$ are each composed of two parts: an electronic part, E_e, that is presumed to be the same for the two cases, and a vibrational part, E_{vib}, that is different in the way we have described above. If the preexponential factors A_{H} and A_{D} are identical,[109] then the ratio $k_{\mathrm{H}}/k_{\mathrm{D}}$ can be calculated to be

$$k_{\mathrm{H}}/k_{\mathrm{D}} = \exp[(1.15 \times 10^3\ \mathrm{cal\ mol}^{-1})/(300\ \mathrm{deg} \times 1.98\ \mathrm{cal\ deg}^{-1}\ \mathrm{mol}^{-1})] \qquad (6.50)$$

so

$$k_{\mathrm{H}}/k_{\mathrm{D}} = \exp(1.94) \approx 7 \qquad (6.51)$$

Thus we predict that in a purely unimolecular, thermal reaction a C—H bond should dissociate seven times as fast as a C—D bond.[101(c)]

Most organic processes of interest are not unimolecular thermolysis reactions.[110] More often, we are interested in reactions in which more than one bond is broken or formed in the same elementary step. Based on the Hammond postulate, we can envision three scenarios for a reaction in which

---

[108]For a discussion of the infrared spectra of deuterium-labeled compounds, see Pinchas, S.; Laulicht, I. *Infrared Spectra of Labelled Compounds*; Academic Press: London, 1971; pp. 65 *ff*.

[109]This assumption is adequate for our purposes, but it is not always rigorously correct. See, for example, reference 100(a), p. 146 for a discussion of $A_{\mathrm{H}}/A_{\mathrm{D}}$ values in studies of bimolecular elimination reactions.

[110]Some reactions of interest are unimolecular dissociations. For example, the shock tube dissociation of toluene to benzyl radical and hydrogen atom has been studied by Hippler, H.; Troe, J. *J. Phys. Chem.* **1990**, *94*, 3803.

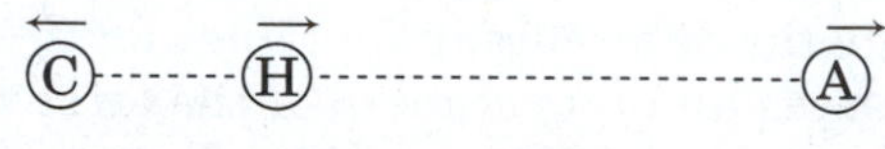

Figure 6.34
C · H · · · A vibration in an early transition state.

an atom A abstracts a hydrogen atom from a carbon atom.[111] In the first, there is an exothermic reaction and the transition state occurs early (to the left). In the second there is an endothermic reaction, so the transition state occurs late (to the right). In the third, there is a more thermoneutral reaction, so the transition state occurs near the center of a reaction coordinate diagram.

If the transition state occurs early, the C—H or C—D bond will be only slightly broken in the transition structure. The stretching vibrational mode shown in Figure 6.34 will be affected by the mass of the hydrogen (or deuterium) atom, so that the difference in transition state energy for the C—H

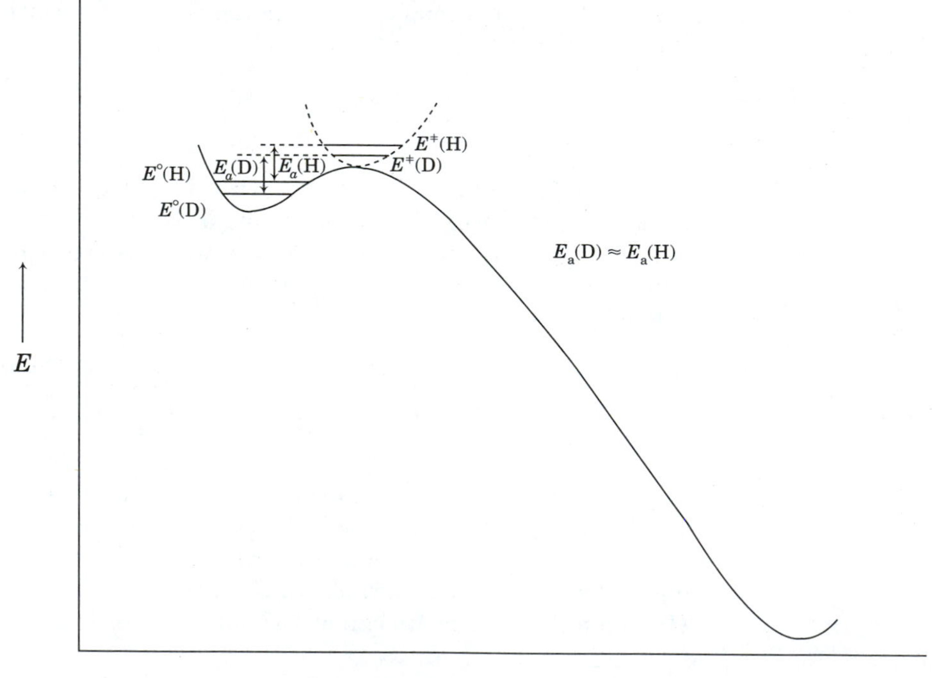

Figure 6.35
Origin of low k_H/k_D values for an exothermic reaction. (Modified from reference 101(b).)

[111]Examples of such reactions include (a) the transfer of a hydrogen atom from diphenylmethane to a benzyl radical, reported by Bockrath, B. C.; Bittner, E. W.; Marecic, T. C. *J. Org. Chem.* **1986**, *51*, 15 and (b) the cytochrome P-450 catalyzed oxidation of octane reported by Jones, J. P.; Trager, W. F. *J. Am. Chem. Soc.* **1987**, *109*, 2171.

or C—D species will be nearly the same in the transition structure as in the reactant. As shown on the hypothetical potential energy surface for the abstraction in Figure 6.35, there will be only a small difference in activation energies for the C—H or C—D reactions. Therefore, k_H/k_D will be close to unity for such a process.

In a very endothermic reaction (Figure 6.36), the transition structure closely resembles the product. Although the original C—H or C—D bond will be almost fully broken, the bond between H or D and the abstracting atom (A) will be nearly fully developed. In the transition structure there will be nearly the same difference in zero point energies for the developing A—H or A—D bond as there is in the C—H or C—D bond in the reactant. As a result, the activation energy for abstracting a deuterium will be very similar to that for abstracting a hydrogen atom, so there will be only a very small primary kinetic isotope effect.

In a thermoneutral reaction the transition state is symmetrically located along the reaction coordinate. Now the hydrogen atom does not move in the vibrational mode illustrated in Figure 6.37 (page 356), so the frequency of the vibration does not depend on its mass. Therefore $E^{\ddagger}(D) = E^{\ddagger}(H)$; that is, the transition state energy is the same, whether it is a hydrogen or deuterium atom that is being abstracted. As shown in Figure 6.38 (page 356), the activation energies for the abstraction of H and D will differ by an amount equal to the difference in the C—H and C—D bond zero point

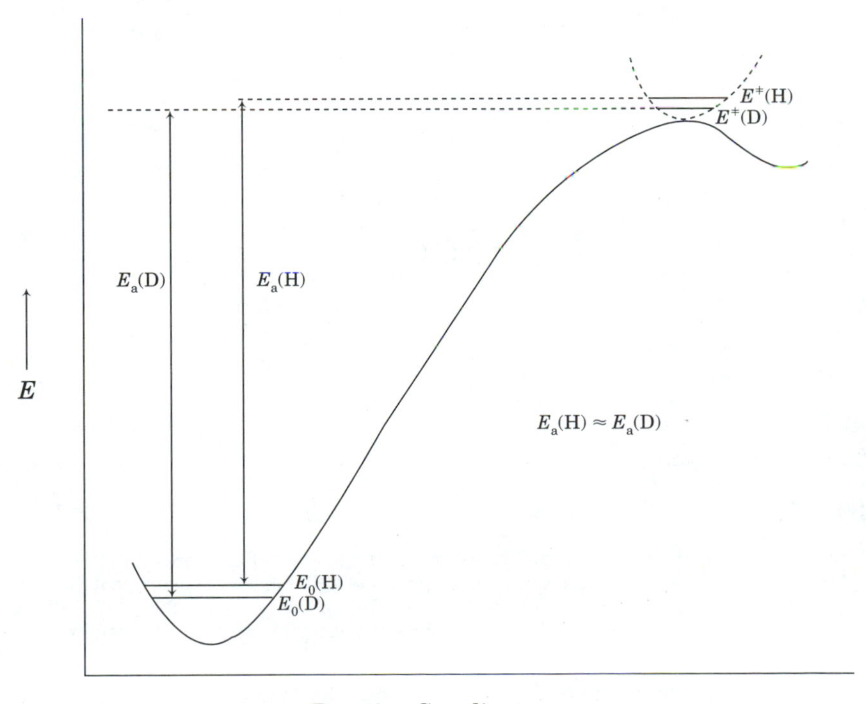

Figure 6.36
Origin of low k_H/k_D values for an endothermic reaction. (Modified from reference 101(b).)

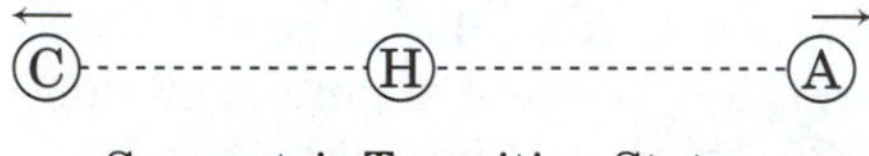

Figure 6.37
C··H··A vibrational mode in a symmetric transition state.

energies, so there should be a significant kinetic isotope effect.[112,113] This situation appears to describe the oxidation of $CH_3CDOHCH_3$ shown in Figure 6.31. Westheimer and Nicolaides found that the value of k_H/k_D for the reaction was approximately 6, thereby confirming that the H—C bond is broken in the rate-limiting step of the reaction sequence.[114]

In view of the discussion here, we might expect that the maximum hydrogen primary kinetic isotope effect for a reaction in which a C—H bond is

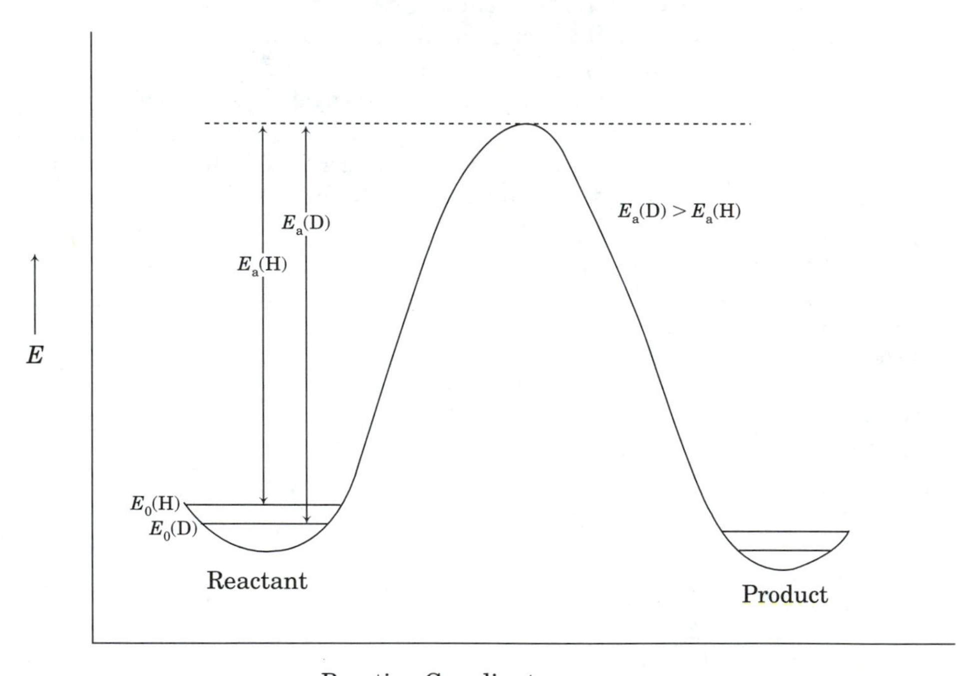

Figure 6.38
Origin of large k_H/k_D value for a nearly thermoneutral reaction. (Adapted from reference 101(b), p. 732.)

---

[112]The theoretical basis for kinetic isotope effects presented here is quite simplified. (See, for example, reference 101(c), pp. 268–269). For more theoretical discussions, see the sources cited in references 100 and 101. See also reference 46, pp. 296–297; Gilliom, R. *Introduction to Physical Organic Chemistry*; Addison-Wesley Publishing Co.: Reading, MA, 1970.

[113]The effect of transition state location on kinetic isotope effects has been analyzed by Garrett, B. C.; Truhlar, D. G. *J. Am. Chem. Soc.* **1980**, *102*, 2559.

[114]Westheimer, F. H.; Nicolaides, N. *J. Am. Chem. Soc.* **1949**, *71*, 25.

broken would be $k_H/k_D = 7$, while we expect k_H/k_D to be less than the maximum for transition states that are not exactly midway between reactants and products. The usual range for primary kinetic isotope effects is about 5–8, although a ratio of k_H/k_D of 25 has been reported in one case,[115] and a value of 13,000 in one unusual situation.[116] Obviously, our analysis of the possible magnitude of k_H/k_D has been oversimplified. Two factors were specifically ignored:

1. We have considered C—H or C—H stretching vibrations but not C—H or C—D bending vibrations. If one adds these vibrations to the theoretical determination of k_H/k_D values, the maximum value is larger. For example, Wiberg calculated the C—H/C—D energy difference to be 2.3 kcal/mol, so that $(k_H/k_D)_{max} \approx 48$.[117,118,119]
2. Our analysis has ignored the possibility of tunneling, through which a proton can cross an energy barrier with less energy than the E_{act}, whereas a more massive deuterium or tritium cannot easily do so.[116,120,121,122]

The discussion to this point has considered only *intermolecular* effects, that is the comparison of reaction rates for C—H or C—D bonds in different molecules. Let us now consider *intramolecular* kinetic isotope effects resulting from competition for reaction between C—H and C—D bonds in the same molecule.[123] Our example is the free radical abstraction of a H (or D) atom from α-*d*-toluene (Figure 6.39).[124] The rate constants k_H and k_D represent the processes for abstracting hydrogen and deuterium, respectively.

---

[115]Reference 95, p. 129.

[116]Brunton, G.; Griller, D.; Barclay, L. R. C.; Ingold, K. U. *J. Am. Chem. Soc.* **1976**, *98*, 6803.

[117]Wiberg, K. B. *Physical Organic Chemistry*; Wiley: New York, 1964; pp. 352–353.

[118]Since tritium is even more massive than deuterium, k_H/k_T values can be even larger, in the range of 50–75. See reference 101(a), p. 130.

[119]KIE values can be calculated from models that use molecular mechanics to determine either or both rotational and translational contributions and vibrational contributions: Canadell, E.; Olivella, S.; Poblet, J. M. *J. Phys. Chem.* **1984**, *88*, 3545.

[120]For a review see Bell, R. P. *The Tunnel Effect in Chemistry*; Chapman and Hall: New York, 1980.

[121]Lewis, E. S.; Robinson, J. K. *J. Am. Chem. Soc.* **1968**, *90*, 4337.

[122]It has been calculated that 80% of the hydrogen atom transfer reaction O + HD → OH + D occurs by tunneling. Garrett, B. C.; Truhlar, D. G.; Bowman, J. M.; Wagner, A. F.; Robie, D.; Arepalli, S.; Presser, N.; Gordon, R. J. *J. Am. Chem. Soc.* **1986**, *108*, 3515.

[123]The magnitude of the secondary isotope is assumed to be small in comparison with the primary isotope effect for this reaction (reference 128). A method for the separation of primary and secondary isotope effects was reported by Hanzlik, R. P.; Hogberg, K.; Moon, J. B.; Judson, C. M. *J. Am. Chem. Soc.* **1985**, *107*, 7164, and a detailed analysis of primary and secondary isotope effects in this system has been given by Hanzlik, R. P.; Schaefer, A. R.; Moon, J. B.; Judson, C. M. *J. Am. Chem. Soc.* **1987**, *109*, 4926.

[124]For a discussion of the abstraction of hydrogen atoms by radicals, see Tedder, J. M. *Quart. Rev. Chem. Soc.* **1960**, *14*, 336. For a discussion of the photochlorination of alkanes in solution, see Ingold, K. U.; Lusztyk, J.; Raner, K. D. *Acc. Chem. Res.* **1993**, *23*, 219.

Note that the potential energy well of the reactant shows only one vibrational energy level, since any one molecule can have only one minimum vibrational energy level (Figure 6.40). However, abstraction of the hydrogen will leave behind a C—D bond, so the transition state will be lower in energy than that for abstraction of a deuterium, which would leave behind a C—H bond instead.[125] As before, the removal of deuterium will require more energy, and thus be kinetically disfavored relative to the removal of hydrogen.

This analysis is consistent with experimental data for free radical halogenation of toluene. If the abstracting radical is a chlorine atom, k_H/k_D is 1.3 for reaction in CCl_4 solution at 77°C. If the abstracting atom is bromine, however, the ratio is 4.6.[126] We interpret the results to mean that the transition state lies further to the right for the bromine abstraction, so that more radical character is developed at the transition state and there is a greater difference between the energies of the benzyl radical with deuterium and the radical without deuterium substitution (Figure 6.41).

Kinetic isotope effects generally decrease with increasing reaction temperatures, since the small difference between C—H and C—D dissociation pathways becomes less significant as more energy is available to the system.[127] Table 6.3 shows results for gas-phase bromine-mediated free radical bromination of α-deuteriotoluene to give benzyl bromide and α-deuteriobenzyl bromide.[128]

It may be obvious that primary kinetic isotope effects may be most pronounced for reactions that involve dissociation of C—H bonds, since the mass of the leaving group doubles on going from hydrogen to deuterium and

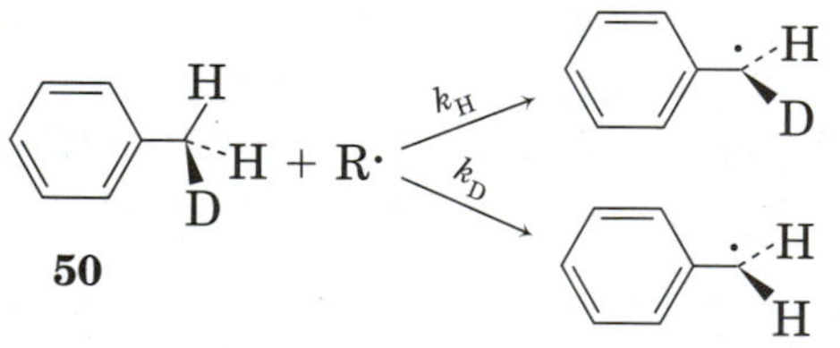

Figure 6.39
The intramolecular primary hydrogen isotope effect.

---

[125]It is the vibrational zero point energy of the C-H or C-D bond remaining in the molecule, not the vibrational zero point energy of the bond that is being broken, that makes the difference in transition state energy.

[126]Wiberg, K. B.; Slaugh, L. H. *J. Am. Chem. Soc.* **1958**, *80*, 3033 and references therein.

[127]This statement is correct for PKIE measurements in which a single reaction step determines the observed reaction rate. In more complex cases, however, the observed PKIE can increase with increasing temperature. For example, see Koch, H. F.; Koch, A. S. *J. Am. Chem. Soc.* **1984**, *106*, 4536.

[128]Timmons, R. B.; de Guzman, J.; Varnerin, R. E. *J. Am. Chem. Soc.* **1968**, *90*, 5996. Product ratios were determined from ratios of H_2, HD, and D_2 produced from the HBr and DBr byproducts of the reaction. Because $C_6H_5CH_2D$ contains two abstractable hydrogen atoms and one deuterium atom, the ratio of the rate constant for hydrogen abstraction divided by the rate constant for deuterium abstraction was divided by two to give a k_H/k_D value for the reaction.

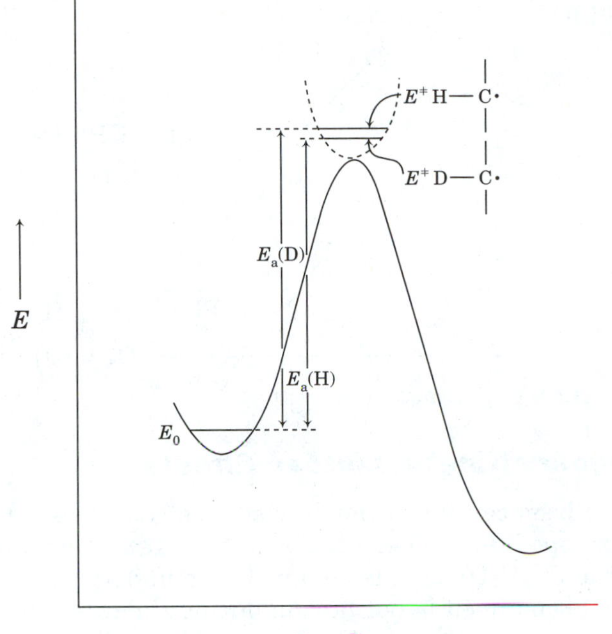

Figure 6.40 Reaction coordinate diagram for intramolecular primary hydrogen kinetic isotope effect. (Adapted from reference 101(b).)

triples on going from hydrogen to tritium.[129] Nevertheless, kinetic investigations using isotopes of other atoms have been found to give useful information about reaction mechanisms.[130] In addition, isotope effects can have surprising effects in spectroscopy. For example, replacement of hydrogen by deuterium can alter ^{13}C chemical shifts up to six bonds away from the site of the substitution.[131]

---

[129]The relationship between deuterium and tritium primary kinetic isotope effects is given by the Swain-Schaad equation:

$$\frac{k_H}{k_D} = \left(\frac{k_D}{k_T}\right)^x = \left(\frac{k_H}{k_T}\right)^{x/(x+1)}$$

where $x = 2.344$. Swain, C. G.; Stivers, E. C.; Reuwer, Jr., J. F.; Schaad, L. J. *J. Am. Chem. Soc.* **1985**, *80*, 5885. Also see Streitwieser, Jr., A.; Kaufman, M. J.; Bors, D. A.; Murdoch, J. R.; MacArthur, C. A.; Murphy, J. T.; Shen, C. C. *J. Am. Chem. Soc.* **1985**, *107*, 6983.

[130]Buncel, E.; Saunders, Jr., W. H., eds., *Isotopes in Organic Chemistry*, Vol. 8. *Heavy Atom Isotope Effects*; Elsevier: Amsterdam, 1992. For an example of the application of both ^{13}C and ^{18}O isotope effects in the study of one reaction, see Jacober, S. P.; Hanzlik, R. P. *J. Am. Chem. Soc.* **1986**, *108*, 1594; for the application of ^{14}C, ^{37}Cl and ^{2}H isotope effects to a solvolysis reaction, see Burton, G. W.; Sims, L. B.; Wilson, J. C.; Fry, A. *J. Am. Chem. Soc.* **1977**, *99*, 3371; for the application of $^{11}C/^{14}C$ kinetic isotopes to study an enzymatic reaction, see Axelsson, B. S.; Bjurling, P.; Matsson, O.; Långström, B. *J. Am. Chem. Soc.* **1992**, *114*, 1502.

[131]Berger, S.; Künzer, H. *Tetrahedron* **1983**, *39*, 1327; Servis, K. L.; Domenick, R. L. *J. Am. Chem. Soc.* **1986**, *108*, 2211.

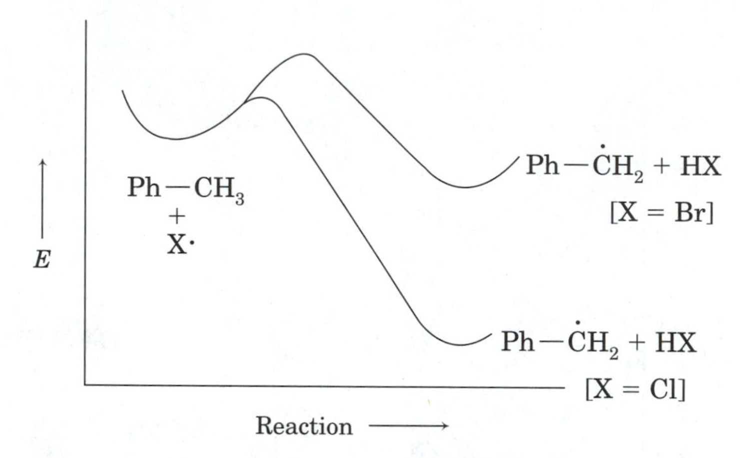

Figure 6.41 Reaction coordinate diagram for abstraction of H by Br · and by Cl · . (Adapted from reference 101(b), p. 733.)

Secondary Kinetic Isotope Effects

We have been careful in the discussion above to use the term *primary* kinetic isotope effect. There can also be a secondary kinetic isotope effect when the C—H(D) bond is on a molecule undergoing bond change, but the C—H(D) bond itself is not broken during the rate-limiting step. Generally secondary kinetic isotope effects are much smaller in magnitude than most examples of primary kinetic isotope effects.[132] On the other hand, the magnitude of secondary kinetic isotope effects are usually thought to increase as the transition structure changes from reactant-like to product-like, so secondary kinetic isotope effects can serve as a probe of transition state structure.[42,133]

We divide secondary kinetic isotope effects into subgroups depending upon the location of the isotope with respect to the reaction site. If the isotope is on the atom undergoing reaction, the effect is said to be an α sec-

Table 6.3 k_H/k_D values for gas phase bromination of α-deuteriotoluene.

T (°C)	k_H/k_D
121	6.69
130	6.53
142	6.17
150	5.93
160	5.69

(Data from reference 128.)

[132]For a review of secondary isotope effects, see Halevi, E. A. *Prog. Phys. Org. Chem.* **1963**, *1*, 109.

[133]Kovach, I. M.; Elrod, J. P.; Schowen, R. L. *J. Am. Chem. Soc.* **1980**, *102*, 7530.

ondary isotope effect. We further divide α effects into the following three categories:[134]

1. **Normal** ($k_H/k_D > 1$) α secondary kinetic isotope effects are seen for hybridization changes of the type $sp^3 \rightarrow sp^2$ or $sp^2 \rightarrow sp$. Consider the ionization of an alkyl halide, Figure 6.42(a).[135] The carbon orbital used for C—H bonding changes from an sp^3 hybrid to an sp^2 hybrid. The energy associated with that change will vary depending on whether the carbon atom is bonded to a hydrogen or deuterium atom, and the reaction will be found to be slightly faster for the molecule with the C—H bond. Due to the anharmonicity of the C—H bond vibration, higher vibrational energy means a greater C—H bond length than C—D bond length (see Figure 6.33). We might therefore expect that reactions that involve a change in hybridization from $sp^3 \rightarrow sp^2$ or $sp^2 \rightarrow sp$ would occur faster with D than with H, since the bond length should decrease with more *s* character in the hybrid orbital and since the C—D bond is already shorter than the C—H bond. In fact, the reactions are faster with H than with D. So far our analysis of isotope effects has considered only stretching vibrations, but bending vibrations become important for reactions involving hybridization change, and the α secondary hydrogen isotope effect is best explained on the basis of bending vibrations of the C—H and C—D bonds.[136] The proton-induced elimination of water from *p,p'*-dimethoxybenzhydrol (**52**) in Figure 6.43(a) has a $k_H/k_D = 1.18$,[137] and a k_H/k_D of 1.15 per D is typical for normal α secondary hydrogen kinetic isotope effects.

2. **Inverse** ($k_H/k_D < 1$) secondary kinetic isotope effects are seen for hybridization changes in the opposite direction ($sp \rightarrow sp^2$ or $sp^2 \rightarrow sp^3$). One process leading to an inverse secondary kinetic isotope effect would be the protonation of an alkene, Figure 6.42(b). Similarly, the cis-trans isomerization of maleic acid occurring according to the mechanism illustrated in Figure 6.43(b) has a k_H/k_D of 0.86 at 25°C.[138]

3. Reactions that convert sp^3-hybridized atoms to trigonal bipyramidal transition structures, as in S_N2 reactions, are a third category of α effects. Even though we commonly describe such trigonal bipyramidal transition structures as sp^2-hybridized, the isotope effects are not the normal effects expected; in many cases, α k_H/k_D values are unity or very slightly inverse for these kinds of reactions.[139] For the reaction shown in Figure 6.43(c), $k_H/k_D = 0.96$.[140]

---

[134]Reference 132, p. 145.

[135]For a discussion of secondary deuterium isotope effects on reactions that involve carbocation formation, see Sunko, D. E.; Hehre, W. J. *Prog. Phys. Org. Chem.* **1983**, *14*, 205.

[136]Streitwieser, Jr., A.; Jagow, R. H.; Fahey, R. C.; Suzuki, S. *J. Am. Chem. Soc.* **1958**, *80*, 2326.

[137]Stewart, R.; Gatzke, A. L.; Mocek, M.; Yates, K. *Chem. Ind. (London)* **1959**, 331.

[138]Seltzer, S. *J. Am. Chem. Soc.* **1961**, *83*, 1861.

[139]Reference 132, p. 173; Westaway, K. C. in reference 100(c), p. 275; McLennan, D. J. in reference 100(c), p. 393.

[140]Johnson, R. R.; Lewis, E. S. *Proc. Chem. Soc.* **1958**, 52.

Figure 6.42 Types of secondary kinetic hydrogen isotope effects.

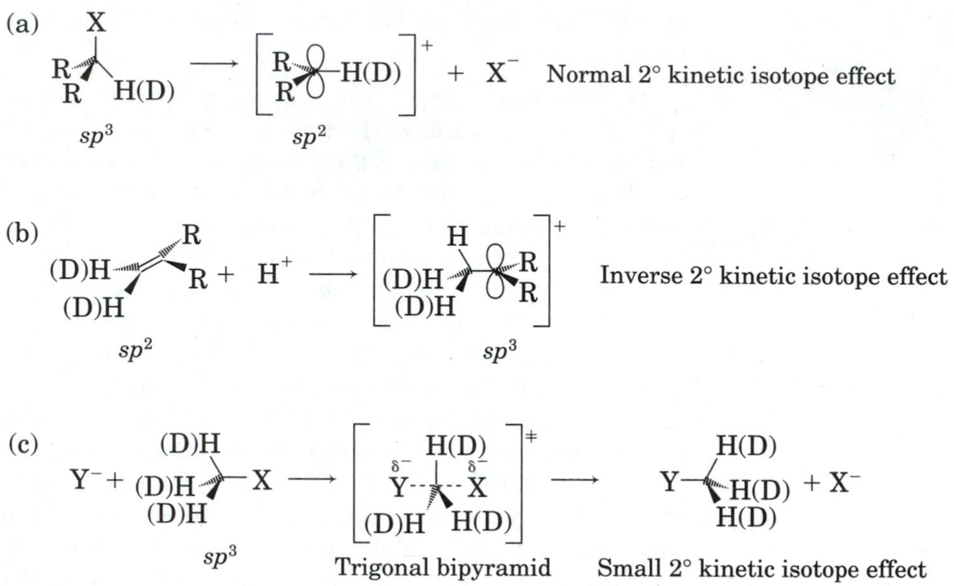

Figure 6.43 Examples of secondary hydrogen kinetic isotope effects.

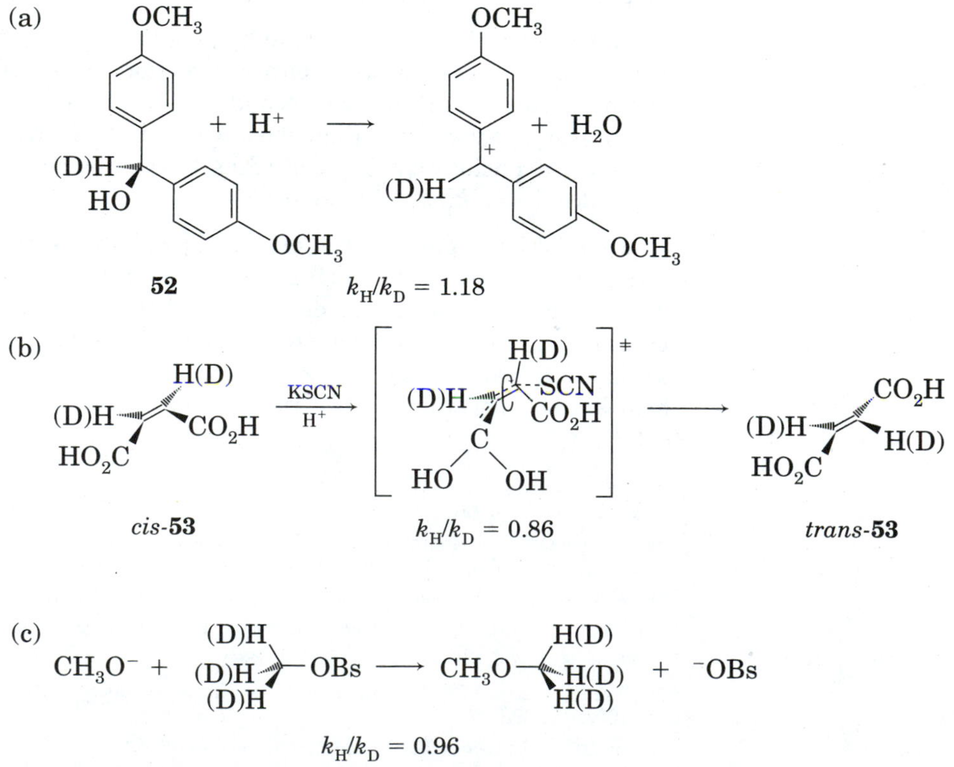

If the deuterium is bonded to an atom that is adjacent to the site of reaction, then a **β secondary isotope effect** may be observed. We might expect β secondary kinetic isotope effects to be smaller than α effects because the isotopic substituent is removed further from the reaction site, but this is not necessarily the case. The acidities of several carboxylic acids substituted with deuterium on the β carbon atom have equilibrium isotope effects, $K_{a(H)}/K_{a(D)}$ greater than 1.0 (Table 6.4).[141] The simplest explanation of the data is that CD_3 or CD_2 groups are electron-releasing with respect to CH_3 and CH_2 groups,[142] making the deuterated carboxylates less stable than their protonated analogues. If that is the case, such an effect can be rationalized by noting that the C—D bond is shorter than the C—H bond (Figure 6.33), so we would expect more electron density near the carbon nucleus. It is possible to see effects still further removed, but γ kinetic isotope effects are usually very small.[143]

Let us consider another case. S_N1 substitutions with β-D generally show $k_H/k_D > 1$, as indicated by the series of deuterated cyclopentyl tosylates in Table 6.5.[144] Numerous explanations have been advanced to explain results of this kind, but the most generally accepted view is that isotope effects on vibrations of β C—H (C—D) bonds dominate any electron donation effects in reactions of this kind. In short, the donation of electron density to a center of positive charge from β carbon atoms weakens the bond strength

Table 6.4 Isotope effects on acid strengths.

Deuterated Acid	$K_{a(H)}/K_{a(D)}$
C_6D_5OH	1.12 ± .02
C_6D_5COOH	1.024 ± .006
$C_6H_5CD_2COOH$	1.12 ± .02
DCOOH	1.06 ± .03
CD_3COOH	1.033 ± .002

(Data from the compilation in reference 141.)

[141]Data from Streitwieser, Jr., A.; Van Sickle, D. E. *J. Am. Chem. Soc.* **1962**, *84*, 254 and references therein. The values of $K_{a(H)}/K_{a(D)}$ in Table 6.4 reflect the contribution from all of the deuterium atoms in the structure indicated (i.e., the values shown are not on a per deuterium basis).

[142]See the discussion by Servis, K. L.; Domenick, R. L. *J. Am. Chem. Soc.* **1985**, *107*, 7186; Halevi, E. A.; Nussim, M.; Ron, A. *J. Chem. Soc.* **1963**, 866.

[143]Leffek, K. T.; Llewellyn, J. A.; Robertson, R. E. *J. Am. Chem. Soc.* **1960**, *82*, 6315.

[144]Data from reference 136.

Table 6.5 Isotope effects on cyclopentyl tosylate solvolysis (relative to rate for cyclopentyl tosylate).

Compound	k_H/k_D
cyclopentyl tosylate with D on C bearing OTs (OTs, D)	1.15
cyclopentyl tosylate with H/D on β carbon (OTs, H; H, D)	1.16
cyclopentyl tosylate with D/H on β carbon (OTs, H; D, H)	1.22

of C_β—H or C_β—D bonds. Thus there is a smaller difference between the ZPE of C—H and C—D in the transition structure than in the reactant, so the activation energy increases for the deuterated compound.[145] In addition to stretching vibrations, bending modes of the β-substituted atoms may also be important.[136,146]

We may also see isotope effects for processes that do not lead to bond breaking. Again recall from Figure 6.33 that the lower ZPE of C—D than C—H and the anharmonicity of the electronic potential energy curve cause the C—D bond to be slightly shorter than the C—H bond, suggesting the possibility of a steric isotope effect.[147] In molecular mechanics calculations, for example, deuterium is taken to have a smaller van der Waals radius than hydrogen.[148] Cyclohexane-d_1 (**54**) is reported to prefer the conformation with the D equatorial over that with the D axial by 6.3 *cal*/mol (0.0063 kcal/mol),[149] a result that agrees with theoretical calculations.[150] Another

[145]Boozer, C. E.; Lewis, E. S. *J. Am. Chem. Soc.* **1954**, *76*, 794.

[146]For a theoretical study of the relationship between the structure of S_N2 transition states and secondary α-deuterium kinetic isotope effects, see Poirier, R. A.; Wang, Y.; Westaway, K. C. *J. Am. Chem. Soc.* **1994**, *116*, 2526.

[147]For a more detailed description, see reference 101(a), pp. 189 *ff*; Carter, R. E.; Melander, L. *Adv. Phys. Org. Chem.* **1973**, *10*, 1; Bartell, L. S. *J. Am. Chem. Soc.* **1961**, *83*, 3567.

[148]Allinger, N. L.; Flanagan, H. L. *J. Comput. Chem.* **1983**, *4*, 399–403.

[149]Anet, F. A. L.; Kopelevich, M. *J. Am. Chem. Soc.* **1986**, *108*, 1355, 2109; a value of 8.3 ± 1.5 cal/mol was reported later by Anet, F. A. L.; O'Leary, D. J. *Tetrahedron Lett.* **1989**, *30*, 1059.

[150]Williams, I. H. *J. Chem. Soc., Chem. Commun.* **1986**, 627; Aydin, R.; Günther, H. *Angew. Chem., Int. Ed. Engl.* **1981**, *20*, 985.

manifestation of steric isotope effects is seen in the atropisomerization of the biphenyl **55**, in which k_H/k_D is found to be 0.84.[151,152]

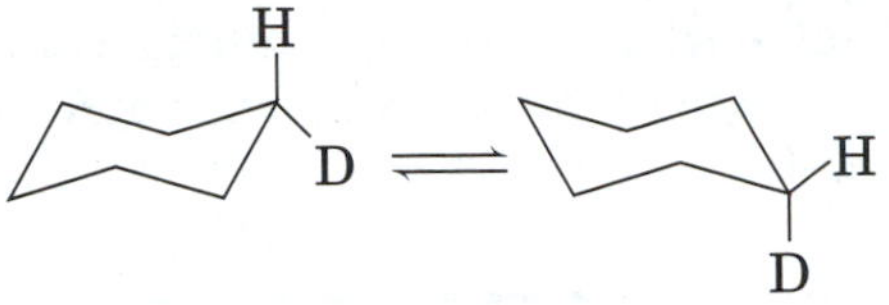

54e **54a** 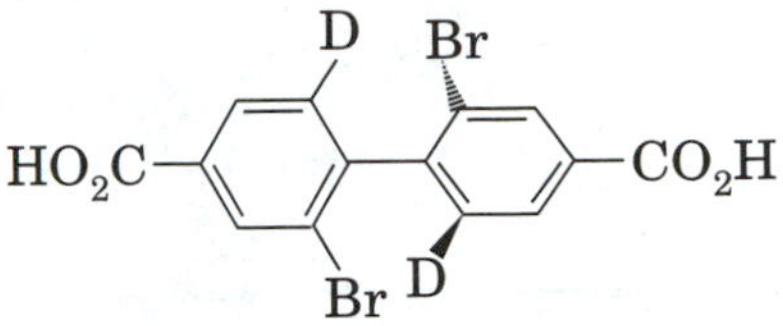 **55**

Solvent Isotope Effects

A ***solvent isotope effect*** results when the rate constant or the equilibrium constant for a process changes when a solvent is replaced with an isotopically substituted solvent.[153] Most often solvent isotope effect studies involve using a hydroxylic solvent in which the OH group is replaced with an OD group. For example, the value of K_H/K_D for ionization of phenylacetic acid in H_2O versus ionization of phenylacetic acid-O-*d* in D_2O is 3.[132] Schowen has indicated three ways for solvent isotope effects to occur:[153a]

1. The solvent may be a reactant; as with other reactions, there may be a 1° solvent isotope effect if proton is transferred in the rate-limiting step, otherwise there may be a 2° solvent isotope effect.
2. Solvent protons (or deuterons) on solvent molecules may exchange with reactant protons, leading to labeled reactant and a kinetic isotope effect. The isotopic fractionation factor, denoted ϕ, indicates the equilibrium distribution of the isotopic label between solute (R—H) and solvent (S—D). For the reaction

$$\text{R—H} + \text{S—D} \rightleftharpoons \text{R—D} + \text{S—H} \qquad (6.52)$$

ϕ is defined as

$$\phi = \frac{[\text{R—D}][\text{S—H}]}{[\text{R—H}][\text{S—D}]} \qquad (6.53)$$

3. Interactions of the transition structure with solvent molecules may be different in an isotopically labeled solvent.

Interpreting solvent kinetic isotope effects in terms of these interactions can provide information about the transition structure, its solvation, or

---

[151]Melander, L.; Carter, R. E. *J. Am. Chem. Soc.* **1964**, *86*, 295.

[152]A small steric isotope effect may also be seen in the flattening of the "butterfly" shape exhibited by the parent structure in perdeuterated tetracyanoanthraquinodimethane, which was reported by Heimer, N. E.; Mattern, D. L. *J. Am. Chem. Soc.* **1993**, *115*, 2217.

[153]For an introduction, see (a) Schowen, R. L. *Prog. Phys. Org. Chem.* **1972**, *9*, 275; (b) Kresge, A. J.; More O'Ferrall, R. A.; Powell, M. F. in *Isotopes in Organic Chemistry, Volume 7, Secondary and Solvent Isotope Effects*, Buncel, E.; Lee, C. C., eds.; Elsevier, Amsterdam, 1987; pp. 177 *ff*.

both. With one technique, known as the proton inventory technique, the rate of a reaction is studied in a series of mixtures of H_2O and D_2O. The isotope effect, k_n/k_{H_2O}, is plotted versus the parameter n (the mole fraction of deuterium in the solvent) and the slope and curvature of the resulting plot is analyzed for information about the number and role of solvent molecules involved in the rate-limiting step.[154,155]

6.6 Substituent Effects and Linear Free Energy Relationships

Substituent Effects

In studying the mechanism of an organic reaction, our goal is not just to be able to write down the steps that occur in a reaction. We also want to use that information to predict properties of a different reaction. If we carry out a chemical reaction such as

$$R_1{-}X \longrightarrow R_1{-}Y$$

and then modify the reactant in some way so that now the reaction is

$$R_2{-}X \longrightarrow R_2{-}Y$$

we want to be able to predict the effect of the structural change on the rate (or equilibrium) of the reaction. One useful approach is to change from R_1 to R_2 by making a substitution on the group R. Traditionally, the effect of such a substituent has been ascribed to one of three factors: induction, resonance, and steric hindrance.[156] Let us first look more closely at inductive and resonance effects.

Substituents that can donate or withdraw electrons by resonance are categorized as **+R** and **−R**, respectively.[157,158] The substituent must have a *p* orbital (alone or as part of a π system) that can either donate one or more electrons to, or accept one or more electrons from, another part of the molecule. Some +R substituents include the halogens as well as OR and NR_2 groups (where R can be H). Methyl and other alkyl groups are also considered +R substituents, which is consistent with the PMO model of stabilization of carbocations and radicals. Some −R groups include NO_2, CN, CO_2R, and C_6H_5.[156]

Similarly, substituent groups have been categorized according to their inductive effects as being either electron donating (**+I**) or electron with-

---

[154]Batts, B. D.; Gold, V. *J. Chem. Soc. A* **1969**, 984.

[155]For an example of the use of the proton inventory technique in the study of a reaction mechanism, see Bokser, A. D.; York, K. A.; Hogg, J. L. *J. Org. Chem.* **1986**, *51*, 92.

[156]Additional effects can be classified as statistical considerations (the effect of an additional carboxylic acid group on the pK_a of a carboxylic acid, for example), tautomeric, solvation, polarizability and internal hydrogen bonding effects. For a discussion, see Clark, J.; Perrin, D. D. *Quart. Rev. Chem. Soc.* **1964**, *18*, 295.

drawing (**−I**). The +I substituents are limited to alkyl groups such as methyl and to negatively charged substituents such as O^- and NH^- groups. The −I substituents include all those groups that are deactivating in electrophilic aromatic substitution, as well as other groups that have electronegative atoms such as OR, SR, and others at the point of attachment.

The effect that we commonly call induction is represented by designation of partial charges on atoms or by using arrows to represent the polarization of σ bonds, Figure 6.44(a). For example, we could ascribe the greater acidity of 2-fluoroethanol over that of ethanol to the inductive effect of the electronegative fluorine, which stabilizes the alkoxide ion by reducing the extent of negative charge localization on the oxygen.[159] However, induction may actually be more complex than is suggested by a simple model involving shifts of electrons in one σ bond after another through a molecule. We also have to consider the possibility of ***field effects*** (electrostatic effects) in which the substituent influences a remote site by an electrostatic potential operating through space, Figure 6.44(b). However, distinguishing between the two effects is not easy. In most cases, increasing the number of bonds between a substituent and a probe site also increases the distance between them, so that it is difficult to assign an effect to one or the other mode of transmission. Moreover, when model systems are chosen, the molecular skeletons must be fixed, otherwise conformational changes will allow a variety of orientations between the substituent and the reactant dipoles.

Several studies have been designed to determine the importance of through-space (field) and through-bond (inductive) effects. An electron withdrawing substituent that would be expected to increase the acidity of a model carboxylic acid by a through-bond effect could theoretically decrease the acidity by a through-space effect if the negative end of the C—X dipole is closer to the carboxylate group than is the positive end of the dipole. In fact, for the carboxylic acids **56** and **57**, the electron withdrawing chlorine substituent decreases the acidity of the parent (X = H) compound.[160] Similarly, in a series of 3-(8-substituted-1-naphthyl)propiolic acids (**58–60**), acidity decreases as the substituent X is changed from H to Br to Cl.[161] The

---

[157]See the listing in reference 2, p. 218. Resonance effects are sometimes referred to as **mesomeric effects**. Substituents that interact through resonance interactions can also be described as +**M** (electron donating through resonance) or −**M** (electron withdrawing through resonance) substituents. See, for example, *Compendium of Chemical Terminology. IUPAC Recommendations*; Gold, V.; Loening, K. L.; McNaught, A. D.; Sehmi, P., compilers; Blackwell Scientific Publications: Oxford, England, 1987; p. 250.

[158]For a more detailed analysis of possible interactions between substituents and a reaction site, see Katritzky, A. R.; Topsom, R. D. *J. Chem. Educ.* **1971**, *48*, 427.

[159]It might appear that the acid-strengthening effect of the fluorine is enthalpic, but there can also be an entropic effect due to differences in solvent reorganization upon deprotonation of ethanol and 2-fluoroethanol. This topic will be addressed in Chapter 7.

[160]Golden, R.; Stock, L. M. *J. Am. Chem. Soc.* **1966**, *88*, 5928.

[161]Bowden, K.; Hojatti, M. *J. Chem. Soc., Chem. Commun.* **1982**, 273; *J. Chem. Soc. Perkin Trans. 2* **1990**, 1197. See also Roberts, J. D.; Carboni, R. A. *J. Am. Chem. Soc.* **1955**, *77*, 5554.

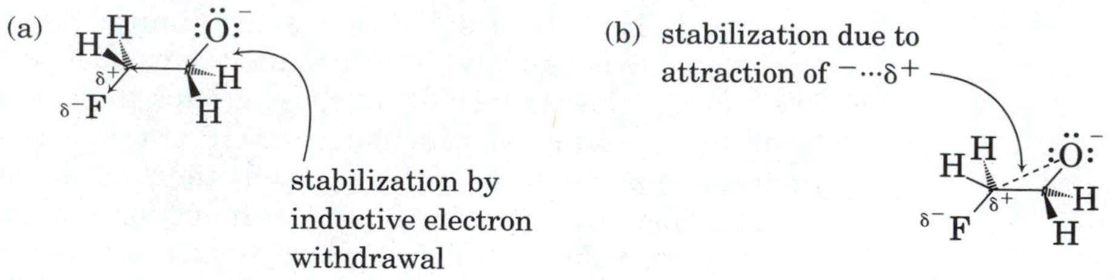

Figure 6.44
(a) Inductive (through bond) effect; (b) field (through space) effect.

(a) stabilization by inductive electron withdrawal

(b) stabilization due to attraction of $^{-}\cdots\delta^{+}$

effect of angularity is even more clearly seen with compounds **61** and **62**. Even though the electron-withdrawing chlorine is one bond closer to the carboxylic acid functionality in **62** than in **61**, the acidity of **61** is greater. The simplest explanation is that the negative end of the C—Cl dipole is aligned toward the carboxylate ion in **62** but away from it in **61**, suggesting that through-space interactions are more important than through-bond interactions.[162] However, alternative explanations are possible. For example, any substitution may alter solvation shells so as to reduce the stability of the carboxylate ion.

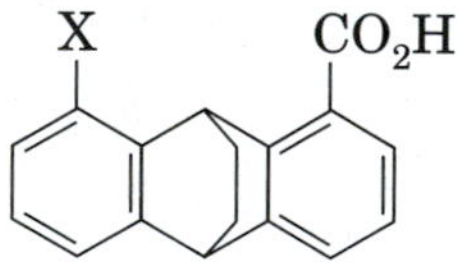

56 X = H pk_a = 6.04 ± .03
57 X = Cl pk_a = 6.25 ± .02

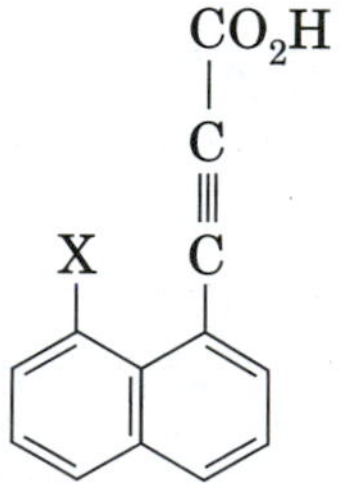

58 X = H pk_a = 4.42
59 X = Br pk_a = 4.70
60 X = Cl pk_a = 4.90

[162]Grubbs, E. J.; Wang, C.; Deardurff, L. A. *J. Org. Chem.* **1984**, *49*, 4080.

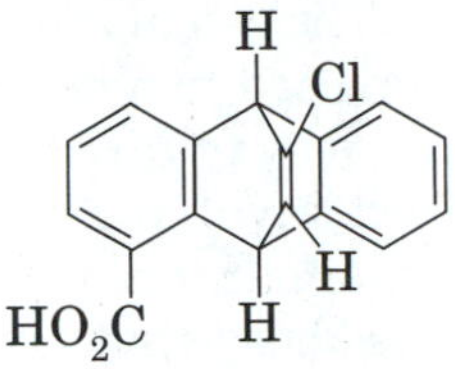

61
$pk_a = 5.72 \pm 0.01$

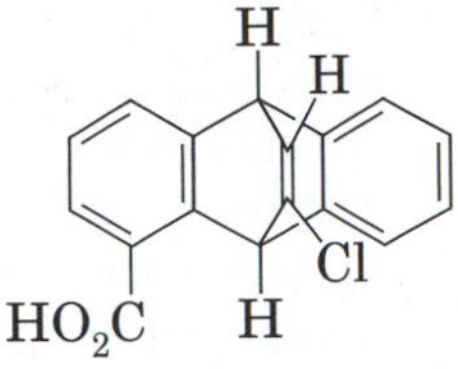

62
$pk_a = 5.90 \pm 0.01$

Another approach is to study compounds based on derivatives of bicyclo[2.2.2]octane and cubane.[163,164] Carbon atoms 1 and 4 are essentially the same distance apart in the two parent compounds (± 0.01 Å). While there are three different three-bond paths connecting these two atoms in bicyclo[2.2.2]octane, however, there are six three-bond paths in cubane. Because a series of 4-substituted cubanecarboxylic acids (**63**)[165] and 4-substituted bicyclo[2.2.2]octane-1-carboxylic acids (**64**)[166] showed similar effects of a substituent X on the pK_a of the parent compound, it appears that field effects and not inductive effects are dominant.

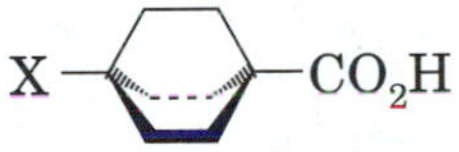

63
A derivative of
bicyclo[2.2.2]octanecarboxylic acid

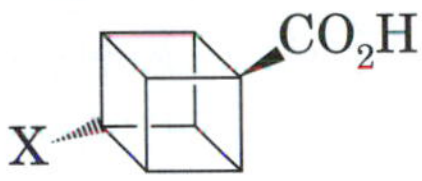

64
A derivative of
1-cubanecarboxylic acid

Whenever conclusions about the magnitude of some nonobservable property are based on comparison of an experimentally determined value and the value expected for some reference system, it is essential to examine the assumptions made about the reference system. In this case, our expectation that inductive effects in compounds **63** and **64** should be different because of a different number of bond pathways may be invalid if we are intuitively viewing the molecules **63** and **64** as atoms held together by lines (perhaps thin metallic wires) that conduct electric charge. Davidson and Williams used CNDO calculations to study the transmission of substituent electronic effects through both cubyl and bicyclo[2.2.2]octyl systems in detail.[163] The CNDO calculations agreed with the experimental results in that both systems were seen to transmit substituent electronic effects equally,

---

[163]Davidson, R. B.; Williams, C. R. *J. Am. Chem. Soc.* **1978**, *100*, 2017 and references therein.
[164]A study of 6-substituted derivatives of spiro[3.3]heptane-2-carboxylic acids also indicated that field and not inductive effects were dominant: Liotta, C. F.; Fisher, W. F.; Greene, Jr., G. H.; Joyner, B. L. *J. Am. Chem. Soc.* **1972**, *94*, 4891.
[165]Cole, Jr., T. W.; Mayers, C. J.; Stock, L. M. *J. Am. Chem. Soc.* **1974**, *96*, 4555 and references therein.
[166]Wilcox, C. F.; Leung, C. *J. Am. Chem. Soc.* **1968**, *90*, 336.

that is, equally poorly. The calculations revealed that the electronic effect of most substituents decreased to a negligible value through either molecular framework, and thus it was difficult to ascribe the effects of the substituents to any effect involving transmission of electronic properties through the σ framework. Similar results have been obtained in other investigations, such as those that determine the effect of substituents on NMR chemical shifts of atoms at various points of attachment on a molecular skeleton.[167] This conclusion is reinforced by results of theoretical calculations in which the effect of a substituent on a molecular framework was found to be essentially the same, even when actual chemical bonds between the substituent and reacting sites were omitted (Figure 6.45).[168]

The cumulative evidence from the studies cited here and from other studies on the acidities of carboxylic acids suggests that inductive effects generally are not transmitted through many carbon-carbon σ bonds and that simple electrostatic effects can satisfactorily describe the experimental results in most cases.[169,170] There are other analyses, however, that seem more consistent with the dominance of an inductive effect for studies involving dissociation of positively charged ions or for investigations of the electrostatic theory in solvents of varying polarity.[171] Exner and Friedl concluded that the field effects and induction were "two opposite approximations," both of which are capable of being extended to produce more accurate results for most compounds.[171] The two explanations are, therefore, yet another example of complementary models that provide useful beginning points for the discussion of organic chemistry.

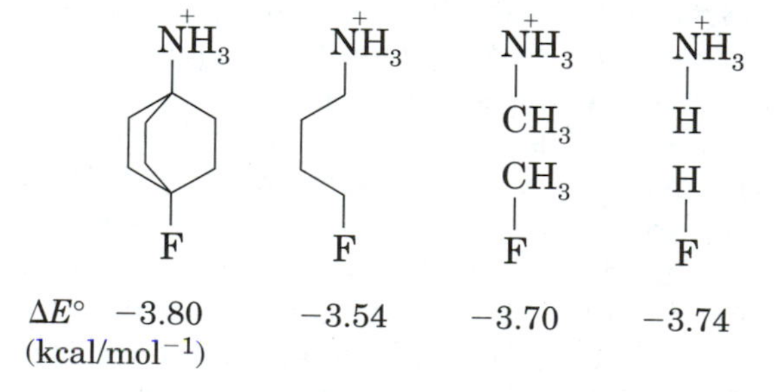

Figure 6.45 Stabilization energies calculated for model compounds. (Reproduced from reference 168.)

---

[167] *Cf.* Adcock, W.; Butt, G.; Kok, G. B.; Marriott, S.; Topsom, R. D. *J. Org. Chem.* **1985**, *50*, 2551.

[168] Marriott, S.; Topsom, R. D. *J. Am. Chem. Soc.* **1985**, *107*, 2253 and references therein.

[169] (a) Stock, L. M. *J. Chem. Educ.* **1972**, *49*, 400; (b) Bowden, K.; Grubbs, E. J. *Prog. Phys. Org. Chem.* **1993**, *19*, 183.

[170] However, Bianchi, G.; Howarth, O. W.; Samuel, C. J.; Vlahov, G. *J. Chem. Soc., Chem. Commun.* **1994**, 627 have analyzed NMR shifts of long-chain esters and acids in terms of inductive effects operating through up to fourteen C-C σ bonds.

[171] Exner, O.; Friedl, Z. *Prog. Phys. Org. Chem.* **1993**, *19*, 259.

Linear Free Energy Relationships

Understanding how the electronic effect of a substituent is transmitted through a molecule offers some utility. However, we would like to be able to extend the understanding of the effect of a substituent in one reaction into the ability to predict how the substituent will affect some other reaction. Moreover, it would be helpful to have a procedure for using the effect that different substituents have on a reaction to help us understand the mechanism of that reaction. Because it is more difficult to treat both steric and electronic effects simultaneously, we will first consider systems in which the effects of a substituent are primarily induction/field effects or resonance effects but not steric effects. The approach that we will take illustrates the use of a **linear free energy relationship** to relate values of ΔG and $\Delta G^{\ddagger}$ from one reaction to another.[172,173,174]

Consider a molecular system composed of three parts: a reactive site (R) is attached through a molecular framework (□) to a substituent (S), Figure 6.46. The product of the reaction has the same structure except for a functional group change (P) associated with the reaction. We assume that S and R (or P) are sufficiently far apart that they do not interact with each other sterically; that is, a change in the size of S does not affect the R → P transformation. The effect of the connecting framework is to transmit the electronic effect of the substituent (either induction or resonance or both) to the site of the R → P transformation.

Hammett proposed that for systems of this kind it should be possible to correlate the position of the equilibrium with the substituent effect as shown in equation 6.54:[175,94,176,177]

$$\log (K/K_0) = \sigma\rho \qquad \textbf{(6.54)}$$

Figure 6.46 Schematic representation of a model system for studying substituent effects.

[172]Reference 78, pp. 172–235.

[173]Fuchs, R.; Lewis, E. S. in *Investigation of Rates and Mechanisms of Reactions*, 3rd Ed., Part I; Lewis, E. S., ed.; Wiley-Interscience: New York, 1974; pp. 777–824.

[174]Buncel and Wilson have discussed the relationship of linear free energy relationships to the larger problem of structure-activity relationships in organic chemistry: Buncel, E.; Wilson, H. *J. Chem. Educ.* **1987**, *64*, 475. Their conclusion—that the concept of structure-activity relationships is an oversimplification, but one that provides a model that is complementary to other approaches to studying organic chemistry—is entirely consistent with the major theme of this text.

[175]Hammett, L. P. *J. Am. Chem. Soc.* **1937**, *59*, 96; reference 95, pp. 347 *ff.* and references therein; See also Hammett, L. P. *Chem. Rev.* **1935**, *17*, 125.

[176]Jaffé, H. H. *Chem. Rev.* **1953**, *53*, 191.

[177]Johnson, C. D. *The Hammett Equation*; Cambridge University Press: Cambridge, England, 1973.

where K is the equilibrium constant for the reaction with a substituent S in place and K_0 is the equilibrium constant for a reference compound having a hydrogen atom as a substituent at that same position.

The symbol σ in these equations is the *substituent constant*, and is a measure of the electronic effect of the substituent S at that particular position. The symbol ρ is the *reaction constant*, which measures the sensitivity of the reaction to the electronic effect of a substituent at a particular position in the molecular framework.

For the reaction of the unsubstituted compound (where the substituent is H) and the substituted analogue, respectively, we can draw the reaction coordinate diagrams in Figure 6.47, where ΔG_0 is the ΔG for the conversion of reactant to product for the unsubstituted reactant, while ΔG is the free energy change for the reaction with a substituent S in place. In this example, the effect of the substituent is such that ΔG is less than ΔG_0, but in some cases it might be larger.

Rewriting equation 6.54 as $$\log (K/K_0) = \sigma\rho = \log K - \log K_0 \quad \textbf{(6.55)}$$

We calculate that $$-\Delta G/2.3RT + \Delta G_0/2.3RT = \sigma\rho \quad \textbf{(6.56)}$$

So $$\Delta G = \Delta G_0 - (2.3RT)\sigma\rho \quad \textbf{(6.57)}$$

Equation 6.57 makes it clear that equation 6.54 is a linear free energy relationship. That is, the free energy change for one reaction is directly proportional to the product of two terms, one that depends on the nature of the reaction (ρ) and one that depends on the nature of the substituent (σ) in the new reactant. Similarly, we can replace the equilibrium constant K by the reaction rate k in the discussion above and can substitute the transition-state theory definition of k to arrive at the corresponding kinetic relationship

$$\Delta G^{\ddagger} = \Delta G_0^{\ddagger} - (2.3RT)\sigma\rho \quad \textbf{(6.58)}$$

The Hammett equation (equation 6.54) is one equation with four unknowns. Therefore, we cannot determine K for a particular reaction unless we know K_0 and ρ for that reaction and σ for the particular substituent that is present. Hammett solved that problem by adopting as a standard system the ionization of benzoic acid derivatives in water and by arbitrarily setting

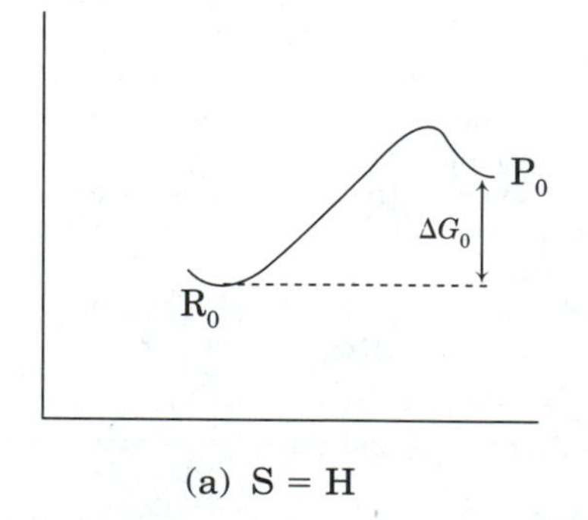

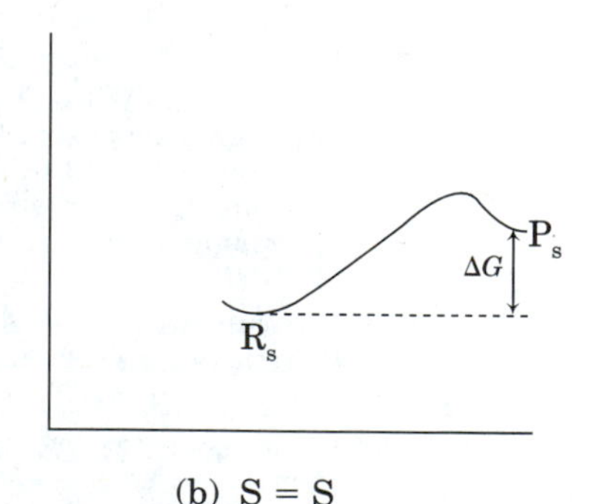

Figure 6.47 Reaction coordinate diagrams for (a) reference and (b) substituent reaction.

the value of ρ for that equilibrium to 1. Therefore, K_0 is the ionization constant for benzoic acid itself. For benzoic acid, with hydrogen as the substituent, the substituent constant σ was set equal to 0. Then the value of K was determined for a substituted benzoic acid, allowing the value of σ for *that substituent in that position* to be determined from the relation

$$\sigma = \log (K/K_H) \text{ for } \rho \equiv 1 \tag{6.59}$$

Some data obtained in this manner are listed in Table 6.6.[178]

These are results that we can rationalize easily and which we might have predicted qualitatively, even if not quantitatively. The Cl, CN, and NO_2 substituents are electron withdrawing by induction and/or resonance, so they would be expected to increase the stability of the benzoate ion, thus increasing the acidity of the corresponding acid. The methyl and amino groups are electron donating and thus destabilize the benzoate ion, decreasing the acidity of the corresponding acid.

Using the values of σ for each substituent determined from this approach, we may analyze quantitatively the effect of substituents on other reactions, either equilibrium or kinetic rate processes. Our approach is to measure the K (or k) for each substituted compound and to plot the value of log K/K_0 (or log k/k_0 for rates) versus σ. If we observe a straight line, we say that **a linear Hammett correlation** is obtained, and the slope of the straight line is ρ. An example is the correlation of log k_{obs} with σ in the reaction of benzaldehyde with semicarbazide in 25% ethanol at pH 1.75 (Figure 6.48), where ρ = + 0.91.[179] A linear correlation tells us that the reaction or equilibrium is affected in a consistent way by the electron donating or with-

Table 6.6 Some σ values.

Benzoic Acid Substituents	log $(K/K_0) \equiv \sigma$
p-NH_2	−0.66
p-CH_3	−0.17
H	0*
p-Cl	0.23
p-CN	0.66
p-NO_2	0.78

(Data from reference 178.)
*By definition.

[178] McDaniel, D. H.; Brown, H. C. *J. Org. Chem.* **1958**, *23*, 420.

[179] In this plot, the y-axis is log k, not log (k/k_0). Problem 15 (page 390) asks the reader to demonstrate that this approach is equivalent to plotting log (k/k_0) versus σ . The units of the rate constant, min^{-1}, suggest a first order rate constant. The study was conducted with a large excess of semicarbazide, so that the kinetics are pseudo-first order. For details, see reference 193.

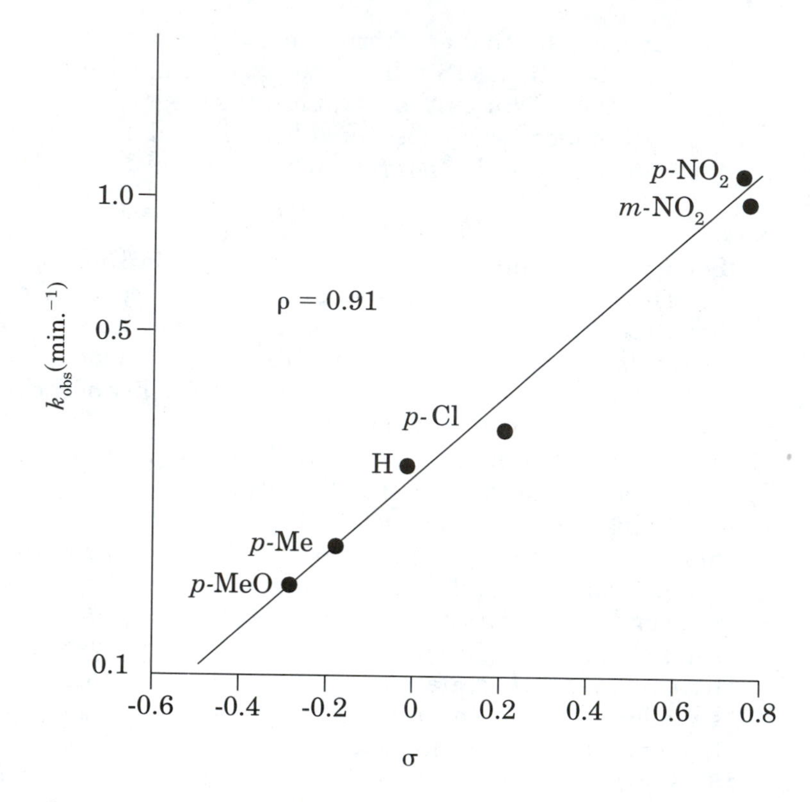

Figure 6.48 Linear Hammett correlation observed in the formation of benzaldehyde semicarbazone at pH 1.75. (Reproduced from reference 193.)

drawing ability of substituents. A positive ρ means that the equilibrium or reaction rate increases with electron withdrawing substituents. Conversely, a negative ρ indicates that the equilibrium or reaction rate is favored by electron donating substituents.

Frequently we seek further meaning from ρ values. For example, we may reason that a positive ρ indicates the development of negative charge in the transition structure (for a rate) or product (for an equilibrium) of a reaction. Conversely, we often conclude that there is an accumulation of positive charge in the transition structure or product for reactions with a negative value of ρ. It is more difficult to interpret the magnitude of ρ. It is technically true to say only that the greater the magnitude of ρ, the more sensitive is the reaction (equilibrium) to the electron donating or withdrawing power of substituents. However, we sometimes *infer* that the magnitude of ρ can tell us about the extent to which charge develops in the product or transition structure during the reaction.[180] For example, anionic polymerization involving benzylic anions produced a ρ value of +5.0, but the corresponding radical polymerization of the same compounds produced a ρ of +0.5.[181]

---

[180]Theoretical calculations suggest that σ values correlate with the accumulation of partial charge at the oxygen and hydrogen of the OH group in benzoic acids: Kim, K. H.; Martin, Y. C. *J. Org. Chem.* **1991**, *56*, 2723 and references therein.

[181]Shima, M.; Bhattacharyya, D. N.; Smid, J.; Szwarc, M. *J. Am. Chem. Soc.* **1963**, *85*, 1306.

Relating the magnitude of ρ to the extent of charge development must be done with caution. The ionization of benzoic acids, for which ρ was defined to be 1 in water, has a ρ of 1.7 in 40% aqueous ethanol.[182] We presume that the charge on the carboxylate ion is essentially the same in both solvent systems, but the differing polarity of the solvents causes the sensitivity of the equilibrium to substituent effects to be different. Values of ρ can also vary among members of a series of acids in the same solvent. Ionization of substituted phenylacetic acids in water gives a ρ of 0.49, while ionization of β-phenylpropionic acids in water gives a ρ of 0.212.[183] Thus the magnitude of ρ falls off with increasing distance between the substituent and the reaction site. Again, the extent of charge developing at the reaction site is not changing, just the sensitivity of the reaction to the effect of remote substituents.

One important application of Hammett correlations is in the determination of charge development in transition structures, which can help us determine reaction mechanisms.[184] Let us consider first the alkaline hydrolysis of methyl benzoate, which has a ρ value of 2.2 (Figure 6.49).[185,186] If the reaction were a direct S_N2-type displacement, as shown in Figure 6.50, we would not expect ρ to be so large because little charge develops on the benzoyl system. In fact, small values of ρ (either positive or negative)

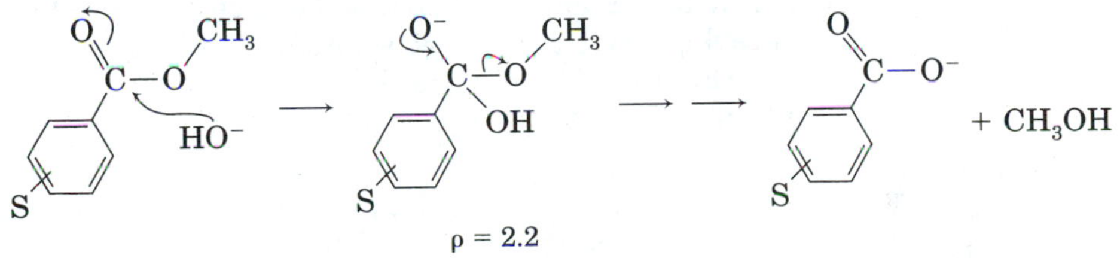

Figure 6.49 Hydrolysis of an ester.

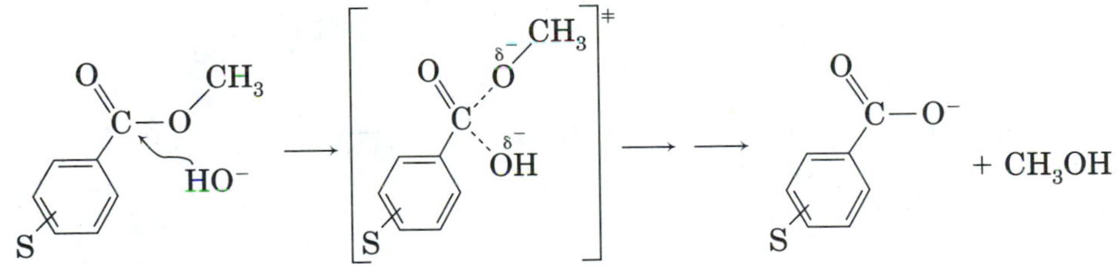

Figure 6.50 Hypothetical mechanism for ester hydrolysis.

[182]Reference 176, based on data reported by Bright, W. L.; Briscoe, H. T. *J. Phys. Chem.* **1933**, *37*, 787.

[183]Reference 176, p. 199.

[184]Indeed, it has been noted that "Almost every kind of organic reaction has been treated via the Hammett equation, or its extended form." Hansch, C.; Leo, A.; Taft, R. W. *Chem. Rev.* **1991**, *91*, 165.

[185]The following examples are taken from the compilation by Jaffé (reference 176).

[186]Reference 176, based on data reported by Tommila, E.; Hinshelwood, C. N. *J. Chem. Soc.* **1938**, 1801.

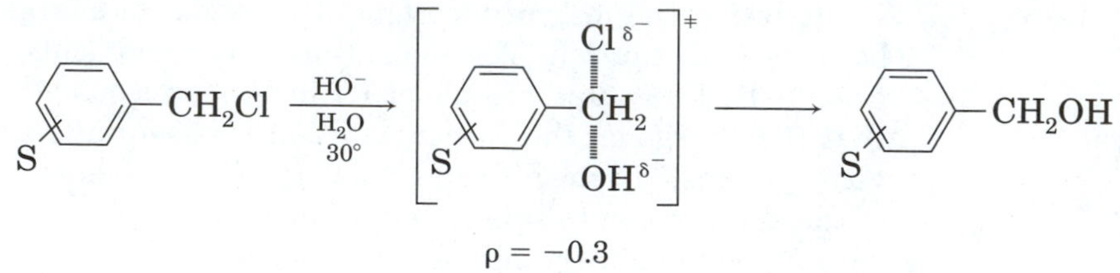

Figure 6.51
S_N2 reaction of benzyl chlorides with hydroxide ion.

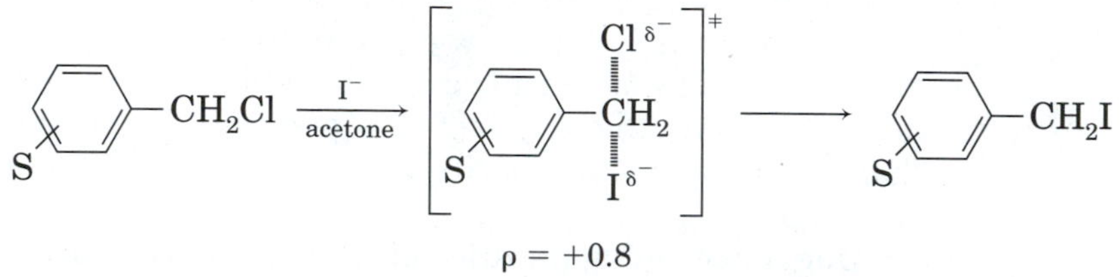

$\rho = -0.3$

Figure 6.52
S_N2 reaction of benzyl chlorides with iodide ion.

$\rho = +0.8$

are observed in S_N2 reactions. The reaction of benzyl chloride with hydroxide ion by an S_N2 pathway (Figure 6.51) has a ρ of -0.3.[187] On the other hand, the reaction of benzyl chloride with iodide in acetone (Figure 6.52) has a ρ of $+0.8$.[188,189]

Larger ρ values may be observed in S_N2 reactions if the substituent is on the nucleophile or the leaving group. For the reaction of phenoxide with ethylene oxide (Figure 6.53), $\rho = -0.95$.[190] For phenoxide reacting with ethyl iodide (Figure 6.54), ρ was found to be -1.[191] Reaction of aniline with benzoyl chloride in benzene solution gave a Hammett correlation with $\rho = -2.8$ (Figure 6.55).[192] Here the nonpolar solvent may enhance the magnitude of ρ.

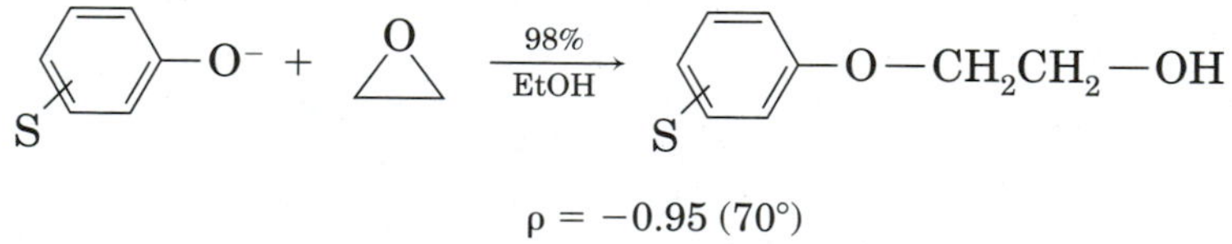

Figure 6.53
Reaction of phenoxides with ethylene oxide.

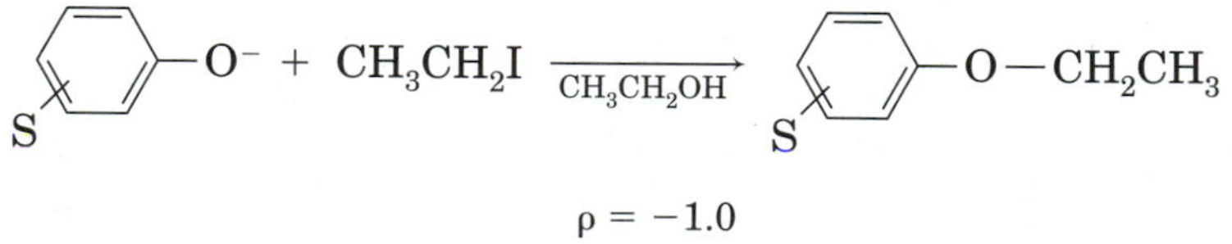

$\rho = -0.95$ (70°)

Figure 6.54
Reaction of phenoxides with ethyl iodide.

$\rho = -1.0$

---

[187]Reference 176, based on data of Olivier, S. C. J.; Weber, A. P. *Rec. Trav. Chim. Pays-Bas* **1934**, *53*, 869.

[188]Reference 176, based on data of Bennett, G. M.; Jones, B. *J. Chem. Soc.* **1935**, 1815.

[189]This contrast indicates that small positive or negative ρ values can appear in S_N2 reactions, depending on the degree to which bond forming or bond breaking is greater in the transition structure. Still, the magnitude of ρ is much smaller for this kind of reaction than for the hydrolysis of methyl benzoate.

[190]Reference 176, based on data of Boyd, D. R.; Marle, E. R. *J. Chem. Soc.* **1914**, *105*, 2117.

[191]Reference 176, based on data by Goldsworthy, L. J. *J. Chem. Soc.* **1926**, 1254.

[192]Reference 176, based on data reported by Williams, E. G.; Hinshelwood, C. N. *J. Chem. Soc.* **1934**, 1079.

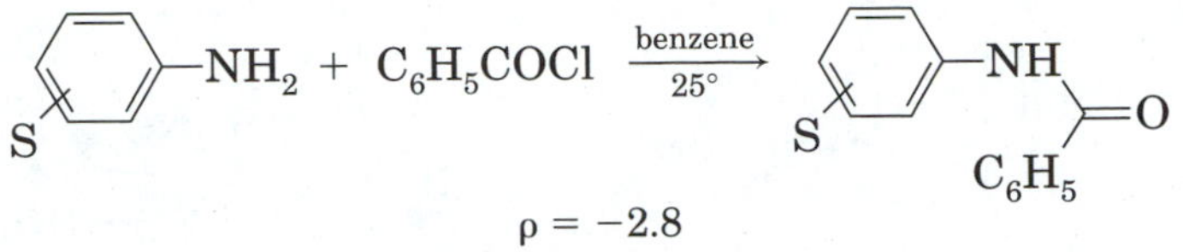

Figure 6.55 Reaction of anilines with benzoyl chloride.

$$\rho = -2.8$$

In some cases nonlinear, even bent Hammett correlation plots are found. These situations arise when a reaction mechanism involves more than one step, either of which can be rate-limiting under certain conditions. In such a case, changing a substituent can change the step that is rate limiting, thus changing the apparent ρ for the reaction. Figure 6.56 shows the two steps involved in the formation of a semicarbazone from reaction of benzaldehyde and semicarbazide.[193] Both the rate and equilibrium of the first step are favored by electron withdrawing substituents on the aldehyde, while the rate of the second step is favored by electron donating substituents on the aldehyde. The effects of a substituent oppose each other at neutral pH, and there is little variation of rate with σ. At low pH (1.75), the first step is rate-limiting, so a positive value of ρ is observed (Figure 6.48). However, at pH 3.9, the first step is rate-limiting for electron withdrawing substituents but not for electron donating substituents. Therefore the Hammett correlation is nonlinear (Figure 6.57, page 378).[194,179]

Similarly, two steps (addition of the amine, elimination of water) occur in the uncatalyzed formation of the *n*-butyl imine of a series of benzaldehydes (Figure 6.58, page 379). Again, the two steps of the reaction are favored by different substituent effects. In this case the maximum rate constant is observed for benzaldehyde itself, so the Hammett plot shows a maximum.[195]

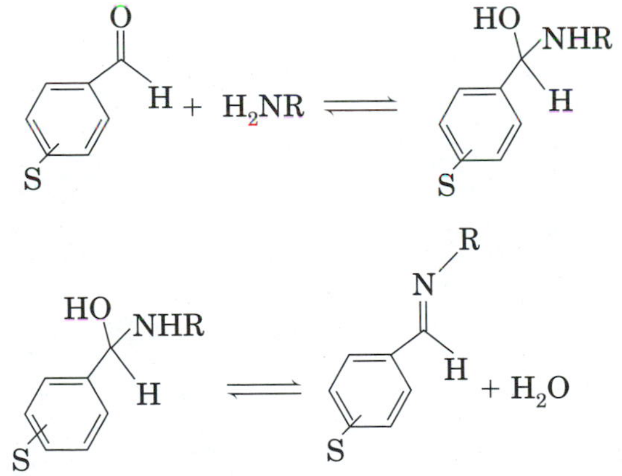

Figure 6.56 Two steps in the mechanism of formation of benzaldehyde semicarbazone.

---

[193]Anderson, B. M.; Jencks, W. P. *J. Am. Chem. Soc.* **1960**, *82*, 1773.

[194]For a discussion of nonlinear Hammett correlations, see Schreck, J. O. *J. Chem. Educ.* **1971**, *48*, 103. See also Henri-Rousseau, O.; Texier, F. *J. Chem. Educ.* **1978**, *55*, 437; Jencks, W. P. *Chem. Soc. Rev.* **1981**, *10*, 345.

[195]The reaction was carried out in methanol solution. The rate expression for the reaction was rate = $(k_0 + k_{HOAc}[HOAc])[RCHO][BuNH_2]$. Here log k_0, the rate constant for the reaction in the absence of acid catalysis, is plotted versus σ . The constant factor of 2 added to log k_0 makes the unit of the *y*-axis easier to read but does not affect the slope of the line. Santerre, G. M.; Hansrote, Jr., C. J.; Crowell, T. I. *J. Am. Chem. Soc.* **1958**, *80*, 1254.

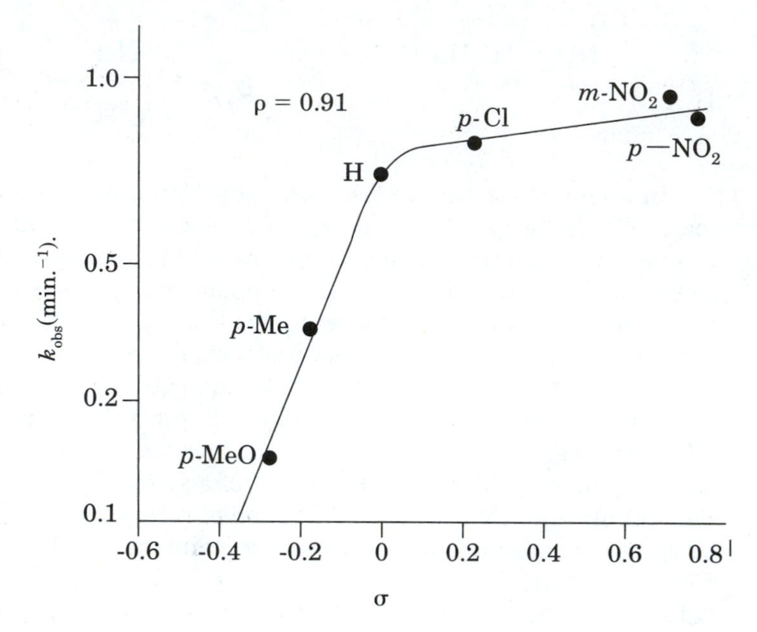

Figure 6.57 Nonlinear Hammett correlation observed for formation of benzaldehyde semicarbazone at pH 3.9. (Reproduced from reference 193.)

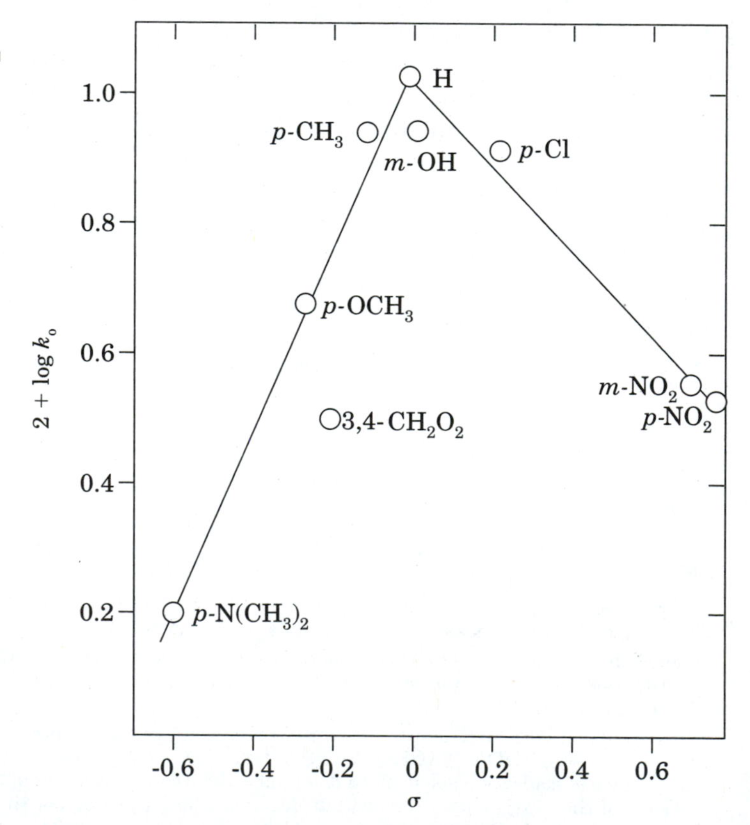

Figure 6.58 Hammett correlation for uncatalyzed reaction of benzaldehydes with *n*-butylamine. (Reproduced from reference 195.)

Having discussed the interpretation of ρ values, let us look more closely at some trends in σ values. Consider the difference between values of σ for the same substituents in para and meta positions, as shown in Table 6.7.[178] For the first two, $|\sigma_m| > |\sigma_p|$. The interactions expected for the trimethylammonium substituent are represented in Figure 6.59 (page 380). Both the inductive effect (represented by arrows superimposed on the σ bonds) and the field effect (represented by a dashed line) are expected to decrease with increasing distance between the positively charged nitrogen atom and the site of the reaction (R). The decreasing ability of the electronic effect of the substituent to be transmitted to the reaction site accounts for the lower magnitude of σ_p than σ_m.

For the next two substituents in Table 6.7, $|\sigma_p| > |\sigma_m|$. The NH_2 group is able to donate electron density by resonance (Figure 6.60, page 380), while $COCH_3$ group is able to accept electron density by resonance (Figure 6.61, page 380). Since resonance effects are more pronounced in para (and ortho) positions than meta positions, resonance makes the magnitude of σ greater when the substituents are para to the carbon atom attached to the reaction site than when they are meta.

In the previous examples it has been possible to discuss the effect of a substituent in terms of only the induction/field effect or the resonance effect, but many substituents exhibit both types of electronic interactions, and the orientation of some substituents with respect to the site of the reaction can alter their effect. Methoxy, for example, functions primarily as an electron withdrawing substituent by induction when it is meta to the reaction site (Figure 6.62, page 381). In the para position, inductive withdrawal of electron density still operates, but it is weakened by the greater distance to the reaction site. Resonance donation of electron density now overcomes the inductive effect, and the substituent is overall electron donating (at least, as judged by the ionization of benzoic acids).

The much larger difference in σ_m and σ_p values for fluoro than trimethylammonium suggests that a considerable electron donating resonance effect may be operating with *p*-F also, but that the electron withdrawing field/induction effect is even stronger (Figure 6.63, page 381).

Table 6.7 Comparison of *p* and *m* σ values.

Substituent	σ_p	σ_m
F	.06	.34
NMe_3^+	.82	.88
NH_2	−.66	−.16
$COCH_3$	.50	.38
OCH_3	−.27	.12

Figure 6.59 Illustration of inductive (arrows) and field (dashed line) effects for *p*- (left) and *m*- (right) trimethylammonium substituent. R = site of reaction.

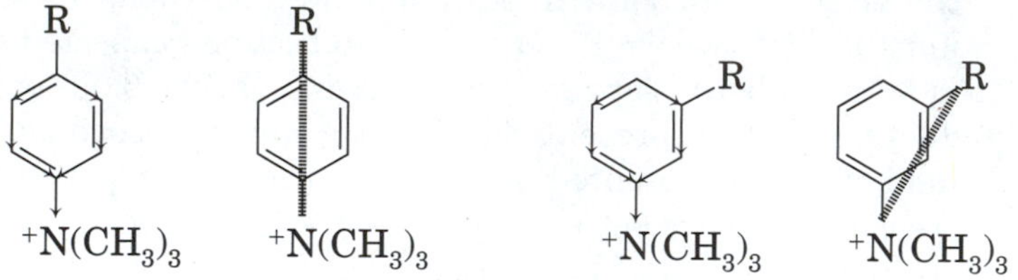

Figure 6.60 Resonance effect of para (left) and meta (right) NH_2 substituent.

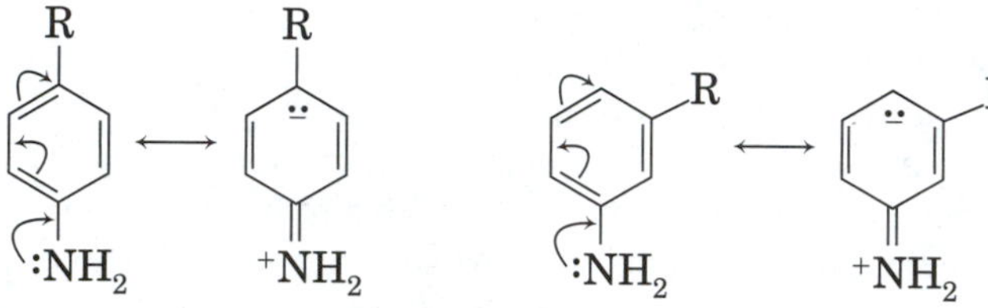

It is important to remember that the initial definition of σ values was based on substituent effects for the ionization of benzoic acids. For these reactions we can imagine the stability of the resulting benzoate ion to be influenced by charge density accumulating on the benzene carbon atom attached to the carboxylate, whether that charge density results from field or resonance effects. For other reactions, however, the relative effects of induction and resonance might be quite different. Solvolysis of a series of cumyl chlorides by a mechanism involving carbocation formation (equation 6.60) should proceed much faster when a para substituent is capable of direct resonance interaction with the carbocation center. Thus Brown and Okamoto suggested the designation of a new substituent constant, σ^+, for this kind of reaction.[196]

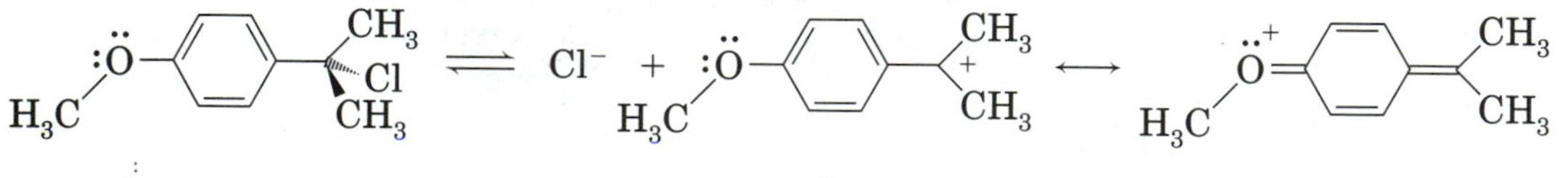

(6.60)

Figure 6.61 Resonance effect of para (left) and meta (right) $COCH_3$ substituent.

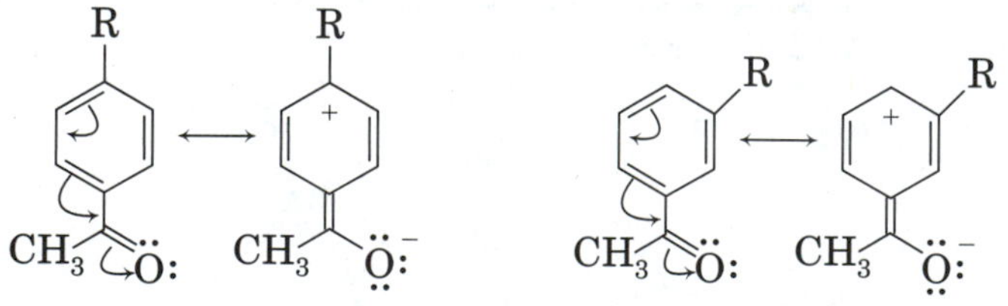

[196] Okamoto, Y.; Brown, H. C. *J. Org. Chem.* **1957**, *22*, 487; Brown, H. C.; Okamoto, Y. *J. Am. Chem. Soc.* **1958**, *80*, 4979.

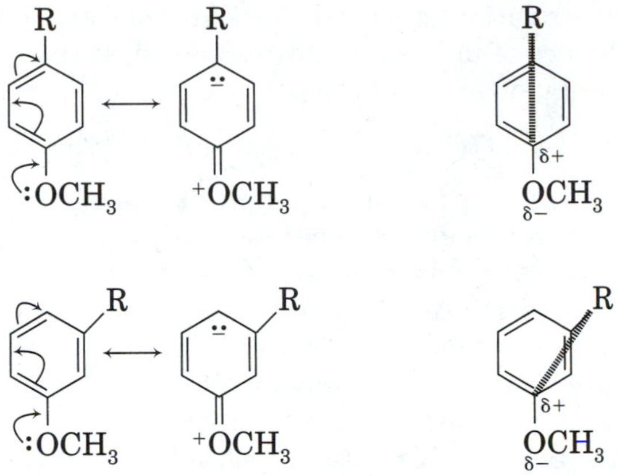

Figure 6.62 Resonance (left) and field (right) interactions for a methoxy substituent in the para (top) and meta (bottom) positions.

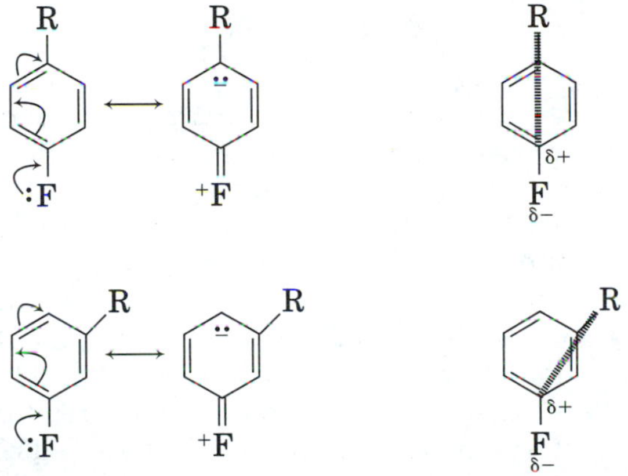

Figure 6.63 Resonance (left) and field (right) interactions for a fluoro substituent in the para (top) and meta (bottom) positions.

Similarly, the ionization of a substituted phenol should be enhanced if there is a *p*-substituent that can directly stabilize the resulting negatively charged oxygen. Thus we can also define σ^- values for substituents such as *p*-nitro, as shown in equation 6.61.[197]

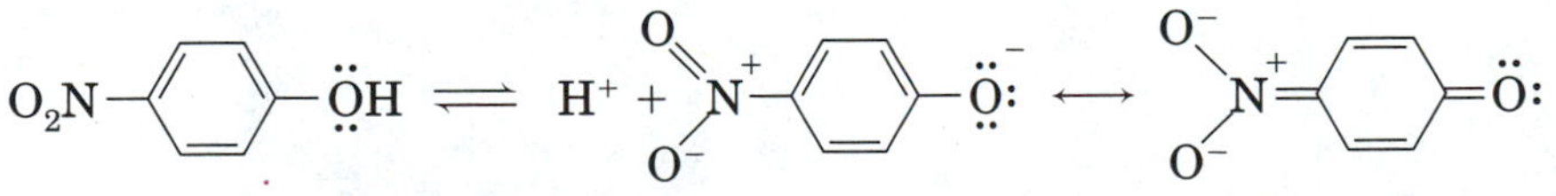

$$O_2N\text{-}C_6H_4\text{-}\ddot{O}H \rightleftharpoons H^+ + O_2N^+\text{-}C_6H_4\text{-}\ddot{O}:^- \longleftrightarrow (O^-)_2N^+{=}C_6H_4{=}\ddot{O}: \quad \textbf{(6.61)}$$

Unless we already know something about the mechanism of a reaction, we may not know ahead of time whether log (k_x/k_0) values are more likely to correlate with σ, σ^+, or σ^-. Consider a reaction in which electron-donating substituents are observed to increase the reaction rate. One way to choose between σ and σ^+ is to plot the $\log(k_x/k_0)$ versus both σ and σ^+ and deter-

[197]Reference 95, p. 360; Cohen, L. A.; Jones, W. M. *J. Am. Chem. Soc.* **1963**, *85*, 3397.

mine which correlation is more linear. Another approach is to try to separate the influence of induction and resonance on the reaction. Yukawa and Tsuno proposed using the equation

$$\log (k_x/k_0) = \rho[\sigma + r(\sigma^+ - \sigma)] \tag{6.62}$$

where the *r* parameter measures the influence of resonance relative to the ionization of benzoic acids.[198] Experimentally one first determines ρ by plotting (log k_x/k_0) versus σ for those substituents that do not act as strong electron donors by resonance. Then one plots $[(\log k_x/k_0)/\rho - \sigma]$ versus $(\sigma^+ - \sigma)$ for those substituents for which $\sigma^+ \neq \sigma$. The slope of the correlation is *r*.

As we have seen, σ values may involve some resonance contribution, so the use of σ and σ^+ terms does not totally distinguish between field and resonance effects. Swain and Lupton proposed using discrete field (*F*) and resonance (*R*) substituent constants.[199] Instead of a single reaction constant ρ, two terms measured the influence of field effects (*f*) and resonance (*r*) on the reaction.

$$\log (k_x/k_0) = fF + rR \tag{6.63}$$

A comprehensive treatment of the dual parameter treatment has been given by Ehrenson, Brownlee, and Taft.[200] Several attempts have been made to define σ values for reactions in which only field effects could be important. These have involved the ionization of phenylacetic acids to generate a series of σ^0 values,[201] of acetic acids to generate σ_I values,[202] or other reactions in which resonance was considered unimportant to generate σ^n values.[203]

Taft and Topsom developed a detailed analysis of the role that substituent effects play in properties such as acidity and basicity.[204] The four primary kinds of substituent effects were considered to be electronegativity (induction), field, resonance, and polarizability effects. As in the Hammett equation, the σ values are substituent properties, while the ρ values represent the sensitivity of the reaction to each of these properties. The general form of such an equation is given in equation 6.64,[204] where the symbols *F*, χ, α, and *R* represent field, electronegativity, polarizability, and resonance effects, respectively.

$$\delta\Delta E^\circ = \sigma_F\rho_F + \sigma_\chi\rho_\chi + \sigma_\alpha\rho_\alpha + \sigma_R\rho_R \tag{6.64}$$

---

[198]Yukawa, Y.; Tsuno, Y. *Bull. Chem. Soc. Jpn.* **1959**, *32*, 965, 971.

[199]Swain, C. G.; Lupton, Jr., E. C. *J. Am. Chem. Soc.* **1968**, *90*, 4328.

[200]Ehrenson, S.; Brownlee, R. T. C.; Taft, R. W. *Prog. Phys. Org. Chem.* **1973**, *10*, 1. See also the treatment of substituent effects in nonaromatic unsaturated systems by Charton, M. *ibid.*, p. 81.

[201]Taft, Jr., R. W. *J. Phys. Chem.* **1960**, *64*, 1805.

[202]Charton, M. *J. Org. Chem.* **1964**, *29*, 1222.

[203]van Bekkum, H.; Verkade, P. E.; Wepster, B. M. *Rec. Trav. Chim. Pays-Bas* **1959**, *78*, 815.

[204]Taft, R. W.; Topsom, R. D. *Prog. Phys. Org. Chem.* **1987**, *16*, 1.

In this study inductive effects were deemed to be important primarily as a source of electric dipoles that can cause field effects, so the effect of a substituent usually could be described in terms of the other three effects. Just as with the Hammett equation, direct conjugation between substituents and reaction sites can cause resonance effects that are different from the effects when direct conjugation is not possible. Therefore the σ_R term may be replaced by σ_R^- for substituents with available electron pairs for donation by resonance, or the term σ_R^+ may be used for substituents that may accept electrons by resonance. Some values of these σ parameters for common substituents are given in Table 6.8.

All of the approaches to linear free energy relationships discussed to this point have involved only meta and para substituents because ortho substituents may provide steric hindrance in addition to electronic effects. Taft proposed a substituent constant, σ^*, to measure the polar effect of alkyl substituents in aliphatic systems.[205] This method is based on the idea that resonance is unimportant in aliphatic systems and that steric effects are the same for ester hydrolysis in acid and base, so only the polar effect of the substituent is different under the two reaction conditions.[206] The value of σ^* for a substituent, R, was based on the relative rates for acid-catalyzed and base-promoted hydrolysis of the ester RCO_2R'. A factor of 2.48 was used to relate σ^* values to the Hammett σ values. Thus

$$\sigma^* = [\log(k/k_0)_B - \log(k/k_0)_A]/2.48 \qquad \textbf{(6.65)}$$

where k is the rate for hydrolysis with the substituent R, k_0 is the rate of hydrolysis of the ester with the reference substituent (CH_3), and A and B refer to acid and base conditions, respectively.[207,208]

Taft was also able to deduce a steric substituent constant, E_s, from the relationship

$$E_s = \log\,(k/k_0)_A \qquad \textbf{(6.66)}$$

on the assumption that polar effects did not affect acid-catalyzed hydrolysis, so only steric effects contributed to differences in rates. The Taft equation (equation 6.67) is a linear free energy relationship that includes the influence of steric effects for certain reactions.

$$\log k_x/k_0 = \rho^*\sigma^* - SE_s \qquad \textbf{(6.67)}$$

---

[205]Taft, Jr., R. W. in *Steric Effects in Organic Chemistry*; Newman, M. S., ed.; John Wiley & Sons: New York, 1956; p. 556 and references therein.

[206]For a discussion, see Shorter, J. *Quart. Rev. Chem. Soc.* **1970**, *24*, 433.

[207]In Table 6.8 the values of σ^* for H, CH_3 and CH_3CH_2 are +0.490, 0.000, and −0.100, respectively. However, Bordwell, F. G.; Bartmess, J. E.; Hautala, J. A. *J. Org. Chem.* **1978**, *43*, 3095, noted that while these values correlate well with rates of some reactions, they do not correlate well with many other reactions. Instead, better correlations may be obtained if the σ^* values for hydrogen and for alkyl groups in general are taken to be 0.000.

[208]Ritchie, C. D.; Sager, W. F. *Prog. Phys. Org. Chem.* **1964**, *2*, 323 have advocated the use of the σ_I scale instead of the σ^* scale. This source provides an extensive listing of σ_I values on pp. 334–337.

Table 6.8

Selected substituent constants*

Substituent	σ_p	σ_m	σ_p^+	F	R	σ^*	E_s	σ_χ	σ_α	σ_F	σ_R	σ_R^-
H	0.000	0.000	0.000	0.000	0.000	0.490		0.000	0.000	0.000	0.000	0.000
CH_3	−0.170	−0.069	−0.256	−0.052	−0.141	0.000	0.00	0.00	−0.35	0.00	−0.08	
CH_2CH_3	−0.151	−0.07	−0.218	−0.065	−0.114	−0.1000	−0.07					
$C(CH_3)_3$	−0.197	−0.10	−0.275	−0.104	−0.138	−0.3000	−1.54	−0.02	−0.75	0.00	−0.07	
C_6H_5	−0.01	0.06	−0.085	0.139	−0.088	0.600						
CO_2H	0.45	0.37	0.472	0.552	0.140							
CO_2^-	0.0	−0.1	0.109	−0.221	0.124							
$COCH_3$	0.502	0.376	0.567	0.534	0.202			−0.04	−0.55	0.26	0.17	0.00
$CO_2CH_2CH_3$	0.45	0.37	0.472	0.552	0.140							
CO_2CH_3								0.04	−0.49	0.24	0.16	0.00
CN	0.660	0.56	0.674	0.847	0.184			0.30	−0.46	0.60	0.10	0.00
CF_3	0.54	0.43	0.582	0.631	0.186		−1.16	0.02	−0.25	0.44	0.07	0.00
NH_2	−0.66	−0.16	−1.111	0.037	−0.681			0.33	−0.16	0.14	−0.52	−0.28
$N(CH_3)_3^+$	0.82	0.88	0.636	1.460	0.000			0.46	−0.26	0.65	0.18	0.00
NO_2	0.778	0.710	0.740	1.109	0.155	0.97						
OH	−0.37	0.121	−0.853	0.487	−0.643			0.54	−0.03	0.30	−0.38	−0.28
OCH_3	−0.268	0.115	−0.648	0.413	−0.500	−0.22		0.55	−0.17	0.25	−0.42	−0.27
F	0.062	0.337	−0.247	0.708	−0.336	0.41		0.70	0.13	0.44	−0.25	
Cl	0.227	0.373	0.035	0.690	−0.161	0.37		0.16	−0.43	0.45	−0.17	
Br	0.232	0.391	0.025	0.727	−0.176	0.38						
I	0.18	0.352	−0.034	0.672	−0.197	0.38						
SCH_3	0.00	0.15	−0.164	0.332	−0.186			−0.15	−0.68	0.25	−0.27	
$S(CH_3)_2^+$	0.90	1.00	0.660	1.687	−0.042							
$Si(CH_3)_3$	−0.070	−0.040	−0.040	−0.093	−0.047	−0.044			−0.072	−0.02	0.02	0.00

*Values of σ_p and σ_m are taken from reference 178. Values of σ_p^+, F, and R are taken from reference 199. Values of σ^* and E_s are taken from reference 205. Values of σ_χ, σ_α, σ_F, σ_R and σ_R^- are taken from reference 204.

The σ^* values measure polar effects in this series and the E_s values measure steric effects. The terms ρ^* and S are the corresponding reaction constants. Table 6.8 collects substituent parameters for a number of common substituents.[209,210]

In discussions of the effects of various substituents, it is easy to begin thinking primarily in terms of energy or enthalpy effects. We must not forget that rates and equilibria depend on $\Delta G^{\ddagger}$ and ΔG, not just on $\Delta H^{\ddagger}$ and ΔH, respectively. One effect of entropy in these relationships can be to generate a temperature dependence with a surprising result. If $\Delta\Delta H$ and $\Delta\Delta S$ are linearly related for a series of substituents, then equations 6.68 and 6.69 hold:[211,212,213]

$$\Delta\Delta H^{\ddagger} = \beta\Delta\Delta S^{\ddagger} \tag{6.68}$$

$$\Delta\Delta G^{\ddagger} = (1 - T/\beta)\Delta\Delta S^{\ddagger} \tag{6.69}$$

Equation 6.69 suggests that at a certain temperature, β , known as the ***isokinetic temperature***, $\Delta\Delta G^{\ddagger}$ will be zero.[214] That is, there will be no change in the $\Delta G^{\ddagger}$ for the reaction with a change in substituents, so the apparent value of ρ will be zero at that temperature.[215] As an example, Meloche and Laidler found a linear correlation between E_a and $T\Delta S^{\ddagger}$ for the acid-catalyzed hydrolysis of a series of benzamides with *p*-nitro, *p*-chloro, *p*-methyl, and *p*-hydrogen substituents.[216] The slope of the line gave a value of β, the isokinetic temperature, of 400 K.[78] Although isokinetic temperatures are often outside the range of experimentally accessible conditions and cannot be practicably observed, equation 6.69 does serve to remind us that ρ values are temperature-dependent. Moreover, if a series of related reactions can be shown to have the same isokinetic temperature (by extrapolation of temperature effect data), then there is strong evidence that the reactions all proceed by the same mechanism.[217]

In later chapters we will see other uses of the Hammett correlation as a tool to understand reaction mechanisms and the effect of substituents on them. We will also consider free energy relationships in the study of acid- and base-catalyzed reactions (Brønsted equation) and substitution reac-

---

[209]Adapted from reference 199.

[210]For an extensive listing of substituent constants, see reference 184.

[211]Leffler, J. E. *J. Org. Chem.* **1955**, *20*, 1202.

[212]See also reference 78, pp. 324–342; reference 54, p. 412; reference 208, pp. 352–378.

[213]The terms $\Delta\Delta H^{\ddagger}$ and $\Delta\Delta S^{\ddagger}$ are often written as $\delta\Delta H^{\ddagger}$ and $\delta\Delta S^{\ddagger}$, respectively.

[214]Linert, W.; Jameson, R. F. *Chem. Soc. Rev.* **1989**, *18*, 477; also see Linert, W.; *Chem. Soc. Rev.* **1994**, *23*, 429.

[215]A similar relationship involving ΔG, ΔH, and ΔS would be known as an isoequilibrium relationship.

[216]Meloche, I.; Laidler, K. J. *J. Am. Chem. Soc.* **1951**, *73*, 1712.

[217]Exner, O. *Prog. Phys. Org. Chem.* **1973**, *10*, 411.

tions (Swain-Scott, Edwards, and Winstein-Grunwald equations).[218] What should be emphasized at this point is that a linear free energy correlation can be developed from the study of almost any reaction. When we try to apply that equation (and the same substituent constants) to study a new reaction, we are implicitly modeling the new reaction on the previous one. A linear free energy relationship must be seen, therefore, as one of the many conceptual models used to describe organic chemistry.

Problems

1. The Criegee mechanism[219] for the ozonolysis of alkenes is shown in Figure 6.64. As shown in Figure 6.65, ozonolysis of 3-heptene (**65**) was found to give only the ozonide **66** and not **67** or **68**. Does this finding rule out a mechanism for ozonolysis that requires dissociation of one molecule into two fragments that recombine? What other experiments can you suggest to determine whether the Criegee mechanism might be occurring?

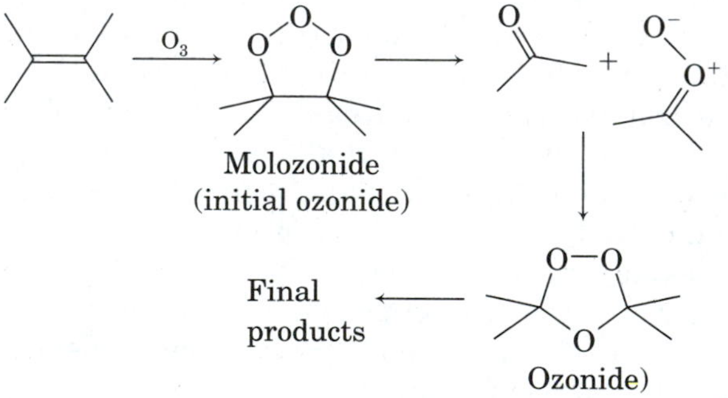

Figure 6.64 The Criegee mechanism.

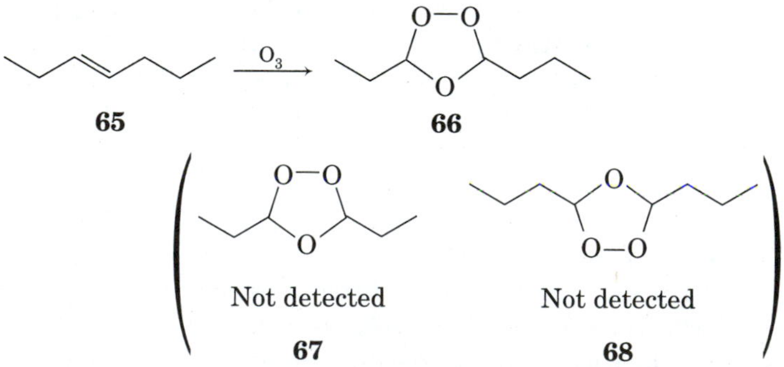

Figure 6.65 Ozonolysis of 3-heptene.

[218]Similar concepts are important in medicinal chemistry, in which the effect of an unsynthesized drug can be predicted on the basis of the effect of a series of known drugs. A leading researcher in this area is Corwin Hansch, for whom the Hansch correlation is named. See reference 184 for a review of this field by Hansch and others.

[219]Criegee, R. in *Peroxide Reaction Mechanisms*; Edwards, J. O., ed.; Wiley-Interscience: New York, 1962 and references therein.

2. The reactions of propylene oxide and isobutylene oxide with ^{18}O-labeled hydroxide in ^{18}O-labeled water lead to the corresponding glycols in which the oxygen label is distributed between the two oxygen atoms of each product as shown in equations 6.70 and 6.71. O* = ^{18}O.

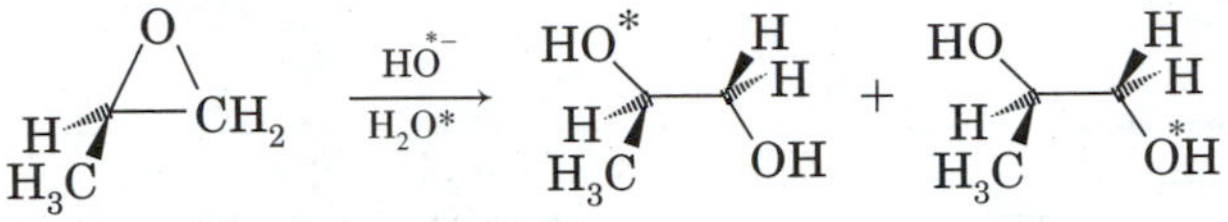

20% 80% **(6.70)**

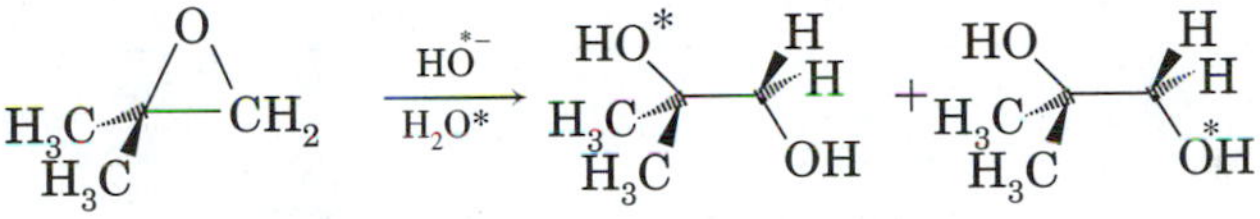

9% 91% **(6.71)**

What do these results suggest about the regiochemistry of the hydrolysis of epoxides under basic conditions?

3. Derive the steady state equation for $A + B \underset{k_{-1}}{\overset{k_1}{\rightleftharpoons}} C \xrightarrow{k_2} D$ shown in equation 6.25 on page 339.
4. Propose a qualitative explanation for the observation that compound A reacts 9000 times faster than compound B at 0°, but the two compounds react at the same rate at a temperature of 103°.
5. Rationalize the fact that $\Delta S^{\ddagger}$ for the dimerization of cyclopentadiene to dicyclopentadiene in the gas phase is −26 eu, whereas for the pyrolysis of dicyclopentadiene to cyclopentadiene the $\Delta S^{\ddagger}$ is 0 eu.
6. The thermal rearrangement of dideuteriomethylenecyclopropane (**69**) to methylenecyclopropane-2,2-d_2 (**70**) was found to have $\Delta H^{\ddagger}$ of 40.5 kcal/mol and $\Delta S^{\ddagger}$ of 1.5 eu. However, the rearrangement of **71** to **72** has a $\Delta H^{\ddagger}$ of 23.8 kcal/mol and $\Delta S^{\ddagger}$ of −6.0 eu. Explain these results, especially the appearance of a negative activation entropy for a reaction that apparently involves bond breaking.

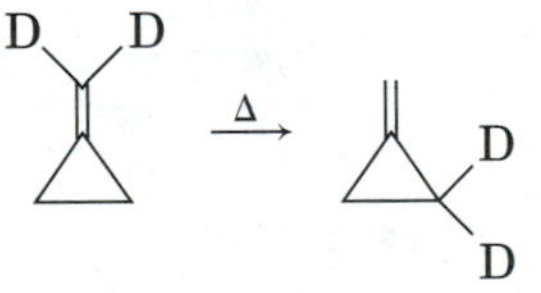

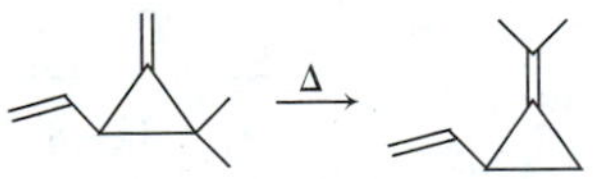

69 **70** **71** **72**

7. Data for the thermal decarboxylation of the β-keto acid 2,2-dimethylbenzoylacetic acid (**73**, Figure 6.66) in 0.1 M HCl are shown in Table 6.9. (Note that the rates are multiplied by 10^5. At 25.0° the rate is 0.000037 sec^{-1}.) Calculate $\Delta H^{\ddagger}$ and $\Delta S^{\ddagger}$ for the reaction and propose a possible mechanism for the decomposition.

Figure 6.66 Decarboxylation of 2,2-dimethylbenzoylacetic acid.

$$\underset{\mathbf{73}}{Ph-\overset{O}{\overset{\|}{C}}-\underset{CH_3}{\overset{CH_3}{\overset{|}{\underset{|}{C}}}}-\overset{O}{\overset{\|}{C}}-CO_2H} \xrightarrow{\Delta} \underset{\mathbf{74}}{Ph-\overset{O}{\overset{\|}{C}}-\underset{CH_3}{\overset{CH_3}{\overset{|}{\underset{|}{C}}}}-H} + CO_2$$

Table 6.9 Kinetic data for the thermal decarboxylation of **73**.

T (°C)	Rate ($10^5\ k$, sec^{-1})
25.0	3.7
48.5	69
63.0	300
79.0	1,400

8. (Z)-2,2,3,4,5,5,-hexamethyl-3-hexene undergoes cis to trans isomerization with rate constants that vary with temperature as shown in Table 6.10. Determine the E_a value for the isomerization. The corresponding value for the isomerization of 2-butene is E_a = 65 kcal/mol. Explain any difference between the activation energies for the two different compounds.

Table 6.10 Kinetic data for isomerization of (Z)-2,2,3,4,5,5-hexamethyl-3-hexene.

T (°C)	Rate ($10^5\ k$, sec^{-1})
195.2	4.3 ± 0.16
205.2	10.7 ± 0.24
209.6	13.7 ± 0.49
214.8	27.5 ± 0.63
225.1	65.4 ± 1.6

9. The rate constants for the pyrolysis of an organic compound varied with temperature as shown in Table 6.11. Determine E_a, A, $\Delta H^‡$, and $\Delta S^‡$ for the reaction.

Table 6.11 Kinetic data for a pyrolysis reaction.

T (°C)	Rate ($10^3\ k$, sec^{-1})
324.8	1.27 ± 0.09
335.4	3.14 ± 0.16
345.9	4.75 ± 0.22
355.6	7.10 ± 0.25
367.2	15.0 ± 0.4
376.6	29.2 ± 1.1
383.8	44.8 ± 1.3
394.8	84.1 ± 2.2

10. Compound **75** reacts thermally to form **76** in *sec*-butylbenzene. A study of the effect of temperature on reaction rate produced the following data: log A = 13.12, E_a = 30.6 kcal/mol; $\Delta H^\ddagger$ = 29.8 kcal/mol, $\Delta S^\ddagger$ = −11.3 eu. Based on these values, would you favor a mechanism in which **75** dissociates into two molecules that recombine in a different way to form the product, or would you favor a mechanism in which the reactant rearranges without dissociation?

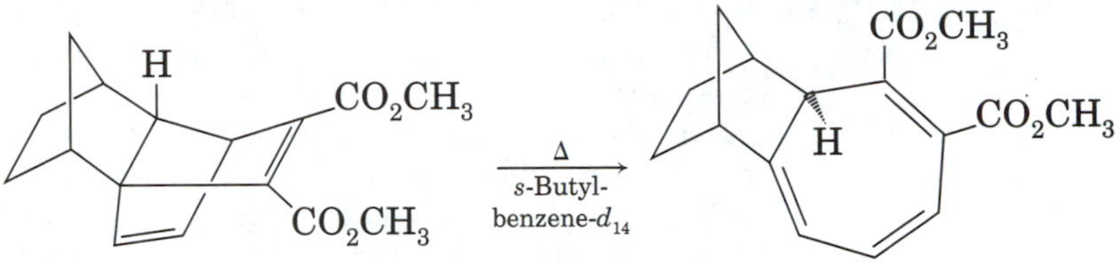

H CO_2CH_3 CO_2CH_3 Δ *s*-Butylbenzene-d_{14} CO_2CH_3 H CO_2CH_3

75 **76**

11. The relative rates for the reaction C—H + X· ⟶ C· + X—H vary according to the halogen, X, and according to the nature of the carbon atom (1°, 2°, 3°). For X = Cl, the relative rates are 3° : 2° : 1° = 5.1 : 3.9 : 1.0. For X = Br, the rates are 3° : 2° : 1° = 1600 : 82 : 1.[220] (Note that the relative scales are based on the reactivity of 1° hydrogen atoms in each case alone. Cl and Br do not react with a 1° hydrogen atom at the same rate.) Construct reaction coordinate diagrams for the hydrogen abstraction steps for Cl· reacting with all three types of hydrogen atoms; do the same for reaction of Br· with all three types of hydrogen atoms. Using the Hammond postulate and the ΔH values for the reaction of Cl· and Br· with various alkanes, explain
    a. why the selectivity is greater for bromination than for chlorination,
    b. why the ratio k_H/k_D is greater for bromination than chlorination, and
    c. if the value of k_H/k_D for the bromination of toluene in CCl_4 at 77° is 4.86, what would you expect the *approximate* value to be for the bromination of ethylbenzene and for bromination of cumene? (Note that for reaction of cumene the observed isotope effect is necessarily an intermolecular isotope effect.)
12. Indicate the kind of isotope effect one would expect (e.g., primary or secondary, α or β, etc.) and predict the value of the isotope effect for the equilibrium constant for the dissociation of cyclopentanone methyl hemiketal (**77** in Figure 6.67).

Figure 6.67 Dissociation of cyclopentanone methyl hemiketal.

(D)H H(D) OH OCH_3 (D)H H(D) ⟶ (D)H H(D) =O + CH_3OH (D)H H(D)

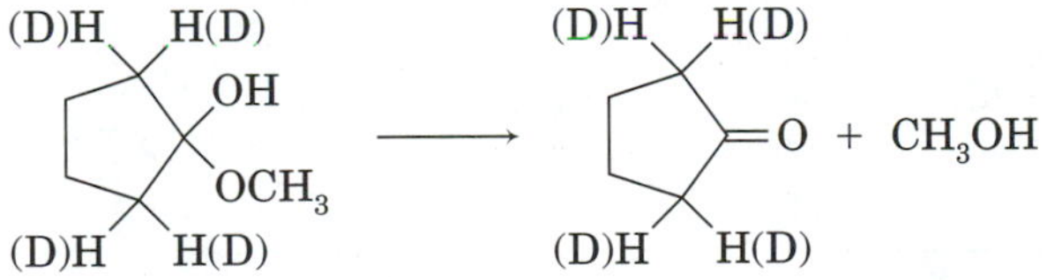

77 **78**

13. Indicate the kind of isotope effect one would expect (e.g., primary or secondary, α or β, etc.) and predict the value of the isotope effect for the rate of hydration of acetaldehyde-d_1 (**79**, Figure 6.68).

[220] Data summarized by Tedder, J. M. (reference 124).

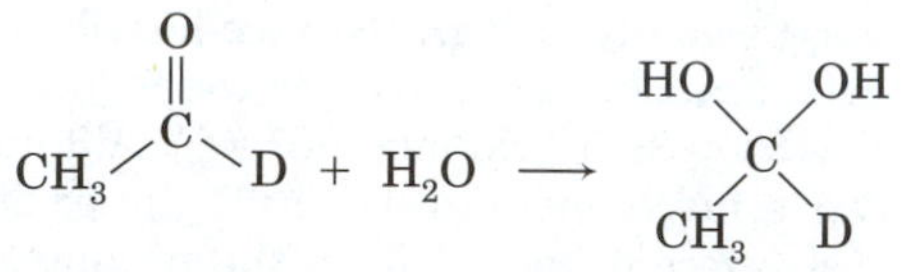

Figure 6.68
Hydration of acetaldehyde-d_1.

$$CH_3C(=O)D + H_2O \longrightarrow CH_3C(OH)_2D$$

79 → **80**

14. Suggest an explanation for the fact that the ρ value for the ionization of benzoic acids in ethanol is 1.96, whereas in water it is 1.0.
15. Some investigators try to observe Hammett correlations without measuring k_0 values for the unsubstituted (substituent is hydrogen) compound. Show that a plot of log k versus σ should give a linear Hammett correlation with slope ρ that is the same slope one would obtain by plotting log (k/k_0) versus σ.
16. In a study of the reaction between ethyl (substituted)aryl disulfides with triphenylphosphine and water, the investigators proposed the mechanism shown in Figure 6.69, with step 1 being rate-limiting. They found that the rate constant for the first reaction varied according to the substituent (on the Ar of Ar-S-S-Et) as indicated in Table 6.12. Show that these data give a linear Hammett correlation. Calculate the value of ρ . Do you think this value of ρ supports the authors' conclusion that the transition structure for step 1 is like the arrangement indicated in Figure 6.69(b)?

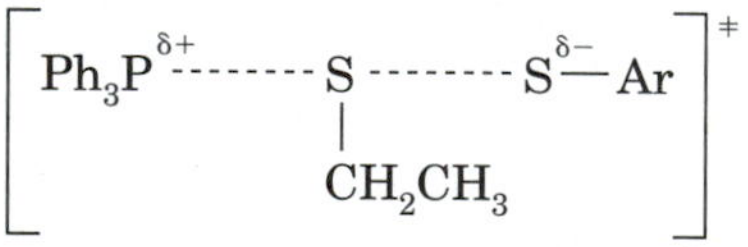

Figure 6.69
(a) Two steps in proposed mechanism; (b) proposed transition structure for the first step.

(a) $$Ar-S-S-CH_2CH_3 + Ph_3P \xrightleftharpoons{\text{slow}} Ph_3P^+-S-CH_2CH_3 + Ar-S^-$$

$$Ph_3P^+-S-CH_2CH_3 + HO^- \xrightarrow{\text{fast}} Ph_3PO + CH_3CH_2SH$$

(b) $$\left[Ph_3\overset{\delta+}{P}\cdots\cdots \underset{CH_2CH_3}{S} \cdots\cdots \overset{\delta-}{S}-Ar \right]^{\ddagger}$$

Table 6.12 Rate data for the reaction of ethyl (substituted)aryl disulfides with triphenylphosphine and water.

Substituent	Rate ($10^4\ k_1$, $M^{-1}\ min^{-1}$)	Substituent	Rate ($10^4\ k_1$, $M^{-1}\ min^{-1}$)
4-NO_2	6.54	3-OCH_3	0.355
3-NO_2	6.19	4-OCH_3	0.097
3-Cl	1.08	4-NH_2	0.0219
4-Cl	0.684		

17. Consider the data in Table 6.13 for the reaction of a series of substituted phenoxides with *N*-chloroacetanilide (equation 6.72) and for the acidities of the phenols in water.
    a. Determine whether the acidities correlate better with σ or with σ^- and rationalize the results.
    b. Determine whether the rate constants correlate better with σ or with σ^-. From the value of ρ you obtain, suggest a likely transition structure for the reaction of phenoxide with *N*-chloroacetanilide.

Table 6.13 Acidities and reaction rate constants for reaction of substituted phenoxides with *N*-chloroacetanilide.

Substituent	pK_a^*	Rate (M^{-1} min^{-1})
p-CH_3	10.28	1.1×10^3
H	9.95	5.3×10^2
m-Cl	9.42	3.6×10^2
m-NO_2	8.52	8.5
p-NO_2	7.13	4.1×10^{-1}

*Under the reaction conditions.

$$C_6H_5N(Cl)C(=O)CH_3 + C_6H_5OH \longrightarrow C_6H_5N(H)C(=O)CH_3 + C_6H_5OCl \quad (6.72)$$

18. Rate data for the ozonolysis of a series of substituted styrenes in CCl_4 solution at 25° are summarized in Table 6.14. Do the data show a linear Hammett correlation with σ, σ^+, or σ^-? If so, what is the value of ρ at 25°? What do these results suggest about the reaction mechanism? Specifically, is the addition of ozone electrophilic or nucleophilic in nature?

Table 6.14 Rate constants for the reaction of ozone with $XC_6H_4CH{=}CH_2$ in CCl_4 solution.

X	Rate (k, M^{-1} sec^{-1}) at 25°
p-CH_3	5.29
H	3.64
p-Cl	2.25
m-Cl	1.70
m-NO_2	0.84

19. Ozonolysis of propene, propene-1-*d* and propene-2-*d* gave the k_H/k_D values shown in Figure 6.70. What do these results suggest about the symmetry of the pathway for the addition of ozone?

Figure 6.70 Kinetic isotope effects in addition of ozone to propene.

$CH_3CH{=}CHD$	$CH_3CD{=}CH_2$
propene-1-*d*	propene-2-*d*
$k_H/k_d = 0.88$	$k_H/k_d = 0.88$

20. In the Beckmann rearrangement (equation 6.73), oximes of ketones are converted to amides by acid catalysts.

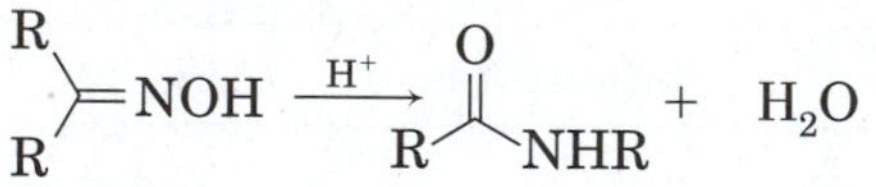

$$\mathrm{R_2C{=}NOH} \xrightarrow{H^+} \mathrm{RC(=O)NHR} + H_2O \qquad (6.73)$$

In principle, two mechanisms can be considered for the rearrangement (Figure 6.71). Mechanism (a) involves a migration of an alkyl group that is concerted with dissociation of water. Mechanism (b) involves dissociation instead of migration of the alkyl group, leading to a nitrile and carbocation that can then undergo a Ritter reaction.[221]

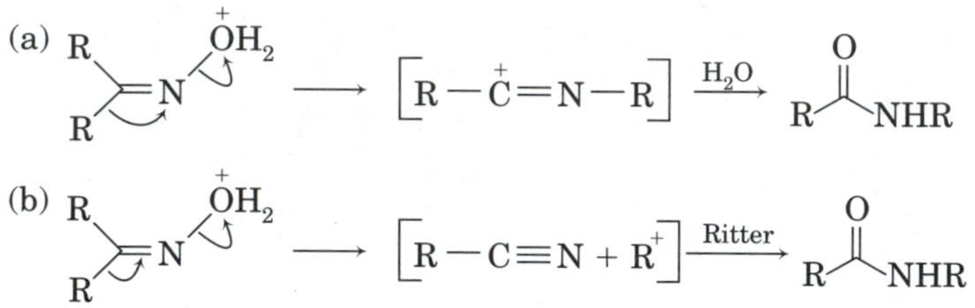

$$\text{(a)}\quad \mathrm{R_2C{=}N{-}\overset{+}{O}H_2} \longrightarrow \left[\mathrm{R{-}\overset{+}{C}{=}N{-}R}\right] \xrightarrow{H_2O} \mathrm{RC(=O)NHR}$$

$$\text{(b)}\quad \mathrm{R_2C{=}N{-}\overset{+}{O}H_2} \longrightarrow \left[\mathrm{R{-}C{\equiv}N} + \mathrm{R^+}\right] \xrightarrow{\text{Ritter}} \mathrm{RC(=O)NHR}$$

Figure 6.71 (a) Nondissociative and (b) dissociative mechanisms for the Beckmann rearrangement.

Most experimental studies have supported the nondissociative mechanism (a) for this reaction,[222] but one study reported the following results: The polyphosphoric acid (PPA) catalyzed Beckmann rearrangement of pinacolone oxime produced *N-t*-butylacetamide, while the PPA catalyzed rearrangement of 2-methyl-2-phenylpropiophenone oxime produced *N*-benzoyl-α,α-dimethylbenzylamine and benzamide. Carrying out the rearrangement on a mixture of pinacolone oxime and 2-methyl-2-phenylpropiophenone oxime produced the products expected from each reactant, plus *N-t*-butylbenzamide and *N*-acetyl-α,α-dimethylbenzylamine.

a. What does the observation of the last two products suggest about the mechanism of the Beckmann reactions in this study?
b. Can you propose a stereochemical study to further distinguish between the two possible pathways for the reaction?
c. For what types of oximes would a dissociative pathway be most likely?

21. The hydrolysis of a series of substituted benzoic anhydrides in a solvent composed of 75 parts dioxane and 25 parts water yielded the activation parameters listed in Table 6.15. Do these data show an isokinetic relationship? If so, determine the isokinetic temperature.

---

[221] Ritter, J. J.; Minieri, P. P. *J. Am. Chem. Soc.* **1948**, *70*, 4045; Ritter, J. J.; Kalish, J. *J. Am. Chem. Soc.* **1948**, *70*, 4048.

[222] For a review and leading references, see Donaruma, L. G.; Heldt, W. Z. *Org. React.* **1960**, *11*, 1.

Table 6.15 Activation parameters for hydrolysis of substituted benzoic anhydrides.

Substituent	E_a (kcal/mol)	$\Delta S^{\ddagger}$ (eu)
p-Methoxy	20.1 ± 0.4	-27.8
m-Methyl	17.8 ± 0.4	-31.4
m-Nitro	11.6 ± 0.3	-38.8
p-Nitro	10.7 ± 0.4	-40.5

CHAPTER 7

Acid-Base Catalyzed Reactions

7.1 Acidity and Basicity of Organic Compounds

Acid-Base Measurements in Solution

Acidity and basicity are fundamental properties of organic compounds, and acid-base reactions are essential steps in many organic transformations. Although there are several definitions of acidity and basicity,[1,2,3,4] the Brønsted theory and the Lewis theory are used most often in organic chemistry. In Lewis theory, an **acid** is an electron pair acceptor and a **base** is an electron pair donor, as in the reaction of a trialkylamine with boron trifluoride, equation 7.1.

$$R_3N: + BF_3 \rightleftharpoons R_3N^{+}\text{—}B^{-}F_3 \qquad (7.1)$$

[1]The earlier Arrhenius theory held that an acid was a proton donor and a base was a hydroxide donor. For a discussion, see reference 3. For a discussion of the merits of teaching this definition, see reference 8.

[2]The hard-soft acid-base theory was developed by Pearson. For leading references, see Pearson, R. G. *J. Chem. Educ.* **1987**, *64*, 561.

[3]For a discussion of acid-base theories, see Finston, H. L.; Rychtman, A. C. *A New View of Current Acid-Base Theories*; Wiley-Interscience: New York, 1982.

[4]A history of acid-base theories has been given by Jensen, W. B. *The Lewis Acid-Base Concepts: An Overview*; Wiley-Interscience: New York, 1980.

In Brønsted theory, an acid is a proton donor and a base is a proton acceptor, as in the reaction of an amine with HCl in equation 7.2.[5,6,7] There is a conceptual error in the latter definition, however. As Hawkes has noted, "it makes no more sense to speak of HCl 'donating' a proton than of 'donating' your purse to a mugger." It is more appropriate to speak of a Brønsted acid as being a substance "from which a proton can be removed" and of a Brønsted base as a substance "that can remove a proton from an acid."[8]

$$R_3N{:} + HCl \rightleftharpoons R_3N^{+}{-}H + Cl^{-} \quad (7.2)$$

The equilibrium constant for the Brønsted acid-base reaction shown in equation 7.3 is the acidity constant, K_a, for the acid A—H. K_a is calculated as indicated in equation 7.4. It is convenient to indicate the acidity of a substance with its pK_a value, which is defined in equation 7.5.[9,10]

$$A{-}H \rightleftharpoons A^{-} + H^{+} \quad (7.3)$$

$$K_a = [A^{-}][H^{+}]/[A{-}H] \quad (7.4)$$

$$pK_a = -\log_{10}K_a = pH + \log([A{-}H]/([A^{-}]) \quad (7.5)$$

Traditionally, the concentrations of A—H and A^- have been measured by UV-vis spectrophotometry or potentiometry, although NMR and IR methods

---

[5]Brønsted, J. N. *Recl. Trav. Chim. Pays-Bas* **1923**, *42*, 718.

[6]More precisely, a Brønsted acid is a hydron donor and a Brønsted base is a hydron acceptor. (Commission on Physical Organic Chemistry, IUPAC, *Pure Appl. Chem.* **1994**, *66*, 1077). Hydron is a general term for H^+, irrespective of isotope, and includes the proton ($^1H^+$), deuteron ($^2H^+$), and triton ($^3H^+$). (For a discussion of the recommended terminology for hydrogen atoms and ions, see Commission on Physical Organic Chemistry, IUPAC, *Pure Appl. Chem.* **1988**, *60*, 1115.) This distinction is particularly useful in discussion of isotope effects. However, the term *proton* has long been used to represent H^+ in general as well as to represent the specific isotope $^1H^+$. Because of this familiar usage, a mechanistic discussion that uses the terms *protonation* and *deprotonation* may be clearer than a discussion that uses the terms *hydronation* and *dehydronation*. Therefore, the presentation here will retain the term *proton* as a general term for H^+. Instances of the use of the term *proton* to mean the specific isotope $^1H^+$ can be understood in context.

[7]A proton donor is often defined as a Brønsted-Lowry acid. For a discussion of the degree to which Lowry should share credit with Brønsted, see the discussion in Bell, R. P. *The Proton in Chemistry*, 2nd Ed.; Cornell University Press: Ithaca, New York, 1973.

[8]Hawkes, S. J. *J. Chem. Educ.* **1992**, *69*, 542.

[9]The symbol pX means the negative of the logarithm (base 10) of the quantity X. Thus pH is $-\log_{10} a_{H^+}$, which we often approximate as $-\log_{10}[H^+]$. (Here and in equation 7.6, a_{H^+} is the activity of H^+ and γ_{H^+} is the activity coefficient of H^+.) For a discussion, see Albert, A.; Serjeant, E. P. *The Determination of Ionization Constants: A Laboratory Manual*, 3rd Ed.; Chapman and Hall: London, 1984; p. 203.

[10]For a thorough discussion of pH measurements, see Bates, R. G. *Determination of pH: Theory and Practice*; John Wiley & Sons, Inc.: New York, 1964.

have also been utilized.[16] There are a number of tabulations of pK_a values in aqueous solution,[11,12] and correlations of pK_a values with molecular structure have been discussed.[13,14]

The expression of pK_a values in terms of concentrations (equation 7.5) is not rigorously correct. Instead, the acidity of a solution at equilibrium should be defined by the *activity* (a) of each species or by the product of the concentration of each species and its *activity coefficient* (γ). Thus the **thermodynamic acidity constant**, K_a^T, is defined as

$$K_a^T = a_{H^+} + a_{H^-}/a_{H-A} = \gamma_{H^+}[H^+]\gamma_{A^-}[A^-]/\gamma_{A-H}[A-H]$$

$$= K_a(\gamma_{H^+})(\gamma_{A^-})/\gamma_{A-H} \qquad \textbf{(7.6)}$$

Since the activity coefficients are concentration-dependent and approach unity as the solution becomes more dilute, K_a^T is approximately the same as K_a in very dilute solution.[15]

Equation 7.3 does not explicitly consider the role of the reaction medium. We have represented the product of the ionization as H^+, but in solution the product involves solvent molecules as participants in the reaction. In water the actual products are solvated protons, $H(H_2O)_n^+$. Furthermore, ion pairing effects may be important in acid-base reactions. Therefore, a more complete description of the reaction may be given by equations 7.7 and 7.8, in which the acid and base initially form an ion pair (with equilibrium constant K_i) and then separate (with an equilibrium constant K_d). Changes in the dielectric constant of the medium are thought to influence K_d more than K_i.[16,17] Ordinarily, however, we use the simpler description of equation 7.3, in which the role of the solvent is considered only implicitly. Table 7.1 lists pK_a values for selected carboxylic acids, alcohols, phenols, and other compounds.

$$S{:} + A-H \overset{K_i}{\rightleftharpoons} A^-SH^+ \qquad \textbf{(7.7)}$$

$$A^-SH^+ \overset{K_d}{\rightleftharpoons} A^- + SH^+ \qquad \textbf{(7.8)}$$

---

[11]Serjeant, E. P.; Dempsey, B. *Ionisation Constants of Organic Acids in Aqueous Solution*; Pergamon Press: Oxford, England, 1979.

[12]Körtum, G.; Vogel, W.; Andrussow, K. *Dissociation Constants of Organic Acids in Aqueous Solution*; Butterworths: London, 1961.

[13]Perrin, D. D.; Dempsey, B.; Serjeant, E. P. *pK_a Prediction for Organic Acids and Bases*; Chapman and Hall: London, 1981.

[14]Barlin, G. B.; Perrin, D. D. *Quart. Rev. Chem. Soc.* **1966**, *20*, 75.

[15]The K_a from equation 7.4 might be called the "concentration-dependent acidity constant," since it varies with concentration. For a discussion, see Albert, A.; Serjeant, E. P. (reference 9), p. 4.

[16]Cookson, R. F. *Chem. Rev.* **1974**, *74*, 5 and references therein.

[17]In this formulation K_a is defined as

$$K_a = K_iK_d/(1 + K_d)$$

See King, E. J. *Acid-Base Equilibria*; Pergamon Press: Oxford, England, 1965; reference 16.

Even if literature acidity data are not readily available for a particular compound, empirical correlations may allow an estimation of its pK_a value. For example, a very good prediction of the pK_a value for a multiply substituted benzoic acid can be obtained from the relationship[14]

$$pK_a = 4.20 - \Sigma\sigma \tag{7.9}$$

in which $\Sigma\sigma$ is the sum of the Hammett σ values for the individual substituents on the benzene ring. Interestingly, almost any ortho substituent increases the acidity of benzoic acid (see Table 7.1), so almost all σ values for ortho substituents are positive.[19] For example, the Hammett σ value for a methyl group at the ortho position in benzoic acid is +0.29. (In contrast, the σ value for a methyl group in the para position is −0.14, while that for a methyl group in the meta position is −0.06.) It has been suggested that ortho substituents may reduce resonance interaction of the carboxyl group with the benzene ring through a steric effect. That is, if the phenyl group is acid-strengthening by induction but acid-weakening by resonance, then the steric effect of the ortho substituent would increase the acidity of the benzoic acid. However, results of gas phase acidity studies by Exner, Gal, and co-workers were more consistent with an alternative explanation, which is that ortho substituents perturb the solvation of the protonated carboxyl group more than they perturb the solvation of the carboxylate ion.[20]

The acidities of the aliphatic carboxylic acids (RCO_2H) in Table 7.1 correlate well with the Taft σ^* values of the substituents,[21] with the correlation being given by the equation[14]

$$pK_a = 4.66 - 1.62\sigma^* \tag{7.10}$$

The corresponding equation for derivatives of acetic acid, RCH_2CO_2H, is[14]

$$pK_a = 5.16 - 0.73\sigma^* \tag{7.11}$$

For alcohols, RCH_2OH, the relationship is[14,22]

$$pK_a = 15.9 - 1.42\sigma^* \tag{7.12}$$

As the last two equations indicate, there is a large difference between the acidities of alcohols and carboxylic acids. Organic chemists have traditionally ascribed the greater acidity of acetic acid than that of isopropyl alcohol to resonance stabilization of the negative charge in the acetate ion, which makes it more stable than an isopropoxide ion. However, Siggel, Streitwieser, and

---

[18]Stewart, R. *The Proton: Applications to Organic Chemistry*; Academic Press: New York, 1985.

[19]For a discussion, see reference 18, pp. 26–27; for listing of ortho σ values, see reference 13, pp. 137–138.

[20]Decouzon, M.; Ertl, P.; Exner, O.; Gal, J.-F.; Maria, P.-C. *J. Am. Chem. Soc.* **1993**, *115*, 12071.

[21]Taft, Jr., R. W. in *Steric Effects in Organic Chemistry*, Newman, M. S., ed.; John Wiley & Sons: New York, 1956; p. 556 and references therein.

[22]For an extensive list of Taft relationships for predicting the acidities of organic acids and protonated bases, see reference 13, pp. 126–135.

Table 7.1 pK_a data for selected organic compounds in aqueous solution.

Compound	pK_a	Compound	pK_a
Carboxylic acids		Cyclohexanecarboxylic acid	4.90[a]
Formic acid	3.75[a]	Benzoic acid	4.20[a]
Performic acid	7.1[b]	2-Methylbenzoic acid	3.91[d]
Acetic acid	4.76[c]	3-Methylbenzoic acid	4.27[d]
Fluoroacetic acid	2.59[a]	4-Methylbenzoic acid	4.37[d]
Chloroacetic acid	2.87[a]	2-*t*-Butylbenzoic acid	3.54[d]
Bromoacetic acid	2.90[a]	2-Bromobenzoic acid	2.85[d]
Iodoacetic acid	3.18[a]	3-Bromobenzoic acid	3.81[a]
Cyanoacetic acid	2.47[a]	4-Bromobenzoic acid	4.00[a]
Methoxyacetic acid	3.57[a]	2-Chlorobenzoic acid	2.91[d]
Nitroacetic acid	1.48[d]	3-Chlorobenzoic acid	3.83[a]
Mercaptoacetic acid	3.56[d]	4-Chlorobenzoic acid	3.99[a]
2-Hydroxyacetic acid (Glycolic acid)	3.38[a]	2-Fluorobenzoic acid	3.27[d]
Phenylacetic acid	4.31[b]	3-Fluorobenzoic acid	3.86[d]
Phenoxyacetic acid	3.16[b]	4-Fluorobenzoic acid	4.14[d]
Difluoroacetic acid	1.34[d]	2-Iodobenzoic acid	2.86[d]
Dichloroacetic acid	1.35[d]	3-Iodobenzoic acid	3.85[d]
Dibromoacetic acid	1.48[b]	4-Iodobenzoic acid	4.00[d]
Trifluoroacetic acid	0.52[b]	2-Hydroxybenzoic (Salicylic) acid	2.97[a]
Trichloroacetic acid	0.51[b]	3-Hydroxybenzoic acid	4.07[d]
Tribromoacetic acid	0.72[b]	4-Hydroxybenzoic acid	4.58[a]
Propanoic acid	4.87[a]	2-Cyanobenzoic acid	3.14[d]
Acrylic acid ($H_2C{=}CH{-}CO_2H$)	4.25[b]	3-Cyanobenzoic acid	3.60[d]
Propiolic acid ($HC{\equiv}C{-}CO_2H$)	1.89[d]	4-Cyanobenzoic acid	3.55[a]
Pyruvic acid (CH_3COCO_2H)	2.39[b]	2-Nitrobenzoic acid	2.21[d]
2,2,3,3,3-Pentafluoropropanoic acid	−0.41[b]	3-Nitrobenzoic acid	3.49[d]
Butanoic acid	4.82[c]	4-Nitrobenzoic acid	3.44[a]
cis-2-Butenoic acid	4.42[b]	Acetylsalicylic acid	3.38[b]
trans-2-Butenoic acid	4.70[b]	Pentafluorobenzoic acid	1.75[b]
2-Butynoic acid ($CH_3{-}C{\equiv}C{-}CO_2H$)	2.59[b]	1-Naphthoic acid	3.60[b]
3-Butynoic acid ($HC{\equiv}C{-}CH_2CO_2H$)	3.32[b]	2-Naphthoic acid	4.14[b]
Cyclopentanecarboxylic acid	4.99[e]	**Alcohols and Phenols**	
		(Water	15.74[d])
		Methanol (CH_3OH)	15.5[d]
		Ethanol	15.9[d]

Table 7.1 *Continued*

Compound	pK_a	Compound	pK_a
2-Chloroethanol	14.3[d]	1-Naphthol	9.39[f]
2-Methoxyethanol	14.8[d]	2-Naphthol	9.59[f]
2,2,2-Trifluoroethanol	12.4[d]		
1-Propanol	16.1[d]	**Other Acids**	
2-Propanol	17.1[d]	Methanethiol CH_3SH	10.33[d]
Allyl alcohol	15.5[d]	Ethanethiol	10.61[d]
Propargyl alcohol ($H\text{-}C{\equiv}C\text{-}CH_2OH$)	13.6[d]	Thiophenol	6.52[f]
		p-Toluenesulfonic acid	−1.34[b]
1-Butanol	16.1[d]	Nitric acid	−1.44[e]
2-Butanol	17.6[d]	HBr	−8.[e]
t-Butyl alcohol	19.2[d]	HCl	−6.1[e]
Benzyl alcohol	15.4[d]	HF	3.18[e]
Phenol	10.0[f]	HCN	9.22[e]

[a]Data from the compilation in reference 75. [b]Data from reference 11. [c]Data from the compilation in reference 76. [d]Data from reference 18. [e]Data from reference 9. [f]Data from the compilation in reference 14.

Thomas concluded from *ab initio* calculations that resonance contributes only about 2–5 kcal/mol to the stabilization of the carboxylate ion. Instead, about 80% of the difference in acidity between acetic acid and isopropyl alcohol was determined to result from the inductive effect of the carbonyl group on the adjacent OH function in acetic acid.[23,24,25] This conclusion was supported[26] by some investigators but challenged[27] by others. In particular, Perrin reported that the theoretical basis for the conclusion that resonance is unimportant proved to be unreliable in model calculations. He emphasized, however, that finding a flaw in an argument against resonance effects does not allow the conclusion that the resonance explanation is correct and

---

[23]Siggel, M. R.; Thomas, T. D. *J. Am. Chem. Soc.* **1986**, *108*, 4360.

[24]Siggel, M. R. F.; Streitwieser, Jr., A.; Thomas, T. D. *J. Am. Chem. Soc.* **1988**, *110*, 8022.

[25]McClard, R. W. *J. Chem. Educ.* **1987**, *64*, 416, has noted that there should be a relatively flat potential energy surface between the transition state and the anion (ethoxide or acetate) in these acid-base reactions. For a discussion of the rate constants of proton transfer reactions, see reference 7, pp. 111–132.

[26]Taft and co-workers made quantitative estimates of the effects of field/inductive and resonance effects on the acidity of acetic acid versus ethanol. They concluded that the $10^{22.5}$ greater acidity of acetic acid over ethanol in the gas phase arises primarily (by a factor of 10^{14}) from a field or induction interaction, not primarily from resonance. On the other hand, they also concluded that the greater acidity of phenol over cyclohexanol arises primarily through resonance. Taft, R. W.; Koppel, I. A.; Topsom, R. D.; Anvia, F. *J. Am. Chem. Soc.* **1990**, *112*, 2047.

[27]Exner, O. *J. Org. Chem.* **1988**, *53*, 1810.

that there remain valid questions about the role that resonance plays in organic structure and reactivity.[28] Thus, the familiar resonance explanation of carboxylic acid acidity is a useful conceptual model, but it—like all of our other models—is subject to continual questioning and, perhaps, reinterpretation.[29]

Basicities in water can be represented by K_b values for the equilibrium

$$B + H_2O \underset{}{\overset{K_b}{\rightleftharpoons}} BH^+ + OH^- \tag{7.13}$$

but they usually are represented by K_{BH^+} values. K_{BH^+} is the equilibrium constant for deprotonation of the protonated base, BH^+

$$BH^+ \overset{K_{BH^+}}{\rightleftharpoons} B + H^+ \tag{7.14}$$

in which H^+ is again the protonated solvent.[30] This approach allows equilibrium constants for acids and bases to be written on the same scale, in which a smaller pK value corresponds to a stronger acid (or a weaker base), and a larger pK value corresponds to a weaker acid (or a stronger base).[13] In aqueous solution, values of pK_b and pK_{BH^+} are related by equation 7.30.

$$pK_b + pK_{BH^+} = 14 \tag{7.15}$$

Values of pK_{BH^+} for the conjugate acids of selected nitrogen and oxygen bases in aqueous solution are shown in Table 7.2.[31,32,33]

The range of pK_a values that can be measured in water is limited by the fact that water is itself both an acid (pK_a = 15.75) and a base (pK_{BH^+} = −1.75). Acid-base reactions can be carried out in other media, but the pK_a of a substance in a nonaqueous solvent is usually very different from the pK_a in water. In such studies, therefore, it is usually necessary to determine the pK_a of a substance (A_1—H) indirectly by relating its pK_a to that of some other substance (A_2—H) with a known pK_a. The acid-base reaction used for

---

[28] Perrin, C. L. *J. Am. Chem. Soc.* **1991**, *113*, 2865.

[29] Similar questions have been raised about the role of resonance in determining the structure and properties of the amide group. For leading references and an experimental study related to this question, see Bennet, A. J.; Wang, Q.-P.; Ślebocka-Tilk, H.; Somayaji, V.; Brown, R. S.; Santarsiero, B. D. *J. Am. Chem. Soc.* **1990**, *112*, 6383; also see reference 28.

[30] More rigorously, the acidity constants of protonated bases are calculated from activities instead of concentrations:

$$K_{BH^+} = a_B a_{H^+}/a_{BH^+}$$

[31] Arnett, E. M.; Wu, C. Y. *J. Am. Chem. Soc.* **1960**, *82*, 4999.

[32] Levy, G. C.; Cargioli, J. D.; Racela, W. *J. Am. Chem. Soc.* **1970**, *92*, 6238.

[33] Literature pK_{BH^+} values for several types of organic bases differ considerably. For example, Albert and Serjeant (reference 9, p. 160) cite values for primary alcohols ranging from −2.2 to −4.8.

Table 7.2 Values of pK_{BH^+} for the conjugate acids of selected organic compounds in aqueous solution.

Protonated Base, BH^+	pK_{BH^+}
Methylamine·H^+	10.66[a]
Dimethylamine·H^+	10.73[a]
Trimethylamine·H^+	9.80[a]
Pyridine·H^+	5.23[a]
Aniline·H^+	4.87[a]
m-Nitroaniline·H^+	2.46[a]
p-Nitroaniline·H^+	1.02[a]
Tetrahydrofuran·H^+	−2.08[b]
Diethyl ether·H^+	−3.59[b]
Anisole·H^+	−6.54[b]
Acetophenone·H^+	−6.3[c]
Acetone·H^+	−7.5[c]
Cyclobutanone·H^+	−9.5[c]
1-Fluoroacetone·H^+	−10.8[c]
1,3-Difluoroacetone·H^+	−12.9[c]
1,1,1-Trifluoroacetone·H^+	−14.9[c]
1,1,3,3-Tetrafluoroacetone·H^+	≈ −17[c]

[a]Reference 9. [b]Reference 31. Uncertainties are ±0.18 for tetrahydrofuran, ±0.10 for diethyl ether, and ±0.02 for anisole. [c]Reference 32. Values were determined by NMR from H_0 values (discussed beginning on page 27) at half-protonation in H_2SO_4-H_2O mixtures.

the measurement is indicated in equation 7.16, and the pK_a is given by equation 7.17.[34]

$$A_1{-}H + A_2^- \underset{}{\overset{K_{ion}}{\rightleftharpoons}} A_1^- + A_2{-}H \qquad (7.16)$$

$$pK_{A_1H} = pK_{A_2H} - \log K_{ion} \qquad (7.17)$$

In equation 7.16, the symbol for the equilibrium (K_{ion}) indicates that free carbanions are involved in the reaction. If the nonaqueous solvent is

[34]For a review, see Streitwieser, Jr., A.; Juaristi, E.; Nebenzahl, L. L. in *Comprehensive Carbanion Chemistry, Part A: Structure and Reactivity*, Buncel, E.; Durst, T., eds.; Elsevier Scientific Publishing Company: Amsterdam, 1980; p. 323 and references therein. Some of the symbols used in this reference have been changed to avoid duplication of symbols used in other parts of the present discussion.

not sufficiently polar, the anions may be closely associated with cations, so that the equilibria (K_{ip}) involve ion pairs (equations 7.18 and 7.19).

$$A_1\text{—}H + A_2^-M^+ \overset{K_{ip}}{\rightleftharpoons} A_1^-M^+ + A_2\text{—}H \quad (7.18)$$

$$pK_{A_1H} = pK_{A_2H} - \log K_{ip} \quad (7.19)$$

Streitwieser and co-workers have reported extensive studies of measurements of pK_a values of weak acids in cyclohexylamine (CHA) solution (designated as pK_{CsCHA} values) using the cesium salt of cyclohexylamine as the base. They have also established a pK_a scale involving Li^+ and Cs^+ counter ions in tetrahydrofuran (THF) solution. (pK_a values determined in THF are useful because that solvent is often used for synthetic reactions involving carbanions.) The lithium salts behaved as solvent separated ion pairs, while the cesium salts appeared to be contact ion pairs.[35,36] The p$K_{Cs/THF}$ values of *p*-methylbiphenyl, fluorene, and 9-biphenylfluorene were found to be 38.73, 22.90, and 17.72, respectively.[37,38]

Bordwell has presented extensive data on equilibrium acidity measurements in dimethyl sulfoxide (DMSO) solutions, in which ion pairing was found to be essentially unimportant.[39] As illustrated in Table 7.3 by selected examples taken from reference 39, the pK_a values measured in DMSO differ from aqueous pK_a values in several important respects. First, values of pK_a as large as 32 can be determined without complications due to the leveling effect of the solvent (which has a pK_a of 35). In contrast, adding the same molar quantity of any one of the first four substances listed in Table 7.3 to water would produce the same concentration of H^+ because they would all be fully dissociated.[40] Second, pK_a values for a substance can vary dramatically with solvent because of differing solvation energies, particularly of the ions. For example, the two pK_a values for water shown in Table 7.3 differ by 16, and there are also large differences in the pK_a values for other acids that form oxyanions capable of hydrogen bonding strongly with water. The delocalized anions formed by stronger acids do not hydrogen bond so strongly to water, however, so differences in the pK_a values of these compounds are smaller. For example, the pK_a value of 2,4,6-trinitrophenol (picric acid) is

---

[35]Kaufman, M. J.; Gronert, S.; Streitwieser, Jr., A. *J. Am. Chem. Soc.* **1988**, *110*, 2829.

[36]The pK_a values were sometimes found to be concentration-dependent because of aggregation of the ion pairs, primarily in the case of localized carbanions. Kaufman, M. J.; Streitwieser, Jr., A. *J. Am. Chem. Soc.* **1987**, *109*, 6092; Gronert, S.; Streitwieser, Jr., A. *J. Am. Chem. Soc.* **1988**, *110*, 2836.

[37]Streitwieser, A.; Ciula, J. C.; Krom, J. A.; Thiele, G. *J. Org. Chem.* **1991**, *56*, 1074.

[38]For those compounds that do not establish acid-base equilibria rapidly, studies of rates of isotopic exchange can be used to determine relative kinetic acidities of carbon acids. For a discussion of the correlation of kinetic acidities with equilibrium acidities, see Streitwieser, Jr., A.; Kaufman, M. J.; Bors, D. A.; Murdoch, J. R.; MacArthur, C. A.; Murphy, J. T.; Shen, C. C. *J. Am. Chem. Soc.* **1985**, *107*, 6983 and references therein.

[39]Bordwell, F. G. *Acc. Chem. Res.* **1988**, *21*, 456.

[40]The aqueous pK_a values shown are estimated from measurements using the acidity function, H_0, but there are uncertainties with these measurements.

Table 7.3 Equilibrium acidities in DMSO and H_2O.

Acid	pK_a (H_2O)	pK_a (DMSO)	Acid	pK_a (H_2O)	pK_a (DMSO)
F_3CSO_3H	−14.*	0.3	$(CH_3CO)_2CH_2$	8.9	13.3
HBr	−9.*	0.9	HCN	9.1	12.9
HCl	−8.*	1.8	CH_3NO_2	10.0	17.2
CH_3SO_3H	−0.6*	1.6	C_6H_5OH	10.0	18.0
2,4,6-$(NO_2)_3C_6H_2OH$	0	≈0	$CH_2(CN)_2$	11.0	11.0
HF	3.2	15 ± 2	CH_3CONH_2	15.1	25.1
$C_6H_5CO_2H$	4.25	11.1	CH_3OH	15.5	29.0
CH_3CO_2H	4.75	12.3	H_2O	15.75	32

*These values are estimated by the H_0 method. (Reproduced in part from reference 39.)

essentially the same in water and in DMSO. pK_a (DMSO) values are of wide interest because they have been shown to correlate with chemical reactivity and to allow estimates of bond dissociation energies, relative radical stabilities, and the acidities of radical cations.[39,41]

Acid-Base Reactions in the Gas Phase

Gas Phase Acidity and Basicity Measurements

As was noted in the previous section, solvent effects play an important role in the determination of pK_a values in solution. Indeed, there are pairs of acids for which the relative order of acidity can be reversed by a change of solvent.[42,43] In order to study the effects of structure on acidity and basicity in the absence of solvent, a variety of experimental techniques have been developed to study acid-base reactions in the gas phase. Three of the major experimental techniques are high pressure mass spectrometry (HPMS), flowing afterglow (FA or FA-SIFT[44]) studies, and pulsed ion cyclotron resonance (ICR) spectrometry.[45,46]

[41]See also the discussion of structural and solvent effects on pK_a values measured in DMSO and in the gas phase by Taft, R. W.; Bordwell, F. G. *Acc. Chem. Res.* **1988**, *21*, 463.

[42]Allen, C. R.; Wright, P. G. *J. Chem. Educ.* **1964**, *41*, 251.

[43]The order of acidity of a pair of compounds may also change with temperature, and Edward has discussed the effect of ion size on the enthalpy and entropy changes associated with ionization in solution. Edward, J. T. *J. Chem. Educ.* **1982**, *59*, 354.

[44]Van Doren, J. M.; Barlow, S. E.; DePuy, C. H.; Bierbaum, V. M. *Int. J. Mass Spectrom. Ion Proc.* **1987**, *81*, 85.

[45]For a discussion of each of these techniques, see Pellerite, M. J.; Brauman, J. I. in *Comprehensive Carbanion Chemistry, Part A: Structure and Reactivity*; Buncel, E.; Durst, T., eds.; Elsevier Scientific Publishing Company: Amsterdam, 1980; pp. 55 *ff*.

[46]Also see Aue, D.; Bowers, M. T. in *Gas Phase Ion Chemistry*, Vol. 2; Bowers, M. T., ed.; Academic Press: New York, 1979; pp. 1–51.

In principle one might try to study the ionic dissociation of an acid (equation 7.3) directly in the gas phase, but ΔH for dissociation of a neutral species into a proton and an anion is usually quite large without solvent stabilization. For example, the ΔH for the gas phase dissociation of methane to methyl anion and a proton (ΔH°_{acid}) was calculated to be +417 kcal/mol.[47] This is much greater than the C—H bond dissociation energy of methane (104 kcal/mol), so thermolysis of methane in the gas phase leads to homolytic instead of heterolytic dissociation. However, the pK_a value of a substance can often be determined by measuring the equilibrium for proton transfer from an acid to a base with a known pK_a, so a scale of gas phase acidity values can be established by "stair-stepping" from one compound to another.

It is particularly difficult to determine the gas phase acidities of alkanes, however, because most alkyl anions cannot be produced as discrete species in the gas phase.[47] Indeed, some alkyl anions are expected to have either a negative or low positive ionization potential, meaning that the ionization of the corresponding hydrocarbon would produce a proton, a radical, and an electron, not a proton and a carbanion.[48] However, DePuy and coworkers used the flowing afterglow method to measure the gas phase acidities of alkanes by determining the ratio of methane and alkane formed from the reactions of hydroxide ion with alkyltrimethylsilanes. A linear free energy relationship between the logarithm of the product ratios and the acidity of the corresponding alkanes, referenced to the known acidities of methane and benzene, allowed calculation of the gas phase acidity values of the alkanes. In these studies the alkyl anions are not produced as discrete carbanions but are thought to be formed as part of an ion-dipole complex in which the carbanion is stabilized by solvation by a trialkylsilanol molecule.[47,49]

ΔH_{acid} values can also be calculated from thermochemical cycles involving measurements of bond dissociation energies, ionization potentials, and electron affinities. Consider the reactions in equations 7.20 through 7.22:

$$\text{A—H} \longrightarrow \text{A}\cdot + \text{H}\cdot \qquad \Delta H = \text{D(A—H)} \tag{7.20}$$

$$\text{H}\cdot \longrightarrow e^- + \text{H}^+ \qquad \Delta H = \text{IP(H)} \tag{7.21}$$

$$\text{A}\cdot + e^- \longrightarrow \text{A}^- \qquad \Delta H = \text{EA(A)} \tag{7.22}$$

in which D(A—H) is the gas-phase homolytic dissociation energy of A—H, IP(H) is the ionization potential of the hydrogen atom, and EA(A) is the

[47]DePuy, C. H.; Gronert, S.; Barlow, S. E.; Bierbaum, V. M.; Damrauer, R. *J. Am. Chem. Soc.* **1989**, *111*, 1968.

[48]Electron transfer from an anionic base to a neutral acid can also compete with proton transfer from the acid to the base in the gas phase. Han, C.-C.; Brauman, J. I. *J. Am. Chem. Soc.* **1988**, *110*, 4048.

[49]DePuy, C. H.; Bierbaum, V. M.; Damrauer, R. *J. Am. Chem. Soc.* **1984**, *106*, 4051.

electron affinity of the radical A·. Summing the reactions in equations 7.20 through 7.22, and taking advantage of the fact that IP(H) is a constant, gives equation 7.23:[50,51]

$$\mathrm{A{-}H} \longrightarrow \mathrm{A^-} + \mathrm{H^+} \qquad \Delta H_{\mathrm{acid}} = \mathrm{D(A{-}H)} - \mathrm{EA(A)} + \mathrm{IP(H)} \qquad \textbf{(7.23)}$$

The standard state for reporting enthalpies and free energies of acid-base reactions in the gas phase is 298 K. Because homolytic dissociation energies are usually reported at 298 K, but electron affinity and ionization potential values derived from spectroscopic data refer to enthalpies at 0 K, the spectroscopic parameters must be corrected.[52,53] ΔH_{acid} values can be used to determine ΔG_{acid} values at 298 K if ΔS_{acid} values are known or can be estimated.[54] Table 7.4 lists gas phase acidity data for several carboxylic acids, alcohols, phenols, and C—H acids.

Similarly, one can measure the basicity of a substance in the gas phase.[60] The **proton affinity (PA)** of species B with charge z is defined as the negative of the ΔH for the protonation of $\mathrm{B^z}$ (equation 7.24).

$$\mathrm{B^z} + \mathrm{H^+} \longrightarrow \mathrm{BH^{z+1}} \qquad -\Delta H = \mathrm{PA(B^z)} \qquad \textbf{(7.24)}$$

Therefore a larger proton affinity corresponds to greater difficulty in removing a proton from $\mathrm{BH^{z+1}}$, that is, to greater basicity of $\mathrm{B^z}$, and a lower proton affinity corresponds to greater acidity of $\mathrm{BH^{z+1}}$. A discussion of the measurement of PA values and an extensive compilation of experimental data

---

[50] Ervin, K. M.; Gronert, S.; Barlow, S. E.; Gilles, M. K.; Harrison, A. G.; Bierbaum, V. M.; DePuy, C. H.; Lineberger, W. C.; Ellison, G. B. *J. Am. Chem. Soc.* **1990**, *112*, 5750.

[51] Alternatively, if the acidity of the compound is known, the same relationship may be used to determine its bond dissociation energy. See, for example, Bordwell, F. G.; Cheng, J.-P.; Harrelson, Jr., J. A. *J. Am. Chem. Soc.* **1988**, *110*, 1229.

[52] For a discussion, see Bartmess, J. E.; McIver, Jr., R. T. in *Gas Phase Ion Chemistry*, Vol. 2; Bowers, M. T., ed.; Academic Press: New York, 1979; pp. 87–121.

[53] Also see Gal, J.-F.; Maria, P.-C. *Prog. Phys. Org. Chem.* **1990**, *17*, 159.

[54] Because $\Delta S°$ is nearly 0 for many proton transfer reactions in the gas phase, the difference in the ΔH_{acid} values for two compounds is essentially the same as the difference in the ΔG_{acid} values for the two compounds. For a discussion, see Majumdar, T. K.; Clairet, F.; Tabet, J.-C.; Cooks, R. G. *J. Am. Chem. Soc.* **1992**, *114*, 2897.

[55] Caldwell, G.; Renneboog, R.; Kebarle, P. *Can. J. Chem.* **1989**, *67*, 611.

[56] Lias, S. G.; Bartmess, J. E.; Liebman, J. F.; Holmes, J. L.; Levin, R. D.; Mallard, W. G. *J. Phys. Chem. Ref. Data* **1988**, *17*, Supplement 1, 1; *Gas-Phase Ion and Neutral Thermochemistry*; American Chemical Society and American Institute of Physics for the National Bureau of Standards: New York, 1988.

[57] Graul, S. T.; Squires, R. R. *J. Am. Chem. Soc.* **1990**, *112*, 2517.

[58] Meot-Ner, M. *J. Am. Chem. Soc.* **1988**, *110*, 3071.

[59] Cumming, J. B.; Kebarle, P. *Can. J. Chem.* **1978**, *56*, 1.

[60] For a discussion, see reference 3, pp. 54 *ff*.

Table 7.4 Gas phase acidity data for selected organic compounds.

Compound	ΔH_{acid} (kcal/mol)	ΔG_{acid} (kcal/mol)
Carboxylic acids		
Formic acid	345.3[a]	338.1[a]
Acetic acid	348.5[a], 348.8[g]	341.5[a]
Propanoic acid	347.4[a]	340.4[a]
Butanoic acid	346.5[a]	339.5[a]
Pentanoic acid	346.2[a]	339.2[a]
2-Methylpropanoic acid	346.0[a]	339.0[a]
2,2-Dimethylpropanoic acid	344.6[a]	337.6[a]
Benzoic acid	340.1[a], 339.9 ± 2.9[b]	333.1[a], 332.9 ± 1.9[b], 333.0[m]
2-Methylbenzoic acid		332.4[m]
3-Methylbenzoic acid		333.7[m]
4-Methylbenzoic acid		334.1[m]
Fluoroacetic acid	338.6[a]	331.6[a]
Chloroacetic acid	336.0[a]	329.0[a]
Bromoacetic acid	335.2[a]	328.2[a]
Iodoacetic acid	334.7[a]	327.7[a]
Methoxyacetic acid	342.3[a]	335.3[a]
Difluoroacetic acid	330.8[a]	323.8[a]
Dichloroacetic acid	328.9[a]	321.9[a]
Dibromoacetic acid	328.3[a]	321.3[a]
Trifluoroacetic acid	324.4[a], 322.9 ± 4.1[b]	317.4[a], 316.2 ± 1.9[b]
Alcohols and Phenols		
(Water	390.8[b]	384.1 ± 0.3[b], 383.9[e])
Methanol	381.7 ± .08[e]	375.1 ± 0.6[e]
Ethanol	378.6 ± .08[e]	372.0 ± 0.6[e]
2-Methoxyethanol	373.8 ± 2.9[b]	366.9 ± 1.9[b]
1-Propanol	376.0 ± 2.2[b]	369.5 ± 1.9[b], 368.1[f,j]
2-Propanol	376.7 ± .08[e]	368.8 ± 1.9[b], 370.1 ± 0.6[e], 367.7[f], 367.5[i]
1-Butanol	375.5 ± 2.4[b]	367.1[f], 368.8 ± 1.9[b]
2-Butanol	374.3 ± 2.4[b]	367.6 ± 1.9[b], 366.3[f]
2-Methylpropanol	374.0 ± 2.1[b]	366.7[i], 368.1 ± 1.9[b]
t-Butanol	375.9 ± .08[e], 374.5 ± 2.1[b]	369.3 ± 0.6[e], 366.6[f], 368.1 ± 1.9[b]
1-Pentanol	372.8 ± 2.4[b]	366.2[f], 367.4 ± 3.1[b]
1-Hexanol	372.1 ± 2.6[b]	366.4 ± 3.1[b]
1-Heptanol	371.7 ± 2.6[b]	365.9 ± 3.1[b]
1-Octanol	371.2 ± 2.4[b]	364.5[f], 365.2 ± 3.1[b]
Benzyl alcohol	370.0 ± 2.9[b]	363.3 ± 1.9[b], 363.0[i]
2-Fluoroethanol	367.0 ± 3.8[b]	363.5 ± 3.6[b]
2,2-Difluoroethanol	366.4 ± 2.9[b]	359.2 ± 1.9[b]
2,2,2-Trifluoroethanol	361.0[g], 361.8 ± 3.6[b]	354.2 ± 1.9[b]

Table 7.4 *Continued*

Compound	ΔH_{acid} (kcal/mol)	ΔG_{acid} (kcal/mol)
cis-2-Methylcyclohexanol		364.1 ± 0.1[f]
trans-2-Methylcyclohexanol		365.1 ± 0.1[f]
cis-4-Methylcyclohexanol		365.1 ± 0.1[f]
trans-4-Methylcyclohexanol		366.2 ± 0.1[f]
Phenol	349.8[l], 349.2 ± 2.4[b]	342.3 ± 1.9[b], 343.4[l]
4-Methylphenol	352.6 ± 2[n]	345.7 ± 2[n]
C—H acids		
Ethane	420.1[d], 421[h], 420.9 ± 1.9[c]	412.3 ± 2.4[b]
Propane (2°H)	419.4[d], 419[h], 419.0 ± 1.9[b]	410.9 ± 2.4[b]
Methane	416.6[d], 417 ± 2[j], 416.8 ± 1.7[b]	408.7 ± 1.7[b]
Propane (1°H)	415.6[d]	
Butane (2°H)	415.7[d]	
Cyclobutane	417.4[d]	
Cyclopentane	416.1[d]	
Isobutane (3°H)	413.1[d], 414[h], 414.0 ± 1.9[b]	406.5 ± 2.4[b]
Fluoromethane	409 ± 4[c]	
Cyclopropane	408 ± 5[c], 411.5[d], 412.0 ± 1.9[b]	403.2 ± 2.6[b]
Ethene	407 ± 3[c], 407.5[d], 409.4 ± .06[e], 406[h]	401.0 ± 0.5[e]
Benzene	401 ± 10[c], 400.7[d], 399[h], 401.7 ± 0.5[o]	329.9 ± 0.4[o]
Chloromethane	396 ± 1[j]	389.1 ± 3.1[b]
Bromomethane	392.7 ± 3.8[b]	385.8 ± 3.1[b]
Propene	390.8 ± 2[n]	384.2 ± 2[n]
Difluoromethane	389 ± 3.5[c], 387 ± 7[j]	
Iodomethane	386.5 ± 5.7[b]	379.3 ± 4.8[b]
Ethyne	377.8 ± 0.6[e]	369.8 ± 0.6[e]
Toluene ($C_6H_5\mathit{CH_3}$)	377 ± 3.5[c,k], 380.1 ± 2.4[b]	373.8 ± 1.9[b]
Fluoroform (HCF_3)	376 ± 4.5[c], 377 ± 2[j], 376.9 ± 2.1[b]	369.3 ± 1.9[b]
Dichloromethane	375 ± 3[j], 374.5 ± 3.8[b]	366.9 ± 3.1[b]
Cycloheptatriene	373.9 ± 2[n]	367.9 ± 2[n]
Acetonitrile	369 ± 4.5[c], 373.5[l], 372.8 ± 2.6[b]	365.6[l], 365.2 ± 1.9[b]
Acetone	370.0[l], 369.0 ± 2.6[b]	362.4[l], 361.9 ± 1.9[b]
Acetaldehyde	366.4 ± 2[n]	359.6 ± 2[n]
Chloroform	357 ± 6[j]	349.2 ± 6.0[b]
Nitromethane	357.6[l], 356.4 ± 2.9[b]	350.7[l], 349.7 ± 1.9[b]
Cyclopentadiene	355.5[l]	348.7[l]
Hydrogen Cyanide	351.1 ± 1.9[b]	343.7 ± 1.9[b]

[a]Data from reference 55. [b]Data from reference 56. [c]Data from reference 57. [d]Data from reference 47. [e]Data from reference 50. [f]Data from reference 54 and references therein. The effective temperature of the measurement varied from species to species. [g]Data from reference 58. [h]Data from reference 49. [i]Data from reference 69. [j]Data from reference 56. [k]The acidic hydrogen atoms are italicized. [l]Data from reference 59. [m]Data from reference 20. [n]Data from reference 52. [o]Data from Davico, G. E.; Bierbaum, V. M.; DePuy, C. H.; Ellison, G. B.; Squires, R. R. *J. Am. Chem. Soc.* **1995**, *117*, 2590.

have been given by Szulejko and McMahon.[61] It is also possible to determine the **gas phase basicity (GB)**, which is defined as

$$GB(B^z) = -\Delta G \tag{7.25}$$

for the reaction in equation 7.24. GB and PA are interconverted with the relationship[62]

$$PA = GB - T\Delta S \tag{7.26}$$

Selected values of gas phase basicity are shown in Table 7.5.[63]

Comparison of Gas Phase and Solution Acidities

Measurements of acidity and basicity in the gas phase allow a re-examination of the relationship of structure to acidity and basicity. The order of acidity of alcohols in the gas phase (Table 7.4) is *t*-butyl > isopropyl > ethyl > methyl, with methyl alcohol being more acidic than water, which is entirely the *opposite* of the order of acidity of these compounds in aqueous solution (Table 7.1).[68] The decrease of acidity in solution with increasing alkyl substitution has traditionally been ascribed to the electron donating property of alkyl groups, which was said to decrease the stability of alkoxide ions. However, the increase of acidity of an alcohol with alkyl substitution in the gas phase suggests that alkyl groups are able to polarize electrons away from a center of negative charge to stabilize an anion, just as they are able to polarize electrons toward a center of electron deficiency to increase the stability of a cation.[69] Theoretical calculations provide support for this view. Silla and co-workers determined from *ab initio* calculations that the charge associated with the carbon and oxygen atoms of the C—O group of a series of alkoxides in the gas phase decreased along the series CH_3O^-, $CH_3CH_2O^-$, $(CH_3)_2HCO^-$, and $(CH_3)_3CO^-$.[70] The effect of this stabilization is to reduce

---

[61]Szulejko, J. E.; McMahon, T. B. *J. Am. Chem. Soc.* **1993**, *115*, 7839.

[62]For a discussion, see Bouchoux, G.; Djazi, F.; Houriet, R.; Rolli, E. *J. Org. Chem.* **1988**, *53*, 3498.

[63]Except as noted, ΔG values are measured at 300 K.

[64]Brickhouse, M. D.; Squires, R. R. *J. Am. Chem. Soc.* **1988**, *110*, 2706.

[65]Lias, S. G.; Liebman, J. F.; Levin, R. D. *J. Phys. Chem. Ref. Data* **1984**, *13*, 695.

[66]Abboud, J.-L. M.; Elguero, J.; Liotart, D.; Essefar, M.; El Mouhtadi, M.; Taft, R. W. *J. Chem. Soc. Perkin Trans. 2* **1990**, 565.

[67]Santos, I.; Balogh, D. W.; Doecke, C. W.; Marshall, A. G.; Paquette, L. A. *J. Am. Chem. Soc.* **1986**, *108*, 8183.

[68]For a discussion of the gas phase acidities of these and other alcohols, see Brauman, J. I.; Blair, L. K. *J. Am. Chem. Soc.* **1968**, *90*, 6561.

[69]For a more detailed discussion of this point, see the discussion by Boand, G.; Houriet, R.; Gäumann, T. *J. Am. Chem. Soc.* **1983**, *105*, 2203.

[70]The sum of the fractional charges for the carbon and oxygen atoms in each case was calculated to be CH_3O^-, − 0.957; $CH_3CH_2O^-$, −0.801; $(CH_3)_2HCO^-$, −0.664; $(CH_3)_3CO^-$, −0.548. In each case the remaining negative charge was calculated to be dispersed among the atoms of the alkyl groups.

Table 7.5 Selected values of gas phase basicity of organic compounds.

Compound	PA (kcal/mol)	GB (kcal/mol)
Anions		
Amide ion (H_2N^-)	403.6[a]	
Allyl anion	390.7[a]	
Hydroxide ion	390.7[a]	
Methoxide ion	380.6[a]	
$C_6H_5CH_2^-$	380.8[a]	
$CH_3COCH_2^-$	369.1[a]	
$O_2NCH_2^-$	356.4[a]	
Amines		
Trimethylamine	224.3[b], 225.1[c]	216.5[b]
Dimethylamine	220.6[c], 222.5[g]	212.3[b]
Methylamine	214.1[c], 215.4[g]	205.7[b]
Ammonia	204.0[c], 203.5[g]	196.4[b]
Pyridine	220.8[c]	212.6[b]
4-Cyanopyridine	209.4[c]	202.9[b]
Aniline	211.5[b]	203.1[b]
Oxygen bases		
Isopropyl ether	206.0[b]	198.7[b], 199.4[f]
2,4-Dimethyl-3-pentanone		197.3[f]
Methyl benzoate		197.0[f]
Propyl acetate	202.0[b]	194.2[b]
Diethyl ether	200.2[c], 200.2[g]	193.1[b]
Cyclopentanone	199.8[b]	192.5[b]
Tetrahydrofuran	198.8[c]	192.4[b]
Acetone	197.2[b], 196.1[c], 193.7[g]	189.9[b]
t-Butyl alcohol	195.[b]	187.[b]
Acetic acid	190.7[b], 188.1[g]	182.5[b]
Methanol	182.5[e], 181.7[g]	
Water	167.3[e], 165.0[g]	
Other bases		
Cubane		200.7[f]
Dodecahedrane		196.6[f]
Cyclohexene		181.5[d]
Cyclohexane		161.[d]
Ethene	162.6[g]	
Ethane	142.7[g]	
Methane	130.2[g]	
CO_2	129.4[g]	
Xe	120.3[g]	
N_2	118.7[g]	

[a]Data from the compilation of literature values in reference 64. [b]Data from reference 46. [c]Data from Drago et al. (reference 81). [d]Data from reference 65. [e]Data from reference 66. [f]Data from reference 67 for equilibrium measurements at 155°. The relatively high basicity of cubane and of dodecahedrane were ascribed primarily to bond strain and to polarizability effects, respectively. [g]Data from reference 61.

the calculated ΔG for gas phase deprotonation of methanol by 3.01, 4.98, and 6.21 kcal/mol for ethanol, 2-propanol, and *t*-butyl alcohol, respectively.[71]

If a substituent has the same electronic effect in solution as in the gas phase, then the apparent acid-weakening effect of alkyl substituents in solution must arise because of a solvent effect that masks the charge-dispersing properties of the substituent.[43,72] The order of the calculated electrostatic energies of alkoxide ions in solution varies in the opposite direction from the order calculated for the gas phase. (For example, methoxide anion is stabilized by 9.5 kcal/mol more than the *t*-butoxide anion.) In solution, therefore, solvation of the anion overcomes the electronic effect, and methanol is the most acidic member of the series.[71,73,74]

One hypothesis for the acid-weakening effect of alkyl substituents in solution might be an increase in enthalpy of the solvated anion due to an increase in the distance from the negative charge to polar solvent molecules. However, entropy changes associated with increased alkyl substitution cannot be ignored. It has long been known that the absolute values of $T\Delta S$ are greater than values of ΔH for ionization of many carboxylic acids in aqueous solution. For example, ΔH values for ionization of formic acid and acetic acid in water at 25° are reported to be +0.01 and −0.02 kcal/mol, respectively, but the corresponding values of ΔS for the two ionizations are −17.1 and −21.9 eu.[75,76] Furthermore, Bartmess and co-workers concluded from thermodynamic calculations that the enthalpy of ionization in solution is

---

[71] Tuñón, I.; Silla, E.; Pascual-Ahuir, J.-L. *J. Am. Chem. Soc.* **1993**, *115*, 2226.

[72] Sebastian has summarized some other experimental evidence that supports this view of the electronic effect of alkyl groups. Sebastian, J. F. *J. Chem. Ed.* **1971**, *48*, 97.

[73] Solvent effects also determine the relative acidity of carboxylic acids. In the gas phase the order of acidity is acetic acid < propionic acid < butanoic acid, but the opposite order is found in aqueous solution. Interestingly, formic acid would be predicted by this explanation to be less acidic in the gas phase than any of the three acids mentioned above, but formic acid is more acidic than butanoic acid. Siggel, M. R. F.; Thomas, T. D. *J. Am. Chem. Soc.* **1992**, *114*, 5795, have reported *ab initio* calculations that suggest an explanation for this apparent anomaly. In acetic acid the methyl group is bonded to a strongly electron-withdrawing carboxylic acid group and is highly polarized. However, there is less polarization of the methyl group in the acetate ion. As a result, replacing the H on formic acid with CH_3 to make acetic acid stabilizes the carboxylic acid function more than the carboxylate ion, so there is a decrease in the acidity.

[74] There is a similar trend in the order of gas phase acidities of acetylenes, which were found to increase as

$$CH_3C{\equiv}CH < CH_3CH_2C{\equiv}CH < HC{\equiv}CH$$

It has been suggested that dipole moments of hybrid orbitals may play a controlling role in this series. Brauman, J. I.; Blair, L. K. *J. Am. Chem. Soc.* **1971**, *93*, 4315. Additionally, theoretical calculations suggest that the site of protonation of the acetylene anion may not be the anionic carbon atom but may instead be the other carbon atom, followed by rearrangement of the resulting vinylidene to acetylene. Brinck, T.; Murray, J. S.; Politzer, P. *J. Org. Chem.* **1991**, *56*, 5012.

[75] Christensen, J. J.; Izatt, R. M.; Hansen, L. D. *J. Am. Chem. Soc.* **1967**, *89*, 213.

[76] Christensen, J. J.; Oscarson, J. L.; Izatt, R. M. *J. Am. Chem. Soc.* **1968**, *90*, 5949.

actually more favorable for acids with large, bulky R groups than for acids with small R groups, just as it is in the gas phase. They determined that the decrease in acidity that accompanies an increase in the size of the R group in RCH_2CO_2H actually arises from a less favorable *entropy* term. Specifically, the bulkier R group causes the solvent around the anion to become more ordered upon ionization, and this is the determining factor for equilibria in solution.[43,77]

It is also worthwhile to compare the values of ΔH for the ionization of halomethanes in the gas phase with the pK_a values in solution. In the gas phase, the relative acidities of methane and two halomethanes were found to be $CH_4 < CH_3F < CH_3Cl$ (Table 7.4). The greater acidity of methyl chloride than of methyl fluoride cannot be rationalized by any simple prediction based on electronegativity. Instead, the greater acidity of methyl chloride is attributed to the greater polarizability of chlorine than fluorine.[78,79] The observation that the acidity of chloroform in water is 10^7 greater than that of fluoroform (Table 7.1) suggests that the same phenomenon occurs in solution.[80] These results remind us that many of our simple paradigms that correlate molecular structure with physical properties such as electronegativity are, at best, oversimplifications.[81]

Acidity Functions

As noted on page 402, dissolving in water the same molar quantity of two acids that are both fully dissociated will produce the same concentration of H_3O^+. Because of this leveling effect, it is not possible to tell whether one acid is stronger than the other (see equation 7.5).[82,83] This observation led

[77]Wilson, B.; Georgiadis, R.; Bartmess, J. E. *J. Am. Chem. Soc.* **1991**, *113*, 1762.

[78]Schleyer, P. v. R.; Clark, T.; Kos, A. J.; Spitznagel, G. W.; Rohde, C.; Arad, D.; Houk, K. N.; Rondan, N. G. *J. Am. Chem. Soc.* **1984**, *106*, 6467.

[79]For theoretical calculations of the effects of multiple halogen substituents on methane, see Rodriquez, C. F.; Sirois, S.; Hopkinson, A. C. *J. Org. Chem.* **1992**, *57*, 4869.

[80]See also the discussion of kinetics of formation of carbanions from trihalomethanes in Hine, J. *Physical Organic Chemistry*; McGraw-Hill Book Company: New York, 1962; pp. 486–487.

[81]Additional understanding of the role that solvent plays in acid-base reactions comes from analysis of the effects of substituents on acid-base reactions in the gas phase and in solution. An example of this approach is found in a study of the basicity of substituted pyridines by Abboud, J.-L. M.; Catalán, J.; Elguero, J.; Taft, R. W. *J. Org. Chem.* **1988**, *53*, 1137. The order of basicities observed in the gas phase (Table 7.5) for methylamines follows the order trimethylamine > dimethylamine > methylamine > ammonia. In solution, however, the observed order of basicities is methylamine > ammonia > dimethylamine > trimethylamine. (Taft, R. W.; Wolf, J. F.; Beauchamp, J. L.; Scorrano, G.; Arnett, E. M. *J. Am. Chem. Soc.* **1978**, *100*, 1240. For a discussion, see Drago, R. S.; Cundari, T. R.; Ferris, D. C. *J. Org. Chem.* **1989**, *54*, 1042.) In solution, the basicity values correlate well with a linear free energy relationship based on field and resonance terms only. Catalán and Taft concluded that polarization is unimportant in solution, and that the field and resonance parameters are attenuated by a factor of 2.3 from their gas phase values.

[82]Hammett, L. P. *Physical Organic Chemistry: Reaction Rates, Equilibria and Mechanisms*, 2nd Ed.; McGraw-Hill Book Company: New York, 1970; pp. 272–273.

[83]See also Hammett, L. P. *J. Chem. Educ.* **1966**, *43*, 464.

Hammett to consider ways to study acidities of species in more strongly acidic media, such as mixtures of water and sulfuric acid, because an acid that is half-dissociated in 50% H_2SO_4 is a stronger acid than one that is half-dissociated in 10% H_2SO_4. To this end, Hammett suggested the use of an **acidity function**, H_0, to categorize acidities in such environments.[84] In other words, the acidity function can be used to extend the pH scale, which characterizes the acidity of aqueous solutions to highly acidic solutions. As noted in equation 7.14, we can represent the protonation equilibrium for a base by the relationship

$$BH^+ \overset{K_{BH^+}}{\rightleftharpoons} B + H^+ \tag{7.14}$$

in which H^+ represents the protonated solvent. Letting the symbol I stand for the ratio of the concentrations of protonated to unprotonated base, ($[BH^+]/[B]$), we may write

$$pK_{BH^+} = \log I + pH - \log(\gamma_B \gamma_{H^+}/\gamma_{BH^+}) \tag{7.27}$$

in which the γ terms represent the activity coefficients of the indicated species. If the activity coefficients are unity, equation 7.27 reduces to

$$pK_{BH^+} = \log I + pH \tag{7.28}$$

so that if $[BH^+]$ is equal to [B], then $\log I$ is 0 and pK_{BH^+} is equal to pH. In most acidic solutions, however, the activity coefficients cannot be ignored. Therefore, the acidity function H_0 is defined as

$$H_0 = pH - \log(\gamma_B \gamma_{H^+}/\gamma_{BH^+}) \tag{7.29}$$

so that

$$pK_{BH^+} = \log I + H_0 \tag{7.30}$$

Equation 7.30 provides a way to develop a scale of H_0 values. Using a base with a known pK_{BH^+} (measured in aqueous solution), the ratio I is determined spectroscopically in a series of acid solutions. The H_0 values determined for these acid solutions are then used to determine the pK_{BH^+} value for a weaker base. In turn, the second base is used to determine the H_0 value for each of a series of solutions with increasing acid concentration. Those solutions are used to determine the pK_{BH^+} value for a still weaker base, which is then used to extend the H_0 scale even further, and so on. By using a series of bases, H_0 values can be determined for a wide range of acid concentrations.

Hammett used the protonation of a series of nitroanilines in mixtures of water and sulfuric acid to establish the H_0 scale. The protonated and unprotonated bases show different absorption spectra, so UV-vis spectroscopy

[84]Hammett, L. P.; Deyrup, A. J. *J. Am. Chem. Soc.* **1932**, *54*, 2721.

Figure 7.1 Correlation of log ([BH^+]/[B]) with percent H_2SO_4 for a series of nitroanilines. (Reproduced from reference 86.)

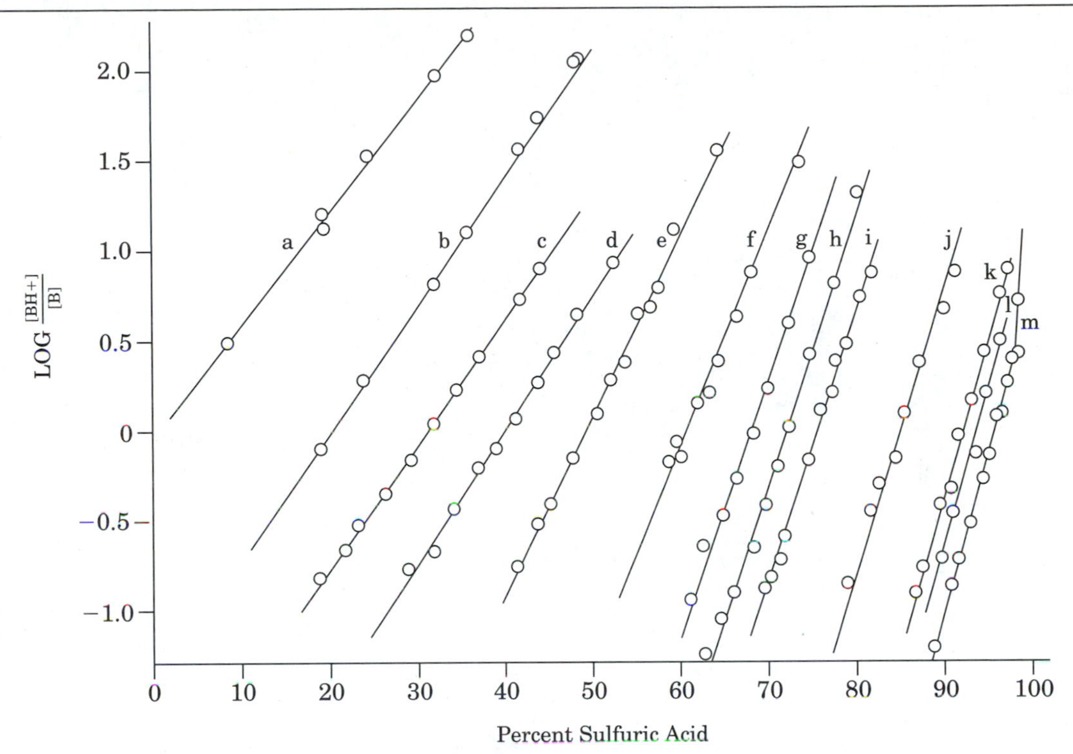

indicates the concentration of each species as a function of solution composition.[85] Figure 7.1 shows the correlation of log I with percent H_2SO_4 for each of a series of nitroanilines.[86,87,88] Figure 7.2 shows how H_0 varies over a concentration range from 0% H_2SO_4 (100% H_2O) to 100% H_2SO_4.

Figure 7.2 also shows several other acidity functions, each of which has been developed with a different series of indicators. H''' is based on *N,N*-dialkylanilines and *N*-alkyldiphenylamines with nitro substituents on the

[85]Although UV-vis spectroscopy has most often been used in acidity function determinations, other techniques are also applicable. For example, an acidity function based on ^{13}C NMR measurements has been reported by Fărcaşiu, D.; Ghenciu, A. *J. Am. Chem. Soc.* **1993**, *115*, 10901.

[86]Jorgenson, M. J.; Hartter, D. R. *J. Am. Chem. Soc.* **1963**, *85*, 878.

[87]The figure also includes data from reference 84. The compounds used are (from left to right) (a) 2-nitroaniline, (b) 4-chloro-2-nitroaniline, (c) 2,5-dichloro-4-nitroaniline, (d) 2-chloro-6-nitroaniline, (e) 2,6-dichloro-4-nitroaniline, (f) 2,4-dinitroaniline, (g) 2,6-dinitroaniline, (h) 4-chloro-2,6-dinitroaniline, (i) 2-bromo-4,6-dinitroaniline, (j) 3-methyl-2,4,6-trinitroaniline, (k) 3-bromo-2,4,6-trinitroaniline, (l) 3-chloro-2,4,6-trinitroaniline, (m) 2,4,6-trinitroaniline.

[88]Note the increasing slope with increasing [H_2SO_4].

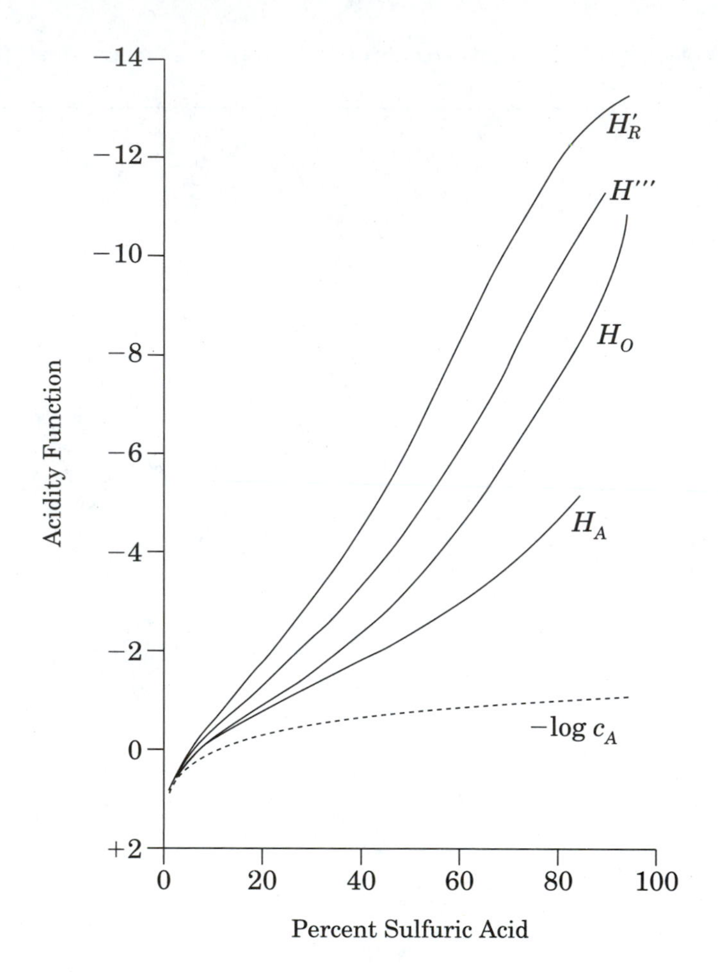

Figure 7.2 Acidity functions of mixtures of water and sulfuric acid. (Reproduced from reference 82, p. 272.)

aromatic rings,[89,90] and H_A is based on a series of amides.[91] Some of these functions change more rapidly with changing acid concentration than does H_0, while others change less rapidly. All of the acidity functions change much more rapidly than does $-\log c_A$, where c_A is the molarity of the acid.

There are some other acidity functions which are defined for processes more complicated than simply protonating a particular type of atom. Consider the result of protonating the alcohol function of a triarylmethanol bearing a substituent (S) on each aromatic ring, as shown in Figure 7.3. The

---

[89]Arnett, E. M.; Mach, G. W. *J. Am. Chem. Soc.* **1964**, *86*, 2671.

[90]The triple prime notation refers to the fact that the protonation takes place on a nitrogen atom in a tertiary aromatic amine. The ordinary H_0 values are sometimes called H' or H_0' because they are based on measurements for the protonation of a primary nitrogen in an aniline. *Cf.* Arnett, E. M.; Mach, G. W. *J. Am. Chem. Soc.* **1966**, *88*, 1177.

[91]Yates, K.; Stevens, J. B.; Katritzky, A. R. *Can. J. Chem.* **1964**, *42*, 1957.

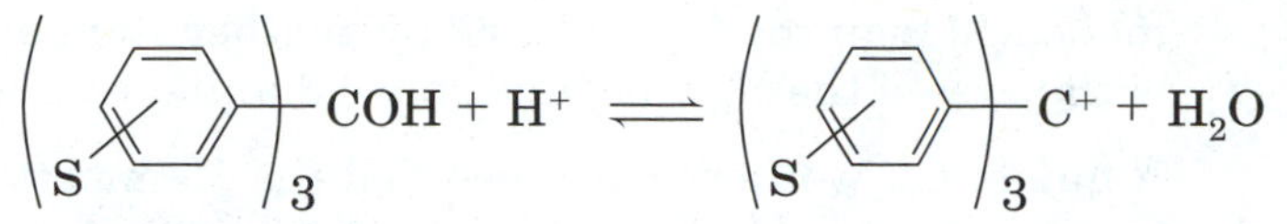

Figure 7.3 Indicator reaction for acidity function H_R.

$$\left(\text{S-}C_6H_4\right)_3\text{COH} + H^+ \rightleftharpoons \left(\text{S-}C_6H_4\right)_3\text{C}^+ + H_2O$$

overall result is the formation of a carbocation and a water molecule. For this reaction, Deno et al. defined an acidity function H_R as shown in equation 7.31, in which a_w is the activity of water.[92] Reagan developed the H_C scale based on the protonation of carbon bases such as azulene,[93] and acidity functions have also been developed for series of indoles (H_I), benzophenones (H_B), azulenes (H_m), and azo compounds (H_{Az}).[94]

$$H_R = -\log a_{H^+}(a^{\circ}_{ROH}/a^{\circ}_{R^+}) + \log a_w \tag{7.31}$$

Although sulfuric acid-water mixtures are often used for the definition of acidity functions, acidity functions have been measured for other water-acid mixtures, including perchloric, hydrochloric, phosphoric, nitric, and toluenesulfonic acids. Similarly, it is possible to measure and compare the acidities of compounds that are not acidic enough to be ionized in water by using a more basic medium. This approach was reported by Stewart and O'Donnell, who established the H_- scale for solutions of tetraalkylammonium hydroxide in sulfolane, DMSO-water, pyridine-water, and water.[95] In essence, the H_- scale extends upward the range for which pK_a values can be measured, which is effectively from 0 to 12 in aqueous solutions, by another 12 pK units of basicity.[39,96]

Hammett suggested that all of the functions are suitable as acidity functions, but noted that

> It is now abundantly evident that a unique operational definition of the acidity of a system is a will-o'-the-wisp. Qualitatively one conveys some significant information by saying for instance that 80% aqueous sulfuric acid is more acid than 5% sulfuric acid. (This represents a semantic change, for 50 years ago one would have said that the 80% acid is less acid because it is less ionized.) But by one standard, that of reaction with bases of the H'_R type, the 80% acid is 11.8 logarithmic units

[92]Deno, N. C.; Jaruzelski, J. J.; Schriesheim, A. *J. Am. Chem. Soc.* **1955**, *77*, 3044. In Figure 7.2, the function H'_R is defined as

$$H'_R = -\log a_{H^+}(a^{\circ}_{ROH}/a^{\circ}_{R^+})$$

For a discussion, see reference 82, pp. 274–275.

[93]Reagan, M. T. *J. Am. Chem. Soc.* **1969**, *91*, 5506.

[94]For a discussion and leading references, see reference 98.

[95]Stewart, R.; O'Donnell, J. P. *J. Am. Chem. Soc.* **1962**, *84*, 493.

[96]For a review of the determination of acidity functions and their application to the study of kinetics and mechanisms, see Paul, M. A.; Long, F. A. *Chem. Rev.* **1957**, *57*, 1; Long, F. A.; Paul, M. A. *Chem. Rev.* **1957**, *57*, 935.

more acid than the 5% acid, and by another standard, that of reaction with bases of the H_A type, it is only 4.3 units more acid.[97]

A number of workers have modified the Hammett approach to acidity functions. In particular, Cox and Yates proposed the **excess acidity function**, X, which represents the difference between the acidity observed for a system and the acidity that would be observed if the system were ideal. In this approach, the activity coefficient ratio in equation 7.27 is taken to be the product of a coefficient m^* times the activity coefficient ratio, X, for a hypothetical standard base, B^*

$$\log(\gamma_B \gamma_{H^+}/\gamma_{BH^+}) = m^* \times \log(\gamma_{B^*}\gamma_{H^+}/\gamma_{B^*H^+}) = m^*X \tag{7.32}$$

in which m^* is a constant characteristic of each base, while X is a constant characteristic of the medium.[98,99,100]

7.2 Acid and Base Catalysis of Chemical Reactions

The properties of acids and bases are important to us not only because organic molecules affect the pH of aqueous solutions, but also because many organic reactions involve proton transfer. For that reason, acid-catalyzed or base-catalyzed reactions may be the "largest single class of organic mechanisms."[101] We will not attempt to survey all possible acid-catalyzed and base-catalyzed reactions. Instead, we will first develop some ideas about how proton transfers are involved in organic reactions and then use these ideas to study selected reactions of carbonyl compounds and carboxylic acid derivatives. More complete discussions of these and related reactions are given in the references.[7,18,102,103,104]

---

[97]Hammett, L. P. (reference 82) p. 278.

[98]Cox, R. A.; Yates, K. *J. Am. Chem. Soc.* **1978**, *100*, 3861.

[99]Cox, R. A.; Yates, K. *Can. J. Chem.* **1980**, *59*, 2116.

[100]For a discussion of different approaches to defining acidity functions and excess acidity, see reference 18, pp. 14–115.

[101]*Cf.* Arnett and Mach (reference 90).

[102]Bunnett, J. F. in *Investigation of Rates and Mechanisms of Reactions. Part I. General Considerations and Reactions at Conventional Rates*, 4th Ed.; Bernasconi, C. F., ed.; Wiley-Interscience: New York, 1986; p. 253.

[103]Jencks, W. P. *Catalysis in Chemistry and Enzymology*; McGraw-Hill Book Company: New York, 1969.

[104]Frost, A. A.; Pearson, R. G. *Kinetics and Mechanism*; John Wiley & Sons: New York, 1953; pp. 231 *ff*.

Specific Acid Catalysis

Let us consider the acid-catalyzed reaction in water of compound **S** to form product **P**, as shown in equations 7.33 and 7.34.[105] In these equations k_1 and k_{-1} are fast relative to k_2, and H^+ represents protonated solvent.

(fast) $$S + H^+ \underset{k_{-1}}{\overset{k_1}{\rightleftharpoons}} SH^+ \qquad (7.33)$$

(rate-limiting) $$SH^+ \xrightarrow{k_2} P \qquad (7.34)$$

Since the second step is the rate-limiting step, the rate of the reaction is:

$$\text{rate} = d[P]/dt = k_2[SH^+] \qquad (7.35)$$

If we express the equilibrium in equation 7.33 as

$$K = [SH^+]/[S][H^+] \qquad (7.36)$$

then $$d[P]/dt = Kk_2[S][H^+] = k_{SH^+}[S][H^+] \qquad (7.37)$$

The concentrations of any acids that might ionize to produce H^+ do not appear in equation 7.37 because the proton has been fully transferred to the substrate before the rate-limiting step of the reaction.[106] Because the reaction rate depends only on the concentrations of S and H^+, the reaction is said to be subject to **specific acid catalysis**. In essence, the term *specific acid catalysis* means that the reaction rate depends only on the concentration of protons in the solvent. In water the term often used is ***specific oxonium ion catalysis***, which refers to the activity of hydrogen ions (H_3O^+ or, more generally, $H(H_2O)_n{}^+$). In other solvents, the more precise term is ***specific lyonium ion catalysis***, referring to the protonated form of solvent in each medium (for example, $CH_3OH_2{}^+$ in methanol).[107,108]

General Acid Catalysis

If partial or complete proton transfer occurs *during* the rate-limiting step of a reaction, then the reaction is subject to **general acid catalysis**. Two processes can give rise to general acid catalysis. In the first, a proton is transferred from an acid, HA, to a substrate, S, in the rate-limiting step of the reaction. The proton transfer may occur as the rate-limiting step in a

---

[105]Although it is not explicitly indicated in equation 7.34 or in equation 7.39, the proton is released in the rate-limiting or a subsequent step.

[106]For a discussion of many variations of the dependence of reaction rates on $[H^+]$, see Gupta, K. S.; Gupta, Y. K. *J. Chem. Educ.* **1984**, *61*, 972.

[107]Bunnett has noted that "It is doubtful whether any major chemical phenomena have been blessed with names more confusing than specific oxonium ion catalysis and general acid catalysis." (reference 102, p. 317.)

[108]A similar treatment for base-catalyzed reactions can be used to develop corresponding equations for **specific base catalysis**. Again, the terms *specific hydroxide ion catalysis* in water or *specific lyate ion catalysis* in another solvent (e.g., CH_3O^- in methanol) may be used for greater precision. (See reference 102.)

two-step reaction (equations 7.38 and 7.39) or the proton may be transferred during a one-step reaction leading directly to protonated product, PH^+ (equation 7.40).[109]

$$\text{(rate-limiting)} \qquad S + HA \underset{k_{-1}}{\overset{k_1}{\rightleftharpoons}} SH^+ + A^- \tag{7.38}$$

$$\text{(fast)} \qquad SH^+ \xrightarrow[\text{fast}]{k_2} P \tag{7.39}$$

or

$$\text{(rate-limiting)} \qquad S + HA \underset{k_{-1}}{\overset{k_1}{\rightleftharpoons}} PH^+ + A^- \tag{7.40}$$

followed by deprotonation of PH^+. In either case the rate law is

$$d[P]/dt = k_1[S][HA] \tag{7.41}$$

The reaction is said to be subject to general acid catalysis because acids in general, not just H^+, catalyze the reaction. If more than one acid is available to transfer protons, then the rate is a summation of the individual rates from the acid catalysis of all of the acids present.[110]

$$\frac{d[P]}{dt} = \sum_i k_i[S][HA_i] \tag{7.42}$$

The second process leading to general acid catalysis involves fast initial proton transfer from HA to S (equation 7.43), followed by rate-limiting proton transfer from the protonated substrate to the *conjugate base* of an acid (equation 7.44). Since the substrate is protonated before the rate-limiting step, this process can also be called **specific acid-general base catalysis**.

$$\text{(fast)} \qquad S + HA \underset{k_{-1}}{\overset{k_1}{\rightleftharpoons}} SH^+ + A^- \tag{7.43}$$

$$\text{(rate-limiting)} \qquad SH^+ + A^- \xrightarrow{k_2} P + HA \tag{7.44}$$

The rate of the reaction is

$$d[P]/dt = k_2[SH^+][A^-] \tag{7.45}$$

Incorporating the relationships in equation 7.46 and 7.47,

$$[A^-] = K_a[HA]/[H^+] \tag{7.46}$$

$$K_{eq}' = [SH^+]/[S][H^+] \tag{7.47}$$

---

[109]In the latter case, the proton transfer may be concerted with some other bonding change in the substrate. For example, see Capon, B.; Nimmo, K. *J. Chem. Soc., Perkin Trans. 2* **1975**, 1113.

[110]A similar treatment for base-catalyzed reactions can be used to develop corresponding equations for **general base catalysis**. For a detailed discussion of general acid-base catalysis of aqueous reactions, see Jencks, W. P. *Chem. Rev.* **1972**, *72*, 705.

we can rewrite 7.45 as

$$d[\mathrm{P}]/dt = k_2[\mathrm{SH^+}][\mathrm{A^-}] = k_2 K_{eq}'[\mathrm{S}][\mathrm{H^+}][\mathrm{A^-}] = k_2 K_a K_{eq}'[\mathrm{S}][\mathrm{HA}] \tag{7.48}$$

The product of k_2 with the two equilibrium constants can be written as k_{HA}, the observed experimental rate constant for catalysis by HA, so that equation 7.48 becomes

$$d[\mathrm{P}]/dt = k_{HA}[\mathrm{S}][\mathrm{HA}] \tag{7.49}$$

If more than one acid is present, then the reaction rate is the sum of the rates of each of the individual general acid-catalyzed rates, and equation 7.42 again holds.

We should not expect that a given reaction will exhibit only general acid or base catalysis or specific acid or base catalysis. In principle, reactions may be subject to more than one kind of catalysis. For example, the catalytic rate expression for the reaction of iodine with acetone in buffer solutions was determined by Dawson and co-workers[111] to be

$$k = k_{H_2O}[\mathrm{H_2O}] + k_{H}[\mathrm{H^+}] + k_{OH}[\mathrm{HO^-}] + k_{A}[\mathrm{A^-}] + k_{HA}[\mathrm{HA}] \tag{7.50}$$

where k_{H2O} is the rate constant for the uncatalyzed or water-catalyzed reaction, and [HA] and [A^-] are the concentrations of the protonated and unprotonated forms of the acid catalyst. Therefore, this reaction was found to be subject to both general and specific acid catalysis and general and specific base catalysis. Usually in such reactions, however, one may choose pH ranges and catalyst concentrations in which either acid or base catalysis is dominant. The procedure for extracting values of k_{HA} has been described by Stewart.[112] It is useful to make the substitutions

$$K_W = [\mathrm{H^+}][\mathrm{HO^-}] \tag{7.51}$$

$$K_A = [\mathrm{H^+}][\mathrm{A^-}]/[\mathrm{HA}] \tag{7.52}$$

and

$$r = [\mathrm{HA}]/[\mathrm{A^-}] \tag{7.53}$$

If the buffer ratio, r, is kept constant for a series of solutions that differ in [HA], then

$$k = k'(r) + [\mathrm{HA}](k_{HA} + k_A/r) \tag{7.54}$$

A plot of k versus [HA] should generate a linear plot with slope $(k_{HA}+k_A/r)$, and analysis of several studies, each with a different value of r, can yield the

---

[111] Dawson, H. M.; Hall, G. V.; Key, A. *J. Chem. Soc.* **1928**, 2844 and references therein. See also the discussion in reference 7, pp. 173 *ff*.

[112] Reference 18, pp. 173 *ff*.

values of k_{HA} and k_A separately. If it can be demonstrated that only general acid catalysis is significant, then keeping a 1 : 1 buffer ratio ($r = 1$) allows k_{HA} to be determined directly from a plot of k versus [HA].[113]

Brønsted Catalysis Law

For general acid catalysis the relationship between the acidity of each acid present and the rate of the reaction is given by the **Brønsted catalysis law**,[114]

$$k_a = G_a(K_a)^\alpha \quad \textbf{(7.55)}$$

which means that there is a linear logarithmic relationship between the rate constants and equilibrium constants for a general acid-catalyzed reaction involving a series of closely related compounds. Taking the logarithm of both sides of equation 7.55 produces equation 7.56.

$$\log k_{HA} = \alpha \log K_a + \text{constant} \quad \textbf{(7.56)}$$

Similarly, for general base catalyzed reactions with a series of bases

$$\log k_B = \beta \log K_b + \text{constant} \quad \textbf{(7.57)}$$

A linear plot of $\log k_{HA}$ versus $\log K_a$ (equivalently, $-\text{p}K_a$) for a series of acids confirms the Brønsted relationship, and the slope of the line is α. For example, Figure 7.4 shows a Brønsted plot for the hydrolysis of di-

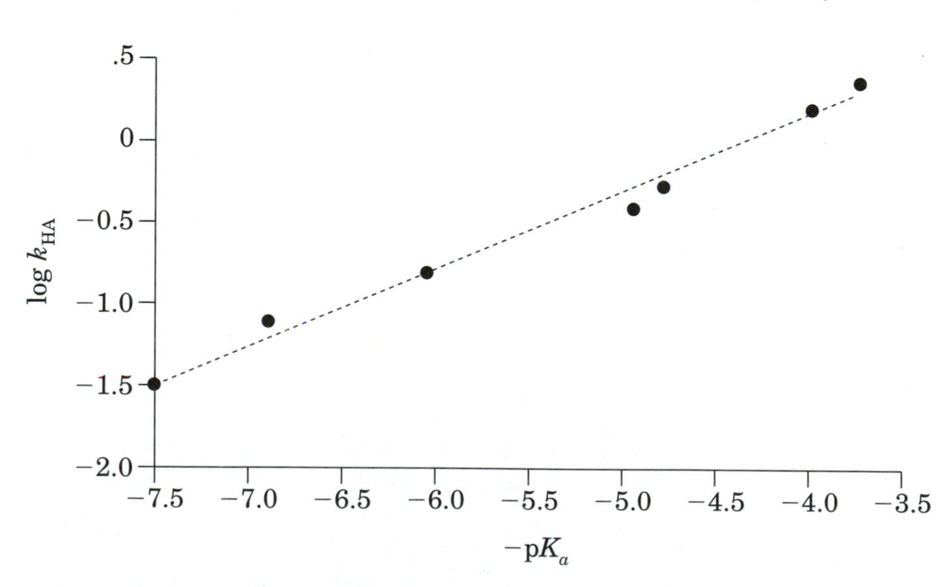

Figure 7.4 A Brønsted catalysis plot for hydrolysis of diethylphenyl orthoformate in 50 : 50 dioxane : water. (Based on data reported in reference 115.)

---

[113]For a discussion of the determination of type(s) of catalysis, see reference 103, pp. 163–167.

[114]Brönsted, J. N. *Chem. Rev.* **1928**, *5*, 231. (The representation of the first vowel in the last name is that used in the publication.)

ethylphenyl orthoformate in 50 : 50 dioxane : water at 25°. There is a linear correlation, and α is 0.47.[115] The value of α is said to represent the "sensitivity of the rate constant to the structural changes in the family of reactions."[116,117] Some investigators relate α to the extent of proton transfer at the transition state of the general acid-catalyzed reaction,[118] but this interpretation is not totally accepted.[119] Moreover, Bordwell has reported α values outside the range of 0–1.[120,121]

7.3 Acid and Base Catalysis of Reactions of Carbonyl Compounds and Carboxylic Acid Derivatives

Addition to the Carbonyl Group

A characteristic reaction of the carbonyl group is addition of polar reagents across the carbon-oxygen double bond.[122,123] Because of the polarity of the carbonyl group, polar additions take place so that the component of the adding reagent that is more acidic/electrophilic adds to the oxygen atom, while the component that is more basic/nucleophilic adds to the carbon atom. Familiar reactions of this type include cyanohydrin formation, bisulfite addition, hemiacetal or hemiketal formation, and addition of amines—the latter usually being followed by elimination of water to give products with carbon-nitrogen double bonds. One of the simplest reactions of aldehydes and some ketones is hydration, the addition of water across the carbon-oxygen double bond. Hydration reactions typically exhibit both general base and general acid catalysis, as illustrated in equation 7.58 and equation 7.59, respectively.[124]

---

[115]Anderson, E.; Fife, T. H. *J. Org. Chem.* **1972**, *37*, 1993.

[116]Bender, M. L. *Chem. Rev.* **1960**, *60*, 53.

[117]For a discussion of the interpretation of α, see Lewis, E. S. *J. Phys. Org. Chem.* **1990**, *3*, 1.

[118]See, for example, Leffler, J. E. *Science* **1953**, *117*, 340.

[119]For leading references, see Streitwieser, Jr., A.; Kaufman, M. J.; Bors, D. A.; Murdoch, J. R.; MacArthur, C. A.; Murphy, J. T.; Shen, C. C. *J. Am. Chem. Soc.* **1985**, *107*, 6983. Also see Wiseman, F.; Kestner, N. R. *J. Phys. Chem.* **1984**, *88*, 4354.

[120]Bordwell, F. G.; Boyle, Jr., W. J.; Yee, K. C. *J. Am. Chem. Soc.* **1970**, *92*, 5926; Bordwell, F. G.; Boyle, Jr., W. J. *J. Am. Chem. Soc.* **1972**, *94*, 3907.

[121]While the equations above suggest that Brønsted plots should be linear, curved plots can arise in some circumstances, as discussed by Kresge, A. J. *Chem. Soc. Rev.* **1973**, *2*, 475. Curved Brønsted plots may also result when a change in the rate-limiting step of a two-step mechanism accompanies the change in base or acid catalyst. For a discussion, see reference 214.

[122]Ingold, C. K. *Structure and Mechanism in Organic Chemistry*, 2nd Ed.; Cornell University Press: Ithaca, New York, 1969; pp. 994 *ff*.

[123](a) For a review of additions in which an equilibrium is established between reactants and products, see Ogata, Y.; Kawasaki, A. in *The Chemistry of the Carbonyl Group*, Vol. 2; Zabicky, J., ed.; Wiley-Interscience: London, 1970. (b) For a theoretical and experimental study of the equilibria of formation of hydrates, hemiacetals and acetals, see Wiberg, K. B.; Morgan, K. M.; Maltz, H. *J. Am. Chem. Soc.* **1994**, *116*, 11067.

[124]Sørensen, P. E.; Jencks, W. P. *J. Am. Chem. Soc.* **1987**, *109*, 4675.

$$B^{-} + H_2O + R_1R_2C{=}O \xrightleftharpoons{\text{slow}} BH + HO\text{–}C(R_1)(R_2)\text{–}O^{-} \xrightleftharpoons{\text{fast}} HO\text{–}C(R_1)(R_2)\text{–}OH + B^{-} \quad (7.58)$$

$$H_2\ddot{O}: + R_1R_2C{=}\ddot{O}{:}\,H\text{–}A \xrightleftharpoons{\text{slow}} H_2\overset{+}{O}\text{–}C(R_1)(R_2)\text{–}OH + {:}A^{-} \xrightleftharpoons{\text{fast}} HO\text{–}C(R_1)(R_2)\text{–}OH + AH \quad (7.59)$$

Although only rarely are the hydrates sufficiently stable to be isolated,[125] the hydration reaction allows an examination of the factors that influence addition to the carbonyl group. It is convenient to represent the equilibrium between an aldehyde or ketone and the corresponding hydrate in terms of a *dissociation constant*, K_d, as shown in equation 7.60.

$$R_1R_2C(OH)_2 \xrightleftharpoons{K_d} R_1C(=O)R_2 + H_2O \quad (7.60)$$

Values of K_d for several hydrates of interest are shown in Table 7.6.[126,127,128,129] The data suggest that the carbonyl side of the equation is favored when R_1 and R_2 are electron donating and is disfavored when R_1 and R_2 are electron withdrawing. In an aromatic aldehyde or ketone, conjugation of the aryl group with the carbonyl group stabilizes the carbonyl side of the equilibrium, with the magnitude of the stabilization estimated at 2.7 kcal/mol.[130,131] Steric factors are also important. The carbonyl side of the equation is favored when R_1 and R_2 are large alkyl groups because a 120° C—C(=O)—C bond angle provides more room for bulky substituents than does the 109° bond angle of the hydrate.[132] In small cycloalkanones, however, steric strain can destabilize the carbonyl side of the equilibrium. In cyclopropanone there is 30° of angle strain at the carbonyl carbon atom because of the difference between the 60° internuclear C—C(=O)—C

[125]The isolation of the dihydrate from hexafluoroacetylacetone was reported by Schultz, B. G.; Larsen, E. M. *J. Am. Chem. Soc.* **1949**, *71*, 3250. The hydrate of *trans*-2,3-di-*tert*-butylcyclopropanone was reported to melt at 105–107° by Pazos, J. F.; Pacifici, J. G.; Pierson, G. O.; Sclove, D. B.; Greene, F. D. *J. Org. Chem.* **1974**, *39*, 1990.

[126]Bell, R. P. *Adv. Phys. Org. Chem.* **1966**, *4*, 1.

[127]Sutton, H. C.; Downes, T. M. *J. Chem. Soc., Chem. Commun.* **1972**, 1.

[128]Hine, J.; Redding, R. W. *J. Org. Chem.* **1970**, *35*, 2769.

[129]The K_d values shown in Table 7.6 are defined (see, for example, reference 132) as

$$K_d = [R_1R_2C{=}O]/[R_1R_2C(OH)_2]$$

and are therefore dimensionless because the concentration of water (55 M, with activity assumed to be unity) is included in the equilibrium constant. Literature values for equilibrium constants in which the concentration of water is not incorporated into the K_d term have been divided by 55.5 to make the values in Table 7.6 comparable.

Table 7.6 Equilibrium constants for dissociation of hydrates of carbonyl compounds.[129]

Compound	K_d[129]
Cyclopropanone[134]	very small
Chloral (α,α,α-Trichloroacetaldehyde)	3.6×10^{-5}
Formaldehyde[127]	4.5×10^{-4}
α-Chloroacetaldehye	2.7×10^{-2}
α-Chlorobutyraldehyde	6.3×10^{-2}
α,α′-Dichloroacetone	0.10
α,α-Dibromobutanal	0.11
α-Chloroheptanal	0.16
2-Chloro-2-methylpropanal	0.19
Methyl pyruvate[132]	0.32
α,α-Dichloroacetone[132]	0.35
α-Bromoheptanal	0.35
Pyruvic acid[132]	0.42
Biacetyl[132]	0.50
Acetaldehyde[124]	0.83
Propanal	1.4
Butanal	2.1
2-Methylpropanal	2.3
Pivaldehyde[132]	4.1
α-Chloroacetone[132]	9.1
Sodium pyruvate[132]	18.5
Benzaldehyde[130]	120
Acetone[128]	720
Acetophenone[130]	1.5×10^5
Benzophenone[130]	8.5×10^6

(Except as noted, data are from reference 126.)

[130]Guthrie, J. P. *Acc. Chem. Res.* **1983**, *16*, 122.

[131]Greenzaid, P. *J. Org. Chem.* **1973**, *38*, 3164.

[132]Bell (reference 126) calculated that the K_d values show a good correlation with Taft's polar and steric constants for alkyl groups, as given by equation 7.60.

$$\log K_d = 2.70 - 2.6\,\Sigma\sigma^* - 1.3\,\Sigma E_s$$

A somewhat better correlation was obtained by Greenzaid et al., who eliminated the need for the E_s term by treating aldehydes separately from ketones in the following equation

$$-\log K_d = 1.70\,\Sigma\sigma^* + 2.03\,\Delta - 2.81$$

in which Δ is the number of aldehyde hydrogen atoms (that is, hydrogen atoms bonded to the carbonyl group) in the compound. Greenzaid, P.; Luz, Z.; Samuel, D. *J. Am. Chem. Soc.* **1967**, *89*, 749.

bond angle and the preferred angle of 120°, for example. However, in cyclopropanone hydrate there is only 24.75° of angle strain because of the smaller difference between the 60° C—C(OH)$_2$—C internuclear bond angle and the preferred angle of 109.5°.[133] Formation of the hydrate relieves angle strain, so cyclopropanone is fully hydrated in the presence of water.[134]

The addition of an alcohol to an aldehyde or ketone produces a hemiacetal or hemiketal. Wheeler reported K_d values for dissociation of the hemiketals produced by reaction of methanol or ethanol with a number of ketones (Table 7.7). The reduced stability of the hemiketals derived from ethanol, in comparison with those derived from methanol, is attributed to the greater steric requirements of the larger ethyl group. It is interesting to note that the methyl hemiketal of cyclobutanone is more stable (smaller K_d) than is that of cyclohexanone, while the trend is the reverse for the hemiketals formed from ethanol. The results can be explained in terms of two conflicting trends. For both ketones, hemiketal formation relieves angle strain in the ketone, but the release of strain is greater for cyclobutanone. Therefore, the hemiketal formed from methanol is more stable for cyclobutanone than for cyclohexanone. Because of the larger size of the ethyl group, there is considerable steric strain in hemiketals formed from ethanol. That strain is greater in a hemiketal formed from cyclobutanone than in a hemiketal formed from cyclohexanone, so the ethyl hemiketal of cyclohexanone is more stable than is the ethyl hemiketal of cyclobutanone.[135]

The addition of a primary amine (H_2NG) to an aldehyde or ketone is the first step of a two-step reaction leading, after dehydration, to a structure with a carbon-nitrogen double bond (equation 7.61). Depending upon

Table 7.7 Equilibrium constants (K_d) for dissociation of hemiketals formed from ketones in methanol or ethanol.

Ketone	K_d (methanol)	K_d (ethanol)
Dipropyl ketone	89.0	
Cyclobutanone	1.11	327
Cyclopentanone	15.1	810
Cyclohexanone	2.16	237
Cycloheptanone	53.5	
Cyclooctanone	268	

(Data from reference 135.)

[133]As was discussed in Chapter 3, angle strain is defined as *half* of the difference between the preferred bond angle and the observed bond angle.

[134]Lipp, P.; Buchkremer, J.; Seeles, H. *Liebigs Ann. Chem.* **1932**, *499*, 1.

[135]Wheeler, O. H. *J. Am. Chem. Soc.* **1957**, *79*, 4191.

the reagents and experimental conditions, either the addition step to form the carbinolamine or the dehydration step can be rate limiting.[136] A particularly important variable is pH, as illustrated in Figure 7.5 for the rate of reaction of hydroxylamine with acetone.[137] From pH 7 to about pH 5, the rate-limiting step in the reaction is the acid-catalyzed dehydration of the addition product, so that the rate increases with decreasing pH. Below pH 5, however, the rate limiting step is the addition of unprotonated hydroxylamine to the carbonyl group, and the concentration of free amine decreases with decreasing pH.[138] Therefore, a bell-shaped pH-rate curve is obtained.

$$—NH_2 + R_1R_2C{=}O \underset{k_{-1}}{\overset{k_1}{\rightleftharpoons}} HO{-}C(R_1)(R_2){-}NHG \underset{k_{-2}}{\overset{k_2}{\rightleftharpoons}} R_1R_2C{=}NG + H_2O \qquad (7.61)$$

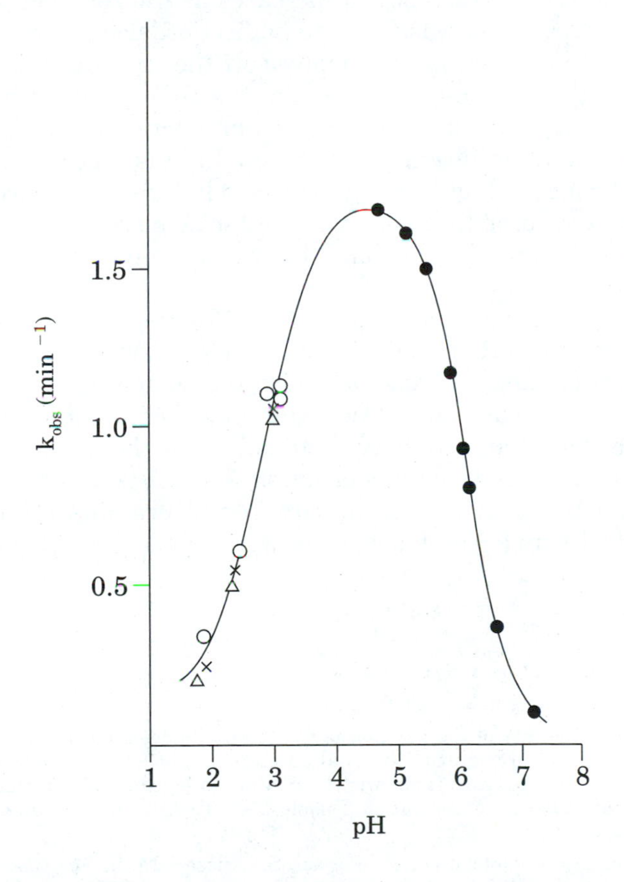

Figure 7.5 Dependence on pH of the rate of reaction of hydroxylamine with acetone. (Adapted from reference 137.)

[136] The two steps have different sensitivities to substituent effects, so that the ρ values for Hammett correlations for the additions can vary according to conditions. (See the discussion in Chapter 6.)

[137] Jencks, W. P. *J. Am. Chem. Soc.* **1959**, *81*, 475.

[138] In addition, there is a specific acid-catalyzed pathway for addition of hydroxylamine to the protonated carbonyl group. Therefore, the rate does not fall off as fast at low pH as might be expected solely on the basis of the concentration of free hydroxylamine.

Enolization of Carbonyl Compounds

In addition to addition to the carbonyl group, ketones with α hydrogen atoms also undergo substitution by halogen.

$$\mathrm{{>}C{-}C({=}O){-}C{<}(H)} \xrightarrow[\text{acid or base}]{X_2} \mathrm{{>}C{-}C({=}O){-}C{<}(X)} + \mathrm{HX} \qquad (7.62)$$

Lapworth found that the substitution reaction is catalyzed by both acids and bases. Since the HX produced as a by-product of the reaction can serve as a catalyst, the acid-catalyzed reaction is said to be **autocatalytic** because the rate of the reaction increases as the reaction proceeds. The rate law for the reaction was found to be first order in ketone but zero order in halogen, so halogen is not involved in the rate-limiting step of the reaction.[139,140,141] The simplest mechanisms consistent with the data involve rate-limiting, acid- or base-catalyzed production of an intermediate, which then reacts with halogen in a fast step. Because species with carbon-carbon double bonds are highly reactive toward halogen, Lapworth proposed mechanisms proceeding through an enol intermediate in the acid-catalyzed reaction (Figure 7.6) and an enolate intermediate in the base-catalyzed process (Figure 7.7).[142]

The mechanisms in Figure 7.6 and Figure 7.7 are supported further by the observations that chiral ketones undergo both acid- and base-catalyzed racemization, and exchange of hydrogen isotopes on the α carbon atoms is also observed. The rates of acid-catalyzed iodination and racemization are identical for phenyl *sec*-butyl ketone,[143] and the same is true for acid-catalyzed bromination and racemization of (+)-2-(2-carboxybenzyl)indanone[144] and for base-catalyzed racemization and bromination of the same compound.[145] Hydrogen exchange rates were shown to be identical to the rates of

---

[139]Lapworth, A. *J. Chem. Soc.* **1904**, 30.

[140]Watson, H. B. *Chem. Rev.* **1930**, *7*, 173.

[141]Obviously the rate of the reaction is not totally independent of halogen concentration, because no reaction can occur in the total absence of halogen. At $[X_2]$ greater than 10^{-8} to 10^{-5} M^{-1}, however, the reaction is apparently 0th order in halogen. Dependence on $[X_2]$ has been observed at low halogen concentration: Dubois, J.-E.; El-Alaoui, M.; Toullec, J. *J. Am. Chem. Soc.* **1981**, *103*, 5393.

[142]For a review of enolization, see Forsén, S.; Nilsson, M. in *The Chemistry of the Carbonyl Group*, Vol. 2; Zabicky, J., ed.; Wiley-Interscience: London, 1970; pp. 157–240.

[143]Bartlett, P. D.; Stauffer, C. H. *J. Am. Chem. Soc.* **1935**, *57*, 2580. However, the rates of racemization and halogenation of a series of dialkyl ketones were found not to be identical in this study.

[144]Ingold, C. K.; Wilson, C. L. *J. Chem. Soc.* **1934**, 773.

[145]Hsü, S. K.; Wilson, C. L. *J. Chem. Soc.* **1936**, 623.

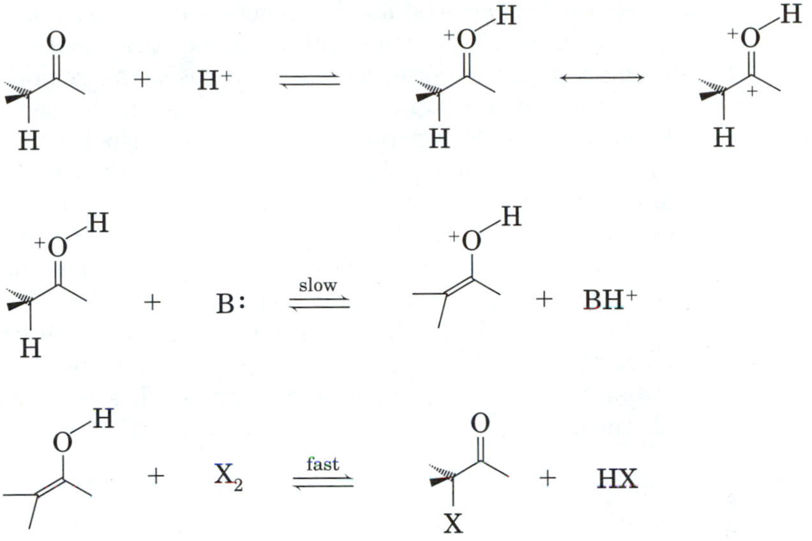

Figure 7.6 Lapworth mechanism for acid-catalyzed α-halogenation of a ketone.

racemization and halogenation with acid catalysis by Reitz[146] and with base catalysis by Ingold and co-workers.[147] The simplest explanation for the experimental results is that halogenation, racemization, and hydrogen exchange occur at identical rates because in each case the rate-limiting step is the deprotonation of the ketone to form an enolate (under base catalysis) or deprotonation of the protonated carbonyl to form an enol (under acid catalysis).[148]

Figure 7.7 Lapworth mechanism for base-catalyzed α-halogenation of a ketone.

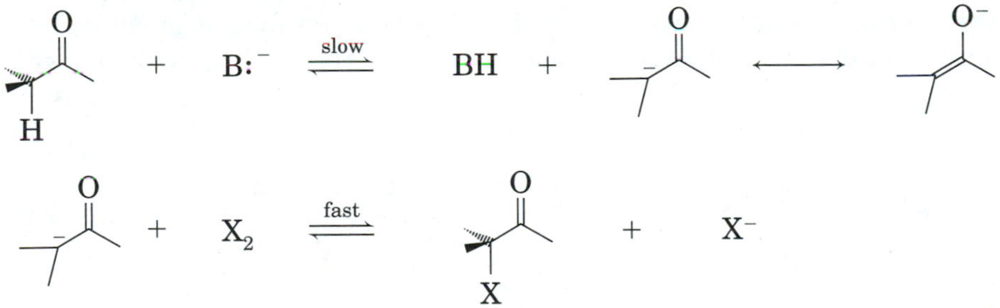

[146]Reitz, O. *Z. phys. Chem.* **1937**, *A179*, 119.

[147]Hsü, S. K.; Ingold, C. K.; Wilson, C. L. *J. Chem. Soc.* **1938**, 78.

[148]Reference 122, p. 822.

If there is at least one hydrogen atom on each of the α carbon atoms of an unsymmetrically substituted ketone, then more than one substitution product might be observed. In such cases the regiochemistry of the halogenation depends upon the substituents on the two α carbon atoms and upon whether the reaction is catalyzed by acid or base. In general, acid-catalyzed halogenation occurs on the α carbon bearing the greater number of alkyl substituents. For example, acid-catalyzed iodination of (−)-menthone (**1**) produced 79% of the product with halogen on the more highly substituted α carbon atom (**2**) and 21% of the product formed by substitution on the less highly substituted α carbon atom (**3**).[149] The results suggest that acid-catalyzed enolization of a ketone leads to preferential formation of the more stable enol, which is the structure with the greater number of alkyl substituents on the carbon-carbon double bond (as is the case with alkenes).[150]

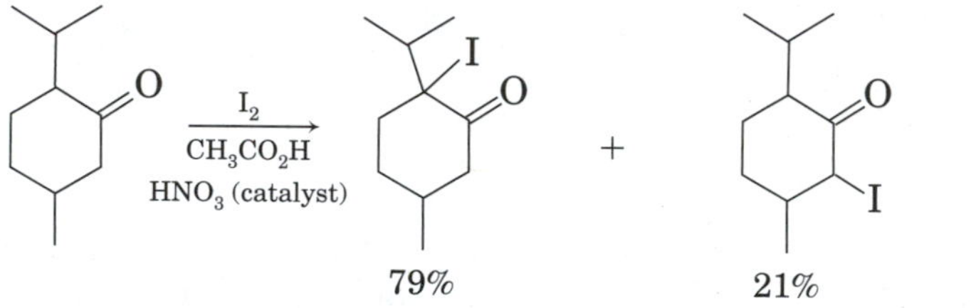

I_2, CH_3CO_2H, HNO_3 (catalyst)

79% 21%

1 **2** **3**

(7.63)

In contrast, base-promoted halogenation tends to occur on the *less*-substituted α carbon atom. For example, base-promoted iodination of methyl alkyl ketones leads initially to the iodomethyl alkyl ketone. Further reaction leads to the α,α-diodo-, then to the α,α,α-triiodoketone and then to iodoform and a carboxylate ion by the haloform reaction (Figure 7.8). Two factors could contribute to the observed regiochemistry of base-promoted halogenation of ketones: one is a kinetic preference for formation of the less-substituted enolate, followed by fast reaction of the enolate with halogen,

Figure 7.8
Iodoform reaction.

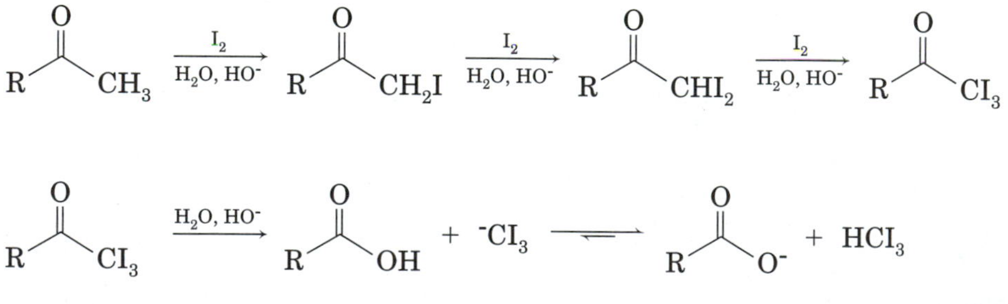

$$RC(=O)CH_3 \xrightarrow[H_2O,\ HO^-]{I_2} RC(=O)CH_2I \xrightarrow[H_2O,\ HO^-]{I_2} RC(=O)CHI_2 \xrightarrow[H_2O,\ HO^-]{I_2} RC(=O)CI_3$$

$$RC(=O)CI_3 \xrightarrow{H_2O,\ HO^-} RC(=O)OH + {}^-CI_3 \rightleftharpoons RC(=O)O^- + HCI_3$$

[149]Bartlett, P. D.; Vincent, J. R. *J. Am. Chem. Soc.* **1933**, *55*, 4992.

[150]For a summary and leading references, see Cardwell, H. M. E.; Kilner, A. E. H. *J. Chem. Soc.* **1951**, 2430.

and the second is a greater thermodynamic stability of the less-substituted enolate ion. The kinetic explanation is consistent with the observation that increasing alkyl substitution decreases the rate of formation of enolate ions. For example, the rate of base-catalyzed iodination of phenyl alkyl ketones with the structure $C_6H_5COCHR_1R_2$ was found to decrease with increasing alkyl substitution, as shown in Table 7.8.[151]

There is also evidence suggesting preferential but not exclusive formation of the less-substituted enolate in equilibrium mixtures of enolate ions. House and Kramar investigated the equilibria between enolates synthesized by reaction of unsymmetrical ketones with triphenylmethylpotassium in 1,2-dimethoxyethane solution (equation 7.64).[152,153]

$$\underset{\mathbf{4}}{(H)(R)C{=}C(O^-)CHR_2} \underset{H_3COCH_2CH_2OCH_3}{\overset{KC(C_6H_5)_3}{\rightleftharpoons}} RH_2C{-}CO{-}CHR_2 \underset{H_3COCH_2CH_2OCH_3}{\overset{KC(C_6H_5)_3}{\rightleftharpoons}} \underset{\mathbf{5}}{RH_2C(O^-)C{=}CR_2} \qquad (7.64)$$

After equilibration, the enolates were allowed to react with D_2O, acetic anhydride or methyl iodide. The results suggested that the less highly substituted enolate (**4**) was present in greater concentration at equilibrium when the starting ketone was monoalkyl substituted on one α carbon and dialkyl substituted on the other (e.g., $R_2CHCOCH_2R$). For methyl ketones (CH_3COCH_2R), however, the equilibrium favored the less substituted enolate if R was a branched alkyl group, but approximately equal concentrations of the two possible enolates were observed if R was a linear alkyl group. Furthermore, the counter ion was found to be a significant factor in

Table 7.8 Substituent effects on base-promoted iodination of phenyl alkyl ketones.

$C_6H_5C(O)CHR_1R_2$	R_1	R_2	Relative Rate
	H	H	238
	H	CH_3	37
	H	CH_2CH_3	29
	CH_3	CH_3	7

(Data from reference 151.)

[151]Literature data summarized by Cardwell, H. M. E. *J. Chem. Soc.* **1951**, 2442.

[152]House, H. O.; Kramar, V. *J. Org. Chem.* **1963**, *28*, 3362.

[153]Synthetic applications of acid- and base-catalyzed α-halogenation of ketones have been discussed by House. House, H. O. *Modern Synthetic Reactions*, 2nd Ed.; W. A. Benjamin, Inc.: Menlo Park, CA, 1972; pp. 459–478.

the equilibrium, with the more highly substituted enolate increasingly favored as the cation varied from sodium or potassium to lithium.[154]

Acid catalysis of enol formation and base catalysis of enolate formation are also key steps in the aldol reaction.[155] The base-catalyzed reaction is shown in Figure 7.9, and the acid-catalyzed reaction is illustrated in Figure 7.10. The analogous reaction involving ketones is usually not synthetically useful because addition of an enolate ion to a ketone is less favorable than is addition to an aldehyde.[156]

In recent years it has become possible to generate enols by processes other than enolization of carbonyl compounds.[157,158] Kresge and co-workers found that the ketonization of photochemically generated acetophenone enol is subject to general acid catalysis and to general base catalysis.[159,160] The results of the study led to the conclusion that the uncatalyzed ketonization (i.e., the reaction without added acid or base catalysts) takes place in

Figure 7.9
Base-catalyzed aldol reaction.

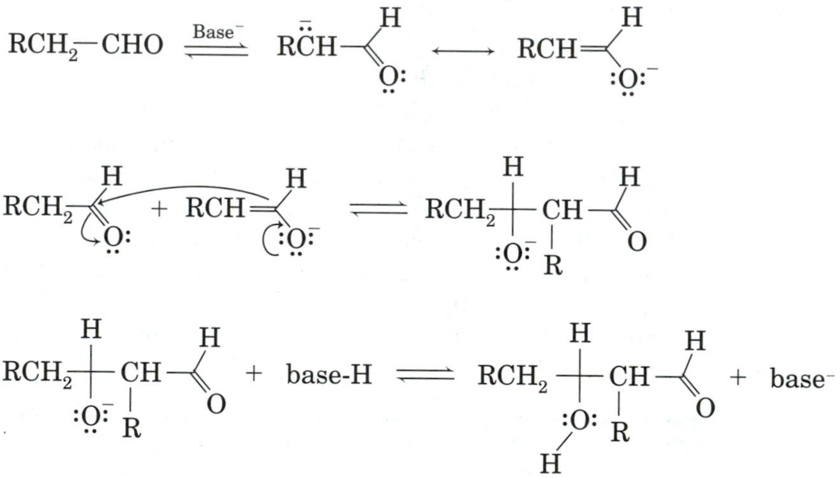

---

[154]Beutelman, H. P.; Xie, L.; Saunders, Jr., W. H. *J. Org. Chem.* **1989**, *54*, 1703, determined that the primary hydrogen kinetic isotope effect for proton removal from alkyl ketones by strong bases such as lithium diisopropylamide in THF or DME was $k_H/k_D = 2.3$ to 5.9 at 0°. The results suggested a very early transition state. However, the reactions appeared to depend on more than one base reacting with each ketone. See also Xie, L.; Saunders, Jr., W. H. *J. Am. Chem. Soc.* **1991**, *113*, 3123.

[155]For leading references and synthetic applications, see chapter 10 of reference 153.

[156]Subsequent elimination of water from the aldol product leads to α,β-unsaturated carbonyl compounds as in the Claisen-Schmidt reaction, for example. For a discussion, see references 153, pp. 632–639.

[157]Capon, B.; Guo, B.-Z.; Kwok, F. C.; Siddhanta, A. K.; Zucco, C. *Acc. Chem. Res.* **1988**, *21*, 135.

[158]Keefe, J. R.; Kresge, A. J.; Schepp, N. P. *J. Am. Chem. Soc.* **1988**, *110*, 1993.

[159]Chiang, Y.; Kresge, A. J.; Santaballa, J. A.; Wirz, J. *J. Am. Chem. Soc.* **1988**, *110*, 5506.

[160]One method of generating enols involves photohydration of phenylacetylene, while another method utilizes the Norrish Type II cleavage of γ-hydroxybutyrophenone. Photochemical reactions will be discussed in Chapter 12.

[161]Phillips, D. C. *Scientific American* **1966** (Nov.), *215*, 78; Vernon, C. A. *Proc. Roy. Soc. (B)* **1969**, *167,* 389. See also Capon, B. *Chem. Rev. 1969*, *69*, 407.

Figure 7.10 Acid-catalyzed aldol reaction.

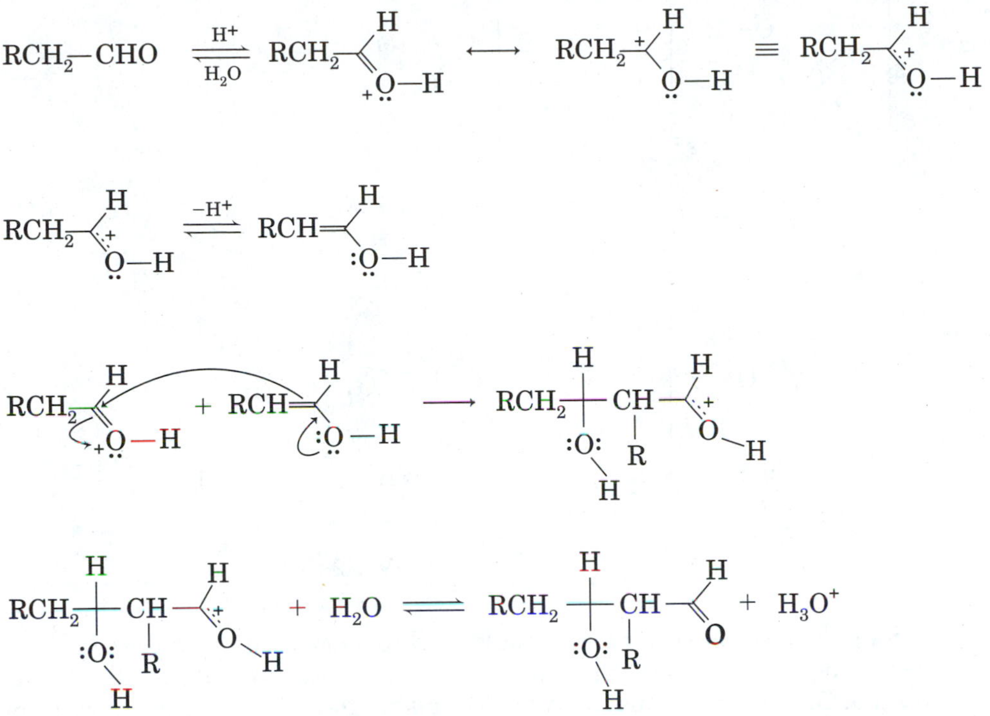

water through a process in which water acts as a base to deprotonate the enol, with subsequent reprotonation of the enolate ion.

Hydrolysis of Acetals

Hydrolysis of acetals is central to the chemistry of carbohydrates,[161] and understanding the operation of enzymes is aided by understanding the mechanisms that can occur in solution. Since acetals are stable in basic solutions, hydrolysis reactions are carried out with acid catalysis. Cordes discussed four possible transition structures for the acid-catalyzed hydrolysis of acetals (Figure 7.11).[162] Structures I and II would be formed in unimolecular decompositions of protonated acetals, so reactions proceeding through these transition structures are termed **A1** (acid-catalyzed, unimolecular) mechanisms. A possible reaction coordinate diagram for a mechanism proceeding through transition structure I is shown in Figure 7.12.[163] Structures III and IV would occur in bimolecular reaction of water with a protonated acetal, so reactions proceeding through these transition structures are termed **A2** (acid-catalyzed, bimolecular) mechanisms.

The possible mechanisms for acetal hydrolysis can be tested with the tools used to study other organic reactions. An A1 mechanism proceeding

[162]Cordes, E. H. *Prog. Phys. Org. Chem.* **1967**, *4*, 1.

[163]Anderson, E.; Capon, B. *J. Chem. Soc. (B)* **1969**, 1033.

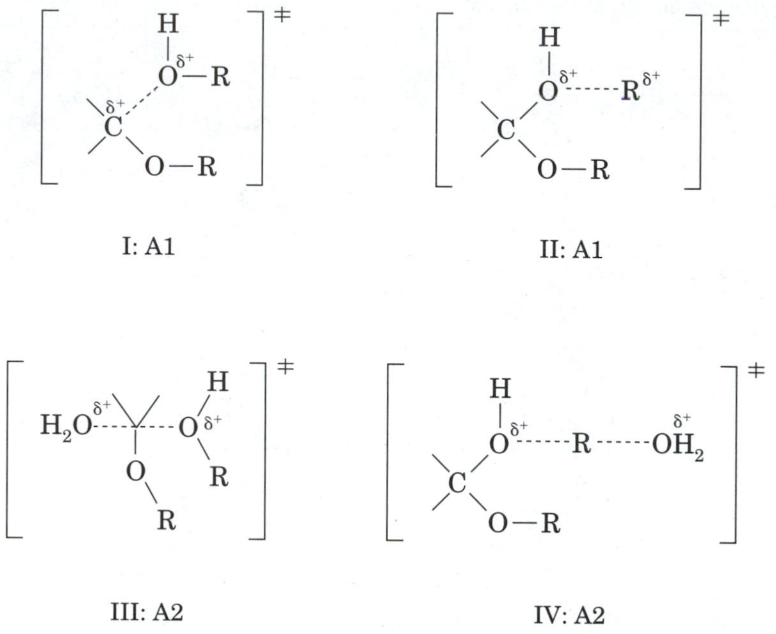

Figure 7.11 Possible transition structures for acid-catalyzed hydrolysis of an acetal. (Adapted from reference 162.)

I: A1

II: A1

III: A2

IV: A2

through transition structure II would lead to a carbocation centered on the departing R group, so racemization of a chiral center would therefore be expected. On the other hand, an A2 mechanism proceeding through transition structure IV would lead to inversion of configuration of a chiral R group. However, Cordes has summarized evidence obtained by a number of workers indicating that—even in the most favorable circumstances—neither inversion nor racemization of a chiral center is observed.[164] Therefore, the acid-catalyzed hydrolysis of acetals appears not to occur through mechanisms involving either transition structure II or transition structure IV.[162]

In transition structure I, there should be appreciable positive charge on the hemiacetal carbon atom, and there should be a negative slope to a Hammett correlation for rates of hydrolysis of acetals formed from aryl aldehydes. Transition structure III, on the other hand, does not show an appreciable positive charge on the acetal carbon atom, so a large Hammett ρ value would not be expected.[165] Fife and Jao found that hydrolysis of *m*-substituted benzaldehyde diethyl acetals gave a good Hammett correlation with σ, with a ρ value of -3.35, which is more consistent with a mechanism involving transition structure I.[166] Observations of near-zero or positive values of $\Delta S^{\ddagger}$ also favor the mechanism shown in Figure 7.12 because a negli-

---

[164]Reference 162, pp. 3 *ff.*

[165]A small positive or negative charge might be present, depending on the relative rates of bond making and bond breaking in the transition structure.

[166]Fife, T. H.; Jao, L. K. *J. Org. Chem.* **1965**, *30*, 1492.

[167]Reference 162, p. 14.

Figure 7.12 Reaction coordinate diagram for A1 hydrolysis of formaldehyde dimethyl acetal. (Adapted from reference 163.)

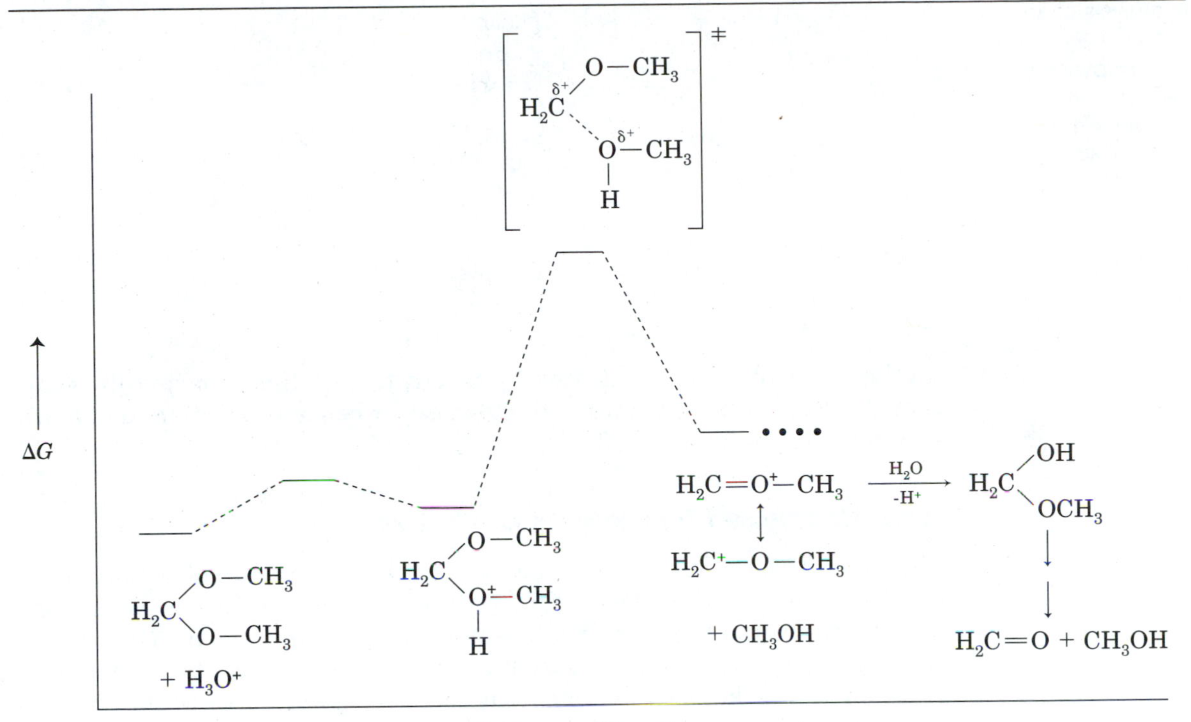

gible or slightly positive $\Delta S^{\ddagger}$ would be expected for an A1 mechanism. However, an A2 mechanism, which requires two species to combine, should have a negative value of $\Delta S^{\ddagger}$.[167] Additional evidence from studies of activation volume and solvent kinetic isotope effects also supports the A1 pathway for acetal hydrolysis.[162,168]

In contrast to acetals, hemiacetals can revert to aldehydes and alcohols through both acid and base catalysis.[169] Some catalysts incorporate both acid and base functions and can serve as bifunctional catalysts. This type of

---

[168]For a discussion of the generality of the A1 mechanism, see Wann, S. R.; Kreevoy, M. M. *J. Org. Chem.* **1981**, *46*, 419.

[169]See, for example, Swain, C. G.; Brown, Jr., J. F. *J. Am. Chem. Soc.* **1952**, *74*, 2534.

[170]Swain, C. G.; Brown, Jr., J. F. *J. Am. Chem. Soc.* **1952**, *74*, 2538.

[171]2-Hydroxypyridine is the tautomer of 2-pyridone:

The equilibrium favors the hydroxy form in the gas phase and the keto form in polar solvents; in nonpolar solvents both tautomers are present in comparable concentrations. For leading references and a theoretical study, see Wong, M. W.; Wiberg, K. B.; Frisch, M. J. *J. Am. Chem. Soc.* **1992**, *114*, 1645.

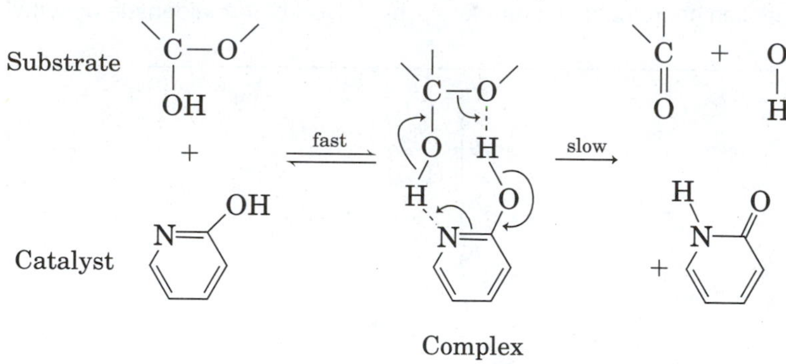

Figure 7.13 Catalysis of the hydrolysis of an acetal by 2-hydroxypyridine. (Adapted from reference 170.)

Substrate

Catalyst

fast

slow

Complex

catalysis was first demonstrated by Swain for the 2-hydroxypyridine-catalyzed mutarotation of tetramethylglucose in benzene solution, as shown schematically in Figure 7.13.[170,171]

Acid-Catalyzed Hydrolysis of Esters

The characteristic reaction of carboxylic acid derivatives is substitution, not addition, because the $-OH$, $-OR$, $-NH_2$, $-X$, $-OCOR$, and $-SR$ groups are less basic (and therefore better leaving groups) than are hydride ions or carbanions that would be produced by the analogous substitution reaction of aldehydes or ketones.[172] A general reaction is shown in equation 7.65, where H—Nu is the protonated form of the nucleophile and L is the leaving group. Both acid and base catalysis of these reactions are possible.[173,174]

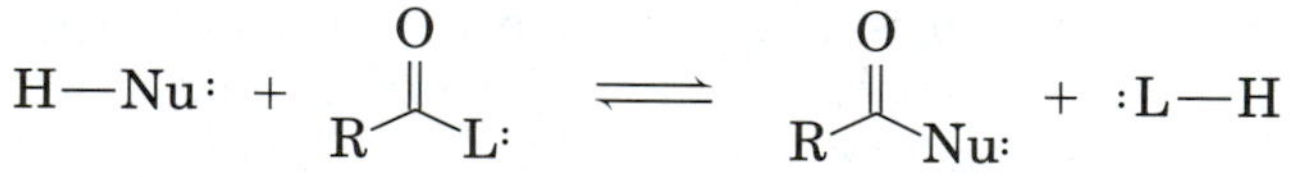

$$\text{H—Nu:} + \text{RC(=O)L:} \rightleftharpoons \text{RC(=O)Nu:} + \text{:L—H} \qquad (7.65)$$

[172]Note the exception to this generalization in the elimination of trihalomethyl anion in the haloform reaction (Figure 7.8).

[173]For reviews, see (a) Euranto, E. K. in *The Chemistry of Carboxylic Acids and Esters*, Patai, S., ed.; Wiley-Interscience: London, 1969; pp. 505–588; (b) Koskikallio, J. in *The Chemistry of Carboxylic Acids and Esters*, Patai, S., ed.; Wiley-Interscience, London, 1969; pp. 103–135; (c) Kirby, A. J. in *Comprehensive Chemical Kinetics. Volume 10. Ester Formation and Hydrolysis and Related Reactions*, Bamford, C. H.; Tipper, C. F. H., eds.; Elsevier Publishing Company: Amsterdam, 1972; pp. 57–207; (d) Talbot, R. J. E. in *Comprehensive Chemical Kinetics. Volume 10. Ester Formation and Hydrolysis and Related Reactions*, Bamford, C. H.; Tipper, C. F. H., eds.; Elsevier Publishing Company: Amsterdam, 1972; pp. 209–293.

[174]The base-catalyzed hydrolysis of a carboxylic acid derivative consumes hydroxide ion and generates a carboxylate ion. To the extent that carboxylate is a less effective base, catalyst is consumed during the reaction. To be precise the process should be called "base-promoted" rather than "base-catalyzed."

The mechanisms of acid-catalyzed hydrolysis of esters can also be categorized as A1 or A2. Two different pathways of each type can be defined, depending upon whether the rate-limiting step involves cleavage of the acyl-oxygen bond ($A_{Ac}1$ or $A_{Ac}2$) or the alkyl-oxygen bond ($A_{Al}1$ and $A_{Al}2$).[175] Only three of these mechanisms are common, however.[176,177] In the $A_{Ac}2$ mechanism (acid-catalyzed, **bimolecular**, with **ac**yl-oxygen cleavage) shown in Figure 7.14, the rate-limiting step involves addition of water to the protonated ester.

Figure 7.14
$A_{Ac}2$ mechanism for ester hydrolysis.

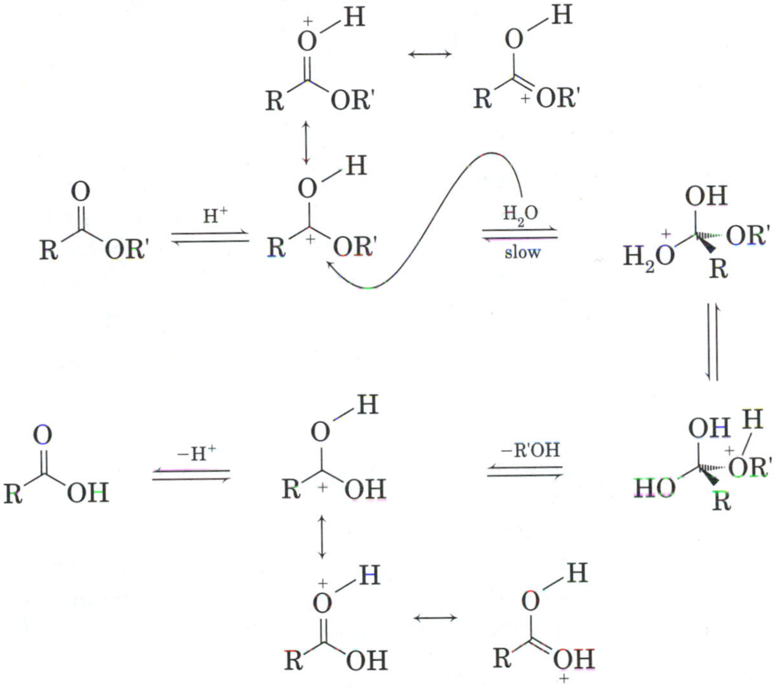

R–C(=O)–OR' ⇌(H⁺) R–C⁺(OH)–OR' [↔ R–C(=O⁺H)–OR' ↔ R–C(OH)=O⁺R'] ⇌(H_2O, slow) H_2O^+–C(OH)(R)–OR' ⇌ HO–C(OH)(R)–O⁺(H)R' ⇌(−R'OH) R–C⁺(OH)–OH [↔ R–C(=O⁺H)–OH ↔ R–C(OH)=O⁺H] ⇌(−H⁺) R–C(=O)–OH

---

[175]According to the 1989 IUPAC Recommendations for the Representation of Reaction Mechanisms, (Commission on Physical Organic Chemistry, IUPAC, *Pure Appl. Chem.* **1989**, *56*, 23; also see Guthrie, R. D.; Jencks, W. P. *Acc. Chem. Res.* **1989**, *22*, 343), bond making (association) processes are denoted A, while bond breaking (dissociation) processes are labeled D. An electrophilic or electrofugic process at a core atom is indicated with a subscript E, and a nucleophilic or nucleofugic process at a core atom is shown with a subscript N. Subscript H indicates hydron as electrophile or electrofuge at a core atom, and subscript h indicates a hydron as electrophile or electrofuge at a peripheral atom. Subscript xh indicates bond breaking or making between hydron and a hydron carrier atom. Stepwise processes are indicated using a plus sign (+). Using this formalism, the $A_{Ac}1$ mechanism is denoted as $A_h + D_N + A_N + D_h$. The $A_{Ac}2$ mechanism is denoted $A_h + A_N + A_hD_h + D_N + D_h$. The $A_{Al}1$ reaction is called $A_h + D_N + A_N + D_h$.

[176]Reference 122, pp. 1128–1164.

[177]Other mechanistic designations are possible. For a discussion of the B-$A_{Ac}3$ mechanism, for example, see Kanerva, L. T.; Euranto, E. K. *J. Chem. Soc. Perkin Trans 2* **1986**, 721.

Figure 7.15
$A_{Ac}1$ mechanism for ester hydrolysis.

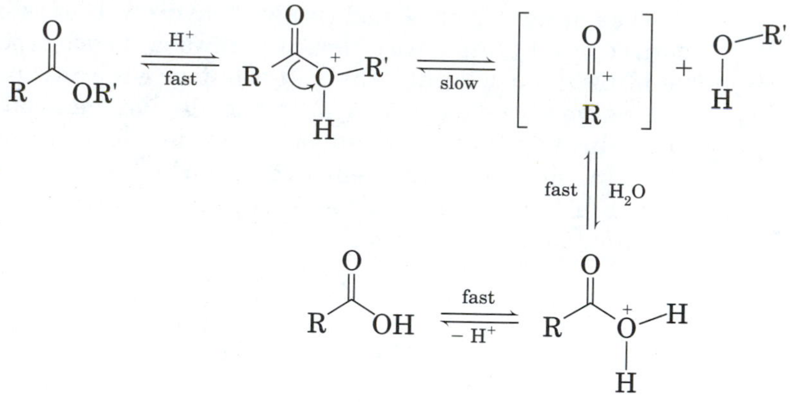

O R OR' $\underset{fast}{\overset{H^+}{\rightleftharpoons}}$ O R $\overset{+}{O}$–R' H $\underset{slow}{\rightleftharpoons}$ [O R +] + O–R' H

fast H₂O

O R OH $\underset{-H^+}{\overset{fast}{\rightleftharpoons}}$ O R $\overset{+}{O}$–H H

The $\mathbf{A_{Ac}1}$ (**a**cid-catalyzed, **unimolecular**, **ac**yl-oxygen cleavage) mechanism is shown in Figure 7.15. The reaction is unimolecular because the rate-limiting step involves dissociation of ROH from the protonated ester.[178] This reaction pathway is most likely with esters of aromatic acids:

1. in which the acylium ion is relatively stable, and
2. in which steric hindrance both destabilizes the ester by reducing resonance interaction of the carbonyl group with the aromatic ring and inhibits nucleophilic addition of water to the carbonyl carbon atom.

For example, acid-catalyzed hydrolysis of methyl mesitoate (**6**) labeled with ^{18}O in the carbonyl oxygen (equation 7.66) led to mesitoic acid (**7**) with no loss of ^{18}O label, even in solutions as low as 3.09 M in H_2SO_4. Had the reaction occurred by the $A_{Ac}2$ pathway, proton exchange and loss of water from the tetrahedral intermediate would probably have resulted in loss of ^{18}O label from starting material and, therefore, from the product.[179]

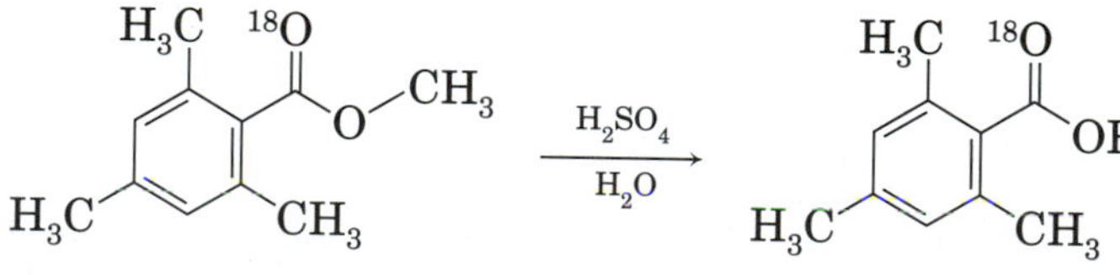

H_3C ^{18}O O CH_3 H_3C CH_3 $\xrightarrow[H_2O]{H_2SO_4}$ H_3C ^{18}O OH H_3C CH_3 **(7.66)**

6 **7**

[178]Reference 122, pp. 145–146.

[179]Bender, M. L.; Ladenheim, H.; Chen, M. C. *J. Am. Chem. Soc.* **1961**, *83*, 123.

[180]Figure 7.16 shows a mechanism proceeding through protonation of the carbonyl oxygen, which is generally agreed to be the predominant site of protonation of an ester. (*Cf.* reference 173(c), p. 59.) Protonation of the alkyl oxygen also occurs to some extent, and an alternative mechanism involving this pathway can be written. See, for example, reference 122, p. 1159.

The **$A_{Al}1$** (acid-catalyzed, **unimolecular**, **al**kyl-oxygen cleavage) mechanism involves formation of a carbocation in a process that is analogous to the ionization step in an S_N1 or E1 reaction (Figure 7.16).[180]

Figure 7.16
$A_{Al}1$ mechanism for ester hydrolysis.

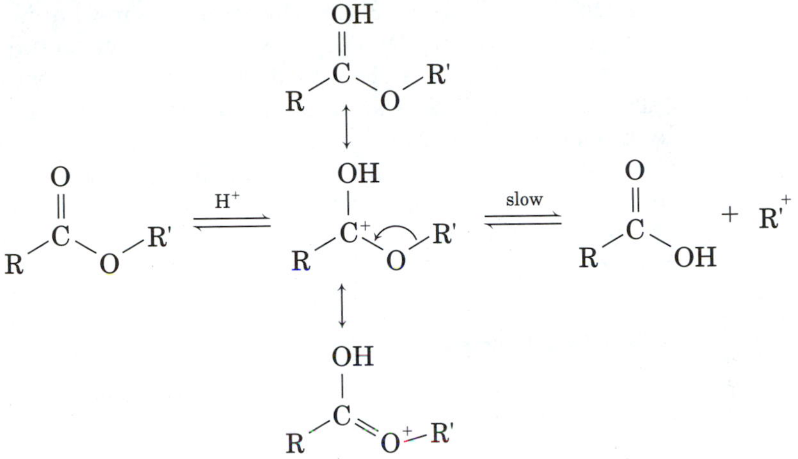

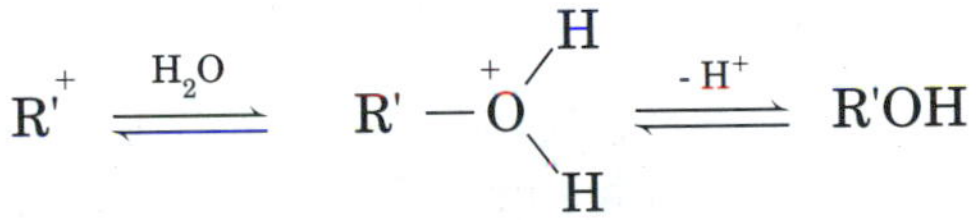

$$RC(=O)OR' \overset{H^+}{\rightleftharpoons} RC^+(OH)OR' \overset{slow}{\rightleftharpoons} RC(=O)OH + R'^+$$

$$R'^+ \overset{H_2O}{\rightleftharpoons} R'-O^+H_2 \overset{-H^+}{\rightleftharpoons} R'OH$$

Unlike acetals, for which one mechanism seems to describe most of the hydrolysis reactions that have been studied, the mechanism of acid-catalyzed hydrolysis of an ester depends on the structure of the ester and on the reaction conditions.[181] For example, Yates has concluded that hydrolysis of primary esters occurs by the $A_{Ac}2$ mechanism below 90% H_2SO_4, but changes to the $A_{Ac}1$ mechanism in solutions with higher concentrations of

Table 7.9 Ester hydrolysis mechanisms in concentrated sulfuric acid solutions.

Acetate	Changeover %	Mechanism
Methyl	75–90	$A_{Ac}2 \rightarrow A_{Ac}1$
sec-Butyl	70–75	$A_{Ac}2 \rightarrow A_{Al}1$
Phenyl	>60	$A_{Ac}2 \rightarrow A_{Ac}1$
t-Butyl		$A_{Al}1$
p-Methoxybenzyl	dilute	$(A_{Ac}2) \rightarrow A_{Al}1$

(Adapted from reference 181.)

[181] Yates, K. *Acc. Chem. Res.* **1971**, *4*, 136.

sulfuric acid (and, therefore, lower concentrations of water). Consistent with this conclusion is the observation that the acid-catalyzed hydrolysis of ethyl acetate in 40.2% H_2SO_4 exhibits a $\Delta H^\ddagger$ of 16.9 kcal/mol and a $\Delta S^\ddagger$ of −15.3 e.u. For hydrolysis of the same compound in 98.4% H_2SO_4, $\Delta H^\ddagger$ is 23.7 kcal/mol and $\Delta S^\ddagger$ is +2.3 e.u. However, the mechanism of hydrolysis of esters from secondary alcohols, as well as benzyl and allyl esters, appears to change from $A_{Ac}2$ to $A_{Al}1$ at an acidity that varies with the stability of the carbocation being generated. For example, the mechanism of benzyl acetate hydrolysis changes from $A_{Ac}2$ to $A_{Al}1$ at a much lower concentration of H_2SO_4 than does the mechanism for hydrolysis of *p*-nitrobenzyl acetate. Table 7.9 (page 437) shows a compilation of the mechanistic assignments for the acid-catalyzed hydrolysis of a series of alkyl and aryl acetates in H_2SO_4 solutions.[182]

Alkaline Hydrolysis of Esters

Now let us consider the hydrolysis of esters under alkaline conditions (equation 7.67).[183,184]

$$\mathrm{RC(=O)OR'} + \mathrm{HO^-} \xrightarrow{\mathrm{H_2O}} \mathrm{RC(=O)O^-} + \mathrm{R'OH} \qquad \textbf{(7.67)}$$

As with acid-catalyzed reactions, the reaction mechanisms can be distinguished by two factors: (1) whether ester cleavage occurs between the oxygen atom and the acyl carbon atom or between the oxygen atom and the alkyl carbon atom, and (2) whether the reaction involves only a single elementary step or involves an intermediate.[185]

---

[182]To facilitate the interpretation of acid-catalyzed hydrolysis reactions, Yates has defined an *r* value:

$$\log k_\psi + H_X = r \log a_{H_2O} + \text{constant}$$

where k_ψ is the pseudo–first order rate constant and H_X is an acidity function appropriate for the type of compound being studied. Here the value of *r* is "the number of water molecules required to convert a protonated substrate molecule to the transition state, or the approximate 'order' of the reaction in water." Values of *r* around 2 are associated with the $A_{Ac}2$ mechanism, since two water molecules are drawn in the step that converts a protonated ester to the transition state. Negative values of *r* result from transition states that are less highly solvated than are the protonated esters, suggesting that acylium ions or carbocations are involved in the reaction. Yates, K.; McClelland, R. A. *J. Am. Chem. Soc.* **1967**, *89*, 2686.

[183]Johnson, S. L. *Adv. Phys. Org. Chem.* **1967**, *5*, 237.

[184]For an overview of reactions of carboxylic acid derivatives, see Bender, M. L. *Chem. Rev.* **1960**, *60*, 53 and references therein.

[185]The terminology used for mechanisms in this section is based on the Ingold system (reference 122). The IUPAC nomenclature (reference 175) for alkaline hydroylses of esters is as follows: The $B_{Ac}1$ mechanism is $D_N + A_N + A_{xh}D_h$. The $B_{Ac}2$ reaction is termed $A_N + D_N + A_{xh}D_h$. The $B_{Al}1$ mechanism is $D_N + A_N + A_{xh}D_h$. The $B_{Al}2$ mechanism is A_ND_N. For further discussion, see Guthrie, J. P. *J. Am. Chem. Soc.* **1991**, *113*, 3941.

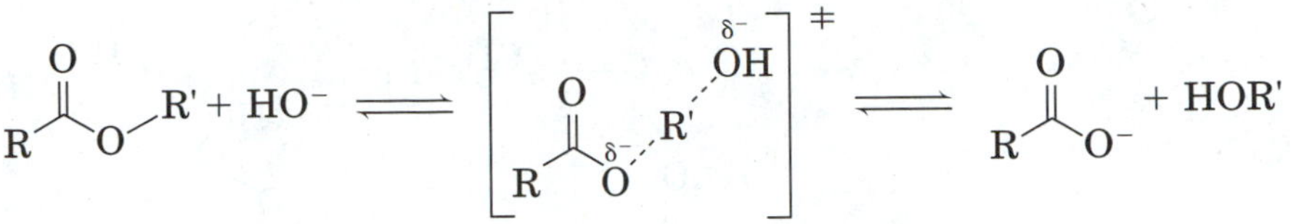

Figure 7.17
$B_{Al}2$ mechanism for ester hydrolysis.

$$RC(=O)OR' + HO^- \rightleftharpoons [RC(=O)O^{\delta-}\cdots R'\cdots {}^{\delta-}OH]^{\ddagger} \rightleftharpoons RC(=O)O^- + HOR'$$

The **B**$_{Al}$**2** (**b**ase promoted, **bimolecular**, with attack on the **al**kyl group) mechanism is essentially an S_N2 reaction, with nucleophilic attack on the alkyl group by hydroxide. Since this mechanism predicts cleavage of the alkyl-carbon—oxygen bond, rather than the acyl-carbon—oxygen bond, isotopic labeling experiments can be used to determine whether this pathway operates. Hydrolysis of ethyl propionate labeled as shown in equation 7.68 produced product in which all of the label was found in the ethanol product, therefore ruling out a $B_{Al}2$ mechanism for this compound.[186] The $B_{Al}2$ mechanism also predicts inversion at a chiral carbon atom bonded to the oxygen, but esters made from optically active alcohols do not show inversion.[187] Therefore, it seems that most base-promoted hydrolysis reactions do not occur by the $B_{Al}2$ mechanism.[188]

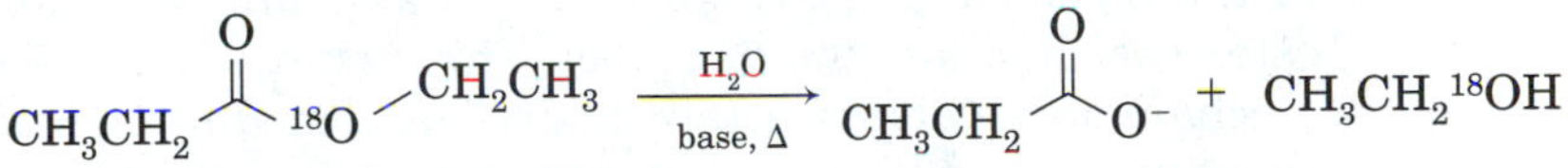

$$CH_3CH_2C(=O){}^{18}OCH_2CH_3 \xrightarrow[\text{base, } \Delta]{H_2O} CH_3CH_2C(=O)O^- + CH_3CH_2{}^{18}OH \quad (7.68)$$

There is evidence for the $B_{Al}2$ pathway in certain cases, however. Olson and Miller found that hydrolysis of β-butyrolactone (**8**) is catalyzed by acid in very acidic solution ($pH < 1$) and by base at pH 9 and above. Between these pH regions the rate of hydrolysis is essentially constant. The product formed upon hydrolysis at neutral pH of the lactone formed from (+)-β-bromobutyric acid was found to be optically active and to have the *opposite* rotation from that produced by the hydrolysis of the same lactone under conditions of acid or base catalysis. The results suggested a mechanism involving back-side attack of water on the alkyl C—O bond in the neutral pH reaction, analogous to the mechanism shown in Figure 7.17.[189] These results were supported by the results of hydrolysis of the lactone in $H_2{}^{18}O$ (equation 7.69), in which the labeled oxygen was found to be in the alcohol moiety of the product.[190]

---

[186] Kursanov, D. N.; Kudryavtsev, R. V. *Zhur. Obscheĭ. Khim.* **1956**, *26*, 1040. See also reference 184 and *Chem. Abstr.* **1956**, *50*, 16666i.

[187] Holmberg, B. *Chem. Ber.* **1912**, *45*, 2997.

[188] However, this pattern of reaction has been identified with methoxide ion as the nucleophile in the formation of dimethyl ether from the reaction of methoxide with methyl benzoate in methanol solution at 100°. Bunnett, J. F.; Robison, M. M.; Pennington, F. C. *J. Am. Chem. Soc.* **1950**, *72*, 2378.

[189] Olson, A. R.; Miller, R. J. *J. Am. Chem. Soc.* **1938**, *60*, 2687.

[190] Olson, A. R.; Hyde, J. L. *J. Am. Chem. Soc.* **1941**, *63*, 2459. See also Olson, A. R.; Youle, P. V. *J. Am. Chem. Soc.* **1951**, *73*, 2468.

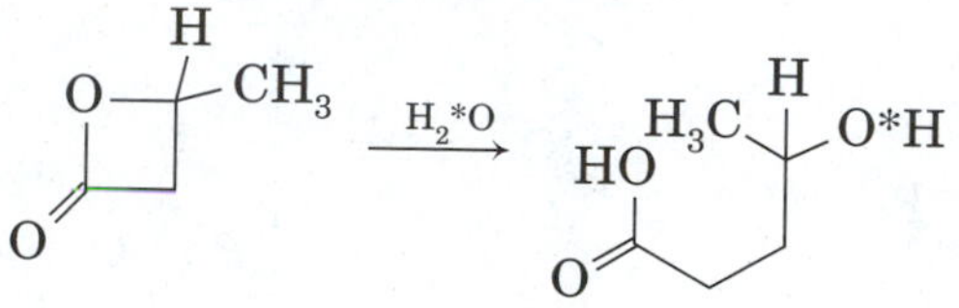

$$\mathbf{8} \xrightarrow{H_2{}^*O} \mathbf{9} \qquad (7.69)$$

The $\mathbf{B_{Ac}2}$ (**b**ase promoted, **bimolecular**, with attack on the **ac**yl group) mechanism is shown in Figure 7.18. In principle either addition of hydroxide to the ester (with rate constant k_1) or decomposition of the tetrahedral intermediate (with rate constant k_3) may be the rate-limiting step in the reaction, so the kinetics of the reaction depend upon the nucleophile and the leaving group.[191] If the leaving group is very poor (as in the case of hydrolysis of amides), then elimination of the leaving group can be rate limiting. For the hydrolysis of esters, however, the attack of hydroxide ion is rate-limiting.[192]

Evidence for the $B_{Ac}2$ mechanism has come from isotopic labeling studies.[193,194] Hydrolysis of ethyl, isopropyl, or *t*-butyl benzoate labeled with ^{18}O in the carbonyl group (Figure 7.19) was found to yield a small amount of ester that had lost the ^{18}O label. This result is most easily explained by a mechanism involving a tetrahedral intermediate, specifically the central structure in brackets as shown in Figure 7.19, that can undergo proton exchange and then loss of labeled water. It must be emphasized that failure to observe isotope exchange would not rule out a mechanism involving a tetra-

Figure 7.18 $B_{Ac}2$ mechanism for ester hydrolysis.

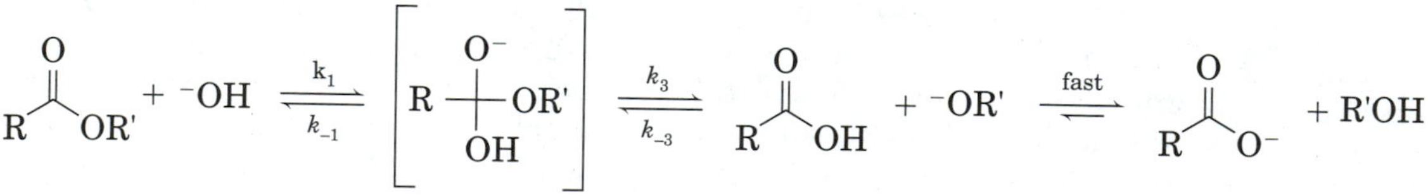

$$RC(=O)OR' + {}^-OH \underset{k_{-1}}{\overset{k_1}{\rightleftharpoons}} \left[R{-}C(O^-)(OH){-}OR' \right] \underset{k_{-3}}{\overset{k_3}{\rightleftharpoons}} RC(=O)OH + {}^-OR' \xrightarrow{\text{fast}} RC(=O)O^- + R'OH$$

[191]The proposed expulsion of an alkoxide ion from the tetrahedral intermediate is consistent with theoretical calculations. However, some studies have suggested that proton transfer to the oxygen may occur prior to breaking of the C-O bond. See Maraver, J. J.; Marcos, E. S.; Bertrán, J. *J. Chem. Soc. Perkin Trans. 2* **1986**, 1323.

[192]For a detailed discussion of the kinetics of the base-promoted hydrolysis of esters, see Johnson, S. L. *Adv. Phys. Org. Chem.* **1967**, *5*, 237.

[193](a) Bender, M. L. *J. Am. Chem. Soc.* **1951**, *73*, 1626; see also reference 184. (b) Bender, M. L.; Thomas, R. J. *J. Am. Chem. Soc.* **1961**, *83*, 4189.

[194]See also Kellogg, B. A.; Tse, J. E.; Brown, R. S. *J. Am. Chem. Soc.* **1995**, *117*, 1731.

[195]Bunton, C. A.; Spatcher, D. N. *J. Chem. Soc.* **1956**, 1079.

[196]Bender, M. L.; Matsui, H.; Thomas, R. J.; Tobey, S. W. *J. Am. Chem. Soc.* **1961**, *83*, 4193.

[197]Bruice, T. C.; Benkovic, S. J. *Bioorganic Mechanisms*, Vol. I; W. A. Benjamin, Inc.: New York, 1966; p. 24.

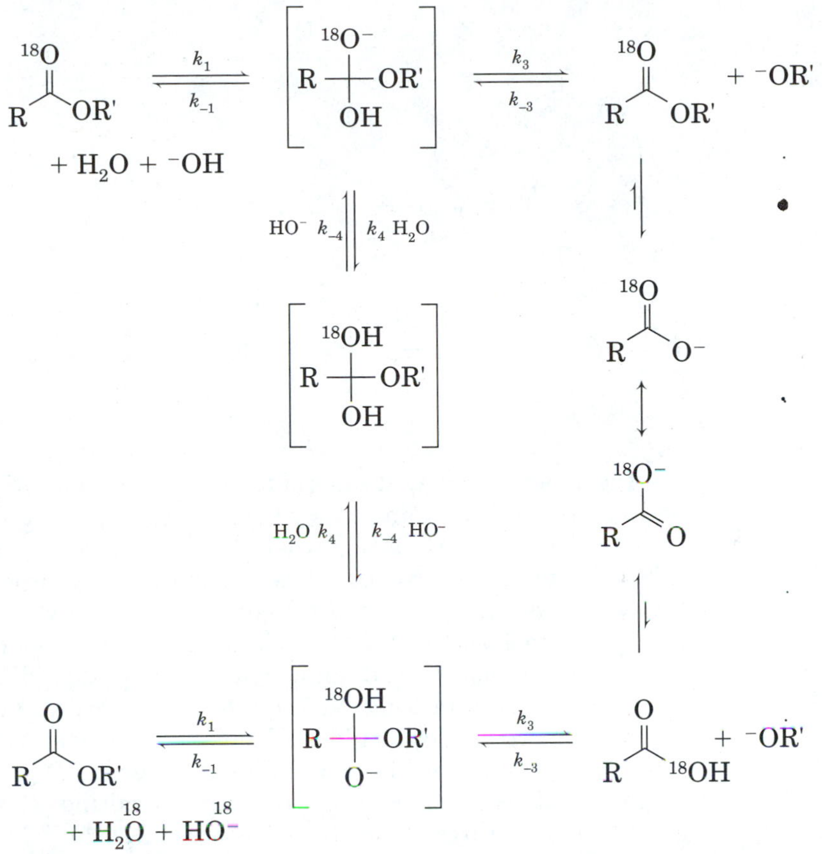

Figure 7.19 Isotopic exchange evidence for the tetrahedral intermediate in ester hydrolysis. (Adapted from reference 193(b).)

hedral intermediate because that situation could result if the proton exchange required to produce the bracketed intermediate is too slow to compete with collapse of the anions with which it is in equilibrium. That is, isotopic exchange would not be observed unless $k_4 \gg k_3, k_{-1}$. Failure to observe isotopic exchange for reaction of phenylbenzoate and some other benzoates has been attributed to larger values of k_3 and k_{-1} than k_4.[195,196,197]

The $\mathbf{B_{Al}1}$ mechanism is illustrated in Figure 7.20. The rate-limiting step in the reaction is a unimolecular dissociation of the ester to a carboxylate ion and a carbocation. The $B_{Al}1$ mechanism is more likely if the ester is derived from a strong acid (making the carboxylate ion more stable), if R′ is a tertiary or benzylic alkyl group (making the carbocation more stable), and if the concentration of base is low enough that the $B_{Ac}2$ mechanism does not compete. For molecules in which the $B_{Al}1$ reaction is rapid, it can be observed (through formation of racemized alcohol product) even in the presence of hydroxide ions. For molecules in which the $B_{Al}1$ reaction is not fast, however, reaction in the presence of aqueous base leads primarily to $B_{Ac}2$ reaction, and the hydrolysis of an ester formed from an optically active alcohol leads to product in which there is retention of configuration in the alcohol. For exam-

Figure 7.20 $B_{Al}1$ mechanism for ester hydrolysis.

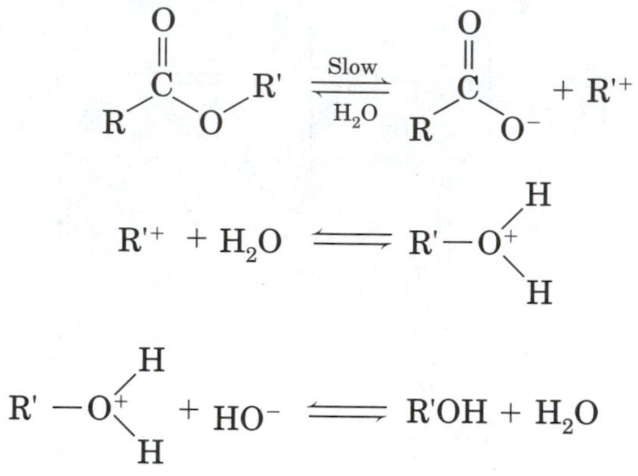

$$RC(=O)OR' \underset{H_2O}{\overset{\text{Slow}}{\rightleftharpoons}} RC(=O)O^- + R'^+$$

$$R'^+ + H_2O \rightleftharpoons R'{-}O^+H_2$$

$$R'{-}O^+H_2 + HO^- \rightleftharpoons R'OH + H_2O$$

ple, Ingold summarized studies of the hydrolysis of a series of alkyl hydrogen phthalates by Kenyon and co-workers indicating that the $B_{Al}1$ process dominates in 10 N NaOH solution when the alkyl group is *p*-methoxybenzhydryl and in dilute aqueous NaOH solution when the alkyl group is *p*-phenoxybenzhydryl. However, only in neutral aqueous solution was racemization of the product alcohol observed when the alkyl group was 1-phenylethyl.[198]

The mechanisms for alkaline hydrolysis of esters discussed to this point have shown the attacking nucleophile as a hydroxide ion. However, there are two processes that could lead to general base-catalyzed reactions: (1) general base catalysis of the addition of water or (2) proton donation to the departing alkoxide ion in the second (rate-limiting) step of a specific base, general acid-catalyzed reaction (Figure 7.21).[199,200,201] Marlier concluded

Figure 7.21 Possible transition states for general base-catalyzed hydrolysis of an ester: deprotonation of water (left) or protonation of the departing alkoxide ion (right).

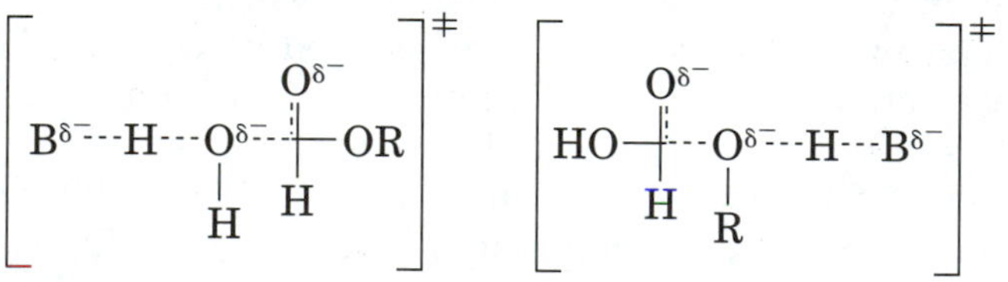

$$\left[B^{\delta-}\cdots H\cdots O^{\delta-}(H)\cdots C(H)(O^{\delta-})-OR\right]^{\ddagger} \qquad \left[HO-C(H)(O^{\delta-})\cdots O^{\delta-}(R)\cdots H\cdots B^{\delta-}\right]^{\ddagger}$$

[198] Reference 122, pp. 1139–1140 and references therein.

[199] Jencks, W. P.; Carriuolo, J. *J. Am. Chem. Soc.* **1961**, *83*, 1743.

[200] See also Dawson, H. M.; Lowson, W. *J. Chem. Soc.* **1929**, 393; Gold, V.; Oakenfull, D. G.; Riley, T. *J. Chem. Soc. B* **1968**, 515.

[201] Sawyer, C. B.; Kirsch, J. F. *J. Am. Chem. Soc.* **1973**, *95*, 7375. Other reaction pathways can be kinetically equivalent to general base catalysis; for a discussion, see reference 203.

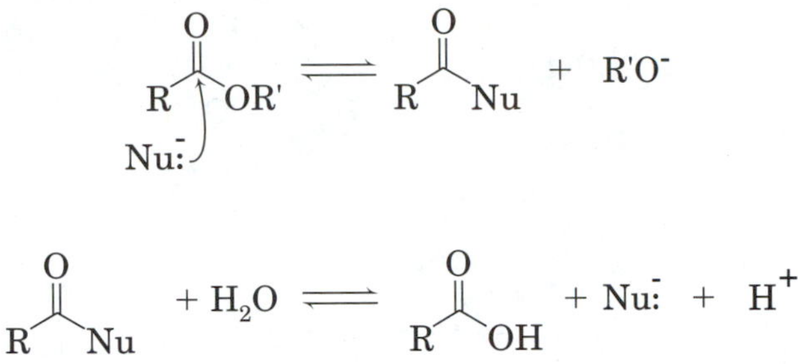

Figure 7.22 Nucleophilic catalysis in ester hydrolysis.

from a study of carbonyl oxygen, carbonyl carbon, and nucleophile oxygen kinetic isotope effects that the hydrolysis of methyl formate in water occurs through a stepwise mechanism in which the rate-limiting step is the formation of the tetrahedral intermediate by addition of a *water molecule*, not hydroxide, with assistance by a general base.[202] Stefanidis and Jencks came to the same conclusion in studies of the hydrolysis of a series of alkyl formates on the basis of the increase in the Brønsted β value for base catalysis with the decrease in the pK_a of the leaving group and on the basis of solvent isotope effects.[203]

An alternative mechanism for base catalysis of ester hydrolysis is nucleophilic catalysis, as illustrated in Figure 7.22. In the first step of the reaction, a nucleophile replaces the alkoxy group. In the second step, that nucleophile is replaced by an OH group.[204] The catalytic effect arises if (at the pH of the experiment) Nu is both a more effective nucleophile than hydroxide ion and is also a better leaving group than the alkoxide ion.

Experimental evidence for nucleophilic catalysis in the hydrolysis of a benzoic acid ester is summarized in Figure 7.23. Hydrolysis of 2,4-dinitrophenyl benzoate catalyzed by acetate ion labeled with ^{18}O on both oxygen atoms led to product in which both benzoic acid and acetic acid were labeled with ^{18}O, but about 75% of the label derived from one ^{18}O in the reactant was found in the benzoate ion.[205] The data were interpreted as evidence for the intermediacy of the mixed anhydride $CH_3COOCOC_6H_5$.

The reaction in Figure 7.23 involves intermolecular nucleophilic catalysis, but *intramolecular* nucleophilic catalysis has also been observed. Fersht and Kirby found incorporation of ^{18}O in the product when 3,5-dinitrosalicylate was hydrolyzed in a solvent containing isotopically enriched water.

[202] Marlier, J. F. *J. Am. Chem. Soc.* **1993**, *115*, 5953.

[203] Stefanidis, D.; Jencks, W. P. *J. Am. Chem. Soc.* **1993**, *115*, 6045.

[204] For examples, see Bender, M. L.; Turnquest, B. W. *J. Am. Chem. Soc.* **1957**, *79*, 1652; Bender, M. L.; Glasson, W. A. *J. Am. Chem. Soc.* **1959**, *81*, 1590.

[205] Bender, M. L.; Neveu, M. C. *J. Am. Chem. Soc.* **1958**, *80*, 5388.

Figure 7.23 Nucleophilic catalysis by acetate ion in hydrolysis of a benzoic acid ester.

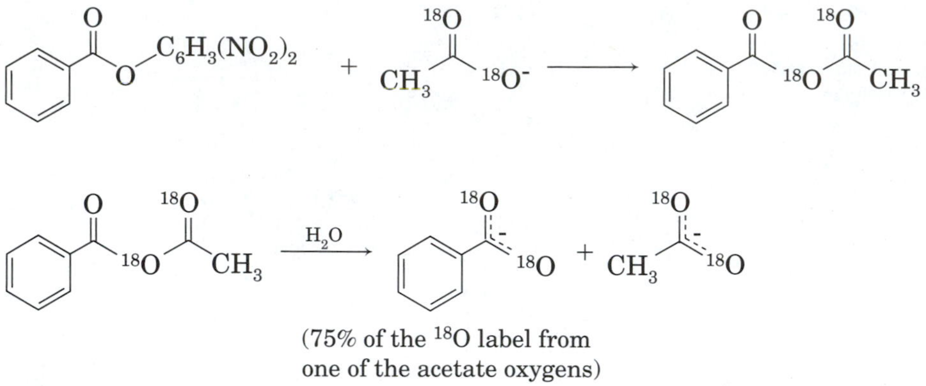

(75% of the ^{18}O label from one of the acetate oxygens)

This result suggested that the mechanism occurs by the process shown in equations 7.70 and 7.71. The reaction in equation 7.70 is the step involving intramolecular nucleophilic catalysis; it is followed by a reaction involving intramoleclar base catalysis (equation 7.71).[206] Intramolecular nucleophilic catalysis was also proposed for the hydrolysis of phthalamic acid,[207] methyl hydrogen phthalate,[208] and nonenolizable β-keto esters.[209]

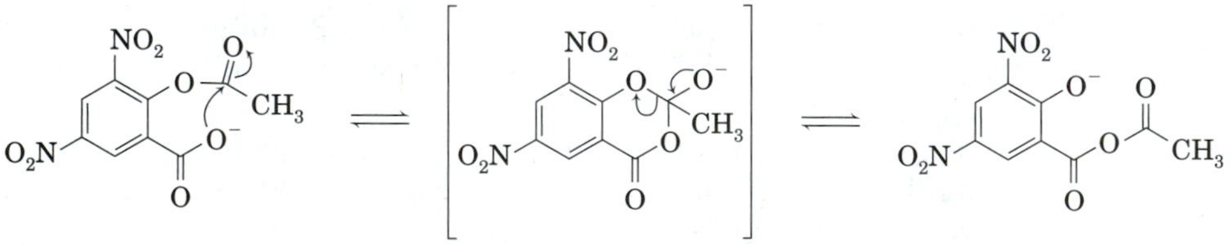

(7.70)

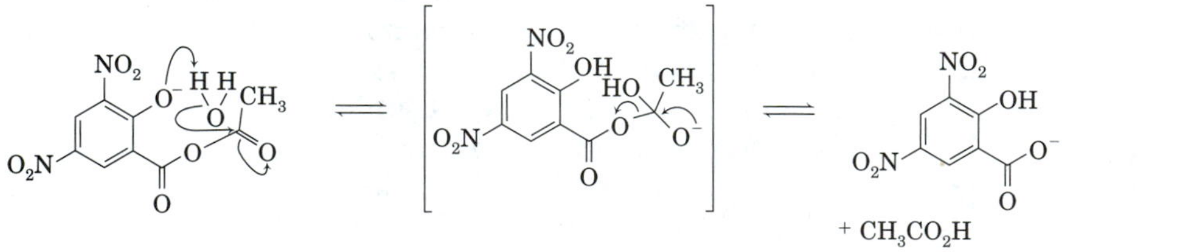

(7.71)

The discussion here has emphasized the role of hydrolysis of alkyl esters. The greater acidity of phenols than of alcohols makes the phenolate ions much better leaving groups in ester hydrolysis. Guthrie has concluded that aryl esters with leaving groups formed from strongly acidic phenols ($pK < 1$) react by way of acylium ions, while esters with leaving groups of

[206]Fersht, A. R.; Kirby, A. J. *J. Am. Chem. Soc.* **1968**, *90*, 5818.

[207]Bender, M. L.; Chow, Y.-L., Chloupek, F. *J. Am. Chem. Soc.* **1958**, *80*, 5380.

[208]Bender, M. L.; Chloupek, F.; Neveu, M. C. *J. Am. Chem. Soc.* **1958**, *80*, 5384.

[209]Washburn, W. N.; Cook, E. R. *J. Am. Chem. Soc.* **1986**, *108*, 5962.

$pK_a > 11$ react by an associative mechanism that may involve a tetrahedral addition intermediate. Esters with pK_a of the leaving group between 1 and 11, however, were determined to react by way of a concerted mechanism.[185,210]

Hydrolysis and other acyl substitution reactions can be catalyzed by molecules that are bifunctional, and these reactions serve as models for enzymatic processes. In addition, metal ions can catalyze some of these reactions.[184] For example, the hydroxide ion-catalyzed hydrolysis of esters of picolinic acid is dramatically enhanced by low concentrations of Ni^{+2} and Cu^{+2} ions. The catalysis appears to arise through formation of a metal chelate, such as **10**, that reduces the barrier for the rate-limiting addition of the nucleophile to the carbonyl group.[211,212] In addition, the hydrolysis of a 1° amide by a catalytic antibody has been reported.[213]

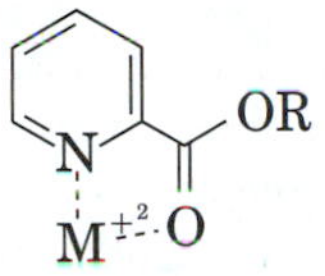

OR
N
M^{+2} O

10

Other Acyl Substitution Reactions

Hydrolysis is only one of many acyl substitution reactions of esters. If the attacking nucleophile is not a hydroxide ion but is an amine, alkoxide or phenoxide ion, halide ion, or other nucleophile, then the mechanism may differ. For example, Jencks and co-workers have made a detailed study of the reaction of substituted phenoxide ions as nucleophiles in substitutions with nitro-substituted phenyl formates and acetates.[214] In such reactions the fate of a tetrahedral intermediate, such as that shown in Figure 7.18, depends on the nucleophile and the leaving group. If the nucleophile is the weaker base, then departure of the leaving group is rate-limiting. If the nucleophile is the stronger base, then attack of the nucleophile is rate-limiting. However, Jencks and co-workers were unable to detect evidence for a change in the rate-limiting step in the reaction of substituted phenoxide nucleophiles with a series of formate and acetate esters. This led to the conclusion that the tetrahedral species is so unstable that it is a transition

[210]Similar conclusions had been reached by Ba-Saif, S.; Luthra, A. K.; Williams, A. *J. Am. Chem. Soc.* **1989**, *111*, 2647.

[211]Fife, T. H.; Przystas, T. J. *J. Am. Chem. Soc.* **1985**, *107*, 1041 and references therein.

[212]Pronounced catalytic activity has been observed with some binuclear metal complexes. See the discussion by Göbel, M. W. *Angew. Chem., Int. Ed. English* **1994**, *33*, 1141.

[213]Martin, M. T.; Angeles, T. S.; Sugasawara, R.; Aman, N. I.; Napper, A. D.; Darsley, M. J.; Sanchez, R. I.; Booth, P.; Titmas, R. C. *J. Am. Chem. Soc.* **1994**, *116*, 6508.

[214]Stefanidis, D.; Cho, S.; Dhe-Paganon, S.; Jencks, W. P. *J. Am. Chem. Soc.* **1993**, *115*, 1650.

structure rather than an intermediate. Similarly, DeTar concluded from a detailed analysis of the kinetics of acyl substitution reactions that aminolysis of aryl esters might not involve a "kinetically significant" tetrahedral intermediate and that a direct displacement should also be considered.[215] Synchronous displacement was also proposed for reactions of acyl halides with nucleophiles such as 2-naphthoxide,[216] the reaction of phenoxide ion with 2-aryloxazolin-5-one,[217] reactions of phenol with chloroacetyl chloride in acetonitrile,[218] alcoholysis of benzoyl chloride,[219] and solvolyses of benzoyl chloride in some water-organic solvent mixtures.[220]

There is also evidence for the role of acylium ions in some acyl substitutions, such as the reactions of acyl halides with nucleophilic reagents in acetonitrile, nitromethane,[221] and ethanol,[222] and the hydrolysis of some acyl fluorides.[223] Acylium ions are more likely candidates for reaction intermediates under acidic conditions. Bender reported that the hydrolysis of a series of substituted methyl 2,6-dimethylbenzoates in 9.7 M sulfuric acid proceeds through an acylium ion intermediate.[224] Hydrolysis of a series of substituted 2,6-dimethylbenzoyl chlorides in 99% CH_3CN - H_2O was determined to involve an acylium intermediate under acidic or neutral conditions, and ρ values (measured by σ^+) were found to be -3.85 and -3.73, respectively. However, the value of ρ under basic conditions was $+1.20$, suggesting that the base-promoted reaction involves a tetrahedral addition intermediate.[225]

The hydrolysis of amides has been the subject of extensive experimental and theoretical investigation. Observations of ^{18}O exchange accompanying the basic hydrolysis of amides such as primary, secondary, or tertiary toluamides suggested that a tetrahedral intermediate is formed along the reaction path.[226] Researchers have concluded from solvent deuterium kinetic isotope effects and other results that the reaction proceeds as shown in Figure 7.24. A significant feature of this proposed mechanism is that the nitrogen departs as a neutral amine and not as an amide ion. Experimental studies led to the conclusion that the rate-limiting step in the hydrolysis is departure of the amine from the tetrahedral intermediate, because greater oxygen exchange was observed with poorer leaving groups.[227,228,229] Oxygen

---

[215]DeTar, D. F. *J. Am. Chem. Soc.* **1982**, *104*, 7205.

[216]Haberfield, P.; Trattner, R. B. *Chem. Commun.* **1971**, 1481.

[217]Curran, T. C.; Farrar, C. R.; Niazy, O.; Williams, A. *J. Am. Chem. Soc.* **1980**, *102*, 6828.

[218]Briody, J. M.; Satchell, D. P. N. *J. Chem. Soc.* **1965**, 168.

[219]Peterson, P. E.; Vidrine, D. W.; Waller, F. J.; Henrichs, P. M.; Magaha, S.; Stevens, B. *J. Am. Chem. Soc.* **1977**, *99*, 7968.

[220]Bentley, T. W.; Carter, G. E.; Harris, H. C. *J. Chem. Soc. Perkin Trans. 2* **1985**, 983.

[221]Reference 218 and references therein.

[222]Kevill, D. N.; Daum, P. H.; Sapre, R. *J. Chem. Soc. Perkin Trans. 2* **1975**, 963.

[223]Song, B. D.; Jencks, W. P. *J. Am. Chem. Soc.* **1987**, *109*, 3160; **1989**, *111*, 8470.

[224]Bender, M. L.; Chen, M. C. *J. Am. Chem. Soc.* **1963**, *85*, 37.

[225]Bender, M. L.; Chen, M. C. *J. Am. Chem. Soc.* **1963**, *85*, 30.

[226]Ślebocka-Tilk, H.; Brown, R. S. *J. Org. Chem.* **1988**, *53*, 1153.

exchange was also observed in the acid-catalyzed hydrolysis of amides.[230] The exchange data, as well as solvent kinetic isotope effect data, are consistent with the mechanism shown in Figure 7.25.

Kollman and co-workers have carried out theoretical calculations for the hydrolysis of formamide in aqueous solution and in the gas phase.[231] Figure 7.26 shows the energies calculated for the gas phase reaction (filled circles) and for the solution reaction (open circles) for several points between reactants and products. The gas phase calculations do include one molecule of water to act as a catalyst by lowering the barrier for transfer of a proton from oxygen to nitrogen during the expulsion of ammonia. Although there is no activation barrier for formation of the tetrahedral addition intermediate in the gas phase, there is a 12 kcal/mol barrier for the expulsion of ammonia. In aqueous solution there are barriers for both the hydroxide addition to the C-O double bond and for the departure of ammonia, with the former (22 kcal/mol) being slightly higher than the latter.[232]

Figure 7.24 Proposed mechanism for the basic hydrolysis of an amide.

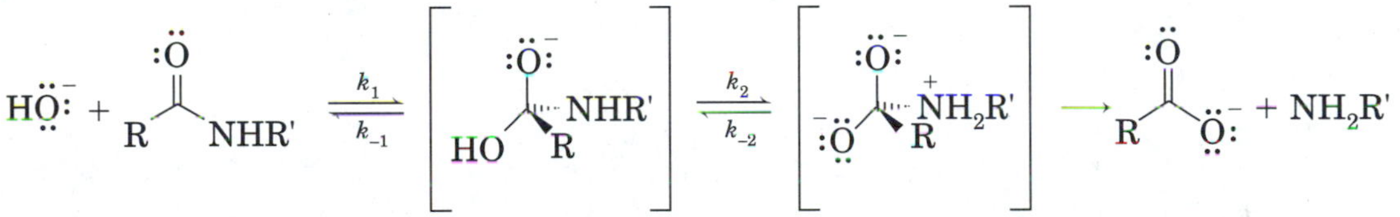

$$HO^- + RC(=O)NHR' \underset{k_{-1}}{\overset{k_1}{\rightleftharpoons}} [RC(O^-)(OH)NHR'] \underset{k_{-2}}{\overset{k_2}{\rightleftharpoons}} [RC(O^-)(O^-)NH_2R'^+] \longrightarrow RCOO^- + NH_2R'$$

[227]Ślebocka-Tilk, H.; Bennet, A. J.; Keillor, J. W.; Brown, R. S.; Guthrie, J. P.; Jodhan, A. *J. Am. Chem. Soc.* **1990**, *112*, 8507.

[228]Ślebocka-Tilk, H.; Bennet, A. J.; Hogg, H. J.; Brown, R. S. *J. Am. Chem. Soc.* **1991**, *113*, 1288.

[229]As the basicity of the leaving amine decreases so that the pK_a of the ammonium ion is less than 6, the nitrogen is less likely to be protonated in the tetrahedral intermediate. Therefore the nitrogen is proportionally more likely to depart as an amide ion, and ^{18}O exchange becomes proportionally less. Brown, R. S.; Bennet, A. J.; Ślebocka-Tilk, H.; Jodhan, A. *J. Am. Chem. Soc.* **1992**, *114*, 3092; Brown, R. S.; Bennet, A. J.; Ślebocka-Tilk, H. *Acc. Chem. Res.* **1992**, *25*, 481.

[230]Ślebocka-Tilk, H.; Brown, R. S.; Olekszyk, J. *J. Am. Chem. Soc.* **1987**, *109*, 4620. The compounds studied in this investigation were acetanilide and *N*-cyclohexylacetamide.

[231]Weiner, S. J.; Singh, U. C.; Kollman, P. A. *J. Am. Chem. Soc.* **1985**, *107*, 2219. The computation involved a combination of *ab initio* quantum mechanics and molecular mechanics calculations.

[232]Krug, J. P.; Popelier, P. L. A.; Bader, R. F. W. *J. Phys. Chem.* **1992**, *96*, 7604 subsequently carried out *ab initio* calculations for the hydrolysis of formamide in the gas phase under four sets of conditions: (i) no acid or base catalysis, but with one molecule of water, (ii) reaction with hydroxide, (iii) reaction with H_3O^+ involving protonation of the carbonyl oxygen, and (iv) similar acid catalysis involving protonation of the amide nitrogen. The detailed model of the structure, charge and energy changes that accompany the hydrolysis reaction provided by these calculations may be relevant to the mechanism of amide hydrolysis in the hydrophobic region of an enzyme.

Figure 7.25 Mechanism proposed for the acid-catalyzed hydrolysis of an amide.

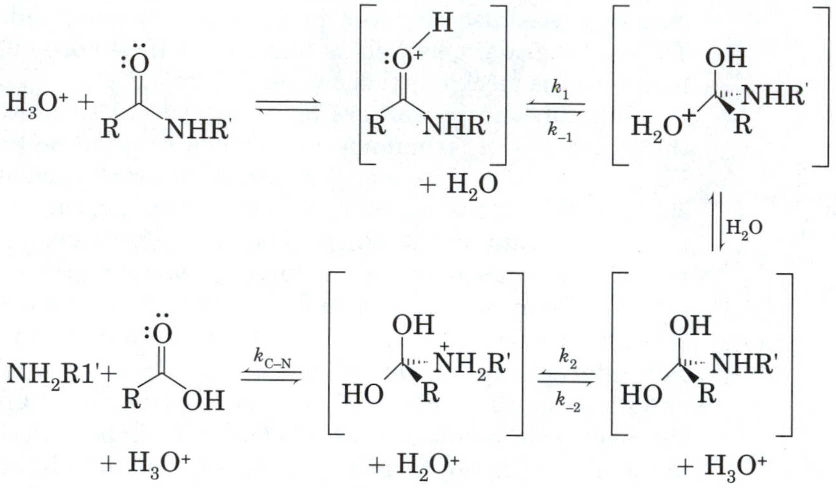

Figure 7.26 Calculated energy profiles for the hydrolysis of formamide in the gas phase (filled circles) and in aqueous solution (open circles). (Reproduced from reference 231.)

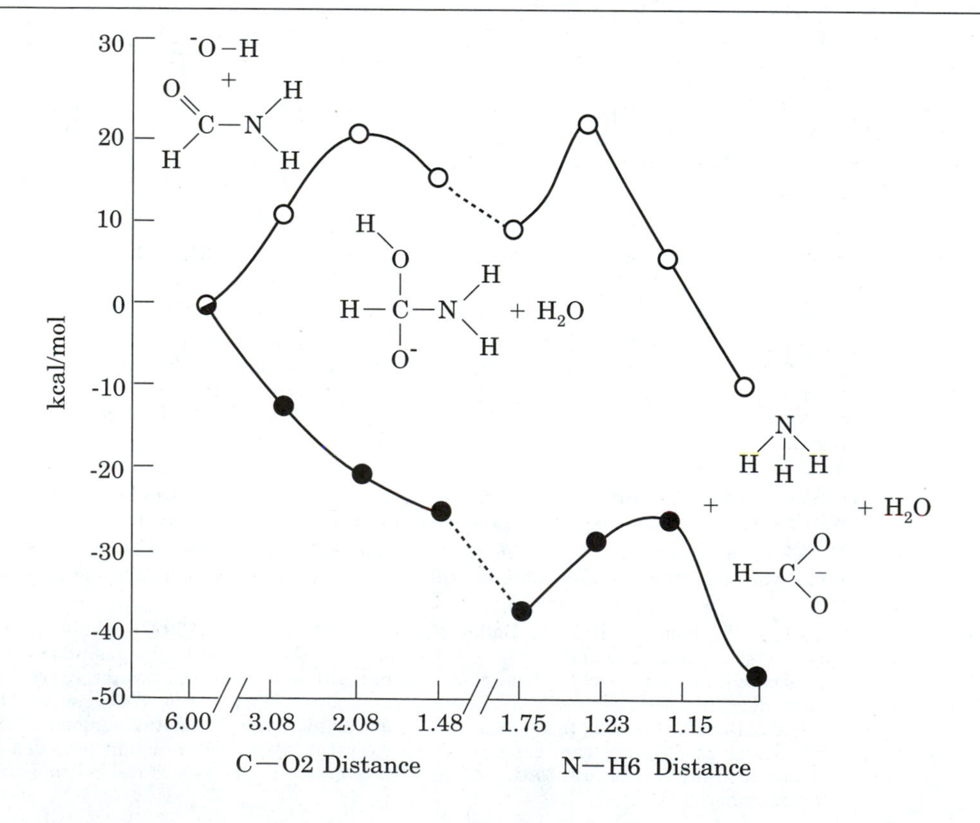

Problems

1. Use equation 7.9 to predict the pK_a of vanillic acid (4-hydroxy-3-methoxybenzoic acid).
2. Convert the ΔG_{acid} values in kcal/mol for acetic and propionic acids in Table 7.4 to pK_a values at 25° by using the relationship $\Delta G = -RT \ln K$. Explain the difference in magnitude of the pK_a values in the gas phase and in solution, and rationalize the different order of acidities of the compounds in the two media.
3. Use the data for *ab initio* calculations of the stabilities of alkoxide ions on page 408 to compare the effect, in kcal/mol, of replacing first one, then two, then three hydrogen atoms on methoxide with methyl groups. Rationalize the trend you observe.
4. Explain the order of the K_d values for hemiketals of cyclobutanone, cyclopentanone, and cyclohexanone listed in Table 7.7.
5. Demonstrate that equation 7.57 can be written in terms of the pK_a value of the protonated base, BH^+, as follows:

$$\log k_B = \beta \, pK_{BH^+} + \text{constant} \qquad \textbf{(7.72)}$$

6. Toxicity has been associated with the coincident ingestion of both the mushroom *Coprinius atramentarius* and ethanol. A compound found to the extent of 0.1% by weight of the dried mushroom has been identifed as coprine (**11**). Although coprine itself is inactive *in vitro*, a derivative of coprine identified as cyclopropanone hydrate was found to be an inhibitor of acetaldehyde dehydrogenase, an enzyme that is involved in the physiological oxidation of ethanol.
   a. Propose a mechanism for the nonenzymatic formation of cyclopropanone hydrate from coprine.
   b. Explain why cyclopropanone hydrate would exist in physiological solution almost entirely as the hydrate and not as the ketone.

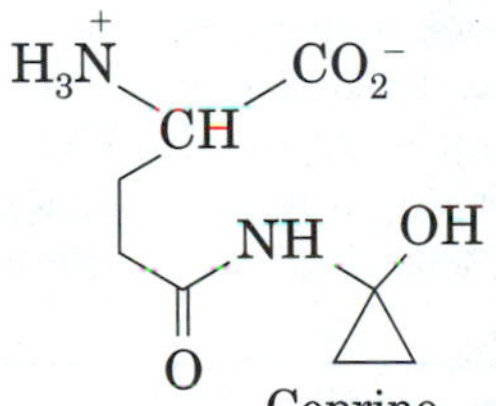

$H_3\overset{+}{N}$ CO_2^- CH NH OH O

Coprine

11

7. Predict the position of acid-catalyzed bromination of each of the following ketones:
   a. 2-butanone
   b. 2-pentanone
   c. 3-methyl-2-butanone
   d. 2-methylcyclohexanone
   e. methyl cyclohexyl ketone
8. Treatment of methyl isobutyl ketone (**12**) with sodium amide and then CO_2 produces, after workup, a carboxylic acid that is converted by diazomethane to methyl isovalerylacetate (**13**). Propose a mechanism for the reaction, and explain the regiochemical preference that leads to the observed product.

$$\underset{\mathbf{12}}{H_3C{-}CH(=CH_3){-}CH_2{-}C(=O){-}CH_3} \xrightarrow[NH_3(\text{liq.})]{NaNH_2} \xrightarrow[EtOEt]{CO_2} \xrightarrow{H_3O} \tag{7.73}$$

$$\xrightarrow{CH_2N_2} \underset{\mathbf{13}}{H_3C{-}CH(CH_3){-}CH_2{-}C(=O){-}CH_2{-}C(=O){-}O{-}CH_3}$$

9. Pyridine bases have been found to catalyze the aldol reaction of D-bglyceraldehyde. Rates for reactions carried out in water at pH 7.0 and 30° with a series of catalysts are shown in Table 7.10. (The observed reaction rates were divided by the molar concentration of unprotonated base at the reaction pH to obtain the specific reaction rates shown.) Determine whether the reaction is subject to general or specific base catalysis. Do the data suggest any role of steric hindrance in the catalysis of the reaction by pyridine bases?
10. Treatment of 5,5-dimethyl-1,3-cyclohexanedione with NaOCl and KOH in aqueous solution at 35°, followed by acidification of the reaction mixture, leads to the formation of 3,3-dimethylpentanedioic acid. Propoes a detailed mechanism to account for the formation of this product.
11. Should a tetrahedral intermediate formed by the addition of a nucleophile to the C=O group of an ester be more stable for phenyl acetate or phenyl formate? Explain your answer.
12. Construct a More O'Ferrall-Jencks diagram for the general base-catalyzed hydration of acetaldehyde. Let the horizontal scale represent proton transfer from water to the general base (to form hydroxide ion), and let the vertical scale represent addition of the oxygen atom of water to the carbonyl carbon atom. The upper right corner of the diagram will represent the products, protonated base, and the oxyanion of the hydrate. What structural features of the base and the

Table 7.10 Rates of aldolization of D-glyceraldehyde catalyzed by pyridine bases.

Catalyst	pK_{BH^+}	Specific Reaction Rate, $M^{-1}\,sec^{-1}$
Pyridine	5.17	1.74×10^{-5}
2-Methylpyridine	5.97	1.90×10^{-5}
3-Methylpyridine	5.68	2.91×10^{-5}
4-Methylpyridine	6.02	3.65×10^{-5}
2,4-Dimethylpyridine	6.63	3.24×10^{-5}
2,5-Dimethylpyridine	6.40	2.74×10^{-5}
3,4-Dimethylpyridine	6.46	7.32×10^{-5}
2,6-Dimethylpyridine	6.72	0.60×10^{-5}

carbonyl compound would affect the location of the transition structure and the reaction coordinate on this projection of the potential energy surface?

13. Propose an explanation for the fact that the apparent molecular weight of freshly prepared 1,3-dihydroxyacetone is 180 g/mol.
14. The extinction coefficient (ϵ) values determined from the UV-vis spectra of 4-methoxybutanal and 4-hydroxybutanal dissolved in 75 : 25 (v : v) dioxane-water were found to be 17.4 and 1.99, respectively. For 5-methoxy- and 5-hydroxypentanal, the corresponding values were 19.2 and 1.17. However, for 6-methoxy- and 6-hydroxyhexanal, the ϵ values were much closer: 19.4 and 16.6. Explain the origin of the difference in the ϵ values for each compound, and explain why the difference varies among members of the series.
15. Hydrolysis of acetals in $H_2{}^{18}O$ produces alcohols with virtually no ^{18}O label. Predict the stereochemistry of the product from the hydrolysis of the acetal formed from acetaldehyde and (−)-α-phenylethyl alcohol.
16. Rationalize the observation that sterically hindered esters are more likely to react via acylium ions than are unhindered esters.
17. Reaction of an aldehyde with *N*-methylhydroxylamine in the presence of an acid catalyst in aqueous solution leads to the formation of a structure known as a nitrone. Experimental evidence indicates that the reaction occurs in two steps, with the second step showing general acid catalysis. Propose a mechanism for the reaction.

$$RCHO + H_3CNHOH \xrightarrow[H_2O]{acid} (R)(H)C{=}N^+(O^-)(CH_3) \qquad (7.74)$$

18. When the hydrolysis of optically active β-butyrolactone is carried out in neutral solution, the product is optically active and has the opposite rotation from the product obtained by hydrolysis of the same lactone under conditions of base catalysis by carbonate ion. However, when acetate ion is used as a base catalyst, the optical activity of the product is identical to that observed at neutral pH. Explain these results.
19. In methanol solution 3-methylbicyclobutanecarbonitrile (**14**) undergoes acid-catalyzed addition of solvent to form as major products the cyclobutanes **15** and **16**. The reaction was investigated by carrying out the addition in a series of buffered methanol solutions at 50° and constant ionic strength. The second order rate constants observed for the acid-catalyzed addition of methanol (and the pK_a values of the buffers under the experimental conditions) were found to be the following: 2.24×10^{-1} (2.75), 1.06×10^{-3} (4.98), 3.52×10^{-5} (6.41), 8.13×10^{-7} (8.35), and 7.8×10^{-8} M^{-1} sec^{-1} (9.42). Demonstrate that the reaction is subject to general acid catalysis, and determine the value of α for the reaction.

$$\mathbf{14} \xrightarrow[H^+]{CH_3OH} \mathbf{15} + \mathbf{16} \qquad (7.75)$$

20. Treatment of 3-(2-bromoethyl)-3-phenyl-2-benzofuranone (**17**) with ammonia in acetonitrile leads to the isolation of a product mixture consisting primarily of **18**, but about 10% of **19** is also formed. Propose a mechanism to account for the formation of the products.

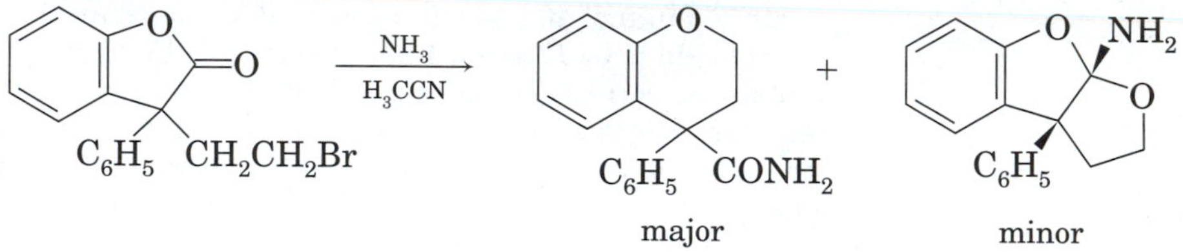

C_6H_5 CH_2CH_2Br ; NH_3 / H_3CCN ; C_6H_5 $CONH_2$ major ; + ; NH_2 C_6H_5 minor

17 **18** **19**

21. The reaction of ketones with acetic anhydride in the presence of BF_3 provides a convenient synthesis of β-diketones. For example, reaction of acetone with AcOAc produces acetylacetone. Two different β-diketones may be formed from unsymmetrical ketones. With a series of methyl ketones having the general structure CH_3COCH_2R, the yield of the two possible products $CH_3COC(COCH_3)HR/CH_3COCH_2COCH_2R$ varied with R as follows: R = methyl (100% / 0%); R = ethyl (90% / 10%); R = isopropyl (45% / 55%). Propose a mechanism for the general reaction, and account for the distribution of products in each of these cases.
22. In contrast to the case with base-promoted hydrolysis of epoxides in ^{18}O-labeled water (problem 2, page 387), the corresponding hydrolysis of unsymmetrical alkene oxides under acid-catalyzed conditions leads to product mixtures in which the majority of the labeled oxygen is on the carbon atom with the greater number of alkyl groups. For example, products from hydrolysis of propylene oxide and isobutylene oxide are shown in equations 7.76 and 7.77 (O* = ^{18}O).

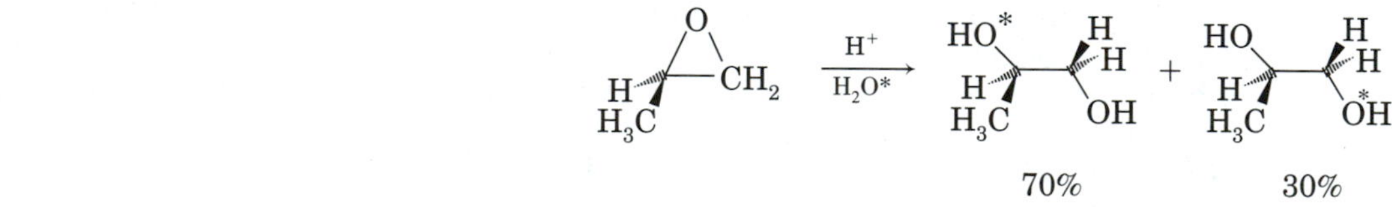

H^+ / H_2O^* ; 70% ; 30% **(7.76)**

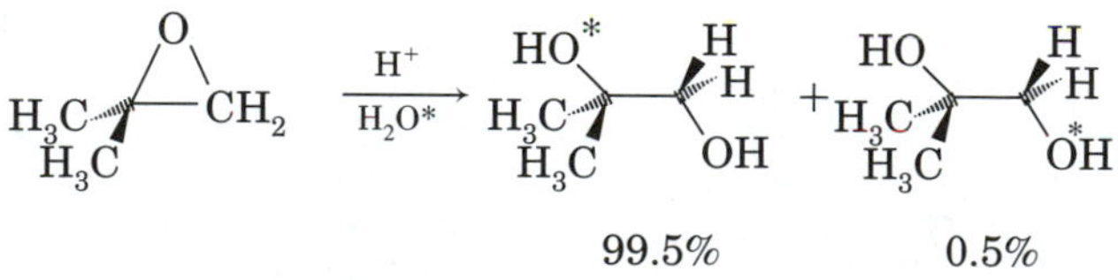

H^+ / H_2O^* ; 99.5% ; 0.5% **(7.77)**

Isobutylene oxide reacts 100 times faster than does propylene oxide under the same conditions. In both cases the log of the rate constant is linear with $-H_0$. Propose an explanation for these results.

CHAPTER 8

Substitution Reactions

8.1 Introduction

In a substitution reaction, an atom or group of atoms (Y) replaces another atom or group of atoms (X) in some molecular entity (RX). In shorthand notation,

$$Y + R{-}X \longrightarrow R'{-}Y + X \tag{8.1}$$

Replacement of one functional group by another may occur without accompanying change in the molecular framework. However, skeletal change may accompany substitution, as in inversion of a chiral center in the S_N2 reaction or allylic rearrangement in the S_N1' and S_N2' reactions (see footnote 11), so R′ may not necessarily be the same as R.

The generalized substitution reaction shown in equation 8.1 does not indicate the source of the electrons that are used to make the R′—Y bond. In nucleophilic substitution, those electrons originate with Y:, which is a **nucleophile** (from the Greek, *philein*, "loving").[1] In electrophilic substitution, an **electrophile** forms a Y—R bond by using both of the electrons in the R—X bond. Leaving groups also are categorized according to the fate of the electrons that bond the substrate to the leaving group. If these electrons depart with the leaving group (typically the case in a nucleophilic substitution), then the leaving group is a **nucleofuge**, as in equation 8.2. If these electrons do *not* depart with the leaving group (typically the case in an electrophilic substitution), then that group is an **electrofuge**, as in equation 8.3.[2] In both cases, the pair of electrons that originally comprises the R—X bond is shown as a bold pair of dots.

---

[1]Commission on Physical Organic Chemistry, IUPAC, *Pure Appl. Chem.* **1994**, *66*, 1077.

[2]These terms are derived from the Latin *fugitivus* ("fleeing"). For a discussion and references to the original use of these terms, see Nickon, A.; Silversmith, E. F. *Organic Chemistry: The Name Game*; Pergamon Press: New York, 1987; pp. 258–259.

$$\underbrace{\mathrm{Y:^-}}_{\text{nucleophile}} + \mathrm{R:}\underbrace{\mathrm{X}}_{\text{nucleofuge}} \longrightarrow \mathrm{RY} + \mathrm{:X^-} \qquad (8.2)$$

$$\underbrace{\mathrm{E^+}}_{\text{electrophile}} + \mathrm{C_6H_5{:}H}\ (\text{electrofuge } \{\mathrm{H}) \longrightarrow \mathrm{C_6H_5{:}E} + \mathrm{H^+} \qquad (8.3)$$

It is convenient to categorize reactions with concise descriptive labels. For substitution reactions we use the notation S_xM, in which the letter S indicates a substitution reaction. The subscript x indicates something of the mechanism, such as N for nucleophilic or E for electrophilic. M usually indicates the molecularity of the reaction, the nature of the reacting species, or additional information.[3,4,5] The most familiar terms for substitution reactions are S_N1 (for Substitution Nucleophilic Unimolecular[6]) as shown in equation 8.4

$$\mathrm{(CH_3)_3CCl} \underset{\mathrm{CH_3CH_2OH}}{\overset{\text{slow}}{\rightleftharpoons}} \mathrm{Cl^-} + \mathrm{(CH_3)_3C^+} \underset{\mathrm{CH_3CH_2OH}}{\rightleftharpoons} \mathrm{(CH_3)_3C{-}\overset{+}{O}(H){-}CH_2CH_3} \overset{-\mathrm{H^+}}{\rightleftharpoons} \mathrm{(CH_3)_3C{-}OCH_2CH_3} \qquad (8.4)$$

and S_N2 (for Substitution Nucleophilic Bimolecular), as illustrated in equation 8.5

$$\mathrm{H_3C\ddot{O}:^-} \ \ \mathrm{H_3C{-}Br} \xrightarrow{\mathrm{H_3COH}} \mathrm{H_3CO{-}CH_3} + \mathrm{:\ddot{B}r:^-} \qquad (8.5)$$

These terms were suggested by Ingold and are familiar to all organic chemists.

Typically the kinetics of simple aliphatic substitutions are overall second order kinetics for S_N2 reactions and overall first order kinetics for S_N1

---

[3]Ingold, C. K.; Rothstein, E. *J. Chem. Soc.* **1928**, 1217.

[4]Ingold, C. K. *Structure and Mechanism in Organic Chemistry*, 2nd Ed.; Cornell University Press: Ithaca, New York, 1969; p. 427.

[5]For a listing of Ingold symbols for other kinds of substitution reactions not considered here, see Orchin, M.; Kaplan, F.; Macomber, R. S.; Wilson, R. M.; Zimmer, H. *The Vocabulary of Organic Chemistry*; Wiley-Interscience: New York, 1980.

[6]The term *unimolecular* does not mean that no other molecules are involved in the rate-limiting step, since the reaction is not observed in the absence of solvent. Therefore the designation *polymolecular* was used by Steigman, J.; Hammett, L. P. *J. Am. Chem. Soc.* **1937**, *59*, 2536, and the term *termolecular,* based on a specific model of solvent interaction, was suggested by Swain, C. G. *J. Am. Chem. Soc.* **1948**, *70*, 1119.

reactions. That is, an S_N2 reaction between Y^- and R—X leads to the rate equation

$$\text{Rate} = -d[\text{R—X}]/dt = k_2[\text{R—X}][\text{Y}^-] \qquad \textbf{(8.6)}$$

whereas for an S_N1 reaction we expect the kinetic expression to be

$$\text{Rate} = -d[\text{R—X}]/dt = k_1[\text{R—X}] \qquad \textbf{(8.7)}$$

It is important to note, however, that the terms S_N1 and S_N2 do not merely identify the kinetic results observed in studies of nucleophilic substitution. Rather, they are intended to designate the mechanisms of those reactions.[7] To put it another way, those terms designate *nonobservable* (mechanistic) and not *observable* (kinetic) properties.

Although the Ingold mechanistic labels for substitution reactions are very familiar, some chemists have recognized the need for new terminology that would designate the details of mechanisms more explicitly. An alternative formalism suggested by the IUPAC Commission on Physical Organic Chemistry[8] designates the steps of a substitution A (attachment) and D (detachment). A suffix N indicates a nucleophilic attachment (A_N) or a nucleofugic detachment (D_N).[9] The placement of the terms A_N and D_N with other symbols can be used to convey the mechanism of the reaction because the two symbols:

1. may be combined without an intervening symbol to indicate that they occur at the same time,
2. may be combined with + to indicate that they occur with enough time between them that any intermediates have time to equilibrate with solvent and other ions in solution, or
3. may be combined with * to indicate that the two steps occur so quickly that equilibration with outside species does not occur.

With this terminology we may not only describe a fully concerted, one step S_N2 reaction (A_ND_N), and a step-wise S_N1 reaction involving intermediate ions that diffuse apart ($D_N + A_N$), but we may also economically describe a step-wise reaction involving a transient ion pair ($D_N{*}A_N$).[10,11] This system

---

[7]Although it is sometimes thought that the 1 and 2 refer to kinetics, Ingold noted in reference 4, p. 427, that "...the numerical indication in the symbolic label, no less than in the verbal name, refers to the *molecularity* of the reaction, and not to its kinetic order."

[8]Commission on Physical Organic Chemistry, IUPAC, *Pure Appl. Chem.* **1989**, *61*, 23; see also Guthrie, R. D.; Jencks, W. P. *Acc. Chem. Res.* **1989**, *22*, 343.

[9]Similarly, a suffix E denotes an electrophilic or electrofugal process.

[10]See the discussion of the roles of ion pairs in nucleophilic substitution reactions (beginning on page 470).

[11]The IUPAC nomenclature system can also be used to describe other substitution reactions. Among those that will not be discussed extensively here are the S_N1' (Substitution Nucleophilic Unimolecular with rearrangement) reaction below (Catchpole, A. G.; Hughes, E. D. *J. Chem. Soc.* **1948**, 1), which is denoted by IUPAC as an ($1/D_N + 3/A_N$) reaction. The numbers before the slash symbols indicate atoms involved in the dissociation and association steps. Thus

for labeling reaction mechanisms has not met with universal acceptance,[12] however, and much of the chemical literature utilizes the well-known Ingold system. It remains to be seen whether the new symbols will supplant the well-entrenched Ingold terminology.

Substitution is such a broad topic in organic chemistry that we cannot begin to cover all of it here. Indeed, we will not be able to cover even one area of substitution in detail.[13] Rather, our primary goal will be to apply to substitution reactions the kinds of analysis we have developed in earlier chapters. In particular, we want to understand the explicit and implicit models we use to describe substitution, and we want to examine carefully the bases for these models.

8.2 Nucleophilic Aliphatic Substitution

Designation of Nucleophilic Aliphatic Substitution Reactions

Nucleophilic substitution is sometimes the first example of chemical reactivity in introductory organic chemistry because classification of substitution reactions either as S_N1 or as S_N2 serves to illustrate the application of chemical kinetics, stereochemical labeling, solvent effects, and structural

---

$1/D_N$ means that the nucleofuge dissociates from one atom (1), while the $3/A_N$ term means that the nucleophile associates at an allylic position (3).

$$\mathrm{Ph{-}CH(X){-}CH{=}CH_2 \rightleftharpoons Ph{-}\overset{+}{C}H{-}CH{=}CH_2\ {+}X^- \longleftrightarrow Ph{-}CH{=}CH{-}\overset{+}{C}H_2\ {+}X^- \rightleftharpoons Ph{-}CH{=}CH{-}CH_2(X)}$$

(X is *p*-nitrobenzoate.)

Similarly, the S_N2' (Substitution Nucleophilic Bimolecular with rearrangement) reaction (Kepner, R. E.; Winstein, S.; Young, W. G. *J. Am. Chem. Soc.* **1949**, *71*, 115. See also DeWolfe, R. H.; Young, W. G. *Chem. Rev.* **1956**, *56*, 753) is a $(3/1/A_ND_N)$ process. For a review of the intramolecular S_N' reaction, see Paquette, L. A.; Stirling, C. J. M. *Tetrahedron* **1992**, *48*, 7383; for a review of the intermolecular S_N2' reaction, see Magid, R. M. *Tetrahedron* **1980**, *36*, 1901.

$$\mathrm{CH_3CH_2{-}CH(Cl){-}CH{=}CH_2 + {:}\overset{-}{C}H(CO_2CH_2CH_3)_2 \longrightarrow CH_3CH_2{-}CH{=}CH{-}CH_2{-}CH(CO_2CH_2CH_3)_2}$$

The S_Ni (Substitution Nucleophilic Internal) reaction below, which was reported by McKenzie, A.; Clough, G. W. *J. Chem. Soc.* **1910**, *97*, 2564, is denoted a $(D_N + D + A_N)$ process.

$$\mathrm{C_6H_5(H_3C)C(CO_2H){-}OH \xrightarrow{SOCl_2} C_6H_5(H_3C)C(CO_2H){-}OSOCl + HCl \xrightarrow{\Delta} C_6H_5(H_3C)C(CO_2H){-}Cl + SO_2}$$

[12]Olah, G. A. *Acc. Chem. Res.* **1990**, *23*, 31. However, see also Jencks, W. P. *Acc. Chem. Res.* **1990**, *23*, 32; Guthrie, R. D. *Acc. Chem. Res.* **1990**, *23*, 33.

[13]In 1963–64 Olah and co-workers summarized the information then known about electrophilic aromatic substitution (S_EAr) in three volumes composed of over 3,900 pages: Olah, G.A., ed. *Friedel-Crafts and Related Reactions*; Wiley-Interscience: New York, 1963–1964.

effects in studying organic reactions. The S_N1 and S_N2 models are familiar to all organic chemists, but such models are two-edged swords. They enable us to assimilate a large volume of diverse material into one conceptual framework; at the same time, they can restrict our ability to envision new mechanistic possibilities. As we look more closely at nucleophilic aliphatic substitution in the sections that follow, we will see that[14]

1. the familiar S_N2 and S_N1 mechanisms describe only two of a possibly infinite number of pathways on a two-dimensional surface describing nucleophilic aliphatic substitution;
2. there is evidence that not all nucleophilic substitutions are heterolytic processes, meaning that both of the electrons from the nucleophile are donated to the substrate at the same time. Instead, some reactions have been found to occur by a process in which only one electron is transferred at a time.

A major mechanistic distinction between S_N2 and S_N1 reactions is the timing of the departure of the leaving group and the arrival of the nucleophile. One tool to help visualize the relationship of these processes is a two-dimensional reaction coordinate diagram, such as that shown in Figure 8.1.[15,16] The horizontal scale represents the extent of R—Y bond formation,

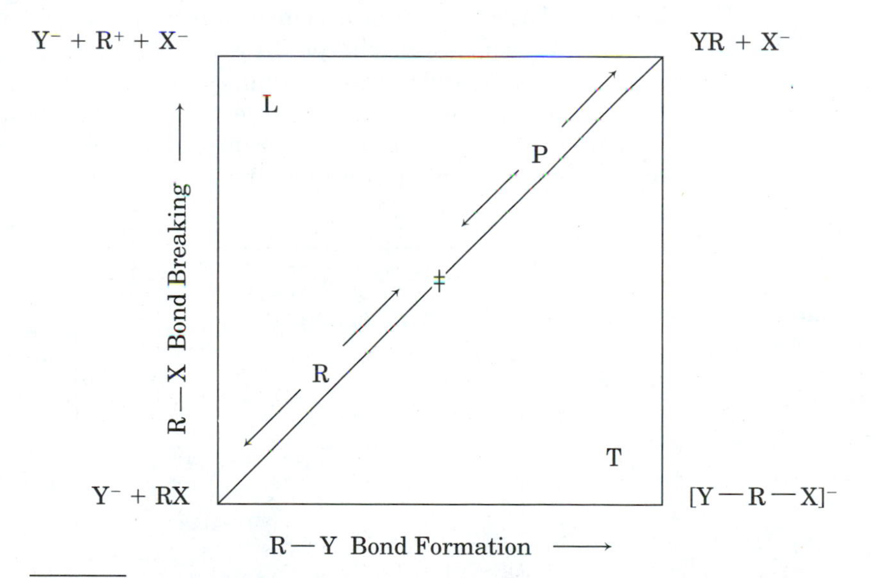

Figure 8.1 Two-dimensional reaction coordinate diagrams for an S_N2 reaction.

[14]These possible complexities of nucleophilic aliphatic substitution only hint at even greater complexity to be found with the entire range of substitution processes. For a review of the mechanisms of nucleophilic aliphatic substitution, see Katritzky, A. R.; Brycki, B. E. *Chem. Soc. Rev.* **1990**, *19*, 83.

[15]*Cf.* More O'Ferrall, R. A. *J. Chem. Soc. B* **1970**, 274; More O'Ferrall, R. A. in *The Chemistry of the Carbon-Halogen Bond*, Vol. 2; Patai, S., ed.; John Wiley & Sons: New York, 1973; Jencks, D. A.; Jencks, W. P. *J. Am. Chem. Soc.* **1977**, *99*, 7848.

[16]The following discussion is drawn in part from a paper by Harris, J. M.; Shafer, S. G.; Moffatt, J. R.; Becker, A. R. *J. Am. Chem. Soc.* **1979**, *101*, 3295.

while the vertical scale represents the extent of R—X bond cleavage. The diagonal pathway corresponds to an S_N2 process in which R—X bond-breaking and R—Y bond-forming occur simultaneously, and the ‡ on the midpoint of that line represents the location of a transition state.

Figure 8.1 helps to define some common terms in substitution mechanisms. If the transition state for an S_N2 reaction occurs on the diagonal but is nearer the reactant than the product, then the transition state is *early* or reactant-like (denoted by **R**). Conversely, if the transition state is nearer the product, the transition state is *late* or product-like (marked by **P**). If the transition state lies off the diagonal near the upper left hand corner, then bond breaking occurs faster than bond forming, so the transition state is *loose* (**L**). If the transition state lies off the diagonal near the lower right hand corner, then bond forming is faster than bond breaking, so the transition state is *tight* (**T**).

In the extreme case of a loose S_N2 transition state, the reaction may become an S_N1 reaction. The curve in Figure 8.2 represents a minimum energy pathway along the same set of nuclear coordinates for an S_N1 process. The upper left corner represents a local energy minimum associated with an intermediate carbocation. The two ‡ marks near it indicate transition states, one for formation of the carbocation and one for its reaction with the nucleophile. The lower right corner of these drawings represents an intermediate with net negative charge. In principle we might propose that such a species might be sufficiently stabilized to be an intermediate in a very tight S_N2 reaction. In fact, such a curve would be expected for an S_NAr reaction (Figure 8.3).[17] Here, the corner represents a local energy minimum associated with a carbanion intermediate.

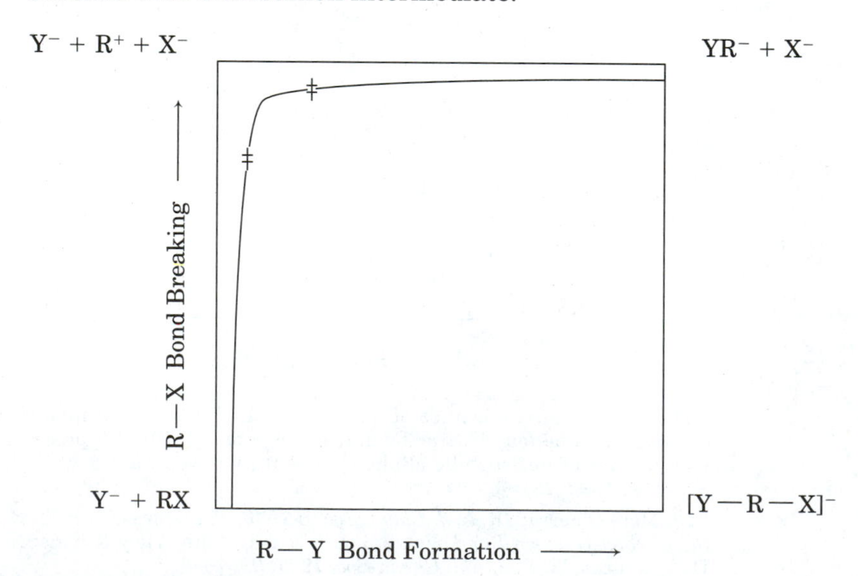

Figure 8.2 Two-dimensional reaction coordinate diagrams for an S_N1 reaction. (Adapted from reference 16.)

[17]Lee et al. have noted in one study that extensive charge transfer from nucleophile to substrate does not always mean a "tight" transition state: Lee, I.; Kang, H. K.; Lee, H. W. *J. Am. Chem. Soc.* **1987**, *109*, 7472.

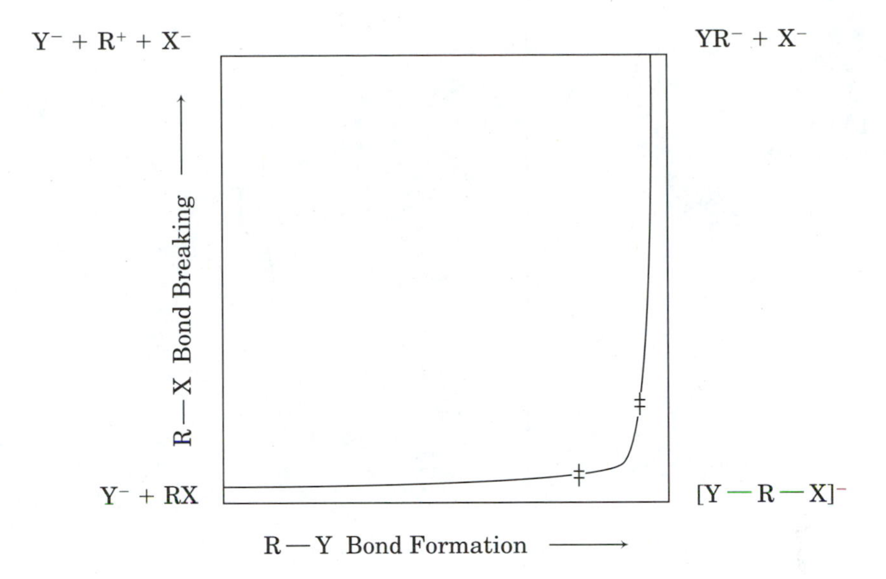

Figure 8.3 Two-dimensional reaction coordinate diagrams for an S_NAr reaction. (Adapted from figure 3, p. 3296 of reference 16.)

These figures can be made three-dimensional by allowing the axis projecting toward the viewer to represent energy, as shown in Figure 8.4 and Figure 8.5. In Figure 8.4 the third dimension is indicated by redrawing Figure 8.1 so that contour lines connect points of equal energy. In Figure 8.5 the coordinate system for the three-dimensional surface is rotated so that the topological surface corresponding to those contour lines is viewed from an angle to the side. With such diagrams we are better able to visualize the energetics of reactions in which bond breaking and bond forming occur at different rates. Now it is clearer that the point indicated by the ‡ in Figure 8.1 is a saddle point on the potential energy surface.

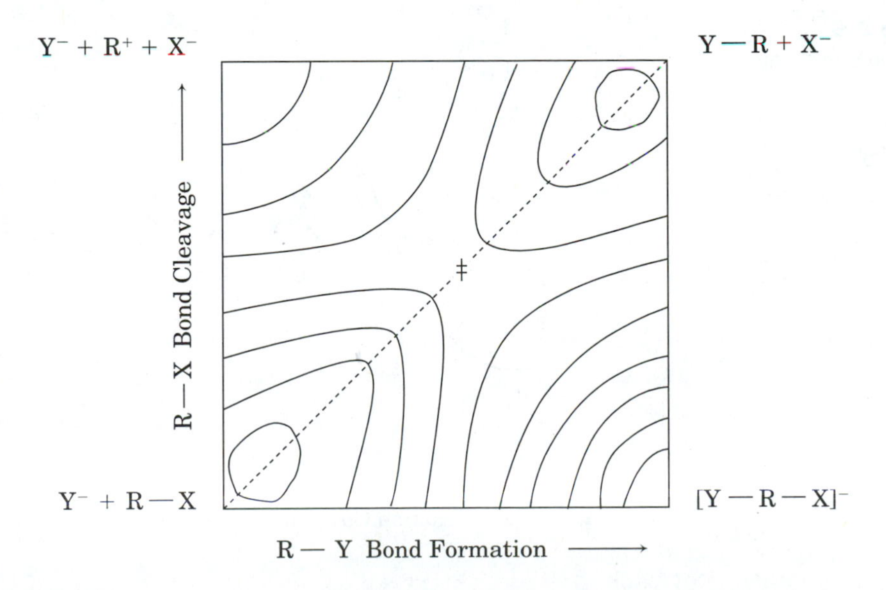

Figure 8.4 Two-dimensional potential energy surface drawn as a contour map. (Adapted from reference 16.)

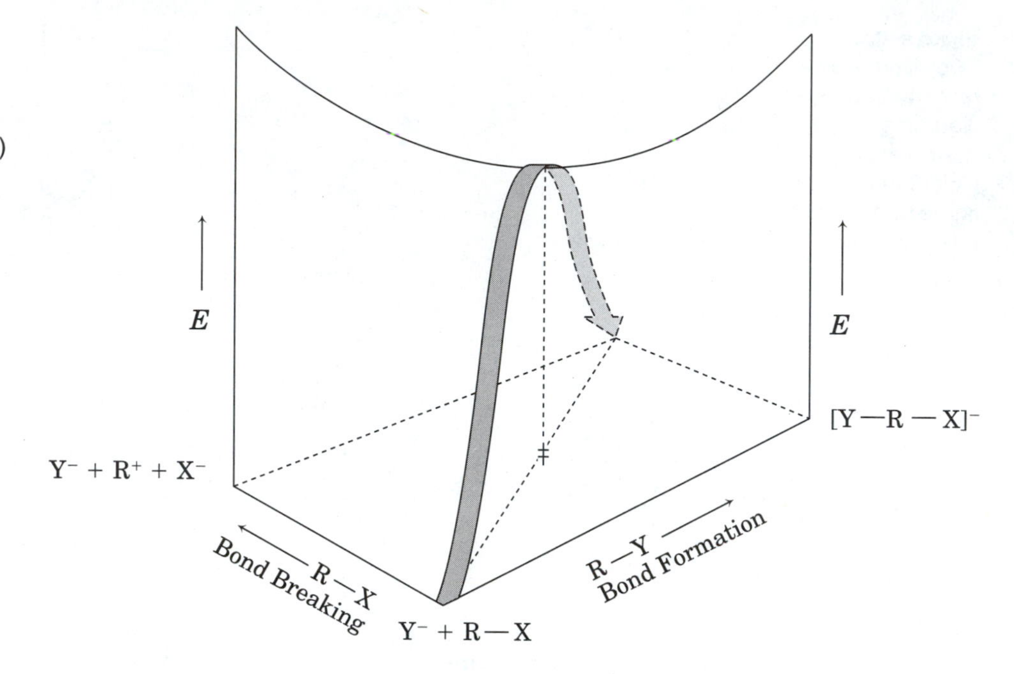

Figure 8.5 A three-dimensional potential energy surface. (Adapted from reference 16.)

Choosing the diagonal line in Figure 8.4 or Figure 8.5 as the x-axis of a two-dimensional reaction coordinate diagram produces Figure 8.6. The wide arrow from reactant to saddle point to product in Figure 8.5 is now seen as the line from reactants to transition state to products in Figure 8.6.

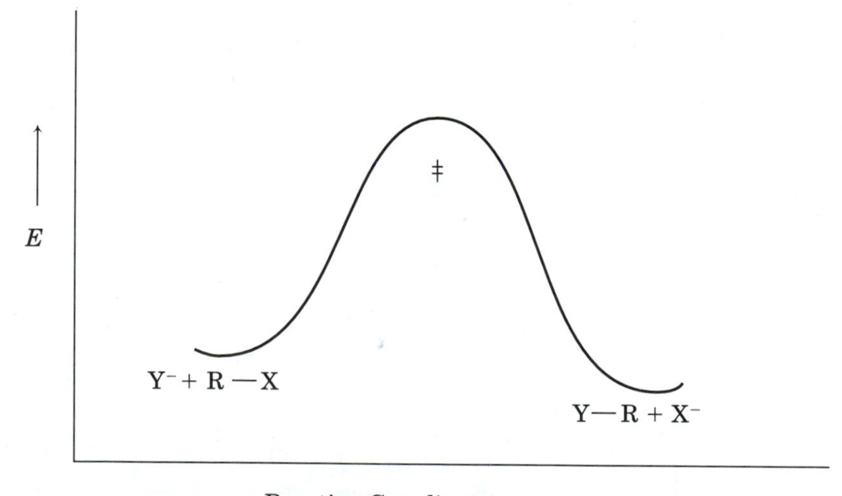

Figure 8.6 Reaction coordinate diagram corresponding to Figure 8.1.

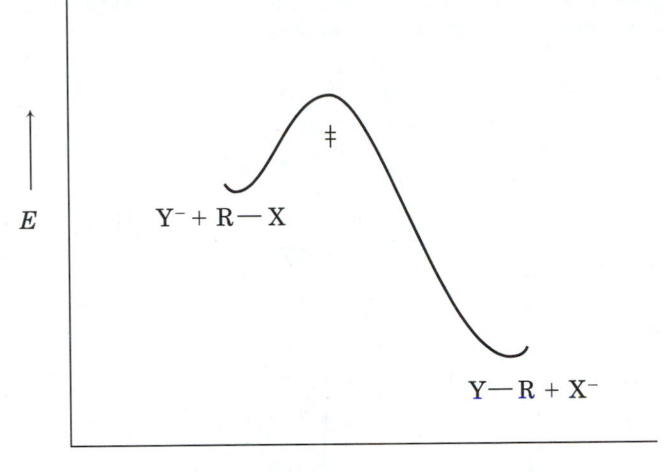

Figure 8.7
Reaction coordinate diagram for exothermic reaction.

According to the Hammond postulate (Chapter 5), increasing the stability of the product in a one-step reaction causes the transition structure to more closely resemble the reactant in both energy and in geometry. The graphical result of increasing product stability, therefore, is a reaction coordinate diagram in which the transition state lies more to the left (Figure 8.7). The net effect of such a change on a contour plot would be to move the transition state *away* from the corner that is stabilized, as shown in Figure 8.8.

Increasing the stability of one the corners that is *not* on the reaction diagonal in Figure 8.5 has a very different effect. Figure 8.9 shows that the ef-

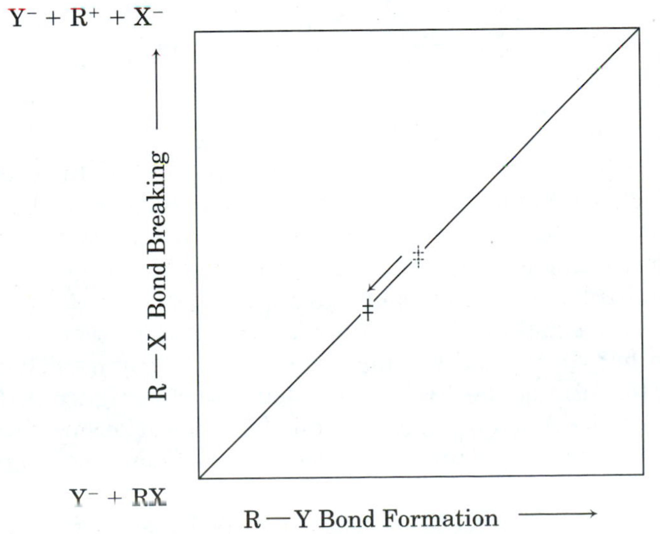

Figure 8.8
Effect of stabilizing upper right corner of a contour plot.

Figure 8.9 Effect of stabilization of one corner of a three-dimensional potential energy surface. (Adapted from reference 16.)

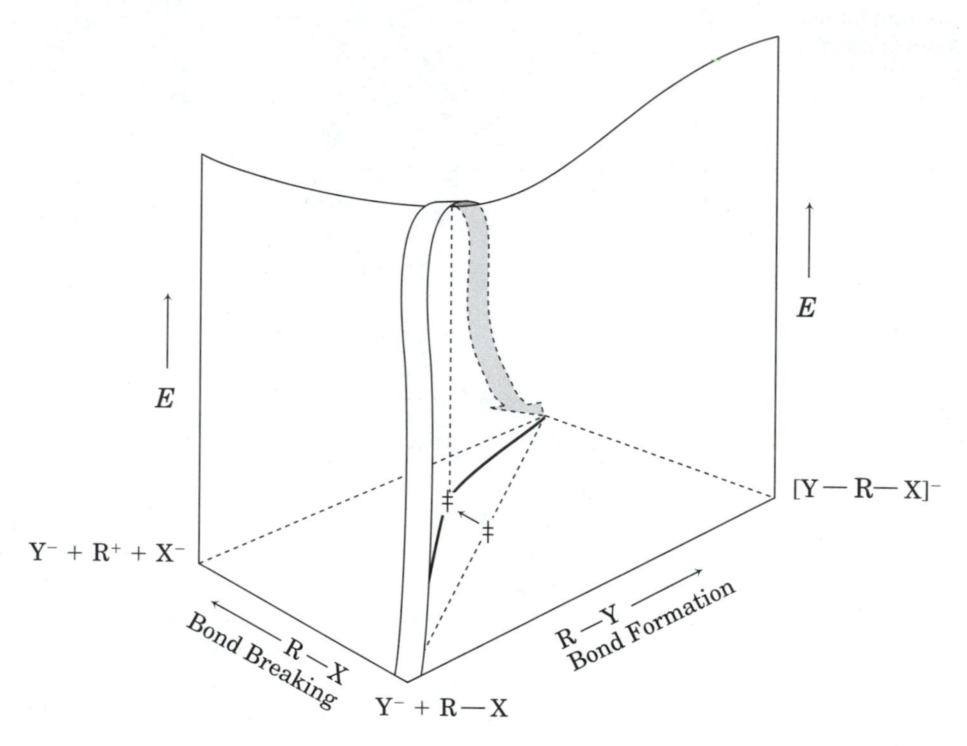

fect of increasing the stability of the species on the upper left corner of Figure 8.5 is to shift the transition state *toward* that corner. This difference in effects on the location of the transition state arises because the transition state is a saddle point—an energy maximum with respect to movement along the diagonal, but an energy minimum with respect to movement perpendicular to the diagonal (see Figure 8.5).

Since making $X:^-$ a better leaving group lowers the energy of both the top left corner and the top right corner of Figure 8.1, the net effect is the vector sum of the two independent effects (Figure 8.10). Figures such as these enable us to correlate the effect of variation of the nucleophile, substrate, leaving group, and solvent on the geometry and energy of the transition structure for substitution reactions.

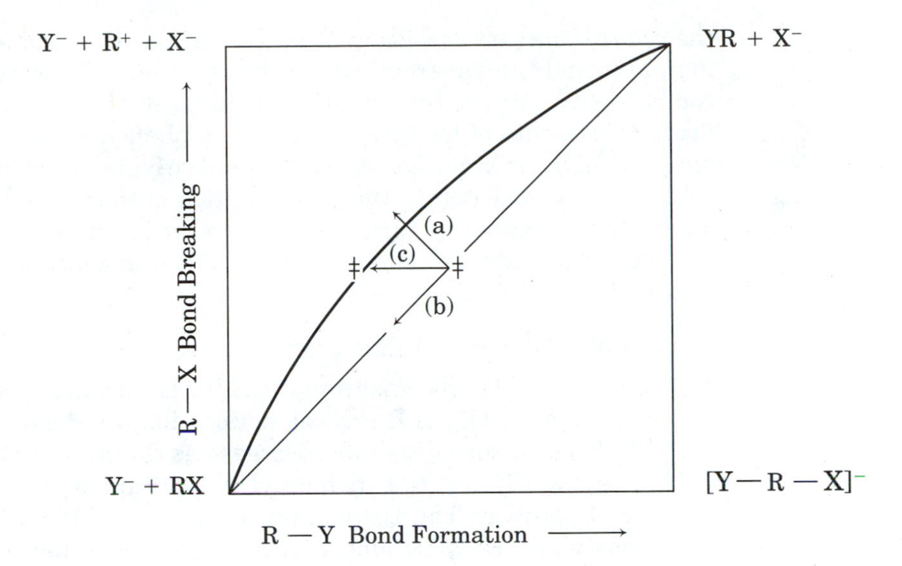

Figure 8.10 Effect (vector c) of stabilizing both top left (vector a) and top right (vector b) corners.

The S_N1 Reaction

Kinetics

In the usual mechanism for the S_N1 reaction (equations 8.8 and 8.9), the rate-limiting step is the ionization of the substrate to form a carbocation and an anion. Therefore, we commonly write that the rate of the reaction is effectively the rate of the unimolecular first step, as shown in equation 8.10. Clearly equation 8.10 cannot be exact, however, because there must be some $Y:^-$ in the solution or the substitution step (equation 8.9) cannot occur.

(slow) $$R{-}X \underset{k_{-1}}{\overset{k_1}{\rightleftharpoons}} R^+ + X:^- \qquad \textbf{(8.8)}$$

(fast) $$R^+ + Y:^- \xrightarrow{k_2} R{-}Y \qquad \textbf{(8.9)}$$

$$\text{Rate} = k_1[R{-}X] \qquad \textbf{(8.10)}$$

A more exact kinetic expression results from applying the steady state approximation to $[R^+]$ to produce equation 8.13.

$$d[R^+]/dt \equiv 0 = k_1[R{-}X] - k_{-1}[R^+][X:^-] - k_2[R^+][Y:^-] \qquad \textbf{(8.11)}$$

and $$[R^+] = k_1[R{-}X]/(k_{-1}[X:^-] + k_2[Y:^-]) \qquad \textbf{(8.12)}$$

so $$\text{Rate} = k_1[R{-}X]k_2[Y:^-]/(k_{-1}[X:^-] + k_2[Y:^-]) \qquad \textbf{(8.13)}$$

In early phases of the reaction $[X:^-]$ is close to zero, so the term $k_1[X:^-]$ is also about zero. Making that assumption allows us to cancel the $k_2[Y:^-]$ terms in the numerator and denominator of equation 8.13, which leads to the *approximate* expression of equation 8.10. As the reaction proceeds, [X:$^-$] increases and the apparent rate of the reaction may decrease if k_1[X:$^-$] becomes significant relative to k_2[Y:$^-$]. Adding X:$^-$ to the solution would also decrease the rate of formation of R—Y, a phenomenon known as the *common ion effect* or *mass law effect.*[18] In **solvolysis reactions**, in which the solvent is the nucleophile, the concentration of the nucleophile is effectively constant and cannot be varied. Therefore the reaction is more properly described as **pseudo–first order**, since only the concentration of the substrate can be varied.

Structural Effects in S_N1 Reactions

The central role of carbocations in the S_N1 reaction is supported by the finding that the reactivity of R—X compounds follows the order of carbocation stability. That is, solvolysis rates decrease as the stability of the carbocation decreases, from 3° to 2° to 1° to methyl, with 1° and methyl being unreactive by the S_N1 pathway. This correlation seems reasonable in view of the Hammond postulate. Because ionization is highly endothermic, the transition structure should resemble the carbocation being formed. Indeed, calculations have suggested that the charge separation in the transition structure for *t*-butyl chloride solvolysis is about 80% of full ionization.[19]

There are also other structural effects on the rates of S_N1 reactions. Consider the reactant and carbocation shown in Figure 8.11. In the reactant the carbon atom bearing the leaving group has bond angles that ordinarily are close to 109.5°. In the carbocation intermediate, the preferred geometry of the trigonal carbon atom is planar with bond angles of 120°. Thus three alkyl groups attached to the reaction site are sterically more crowded in the reactant than they are in the intermediate, so steric relief can accelerate the rate of the ionization step. Such a trend is observed in the rates of hydrolysis of the 3° alkyl halides as shown in Table 8.1.[20]

Figure 8.11 Steric relief accompanying ionization.

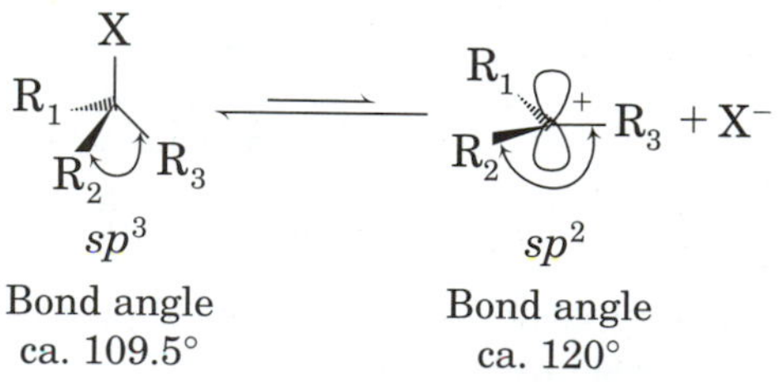

sp^3 Bond angle ca. 109.5°

sp^2 Bond angle ca. 120°

[18]Addition of other ionic species could increase the rate by changing the effective polarity of the microenvironment through a salt effect, and some salts exhibit a "special salt effect." See the discussion on page 474.

[19]Abraham, M. H.; Abraham, R. J. *J. Chem. Soc. Perkin Trans. 2* **1974**, 47; Clarke, G. A.; Taft, R. W. *J. Am. Chem. Soc.* **1962**, *84*, 2295.

[20](a) Brown, H. C.; Fletcher, R. S. *J. Am. Chem. Soc.* **1949**, *71*, 1845; (b) Brown, H. C.; Fletcher, R. S.; Johannesen, R. B. *J. Am. Chem. Soc.* **1951**, *73*, 212.

Table 8.1 Rate constants for the hydrolysis of 3° alkyl halides, $R(CH_3)_2CCl$, in aqueous ethanol at 25°

R	Rate Constant (hr^{-1})	Rel. Rate
CH_3	0.033	1.00
Et	0.055	1.67
Pr	0.052	1.58
i-Pr	0.029	0.88
Bu	0.047	1.42
t-Bu	0.040	1.21
Neopentyl	0.74	22.4

(Data from reference 20(a).)

With some structures, however, the geometry imposed by a molecular skeleton leads to greater steric strain in the carbocation than in the reactant. This effect is particularly notable in 1-substituted bridgehead systems in which the molecular framework forces the carbocation to be nonplanar.[21] In the 1-nortricyclyl cation (**1**) for example, the carbocation is decidedly nonplanar, with the angle of the carbon-carbon-carbon bonds at the carbocation center calculated to be 110°. The strain associated with formation of this carbocation means that 4-tricyclyl derivatives are very resistant to solvolysis.[22] The relative reactivities of several structures substituted at the bridgehead position (Figure 8.12, page 466) confirm that solvolytic reactivity decreases as the size of the bridges decreases.[21]

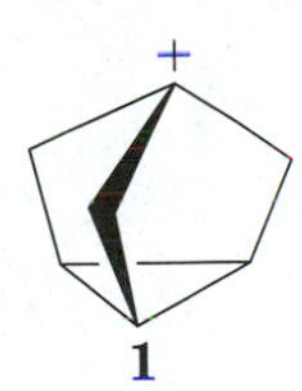

Substituents that can stabilize a carbocation by electron donation can increase the rates of S_N1 reactions. For example, benzyl (**2**), benzhydryl (**3**), and triphenylmethyl (**4**, also known as trityl) cations are stabilized by the aromatic rings, with a resulting dramatic increase in the reactivity of the corresponding chlorides in solvoylsis reactions (Table 8.2).[23]

[21]For a discussion of the reactivity of bridgehead systems in solvolysis reactions, see Bingham, R. C.; Schleyer, P. v. R. *J. Am. Chem. Soc.* **1971**, *93*, 3189.

[22]Sherrod, S. A.; Bergman, R. G.; Gleicher, G. J.; Morris, D. G. *J. Am. Chem. Soc.* **1972**, *94*, 4615.

[23]Nixon, A. C.; Branch, G. E. K. *J. Am. Chem. Soc.* **1936**, *58*, 492.

Figure 8.12 Approximate relative reactivities in solvolysis reactions. (Adapted from reference 21.)

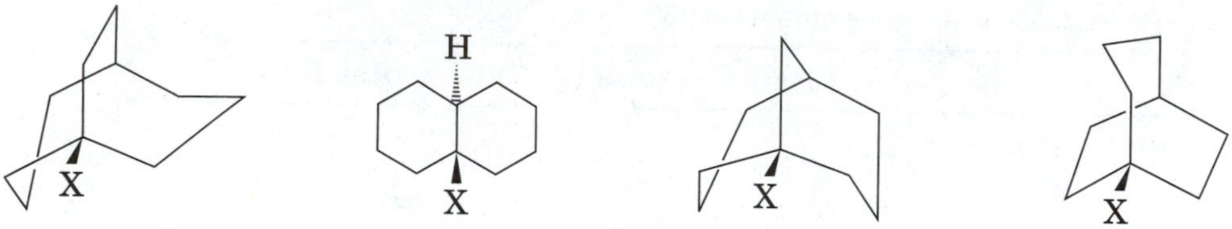

1 | 8×10^{-2} | 3×10^{-3} | 6×10^{-4}

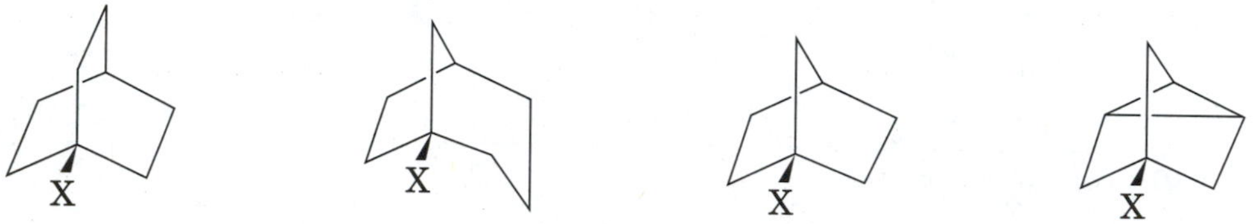

10^{-7} | 10^{-8} | 10^{-15} | 10^{-19}

$$C_6H_5{-}CH_2X \rightleftharpoons C_6H_5{-}CH_2^{+} + X^{-}$$

Benzyl cation

2

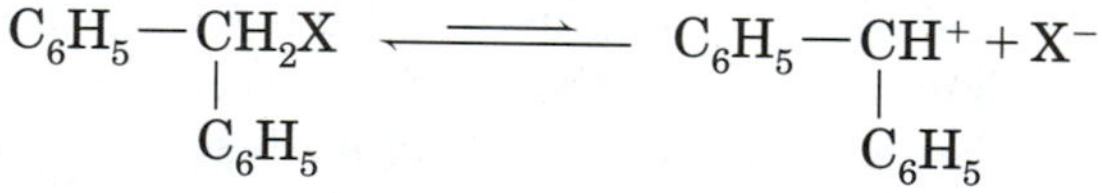

$$C_6H_5{-}CH(C_6H_5)X \rightleftharpoons C_6H_5{-}CH^{+}(C_6H_5) + X^{-}$$

Benzhydryl cation

3

$$(C_6H_5)_3CX \rightleftharpoons (C_6H_5)_3C^{+} + X^{-}$$

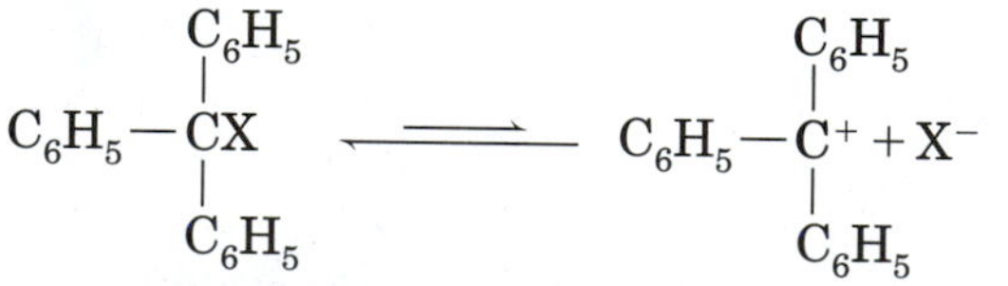

Trityl cation

4

Solvent Polarity and Nucleophilicity

Because the S_N1 reaction involves dissociation of a neutral species to two oppositely charged ions, greater solvent polarity accelerates the rate of reaction.[24] In general, however, solvolysis rates do not correlate well with sol-

[24]For a theoretical study of the effect of solvent polarity on the transition structures for S_N1 reactions, see Mathis, J. R.; Kim, H. J.; Hynes, J. T. *J. Am. Chem. Soc.* **1993**, *115*, 8248.

Table 8.2 Relative reactivities of aryl halides in S_N1 reactions.

Alkyl Halide	Reactivity
$C_6H_5CH_2—Cl$	1
$(C_6H_5)_2CH—Cl$	1.75×10^3
$(C_6H_5)_3C—Cl$	2.5×10^7

(Data for ethanolysis in 60% diethyl ether, 40% ethanol from reference 23.)

vent dipole moment (μ) or dielectric constant (ϵ).[25] Other solvent polarity parameters have been developed, including the Z scale and the E_T scale discussed in Chapter 6, but solvent molecules do more than simply provide a benign polar medium for ion stabilization. Solvent molecules with electrophilic properties may also enhance reactivity by assisting in the solvation of the departing nucleofuge. In addition solvent molecules that can act as nucleophiles can compete with other species for product formation or can assist in the ionization process so that another nucleophile can react to produce a substitution product.

One of the earliest attempts to describe solvent behavior in solvolysis reactions quantitatively was the Grunwald-Winstein equation, which is a linear free energy relationship analogous to the Hammett equation.[26]

$$\log (k/k_0) = mY \tag{8.14}$$

Here, k is the reactivity of a substrate in a given solvent and k_0 is the rate of reaction of the substrate in a standard solvent system. Y is the ionizing power of a solvent, and m is the sensitivity of that substrate to solvent ionizing power. The Y scale was established by considering the relative reactivity of t-butyl chloride in methanol (k_0). That is,

$$Y \equiv \log (k_{t\text{-BuCl, solvent}}/k_{t\text{-BuCl, methanol}}) \tag{8.15}$$

Perhaps the most widely used scale of solvent ionizing power is the Y_{OTs} scale introduced by Schleyer and co-workers.[27] This system is based on the rates of solvolysis of 2-adamantyl tosylate (**5**). The free energy relationship is

$$\log\left(\frac{k}{k_0}\right) = lN + mY \tag{8.16}$$

[25]Fainberg, A. H.; Winstein, S. *J. Am. Chem. Soc.* **1956**, *78*, 2770.

[26]Grunwald, E.; Winstein, S. *J. Am. Chem. Soc.* **1948**, *70*, 846; Winstein, S.; Grunwald, E.; Jones, H. W. *J. Am. Chem. Soc.* **1951**, *73*, 2700.

[27]Schadt, F. L.; Bentley, T. W.; Schleyer, P. v. R. *J. Am. Chem. Soc.* **1976**, *98*, 7667.

where k is the rate constant for ionization in a given solvent, k_0 is the rate constant for ionization in the reference solvent, 80% aqueous ethanol, Y is the solvent ionizing power, and m is the sensitivity of the substrate to solvent ionizing power. Because some substrates show evidence of nucleophilic solvent participation in solvolysis, N is the solvent nucleophilicity and l is the sensitivity of the substrate to solvent nucleophilicity.[28,29] Values of Y_{OTs} and N_{OTs} for selected solvents are shown in Table 8.3. Many other Y scales have been developed for particular applications.[30] As noted in Chapter 6, however, the act of writing a linear free energy relationship implies the modeling of one reaction on another. The ability of the linear free energy equation to correlate experimental data depends, at least in part, on the degree to which the model is appropriately chosen.

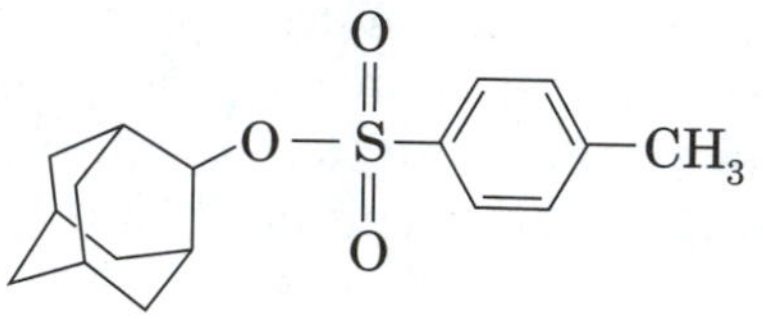

2-Adamantyl tosylate

5

If the solvent for an S_N1 reaction includes added nucleophiles, particularly anions, then not only the rate but also the product of the reaction can be affected. For example, adding 0.05 M sodium azide to a solution of 4,4′-dimethylbenzhydryl chloride in 85% aqueous acetone was seen to increase the rate of the solvolysis reaction by 50%, and the product was found to consist of 66% of 4,4′-dimethylbenzhydryl azide and 34% of the alcohol.[31,32] The rate increase is explained on the basis of a ***normal salt effect***, that is, the

[28]N_{OTs} values are determined from rates of solvolysis of methyl tosylate from the formula:

$$N_{OTs} = \log\left(\frac{k}{k_0}\right)_{CH_3OTs} - 0.3\,Y_{OTs}$$

For a discussion, see references 27 and 30.

[29]Abraham et al. have concluded that at least four solvent parameters are necessary to adequately correlate solvolysis of t-butyl chloride with solvent properties in the most general case. These four properties are solvent dipolarity, solvent hydrogen bond acidity, solvent hydrogen bond basicity, and the cohesive energy density of the solvent. They note, however, that a smaller number of parameters may be adequate for studies in which solvent properties vary in more restricted ways. Abraham, M. C.; Doherty, R. M.; Kamlet, M. J.; Harris, J. M.; Taft, R. W. *J. Chem. Soc. Perkin Trans. 2* **1987**, 913, 1097.

[30]For a discussion of Y_X scales of solvent ionizing power, see Bentley, T. W.; Llewellyn, G. *Prog. Phys. Org. Chem.* **1990**, *17*, 121.

[31]Bateman, L. C.; Church, M. G.; Hughes, E. D.; Ingold, C. K.; Taher, N. A. *J. Chem. Soc.* **1940**, 979.

[32]Huisgen, R. *Angew. Chem., Int. Ed. Engl.* **1970**, *9*, 751.

Table 8.3 Values of Y_{OTs} and N_{OTs} for selected solvents.

Solvent	Y_{OTs}	N_{OTs}
80% Aqueous ethanol	0.00	0.00
Ethanol	−1.96	0.06
50% Ethanol	1.29	−0.09
Methanol	−0.92	−0.04
95% Acetone	−2.95	
50% Acetone	1.26	−0.39
Acetonitrile	−3.21	
50% Acetonitrile	1.2	
30% Acetonitrile	1.9	
10% Acetonitrile	3.6	
Water	4.1	−0.44
2-Propanol	−2.83	0.12
2-Methyl-2-propanol	−3.74	
Acetic acid	−0.9	−2.28
Trifluoroacetic acid	4.57	−5.56
60% (w/w) Sulfuric acid in water	5.29	−2.02
N,N-Dimethylformamide	−4.14	
N,N-Dimethylacetamide	−4.99	

(Data from the compilation in reference 30.)

increased ionic strength of the reaction medium resulting from salt addition increases the effective *polarity* of the solvent. Various theoretical and empirical treatments suggest that the rate constant should increase with the log of the concentration of added salt,[31] with the first power of concentration of added salt,[33] or with the square root of the added salt concentration,[34] depending on the reactants and solvents.

The product distribution resulting from the addition of nucleophiles to a solvolysis reaction depends upon the ability of the added nucleophile and the solvent to compete for attachment of the carbocation intermediate. In contrast to the case in S_N2 reactions,[35] the relative reactivity of nucleophiles toward carbocations often does not vary dramatically with the structure of the carbocation. Instead, the distribution of products of nucleophilic attachment to a cation is a function of the structures of both the nucleophile and

[33]Fainberg, A. H.; Winstein, S. *J. Am. Chem. Soc.* **1956**, *78*, 2763.

[34]Winstein, S.; Klinedinst, Jr., P. E.; Robinson, G. C. *J. Am. Chem. Soc.* **1961**, *83*, 885.

[35]See the discussion beginning on page 498.

the solvent in which the reaction occurs. Ritchie determined that many cation-nucleophile reactions could be described by the equation

$$\log (k_n/k_{H_2O}) = N_+ \tag{8.17}$$

where k_n is the rate constant for the reaction of a nucleophile with a given carbocation, k_{H_2O} is the rate constant for reaction of water as a nucleophile under the same conditions, and N_+ is a nucleophilicity parameter for the nucleophile-solvent system.[36,37] Selected values of N_+ are shown in Table 8.4. The trends in N_+ values, in particular the observation that higher N_+ values are observed for systems in which the nucleophiles are not strongly solvated, suggest that nucleophilic reactivity is a function of a *system* consisting of both a nucleophile and the solvent.[36] Failure to observe a dependence of relative nucleophilicity on the structure of the cation in equation 8.17 suggests that the transition structures strongly resemble the carbocations; that is, there is negligible bond formation, and there has been little disturbance of the cation or its solvation shell in the transition state.[38]

Solvated Ions and Ion Pairs

Discussion of the S_N1 reaction often emphasizes the development of the carbocation, since that is the site of nucleophilic attack and product formation. Yet the roles of the solvent and the nucleofuge are also significant. The sim-

Table 8.4 N_+ Values for nucleophile (solvent) systems at 25°.

Nucleophile (solvent)	N_+	Nucleophile (solvent)	N_+
H_2O (H_2O)	0.0	CH_3O^- (CH_3OH)	7.5
CH_3OH (CH_3OH)	0.5	N_3^- (CH_3OH)	8.5
CN^- (H_2O)	3.8	CN^- (CH_3SOCH_3)	8.6
$C_6H_5SO_2^-$ (CH_3OH)	3.8	CN^- ($HCON(CH_3)_2$)	9.4
HO^- (H_2O)	4.5	N_3^- (CH_3SOCH_3)	10.7
N_3^- (H_2O)	5.4	$C_6H_5S^-$ (CH_3OH)	10.7
CN^- (CH_3OH)	5.9	$C_6H_5S^-$ (CH_3SOCH_3)	13.1

(Reproduced from reference 36.)

[36] Ritchie, C. D. *Acc. Chem. Res.* **1972**, *5*, 348 and references therein.

[37] Scales of nucleophilicity and electrophilicity have been presented by Mayr, H.; Patz, M. *Angew. Chem., Int. Ed. Engl.* **1994**, *33*, 957.

[38] As the carbocation becomes more stable, however, the transition structure would be expected to become somewhat more product like, and some selectivity of the cation for the more nucleophilic species might be expected. For a discussion, see (a) Sneen, R. A.; Carter, J. V.; Kay, P. S. *J. Am. Chem. Soc.* **1966**, *88*, 2594; (b) Raber, D. J.; Harris, J. M.; Hall, R. E.; Schleyer, P. v. R. *J. Am. Chem. Soc.* **1971**, *93*, 4821. Ritchie (reference 36) has discussed the possible origins of such correlations. A discussion of the barrier for carbocation-nucleophile combinations has been given by Richard, J. P. *Tetrahedron*, **1995**, *51*, 1535.

plest model of the S_N1 reaction at a single stereogenic center predicts racemization of an optically active starting material because the nucleophile is able to add equally well to the top lobe or the bottom lobe of an empty *p* orbital. For example, **formolysis** (solvolysis in formic acid) of optically active 1-(α-naphthyl)ethyl acetate (**6**) gave the racemic formate (**7**).[39] Similarly, **methanolysis** of (−)-*p*-methoxybenzhydryl hydrogen phthalate (**8**) produced the totally racemic ether (**9**).[40,41]

CH_3 O CH_3 O — HCO_2H → CH_3 O H O **(8.18)**

optically active **6** racemic **7**

C_6H_5 CH O O CH_3O HO_2C — CH_3OH, warm → C_6H_5 CH O CH_3 CH_3O **(8.19)**

(−) **8** racemic **9**

In some cases, however, solvolysis reactions of chiral substrates show some degree of inversion of configuration. For example, Steigman and Hammett found that **acetolysis** (solvolysis in acetic acid) of α-phenylethyl chloride (**10**) gave the corresponding acetate (**11**) by a strictly first order process, the rate not being affected by added acetate ion.[42,43] The acetate could also be formed by a second order reaction between **10** and tetraethylammonium acetate in acetone. The product formed by both pathways had the same *sign* of rotation of polarized light, although the product obtained through solvolysis had a much smaller rotation. One possible explanation for such behavior is that the reaction takes place by a mixture of S_N1 and S_N2 pathways. However, if the enantiomeric purity of the product(s) is not affected by changes in concentration of nucleophile (meaning that the kinetics are

[39]Balfe, M. P.; Downer, E. A.; Evans, A. A.; Kenyon, J.; Poplett, R.; Searly, C. E.; Tarnoky, A. L. *J. Chem. Soc.* **1946**, 797.

[40]Balfe, M. P.; Doughty, M. A.; Kenyon, J.; Poplett, R. *J. Chem. Soc.* **1942**, 605.

[41]S_N1 reactions do not always lead to complete racemization; see the discussion beginning in the following paragraph.

[42]Determination of first order kinetics is complicated by the need to keep the ionic strength of the medium constant during the reaction.

[43]Steigman, J.; Hammett, L. P. *J. Am. Chem. Soc.* **1937**, *59*, 2536. This paper marks an early use of the word *solvolysis*.

Figure 8.13 Acetolysis of α-phenylethyl chloride.

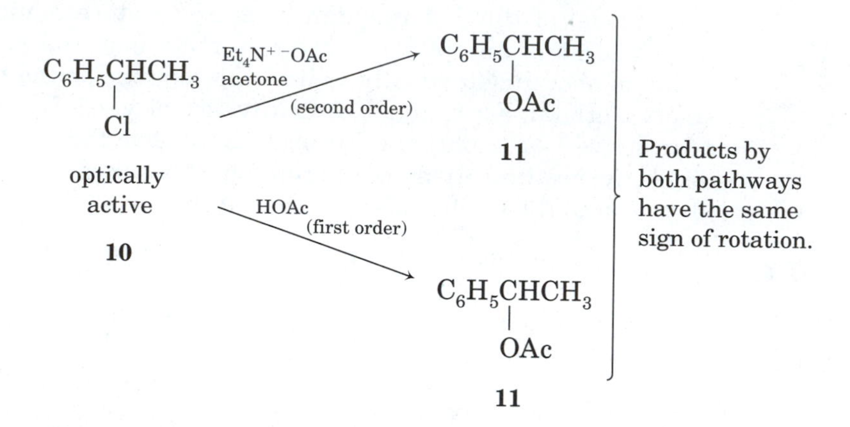

strictly first order), competing S_N1 and S_N2 pathways cannot be occurring, and any observed retention of configuration must be associated with a first order process.

Numerous other cases of partial net inversion in first order solvolysis reactions have led to a variety of proposals for participation of solvent in these reactions. Doering and Zeiss found 54% inversion and 46% racemization in the methanolysis of the phthalate ester **12**.[44] They suggested that the intermediate in solvolyses should not be considered a free carbocation but, instead, a carbocation strongly coordinated by available electron pair donors (Figure 8.15).[45,46] This intermediate (shown as **I** in Figure 8.16) could then undergo complete bond formation with solvent to form inverted product or could undergo exchange of the departing nucleofuge by another sol-

Figure 8.14 Methanolysis of an optically active phthalate.

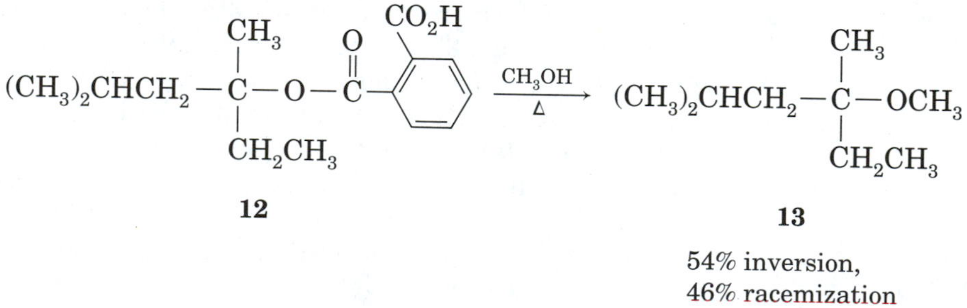

$$(CH_3)_2CHCH_2-\underset{CH_2CH_3}{\overset{CH_3}{\overset{|}{\underset{|}{C}}}}-O-\overset{O}{\overset{\|}{C}}-C_6H_4(CO_2H) \xrightarrow[\Delta]{CH_3OH} (CH_3)_2CHCH_2-\underset{CH_2CH_3}{\overset{CH_3}{\overset{|}{\underset{|}{C}}}}-OCH_3$$

12 **13**

54% inversion,
46% racemization

[44]Doering, W. v. E.; Zeiss, H. H. *J. Am. Chem. Soc.* **1953**, *75*, 4733.

[45]The term *encumbered* has been used to describe carbocations with strong interactions with Lewis base solvents. For a discussion, see Keating, J. T.; Skell, P. S. in *Carbonium Ions. Volume II. Methods of Formation and Major Types*, Olah, G. A.; Schleyer, P. v. R., eds; Wiley-Interscience: New York, 1970; pp. 573–653 and references therein.

[46]Streitwieser and Schaeffer found that solvolysis of 1-butyl-1-*d* *p*-nitrobenzenesulfonate in 75% dioxane–25% acetic acid occurred with 46% net inversion. Furthermore, no products indicating rearrangement of a 1° to a 2° carbocation were found. Streitwieser, Jr., A.; Schaeffer, W. D. *J. Am. Chem. Soc.* **1957**, *79*, 2888.

Figure 8.15 Doering-Zeiss intermediate.

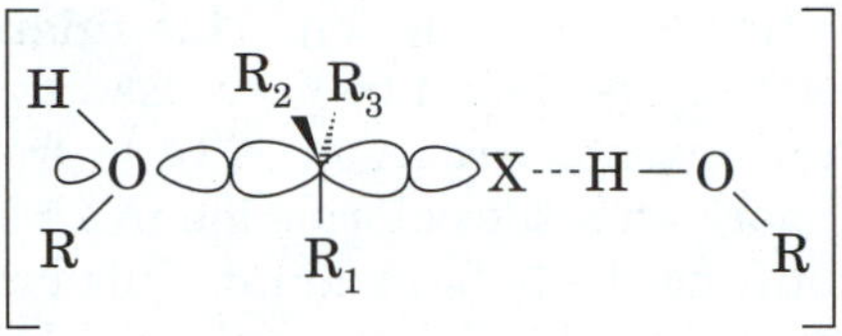

vent molecule to produce intermediate **II**. Products resulting from **II** would be racemic, half showing retention of configuration and half showing inversion (Figure 8.16). The net result would be a slight excess of inversion.

Later studies have extended the view that the departing nucleofuge plays an important role in the S_N1 reaction by suggesting that the carbocation and nucleofuge can form several discrete ion pair intermediates, each of which is characterized by an energy minimum.[47] Early evidence for ion pairing was reported by Winstein from the reaction of α,α-dimethylallyl chloride (**14**) in acetic acid with added acetate. Although the major product of the reaction was the expected acetate, there was rapid isomerization of **14** to the isomeric γ,γ-dimethylallyl chloride (**15**). Formation of **15** can be explained by a mechanism in which the nucleofuge, $Cl{:}^-$, competes as a nucleophile with solvent and rebonds to the allyl cation at the less hindered position. However, the rate of formation of **15** was found not to be a function

Figure 8.16 Doering-Zeiss mechanism.

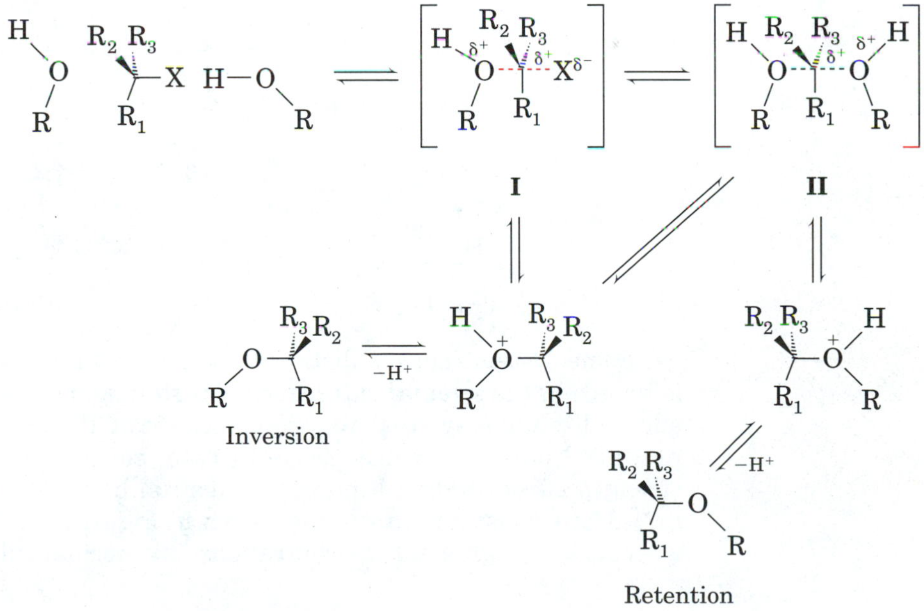

[47]For a discussion, see Hartshorn, S. R. *Aliphatic Nucleophilic Substitution*; Cambridge University Press: Cambridge, England, 1973; pp. 95–127.

of the concentration of added chloride ion. This finding is significant because the rate of the solvolysis should be decreased by addition of chloride ion if the mechanism involves dissociation of **14** to free ions (see equation 8.13). It appears, therefore, that the chloride ion in **15** must originate in the same molecule and not from the bulk medium. This process was characterized as **internal return** of a chloride ion held as part of an intimate ion pair (**16**) within a solvent shell.[48] Similar conclusions have been obtained from the study of other systems. For example, Weinstock found that optically active 3-chloro-3-phenylpropene (**17**) underwent racemization 50% faster than it isomerized to cinnamyl chloride (**18**), which suggested that a tight ion pair was intermediate in the reaction.[49]

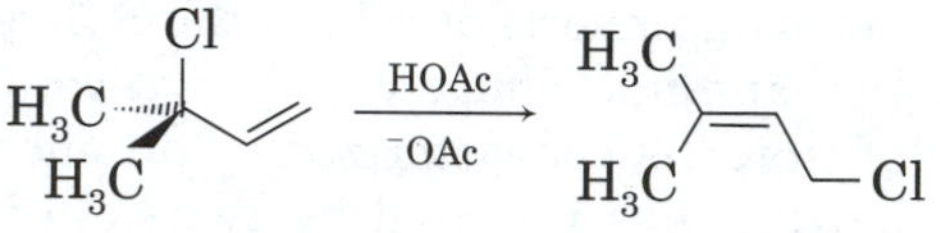

Cl, H_3C, H_3C —(HOAc, ^{-}OAc)→ H_3C, H_3C, Cl **(8.20)**

14 **15**

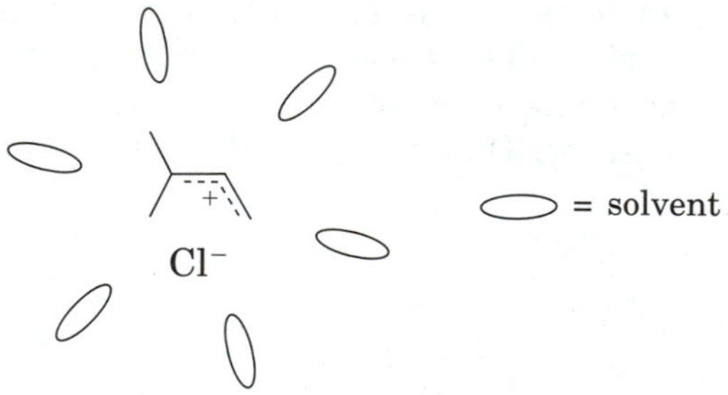

Cl^- ⬭ = solvent

16

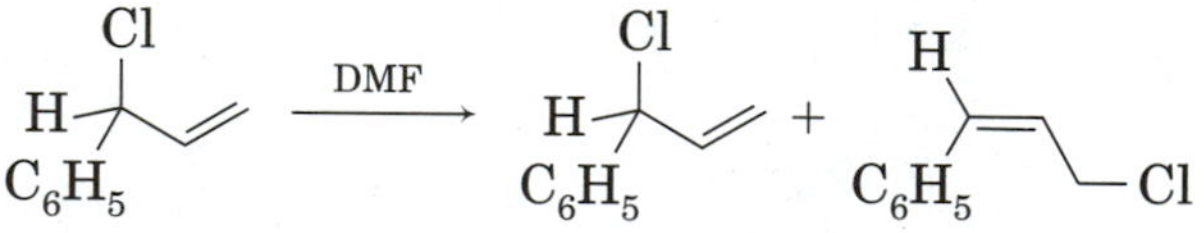

Cl, H, C_6H_5 —(DMF)→ Cl, H, C_6H_5 + H, C_6H_5, Cl **(8.21)**

(−)−17 ⟶ **(+)−17** **18**

Some systems show a distinctive kinetic behavior with added salt that is known as the **special salt effect**. Winstein and co-workers reported that adding lithium tosylate to reaction mixtures of alkyl tosylates in acetic acid produced only a small increase in rate, consistent with a small ionic strength effect. Adding lithium perchlorate, however, caused the observed initial rate constants to vary as shown in Figure 8.17.[34] The linear part of the curve, at higher salt concentrations, is a normal salt effect. The steeply

---

[48]Young, W. G.; Winstein, S.; Goering, H. L. *J. Am. Chem. Soc.* **1951**, *73*, 1958.

[49]Shandala, M. Y.; Waight, E. S., Weinstock, M. *J. Chem. Soc. B* **1966**, 590.

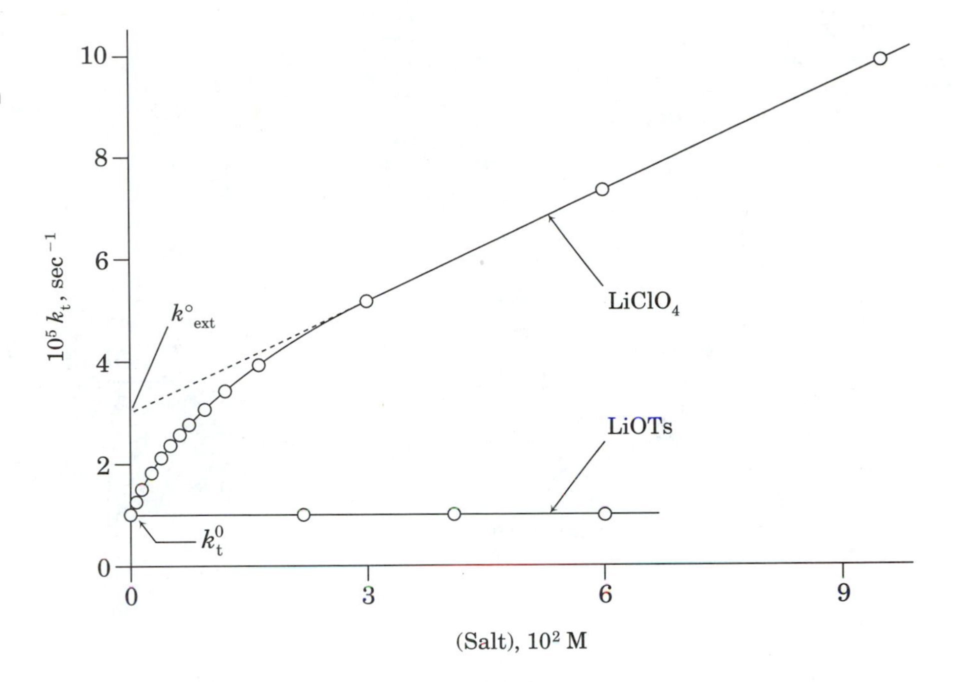

Figure 8.17
An example of the special salt effect in the solvolysis of an alkyl tosylate. (Reproduced from reference 34.)

rising initial portion of the curve for $LiClO_4$, however, is an indication of a special salt effect. Winstein explained the results in terms of two kinds of ion pairs, an intimate ion pair and a solvent-separated ion pair (Figure 8.18). For substrates that undergo internal return only from intimate ion pairs, only a normal salt effect is observed. However, for substrates in which the intimate ion pair separates further to external (solvent-separated) ion pairs, some added salts can trap the nucleofuge and interfere with its return to form the intimate ion pair, thus leading to enhanced rates of solvolysis.[50,51] The external ion pairs can also dissociate further to free ions without added special salt. Depending upon the relative stability of each species and the reactivity of each species with nucleophilic solvent, there may be varying amounts of racemization and inversion of chiral centers.[52]

Theoretical calculations support the existence of discrete ion pair intermediates in solvolysis reactions. Figure 8.19 shows a reaction coordinate diagram, and Figure 8.20(a) through Figure 8.20(c) show calculated

---

[50]Winstein, S.; Clippinger, E.; Fainberg, A. H.; Robinson, G. C. *J. Am. Chem. Soc.* **1954**, *76*, 2597.

[51]Winstein, S.; Klinedinst, Jr., P. E.; Clippinger, E. *J. Am. Chem. Soc.* **1961**, *83*, 4986.

[52]For a review of studies of the role of ions and ion pairs in solvolysis reactions, see Raber, D. J.; Harris, J. M.; Schleyer, P. v. R. in *Ions and Ion Pairs in Organic Reactions*, Vol. 2; Szwarc, M., ed.; Wiley-Interscience: New York, 1974; pp. 247–374.

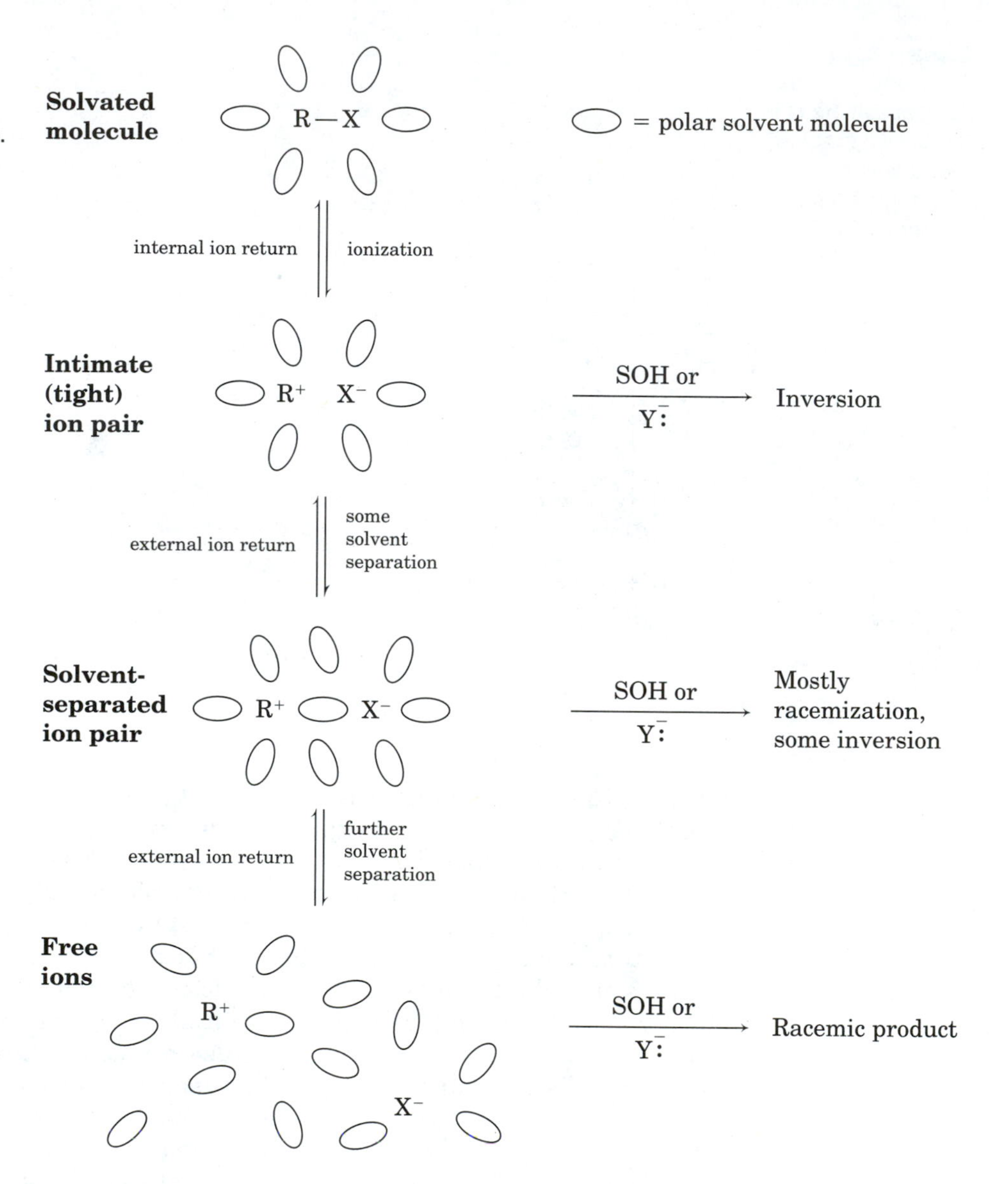

Figure 8.18 Species proposed as intermediates in solvolysis reactions.

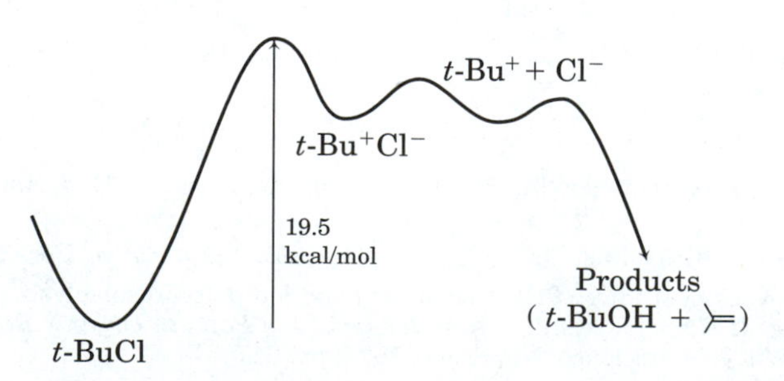

Figure 8.19 Schematic diagram of a possible reaction profile for the solvolysis of *t*-butyl chloride in water. (Adapted from reference 53.)

Figure 8.20 Calculated structures in the solvolysis of *t*-butyl chloride: (a) contact ion pair at 3.0 Å; (b) solvent-separated ion pair at 5.75 Å; (c) tight ion pair at 7.25 Å. (Adapted from reference 53. Notes to the right of each part of the figure are from that reference.)

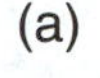
(a)

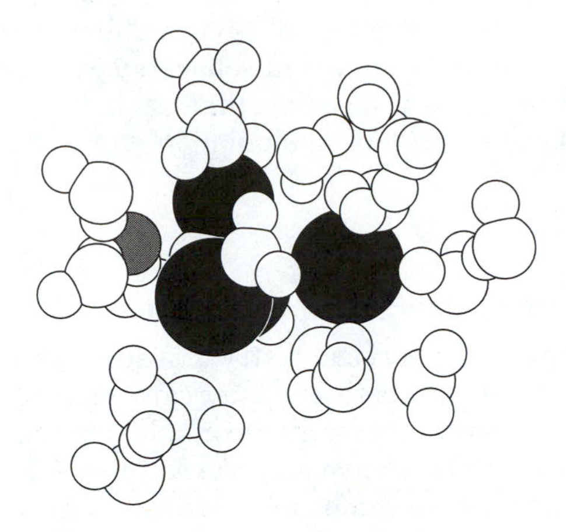

The $(CH_3)_3C^+Cl^-$ contact ion pair at a C—Cl separation of 3.0 Å. Only water molecules with oxygens within 4.5 Å of any atom are shown. The hatched oxygens indicate water molecules solvating the carbenium carbon or between the ions. Methyl groups are shown as single atoms.

(b)

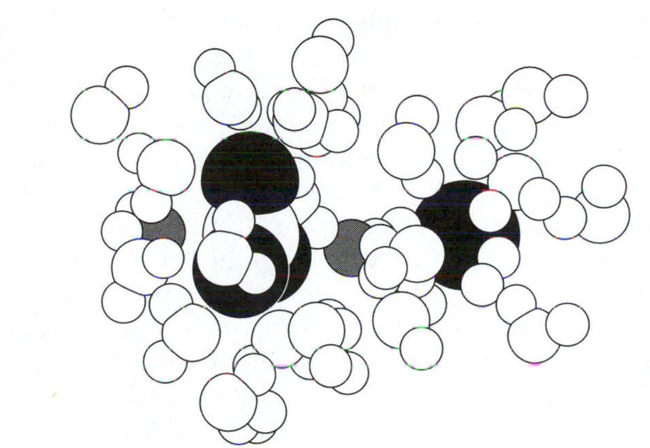

The $(CH_3)_3C^+Cl^-$ solvent-separated ion pair at a C—Cl distance of 5.75 Å.

(c)

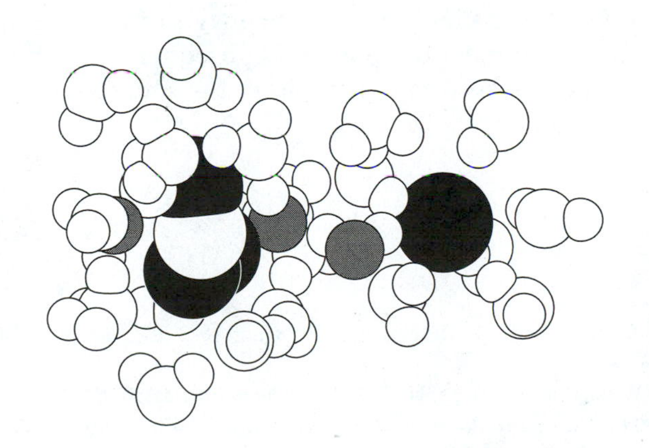

The $(CH_3)_3C^+Cl^-$ ion pair at a C—Cl separation of 7.25 Å. The rightmost hatched oxygen indicates a water molecule that is bridging between the chloride ion and the water molecule coordinated to the carbenium carbon. The latter's hydrogens are now oriented up to accommodate the hydrogen bond; 7-8 hydrogen bonds with the chloride ions are also clearly visible.

structures for the solvolysis of *t*-butyl chloride in water. The $(CH_3)_3C^+Cl^-$ contact ion pair and the solvent separated ion pair are distinct species, with calculated energies indicating that each is a local minimum. The more widely separated ion pair is similar in energy to the solvent-separated ion pair.[53]

Anchimeric Assistance in S_N1 Reactions

Both the rate and stereochemistry of a solvolysis reaction can be affected if the substrate has a substituent that can donate a pair of electrons to the developing carbocation center. For example, treatment of (±)-*threo*-3-bromo-2-butanol (**19**) with HBr gave only the racemic 2,3-dibromobutane (**21**). There was none of the *meso* compound that would have been expected if the reaction involved protonation, loss of water, and formation of an intermediate free carbocation. Similarly, reaction of (±)-*erythro*-3-bromo-2-butanol with HBr gave only *meso*-2,3-dibromobutane. The reaction of **19** seems best explained on the basis of nucleophilic participation of the bromine on the adjacent atom at the time of departure of water. The result is a bridged intermediate (**20**) that is the same bromonium ion expected from the electrophilic addition of molecular bromine to *cis*-2-butene (Figure 8.21).[54] Back-side attack of bromide ion on either carbon atom involved in the three-membered bromonium ring is equally likely, so a racemic mixture results.

There is also kinetic evidence for intramolecular participation in the first steps of some S_N1 reactions. Acetolysis of *trans*-2-acetoxycyclohexyl tosylate (**22**) gives only the trans diacetate (**24**), suggesting the intermediacy of **23**.[55] Furthermore, the rate of the reaction with the *trans*-2-acetoxy compound was found to be nearly 10^3 faster than that of the cis isomer and five times greater than that of cyclohexyl tosylate itself.[56] This evidence strongly supports the view that the acetoxy group participates in the rate-limiting step of the reaction, not in a subsequent step after formation of the intermediate carbocation. Therefore, the ionization of the tosylate is said to be

---

[53]Jorgensen, W. L.; Buckner, J. K.; Huston, S. E.; Rossky, P. J. *J. Am. Chem. Soc.* **1987**, *109*, 1891.

[54]Winstein, S.; Lucas, H. J. *J. Am. Chem. Soc.* **1939**, *61*, 1576.

[55]The literature describing this work uses the term *retention of configuration* to describe the products. Since racemic starting materials were used, the terminology refers to the formation of trans product from trans starting material. Retention of optical activity was not involved.

[56]Winstein, S.; Grunwald, E.; Buckels, R. E.; Hanson, C. *J. Am. Chem. Soc.* **1948**, *70*, 816 and references therein.

Figure 8.21 Anchimeric assistance via bromonium ion intermediate.

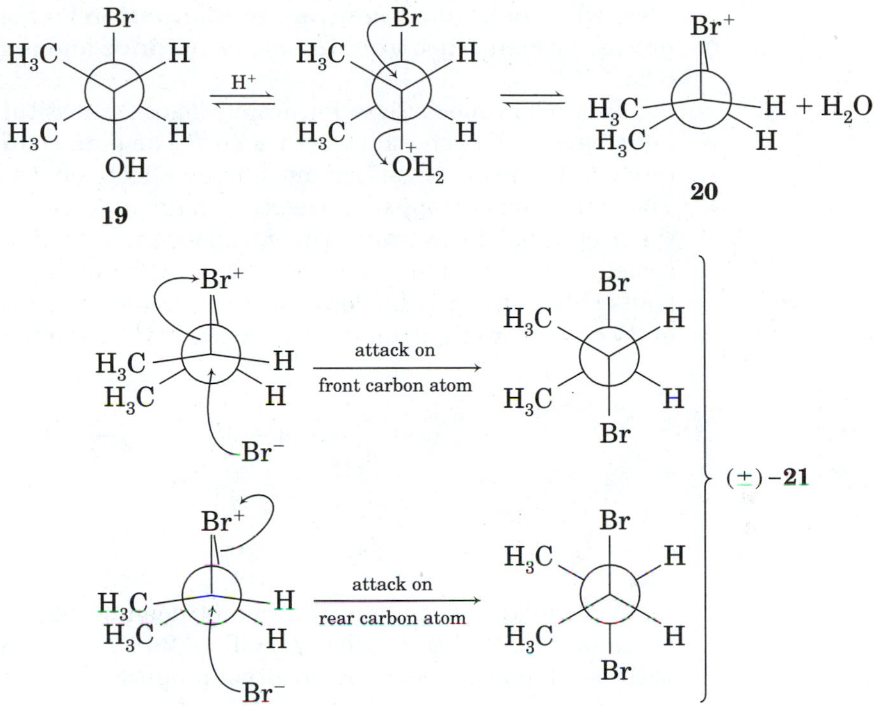

Figure 8.22 Anchimeric assistance in the reaction of *cis*-2-acetoxycyclohexyl tosylate.

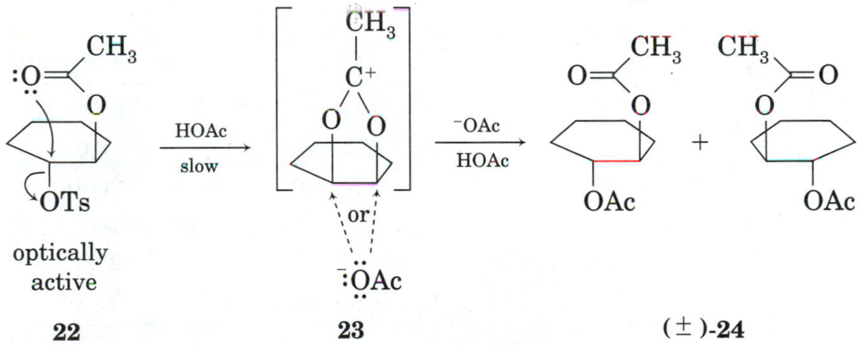

Figure 8.23 Schematic representation of anchimeric assistance.

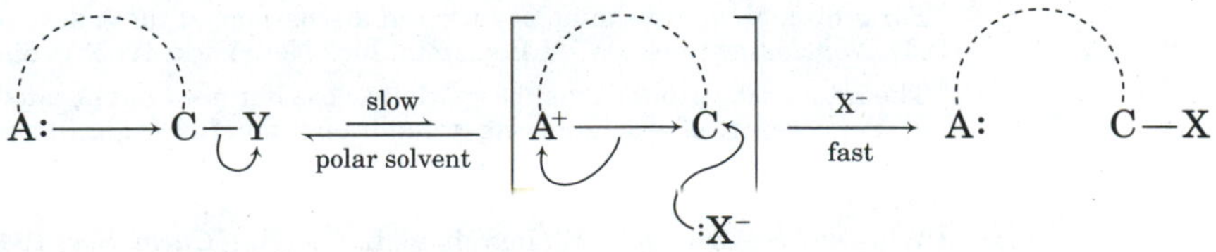

assisted by **neighboring group participation** by the acetoxy group. This process is often called **anchimeric assistance** and is generalized in Figure 8.23.[57,58,59]

Table 8.5 collects data for acetolysis of 2-substituted cyclohexyl brosylates having the general structure **25**.[60] The data confirm that a trans acetate accelerates the reaction much more effectively than does a cis acetate. The activation entropy data suggest a more ordered transition structure for the acetolysis of the trans isomer, consistent with the model of anchimeric assistance. A trans bromine is nearly as effective as a trans acetoxy, but a trans chloro group is far less effective—apparently due to the smaller stability of a chloronium ion in comparison with a bromonium ion.

25

A dramatic example of the role of neighboring group participation is provided by 2,2′-dichlorodiethyl sulfide (**26**). This compound reacts rapidly with nucleophiles to give substitution products, but with rates that are first

Table 8.5 Rate constants for acetolysis of 2-substituted cyclohexyl *p*-bromobenzenesulfonates.

Substituent	$k_{(rel)}$	$\Delta H^{\ddagger}$ (kcal/mol)	$\Delta S^{\ddagger}$ (e.u.)
H	1.00	27.0	+1.5
trans-OAc	0.24	26.0	−4.2
cis-OAc	3.8×10^{-4}	30.9	−3.5
trans-Br	0.1	28.4	+0.8
trans-OCH_3	0.06	27.3	−3.4
trans-Cl	4.6×10^{-4}	33.0	+2.7

(Data from reference 60.)

[57]The term is derived from the Greek *anchi* and *meros*, meaning "neighboring parts:" Winstein, S.; Lindegren, C. R.; Marshall, H.; Ingraham, L. L. *J. Am. Chem. Soc.* **1953**, *75*, 147.

[58]For a discussion of terminology used in discussions of these assisted reactions, see Bartlett, P. D. *Nonclassical ions*; W. A. Benjamin, Inc.: New York, 1965; p. 65.

[59]The interacting groups must be spatially close but need not be adjacent along a carbon skeleton. For a report of a neighboring group interaction involving functional groups that interact through a 17-membered ring that arises due to medium-induced coiling of a linear molecule, see Jiang, X.-K.; Fan, W.-Q.; Hui, Y.-Z. *J. Am. Chem. Soc.* **1984**, *106*, 7202.

[60]Winstein, S.; Grunwald, E.; Ingraham, L. L. *J. Am. Chem. Soc.* **1948**, *70*, 821.

Figure 8.24 Anchimeric assistance in reaction of mustard gas.

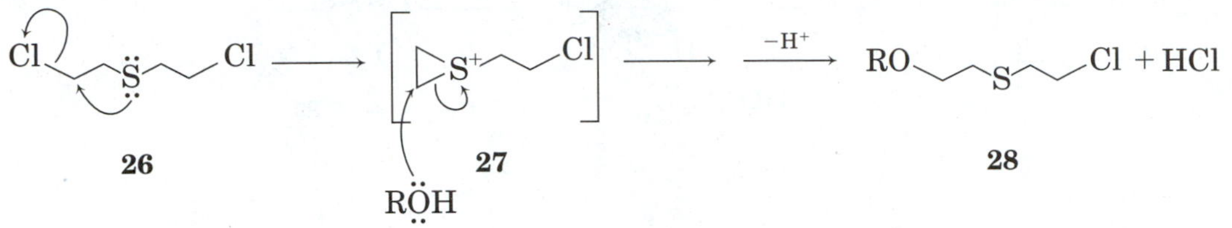

order in **26** and independent of added nucleophile.[61] Thus the kinetics are consistent with the intramolecular formation of the intermediate **27**, which reacts with nucleophiles to give substitution products as shown in Figure 8.24. Alkoxy, hydroxy, or amino groups are rapidly alkylated by **27**, and HCl is liberated as a byproduct, making **26** a powerful vesicant and irritant. In World War I, **26** was used as a poison gas and is still known as mustard gas.

The intramolecular participation of the sulfur in reactions of **26** involves the formation of a three-membered ring in the intermediate. A study of the analogous reaction of nitrogen compounds to form cyclic amines (equation 8.22) led to the conclusion that three-membered ring formation occurs with a greater rate constant than does formation of a four-membered ring (Table 8.6). However, the rate constant for formation of five-membered rings is fastest of all of the compounds studied.[62]

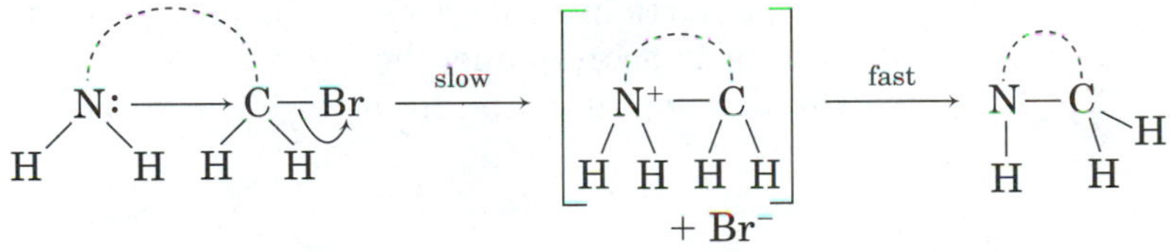

(8.22)

Table 8.6 Rate constants for cyclization of ω-aminoalkyl bromides.

Substrate	Rate constant (min^{-1})	Rel. Rate
Br-$(CH_2)_2$-NH_2	3.6×10^{-2}	0.072
Br-$(CH_2)_3$-NH_2	5.0×10^{-4}	0.001
Br-$(CH_2)_4$-NH_2	ca. 30	60
Br-$(CH_2)_5$-NH_2	0.5	1
Br-$(CH_2)_6$-NH_2	1.0×10^{-3}	0.002

(Data for reaction in water at 25°, reported in reference 62.)

[61] Bartlett, P. D.; Swain, C. G. *J. Am. Chem. Soc.* **1949**, *71*, 1406.

[62] Freundlich, H.; Kroepelin, H. *Z. Phys. Chem.* **1926**, *122*, 39.

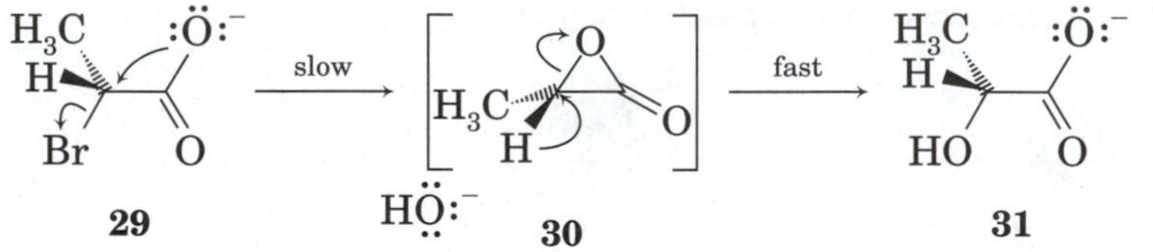

Figure 8.25 Retention of configuration in solvolysis reactions via anchimeric assistance.

The facility of three-membered ring formation is an important factor in other solvolytic reactions. Reaction of α-bromocarboxylates (e.g., **29**) with nucleophiles proceeds with retention of configuration due to the double inversion resulting from two steps: (1) ionization assisted by participation of the carboxylate group to form an α-lactone (**30**), followed by (2) fast attack of a nucleophile to generate a product with the same configuration as the reactant (**31**).[63]

Anchimeric assistance does not require heteroatoms with nonbonded electrons; electrons associated with π bonds on carbon atoms may also accelerate first order substitution reactions. Acetolysis of 2,2,2-triphenylethyl tosylate (**32**) gives products suggesting that rearrangement has accompanied the ionization. Moreover, the reaction takes place 7×10^3 faster (Table 8.7) than with 2,2-dimethyl-2-phenylpropyl tosylate (**33**), even though inductively withdrawing phenyl groups in **32** would be expected to retard a reaction that forms a carbocation.[64] Correcting for the expected inductive effect of the phenyl groups means that **32** reacts 10^7 times as fast as a compound with no participation by one of the phenyl groups.[65] The data suggest that ionization is accompanied by concurrent migration of a phenyl group toward the developing carbocation to form an intermediate **phenonium**

Table 8.7 Acetolysis rate constants and activation parameters for primary tosylates.

Substrate	k (sec^{-1})	$\Delta H^{\ddagger}$ (kcal/mol)	$\Delta S^{\ddagger}$ (e.u.)
2,2-dimethylpropyl tosylate	2.17×10^{-9}	31.5	−1.0
2-methyl-2-phenylpropyl tosylate	9.92×10^{-7}	25.7	−6.4
2,2,2-triphenylethyl tosylate	1.68×10^{-5}	25.2	−2.5

(Data from reference 64.)

[63]Grunwald, E.; Winstein, S. *J. Am. Chem. Soc.* **1948**, *70*, 841 and references therein.

[64]Winstein, S.; Morse, B. K.; Grunwald, E.; Schreiber, K. C.; Corse, J. *J. Am. Chem. Soc.* **1952**, *74*, 1113.

[65]Reference 64. See also Charlton, J. C.; Dostrovsky, I.; Hughes, E. D. *Nature* **1951**, *167*, 986; Brown, F.; Hughes, E. D.; Ingold, C. K.; Smith, J. F. *Nature* **1951**, *168*, 65.

Figure 8.26 Comparison of reactivity of 2,2,2-triphenylethyl tosylate and 2,2-dimethylpropyl tosylate.

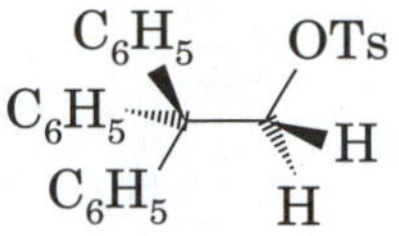

2,2,2-triphenylethyl tosylate (reactivity = 7,700)

32

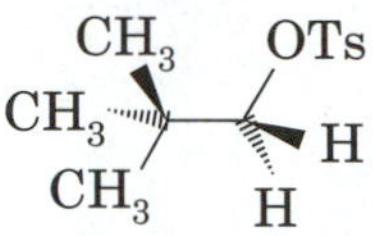

2,2-dimethylpropyl tosylate (reactivity = 1.0)

33

ion, which can be represented by the resonance-stabilized structure **34**.[66,67,68]

Stereochemical evidence for the participation of β-aromatic rings was provided by Cram, who found the acetolysis of D-*erythro*-3-phenyl-2-butyl tosylate (**35**) to proceed with complete retention of configuration. However, acetolysis of the optically active threo isomer (**38**) gave racemic product.[69] The results can most easily be rationalized by the mechanisms shown in Figure 8.28.[70]

Figure 8.27 Resonance structures for a phenonium ion generated by participation of a phenyl group in a solvolysis reaction.

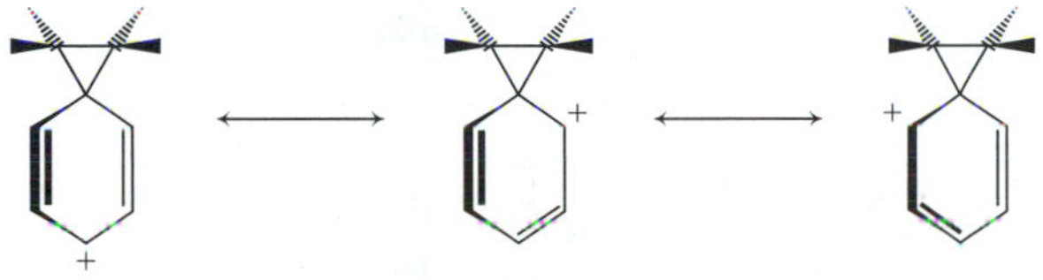

34

[66]For a theoretical study of the phenonium ion, see Sieber, S; Schleyer, P. v. R.; Gauss, J. *J. Am. Chem. Soc.* **1993**, *115*, 6987.

[67]To view the process differently, the carbocation character that develops as the leaving group starts to depart leads to attachment of the incipient carbocation to the aromatic ring in what is essentially the first step in an electrophilic aromatic substitution.

[68]Olah and co-workers used NMR spectroscopy to study the relative stability of phenonium ions, benzylic ions, and phenethyl cations as a function of substituents on the benzene ring. For the unsubstituted benzene ring, the phenonium ion was more stable than the other two possible structures. With a *p*-methoxy substituent, however, the benzylic ion was most stable, while the phenethyl cation was most stable for a *p*-trifluoromethyl substituent. Olah, G. A.; Comisarow, M. B.; Kim, C. J. *J. Am. Chem. Soc.* **1969**, *91*, 1458. Also see Olah, G. A.; Head, N. J.; Rasul, G.; Prakash, G. K. S. *J. Am. Chem. Soc.* **1995**, *117*, 875.

[69]Note that the intermediate phenonium ion formed from the *threo* reactant is a *meso* structure. Optical activity has already been lost at this stage, so the product cannot be optically active.

[70]Cram, D. J. *J. Am. Chem. Soc.* **1949**, *71*, 3863. For a more detailed discussion of this and related papers, see Gould, E. S. *Mechanism and Structure in Organic Chemistry*; Holt, Rinehart and Winston: New York, 1959; pp. 575 *ff.*

Figure 8.28 Stereochemical evidence for phenonium intermediates.

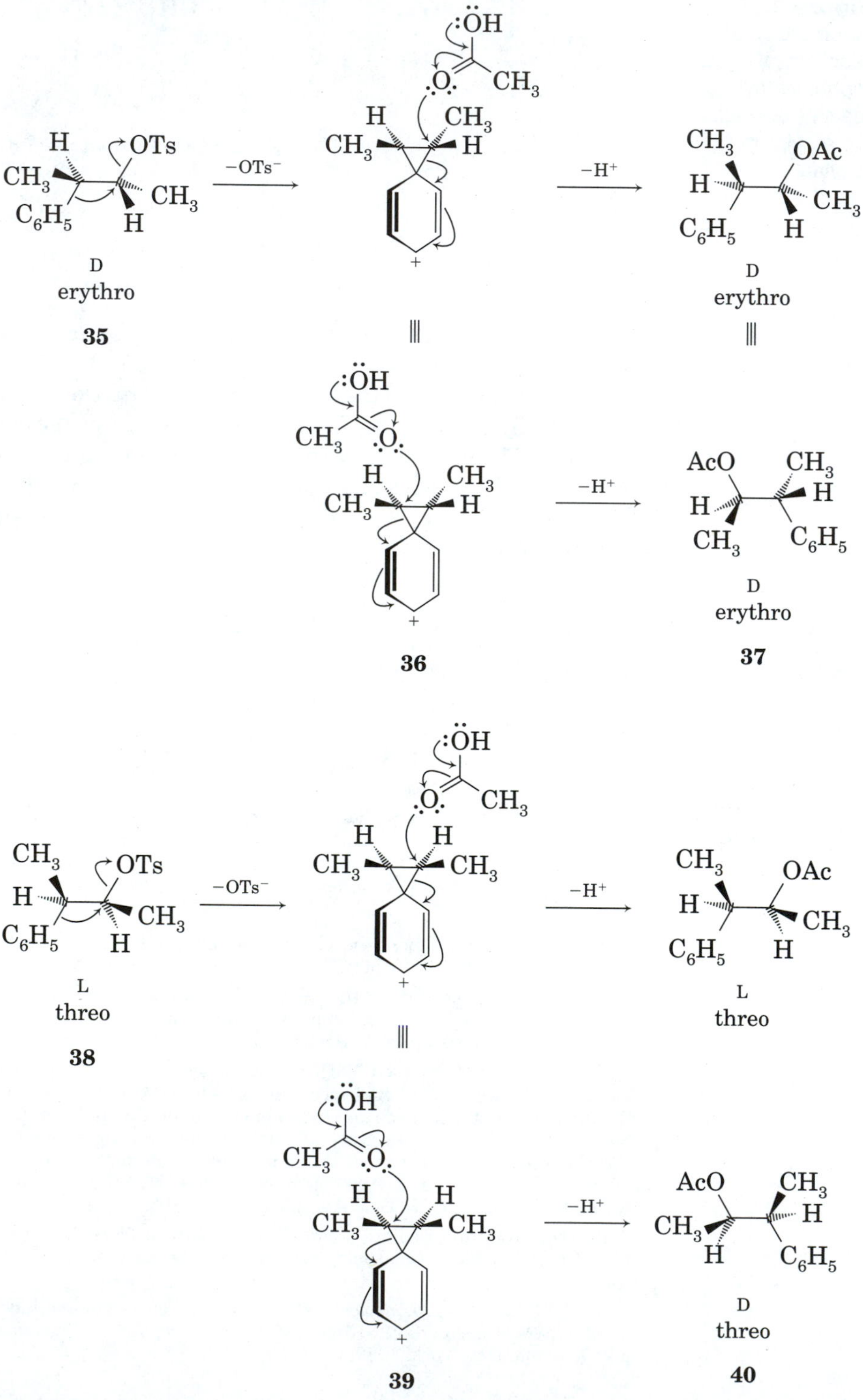

$-OTs^-$ $\quad$ $-H^+$

D erythro **35**

36

D erythro **37**

$-OTs^-$ $\quad$ $-H^+$

L threo **38**

L threo

39

D threo **40**

Nonclassical Carbocations in S_N1 Reactions

A carbon-carbon double bond can also participate in a solvolysis reaction. Acetolysis of cholesteryl tosylate (**41**) gave the β-acetoxy product (**43**), plus some rearrangement product (**44**) with a rate around 100 times greater than the rate of acetolysis of cyclohexyl tosylate.[71] The results were consistent with anchimeric assistance by the double bond concurrent with ionization to give a delocalized carbocation intermediate (shown as the delocalized structure **42** and as the hybrid of the resonance structures, **42**(a) and **42**(b). Nucleophilic attack on **42** by methanol can yield two different products, **43** and **44**.

There is also evidence that σ bonds can participate in the formation of nonclassical carbocation intermediates in solvolysis reactions. In the Wagner-Meerwein rearrangement, camphene hydrochloride (**45**) rearranges to isobornyl chloride (**46**).[72,73] It is possible to write a mechanism in which only classical (carbenium) ions are intermediates in the reaction, as shown in Figure 8.30. However, the major product of the rearrangement is the exo isomer of bornyl chloride (**46**). Very little of the endo isomer (**47**) of the product is seen, even though **47** might be expected to form faster than **46** if the reaction involves a free carbocation because the pathway leading to **47** would not require chloride ion to attack the carbocation near the methyl group on the bridge.

Winstein proposed that the intermediate in the solvolysis of 2-norbornyl (bicyclo[2.2.1]-heptyl) systems is one in which a σ bond participated in the solvolysis reaction to give the nonclassical carbocation **48** (Figure 8.31). As shown in Figure 8.32, nucleophilic attack on the back side of either partial carbon-carbon bond leads to the formation of exo product. Moreover, the presence of the plane of symmetry (passing through carbon atoms 4, 5, and 6) in **48** means that the product must be racemic. As a means of formalizing

---

[71]Winstein, S.; Adams, R. *J. Am. Chem. Soc.* **1948**, *70*, 838. Qualitatively similar results had been reported by Shoppee, C. W. *J. Chem. Soc.* **1946**, 1147 for the methanolysis of cholesteryl tosylate.

[72]The camphene hydrochloride to isobornyl chloride is the prototypical Wagner-Meerwein rearrangement, but the term Wagner-Meerwein rearrangement is a general term for 1,2-alkyl and 1,2-hydride shifts in carbocation rearrangements. Because of the contributions of Whitmore in elucidating these shifts, they are sometimes called Wagner-Meerwein-Whitmore rearrangements. *Cf.* reference 5, p. 414. For a review, see Pocker, Y. in *Molecular Rearrangements*, Part 1, de Mayo, P., ed.; Wiley-Interscience: New York, 1963, 1; Berson, J. A. *ibid.* p. 111.

[73]Isotopic labeling experiments by Roberts and Lee not only confirmed the rearrangement shown in Figure 8.30, but also revealed even more extensive rearrangements involving hydride shifts in this system. Roberts, J. D.; Lee, C. C. *J. Am. Chem. Soc.* **1951**, *73*, 5009; Roberts, J. D.; Lee, C. C.; Saunders, Jr., W. H. *J. Am. Chem. Soc.* **1954**, *76*, 4501. See also the discussion by Bartlett, P. D. (reference 58), pp. 65–67 and references therein.

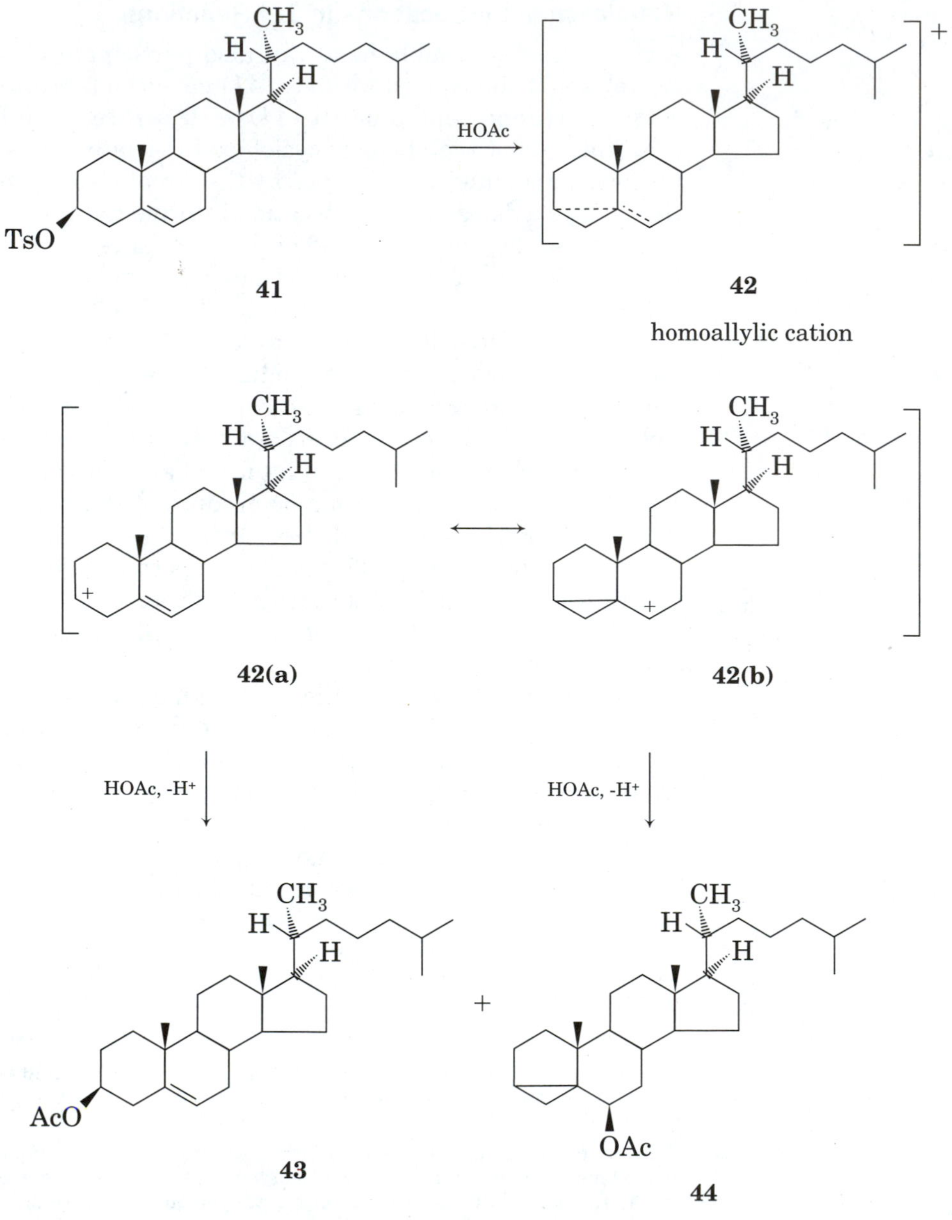

Figure 8.29
Acetolysis of cholesteryl tosylate.

the information in this area, Winstein and others proposed the concepts of ***homoconjugation*** (conjugation involving orbitals on atoms not formally σ-bonded to each other) and of ***homoaromaticity*** (aromaticity achieved through homoconjugation). For example, the intermediate (**42**) in the solvol-

Figure 8.30 Wagner-Meerwein rearrangement.

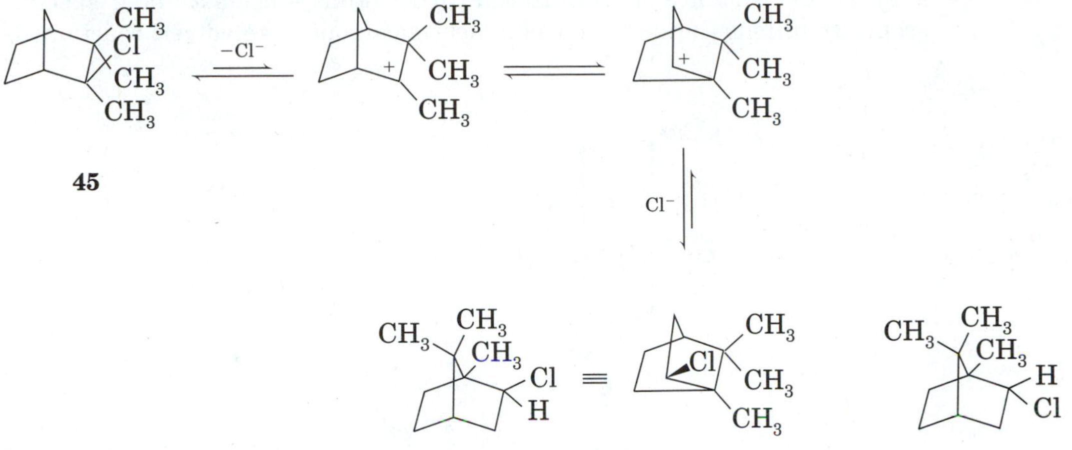

ysis of cholesteryl tosylate would be *homoallylic* since it is allylic with one overlap involving atoms not σ-bonded to each other. Similarly, the norbornyl cation (**48**) would be *bishomocyclopropenyl*, since it is like cyclopropenyl except for two missing σ bonds.[74]

Although there were many proponents of nonclassical ions and anchimeric assistance in solvolysis reactions, there was also resistance to these ideas. In particular, Brown argued that all of the kinetic and stereochemical results cited in support of nonclassical ions could be explained in terms of classical carbocations.[75,76] In the case of the solvolysis of the 2-norbornyl derivatives, he pointed out that the apparent rate acceleration might be the result of steric acceleration of ionization (see Table 8.1), not anchimeric assistance. He suggested that the formation of exclusively exo product from solvolysis of exo 2-norbornyl compounds could result from a higher transition state energy for endo product formation than for exo product formation, and he ascribed the racemization accompanying such solvolyses to the intermediacy of **rapidly equilibrating classical carbocations**

[74]The concept of homoconjugation and homoaromaticity was a topic of great interest in physical organic chemistry, and Winstein summarized much of the work in this area: Winstein, S. *Quart. Rev. Chem. Soc. (London)* **1969**, *23*, 141.

[75]Brown, H. C. *Chem. Soc. (London) Spec. Publ.* **1962**, *16*, 140. This paper is reproduced in reference 58 (pp. 438 *ff.*). See also the commentary by Bartlett, pp. 461 *ff.* in that reference.

[76]See also Brown, H. C. with commentary by Schleyer, P. v. R. *The Nonclassical Ion Problem*; Plenum: New York, 1977.

as shown in Figure 8.33.[77] As noted in Chapter 5, several lines of evidence support the existence of nonclassical carbocations—at least under spectroscopic conditions—and the role of nonclassical ions in solvolysis reactions is

Figure 8.31 Formation of a nonclassical carbocation in solvolysis of a 2-norbornyl derivative.

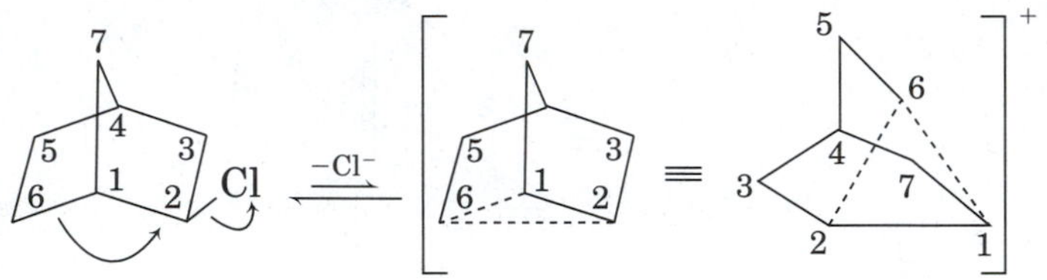

48

Figure 8.32 Formation of racemic product from solvolysis of a 2-norbornyl compound.

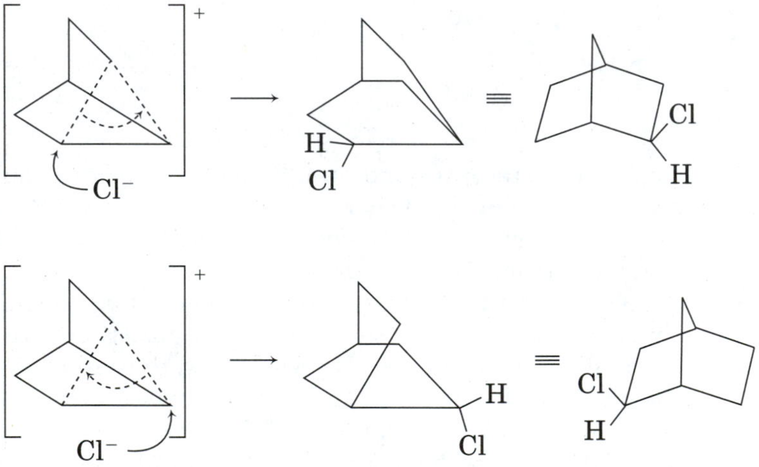

Figure 8.33 Rapidly equilibrating classical carbocation model for the norbornyl cation.

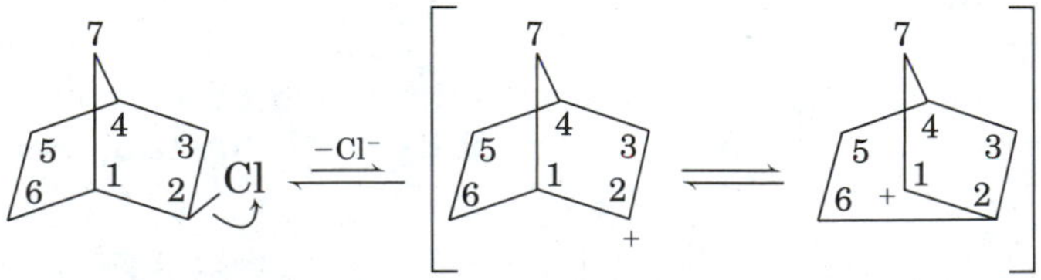

rapidly equilibrating classical carbocations

[77]For a discussion of the arguments against "nonclassical" ions by H. C. Brown and a commentary offering the opposing view, see reference 76.

[78]Eliel, E. L. *Stereochemistry of Carbon Compounds*; McGraw-Hill: New York 1962; p. 116.

widely accepted. Nevertheless, the discussions of classical and nonclassical carbocations in solvolysis reactions prompted physical organic chemists to re-examine fundamental concepts of structure and reactivity.[58]

The S_N2 Reaction

Stereochemistry

Prior to 1895, it was thought that all substitution reactions occur with retention of configuration.[78] In that year, however, Walden reported that inversion of configuration accompanied a substitution reaction.[79] As shown in Figure 8.34, conversion of (−)-malic acid to (+)-malic acid by two substitution reactions indicates that one of the two steps must take place with retention of configuration and the other must take place with inversion. Later, racemization of optically active 2-iodooctane was found to occur at twice the rate of incorporation of radioactive iodine into the structure.[80] The simplest mechanism to explain this result is the reaction pathway shown in Figure 8.35, which is the familiar S_N2 back-side displacement mechanism.

An alternative bimolecular process that would give retention of configuration is the front-side attack mechanism shown in Figure 8.36. One explanation for the failure to observe retention of configuration in S_N2 reactions is that a front-side attack would give a transition structure with two partial

Figure 8.34
The Walden cycle.

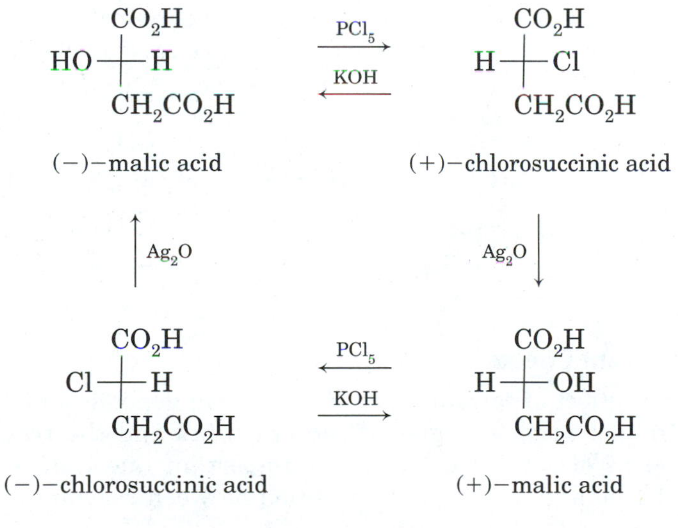

CO_2H / HO—C—H / CH_2CO_2H — (−)−malic acid

PCl_5 → ; ← KOH

CO_2H / H—C—Cl / CH_2CO_2H — (+)−chlorosuccinic acid

Ag_2O (↓)

CO_2H / H—C—OH / CH_2CO_2H — (+)−malic acid

← PCl_5 ; KOH →

CO_2H / Cl—C—H / CH_2CO_2H — (−)−chlorosuccinic acid

Ag_2O (↑)

[79] Walden, P. *Ber. dtsch. chem. Ges.* **1895**, *28*, 1287; *Ber. dtsch. chem. Ges.* **1897**, *30*, 3146.

[80] Hughes, E. D.; Juliusburger, F.; Masterman, S.; Topley, B.; Weiss, J. *J. Chem. Soc.* **1935**, 1525. The rate of racemization was *twice* the rate of inversion. Because inversion both removes one stereoisomer and produces its enantiomer, the rate of radioactive iodine incorporation was presumed to equal the rate of inversion.

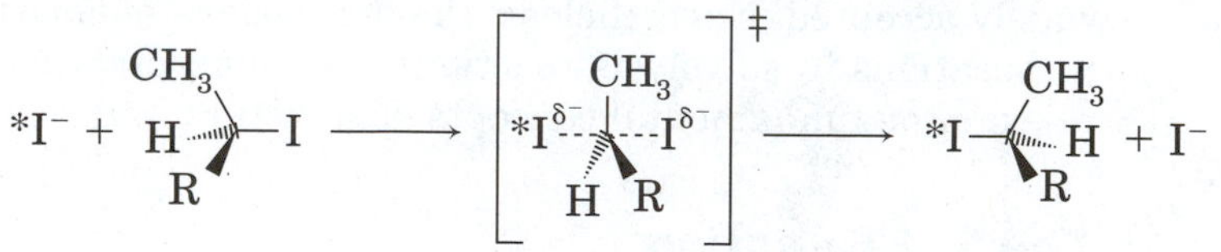

Figure 8.35 Stereochemistry of the back-side pathway in the S_N2 reaction.

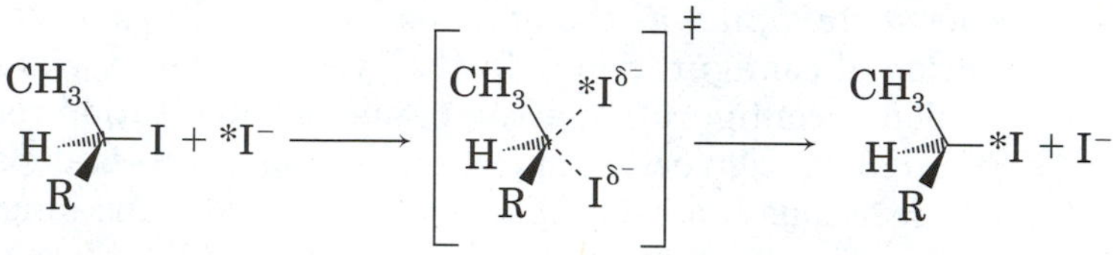

Figure 8.36 Stereochemistry of the hypothetical front-side attack pathway for the S_N2 reaction.

negative charges close to each other, and that geometry should be much less stable than a transition structure with the charges further apart, which is the case with back-side displacement. However, S_N2 reactions with negatively charged nucleophiles and positively charged leaving groups also react by back-side displacement, even though the front-side attack would appear to be more stable due to positive-negative attraction. For example, the reaction of L-(−)-dimethyl-α-phenylethylsulfonium ion (**49**) with bromide ion gives D-(+)-α-phenylethyl bromide (**50**, equation 8.23).[81,82] Therefore it appears that back-side attack generates a transition structure with lower electronic energy than does front-side attack, and theoretical studies utilizing both semi-empirical and *ab initio* calculations have confirmed this explanation.[83]

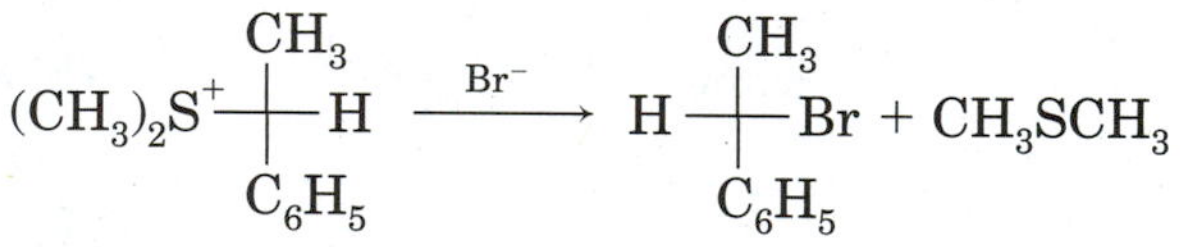

$$(CH_3)_2S^+\text{–}C(CH_3)(H)(C_6H_5)\ \xrightarrow{Br^-}\ H\text{–}C(CH_3)(C_6H_5)\text{–}Br + CH_3SCH_3 \quad (8.23)$$

49   **50**

Solvent Effects

The effect of solvent on an S_N2 reaction depends on the difference between the solvation energies of the reactants and the transition structure. In terms of solvent polarity, the important question is the change in total charge and in charge distribution between reactants and transition struc-

[81] Siegel, S.; Graeffe, A. F. *J. Am. Chem. Soc.* **1953**, *75*, 4521.

[82] Cowdrey, W. A.; Hughes, E. D.; Ingold, C. K.; Masterman, S.; Scott, A. D. *J. Chem. Soc.* **1937**, 1252.

[83] (a) Allinger, N. L.; Tai, J. C.; Wu, F. T. *J. Am. Chem. Soc.* **1970**, *92*, 579; (b) Ritchie, C. D.; Chappell, G. A. *J. Am. Chem. Soc.* **1970**, *92*, 1819; (c) Dedieu, A. Veillard, A. *J. Am. Chem. Soc.* **1972**, *94*, 6730; *Quantum Theory Chem. React.* **1980**, *1*, 69. Also see Hu, W.-P.; Truhlar, D. G. *J. Am. Chem. Soc.* **1994**, *116*, 7797.

tures. Ingold identified four different patterns that might be observed in S_N2 reactions:[84]

1. Negative nucleophile and neutral substrate, such as the substitution of iodide for chloride in methyl chloride.

$$I^- + CH_3Cl \longrightarrow \left[\overset{\delta-}{I} \cdots CH_3 \cdots \overset{\delta-}{Cl} \right]^{\ddagger} \longrightarrow ICH_3 + Cl^- \quad (8.24)$$

2. Neutral nucleophile and neutral substrate, such as the reaction of trimethylamine with methyl chloride.

$$(CH_3)_3N + CH_3Cl \longrightarrow \left[(CH_3)_3\overset{\delta+}{N} \cdots CH_3 \cdots \overset{\delta-}{Cl} \right]^{\ddagger} \longrightarrow (CH_3)_3N^+CH_3 + Cl^- \quad (8.25)$$

3. Negative nucleophile and positive substrate, such as the reaction of hydroxide ion with tetramethylammonium ion.

$$HO^- + CH_3{-}N^+(CH_3)_3 \longrightarrow \left[\overset{\delta-}{HO} \cdots CH_3 \cdots \overset{\delta+}{N}(CH_3)_3 \right]^{\ddagger} \longrightarrow HOCH_3 + N(CH_3)_3 \quad (8.26)$$

4. Neutral nucleophile and positive substrate, such as the reaction of trimethylamine with trimethylsulfonium to produce tetramethylammonium ion and dimethyl sulfide.

$$(CH_3)_3N + CH_3{-}S^+(CH_3)_2 \longrightarrow \left[(CH_3)_3\overset{\delta+}{N} \cdots CH_3 \cdots \overset{\delta+}{S}(CH_3)_2 \right]^{\ddagger} \longrightarrow (CH_3)_4N^+ + CH_3SCH_3 \quad (8.27)$$

Let us illustrate the relationship between reaction type and solvent polarity effects by considering the first two types of reactions. Because all of the charge in a type (i) reaction is initially localized on the nucleophile, there is strong solvation of the nucleophile by a polar solvent and only much weaker solvation of the neutral substrate. The negative charge is dispersed in the transition structure, with some charge on the nucleophile and some on the leaving group, so the solvation energy of the transition structure is diminished because the charge is less localized.[85] Therefore, the activation energy for the reaction is greater in a more polar solvent than in a less polar solvent, and the rate of the reaction decreases with increasing solvent polarity, as is indicated by the data in Table 8.8.

The data in Table 8.8 also indicate that faster rates of reaction are associated with ***aprotic*** (that is, nonhydrogen bonding) solvents that do not strongly solvate anions. For example, the reaction of methyl iodide with

---

[84]Reference 4, pp. 457 *ff.*

[85]For the reaction of chloride ion with methyl chloride in water, Jorgenson has calculated that the heats of hydration of the transition structure and reactants differ by 22 kcal/mol. Chandrasekhar, J.; Smith, S. F.; Jorgensen, W. L. *J. Am. Chem. Soc.* ***1985,*** 107, *154.*

Table 8.8 Rate data for S_N2 reaction of CH_3I with Cl^-.

Solvent	*k* (rel.)
CH_3OH	0.9
H_2O	1.0
$HCONH_2$	14.1
CH_3NO_2	14,100
CH_3CN	35,800
DMF	708,000
Acetone	1,410,000

(Data adapted from tabulation by Parker, reference 86.)

chloride ion increases by a factor of 1.5×10^6 as the solvent is changed from methanol to acetone.[86] Moreover, changing from a protic solvent to acetone reverses the order of nucleophilicity of halide ions[87] ($I^- > Br^- > Cl^-$ in protic solvent), indicating that the energy required to remove hydrogen-bonded solvent molecules from the smaller ions is a significant factor in decreasing nucleophilicity.

Theoretical calculations are also informative in interpreting solvent effects in type (i) S_N2 reactions. Figure 8.37 shows the results obtained with Monte Carlo statistical mechanics calculations by Chandrasekhar and Jorgensen for the reaction of chloride ion with methyl chloride in the gas phase, in dimethylformamide (DMF) and in water. It is interesting that in the gas phase there is an energy minimum both before and after the energy maximum, the minima corresponding to a $[Cl \cdots CH_3 \cdots Cl]^-$ complex. (A simple way to think of the complex is that the polar methyl chloride molecule can partially solvate the chloride ion, since no other solvent is available.) The actual reaction barrier is only slightly higher than the initial energy of the reactants. In DMF solution the complex is less evident and the reaction barrier is higher. In aqueous solution the complexes are not apparent, and the reaction barrier is much higher.[88] Chandrasekhar and Jorgensen calculated the activation energies in water and DMF to be 26.3 and 19.3 kcal/mol, respectively, for the reaction of chloride ion with methyl chloride.

Equation 8.24 explicitly shows only the anionic nucleophile since the counterion does not appear to take part in the reaction. Nevertheless, the counterion can determine the solubility of the nucleophilic salt, and solubil-

---

[86]Parker, A. J. *Chem. Rev.* **1969**, *69*, 1.

[87]Winstein, S.; Savedoff, L. G.; Smith, S.; Stevens, I. D. R.; Gall, J. S. *Tetrahedron Lett.* **1960**, *9*, 24.

[88]Chandrasekhar, J.; Jorgensen, W. L. *J. Am. Chem. Soc.* **1985**, *107*, 2974. The calculation for DMF involved the calculation of the energy of the two reactants plus 180 molecules of DMF.

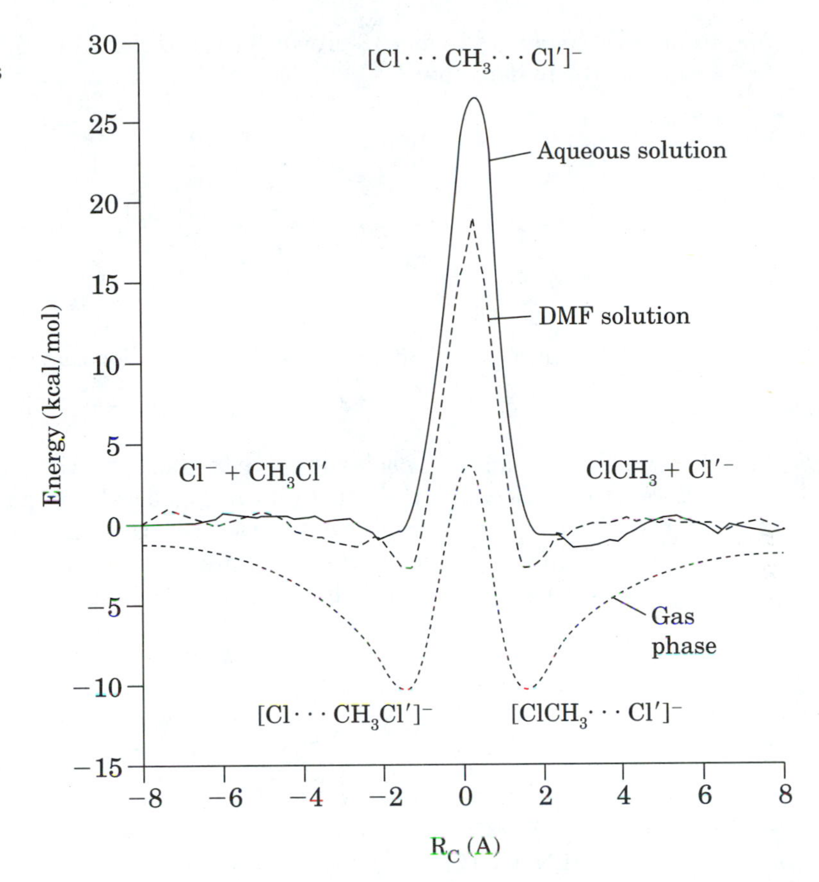

Figure 8.37 Calculated energies for reaction of Cl^- with CH_3Cl. (Reproduced from reference 88.)

ity considerations can influence the polarity of the solvent needed for the reaction. An alternative to the use of a more polar solvent to dissolve a salt for a nucleophilic substitution is to use crown ether additives. Crown ethers are cyclic polyethers that can coordinate with and therefore increase the solubility of cations in organic solvents. However, the nucleophilic anion is not tightly solvated. Thus, the activation energy for substitution does not include a large term for desolvation of the nucleophile, and reaction rates are fast.[89] For example, adding dicyclohexano-18-crown-6 (**53**) to a solution of 1-bromobutane in dioxane was found to increase its rate of reaction with potassium phenoxide by a factor of 1.5×10^4.[90] Liotta and Harris were able

[89]Similarly, S_N2 reactions can be carried out in other solvents if appropriate phase transfer agents (highly soluble organic cations) are used to carry the "naked" nucleophile into solution. For example, Carpino and Sau used tetra-*n*-butylammonium chloride to solubilize KF in acetonitrile for such a reaction. Carpino, L. A.; Sau, A. C. *J. Chem. Soc. Chem. Commun.* **1979**, 514.

[90]Thomassen, L. M.; Ellingsen, T.; Ugelstad, J. *Acta Chem. Scand.* **1971**, *25*, 3024. See also the discussion in de Jong, F., Reinhoudt, D. N. *Stability and Reactivity of Crown-Ether Complexes*; Academic Press: New York, 1981; pp. 35 *ff.*

to use KF solubilized with 18-crown-6 (**52**) to carry out S_N2 reactions on 1-bromooctane in benzene.[91]

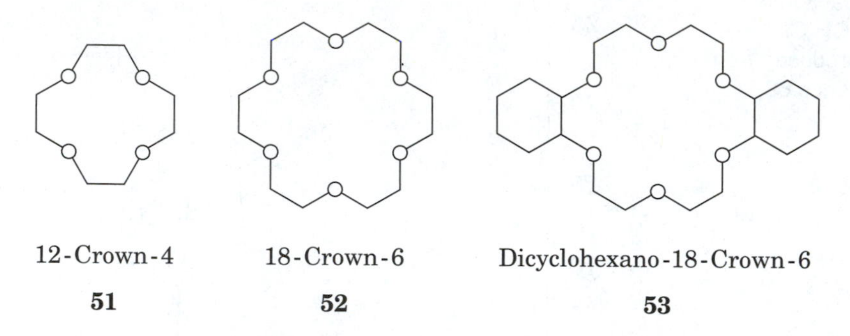

The effect of changing solvent polarity is exactly the opposite in type (ii) S_N2 reactions. Because the reactants are neutral but partial charges develop in the transition structure, type (ii) reactions are facilitated by polar solvent.[92] Figure 8.38 shows the calculated free energy for the reaction of ammonia with methyl chloride in the gas phase and in aqueous solution.[93,94]

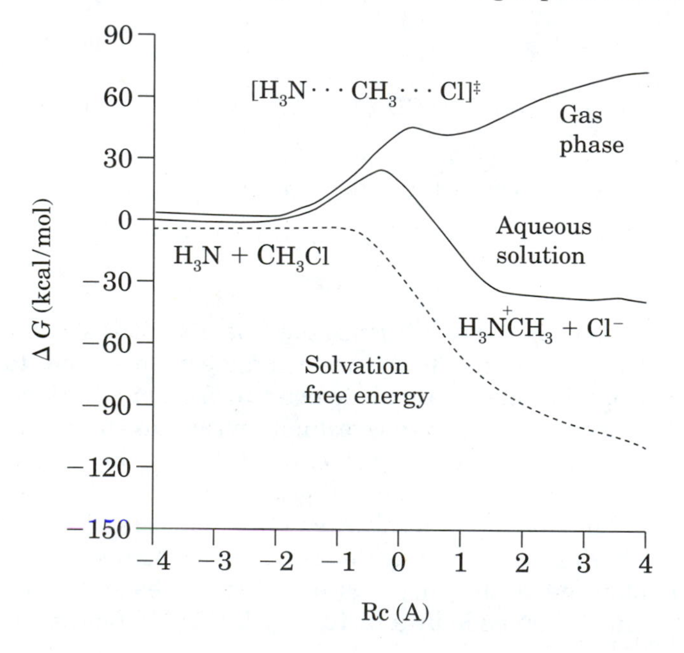

Figure 8.38 Free energy profiles calculated for reaction of ammonia with methyl chloride. (Reproduced from reference 93.)

---

[91]Liotta, C. L.; Harris, H. P. *J. Am. Chem. Soc.* **1974**, *96*, 2250.

[92]For examples, see, Cox, H. E. *J. Chem. Soc.* **1921**, *119*, 142; McCombie, H.; Scarborough, H. A.; Smith, F. F. P. *J. Chem. Soc.* **1927**, 802.

[93]Gao, J. *J. Am. Chem. Soc.* **1991**, *113*, 7796. See also Gao, J.; Xia, X. *J. Am. Chem. Soc.* **1993**, *115*, 9667.

[94]This reaction is a Menshutkin—also spelled Menschutkin—reaction, which is generally described as the S_N2 alkylation of a nitrogen nucleophile, usually by an alkyl halide. For a discussion, see Abboud, J.-L. M.; Notario, R.; Bertràn, J.; Solà, M. *Prog. Phys. Org. Chem.* **1993**, *19*, 1.

The top line shows that the energy increases throughout the reaction in gas phase reactions because there is no solvent to stabilize the developing ions. The middle curve shows the free energy for reaction in aqueous solution. The bottom, dotted curve shows the calculated free energy of solvation for the reaction. A solvent less polar than water should produce somewhat less stabilization, resulting in a curve between those for reaction in water and in the gas phase. Therefore, the activation energy for a type (ii) reaction increases with decreasing solvent polarity.

Substrate Effects

Varying the structure of the substrate can also influence the rate of an S_N2 reaction. As the substrate is varied from methyl to a 1° to a 2° to a 3° alkyl halide, there is greater steric hindrance for attack on the back side of the C—X bond, and the rate decreases. Table 8.9 summarizes similar data for alkyl bromides reacting with bromide.[95] The data show the predicted decrease in reactivity as the halide changes from methyl to 1° to 2° to 3°. It is notable that neopentyl bromide is less reactive than *t*-butyl bromide, even though neopentyl is formally a 1° alkyl halide. Space-filling models indicate that hydrogen atoms on β methyl groups can provide a steric barrier to reaction that is comparable to that of hydrogen atoms on α methyl groups.[95]

This explanation for rate reduction with increasing substitution of the reaction site might be interpreted to mean that the steric effect is primarily enthalpic, but that is not necessarily the case. Table 8.10 summarizes some data for the reaction of chloride with a series of alkyl iodides in acetone.[96] The E_a values for the first five entries in the table differ by only 2 kcal/mol, although that for neopentyl is several kcal/mol higher. Interestingly, however, the log A values are quite different among this series of compounds.

Table 8.9 Relative second order rate constants in acetone at 25° for $^{*}Br^{-}$ + R—Br → *Br—R + Br^{-}.

R	*k* (rel.)	E_a (kcal/mol)	log A
CH_3	76	15.8	10.7
CH_3CH_2	1.0	17.5	10.1
$CH_3CH_2CH_2$	0.65	17.5	9.8
$(CH_3)_2CH$	0.011	19.7	9.7
$(CH_3)_2CHCH_2$	0.033	18.9	9.6
$(CH_3)_3C$	0.003	21.8	10.7
$(CH_3)_3CCH_2$	0.000015	22.0	8.6

(Data from reference 95.)

[95] de la Mare, P. B. D. *J. Chem. Soc.* **1955**, 3180.

[96] Hughes, E. D.; Ingold, C. K.; Mackie, J. D. H. *J. Chem. Soc.* **1955**, 3177.

Table 8.10 Activation enthalpies and entropies for reaction of R—I with chloride in acetone.

R	*k* (rel.)	log *A*	E_a (kcal/mol)
CH_3	11.1	9.4	16.0
CH_3CH_2	1.0	9.1	17.0
$CH_3CH_2CH_2$	0.58	8.8	17.0
$(CH_3)_2CH$	0.032	8.3	18.0
$(CH_3)_2CHCH_2$	0.038	8.3	17.8
$(CH_3)_3CCH_2$	0.000014	7.9	22.0

(Data from reference 96.)

The log *A* values suggest that the activation entropy plays an important role in determining the rates of S_N2 reactions because there is a greater negative entropy of activation for a reaction in which many nuclei must move simultaneously on going from the reactant to the transition state than is the case for a smaller number of nuclei. This decrease in S_N2 reactivity has been called the ponderal effect, meaning that it depends on the mass of the substituents (and not on their space-filling properties) near the carbon atom undergoing substitution.[97]

Not all substituents decrease the rate of S_N2 reactions. Table 8.11 summarizes the rates of reaction of a series of alkyl chlorides with iodide ion in acetone.[98] The much greater reactivity of allyl chloride, benzyl chloride, chloroacetonitrile, and α-chloroacetone than of *n*-butyl bromide is attributed to the π system bonded to the carbon atom undergoing reaction. A sim-

Table 8.11 Relative reactivities of R—Cl with Iodide in acetone at 50°.

R	*k* (rel.)	R	*k* (rel.)
n-butyl	1.0	allyl	79
n-C_7H_{15}	1.30[98(b)]	benzyl	195[98(c)]
cyclohexyl	<0.0001[98(b)]	H_2NCOCH_2	99
$C_6H_5COOCH_2$	59.1[98(d)]	$NCCH_2$	3,070
$C_6H_5COCH_2$	105,000	EtO_2COCH_2	1,720

(Except as noted, data are from reference 98(a).)

[97] de la Mare, P. B. D.; Fowden, L.; Hughes, E. D.; Ingold, C. K.; Mackie, J. D. H. *J. Chem. Soc.* **1955**, 3200.

[98] (a) Conant, J. B.; Kinner, W. R.; Hussey, R. E. *J. Am. Chem. Soc.* **1925**, *47*, 488; (b) Conant, J. B.; Hussey, R. E. *J. Am. Chem. Soc.* **1925**, *47*, 476; (c) Conant, J. B.; Kirner, W. R. *J. Am. Chem. Soc.* **1924**, *46*, 232; (d) Kirner, W. R. *J. Am. Chem. Soc.* **1926**, *48*, 2745.

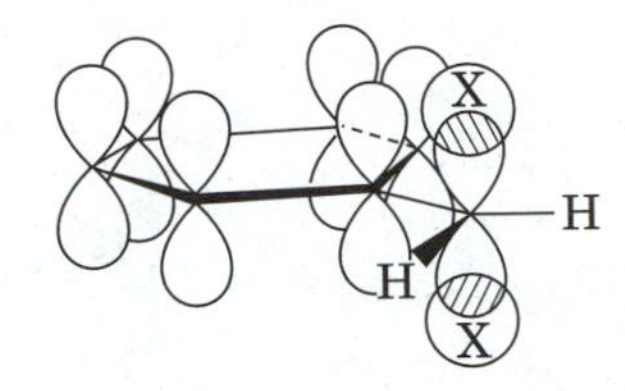

Figure 8.39 Transition state interactions with unsaturated substituents. (Reproduced from reference 99.)

ple model (Figure 8.39) suggests that the sp^2-hybridized carbon atom proposed for the transition structure can interact with the MOs of this π system, stabilizing the transition structure and increasing the rate of the reaction.[99]

The much greater reactivity of α-chloroacetone than of allyl chloride suggests that there is an additional rate accelerating factor in α-halo carbonyl compounds and similar structures. An early rationale for the reactivity of these compounds was that the partial positive charge on the carbonyl carbon atom accentuates the partial charge on the C—X carbon atom, thereby increasing its attraction for a nucleophile.[100] Another explanation suggested that compounds of this type could undergo nucleophilic attack on the carbonyl group, leading to a tetrahedral intermediate that subsequently rearranges to the substitution product (equation 8.28).[101] However, Pearson et al. summarized data inconsistent with either of these mechanisms. In addition, they ruled out a third possibility, that the tetrahedral intermediate formed by attachment of a nucleophile to a carbonyl group might form an epoxide, which could react with another nucleophile to form the substitution product (equation 8.29).[102,103] Instead, Pearson suggested that the increased reactivity of α-halo carbonyl compounds might result from the additional electrostatic attraction of the carbonyl dipole for the approaching nucleophile, a mechanism commonly called *dual attraction* (Figure 8.40).[104,105]

Figure 8.40 Dual attraction model for nucleophilic substitution.

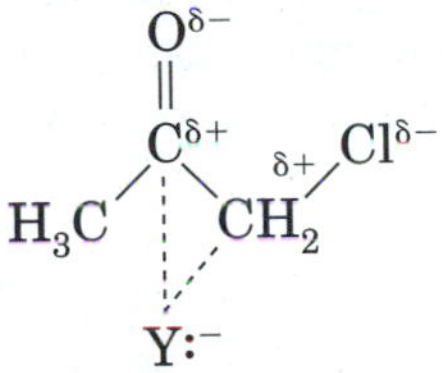

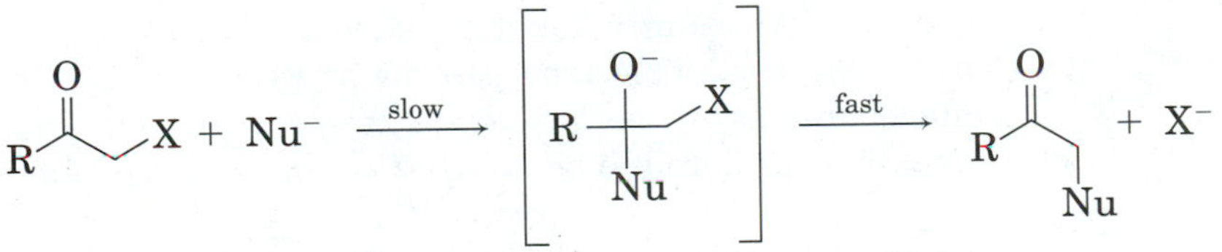

$O^{\delta-}$, $C^{\delta+}$, H_3C, $^{\delta+}CH_2$, $Cl^{\delta-}$, $Y:^-$

R–C(=O)–CH$_2$–X + Nu^- $\xrightarrow{\text{slow}}$ [R–C(O$^-$)(Nu)–CH$_2$–X] $\xrightarrow{\text{fast}}$ R–C(=O)–CH$_2$–Nu + X^- **(8.28)**

[99]King, J. F.; Tsang, G. T. Y.; Abdel-Malik, M. M.; Payne, N. C. *J. Am. Chem. Soc.* **1985**, *107*, 3224 and references therein.

[100]Hughes, E. D. *Trans. Faraday Soc.* **1941**, *37*, 603.

[101]Baker, J. W. *Trans. Faraday Soc.* **1941**, *37*, 632 (see especially the discussion beginning on page 643).

[102]Pearson, R. G.; Langer, S. H.; Williams, F. V.; McGuire, W. J. *J. Am. Chem. Soc.* **1952**, *74*, 5130.

[103]An expoxide was isolated and determined to be an intermediate in the hydrolysis of desyl chloride (2-chloro-2-phenylacetophenone) by Ward, A. M. *J. Chem. Soc.* **1929**, 1541.

[104]Eliel, E. L. in *Steric Effects in Organic Chemistry*, Newman, M. S., ed.; John Wiley & Sons, Inc.: New York, 1956; p. 105.

[105]For a summary and discussion of α- and β-substituent effects on the rates of S_N2 reactions, see Shaik, S. S. *J. Am. Chem. Soc.* **1983**, *105*, 4359. (The theoretical discussion there is based on a model that is introduced on page 506.)

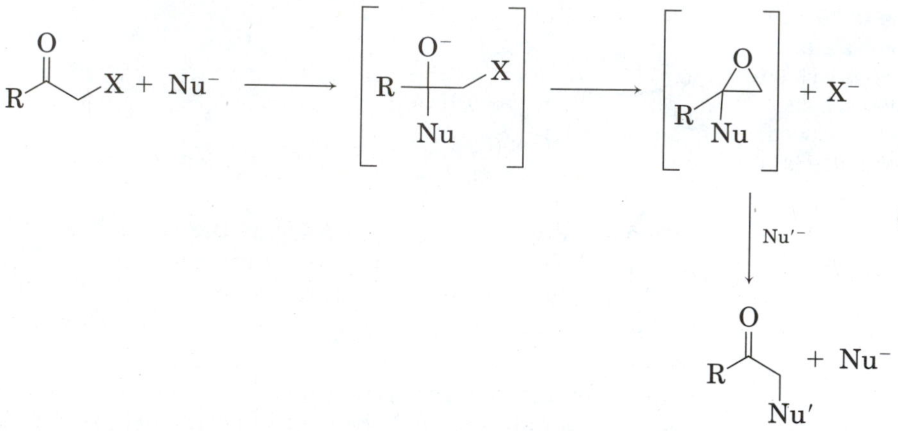

(8.29)

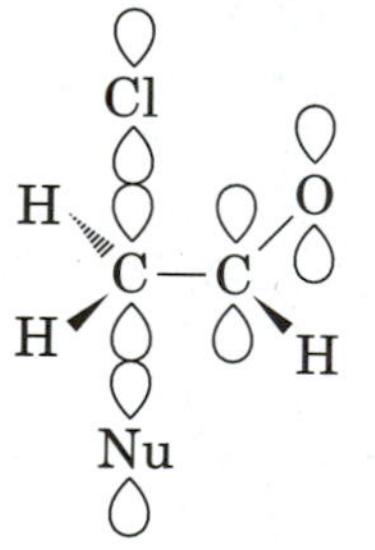

Figure 8.41
Basis set orbitals for the S_N2 reaction of chloroacetaldehyde with a nucleophile. (Adapted from reference 106.)

Bach and co-workers have presented a theoretical analysis that suggests that the enhanced reactivity of α-halocarbonyl compounds arises in part from overlap of the carbon *p* orbital with the π orbitals of the carbonyl group, just as with the benzyl system (Figure 8.39). Figure 8.41 shows a similar representation of the basis set orbitals for the transition structure for the S_N2 reaction of chloracetaldehyde with a nucleophile. The additional reactivity of the α-halocarbonyl system over that of a benzyl system appears to arise because the carbonyl group is more polarizable than is the benzene ring, so that the carbonyl group is better able to delocalize the negative charge in the transition structure.[106]

Quantitative Measures of Nucleophilicity

There is a fundamental relationship between nucleophilicity and basicity, because both require donation of electrons. Although one definition is that a nucleophile is an electron pair donor, i.e., a Lewis base,[107] we often consider nucleophilicity to be a *kinetic concept* (equation 8.30), whereas basicity is usually considered to be an *equilibrium concept* (equation 8.31).[108]

Nucleophilicity: $$Y:^- + R{-}X \xrightarrow{k_2} Y{-}R + X:^- \quad (8.30)$$

Basicity: $$Y:^- + H{-}X \overset{K}{\rightleftharpoons} Y{-}H + X:^- \quad (8.31)$$

Location of the nucleophilic atom in the periodic table is also important. Comparison of nucleophilicities of species in which the nucleophilic atoms bear the same charge and are in the same column of the periodic table gen-

[106] Bach, R. D.; Coddens, B. A.; Wolber, G. J. *J. Org. Chem.* **1986**, *51*, 1030.

[107] Reference 5, p. 283.

[108] This "feeling" is ascribed to Swain, C. G.; Scott, C. B. *J. Am. Chem. Soc.* **1953**, *75*, 141, but it was challenged by Edwards, J. O. *J. Am. Chem. Soc.* **1954**, *76*, 1540.

erally shows that nucleophilicity increases with increasing atomic number. (H_2S is a better nucleophile than is H_2O, for example). This result is consistent with a model in which the polarizability (ease of distortion) of the electron cloud increases as the size of the cloud increases. Greater polarizability is said to enable a nucleophile to form an incipient covalent bond without incurring repulsive electron-electron interactions with substituents near the site of the nucleophilic attack.[109] Furthermore, solvation forces are stronger for small nucleophilic ions than for large ones, especially in protic solvents. In protic solvents the order of nucleophilicities of halide ions is $I^- > Br^- > Cl^- > F^-$, consistent with these factors.[87] Sodium hydride, a strong base but a small, nonpolarizable nucleophile, is unreactive in S_N2 reactions.[110]

Although a qualitative description of nucleophilicity is useful, it is desirable to be able to predict quantitatively the rates of substitution reactions so that empirical data can be assimilated into a coherent framework and so that the best conditions for synthetic reactions can be chosen.[111] There have been many attempts to develop linear free energy relationships so that rates of substitution reactions could be quantitatively predicted and so that the effects of changing reaction conditions on reaction rates could be ascribed to particular aspects of the intermediates involved. However, this has not been easy to accomplish. As Pearson has noted

> While as many as 17 factors have been identified as influencing nucleophilic reactivity, there is still not general agreement on the factors that are most important, except for solvent effects. That is, we are still hard pressed to estimate the rate of an S_N2 reaction that involves a new combination of nucleophile and electrophile.[112]

Wells,[113] Ibne-Rasa,[114] and Harris and McManus[115] have summarized the development of various attempts to quantify nucleophilic reactivity. The most commonly considered approaches are those based on the Brønsted equation, the empirical formulas derived by Swain and Scott and by Edwards, and the hard-soft acid-base approach of Pearson.

Brønsted Correlations One of the earliest correlations of nucleophilicity with basicity was provided by Smith for the reaction of carboxylate ions ($Y^- = RCO_2^-$ or $ArCO_2^-$) with chloroacetate ion.

---

[109]See footnote 9, reference 102.

[110]Cristol, S. J.; Ragsdale, J. W.; Meek, J. S. *J. Am. Chem. Soc.* **1949**, *71*, 1863.

[111]With the advent of artificial intelligence and its application to problems in chemical synthesis, it may someday be possible for computers to design syntheses of interesting molecules. For a discussion of the application of computers to reactions of nucleophiles, see Metivier, P.; Gushurst, A. J.; Jorgensen, W. L. *J. Org. Chem.* **1987**, *52*, 3724.

[112]Pearson, R. G. *J. Org. Chem.* **1987**, *52*, 2131. See also the discussion in reference 37.

[113]Wells, P. R. *Chem. Rev.* **1963**, *63*, 171.

[114]Ibne-Rasa, K. M. *J. Chem. Educ.* **1967**, *44*, 89.

[115]Harris, J. M.; McManus, S. P. in *Nucleophilicity*, Harris, J. M.; McManus, S. P., eds.; American Chemical Society: Washington, D.C., 1987; pp. 1–20.

$$Y^- + ClCH_2CO_2^- \longrightarrow YCH_2CO_2^- + Cl^- \quad (8.32)$$

As shown in Figure 8.42, a linear correlation of log k versus pK_a is obtained for this series of nucleophiles.[116] However, basicity alone cannot be the basis of nucleophilicity, as evidenced by reactions in which the nucleophilic atom is different. For example, phenoxide is more than 10^3 times more basic than thiophenoxide, but thiophenoxide is more than 10^4 times more nucleophilic than phenoxide.[117]

Swain-Scott Equation Swain and Scott developed a linear free energy relationship for nucleophilicity (equation 8.33) in the same form as the Hammett equation.

$$\log (k_n/k_0) = sn \quad (8.33)$$

The term n is the nucleophilicity parameter for a given nucleophile, while s is the measure of the sensitivity of the substrate to the nucleophilicity of the attacking reagent. The rate constants k_n and k_0 are the rate constants for

Figure 8.42 Correlation of nucleophilic reactivity with basicity in hydrolysis of chloroacetate ion. (Figure drawn using data in reference 116.)

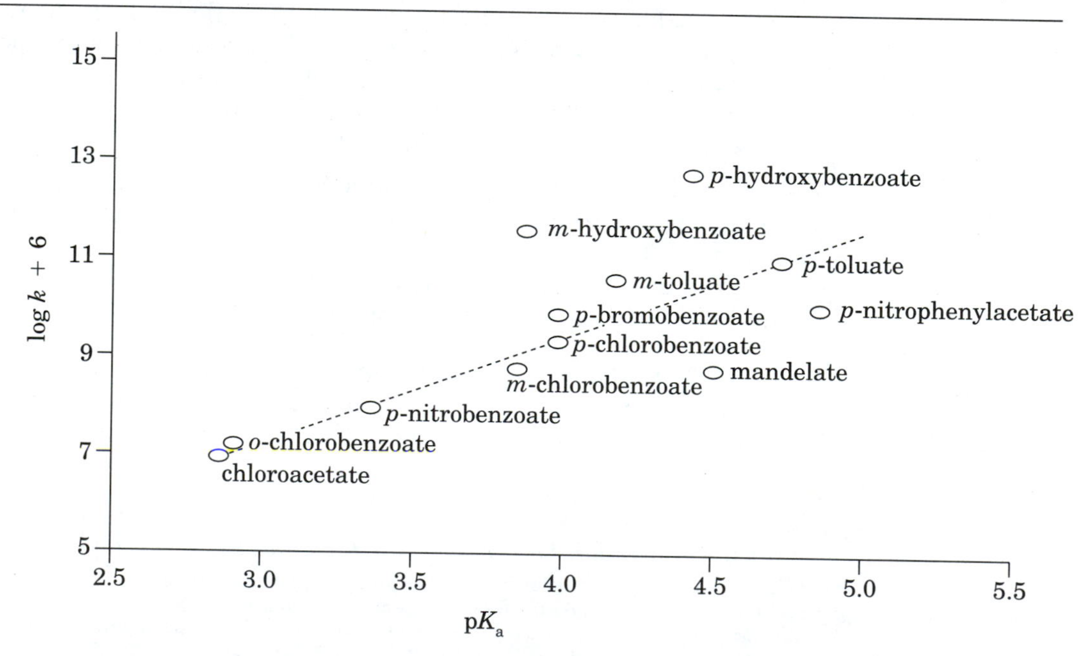

[116]Smith, G. F. *J. Chem. Soc.* **1943**, 521. For the data plotted, the correlation coefficient is 0.7, but a general trend is evident.

[117]Data are for reaction of the two ions with CH_3I in methanol solution. See Pearson, R. G.; Sobel, H.; Songstad, J. *J. Am. Chem. Soc.* **1968**, *90*, 319.

reaction of the nucleophile and water, respectively, and the n value of water was taken to be 1.0.[118] Both n and s values are empirically determined and are not related to other parameters. Table 8.12 shows the n values determined for the reaction of methyl bromide with various nucleophiles, and Table 8.13 collects some values of s.[119] Note that n values are logarithmic values; iodide ion is 100,000 times more reactive than water.[120]

The Edwards Equations Edwards attempted to improve the correlation of nucleophilicity with substrate reactivity by considering separately two different components of nucleophilic reactivity. Two equations were presented, and the distinction between them is not always clear in the literature. In deriving equation 8.34, Edwards noted that basicity is only one measure of a nucleophile's tendency to donate electrons, the other being its oxidation potential.[121] To the extent that the transition structure for an S_N2 reaction resembles a partially oxidized species, then the oxidation potential may be correlated with nucleophilic reactivity. Moreover, he suggested that

Table 8.12 Selected values of Swain nucleophilicity parameter, n.

Nucleophile	n	Nucleophile	n
H_2O	0.00	NO_3^-	1.03
$2,4,6\text{-}(NO_2)_3C_6H_2O^-$	1.9	Cl^-	2.70
HCO_2^-	2.75	$C_6H_5O^-$	3.5
Br^-	3.53	HCO_3^-	3.8
N_3^-	4.00	HO^-	4.20
I^-	5.04	HS^-	5.1
NC^-	5.1		

(Adapted from reference 113.)

[118] Swain, C. G.; Scott, C. B. *J. Am. Chem. Soc.* **1953**, *75*, 141; Swain, C. G.; Dittmer, D. C. *J. Am. Chem. Soc.* **1953**, *75*, 4627; Swain, C. G.; Mosley, R. B. *J. Am. Chem. Soc.* **1955**, *77*, 3727; Swain, C. G.; Mosley, R. B.; Brown, D. E. *J. Am. Chem. Soc.* **1955**, *77*, 3731; Swain, C. G.; Dittmer, D. C.; Kaiser, L. E. *J. Am. Chem. Soc.* **1955**, *77*, 3737.

[119] Data from the compilation by Wells (reference 113) and references therein. A summary is also given by Gordon, A. J.; Ford, R. A. *The Chemist's Companion*; Wiley-Interscience: New York, 1972; p. 151.

[120] The format of the Swain-Scott equation implies that the relative nucleophilicity of a nucleophile is invariant, i.e., that it is not a function of the substrate. As Ibne-Rasa (reference 114) has noted, however, the success of the Swain-Scott equation is greatest for nucleophilic reactions involving attack on a tetrasubstituted carbon atom. The deviations from linearity in the determination of s were severe in other cases. The conclusion may be that the Swain-Scott n values are most properly denoted as n_C values (for attack on a tetravalent carbon atom) and that different n values (n_H for attack on hydron, n_{Ar} for attack on aromatic carbon atom, etc.) would need to be devised for individual systems. For a discussion, see reference 111, p. 01.

[121] Edwards, J. O. *J. Am. Chem. Soc.* **1954**, *76*, 1540.

Table 8.13 Selected values of Swain-Scott *s* value.

Substrate	s
Methyl bromide	1.00
Methyl iodide	1.15
Chloroacetate	1.0
Bromoacetate	1.1
Iodoacetate	1.33
Benzyl chloride	0.87
Benzenesulfonyl chloride	1.25
Benzoyl chloride	1.43
β-Propionolactone	0.77
Ethylene-β-chloroethylsulfonium ion	0.95
Ethyl tosylate	0.66
Trityl fluoride	0.61

(Adapted from reference 113.)

different substrates might require different contributions from basicity and from oxidation of the nucleophile. Therefore Edwards wrote that

$$\log (k/k_0) = \alpha E_n + \beta H \tag{8.34}$$

As before, k and k_0 are the rates of reaction of a nucleophile n and of a standard nucleophile (chosen as H_2O), respectively. H is a measure of the basicity of the nucleophile,

$$H = pK_a + 1.74 \tag{8.35}$$

where the pK_a is the pK_a of the conjugate acid of the nucleophile. E_n is related to the oxidation potential of the nucleophile (X^-) in the reaction

$$2\ X^- \rightleftharpoons X_2 + 2\ e^- \tag{8.36}$$

by the relationship

$$E_n = E_0 + 2.60 \tag{8.37}$$

The parameters α and β then relate the sensitivity of the substrate to each of these measures of nucleophilicity.[122]

Application of equation 8.34 led to situations in which β was found to be negative for some substrates. The most satisfying explanation for this result was that the parameters E_n and H are not totally separable because

[122]Equation 8.34 has been called an *oxibase scale*, since the E_n term is an oxidation term, while the H term is a basicity term: Davis, R. E. in *Organosulfur Chemistry*, Janssen, M. J., ed.; Wiley-Interscience: New York, 1967; pp. 311–328.

those factors that determine basicity can also influence the electrochemical properties of the nucleophile. To preserve the two-parameter nucleophilicity equation, Edwards proposed equation 8.38, in which substrate reactivity is correlated with the basicity (B) and polarizability (P) of the nucleophile.[123,124]

$$\log (k/k_0) = AP + BH \tag{8.38}$$

The polarizability term, P, was determined from the molar refractivity of the nucleophile.

Hard-Soft Acid-Base Theory Pearson and Edwards had noted that some substrates in S_N2 reactions seem most susceptible to the (Brønsted) basicity of the nucleophile, while other substrates seem most susceptible to the polarizability term.[125] Pearson proposed that bases he divided into two categories: *soft* (polarizable) and *hard* (nonpolarizable).[126] On the basis of equilibrium data, Pearson concluded that H^+, Li^+, Na^+, Mg^{2+}, and Ca^{2+} are hard acids, while Cu^+, Ag^+, Hg^{2+}, I^+, Br^+, I_2, and Br_2 are soft. (Some acids, such as Zn^{2+} and Pb^{2+}, are designated as borderline). Pearson noted that hard acids show greater association with hard bases, while soft acids prefer soft bases.[127] One explanation for this behavior is the ionic/covalent bond description: The assumption is that hard acids and hard bases are attracted primarily through ionic interactions, while soft acids and soft bases bond through covalent bonds. In addition, the two types of acids are expected to differ with regard to π-bonding possibilities. Solvation is generally said to soften both hard and soft acids.[128]

More recently Pearson has reexamined the concept of hard and soft acids and bases in the context of the S_N2 reaction by using the concept of *absolute hardness*, η.[129,130]

$$\eta = (I - A)/2 = 1/\sigma \tag{8.39}$$

---

[123]The relationships $A = \alpha a$ and $B = \beta + \alpha b$ may be used to convert equation 8.34 to 8.38: Edwards, J. O. *J. Am. Chem. Soc.* **1956**, *78*, 1819.

[124]This equation is often written as $\log (k/k_0) = \alpha P + \beta H$ (*cf.* reference 125). Thus it is sometimes difficult to know whether the Edwards parameters refer to the original α and β definitions or to the new A and B terms.

[125]Edwards, J. O.; Pearson, R. G. *J. Am. Chem. Soc.* **1962**, *84*, 16.

[126]Pearson, R. G. *J. Am. Chem. Soc.* **1963**, *85*, 3533.

[127]For a discussion of the hard-soft concept and organic reactions, see Pearson, R. G.; Songstad, J. *J. Am. Chem. Soc.* **1967**, *89*, 1827.

[128]Hard and soft acid and base theory has gained wide currency in organic chemistry, but has not been universally accepted. For example, Parker noted the reversal of nucleophilicity of halide ions in protic and nonprotic solvents, adding (reference 86, from page 27) that "one cannot sensibly claim that saturated carbon is 'soft' in protic solvents and 'hard' in dipolar solvents, and yet this is required if the hard acids-soft bases theory is to be applied to this observation."

[129]Pearson, R. G. *J. Org. Chem.* **1987**, *52*, 2131.

[130]Parr, R. G.; Pearson, R. G. *J. Am. Chem. Soc.* **1983**, *105*, 7512.

Table 8.14 Hardness of bases.

Base	$\eta(B)$	Base	$\eta(B)$
F^-	7.0	HO^-	5.6
Cl^-	4.7	H_2O	7.0
Br^-	4.2	HS^-	4.1
I^-	3.7	H_2S	5.3
H^-	6.8	H_2N^-	5.3
CH_3^-	4.0	H_3N	6.9

(Data from reference 130.)

Here A is the electron affinity and I is the gas-phase ionization potential of the acid under consideration. Absolute softness, σ, is the inverse of η. The term $I - A$ is a measure of the energy gap between the LUMO and the HOMO calculated for a π system in MO theory, so that this term is equivalent to the polarizability term used previously to define hardness or softness. This approach is somewhat more successful at predicting relative reactivities (such as the increasing ratio of the reactivity of CH_3I relative to CH_3F as the hardness of the base increases) and in explaining solvent effects on reactivity.[131] One clear example of the utility of the theory is its ability to explain why nucleophiles such as *p*-toluenesulfinate react with the oxygen serving as the nucleophile with hard acids (to give an ester) but with the sulfur serving as the nucleophile with soft acids (to yield a sulfone).[132] Generally, however, the hard-soft acid-base theory has been more widely used to explain the reactions of coordination compounds than to quantitatively describe organic reactions.

The α-Effect Subsequent to the work of Edwards, it was shown that some anions are more nucleophilic than would be expected on the basis of P and H values.[133] These nucleophiles are characterized by the formula :Z—Y:, where :Z represents an electronegative atom with an unshared pair of electrons that is adjacent (α) to a nucleophilic atom, Y:. For example, the basicity of hydroxide (HO^-) is about 16,000 times greater than that of hydroperoxide (HOO^-). However, for addition to carbonyl compounds, hydroperoxide is more than 200 times as nucleophilic as hydroxide.[134] The apparently enhanced nucleophilicity of species such as HOO^- has been called

[131] Reference 129, p. 2135.

[132] Meek, J. S.; Fowler, J. S. *J. Org. Chem.* ***1968***, 33, 3422.

[133] Jencks, W. P.; Carriuolo, J. *J. Am. Chem. Soc.* **1960**, *82*, 1778.

[134] Edwards, J. O.; Pearson, R. G. *J. Am. Chem. Soc.* **1962**, *84*, 16.

the α effect, and its origins have been explored by a number of workers.[135] One theoretical explanation, for example, was the proposal by Ibne-Rasa and Edwards that the presence of α-nonbonded electrons causes destabilization of the nucleophile in the ground state, thus lowering the activation energy.[136] However, DuPuy found experimentally that HOO^- is *less* reactive than HO^- for reaction with CH_3F in the gas phase,[137] and calculations by Jorgensen indicate that HOO^- should be less reactive than HO^- for reaction with methyl chloride in the gas phase.[138] These results suggest that the apparently greater reactivity of HOO^- in solution is due to solvation effects arising from charge delocalization.[139,140,141]

Leaving Group Effects in S_N2 Reactions The more stable the detached leaving group (usually an anion or neutral molecule), the more stable will be the product Y—R + X:⁻ and the greater will be the ***nucleofugality*** (kinetic leaving group ability) of X:⁻ For example, an important principle in the reactions of alcohols is that C—O bonds are broken more easily in substitution reactions when the oxygen departs as a weaker base (such as water or tosylate ion) than when the oxygen departs as a hydroxide ion. Such observations give rise to the general statement that the activation energy for the S_N2 reaction decreases with a decrease in the basicity of the leaving group.[142] Pearson[129] has determined theoretically that the order of nucleofugality should be

$$H_2O > CH_3OH > Br^- \gtrsim NO_3^- > I^- > F^- > Cl^- > SCN^- > (CH_3)_2S >$$

$$C_6H_5O^- > NH_3 > C_6H_5S^- > CH_3O^- > CN^- >> NH_2^- >> H^- > H_3C^-$$

However, just as nucleophiles can interact differently with protic and dipolar aprotic solvents, so can leaving groups, and the solvation energies can shift the location and energies of transition states. The data in Table 8.15 show considerable variation in the experimental values for the relative rate constants for reaction of methyl derivatives with a series of nucleophiles, Y^-.

---

135 One review concluded that "the sources of the α-effect are remarkable in their ambiguity:" Hoz, S.; Buncel, E. *Israel J. Chem.* **1985**, *26*, 313.

136 Ibne-Rasa, K. M.; Edwards, J. O. *J. Am. Chem. Soc.* **1962**, *84*, 763.

137 DuPuy, C. H.; Della, E. W.; Filley, J.; Gragowski, J. J.; Bierbaum, V. M. *J. Am. Chem. Soc.* **1983**, *105*, 2481.

138 Evanseck, J. D.; Blake, J. F.; Jorgensen, W. L. *J. Am. Chem. Soc.* **1987**, *109*, 2349.

139 See also Gao, J.; Garner, D. S.; Jorgensen, W. L. *J. Am. Chem. Soc.* **1986**, *108*, 4784.

140 Hudson, R. F.; Hansell, D. P.; Wolfe, S.; Mitchell, D. J. *J. Chem. Soc. Chem. Commun.* **1985**, 1406.

141 Hudson, R. F. in reference 115, p. 195.

Table 8.15 Relative rate constants of S_N2 reactions of CH_3X.

		log $[k(CH_3X)/k(CH_3I)]$				
Y^-	Solvent	X = Cl	Br	I	OTs	Me_2S^+
N_3^-	CH_3OH	−2.0	−0.2	0.0	+0.8	−3.3
	DMF	−3.3	−0.9	0.0	−1.8	−4.8
Cl^-	CH_3OH		+0.3	0.0	+0.4	
	DMF		−0.8	0.0	−1.7	

(Data from reference 86.)

Aliphatic Substitution and Single Electron Transfer

The discussion to this point has considered only heterolytic processes for substitution reactions. However, there is increasing evidence that some substitution reactions may take place by single electron transfer (SET) pathways. To understand the arguments used to discuss this theory, it is useful to first discuss a conceptual model for describing organic reactions that has been called the **valence bond configuration mixing (VBCM)** model.[143,144] As described by Pross and Shaik, the method is based on an analysis of the energies of all possible valence bond configurations of the reactants and products as a function of the reaction coordinate.

Consider the homolysis of a polar covalent R—X bond in the gas phase and in solution (Figure 8.43). The ground electronic state of R—X is described by equation 8.40, in which the polar bond is considered to result from mixing of two configurations, one nonpolar and the other polar. It is also useful to describe the excited state (S_1) by equation 8.41, in which the polar (zwitterionic) character is dominant, although there is some mixing of nonpolar character as well.[145] Figure 8.43(a) shows the calculated energies of the pure nonpolar and polar configurations (solid lines) as a function of the R—X bond distance as well as the energies of S_0 and S_1 (dashed lines) when mixing is included.

$$(\mathrm{R{-}X})_{S_0} \approx (\mathrm{R{\cdot\cdot}X}) + \lambda(\mathrm{R^+\ {:}X^-}) \tag{8.40}$$

$$(\mathrm{R{-}X})_{S_1} \approx (\mathrm{R^+\ {:}X^-}) - \lambda'(\mathrm{R{\cdot\cdot}X}) \tag{8.41}$$

where

$$(\mathrm{R{\cdot\cdot}X}) = 1/\sqrt{2}\,[(\mathrm{R}\uparrow\downarrow\mathrm{X}) \leftrightarrow (\mathrm{R}\downarrow\uparrow\mathrm{X})] \tag{8.42}$$

---

[142]However, this correlation has been called an "illusion" based on inadequate and perhaps unobtainable data: Stirling, C. J. M. *Acc. Chem. Res.* **1979**, *12*, 198.

[143]Shaik, S. S. *J. Am. Chem. Soc.* **1981**, *103*, 3692.

[144]Pross, A.; Shaik, S. S. *Acc. Chem. Res.* **1983**, *16*, 363 and references therein.

[145]Excited (electronic) states are discussed in chapter 12.

Figure 8.43 Energy diagram for homolytic and heterolytic dissociation: (a) gas phase; (b) solution. (Reproduced from reference 144.)

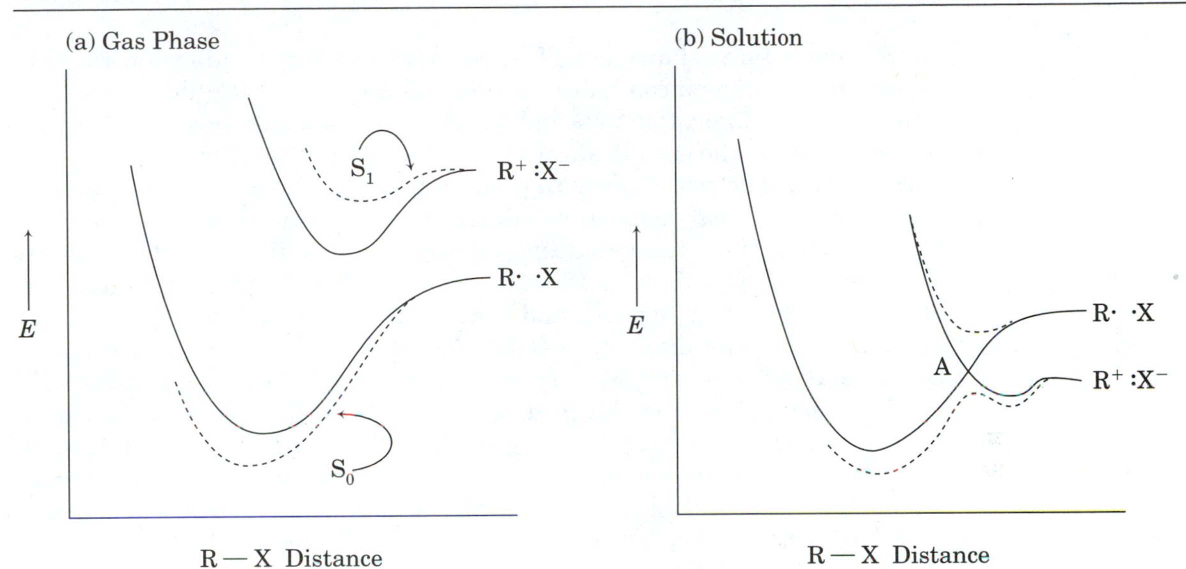

In solution the energy of ionic states is reduced by solvation energies, so the two solid curves cross, Figure 8.43(b). Now, the mixing of the two states produces an *avoided crossing* of the two dotted curves.[146] Figure 8.43(b) thus suggests that stretching of the R—X bond leads at first to a slight decrease in polarity until the energy maximum at the avoided crossing is reached, then to an increase in polarity as ions are formed. This increase in polarity is effectively brought about by a *single electron shift* from R · to · X in the developing R ·· X configuration to form the R^+ $:X^-$ configuration.

Now let us apply this model to a nucleophilic substitution reaction in which the R—X bond is broken and a new bond is formed from R to a nucleophile, Y, which is present initially as the nucleophilic anion, Y^-. We can define valence bond representations **54–57**, where **54** is the reactant configuration, **55** is the product configuration, and **56** and **57** are species with positive and negative charge, respectively, on the carbon atom at the site of the substitution. Pross and Shaik used these descriptions to illustrate their suggestion that S_N2 reactions involve a single-electron shift, as indicated in equation 8.43.

$\bar{Y:}$ R··X	Y··R $:\bar{X}$	$\bar{Y:}$ $\overset{+}{R}$ $:\bar{X}$	Y· $\bar{R:}$ ·X
54	**55**	**56**	**57**

[146]See the references in reference 144 for a discussion of avoided crossings. Also see the discussion in chapter 11.

$$\text{Y:}^- \ \ \text{R}\cdot\cdot\text{X} \longrightarrow \text{Y}\cdot\ \text{R}\cdot\!\cdot\!\cdot\text{X}^- \longrightarrow \text{Y}\cdot\ \text{R}\cdot\ \text{:X}^- \longrightarrow \text{Y}\cdot\cdot\text{R} \ \ \text{:X}^- \tag{8.43}$$

Figure 8.44(a) shows a VBCM reaction coordinate diagram for this process based only on configurations **54** and **55**. (A complete description requires that configurations **56** and **57** also be included.) Now the reaction pathway follows the curve indicated by **1**, the lower dashed avoided crossing curve, and **2**. The net effect is to produce a curve that is overall similar to the two-dimensional reaction coordinate diagram curves often written for S_N2 reactions. There is an analogy to this mechanism in the electrochemical reduction of alkyl halides. Reduction of methyl halide apparently adds an electron to the σ^*_{R-X} orbital, and that leads to dissociation of the radical anion to a methyl radical and a halide ion, R · and X^-.[144] In SET reactions, the R · may rapidly bond with Y · so that no net production of free radicals outside a solvent cage is observed. If configuration **55** is made more stable (i.e., by making X a better leaving group or making Y a better nucleophile), then the curve for configuration 2 will be lowered, leading to an earlier transition state, see Figure 8.44(b). However, this does not mean that X or Y has greater or lesser charge or that free radicals need be formed.[143,144]

Even though long-lived radicals need not be formed in S_N2 reactions that proceed by the SET pathway, evidence for radical intermediates has been found in some cases by using diagnostic tools such as cyclizable radical probes, radical traps, and chiral alky halides. These studies have suggested that the extent of single electron transfer is a function of the substrate, the

Figure 8.44 Reaction coordinate diagrams: (a) VBCM interaction; (b) early and late transition states. (Adapted from reference 144.)

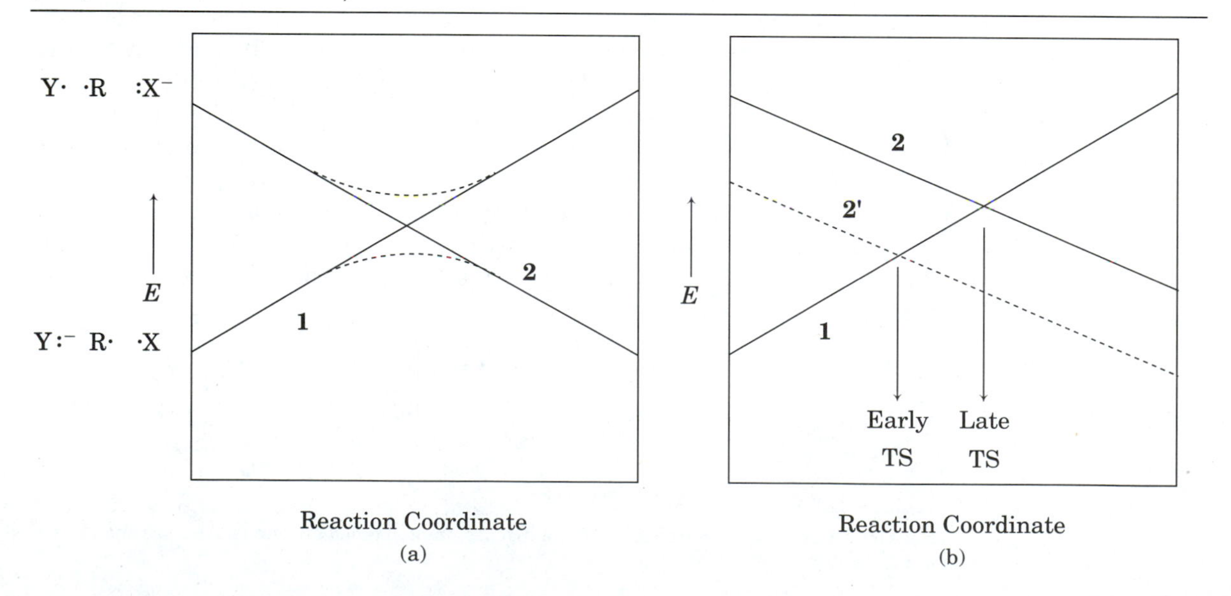

leaving group, the solvent, and the reducing agent.[147] Cyclizable radical probes have been particularly useful in these studies. For example, the reduction of alkyl halides by $LiAlH_4$ had long been considered as S_N2 reaction because inversion of configuration was observed in the reduction of (+)-1-chloro-1-phenylethane with $LiAlD_4$.[148] However, when 5-substituted cyclooctenes, **58(a–d)**, were reduced with $LiAlH_4$, a significant amount of bicyclo[3.3.0]octane (**63**) was formed in reactions of the iodide and bromide, although not with the chloride and tosylate (Figure 8.45). The results suggest that single electron transfer leads to the 4-cyclooctenyl radical (**60**) when the carbon-halogen bond is easily reduced, but that in other cases the reduction is effectively a two-electron transfer that leads to cyclooctene as the major product.[149,150]

Evidence for an SET pathway has also been obtained by studying substrates (R—X) in which the radical R· can undergo a characteristic rearrangement before coupling to form the substitution product. As shown in Figure 8.46, SET reduction of 6-iodo-5,5-dimethyl-1-hexene (**64**) can lead to

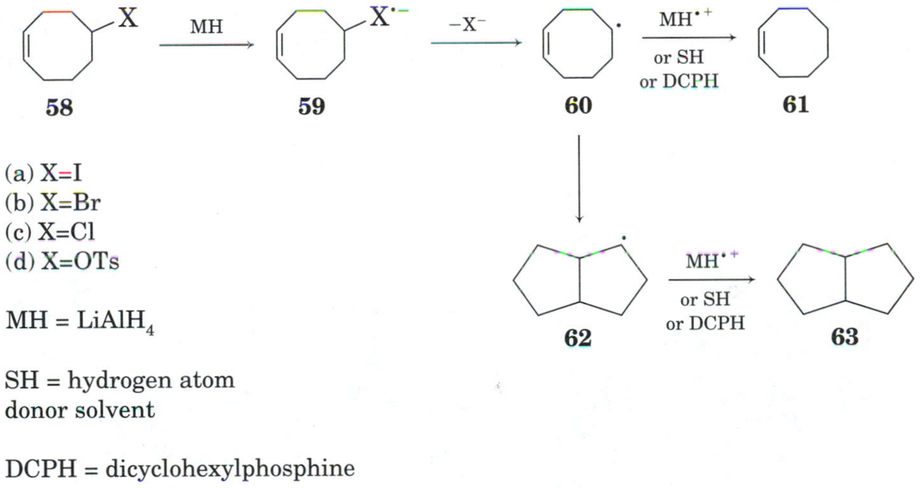

Figure 8.45
Evidence for SET in the metal hydride reduction of 5-halocyclooctenes. (Adapted from reference 149; not all processes are shown.)

(a) X=I
(b) X=Br
(c) X=Cl
(d) X=OTs

MH = $LiAlH_4$

SH = hydrogen atom donor solvent

DCPH = dicyclohexylphosphine

[147] Ashby, E. C.; DePriest, R. N.; Goel, A. B.; Wenderoth, B.; Pham, T. N. *J. Org. Chem.* **1984**, *49*, 3545.

[148] Eliel, E. L. *J. Am. Chem. Soc.* **1949**, *71*, 3970.

[149] Ashby, E. C.; Pham, T. N. *J. Org. Chem.* **1986**, *51*, 3598.

[150] Ashby, E. C.; Argyropoulos, J. N. *J. Org. Chem.* **1978**, *50*, 3274 also found that the enolate of propiophenone reacts with 1-iodo-2,2-dimethyl-5-hexene to give a number of products, including a significant yield of the cyclized products expected if single electron transfer and halide loss produce an alkyl radical intermediate. However, cyclization was not observed with the corresponding bromide or chloride. The results suggest that the S_N2 reaction of enolates with alkyl halides proceeds with some amount of single electron transfer, the degree depending upon the leaving group. Bordwell, F. G.; Wilson, C. A. *J. Am. Chem. Soc.* **1987**, *109*, 5470 found evidence for SET in one system when the redox potentials of the nucleophile and substrate are favorable and when steric hindrance inhibits the "normal" S_N2 reaction.

a radical capable of cyclization to a five-membered ring. Both the initial and cyclized radicals can react with a radical (from the nucleophile) in the solvent cage or can abstract a hydrogen atom from solvent. In the case of reduction of **64** with $LiAlH_4$ in solvents in which only protium is abstractable, products **65** and **66** are identical, as are products **67** and **68**. However, when the reduction is carried out with $LiAlD_4$ in a nondeuterated solvent, those pairs of products are isotopically different and can be distinguished. In an experimental investigation of the reduction of **64** with $LiAlD_4$ in a nondeuterated solvent, it was found that 31% of the noncyclized product was the nondeuterated **66**. Formation of this product is most simply explained by a mechanism involving abstraction of a protium from solvent by the radical formed by single electron transfer.[152]

Leaving group effects can also distinguish between S_N2 and SET processes. For alkyl halides, the relative reactivity in S_N2 reactions is I ≈

Figure 8.46
Product formation in SET reaction with a cyclizable probe.

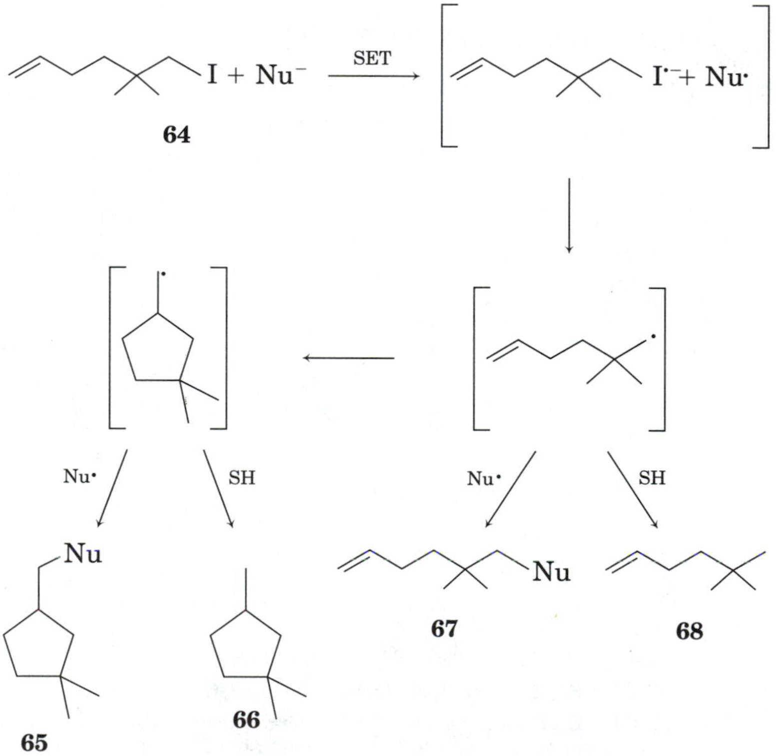

[151]Ashby, E. C.; Park, B.; Patil, G. S.; Gadru, K.; Gurumurthy, R. *J. Org. Chem.* **1993**, *58*, 424.
[152]Ashby, E. C.; Pham, T. N.; Amrollash-Madjdabadi, A. *J. Org. Chem.* **1991**, *56*, 1596.

OTs > Br > Cl, while relative reactivity in SET processes is I > Br > Cl > OTs.[151] For reduction of cyclizable 2° alkyl derivatives, it appears that the tosylates and chlorides undergo reduction only by an S_N2 pathway, that the bromides give mostly S_N2 reduction (but some SET reduction also occurs), and the iodides undergo significant SET reactions. However, with stronger one-electron donors, such as $(CH_3)_3Sn^-$, even alkyl chlorides can undergo SET reaction.[152] Stereochemical studies reinforce the conclusions from leaving group investigations. A study of the reduction of optically active 2-octyl derivatives with $LiAlH_4$ led to the conclusion that reduction of the iodide occurs with racemization but that the tosylate, chloride, and bromide react with inversion of configuration.[147,153] However, the more easily reduced alkyl geminal dichlorides do undergo SET reaction with $LiAlH_4$.[154]

It must be emphasized that while isomerization of a cyclizable probe can be taken as positive evidence for a radical intermediate in a substitution reaction, failure to observe cyclization with such a probe does not rule out a radical intermediate. For example, reaction of 6-bromo-1-hexene with $NaSN(CH_3)_3$ in THF was found not to produce cyclized product.[155] However, cyclized product was observed when the same reaction was carried out in 1:1 THF-pentane.[156] The results led to the conclusions that cyclization of the radical probe occurs outside of the solvent cage in which it is formed and that a lower viscosity solvent allows more radicals to diffuse from the radical cage so that cyclization can occur.

There may be varying degrees of single electron transfer and Y—R bond formation in substitution reactions.[157] One may view the "pure" SET process represented by the solid line marked SET on Figure 8.47 as a limiting case in which one electron is completely transferred from $Y:^-$ to R—X before the R—X bond dissociation occurs. An alternative limiting case is one in which the electron transfer occurs at the same rate as Y—R bond formation (the diagonal line), meaning a "pure" or "classical" two-electron S_N2 process. The dashed line in Figure 8.47 suggests one of a continuum of possible transition states, ranging from pure SET to pure "polar."

Although this discussion has emphasized the competition between SET and S_N2 processes, other processes are also possible in the reaction of alkyl

---

[153]Ashby, E. C.; DePriest, R. N.; Pham, T. N. *Tetrahedron Lett.* **1983**, *24*, 2825.

[154]Ashby, E. C.; Deshpande, A. K. *J. Org. Chem.* **1994**, *59*, 3798.

[155]Park, S.-U.; Chung, S.-K.; Newcomb, M. *J. Org. Chem.* **1987**, *52*, 3275.

[156](a) Ashby, E. C.; Su, W.-Y.; Pham, T. *Organometallics* **1985**, *4*, 1493; (b) Tolbert, L. M.; Sun, X.-J.; Ashby, E. C. *J. Am. Chem. Soc.* **1995**, *117*, 2861.

[157]In addition to references cited above, leading references include (a) Pross, A., *Acc. Chem. Res.* **1985**, *18*, 212; (b) Shaik, S. S. *J. Am. Chem. Soc.* **1984**, *106*, 1227; (c) McLennan, D. J.; Pross, A. *J. Chem. Soc. Perkin Trans. 2* **1984**, 981; (d) Shaik, S. S. *Prog. Phys. Org. Chem.* **1985**, *15*, 197; *Acta Chem. Scand.* **1990**, *44*, 205 and references therein; (e) Shaik, S. S. *Israel J. Chem.* **1985**, *26*, 367; (f) Shaik, S.; Ioffe, A.; Reddy, A. C.; Pross, A. *J. Am. Chem. Soc.* **1994**, *116*, 262; (g) Shaik, S. S.; Schlegel, H. B.; Wolfe, S. *Theoretical Aspects of Physical Organic Chemistry: The S_N2 Mechanism*; Wiley-Interscience: New York, 1992. See also Perrin, C. L. *J. Phys. Chem.* **1984**, *88*, 3611.

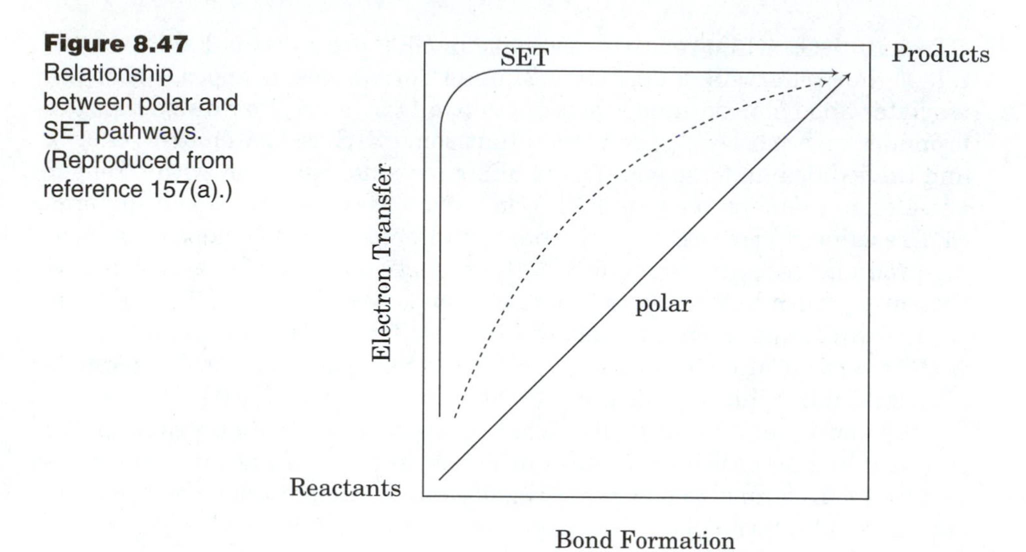

Figure 8.47 Relationship between polar and SET pathways. (Reproduced from reference 157(a).)

Figure 8.48 Competing radical, carbanion, carbene pathways in the reaction of 6-iodo-5,5-dimethyl-1-hexene with LDA in THF. (Adapted from reference 151.)

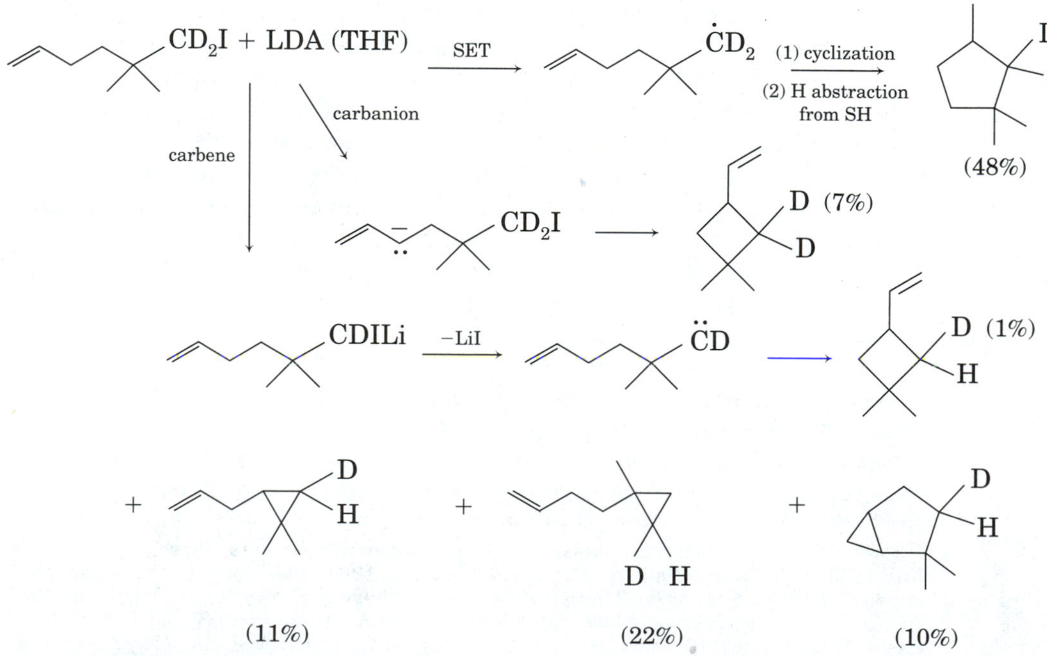

halides with reagents that are strong bases and electron donors. For example, in the case of reduction of 6-iodo-5,5-dimethyl-1-hexene with lithium diisopropylamide (LDA) in THF, products arising from competing radical, carbanion, and carbene pathways were identified (Figure 8.48).[151]

8.3 Electrophilic Aromatic Substitution

The S_EAr Reaction

The S_EAr reaction involves reaction of an electrophile, E^+, with an aromatic compound to produce a substituted aromatic compound and a proton.[158]

$$E^+ + \mathrm{Ar{-}H} \longrightarrow \mathrm{Ar{-}E} + H^+ \tag{8.44}$$

If equation 8.44 represents a one-step process, then a primary hydrogen kinetic isotope effect should be observed for the reaction. In most cases, however, k_H/k_D is near unity for typical S_EAr reactions such as nitration[159] or bromination.[160,161,162] Therefore we surmise that the loss of the C—H bond usually does not occur in the rate-limiting step. Since two essential processes in the reaction are formation of a bond between E^+ and a carbon atom in the aromatic ring and the loss of a proton (hydron) from the same carbon atom, a minimum kinetic formulation consistent with the data is the following two-step process

$$E^+ + \mathrm{Ar{-}H} \underset{k_{-1}}{\overset{k_1}{\rightleftharpoons}} \left[\mathrm{Ar}{<}^{\mathrm{H}}_{\mathrm{E}}\right]^+ \tag{8.45}$$

$$\left[\mathrm{Ar}{<}^{\mathrm{H}}_{\mathrm{E}}\right]^+ \xrightarrow{k_2} \mathrm{Ar{-}E} + H^+ \tag{8.46}$$

---

[158]Although a proton is usually displaced by the electrophile, in *ipso* substitution there is replacement of another substituent, as in the following example:

OCH_3 / I $\xrightarrow{[NO_2^+]}$ OCH_3 / NO_2 (plus other products)

ipso product

Perrin, C. L.; Skinner, G. A. *J. Am. Chem. Soc.* **1971**, *93*, 3389. For an example of a nucleophilic *ipso* substitution, see reference 202.

[159]Lauer, W. M.; Nolan, W. E. *J. Am. Chem. Soc.* **1953**, *75*, 3689.

[160]de la Mare, P. B. D.; Dunn, T. M.; Harvey, J. T. *J. Chem. Soc.* **1957**, 923.

[161]However, a primary kinetic isotope effect is seen for sulfonation: Melander, L.; Saunders, Jr., W. H. *Reaction Rates of Isotopic Molecules*; Wiley-Interscience: New York, 1980; pp. 162 *ff.*

[162]A pronounced kinetic isotope effect can also been seen in diazonium ion coupling reactions (Zollinger, H. *Helv. Chim. Acta* **1955**, *38*, 1617) and mercuration of benzene (Perrin, C.; Westheimer, F. H. *J. Am. Chem. Soc.* **1963**, *85*, 2773).

with the first step being rate-limiting (i.e., $k_2 >> k_{-1}$) for most reactions. This would give rise to the reaction coordinate diagram shown in Figure 8.49.

The simple mechanism represented in equations 8.45 and 8.46 underestimates the complexity of the S_EAr reaction. First, generation of the electrophile may be multi-step process with equilibria that vary according to reagents and experimental conditions.[163] In the case of bromination, for example, the active electrophile may be either Br^+ or HO_2Br^+ when the reagent is HOBr in aqueous dioxane with perchloric acid catalysis, but may be molecular bromine under other conditions.[163(a)] Second, π complexes may also be involved in the reaction. Let us consider as an example one of the most thoroughly studied S_EAr reactions, the nitration of benzene and its derivatives.[164,165,166] Figure 8.50 shows the steps thought to be involved in the generation of the electrophile (nitronium ion, NO_2^+) and its subsequent reaction with benzene. The resonance stabilized intermediate has been called a *σ complex* (because the electrophile is bonded to a carbon atom of the aromatic ring by a σ bond) or a *Wheland intermediate* (after the theoretical contributions of Wheland).[167,168] There is also another intermediate, which is

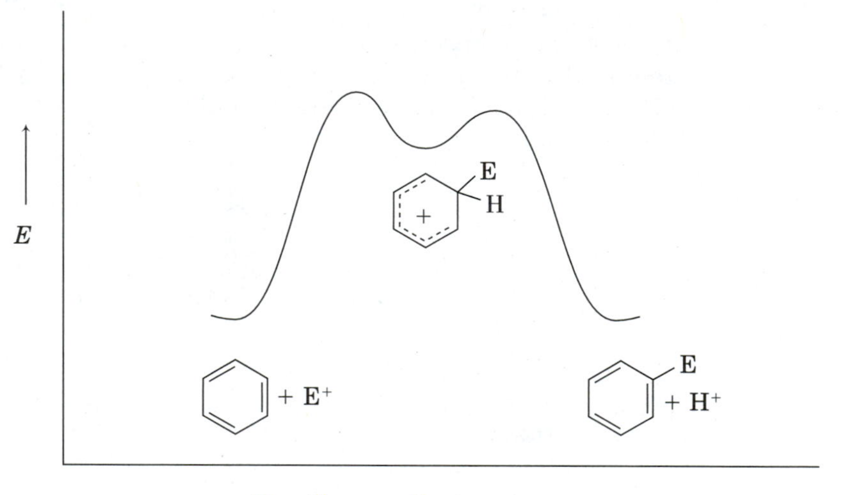

Figure 8.49
Reaction coordinate diagram for electrophilic aromatic substitution.

---

[163](a) de la Mare, P. B. D.; Harvey, J. T. *J. Chem. Soc.* **1956**, 36; (b) Sato, Y.; Yato, M.; Ohwada, T.; Saito, S.; Shudo, K. *J. Am. Chem. Soc.* **1995**, *117*, 3037 have reported that in some cases the electrophile in a Friedel-Crafts acylation reaction may be a dication.

[164]Deno, N. C.; Stein, R. *J. Am. Chem. Soc.* **1956**, *78*, 578.

[165]For a review of work in this area, see reference 175.

[166]For a detailed discussion, see Olah, G. A.; Malhotra, R.; Narang, S. C. *Nitration: Methods and Mechanisms*; VCH Publishers: New York, 1989.

[167]Wheland, G. W. *J. Am. Chem. Soc.* **1942**, *64*, 900.

[168]Wheland, G. W. *Resonance in Organic Chemistry*; John Wiley & Sons: New York, 1955; pp. 476–507.

denoted as an *encounter complex*. This species is considered to be formed by diffusion together of the aromatic and the electrophile to form a complex with some activation barrier for dissociation.

The mechanistic scheme in Figure 8.50 includes many steps, each of which could be the rate-limiting step of the reaction under certain conditions. If formation of the nitronium ion is rate-limiting, the kinetic expression for the reaction is zero order in the aromatic. In other cases the

Figure 8.50 Detailed mechanism for nitration of benzene.

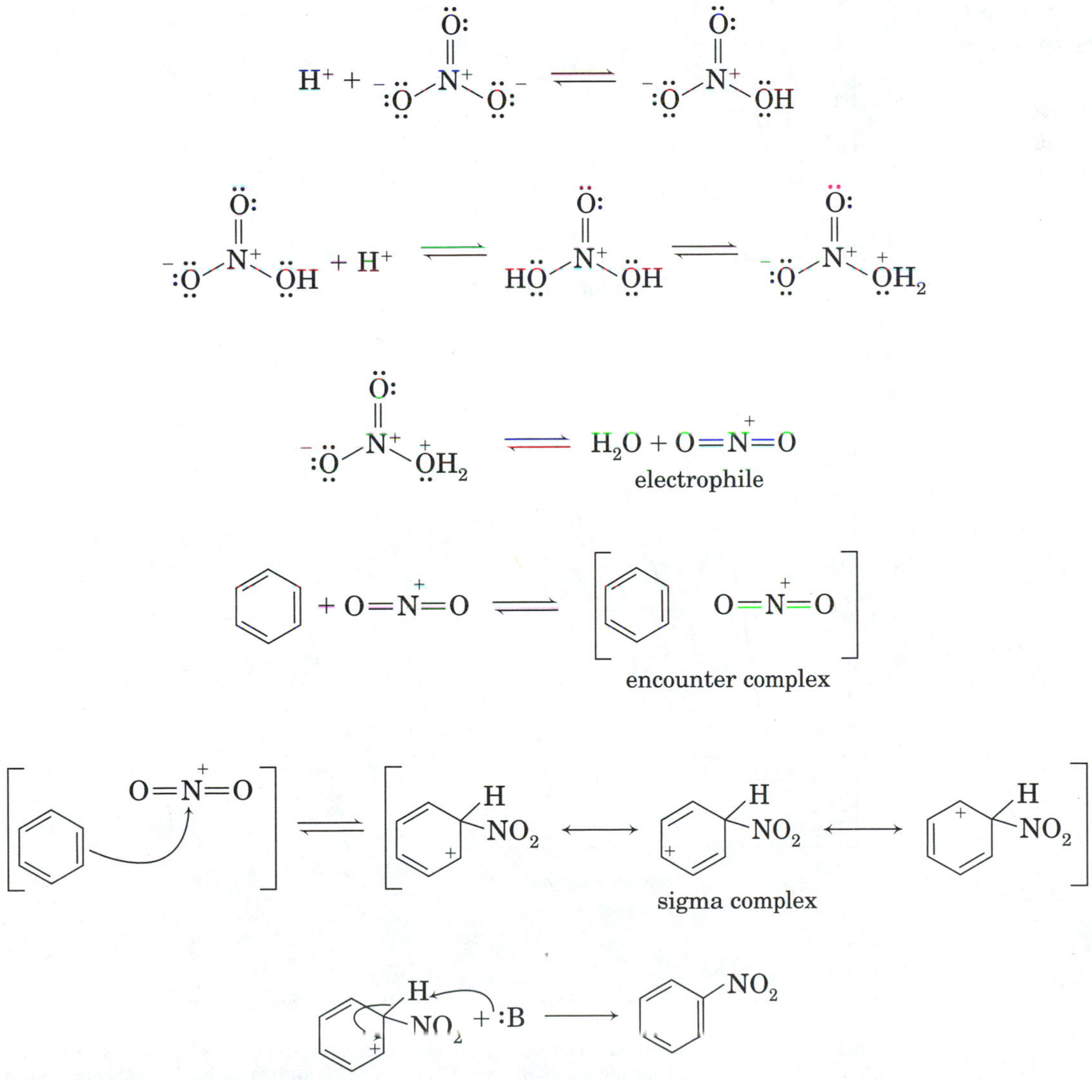

formation of the σ complex can be the rate-limiting step in the reaction, while in still others formation of the encounter complex can be rate-limiting. Figure 8.51 illustrates the latter two situations with reaction coordinate diagrams proposed for reaction of nitronium ion with toluene, Figure 8.51(a), and 1,2,4-trimethylbenzene, Figure 8.51(b).[169,170]

Figure 8.51 Possible reaction coordinate diagrams for nitration of (a) toluene and (b) 1,2,4-trimethylbenzene. (Adapted from reference 175.)

(a)

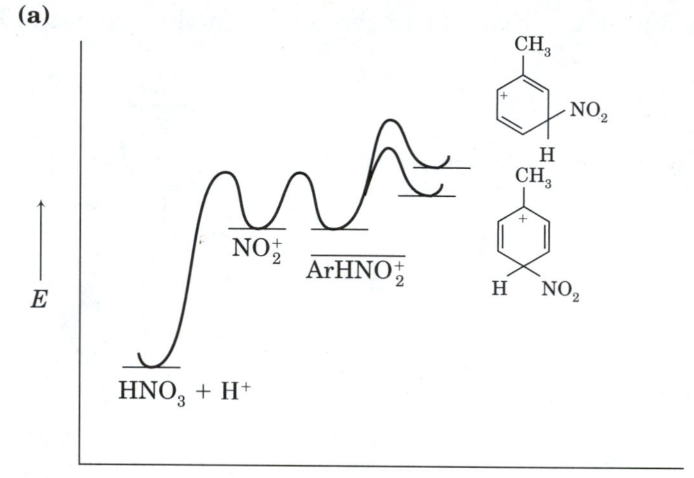

(b)

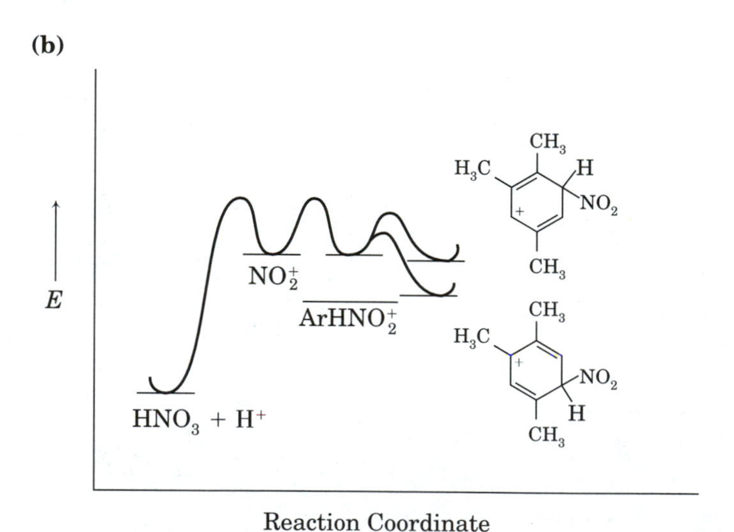

[169]Reference 175, p. 39.

[170]The formation of a complex prior to reaction is well supported in the case of bromination. Molecular bromine forms charge transfer complexes with benzene even in the absence of Lewis acid catalysts. The complexes can be detected spectroscopically and can even be crystallized for structure determination. Complexes of Br^+ with the aromatic may also be formed. For a summary of experimental data in this area, see the discussion in reference 177.

Quantitative Measurement of S_EAr Rates: Partial Rate Factors

More complex than S_EAr reactions of benzene itself are reactions of benzene derivatives. Benzene has six equivalent sites for reaction. However, a monosubstituted benzene has only five, and these five sites are not equivalent because two are ortho to the substituent, two are meta, and one is para. Often it is desirable to accurately compare the reactivity at one site in benzene with the reactivity at one site of each type in a substituted derivative of benzene. The reactivities so determined are called **partial rate factors** (denoted f^Z) and are defined as follows:[171,172,173]

$$f_o^Z = [(k'/2)/(k/6)] \times (\text{percent ortho product}/100) \qquad \textbf{(8.47)}$$

$$f_m^Z = [(k'/2)/(k/6)] \times (\text{percent meta product}/100) \qquad \textbf{(8.48)}$$

$$f_p^Z = [(k'/1)/(k/6)] \times (\text{percent para product}/100) \qquad \textbf{(8.49)}$$

where k is the rate of reaction of benzene and k' is the rate of reaction of the derivative.[174] Thus f_o^Z is the relative reactivity at *one* of the sites ortho to a substituent in a monosubstituted benzene relative to the reactivity of *one* of the six sites on benzene under the same conditions. Equations 8.47 through 8.49 may be rearranged to give equations 8.50 through 8.52.

$$f_o^Z = .03 \times (k'/k) \times \text{percent ortho product} \qquad \textbf{(8.50)}$$

$$f_m^Z = .03 \times (k'/k) \times \text{percent meta product} \qquad \textbf{(8.51)}$$

$$f_p^Z = .06 \times (k'/k) \times \text{percent para product} \qquad \textbf{(8.52)}$$

As an example, the bromination of toluene with HOBr in dioxane-water proceeds 36.2 times as fast as does bromination of benzene, and the product is 70.3% *ortho*-, 2.3% *meta*-, and 27.4% *para*-bromotoluene.[163] Thus the partial rate factors are

$$f_o^Z = .03 \times 36.2 \times 70.3 = 76.3 \qquad \textbf{(8.53)}$$

$$f_m^Z = .03 \times 36.2 \times 2.3 = 2.5 \qquad \textbf{(8.54)}$$

$$f_p^Z = .06 \times 36.2 \times 27.4 = 59.5 \qquad \textbf{(8.55)}$$

Partial rate factors are indicated by the numbers shown near each reaction site (Figure 8.52).

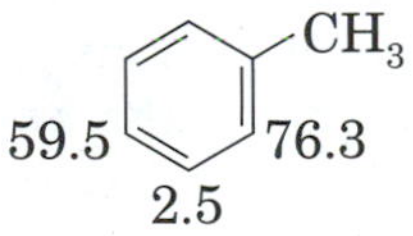

Figure 8.52 Partial rate factors for the S_EAr bromination of toluene. (Data from reference 163.)

CH_3

59.5 76.3

2.5

[171]Reference 4, p. 289.

[172]Taylor, R. *Electrophilic Aromatic Substitution*; John Wiley & Sons: Chichester, England, 1990; pp. 40 *ff*.

[173]The superscript Z refers to a substituent on the aromatic ring. In some sources the partial rate factor is represented with the symbol $f^{\phi Z}$, in which ϕZ more explicitly indicates a monosubstituted derivative of benzene. (See, for example, reference 5, pp. 273–274.) Also used in the literature are symbols in which the position of the letter f and the positional indicator are reversed. For example, f_o^Z may be shown as o_f, as in reference 175.

[174]The statistical factors are included in the equations because there are six equivalent sites for reaction on benzene, while there are two equivalent ortho sites and two equivalent meta sites, but only one para site for reaction in a monosubstituted benzene.

Table 8.16 Partial rate factors for nitration with nitric acid in nitromethane at 25°.

Compound	Relative Rate k/k_a	Isomer Distribution ortho	meta	para	Partial Rate Factor f_o^Z	f_m^Z	f_p^Z
t-Butylbenzene	15	12.2	8.2	79.6	5.5	3.7	71.6
Toluene	21	61.7	1.9	36.4	38.9	1.3	45.8
Chlorobenzene	0.031	29.6	0.9	69.5	0.028	0.00084	0.130
Bromobenzene	0.028	36.5	1.2	62.4	0.030	0.00098	0.103

(Data from reference 175.)

Partial rate factors allow us to quantify the qualitative terms *activating, ortho, para directing, deactivating, meta directing*, and *deactivating, ortho, para directing* that are used to label substituents. If $f^Z > 1$, then that site is activated, since it reacts faster than a site on benzene. Similarly, if $f^Z < 1$, then it is deactivated. If f_o^Z, $f_p^Z > f_m^Z$, the substituent is an ortho, para director. If $f_m^Z > f_o^Z$, f_p^Z, the substituent is a meta director. Although the ortho and para positions are treated together in qualitative discussions of directing effects, they clearly are in different steric environments, and the values of f_o^Z and f_p^Z can be significantly different. Table 8.16 shows the partial rate factors for nitration of four substituted benzenes.[175] Note that the relative reactivity of ortho and para positions is quite different in toluene and *t*-butylbenzene, apparently because of steric hindrance to the approach of the nucleophile in the latter compound.[176,177]

Lewis Structure Models of Reactivity in S_EAr Reactions

The partial rate factor concept suggests that there is a unique rate constant for reaction at each position of the benzene derivative. Therefore, competitive reactions at different sites within one molecule may be described in the same way as are competitive reactions at sites in different molecules. In the case of comparison of ortho and para sites of toluene with the meta position, it is useful to compare the energies of the σ-intermediates for ortho and para reaction to that for meta substitution. The Lewis structure descrip-

[175] Stock, L. M. *Prog. Phys. Org. Chem.* **1976**, *12*, 21.

[176] Product distribution can also be altered by carrying out reactions in heterogeneous environments. For example, the percent yield of *p*-nitrotoluene increases from 34% in homogeneous solution to 55% when the reaction is carried out on a clay support. For a discussion, see Delaude, L.; Laszlo, P.; Smith, K. *Acc. Chem. Res.* **1993**, *26*, 607.

[177] Reaction conditions have strong effects on relative reactivities of benzene derivatives. Under some conditions the relative reactivity of toluene to benzene (k_T/k_B) is nearly 30, but under other conditions the two compounds react at essentially the same rate. Berliner, E. *Prog. Phys. Org. Chem.* **1964**, *2*, 253; see especially the discussion on p. 310.

Figure 8.53 Resonance structures for ortho, meta, and para attack of an electrophile on toluene.

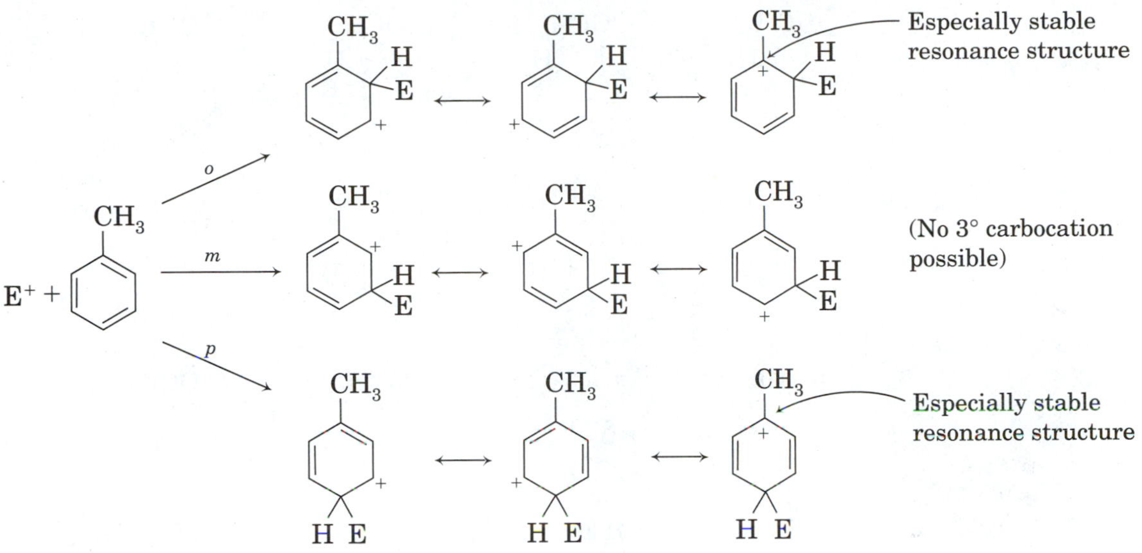

tions of the σ-intermediates formed by reaction of an electrophile with toluene are shown in Figure 8.53. In the case of ortho and para substitution, one of the three resonance structures for the intermediate can be described as a 3° carbocation, while all the contributing resonance structures for the intermediate in meta substitution are 2° carbocations.[168] Based on the Hammond postulate, there should be a lower activation energy for formation of a more stable intermediate, so the rates of formation of the ortho and para substitution products should be greater than the rate of formation of the meta product. For highly activated derivatives such as anisole, the intermediate for ortho and para reaction has a contribution from a fourth resonance structure, but a similar resonance structure is not possible for reaction at the meta position (Figure 8.54).

On the other hand, the intermediates formed upon electrophilic substitution at the ortho or para positions on benzene rings with strongly deactivating substituents (such as nitro or trialkylammonium groups) involve contributions from resonance structures representing unstable carbocations. However, such unstable resonance structures do not contribute to the resonance hybrid for the intermediate for meta reaction (Figure 8.55). Therefore, the activation energy for formation of the σ-intermediate should be lower for meta substitution than for ortho or para substitution, so more meta product should be observed.

Perhaps the most interesting case is that of halogen-substituted aromatics. In intuitive terms, the halogens are electron withdrawing by induction (conveniently viewed as a through-bond interaction) while they are electron donating through the π system. The π donation can be described by

Figure 8.54 Resonance structures for ortho, meta, and para attack of an electrophile on anisole.

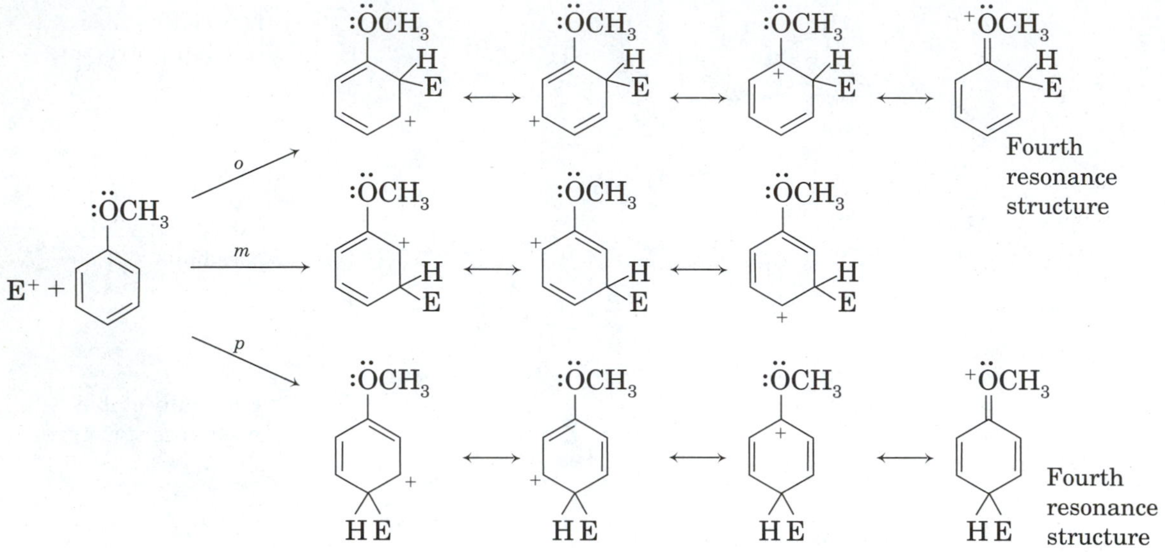

the fourth resonance structure in Figure 8.56. Therefore, the halogens are deactivating but are still ortho, para directing substituents.[178]

Although the mechanism of the S_EAr reaction as described here is widely accepted, alternative explanations have been advanced for some familiar S_EAr reactions. In particular, since single electron transfer processes in S_N2 reactions have been observed, they might be considered possible in S_EAr reactions as well. Perrin suggested that such reactions might occur through the mechanism for nitration shown in equations 8.56 and 8.57.[179,180] Here, the nitronium ion and the aromatic form a radical ion pair in a solvent cage (represented by the bar in the equations below), and the radical ion pair then collapses to the σ complex.[181]

$$^{+}NO_2 + ArH \xrightarrow[\text{controlled}]{\text{encounter}} \overline{NO_2\cdot + {}^{+}ArH\cdot} \tag{8.56}$$

$$\overline{NO_2\cdot + {}^{+}ArH\cdot} \longrightarrow [Ar{<}^{NO_2}_{H}]^{+} \tag{8.57}$$

[178]The analysis of substitution on multiply-substituted benzenes is usually carried out in terms of the same resonance (Lewis structure) approaches that are described here for benzene and its monosubstituted derivatives. For a discussion, see Bures, M. G.; Roos-Kozel, B. L.; Jorgensen, W.L. *J. Am. Chem. Soc.* **1985**, *50*, 4490.

[179]Perrin, C. *J. Am. Chem. Soc.* **1977**, *99*, 5516. Although the conclusions of this paper have been questioned (reference 180), there is experimental evidence for the electron transfer pathway in the nitration of naphthalene: Johnston, J. F.; Ridd, J. H.; Sandall, J. P. B. *J. Chem. Soc., Chem. Commun.* **1989**, 244.

[180]See, however, Eberson, L.; Radner, F. *Acc. Chem. Res.* **1987**, *20*, 53.

[181]Theoretical calculations have reinforced this conclusion: Gleghorn, J. T.; Torossian, G. *J. Chem. Soc. Perkin Trans. 2* **1987**, 1303.

Figure 8.55 Resonance structures for ortho, meta, and para attack of an electrophile on phenyltrialkylammonium ion.

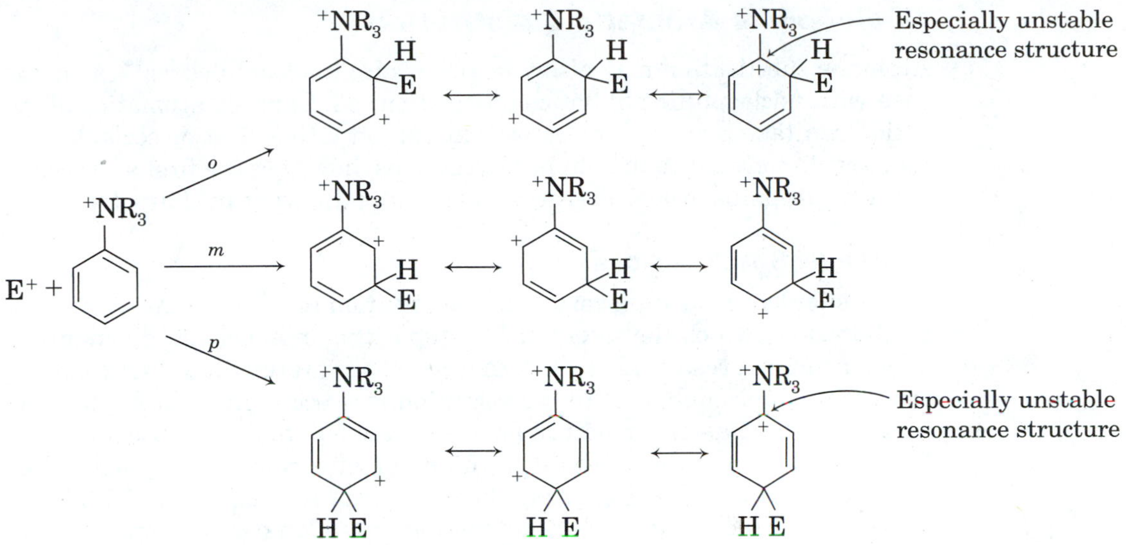

$^{+}NR_3$ $^{+}NR_3$ $^{+}NR_3$

H E ⟷ H E ⟷ H E

Especially unstable resonance structure

$^{+}NR_3$

E^+ +   o   m   p

$^{+}NR_3$ $^{+}NR_3$ $^{+}NR_3$

H E ⟷ H E ⟷ H E

$^{+}NR_3$ $^{+}NR_3$ $^{+}NR_3$

H E ⟷ H E ⟷ H E

Especially unstable resonance structure

Figure 8.56 Resonance structures for ortho, meta, and para attack of an electrophile on chlorobenzene.

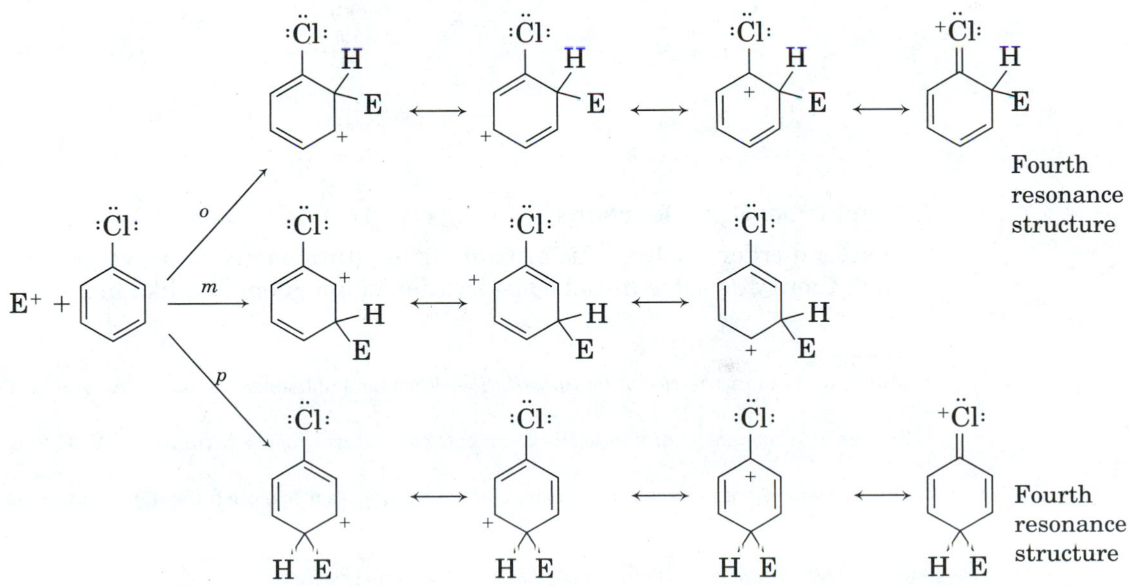

:C̈l: :C̈l: :C̈l: $^{+}$C̈l:

H E ⟷ H E ⟷ H E ⟷ H E

Fourth resonance structure

:C̈l:

E^+ +   o   m   p

:C̈l: :C̈l: :C̈l:

H E ⟷ H E ⟷ H E

:C̈l: :C̈l: :C̈l: $^{+}$C̈l:

H E ⟷ H E ⟷ H E ⟷ H E

Fourth resonance structure

8.4 Nucleophilic Aromatic and Vinylic Substitution

Nucleophilic Aromatic Substitution

Aromatic substitution may also take place with nucleophiles.[182,183] As is the case with nucleophilic aliphatic substitution, nucleophilic aromatic substitution can take place by processes that exhibit either first or second order kinetics. In contrast to the aliphatic reactions, however, the first and second order nucleophilic aromatic reactions are quite different in character.

First Order S_NAr Reactions

The first order nucleophilic aromatic substitution reaction (S_NAr) requires a mechanism in which the rate-limiting step is the unimolecular departure of a leaving group, resulting in the formation of an intermediate that can then react with a nucleophile. If the substitution is to occur on an aromatic ring, then the most feasible structure for the intermediate is a carbocation that has an empty sp^2-hybrid orbital in the plane of a benzene ring formed by dissociation of a C-(leaving group) bond, as shown in equation 8.58. Since aryl carbocations are much less stable than are alkyl carbocations,[184] there must be a strong driving force for the departure of the leaving group. For this reason, first order nucleophilic aromatic substitution reactions essentially all involve decomposition of arenediazonium ions to phenyl carbocations and molecular nitrogen, as illustrated in equation 8.59.[185]

$$\mathrm{Ar-X} \xrightarrow{\text{slow}} \mathrm{Ar^+} + \mathrm{X^-} \xrightarrow[\text{fast}]{\mathrm{Y^-}} \mathrm{Ar-Y} \qquad \textbf{(8.58)}$$

$$\mathrm{C_6H_5-\overset{+}{N}{\equiv}N:} \xrightarrow{\text{slow}} \mathrm{C_6H_5^+} + \mathrm{N_2} \xrightarrow{\mathrm{:Nu^-}} \mathrm{C_6H_5-Nu} \qquad \textbf{(8.59)}$$

Second Order S_NAr Reactions

The second order nucleophilic aromatic substitutions are more varied. Although there was some initial consideration of one-step, S_N2-like bimolecu-

[182] Miller, J. *Aromatic Nucleophilic Substitution*; Elsevier Publishing Company: Amsterdam, 1968.

[183] Terrier, F. *Nucleophilic Aromatic Displacement: The Influence of the Nitro Group*; VCH Publishers: New York, 1991.

[184] Orbitals with more s character are more electronegative, so it is more difficult to produce an empty sp^2-hybrid orbital than an empty sp^3-hybrid orbital.

[185] For another example of a first-order S_NAr reaction, see Himeshima, Y.; Kobayashi, H.; Sonoda, T. *J. Am. Chem. Soc.* **1985**, *107*, 5286.

lar mechanisms for these reactions,[186] Bunnett and Zahler argued against such pathways on theoretical grounds.[187] They suggested that the reactions instead proceed through an attachment-detachment ($A_N + D_N$) mechanism analogous to the mechanism for electrophilic aromatic substitution (equation 8.60). It is important to note a distinction between nucleophilic and electrophilic aromatic substitution, however. In the electrophilic reactions, there are usually several good electrofugal groups (the protons), so substituents can exert an influence on both the overall reactivity of the molecule and on the site of the reaction. (That is, some substituents are activating, ortho, para directors, etc.) In S_NAr reactions, however, there is seldom more than one good nucleofugal group. The detachment of a hydride ion is most unlikely, so the nucleophilic aromatic substitution replaces a good leaving group such as a halide ion. Other substituents on the aromatic ring can influence the reactivity of the molecule, but usually not the site of the reaction.[188,189,190]

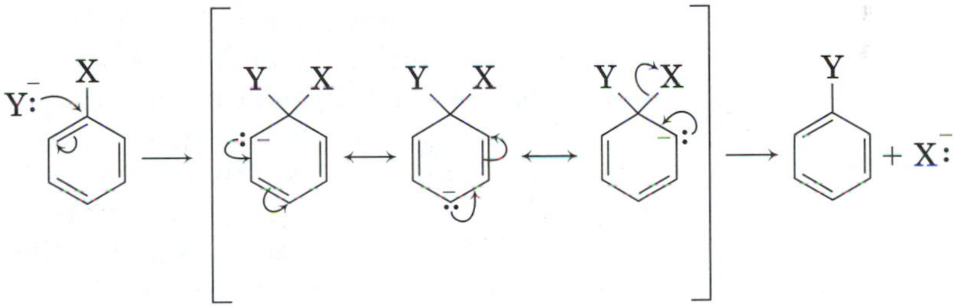

(8.60)

---

[186] *Cf.* Chapman, N. B.; Parker, R. E.; Soanes, P. W. *J. Chem. Soc.* **1954**, 2109 and earlier papers cited therein.

[187] Bunnett, J. F.; Zahler, R. E. *Chem. Rev.* **1951**, *49*, 273.

[188] It is possible, however, to determine the effects of substituents on the regiochemistry of the S_NAr reaction in the case of monosubstituted derivates of pentafluorobenzene because there are now nucleofugal groups in the ortho, meta, and para positions. In molecules with the structure C_6F_5X, for example, the NH_2 group is found to be a deactivating, meta director in S_NAr reactions. For a discussion and leading references, see reference 182, p. 127.

[189] Even though an S_NAr reaction is not likely to displace a hydride ion by the mechanism discussed here, a process that appears to produce that result may occur through a "vicarious" nucleophilic substitution of an aromatic hydrogen. In the equation below, the nucleophile bears a leaving group, L. Attachment of such a nucleophile to an aromatic ring (at a position determined by the electron withdrawing groups on the ring) has been reported to produce an intermediate that can then undergo base-promoted elimination of H (from the benzene ring) and L (from the nucleophile). Subsequent protonation and tautomerization lead to the product shown.

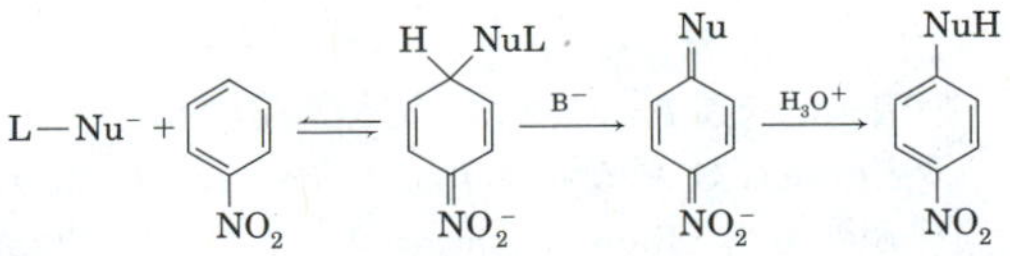

Makosza, M.; Winiarski, J. *J. Org. Chem.* **1984**, *49*, 1494; Makosza, M.; Ludwiczak, S. *J. Org. Chem.* **1984**, *49*, 4562.

[190] A theoretical study has suggested that charge transfer complexes may also be involved in the S_NAr reaction. Dotterer, S. K.; Harris, R. L. *J. Org. Chem.* **1988**, *53*, 777.

Figure 8.57 Resonance interaction of a *p*-nitro substituent in an S_NAr substitution.

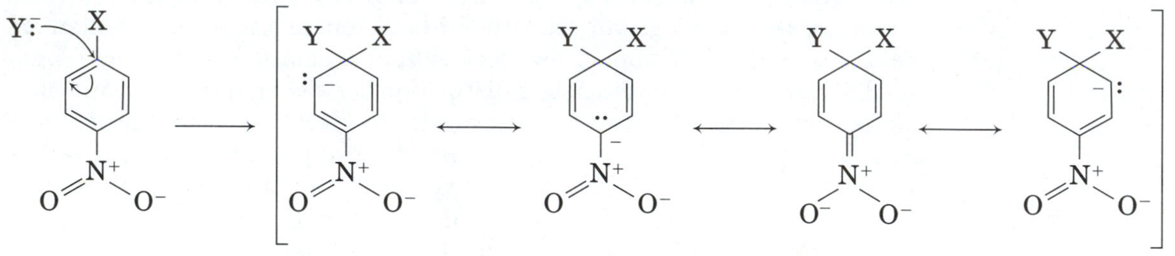

Several lines of evidence support the mechanism shown in equation 8.60.[191,192] Cryoscopic measurements suggest the formation of an adduct between the aromatic compound and a nucleophilic solvent,[193] and adducts known as Jackson-Meisenheimer complexes (analogous to the σ complexes of electrophilic aromatic substitution) can be isolated in some cases.[194] The S_NAr reaction is accelerated by groups such as nitro and cyano that can accept electrons by resonance in the ortho and para positions (Figure 8.57), and a Hammett correlation confirms the role of electron-withdrawing substituents.[195] In addition, there is generally no ***element effect*** (difference in rate constant with different leaving group) in the reaction. For example, reactions of several 1-substituted-2,4-dinitrobenzenes with piperidine in methanol solution show rate constants and activation parameters that are very similar (Table 8.17), indicating that the C—X bond is broken only slightly (if at all) in the transition structures for reaction of these sub-

Table 8.17 Kinetic data for reaction of piperidine with 1-substituted-2,4-dinitrobenzenes in methanol solution at 0°.

Substituent	Rate constant (M^{-1} min^{-1})	E_a (kcal/mol)	$\Delta S^{\ddagger}$ (e.u.)
Br	0.118	11.8	−29.5
Cl	0.117	11.6	−30.2
$C_6H_5SO_2$	0.0860	12.0	−29.3
p-$NO_2C_6H_4O$	0.0812	10.5	−35.3
I	0.0272	12.0	−31.7

(Data from reference 196.)

[191]Ross, S. D. *Prog. Phys. Org. Chem.* **1963**, *1*, 31.

[192]Bunnett, J. F. *Quart. Rev. Chem. Soc.* **1958**, *12*, 1.

[193]Baliah, V.; Ramakrishnan, V. *Rec. Trav. Chim. Pays-Bas* **1959**, *78*, 783.

[194]For a discussion see Crampton, M. R. *Adv. Phys. Org. Chem.* **1969**, *7*, 211; also see Mariella, R. P.; Callahan, J. J.; Jibril, A. O. *J. Org. Chem.* **1955**, *20*, 1721.

[195]Berliner, E.; Monack, L. C. *J. Am. Chem. Soc.* **1952**, *74*, 1574; Bunnett, J. F.; Moe, H.; Knutson, D. *J. Am. Chem. Soc.* **1954**, *76*, 3936.

strates.[196,197,198] An intramolecular S_NAr reaction is the Smiles rearrangement (equation 8.61).[199]

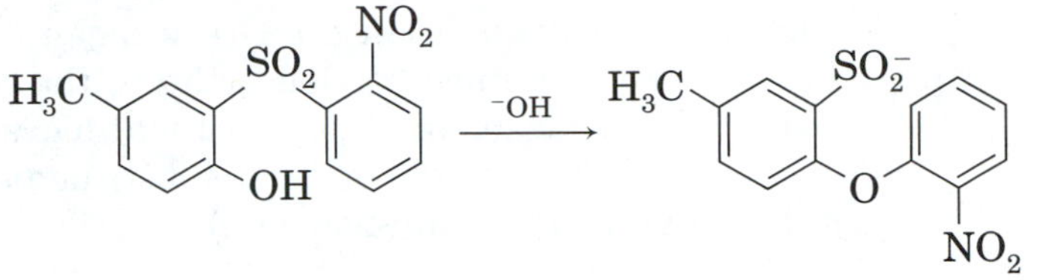

$$\text{(8.61)}$$

There are some notable exceptions to the general rule that nucleophilic aromatic substitution does not lead to replacement of a hydrogen. In one example of the **von Richter reaction**, cyanide and *p*-nitrochlorobenzene produce *m*-chlorobenzoic acid. Results of isotopic labeling experiments have suggested the mechanism shown in Figure 8.58, and this proposed mecha-

Figure 8.58 Mechanism proposed for the von Richter reaction. (Adapted from reference 200(c).)

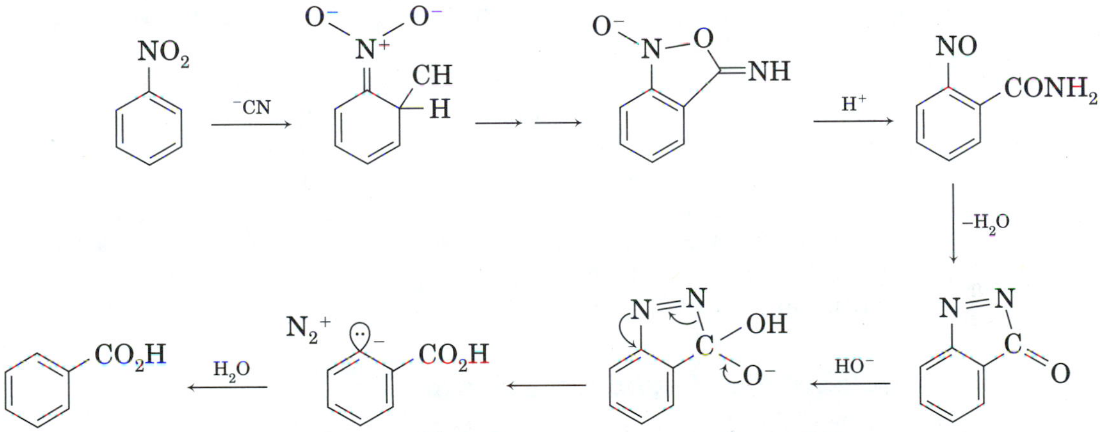

---

[196]Bunnett, J. F.; Garbisch, Jr., E. W.; Pruitt, K. M. *J. Am. Chem. Soc.* **1957**, *79*, 385.

[197]In protic solvents, solvent assistance of the departing nucleofuge makes the attachment of the nucleophile rate-limiting, so no leaving group effect is seen. The rate-limiting step in S_NAr reactions in aprotic solvents such as THF appears to be the detachment of the nucleofuge. Nudelman, N. S.; Mancini, P. M. E.; Martinez, R. D.; Vottero, L. R. *J. Chem. Soc. Perkin Trans. 2* **1987**, 951. In addition, an $^{18}F/^{19}F$ fluorine kinetic isotope effect of 1.0262 has been observed for the reaction of piperidine with 2,4-dinitrofluorobenzene in THF, suggesting that the C—F bond is broken in the rate-limiting step. Matsson, O.; Persson, J.; Axelsson, B. S.; Långström, B. *J. Am. Chem. Soc.* **1993**, *115*, 5288.

[198]A leaving group effect (Br > Cl > F) is observed for the 4-chlorobenzoylCoA dehalogenase-catalyzed substitution of OH for X in *p*-X-halogenobenzoate ion in aqueous solution. One explanation for this result is that the enzyme lowers the activation energy for the first step (attachment of the nucleophile to the aromatic ring), making the second step (detachment of halide ion) rate-limiting. Crooks, G. P.; Copley, S. D. *J. Am. Chem. Soc.* **1993**, *115*, 6422.

[199]Levy, A. A.; Rains, H. C.; Smiles, S. *J. Chem. Soc.* **1931**, 3264. For an extensive discussion, see Truce, W. E.; Kreider, E. M.; Brand, W. W. *Org. React.* ***1970***, 18, *99*.

nism has been supported by studies demonstrating that both the *o*-nitrosobenzamide and the subsequent indazolone produce benzoic acid when subjected to the reaction conditions.[200] This reaction can be called a **cine substitution** (from the Greek meaning "to move"), since the nucleophilic atom becomes bonded ortho to the leaving group.[201] Substitution leading to replacement of a substituent more than one atom removed (usually meta or para to the original substituent on an aromatic ring) is called **tele substitution** (equation 8.62).[202]

CH_3, NO_2, CH_3, NO_2 $\xrightarrow[\text{DMSO}]{\text{ArSNa}}$ CH_2SAr, NO_2, CH_3 **(8.62)**

Single Electron Transfer in S_NAr Reactions

We have noted that both S_N2 and S_EAr reactions may occur through single electron transfer (SET) processes. There is good evidence that the S_NAr reaction may involve such intermediates also. Figure 8.59 shows species identified by Bacaloglu, Bunton, and Cerichelli in a fast kinetic spectroscopy study of the reaction of hydroxide ion with 1-chloro-2,4,6-trinitrobenzene (picryl chloride, **69**).[203] Depending on reaction conditions, these workers could see transients ascribed to the π complex **70**, a species produced by single electron transfer intermediate in the reaction (**71**), and one or more σ complexes (**72**, **74**). In addition, evidence was obtained for the reversible formation of a phenyl carbanion (**75**) that probably does not lead directly to substitution product.

Nucleophilic Vinylic Substitution

Analogous to nucleophilic aromatic substitution is nucleophilic vinylic substitution (equation 8.63), in which a nucleophile replaces a leaving group on

[200](a) Rauhut, M. M.; Bunnett, J. F. *J. Org. Chem.* **1956**, *21*, 939; Bunnett, J. F.; Rauhut, M. M. *J. Org. Chem.* **1956**, *21*, 944 and references therein; (b) Samuel, D. *J. Chem. Soc.* **1960**, 1318; (c) Rosenblum, M. *J. Am. Chem. Soc.* **1960**, *82*, 3796; (d) Ibne-Rasa, K. M.; Koubek, E. *J. Org. Chem.* **1963**, *28*, 3240. For further discussion, see Jones, R. A. Y.; *Physical and Mechanistic Organic Chemistry*, 2nd ed.; Cambridge University Press: Cambridge, 1984; pp. 2–5.

[201]The initial meaning of *cine* was loss of an original substituent and replacement with another group at **any** other site on the molecular framework. However, *cine* has come to mean that the replacement occurs on the carbon atom adjacent (ortho on an aromatic ring) to the site of the original substituent, while *tele* has come to mean substitution at some other site. For an account of the origin of these terms, see Nickon and Silversmith (reference 2), pp. 249–250.

[202]Novi, M.; Dell'Erba, C.; Sancassan, F. *J. Chem. Soc. Perkin Trans. 1* **1983**, 1145.

[203]Bacaloglu, R.; Bunton, C. A.; Cerichelli, G. *J. Am. Chem. Soc.* **1987**, *109*, 621. See also Bacaloglu, R.; Blaskó, A.; Bunton, C. A.; Ortega, F.; Zucco, C. *J. Am. Chem. Soc.* **1992**, *114*, 7708.

Figure 8.59 Evidence for single electron transfer in an S_NAr reaction. (Adapted from reference 203.)

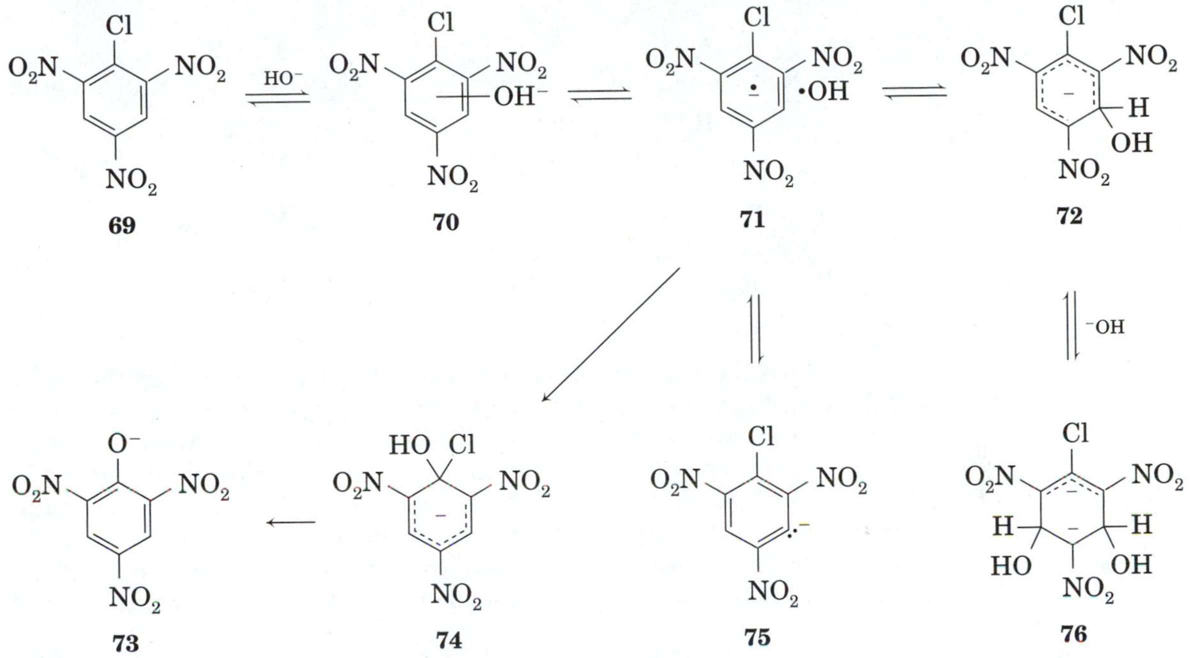

an olefinic carbon atom. In principle, reactions of the type shown in equation 8.63 may take place by a variety of mechanisms. Rappoport has identified sixteen different processes for nucleophilic substitution of a leaving group attached to olefins,[205] including an S_N1 reaction involving a vinyl cation,[204] a one-step substitution similar to an S_N2 reaction, an elimination-addition process (analogous to the benzyne process discussed on page 530), and an attachment-detachment pathway analogous to the S_NAr reaction.[205,206,207] The reaction in this area that has received most interest, and

[204]Stang, P. J.; Rappoport, Z.; Hanack, M.; Subramanian, L. R. *Vinyl Cations*; Academic Press: New York, 1979.

[205]Rappoport, Z. *Adv. Phys. Org. Chem.* **1969**, *7*, 1.

[206]Modena, G. *Acc. Chem. Res.* **1971**, *4*, 73.

[207]Rappoport, Z. *Acc. Chem. Res.* **1981**, *14*, 7.

[208]However, an *ab initio* study has suggested that a concerted S_N2 substitution reaction is possible in the case of *unactivated* substrates in which the attachment-detachment pathway is not favored by electron-withdrawing substituents, and some experimental data are consistent with this prediction: Glukhovtsev, M. N.; Pross, A.; Radom, L. *J. Am. Chem. Soc.* **1994**, *116*, 5961.

the process that most closely resembles the S_NAr mechanism, is the attachment-detachment pathway (equation 8.64).[208]

$$\mathrm{Nu^-} + \mathrm{(Y)(R_1)C_\beta{=}C_\alpha(X)(R_2)} \longrightarrow \mathrm{(Y)(R_1)C_\beta{=}C_\alpha(Nu)(R_2)} + \mathrm{X^-} \quad \textbf{(8.63)}$$

$$\mathrm{(Y)(R_1)C{=}C(X)(R_2)} + \mathrm{Nu^-} \underset{k_{-1}}{\overset{k_1}{\rightleftharpoons}} \left[\mathrm{(Y)(R_1)\overset{-}{C}{-}C(Nu)(X)(R_2)} \right] \xrightarrow{k_2} \mathrm{(Y)(R_1)C{=}C(Nu)(R_2)} + \mathrm{X^-} \quad \textbf{(8.64)}$$

Like the S_NAr reaction, the rate-limiting step is usually thought to be the attachment of the nucleophile to the unsaturated system, so the presence of groups on the β carbon atom that can stabilize negative charge lowers the energy of the intermediate in brackets and increases the rate of the reaction.[209] For example, Modena has noted the enhancement of reactivity achieved by replacing a methyl group with a 2,4-dinitrophenyl group at the β position of a vinyl halide. Note especially that the reaction in equation 8.65 requires temperatures around 130°, while the reaction in equation 8.66 occurs at − 20°.[206]

$$\mathrm{CH_3CH{=}CHCl} + \mathrm{C_2H_5S^-} \xrightarrow[\text{ca. 130°}]{\mathrm{CH_3CH_2OH}} \mathrm{CH_3CH{=}CHSC_2H_5} + \mathrm{Cl^-} \quad \textbf{(8.65)}$$

$$\mathrm{2,4\text{-}(NO_2)_2C_6H_3CH{=}CHBr} + \mathrm{C_6H_5S^-} \xrightarrow[\text{-20°}]{\mathrm{CH_3OH}} \mathrm{2,4\text{-}(NO_2)_2C_6H_3CH{=}CHSC_6H_5} + \mathrm{Br^-} \quad \textbf{(8.66)}$$

The attachment-detachment mechanism shown in equation 8.64 is supported by several lines of evidence. The reactions typically show kinetics that are overall second order, first order in the vinyl halide and first order in the nucleophile.[210] Particularly significant is the lack of an element effect. That is, for most vinyl halides, the rates of substitution are about the same for the chloride as for the bromide, which suggests that the halide is not eliminated in the rate-limiting step for the reaction. Moreover, the intermediate shown in brackets in equation 8.64 has been observed spectroscopically in a few cases.[211] However, the lifetime of the intermediate depends on

---

[209]In discussions of nucleophilic vinylic substitution, the carbon atom bearing the leaving group is commonly denoted the "α" carbon atom, while the other olefinic carbon atom is the "β" position. For a discussion, see reference 206.

[210]See, for example, Silversmith, E. F.; Smith, D. *J. Org. Chem.* **1958**, *23*, 427.

[211]Bernasconi, C. F.; Killion, Jr., R. B.; Fassberg, J.; Rappoport, Z. *J. Am. Chem. Soc.* **1989**, *111*, 6862; Bernasconi, C. F.; Fassberg, J.; Killion, Jr., R. B.; Rappoport, Z. *J. Am. Chem. Soc.* **1990**, *112*, 3169; also see Bernasconi, C. F.; Fassberg, J.; Killion, Jr., R. B.; Schuck, D. F.; Rappoport, Z. *J. Am. Chem. Soc.* **1991**, *113*, 4937.

the relative magnitude of the rate constants for the attachment and detachment steps, which are complex functions of the activating group(s) on the β carbon atom as well as the other substituent on the α carbon atom, the strength of the nucleophile, and the nucleofugality of the leaving group. In some cases the mechanism may approach a one-step process involving a transition state and not a true intermediate.[207]

One of the interesting aspects of nucleophilic vinylic substitution is the observation that the product often has the same stereochemistry as the reactant, as illustrated by the reactions of (*Z*)- and (*E*)-β-bromoethyl crotonate in equations 8.67 and 8.68.[212] (This stereochemical outcome is termed *retention of configuration.*) Not all reactions are stereospecific, however, and the stereochemical outcome depends on the nature of the nucleophile, leaving group, and especially on the activating group(s) on the β carbon atom.[213]

$$\mathrm{(Br)(H_3C)C{=}C(CO_2CH_2CH_3)(H)} \xrightarrow[\mathrm{CH_3CH_2OH},\ 25^\circ]{\mathrm{C_6H_5S^-}} \mathrm{(C_6H_5S)(H_3C)C{=}C(CO_2CH_2CH_3)(H)} \tag{8.67}$$

$$\mathrm{(H_3C)(Br)C{=}C(CO_2CH_2CH_3)(H)} \xrightarrow[\mathrm{CH_3CH_2OH},\ 25^\circ]{\mathrm{C_6H_5S^-}} \mathrm{(H_3C)(C_6H_5S)C{=}C(CO_2CH_2CH_3)(H)} \tag{8.68}$$

The tendency for nucleophilic vinylic substitution to give retention of configuration can be explained on the basis of the scheme in Figure 8.60,[214] with the following additional restrictions.[205,206]

1. The nucleophile approaches the α carbon atom on a line perpendicular to the plane of the olefin.
2. The α carbon atom is nearly tetrahedral in the transition structure because bond making from the nucleophile to the carbon atom is more advanced than is C_α—X bond breaking.
3. The β carbon atom remains relatively planar (especially if there is resonance delocalization of the negative charge) or is pyramidal but inverts so rapidly that its time average geometry is planar.
4. The leaving group departs on a line that is perpendicular to the plane of the developing double bond.

This last restriction means that some rotation about the olefinic carbon-carbon bond must occur in order for the leaving group to depart. A 60° rotation followed by detachment gives a substitution product in which the original stereochemistry is retained. A 120° rotation followed by detachment gives the diastereomer of that product. Several studies suggest that the 60° rota-

[212]Théron, F. *Bull. Soc. Chim. France* **1969**, 278. See also the discussion in Peishoff, C. E.; Jorgensen, W. L. *J. Org. Chem.* **1985**, *50*, 1056.

[213]Rappoport, Z.; Gazit, A. *J. Org. Chem.* **1956**, *51*, 4112.

[214]This figure is a considerably simplified version of more detailed schemes provided in the references.

Figure 8.60 Inversion (top) and retention (bottom) of configuration in nucleophilic vinylic substitution reactions.

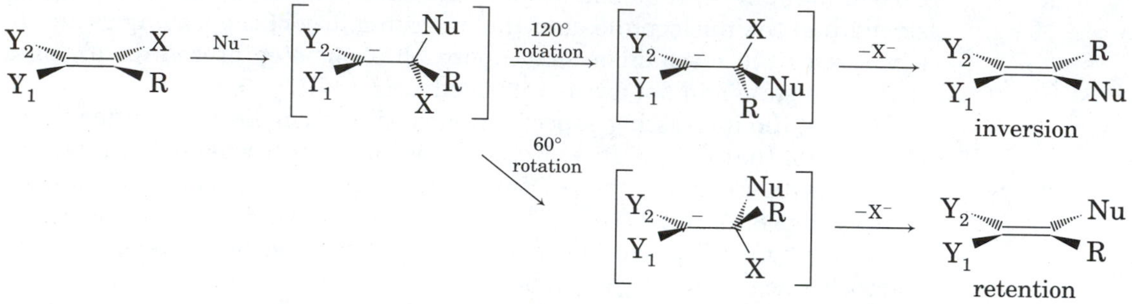

tion is faster than the 120° rotation, so retention is most likely with short-lived carbanionic intermediates.[215]

Nucleophilic Substitution Involving Benzyne Intermediates

Introduction

Because of the requirement for S_NAr reactions to have nitro or other electron-withdrawing groups ortho or para to the leaving group, unsubstituted benzenes generally react only slowly with nucleophiles. In an investigation of the effect of activing, ortho, para-directing substituents on such reactions, however, Bergstrom and co-workers found that chlorobenzene and other unactivated aromatic halides do react to give aniline (plus diphenylamine and other products) when treated with amide ion in liquid ammonia at − 33°. However, they do not react with sodium amide in diethyl ether at room temperature.[216,217]

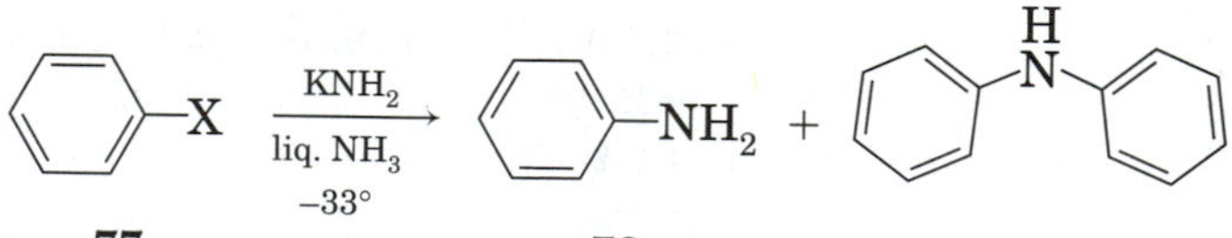

KNH_2, liq. NH_3, −33° (8.69)

77 (a) X = Cl; (b) X = Br; (c) X = 1 (X ≠ F) → **78** + **79** (plus other products)

Later Gilman and Avakian reported an unusual substitution reaction of 4,6-diiododibenzofuran (**80**) when they attempted to prepare the 4,6-

[215] See the discussion in reference 207 and references therein.

[216] Bergstrom, F. W.; Wright, R. E.; Chandler, C.; Gilkey, W. A. *J. Org. Chem.* **1936**, *1*, 170.

[217] The diphenylamine and triphenylamine were found to be formed as secondary products of the reaction of aniline and diphenylamine, respectively. Wright, R. E.; Bergstrom, F. W. *J. Org. Chem.* **1936**, *1*, 179.

diamino derivative by substitution of **80** with sodium amide in liquid NH_3. Instead of the expected product (4-amino-6-iododibenzofuran), they observed formation of 3-amino-6-iododibenzofuran (**81**) plus other products. They also found that 4-iododibenzofuran (**82**) gave the 3-amino derivative (**83**), but that 2-iododibenzofuran gave the 2-amino derivative.[218] Extending the study to derivatives of benzene, Gilman and Avakian found that under the same conditions *m*-anisidine was obtained from reaction of *o*-iodoanisole, *o*-bromoanisole, or *o*-chloroanisole. They concluded that the apparent rearrangement of *o*-halo aromatic ethers to yield *m*-amino aromatic ethers was characteristic of molecules with a halogen ortho to an aromatic ether oxygen.

$NaNH_2$, liq. NH_3 **(8.70)**

80 → **81**

plus other products

$NaNH_2$, liq. NH_3 **(8.71)**

82 → **83**

X = Br, I

In later investigations, Gilman and co-workers observed rearranged product from the reaction of *o*-bromodimethylaniline under similar conditions.[219] Further studies showed that *p*-bromoanisole reacted to give *m*-methoxyaniline.[220] Reaction of *o*-chlorotrifluoromethylbenzene gave only the *m*-trifluoromethylaniline, but reaction of *m*-chlorotrifluoromethylbenzene gave only the expected *m*-trifluoromethylaniline. The language in these papers suggests that the authors believed the "normal" reaction (taking place without apparent rearrangement) to be an S_NAr type reaction, while the rearrangement was believed to take place by a different mechanism. For

[218] Gilman, H.; Avakian, S. *J. Am. Chem. Soc.* **1945**, *67*, 349.

[219] Gilman, H.; Kyle, R. H.; Benkeser, R. A. *J. Am. Chem. Soc.* **1946**, *68*, 143.

[220] Gilman, H.; Kyle, R. H. *J. Am. Chem. Soc.* **1948**, *70*, 3945.

simplicity, the "normal" mechanism (equation 8.72) and a possible "abnormal" mechanism (equation 8.73) are illustrated for benzene:

Resonance structures

(8.72)

Resonance structures

(8.73)

Experimental Evidence for Benzyne Intermediates

The benzyne mechanism for these reactions was established through work of Roberts and co-workers.[221,222,223] Summarizing the various reactions reported in the literature, Roberts noted that

1. The amino group was always found either on the carbon atom from which the leaving group departed or, at most, one carbon atom away;
2. Neither the starting materials (aryl halides) nor the products (arylamines) appeared to isomerize under the reaction conditions; and
3. The reactivities of the halogens were Br > I > Cl >> F.

Two mechanistic alternatives could explain these results. In the first, the aryl halide reacts with the incoming amino group by competitive "normal" (equation 8.72) and "abnormal" (equation 8.73) pathways. In the second, an

---

[221] Roberts, J. D.; Semenow, D. A.; Simmons, Jr., H. E.; Carlsmith, L. A. *J. Am. Chem. Soc.* **1956**, *78*, 601.

[222] Particularly noteworthy with regard to our continuing emphasis on the roles of mechanisms as models in organic chemistry is the use of quotation marks around the word "prove" in the statement by Roberts (reference 221) that "The above facts strongly indicate but do not 'prove' that benzyne is the intermediate in the [reaction]. Therefore, other reaction mechanisms will be considered in order to determine whether a more satisfactory formulation can be found."

[223] The benzyne mechanism can be denoted as an $\mathbf{S_N(EA)}$ (Substitution Nucleophilic Elimination Addition) reaction.

elimination-addition mechanism produces as an intermediate a benzyne, an aromatic ring with a formal triple bond (equation 8.74).

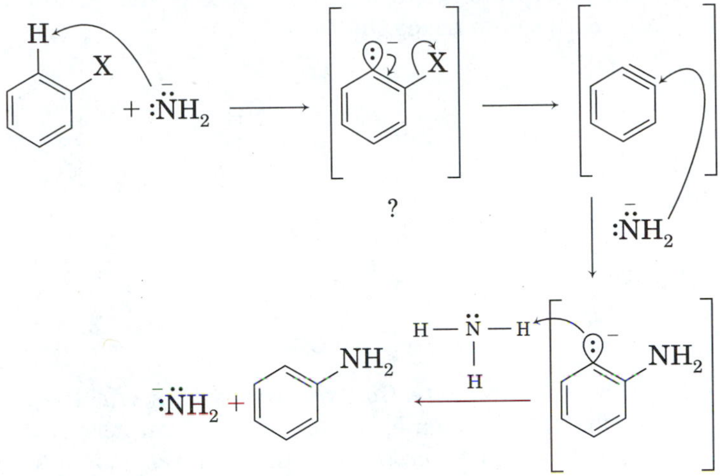

(8.74)

To determine the regiochemistry of halide replacement by amide ion, Roberts and co-workers studied the reaction of chlorobenzene-^{14}C and iodobenzene-^{14}C with KNH_2 in liquid NH_3, as shown in Figure 8.61. The relative yield of the two products is consistent with the mechanism shown in Equation 8.74.[224] Although it is conceivable that the same products might be formed through competing "normal" and "abnormal" S_NAr pathways (equations 8.72 and 8.73), it would seem to be highly unlikely that the two halides would give the same product distribution in that case, since the proportion of S_N2 and E2 reactions in alkyl halides is strongly dependent on the leaving group.

The mechanism shown in equation 8.74 requires that both a C—H and C—X bond be broken in proceeding from the reactant to the benzyne intermediate. Roberts and co-workers considered that these two groups might be lost in a concerted reaction (equation 8.75) or in consecutive processes (equation 8.76). Use of halobenzenes labeled with deuterium on one of the

Figure 8.61 Distribution of isotopic label after substitution.

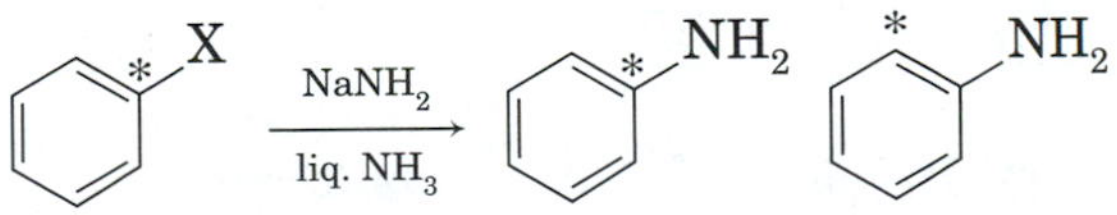

$NaNH_2$ / liq. NH_3

	(label at C bearing NH_2)	(label ortho to NH_2)
X = I	46%	52%
X = Cl	43%	53%

[224]The relative yields are not exactly 50 : 50 due to a small carbon isotope effect favoring attachment of the nucleophile to ^{12}C.

positions ortho to the halogen gave further insight into the reaction. If the reaction occurs in step-wise fashion, then the initially formed carbanion might be reprotonated, leading to loss of deuterium label in the reactant. Table 8.18 summarizes the results.

(8.75)

(8.76)

The data indicate that the reaction is effectively concerted (equation 8.75) for bromobenzene, because deuterium loss in the reactant is negligible. However, chlorobenzene appears to react by a stepwise process (equation 8.76). During a reaction period sufficient to convert 28% of reactant to product, 13% of the reactant underwent proton exchange (deuterium loss). The fluorobenzenes also underwent rapid proton exchange, but the reaction did not lead to anilines. The ease with which an ortho hydrogen is abstracted by the carbanion formed by proton loss from a halobenzene varies with the electronegativity of the halogen, F > Cl > Br > I; however, the ease with which the halide ion leaves is I > Br > Cl > F. For bromobenzene and iodobenzene, the rate-limiting step is hydron removal; for fluorobenzene and chlorobenzene the rate-limiting step is halide ion departure. Thus the overall reactivity of aryl halides varies as Br > I > Cl >> F.[221]

Product Ratios in Benzyne Reactions

The near-equivalence of the rate constants for protonation and for departure of halide ion from the aryl carbanion derived from chlorobenzene sug-

Table 8.18 Deuterium loss and aniline formation in the reaction of *o*-deuterioaryl halides with KNH_2 in liquid ammonia.

Aryl Halide	Deuterium Loss from Reactant	% Aniline Formation
Fluorobenzene-2,4,6-$^2H_3$	100%	0
Fluorobenzene-3,5-$^2H_2$	100%	0
Fluorobenzene-2-^2H	100%	0
Chlorobenzene-2-^2H	13%	28
Bromobenzene-2-^2H		72

(Data from reference 221.)

Table 8.19 Product distributions for reaction of substituted halobenzenes with amide in liquid ammonia.

R	X	Yield (%)	% Ortho	% Meta	% Para
o-CF_3	Cl	28		100	
o-CH_3	Cl	66	45	55	
o-CH_3	Br	64	48.4	51.5	
o-OCH_3	Br	33		100	
p-CF_3	Cl	25		50	50
p-CH_3	Cl	35		62	38
p-OCH_3	Br	31		49	51
m-CF_3	Cl	16		100	
m-CH_3	Cl	66	40	52	8
m-CH_3	Br	61	22	56	22
m-OCH_3	Br	59		100	

(Data from reference 225.)

gests an explanation for the effects of substituents on the rates and product distributions observed in benzyne reactions. Table 8.19 summarizes the distribution of substituted aniline products observed in the reaction of substituted bromo- and chlorobenzenes with amide ion in liquid ammonia.[225] For the ortho-substituted compounds, benzyne formation can only occur so as to give the intermediate with the triple bond located between carbon atoms 2 and 3 (relative to the substituent), as shown in Figure 8.62. Attach-

Figure 8.62 Effect of ortho substituents on patterns of benzyne substitution.

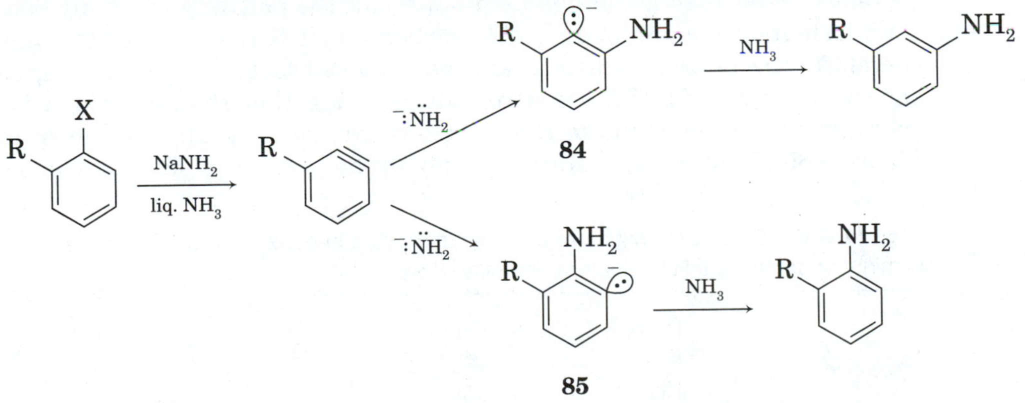

[225] Roberts, J. D., Vaughan, C. W., Carlsmith, L. A., Semenow, D. A. *J. Am. Chem. Soc.* **1956**, *78*, 611.

Figure 8.63
Effect of para substituents on patterns of benzyne substitution.

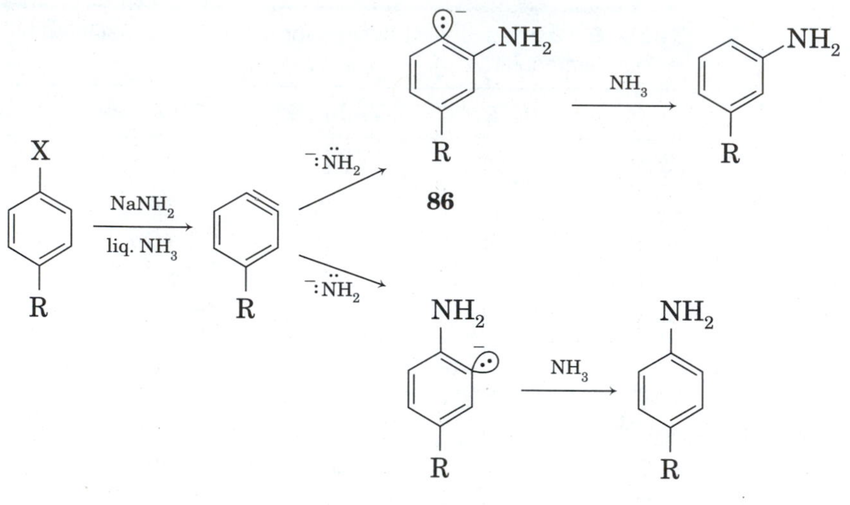

ment of a nucleophile to the triple bond of such a benzyne could yield only the ortho and meta products. For R = *o*-CF_3, only the meta product is formed, however. We infer that the transition state leading to phenyl carbanion **84** is lower than that leading to **85** because of the greater stability of the carbanion in which the negative charge is closer to the electron-withdrawing CF_3 group. When R = *o*-CH_3, which we traditionally consider to be weakly electron donating in electrophile aromatic substitution, approximately equal amounts of ortho and meta products are formed. Similar arguments apply to the intermediates (**86** and **87**) when *p*-substituted halobenzenes react.

Two different benzyne intermediates can be formed by reaction of meta-substituted halobenzenes, but we expect a lower activation energy (and therefore faster reaction and more product) for the pathway involving benzyne formation via the more stable carbanion. If R is electron donating, therefore, the proton para to R will be removed faster by amide ion, as shown in Figure 8.64. If R is electron withdrawing, then the carbanion with the negative charge ortho to R will be formed preferentially, as shown in Figure 8.65. Similar considerations apply to the prediction of the site of re-

Figure 8.64 Effect of substituent on elimination step in benzyne mechanism: favored elimination pathway if R is electron donating.

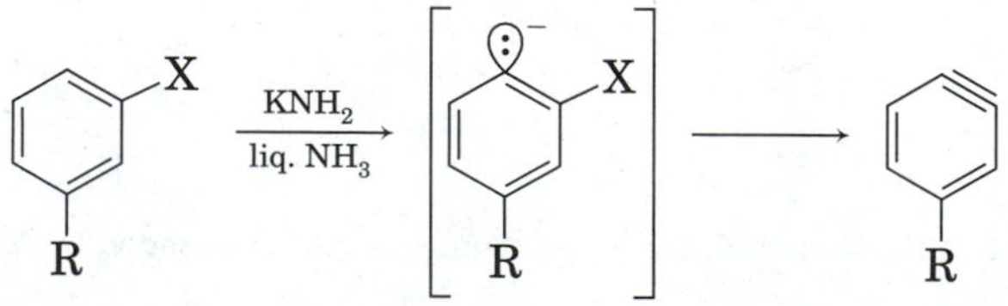

Figure 8.65 Effect of substituent on elimination step in benzyne mechanism: favored elimination pathway if R is electron withdrawing.

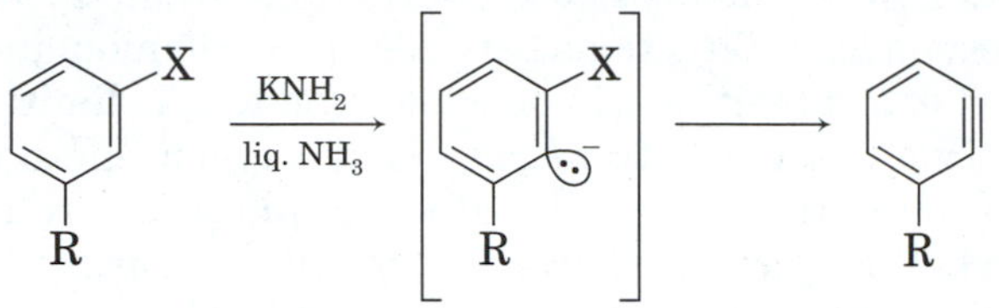

action of the resulting benzyne with amide ion, as shown in Figure 8.66 and Figure 8.67.

Benzyne Substituent Effects and Implicit Models

The preceding explanation would seem to explain most of the data in Table 8.19, but there is one apparent discrepancy. We might have expected the methoxy substituent to be electron donating, but it gives the same product orientation as does trifluoromethyl. However, this intuitive expectation of the substituent effect of methoxy is based primarily on its influence on either electrophilic aromatic substitution (S_EAr) or on nucleophilic aromatic substitution (S_NAr), both of which involve attachment of a species to an aromatic ring with subsequent formation of an intermediate having a tetrasubstituted carbon atom in the six-membered ring. In contrast, the carbanionic intermediates presumed to be formed in the reaction have the nonbonded pair of electrons in an sp^2-hybrid orbital in the plane of the aromatic ring.

Figure 8.66 Effect of substituents on addition step of benzyne mechanism: favored pathway if R is electron donating.

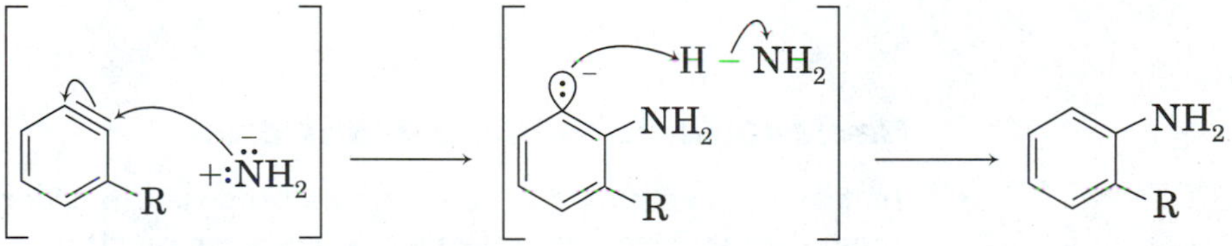

Figure 8.67 Effect of substituents on addition step of benzyne mechanism: favored pathway if R is electron withdrawing.

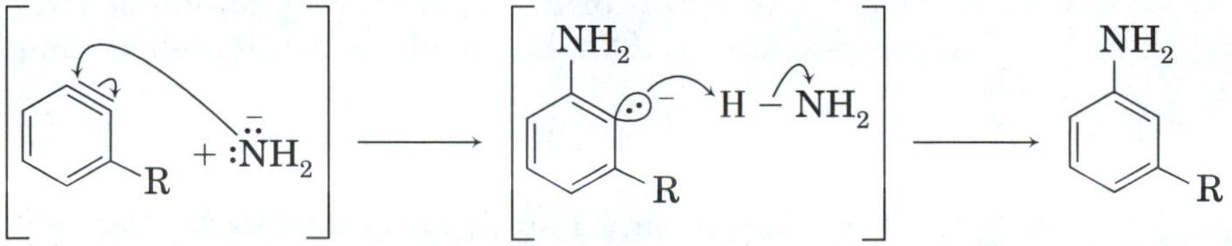

This sp^2-hybrid orbital is therefore orthogonal to the aromatic π system, so interaction of this orbital with the π system and with p orbitals on substituents is precluded. In other words, the geometry of the carbanion prevents resonance interactions between the carbanion and the substituent, so the effect of a substituent in benzyne reactions is limited to inductive (field) effects. Due to the electronegativity of oxygen, methoxy is electron-withdrawing by induction, so its effect is qualitatively the same as trifluoromethyl. This example shows the potential hazards in transferring models from one reaction to another without explicitly considering the underlying bases of those models.

One other aspect of benzyne chemistry deserves mention. We have drawn the benzyne structure as **88**, which might be called the aryne representation. However, we cannot *a priori* state that a structure with what is formally one triple bond and two double bonds in a six-membered ring is more stable than a structure with four double bonds, which is the cumulene representation of benzyne (**91**). We should also consider the possibility of some biradicaloid nature of benzyne, as suggested by the resonance structures **89** and **90**. There have been several attempts to determine experimentally and theoretically which drawing is the best graphical representation of benzyne, and Liang and Berry found that analysis of infrared absorption bands of matrix-isolated benzyne suggested the aryne structure.[226] Nevertheless, our structural models should not be limited by our desire to draw simple graphical models in synthetic or mechanistic schemes, and there may be utility in remembering the possible contribution of the other resonance structures also.

88 **89** **90** **91**

Radical-Nucleophilic Substitution

Just as the benzyne mechanism was proposed in order to explain apparent anomalies in reactions intended to be S_NAr reactions, the $S_{RN}1$ mechanism for nucleophilic aromatic substitution was suggested by the results of studies under conditions in which the benzyne mechanism was expected to operate. Kim and Bunnett investigated the reaction of halogen-substituted isomers of pseudocumene (1,2,4-trimethylbenzene, **92**) with KNH_2 in liquid NH_3.[227] As shown in Figure 8.68, elimination of HX from both the 5-halopseudocumenes (**93a, b, c**) and the 6-halopseudocumenes (**94a, b, c**) should

[226]Liang, J. W.; Berry, R. S. *J. Am. Chem. Soc.* **1976**, *98*, 660.

[227]Kim, J. K.; Bunnett, J. F. *J. Am. Chem. Soc.* **1970**, *92*, 7463.

produce the same aryne intermediate, **95**. As a result, the distribution of 5-(**96**) and 6-pseudocumidine (**97**) formed via a benzyne reaction should be the same, no matter whether the reactant was **93a, 93b, 93c, 94a, 94b**, or **94c**.

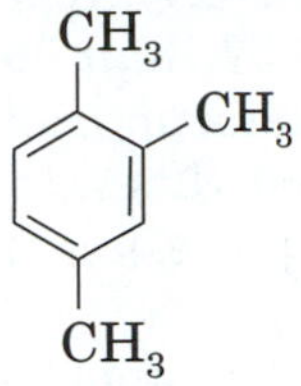

92

The ratio of products from reaction of **93a, 93b, 94a**, and **94b** were found to be identical within experimental error. (Ratios of **97** to **96** ranged from 1.45 to 1.55.) There was a wide variation in the product distribution formed from the iodopseudocumenes, however. When the reactant was **93c**, the ratio of **97** to **96** was 0.63; when the reactant was **94c**, the ratio of **97** to **96** was 5.86. In addition, significant yields of **92** were observed from reactions with **93c** and **94c**.

Figure 8.68 Aryne mechanism for reaction of isomeric halo-pseudocumenes with KNH_2 in liquid NH_3. (Adapted from reference 227.)

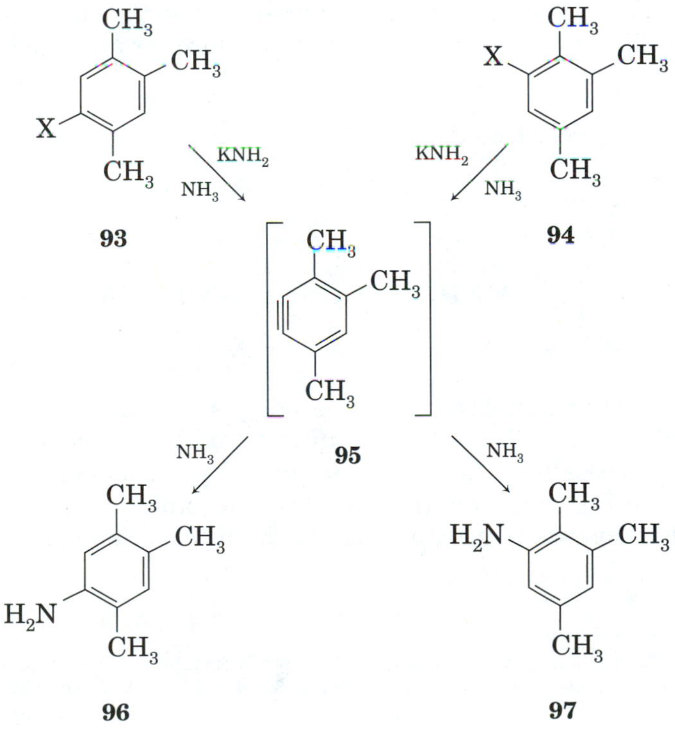

a, X = Cl; b, X = Br; c, X = I

The observation that some **97** formed in the reaction of **93c** and some **96** from reaction of **94c** suggested that at least some product must arise through the aryne mechanism shown in Figure 8.68. However, the observation of relatively greater amounts of **96** than **97** from **93c** and of greater amounts of **97** than **96** from **94c** indicated that there must also be a competing mechanism leading to the unrearranged products. It was considered unlikely that this alternative mechanism could be an S_NAr pathway, because that process is more favorable for aryl bromides and chlorides than for iodides.

Kim and Bunnett also found that the product ratios from reactions **93c** and **94c** could be shifted toward the ratios observed with other halogens by adding a radical trap such as tetraphenylhydrazine to the solvent (liquid ammonia) or by carrying out the reaction in a solvent composed of 50% ammonia and 50% diethyl ether. Therefore, the authors proposed a radical chain mechanism, shown in equations 8.77 through 8.81, as a mechanism for formation of unrearranged products. In the initiation step, an unidentified electron donor transfers an electron to either **93c** or **94c** (both denoted as Ar). In the first propagation step of the chain reaction, the radical anion of the aryl iodide detaches an iodide ion and forms an aryl radical. The aryl radical reacts with amide ion to form the radical anion of the product, **96** or **97**. The product radical anion then transfers an electron to the aryl iodide, and the chain continues. Termination steps, such as the reaction of the aryl radical with a hydrogen donor to produce the arene (**92**). interrupt the chain reaction.[228]

$$\text{electron donor} + \text{ArI} \longrightarrow [\text{ArI}]\cdot^- + \text{electron donor residue} \quad (8.77)$$

$$[\text{ArI}]\cdot^- \longrightarrow \text{Ar}\cdot + \text{I}^- \quad (8.78)$$

$$\text{Ar}\cdot + \text{NH}_2^- \longrightarrow [\text{ArNH}_2]\cdot^- \quad (8.79)$$

$$[\text{ArNH}_2]\cdot^- + \text{ArI} \longrightarrow \text{ArNH}_2 + [\text{ArI}]\cdot^- \quad (8.80)$$

$$\text{Ar}\cdot + \text{H-donor} \longrightarrow \text{Ar-H} + \cdot\text{donor} \quad (8.81)$$

This chain reaction is analogous to radical chain mechanisms for aliphatic nucleophilic substitution that had been suggested independently by Russell and by Kornblum and their co-workers.[229,230] The descriptive title $S_{RN}1$ (substitution radical-nucleophilic unimolecular) was suggested for this reaction by analogy to the S_N1 mechanism for aliphatic substitu-

[228]Under some conditions, termination by dimerization of aryl radicals may lead to appreciable yields of Ar-Ar. Ettayeb, R.; Savéant, J.-M.; Thiébault, A. *J. Am. Chem. Soc.* **1992**, *114*, 10990.

[229]Kornblum, N.; Michel, R. E.; Kerber, R. C. *J. Am. Chem. Soc.* **1966**, *88*, 5662.

[230]Russell, G. A.; Danen, W. C. *J. Am. Chem. Soc.* **1966**, *88*, 5663; **1968**, *90*, 347.

tion.[227,231,232] When the reaction was carried out in the presence of solvated electrons formed by adding potassium metal to the ammonia solution, virtually no aryne (rearranged) products were observed. Instead, reaction of **93c** produced only **96** (40%) and **92** (40%) but no **97**, and reaction of **94c** produced **97** (54%) and **92** (30%) with only a trace of **96**.[233]

Subsequent research has shown the $S_{RN}1$ mechanism to occur with many other aromatic substrates.[234] The reaction has been shown to be initiated by solvated electrons, by electrochemical reduction, and by photoinitiated electron transfer.[235] Not only I but also Br, Cl, F, SC_6H_5, $N(CH_3)_3^+$, and $OPO(OCH_2CH_3)_2$ have been found to serve as electrofuges. In addition to amide ion, phosphanions, thiolate ions, benzeneselenolate ion ($C_6H_5Se^-$), ketone, and ester enolate ions as well as the conjugate bases of some other carbon acids have been identified as nucleophiles. The $S_{RN}1$ reaction has been observed with naphthalene, phenanthrene, and other polynuclear aromatic systems, and the presence of alkyl, alkoxy, phenyl, carboxylate and benzoyl groups on the aromatic ring does not interfere with the reaction.[236]

The $S_{RN}1$ mechanism is also important for many aliphatic structures that are unreactive in either S_N2 or S_N1 reactions, including perfluoroalkyl iodides, 1-substituted bridgehead compounds, substituted cyclopropanes, and neopentyl halides.[234,237,238] As noted on page 465, the 1-substituted bridgehead compounds are unreactive in S_N1 reactions because the resulting carbocations are nonplanar.[21] For example, 4-tricyclyl triflate (**98**) has been calculated to undergo solvolysis in 60% aqueous ethanol at 25° with a

---

[231]By analogy to the S_N2 mechanism, an $S_{RN}2$ mechanism was also considered by Galli, C.; Bunnett, J. F. *J. Am. Chem. Soc.* **1979**, *101*, 6137. In this reaction the two propagation steps would be

$$[ArX]\cdot^- + Y^- \longrightarrow [ArY]\cdot^- + X^-$$
$$[ArY]\cdot^- + ArX \longrightarrow ArY + [ArX]\cdot^-$$

For discussions of this mechanistic possibility, see (a) Rossi, R. A.; Palacios, S. M. *Tetrahedron* **1993**, *49*, 4485; (b) Savéant, J.-M. *Tetrahedron* **1994**, *50*, 10117.

[232]The IUPAC notation (reference 8) for the $S_{RN}1$ reaction is $(T + D_N + A_N)$, in which the symbol T refers to an electron transfer.

[233]Kim, J. K.; Bunnett, J. F. *J. Am. Chem. Soc.* **1970**, *92*, 7464.

[234]For a detailed discussion, see Rossi, R. A.; de Rossi, R. H. *Aromatic Substitution by the $S_{RN}1$ Mechanism*, ACS Monograph 178; American Chemical Society: Washington, DC, 1983.

[235]For a review of photochemically stimulated $S_{RN}1$ reactions, see Bowman, W. R. in *Photoinduced Electron Transfer*., Part C, Fox, M. A.; Chanon, M., eds.; Elsevier: Amsterdam, 1988; pp. 487–552.

[236]Bunnett, J. F. *Acc. Chem. Res.* **1978**, *11*, 413.

[237]Rossi, R. A.; Pierini, A. B.; Palacios, S. M. *J. Chem. Educ.* **1989**, *66*, 720.

[238]Kornblum and co-workers have found evidence for a radical chain mechanism initiated by electron transfer from the nucleophile to the nitroaromatic ring in the reaction of *p*-nitrocumyl chloride with nucleophiles such as azide ion in hexamethylphosphoramide (HMPA) solution. Kornblum, N.; Wade, P. A. *J. Org. Chem.* **1987**, *52*, 5301. See also Kornblum, N.; Cheng, L.; Davies, T. M.; Earl, G. W.; Holly, N. L.; Kerber, R. C.; Kestner, M. M.; Manthey, J. W.; Musser, M. T.; Pinnick, H. W.; Snow, D. H.; Stuchal, F. W.; Swiger, R. T. *J. Org. Chem.* **1987**, *52*, 196.

half-life of 4.6×10^9 years.[22] However, 4-iodotricyclene (**99**) readily undergoes the $S_{RN}1$ reaction with $(C_6H_5)_2P^-$ ions in liquid ammonia at $-33°$.[239]

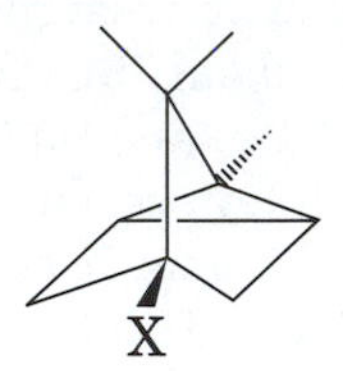

98 (X = OTf) **99** (X = I)

Concluding Remarks

Substitution processes are diverse both in scope and in mechanism. Upon detailed investigation, even those reactions that seem most familiar to us offer surprising degrees of complexity and variation that belie our attempts to fit all of organic chemistry into a few distinct compartments. As one author put it, "In summary, nitration—the classic example of electrophilic aromatic substitution—is not always an electrophilic aromatic substitution . . ."[240]

We are reminded again of the need to consider all mechanisms in chemistry as nothing more than temporary descriptions that we adopt because they are useful to us, but with the understanding that our ideas may change as new theories and experimental results become available. There is truth in the statement of Coulson that "Reaction mechanisms in general are elucidated in successive approximation."[241] In this regard it may be useful to remember the words of Lucretius:[242]

No single thing abides, but all things flow.
Fragment to fragment clings, and thus they grow
Until we know and name them.
Then by degrees they change and are no more
The things we know.

---

[239]The corresponding chloride is not reactive under the same conditions, however, apparently because of the less favorable reduction potential of the chloride. Santiago, A. N.; Morris, D. G.; Rossi, R. A. *J. Chem. Soc., Chem. Commun.* **1988**, 220.

[240]Perrin, C. *J. Am. Chem. Soc.* **1977**, *99*, 5516.

[241]Coulson, C. A. *Proc. Chem. Soc.* **1962**, 265.

[242]This quotation is taken from the introduction to chapter 4 of Leffler, J. E.; Grunwald, E. *Rates and Equilibria of Organic Reactions*; John Wiley & Sons, Inc.: New York, 1963; p. 57.

Problems

1. Predict the products and effect of increasing solvent polarity on the rate of each of the following reactions:
   a. $I^- + (CH_3)_4N^+ \longrightarrow$
   b. $NH_3 + (CH_3)_3S^+ \longrightarrow$
2. Complete the diagram shown in Figure 8.69 for nucleophilic aromatic substitution by drawing in the structures corresponding to the lower left and upper right corners. Identify the reaction pathway corresponding to the dashed lines labeled path 1 and path 2.

Figure 8.69

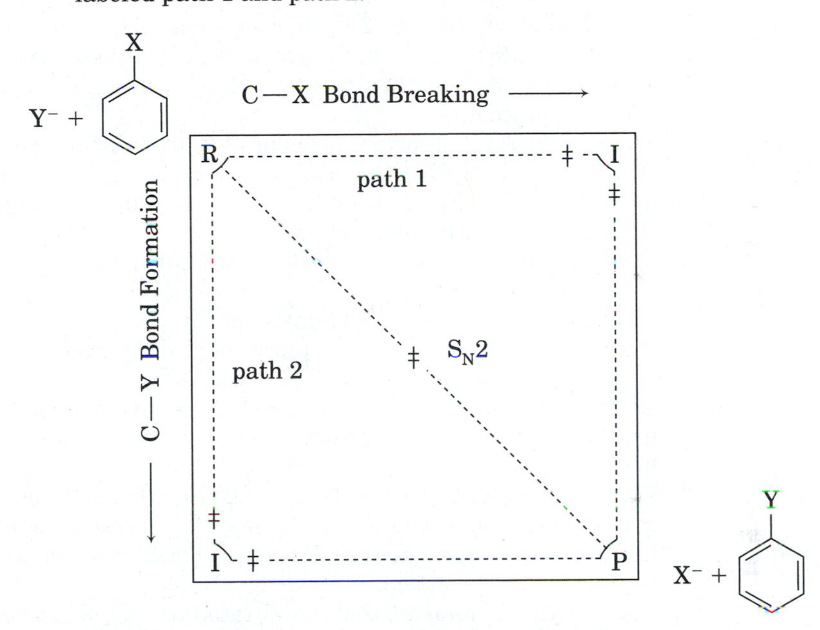

3. Reaction of allyl-1,1-d_2 chloride with phenyllithium produces allyl benzene in which about 25% of the product is allylbenzene-1,1-d_2 and about 75% is allylbenzene-3,3-d_2.

D D (allyl-1,1-d_2 chloride, —Cl) $\xrightarrow{\text{Ph—Li}}$ D, —D (with Ph) + Ph D, —D + LiCl

ca. 75% ca. 25%

What do these results suggest about the relative contribution of S_N2 and S_N2' processes in this reaction?

4. Cyclopropylcarbinyl halides react with piperidine to give products as shown in Figure 8.70. Both products are formed by second order processes. The distribution of the two products varies with the nature of the R group: the larger the R group, the greater is the yield of rearranged product.
   a. Propose a mechanism for the formation of each product.
   b. Explain why the size of the R group affects the product distribution.

Figure 8.70

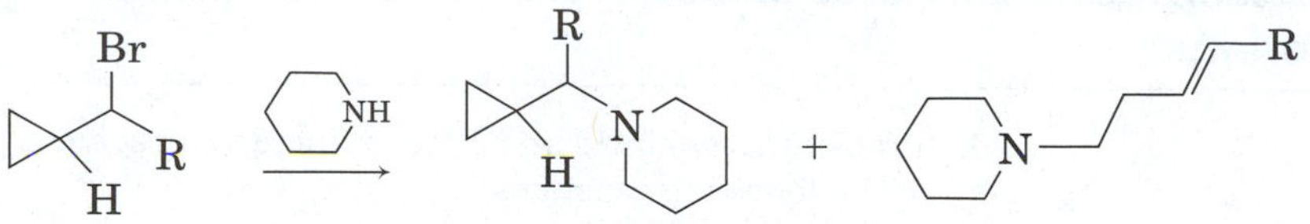

5. (+)-D-α-bromopropionate reacts in aqueous NaOH with second order kinetics to give (−)-L-α-hydroxypropionate. However, in water it undergoes first order reaction to give (+)-D-hydroxypropionate. Propose an explanation for these results.
6. Treatment of (−)-2-hydroxy-2-phenylpropanoic acid with $SOCl_2$ leads to the formation of (−)-2-chloro-2-phenylpropanoic acid. However, when the same reactant is treated with PCl_5, the product that is isolated is (+)-2-chloro-2-phenylpropanoic acid. Explain the different stereochemical outcomes of these two synthetic procedures.
7. The relative rates of solvolysis in buffered acidic solution of compounds with the structure $H_2NCH_2CR_2CH_2CH_2Br$ vary with the structure of the R group as follows: R = H, 1.0; R = methyl, 158; R = ethyl, 594; R = isopropyl, 9190. Explain this variation in rates of reaction.
8. a. Verify that the stereochemical assignments for each structure in Figure 8.28 are correct.
   b. What stereoisomer of 3-phenyl-2-pentyl tosylate gives acetolysis products that are the mirror image of those obtained by acetolysis of L-*threo*-2-phenyl-3-pentyl tosylate?
9. Sketch the expected shape of plots of ΔG versus reaction coordinate (as in Figure 8.38) for the reaction of ammonia with methyl chloride in (i) acetone and (ii) hexane solution.
10. When treated with fuming HBr, (±)-2,3-diacetoxybutane is converted to *meso*-2,3-dibromobutane. There is evidence that (±)-*erythro*-3-bromo-2-butanol is an intermediate in the reaction. Propose a mechanism to account for the stereospecificity of the reaction.
11. For nitration of toluene, the partial rate factors under one set of experimental conditions are $f_o^Z = 41$, $f_m^Z = 2.1$, $f_p^Z = 51$. What is the total relative rate of reaction of toluene to benzene under these conditions?
12. The crown ether 21-crown-7 was found to influence the rate and selectivity of nitration of anisole by tetrabutylammonium nitrate and trifluoroacetic acid in methylene chloride solution. Table 8.20 shows the relative reactivity and product distribution for various concentrations of the crown ether.[243]
    a. Calculate the partial rate factor for the ortho and para positions for each [21C7].
    b. Rationalize the change (if any) seen in the partial rate factors as a function of [21C7].
13. Rate constants reported for the $AlCl_3$-catalyzed benzoylation of benzene and its derivatives in benzoyl chloride solution are shown in Table 8.21.
    a. Calculate the relative rate of reaction (with the rate of reaction of benzene set equal to 1.0) for each compound. Are the results consistent with your expectations based on the activating or deactivating nature of the substituent groups?

---

[243] Masci, B. *J. Org. Chem.* **1985**, *50*, 4081.

Table 8.20 Effect of [21C7] on nitration of anisole

	[21C7]			
	0.0 M	**0.0082 M**	**0.032 M**	**0.131 M**
k/k_B	2220	4940	10800	27600
%-ortho	78.1	33.8	13.2	4.8
%-para	21.9	66.2	86.8	95.2

b. The relative isomer distribution of products from benzoylation of toluene is 9.3% ortho, 1.45% meta and 89.3% para. Calculate the partial rate factor for reaction of toluene at each position.

c. Use the data in Table 8.21 to determine whether a Hammett correlation is observed in this reaction. Do the data fit better when the log of the rates is plotted vs. σ or σ^+? Interpret the meaning of the sign and magnitude of the value of ρ you obtain.

Table 8.21 Rate data for benzoylation and benzene and derivatives in benzoyl chloride solution at 25°

Compound	**Rate (M^{-1} min^{-1})**
Benzene	0.00297
Toluene	0.326
t-Butylbenzene	0.215

14. a. When heated in 2,2,2-trifluoroethanol (TFE) buffered with 2,6-lutidine at 120° for 5 hours, 2,6-bis(trimethylsilyl)phenyl triflate (**100**) gave a quantitative yield of the trifluormethyl ether **101**. The rate of the reaction was found to be first order in **100** and zero order in 2,6-lutidine. The $\Delta H^{\ddagger}$ and $\Delta S^{\ddagger}$ values for the reaction were found to be 26.5 kcal/mol and −5.5 eu, respectively. Propose a mechanism and give a mechanistic label for the reaction.

OSO_2CF_3 / $(CH_3)_3Si$ / $Si(CH_3)_3$ $\xrightarrow[120°]{HOCH_2CF_3}$ OCH_2CF_3 / $(CH_3)_3Si$ / $Si(CH_3)_3$ **(8.82)**

100 **101**

b. Trifluoroethanolysis of the triflate **102** at 100° produced the two ethers **103** (66%) and **104** (27%). Propose a mechanism for the formation of **104** that involves both a 1,4-hydride shift and a 1,2-aryl shift.

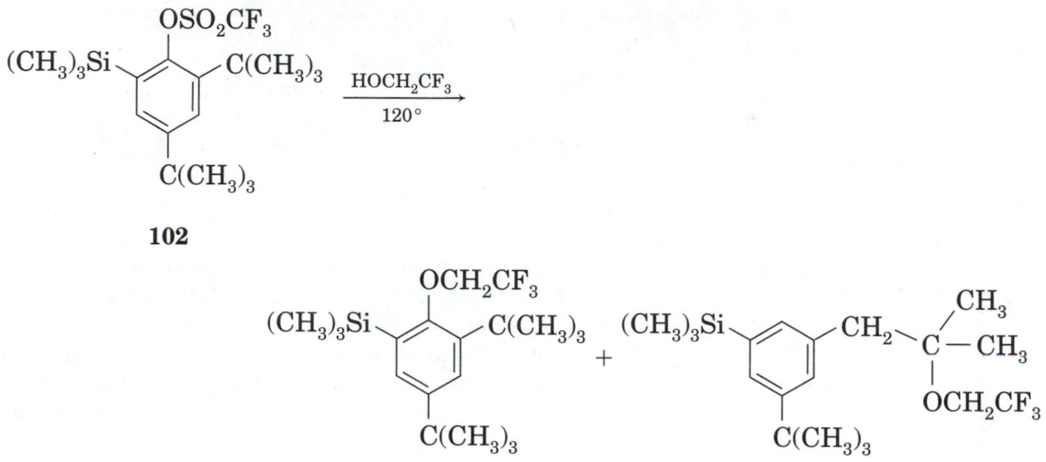

OSO_2CF_3 $(CH_3)_3Si$ $C(CH_3)_3$ $\xrightarrow[120°]{HOCH_2CF_3}$ $C(CH_3)_3$

102

(8.83)

OCH_2CF_3 $(CH_3)_3Si$ $C(CH_3)_3$ $C(CH_3)_3$ + $(CH_3)_3Si$ CH_2–$C(CH_3)_2$–OCH_2CF_3 $C(CH_3)_3$

103 **104**

15. In a study of the iodination of phenol, the second order rate constant for the reaction of 2,4,6-trideuteriophenol with iodine in aqueous solution[244] was found to be $3.05 \times 10^{-4}\ M^{-1}\ sec^{-1}$, while the corresponding rate constant for reaction of undeuterated phenol was found to be $1.21 \times 10^{-3}\ M^{-1}\ sec^{-1}$.
   a. Calculate the 1° hydrogen kinetic isotope effect for the reaction.
   b. Propose a mechanism involving the formation of an intermediate that will account for both the observed second order kinetics and the kinetic isotope effect.
   c. A 1° hydrogen kinetic isotope effect has also been observed in the sulfonation of benzene and in the cyclodehydration of 2-anilino-2-pentene-4-one. Based on the intermediates expected in each case, can you suggest why 1° hydrogen kinetic isotope effects are observed in these reactions but not in most other S_EAr reactions?
16. Propose a detailed mechanism for the Smiles rearrangement shown as equation 8.61, page 525.
17. Summarize the types of labeling experiments that could be used to study the proposed mechanism for the von Richter reaction (Figure 8.58), page 525.
18. Propose an explanation for the observation that potassium amide enhances the rate of formation of tetraphenylmethane in the reaction of chlorobenzene with potassium triphenylmethide in liquid ammonia.
19. Consider an alternative mechanism (Figure 8.71) for the formation of *m*-trifluoromethylaniline by reaction of *o*-chlorotrifluoromethylbenzene with KNH_2 in liquid ammonia in which a phenyl carbanion isomerizes prior to an S_NAr reaction. Suggest experiments to determine whether such a mechanism occurs in this case or in other reactions of aryl halides with KNH_2 in liquid ammonia.

---

[244]Initial concentrations were 8×10^{-3} M phenol, 1.96×10^{-3} M iodine, 0.5 M acetic acid, 0.05 M sodium acetate, 0.23 M sodium perchlorate and 0.02 M sodium iodide.

Figure 8.71

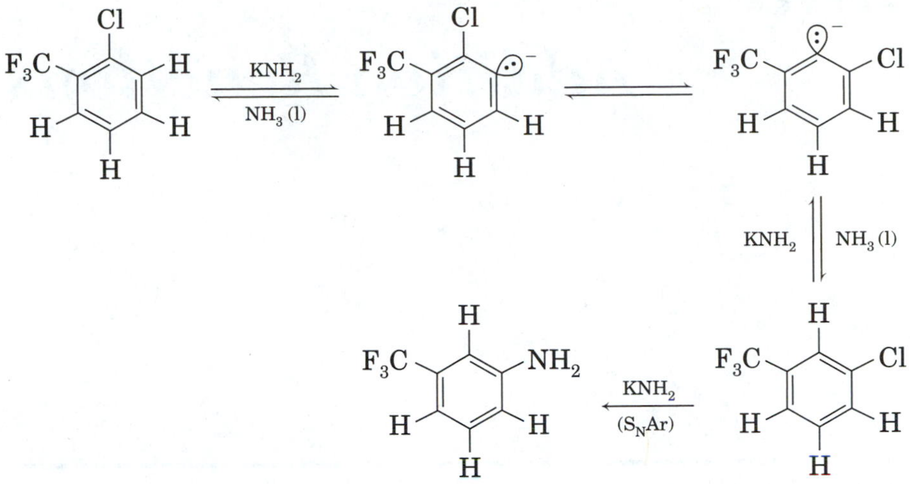

20. Propose a detailed mechanism for the transformation of nitrobenzene to *p*-nitroaniline with 4-amino-1,2,4-triazole shown in equation 8.84.

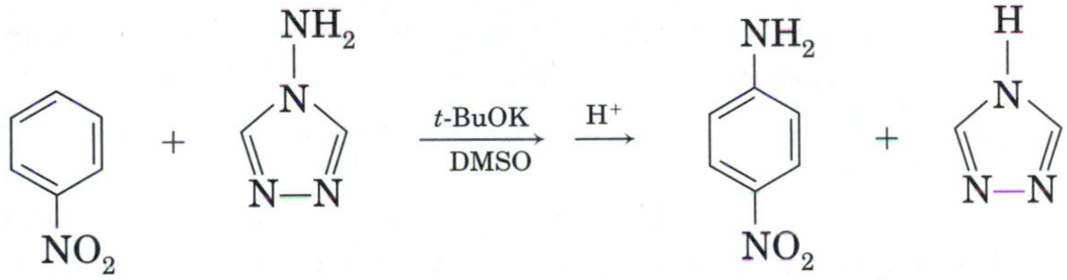

(8.84)

21. Based on the data in Table 8.17 (page 524), draw a reaction coordinate diagram for the reaction of piperidine with 1-bromo-2,4-dinitrobenzene.
22. Rationalize the distribution of products from the reaction of *o*-bromotoluene with KNH_2 in liquid ammonia (Table 8.19, page 535.)
23. One of the observations that led to the formulation of the benzyne mechanism was the failure of either bromomesitylene or bromodurene to react with sodium amide in liquid ammonia.
    a. Why are these compounds unreactive by the benzyne mechanism?
    b. Why is the lack of reaction of these compounds alone insufficient to establish the benzyne mechanism for reaction of compounds such as bromobenzene with amide ion in liquid ammonia? That is, what factors might make bromomesitylene and bromodurene unreactive with amide ion in liquid ammonia whether or not the reaction of bromobenzene involves a benzyne mechanism?

CHAPTER 9

Addition Reactions

9.1 Introduction

An addition reaction occurs when two or more molecular entities combine to form a single reaction product containing all of the atoms of the reactants. In the process two bonds are formed, one bond breaks in the adding reagent, and there is a net reduction in bond multiplicity in the other reactant.[1] The structure undergoing addition can be any species with a multiple bond, but our focus will be on alkenes, alkynes, and carbonyl compounds. While addition to other functional groups is certainly important, alkenes exhibit a wide range of electrophilic addition reactions, and additions to the carbonyl group will illustrate some theoretical aspects of nucleophilic addition.

Addition mechanisms are broadly defined to be heterolytic, homolytic, or cyclic, depending upon whether the process involves ionic or radical intermediates or is concerted.[2] Concerted reactions will be discussed in Chapter 11. Our emphasis here will be heterolytic (ionic) additions, although we will also consider some aspects of radical reactions. Addition reactions may be categorized further as being electrophilic or nucleophilic. In an electrophilic addition, a compound with a multiple bond reacts with an electrophilic reagent to produce an intermediate that subsequently reacts with a nucleophile. In nucleophilic addition, the unsaturated substrate reacts with a nucleophile to produce an intermediate that subsequently reacts with an electrophile to produce the final product.[3]

---

[1]Commission on Physical Organic Chemistry, IUPAC, *Pure Appl. Chem.* **1994**, *66*, 1077.

[2]de la Mare, P. B. D.; Bolton, R. *Electrophilic Additions to Unsaturated Systems*, 2nd Ed.; Elsevier Scientific Publishing Company: New York, 1982; p. 2.

[3]While we generally think of electrophilic and nucleophilic additions as being heterolytic processes, there can be electrophilic and nucleophilic character to radical additions also. Alkyl or aryl substituents on a carbon-centered radical make the radical more nucleophilic than a methyl radical, while electron withdrawing substituents make it more electrophilic. For a discussion, see Zipse, H.; He, J.; Houk, K. N.; Giese, B. *J. Am. Chem. Soc.* **1991**, *113*, 4324 and references therein. Perfluoroalkyl radicals are strongly electrophilic; see, for example, Avila, D. V.; Ingold, K. U.; Lusztyk, J.; Dolbier, W. R.; Pan, H.-Q. *J. Am. Chem. Soc.* **1993**, *115*, 1577.

One important stereochemical distinction to be made in additions to alkenes is that syn addition and anti addition to a carbon-carbon double bond may produce stereoisomeric products.[4] The term ***anti*** means that one group adds from the top of the molecule (as defined by the plane of the substituents on the double bond), while the other adds from the bottom. In ***syn*** addition, both substituents add from the same face. As noted in Figure 9.1, the syn and anti pathways can produce different configurations at one of the two tetrahedral centers created by the addition to a carbon-carbon double bond.[5]

Many different mechanisms have been proposed for addition reactions, and they have long been described in terms of the Ingold formalism.[6] In this system, addition reactions are labeled with symbols incorporating Ad for addition, E or N for electrophilic or nucleophilic, and a number indicating the molecularity of the reaction.[7] For example, an Ad_E2 mechanism involves attachment of an electrophile to the substrate, followed by attachment of a nucleophile to the resulting intermediate. The corresponding IUPAC term for the same reaction would be $(A_E + A_N)$.[8] Because most of the literature of organic chemistry has been written using the Ingold terminology, we will note the IUPAC nomenclature but will usually discuss reactions in terms of the more familiar Ingold designations.

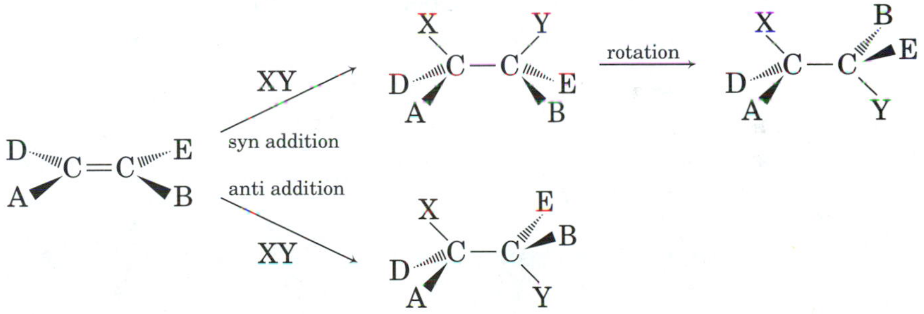

Figure 9.1 Diastereomeric products of syn and anti addition to an alkene. (In this example, the products are diastereomers.)

[4]Whether stereoisomeric products are formed and, if so, whether the products are enantiomers or diastereomers, depends on the substitution pattern of the olefin and on whether the adding reagent is symmetric (X—X) or not (X—Y).

[5]In older terminology, these processes were referred to as trans or cis additions. Current preference uses the terms *syn* and *anti* for *mechanisms* and reserves the terms *cis* and *trans* for *structures*. Eliel, E.; editor's comment, p. 328, in Fahey, R. C. *Top. Stereochem.* **1968**, 3, 237.

[6]Ingold, C. K. *Structure and Mechanism in Organic Chemistry*, 2nd Ed.; Cornell University Press: Ithaca, NY, 1969. See also Fahey, R. C.; Lee, D.-J. *J. Am. Chem. Soc.* **1968**, *90*, 2124.

[7]Wilson, M. A. *J. Chem. Educ.* **1975**, *52*, 495, has summarized the Ingold terminology for many of the mechanistic possibilities.

[8]Commission on Physical Organic Chemistry, IUPAC, *Pure Appl. Chem.* **1989**, *61*, 23. For a discussion, see Guthrie, R. D.; Jencks, W. P. *Acc. Chem. Res.* **1989**, *22*, 343.

9.2 Addition of Halogens to Alkenes

Electrophilic Addition of Bromine to Alkenes[9,10,11]

Introduction

Let us begin our discussion of addition reactions by considering one of the classic reactions of organic chemistry, the addition of bromine to an alkene. We will first summarize the evidence that led to the textbook mechanism (shown in Figure 9.2), then we will survey more recent studies that reveal the complexities that underlie this simple mechanistic representation.

Two important features of the mechanism shown in Figure 9.2 are:

1. The intermediacy of a "bromonium" ion, a three-membered ring containing a bromine atom bearing a formal charge of +1. The structure is also called a σ *complex* because the valence bond representation implies a full or partial σ bond between the olefinic carbon atoms and the bromonium atom.[12]
2. The *anti* attachment of the two bromine atoms to the carbon-carbon double bond. This stereochemistry results because in the second step a bromide ion attacks the backside of one of the bromonium ion C—Br bonds in a process much like an S_N2 reaction.[13,14,15]

The bromonium ion in Figure 9.2 was proposed by Roberts and Kimball to explain the stereospecificity of the addition to alkenes.[16] Independent evidence for the existence of bromonium ions was provided by Winstein and

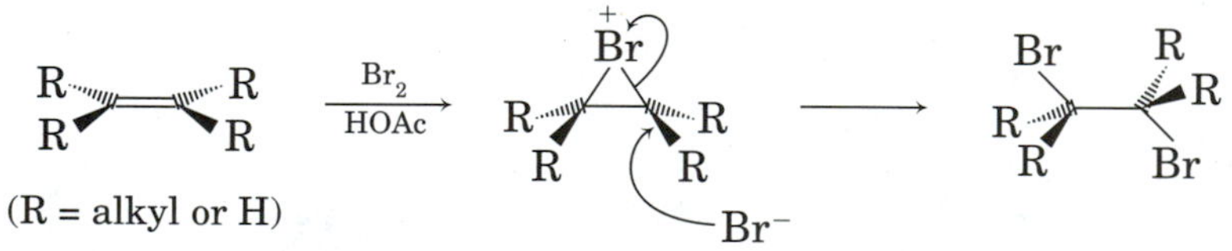

Figure 9.2 Generalized mechanism for addition of bromine to an alkene.

[9]de la Mare, P. B. D.; Bolton, R. *Electrophilic Addition to Unsaturated Systems*; Elsevier Publishing Company; Amsterdam, 1966; p. 32.

[10]Schmid, G. H.; Garratt, D. G. in *The Chemistry of Double-bonded Functional Groups*, Supplement A, Part 2; Patai, S., ed.; John Wiley & Sons: London, 1977; pp. 725–912.

[11]Schmid, G. H. in *The Chemistry of Double-bonded Functional Groups*, Supplement A, Volume 2, Part 1; Patai, S., ed.; John Wiley & Sons, Inc.: London, 1989; pp. 679–731.

[12]The results of *ab initio* calculations (reference 73) have suggested, however, that it might be more correctly described as a strong π-complex.

[13]Meer, N.; Polanyi, M. *Z. phys. Chem.* **1932**, *B19*, 164; Bergmann, I. E.; Polanyi, M.; Szabo, A. *Z. phys. Chem.* **1933**, *B20*, 161.

[14]Olson, A. R. *J. Chem. Phys.* **1933**, *1*, 418; Olson, A. R.; Voge, H. H. *J. Am. Chem. Soc.* **1934**, *56*, 1690.

[15]A Frontier MO rationalization for *anti*-1,2 addition (as well as for *syn*-1,4 addition and *anti*-1,6 addition) was provided by Fukui, K. *Tetrahedron Lett.* **1965**, 2427.

[16]Roberts, I.; Kimball, G. E. *J. Am. Chem. Soc.* **1937**, *59*, 947.

Lucas, who found that (±)-*erythro*-3-bromo-2-butanol was converted to *meso*-2,3-dibromobutane upon reaction with fuming HBr (equation 9.1), but the (±)-*threo* reactant formed (±)-2,3-dibromobutane. These authors suggested that the bromine provides anchimeric assistance to the departure of the protonated OH group and forms a bromonium ion which, upon nucleophilic attack by bromide ion, gives a dibromide with retention of configuration of the original bromohydrin (Figure 9.3).[17]

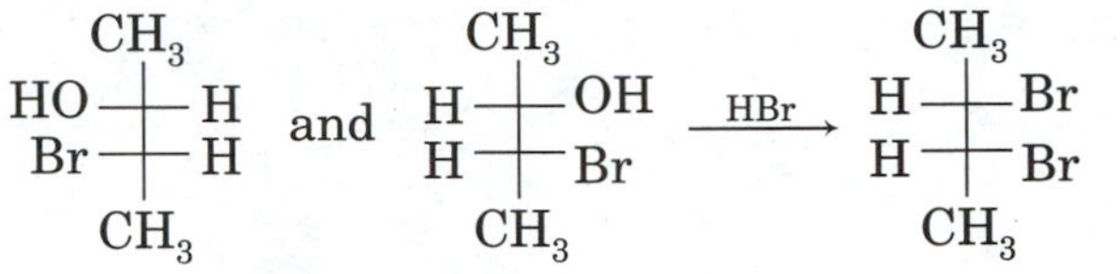

racemic erythro $\xrightarrow{HBr}$ meso **(9.1)**

The mechanism of bromine addition shown in Figure 9.2 is supported by several additional lines of evidence:

1. Addition of bromine to an alkene is highly stereospecific and gives products with stereochemistry consistent with an anti addition mechanism.[18] For example, addition of bromine to cyclohexene produces *trans*-1,2-

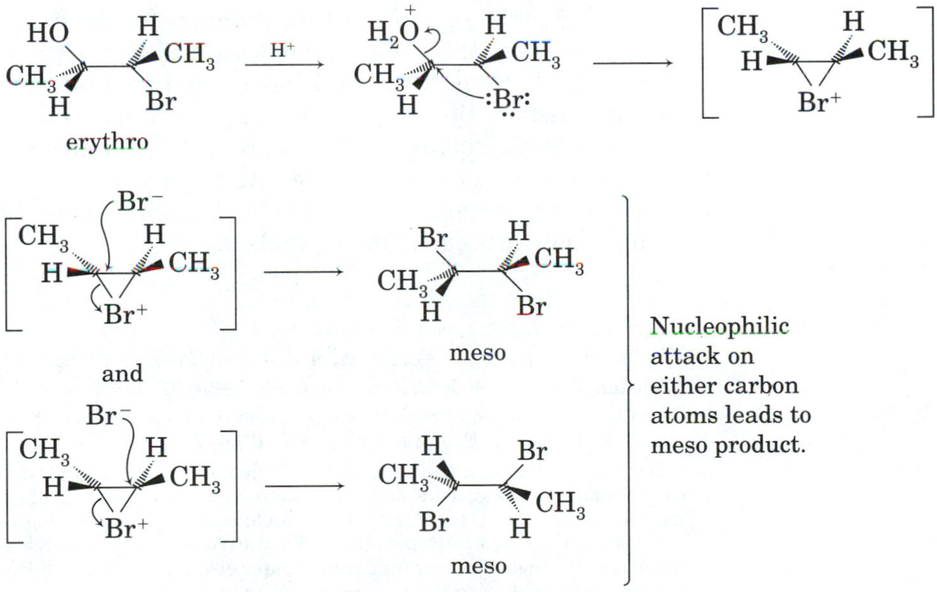

Figure 9.3 Proposed mechanism for the conversion of *erythro*-3-bromo-2-butanol into *meso*-2,3-dibromobutane.

erythro $\xrightarrow{H^+}$ … ⟶ meso

Nucleophilic attack on either carbon atoms leads to meso product.

[17]Winstein, S.; Lucas, H. J. *J. Am. Chem. Soc.* **1939**, *61*, 1576, 2845.

[18]For a summary of reactions illustrating the stereochemistry, see Terry, E. M.; Eichelberger, L. *J. Am. Chem. Soc.* **1925**, *47*, 1067 and references therein. See also the discussion by Fahey (reference 5).

dibromocyclohexane.[19] Addition of bromine to *cis*-2-butene produces racemic 2,3-dibromobutane (equation 9.2), while addition to the trans isomer gives the meso product (equation 9.3).[20,21,22]

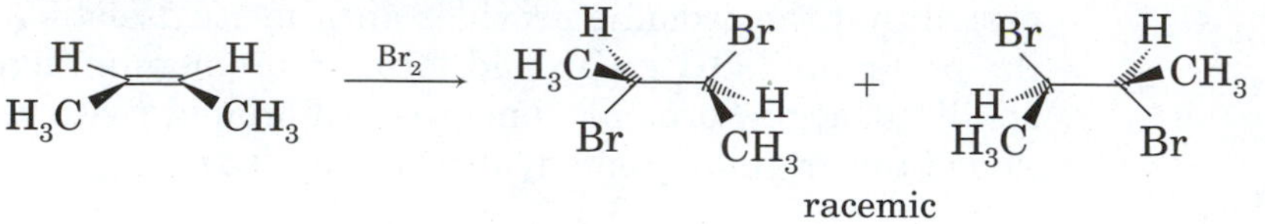

Br_2 → + (9.2)

racemic

Br_2 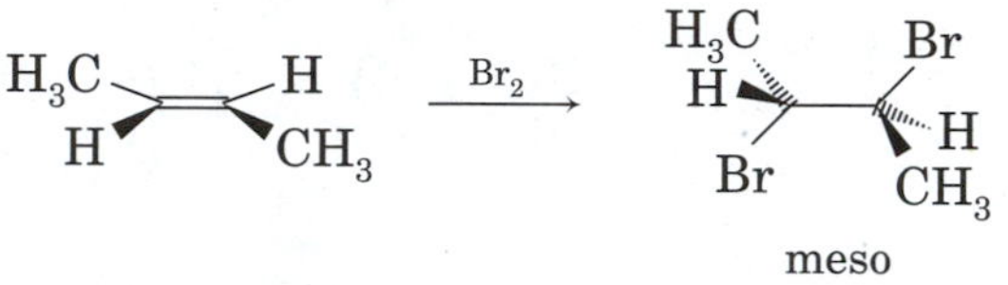 → (9.3)

meso

2. The reaction is faster with electron rich alkenes.[23] The relative rates of addition of bromine to a series of alkenes was found to increase in the order ethene < propene < 2-butene ≈ isobutene < 2-methyl-2-butene.[24] In other words, each methyl group that replaces a hydrogen atom on ethene increases the reactivity. The addition of bromine to substituted ethenes in methanolic sodium bromide can be correlated to the equation

$$\log k_2 = -3.10\Sigma\sigma^* + 7.02 \tag{9.4}$$

where $\Sigma\sigma^*$ is the sum of the Taft σ^* values for the four substituents.[25] Since the polar effects of the alkyl groups are additive, regardless of their position on the double bond, the results have been interpreted as evidence for a symmetrical distribution of positive charge on the two alkenyl carbon atoms in the transition structure for the addition.[26,27] Moreover, as would be expected for an electrophilic reaction, the rate of the reaction is decreased if electron withdrawing groups are near the double bond, as is shown by the series of halogen-substituted olefins in Table 9.1.[28]

---

[19]Winstein, S. *J. Am. Chem. Soc.* **1942**, *64*, 2792.

[20]Dillon, R. T.; Young, W. G.; Lucas, H. J. *J. Am. Chem. Soc.* **1930**, *52*, 1953.

[21]Rolston, J. H.; Yates, K. *J. Am. Chem. Soc.* **1969**, *91*, 1469.

[22]Addition of bromine to cholesterol produced the 5α,6β isomer of cholesterol dibromide. Barton, D. H. R.; Miller, E. *J. Am. Chem. Soc.* **1950**, *72*, 1066.

[23]The term *electron rich* has been applied to alkenes with several alkyl substituents because these compounds typically react rapidly with electrophiles. However, Vardhan and Bach (reference 221) have determined that the HOMO of such an alkene has a smaller degree of π character than does the HOMO of a less-substituted alkene. They attribute the greater reactivity of highly substituted alkenes to the higher energy (and thus greater electron donating ability) of the HOMO.

[24]Davis, H. S. *J. Am. Chem. Soc.* **1928**, *50*, 2769.

[25]Linear free energy relationships, including the use of σ^* and σ^+ constants, are discussed in Chapter 6.

[26]Ruasse, M.-F. *Acc. Chem. Res.* **1990**, *23*, 87 and references therein.

[27]However, the steric effects of alkyl substituents can retard the rates of bromine addition reactions, and this steric hindrance can be large enough to mask the electronic effect of alkyl substituents. Ruasse, M.-F.; Motallebi, S.; Gal, B.; Lomas, J. S. *J. Org. Chem.* **1990**, *55*, 2298.

[28]Swedlund, B. E.; Robertson, P. W. *J. Chem. Soc.* **1947**, 630.

Table 9.1 Relative reactivity of alkenes in bromine addition.

Alkene	Relative Reactivity
Ethene	1
Allyl bromide	0.3
Vinyl bromide	3.0×10^{-4}
cis-1,2-Dichloroethene	1.0×10^{-7}
Trichloroethene	3×10^{-10}
Tetrachlorethene	too low to measure

(Data from reference 28.)

3. The cationic intermediate suspected in the reaction can be trapped by added nucleophiles (including nucleophilic solvent molecules) if its lifetime is long enough for diffusion to bring the two species together. For example, the addition of bromine to *trans*-2-butene in acetic acid gave 98% of the expected *meso*-2,3-dibromobutane, but 2% of 1-acetoxy-2-bromobutane was also observed.[21] Addition of bromine to 1-hexene in methanol was found to produce 31% of 1,2-dibromohexene (**2**), and 59% of a 4 : 1 ratio of 1-bromo-2-methoxyhexane (**3**) and 2-bromo-1-methoxyhexane (**4**), as shown in equation 9.5.[29] The formation of **4** implies that the cationic intermediate is not simply a secondary carbocation.

$$\mathbf{1} \xrightarrow[CH_3OH]{Br_2} \mathbf{2}\ (\text{Br, Br};\ 31\%) + \mathbf{3}\ (\text{Br, }OCH_3;\ \text{ca. } 48\%) + \mathbf{4}\ (OCH_3\text{, Br};\ \text{ca. } 11\%) \qquad (9.5)$$

4. Olah obtained direct evidence for the existence of the tetramethylethylene bromonium ion (and the chloronium and the iodinium ion but not the fluoronium ion) by PMR spectra of halogenated precursors in antimony pentafluoride–sulfur dioxide solution at $-60°$.[30,31,32]

[29]Puterbaugh, W. H.; Newman, M. S. *J. Am. Chem. Soc.* **1957**, *79*, 3469. See also Ecke, G. G.; Cook, N. C.; Whitmore, F. C. *J. Am. Chem. Soc.* **1950**, *72*, 1511 and references cited therein.

[30]Olah, G. A.; Bollinger, J. M. *J. Am. Chem. Soc.* **1967**, *89*, 4744.

[31]Also see Olah, G. A. *Halonium Ions*; Wiley-Interscience: New York, 1975.

[32]Evidence for the existence of the fluoronium ion in the gas phase was reported by Nguyen, V.; Cheng, X.; Morton, T. H. *J. Am. Chem. Soc.* **1992**, *114*, 7127.

5. A yellow solid identified as a bromonium ion was isolated by Wynberg and co-workers from the reaction of bromine with adamantylideneadamantane.[33] Brown and co-workers obtained an X-ray crystal structure of the species **5** as a tribromide salt.[34,35,36]

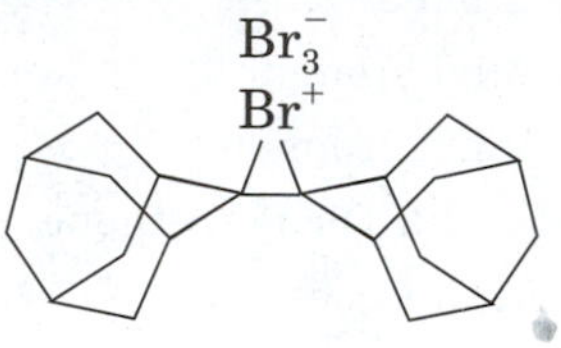

Br_3^-
Br^+

5

All of the data cited so far are consistent with the mechanism suggested in Figure 9.2. However, additional experimental results suggest a much more complex range of interactions among alkene, halogen, solvent, and added nucleophiles than is apparent from that figure.

Role of Charge Transfer Complexes in Bromine Addition Reactions

One of the earliest observations about bromine addition reactions is that mixing bromine and alkenes leads to a new UV-vis absorption band, suggesting that a bromine-alkene charge transfer (CT) complex is formed.[37] Quantifying the extent of weak complex formation is a difficult task, however.[38] Furthermore, observation of a complex between two reactants does not necessarily mean that the complex is a necessary step on the reaction path, because it might be a nonproductive "detour" instead. Because there was a correlation between the CT complex stabilization energies (determined by spectroscopy) and the rates of bromine addition, Dubois and Garnier concluded that complex formation may be a fast step that precedes the formation of the bromonium ion.[39] Moreover, spectroscopic studies reported

[33]Strating, J.; Wieringa, J. H.; Wynberg, H. *J. Chem. Soc., Chem. Commun.* **1969**, 907.

[34]Slebocka-Tilk, H.; Ball, R. G.; Brown, R. S. *J. Am. Chem. Soc.* **1985**, *107*, 4504.

[35]Bennet, A. J.; Brown, R. S.; McClung, R. E. D.; Klobukowski, M.; Aarts, G. H. M.; Santarsiero, B. D.; Bellucci, G.; Bianchini, R. *J. Am. Chem. Soc.* **1991**, *113*, 8532.

[36]An X-ray structure of the corresponding triflate salt confirmed the symmetrical geometry of the three-membered ring containing bromine. Brown, R. S.; Nagorski, R. W.; Bennet, A. J.; McClung, R. E. D.; Aarts, G. H. M.; Klobukowski, M.; McDonald, R.; Santarsiero, B. D. *J. Am. Chem. Soc.* **1994**, *116*, 2448.

[37]Because of the evidence for charge-transfer complexation, Garnier and Dubois proposed that the mechanistic description should be Ad_EC1 (unimolecular electrophilic addition proceeding through a CT complex) for the process in which progression from the CT complex to the bromonium ion is the rate-limiting step and Ad_EC2 (bimolecular electrophilic addition involving reaction of a CT complex with a nucleophile) for processes in which the CT complex reacts with another species in the rate-limiting step. Garnier, F.; Dubois, J.-É. *Bull. Soc. Chim. France* **1968**, 3797.

[38]Person, W. B. *J. Am. Chem. Soc.* **1965**, *87*, 167.

[39]Dubois, J. E.; Garnier, F. *Tetrahedron Lett.* **1966**, 3047; *Chem. Commun.* **1968**, 241.

by Bellucci and co-workers indicated that the decay of the absorption (λ_{max} near 287 nm) by the CT complex from bromine and cyclohexene corresponded to the rate of formation of the bromine addition product. The kinetic data pointed to an apparent negative activation energy, −7.8 kcal/mol, for the bromination, while the formation of the CT complex in 1,2-dichloroethane was exothermic, with $\Delta H = -4.6$ kcal/mol and $\Delta S = -17.0$ eu in the temperature range of 15° to 35°. The results of the detailed kinetic and thermodynamic analysis were consistent with a mechanism in which the CT complex is an essential step in the addition of bromine to cyclohexene, but were not consistent with a mechanism in which complex formation is an unproductive pathway that serves only to reduce the concentrations of the reactants.[40]

Not only does the experimental evidence support the role of *a* CT complex as an intermediate in bromine addition reactions, but it appears that *many* different CT complexes may occur in mixtures of alkenes and bromines. Bellucci and co-workers reported evidence for the interaction of adamantylideneadamantane and bromine to form complexes with the stoichiometry of 2 : 1, 1 : 1, 1 : 2 and 1 : 3 alkene : Br_2 ratios.[41] Conductivity experiments suggested that the 1 : 2 and 1 : 3 complexes are tribromide and pentabromide salts, respectively, of the adamantylideneadamantane bromonium ion.

Kinetics of Bromine Addition Reactions

The kinetics of bromine addition to alkenes can be more complex than the simple model in Figure 9.2 would suggest.[42] Early studies revealed considerable kinetic variation with solvent, temperature, and the presence of additives such as water. The kinetic investigations can also be complicated because of the involvement of HBr_3 (from Br_2 and HBr) or of intermolecular processes involving alkene, Br_2, and Br^-.[43] The Br^- can be produced if the bromonium ion is intercepted by solvent or solvent anion (e.g., acetate ion). The equilibrium constant for dissociation of Br_3^- to Br_2 and Br^- was determined to be 5×10^{-2} in water and 2×10^{-3} in methanol.[44] Thus a general kinetic expression for the addition of bromine to alkenes can be represented as:[21,45]

$$-d[Br_2]/dt = [\text{alkene}](k_2[Br_2] + k_3[Br_2]^2 + k_{Br_3^-}[Br_3^-]) \qquad \textbf{(9.6)}$$

---

[40]Bellucci, G.; Bianchini, R.; Ambrosetti, R. *J. Am. Chem. Soc.* **1985**, *107*, 2464.

[41]Bellucci, G.; Bianchini, R.; Chiappe, C.; Marioni, F.; Ambrosetti, R.; Brown, R. S.; Slebocka-Tilk, H. *J. Am. Chem. Soc.* **1989**, *111*, 2640. The formation constants observed for these species in 1,2-dichloroethane were $K_{2:1} = 1.11 \times 10^3\ M^{-2}$, $K_{1:1} = 2.9 \times 10^2\ M^{-1}$, $K_{1:2} = 3.2 \times 10^5\ M^{-2}$, and $K_{1:3} = 7.2 \times 10^6\ M^{-3}$, respectively.

[42]For a discussion and compilation of kinetic parameters for electrophilic addition to unsaturated compounds, see Bolton, R. in *Comprehensive Chemical Kinetics. Volume 9. Addition and Elimination Reactions of Aliphatic Compounds*; Bamford, C. H.; Tipper, C. F. H., eds.; Elsevier: Amsterdam, 1973; pp. 1–86.

[43]Morton, I. D.; Robertson, P. W. *J. Chem. Soc.* **1945**, 129 and references therein.

[44]Bartlett, P. D.; Tarbell, D. S. *J. Am. Chem. Soc.* **1936**, *58*, 466.

[45]Schmid, G. H.; Toyonaga, B. *J. Org. Chem.* **1984**, *49*, 761 and references therein.

Some early studies suggested that only the k_2 process should occur in polar solvents at low $[Br_2]$, that the k_3 process alone should be seen in nonpolar solvents, and that both the k_2 and k_3 processes can occur in polar solvents at higher $[Br_2]$. However, Fukuzumi and Kochi demonstrated that both second-order and third-order reactions could occur in a nonpolar solvent, CCl_4.[46] Schmid and Toyonaga determined that the rate constants for the second-order and third-order processes for addition of bromine to *cis*-2-butene in CCl_4 at 25° are $k_2 = 9 \times 10^{-4}\ M^{-1}\ s^{-1}$ and $k_3 = 4\ M^{-2}\ s^{-1}$, respectively. Thus, for *cis*-2-butene there is a concentration range ($<10^{-4}$ M $[Br_2]$) in which the second-order process dominates, a region ($>10^{-2}$ M $[Br_2]$) in which the third-order process dominates, and a range between in which both processes are significant.[45] The Br_3^- process is thought to be unimportant in most cases unless bromide ion has been added as a reagent. Because of the kinetic complexity, mechanistic discussions based on kinetics studies must take account of the concentration of the brominating agent(s).

The kinetics of bromine addition can also be affected by the involvement of species such as tribromide and pentabromide ions.[47] One possible mechanism for these reactions involves the dissociation of Br_3^- to Br_2 and Br^- before reaction of Br_2 with the alkene.[48] Alternatively, Br_3^- might act as an electrophile and add directly to the alkene.[44,49] Bellucci and co-workers[50] used the stopped-flow technique to study the kinetics of addition of bromine to cyclohexene in 1,2-dichloroethane solution. With molecular bromine the kinetics were overall third order, first order in [cyclohexene] and second order in $[Br_2]$.

$$-d[Br_2]/dt = k_3[Br_2]^2[\text{alkene}] \tag{9.7}$$

With tetrabutylammonium tribromide, the reaction was second order, first order in [cyclohexene] and first order in $[Br_3^-]$.

$$-d[Br_3^-]/dt = k_2[Br_3^-][\text{alkene}] \tag{9.8}$$

Added tetrabutylammonium bromide did not appreciably affect the rate of the latter reaction. There was a solvent kinetic isotope effect on the tribromide reaction in $CHCl_3/CDCl_3$, $k_H/k_D = 1.175$, but there was no solvent isotope effect for the addition of Br_2. Reaction with Br_2 (the third-order reaction, equation 9.7) gave a $\Delta H^{\ddagger}$ value of -8.4 kcal/mol,[40] while the reaction with tribromide gave a $\Delta H^{\ddagger}$ value of $+6.0$ kcal/mol. The investigators suggested that for reactions exhibiting third-order kinetics, the rate-limiting step involves formation of a bromonium ion–tribromide ion pair from a complex of one alkene and two bromine molecules. With tetra-*n*-butylam-

---

[46]Fukuzumi, S.; Kochi, J. K. *J. Am. Chem. Soc.* **1982**, *104*, 7599.

[47]Organic tribromide salts such as tetraethylammonium and pyridinium tribromide can also be used for S_EAr reactions with aromatics and for electrophilic additions to alkenes. For a leading reference, see Giordano, C.; Coppi, L. *J. Org. Chem.* **1992**, *57*, 2765.

[48]Wolf, S. A.; Ganguly, S.; Berliner, E. *J. Org. Chem.* **1985**, *50*, 1053 and references therein.

[49]DeYoung, S.; Berliner, E. *J. Org. Chem.* **1979**, *44*, 1088 and references therein.

monium tribromide as the source of bromine, the rate-limiting step is thought to be a backside nucleophilic attack at an olefin-Br_2 charge-transfer complex in equilibrium between Br_3^- and the olefin by the ammonium bromide ion pair that has become detached from Br_2 at the moment of formation of the CTC or that is present as added salt.[51]

Kinetic isotope effects also indicate a difference in mechanism when the brominating agent is Br_2 and Br_3^-. Brown and co-workers[52] measured rate constants for addition of molecular bromine to cyclohexene, to 3,3,6,6-tetradeuteriocyclohexene and to cyclohexene-d_{10} with Br_2 and with Br_3^- in acetic acid. For bromination with Br_2, the β secondary isotope effect (found by comparison of the rate for cyclohexene and 3,3,6,6-tetradeuteriocyclohexene) was 1.0 (Figure 9.4). Addition of bromine to cyclohexene and cyclohexene-d_{10}, however, gave an inverse secondary kinetic isotope effect, k_H/k_D = 0.53. The results suggested that in the transition structure there is significant rehybridization of the alkene carbon atoms from sp^2 toward sp^3. For addition of bromine with Br_3^-, however, the value of k_H/k_D calculated for cyclohexene/cyclohexene-d_{10} was 0.78, suggesting less extensive rehybridization of the alkene carbon atoms in the transition structure.

In addition to *trans*-1,2-dibromocyclohexane, *trans*-1-acetoxy-2-bromocyclohexane is formed by the reaction of bromine with cyclohexene in acetic acid.[52] The product distribution varies with the concentration of added LiBr (with ionic strength held constant). With [LiBr] = 0, Brown and co-workers found that the product consisted of 27% 1,2-dibromo and 73% bromoacetoxy adducts. At [LiBr] = 0.1 M, 90% of the product was 1,2-dibromocyclohexane and only 10% was the bromoacetoxy derivative. Based on the conclusion that the low α-secondary deuterium KIE requires a nucleophilic and not an

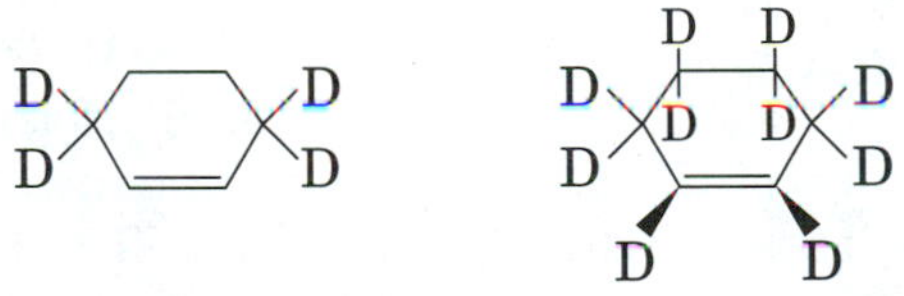

Figure 9.4
Kinetic isotope effects in the bromination of cyclohexene with Br_2 and Br_3^-.

Bromination agent	3,3,6,6–Tetradeuterio-cyclohexene	Cyclohexene–d_{10}
Br_2	k_H/k_D = 1.0	k_H/k_D = 0.53
Br_3^-		k_H/k_D = 0.78

[50]Bellucci, G.; Bianchini, R.; Ambrosetti, R.; Ingrosso, G. *J. Org. Chem.* **1985**, *50*, 3313.

[51]Further support for the difference in mechanism for addition with Br_2 and Br_3^- comes from the observation that the two reagents give different stereochemistry upon addition of bromine to 3-substituted cyclohexenes. Bellucci, G.; Bianchini, R.; Vecchiani, S. *J. Org. Chem.* **1986**, *51*, 4224. Furthermore, the two reagents yield diastereomeric products upon reaction with chiral ketals (reference 47).

[52]Slebocka-Tilk, H.; Zheng, C. Y.; Brown, R. S. *J. Am. Chem. Soc.* **1993**, *115*, 1347 and references therein.

Figure 9.5 Mechanism proposed for bromination of cyclohexene in acetic acid. (Modified from reference 52.)

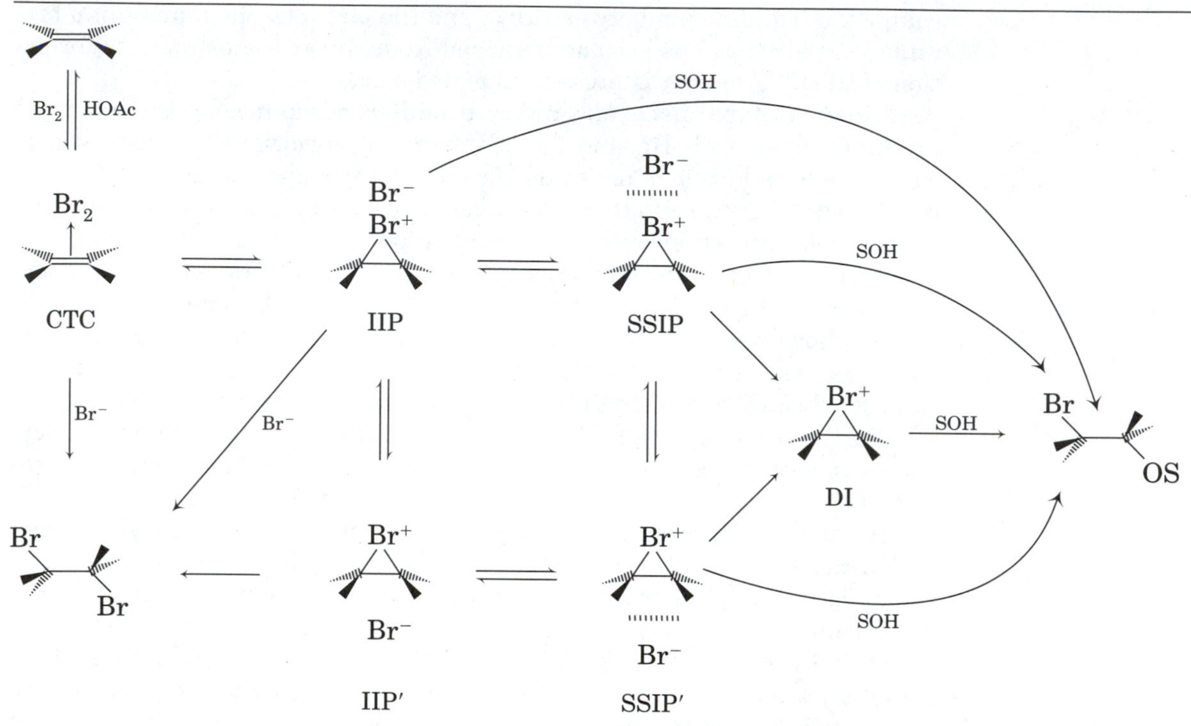

electrophilic role for Br^-, Brown et al. proposed the detailed mechanism shown in Figure 9.5. Here, **CTC** is a charge-transfer complex; **IIP** and **IIP′** are intimate ion pairs; **SSIP** and **SSIP′** are solvent-separated ion pairs; **DI** is a dissociated ion; and **SOH** is a hydroxylic solvent. The key feature of the mechanism is the necessity for Br^- migration to occur in the rearrangement of IIP to IIP′ so that backside attack can produce the dibromo product. The SSIP can rearrange to SSIP′, but the latter must then reorganize to form IIP′ (so that the Br^- is inside the solvent shell) before nucleophilic attack can occur. Added Br^- can react with the CTC or with IIP to produce the dibromo product.

The complex mechanism in Figure 9.5 can rationalize some observations about yields of the solvent incorporation products. In acetic acid, the bromoacetoxy product can arise by nucleophilic backside attack by solvent on IIP, IIP′, and SSIP′, and all of these processes can compete with pathways leading to the dibromo product. Once the SSIP or SSIP′ has dissociated to free ions (DI), however, only nucleophilic attack by solvent is likely, so DI leads only to the bromoacetoxy product. It should be emphasized that the pathways shown in Figure 9.5 are suggested by the observed kinetics and product distributions, but they are not the only possibilities.[53] Any one

[53]Because the mechanistic scheme has many more rate constants than there are observables, consistency between a calculated rate expression and experimental results does not ensure that the mechanism is correct.

of the reaction pathways might be replaced by another process that is its kinetic equivalent. For example, an alternative pathway for Br^- participation in the reaction would be one in which it acts as a base to facilitate removal of a proton from acetic acid in the transition structure for nucleophilic attack by the acetoxy oxygen.[52] In any case, the mechanism in Figure 9.5 shows considerably more detail than the simplified form in Figure 9.2.

Solvent Effects in Bromine Additions

The mechanism in Figure 9.5 also provides a rationalization for some observations of solvent effects in bromine addition reactions. The major product of the reaction of bromine with cyclohexene in methanol is *trans*-2-bromo-1-methoxycyclohexane. *trans*-1,2-Dibromocyclohexane is observed if Br^- is added to the solution, but the yield of the dibromo adduct approaches 0 as $[Br^-]$ approaches 0.[54] This result stands in contrast to the 27% of 1,2-dibromo adduct obtained from the corresponding reaction in acetic acid. It appears that the greater polarity of methanol accelerates the dissociation of IIP to DI, and the greater nucleophilicity of methanol enhances the reaction of solvent with IIP, SSIP, DI and SSIP′.[52]

In addition to polarity and nucleophilicity toward the bromonium ion, other solvent properties may also come into play. The rate-limiting step in electrophilic bromine addition is thought to be the conversion of a CT complex to a cation/bromide ion pair, which is consistent with the observation that the rate of bromine addition increases as solvent polarity increases.[37] Solvent may also electrophilically assist the removal of the bromide ion from the alkene-Br_2 CT complex by hydrogen bonding to a developing bromide ion (Figure 9.6).[55,56] It has been estimated that this electrophilic solvent participation can lower the energy of bromonium ion formation by 60 kcal/mol.[57,58] Additional support for the role of electrophilic participation of

Figure 9.6 Possible role for electrophilic properties of solvent in addition of bromine to alkenes.

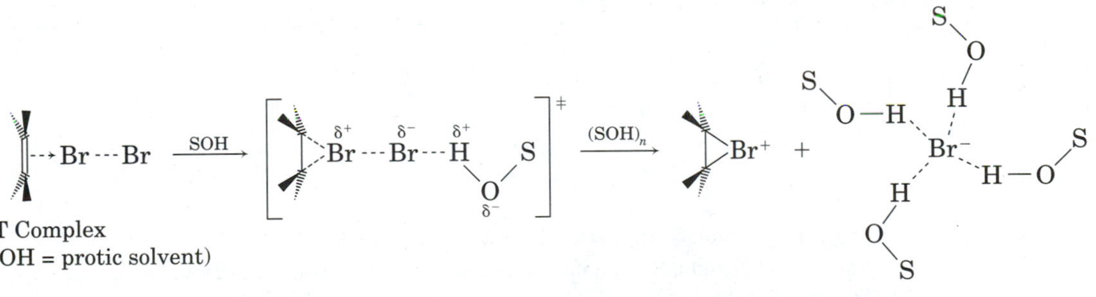

CT Complex
(SOH = protic solvent)

[54]Nagorski, R. W.; Brown, R. S. *J. Am. Chem. Soc.* **1992**, *114*, 7773.

[55](a) Garnier, F.; Donnay, R. H.; Dubois, J.-E. *J. Chem. Soc. D, Chem. Commun.* **1971**, 829; (b) Modro, A.; Schmid, G. H.; Yates, K. *J. Org. Chem.* **1979**, *44*, 4221.

[56]Modro, A.; Schmid, G. H.; Yates, K. *J. Org. Chem.* **1977**, *42*, 3673.

[57]Ruasse, M.-F.; Zhang, B.-L. *J. Org. Chem.* **1984**, *49*, 3207.

[58]Ruasse, M.-F.; Motallebi, S.; Galland, B. *J. Am. Chem. Soc.* **1991**, *113*, 3440 and references therein.

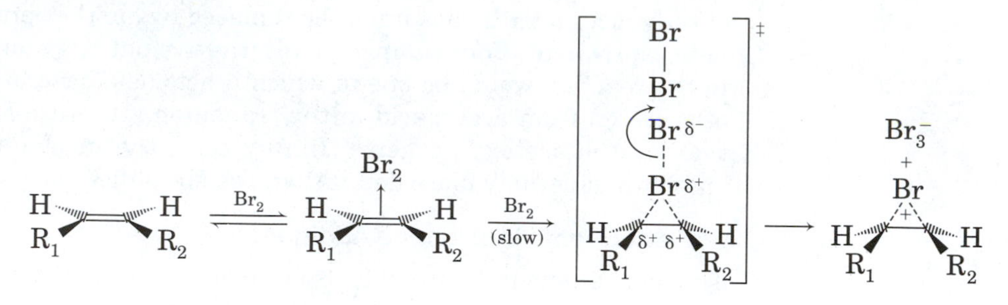

Figure 9.7 Model for electrophilic assistance of departing Br^- by Br_2 in a nonprotic solvent.

protic solvent comes from solvent kinetic isotope effect studies. The value of k_H/k_D for addition of bromine to 1-pentene in methanol/methanol-*d* was found to be 1.21, while for addition in acetic acid/acetic acid-d_4 it was 1.25.[59,55a] In an aprotic solvent, a second Br_2 molecule may assist collapse of the charge transfer complex by removing the departing Br^- as a Br_3^- ion (Figure 9.7).[60,58]

Based on solvent isotope effects and on the sensitivity of reaction rates to solvent, it has also been proposed that solvent molecules can provide nucleophilic assistance to the conversion of the CT complex to the bromonium ion, as shown in Figure 9.8.[56,58] This process is analogous to that proposed for one possible mechanism (Figure 9.9) for the bromide catalyzed addition to highly deactivated alkenes such as cinnamic acid.[49,61]

Reversibility of the Addition of Bromine

Most studies of the bromine addition to alkenes have presumed that the intermediate proceeds to product and does not revert to alkene and bromine.[62] However, Brown and co-workers determined that bromonium ions generated from the solvolysis of the *trans*-2-bromo-1-brosylates of cyclohexene or

Figure 9.8 Proposed nucleophilic role of solvent in converting CT complex to bromonium ion.

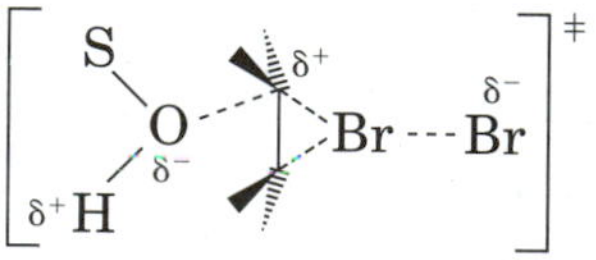

[59]Modro, A.; Schmid, G. H.; Yates, K. *J. Org. Chem.* **1979**, *44*, 4221.

[60]Yates, K.; McDonald, R. S.; Shapiro, S. A. *J. Org. Chem.* **1973**, *38*, 2460.

[61]The role of nucleophilic solvent participation is more clearly developed in the case of bromine addition to alkynes: Modena, G.; Rivetti, F.; Tonellato, U. *J. Org. Chem.* **1978**, *43*, 1521.

[62]Barton and Miller (reference 22) had reported that the 5α,6β-dibromocholestan-3β-ol produced initially by addition of bromine to cholesterol is converted, upon standing in methanol solution at room temperature for three days, to a 4 : 1 mixture of 5β,6α-dibromocholestan-3β-ol and the 5α,6β-isomer. However, the rate of the rearrangement was not affected by a 30-fold molar excess of cyclohexene, so it appeared that the rearrangement did not occur by dissociation of the cholesterol dibromide to cholesterol and bromine.

Figure 9.9 Proposed nucleophilic role of bromide ion in reaction of CT complex.

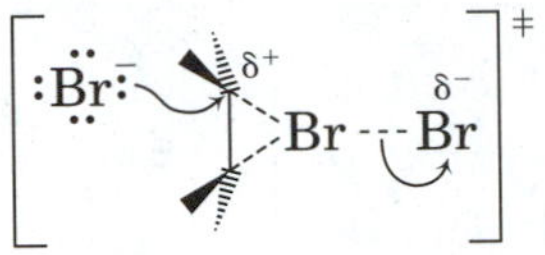

cyclopentene could react with added Br^- to produce Br_2.[63] Furthermore, *erythro*-2-bromo-1,2-diphenylethanol was found to react with anhydrous HBr (in 1,2-dichloroethane or in chloroform) to produce both *trans*-stilbene and *meso*-1,2-dibromo-1,2-diphenylethane. The reaction of the erythro diastereomer can be explained by a mechanism involving anchimeric assistance in departure of water, which leads to a bromonium ion that reverts to the stilbene, as shown in Figure 9.10.[64,65]

threo-2-Bromo-1,2-diphenylethanol was also found to produce *trans*-stilbene and *meso*-stilbene dibromide under the same reaction conditions. It appears that the internal strain of the bromonium ion initially formed by loss of water can be relieved by isomerization, via a β-bromocarbocation, to the isomeric bromonium ion having the same stereochemistry as that in Figure 9.10. As a result, both diastereomers of starting material lead to the same stereoisomeric products.

Figure 9.10 Evidence for reversibility of bromonium ion formation from an erythro bromohydrin. (Adapted from reference 64.)

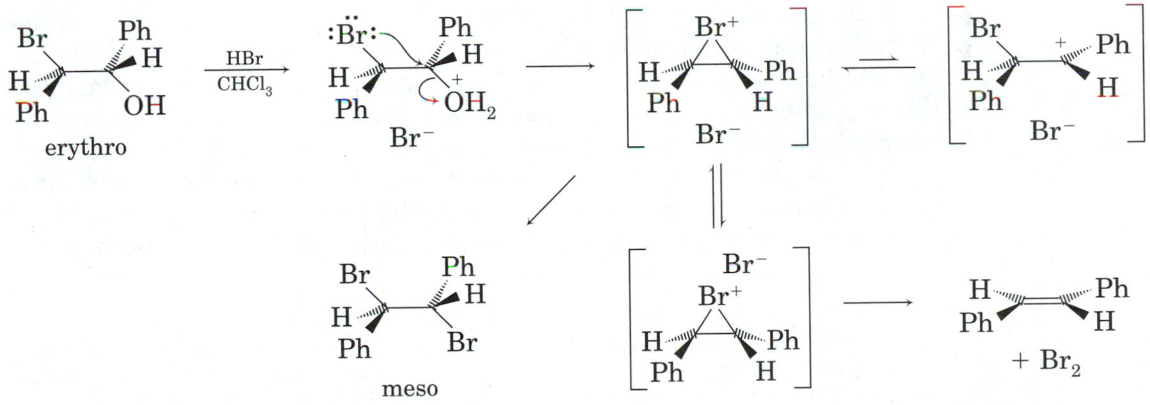

[63]Brown, R. S.; Gedye, R.; Slebocka-Tilk, H.; Buschek, J. M.; Kopecky, K. R. *J. Am. Chem. Soc.* **1984**, *106*, 4515.

[64]Bellucci, G.; Chiappe, C.; Marioni, F. *J. Am. Chem. Soc.* **1987**, *109*, 515.

[65]Also see Brown, R. S.; Slebocka-Tilk, H.; Bennet, A. J.; Bellucci, G.; Bianchini, R.; Ambrosetti, R. *J. Am. Chem. Soc.* **1990**, *112*, 6310 and references therein.

Figure 9.11 Evidence for reversibility of bromonium ion formation from a threo bromohydrin. (Adapted from reference 64.)

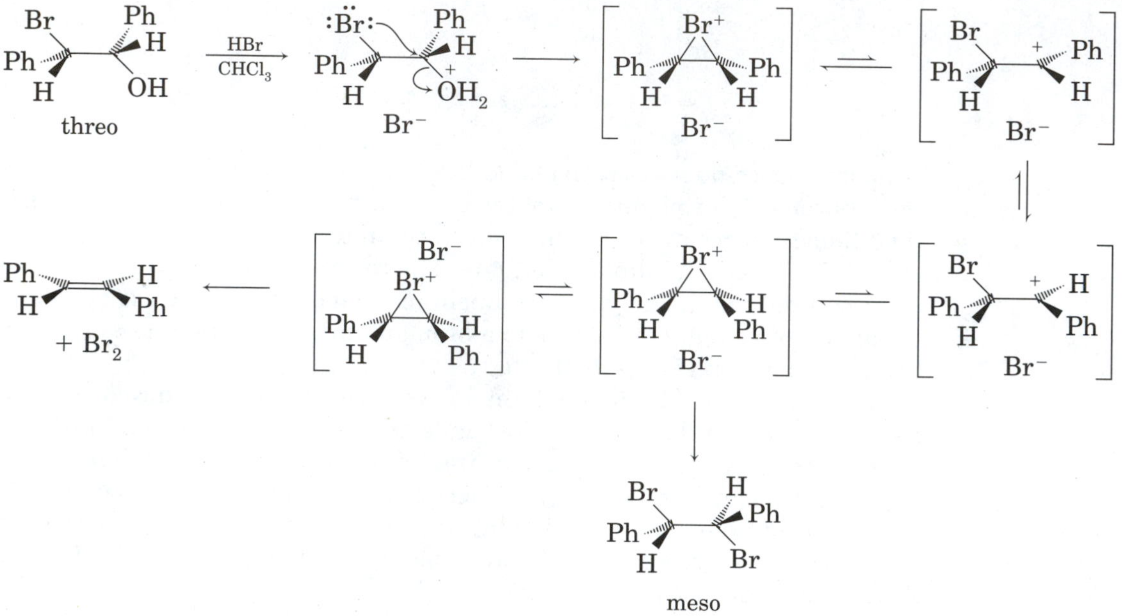

threo

$\xrightarrow[\text{CHCl}_3]{\text{HBr}}$

meso

Reversibility of bromonium ion formation has been directly observed in the regeneration of adamantylideneadamantane from the bromonium ion salt (Figure 9.12).[66,67] Further evidence for reversibility is the observation that the bromonium ion from adamantylideneadamantane can transfer Br^+ to cyclohexene in CH_2Cl_2 solution. Since a sterically hindered bromonium ion can transfer Br^+, it seems reasonable that bromonium ions that are not sterically hindered should also be capable of transferring Br^+. Therefore, reversibility of bromonium ion formation could be a general process.[35] However, reversibility may be less likely with bromonium ions formed from sterically unhindered alkenes in nucleophilic protic solvents because the

[66]Reversibility is also not anticipated for unsymmetrical alkenes such as *gem*-disubstituted alkenes. As is discussed in the following section, the cationic intermediates in such reactions are highly asymmetric, with much of the positive charge on the carbon atom and relatively little on the bromine. Since reversibility implies nucleophilic attack of bromide ion on the bromine in the intermediate, the smaller the charge on bromine, the less likely is that nucleophilic attack. For further discussion, see reference 58.

[67]Furthermore, from the solvolysis of the *trans*-2-bromo-1-trifluoromethane sulfonate of cyclohexane in the presence of Br^- in both acetic acid and in methanol, it was determined that the solvolytic bromonium ions actually react with Br^- preferentially on the Br^+, thus generating free Br_2. Zheng, C. Y.; Slebocka-Tilk, H.; Nagorski, R. W.; Alvarado, L.; Brown, R. S. *J. Org. Chem.* **1993**, *58*, 2122.

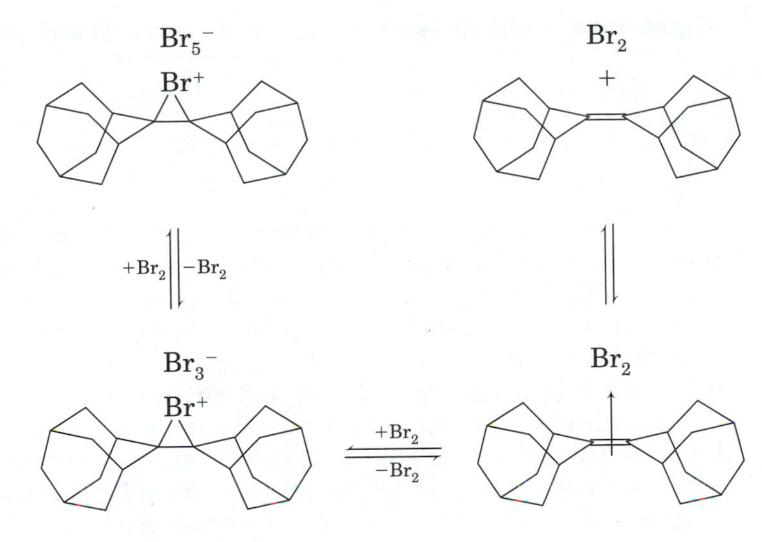

Figure 9.12 Multiple equilibria in the formation of bromine from the adamantylidene-adamantane bromonium ion. (Adapted from reference 41.)

solvent capture is very rapid. As we noted, the protic solvent provides about 60 kcal/mol of electrophilic assistance in removal of bromide ion to form the bromonium ion (or β-bromocarbocation), so this barrier would have to be overcome in the reverse reaction. Moreover, lack of steric hindrance would lower the barrier to attack of the bromide ion on the bromonium ion, so the lifetime of the bromonium ion should be decreased.

Intermediates in the Addition of Bromine to Alkyl-Substituted Alkenes

The Lewis structure of the bromonium ion in Figure 9.2 shows all of the positive charge on the bromine atom. However, polarization of the C—Br bond might be expected to result in some positive charge on the two carbon atoms. One way to indicate the charge distribution is to depict the bromonium ion as a resonance hybrid of three Lewis structures, as shown in Figure 9.13.[68] Still, we often find it convenient to represent the bromonium ion with just the Lewis structure having the positive charge on the bromine atom, just as we frequently represent benzene with only one of the Kekulé structures.

[68]Dewar proposed that the intermediate might be a π complex: Dewar, M. J. S. *J. Chem. Soc.* **1946**, 406.

Figure 9.13 Resonance structures contributing to the "bromonium ion".

Even though there is experimental evidence for the existence of bromonium ions, it need not be the case that all bromine addition reactions involve identical intermediates. The totally symmetric structure represented in Figure 9.13 may be only one member (**I** in Figure 9.14) of a range of possible intermediates for electrophilic halogen addition reactions. At the other extreme would be the bromosubstituted carbocation **III**, a species capable of rotation about the former carbon-carbon double bond. Since carbocation stability is strongly affected by substituents on the positively charged carbon atom, we might expect structure **III** to be more likely with structures such as *gem*-dialkyl olefins. In between these two extremes could be intermediates that are formally described as carbocations, but in which stabilization of the ion and restricted rotation about the C-C bond results from some interaction between the bromine atom and the more distant olefinic carbon atom.[69,70] The stabilizing interaction could be described either as partial bonding (**IIa**) or as an ion-dipole interaction involving the positive charge and the carbon-bromine bond dipole (**IIb**). Structures such as **IIa** and **IIb** are said to be *partially bridged*.

Figure 9.14 Spectrum of possible intermediates in the bromination of alkenes: I: bromonium ion; II(a) and II(b): partially bridged ion; III: bromocarbocation.

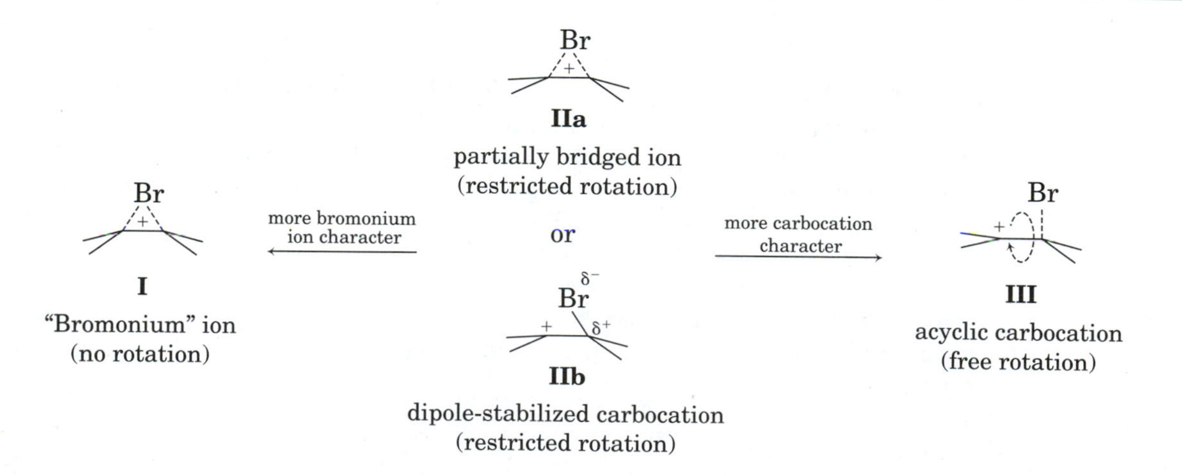

[69]Yates, K.; McDonald, R. S. *J. Org. Chem.* **1973**, *38*, 2465.

[70]The structures in Figure 9.14 are modified from those presented in reference 69.

One test for the intermediacy of carbocations in reaction mechanisms is to look for rearrangements, for example, from a 2° carbocation to a 3° carbocation. However, in the addition of bromine to *t*-butylethylene (3,3-dimethyl-1-butene, **6**) in methanol, the only products observed were 1,2-dibromo-3,3-dimethylbutane (**7**), 45%, and 2-bromo-1-methoxy-3,3-dimethylbutane (**8**), 44%.[29] There was no evidence for products such as **9**, which might have been expected if a free 2° carbocation were formed and then underwent a methyl shift to yield a 3° carbocation, so the intermediate in the addition of bromine to alkyl-substituted alkenes appears not to behave like a carbocation.

H_3C, H_3C, CH_3 — **6** —(Br_2 / CH_3OH)→ Br, H_3C, Br, H_3C, CH_3; 45% **7** + Br, H_3C, OCH_3, H_3C, CH_3; 44% **8** (H_3C, H_3C, Br, H_3C, Br; not observed) **9** **(9.9)**

There is also kinetic evidence that the intermediate in the addition of bromine to aliphatic alkenes is not a carbocation. Nagorski and Brown studied the addition of bromine to alkenes in methanol containing varying concentrations of added bromide ion or azide ion. The ratios of the rate constants for formation of addition products incorporating azide ion or methanol (k_{N_3}/k_{CH_3OH}) were found to be 5.9 and 4.9 M^{-1}, respectively, for cyclopentene and cyclohexene. Since azide is a much stronger nucleophile than is methanol, the relatively small ratios suggested that the intermediate is a highly reactive (nonselective) species. Assuming that the reaction of the intermediate with N_3^- is diffusion-limited (meaning a rate constant of about 10^{10} M^{-1} sec^{-1}), the lifetimes of the intermediates from these alkenes were calculated to be 5.9×10^{-10} and 5.0×10^{-10} sec, respectively.[54] These lifetimes are two orders of magnitude greater than the lifetime expected for a secondary carbocation.[71,72] The longer lifetimes and the observation of exclusive anti addition support the view that the intermediates are bromonium ions and not β-bromocarbocations.[54]

The bromonium ion derived from ethene has been studied by *ab initio* methods. The results of one study found the bromonium ion to be an energy minimum on the $C_2H_4Br^+$ hypersurface, with the open β-bromocarbocation being higher in energy.[73] Contrary to the depiction in Figure 9.2, however,

---

[71] Chiang, Y.; Chwang, W. K.; Kresge, A. J.; Powell, M. F.; Szilagyi, S. *J. Org. Chem.* **1984**, *49*, 5218.

[72] Chiang, Y.; Kresge, A. J. *J. Am. Chem. Soc.* **1985**, *107*, 6363.

[73] Hamilton, T. P.; Schaefer, III, H. F. *J. Am. Chem. Soc.* **1990**, *112*, 8260.

the bromine atom in the bromonium ion was calculated by one theoretical method to be almost neutral, with the charge residing primarily on the carbon atoms.[73,74] However, a study of the bromonium ion derived from ethene by another theoretical approach led to the conclusion that there is appreciable positive charge on the bromine, but that the magnitude of the charge is sensitive to solvent, to substituents, and to the orientation of the bromide ion with respect to the bromonium ion.[75]

An MNDO study of methyl-substituted bromonium ions indicated that *symmetrical* bromonium ions are energy minima only when the two carbon atoms in the bromonium ion ring *each* have 0, 1, or 2 methyl groups. When one of the carbon atoms has a greater number of alkyl groups than does the other one, the structures corresponding to energy minima were found to be highly asymmetric, with a longer distance from the bromine atom to the more substituted carbon atom (Figure 9.15). The resulting picture is more nearly consistent with a structure in which there is only a small bonding interaction between the bromine and the more highly substituted carbon atom.[76] However, Klobukowski and Brown found that the geometry calculated for the bromonium ion derived from isobutene depends strongly upon the level of the calculation. Higher order calculations predicted a more symmetric ion but with a rather flat potential energy surface for variation of the $C-CH_2-Br$ angle (10–15° variation from an angle near 85° causing an energy change of only about 2 kcal/mol). Although the resulting geometry differs somewhat from that shown in Figure 9.15, the higher order calculations do indicate greater positive charge on the carbon atom bearing the methyl substituents. Furthermore, Klobukowski and Brown suggested that the flat potential energy surface for variation of the $C-C-Br$ angle may be perturbed by an incoming nucleophile.[77]

An additional perspective on the reactivity of these bromonium ions comes from comparison of the square of the LUMO coefficient at the two

Figure 9.15 MNDO structures for intermediates in the bromination of ethene (left) and 2-methylpropene (right). (Data from reference 76.)

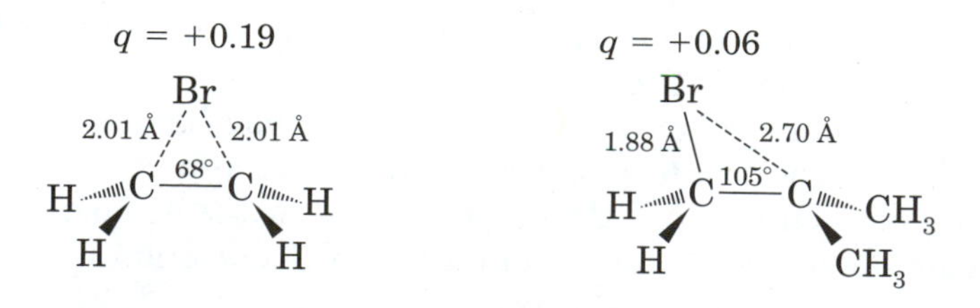

[74]Cioslowski, J.; Hamilton, T.; Scuseria, G.; Hess, B. A.; Hu, J.; Schaad, L. J.; Dupuis, M. *J. Am. Chem. Soc.* **1990**, *112*, 4183.

[75]Cossi, M.; Persico, M.; Tomasi, J. *J. Am. Chem. Soc.* **1994**, *116*, 5373.

[76]Galland, B.; Evleth, E. M.; Ruasse, M.-F. *J. Chem. Soc., Chem. Commun.* **1990**, 898.

[77]Klobukowski, M.; Brown, R. S. *J. Org. Chem.* **1994**, *59*, 7156.

carbon atoms in the three-membered ring in bromonium ions. In the bromonium ion from ethene, the square of the LUMO coefficient is 0.22 on each carbon atom. In the bromonium ion from isobutene, however, the square of the LUMO coefficient on the carbon atom with two hydrogen substituents is 0.035, while that on the carbon atom bearing the two methyl groups is 0.53.[78,79]

Further insight into the nature of bromonium ions comes from studies of the competitive reactions of Br^- and methanol with bromonium ions in methanol solution. The reaction can be categorized with regard to the chemoselectivity (relative reactivity with bromide ion and with methanol) and regioselectivity (Markovnikov or anti-Markovnikov orientation of the bromomethoxy product). In this context, ***Markovnikov*** is defined as the formation of product in which the bromine is bonded to the carbon atom having the fewer alkyl or aryl substituents, which is the orientation expected upon attachment of Br^+ to give the more stable carbocation. The carbon atom with the greater number of alkyl substituents is designated as C_α in Figure 9.16. The results, some of which are shown in Table 9.2, are that the more unsymmetrical bromonium ions were found to produce greater amounts of methanol addition products relative to dibromide. Moreover, the more unsymmetrical alkenes also gave relatively greater amounts of Markovnikov bromomethoxy compounds. On the other hand, the more symmetrical

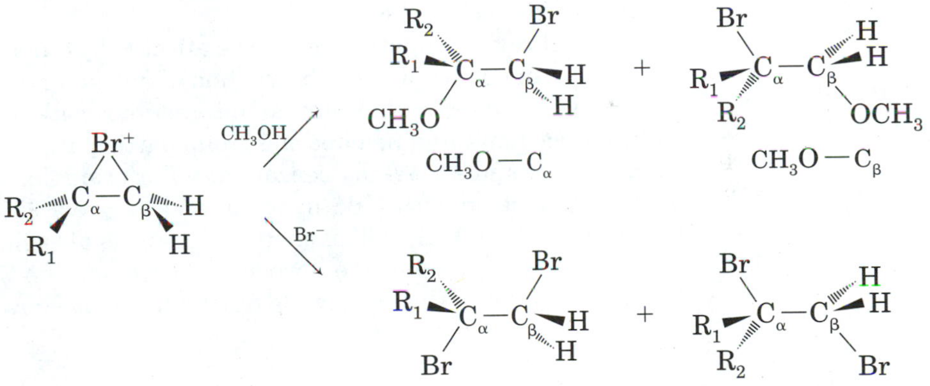

CH_3OH; Br^-; CH_3O-C_α; CH_3O-C_β

Figure 9.16 Chemo- and regioselectivity in bromination of alkenes. (Reproduced from reference 80.)

[78]PM3 calculations were conducted using MOPAC version 3.6 on a CAChe worksystem.

[79]The cover shows an artist's rendering of a portion of the electron density surface (representing a surface having an electron density of 0.01 e/Å^3) of the bromonium ion derived from isobutene. In this calculation the $C-CH_2-Br$ bond angle is fixed at 85° (in accordance with the results reported in reference 77). The color contours illustrate regions of similar nucleophilic susceptibility, which is calculated on the basis of the density of frontier MOs (Fukui, K.; Yonezawa, T.; Nagata, C.; Shingu, H. *J. Chem. Phys.* **1954**, *22*, 1433). The nucleophilic susceptibility on the carbon atom with the two hydrogen atoms is 0.04, while that for the carbon atom with two methyl substituents is 1.18. (By comparison, nucleophilic susceptibility was found to be 0.46 for each of the two carbon atoms in the bromonium ion formed from ethene.) These results are consistent with the tendency for the bromonium ion from isobutene to react with nucleophiles at the more highly substituted position, but the calculated structures implicitly represent gas phase species. Structures in solution may be affected by solvent or counterions.

Table 9.2 Chemoselectivity and regioselectivity in addition of bromine to alkenes.

Alkene	R_1	R_2	Bromomethoxy-alkane	CH_3O-C_α	CH_3O-C_β
Ethene	H	H	38 %	(not defined)	(not defined)
Propene	CH_3	H	61 %	50 %	11 %
2-Methylpropene	CH_3	CH_3	85 %	85 %	0 %

(Data from reference 80.)

alkenes gave higher yields of dibromo products, and the bromomethoxy products that were produced were composed of relatively more of the anti-Markovnikov (CH_3O-C_β) products.

These results are also consistent with a model in which the bromonium ion from a more symmetrically substituted alkene behaves more like a bromonium ion and less like a carbocation. That is, with a symmetrical ion there is relatively little solvent incorporation, and solvent is relatively indiscriminant in attacking the two carbon atoms of the bromonium ion. On the other hand, there is more solvent incorporation product formed from the more unsymmetrical ions, and the solvent is more likely to add to the more highly substituted carbon atom than to the less highly substituted carbon atom. This behavior suggests that solvent is attracted to the center of greater partial positive charge.[80]

Product regiochemistry can also be affected by other functional groups in the alkene. In particular, the observation of anchimeric assistance in solvolysis reactions suggests that similar interactions could occur in electrophilic addition reactions, and neighboring group participation has been noted in the addition of bromine to *N*-allylbenzamides. For example, *N*-*p*-methoxybenzoyl-allylamine reacted with NBS in acetic acid to give a 95% yield of the oxazoline **10** (equation 9.10).[81] This methodology was also shown to be useful for the synthesis of cyclohexane derivatives with specific 2-*cis*,3-*trans* orientation with respect to the benzamide functionality, as shown in equation 9.11.

OCH$_3$ … HN … O … Br—X $\xrightarrow[-H^+]{-X^-}$ OCH$_3$ … N … O … CH$_2$Br

10

(BrX = Br_2 or *N*-bromosuccinimide) **(9.10)**

[80]Dubois, J.-E.; Chretién, J. R. *J. Am. Chem. Soc.* **1978**, *100*, 3506. Regioselectivity could be correlated with the difference in partial positive charge at the two carbon atoms as determined by NMR.

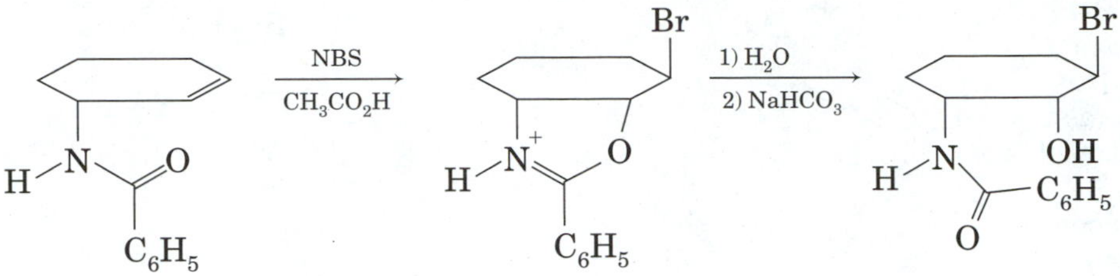

(9.11)

Intermediates in the Addition of Bromine to Aryl-Substituted Alkenes

The discussion to this point has focused on the addition of bromine to alkenes bearing alkyl substituents. There are significant differences in the addition of bromine to alkenes with aryl substituents. For example, the addition of bromine to aryl-substituted alkenes is not stereospecific. Reaction of *cis*-β-methylstyrene with bromine in CCl_4 led to the formation of 17% of *erythro* and 83% of *threo*-(1,2-dibromopropyl)benzene (equation 9.12). Reaction of *trans*-β-methylstyrene under the same conditions yielded 88% of the erythro and 12% of the threo product. When the benzene ring in the trans reactant was substituted with a 4-methoxy group (*trans*-anethole), the reaction became even less stereoselective, giving 63% erythro and 37% threo product.[82,83]

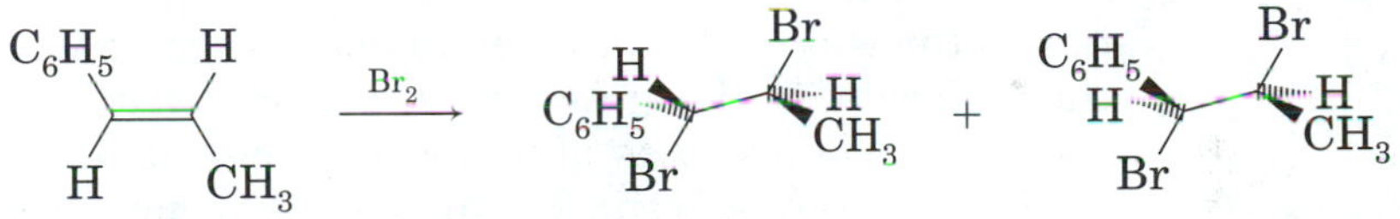

trans-β-methylstyrene **11** $\xrightarrow{Br_2}$ (±)-*erythro* **12** + (±)-*threo* **13** (9.12)

Yates, McDonald, and Shapiro studied the addition of bromine in acetic acid to ring-substituted styrenes under concentration conditions ($< 10^{-3}$ M) in which second order kinetics were observed.[84] The reaction rate was decreased by electron-withdrawing substituents on the aromatic ring, and the rates correlated ($r = .997$) with σ^+ to give a ρ of -4.8.[25] That ρ value is very

[81]Winstein, S.; Goodman, L.; Boschan, R. *J. Am. Chem. Soc.* **1950**, *72*, 2311.

[82]Fahey, R. C.; Schneider, H.-J. *J. Am. Chem. Soc.* **1968**, *90*, 4429.

[83]The reaction of bromine with the stryenes is not said to be stereospecific because it is not observed that stereoisomeric reactants give different stereoisomeric products. Rather, the two reactions can only be said to be stereoselective, meaning that each of the two reactants gives more of one steroisomer than of another.

[84]Yates, K.; McDonald, R. S.; Shapiro, S. A. *J. Org. Chem.* **1973**, *38*, 2460.

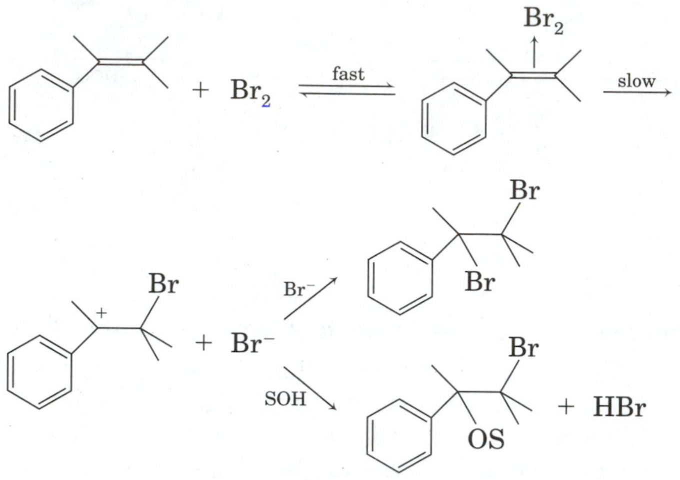

Figure 9.17 Mechanism proposed for second order bromination of styrenes. (Reproduced from reference 84)

nearly the same as the ρ value observed for the solvolysis of cumyl chloride in ethanol (−4.67) or 90% aqueous acetone (−4.54),[85] suggesting that the addition of bromine involves a species that closely resembles a benzylic carbocation (Figure 9.17).[86]

It is unlikely that a free carbocation (or a solvent separated ion pair) is the intermediate in these addition reactions, however. Rolston and Yates studied the addition of bromine to derivatives of styrene in acetic acid and other solvents.[21,87] In acetic acid the products were found to be the 1,2-dibromo addition products plus variable amounts of the bromoacetoxy products derived from solvent. For all of the styrene derivatives studied except for β,β-dimethylstyrene, the solvent incorporation product in each case was exclusively the 1-acetoxy-2-bromo compound—the product expected from nucleophilic attack of solvent on a benzylic carbocation. The dibromo and bromoacetoxy products were formed with different stereoselectivity, however. The reaction of *cis*-β-methylstyrene produced (±)-*threo*-1,2-dibromo-1-phenylpropane (**13**, 58%), (±)-*erythro*-1,2-dibromo-1-phenylpropane (**12**, 22%), and (±)-*threo*-1-acetoxy-2-bromo-1-phenylpropane (**15**, 20%).

---

[85] Brown, H. C.; Okamoto, Y. *J. Am. Chem. Soc.* **1958**, *80*, 4979.

[86] The second- and third-order reactions showed interesting differences. For the second-order reactions, $\Delta H^{\ddagger}$ values were 9.0 and 4.7 kcal/mol for addition of bromine to *p*-nitrostyrene and to styrene, while $\Delta S^{\ddagger}$ values were −39.5 and −37.6 eu, respectively, for these two compounds. For the third-order reaction, however, the values of $\Delta H^{\ddagger}$ were 0.9 and 0.01 kcal/mol, and the $\Delta S^{\ddagger}$ values were −50.5 and −37.6 eu, respectively, for addition to *m*-nitrostyrene and to styrene. Thus, in reference 84 the second-order reaction was said to be "enthalpy controlled," while the third-order reaction was said to be "entropy controlled."

Br_2, HOAc

cis-β-Methylstyrene **14** → (±)-*threo* 58% **13** + (±)-*erythro* 22% **12** + (±)-"*threo*" 20% **15** **(9.13)**

Reaction of *trans*-β-methylstyrene produced 64% of **12**, 13% of **13**, and 23% of (±)-*erythro*-1-acetoxy-2-bromo-1-phenylpropane (**16**).

Br_2, HOAc

trans-β-Methylstyrene **11** → (±)-*erythro* 64% **12** + (±)-*threo* 13% **13** + (±)-"*erythro*" 23% **16** **(9.14)**

In both cases, therefore, the bromoacetoxy products were those expected from anti addition, while formation of the dibromo products was much less stereoselective.[21,88,89] Furthermore, the fraction of dibromide formed by apparent syn addition to *cis*-β-methylstyrene varied with solvent polarity,

---

[87]Rolston, J. H.; Yates, K. *J. Am. Chem. Soc.* **1969**, *91*, 1477.

[88]In this discussion, the words *erythro* and *threo* are shown in quotation marks for those structures in which the labels would apply if the two groups that added to the olefinic carbon atoms were identical, even though they are not the same. In the structures below, the erythro structure would become a meso structure if the phenyl and methyl were made equivalent. In the "erythro" structure to its right, however, the bromine and acetoxy groups would also have to be made identical in order for the compound to become a meso structure.

erythro   "erythro"   threo   "threo"

The "erythro" and "threo" labels provide a convenient shorthand notation for unsymmetrical adducts. For further discussion, see footnote 18 of Ruasse, M. F.; Argile, A.; Dubois, J. E. *J. Am. Chem. Soc.* **1978**, *100*, 7645.

[89]Similar results were obtained in a study of the reaction of *trans*-β-methylstyrene and its derivatives with bromine in methylene chloride. Only when the aromatic ring was substituted with trifluoromethyl groups at the 3 and 5 positions was the reaction 100% anti stereospecific. With one trifluoromethyl group at the three position, 91% of the product was the "erythro" product, while 9% was the "threo" product. With no substituent other than hydrogen on the aromatic ring, 81% of the product was erythro, while with a 4-methoxy substituent, only 63% was erythro. See reference 88.

from 27% in acetic acid to 55% in nitrobenzene.[87] The results were discussed in terms of initial formation of an intimate bromocarbocation–bromide ion pair that can undergo attachment of acetic acid from the face of the ion away from the bromide ion, thus leading to anti formation of bromoacetoxy compounds. Reorientation of the bromide ion within the solvent shell for backside attachment to the bromonium ion provides time for rotation about the carbon-carbon single bond of the bromocarbocation, so the dibromo products are formed with less stereoselectivity.

There is also kinetic evidence that the addition of bromine to styrene involves an intermediate that is neither a bromonium ion nor a carbocation but which does have some interaction between the bromine atom and the phenyl-substituted carbon atom. The lifetime of a 1-phenylethyl cation was determined to be 10^{-11} sec,[72] but the lifetime estimated by Nagorski and Brown for the intermediate in the addition of bromine to styrene is 2.7×10^{-10} sec. Thus, the longer lifetime of the intermediate in the addition reaction would support the view of some kind of stabilization, such as a dipolar interaction or weak bridging.[54,90] That would mean that the intermediate is somewhat closer to the center of the range in Figure 9.14.

It is also possible that there is not one intermediate, but several intermediates that are formed and react in parallel. In the case of addition of bromine to stilbene, Ruasse and co-workers have argued for the competitive formation of three discrete intermediates: two different carbocations and a bromonium ion (Figure 9.18).[26] If formation of the intermediates is irreversible, then the rate constant for the reaction, k, is the sum of the rate constants for the three processes. That is,

$$k = k_{\alpha} + k_{\beta} + k_{Br} \tag{9.15}$$

If X is a strongly electron-donating group, then a carbocation adjacent to the substituted ring will be most stable. Therefore, k will essentially be k_{α}, and a Hammett correlation will give a correlation of log k with σ^+. If X is a strongly electron-withdrawing group, however, then the carbocation with positive charge further away from the substituted ring will be most stable. In this case, k_{β} will effectively determine the value of k, and log k will correlate with σ.[25] If X is neither strongly electron donating nor withdrawing, then the intermediate may be a bromonium ion, and there should again be a correlation of log k with σ. The curved $\sigma\rho$ plot (Figure 9.19) shows a change of slope as X varies from electron donating to electron withdrawing,

---

[90]The intermediacy of a species with appreciable charge on the benzylic carbon atom is also supported by the results of Rolston and Yates, who observed a Hammett correlation (with σ^+) with a ρ value of -4.21 for the addition of bromine to styrene derivatives in acetic acid: Rolston, J. H.; Yates, K. *J. Am. Chem. Soc.* **1969**, *91*, 1483.

[91]For further discussion and results for styrene derivatives, see reference 26.

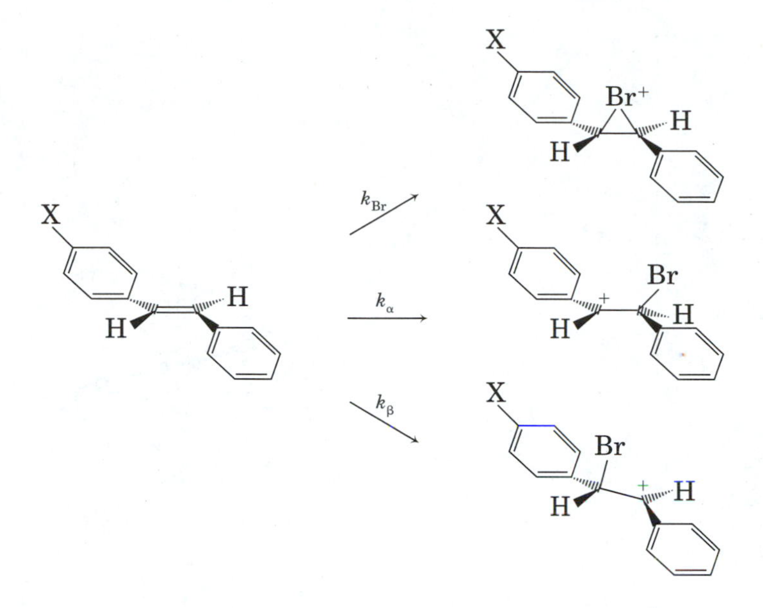

Figure 9.18
Competitive pathways proposed for the electrophilic addition of bromine to stilbene derivatives.

suggesting a mechanistic changeover consistent with the role of discrete intermediates.[91,92]

Another report suggested that a single intermediate with a variable degree of bridging might be involved in the addition of bromine to stilbene derivatives. Bellucci and co-workers found evidence for reversibility in the addition of bromine to stilbene derivatives in 1,2-dichloroethane solution (Figure 9.20).[93] The cis isomer of *p*-methylstilbene, stilbene, *p*-trifluoromethylstilbene and *p,p'*-bis(trifluoromethyl)stilbene each showed some conversion to the corresponding trans isomer during the reaction. Isomerization of starting material was greatest with the *p,p'*-bis(trifluoromethyl)isomer and least with the *p*-methyl isomer. The ratio of meso to racemic product was

---

[92]Bellucci, G.; Bianchini, R.; Chiappe, C.; Marioni, F. *J. Org. Chem.* **1990**, *55*, 4094, however, described the addition of bromine to stilbene derivatives in terms of a spectrum of intermediates with variable bridging. One study involved the addition of bromine to *cis*- and *trans*-stilbene in 1,2-dichloroethane. With the cis isomer and bromine both at concentrations greater than 10^{-3} M, the addition was essentially nonstereoselective; roughly a 50 : 50 mixture of the meso and (±) dibromides was obtained. As the concentrations were reduced, however, the meso isomer became dominant, amounting to 96% of the product when the initial concentration of *cis*-stilbene was 10^{-5} M and that of Br_2 was 10^{-3} M. A similar trend was observed for reaction of *trans*-stilbene. The ratio of meso to racemic products was about 70 : 30 at higher reagent concentrations but was 96 : 4 at the lowest concentrations studied. These results were explained by a scheme involving multiple equilibria between intimate and solvent separated ion pairs of bromonium ions and β-bromocarbocations.

[93]Bellucci, G.; Bianchini, R.; Chiappe, C.; Brown, R. S.; Slebocka-Tilk, H. *J. Am. Chem. Soc.* **1991**, *113*, 8012.

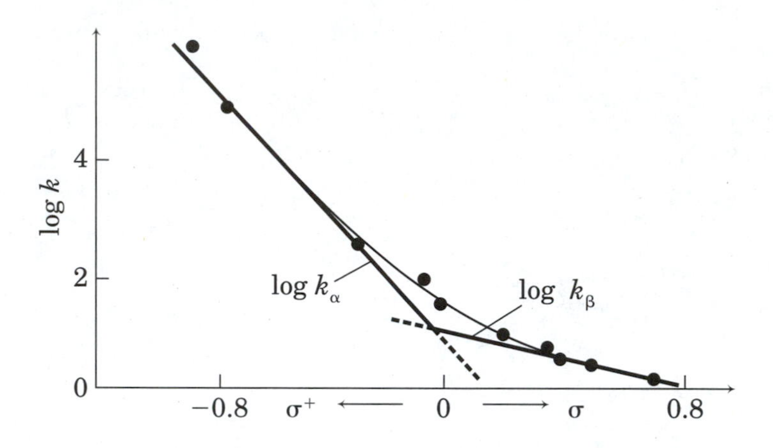

Figure 9.19 Curved σρ plot observed for bromination of monosubstituted stilbenes. (Reproduced from reference 26.)

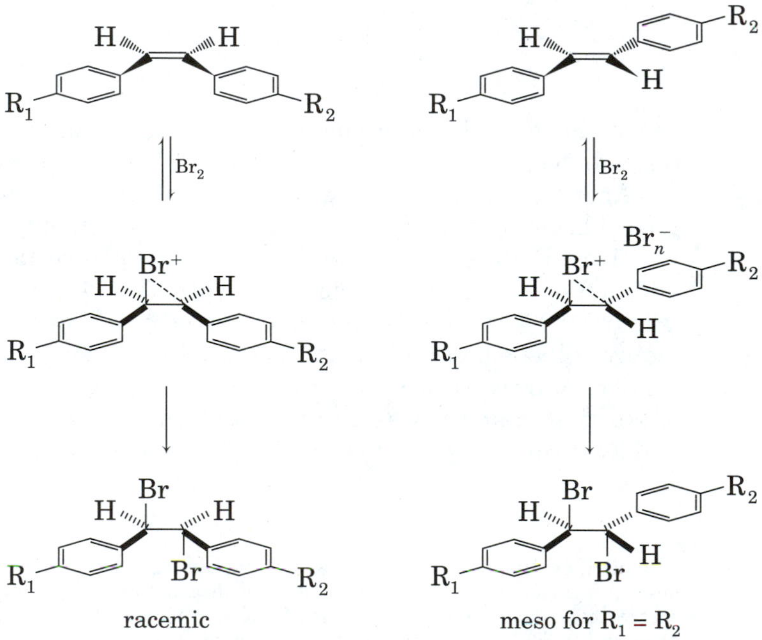

Figure 9.20 Proposed mechanism for bromine addition to stilbenes. (Note the process leading to cis-trans isomerization.)

racemic

meso for $R_1 = R_2$

essentially constant at about 70 : 30 for the *p*-methyl isomer no matter whether the cis or trans isomer was used as starting material. This result suggested that the intermediate resembles an open bromocarbocation. For the other three compounds the product ratio for the cis reactant was different from that obtained when the trans isomer was the starting material. For the trans isomer of *p,p′*-bis-(trifluoromethyl)stilbene, the formation of almost exclusively meso product suggested stereospecific anti addition. The investigators concluded that the cationic intermediates formed from these stilbenes have properties similar to those of β-bromocarbocations but with some stabilization or bridging involving the bromine (as shown by the dotted line in the central structures in Figure 9.20) and that the degree of bridging increases as the electron-withdrawing properties of the substituents increase.[94]

Summary

It appears that addition of bromine to an alkene is much more complicated than the simple representation in Figure 9.2 would suggest. Therefore, the classical bromonium ion description of electrophilic addition of bromine to an alkene is useful only as a beginning point to describe the mechanistic options. The structure of the intermediate, the kinetics of the reaction, and both the stereochemistry and the regiochemistry of the products are all complex functions of the nature and concentration of the brominating agent, the solvent and any added nucleophiles, and the structure of the alkene.

Only symmetrical alkenes react by means of true bromonium ions (that is, strongly bridged, symmetrical structures). If the alkene is unsymmetrically substituted with alkyl groups, the bromonium ion takes on the character of a β-bromocarbocation with restricted rotation about the C^+—C bond. If an aryl group is present on one or both of the olefinic carbon atoms, the reaction takes on character suggesting an intermediate that is even more nearly like a benzylic carbocation.

Preceding the cationic intermediate is at least one charge transfer complex, with the rate-limiting step usually being the conversion of the CT complex to a bromonium ion or bromocarbocation intermediate. Protic solvents can assist this ionization process by acting as an electrophile to remove a Br^- through hydrogen bonding. Nonpolar aprotic solvents cannot assist in that way, but a second Br_2 molecule may participate in the reaction by removing Br^- as a Br_3^- ion (and Br_5^- ions may be involved at high bromine concentration). Formation of the cationic intermediate is reversible, although the reverse reaction is usually not seen unless the solvent is aprotic or the alkene provides a steric barrier to the nucleophilic attack needed to complete the addition reaction.

---

[94]The authors further concluded that for *p*-methylstilbene the activation energy for formation of the β-bromocarbocation intermediate is higher than the activation energy for formation of the dibromide. However, for the reaction of *trans-p,p'*-bis(trifluoromethyl)stilbene, the activation energy appeared to be greater for addition of the nucleophile to the cation. This conclusion is consistent with the hypothesis that reversibility is most likely when the intermediate is a bromonium ion (so that significant positive charge is localized on the bromine), and it is least likely when the intermediate is a β-bromocarbocation (with little charge on bromine). For a discussion, see reference 93.

The cationic intermediate probably is formed as a tight ion pair; its subsequent reaction depends on the nucleophilicity and polarity of the solvent. In a less polar solvent, the initially formed tight ion may rearrange to an ion pair in which the bromide ion is on the backside of the C — Br bonds, which facilitates the second step of the addition reaction. In a more polar solvent, the bromide ion is more likely to diffuse away from the cation, so nucleophilic attack by solvent becomes more likely. Regardless of whether the intermediate is a bromonium ion or a β-bromocarbocation and whether the product is the dibromide or solvent incorporation product, the stereochemistry of the addition is anti with alkyl-substituted alkenes. If the alkene is unsymmetrically substituted with aryl groups, the tendency for the cationic intermediate to behave as a β-bromocarbocation is even greater, and both syn and anti addition are observed.

Addition of Other Halogens to Alkenes

Chlorine

The addition of bromine may be the most-studied addition of molecular halogen to alkenes, but reactions that affect addition of other halogens are also important. The study of the addition of chlorine to alkenes is even older than the study of bromine addition. The synthesis of 1,2-dichloroethane was first reported in 1795, and Michael Faraday studied the addition of chlorine and iodine to ethene.[95] In many ways the addition of chlorine to alkenes is similar to the addition of bromine to alkenes. For example, each additional alkyl group increases the relative reactivity of an olefin by a factor of 50 to 100. (Reactivity data are given in a column on the left of Table 9.3). A Taft $\rho^*\sigma^*$ correlation is observed, and the ρ^* for chlorine addition is -2.9, which is similar to the value of -3.1 for addition of bromine. This result suggests that charges on the alkenyl carbon atoms in the transition structures for addition of chlorine are similar to those in the addition of bromine.[96]

The chloronium ion (the chlorine equivalent of the bromonium ion) was proposed by Lucas and Gould,[97] and Olah reported evidence for a stable chloronium ion by NMR.[30] Thus we would expect the stereochemistry of chlorine addition to be quite similar to that for bromine. Lucas and Gould reported that addition of chlorine to 2-butene occurs by an anti pathway, with *trans*-2-butene giving the meso adduct, while the cis isomer yields racemic product.[97]

---

[95] Anantakrishnan, S. V.; Venkataraman, R. *Chem. Rev.* **1943**, *33*, 27 and references therein.

[96] Bienvenüe-Goëtz, E.; Ratsimandresy, B.; Ruasse, M. F.; Dubois, J. E. *Tetrahedron Lett.* **1982**, 3273.

[97] Lucas, J. J.; Gould, C. W. *J. Am. Chem. Soc.* **1941**, *63*, 2541.

There are significant differences between the bromine addition and chlorine addition reactions, however. The addition reaction of ethene and chlorine is exothermic by 44 kcal/mol, which is 15 kcal/mol more exothermic than the addition of ethene and bromine.[98] Poutsma noted that addition of chlorine to alkenes in nonpolar solvents can occur by either radical or ionic pathways, but that oxygen inhibits the radical reaction.[99] For example, addition of chlorine to cyclohexene (neat) gave *trans*-1,2-dichlorocyclohexane (**18**), 3-chlorocyclohexene (**19**), and 4-chlorocyclohexene (**20**) in a 1.95 : 1.00 : 0.60 ratio when the reaction was carried out under a nitrogen atmosphere. In the presence of oxygen, however, the ratio was 3–4 : 1.00 : 0.[100]

Cl_2 / N_2 → Cl, Cl + Cl, H + Cl, H **(9.16)**

17 **18** **19** **20**

Poutsma found that the tendency to give substitution as well as addition with Cl_2 is general for the reaction of neat alkyl-substituted alkenes and that this tendency increases with increasing alkyl substitution. Table 9.3 shows the yields of substitution products obtained from a series of alkenes.[101] The formation of substitution products in chlorine addition reactions had been explained by Taft as resulting from a β-chlorocarbocation.[102] However, addition of chlorine to *cis*-1,2-di-*t*-butylethylene in carbon tetrachloride solvent with oxygen present gave racemic 3,4-dichloro-2,2,5,5-tetramethylhexane. Similar reaction of the trans isomer gave the meso dichloride, but the substitution product was also formed.[103] The results suggest that the isomerizational stability of the intermediate ion must be at least enough to overcome the greater steric strain of the cis alkene than that of the trans, which is about 9 kcal/mol. An intermediate carbocation would be expected to relax to a conformation with less steric strain, but a chloronium ion could provide isomerizational stabilization. Therefore, Fahey concluded that the

---

[98]Conn, J. B.; Kistiakowsky, G. B.; Smith, E. A. *J. Am. Chem. Soc.* **1938**, *60*, 2764.

[99]The addition of bromine to alkenes can also occur by a free radical pathway if the reaction is carried out in a nonpolar solvent such as CCl_4 and if the reaction mixture is irradiated with light. Allylic substitution can be a competing reaction if the alkene contains allylic hydrogens. For a discussion of the competing substitution and addition reactions, see McMillen, D. W.; Grutzner, J. B. *J. Org. Chem.* **1994**, *59*, 4516.

[100]Poutsma, M. L. *J. Am. Chem. Soc.* **1965**, *87*, 2161.

[101]The material balance not accounted for by the sum of the percent yields was addition product.

[102]Taft, Jr., R. W. *J. Am. Chem. Soc.* **1948**, *70*, 3364.

[103]Fahey, R. C. *J. Am. Chem. Soc.* **1966**, *88*, 4681.

Table 9.3 Reactivities of alkenes and yields of substitution products in reactions of Cl_2 with neat alkenes.

Reactivity*	Reagents	Substitution Products (% Yields)
1.00	Cl_2 / O_2	$ClCH_2$ 3%
1.15	Cl_2 / O_2	CH_2Cl 10%
50	Cl_2 / O_2	Cl 20%
58	Cl_2 / O_2	CH_2Cl 87%
160	Cl_2 / O_2	CH_2Cl 58% + CH_2Cl 29% + CH_2Cl 8%
1,100	Cl_2 / O_2	Cl 85%
43,000	Cl_2 / O_2	Cl >99%

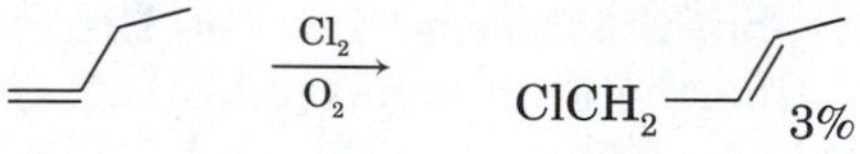

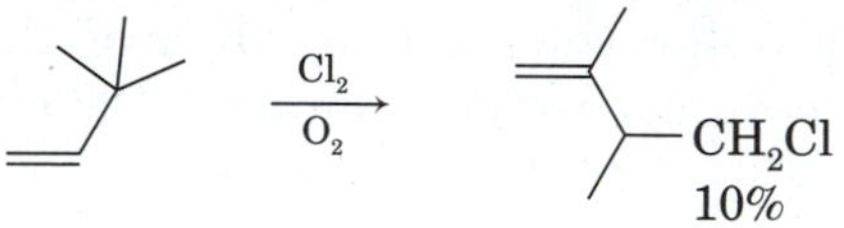

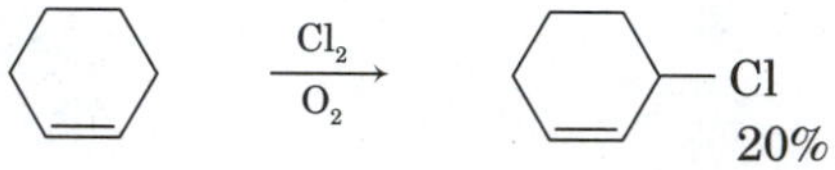

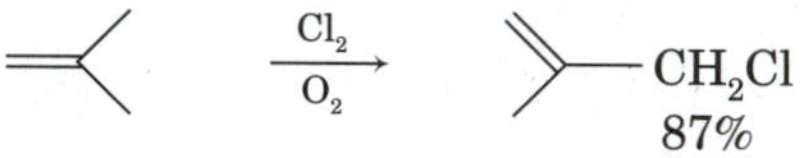

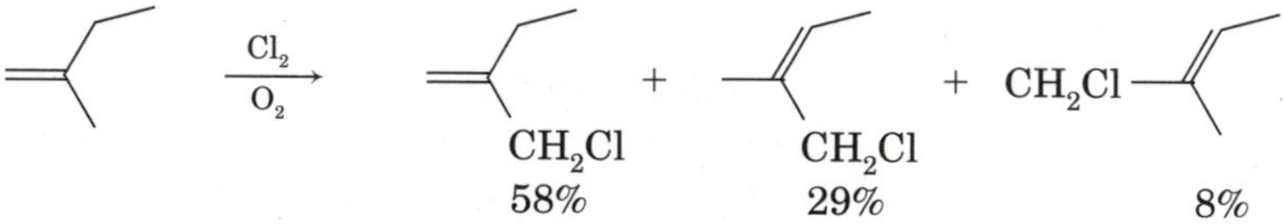

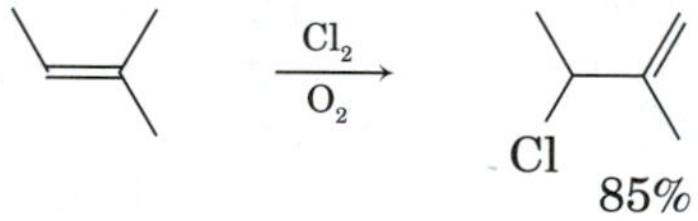

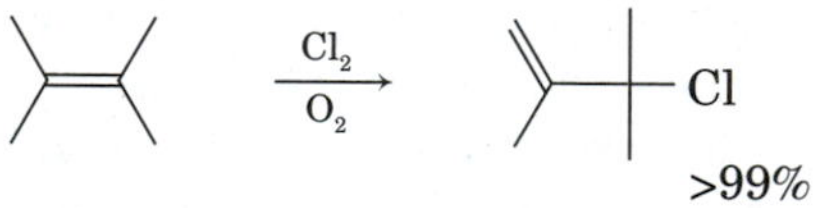

(Adapted from Reference 105.)
*Substitution and addition.

reactions must proceed either through a bridged intermediate or through an open chlorocarbocation with sufficient interaction between the chlorine atom and the carbocation carbon atom to provide significant stabilization. A mech-

Figure 9.21 Possible mechanism for substitution via a chloronium ion intermediate.

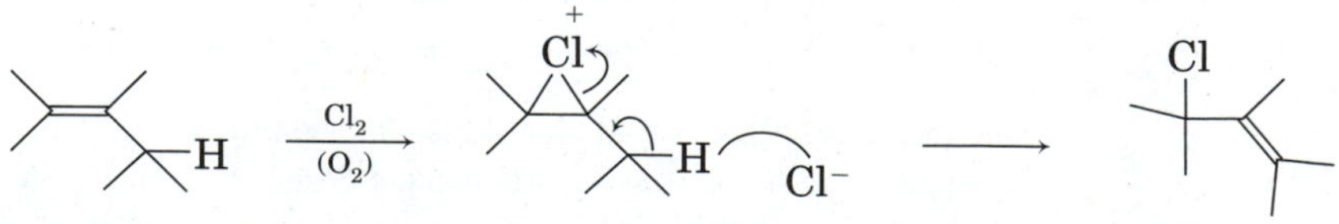

anism that can account for the formation of substitution product by way of a chloronium ion is shown in Figure 9.21.[104,105]

Addition of chlorine to the isomers of 2-butene gave 97–98% anti addition product and 2–3% substitution product if the alkene was neat or was dissolved in a nonpolar solvent with oxygen present to suppress radicals. Under a nitrogen atmosphere, however, 18% of the product was substitution, and the 82% of the product that was derived from adding Cl_2 was not formed stereospecifically (Figure 9.22). Poutsma concluded that radical reactions are difficult to suppress in chlorine addition because they do not require an initiator and seem to begin spontaneously.[106]

Electrophilic addition of Cl_2 to alkenes in a polar solvent seems not to be affected as severely by radical reactions as does addition in a nonpolar solvent. Figure 9.23 shows results of Poutsma and Kartch for the addition of Cl_2 to *trans*-2-pentene in acetic acid.[107] In addition to a small amount of substitution product, both the dichloride and two chloroacetoxy adducts were observed. Because the dichloride and the chloroacetoxy products were formed by anti addition, a mechanism involving a chloronium ion—chloride ion pair (similar to the mechanism for dibromide formation) was proposed for the reaction. Formation of both the 2-acetoxy-3-chloro and 3-acetoxy-2-

Figure 9.22 Products of reaction of *trans*-2-butene with Cl_2 in pentane.

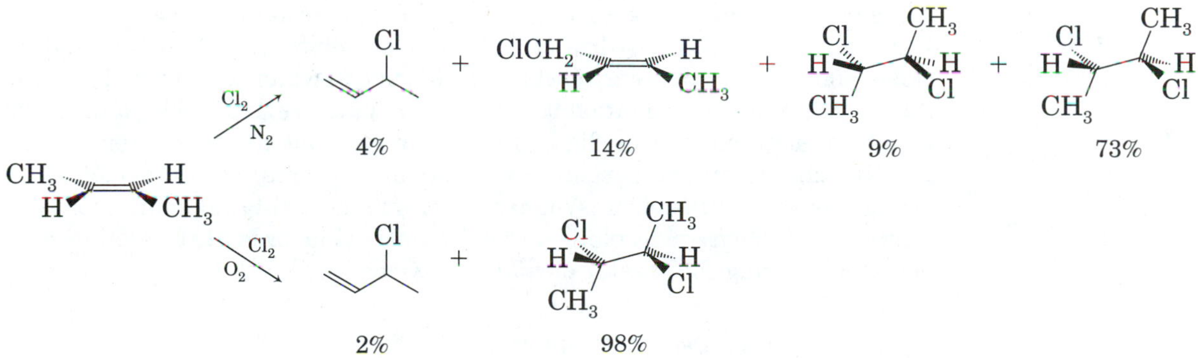

[104]Except for the complication of bridging by chlorine, the competition between addition and elimination in the second step to each product is reminiscent of the E1/S_N1 competition exhibited by other carbocations.

[105]Addition of chlorine to neat *t*-butylethylene gave not only the expected dichloride but also 10% of 4-chloro-2,3-dimethyl-1-butene, indicating that a rearrangement consistent with a methyl shift in a carbocation had occurred. Poutsma, M. L. *J. Am. Chem. Soc.* **1965**, *87*, 4285. Small amounts of substitution products accompanied the addition. Addition of chlorine to linear alkenes gave mostly addition, but addition of chlorine to branched alkenes gave mostly allylic substitution products. Addition is favored at lower temperatures, as is the case with the competition between S_N1 and E1 reactions.

[106]Poutsma, M. L. *J. Am. Chem. Soc.* **1965**, *87*, 2172.

[107]Poutsma, M. L.; Kartch, J. L. *J. Am. Chem. Soc.* **1967**, *89*, 6595.

Figure 9.23 Addition of Cl_2 to *trans*-2-pentene in acetic acid. (All products are racemic.)

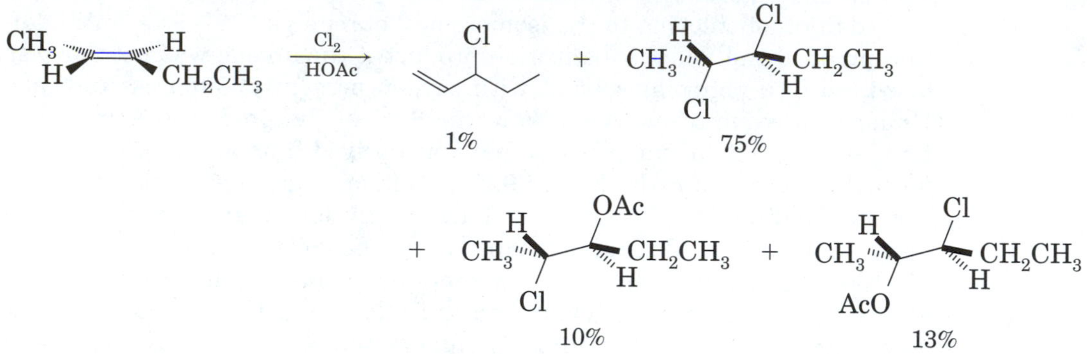

chloro products was ascribed to nucleophilic attack on the chloronium ion by solvent. There appeared to be a slight preference for solvent attack on the carbon atom with the smaller alkyl group, perhaps for steric reasons.

Reaction of an unhindered alkene with chlorine in aqueous solution results in the addition of the elements of hypochlorous acid, HOCl, to produce a chlorohydrin (β-hydroxyalkyl chloride).[108] The stereochemistry of chlorohydrin formation is anti. For example, reaction of *cis*-2-butene with chlorine and water produces (±)-*threo*-2-chloro-2-butanol (equation 9.17).[109] Chlorohydrin formation is generally highly regioselective, with the OH group becoming attached to the alkene carbon atom bearing the greater number of alkyl groups. For example, the chlorohydrin of methylcyclohexene is 2-chloro-1-methylcyclohexanol,[110] and isobutene reacts with chlorine in water to produce 1-chloro-2-methyl-2-propanol (equation 9.18).[111] The results are explicable in terms of an unsymmetrical intermediate chloronium ion that has a greater partial positive charge on the carbon atom bearing the greater number of alkyl groups, as is the case for unsymmetrically substituted bromonium ions (see Figure 9.15). Then nucleophilic attack by water is favored at the olefinic carbon atom bearing the greater partial positive charge.[112,113]

---

[108]Gomberg, M. *J. Am. Chem. Soc.* **1919**, *41*, 1414.

[109]Collis, M. J.; Merrall, G. T. *Chem. Ind. (London)* **1964**, 711. The stereochemistry of the chlorohydrin was determined indirectly by noting that reaction of the chlorohydrin with aqueous base produced *cis*-2,3-epoxybutane. Bartlett, P. D. *J. Am. Chem. Soc.* **1935**, *57*, 224, had earlier concluded that addition of HOCl to cyclohexene produces *trans*-2-chlorocyclohexanol.

[110]Bartlett, P. D.; Rosenwald, R. H. *J. Am. Chem. Soc.* **1934**, *56*, 1990.

[111]de la Mare, P. B. D.; Salama, A. *J. Chem. Soc.* **1956**, 3337.

[112]The argument is closely analogous to that used to explain the regioselectivity of formation of bromoacetoxy compounds (Table 9.2) formed in the addition of bromine to alkenes in acetic acid.

[113]Similarly, addition of bromine to alkenes in water produces bromohydrins. Although they are more difficult to synthesize, iodohydrins and fluorohydrins are also known. For a review of the synthesis and reactions of halohydrins, see Rosowsky, A. in *Heterocyclic Compounds with Three- and Four-Membered Rings*, Part One; Weissberger, A., ed.; Wiley-Interscience: New York, 1964; p. 1.

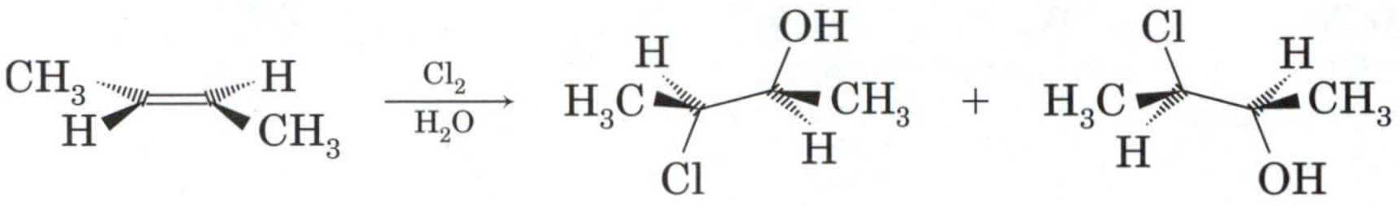

(9.17)

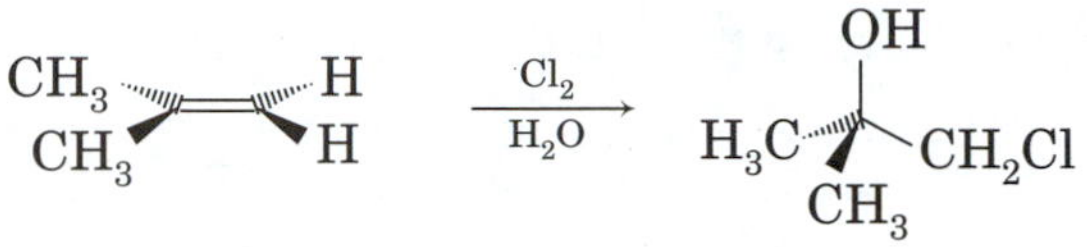

(9.18)

In contrast to the anti addition of chlorine to 2-butene, addition of chlorine to 1-phenylpropene (β-methylstyrene) is nonstereospecific in both polar and nonpolar solvents. While 2-butene reacts via a chloronium ion, 1-phenylpropene probably reacts through an open (benzylic) carbocation.[114] *cis*-Stilbene was found to react with chlorine to give a 9 : 1 mixture of meso to racemic product, while *trans*-stilbene gave a 35 : 65 mixture of those two products.[115] Thus the addition to stilbene must have proceeded through intermediate(s) similar to chlorocarbocations so that appreciable syn reaction can take place. Similarly, addition of chlorine to phenanthrene gave appreciable *cis*-9,10-dichloro-9,10-dihydrophenanthrene,[116] and acenaphthylene (**21**) reacted with chlorine in CCl_4 to give the *cis*-1,2-dichloroacenaphthene (**22**) exclusively.[117]

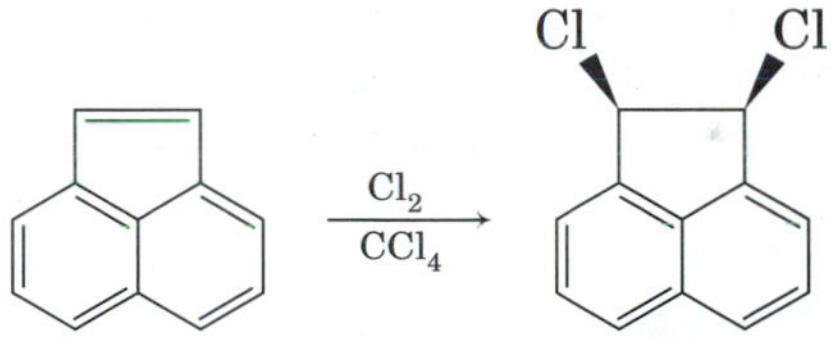

(9.19)

An attractive mechanism to explain these results is shown in Figure 9.24.[114] The addition of chlorine to the double bond of *cis*-stilbene produces a benzylic carbocation–chloride ion pair. Attachment of Cl^- to the carbocation before reorganization gives net syn addition. Either rearrangement of the

[114]Fahey, R. C.; Schubert, C. *J. Am. Chem. Soc.* **1965**, *87*, 5172.

[115]Buckles, R. E.; Knaack, D. F. *J. Org. Chem.* **1960**, *25*, 20, reported that tetrabutylammonium iodotetrachloride gives stereospecific anti addition of chlorine to stilbene and other compounds. The stereospecificity of the reaction is attributed to an unspecified role for the iodotetrachloride or iododichloride ions present in the reaction mixture.

[116]de la Mare, P. B. D.; Koenigsberger, R. *J. Chem. Soc.* **1964**, 5327.

[117]Cristol, S. J.; Stermitz, F. R.; Ramey, P. S. *J. Am. Chem. Soc.* **1956**, *78*, 4939. Interestingly, acenaphthylene was found to react with iodobenzene dichloride in chloroform to give the trans isomer.

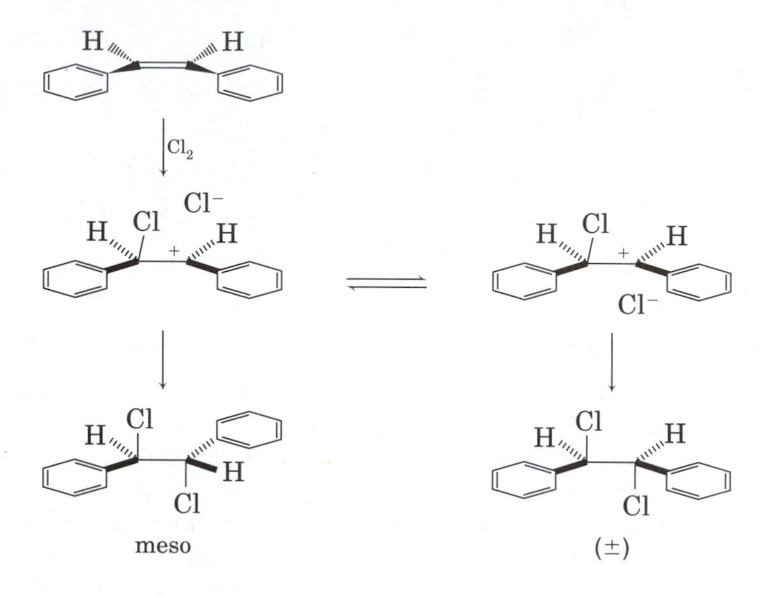

Figure 9.24 Ion pair intermediates proposed for the addition of chlorine to *cis*-stilbene.

ion pair to one with the chloride ion on the other face of the molecule (shown) or rotation about the C—C bond (not shown) could produce apparent anti addition. Since rotation about the single bond cannot occur in the case of chlorine addition to phenanthrene or acenaphthylene, reaction through rearrangement of the tight ion pair must be the main route to cis product formation for these compounds, and it could be the main route for formation of racemic dichlorostilbenes also.[118]

Fluorine

The exothermicity of the addition of F_2 to an alkene both makes it difficult to study the reaction and makes the addition of limited synthetic utility. Merritt suggested that carrying out the addition of fluorine at low temperature would prevent some secondary reactions that result from the heat gen-

---

[118]Theoretical calculations also support the intermediacy of chloronium ions in addition reactions. *Ab initio* calculations indicated that the chloronium ion is a minimum on the $C_2H_4Cl^+$ potential energy surface. The 2-chloroethyl carbocation was calculated not to be a minimum, however; instead, it was found to isomerize without an energy barrier to the 1-chloroethyl carbocation, which was a global minimum (4.3 kcal/mol more stable than the chloronium ion). The activation barrier for conversion of the chloronium ion to the 1-chloroethyl carbocation was about 28 kcal/mol. Rodriquez, C. F.; Bohme, D. K.; Hopkinson, A. C. *J. Am. Chem. Soc.* **1993**, *115*, 3263. Also see Reynolds, C. H. *J. Am. Chem. Soc.* **1992**, *114*, 8676.

There was also experimental evidence for the existence of chloronium ions from ion cyclotron resonance spectrometry experiments. Berman, D. W.; Anichich, V.; Beauchamp, J. L. *J. Am. Chem. Soc.* **1979**, *101*, 1239, reported that the stability of cyclic ions varies with the atom or group that is bonded to two carbon atoms in a three-membered ring. For the following species as the bridging group, relative stability is OH $<$ Cl $<$ Br $<$ SH, PH_2 $<$ NH_2.

erated by addition of fluorine to alkenes, and reactions at −78° were found to be synthetically useful.[119]

Because fluorine is inefficient as a bridging ion,[30] we might consider it likely that addition of fluorine would occur through a two-step mechanism involving a β-fluorocarbocation. However, the stereochemistry of the fluorine addition reaction has generally been found to be that resulting from syn addition, which suggests that the reaction does not proceed through a long-lived carbocation that can undergo nucleophilic attachment from either the top or the bottom lobe of the empty *p* orbital. The results could be explained if the reaction proceeds through a short-lived carbocation-fluoride ion pair that quickly collapses to syn product. An alternative possibility is a concerted, four-center pathway.[120] Results of *ab initio* calculations for fluorine addition to ethene suggest a four-centered transition structure leading to syn addition (Figure 9.25) is feasible.[121]

The four-center concerted pathway has been criticized, however, both on theoretical grounds and on the basis of experimental data from an alternative method of fluorine addition. Rozen and Brand[122] carried out the direct addition of fluorine to alkenes by bubbling a gas mixture composed of 1% fluorine in nitrogen into a solution of the alkene dissolved in a mixture of $CFCl_3$, $CHCl_3$, and ethanol, a solvent system chosen to diminish the proba-

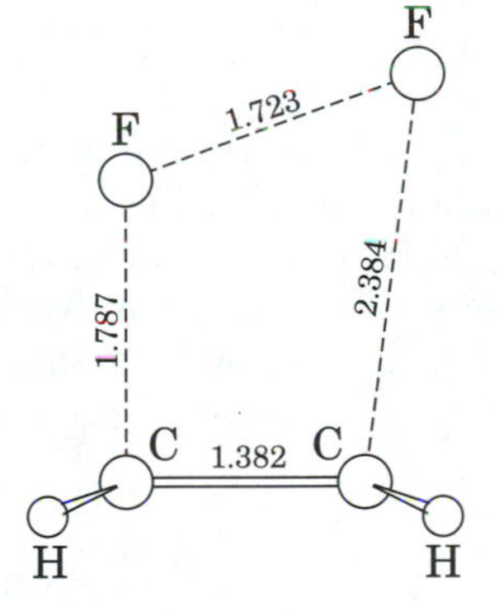

Figure 9.25
Transition structure calculated for the addition of F_2 to alkenes. (Distances in Å. Adapted from reference 121.)

[119]Merritt, R. F.; Johnson, F. A. *J. Org. Chem.* **1966**, *31*, 1859. See also Merritt, R. F.; *J. Org. Chem.* **1966**, *31*, 3871; Merritt, R. F. *J. Am. Chem. Soc.* **1967**, *89*, 609.

[120]The concerted mechanism is termed an Ad_E2M pathway.

[121]Yamabe, S.; Minato, T.; Inagaki, S. *J. Chem. Soc., Chem. Commun.* **1988**, 532. The same calculations identified three-centered transition structures for the addition of chlorine and bromine; such transition structures should lead to chloronium and bromonium ions, respectively.

[122]Rozen, S.; Brand, M. *J. Org. Chem.* **1986**, *51*, 3607.

bility of radical reactions competing with the ionic addition process. For alkenes substituted with alkyl groups (e.g., cycloheptene) or with electron-withdrawing groups (*cis-* or *trans*-ethyl cinnamate), the products isolated were exclusively from those of syn addition. When *cis*-stilbene was treated under the same conditions, the major product was the *meso*-1,2-difluoro-1,2-diphenylethane isomer, but an appreciable yield of the racemic product was also found. The results were interpreted to mean that in most cases the additions take place through electrophilic addition of fluorine to the alkene to produce a tight β-fluorocarbocation—fluoride tight ion pair. The ion pair can collapse to give syn addition before rotation about a carbon-carbon single bond or reorientation of the fluoride ion in the ion pair can result in anti addition. When stilbene reacts, however, the β-fluorocarbocation is a benzylic carbocation and is sufficiently stable for some rotation or reorientation to occur before attack of the fluoride ion to the cation. In no case, however, is there evidence for a long-lived carbocation or even for a solvent-separated ion pair, since no ethyl ethers are observed in the product mixture.[123]

Because of the danger and difficulty of working with molecular fluorine, researchers have sought other reagents for addition of fluorine to alkenes. It has been found that XeF_2 reacts with ethene to give a 35% yield of 1,1-difluoroethane and a 45% yield of 1,2-difluoroethane. However, propene reacts to give 1,1-difluoropropane as the main product.[124] The investigators suggested that 1,2-difluoropropene is formed as a *kinetic* (rate-controlled) product, but that it rearranges to the thermodynamically more stable 1,1-difluoropropane (equilibrium-controlled product).[125]

A significant complication in understanding the mechanism of addition of fluorine to alkenes with XeF_2 is uncertainty about the possible role of single electron transfer in the reaction. Figure 9.26 shows that the hydrogen fluoride–catalyzed reaction of XeF_2 with a styrene derivative could produce the same carbocation by either a one-step transfer of F^+ to the alkene or by a two-step route involving single electron transfer to form the radial cation of the alkene, which then abstracts a fluorine atom from FXe·. Stavber and co-workers studied the HF-catalyzed reaction of a series of phenyl-substituted alkenes with XeF_2 in CH_2Cl_2 at room temperature. The logarithms of

[123]The observation of increasing proportion of syn addition of X_2 with increasing electronegativity of X is also consistent with the suggestion by Phillips, L.; Wray, V. *J. Chem. Soc., Chem. Commun.* **1973**, *90*, that there is a parallel between the vicinal relationship of the halogen atoms in syn addition to an alkene and the relationship of vicinal halogen atoms in molecules that exhibit a gauche effect. These authors suggest that the electronic factors responsible for the gauche effect may also be reflected in the potential energy surface for electrophilic addition of X_2, even if the reaction proceeds through an unsymmetrical carbocation.

[124]Shieh, T.-C.; Yang, N. C.; Chernick, C. L. *J. Am. Chem. Soc.* **1964**, *86*, 5021.

[125]For a discussion of the origins of the greater stability of *gem*-difluorides than of *vic*-difluorides, see Hine, J. *J. Am. Chem. Soc.* **1963**, *85*, 3239.

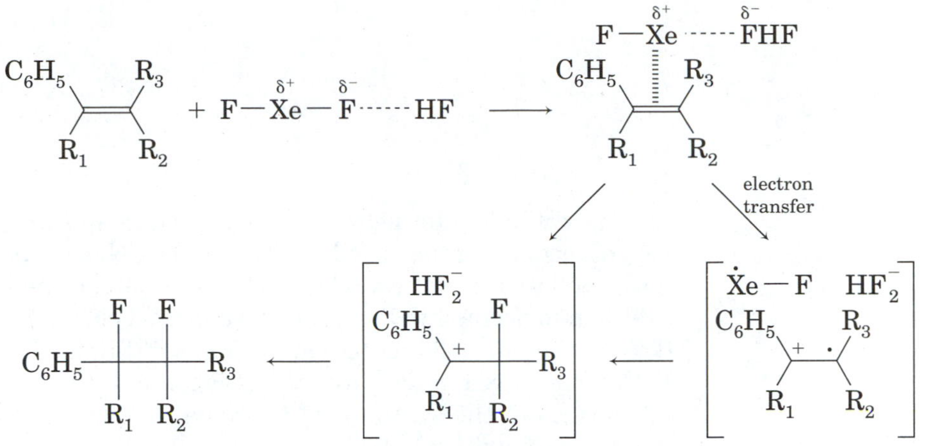

Figure 9.26 Competitive pathways for reaction of XeF_2 with alkenes. (Adapted from reference 126.)

the relative rates of addition across the carbon-carbon double bond correlate with the ionization potentials of the alkenes, with a lower IP being associated with a faster rate of addition.[126] The authors concluded that the experimental data indicate that the rate-limiting step in the addition is disruption of the π bond, but that they do not allow exclusion of either of the two pathways shown in Figure 9.26. The investigators suggested that it might be the case that not all addition reactions occur by the same mechanism. The operation of the ionic or electron transfer mechanism of addition in a particular case might be a function of the structure of the alkene, the concentration of the reagents, the solvent, and the temperature.[126]

If a synthetic scheme requires that only one fluorine be attached to the carbon skeleton, then other reagents are available. Appelman and Rozen and their co-workers have described the synthesis and reactions of *t*-butyl hypofluorite, $(CH_3)_3COF$.[127] Reaction of $(CH_3)_3COF$ with styrene (**23**) in CH_3CN solution produced 1-fluoro-1-phenyl-2-*t*-butoxyethane (**24**) as the major product.[128] This regiochemistry suggests that the reaction involves attachment of the "*t*-butoxylium" cation ($((CH_3)_3CO^+)$) to the alkene, followed by subsequent attachment of F^- to the cationic intermediate.

[126]Stavber, S.; Sotler, T.; Zupan, M.; Popovič, A. *J. Org. Chem.* **1994**, *59*, 5891.

[127]Appelman, E. H.; French, D.; Mishani, E.; Rozen, S. *J. Am. Chem. Soc.* **1993**, *115*, 1379.

[128]Small amounts of 1-fluoro-1-phenyl-3-cyanopropane were also observed, perhaps as a result of radical ion reactions involving solvent.

F H

$(CH_3)_3COF$ / CH_3CN → $OC(CH_3)_3$

H H

23 **24**

(9.20)

The cationic intermediate in this reaction appears to have some bridged ion character. Addition of $(CH_3)_3COF$ in CH_3CN to *trans*-β-methylstyrene produced a 3 : 1 ratio of *erythro*- to *threo*-1-fluoro-1-phenyl-2-*t*-butoxypropane. Addition to acenaphthylene (**21**) gave *threo*-1-fluoro-2-*t*-butoxyacenaphthene (**25**) as the only adduct (equation 9.21).[129] Similarly, methyl hypofluorite, H_3COF, prepared *in situ* from the reaction of F_2 with methanol, also adds to alkenes, as in the synthesis of (±)-*trans*-1-fluoro-2-methoxyindan (**27**) from indene (**26**, equation 9.22).[130]

$C(CH_3)_3$ F^- O^+ F $OC(CH_3)_3$

t-BuOF / CH_3CN →

21 **25**

(9.21)

F

H_3COF / H_3CCN → OCH_3

26 **27**

(9.22)

Iodine

Because iodine is the best of the halogens at bridging, we might expect the iodinium ion to be the intermediate in the addition of iodine to alkenes. Indeed, an iodinium ion has been isolated from the reaction of iodine with adamantylideneadamantane.[36] However, the addition of iodine to alkenes is generally not synthetically useful, in large part because the formation of 1,2-diiodo compounds from I_2 and alkenes is not as exothermic as are the additions of other halogens to alkenes. For example, the ΔH for addition of iodine to ethene is exothermic by only about 11 kcal/mol in the gas phase. Since ΔS is -31.5 eu,[131] adduct stability decreases with increasing temperature. In one study it was found that a mixture of styrene and iodine pro-

[129] The addition of *t*-BuOF stands in contrast to the pattern of reaction of other alkyl hypofluorites, such as CF_3OF and CF_3CF_2OF, since they add in a syn process and with the opposite regiochemistry for fluorine addition: Lerman, O.; Rozen, S. *J. Org. Chem.* **1980**, *45*, 4122; Rozen, S.; Lerman, O.; Kol, M.; Hebel, D. *J. Org. Chem.* **1985**, *50*, 4753.

[130] Rozen, S.; Mishani, E.; Kol, M.; Ben-David, I. *J. Org. Chem.* **1994**, *59*, 4281.

[131] Benson, S. W.; Amano, A. *J. Chem. Phys.* **1962**, *36*, 3464.

duced styrene diiodide; the product could be collected by filtration at 0°, but it decomposed at room temperature to styrene and iodine.[132]

The addition of iodine to alkenes takes place through radical as well as ionic pathways. Addition of I_2 to 1-butene occurred photochemically to give a product, presumed to be 1,2-diiodobutane, that was stable below −15°. Addition of iodine to *trans*-2-butene produced 2,3-diiodobutane, which melted with decomposition at −11°.[133] Alkenes were also found to add iodine photochemically in refluxing propane at −42° to produce crystalline diiodides that decomposed at room temperature.[134] The stereochemistry of the addition was found to be anti, with *cis*-2-butene producing racemic diiodide and *trans*-2-butene producing *meso*-2,3-diiodobutane. For these additions, the authors proposed a radical chain mechanism involving a bridged radical intermediate (with propagation steps 9.23 and 9.24) and suggested that the relative bridging power of β-halogenated radicals is[135]

$$\mathrm{F} \ll \mathrm{Cl} < \mathrm{Br} < \mathrm{I}$$

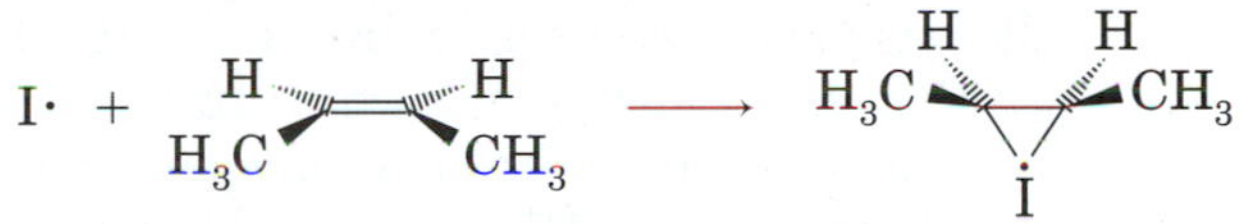

$$\mathrm{I}\cdot + \text{H(H}_3\text{C)C=C(H)CH}_3 \longrightarrow \text{H}_3\text{C(H)C}\overset{\dot{\text{I}}}{\triangledown}\text{C(H)CH}_3 \quad \textbf{(9.23)}$$

$$\text{H}_3\text{C(H)C}\overset{\dot{\text{I}}}{\triangledown}\text{C(H)CH}_3 + \mathrm{I_2} \longrightarrow \text{H}_3\text{C(H)(I)C–C(I)(H)CH}_3 + \mathrm{I}\cdot \quad \textbf{(9.24)}$$

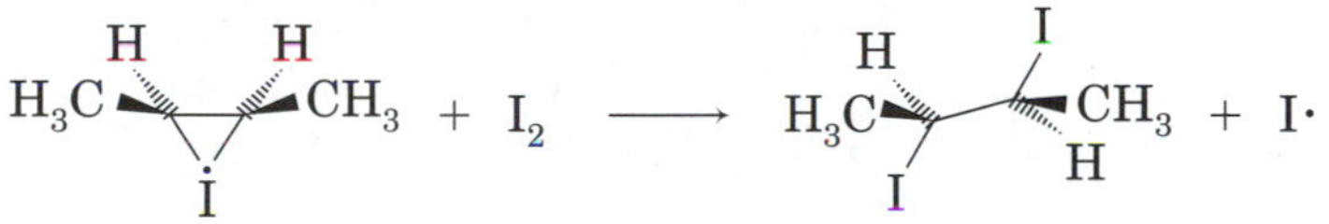

(and enantiomer)

Addition of Mixed Halogens

The mixed halogens ICl, IBr, and BrCl also add to alkenes. White and Robertson found third-order kinetics and determined that the relative reactivities varied as follows:[136]

$$\mathrm{BrCl} > \mathrm{ICl} > \mathrm{Br_2} > \mathrm{I_2}$$

The large reactivity of BrCl means that addition of BrCl occurs when alkenes react with mixtures of Br_2 and Cl_2.[137] The regioselectivity of addition of mixed halogens was reported by Ingold and Smith. Addition of ICl to propene gave 69% of 2-chloro-1-iodopropane (**29**) and 31% of 1-chloro-2-iodopropane (**30**, equation 9.25), and addition to styrene gave more than 95% of 1-chloro-2-iodo-1-phenylethane.[138]

---

[132]Fraenkel, G.; Bartlett, P. D. *J. Am. Chem. Soc.* **1959**, *81*, 5582.

[133]Forbes, G. S.; Nelson, A. F. *J. Am. Chem. Soc.* **1937**, *59*, 693.

[134]Skell, P. S.; Pavlis, R. R. *J. Am. Chem. Soc.* **1964**, *86*, 2956.

[135]Investigators have also concluded that both radical chain and nonradical chain mechanisms are possible: (a) reference 132; (b) Trifan, D. S.; Bartlett, P. D. *J. Am. Chem. Soc.* **1959**, *81*, 5573.

[136]White, E. P.; Robertson, P. W. *J. Chem. Soc.* **1939**, 1509.

[137]Buckles, R. E.; Forrester, J. L.; Burham, R. L.; McGee, T. W. *J. Org. Chem.* **1960**, *25*, 24.

[138]Ingold, C. K.; Smith, H. G. *J. Chem. Soc.* **1931**, 2742.

$$\text{28} \xrightarrow{\text{ICl}} \text{29 (69\%)} + \text{30 (31\%)} \quad (9.25)$$

The regiochemistry can be rationalized if one imagines that the XY molecule adds as X^+ and Y^-, with Y being the more electronegative of the two atoms.[139] Furthermore, attachment of X^+ may result in a cyclic "onium" ion if such an intermediate is formed in the addition of X_2 to the same alkene. For example, *cis*-stilbene reacted in a mixture of Br_2 and Cl_2 to produce 68% of *threo*-1-bromo-2-chloro-1,2-diphenylethane and only about 5% of the erythro diastereomer.[137]

Addition of XY to alkenes can also be accomplished by indirect means. Buckels and Long[140] used *N*-bromoacetamide and HCl to add BrCl to alkenes. The stereochemistry of the reaction suggested an anti addition, which was consistent with attachment of Br^+ from protonated *N*-bromoacetamide to give a bromonium ion (or β-bromocarbocation), followed by attachment of Cl^-. Reaction of *N*-bromoacetamide and HF in a proton-accepting solvent such as THF (which increases solubility of HF) gave anti addition of FBr to cyclohexene. Similarly, reaction of cyclohexene with *N*-iodoacetamide and HF produced *trans*-1-fluoro-2-iodocyclohexane.[141,142]

9.3 Other Addition Reactions

Addition of Hydrogen Halides to Alkenes

Regiochemistry of Addition: Markovnikov's Rule

The addition of hydrogen halides to alkenes is also one of the paradigms of mechanistic organic chemistry. Specifically, the addition of HCl to propene to give isopropyl chloride is the classic example of *Markovnikov's rule*.[143,144]

---

[139] Ingold, C. K. *Chem. Rev.* **1934**, *15*, 225 (see especially pp. 270–271).

[140] Buckels, R. E.; Long, J. W. *J. Am. Chem. Soc.* **1951**, *73*, 998.

[141] Bowers, A.; Cuéllar Ibáñez, L.; Denot, E.; Becerra, R. *J. Am. Chem. Soc.* **1960**, *82*, 4001.

[142] These reactions are useful for the synthesis of steroids, where fluorine substitution has been found to give useful pharmacological properties. Bowers, A.; Denot, E.; Becerra, R. *J. Am. Chem. Soc.* **1960**, *82*, 4007.

[143] The spelling currently popular is "Markovnikov", although different spellings were used in the first publications in German and French: (a) Markownikoff, W. *Liebigs Ann. Chem.* **1870**, *153*, 228 (particularly the discussion beginning on page 256); (b) Markovnikoff, V. *Compt. rend.* **1875**, *81*, 668.

[144] A brief autobiography of Markovnikov was given by Leicester, H. M. *J. Chem. Educ.* **1941**, *18*, 53. Several authors have discussed the history of Markovnikov's rule and its various forms and spellings. Jones, G. *J. Chem. Educ.* **1961**, *38*, 297 noted the contribution of Henry in writing a more general form of "Markovnikov's rule" in which the more negative group in a molecule XY adds to the carbon atom with the fewer number of hydrogen atoms (that is, the carbon atom with the greater number of alkyl substituents). See also Tierney, J. *J. Chem. Educ.* **1988**, *65*, 1053.

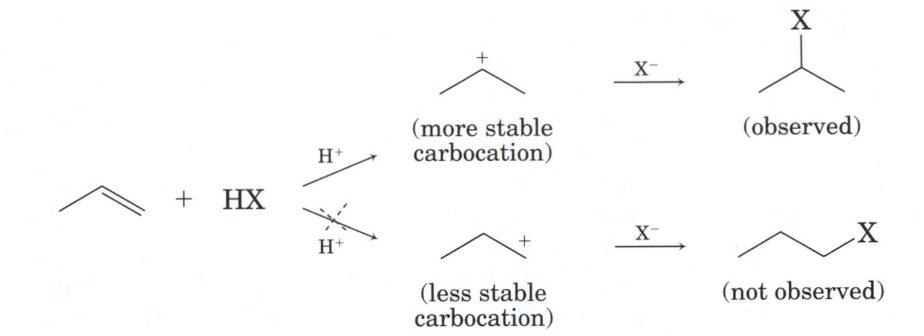

Figure 9.27 Rationale for Markovnikov regiochemistry in terms of carbocation stability.

As translated by Kharasch and Reinmuth,[145] Markovnikov wrote that the addition of halogen acids to unsymmetrical alkenes occurs so that the halogen becomes attached to the less-hydrogenated carbon atom (Figure 9.27).[146] In the case of vinyl chloride and other haloalkenes, the halogen of the halogen acid adds to the carbon atom that already has a halogen substituent.[145] Both statements are easily seen as manifestations of ionic addition proceeding through the more stable of two possible carbocations.

Just as Figure 9.2 is an extremely simplified mechanism for bromine addition, so it appears that the simple picture in Figure 9.27 understates the complexity of the addition of HX to an alkene, particularly in nonpolar solvents. Figure 9.27 suggests that the addition should be overall second order, first order in alkene and first order in hydrogen halide, but this is generally not the case. In nonpolar solvents the addition of HBr to an alkene is first order in alkene and either second order or third order in HBr. In a study of the addition of HBr to propene in pentane, Mayo and Savoy found that it was quite difficult to eliminate all traces of the abnormal addition (i.e., radical reaction; see page 593). The kinetic expression most closely resembled

$$\text{rate} = k[\text{alkene}][\text{HBr}]^3$$

When hydroxylated solvents were added, the rate increased, and the rate expression appeared to be of lower order.[147] Similarly, the addition of HCl to isobutylene indicated overall fourth-order kinetics, again third order in the HX.[148] The addition of HBr to cyclohexene in acetic acid was found to give both cyclohexyl bromide and cyclohexyl acetate. Through deuterium labeling of the cyclohexene, the authors demonstrated that the predominant pattern of addition was anti, with the proportion of syn HBr addition varying

[145]Kharasch, M. S.; Reinmuth, O. *J. Chem. Educ.* **1931**, *8*, 1703.

[146]Some current textbooks state that the addition of HX to an unsymmetrical alkene occurs so that the proton becomes bonded to the olefinic carbon atom that already had the greater number of hydrogen atoms. Although this statement may be easier to remember, it is not precisely the original rule.

[147]Mayo, F. R.; Savoy, M. G. *J. Am. Chem. Soc.* **1947**, *69*, 1348.

[148]Mayo, F. R.; Katz, J. J. *J. Am. Chem. Soc.* **1947**, *69*, 1339.

from 4 to 6% over the temperature range 15–60°. The formation of cyclohexyl acetate was greater at the higher temperature.[149,150,151] Instead of a bare H^+, therefore, the electrophile may be one or more molecules of HBr, and there may also be one or more HBr molecules required to solvate the departing Br^- ion.[152]

Addition of HCl shows similar complexity. Studies by Fahey and coworkers have suggested that anti addition of HCl to cyclohexene occurs by a termolecular process depending on [alkene], [HCl], and [Cl^-] (an Ad_E3 mechanism) along with syn addition through a carbocation-chloride ion pair (Ad_E2 mechanism).[153,154,155] The mechanism of the addition also depends on the structure of the alkene. The addition of HCl to styrene in acetic acid has been reported to occur by an Ad_E2 pathway (Figure 9.28),[156] but the addition of HCl to cyclohexene has been reported to occur at least partly by an Ad_E3 mechanism (Figure 9.29).[157,158]

The addition of HCl to alkenes is slower than is the addition of HBr, and addition of HCl is not observed unless the alkene is substituted with many

Figure 9.28
Ad_E2 addition of HCl to styrene

HCl / HOAc (slow); CH_3; + Cl^-; Cl; CH_3

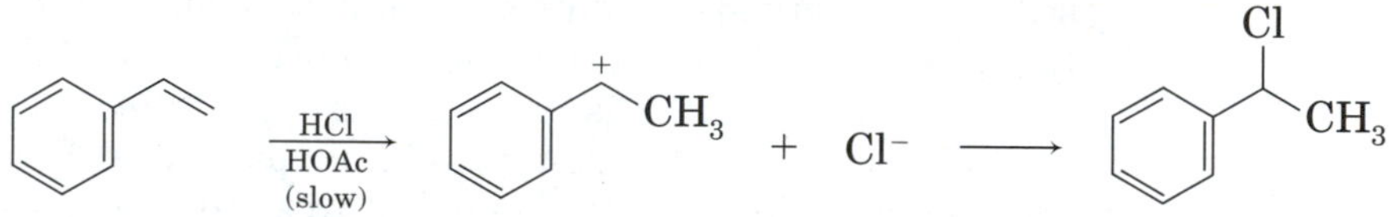

[149]Fahey, R. C.; Smith, R. A. *J. Am. Chem. Soc.* **1964**, *86*, 5035.

[150]Addition of HI to propene gave only 2-iodopropane, whether antioxidants or peroxides were added. Addition of HI to allyl bromide produced only 1-bromo-2-iodopropane, and addition of HI to allyl chloride gave only 1-chloro-2-iodopropane. The reason for the Markovnikov orientation may be that HI reacted with peroxides to give iodine, thus stopping initiation of radical chain reaction. Kharasch, M. S.; Norton, J. A.; Mayo, F. R. *J. Am. Chem. Soc.* **1940**, *62*, 81.

[151]Addition of HF to alkenes has been shown to follow Markovnikov's rule (Sharts, C. M.; Sheppard, W. A. *Org. React.* **1974**, *21*, 125 and references therein). However, direct addition of HF is seldom carried out because of the extreme danger of working with that reagent and because of the development of newer synthetic reagents. It is not only safer but also more convenient to use KF to synthesize a 2-fluoroalkane from the corresponding 2-chloroalkane than to add HF to the corresponding 1-alkene.

[152]Solvent molecules might also serve to facilitate the reaction. Isenberg, N.; Grdinic, M. *J. Chem. Educ.* **1969**, *46*, 601.

[153]Fahey, R. C.; McPherson, C. A. *J. Am. Chem. Soc.* **1971**, *93*, 2445.

[154]Fahey, R. C.; Monahan, M. W.; McPherson, C. A. *J. Am. Chem. Soc.* **1970**, *92*, 2810; Fahey, R. C.; Monahan, M. W. *J. Am. Chem. Soc.* **1970**, *92*, 2816.

[155]In one case it was also suggested that the addition of HCl could occur through a concerted, four-center pathway (Ad_E2M) in solution. Freeman, P. K.; Raymond, F. A.; Grostic, M. F. *J. Org. Chem.* **1967**, *32*, 24.

[156]Fahey, R. C.; McPherson, C. A. *J. Am. Chem. Soc.* **1969**, *91*, 3865; see also the discussion in reference 7.

[157]See reference 154.

[158]The intermediate has been proposed as a nonclassical carbocation in some cases. See, for example, Cristol, S. J.; Caple, R. *J. Org. Chem.* **1966**, *31*, 2741; Cristol, S. J.; Sullivan, J. M. *J. Am. Chem. Soc.* **1971**, *93*, 1967.

Figure 9.29 Ad_E3 addition of HCl to cyclohexene.

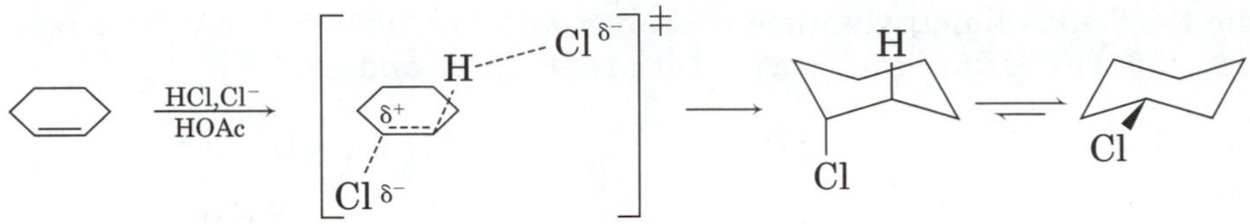

alkyl groups, is substituted with at least one aryl group, or is strained. For example, α-pinene reacts with HCl in $CHCl_3$ solution, but 1-octene does not.[159,160] Solvent effects are also important in HC1 addition. The rate of addition of HCl to α-pinene was found to vary with solvent as follows: $CHCl_3$ > xylene> nitrobenzene >> methanol > dioxane > diethyl ether (no apparent reaction).[161,162] Investigators have concluded that the rate of the addition reaction decreases with an increasing tendency of HX to coordinate with an electron pair donor of the solvent.[163] However, Kropp and co-workers have reported that the addition of HCl to 1-octene can be effected in $CHCl_3$ solution under surface-mediated conditions involving the generation of HCl *in situ* in the presence of silica gel or alumina. The surface-mediated reaction appears to involve a longer-lived carbocation intermediate, since 3-chlorobutane is isolated along with 2-chlorobutane, and both (*E*)- and (*Z*)-2-octene can also be recovered from the reaction mixture.[160,164]

Although a carbocation is not the only possible intermediate in the addition of hydrogen halides to alkenes,[165] the observation of characteristic carbocation rearrangement products strongly supports the role of a carbocation-like species in this reaction under these surface-mediated conditions. Addition of HCl to neat 3,3-dimethyl-1-butene (*t*-butylethylene, **6**) produced 60% of the rearranged product (2-chloro-2,3-dimethylbutane, **31**) and about

[159] Kropp, P. J.; Daus, K. A.; Crawford, S. D.; Tubergen, M. W.; Kepler, K. D.; Craig, S. L.; Wilson, V. P. *J. Am. Chem. Soc.* **1990**, *112*, 7433.

[160] The greater reactivity of HBr in addition reactions, however, means that HBr does add to 1-octene in $CHCl_3$. Kropp, P. J.; Daus, K. A.; Tubergen, M. W.; Kepler, K. D.; Wilson, V. P.; Craig, S. L.; Baillargeon, M. M.; Breton, G. W. *J. Am. Chem. Soc.* **1993**, *115*, 3071.

[161] α-Pinene is 2,6,6-trimethylbicyclo[3.1.1]hept-2-ene.

[162] Hennion, G. F.; Irwin, C. F. *J. Am. Chem. Soc.* **1941**, *63*, 860. (Products from the reaction of pinene with HCl in methanol were not reported.)

[163] O'Connor, S. F.; Baldinger, L. H.; Vogt, R. R.; Hennion, G. F. *J. Am. Chem. Soc.* **1939**, *61*, 1454.

[164] The surface-mediated reaction enables the addition of HBr to 1-octene to proceed at room temperature without the formation of 1-bromooctane by free radical chain reaction addition.

[165] Brown, H. C.; Liu, K.-T. *J. Am. Chem. Soc.* **1975**, *97*, 600.

40% of the unrearranged 2-chloro-3,3-dimethylbutane (**32**). Less rearrangement was observed with addition of HI, with only 10% of the product being the 2-iodo-3,3-dimethylbutane.[166] Fahey and McPherson found that reaction of **6** with HCl in acetic acid also gave both **31** and **32**.[167,168]

$$\mathbf{6} \xrightarrow{HCl} \underset{\mathbf{31}\ (60\%)}{(CH_3)_2CH\text{-}C(Cl)(CH_3)_2} + \underset{\mathbf{32}\ (40\%)}{(CH_3)_3C\text{-}CHCl\text{-}CH_3} \quad (9.26)$$

The strongly electron-withdrawing trifluoromethyl group destabilizes an adjacent carbocation and retards the rate of addition of hydrogen halides to an alkene. For example, adding HBr to 3,3,3-trifluoromethylpropene (**33**) produced the anti-Markovnikov product, 3-bromo-1,1,1-trifluoropropane (**34**), but only when the reactants were heated in a sealed tube at 100° with $AlBr_3$ catalyst (equation 9.27).[169]

$$\underset{\mathbf{33}}{CH_2{=}CH\text{-}CF_3} \xrightarrow[AlBr_3]{HBr} \underset{\mathbf{34}}{Br\text{-}CH_2CH_2\text{-}CF_3} \quad (9.27)$$

If this addition occurs through a carbocation intermediate, as in Figure 9.27, then the regiochemistry might be attributed to the greater stability of a 1° cation in which the trifluoromethyl group is further removed ($CF_3CH_2CH_2^+$) than a 2° carbocation with an adjacent trifluoromethyl group ($CF_3CH^+CH_3$). However, Myhre and Andrews found that the reaction of 3,3,3-trifluoromethylpropene with HCl and $AlCl_3$ produced nearly equal amounts of $CF_3CH_2CH_2Cl$ (**35**) and $CF_2ClCH_2CH_2Cl$ (**36**, equation 9.28).

---

[166]Ecke, G. G.; Cook, N. C.; Whitmore, F. C. *J. Am. Chem. Soc.* **1950**, *72*, 1511.

[167]Fahey, R. C.; McPherson, C. A. *J. Am. Chem. Soc.* **1969**, *91*, 3865.

[168]This result confirmed the earlier work of Whitmore, F. C.; Johnston, F. *J. Am. Chem. Soc.* **1933**, *55*, 5020, who had found that the addition of HCl to 3-methyl-1-butene (no solvent, sealed reaction tube for 7 weeks) gave both 2-chloro-3-methylbutane and 2-chloro-2-methylbutane (*t*-amyl chloride). This result contradicted the suggestion of earlier investigators that the 3° alkyl halide was formed by rearrangement of 2° alkyl halide formed from the addition reaction. Hammond, G. S.; Collins, C. H. *J. Am. Chem. Soc.* **1960**, *82*, 4323, found that addition of HCl to 1,2-dimethylcyclopentene produced 1-chloro-*trans*-1,2-dimethylcyclopentane as the major (perhaps only) addition product. However, the product isomerized to 1-chloro-*cis*-1,2-dimethylcyclopentane.

[169]Henne, A. L.; Kaye, S. *J. Am. Chem. Soc.* **1950**, *72*, 3369.

$$CH_2{=}CHCF_3 \ (\mathbf{33}) \xrightarrow[AlCl_3]{HCl} ClCH_2CH_2CF_3 \ (\mathbf{35}) + ClCH_2CH_2CF_2Cl \ (\mathbf{36}) \qquad (9.28)$$

They proposed that $AlCl_3$-promoted removal of fluoride from the starting material would give 1,1-difluoroallyl cation, which could add Cl^- to give 1,1-difluoro-3-chloropropene. That product could then add either HCl or HF to give the two products observed (equation 9.29). A similar mechanism might be expected in the reaction with $HBr/AlBr_3$, even though Myhre and Andrews found 3-bromo-1,1,1-trifluoropropane as the only product of the reaction.[170] As Myhre and Newton have discussed, therefore, one should resist the temptation to rationalize the HBr addition product in terms of relative stabilities of $CF_3CH_2CH_2^+$ and $CF_3CH^+CH_3$ carbocations.[171]

$$CH_2{=}CHCF_3 \ (\mathbf{33}) \xrightarrow{AlCl_3} \left[[AlCl_3F]^- + CH_2{\cdots}CH{\cdots}CF_2^{+}\right] \longrightarrow \left[ClH_2C{-}CH{=}CF_2\right] \qquad (9.29)$$

$$\left[ClH_2C{-}CH{=}CF_2\right] \xrightarrow{HF} ClCH_2CH_2CF_3 \ (\mathbf{35}) \qquad \left[ClH_2C{-}CH{=}CF_2\right] \xrightarrow{HCl} ClCH_2CH_2CF_2Cl \ (\mathbf{36})$$

Anti-Markovnikov Addition of HBr to Alkenes

For some time the stereochemistry of the addition of HBr to alkenes was the subject of controversy because the results did not appear to be the same from laboratory to laboratory. Sometimes even the same researchers found different results under apparently similar conditions.[147] In 1933 Kharasch and Mayo distinguished between *normal* (now called Markovnikov) addition of HBr to allyl bromide in the presence of radical inhibitors and the *abnormal* or *unnormal* reaction observed in the presence of peroxides, air, and

---

[170] Myhre, P. C.; Andrews, G. D. *J. Am. Chem. Soc.* **1970**, *92*, 7596.

[171] Newton, T. A. *J. Chem. Educ.* **1987**, *64*, 531.

some other reagents.[172,173] Addition of HBr to propene gave 2-bromopropane if antioxidants were present but 1-bromoproprane if peroxides were added.[174] Similar results were obtained in the addition of HBr to pentene.[175,176] In these reactions the dielectric constant of the solvent did not appear to influence the *orientation* of the addition, but it did affect the *rate* of the normal addition.

The mechanism proposed for the *peroxide effect* involves a radical chain reaction.[177] The initiation step (equation 9.30) produces a bromine atom, which then attaches to the less alkyl-substituted carbon atom of a carbon-carbon double bond (equation 9.31). Then the resulting alkyl radical abstracts a hydrogen atom from HBr to produce the anti-Markovnikov product and regenerate a bromine atom in the second propagation step (equation 9.32). Termination steps, not shown, interrupt the chain reaction.

$$\mathrm{RO\cdot + HBr \longrightarrow ROH + Br\cdot} \qquad \textbf{(9.30)}$$

Br· + [propene] ⟶ [2-radical 1-bromopropane] Br — propagation step 1 (attachment) **(9.31)**

[2-radical 1-bromopropane] Br + HBr ⟶ [1-bromopropane] Br + Br· — propagation step 2 (transfer) **(9.32)**

Among the hydrogen halides, only HBr is known to undergo radical addition readily, even though the addition of HX to a carbon-carbon double bond is exothermic for all the hydrogen halides.[178] Calculation of the ΔH values for the two propagation steps in the chain reaction suggest a reason for the singular reactivity of HBr: only for HBr are *both* propagation steps exothermic (Table 9.4).[179] The endothermic radical attachment step precludes radical chain addition of HI, and the highly endothermic transfer step prevents the addition of HF. For radical addition of HCl to an alkene, the

---

[172]Credit has also been given to Hey and Waters for independent discovery of the radical nature of anti-Markovnikov addition of HBr to alkenes: Hey, D. H.; Waters, W. A. *Chem. Rev.* **1937**, *21*, 169.

[173]Kharasch, M. S.; Mayo, F. R. *J. Am. Chem. Soc.* **1933**, *55*, 2468.

[174]Kharasch, M. S.; McNab, M. C.; Mayo, F. R. *J. Am. Chem. Soc.* **1933**, *55*, 2531.

[175]Kharasch, M. S.; Hinckley, Jr., J. A.; Gladstone, M. M. *J. Am. Chem. Soc.* **1934**, *56*, 1642.

[176]Vaughan, W. E.; Rust, F. F.; Evans, T. W. *J. Org. Chem.* **1942**, *7*, 477.

[177]Kharasch, M. S.; Engelmann, H.; Mayo, F. R. *J. Org. Chem.* **1938**, *2*, 288.

[178]The addition of HI to propene, 1-bromopropene, allyl chloride or allyl bromide gave only the normal addition products, and antioxidants did not inhibit the reaction. Moreover, HI inhibited the radical addition of HBr to alkenes. One possible route for this inhibition would be the reaction of peroxides with HI to produce I_2. Kharasch, M. S.; Norton, J. A.; Mayo, F. R. *J. Am. Chem. Soc.* **1940**, *62*, 81.

[179]For a discussion see reference 152. The values in Table 9.4 and those in that reference differ slightly due to the use here of more recent bond strength data. See also Pryor, W. A. *Chem. Eng. News* **1968** (Jan. 15), *46*, 70.

Table 9.4 Enthalpies of the propagation steps in the radical addition of HX to ethene.

HX	Attachment (equation 9.31)	Transfer (equation 9.32)
HF	−46 kcal/mol	36 kcal/mol
HBr	−3 kcal/mol	−11 kcal/mol
HCl	−17 kcal/mol	4 kcal/mol
HI	12 kcal/mol	−27 kcal/mol

(Adapted from reference 152.)

slightly endothermic transfer step minimizes the importance of radical chain addition, although the photochemical addition of HCl to ethene in the gas phase was considered to be a radical chain addition process.[180] Furthermore, addition of anhydrous HCl to neat 3,3-dimethyl-1-butene in the presence of dibenzoyl peroxide produced up to 24% of 1-chloro-3,3-dimethylbutane, apparently as a result of radical chain addition. This anti-Markovnikov product was only seen at low concentrations of HCl, however.[166]

The regiochemistry of anti-Markovnikov addition can be rationalized conveniently by the argument that the reaction proceeds through the intermediacy of the more stable free radical, a statement that parallels the proposal that Markovnikov orientation involves a mechanism in which the more stable carbocation is an intermediate in the reaction. However, Tedder noted an important limitation of this rationalization.[181] Since the attachment of a bromine atom to an alkene (equation 9.31) is strongly exothermic, the transition state should be early, and differences in product stabilities of the developing radicals should not be manifest to a large degree in differences in transition state energies. A more satisfying explanation for the tendency of the bromine atom to bond to the less alkyl-substituted carbon atom is that steric effects are minimized.[182] Thus, in equation 9.31 the observed pathway not only passes through the more stable radical, it also proceeds through the less sterically hindered addition pathway.[183]

As in other cases, the reactive intermediates in additions to alkenes are often more complex than our simple models suggest. That is apparently true also for the intermediate in the free radical addition of HBr to alkenes. Peroxide-promoted addition of HBr to 1-bromocyclohexene in pentane solution

[180]Raley, J. H.; Rust, F. F.; Vaughan, W. E. *J. Am. Chem. Soc.* **1948**, *70*, 2767.

[181]Tedder, J. M. *J. Chem. Educ.* **1984**, *61*, 237.

[182]Tedder (reference 181) also discussed the importance of polar effects, for example in the case of attachment of methyl or trifluoromethyl radicals to vinyl fluoride.

[183]Goering and Larsen reported that the radical chain addition of HBr (or DBr) to 2-bromo-2-butene is stereospecific anti addition at −80°, but not stereospecific at room temperature. Goering, H. L.; Larsen, D. W. *J. Am. Chem. Soc.* **1959**, *81*, 5937.

was reported to give *cis*-1,2-dibromocyclohexane (equation 9.33). Similarly, addition to 1-methylcyclohexene gave *cis*-1-bromo-2-methylcyclohexane. In both cases, the trans products would have been thermodynamically more stable. Therefore the investigators concluded that a planar carbon free radical is not the intermediate. Rather, they proposed that the intermediate radical is a bromine-bridged species analogous to a bromonium ion. As shown in Figure 9.30, a bridging bromine atom effectively blocks one face of the cyclohexane ring. This interaction causes the reaction of the radical with HBr to occur from the other face of the structure, thus accounting for the observed stereochemistry.[184]

HBr / ROOR (9.33)

Although this discussion has focused on the addition of HBr, radical addition pathways have been elucidated for other reagents also, and some of these reactions effect the formation of carbon-carbon bonds.[185] For example, ethanol adds to 1-hexene to give 2-octanol.[186] In the presence of diacetyl peroxide initiator, CCl_3COCl and $CHCl_2COOCH_3$ add to alkenes by a free radical mechanism.[187] Other free radical chain reaction additions to alkenes include mercaptans;[188] CCl_4, CBr_4, $CHCl_3$, and $CHBr_3$;[189] CBr_2Cl_2 and

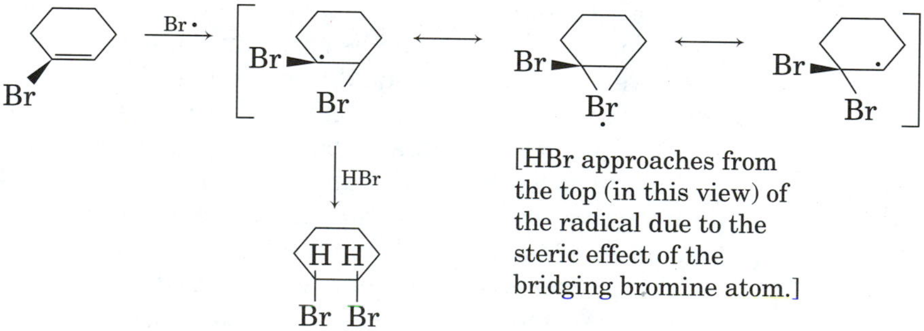

Figure 9.30 Mechanism proposed for formation of *cis*-1,2-dibromocyclohexane by free radical addition of HBr to 1-bromocyclohexene.

[HBr approaches from the top (in this view) of the radical due to the steric effect of the bridging bromine atom.]

[184]Goering, H. L.; Abell, P. I.; Aycock, B. F. *J. Am. Chem. Soc.* **1952**, *74*, 3588.

[185]Walling, C.; Huyser, E. S. *Org. React.* **1963**, *13*, 91.

[186]Urry, W. H.; Stacey, F. W.; Huyser, E. S.; Juveland, O. O. *J. Am. Chem. Soc.* **1954**, *76*, 450.

[187]Kharasch, M. S.; Urry, W. H.; Jensen, E. V. *J. Am. Chem. Soc.* **1945**, *67*, 1626.

[188]Jones, S. O.; Reid, E. E. *J. Am. Chem. Soc.* **1938**, *60*, 2452.

[189]With a generalized unsymmetrical alkene, $RCH{=}CH_2$, the products from these reagents are $RCHClCH_2CCl_3$, $RCHBrCH_2CBr_3$, $RCH_2CH_2CCl_3$, and $RCHBrCH_2CCl_3$, respectively. Kharasch, M. S.; Jensen, E. V.; Urry, W. H. *J. Am. Chem. Soc.* **1947**, *69*, 1100.

$CHBrCl_2$;[190] CF_2I_2[191] and $CBrCl_3$.[192] A radical chain mechanism has also been reported for the addition of perfluoroalkyl iodides to alkenes.[193]

Hydration of Alkenes

Hydration of alkenes is the addition of the elements of water (H, OH) across the carbon-carbon double bond. There is substantial evidence that acid-catalyzed addition of water to an alkene involves a cationic intermediate. Rate constants for hydration increase with the electron-donating ability of the substituents on the double bond, and rates of hydration of unsymmetrical alkenes with the general formula $R_1R_2C{=}CH_2$ give a good correlation with σ^+ values,

$$\log k_2 = \rho^+\Sigma\sigma^+ + C \tag{9.34}$$

where $\rho^+ = -12.3$ and $C = -10.1$.[194] Because of the lower acidity of water in comparison with hydrogen halides, addition of water is carried out with acid catalysis. For hydration of 2-methyl-2-butene in aqueous nitric acid solutions, the reaction was found to be overall second order, first order in alkene and first order in H^+, with $E_a = 18.9$ kcal/mol.[195]

Acid-catalyzed hydration reactions occur with Markovnikov orientation; for example, hydration of 2-methyl-2-butene gives 2-methyl-2-butanol, consistent with the intermediacy of the more stable carbocation. Hydration of both 2-methyl-2-butene and 2-methyl-1-butene (*asym*-methylethylethylene) was found to produce *t*-butyl alcohol, with no indication of isomerization during hydration.[196] These results suggest, but do not confirm,[9] that the cationic intermediate undergoes nucleophilic attack by water faster than it loses a proton to revert to starting material.[197]

Early studies of acid-catalyzed hydration were interpreted in terms of a π complex as an intermediate,[198] and *ab initio* calculations do suggest that a

---

[190]With a generalized unsymmetrical alkene, $RCH{=}CH_2$, the products from these reagents are $BrCHRCH_2CBrCl_2$ and $BrCHRCH_2Cl_2$, respectively. Kharasch, M. S.; Kuderna, B. M.; Urry, W. *J. Org. Chem.* **1948**, *13*, 895.

[191]With a generalized unsymmetrical alkene, $RCH{=}CH_2$, the product from this reagent is $RCHICH_2CF_2I$. Elsheimer, S.; Dolbier, Jr., W. R.; Murla, M.; Seppelt, K.; Paprott, G. *J. Org. Chem.* **1984**, *49*, 205.

[192]With a generalized unsymmetrical alkene, $RCH{=}CH_2$, the product from this reagent is $RCHBrCH_2CCl_3$. Kharasch, M. S.; Reinmuth, O.; Urry, W. H. *J. Am. Chem. Soc.* **1947**, *69*, 1105.

[193]Feiring, A. E. *J. Org. Chem.* **1985**, *50*, 3269 and references therein.

[194]Oyama, K.; Tidwell, T. T. *J. Am. Chem. Soc.* **1976**, *98*, 947.

[195]Lucas, H. J.; Liu, Y.-P. *J. Am. Chem. Soc.* **1934**, *56*, 2138.

[196]Levy, J. B.; Taft, Jr., R. W.; Hammett, L. P. *J. Am. Chem. Soc.* **1953**, *75*, 1253.

[197]Lucas cited unpublished data of Welge indicating that the hydration is reversible but that the equilibrium lies heavily in favor of the alcohol in aqueous solution, $K = 7.5 \times 10^3$ at 25°. The dehydration step has an activation energy of 34.8 kcal/mol. Eberz, W. F.; Lucas, H. J. *J. Am. Chem. Soc.* **1934**, *56*, 1230.

[198]Purlee, E. L.; Taft, Jr., R. W. *J. Am. Chem. Soc.* **1956**, *78*, 5807.

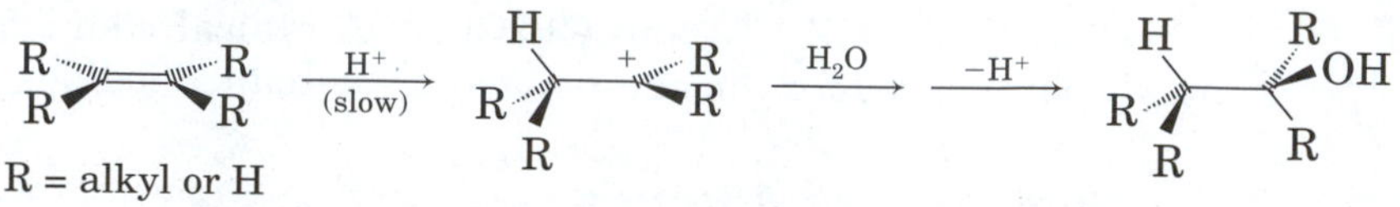

Figure 9.31
$A_{SE}2$ mechanism for hydration of alkenes.

hydrogen-bridged structure is more stable than an open cation structure for both the 2-butyl cation and for the ethyl cation.[199,200] However, it is not clear that structures that are lowest in energy in the gas phase are also lowest in energy in solution,[200] and there is no compelling evidence for bridged carbocations as significant intermediates in the hydration of alkenes. The available experimental data are consistent with the conclusion that hydration and other electrophilic reactions involving proton attachment to alkenes proceed directly through formation of carbocations (by an $A_{SE}2$ mechanism,[201] Figure 9.31), and that π complexes are not required intermediates.[202,203] Currently, hydration of alkenes is taken as the paradigm for electrophilic additions that do not involve 'onium ion intermediates.[204]

The observation of general acid catalysis in the hydration of both *trans*-cyclooctene and 2,3-dimethyl-2-butene established the $A_{SE}2$ mechanism for hydration of alkenes.[205] Furthermore, 1,2-dicyclopropylethene was found to react much faster than either *cis*- or *trans*-1,2-dicyclopropylethene, which indicates that substituent location (and not just the total electron donating ability of the substituents) determines the rate of hydration reactions. Therefore, it appears that in the transition structure the approaching proton is undergoing bond formation to one of the olefinic carbon atoms and not to the other, so the regiochemistry of the hydration reaction is determined by the approach that leads to the development of the more stable carbocation.[206]

Several lines of evidence have led to the acceptance of the $A_{SE}2$ mechanism for the acid-catalyzed hydration of styrenes also. In a study of the rate

[199]Carneiro, J. W. de M.; Schleyer, P. v. R.; Koch, W.; Raghavachari, K. *J. Am. Chem. Soc.* **1990**, *112*, 4064.

[200]Klopper, K.; Kutzelnigg, W. *J. Phys. Chem.* **1990**, *94*, 5625.

[201]The mechanism shown can also be described as an Ad_E2 reaction. $A_{SE}2$ (or A-S_E2) has often been used in the literature to describe reactions, including the hydration of alkenes by the mechanism shown in Figure 9.31, in which a rate-limiting proton transfer from a general acid leads to product formation. Stewart, R. *The Proton: Applications to Organic Chemistry*; Academic Press: New York, 1985; pp. 259–261, indicated that the terminology is derived from the prototypical A-S_E2 reaction, the exchange of ring protons on aromatic compounds, with the S_E standing for electrophilic substitution.

[202]For a discussion and leading references, see Nowlan, V. J.; Tidwell, T. T. *Acc. Chem. Res.* **1977**, *10*, 252.

[203]See also footnote 4 in reference 207.

[204]Schmid, G. H.; Tidwell, T. T. *J. Org. Chem.* **1978**, *43*, 460.

[205]Kresge, A. J.; Chiang, Y.; Fitzgerald, P. H.; McDonald, R. S.; Schmid, G. H. *J. Am. Chem. Soc.* **1971**, *93*, 4907.

[206]Knittel, P.; Tidwell, T. T. *J. Am. Chem. Soc.* **1977**, *99*, 3408.

constants and equilibria, Schubert and Keeffe noted the following conclusions:[207]

1. Rate constants for hydration (k_{hyd}) correlate well with $-H_0'$ and $-H_R'$. The correlation of k_{hyd} with acidity of the medium indicates that a proton is transferred in the rate-limiting step of the reaction.
2. Observation of general acid catalysis for the reaction suggests that a base is also present in the transition structure.
3. Proton transfer apparently does not occur prior to the rate-limiting step. When β,β-dideuteriostyrene was subjected to hydration conditions, loss of deuterium label in the starting material was not observed in the initial stages of the reaction.[208] If proton transfer occurred prior to the rate-limiting step, then the reverse of such a fast step should have led to deuterium loss.
4. Observation of a good Hammett correlation ($\rho^+ = -3.58$) for the hydration of *p*-substituted styrenes suggests that appreciable positive charge is developing on the incipient benzylic carbon atom in the transition structure. A ρ^+ value of -3.2 was found for acid-catalyzed hydration of 2-arylpropenes.[209] However, both values are somewhat smaller in magnitude than the value expected for a fully developed benzylic carbocation.[210]
5. A primary solvent isotope effect suggests that a proton is undergoing bonding change in the transition structure. The value of $k_{H_3O^+}/k_{D_3O^+}$ was found to vary from 2 to 4, depending on reactant and conditions, for reactions in aqueous $HClO_4$ or $DClO_4$. A somewhat smaller solvent isotope effect of 1.45 was found for the hydration of isobutene in aqueous perchloric acid solutions,[211] suggesting that the nature of substituents (alkyl or aryl) on the olefinic carbon atoms can affect the degree of proton transfer in the transition structure of hydration reactions. With other data, the results suggest that proton transfer is more nearly complete in the transition structure for hydration of alkyl-substituted alkenes than for hydration of aryl-substituted alkenes.[207,211]

Before leaving this section, we should note that the simple carbocation structure in Figure 9.31 may camouflage a complex ion-solvent species with considerable variation possible from one alkene-solvent system to another.[212] Chiang and Kresge determined that the 3° carbocation intermedi-

---

[207]Schubert, W. M.; Keeffe, J. R. *J. Am. Chem. Soc.* **1972**, *94*, 559 and references therein.

[208]Because of the equilibrium between styrene and α-phenylethanol under the reaction conditions, some loss of deuterium label in the starting material can occur as the hydration reaction nears completion. When the concentration of alcohol is minimal in the early stages of the reaction, however, the return path is negligible, and loss of label is not detected.

[209]Deno, N. C.; Kish, F. A.; Peterson, H. J. *J. Am. Chem. Soc.* **1965**, *87*, 2157.

[210]Brown, H. C.; Okamoto, Y. *J. Am. Chem. Soc.* **1957**, *79*, 1913.

[211]Gold, V.; Kessick, M. A. *J. Chem. Soc.* **1965**, 6718.

[212]Another complication not discussed here is that the cyclooctyl carbocation may have a bridged structure, not the simple 2° structure shown. For discussions on this point, see reference 72 and references therein.

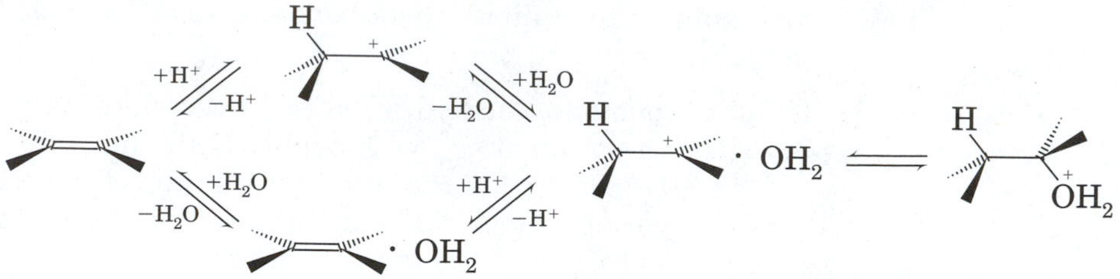

Figure 9.32 Hydration of an alkene through an equilibrated carbocation (upper pathway) or by a preassociation mechanism (lower pathway).

ate formed by protonation of 2,3-dimethyl-2-butene has a lifetime (τ) of about 10^{-10} sec in aqueous solution and therefore is a *conformationally equilibrated* carbocation. The 2° cyclooctyl carbocation formed by protonation of the cis and trans isomers of cyclooctene was found to have a lifetime that varies with the solvent; τ was found to be on the order of 10^{-8} sec in highly acidic solutions (about 50% $HClO_4$) such as those used to study the reaction of *cis*-cyclooctene, but τ was around 5×10^{-12} sec in dilute acid solutions used to study the reaction of *trans*-cyclooctene.[72] Chiang and Kresge concluded that the hydration of *trans*-cyclooctene probably occurs through an early transition state in which there is still appreciable double bond character between the olefinic carbon atoms in the transition structure. For 2° carbocations formed in dilute aqueous acid solutions from unstrained alkenes, however, such a short lifetime would be less than the rotational correlation time of water, so the water molecule that attacks the carbocation must already be oriented properly before protonation occurs. Therefore, the reaction of *trans*-cyclooctene was said to occur by a *preassociation mechanism*, as illustrated in Figure 9.32.[72,213]

Oxymercuration

Alkenes can also be converted to alcohols through a pair of reactions known collectively as oxymercuration-demercuration.[214] In the oxymercuration reaction, an alkene combines with a salt of Hg^{+2} and a nucleophile (usually protic solvent) to give an organomercury compound. The demercuration reaction involves reduction of the organomercury compound with $NaBH_4$ or another reducing agent to produce the final product. The synthetic utility of the oxymercuration-demercuration reaction lies primarily in the mild con-

[213]If the lifetime of the carbocation is shorter than the solvent reorientation time, then proton removal to reform the alkene will be faster than nucleophilic attack by water, so the reaction cannot proceed by the top pathway. Reaction will only occur by the bottom pathway, in which a water molecule is already properly oriented for nucleophilic attack before protonation of the alkene. In that situation the water orientation becomes the rate-limiting step in the reaction.

[214]For a review, see Chatt, J. *Chem. Rev.* **1951**, *48*, 7.

ditions that can hydrate an alkene with Markovnikov orientation (that is, with attachment of the OH group to the alkene carbon atom bearing the greater number of alkyl groups). For example, treatment of 3,3-dimethyl-1-butene with a suspension of mercuric acetate in water-tetrahydrofuran for 10 minutes produced an organomercurial that was reduced *in situ* by $NaBH_4$ in aqueous NaOH to produce 3,3-dimethyl-2-butanol (**38**) in 94% yield.[216]

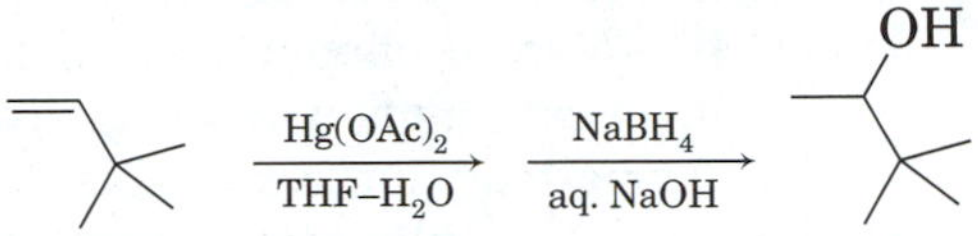

$$\mathbf{37} \xrightarrow[\text{THF–H}_2\text{O}]{\text{Hg(OAc)}_2} \xrightarrow[\text{aq. NaOH}]{\text{NaBH}_4} \mathbf{38} \qquad (9.35)$$

This reaction produced none of the rearrangement product (**39**) that would be expected from acid-catalyzed hydration of the reactant via an intermediate carbocation.

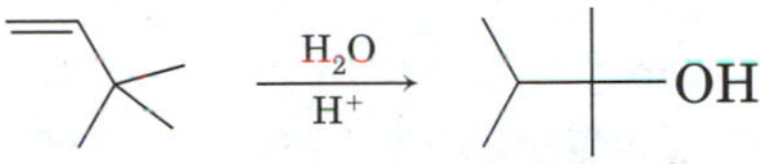

$$\mathbf{37} \xrightarrow[\text{H}^+]{\text{H}_2\text{O}} \mathbf{39} \qquad (9.36)$$

Furthermore, this method offers complementary regiochemistry to that observed from carrying out the hydroboration-oxidation procedure (see page 606) on the same reactant.

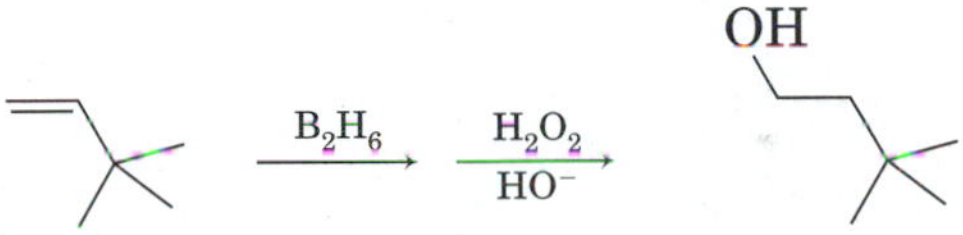

$$\mathbf{37} \xrightarrow{\text{B}_2\text{H}_6} \xrightarrow[\text{HO}^-]{\text{H}_2\text{O}_2} \mathbf{40} \qquad (9.37)$$

In addition to avoiding the carbocation rearrangements common with hydration reactions that involve carbocations, the oxymercuration-demercuration procedure offers another synthetic advantage. As shown in Figure 9.33 and in Figure 9.34, hydration of the alkene proceeds more easily from the less hindered side of the molecule, which is the same side from which attack of a Grignard reagent on the corresponding ketone would occur. There-

Figure 9.33 Stereochemistry of oxymercuration-demercuration and Grignard alcohol synthesis.

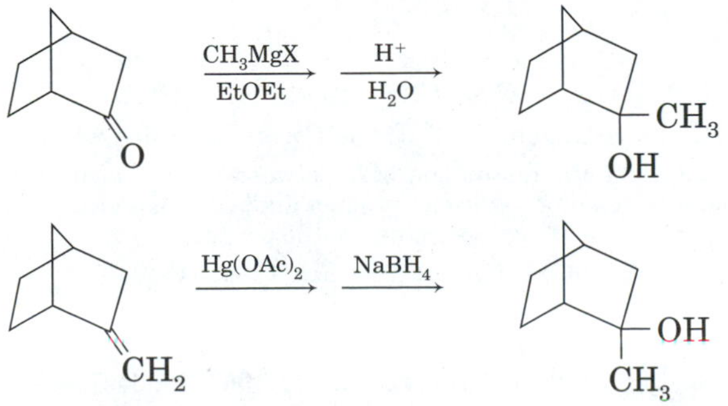

$$\xrightarrow[\text{EtOEt}]{\text{CH}_3\text{MgX}} \xrightarrow[\text{H}_2\text{O}]{\text{H}^+}$$

$$\xrightarrow{\text{Hg(OAc)}_2} \xrightarrow{\text{NaBH}_4}$$

Figure 9.34 Stereoselectivity in the oxymercuration-demercuration of 2,3-dimethylcyclopentene.

$$\text{2,3-dimethylcyclopentene} \xrightarrow{Hg(OAc)_2} \xrightarrow{NaBH_4} \text{(HO, } CH_3\text{, } CH_3\text{) } 73\% + \text{(} CH_3\text{, OH, } CH_3\text{) } 27\%$$

fore, oxymercuration-demercuration provides stereochemistry that is complementary to stereochemistry available through the synthesis of alcohols by reaction of carbonyls with Grignard reagents.[215]

The oxymercuration reaction is thought to be a two-step process. In the first step, electrophilic attachment of the mercury ion to the alkene produces a positively charged intermediate (equation 9.38). In the second step of oxymercuration, a nucleophile (most likely a solvent molecule, SOH) reacts with the intermediate to produce the organomercury compound (equation 9.39). For reactions in water, both the organomercury compound and the final product are alcohols.[216] If the hydroxylic solvent is an alcohol, the reaction produces an ether as the product and is often called *solvomercuration*.[217] Better yields are obtained if the anion of the mercuric salt is not a better nucleophile than is the solvent. For this reason, mercuric trifluoroacetate is often used instead of mercuric acetate for reactions in tertiary alcohols.[218,219]

$$R_1R_2C{=}CR_3R_4 + HgX_2 \underset{k_{-1}}{\overset{k_1}{\rightleftharpoons}} [\text{Intermediate}]^+ + X^- \quad \text{(Step 1)} \qquad \textbf{(9.38)}$$

$$[\text{Intermediate}]^+ + SOH \underset{k_{-2}}{\overset{k_2}{\rightleftharpoons}} SO\text{–}CR_1R_2\text{–}CR_3R_4\text{–}HgX + H^+ \quad \text{(Step 2)} \qquad \textbf{(9.39)}$$

(SOH = protic solvent)

Typical kinetic results are that the oxymercuration reaction is second order overall, first order in alkene and first order in mercuric salt.[220] Vardhan and Bach concluded that the oxymercuration reaction involves fast equilibrium formation of an intermediate (with $K_1 = k_1/k_{-1}$), followed by

[215]Brown, H. C.; Hammar, W. J. *J. Am. Chem. Soc.* **1967**, *89*, 1524.

[216]Brown, H. C.; Geoghegan, Jr., P. *J. Am. Chem. Soc.* **1967**, *89*, 1522.

[217]We will not make the distinction here between *oxymercuration* (in which water is the solvent/nucleophile) leading to alcohol product and *solvomercuration* (in which an alcohol is the solvent/nucleophile) leading to an ether as the product.

[218]Brown, H. C.; Rei, M.-H. *J. Am. Chem. Soc.* **1969**, *91*, 5646.

[219]Solvomercuration in acetonitrile can be used as a route to amines. Brown, H. C.; Kurek, J. T. *J. Am. Chem. Soc.* **1969**, *91*, 5647.

[220]Halpern, J.; Tinker, H. B. *J. Am. Chem. Soc.* **1967**, *89*, 6427.

rate-limiting attack of the nucleophile on this species (with rate constant k_2).[221] Thus, the rate law for the reaction would be

$$k_{obs} = (k_1 k_2 / k_{-1})[\text{alkene}][\text{Hg salt}] = K_1 k_2[\text{alkene}][\text{Hg salt}] \quad \textbf{(9.40)}$$

Since *formation* of the intermediate is rate-limiting in addition of bromine, but *reaction* of the intermediate is rate-limiting in oxymercuration, there is a large difference in substituent effects between these two reactions.[221] Alkyl groups greatly enhance the rate of bromine addition,[24] but their effect on the rate of oxymercuration is more complex. For example, reactivity in oxymercuration increases from ethene to propene to isobutene, but the rate of reaction of either *cis*- or *trans*-2-butene is much less than that of isobutene, and alkenes with four alkyl substituents undergo oxymercuration quite slowly.[220] These relative reactivities have been attributed to a much larger role of steric effects of alkyl substituents in the case of oxymercuration than for bromine addition.[221]

Steric effects may play a role in the regiochemistry of the reaction as well. While the oxymercuration-demercuration reaction is generally considered to give only Markovnikov hydration, it should be noted that there are exceptions. For example, methoxymercuration of 3,3-dimethyl-1-butene produced 2% of 3,3-dimethyl-1-butyl methyl ether (**42**).[218]

$Hg(OAc)_2$ / CH_3OH → $NaBH_4$ →

OCH_3 + OCH_3 **(9.41)**

98% 2%

37 **41** **42**

We have not yet defined the structure of the intermediate in equations 9.38 and 9.39. Based on analogies with other electrophilic additions, it seems reasonable that the intermediate might be a cyclic ion analogous to the bromonium ion.[222] Lucas, Hepner, and Winstein proposed a symmetrically bridged "mercurinium" ion, which might be described as a resonance hybrid of valence bond structures having positive charge on the mercury and on each of the olefinic carbon atoms (Figure 9.35).[223,224] Experimental evidence for a symmetrical mercurinium ion formed from ethene was found by Olah and Clifford by NMR[225] as well as by Bach and co-workers by ion cy-

---

[221]Vardhan, H. B.; Bach, R. D. *J. Org. Chem.* **1992**, *57*, 4948.

[222]The positively charged intermediate in the oxymercuration reaction has also been called a *mercuronium ion* (see reference 221, for example). However, because the term *mercurinium* has been used so extensively in the literature of organic chemistry, that is the term that will be used in the present discussion.

[223]Lucas, H. J.; Hepner, F. R.; Winstein, S. *J. Am. Chem. Soc.* **1939**, *61*, 3102.

[224]Evidence for such an intermediate was deduced from steric effects by Pasto, D. J.; Gontarz, J. A. *J. Am. Chem. Soc.* **1970**, *92*, 7480.

[225]Olah, G. A.; Clifford, P. R. *J. Am. Chem. Soc.* **1971**, *93*, 1261, 2320.

Figure 9.35 Resonance description of mercurinium ion. (Compare with Figure 9.13.)

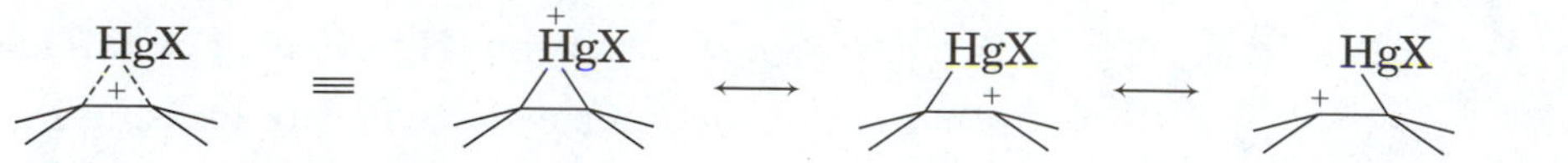

clotron resonance (ICR) spectroscopy[226] and by NMR.[227] Molecular orbital (CNDO/2) calculations also indicated an energy minimum for a symmetrically bridged ion composed of Hg^{+2} ion and ethene, but Bach and Henneike concluded that the bridged ion could be described as a π complex instead of a σ-bonded three-membered ring.[228]

In Figure 9.35 the mercurinium ion is depicted as a structure in which the mercury ion is symmetrically placed over the olefinic carbon atoms. As with other additions, however, the bridging atom may be unsymmetrically located between the two carbon atoms if the reacting olefin is unsymmetrically substituted. Figure 9.36 shows four structures that span a spectrum of possible structures for the organomercury intermediate.[229] CNDO/2 calculations revealed that there is only a shallow energy minimum on the potential energy surface associated with shifting the mercury along the C—C axis toward an unsymmetrical ion, so that the unsymmetrical ion might be lower in energy for unsymmetrically substituted alkenes.[228] As was the case with bromonium and chloronium ions, backside nucleophilic attack by solvent (leading to anti addition) is thus favored at the mercurinium carbon atom bearing the greater number of alkyl groups because that site bears the greater positive charge.

The theoretical results are consistent with experimental data. Oxymercuration of ethene appears to proceed through a symmetrical mercurinium ion with most of the positive charge on mercury, not on the olefinic carbon atoms.[229] If one of the olefinic carbon atoms has an aryl substituent (partic-

Figure 9.36 Possible geometries for the intermediate in the oxymercuration reaction.

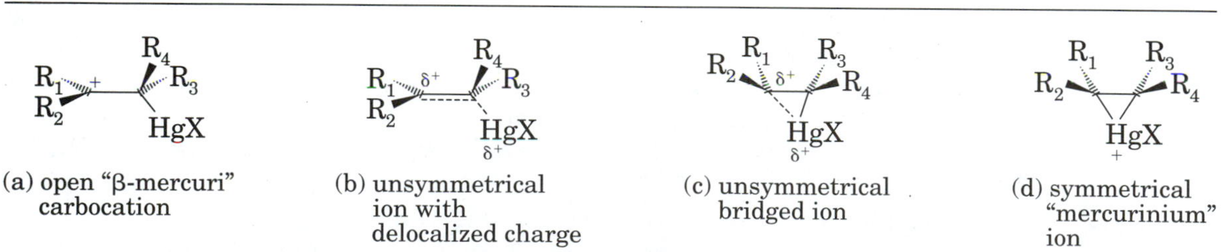

(a) open "β-mercuri" carbocation

(b) unsymmetrical ion with delocalized charge

(c) unsymmetrical bridged ion

(d) symmetrical "mercurinium" ion

[226] Bach, R. D.; Gauglhofer, J.; Kevan, L. *J. Am. Chem. Soc.* **1972**, *94*, 6860.

[227] Bach, R. D.; Richter, R. F. *J. Org. Chem.* **1973**, *38*, 3442.

[228] Bach, R. D.; Henneike, H. F. *J. Am. Chem. Soc.* **1970**, *92*, 5589.

[229] Ambidge, I. C.; Dwight, S. K.; Rynard, C. M.; Tidwell, T. T. *Can. J. Chem.* **1977**, *55*, 3086.

ularly a benzene ring substituted with an electron-donating substituent), then an unsymmetrical mercurinium ion or partially bridged mercury-substituted carbocation is proposed as the intermediate.[230] With substituents such as cyclopropyl groups that are very effective at stabilizing carbocations, the intermediate appears to be an open carbocation.[229]

Although the oxymercuration reaction is stereospecific, the demercuration reaction often is not. Reduction of the *trans*-2-hydroxycyclopentylmercuric acetate (from oxymercuration of cyclopentene) with $NaBD_4$ led to 2-deuteriocyclopentanol, which was determined to be at least 95% pure trans isomer (**44**).

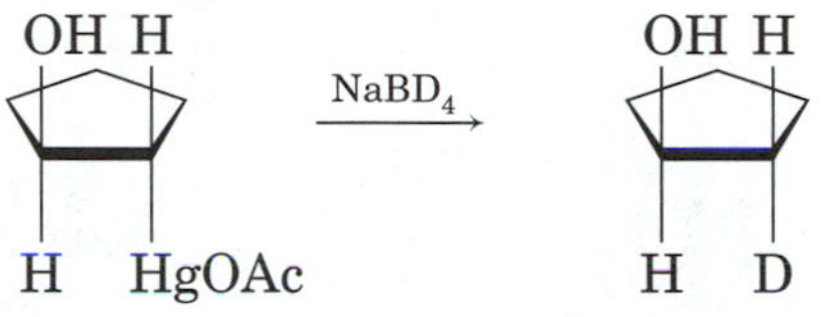

OH H; H HgOAc — $NaBD_4$ → OH H; H D

(>95% trans)

43 (±)-**44**

(9.42)

However, oxymercuration of either *cis*-2-butene or *trans*-2-butene, followed by reduction with $NaBD_4$, led to a 50 : 50 mixture of *erythro*- and *threo*-3-deuterio-2-butanols (Figure 9.37).[231]

These results can be rationalized with a radical mechanism in which an organomercury hydride, produced by reduction of R—Hg—X with borohydride, dissociates to alkyl and ·Hg—H radicals within a solvent cage. Abstraction of the hydrogen atom from ·Hg—H by the alkyl radical would produce the carbon-hydrogen bond and elemental mercury.[231] With a substrate derived from an acyclic alkene, rotation about the central carbon-carbon single bond can occur faster than hydrogen abstraction, leading to loss

Figure 9.37 Loss of stereochemistry in the demercuration reaction. All products are racemic. (Adapted from reference 231.)

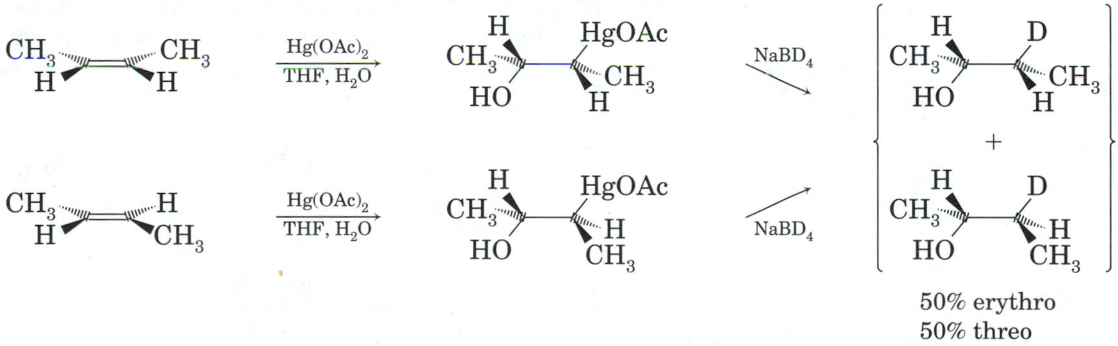

$Hg(OAc)_2$, THF, H_2O → ; $NaBD_4$ →

50% erythro
50% threo

[230]Lewis, A.; Azoro, J. *J. Org. Chem.* **1981**, *46*, 1764 and references therein. See also Lewis, A. *J. Org. Chem.* **1984**, *49*, 4682.

[231]Pasto, D. J.; Gontarz, J. A. *J. Am. Chem. Soc.* **1969**, *91*, 719.

of stereochemistry of the reactant (equation 9.43). With a reactant derived from a cyclic alkene, however, it is more likely that the alkyl radical can abstract a hydrogen atom from ·Hg—H before reorientation can occur within the solvent cage, so the carbon-hydrogen bond is more likely to have the same stereochemistry as the carbon-mercury bond it replaces (equation 9.42). Therefore the stereochemistry of the demercuration reaction depends strongly on the structure of the organomercury compound.[231]

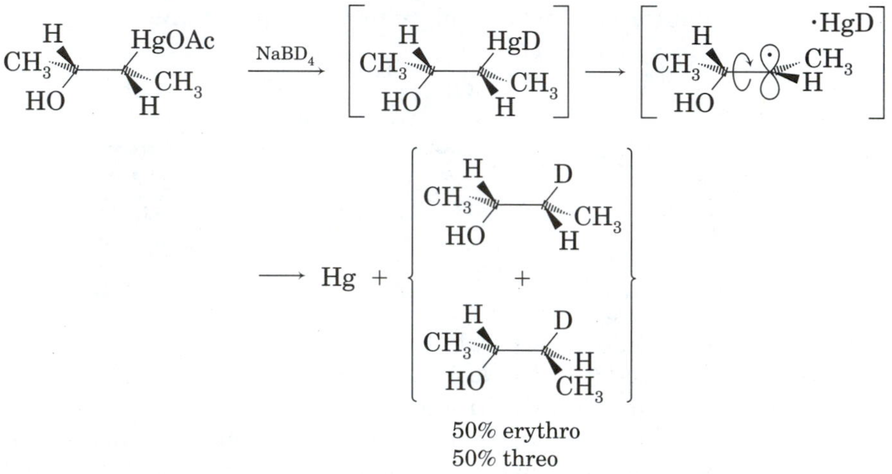

(9.43) $NaBD_4$; Hg +; 50% erythro; 50% threo

Hydroboration

The hydroboration-oxidation procedure is a valuable method to hydrate an alkene with anti-Markovnikov orientation and with syn addition of the H and OH groups.[232,233,234,235] Addition of BH_3 (which may be added to the reaction as diborane, B_2H_6) to an alkene occurs readily in diethyl ether, THF, or similar solvent. The hydroboration is strongly exothermic, with a ΔH of −33 kcal/mol per B—H bond that reacts.[254] If stoichiometry and the steric requirements of the alkyl substituents on the boron atom permit, the reaction proceeds until three alkyl groups are attached to each boron atom. The trialkylborane can then be oxidized with hydrogen peroxide in aqueous base to produce the alcohol.

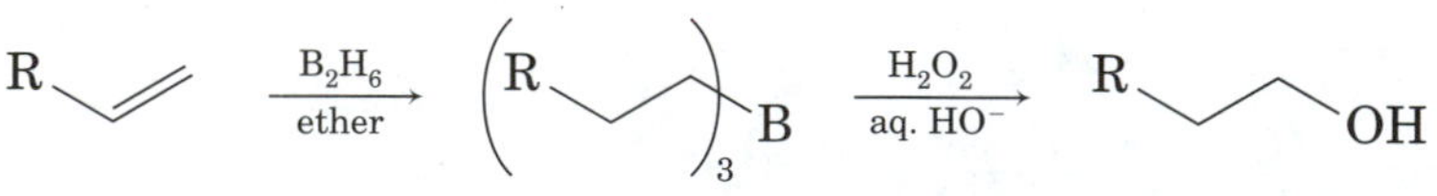

(9.44) B_2H_6, ether; H_2O_2, aq. HO^-

---

[232]Brown, H. C.; Subba Rao, B. C. *J. Am. Chem. Soc.* **1956**, *78*, 5694; **1959**, *81*, 6423.

[233]Brown, H. C. *Hydroboration*; W. A. Benjamin, Inc.; New York, 1962.

[234]Brown, H. C. *Boranes in Organic Chemistry*; Cornell University Press: Ithaca, NY, 1972.

[235]Numerous other reactions include the synthesis of trialkylcarbinols from trialkylboranes (Brown, H. C. *Acc. Chem. Res.* **1969**, *2*, 65) and hydroboration-protonolysis leading to alkanes (Brown, H. C.; Murray, K. *J. Am. Chem. Soc.* **1959**, *81*, 4108). Other features include isomerization and disproportionation of the alkylboranes. For details see reference 233.

The hydroboration reaction is generally highly, but not completely, regioselective. For example, reaction of 1-hexene with diborane produces 1-hexanol in high yield, with only a small amount of 2-hexanol (equation 9.45).[236,237] Brown determined that the preference for the boron atom to add to the less substituted carbon atom is about 94% for monosubstituted alkenes such as 1-pentene, 99% for *gem*-disubstituted alkenes such as 2-methyl-1-butene, and 98% for trisubstituted alkenes such as 2-methyl-2-butene.[236]

$$\mathbf{45}\ \xrightarrow[\text{ether}]{B_2H_6}\ \xrightarrow[\text{aq. } HO^-]{H_2O_2}\ \underset{\substack{94\% \\ \mathbf{46}}}{CH_3(CH_2)_5OH} + \underset{\substack{6\% \\ \mathbf{47}}}{CH_3(CH_2)_3CH(OH)CH_3} \quad \mathbf{(9.45)}$$

The addition of boron and hydrogen across the carbon-carbon double bond appears to occur by stereospecific syn addition. The replacement of the C—B bond by a C—OH bond in the oxidation step is also stereospecific. The mechanism proposed for the oxidation of one of the carbon-boron bonds is shown in equations 9.46 through 9.50,[237,238] so the configuration of the carbon-boron bond is retained in the product. As an example, hydroboration-oxidation of 1-methylcyclopentene (equation 9.51) gave only the *trans*-2-methylcyclohexanol (**49**).

$$H_2O_2 + HO^- \rightleftharpoons HO_2^- + H_2O \quad \mathbf{(9.46)}$$

$$HO_2^- + R_3B \longrightarrow \left[R_3\overset{-}{B}-O-OH \right] \quad \mathbf{(9.47)}$$

$$\left[R_3\overset{-}{B}-O-OH \right] \longrightarrow R_2B-OR + HO^- \quad \mathbf{(9.48)}$$

[236] Brown, H. C.; Zweifel, G. *J. Am. Chem. Soc.* **1960**, *82*, 4708.

[237] Synthetic applications have been discussed: Zweifel, G.; Brown, H. C. *Org. React.* **1963**, *13*, 1.

[238] House, H. O. *Modern Synthetic Reactions*, 2nd Ed.; W. A. Benjamin, Inc.: Menlo Park, CA, 1972; pp. 106–144.

$$R{-}\underset{\underset{R}{|}}{B}{-}OR + HO^- \longrightarrow R{-}\overset{\overset{OH}{|}}{\underset{\underset{R}{|}}{\bar{B}}}{-}OR \qquad (9.49)$$

$$R{-}\overset{\overset{OH}{|}}{\underset{\underset{R}{|}}{\bar{B}}}{-}OR \xrightarrow{H_2O} R{-}\underset{\underset{R}{|}}{B}{-}OH + HOR + HO^- \qquad (9.50)$$

$$\textbf{48} \xrightarrow[\text{ether}]{B_2H_6} (\text{H, } CH_3\text{, H, } BR_2) \xrightarrow[\text{aq. } HO^-]{H_2O_2} (\pm)\text{-}\textbf{49} \qquad (9.51)$$

The reaction in equation 9.51 does not provide a convincing demonstration of the stereochemistry of hydroboration-oxidation, since **49** is thermodynamically more stable than the corresponding cis isomer. Therefore Brown and Zweifel also carried out the procedure on 1,2-dimethylcyclopentene (equation 9.52). The product of that reaction was the less thermodynamically stable isomer (with the two methyl groups cis to each other), confirming the syn pathway for the addition.[239,240]

$$\textbf{48} \xrightarrow[\text{ether}]{B_2H_6} \xrightarrow[\text{aq. } HO^-]{H_2O_2} (\pm)\text{-}\textbf{51} \qquad (9.52)$$

The observation of consistent syn addition in the hydroboration step suggests that the mechanism for addition of BH_3 is different from all of the addition mechanisms discussed so far except one. As discussed on page 581, a one-step, four-center mechanism was considered in the syn addition of F_2 to an alkene, and a similar mechanism can explain the stereochemistry of BH_3 addition. Moreover, the regiochemistry of addition to unsymmetrical alkenes can be rationalized in terms of the steric and electronic effects present in such a transition structure.

---

[239]The production of *exo*-norborneol from hydroboration-oxidation of norbornene further supported the syn pathway. Brown, H. C.; Zweifel, G. *J. Am. Chem. Soc.* **1959**, *81*, 247.

[240]Syn addition of molecular hydrogen to alkenes can be effected by protonolysis of organoboranes by carboxylic acids. For a discussion of the mechanism of the reaction and examples of its use in synthesis, see reference 238.

Figure 9.38 Transition structure models for hydroboration of an alkene with one alkyl substituent.

Attachment of boron to the carbon atom with fewer alkyl substituents

CH_3, δ^+ CH $=$ CH_2 δ^-, H δ^- ····· B δ^+, R(H), R(H)

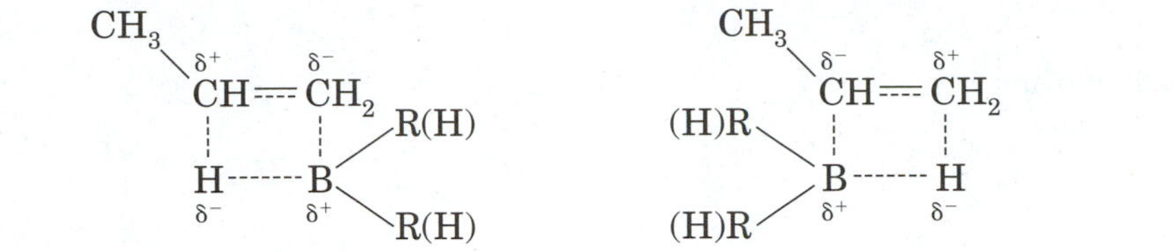

Sterically favored
Electronically favored

Attachment of boron to the carbon atom with more alkyl substituents

CH_3, δ^- CH $=$ CH_2 δ^+, (H)R, (H)R, B δ^+ ····· H δ^-

Sterically disfavored
Electronically unfavored

The two transition structures in Figure 9.38 show two possible orientations for concerted addition of borane to an unsymmetrical alkene. Because of the presumed polarization of the B — H bond so that the hydrogen atom has a slight negative charge and the boron atom has a slight positive charge, association with the alkene would be expected to drain electron density from the double bond, with the charge being more pronounced on the carbon atom that will become bonded to hydrogen. As suggested in Figure 9.38, this interaction could result in a slight electronic effect favoring the attachment of boron to the carbon atom having fewer alkyl substituents because this approach places a partial positive charge on the more highly substituted carbon atom. However, this explanation may be incomplete. The partial charge development may be small, especially in the case of an early transition state.[241] Moreover, there is also the possibility of steric hindrance between the other substituents on boron with the substituents on the alkene. Such a steric effect also correctly predicts the attachment of the boron atom to the less substituted carbon atom of the alkene.

Analogous transition structures for addition of BH_3 to styrene and its derivatives are shown in Figure 9.39. Again, the steric factor should favor attachment of boron to the carbon atom with fewer alkyl substituents. Now, however, the ability of the phenyl group to offer some electronic stabilization to a partial negative charge on the α carbon atom means that there can be some electronic stabilization for the addition pathway leading to attachment of boron to the carbon atom bearing the phenyl group. This stabilization would be more important for compounds such as *p*-nitrostyrene than for styrene, and less important for styrenes with electron-donating groups.

To investigate the roles of charge and steric effects on the regiochemistry of hydroboration, Vishwakarma and Fry studied the relative rates of reaction of *p*-substituted styrenes with 9-borabicyclo[3.3.1]nonane (9-BBN, **52**) in THF at 25°.[242] In all cases the hydroboration-oxidation was highly re-

---

[241]The partial charges are not indicated in the transition structures drawn by all investigators. See, for example, reference 256.

[242]Vishwakarma, L. C.; *Fry, A. J. Org. Chem.* ***1980***, 45, *5306.*

Figure 9.39 Transition structure models for hydroboration of an alkene with one aryl substituent.

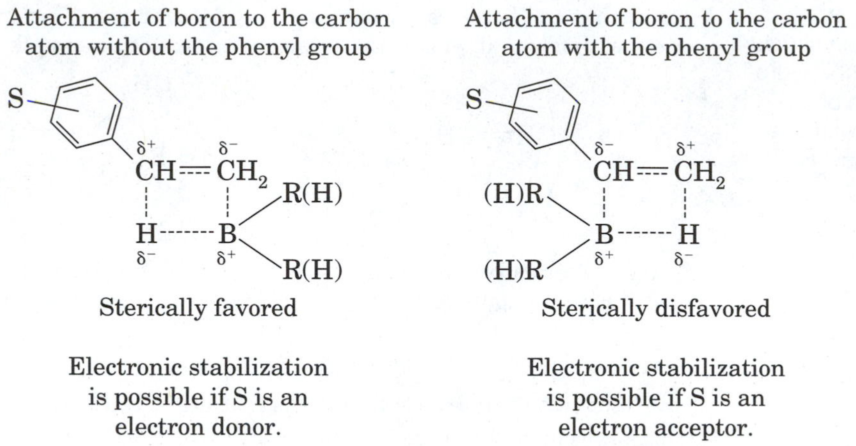

Attachment of boron to the carbon atom without the phenyl group

Sterically favored

Electronic stabilization is possible if S is an electron donor.

Attachment of boron to the carbon atom with the phenyl group

Sterically disfavored

Electronic stabilization is possible if S is an electron acceptor.

gioselective, with the β-phenylethanol comprising more than 97% of the product. (The remaining product was the α-phenylethanol.) Apparently the steric requirements of the 9-BBN were able to overcome any electronic stabilization of the alternative addition pathway. The rate constants for the addition reactions gave a Hammett correlation with σ^+ ($r = 0.94$), with a ρ value of -0.49. The results suggest that there is only a small amount of charge developed on the α carbon atom in the transition structure for the reaction.[243] Vishwakarma and Fry concluded that substituents on the styrene moiety exert electronic influences that change the *rate* of hydroboration, but that steric factors are the primary determinant of *regioselectivity*.

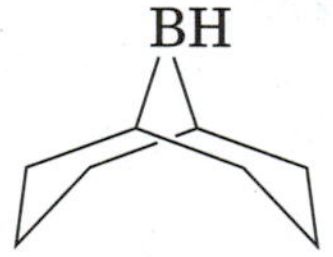

BH

52

While the four-center transition structure for BH_3 addition is a widely used model, other reaction pathways have also been considered. In a synthesis of optically active (−)-1-butanol-1-*d*, Streitwieser and co-workers used the optically active borane formed from diborane and (+)-α-pinene (R_2BH) to carry out the hydroboration-oxidation of (*Z*)-1-butene-1*d*.[244] To explain the observed stereochemistry of the reduction, they proposed that

[243] These results differed somewhat from the results of hydroboration of substituted styrenes reported by Brown, H. C.; Sharp, R. L. *J. Am. Chem. Soc.* **1966**, *88*, 5851, who had found that the Markovnikov hydration product was formed in nearly 20% yield. With electron donating substituents, the yield of anti-Markovnikov product was increased, while for electron withdrawing substituents the yield of Markovnikov products was increased. See also the discussion in reference 236.

[244] Streitwieser, Jr., A.; Verbit, L.; Bittman, R. *J. Org. Chem.* **1967**, *32*, 1530.

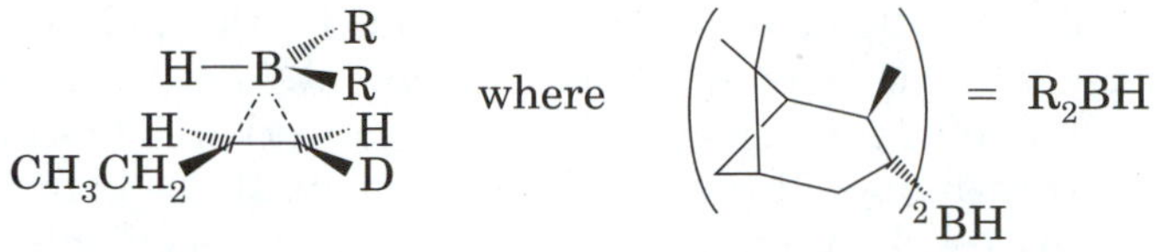

Figure 9.40
Proposed triangular complex for hydroboration of an alkene.

hydroboration involves a π complex between R_2BH and the alkene (Figure 9.40). There is a close resemblance of such a complex to the 'onium complexes proposed for halogen addition or oxymercuration. Streitwieser proposed that the transition structure for hydroboration involves only a small perturbation from the triangular π complex.[245] It is important to note that the structure shown in Figure 9.40 is proposed as an energy minimum and therefore might be an intermediate in the reaction, while the four-center structure shown on the left in Figure 9.38 is proposed to be a transition structure. It is conceivable that the π complex might be an intermediate on the path to the four-center transition structure.[246,247]

The nature of the transition structure in hydroboration has been the subject of extensive theoretical investigation. Nelson found that the rates of hydroboration of alkenes by 9-BBN correlate with the HOMO energy of the alkenes (from an MNDO calculation) and that the regiochemical preference for C — B bond formation is predicted by the larger olefinic carbon atomic orbital in the alkene HOMO.[248] Dewar and McKee used the MNDO method to study the hydroboration of alkenes and alkynes.[249] In the cases of ethene, propene, and isobutene, there was a transition state leading to a loose π-type adduct along the reaction coordinate and another transition state leading away from the complex, but in each case the complex was marginally stable and required little or no activation to proceed to product.[250] Lipscomb and co-workers also studied hydroboration with semi-empirical MO theory and concluded that the mechanism involves a loosely bound complex that is formed prior to a transition state for the reaction.[251]

---

[245] For further discussion on this subject, see Pasto, D. J.; Klein, F. M. *J. Org. Chem.* **1968**, *33*, 1468; Jones, P. R. *J. Org. Chem.* **1972**, *37*, 1886.

[246] Klein, J.; Dunkelblum, E.; Wolff, M. A. *J. Organometal. Chem.* **1967**, *7*, 377.

[247] Sundberg, K. R.; Graham, G. D.; Lipscomb, W. N. *J. Am. Chem. Soc.* **1979**, *101*, 2863.

[248] Nelson, D. J.; Cooper, P. J. *Tetrahedron Lett.* **1986**, *27*, 4693.

[249] Dewar, M. J. S.; McKee, M. L. *Inorg. Chem.* **1978**, *17*, 1075.

[250] These authors noted that the synchronous cycloaddition is not a *forbidden* process because the empty *p* orbital on boron makes the conjugated system more nearly resemble pentadienyl than cyclobutadiene. The distinction between *allowed* and *forbidden* concerted reactions will be discussed in Chapter 11.

[251] Reference 247. The overall reaction was described as a two-step donation-back-donation mechanism, with the boron having considerable negative charge in the transition structure, but releasing this charge as the reaction proceeds. The calculation also indicated considerable charge separation involving the two olefinic carbon atoms, with the carbon atom nearest boron acquiring considerable negative charge, while the other olefin carbon atom becomes more positive.

The results of *ab initio* calculations for the addition of borane to ethene depend on the level of the calculation.[252,253] For the reaction of borane with propene in the gas phase, Houk and co-workers identified a π-complex and two transition states (one for attachment of the boron atom to the carbon bearing two hydrogen atoms and one for attachment of the boron atom to the methyl-substituted carbon atom) in a calculation at the 3-21G level. As in the case of ethene itself, the π-complex was found to be lower in energy than the reactants. The transition state for attachment of the boron atom to the CH_2 position was found to be 3.7 kcal/mol lower in energy than the alternative transition state, and the difference in energy was attributed to a combination of electronic and steric factors.[252]

The calculations discussed so far are for reaction of monomeric BH_3 with alkenes in the gas phase. In solution the borane is most likely to be a dimer or, in ether solvents such as THF, a borane-solvent complex. Activation energies measured in solution are much higher than those in the gas phase. As Pasto and co-workers have reported, however, it is difficult to study the kinetics of borane addition because the reaction is complicated by three addition steps (one for each B—H bond), three redistribution equilibria (in which borane and the alkyl boranes exchange substituents), and five different monomer-dimer equilibria involving all the species with at least one B—H bond.[254] These investigators studied the hydroboration of 2,3-dimethyl-2-butene with diborane in THF, in which the reacting species is most likely a borane-THF complex. The reaction was found to be second order overall, first order in alkene and first order in BH_3-THF. The E_a was found to be 9.2 kcal/mol, while the activation entropy was -27 eu. These results stand in contrast to the value of 2 kcal/mol determined for $\Delta H^{\ddagger}$ for the reaction of BH_3 with ethene in the gas phase.[255]

Pasto and Kang studied the addition of monochloroborane ($BClH_2$) to olefins in THF solution.[256] The advantages of this reagent are (1) it is monomeric (as the THF complex); (2) it reacts only to the monoalkylmonochloroborane stage; (3) the addition product is monomeric in THF and does not disproportionate. Therefore the kinetics are somewhat easier to study than is the case for BH_3—THF. The rates of addition to *p*-nitrostyrene, styrene, and *p*-methoxystyrene were found to be 3.24, 7.31, and 27.8×10^{-3} M^{-1} min^{-1}, respectively. The regioselectivity of product formation varied among these compounds, with 67%, 90%, and 93.3%, respectively, of the product formed through attachment of the boron atom to the $\beta(CH_2)$ carbon atom. Separation of the overall rate constant data into rates for attachment

[252] Wang, X.; Li, Y.; Wu, Y.-D.; Paddon-Row, M. N.; Rondan, N. G.; Houk, K. N. *J. Org. Chem.* **1990**, *55*, 2601.

[253] van Eikema Hommes, N. J. R.; Schleyer, P. v. R. *J. Org. Chem.* **1991**, *56*, 4074.

[254] Pasto, D. J.; Lepeska, B.; Cheng, T.-C. *J. Am. Chem. Soc.* **1972**, *94*, 6083.

[255] Fehlner, T. P. *J. Am. Chem. Soc.* **1971**, *93*, 6366.

[256] Pasto, D. J.; Kang, S.-Z. *J. Am. Chem. Soc.* **1968**, *90*, 3797.

to α and β carbon atoms gave a Hammett correlation for each, with ρ values of −0.65 and −1.43, respectively. Reaction with $BClD_2$ gave isotope effects of 1.78 and 1.87, respectively, for the α and β reactions. Pasto concluded from these data that the addition is electrophilic in nature, but with a very early transition state such that there is little charge development on either carbon atom of the double bond in the transition structure. The results support the four-center model for the transition structure advanced by Brown.

Epoxidation

Epoxidation of an alkene produces an epoxide (oxirane), a three-membered ring containing one oxygen atom. Reagents that are peroxides, either H_2O_2 itself or a peroxy acid (peracid) with the general formula RCO_3H, such as peroxyacetic acid, are often used to effect the transformation.[257,258,259]

$$R_2C{=}CR_2 + RC(=O)O_2H \longrightarrow R_2C\overset{O}{\frown}CR_2 + RC(=O)OH \qquad (9.53)$$

(R = hydrogen or alkyl)

Epoxidation appears to involve electrophilic addition to the alkene, since the reaction is favored by electron-withdrawing groups on the peracid and electron donating groups on the alkene.[260] The epoxidation reaction is highly exothermic, with an experimental heat of reaction of −38 kcal/mol.[266] The kinetic expression is overall second order, first order in the alkene and first order in the peracid.[261] Steric effects do not appear to be important. The rate of the reaction increases with the number of alkyl substituents on the alkene, and *cis*-2-butene, *trans*-2-butene, and isobutene react at nearly the same rate.[258] However, the rate of the reaction is sensitive to strain, with faster rates observed for alkenes that produce greater relief of strain upon epoxidation.[262]

---

[257]Epoxides may also be formed by oxidation of alkenes with other reagents (for a summary, see Hudlický, M. *Oxidations in Organic Chemistry*, ACS Monograph No. 186; American Chemical Society: Washington, D.C., 1990) and by dehydrohalogenation of halohydrins.

[258]Swern, D. *Org. React.* **1953**, *7*, 378. Credit for the discovery of the peracid reaction is given to N. Prileschajew in 1908.

[259]*m*-Chloroperbenzoic acid is often the reagent of choice for epoxidation of alkenes because it is soluble in CH_2Cl_2 but the *m*-chlorobenzoic acid byproduct is not. Precipitation of the acid from solution reduces the possibility that the epoxide will undergo subsequent acid-catalyzed ring-opening reaction.

[260]Lynch, B. M.; Pausacker, K. H. *J. Chem. Soc.* **1955**, 1525.

[261]In most cases peroxidation is found not to be acid catalyzed, although catalysis by trichloroacetic acid was reported by Berti, G., Bottari, F. *J. Org. Chem.* **1960**, *25*, 1286.

[262]Shea, K. J.; Kim, J.-S. *J. Am. Chem. Soc.* **1992**, *114*, 3044.

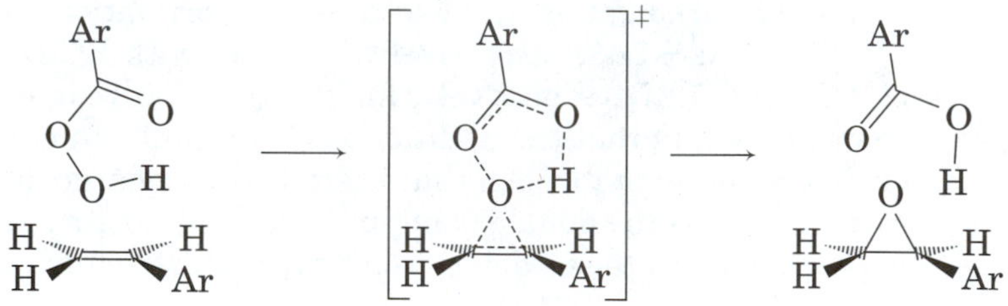

Figure 9.41
Bartlett mechanism for epoxidation.

An early suggestion for the mechanism of peracid epoxidation of alkenes was the butterfly mechanism proposed by Bartlett (Figure 9.41).[263,264] The transition structure shown there is symmetric, but there are variations of the Bartlett mechanism in which the three-membered ring of the developing epoxide is asymmetric, with the incoming oxygen atom over one of the carbon atoms.[265] Plesničar and co-workers studied the reaction with *ab initio* calculations and identified several possible transition structures, including the butterfly structure in Figure 9.41, as well as a spiro structure with the oxygen centered over the middle of the two alkene carbon atoms, Figure 9.42(a), a spiro structure with the oxygen centered over one of the carbon atoms, Figure 9.42(b), and a planar structure with the oxygen centered over one of the alkene carbon atoms, Figure 9.42(c).[266,267]

Hanzlik and Shearer found a ρ of −1.1 for epoxidation of substituted *trans*-stilbenes with perbenzoic acid, but a value of ρ of +1.4 for epoxidation of *trans*-stilbene with substituted peroxybenzoic acids. A k_H/k_D value of 1.17 was observed for reaction with $C_6H_5CO_3D$, suggesting that the proton on the peracid was either only slightly (or alternatively, nearly fully) trans-

[263]Bartlett, P. D. *Rec. Chem. Progr.* **1950**, *11*, 47. See also Mimoun, H. *Angew. Chem., Int. Ed. Engl.* **1982**, *21*, 734.

[264]See also the discussion in Rebek, Jr., J.; Marshall, L.; McManis, J.; Wolak, R. *J. Org. Chem.* **1986**, *51*, 1649.

[265]Ogata, Y.; Tabushi, I. *J. Am. Chem. Soc.* **1961**, *83*, 3440.

[266]Plesničar, B.; Tasevski, M.; Ažman, A. *J. Am. Chem. Soc.* **1978**, *100*, 743.

[267]An alternative mechanism involving a rather different transition structure geometry was proposed by Kwart, H.; Hoffman, D. M. *J. Org. Chem.* **1966**, *31*, 419; Kwart, H.; Starcher, P. S.; Tinsley, S. W. *Chem. Commun.* **1967**, 335. Woods, K. W.; Beak, P. *J. Am. Chem. Soc.* **1991**, *113*, 6281, used the **endocyclic restriction test** to study the epoxidation of alkenes. In this test two functional groups that normally react in an intermolecular process are tethered together with a carbon skeleton that varies in size and, therefore, that restricts the possible orientations of the two functional groups. By determining which chain lengths produce the same reaction seen in the intermolecular reaction and which do not, it is possible to put limits on the orientation of the two reacting groups in the intermolecular reaction. Woods and Beak concluded that this mechanistic test strongly supported the Bartlett mechanism for the intermolecular epoxidation of alkenes. However, the endocyclic restriction test cannot distinguish between the several butterfly and spiro (page 68) variants of the Bartlett mechanism that differ in the orientation of the alkenyl and peracid moieties. For leading references to the development and use of the endocyclic restriction test, see Tenud, L.; Farooq, S.; Seibl, J.; Eschenmoser, A. *Helv. Chim. Acta* **1970**, *53*, 2059; Hogg, D. R.; Vipond, P. W. *J. Chem. Soc. C* **1970**, 2142; Beak, P.; Basha, A.; Kokko, B.; Loo, D. *J. Am. Chem. Soc.* **1986**, *108*, 6016.

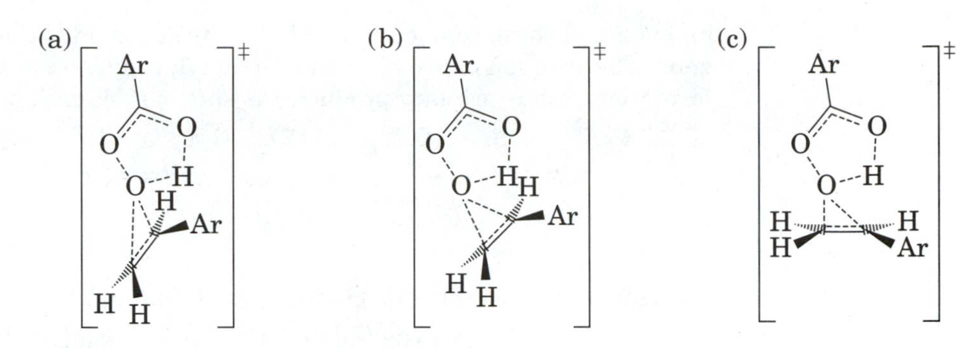

Figure 9.42 Alternative transition structures considered for peracid epoxidation of alkenes.

ferred in the transition structure. A secondary deuterium kinetic isotope effect was observed for the alkenyl hydrogen atoms on the β carbon atom of styrene ($k_H/k_D = 0.82$) as well as the α carbon atom ($k_H/k_D = 0.99$), indicating that rehybridization from sp^2 to sp^3 is further advanced for the β carbon atom than for the α carbon atom in the transition structure. These observations led to the proposal of the asymmetric transition structure shown in Figure 9.43.[268] However, substituent effects on the epoxidation of ring-substituted α-methylstilbenes with peroxybenzoic acid suggested that the nature of the transition structure varies according to the electron-donating ability of the alkene substituents and the electron-withdrawing ability of substituents on the peroxybenzoic acid ring.[265] Thus, the transition structure may be symmetric for epoxidation of ethene or symmetrically substituted alkenes, but it may be highly asymmetric when one carbon atom is substituted with a group that can stabilize a carbocation (such as in Figure 9.43).[269]

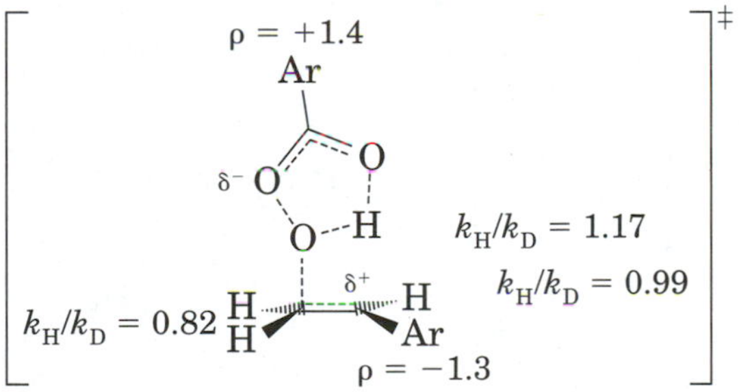

$\rho = +1.4$ Ar; δ^- O; $k_H/k_D = 1.17$; $k_H/k_D = 0.99$; δ^+; $k_H/k_D = 0.82$; $\rho = -1.3$

Figure 9.43 Model for the transition structure in the epoxidation of styrene. (See reference 268.)

Electrophilic Addition to Alkynes and Cumulenes

The discussion to this point has emphasized the addition of electrophilic reagents to alkenes. All of these reagents also add to alkynes and cumulenes, although there are often significant differences in reactivity between

[268] Hanzlik, R. P.; Shearer, G. O. *J. Am. Chem. Soc.* **1975**, *97*, 5231.

[269] For arguments against this point of view, see the discussion in reference 268.

alkenes and these compounds.[270] Let us first consider the addition of halogens. The reaction of bromine and 1-phenylpropyne in acetic acid led to the formation of four primary products, as shown in equation 9.54.[271,272]

$$\underset{\mathbf{53}}{C_6H_5-C\equiv C-CH_3} \xrightarrow[HOAc]{Br_2} \underset{\mathbf{54}}{(C_6H_5)(Br)C=C(Br)(CH_3)} + \underset{\mathbf{55}}{(C_6H_5)(Br)C=C(CH_3)(Br)} + \underset{\mathbf{56}}{(C_6H_5)(AcO)C=C(Br)(CH_3)} + \underset{\mathbf{57}}{(C_6H_5)(AcO)C=C(CH_3)(Br)} \quad (9.54)$$

The rate expression for this reaction was found to be

$$\text{rate} = [\text{alkyne}](k_2[Br_2] + k_3[Br_2]^2 + k_{Br^-}[Br_2][Br^-]) \quad (9.55)$$

In the absence of Br^- and at low concentrations of Br_2, the kinetic expression is first order in alkyne and first order in bromine. Moreover, a ρ value of -5.17 for bromine addition to aryl-substituted derivatives is consistent with the intermediacy of a vinyl cation–bromide ion pair that can combine to form both cis and trans dibromide adducts or can react with solvent to form *cis*- and *trans*-1-acetoxy-2-bromo-1-phenylpropene, as shown in equations 9.56 through 9.58.[273,274]

$$C_6H_5-C\equiv C-CH_3 \underset{\text{fast}}{\overset{Br_2}{\underset{HOAc}{\rightleftharpoons}}} \left[C_6H_5-C\overset{Br_2}{\overset{\uparrow}{\equiv}}C-CH_3\right] \xrightarrow{\text{slow}} \left[C_6H_5-\overset{+}{C}=C(Br)(CH_3)\;\; Br^-\right] \quad (9.56)$$

$$\left[C_6H_5-\overset{+}{C}=C(Br)(CH_3)\;\; Br^-\right] \longrightarrow (C_6H_5)(Br)C=C(Br)(CH_3) + (C_6H_5)(Br)C=C(CH_3)(Br) \quad (9.57)$$

$$\left[C_6H_5-\overset{+}{C}=C(Br)(CH_3)\;\; Br^-\right] \xrightarrow{HOAc} (C_6H_5)(AcO)C=C(Br)(CH_3) + (C_6H_5)(AcO)C=C(CH_3)(Br) \quad (9.58)$$

---

[270]Schmid, G. H. in *The Chemistry of the Carbon-Carbon Triple Bond*, Patai, S., ed.; Wiley-Interscience: Chichester, England, 1978; pp. 275–341.

[271]Pincock, J. A.; Yates, K. *J. Am. Chem. Soc.* **1968**, *90*, 5643. In addition to the products shown, 1,1-dibromoethyl phenyl ketone was formed as a result of subsequent reaction of products **56** and **57**.

[272]Pincock, J. A.; Yates, K. *Can. J. Chem.* **1970**, *48*, 3332.

[273]For a discussion of vinyl cations and electrophilic additions to alkynes, see Stang, P. J.; Rappoport, Z.; Hanack, M.; Subramanian, L. R. *Vinyl Cations*; Academic Press: New York, 1979.

[274]Apparently the 1-phenyl carbocation is the intermediate in the nonstereospecific addition of trifluoroacetic acid to 1-phenylpropene, which produces approximately equal yields of the cis and trans adducts. Peterson, P. E.; Duddey, J. E. *J. Am. Chem. Soc.* **1966**, *88*, 4990. In addition to the vinyl esters, hexaethylbenzene was also formed, with the proportion of this product increasing with increasing initial concentration of 1-phenylpropyne.

The k_{Br^-} term in equation 9.55 is significant only in the presence of added Br^-. The ratio of **54** to **55** was found to be 4.2 : 1 in the absence of added bromide ion, but the ratio increased with increasing bromide ion concentration, and essentially 100% anti addition was observed at the highest concentrations of bromide ion studied.[275] These observations led the investigators to conclude that there could be an Ad_E3 mechanism (equation 9.59) in the presence of bromide ion.

$$C_6H_5{-}C{\equiv}C{-}CH_3 \xrightleftharpoons[\text{HOAc, fast}]{Br_2} \left[C_6H_5{-}C{\equiv}C{-}CH_3 \cdot Br_2 \right] \xrightarrow[\text{slow}]{Br^-} \left[C_6H_5{-}C{\equiv}C{-}CH_3 \ (Br^{\delta-}\cdots Br \cdots Br^{\delta-}) \right]^{\ddagger} \longrightarrow (C_6H_5)(Br)C{=}C(Br)(CH_3) + Br^- \quad (9.59)$$

In contrast to the results for 1-phenylpropyne, both 1-hexyne and 3-hexyne formed only trans bromine addition products. The reaction appears to occur through the intermediacy of a bromonium ion (as shown in equation 9.60) in the absence of bromide ion, or an Ad_E3 mechanism in the presence of bromide ion.[272]

$$\underset{\mathbf{58}}{C_2H_5{-}C{\equiv}C{-}C_2H_5} \xrightleftharpoons{Br_2} \left[C_2H_5{-}C{\equiv}C{-}C_2H_5 \cdot Br_2\ (Br^-) \right] \rightleftharpoons \left[C_2H_5{-}\overset{Br^+}{C{=}C}{-}C_2H_5 \ \ Br^- \right] \longrightarrow \underset{\mathbf{59}}{(C_2H_5)(Br)C{=}C(Br)(C_2H_5)} \quad (9.60)$$

In general, alkynes with alkyl substituents on the acetylenic carbons are more likely to react through cyclic 'onium ions than are those with phenyl groups capable of stabilizing vinyl cations. The tendency for the reaction to proceed through an open cation instead of a cyclic 'onium ion also

[275] Similarly, the addition of bromine to phenylacetylene in acetic acid was found to give a mixture of (*E*)- to (*Z*)-1,2-dibromo-1-phenylethene that varied from 70 : 30 without added bromide to 97 : 3 with 0.1 M LiBr. The ratio was found to be 82 : 18 in chloroform solution. König, J.; Wolf, V. *Tetrahedron Lett.* **1970**, 1629.

increases along the series I, Br, Cl, as the bridging ability of the halogen decreases.[276,277] Mixed halogens add to alkynes also, and the kinetics of the reactions can be complex, especially when the alkene or alkyne reacts slowly and considerable assistance by other molecules of mixed halogen is necessary to evoke reaction.[278]

Hydrogen halides also add to alkynes. The addition of HBr to alkynes can be difficult to interpret because (as with alkenes) both ionic and free radical mechanisms occur, and the free radical process can be difficult to suppress.[279] Reaction of HBr with propyne in the liquid phase at −78° led to the formation of *cis*-1-bromopropene (equation 9.61), indicating stereoselective anti addition.[280] When the reaction was carried out at room temperature, however, a mixture of cis and trans isomers was obtained (equation 9.62). The results suggested that the intermediate vinyl radical (from attachment of a bromine atom to the terminal carbon atom of the 1-alkyne) undergoes isomerization with a half-life on the order of 10^{-7} sec.[281]

$$CH_3{-}C{\equiv}CH \xrightarrow[-78°]{HBr} \text{(H}_3\text{C)(H)C=C(Br)(H)} \tag{9.61}$$

60 → **61**

$$CH_3{-}C{\equiv}CH \xrightarrow[\text{room temperature}]{HBr} \text{(H}_3\text{C)(H)C=C(Br)(H)} + \text{(H}_3\text{C)(Br)C=C(H)(H)} \tag{9.62}$$

60 → **62** + **63**

The addition of HCl to alkynes in acetic acid solution does take place by a heterolytic pathway. By analogy to the addition of electrophiles to alkenes, one might write a mechanism involving a vinyl cation, which can then combine with the nucleophile anion to produce the adduct (equation 9.63).[273] Vinyl cations are much less stable than are 2° or 3° alkyl cations and allyl or benzyl cations.[282] Therefore, mechanisms involving electrophilic addition to

---

[276]Bassi, P.; Tonellato, U. *J. Chem. Soc., Perkin Trans. 1* **1973**, 669.

[277]Addition of I_2 appears to occur by anti addition also. (See reference 270.) Addition of fluorine to alkynes occurs to give the tetrafluoroalkane, even when the fluorinating reagent is XeF_2: Zupan, M.; Pollak, A. *J. Org. Chem.* **1974**, *39*, 2646.

[278]For example, the addition of ICl to ethyl butyne-3-oate in propionic acid has been reported to occur by a combination of Ad_E3 and Ad_E4 pathways. Tendil, J.; Verney, M.; Vessiere, R. *Tetrahedron* **1974**, *30*, 579.

[279]Fahey, R. C.; Lee, D.-J. *J. Am. Chem. Soc.* **1968**, *90*, 2124 and references therein.

[280]Skell, P. S.; Allen, R. G. *J. Am. Chem. Soc.* **1958**, *80*, 5997.

[281]Skell, P. S.; Allen, R. G. *J. Am. Chem. Soc.* **1964**, *86*, 1559.

[282]Some authors have estimated that vinyl cations are comparable in stability to methyl or, perhaps, primary carbocations. For a discussion, see Weiss, H. M. *J. Chem. Educ.* **1993**, *70*, 873. See also Tidwell, T.T. *J. Chem. Educ.* **1996**, 73, 1081; Weiss, H.M. *J. Chem. Educ.* **1996**, *73*, 1082.

alkynes are most feasible when one of the acetylenic substituents is capable of stabilizing the incipient carbocation by resonance.

$$R-C\equiv C-R \xrightarrow{E-Nu} \left[(E)(R)C=C^{+}-R \quad Nu^{-} \right] \longrightarrow (E)(R)C=C(R)(Nu) \qquad (9.63)$$

The reaction of 1-phenylpropyne with 0.1 M HCl in acetic acid solution produced five products, as shown in equation 9.64.[283] The major product (**64**) is the result of syn addition of HCl, with attachment of the proton to C2 of the reactant. There is also a stereoisomer (**65**) with the same regiochemistry, indicating some anti addition of HCl. Both of these products are consistent with the intermediacy of a vinyl cation–chloride ion pair. There are trace amounts of the regioisomers **66** and **67**, which apparently are formed by electrophilic addition with initial protonation of Cl of the alkyne.[284]

$$\underset{\mathbf{53}}{C_6H_5-C\equiv C-CH_3} \xrightarrow[HOAc]{HCl} \underset{\substack{\mathbf{64}\\ 81\%}}{(C_6H_5)(Cl)C=C(CH_3)(H)} + \underset{\substack{\mathbf{65}\\ 8\%}}{(C_6H_5)(Cl)C=C(H)(CH_3)} + \underset{\substack{\mathbf{66}\\ <0.3\%}}{(C_6H_5)(H)C=C(Cl)(CH_3)} \qquad (9.64)$$

$$+ \underset{\substack{\mathbf{67}\\ <0.3\%}}{(C_6H_5)(H)C=C(CH_3)(Cl)} + \underset{\substack{\text{(from HOAc addition)}\\ \mathbf{68}\\ (11\%)}}{C_6H_5-C(=O)-CH_2CH_3}$$

If 0.5 M chloride ion (tetramethylammonium chloride, TMAC) is added to the reaction mixture, the rate of the reaction shown in equation 9.64 increases by a factor of five, and the product distribution is altered, as shown in equation 9.65. Now there is a much greater proportion of **65** and **66**.

---

[283]Fahey, R. C.; Payne, M. T.; Lee, D.-J. *J. Org. Chem.* **1974**, *39*, 1124.

[284]Propiophenone is also formed in the reaction. Apparently solvent reacts with the vinyl cation intermediate to produce the vinyl acetate. The vinyl acetates are thought to react with trace water to hydrolyze to unstable enols, which tautomerize to the ketone.

$$(C_6H_5)(AcO)C=C(CH_3)(H) \text{ and } (C_6H_5)(AcO)C=C(H)(CH_3) \xrightarrow[\text{trace}]{H_2O} C_6H_5-C(=O)-CH_2CH_3$$

$$C_6H_5{-}C{\equiv}C{-}CH_3 \xrightarrow[\text{HOAc, 0.5 M TMAC}]{\text{HCl}} (C_6H_5)(Cl)C{=}C(CH_3)(H) + (C_6H_5)(Cl)C{=}C(H)(CH_3) + (C_6H_5)(H)C{=}C(Cl)(CH_3) \tag{9.65}$$

53 → **64** (36%) + **65** (46%) + **66** (10.6%)

$$+ (C_6H_5)(H)C{=}C(CH_3)(Cl) + C_6H_5{-}C({=}O){-}CH_2CH_3$$

67 (<0.3%) + **68** (7%) (from HOAc addition)

To account for the rates of reaction and product distribution observed in these and other studies, Fahey and co-workers suggested that two competitive processes are involved.[283] As shown in equations 9.66 through 9.68, both processes begin with the formation of an alkyne-HCl complex or its kinetic equivalent, an association of the alkyne and HCl in a solvent cage.[285] If the developing carbocation is sufficiently stable, the complex can proceed to a vinyl cation–chloride ion pair, leading to product formation by an Ad_E2 mechanism (equation 9.67). Alternatively, the complex can react with added chloride ion to form product by an Ad_E3 mechanism (equation 9.68). In the absence of added chloride ion, another molecule of HCl may serve the same purpose.[286]

$$-C{\equiv}C- \underset{\text{HOAc, fast}}{\overset{\text{HCl}}{\rightleftharpoons}} \left[-C{\equiv}C- \cdots HCl \right] \tag{9.66}$$

$$\left[-C{\equiv}C- \cdots HCl \right] \rightleftharpoons \left[Cl^- \; -\overset{+}{C}{=}C(H)- \right] \rightleftharpoons \text{products} \tag{9.67}$$

---

[285]Evidence for the existence of alkyne-HCl complexes has been reported by Mootz, D.; Deeg, A. *J. Am. Chem. Soc.* **1992**, *114*, 5887, who obtained an X-ray structure for a 1 : 1 complex of HCl and 2-butyne. The complex was found to be a T-shaped structure, with the H—Cl bond axis perpendicular to the axis of the carbon-carbon triple bond. The authors also obtained an X-ray structure for a 2 : 1 HCl : 2-butyne complex.

[286]Fahey and Lee (reference 279) found the addition of HCl to 3-hexyne in acetic acid to be termolecular, first order in the alkyne and second order in HCl. However, the reaction showed primarily anti addition at 25°, although there was more syn addition at higher temperatures. Similar considerations apply in the addition of other electrophiles to alkynes. For a discussion, see reference 273.

$$\left[\begin{array}{c}\text{HCl}\\ \uparrow\\ -\text{C}\equiv\text{C}-\end{array}\right] \xrightleftharpoons{\text{Cl}^-} \left[\begin{array}{c}\text{H}\cdots\text{Cl}\\ \vdots\\ -\text{C}\equiv\text{C}-\\ \vdots\\ \text{Cl}^-\end{array}\right] \rightleftharpoons \text{products} \qquad \textbf{(9.68)}$$

The complexities associated with the addition of hydrogen halides to alkynes in homogeneous solution limit the synthetic utility of these reactions. However, Kropp and Crawford found that the surface-mediated addition of hydrogen halides to alkynes proceeds more rapidly and gives less complex product mixtures.[287] For example, addition of HCl to 1-phenylpropyne on a suspension of alumina in CH_2Cl_2 gave a high yield of (*E*)-1-chloro-1-phenylpropene at short reaction times, but at longer reaction times, product equilibration produced a high yield of the (*Z*)-diastereomer (equation 9.69). Furthermore, surface-mediated addition of HBr to alkynes was unaccompanied by products of radical reaction.[287]

$$\underset{\textbf{53}}{C_6H_5-C\equiv C-CH_3} \xrightarrow[\text{alumina, } CH_2Cl_2]{\text{HCl}} \underset{\textbf{64}}{(C_6H_5)(Cl)C=C(CH_3)(H)} \rightleftharpoons \underset{\textbf{65}}{(C_6H_5)(Cl)C=C(H)(CH_3)} \qquad \textbf{(9.69)}$$

Product ratio, t = 0.08 hr 12 : 1
Product ratio, t = 3 hr 0.03 : 1

Dialkylacetylenes and phenylalkylacetylenes also add mercuric acetate in acetic acid solution. The products point to exclusively anti addition, but with a regiochemistry that depends upon the nature of the alkyl group.[288,289] The results suggest that the solvomercuration of both alkynes and alkenes appears to take place through mercurinium ions (**69**).[290]

$$C_6H_5-C\equiv C-R \xrightarrow[\text{HOAc}]{Hg(OAc)_2} \xrightarrow{\text{aq. KCl}} \underset{\textbf{56}}{(C_6H_5)(AcO)C=C(HgCl)(R)} + \underset{\textbf{57}}{(C_6H_5)(ClHg)C=C(OAc)(R)} \qquad \textbf{(9.70)}$$

---

[287] Kropp, P. J.; Crawford, S. D. *J. Org. Chem.* **1994**, *59*, 3102.

[288] See, for example, Uemura, S.; Miyoshi, H.; Okano, M. *J. Chem. Soc., Perkin Trans.* **1980**, *1*, 1098.

[289] The kinetics of the addition are first order in alkyne and first order in the mercurating agent. Bassetti, M.; Floris, B. *J. Org. Chem.* **1986**, *51*, 4140.

[290] Diphenylacetylene, however, gives exclusively syn addition. See the discussion in reference 288; also see Bach, R. D.; Woodard, R. A.; Anderson, T. J.; Glick, M. D. *J. Org. Chem.* **1982**, *47*, 3707.

$$C_6H_5-C\equiv C-R \xrightarrow[HOAc]{Hg(OAc)_2} \mathbf{69} \xrightarrow{aq.\ KCl} (C_6H_5)(AcO)C=C(HgCl)(R) + (C_6H_5)(ClHg)C=C(OAc)(R) \quad (9.71)$$

Alkynes also undergo acid-catalyzed hydration to yield ketones. Alkynes substituted with alkoxy or phenoxy groups undergo hydration readily, and butyl acetate was obtained by heating butoxyacetylene in distilled water.[291] Phenylacetylene and 1-phenylpropyne hydrate in solutions containing about 50% sulfuric acid in water,[292,293] and even acetylene can be hydrated to acetaldehyde in sufficiently acidic solutions.[294] The hydrations are acid-catalyzed, and a solvent deuterium isotope effect on the hydration of 1-phenylpropyne indicates that proton transfer occurs during the rate-limiting step. Therefore, the data are consistent with the intermediacy of a vinyl cation, as illustrated for 1-phenylpropyne in equation 9.72.

$$C_6H_5-C\equiv C-CH_3 \xrightarrow[H_2O,\ slow]{H_2SO_4} [C_6H_5-\overset{+}{C}=C(H)(CH_3)] \xrightleftharpoons{H_2O} [(C_6H_5)(H_2O^+)C=C(H)(CH_3)] \xrightleftharpoons{-H^+} [(C_6H_5)(HO)C=C(H)(CH_3)] \rightleftharpoons C_6H_5C(=O)C_2H_5 \quad (9.72)$$

The hydration of alkynes can be carried out in less acidic solutions if catalysts such as Hg^{2+} are used.[295,296] The kinetics of the reaction in aqueous sulfuric acid are complex,[296] but there is evidence for the formation of a bis-(alkyne)-Hg^{2+} complex in aqueous dioxane solutions.[297] Many textbooks do not indicate a mechanism for the mercury-catalyzed hydration, and

[291] Jacobs, T. L.; Searles, Jr., S. *J. Am. Chem. Soc.* **1944**, *66*, 686 and references therein.

[292] Noyce, D. S.; Schiavelli, M. D. *J. Org. Chem.* **1968**, *33*, 845.

[293] Noyce, D. S.; Matesich, M. A.; Schiavelli, M. D.; Peterson, P. E. *J. Am. Chem. Soc.* **1965**, *87*, 2295.

[294] Lucchini, V.; Modena, G. *J. Am. Chem. Soc.* **1990**, *112*, 6291.

[295] Hess, K.; Munderloh, H. *Chem. Ber.* **1918**, *51*, 377.

[296] For a discussion, see Rutledge, T. F. *Acetylenes and Allenes. Addition, Cyclization, and Polymerization Reactions*; Reinhold Book Corporation: New York, 1969; pp. 124–125.

[297] Budde, W. L.; Dessy, R. E. *J. Am. Chem. Soc.* **1963**, *85*, 3964; *Tetrahedron Lett.* **1963**, 651.

those that do may differ in the mechanisms presented. In particular, the intermediate has been shown as a π-complex, a "mercurinium" ion, and a vinyl cation with a mercury ion bonded to C2. By analogy with bromonium ions and the mercurinium ion in oxymercuration, it seems reasonable that the distinction between a π-complex and a mercurinium ion may be in part a function of the valence bond structures with which we draw a structure—for an asymmetrically substituted alkyne, at least. Therefore the intermediate in equation 9.73 for the hydration of a 1-alkyne is represented as an asymmetric bridged ion. Attack by water on the alkyne carbon likely to have the greater partial positive charge[298] leads to a vinyl mercury species that decomposes in acid to the enol, which in turn tautomerizes to the ketone.[299,300]

$$R{-}C{\equiv}C{-}H \xrightarrow[HgSO_4,\ H_2O]{H_2SO_4} \left[R{-}C{\equiv}C{-}H \cdots HgX^{+},\ H_2O \right] \xrightarrow[slow]{H_2O} \left[(H_2O^{+})(R)C{=}C(H)(HgX) \right] \xrightleftharpoons{-H^{+}} \left[(HO)(R)C{=}C(H)(HgX) \right] \xrightleftharpoons[-Hg^{++}]{H_3O^{+}} \left[(HO)(R)C{=}CH_2 \right] \rightleftharpoons R{-}CO{-}C_2H_5 \qquad (9.73)$$

Alkynes readily undergo hydroboration, and product stereochemistry indicates syn addition. It thus appears that the reaction involves a four-center transition structure, as does the hydroboration of alkenes. Dihydroboration occurs readily in the case of terminal alkynes, so sterically hindered hydroborating agents such as disiamylborane are used to effect monohydroboration. The hydroboration reaction appears to give the regiochemical result analogous to that observed with alkenes, in which the boron atom becomes attached to the carbon atom bearing fewer alkyl substituents. Therefore, oxidation of the product from reaction of a terminal alkyne with disiamylborane is an efficient means for the synthesis of aldehydes.

---

[298] More specifically, the greater nucleophilic susceptibility.

[299] Myddleton, W. W.; Barrett, A. W.; Seager, J. H. *J. Am. Chem. Soc.* **1930**, *52*, 4405.

[300] In this equation, the representation of the mercury species bonded to carbon as HgX is a convenience so that the coordination of anions about the atom can be ignored.

$$CH_3(CH_2)_5{-}C{\equiv}C{-}H \xrightarrow[\text{Diglyme}]{\text{Disiamylborane}} \xrightarrow[\substack{\text{aq. NaOH} \\ \text{(pH 7–8)}}]{15\%\ H_2O_2} CH_3(CH_2)_6{-}CHO \quad (9.74)$$

70 → **71**

Hydroboration of unsymmetrical internal alkynes leads to a mixture of regioisomers, but hydrogenolysis of the hydroboration product provides a route to cis alkenes.[301]

$$CH_3{-}C{\equiv}C{-}CH_2CH_3 \xrightarrow[\text{Diglyme}]{\text{Disiamylborane}} \xrightarrow{CH_3CO_2H} \text{cis-}CH_3CH{=}CHCH_2CH_3 \quad (9.75)$$

72 → **73** ($\geq$ 99% cis)

Alkynes also undergo peroxidation reactions. Depending upon the structure of the alkyne, the peracid, solvent, and other conditions, a wide variety of products can be formed. The reaction products are consistent with the formation of an oxirene intermediate (**74**), which can then undergo further oxidation by peracid, addition of solvent, or rearrangement to a ketene.[302,303] For example, reaction of diphenylacetylene with pertrifluoroacetic acid in methylene chloride produced mostly benzil (**76**) and benzoic acid (equation 9.76),[302] but peroxidation with *m*-chloroperoxybenzoic acid in CCl_4 solution produced only a 5.8% yield of benzil, no benzoic acid, and a total of 53% yield of five other products (equation 9.77).[303]

R–C(=C)–R oxirene (O bridging)

74

$$C_6H_5{-}C{\equiv}C{-}C_6H_5 \xrightarrow[\substack{CH_2Cl_2 \\ Na_2HPO_4}]{CF_3CO_3H} C_6H_5{-}C(=O){-}C(=O){-}C_6H_5 + C_6H_5CO_2H \quad (9.76)$$

75 → **76** (76%) + $C_6H_5CO_2H$ (17%)

[301]Brown, H. C.; Zweifel, G. *J. Am. Chem. Soc.* **1959**, *81*, 1512; **1961**, *83*, 3834.

[302]McDonald, R. N.; Schwab, P. A. *J. Am. Chem. Soc.* **1964**, *86*, 4866.

[303]Stille, J. K.; Whitehurst, D. D. *J. Am. Chem. Soc.* **1964**, *86*, 4871.

$$C_6H_5{-}C{\equiv}C{-}C_6H_5 \xrightarrow[CCl_4]{MCPBA} \underset{5.8\%}{C_6H_5{-}\overset{O}{\overset{\|}{C}}{-}\overset{O}{\overset{\|}{C}}{-}C_6H_5} + \underset{10.7\%}{C_6H_5{-}\overset{O}{\overset{\|}{C}}{-}O{-}\underset{C_6H_5}{\underset{|}{CH}}{-}O{-}\overset{O}{\overset{\|}{C}}{-}C_6H_4Cl} + \underset{17.1\%}{C_6H_5{-}\underset{C_6H_5}{\underset{|}{\overset{OH}{\overset{|}{C}}}}{-}CO_2H} \quad \textbf{(9.77)}$$

$$+ \underset{19.2\%}{C_6H_5{-}\overset{O}{\overset{\|}{C}}{-}O{-}O{-}\overset{O}{\overset{\|}{C}}{-}C_6H_5} + \underset{6.4\%}{C_6H_5{-}\overset{O}{\overset{\|}{C}}{-}C_2H_5} + \underset{(trace)}{C_6H_5{-}\overset{O}{\overset{\|}{C}}{-}\underset{C_6H_5}{\underset{|}{CH}}{-}O{-}\overset{O}{\overset{\|}{C}}{-}C_6H_4Cl}$$

$$\left(MCPBA = \text{3-}ClC_6H_4{-}CO_3H\right)$$

Electrophilic additions also occur to cumulated dienes (allenes).[304] Caserio and co-workers established that the mechanisms of both the oxymercuration reaction and the electrophilic addition of bromine to allenes are similar to the corresponding additions to olefins.[305] The oxymercuration of (*R*)-(−)-1,3-dimethylallene (**77**) in methanol produced an 83% yield of (*S*)-*trans*-3-acetoxymercuri-4-methoxy-2-pentene (**78**), confirming the anti pathway for the addition. Also formed was 17% of *cis*-3-acetoxymercuri-4-methoxy-2-pentene (**79**), which was presumed to have the (*R*) configuration.[306] As shown in Figure 9.44, the product ratios can be explained on the basis of preequilibrium formation of mercurinium ions formed by attachment of mercury to either the top or the bottom of one of the double bonds. Subsequent rate-limiting attack of methanol on the mercurinium ions is easier for the pathway in which solvent can avoid the steric hindrance due to the remaining vinyl methyl group, so formation of (*S*)-**78** predominates.

Addition of bromine to the same allene in methanol led to the formation of 83% of the (*S*)-*trans*-3-bromo-4-methoxy-2-pentene (**80**), confirming the anti pathway for the addition of bromine under these conditions. Also formed was 17% of *cis*-3-bromo-4-methoxy-2-pentene (**81**), which was again presumed to have the (*R*) configuration. As shown in Figure 9.45, the same ratio of diastereomers as was observed in the oxymercuration reaction is consistent with an anti addition mechanism that is closely analogous to the mechanism of oxymercuration.

---

[304] See, for example, Fischer, H. in *The Chemistry of Alkenes*; Patai, S., ed.; Wiley-Interscience: London, 1964; pp. 1025–1159.

[305] Waters, W. L.; Linn, W. S.; Caserio, M. C. *J. Am. Chem. Soc.* **1968**, *90*, 6741.

[306] Although the acetoxymercuri compound is shown in Figure 9.44, the investigators treated the initial oxymercuration product with aqueous sodium chloride and isolated the corresponding chloromercuri compounds.

Figure 9.44 Addition of $Hg(OAc)_2$ to (*R*)-(−)-1,3-dimethylallene in methanol.

H_3C C=C=C CH_3 H H
(*R*)-**77**
$Hg(OAc)_2$ / CH_3OH

[H_3C HgOAc CH_3 H H OH H_3C] less steric hindrance → AcOHg CH_3 H_3C H H OCH_3
(*S*)-**78**
83%

[(H_3C) OH H_3C CH_3 H H HgOAc] more steric hindrance → H_3C OCH_3 H CH_3 AcOHg H
(*R*)-**79**
17%

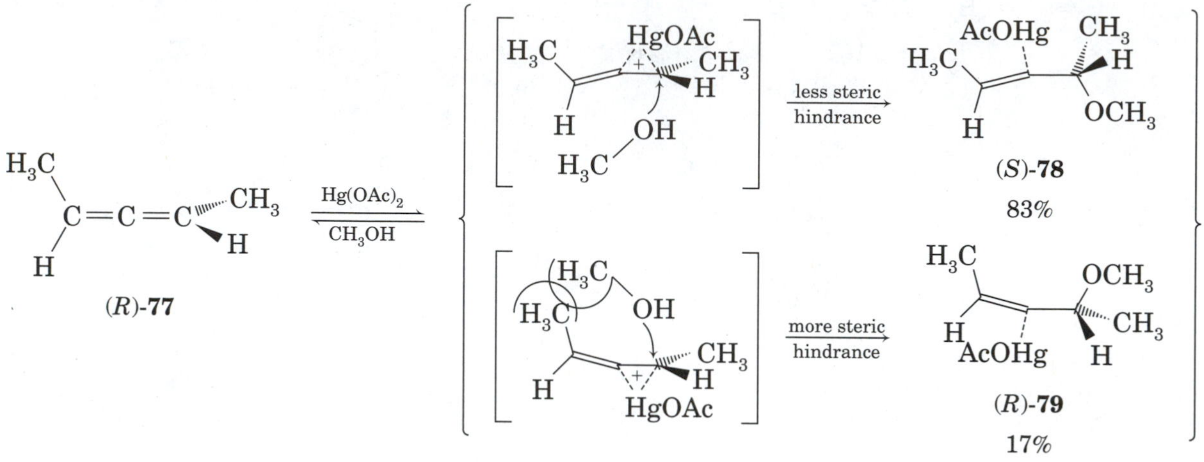

Figure 9.45 Addition of bromine to (*R*)-(−)-1,3-dimethylallene in methanol.

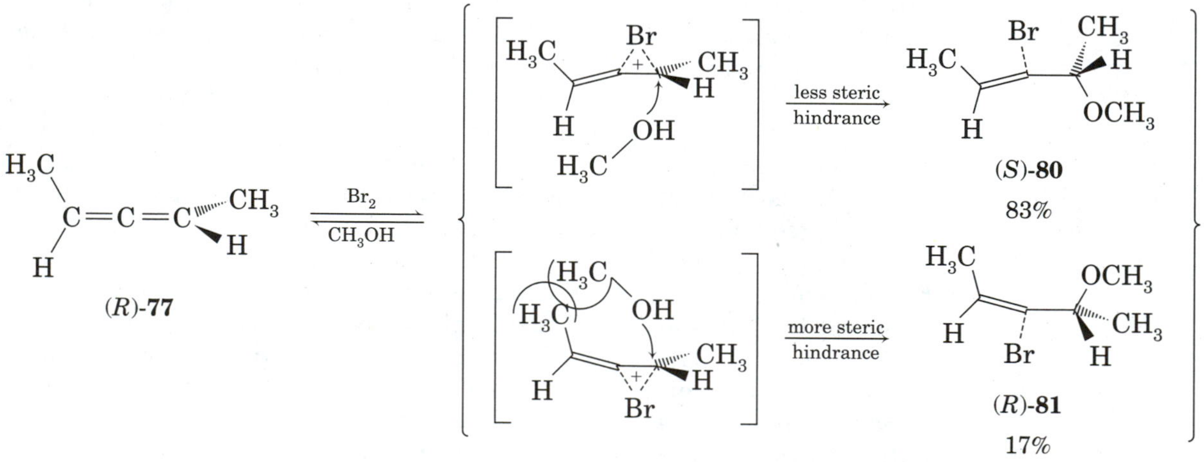

H_3C C=C=C CH_3 H H
(*R*)-**77**
Br_2 / CH_3OH

[H_3C Br CH_3 H H OH H_3C] less steric hindrance → Br CH_3 H_3C H H OCH_3
(*S*)-**80**
83%

[(H_3C) OH H_3C CH_3 H H Br] more steric hindrance → H_3C OCH_3 H CH_3 Br H
(*R*)-**81**
17%

Nucleophilic Addition to Alkenes and Alkynes

To this point we have discussed only electrophilic addition to compounds with carbon-carbon double or triple bonds. Nucleophilic additions to alkenes and alkynes are also possible, but these reactions generally require that the substrate have substituents that can stabilize the resulting carbanionic intermediate. Therefore nucleophilic additions are most likely for compounds with carbon-heteroatom multiple bonds, such as carbonyl compounds, imines and cyano compounds. We will not discuss the synthetic aspects of these reactions at length here, but we will address some mechanistic questions about these reactions.

Two main types of substituents activate alkenes and alkynes for nucleophilic attack. The first type consists of those activating groups (labeled AG in equation 9.78) that can stabilize an adjacent carbanion by induction.[307]

$$Nu^- + R_1R_2C{=}CR_3(AG) \longrightarrow Nu{-}CR_1R_2{-}C^-R_3(AG) \xrightarrow{HNu} Nu{-}CR_1R_2{-}CR_3(AG){-}H + Nu^- \quad (9.78)$$

The synthesis of fluoroalkyl and fluorochloroalkyl ethers occurs readily by base-catalyzed nucleophilic addition of alcohols to haloalkenes. An example is the synthesis of the alkyl 2,2-dichloro-1,1-difluoroethyl ethers (where R is methyl or 1° or 2° alkyl) by Tarrant and Brown.[308]

$$F_2C{=}CCl_2 \xrightarrow{RO^-} \left[RO{-}CF_2{-}C^-Cl_2 \right] \xrightarrow{ROH} RO{-}CF_2{-}CHCl_2 \quad (9.79)$$

Nucleophilic addition can also occur with alkynes that are activated by the presence of groups such as trifluoromethyl. An example is the addition of methoxide to hexafluoro-2-butyne (equation 9.80).[309] Note that in this reaction there is anti addition to the alkyne, a pattern that is frequently (but not always) observed.[310,311]

$$\underset{\mathbf{82}}{F_3C{-}C{\equiv}C{-}CF_3} \xrightarrow[(CH_3CH_2)_3N]{CH_3OH} \underset{\mathbf{83}}{\underset{>97\%\ trans}{(F_3C)(CH_3O)C{=}CH(CF_3)}} \quad (9.80)$$

[307]In some cases, the addition is followed by elimination of one of the olefinic substituents, so the result of addition-elimination is nucleophilic vinylic substitution. Rappoport, Z. *Adv. Phys. Org. Chem.* **1969**, *7*, 1.

[308]Tarrant, P.; Brown, H. C. *J. Am. Chem. Soc.* **1951**, *73*, 1781. For related examples, see Miller, Jr., W. T.; Fager, E. W.; Griswold, P. H. *J. Am. Chem. Soc.* **1948**, *70*, 431; Park, J. D.; Sharrahm, M. L.; Breen, W. H.; Lacher, J. R. *J. Am. Chem. Soc.* **1951**, *73*, 1329.

[309]Raunio, E. K.; Frey, T. G. *J. Org. Chem.* **1971**, *36*, 345.

[310]See, for example, (a) Truce, W. E.; Simms, J. A. *J. Am. Chem. Soc.* **1956**, *78*, 2756; (b) Miller, S. I. *J. Am. Chem. Soc.* **1956**, *78*, 6091.

[311]For discussion of nucleophilic addition to alkynes, see Miller, S. I.; Tanaka, R. in *Selective Organic Transformations*, Vol. I; Thyagarajan, B. S., ed.; Wiley-Interscience: New York, 1970; pp. 143–238; Dickstein, J. I.; Miller, S. I. in *The Chemistry of the Carbon-Carbon Triple Bond*, Part 2; Patai, S., ed.; Wiley-Interscience: Chichester, England, 1978; pp. 843–911.

Some activating groups are able to stabilize the carbanion primarily through resonance, as illustrated in equation 9.81.

$$Nu^- + R_1R_2C{=}C(R_3)C(R_4){=}Z \longrightarrow \left[Nu{-}CR_1R_2{-}\bar{C}(R_3){-}C(R_4){=}Z \longleftrightarrow Nu{-}CR_1R_2{-}C(R_3){=}C(R_4){-}Z^- \right] \xrightarrow{HNu} Nu{-}CR_1R_2{-}CH(R_3){-}C(R_4){=}Z + Nu^- \quad (9.81)$$

For example, addition of the diphenylmethide ion to 1,1-diphenylethene has been shown to occur by nucleophilic addition to the carbon-carbon double bond (equations 9.82 through 9.84).[312] In this case, delocalization of the negative charge over the two phenyl groups stabilizes the intermediate carbanion, thus lowering the activation energy for the addition.

$$(C_6H_5)_2CH_2 \xrightarrow{^-NH_2} (C_6H_5)_2\bar{C}H \quad (9.82)$$

$$(C_6H_5)_2\bar{C}H + H_2C{=}C(C_6H_5)_2 \longrightarrow \left[(C_6H_5)_2CHCH_2\bar{C}(C_6H_5)_2 \longleftrightarrow (C_6H_5)_2CHCH_2C(C_6H_5){=}C_6H_5^- \right] \quad (9.83)$$

[and other resonance structures]

$$(C_6H_5)_2CHCH_2\bar{C}(C_6H_5)_2 \xrightarrow{H^+} (C_6H_5)_2CHCH_2CH(C_6H_5)_2 \quad (9.84)$$

Substituents that are particularly effective at stabilizing nucleophilic addition to alkenes and alkynes are those in which the group C=Z is a C=O (including aldehydes, ketones, esters, acid halide, and other car-

---

[312]Kofron, W. G.; Goetz, J. M. *J. Org. Chem.* **1973**, *38*, 2534.

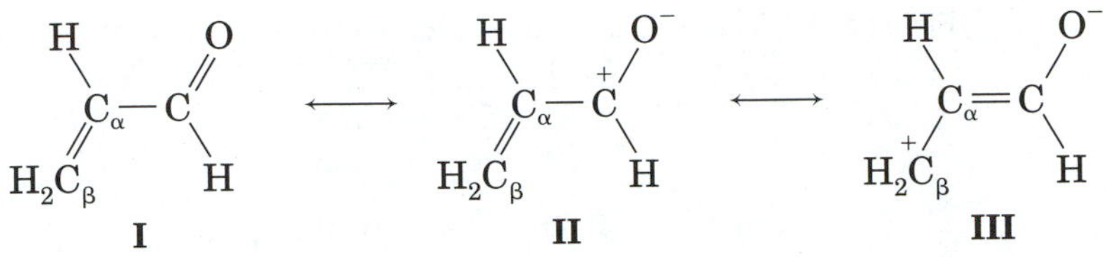

Figure 9.46 Resonance structures for an α,β-unsaturated carbonyl compound.

boxylic acid derivatives except amides), C=NR, S=O and P=O groups.[313] One significant complication with compounds of this type, particularly with carbonyl compounds, is that two different pathways for addition are possible. The resonance structures for the α,β-unsaturated carbonyl compound acrolein (propenal) shown in Figure 9.46 suggest that there should be appreciable positive charge on both the carbonyl carbon atom and the β carbon atom. Therefore, a nucleophile might add directly to the carbonyl carbon atom in a 1,2 addition process, or it could add to the β carbon atom of the α,β-unsaturated carbonyl compound in a 1,4 or conjugate addition reaction, also known as Michael addition.[314,315]

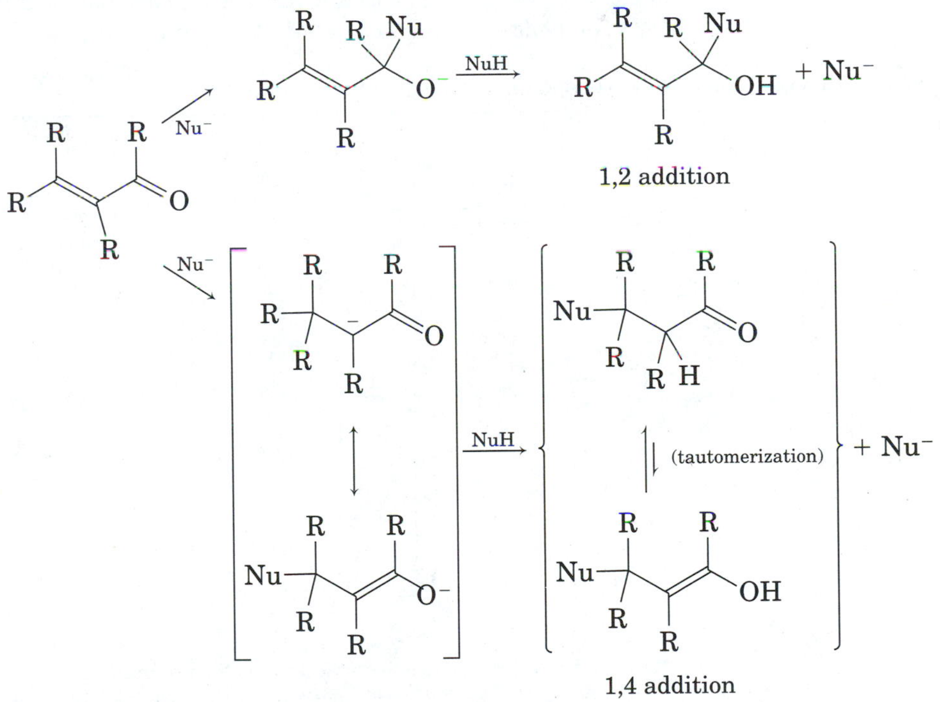

Figure 9.47 1,2 (top) and 1,4 (bottom) addition to an α,β-unsaturated carbonyl compound.

1,2 addition

1,4 addition

[313]Perlmutter, P. *Conjugate Addition Reactions in Organic Synthesis*; Pergamon Press: Oxford, England, 1992.

[314]For a discussion of the terminology of conjugate addition reactions, see reference 313.

[315]For a review of Michael addition in synthesis, see Bergmann, E. D.; Ginsburg, D.; Pappo, R. *Org. React.* **1959**, *10*, 179. For a review of the stereochemistry of the base-promoted Michael addition, see Oare, D. A.; Heathcock, C. H. *Top. Stereochem.* **1989**, *19*, 227–407.

Figure 9.48 LUMO coefficients for acrolein. (Data from reference 319.)

	$H_2C{=}CH{-}CH{=}O$		
	β	α	carbonyl
LUMO coefficient	0.59	-0.48	-0.30

The regiochemistry of nucleophilic addition to an alkene or alkyne stabilized by a carbonyl group is a function of both the substrate and the nucleophile. Hard nucleophiles, such as Grignard reagents, tend to add preferentially at the carbonyl carbon atom.[316] However, soft nucleophiles, such as organocopper compounds, tend to give 1,4 addition. For example, Kharasch and Tawney found that adding a catalytic amount of CuCl to a CH_3MgBr solution in diethyl ether changed the regiochemistry from 1.5% to more than 80% conjugate addition.[317,318] The origin of this change in regiochemistry has been ascribed to the greater LUMO orbital coefficient on the β carbon atom than on the carbonyl carbon atom of an α,β-unsaturated carbonyl compound (Figure 9.48), which makes the softer nucleophile more reactive with the β site.[319] Steric hindrance can also be a factor in determining the preference for 1,2 or 1,4 addition, and a large substituent on the car-

Figure 9.49 Conjugate addition in the Robinson annulation reaction.

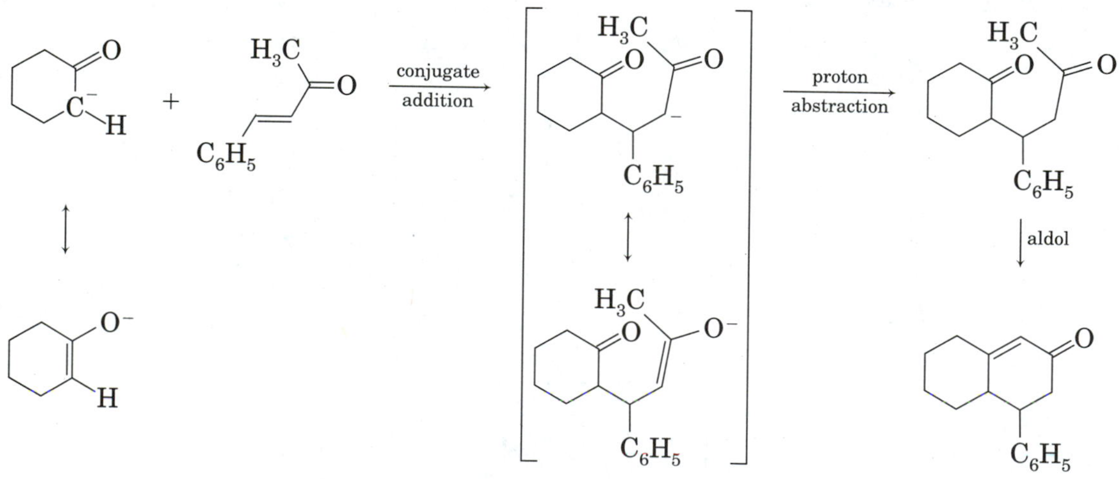

---

[316]Organolithium reagents show an even greater preference for 1,2 addition. Wakefield, B. J. *The Chemistry of Organolithium Compounds*; Pergamon Press: Oxford, England, 1974.

[317]Kharasch, M. S.; Tawney, P. O. *J. Am. Chem. Soc.* **1941**, *63*, 2308.

[318]Organocopper reagents are widely used to effect addition to the β carbon atom of α,β-unsaturated carbonyl compounds. For a discussion, see Posner, G. H *Org. React.* **1972**, *19*, 1. The nature of the organocopper compound is a significant factor in these reactions. See, for example, Lipshutz, B. H.; Wilhelm, R. S.; Kozlowski, J. A. *J. Org. Chem.* **1984**, *49*, 3938.

[319]Fleming, I. *Frontier Orbitals and Organic Chemical Reactions*; Wiley-Interscience: London, 1976; pp. 70, 163.

bonyl carbon atom can greatly increase the yield of conjugate addition product.[320] One of the best-known examples of a conjugate addition is the first step of the Robinson annulation. In an example reported by Robinson (Figure 9.49), conjugate addition of the enolate of cyclohexanone to methyl styryl ketone is followed by an aldol reaction.[321]

As a general rule, nucleophiles add to α,β-unsaturated carbonyl compounds only at the carbonyl carbon atom or the β carbon atom. However, the addition of phenoxide and thiophenoxide to benzoyl(trifluoromethyl)acetylene was found to occur at the α carbon atom (equation 9.85). In this case the anti-Michael orientation is consistent with the observation that both the partial positive charge and the coefficient of the LUMO of the reactant are greater at C_α than at C_β. Also the intermediate carbanion resulting from nucleophilic attachment to C_α in this case is believed to be more stable than the carbanion intermediate generated by nucleophilic attachment to C_β.[322]

$$C_6H_5C(=O)-C_\alpha\equiv C_\beta-CF_3 \xrightarrow[\text{25°}]{C_6H_5O^- \;/\; (CH_3)_3COH} (C_6H_5)C(=O)-C(OC_6H_5)=CH(CF_3) \qquad (9.85)$$

Nucleophilic Addition to Carbonyl Compounds

Regiochemistry is ordinarily not a concern in the addition of nucleophiles to structures having carbon-heteroatom multiple bonds that are not conjugated with carbon-carbon double or triple bonds, because only 1,2-addition is expected.[323] However, if the addition creates a new chiral center, then stereoisomeric addition products can be formed. Considerable interest in this area was sparked by a report by Cram that nucleophilic addition to ketones with chiral α carbon atoms gave unequal yields of the possible diastereomeric adducts. For example, addition of ethyllithium to the methyl ketone (+)-**84** gave primarily the erythro product (+)-**85**.

$$\underset{(+)\text{-}\mathbf{84}}{(C_2H_5)(H_3C)(C_6H_5)C-C(=O)CH_3} \xrightarrow[2.\ H^+]{1.\ CH_3CH_2Li} \underset{\substack{(+)\text{-}\mathbf{85}\\ \text{erythro}\\ \text{major product}}}{(C_2H_5)(H_3C)(C_6H_5)C-C(OH)(CH_3)(CH_2CH_3)} \qquad (9.86)$$

---

[320]See, for example, (a) DeMeester, W. A.; Fuson, R. C. *J. Org. Chem.* **1965**, *30*, 4332; (b) Cooke, Jr., M. P. *J. Org. Chem.* **1986**, *51*, 1638.

[321]Rapson, W. S.; Robinson, R. *J. Chem. Soc.* **1935**, 1285.

[322]Bumgardner, C. L.; Bunch, J. E.; Whangbo, M.-H. *J. Org. Chem.* **1986**, *51*, 4082.

[323]The most important reactions of this type involve addition to the carbonyl group, but reactions involving imines, iminium ions, and nitriles are also synthetically important.

However, addition of methyllithium to the ethyl ketone (+)-**86** produced the threo diastereomer (+)-**87**.[324]

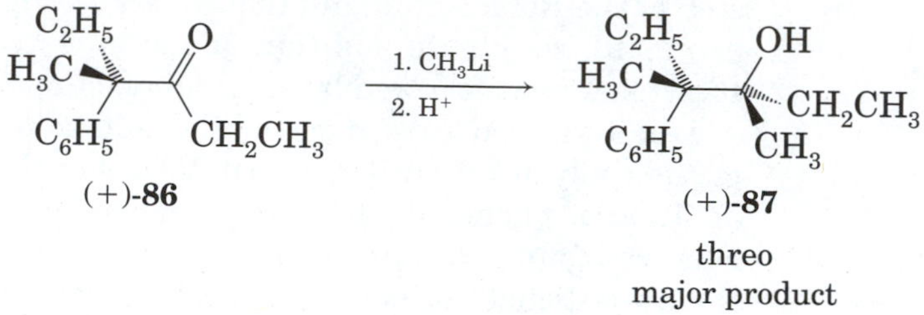

C_2H_5, H_3C, C_6H_5, O, CH_2CH_3 — (+)-**86** → (1. CH_3Li; 2. H^+) → C_2H_5, H_3C, C_6H_5, OH, CH_2CH_3, CH_3 — (+)-**87** threo major product **(9.87)**

To explain this diastereoselectivity, Cram proposed that the asymmetric induction at the carbonyl group is determined by steric effects in the encounter between the ketone and the nucleophilic reagent.[325,326] As shown in Figure 9.50, the preferred conformation of a chiral ketone was considered to be one in which the large substituent (L) is *anti-periplanar* with regard to the carbonyl group, especially if coordination of the oxygen with the metal atom (Z) of the nucleophilic reagent R′—Z increases the effective size of the oxygen atom. The other two substituents on the α carbon atom were designated S and M (for small and medium). The sterically less hindered route of attack, therefore, was thought to bring the incoming nucleophile to the back of the carbonyl on the side of the small substituent. This *rule of steric control of asymmetric induction* was found to correctly predict the major diastereomer in the synthesis of 3-cyclohexyl-2-butanol with methylmagnesium iodide, methyllithium in ether or pentane, lithium aluminum hydride, or sodium borohydride.[327]

For those ketones in which one of the substituents on the α carbon atom is also capable of coordinating with the metal (Z in R′—Z), a second, rigid

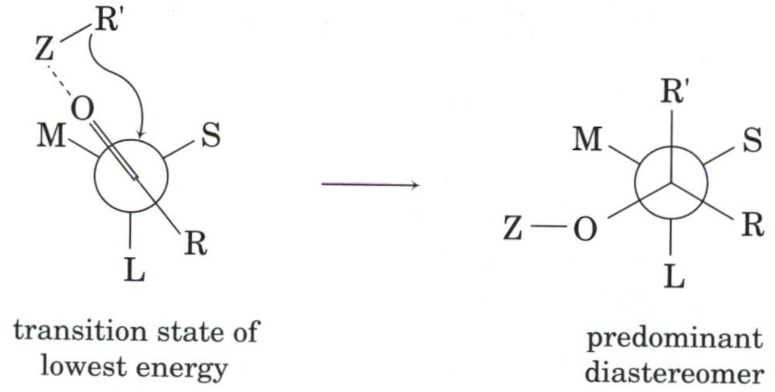

Figure 9.50 Steric model for asymmetric induction in chiral ketones. (Reproduced from reference 326.)

[324] Cram, D. J.; Knight, J. D. *J. Am. Chem. Soc.* **1952**, *74*, 5835.

[325] Cram, D. J.; Abd Elhafez, F. A. *J. Am. Chem. Soc.* **1952**, *74*, 5828.

[326] Cram, D. J.; Kopecky, K. R. *J. Am. Chem. Soc.* **1959**, *81*, 2748.

[327] However, it did not predict the correct product when 3-cyclohexyl-2-butanone was reduced by aluminum isopropoxide. Cram, D. J.; Greene, F. D. *J. Am. Chem. Soc.* **1953**, *75*, 6005.

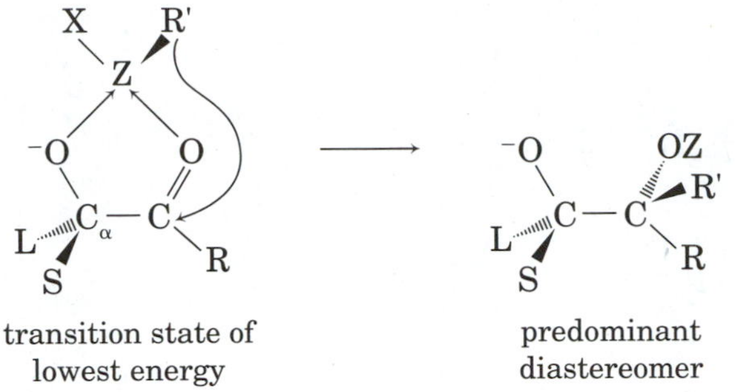

Figure 9.51
Rigid model for diastereoselectivity in nucleophilic addition to carbonyl compounds. (Reproduced from reference 326.)

transition state of lowest energy

predominant diastereomer

model, was proposed.[326] A five-membered ring was said to be formed by the carbonyl carbon atom, the α carbon atom, the electron-donating α substituent, the metal Z, and the carbonyl oxygen atom (Figure 9.51). Again, the nucleophile was considered more likely to attack from the side of the smaller substituent (S) rather than the larger one (L), leading to the diastereoselectivity. This rule correctly predicts the addition of methyllithium to the ketone **88** in equation 9.88,[326] and Eliel and co-workers obtained evidence for the role of chelates in the addition of nucleophiles to α-alkoxy ketones.[328,329,330]

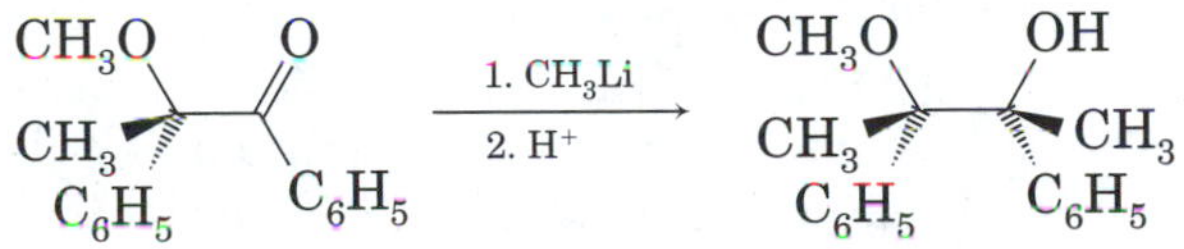

1. CH_3Li
2. H^+

88

89
erythro
major product

(9.88)

The Cram model was questioned by Karabatsos, however, who noted that results for some compounds suggested that the apparent size of methyl was greater than that of isopropyl. Furthermore, he noted that the Curtin-Hammett principle requires knowledge of transition state energies, not just the energies of initial conformations. He proposed an alternative model in which there is a very early transition state and in which the incoming nucleophile prefers to be closest to the smallest group, as shown in the two transition structures in Figure 9.52. Then the *sum* of the steric interactions

---

[328] Frye, S. V.; Eliel, E. L. *J. Am. Chem. Soc.* **1988**, *110*, 484.

[329] Chen, X.; Hortelano, E. R., Eliel, E. L.; Frye, S.V. *J. Am. Chem. Soc.* **1992**, *114*, 1778. A key piece of evidence suggesting that the chelates are true intermediates (and not structures in equilibrium with the reactants but not along the reaction path) was the fact that the most reactive compounds also gave the greatest diastereoselectivity. As the authors noted, this result also suggests an analogy between chelates and enzymatic processes.

[330] Further evidence supporting the role of chelates was reported by Reetz, M. T. *Acc. Chem. Res.* **1993**, *26*, 462 and references therein.

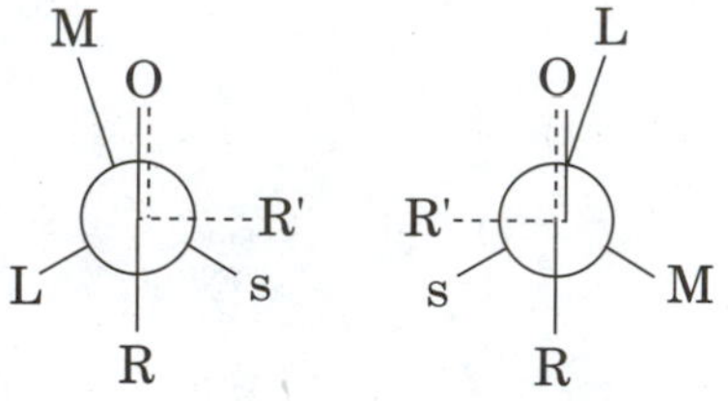

Figure 9.52 Transition structures proposed for asymmetric induction by Karabatsos. (Adapted from reference 331.)

between the other substituents (particularly between the carbonyl oxygen and the substituent M or L on the eclipsed bond) was analyzed to predict correctly the diastereomeric preference of many addition reactions.[331]

Still another model was proposed by Felkin and co-workers, who suggested that the most important steric interaction is between the large group L and the incoming nucleophile.[332] If that is the case, then the lowest energy transition structure is one in which L and Nu are *anti-periplanar*, which means that the nucleophile should approach between the medium and small substituents. The preferred path according to this model, therefore, is the one in which the *small* substituent is closer to the substituent R on the carbonyl carbon atom (see Figure 9.53). The authors also suggested the possibility that there might be some torsional energy associated with interaction of the incipient σ bond with σ bonds already present in the molecule.

The consideration of the development of torsional strain in the transition state for nucleophilic addition afforded an explanation for product distributions observed in the addition of nucleophiles to substituted cyclohexanones. Dauben and co-workers found that reduction of 4-methylcyclohexanone with $LiAlH_4$ produced 81% of the trans alcohol and 19% of the cis isomer.[333] With 4-*t*-butylcyclohexanone, a conformationally more rigid structure, a similar product distribution was obtained.[334] Initially the

Figure 9.53 Favored (left) and disfavored (right) pathways for attachment of nucleophile according to the Felkin model. (Reproduced from reference 339.)

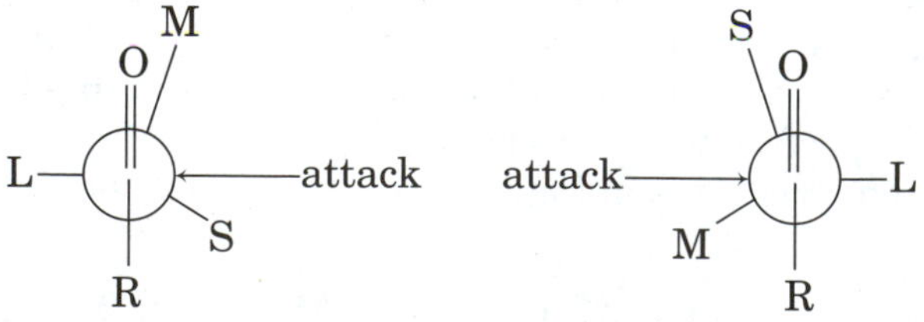

[331] Karabatsos, G. J. *J. Am. Chem. Soc.* **1967**, *89*, 1367.

[332] Chérest, M.; Felkin, H.; Prudent, N. *Tetrahedron Lett.* **1968**, 2199.

[333] Dauben, W. G.; Fonken, G. J.; Noyce, D. S. *J. Am. Chem. Soc.* **1956**, *78*, 2579 and references therein.

[334] Cieplak, A. S. *J. Am. Chem. Soc.* **1981**, *103*, 4540.

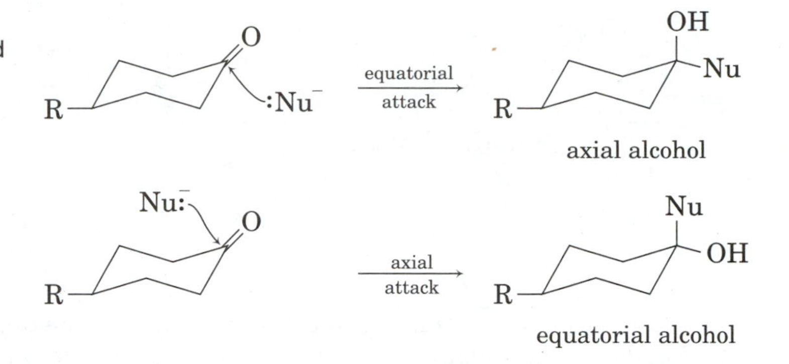

Figure 9.54 Equatorial (top) and axial (bottom) attack of a nucleophile on a cyclohexanone carbonyl group.

results were explained in terms of *product development control* in which the more stable (equatorial) alcohol is formed by axial attack of the nucleophile on the carbon-oxygen double bond (Figure 9.54). With larger nucleophiles, relatively larger yields of product from equatorial attack could then be attributed to a greater steric barrier for axial attack of the nucleophile.[333]

Most theoretical models now indicate that the transition state in nucleophilic addition to the carbonyl group is very early, so product distributions need not reflect the thermodynamic stability of products. As an alternative explanation, Felkin proposed that equatorial attack of a nucleophile (leading to axial alcohol) involves torsional strain between the incipient Nu···C bond and the C_α—H bond. Axial attack of the nucleophile (leading to equatorial alcohol) should also involve steric interaction with ring substituents. The product observed in a particular case, then, is expected to be the product for which the path of attack is lower in energy.[335]

Although the models considered to this point emphasized steric interactions, Nguyên Trong Anh and Eisenstein and co-workers proposed that stereoelectronic factors are comparable in importance.[336] *Ab initio* calculations suggested that the electron cloud around the carbonyl group is dissymmetric in chiral aldehydes and ketones and that this dissymmetry might be a factor in nucleophilic attack. *Ab initio* calculations at the STO-3G level led to the conclusion that the approach of the nucleophile is not along a path perpendicular to the carbon-oxygen double bond but instead is closer to an O—C···Nu bond angle of 109.5°.[337] Nguyên Trong Anh and Eisenstein proposed that the group *anti-periplanar* to the incoming nucleophile is not necessarily the largest group but is, instead, the group with the lowest energy σ* orbital.[338] Now the preferred pathway for attack of the nu-

---

[335]Chérest, M.; Felkin, H. *Tetrahedron Lett.* **1968**, 2205.

[336]Nguyên Trong Anh; Eisenstein, O.; Lefour, J.-M.; Trân Huu Dâu, M-E. *J. Am. Chem. Soc.* **1973**, *95*, 6146.

[337]Bürgi, H. B.; Dunitz, J. D.; Shefter, E. *J. Am. Chem. Soc.* **1973**, *95*, 5065; Bürgi, H. B.; Lehn, J. M.; Wipff, G. *J. Am. Chem. Soc.* **1974**, *96*, 1956.

[338]Nguyên Trong Anh; Eisenstein, O. *Nouv. J. Chem.* **1977**, *1*, 61.

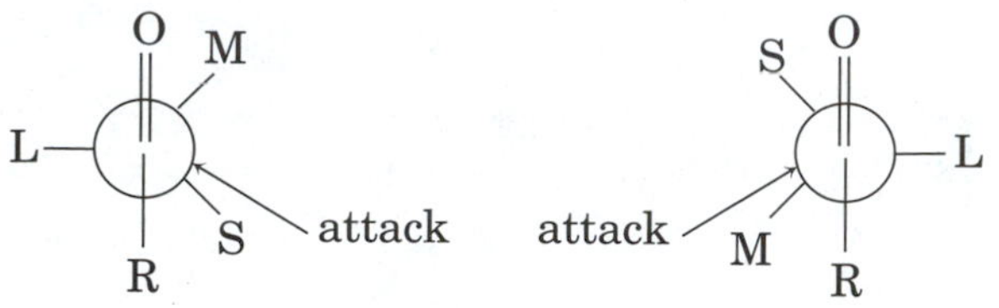

Figure 9.55 Lower (left) and higher (right) energy transition structures predicted by Nguyên Trong Anh and Eisenstein model. (Reproduced from reference 339.)

cleophile is near the small group rather than the medium group (Figure 9.55). Also, some stabilization may arise from mixing of the developing σ orbital with a σ* orbital on the α carbon atom. The more electronegative the group *anti-periplanar* to the incoming nucleophile, the lower the energy of that σ* orbital, and the more stable the transition structure (Figure 9.56).[339]

A rather different model for predicting the addition of nucleophiles to carbonyl compounds was later advanced by Cieplak.[334] According to this model, the preferred transition structure is not the one that is less *disfavored* by steric interactions but is instead the one that is the *more favored* by electronic interaction between the σ* orbital of the *developing* nucleophile-carbon bond with a σ bond in the molecule (Figure 9.57). Cieplak suggested that the energy associated with such an interaction should be comparable to that of steric interactions, that is, from a few tenths of a kcal/mol to a few kcal/mol. Cieplak and Johnson concluded that this model best fits the known data for nucleophilic addition to carbonyl groups substituted with electron-withdrawing groups.[340] The Cieplak model has been supported by results of le Noble,[341] Meyers,[342] and Danishefsky.[343] However, electrostatic effects, and not electronic effects (on the energy of the C_α—X σ

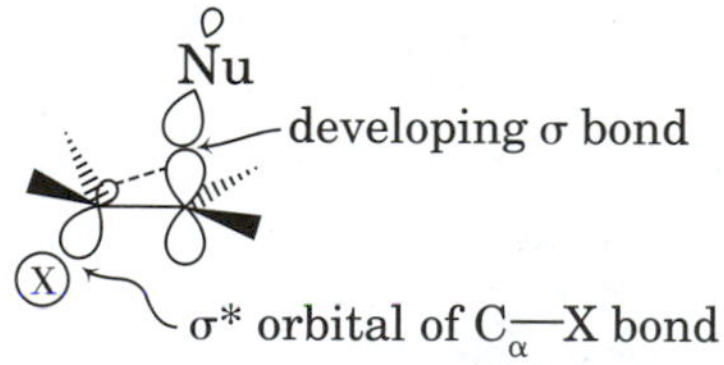

Figure 9.56 Stabilization of developing σ bond by adjacent σ* orbital. (Adapted from reference 340.)

---

[339]Lodge, E. P.; Heathcock, C. H. *J. Am. Chem. Soc.* **1987**, *109*, 3353 tested the theoretical models by determining diastereoselectivity in the addition of a lithium enolate to a series of chiral aldehydes. Their results were in general agreement with the Nguyên Trong Anh-Eisenstein adaptation of the Felkin model, but they concluded that steric effects were as important as electronic effects (energy of the σ* orbital) in determining the transition state for attack of the nucleophile on the carbonyl carbon atom.

[340]Cieplak, A. S.; Tait, B. D.; Johnson, C. R. *J. Am. Chem. Soc.* **1989**, *111*, 8447.

[341]Cheung, C. K.; Tseng, L. T.; Lin, M.-H.; Srivastava, S.; le Noble, W. J. *J. Am. Chem. Soc.* **1986**, *108*, 1598; Srivastava, S.; le Noble, W. J. *J. Am. Chem. Soc.* **1987**, *109*, 5874.

[342]Meyers, A. I.; Romine, J. L.; Fleming, S. A. *J. Am. Chem. Soc.* **1988**, *110*, 7245.

[343]Danishefsky, S.; Langer, M. E. *J. Org. Chem.* **1985**, *50*, 3674.

Figure 9.57 Stabilization of transition structure by interaction of developing σ* orbital with σ orbital on α carbon atom. (Adapted from reference 340.)

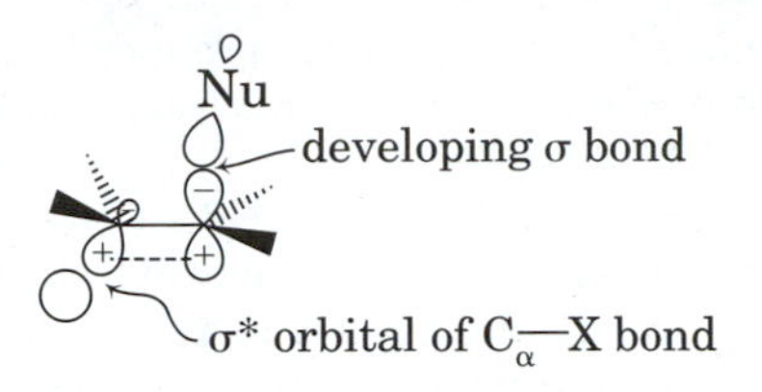

bond) have been proposed as an explanation for the stereoselectivity of addition of nucleophiles to 2,3-disubstituted 7-norbornanones.[344,345,346]

There has continued to be interest in diastereoselective nucleophilic addition in recent years. Cieplak and Wiberg suggested that theoretical calculations representing gas phase addition of nucleophiles may not accurately represent solution processes.[347] However, Squires reported that addition of hydride ions to carbonyls in the gas phase gave the same patterns of diastereoselectivity observed in solution reductions.[348] There has also been some consideration that single electron transfer might play a role in nucleophilic additions, particularly those in which high diastereoselectivity is not observed.[349] Dannenberg and co-workers have discussed the use of polarized π-frontier molecular orbitals to predict the diastereofacial selectivities of addition reactions.[350] All of these discussions about the stereochemistry of nucleophilic addition to carbonyl compounds provide contemporary examples of the utility of alternative models in the theory and practice of organic chemistry.[351]

Problems

1. In the addition of bromine to 1,2-diarylethenes, why is the observation that addition of Br_2 to *trans*-stilbene produces almost exclusively meso product not sufficient to show that the mechanism of the addition is anti? That is, why is it necessary to study addition to the cis isomer also?

---

[344]Wong, S. S.; Paddon-Row, M. N. *Aust. J. Chem.* **1991**, *44*, 765.

[345]Paddon-Row, M. N.; Wu, Y.-D., Houk, K. N. *J. Am. Chem. Soc.* **1992**, *114*, 10638.

[346]Adcock, W.; Cotton, J.; Trout, N. A. *J. Org. Chem.* **1994**, *59*, 1867.

[347]Cieplak, A. S.; Wiberg, K. B. *J. Am. Chem. Soc.* **1992**, *114*, 9226.

[348]Ho, Y.; Squires, R. R. *J. Am. Chem. Soc.* **1992**, *114*, 10961.

[349]Rein, K. S.; Chen, Z.-H.; Perumal, P. T.; Echegoyen, L.; Gawley, R. E. *Tetrahedron Lett.* **1991**, *32*, 1941. See also Rychnovsky, S. D. *Chemtracts-Org. Chem.* **1991**, *4*, 456 and references therein.

[350]Huang, X. L.; Dannenberg, J. J.; Duran, M.; Bertrán, J. *J. Am. Chem. Soc.* **1993**, *115*, 4024; Huang, X. L.; Dannenberg, J. J. *J. Am. Chem. Soc.* **1993**, *115*, 6017. For a comment, see reference 340.

[351]For a computer-assisted method for predicting the outcome of diastereoselective additions, see Fleischer, J. M.; Gushurst, A. J.; Jorgensen, W. L. *J. Org. Chem.* **1995**, *60*, 490.

2. Predict the product of the reaction of *trans*-cinnamic acid with a mixture of Br_2 and Cl_2 in $CHCl_3$ solution.
3. A reaction mixture composed of 5×10^{-2} M Br_2 and 2.5×10^{-2} M cyclohexene in acetonitrile (containing less than 0.5% water) produced a 90% yield of a product identified as *trans*-1-(*N*-acetylamino)-2-bromocyclohexane. Propose a mechanism for the formation of this compound.
4. Refluxing phenylacetaldehyde in acetic anhydride with some potassium acetate leads to the formation of a product, **A**, with molecular formula $C_{10}H_{10}O_2$. Treatment of **A** with bromine in CCl_4 leads to a solution containing compound **B**. Adding methanol and allowing the mixture to sit at room temperature leads to a solution from which can be isolated compound **C**, which has the molecular formula $C_{10}H_{13}O_2Br$. Identify compounds **A**, **B**, and **C**.
5. When the disodium salt of (*E*)-2,3-dimethyl-2-butenedioic acid is allowed to react with chlorine water, a product described as a chloro β-lactone is isolated. When the (*Z*) diastereomer reacts under the same conditions, an isomeric product is obtained. What are the structures of the products, and how are they formed?
6. Addition of bromine to *cis*-β-methylstyrene has been found to give primarily anti addition in solvents of low polarity (e.g., acetic acid, $\mu = 6$) but primarily syn addition in solvents of higher polarity (e.g., nitrobenzene, $\mu = 35$). In dioxane ($\mu = 2.2$) however, the bromine adduct is formed with only 20% anti addition. Explain this apparent contradiction.
7. Propose a detailed mechanism to account for each of the following transformations:

a. 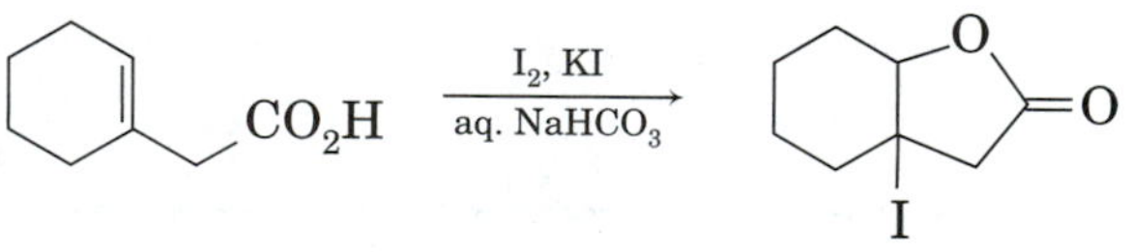 CO_2H $\xrightarrow[\text{aq. NaHCO}_3]{I_2,\ KI}$ O, =O, I

b. 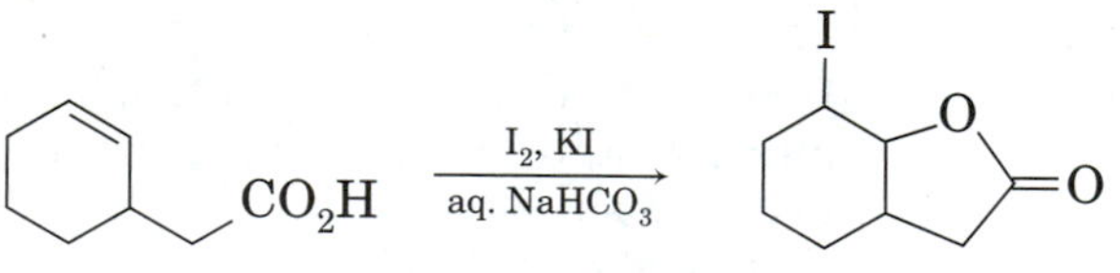 CO_2H $\xrightarrow[\text{aq. NaHCO}_3]{I_2,\ KI}$ I, O, =O

c. 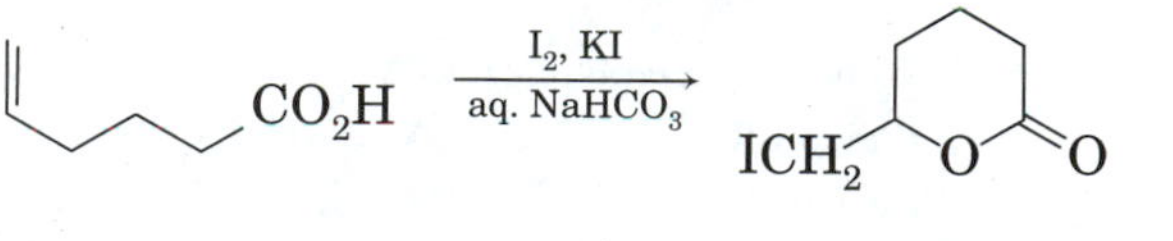 CO_2H $\xrightarrow[\text{aq. NaHCO}_3]{I_2,\ KI}$ ICH_2, O, O

d. 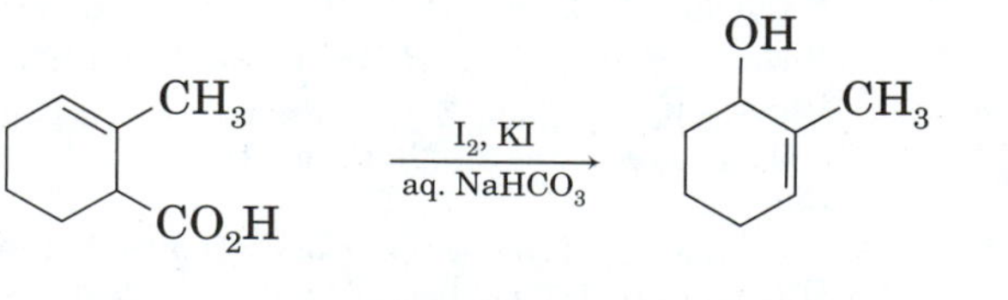 CH_3, CO_2H $\xrightarrow[\text{aq. NaHCO}_3]{I_2,\ KI}$ OH, CH_3

8. Reaction of 5-chloro-5-deuterio-1-hexene in trifluoroacetic acid leads to the formation of 5-chloro-5-deuterio-2-hexyl trifluoroacetate (60%) and 5-chloro-2-deu-

terio-2-hexyl acetate (40%). Propose a mechanism for the formation of these products.

9. One of the products isolated from the photochemical reaction of iodine with styrene is 1,4-diiodo-2,3-diphenylbutane. Account for the formation of this material.
10. Treatment of 2-ethoxy-1,3-butadiene with acid in 80% aqueous acetone yields methyl vinyl ketone. Propose a mechanism to account for the formation of this product.
11. The addition of bromine to 2-cyclohexenyl benzoate in 1,2-dichloroethane produces four dibromide derivatives. Propose a mechanism to account for the formation of each of these compounds.

12. Rationalize the differences in percentage of product formed by syn addition (shown in parentheses) when 1-phenylpropene reacts with each of the following electrophiles: DBr (88%), F_2 (78%), Cl_2 (62%), Br_2 (17%).
13. In a study of the addition of bromine to styrene derivatives in acetic acid solution, it was found that reaction of β,β-dimethylstyrene produces both 1-acetoxy-2-bromo and 2-acetoxy-1-bromo derivatives (in roughly a 3 : 1 yield) as well as the 1,2-dibromo product. However, only the 1,2-dibromo adduct and the 1-acetoxy-2-bromo compound were observed in the reaction of styrene, 3-chlorostyrene, 3-nitrostyrene, *cis*-β-methylstyrene or *trans*-β-methylstyrene under the same conditions. Explain why β,β-dimethylstryene gives different results from those of other styrene derivatives, and suggest other compounds that might give the same or a greater amount of the 2-acetoxy-1-bromo derivative.
14. Reaction of chlorine with methyl *trans*-cinnamate in acetic acid leads to the formation of both methyl *erythro*- and methyl *threo*-3-acetoxy-2-chloro-3-phenylpropionate and both methyl *erythro*- and methyl *threo*-2,3-dichloro-3-phenylpropionate in a ratio of 41 : 7 : 12 : 40. Explain the origin of each of these products, and discuss their relative yields.
15. Give the structure and stereochemistry of the major product expected from reaction at $-78°$ of *trans*-3-hexen-1-ol acetate (dissolved in a mixture of $CFCl_3$, $CHCl_3$, and ethanol) with a mixture of 1% fluorine in nitrogen that is bubbled into the reaction mixture.
16. Propose a detailed mechanism for the specific acid-catalyzed formation of butyl acetate when butoxyacetylene is heated in water, and explain why this compound is more reactive toward hydration in water than is butylacetylene.
17. Oxymercuration of the following pentenols has been found to occur with the rate constants indicated (units of $M^{-1}\ sec^{-1}$) at 25°: 1 penten-3-ol, 1.5×10^2; 1-penten-4-ol, 6.1×10^3; 1-penten-5-ol, $> 10^6$.
    a. Rationalize the rate constants for these three compounds.
    b. Draw the structure of the product(s) expected in each case.
18. Predict the stereochemistry (erythro or threo) of the products resulting from the reaction of (*E*)- and (*Z*)-1-hexene-*1,2-d*$_2$ with dicyclohexylborane.

19. Predict the products (specifying both regiochemistry and stereochemistry) of the following reactions:

a. 2,4,4-Trimethyl-1-pentene $\xrightarrow[\text{THF–H}_2\text{O}]{\text{Hg(OAc)}_2}$ $\xrightarrow{\text{NaBH}_4}$

b. 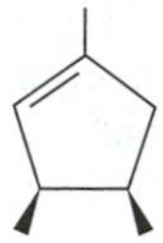 $\xrightarrow[\text{THF–H}_2\text{O}]{\text{Hg(OAc)}_2}$ $\xrightarrow{\text{NaBH}_4}$

20. Solvomercuration of cyclohexene with mercuric nitrate in acetonitrile, followed by reduction with sodium borohydride, leads to a product with the molecular formula $C_8H_{15}NO$.
   a. What is the structure of the product?
   b. Propose a mechanism to account for its formation.
   c. Why would mercuric nitrate be used in this reaction instead of mercuric acetate?
21. Predict which double bond will be epoxidized faster in each of the following compounds:
   a. isoprene, b. methyl 2,4-hexadienoate, c. 2-methyl-2,3-butadiene.
22. Propose an explanation for the observation that in the addition of bromine to alkenes, the relative reactivity of propene to ethene is 550 in 1,1,2,2-tetrachloroethane, 80 in acetic acid, 61 in methanol, and 26 in water.
23. It has been proposed that there is a close analogy between the S_N1 reaction and the electrophilic addition of bromine to an alkene. Compare and contrast the two reactions in terms of (a) gross mechanistic features (especially charge development), (b) substituent effects, and (c) solvent effects.
24. Respond to the following statement by discussing the similarities and differences between conceptual models used in mechanistic chemistry and those used in synthetic chemistry:

> . . . halogenation in polar solvents now assumes a formal resemblance to general acid catalysis initiated by proton donors, which later can be arranged in a series of decreasing efficacy which parallels their decreasing acid strength. Although we regard positive halogen ions as a fiction, it is not to be ignored that free hydrogen ions in solution are also a fiction, and it may develop that for purposes of classification, positive halogen ions may be just as useful a fiction as free halogen ions.[352]

[352]Bartlett, P. D.; Tarbell, D. S. *J. Am. Chem. Soc.* **1936**, *58*, 466.

CHAPTER 10

Elimination Reactions

10.1 Introduction

Elimination is the reverse of addition. In an elimination reaction, a substrate loses two groups, with a resulting increase in its number of units of unsaturation.[1] Elimination reactions, particularly E1 and E2 reactions of alkyl halides, provide some of the fundamental conceptual models for understanding and categorizing the reactions of organic compounds. As with substitution and addition reactions, however, simple mechanistic labels serve only as a beginning point for the discussion of a wide range of elimination reactions.

The most familiar elimination reactions are 1,2-eliminations, which are also known as β-eliminations. In such reactions, a leaving group departs from one (α) carbon atom, while a proton or other group leaves from the adjacent (β) atom (Figure 10.1).[2,3] Many fundamental aspects of 1,2-elimination reactions were developed through the work of Hughes, Ingold, and their co-workers and reported in 1948.[4,5] The common mechanistic designations E1 (Elimination unimolecular) and E2 (Elimination bimolecular) were also suggested by Ingold.[6,7]

---

[1]Commission on Physical Organic Chemistry, IUPAC, *Pure Appl. Chem.* **1994**, *66*, 1077.

[2]The equation is not balanced because the byproducts depend on the reaction conditions.

[3]The IUPAC nomenclature system (reference 1) refers to a 1,2-elimination as a 1/2/elimination.

[4]A series of papers culminated in a discussion of the mechanisms of elimination reactions by Dhar, M. L.; Hughes, E. D.; Ingold, C. K.; Mandour, A. M. M.; Maw, G. A.; Woolf, L. I. *J. Chem. Soc.* **1948**, 2093.

[5]An earlier series of investigations was reported by Hughes, E. D.; MacNulty, B. J. *J. Chem. Soc.* **1937**, 1283, and foregoing papers in that volume.

[6]Hughes, E. D.; Ingold, C. K.; Scott, A. D. *J. Chem. Soc.* **1937**, 1271.

[7]In the nomenclature system recommended by the Commission on Physical Organic Chemistry, IUPAC, *Pure Appl. Chem.* **1989**, *61*, 23, the E2 reaction is denoted $A_nD_ED_N$, and the E1 reaction is termed $D_N + D_E$.

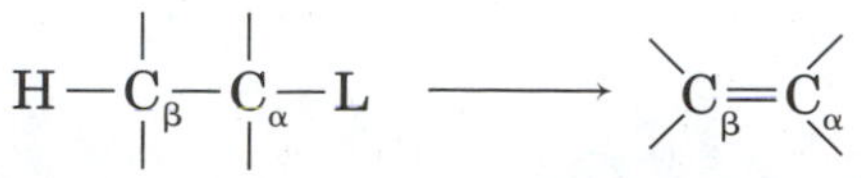

Figure 10.1
A 1,2-elimination reaction.

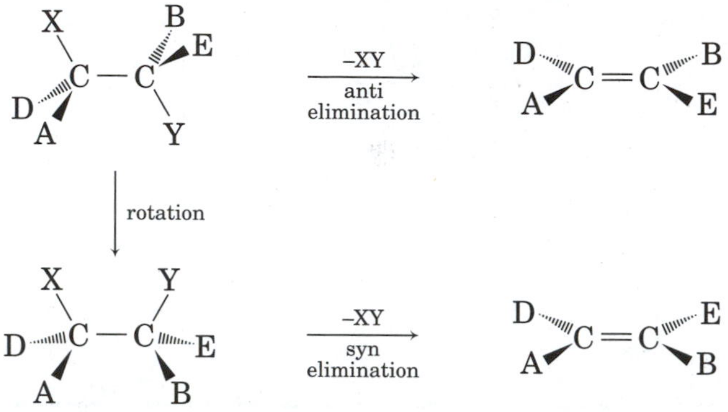

$$\mathrm{H{-}C_{\beta}{-}C_{\alpha}{-}L \longrightarrow C_{\beta}{=}C_{\alpha}}$$

Figure 10.2
Diastereomeric products of syn and anti elimination from a 1,2-disubstituted compound.

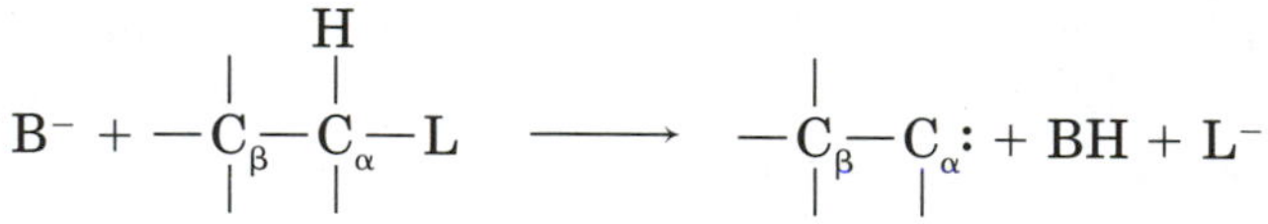

Figure 10.3
A 1,1-elimination.

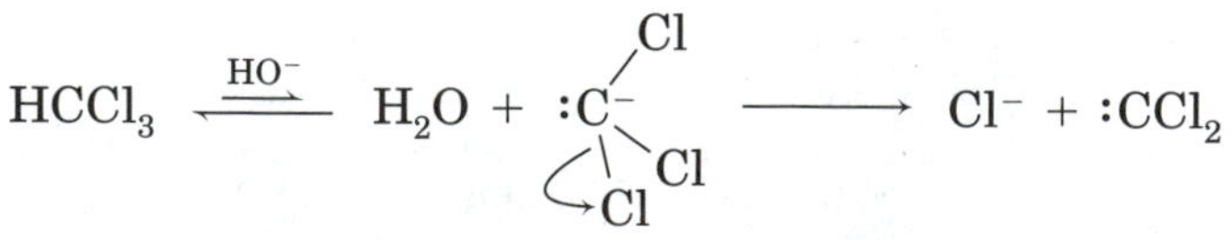

$$\mathrm{B^- + {-}C_{\beta}{-}C_{\alpha}(H){-}L \longrightarrow {-}C_{\beta}{-}C_{\alpha}{:} + BH + L^-}$$

As with addition reactions, an important stereochemical distinction to be made in elimination reactions is that of syn and anti pathways.[8] The term *anti* means that one group detaches from the top of the molecule (as defined by the developing olefinic unit) while the other group detaches from the bottom. In syn elimination, both groups detach from the same face. As noted in Figure 10.2, the syn and anti pathways may be distinguished by the formation of diastereomeric products from appropriately substituted reactants.

There are also elimination reactions in which the two departing groups are not located on adjacent atoms. In a 1,1-elimination (α-elimination), for example, the two leaving groups are bonded to the same atom, so carbenes are produced (Figure 10.3).[9] A specific example is the base-promoted synthesis of dichlorocarbene from chloroform (equation 10.1).

$$\mathrm{HCCl_3 \underset{}{\overset{HO^-}{\rightleftharpoons}} H_2O + {:}CCl_3^- \longrightarrow Cl^- + {:}CCl_2} \qquad \textbf{(10.1)}$$

[8]In older terminology, these pathways were referred to as "trans" or "cis" additions. As was noted in chapter 9, current preference uses the terms *syn* and *anti* for *mechanisms* and reserves the terms *cis* and *trans* for *structures*.

[9]The base-promoted 1,1-cleavage of chloroform to dichlorocarbene was reported by Hine, J.; Dowell, Jr., A. M. *J. Am. Chem. Soc.* **1954**, *76*, 2688.

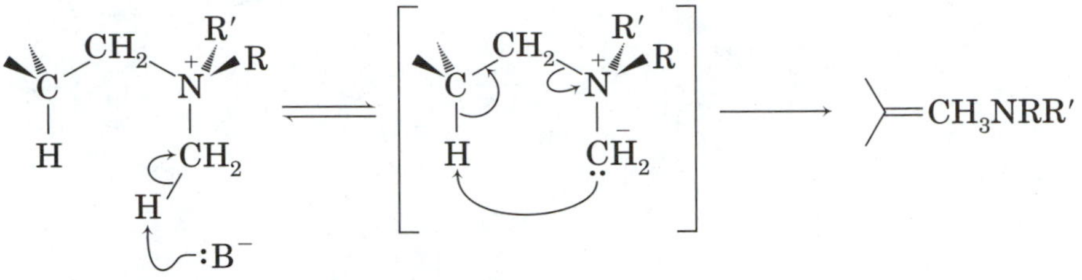

Figure 10.4
An α′,β-elimination pathway.

A 1,1-elimination pathway has also been considered as a possible mechanism for the elimination of trialkylamines from tetraalkylammonium ions. Rearrangement of the carbene produces an alkene, so such a pathway can be distinguished from 1,2-elimination only through isotopic labeling (equation 10.2). An alternative pathway for the elimination of tetraalkylammonium ions is the ylid (α′β) pathway shown in Figure 10.4.[10] Evidence for the ylid mechanism in the reaction of **1** was reported by Cope and Mehta, who found that the deuterium label appeared in one of the methyl groups of trimethylamine (**3**) and not in the alkene (equation 10.3).[11]

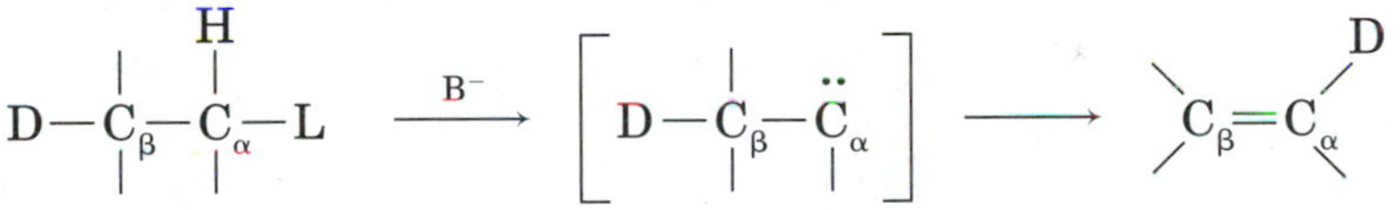

$$\mathrm{D{-}C_\beta{-}C_\alpha(H){-}L} \xrightarrow{B^-} \left[\mathrm{D{-}C_\beta{-}\ddot{C}_\alpha}\right] \longrightarrow \mathrm{C_\beta{=}C_\alpha(D)} \qquad (10.2)$$

$$\mathrm{((CH_3)_3C)_2CD{-}CH_2\overset{+}{N}(CH_3)_3\ {}^{-}OH} \longrightarrow \mathrm{((CH_3)_3C)_2C{=}CH_2 + DCH_2N(CH_3)_2} \qquad (10.3)$$

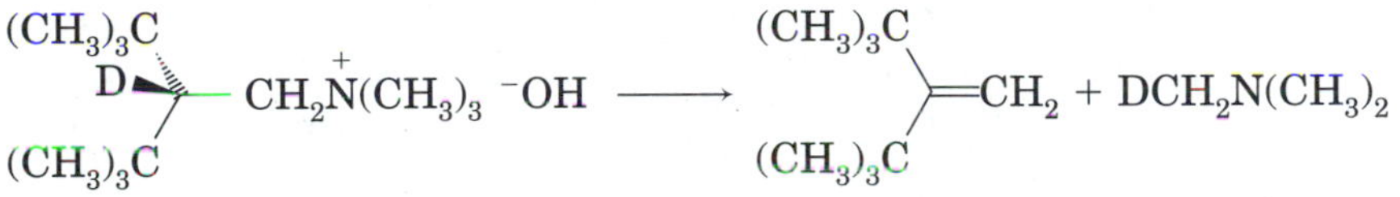

1 **2** **3**

Reactions that are formally 1,3-eliminations (γ-eliminations) have also been observed, as in the example shown in equation 10.4.[12] However, reactions of this type could also be termed intramolecular S_N2 reactions. 1,4- or δ-eliminations can occur when there is a double bond between the atoms bearing the leaving groups, as in the reaction shown in equation 10.5.[13,14] In addition, 1,6-, 1,8- and 1,10-eliminations have been reported.[15]

---

[10]The mechanism of the ylid reaction is very similar to the concerted mechanism proposed for pyrolytic eliminations. See the discussion beginning on page 695.

[11]Cope, A. C.; Mehta, A. S. *J. Am. Chem. Soc.* **1963**, *85*, 1949.

[12]Bumgardner, C. L. *Chem. Commun.* **1965**, 374. 1,2-Dehydrohalogenation (and subsequent reaction with sodium amide and additional starting material to produce diphenylhexenes) predominates when L is bromine or chlorine. The results suggest that the rate constant for 1,2-elimination is more sensitive to leaving group ability than is the rate constant for 1,3-elimination.

[13]Moss, R. J; Rickborn, B. *J. Org. Chem.* **1986**, *51*, 1992.

[14]These elimination reactions can be synthetically useful. See, for example, Tobia, D.; Rickborn, B. *J. Org. Chem.* **1986**, *51*, 3849; Banwell, M. G.; Papamihail, C. *J. Chem. Soc., Chem. Commun.* **1981**, 1182.

$$C_6H_5{-}CH_2CH_2CH_2{-}L \xrightarrow[NH_3\ (liq.)]{NaNH_2} \left[C_6H_5{-}\bar{C}H\,CH_2CH_2{-}L\right] \longrightarrow C_6H_5\text{-cyclopropane} \quad (10.4)$$

L = F, OTs, or $\overset{+}{N}(CH_3)_3$

$$\text{1-methoxy-4-deuterio-1,4-dihydronaphthalene} \xrightarrow[hexane]{LDA} \text{1-deuterionaphthalene} \quad (10.5)$$

In all of the reactions above, the carbon skeleton remains intact. There is also a class of elimination reactions, elucidated by Grob, that are known as *fragmentation reactions* because skeletal bonds are broken during the reaction.[16] In the generalized fragmentation process in equation 10.6,

$$a{-}b{-}c{-}d{-}L \longrightarrow a{-}b + c{=}d + L \quad (10.6)$$

a, b, c, and d generally represent carbon, nitrogen, or oxygen atoms, and L is a leaving group.[17] Fragmentation reactions may occur by one-step or by multi-step mechanisms.[16(c),18] These reactions are useful both for structure elucidation and for synthesis.[16(b)] One example of a fragmentation reaction is the reaction of *cis*-1,4-dibromocyclohexane with zinc in hot dioxane solution.[19]

$$\text{Br-C}_6\text{H}_{10}\text{-Br} \xrightarrow{Zn} \text{Br-C}_6\text{H}_{10}\text{-ZnBr} \xrightarrow[\Delta]{dioxane} 2\ CH_2{=}CH{-}CH{=}CH_2\ \text{fragments} + ZnBr_2 \quad (10.7)$$

Another example is the fragmentation that accompanies solvolysis of γ-aminoalkyl compounds.

[15]For examples of 1,6-, 1,8-, and 1,10-eliminations, respectively, see van Boom, J. H.; Brandsma, L.; Arens, J. F. *Recl. Trav. Chim. Pays-Bas* **1966**, *85*, 952; Rudolf, K.; Koenig, T. *Tetrahedron Lett.* **1985**, *26*, 4835; Rappoport, Z.; Greenblatt, J.; Apeloig, Y. *J. Org. Chem.* **1979**, *44*, 3687.

[16](a) Grob, C. A. in *Theoretical Organic Chemistry*; Butterworths Scientific Publications: London, 1959; pp. 114–126; (b) Grob, C. A.; Schiess, P. W. *Angew. Chem., Int. Ed. Engl.* **1967**, *6*, 1; (c) Grob, C. A. *Angew. Chem., Int. Ed. Engl.* **1969**, *8*, 535.

[17]For the purposes of generality, charges are deliberately omitted.

[18]As noted in reference 21, p. 597, fragmentation reactions can be described as F1, F2, and F1cb by analogy to E1, E2, and E1cb reactions.

[19]The mechanism shown is intended to illustrate a possible pattern of electron movement leading to bond breaking. Grob, C. A.; Baumann, W. *Helv. Chim. Acta* **1955**, *38*, 594, suggest a mechanism involving decomposition of an intermediate with a carbon-zinc bond.

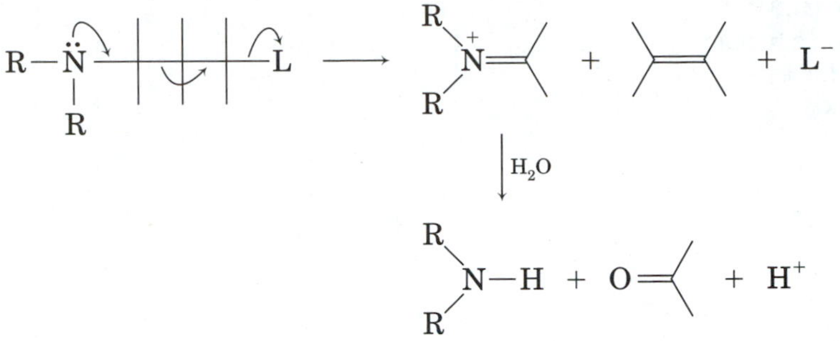

R—N̈—L ⟶ R_2N^+= + (alkene) + L^-

↓ H_2O

R_2N—H + O= + H^+ (10.8)

Fragmentation reactions also formally include the reverse aldol reaction

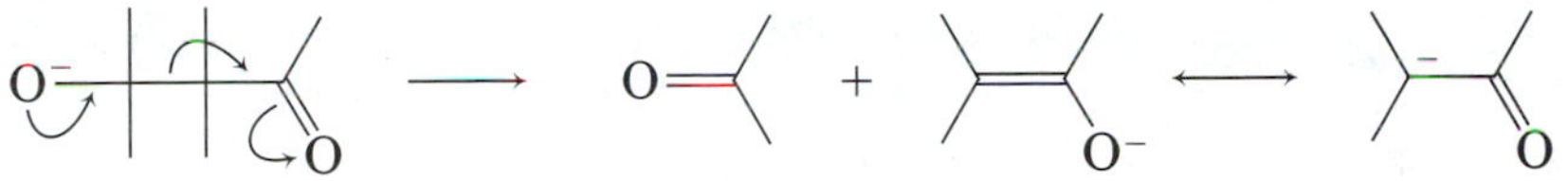

O^- … O ⟶ O= + (alkene)—O^- ⟷ (carbanion)—O (10.9)

and reverse Michael addition.

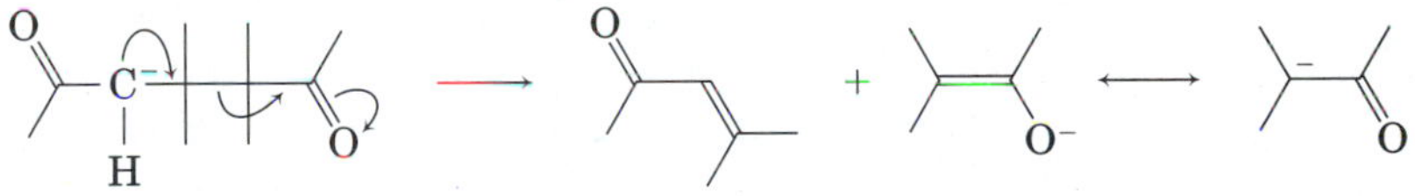

O … C$^-$(H) … O ⟶ O … + (alkene)—O^- ⟷ (carbanion)—O (10.10)

Extensive discussion of these and many other types of elimination reactions can be found in several excellent reviews of elimination reactions.[20,21,22,23] We will not attempt to survey all categories of elimination reactions here but will focus instead on 1,2-eliminations, especially dehydrohalogenation, dehydration, dehalogenation, deamination reactions, and pyrolytic eliminations.

10.2 Dehydrohalogenation and Related 1,2-Elimination Reactions

Potential Energy Surfaces for 1,2-Elimination

There are three essential elements in a 1,2-elimination reaction:

1. the bond between C_β and a hydrogen atom is lost,
2. the bond between C_α and the leaving group, L, is lost, and
3. an additional bond forms between C_α and C_β.

---

[20] Banthorpe, D. V. *Elimination Reactions*; Elsevier Publishing Company: Amsterdam, 1963.

[21] Saunders, Jr., W. H.; Cockerill, A. F. *Mechanisms of Elimination Reactions*; Wiley-Interscience: New York, 1973.

[22] Bartsch, R. A.; Závada, J. *Chem. Rev.* **1980**, *80*, 453.

[23] Cockerill, A. F.; Harrison, R. G. in *The Chemistry of Double-Bonded Functional Groups*, Part I, Patai, S., ed.; Wiley-Interscience: New York, 1977; pp. 155–189.

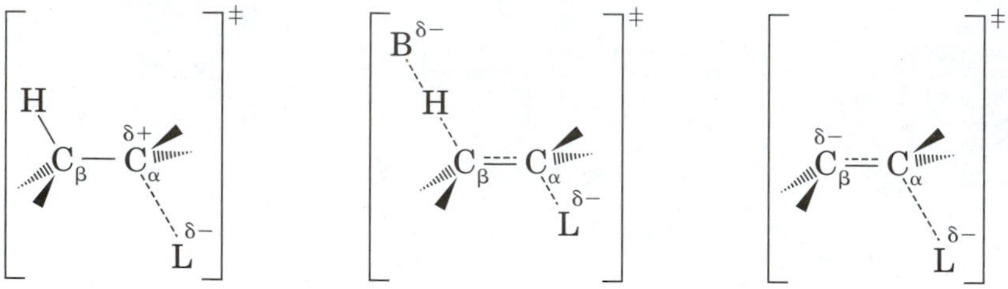

Figure 10.5
Transition structure models for 1,2-elimination reactions.

We can distinguish three general classifications of elimination reactions on the basis of the timing of the first two steps, and transition structures for each are shown in Figure 10.5.[24] If the C_α—L bond dissociates first, the reaction is termed an E1 reaction. If both the C_α—L and the C_β—H bonds are lost at the same time, the reaction is an E2 mechanism. If the C_β proton is abstracted first, then we call the reaction an **E1cb** (**e**limination **u**nimolecular **c**onjugate **b**ase), since proton removal leaves a carbanion that is the conjugate base of the original substrate.[25,26]

The three categories of 1,2-eliminations can be represented as lines on a More O'Ferrall-Jencks diagram, a two-dimensional projection of a three-di-

---

[24]All of the 1,2-elimination mechanisms discussed here have assumed that, at some point, a base abstracts a proton β to the leaving group by directly attacking that proton. However, some authors have distinguished between the **E2H** ("normal E2") pathway and the **E2C** pathway, in which the base interacts with the α-carbon atom attached to the leaving group prior to removal of the β-hydrogen atom.

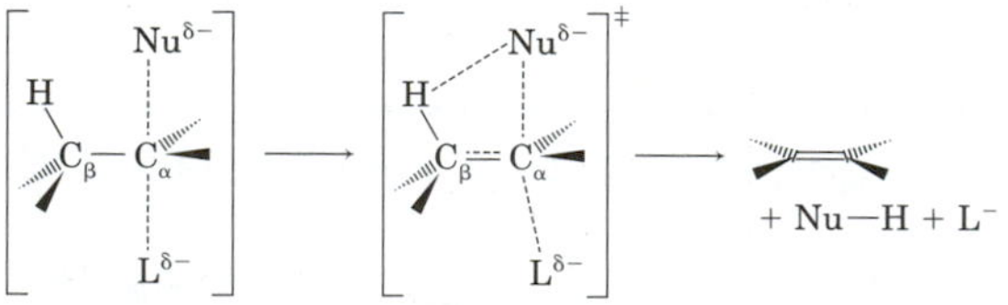

It was thought that this model could provide for a weaker C_β-H bond that could more easily be removed by the weak base by allowing the base to act first as a nucleophile to partially loosen the C_α-L bond. The degree to which the E2C mechanism might compete with the E2H mechanism was thought to be a function of the substitution of the substrate and the polarity of the medium. However, elimination mechanisms are not currently discussed in terms of the E2C pathway. For leading references to literature discussions of this issue, see Ford, W. T. *Acc. Chem. Res.* **1973**, *6*, 410; Biale, G.; Cook, D.; Lloyd, D. J.; Parker, A. J.; Stevens, I. D. R.; Takahashi, J.; Winstein, S. *J. Am. Chem. Soc.* **1971**, *93*, 4735; Parker, A. J. *Chem. Tech.* **1971**, 297; Kwart, H.; Wilk, K. A. *J. Org. Chem.* **1985**, *50*, 3038; McLennan, D. J. *Annual Reports on the Progress of Chemistry* **1970**, *B*, 59; Bunnett, J. F.; Midgal, C. A. *J. Org. Chem.* **1989**, *54*, 3041.

[25]In Figure 10.5 the transition structure for the E1cb reaction indicates that the β-proton has already been abstracted by a base, so the rate-limiting step is detachment of the leaving group. As discussed on page 649, there are many variations of the E1cb mechanism.

[26]In the IUPAC nomenclature system (reference 7), the E1cb reaction is termed $A_nD_E + D_N$.

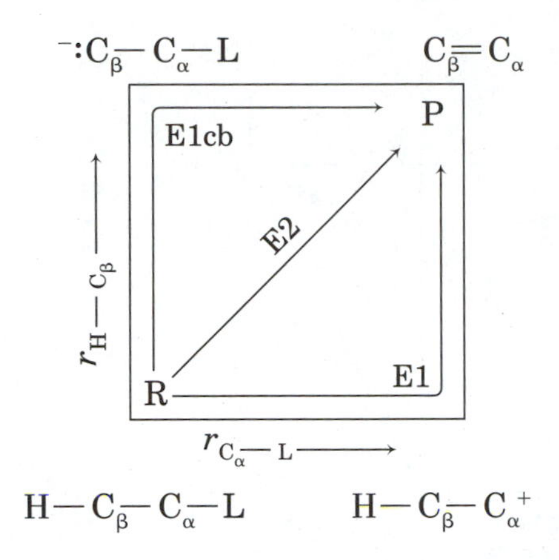

Figure 10.6 Structure-energy surface for 1,2-elimination reactions.

mensional potential energy surface for a reaction.[27,28] In Figure 10.6 the vertical axis (on the printed page) represents the dissociation of the $H—C_\beta$ bond, while the horizontal axis represents the dissociation of the $C_\alpha—L$ bond. The energy of the species at any point on the surface would be represented in the third dimension, coming out of the page toward the viewer. Each reaction would have its own three-dimensional potential energy surface, with the line in Figure 10.6 being the projection onto the plane of a line that follows the lowest energy pathway from reactant to product on the surface above.

Figure 10.6 indicates graphically that the key differences among the elimination pathways are the timing of the two bond-breaking steps and the possible existence of an intermediate along the reaction coordinate. If the proton and leaving group depart to the same extent at any point along the reaction coordinate, then there is a concerted E2 reaction, and the reaction follows the diagonal line from reactant to product. The E1 reaction, in which departure of the leaving group produces a cation that later undergoes proton removal, is represented by the line proceeding through the lower right corner. The E1cb pathway, in which proton removal leaves a carbanion intermediate, proceeds through the upper left corner. Note that this figure only represents the bonding changes for the reacting substrate, not for any bases or solvent molecules that may be important to the reaction.

The idea that Figure 10.6 is a projection of a three-dimensional surface may be made clearer by Figure 10.7, which represents a portion of that surface from an oblique angle. Although it is difficult to represent the potential energy surface itself in such a figure,[29] the shaded portion of the drawing rep-

[27] *Cf.* More O'Ferrall, R. A. *J. Chem. Soc. B* **1970**, 274; More O'Ferrall, R. A. in *The Chemistry of the Carbon-Halogen Bond*, Vol. 2, Patai, S., ed.; John Wiley & Sons: New York, 1973.; Jencks, D. A.; Jencks, W. P. *J. Am. Chem. Soc.* **1977**, *99*, 7848.

[28] Bunnett, J. F. *Angew. Chem., Int. Ed. Engl.* **1962**, *1*, 225.

[29] These figures are analogous to the drawings representing three-dimensional potential energy surfaces for substitution pathways in Chapter 8.

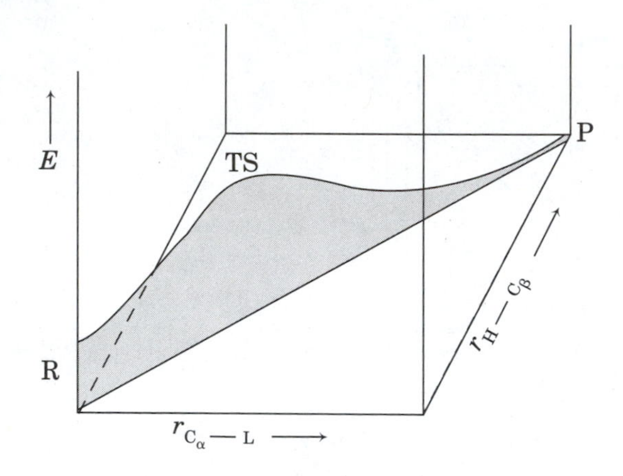

Figure 10.7 Cross section through potential energy surface for E2 reaction.

resents a cross section through the surface from reactant to product. Any deviation perpendicular to the curve shown in Figure 10.7 requires an increase in energy of the system. Thus the transition state is only an energy maximum for the diagonal pathway from $H—C_\beta—C_\alpha—L$ to $C_\beta{=}C_\alpha$; it is an energy minimum on another cross section, from $^{-}{:}C_\beta—C_\alpha—L$ to $H—C_\beta—C_\alpha^{+}$. In other words, the transition state is a saddle point on the three-dimensional curve in space that is the energy surface for this generalized reaction. The diagonal cross section can be redrawn as Figure 10.8, which is the diagram that is commonly used to represent the lowest energy pathway from reactants to products in an E2 reaction.

An oblique view of the three-dimensional potential energy for an E1 reaction is shown in Figure 10.9. Now the lowest energy pathway appears to hug the boundaries of the drawing. Again, any perpendicular deviation from this pathway represents an increase in energy. We see that the usual reaction coordinate diagram for the E1 reaction (Figure 10.10) has a horizontal scale labeled progress of reaction that can conveniently be viewed as a combination of two different sets of nuclear coordinates. From the reactant (R) to the carbocation intermediate, the reaction coordinate is approximated by

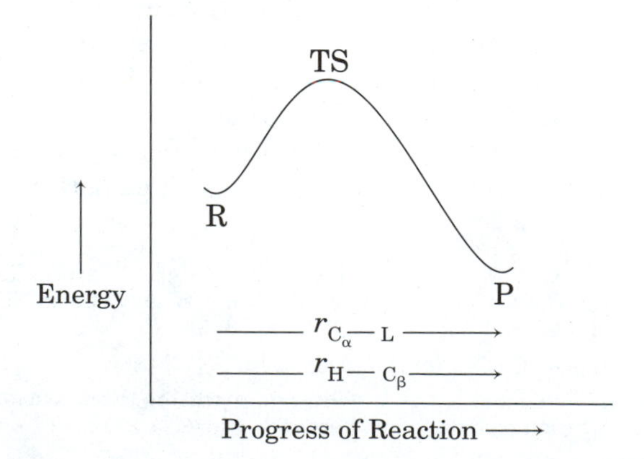

Figure 10.8 Reaction coordinate diagram for E2 reaction. (Redrawn from reference 29.)

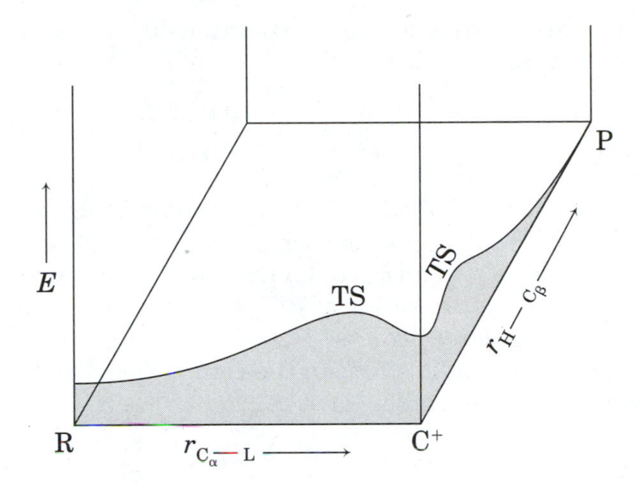

Figure 10.9 Potential energy surface for E1 reaction.

the distance between C_α and L, while from the intermediate to the product (P), the reaction coordinate is essentially the change in the separation of C_β and a proton.

Analogous drawings can be made for the E1cb reaction, but the situation is more complicated because there are many variations of this type of reaction. It is useful to differentiate among several categories of E1cb reactions with reference to the rate constants defined as shown in equation 10.11.[30,31]

$$\mathrm{B:^-} + \mathrm{H{-}C{-}C{-}L} \underset{k_{-1}}{\overset{k_1}{\rightleftharpoons}} \mathrm{B{-}H} + \left[\mathrm{^-:C{-}C{-}L}\right] \xrightarrow{k_2} \mathrm{C{=}C} \quad \textbf{(10.11)}$$

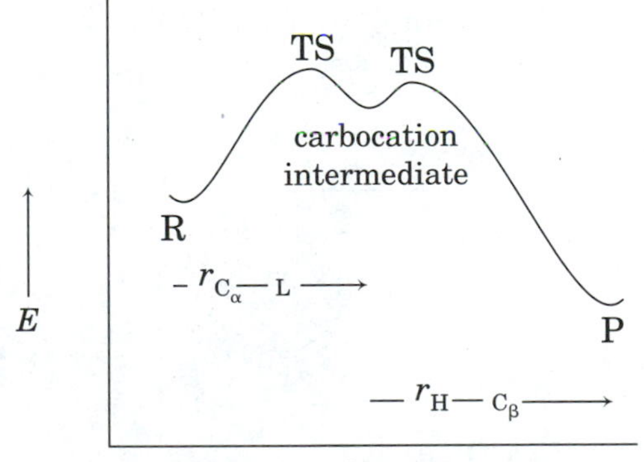

Figure 10.10 Reaction coordinate diagram for E1 reaction.

[30] Bordwell, F. G. *Acc. Chem. Res.* **1972**, *5*, 374, has analyzed a range of mechanisms for elimination reactions.

[31] See also (a) Rappoport, Z. *Tetrahedron Lett.* **1968**, 3601; (b) reference 22.

Applying the steady state approximation to the carbanion intermediate leads to the rate expression

$$\text{Rate} = \frac{k_2 k_1[\text{RL}][\text{B}^-]}{k_{-1}[\text{BH}] + k_2} \tag{10.12}$$

The rate law for a particular E1cb reaction depends on the relative magnitudes of k_1, k_{-1}, and k_2 and the concentrations of B^- and BH. If k_1 is much greater than both k_{-1} and k_2, and if the initial concentration of base is larger than the initial concentration of RL, then essentially all of RL is converted to the carbanion intermediate. Therefore, changes in the concentration of B^- have no appreciable effect on the rate of the reaction, and the reaction appears to follow first order kinetics,

$$\text{rate} \approx k_{\text{exp}}[\text{RL}] \tag{10.13}$$

This type of reaction, termed $\mathbf{E1_{(anion)}}$ or $\mathbf{E1cb_{(anion)}}$, is characterized by rapid isotopic exchange of C_β—D with protons from protic solvents, a β-hydrogen 1° kinetic isotope effect (k_H/k_D) of 1.0, and an appreciable element effect.[32] An example of an $E1_{(anion)}$ mechanism is the elimination of methanol from the 2-nitroethyl methyl ether by the mechanism shown in Figure 10.11.[33]

If $k_{-1}[\text{BH}]$ is much greater than k_2 but is less than $k_1[\text{RL}][\text{B}^-]$, the equilibrium for the first step in equation 10.11 does not lie completely in favor of the anion. Now equation 10.12 becomes

$$\text{Rate} = \frac{k_2 k_1[\text{RL}][\text{B}^-]}{k_{-1}[\text{BH}]} \tag{10.14}$$

If the solvent is BH, then its concentration effectively does not vary during the reaction, so the rate expression becomes

$$\text{Rate} = k_{\text{obs}}[\text{B}^-][\text{RL}] \tag{10.15}$$

Figure 10.11 E1 (anion) mechanism for 1,2-elimination.

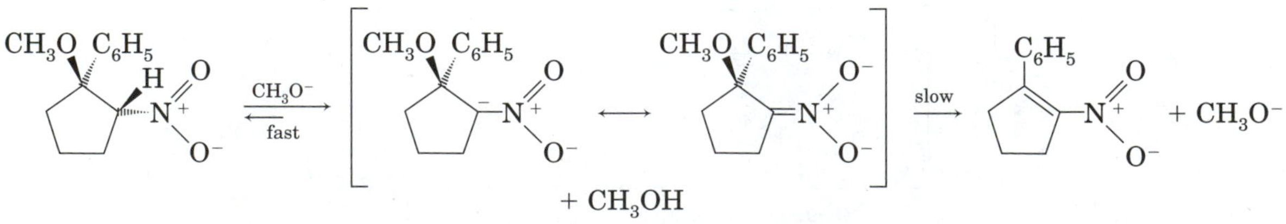

[32]That is, the rate of the reaction depends on the leaving group ability of L.

[33]Bordwell, F. G.; Yee, K. C.; Knipe, A. C. *J. Am. Chem. Soc.* **1970**, *92*, 5945; see also Bordwell, F. G.; Vestling, M. M.; Yee, K. C. *J. Am. Chem. Soc.* **1970**, *92*, 5950.

Figure 10.12
Reaction coordinate diagram for $(E1cb)_R$ reaction.

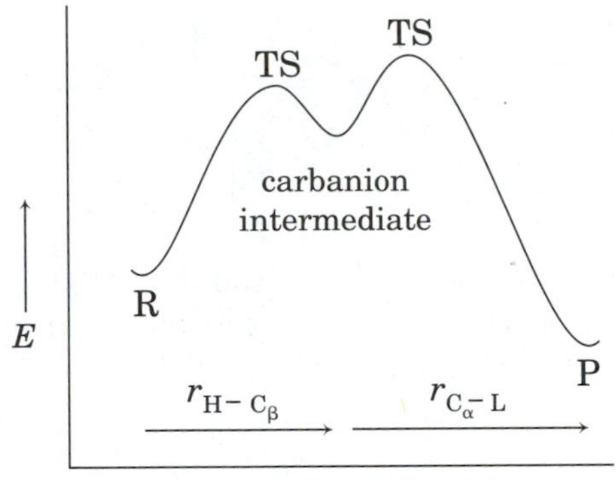

This type of elimination is known as an $\mathbf{E1cb_R}$ (**e**limination, **u**nimolecular, **c**onjugate **b**ase, **r**eversible) reaction, and a generalized reaction coordinate diagram is shown in Figure 10.12. Such reactions exhibit C_β—H exchange, a 1° hydrogen kinetic isotope effect (k_H/k_D) of 1.0, and small element effects. An $E1cb_R$ mechanism is illustrated by the reaction in Figure 10.13.[34]

Just as ion pair intermediates are important in substitution reactions, they may also play an important role in elimination reactions.[35] Figure 10.14 shows an example of an $\mathbf{E1cb_{ip}}$ (**e**limination, **u**nimolecular, **c**onjugate **b**ase, **i**on **p**air) mechanism.[34b] Here again there is a fast preequilibrium formation of a carbanion. In this case, however, the carbanion and the cation are held together as an ion pair due to Coulombic forces that are not overcome by solvation.

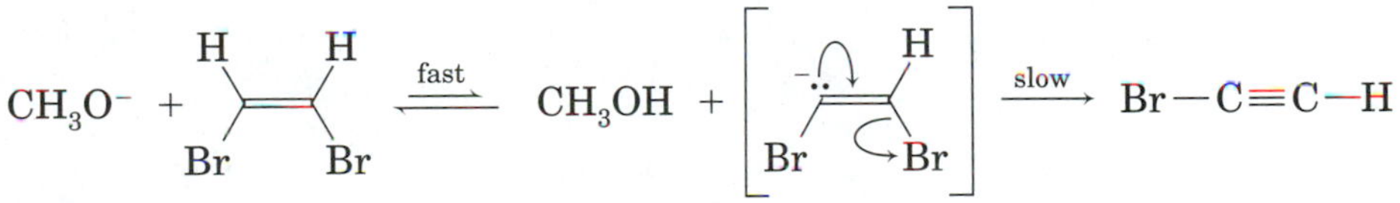

Figure 10.13
$E1cb_R$ mechanism. (Adapted from reference 34(b).)

CH_3O^- + (H)(Br)C=C(H)(Br) ⇌ (fast) CH_3OH + [$^-$C(Br)=C(H)(Br)] → (slow) Br—C≡C—H

---

[34](a) Miller, S. I.; Lee, W. G. *J. Am. Chem. Soc.* **1959**, *81*, 6313; (b) Kwok, W. K.; Lee, W. G.; Miller, S. I. *J. Am. Chem. Soc.* **1969**, *91*, 468.

[35]As another example, the $\mathbf{E2_{ip}}$ (**e**limination, **b**imolecular, **i**on **p**air), mechanism is characterized by dissociation of the C_α—L bond to form a tight carbocation-L^- ion pair, which then undergoes rate-limiting proton abstraction by a base

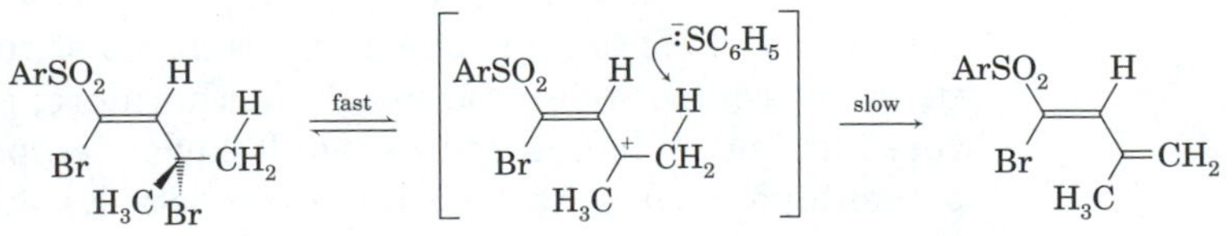

$ArSO_2$(Br)C=C(H)–C(H)(Br)(H_3C)–CH_2H ⇌ (fast) [$ArSO_2$(Br)C=C(H)–C^+(H_3C)–CH_2H :SC_6H_5] → (slow) $ArSO_2$(Br)C=C(H)–C(H_3C)=CH_2

For a discussion, see reference 30. Also see Bordwell, F. G.; Mecca, T. G. *J. Am. Chem. Soc.* **1972**, *94*, 2119. It may also be possible that the ion pair forms during, and not before, attack of the base on the substrate. Saunders, Jr., W. H. *Acc. Chem. Res.* **1976**, *9*, 19, has discussed the approaches used to distinguish concerted and nonconcerted elimination mechanisms.

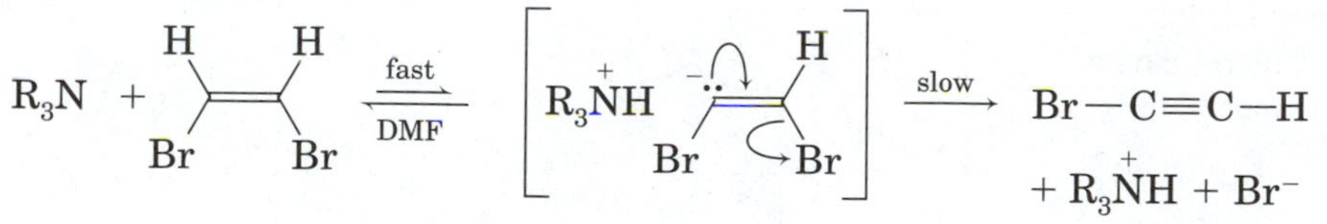

Figure 10.14
$E1cb_{ip}$ mechanism. (Adapted from reference 34(b).)

$$R_3N + \text{HBrC=CHBr} \xrightleftharpoons[\text{DMF}]{\text{fast}} \left[R_3\overset{+}{N}H \;\; \overset{-}{C}(Br)\text{=}C(H)Br \right] \xrightarrow{\text{slow}} Br-C\equiv C-H + R_3\overset{+}{N}H + Br^-$$

If k_2 is much greater than k_{-1}[BH] and if k_{-1}[BH] is much greater than k_1 (which may occur if the intermediate carbanion is strongly hydrogen bonded in a protic solvent), then equation 10.12 becomes

$$\text{Rate} \approx k_1[\text{RL}][\text{B}^-] \qquad \textbf{(10.16)}$$

That is, the reaction exhibits second order kinetics even though the actual loss of the leaving group occurs by a unimolecular decomposition of the conjugate base of the reactant.[36] Such a mechanism is termed $\mathbf{E1cb_I}$ (**e**limination, **u**nimolecular, **c**onjugate **b**ase, **i**rreversible) and is characterized by negligible C_β — H isotope exchange with solvent and with a substantial 1° hydrogen kinetic isotope effect (k_H/k_D in the range of 2 to 8). Inoue and Bruice found that the hydrolysis of *p*-nitrophenyl 2-cyano-3,3-dimethylbutanoate (**4**) occurs by an $E1cb_I$ mechanism at pH values at which *p*-nitrophenyl 2-cyanoacetate reacts by an $E1cb_R$ mechanism. In this case it appears that the deprotonation is effectively irreversible because of steric hindrance to reprotonation of the intermediate anion.[37]

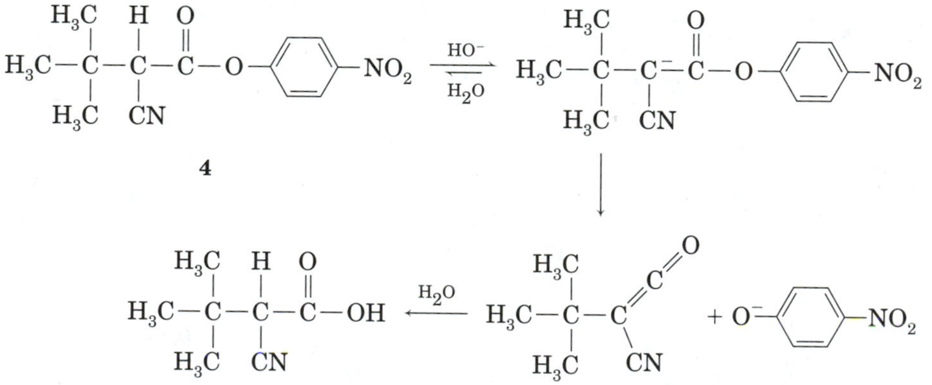

$$(H_3C)_3C-CH(CN)-C(=O)-O-C_6H_4-NO_2 \;(\mathbf{4}) \xrightleftharpoons[H_2O]{HO^-} (H_3C)_3C-\overset{-}{C}(CN)-C(=O)-O-C_6H_4-NO_2$$

$$\downarrow$$

$$(H_3C)_3C-CH(CN)-C(=O)-OH \xleftarrow{H_2O} (H_3C)_3C-C(CN)=C=O \; + \; {}^-O-C_6H_4-NO_2$$

Competition between Substitution and Elimination

It is an axiom of organic chemistry that substitution and elimination reactions are competitive processes.[6] Furthermore, elimination is generally favored at higher temperatures, while lower temperatures result in a higher percentage of substitution product in both E1 and E2 reactions. For exam-

---

[36] Bordwell, F. G.; Weinstock, J.; Sullivan, T. F. *J. Am. Chem. Soc.* **1971**, *93*, 4728.

[37] Inoue, M.; Bruice, T. C. *J. Org. Chem.* **1986**, *51*, 959.

Table 10.1 Activation parameters for substitution and elimination reactions.

Substrate	Solvent	log *A* (S_N2)	E_a (S_N2) (kcal/mol)	log *A* (E2)	E_a (E2) (kcal/mol)
2-Bromopropane	60% ethanol	9.4	20.8	10.4	22.1
2-Bromopropane	80% ethanol	10.1	21.7	10.9	22.6
2-Iodopropane	60% ethanol	10.1	20.7	11.1	22.2
2-Chloropropane	80% ethanol	9.4	23.1	10.7	24.8
t-Butyl bromide	100% ethanol			10.1	19.7
t-Butyldimethylsulfonium	100% ethanol			14.9	24.0
(2-Phenylethyl)dimethylsulfonium	100% ethanol			15.0	23.9

(Data from Reference 38.)

ple, in the solvolysis of *t*-butyl chloride in 80% aqueous ethanol, the percentage of isobutene increased from 16.8% to 36.3% as the temperature was increased from 25° to 65°.[38,39,40] Kinetic data for competition between S_N2 and E2 pathways for reactions of several substrates with sodium ethoxide shown in Table 10.1 indicate that the activation energy for elimination is greater than that for substitution. In addition, reactions of the neutral substrates are seen to have higher activation energies and smaller A values for E2 reaction than do the corresponding reactions of ionic substrates. The smaller A values (i.e., more negative values of $\Delta S^{\ddagger}$) for neutral reactants may be related to the change in solvent order between the reactants and the transition structures for these species. The change in solvent entropy on going from a charged reactant to a transition structure with dispersed charge would be expected to be less than would be the change in solvent entropy for the conversion of a neutral substrate to a transition structure with developing charges.[38]

Hughes and Ingold also determined that elimination becomes more competitive with substitution as the number of alkyl substituents on the substrate increases. Table 10.2 shows rate constants for both S_N2 and E2 reactions of a series of alkyl bromides with sodium ethoxide in ethanol at 55°.[41] The rate constant for elimination increases from ethyl bromide to

[38]Cooper, K. A.; Hughes, E. D.; Ingold, C. K.; Maw, G. A.; MacNulty, B. J. *J. Chem. Soc.* **1948**, 2049. Kinetic studies revealed an E_a of 23.2 kcal/mol and an Arrhenius log *A* value of 11.9.

[39]The effect of temperature on the ratio of substitution to elimination in a 2° substrate was smaller: the percentage of elimination product from the reaction of 2-bromopropane with ethoxide in ethanol was 53% at 45° and increased to 57% at 75°.

[40]In many early papers the solvent composition is reported in quotation marks, as in "80% ethanol", meaning a solvent mixture composed by mixing 80 volumes of absolute ethanol and 20 volumes of water. Due to volume changes on mixing, the volume of ethanol added is not necessarily 80% of the final total volume of the mixture.

[41]Dhar, M. L.; Hughes, E. D.; Ingold, C. K.; Masterman, S. *J. Chem. Soc.* **1948**, 2055.

Table 10.2 Structural effects on rate constants of E2 and S_N2 reactions with sodium ethoxide in ethanol solution at 55°.

Alkyl Bromide	$10^4\ k_{S_N2}$ (M^{-1} sec^{-1})	$10^5\ k_{E2}$ (M^{-1} sec^{-1})
CH_3CH_2Br	17.2	1.6
$CH_3CH_2CH_2Br$	5.5	5.3
$CH_3CH_2CH_2CH_2Br$	4.0	4.3
$CH_3CH_2CH_2CH_2CH_2Br$	3.6	3.5
$(CH_3)_2CHCH_2Br$	0.6	8.5

(Data from reference 41.)

propyl bromide to isobutyl bromide. The more highly substituted alkenes formed with propyl bromide and (to a greater extent) with isobutyl bromide are more stable than the alkene formed from ethyl bromide. Because the transition structures have some double bond character, the Hammond postulate predicts lower transition state energies and faster rates of formation of the more highly substituted products. Steric effects can also be seen in elimination reactions. For example, the rate constants for the E2 reactions decrease slightly on going from propyl to *n*-butyl to *n*-pentyl bromides. The trend closely parallels the decrease in the rate constants of S_N2 reactions along the same series, a trend that is attributed to the dominance of steric effects on the rates of S_N2 reactions.

Only minor variations with increasing size of the alkyl group are seen in the rates of E2 and S_N2 reaction of 2° alkyl halides. With 4 N KOH in ethanol at 80°, the rate constants for the E2 reaction of isopropyl, isobutyl and isoamyl bromides were found to be 1.41, 1.91, and 1.5×10^{-3} M^{-1} sec^{-1}, respectively. The corresponding rate constants for the S_N2 reactions of these compounds were 2.1, 1.8, and 1.6×10^{-4} M^{-1} sec^{-1}. Again, the rate constants for the S_N2 reaction are seen to decrease slightly with the increasing size of the alkyl group. However, the rate constants for the E2 reaction do not show a monotonic trend. The variation in the E2 rate constants was attributed to competing factors, one being the formation of a more highly substituted alkene and the other being an increasing steric barrier for the reaction.[42,43]

[42]Dhar, M. L.; Hughes, E. D.; Ingold, C. K. *J. Chem. Soc.* **1948**, 2058.

[43]The substitution and elimination reactions of 2° alkyl halides are more complicated than those of 1° alkyl halides because of the possibility that S_N1, S_N2, E1 and E2 reactions may occur concurrently. The rates of S_N2 and E2 reactions may be determined under conditions of modest solvent polarity (in ethanol, for example) and relatively large concentrations of strong base, such as ethoxide or hydroxide. However, it is not always possible to study the rates of S_N1 and E1 processes without competition from the bimolecular reactions, even in very polar solvents.

Stereochemistry of 1,2-Elimination Reactions

It is useful to discuss the stereochemistry of bimolecular elimination reactions in terms of the H—C—C—L dihedral angle (Figure 10.15). In the *anti*-periplanar conformation, the dihedral angle H—C_β—C_α—L is 180°, while it is 0° in the *syn*-periplanar conformation. The *anti*-clinal conformation has a dihedral angle of approximately 120°, while the *syn*-clinal conformation has a dihedral of about 30°.

Figure 10.15 Conformational designations.

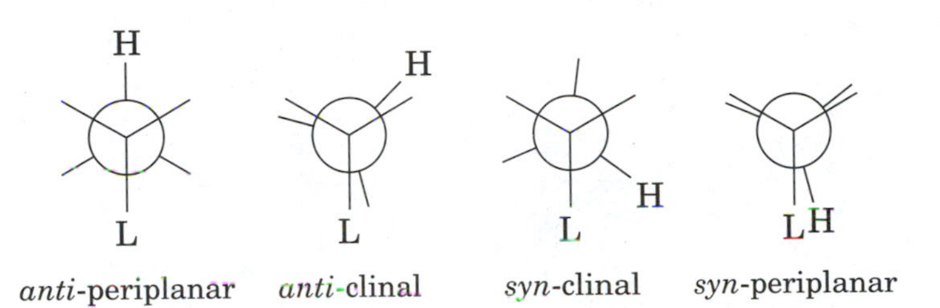

Representing the C_α— L and C_β— H bonds with localized orbitals, as shown in Figure 10.16, makes it apparent that both the *anti*-periplanar and *syn*-periplanar conformations allow the development of π bonding without the necessity of rotation about the carbon-carbon double bond. With the *anti*-clinal and *syn*-clinal conformations, however, rotation of 60° must accompany the elimination in order for there to be parallel *p* orbitals on the carbon atoms involved in the developing double bond. Thus, we would expect the electronic energies to be lower for transition structures having *syn*-periplanar and *anti*-periplanar conformations than for other conformations. A qualitative representation of the electronic barrier expected for H — C_β— C_α— L dihedral angles ranging from 0° to 180° is shown in Figure 10.17.[44]

There should be an additional electronic preference favoring anti elimination over syn elimination because the anti E2 reaction involves a favorable S_N2-like attack of electrons from the C_β — H bond on the back side of the C — L bond.[51,45,46] Bach suggested that the anti orientation allows the

Figure 10.16 Orbital model for *anti*-periplanar (left) and *syn*-periplanar (right) elimination.

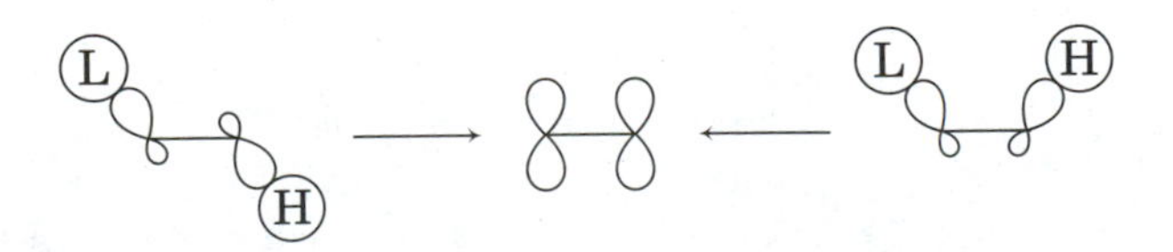

[44]For a discussion, see reference 22 and references therein.

[45]See also Ingold, C. K. *Proc. Chem. Soc.* **1962**, 265.

[46]This intuitive view has received theoretical support from a number of investigators. See, for example, Lowe, J. P. *J. Am. Chem. Soc.* **1972**, *94*, 3718.

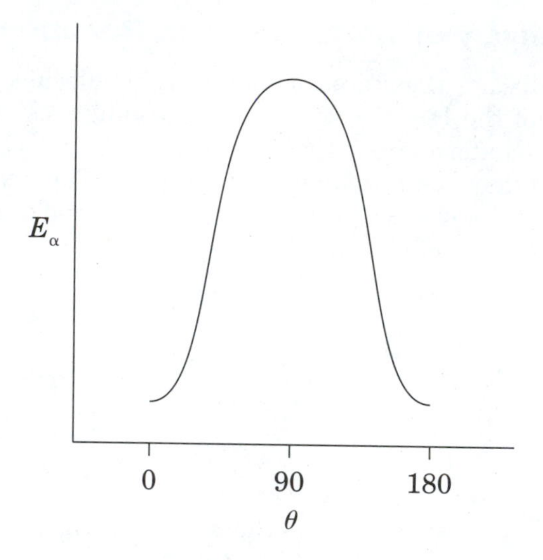

Figure 10.17 Electronic barrier to 1,2-elimination as a function of $H—C_\beta—C_\alpha—L$ dihedral angle. (Reproduced from reference 22.)

electrons in the $C_\beta—H$ σ bonding orbital (the HOMO as the proton removal begins) to interact with the σ* orbital, which is essentially localized along the $C_\alpha—L$ axis (Figure 10.18). Population of the LUMO effectively expels the leaving group, with concomitant production of the carbon-carbon double bond. On the other hand, Bach suggested that syn elimination arises from a transition structure with considerable E1cb character, so that inversion of the developing carbanion also leads to population of the C—L σ* orbital, resulting in expulsion of L and formation of the double bond.[47,48] DePuy and co-workers found evidence for this view in the base-promoted elimination of

Figure 10.18 Interaction of the C-H bonding electrons with the vacant σ* orbital of the C-L bond in (left) concerted *anti*-periplanar pathway and (right) E1cb-like *syn*-periplanar pathway. (Adapted from reference 48.)

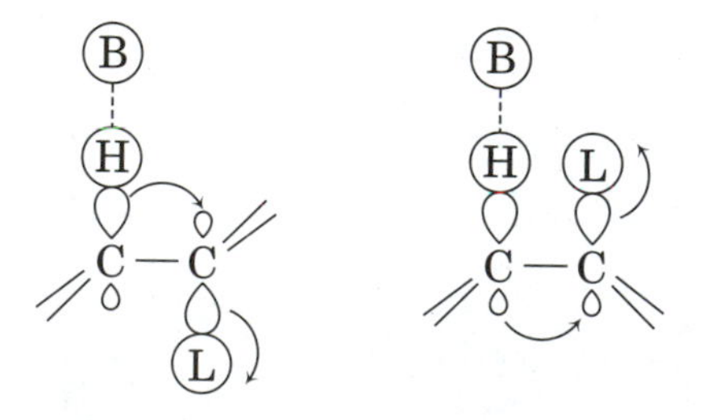

[47](a) Bach, R. D.; Badger, R. C.; Lang, T. J. *J. Am. Chem. Soc.* **1979**, *101*, 2845. (b) This conclusion was supported by calculations of Minato, T.; Yamabe, S. *J. Am. Chem. Soc.* **1988**, *110*, 4586.

[48]Dohner, B. R.; Saunders, Jr., W. H. *J. Am. Chem. Soc.* **1986**, *108*, 245 also concluded that syn elimination has more carbanion character than does anti elimination.

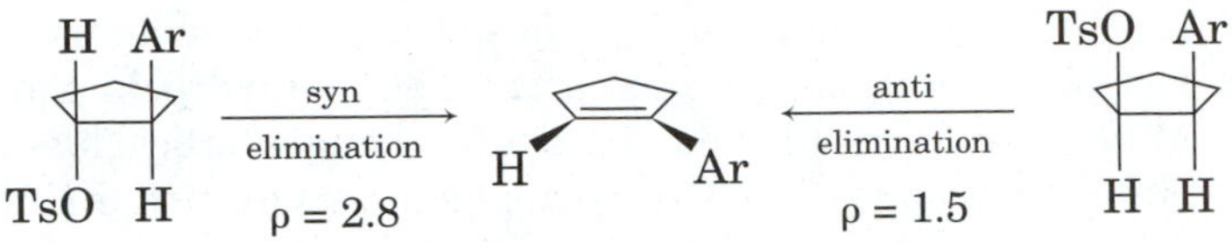

Figure 10.19 ρ values for syn and anti elimination from 2-arylcyclopentyl tosylates.

cis- and *trans*-2-arylcyclopentane tosylates.[49] The observation of a Hammett ρ value of +2.8 for syn elimination (in contrast to a ρ of +1.5 for anti elimination) supported the view that syn elimination occurs by a mechanism with considerable E1cb character.

Electronic energy due to orbital interactions (Figure 10.18) is not the only factor to consider in predicting the stereochemistry of 1,2-elimination reactions, and steric interactions can also be significant. There is some torsional strain in the transition structure for syn 1,2-elimination because the bonds from the C_β and C_α carbon atoms to their substituents are eclipsed (Figure 10.20). On the other hand, such torsional strain is not present in the transition structure for anti 1,2-elimination because those bonds are staggered. Torsional strain in a conformation leading to syn elimination would thus be expected to raise the barrier for syn elimination in a conformationally mobile substrate.[50] The syn pathway also involves greater steric interactions between the base and the leaving group, but these steric interactions are avoided in the anti pathway. Moreover, there could be electrostatic repulsion between a base (bearing a negative charge or a nonbonded pair of electrons) and a leaving group that is becoming negatively charged (or developing a nonbonded pair of electrons) as shown in Figure 10.20.

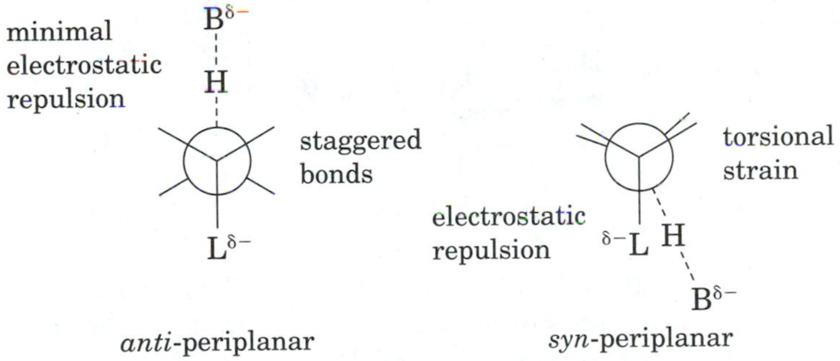

Figure 10.20 *Anti*-periplanar and *syn*-periplanar conformations for 1,2-elimination.

[49]DePuy, C. H.; Morris, G. F.; Smith, J. S.; Smat, R. J. *J. Am. Chem. Soc.* **1965**, *87*, 2421. Cyclopentyl derivatives were studied because coplanarity is achieved more easily in five-membered rings than in six-membered rings.

[50]The interplay of steric and electronic factors can lead to a very shallow potential energy surface with regard to the H-C_β-C_α-L bond angle. Gronert found a slight preference for syn-clinal transition structures (20°–60°) in calculations of the gas phase reaction of F^- with alkyl chlorides: Gronert, S. *J. Am. Chem. Soc.* **1992**, *114*, 2349; **1993**, *115*, 652; *J. Org. Chem.* **1994**, *59*, 7046.

In one of the earliest studies of the stereochemistry of the E2 reaction, Cristol found the rate constants for the dehydrochlorination of the β isomer of benzene hexachloride (1,2,3,4,5,6-hexachlorocyclohexane), in which each chlorine atom is cis to the hydrogen atoms on either side of it (**5**), to be about 10^{-4} that of the other benzene hexachloride isomers. Since each of the other isomers has at least one hydrogen atom trans to a chlorine atom on an adjacent carbon atom, the low reactivity of **5** suggested that the E2 reaction occurs preferentially when there is a trans relationship for the hydrogen atom and chlorine atom on the cycloalkane.[51]

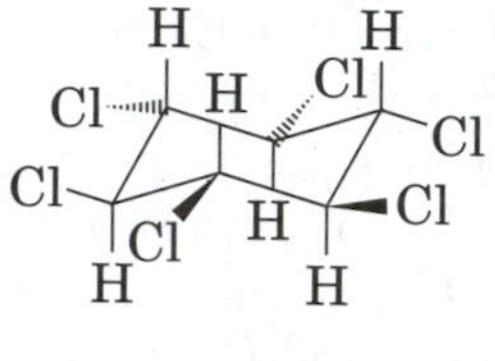

5

An indication for a similar preference for an *anti*-periplanar relationship in acyclic compounds was detected by Cram and co-workers. *threo*-1,2-Diphenyl-1-propyl chloride was found to undergo E2 elimination to give exclusively (*E*)-1,2-diphenylpropene (equation 10.17), while the erythro diastereomer produced only the (*Z*)-1,2-diphenylpropene (equation 10.18).[52,53]

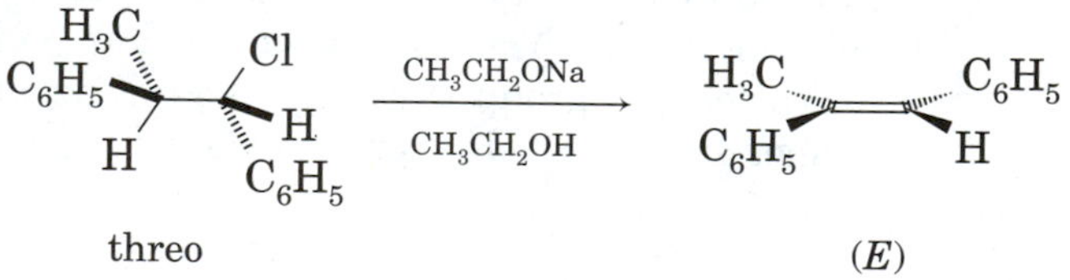

threo $\xrightarrow[CH_3CH_2OH]{CH_3CH_2ONa}$ (*E*) **(10.17)**

erythro $\xrightarrow[CH_3CH_2OH]{CH_3CH_2ONa}$ (*Z*) **(10.18)**

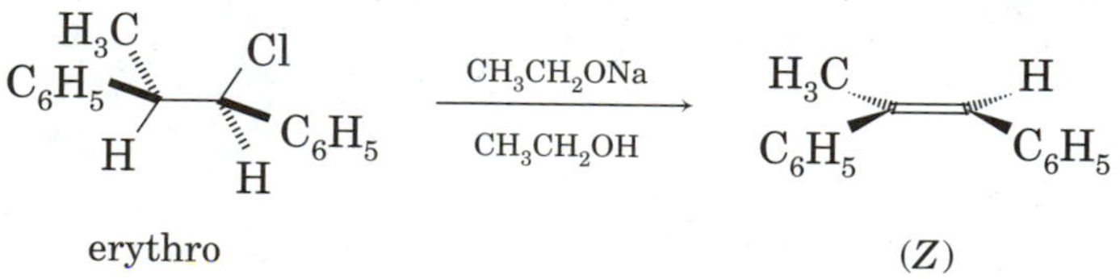

In spite of the electronic and steric factors favoring anti elimination, syn elimination does occur. One situation that gives rise to syn elimination is a molecular structure that precludes an *anti*-periplanar transition structure geometry.[54] For example, Cristol and Hoegger found that *trans*-2,3,-

[51]Cristol, S. J. *J. Am. Chem. Soc.* **1947**, *69*, 338. See also Cristol, S. J.; Hause, N.L.; Meek, J. S. *J. Am. Chem. Soc.* **1951**, *73*, 674.

[52]Cram, D. J.; Greene, F. D.; DePuy, C. H. *J. Am. Chem. Soc.* **1956**, *78*, 790.

[53]Winstein, S.; Pressman, D.; Young, W. G. *J. Am. Chem. Soc.* **1939**, *61*, 1645, had reported earlier that meso-2,3-dibromobutane reacts with sodium iodide in propanol to give only trans-2-butene, while (±)-2,3-dibromobutane produces only the cis isomer.

[54]Cristol, S. J.; Arganbright, R. P. *J. Am. Chem. Soc.* **1957**, *79*, 3441 and references therein.

dichloronorbornane (**6**) undergoes dehydrohalogenation to form 2-chloronorbene 85 times faster than does the *endo-cis* isomer (**7**). This rate difference was attributed to the fact that a *syn*-periplanar relationship is feasible in the transition structure for reaction of **6**, but an *anti*-periplanar relationship is not feasible for the transition structure in the reaction of **7** (Figure 10.20).[55]

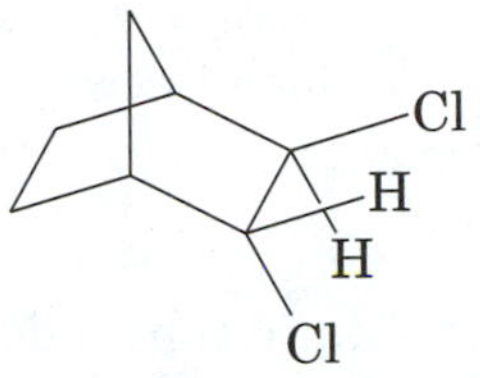

Cl H H Cl **6**   H H Cl Cl **7**

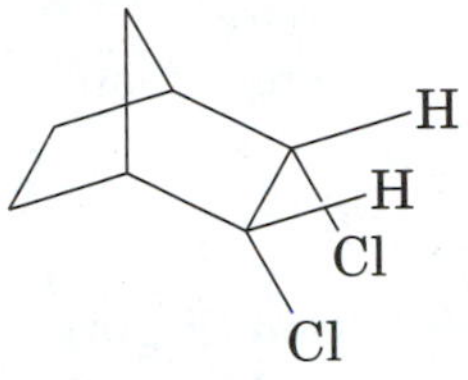

Similarly, the rate of dehydrochlorination of **9**, which must react by syn elimination, was determined to be about eight times greater than the rate of reaction of **8**, which can only react from an *anti*-clinal conformation (Figure 10.21).[56]

In some cases, steric factors may favor syn elimination in acyclic compounds.[57] Tao and Saunders found considerable syn elimination in reactions of acyclic compounds having the general formula $R_1R_2CHCHD(CH_3)_3N^+$ with hydroxide ion in 50 : 50 $(CH_3)_2SO$-H_2O at 80°.[58] When R_1 was phenyl, the percent of syn elimination was found to be 68.5% when R_2 was isopropyl and 26.5% when R_2 was CH_3. The results were rationalized in terms of a greater steric barrier between the leaving group and the substituents R_1 and R_2 in the transition structure for anti elimination than for syn elimina-

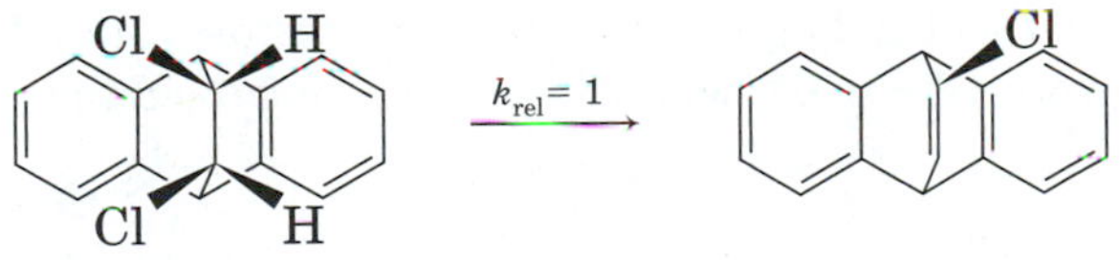

Figure 10.21 Relative rates of syn and anti elimination.

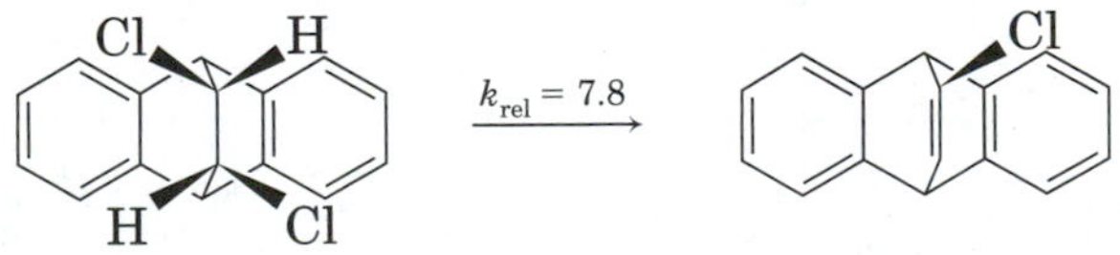

Cl H Cl H **8** $k_{rel} = 1$ Cl

Cl H H Cl **9** $k_{rel} = 7.8$ Cl

[55]Cristol, S. J.; Hoegger, E. F. *J. Am. Chem. Soc.* **1957**, *79*, 3438.

[56]Cristol, S. J.; Hause, N. L. *J. Am. Chem. Soc.* **1952**, *74*, 2193.

[57]Sicher and Závada and co-workers (reference 22 and references therein) noted that with some cyclic systems there is a marked preference for formation of *cis*-cycloalkenes through anti elimination but for formation of *trans*-cycloalkenes by syn elimination.

[58]Tao, Y.-T.; Saunders, Jr., W. H. *J. Am. Chem. Soc.* **1983**, *105*, 3183.

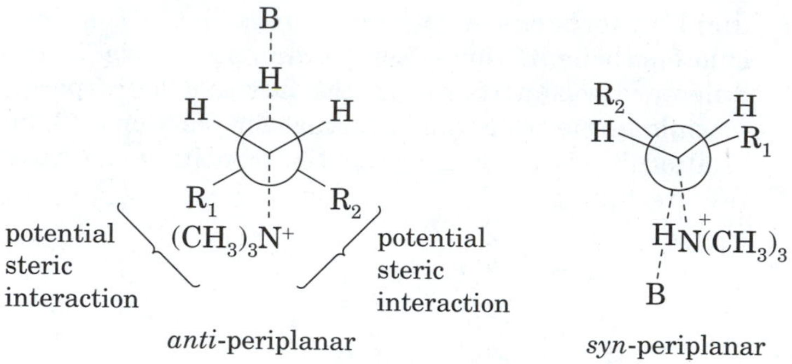

Figure 10.22 Greater steric interactions in anti (left) than in syn (right) transition structures.

tion (Figure 10.22). For these compounds both the syn and anti elimination pathways appear to have some carbanion character, although still the carbanion character is greater for syn elimination. For example, the ρ values for syn and anti elimination of substituted aryl derivatives were found to be 3.02 ± 0.22 for anti elimination and 3.69 ± 0.20 for syn elimination.[48]

Syn elimination is also seen in the dehydrohalogenation of β-halogen activated *trans*-1,2-dichlorocycloalkanes with heterogeneous mixtures of $NaNH_2$ and *t*-BuONa in THF.[59] For instance, the ratio of the rate constants for anti and syn elimination to form 1-chlorocycloheptene from *cis*- and *trans*-1,2-dichlorocycloheptane, respectively, was found to be about 9. This relatively small ratio was attributed to enhancement of the rate of syn elimination as a result of a cyclic, six-membered transition structure (Figure 10.23) in which both the halogen and the β-hydrogen atom interact with the surface of the complex base. In some cases, interaction between the base counterion and the leaving group may also be a factor in syn elimination in homogeneous solution.[60] For example, the percent of syn elimination by reaction of *meso*-1,2-dichloro-1,2-diphenylethane with *t*-BuOK in *t*-BuOH decreases from 13% to 0% when 18-crown-6 is added to the reaction mixture.[61]

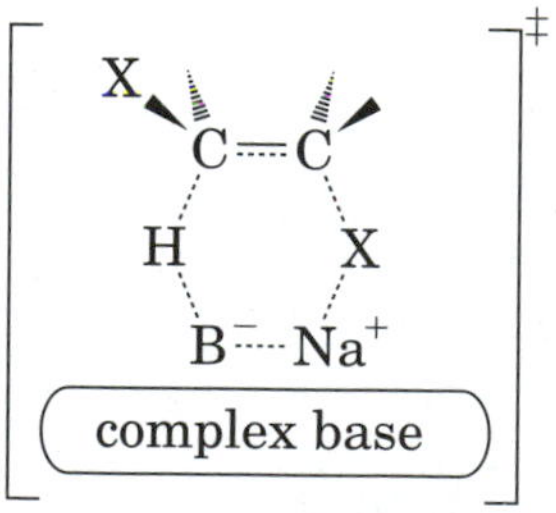

Figure 10.23 Cyclic transition structure suggested for syn dehydrohalogenation with heterogeneous bases. (Adapted from reference 59.)

[59]Croft, A. P.; Bartsch, R. A. *J. Org. Chem.* **1994**, *59*, 1930 and references therein.

[60]Hunter, D. H.; Shearing, D. J. *J. Am. Chem. Soc.* **1973**, *95*, 8333.

[61]Baciocchi, E.; Ruzziconi, R. *J. Org. Chem.* **1984**, *49*, 3395 and references therein.

The dehydrohalogenation of vinyl halides in the synthesis of alkynes shows conformational preferences analogous to those of alkyl halides.[62] Cristol and Norris found the reaction of (*Z*)-β-bromostyrene with hydroxide ion in isopropyl alcohol at 43° to be 2.1×10^5 faster than the reaction of the (*E*) isomer. The results were interpreted in terms of differing mechanisms for the eliminations of the two compounds. As shown in equation 10.19, the (*Z*) isomer can undergo concerted elimination of hydrogen and bromine because they are in the proper orientation for *anti*-periplanar elimination.

$$(Z)\text{-}C_6H_5CH{=}CHBr + HO^- \longrightarrow C_6H_5{-}C{\equiv}C{-}H + Br^- \qquad \textbf{(10.19)}$$

The *anti*-periplanar relationship is not possible with the (*E*) isomer, however, so an E1cb mechanism involving formation of a vinyl carbanion and subsequent elimination of the bromide ion was proposed (equation 10.20).[63] The observation that the ratio of the rate constants for reaction of (*Z*)-*p*-nitro-β-bromostyrene to that for reaction of the (*E*) isomer (1.6×10^4) was an order of magnitude smaller than the ratio of rate constants for (*Z*)- and (*E*)-β-bromostilbenes was attributed to the stabilization of the carbanion intermediate by the *p*-nitro group.[64]

$$(E)\text{-}C_6H_5CH{=}CHBr + HO^- \longrightarrow \left[C_6H_5\ddot{C}^-{=}CHBr \right] \longrightarrow C_6H_5{-}C{\equiv}C{-}H + Br^- \qquad \textbf{(10.20)}$$

With a less acidic $C_\beta{-}H$ proton, the carbanion mechanism in equation 10.20 is slower, but the concerted mechanism in equation 10.19 is not affected as significantly. Therefore the ratio of rates is much greater for elimination of HCl from isomeric chloroalkenes. Cristol and Helmreich were able to determine that the product of a synthesis of β-chlorostyrene was composed of 70% of the (*Z*) isomer and 30% of the (*E*) isomer because only 70% of the product mixture underwent dehydrochlorination with KOH-ethanol.[65]

Still another reaction pathway may become operative if the base is very strong, such as amide ion or an alkyllithium. When treated with phenyllithium in diethyl ether, for example, both (*Z*)- and (*E*)-β-bromostyrene are

---

[62] For a discussion of elimination reactions used for the synthesis of alkynes, see Jacobs, T. L. *Org. React.* **1949**, *5*, 1.

[63] The electronic reorganization associated with the formation of the alkyne is similar to the inversion of vinyl carbanion. Bach, R. D., Evans, J. C. *J. Am. Chem. Soc.* **1986**, *108*, 1374.

[64] Cristol, S. J.; Norris, W. P. *J. Am. Chem. Soc.* **1954**, *76*, 3005.

[65] Cristol, S. J.; Helmreich, R. F. *J. Am. Chem. Soc.* **1955**, *77*, 5034.

converted to phenylacetylene. However, because the ratio of rate constants for reaction of the (*E*) and (*Z*) isomers differed only by a factor of two in this case, the results seemed more consistent with a 1,1-elimination mechanism:[65]

$$(C_6H_5)(H)C{=}C(H)(Br) + C_6H_5Li \xrightarrow{\text{ether}} (C_6H_5)(H)C{=}C(Li)(Br) + C_6H_6 \quad (10.21)$$

$$(C_6H_5)(H)C{=}C(Li)(Br) \longrightarrow (C_6H_5)(H)C{=}C\colon + LiBr \quad (10.22)$$

$$(C_6H_5)(H)C{=}C\colon \longrightarrow C_6H_5{-}C{\equiv}C{-}H \xrightarrow{C_6H_5Li} C_6H_5{-}C{\equiv}C{-}Li \quad (10.23)$$

$$C_6H_5{-}C{\equiv}C{-}Li \xrightarrow[\text{up}]{\text{work}} C_6H_5{-}C{\equiv}C{-}H \quad (10.24)$$

Extending the intuitive idea that a concerted 1,2-elimination reaction is an intramolecular S_N2 process suggests that a concerted 1,4-elimination can be considered an intramolecular analogue of the S_N2' reaction.[66] Therefore it should occur with net syn stereochemistry, and most concerted 1,4-eliminations have been found to occur with syn stereochemistry.[67,68] For example, in the reaction shown in equation 10.5, the elimination was found to be more than 99% syn, consistent with the mechanism shown in Figure 10.24.[13] In some cases, however, it appears that 1,4-eliminations occur not by a concerted pathway but by a mechanism with considerable carbanion character.[69] In the gas phase, 1,4-elimination of methanol from 3-methoxycyclohexene has been found to occur by a concerted syn pathway when the base is weak (such as fluoride), but by a nonstereoselective carbanion pathway when the base is stronger (such as amide or hydroxide).[70]

---

[66]This intuitive idea is consistent with the results of theoretical calculations: (a) Fukui, K. *Tetrahedron Lett.* **1965**, 2427; (b) Nguyen Trong Anh, *Chem. Commun.* **1968**, 1089; (c) Tee, O. S.; Altmann, J. A.; Yates, K. *J. Am. Chem. Soc.* **1974**, *96*, 3141.

[67]See, for example, Hill, R. K.; Bock, M. G. *J. Am. Chem. Soc.* **1978**, *100*, 637; Cristol, S. J.; Barasch, W.; Tieman, C. H. *J. Am. Chem. Soc.* **1955**, *77*, 583.

[68]Failure to observe syn stereochemistry in an enzyme-catalyzed 1,4-elimination has been used to support the conclusion that the enzyme-catalyzed reaction is a two-step process and not a concerted reaction: (a) Hill, R. K.; Newkome, G. R. *J. Am. Chem. Soc.* **1969**, *91*, 5893; (b) Onderka, D. K.; Floss, H. G. *J. Am. Chem. Soc.* **1969**, *91*, 5894.

[69]Cristol, S. J. *Acc. Chem. Res.* **1971**, *4*, 393.

[70]Rabasco, J. J.; Kass, S. R. *J. Org. Chem.* **1993**, *58*, 2633.

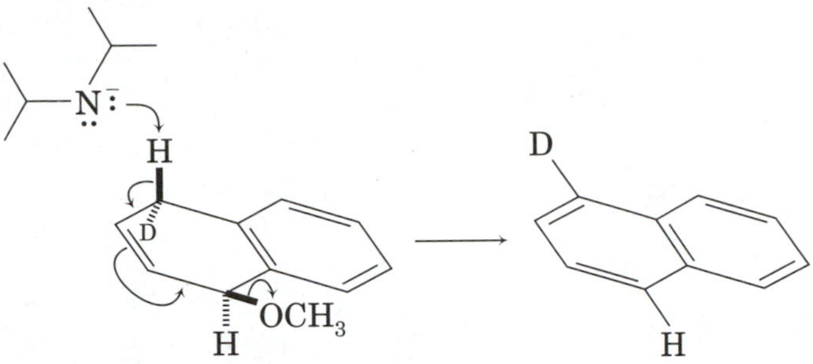

Figure 10.24 Possible mechanism for syn 1,4-elimination.

Regiochemistry of 1,2-Elimination Reactions

Many substrates have nonequivalent β-protons,[71] so a 1,2-elimination may produce more than one alkene. For example, ethoxide-promoted elimination of HI from 2-iodo-3-methylbutane produced 82% of 2-methyl-2-butene and 18% of 3-methyl-1-butene (equation 10.25).[72] The generalization that 1,2-elimination reactions of alkyl halides usually give the more highly substituted alkene is known as the ***Saytzeff rule***. Saytzeff observed that the regiochemistry of elimination could be correlated with removal of a hydrogen atom from that β-carbon atom of an alkyl halide that has the smaller number of hydrogen atoms.[73] Hughes and Ingold generalized this observation to mean that elimination from alkyl halides generally produces in higher yield the alkene with the greater number of alkyl substituents on the carbon-carbon double bond.[4]

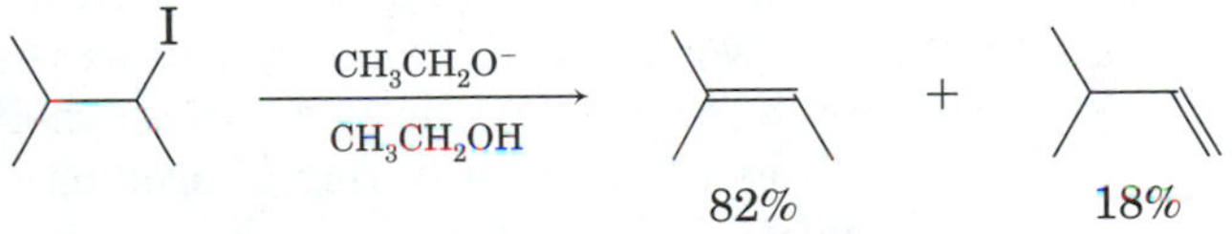

$$\xrightarrow[CH_3CH_2OH]{CH_3CH_2O^-} \quad 82\% \quad + \quad 18\% \qquad (10.25)$$

The Saytzeff rule applies to both E1 and E2 reactions. For example, the reaction of *t*-amyl bromide with 0.05 M sodium ethoxide in ethanol at 25° was found to occur by competing E1 and E2 pathways.[74] Both the E1 and E2 processes produced higher yields of 2-methyl-2-butene (**11**) than of 2-methyl-1-butene (**12**). The percent yield of **11** was found to be 71% by the E2 pathway and 82% by the E1 pathway, so the percent of 2-methyl-2-butene decreases with increasing concentration of ethoxide in the reaction mixture.[75]

---

[71] That is, β-protons that are constitutionally heterotopic.

[72] Hughes, E. D.; Ingold, C. K.; Mandour, A. M. M. *J. Chem. Soc.* **1948**, 2090.

[73] Saytzeff, A. *Liebigs Ann. Chem.* **1875**, *179*, 296. See also the discussion in reference 4.

[74] The reaction produced a total of 56% elimination accompanied by 44% substitution.

[75] Dhar, M. L.; Hughes, E. D.; Ingold, C. K. *J. Chem. Soc.* **1948**, 2065.

$$\underset{\textbf{10}}{H_3C{-}CH_2{-}C(CH_3)_2{-}Br} \xrightarrow{E2} \underset{82\%}{(H_3C)(H)C{=}C(CH_3)(CH_3)} + \underset{18\%}{(CH_3CH_2)(H_3C)C{=}C(H)(H)} \qquad \textbf{(10.26)}$$

$$\underset{\textbf{10}}{H_3C{-}CH_2{-}C(CH_3)_2{-}Br} \xrightarrow{E1} \underset{\textbf{11}}{\underset{71\%}{(H_3C)(H)C{=}C(CH_3)(CH_3)}} + \underset{\textbf{12}}{\underset{29\%}{(CH_3CH_2)(H_3C)C{=}C(H)(H)}} \qquad \textbf{(10.27)}$$

The tendency for alkyl halides to eliminate with Saytzeff orientation can be rationalized with the Hammond postulate. Figure 10.25 is a modification of Figure 10.8 to show two competing E2 reactions. Because the more highly substituted alkene is more stable than the less highly substituted isomer and because the transition structure for each pathway has some double bond character, the transition structure leading to the more highly substituted alkene is more stable than the transition structure leading to the less highly substituted isomer.[4,76] If the eliminations are effectively irreversible under the reaction conditions, the product produced by the pathway with lower activation energy will be the major product (assuming that activation entropies for the two pathways are similar). An analogous argument applies to the distribution of products formed by proton removal from a carbocation intermediate in an E1 reaction (Figure 10.26).

Not all eliminations follow the Saytzeff rule. The Hofmann elimination involves heating a tetraalkylammonium hydroxide until reaction occurs to produce an alkene along with a trialkylamine and water as byproducts. Hofmann[77] reported that elimination of a quaternary ammonium hydroxide containing differing alkyl groups on the nitrogen always produced ethene if an ethyl group was one of the substituents. An example of the Hofmann

---

[76]The stabilization of the incipient double bond in the transition structure is referred to as the "electromeric effect" in the early papers. The term *electromeric* refers to hyperconjugation involving hydrogen atoms γ to the leaving group. However, it is not necessary to ascribe the effect of alkyl substituents on the β-carbon atom to any specific interaction. On the basis of the Hammond postulate, one can simply say that because alkenes with greater number of alkyl groups on the carbon-carbon double bond are more stable than isomeric alkenes with fewer alkyl substituents on the carbon-carbon double bond, the forces that affect alkene stability will—to some extent—affect the transition structures leading to the formation of the alkenes.

[77]This is one spelling of a famous name in chemistry. There are many other Hofmann "name" rules and reactions in chemistry, and some of them are spelled differently.

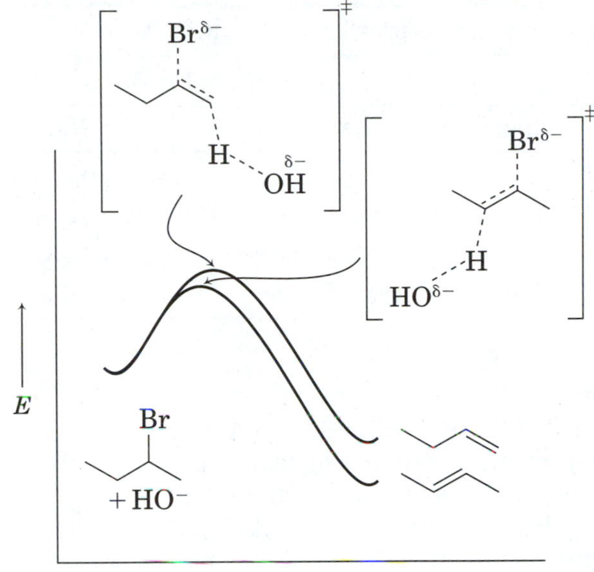

Figure 10.25 Reaction coordinate diagram for dehydrobromination of 2-bromobutane to 1-butene and 2-butene with Saytzeff orientation.

elimination is shown in equation 10.28[78] Since the quaternary ammonium ion has three β-protons on the ethyl group and two β-protons on the propyl group, reaction of hydroxide ion with the alkyl groups on a purely statistical basis would lead to 60% ethene and 40% propene. However, the observed

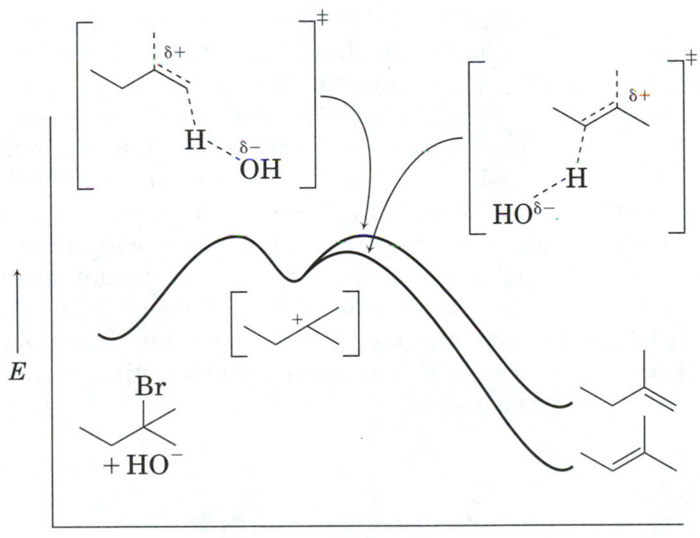

Figure 10.26 Reaction coordinate diagram for E1 elimination of HBr from 2-bromo-2-methylbutane.

[78]Cope, A. C.; LeBel, N. A.; Lee, H.-H.; Moore, W. R. *J. Am. Chem. Soc.* **1957**, *79*, 4720.

Table 10.3 Product distributions in Hofmann eliminations of $R_1R_2(CH_3)_2N^+$ HO^-.

Alkyl Groups		Olefinic Products			
		Experimental Results		Statistical Prediction	
R_1	**R_2**				
ethyl	propyl	97.6% ethene	2.4% propene	60% ethene	40% propene
ethyl	isopropyl	41.2% ethene	58.8% propene	33.3% ethene	67.7% propene
ethyl	isobutyl	99.1% ethene	0.9% isobutene	75% ethene	25% isobutene
n-butyl	isobutyl	64% 1-butene	36% isobutene	67.7% 1-butene	33.3% isobutene

(Data from a tabulation in reference 78.)

product distribution is quite different. Furthermore, the data in Table 10.3 show that product distributions from Hofmann elimination reactions are usually very different from a statistical prediction based only on the number of β-hydrogen atoms on each alkyl group. The observation that the Hofmann elimination usually produces mostly the less highly substituted alkene (ethene instead of propene in the example shown) is known as the ***Hofmann rule***.[79]

$$(H_3C)_2N^+(CH_2CH_3)(CH_2CH_2CH_3) \xrightarrow[\Delta]{HO^-} \underset{97.6\%}{CH_2{=}CH_2} + \underset{2.4\%}{CH_2{=}CHCH_3} \qquad (10.28)$$

Hughes, Ingold, and co-workers generalized Hofmann's observation to mean that elimination of an alkene from a quaternary ammonium ion bearing only alkyl groups will produce as the major product that alkene with the fewer number of alkyl substituents on the carbon-carbon double bond.[4,80] This generalization was not limited to the Hofmann elimination itself, but it was also applied to any bimolecular elimination in which the leaving group bears a positive charge when bonded to the substrate (and thus leaves as a neutral species). As an example, Ingold and co-workers noted that ethoxide-promoted elimination of dimethyl-*sec*-butylsulfonium ion produced 26% of the 2-butenes (predominantly trans) and 74% of 1-butene (equation 10.29).[81] Other molecules that give similar selectivity include tetraalkylphosphonium salts, so Hofmann orientation is said to be a characteristic of the 'onium compounds.

[79]Hofmann, A. W. *Liebigs Ann. Chem.* **1851**, *78*, 253; **1851**, *79*, 11.

[80]Hanhart, W.; Ingold, C. K. *J. Chem. Soc.* **1927**, 997.

[81]Hughes, E. D.; Ingold, C. K.; Maw, G. A.; Woolf, L. I. *J. Chem. Soc.* **1948**, 2077. The product distributions in equation 10.29 reflect only the elimination products. More than 30% substitution product was also obtained.

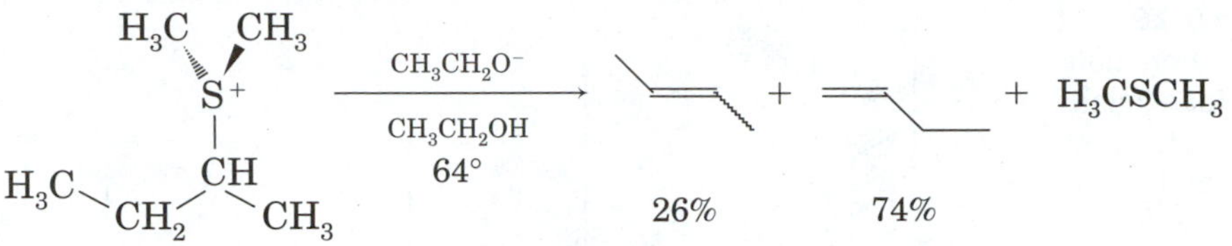

$$\ce{(CH3)2S^+-CH(CH3)CH2CH3} \xrightarrow[\text{CH}_3\text{CH}_2\text{OH, }64^\circ]{\text{CH}_3\text{CH}_2\text{O}^-} \underset{26\%}{\text{2-butene}} + \underset{74\%}{\text{1-butene}} + \text{H}_3\text{CSCH}_3 \quad \textbf{(10.29)}$$

The *anti*-periplanar reaction pathways that lead to Saytzeff or Hofmann orientation are shown in Figure 10.27. Why should one kind of molecule react preferentially by one pathway, while another reacts primarily by a different pathway? If the elimination reaction is not reversible under the experimental conditions, then a higher yield of Hofmann product than Saytzeff product suggests that the transition state leading to the Saytzeff product must be higher in energy than that leading to Hofmann product. Since the more highly substituted alkene is more stable than the less highly substituted alkene, no matter the pathways by which these products are formed, then the transition states in a reaction producing Hofmann orientation must not reflect product stabilities. That conclusion suggests a reaction coordinate diagram such as the one shown in Figure 10.28, in which either some interaction raises the energy of the transition state leading to the more highly substituted alkene or some interaction lowers the energy of the transition state leading to the less highly substituted alkene.

Hughes and Ingold ascribed Hofmann orientation to an inductive effect of the positively charged leaving group, which was said to lower the energy of the transition structure leading to the less highly substituted alkene.[4] As

Figure 10.27 Two views of the generalized Saytzeff and Hofmann eliminations.

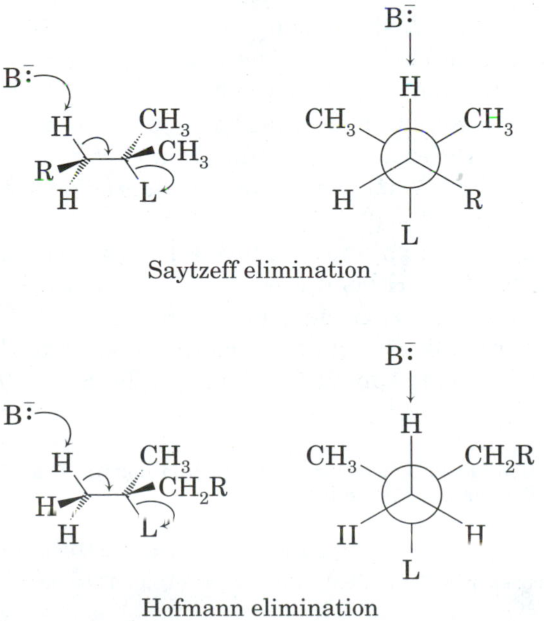

Saytzeff elimination

Hofmann elimination

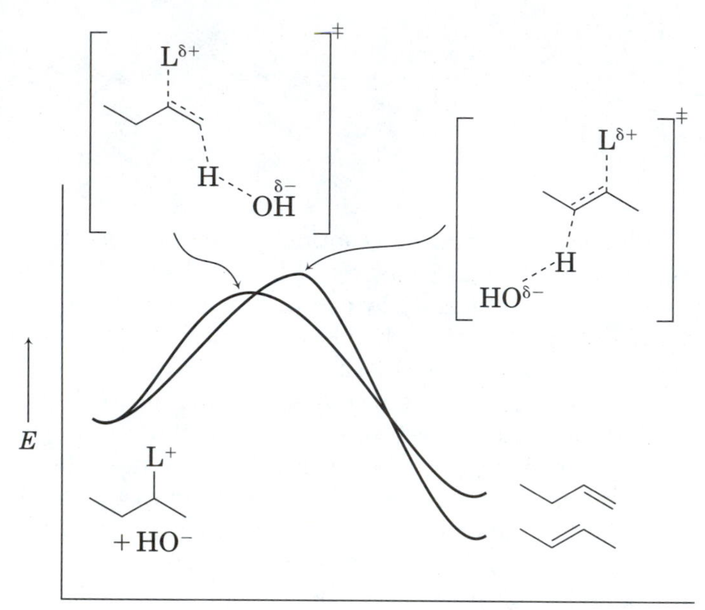

Figure 10.28 Reaction coordinate diagram for elimination leading to 1-butene and 2-butene with Hofmann orientation.

shown in Figure 10.29, Ingold suggested that a cationic substituent would withdraw electron density from the α-carbon atom, which in turn would withdraw electron density from the other atoms in the structure. This electron withdrawal was thought to increase the acidity of a β-hydrogen atom, thus lowering the energy of a transition structure involving removal of the β-proton by base. However, a more highly substituted β-carbon atom was said to have less acidic hydrogen atoms due to the electron donating ability of the attached alkyl group(s). The more acidic protons on the less substituted β-carbon atom would, therefore, be expected to react faster with a base, leading to Hofmann orientation.[82] Calculations suggested that the activation energy differences resulting from such electron withdrawal might be as much as 1 kcal/mol, which was considered adequate to explain the experimental results.[83]

The induction explanation for Hofmann orientation suggests that breaking of the C_β—H bond precedes C_α—L dissociation in the transition structure, so the reaction must have some E1cb character,[84] and the Hofmann elimination does seem to be more sensitive to the electronic effect of electron withdrawing substituents than is the Saytzeff elimination. For ex-

---

[82]This discussion is couched in terms of "electron donating" alkyl groups and in terms of "through-bond" induction. As we have seen, these may not be the most generally applicable models for these phenomena.

[83]Banthorpe, D. V.; Hughes, E. D.; Ingold, C. *J. Chem. Soc.* **1960**, 4054.

[84]See the discussion in Ford (cited in reference 24), p. 411 and in Wolfe, S. *Acc. Chem. Res.* **1972**, *5*, 102.

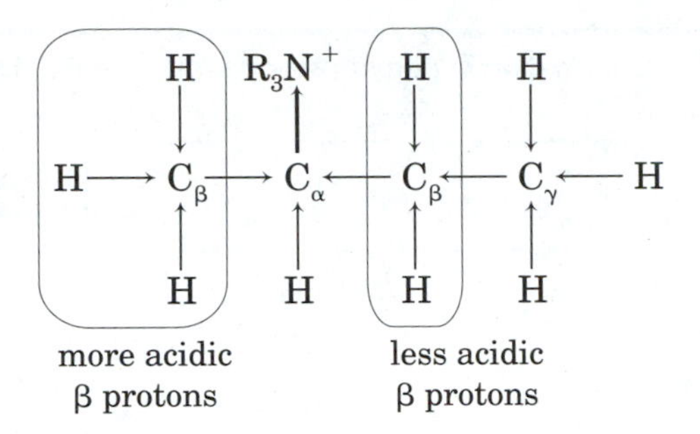

Figure 10.29
An inductive model for Hofmann elimination.

ample, a Hammett correlation for the bimolecular elimination of a series of 2-phenylethyl bromides gave a ρ of 2.1, while elimination of dimethyl sulfide from a series of 2-phenylethyldimethylsulfonium bromides gave a ρ value of 2.6.[85]

Steric effects must also be considered as an explanation for Hofmann orientation in bimolecular elimination reactions of 'onium compounds, however. The leaving group in 'onium compounds is usually much larger than the leaving group in substrates that exhibit Saytzeff orientation.[86] As shown in Figure 10.30, approach of a base to a molecule with a bulky leaving group will be more difficult on a carbon atom with many substituents than on one with few. This interaction could cause the transition state energy for the Hofmann pathway to be lower in energy than the transition state energy for the Saytzeff pathway.

In a series of papers in 1956, Brown and co-workers reported evidence that increasing the size of any one or all of the alkyl substituents on the β-carbon atom(s), the leaving group, and the attacking base can shift the proportion of products from Saytzeff orientation toward Hofmann orientation. For example, Brown investigated the elimination of HBr from a series of 3°

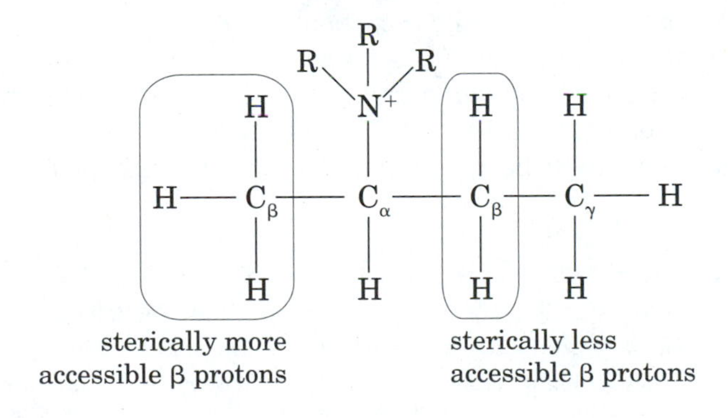

Figure 10.30
Steric explanation for Hofmann elimination.

[85]DePuy, C. H.; Froemsdorf, D. H. *J. Am. Chem. Soc.* **1957**, *79*, 3710; Saunders, Jr., W. H.; Williams, R. A. *J. Am. Chem. Soc.* **1957**, *79*, 3712.

[86]This idea was suggested by Schramm, C. H. *Science* **1950**, *112*, 367, but experimental evidence was not provided.

Table 10.4 1-Alkene formation in dehydrohalogenation of $RCH_2C(CH_3)_2Br$ with pyridine or ethoxide bases.

3° Bromide	R	% 1-Olefin (Pyridine[a])	% 1-Olefin (Ethoxide[b])
2-bromo-2-methylbutane	Methyl	25	30
2-bromo-2-methylpentane	Ethyl	32	50
2-bromo-2,4-dimethylpentane	*i*-Propyl	44	54
2-bromo-2,4,4-trimethylpentane	*t*-Butyl	70	86

[a]Neat pyridine, 70°
[b]1 M $KOCH_2CH_3$ in ethanol, 70°

bromides, $RCH_2C(CH_3)_2Br$ by pyridine and by potassium ethoxide (equation 10.30).[87] The results, shown in Table 10.4, indicate that increasing the number of alkyl substituents on the β-carbon atom increases the percentage of Hofmann orientation.

$$RCH_2-C(Br)(CH_3)CH_3 \xrightarrow{\text{base}} \underset{\text{1-Olefin}}{RCH_2-C(=CH_2)CH_3} + \underset{\text{2-Olefin}}{(R)(H)C=C(CH_3)_2} \tag{10.30}$$

Brown and Wheeler studied the effect of the leaving group on product distribution in ethoxide-promoted E2 reactions of a series of 2-pentyl compounds in ethanol solution (Table 10.5).[88] The data are consistent with a model in which a larger leaving group produces a greater yield of the less substituted alkene, whether or not the leaving group has a positive charge on the atom bonded to the α-carbon atom.

To study the effect of base size, Brown and co-workers determined the product distribution for dehydrobromination of 2-bromo-2-methylbutane by a series of alkoxides.[89,90,91] The percent 1-alkene (2-methylbutene) formed with each base was found to be 30% with $CH_3CH_2O^-$, 72.5% with

[87]Brown, H. C.; Moritani, I.; Nakagawa, M. *J. Am. Chem. Soc.* **1956**, *78*, 2190.

[88]Brown, H. C.; Wheeler, O. H. *J. Am. Chem. Soc.* **1956**, *78*, 2199.

[89]Brown, H. C.; Moritani, I.; Okamoto, Y. *J. Am. Chem. Soc.* **1956**, *78*, 2193.

[90]The tendency for highly hindered bases to give more Hofmann orientation has been used synthetically, as in the isomerization of 1-methylcyclohexene to methylenecyclohexane by first adding HCl and then eliminating HCl with a sterically hindered base. Acharya, S. P.; Brown, H. C. *Chem. Commun.* **1968**, 305.

[91]Gould, E. S. *Mechanism and Structure in Organic Chemistry*; Holt, Rinehart and Winston: New York, 1959; p. 485, noted that this result could occur because of steric effects between the base and the substrate or because of intramolecular strain in the substrate in conformations required for E2 elimination to occur.

Table 10.5 Dependence of elimination products on leaving group in elimination of HL from 2-pentyl compounds.

2-Pentyl Derivative, RL	L	% 1-Pentene	% *cis*-2-Pentene	% *trans*-2-Pentene
Bromide[a]	Br	31	18	51
Iodide[a]	I	30	16	54
Tosylate[a]	OTs	48	18	34
Dimethylsulfonium[a]	$(CH_3)_2S^+$	87	5	8
Methyl sulfone[b]	SO_2CH_3	89	2.5	9
Trimethylammonium[b]	$(CH_3)_3N^+$	98	1	1

(Data from Reference 88.)
[a]1 M $KOCH_2CH_3$, 80°
[b]4 M $KOCH_2CH_3$, 130°

$(CH_3)_3CO^-$, and 88.5% with $(CH_3CH_2)_3CO^-$, indicating greater Hofmann orientation with larger bases. Brown and Nakagawa also investigated the effects of base strength and size on the products of dehydrobromination of 2-bromo-2-methylbutane by a series of pyridine bases.[92] Here the distribution of 1-alkene in the products was found to vary with base as follows: 4-picoline, 25%; pyridine, 25%; 2-picoline, 30%; and 2,6-lutidine, 44.5%. Since 4-picoline and pyridine differ in base strength but should have similar steric requirements in elimination, while 2- and 4-picoline have similar base strengths but different steric requirements, the results were consistent with the view that it is the steric requirement of the larger base in the transition structure for E2 elimination that leads to a greater proportion of Hofmann elimination.

In some cases changing the strength of the base in a series of alkoxides can affect product distributions, however. Froemsdorf and Robbins found that in the elimination of *sec*-butyl tosylate in DMSO, 31% of 1-butene was formed when the base was phenoxide, but only 16% of the 1-butene was formed when *p*-nitrophenoxide was the base. The two anions have different basicities, but the steric environments near the oxygen anions should be similar. This result suggests that change in base strength does influence the regioselectivity of the E2 elimination, which is consistent with the induction model of Ingold.[93,94]

[92]Brown, H. C.; Nakagawa, M. *J. Am. Chem. Soc.* **1956**, *78*, 2197.

[93]Froemsdorf, D. H.; Robbins, M. D. *J. Am. Chem. Soc.* **1967**, *89*, 1737.

[94]The formation of a slightly larger yield of 1-alkene with ethoxide than with pyridine for a given alkyl bromide in Table 10.4 is consistent with a tendency toward more Hofmann product with increasing base strength. However, this effect need not be related to the acidities of the different β-protons. Instead, it may reflect an earlier transition state and, as a result, a smaller dependence of the difference in $\Delta G^\ddagger$ values for the two elimination pathways ($\Delta\Delta G^\ddagger$) on the thermodynamic stability of the two alkenes.

The results of one experiment strongly suggest that steric effects can overcome any inductive effect by an 'onium group. As shown in Figure 10.31, neomenthyl chloride (**13**) underwent E2 reaction with ethoxide ion in ethanol at 100° to give 78% of 3-menthene (**14**) and 22% of 2-menthene (**15**). With trimethylamine as the leaving group instead of chloride (in **16**), the product of elimination with hydroxide as base in water at 156° consisted of 88% of 3-menthene.[95] In other words, this Hofmann elimination gives Saytzeff orientation. A possible explanation for this seemingly contradictory result may be that the molecule adopts a conformation in which steric effects are less important, so the transition states for the two elimination pathways more closely reflect stabilities of the products. However, it may also be true that steric effects are actually more important than in most Hofmann eliminations because steric strain of the axial $(CH_3)_3N^+$ group weakens the C-N bond and changes its leaving group ability.[96,97]

Figure 10.31
Product distributions from second-order eliminations from neomenthyl chloride and neomenthyltrimethylammonium hydroxide.

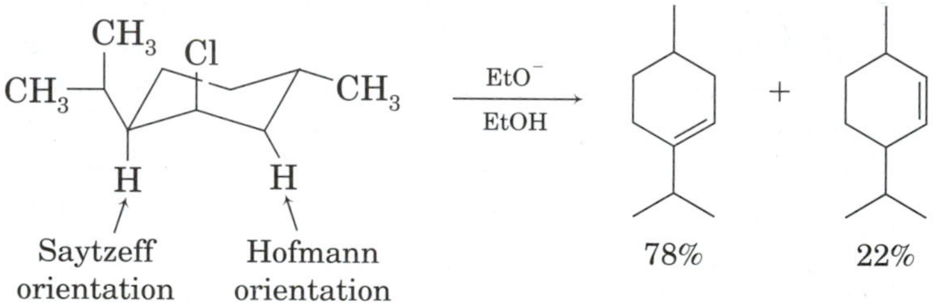

CH_3 CH_3 Cl CH_3 H H Saytzeff orientation Hofmann orientation **13** $\xrightarrow[\text{EtOH}]{\text{EtO}^-}$ 78% **14** + 22% **15**

CH_3 CH_3 $\overset{+}{N}(CH_3)_3$ CH_3 H H Saytzeff orientation Hofmann orientation **16** $\xrightarrow{\text{OH}^-}$ 88% + 12%

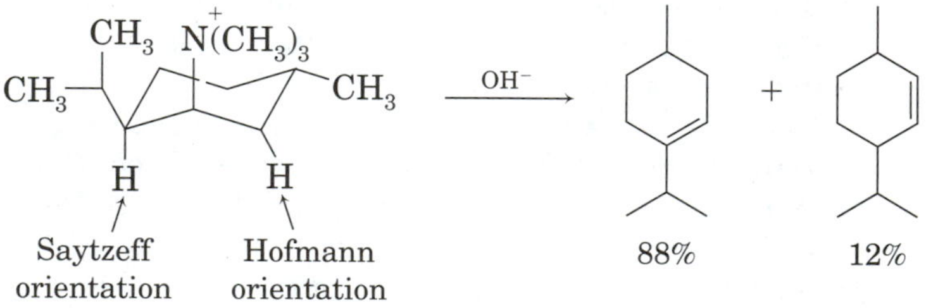

[95]Hughes, E. D.; Wilby, J. *J. Chem. Soc.* **1960**, 4094. The product distributions reflect second order elimination. Some first order elimination accompanied the reaction.

[96]Reference 21, p. 192 and references therein.

[97]It is interesting to consider the possibility that the theory of Hofmann orientation as presented by Hughes and Ingold was based in part on a typographical error. Hughes and Ingold cited the report by Hückel, W.; Tappe, W.; Legutke, G. *Liebigs Ann. Chem.* **1940**, *543*, 191, that neomenthyl chloride underwent base elimination to give 75% 3-methene, while neomenthyltrimethylammonium ion gave 80% 2-methene. However, Brown, H. C.; Moritani, I. *J. Am. Chem. Soc.* **1956**, *78*, 2203, noted that Hückel's experimental data indicate that the trimethylammonium ion actually gave 80% 3-menthene—the report of 80% 2-menthene was a typographical error in the discussion section of the paper. A repetition of the experiment by McNiven, N. L.; Read, J. *J. Chem. Soc.* **1952**, *153*, confirmed that 3-menthene is the major product.

Table 10.6 Product distribution from reaction of 2-halohexanes with $NaOCH_3$.[a]

Leaving Group	% 1-hexene	% *trans*-2-hexene	% *cis*-2-hexene	Saytzeff/Hofmann Ratio	*trans/cis* 2-hexene Ratio
F	69.9	21.0	9.1	0.43	2.3
Cl	33.3	49.5	17.1	2.0	2.9
Br	27.6	54.5	17.9	2.6	3.0
I	19.3	63.0	17.6	4.2	3.6

(Data from reference 99.)
[a]Yields for reaction of 2-halohexanes with $NaOCH_3$ in CH_3OH at 100°.

The idea that leaving group ability, and not just leaving group size or charge, can determine the orientation of E2 reactions is supported by detailed studies of dehydrohalogenation of alkyl halides. Saunders and coworkers found that the major product from reaction of 2-fluoropentane with sodium ethoxide in ethanol is 1-pentene, even though the fluorine atom is neither large nor positively charged.[98] Furthermore, Bartsch and Bunnett found a definite trend toward more Hofmann orientation with a *smaller* leaving group (Table 10.6) in the reactions of 2-halohexanes with methoxide ion in methanol, which is exactly opposite the trend expected on the basis of size of the leaving group.[99,100]

Bartsch and Bunnett calculated the rate constants for formation of each of the products at several temperatures and determined the values of $\Delta H^{\ddagger}$ and $\Delta S^{\ddagger}$ for each product and for each halide. As shown in Table 10.7, the magnitudes of both $\Delta H^{\ddagger}$ and $\Delta S^{\ddagger}$ increase along the series I < Br < Cl < F. With all of the halogens, the $\Delta S^{\ddagger}$ value for formation of 1-hexene is less negative than the $\Delta S^{\ddagger}$ value for formation of either 2-hexene. Bartsch and Bunnett suggested that this trend may reflect some difference in entropies of the products being formed, or it may be a measure of the entropies of the transition structures themselves.[99] In either case, the $T\Delta S^{\ddagger}$ term tends to favor 1-hexene formation.[101] The more negative values of $\Delta S^{\ddagger}$ for Cl and—to an even greater degree—F than for Br and I were attributed to greater hydrogen bonding between the departing anion and solvent in the transition

[98]Saunders, Jr., W. H.; Fahrenholtz, S. R.; Caress, E. A.; Lowe, J. P.; Schreiber, M. *J. Am. Chem. Soc.* **1965**, *87*, 3401.

[99]Bartsch, R. A.; Bunnett, J. F. *J. Am. Chem. Soc.* **1968**, *90*, 408.

[100]The tendency for the Saytzeff elimination to give *trans*-2-hexene followed the trend F < Cl < Br < I. Some substitution product, 2-hexyl methyl ether, was also formed in the reactions.

[101]Among other factors to be considered in predicting the products of bimolecular elimination reactions are the role of aggregation of the base, which can be particularly significant for *t*-BuOK in *t*-BuOH solution. A variety of different models have been proposed to account for the interaction of aggregated bases with substrates in E2 reactions. For a discussion, see reference 22.

Table 10.7 Activation parameters for reactions of 2-halohexanes with methoxide ion.[a]

Leaving Group	1-Hexene		*trans*-2-Hexene		*cis*-2-Hexene	
	$\Delta H^{\ddagger}$	$\Delta S^{\ddagger}$	$\Delta H^{\ddagger}$	$\Delta S^{\ddagger}$	$\Delta H^{\ddagger}$	$\Delta S^{\ddagger}$
F	30.2	−13.5	29.1	−16.6	29.0	−18.7
Cl	27.1	−9.0	25.1	−11.2	26.2	−10.6
Br	25.3	−6.2	23.7	−7.1	24.1	−8.5
I	24.7	−6.0	22.1	−7.8	23.1	−7.5

(Data from reference 99.)
[a]Values of $\Delta H^{\ddagger}$ in kcal/mol; values of $\Delta S^{\ddagger}$ in eu.

structures for reactions of the chloride and fluoride. However, the authors concluded that this additional solvation could not produce a steric effect sufficient to make the tendency toward more Hofmann orientation with decreasing atomic number a result of the size of the leaving group. They also concluded that the polar effects of the halogens could not be the basis for the tendency toward Hofmann orientation because the Taft σ^* values for XCH_2 groups vary only slightly along the series.

Bartsch and Bunnett explained the results of their study in terms of a variable transition state theory that had earlier been advanced by Bunnett.[28] According to this model, the E2 mechanism can be synchronous, or it can have more E1 character or more E1cb character. That is, the reaction path could follow the diagonal line from reactant to product in Figure 10.6, or it could follow a path that curves toward one or the other of the corners corresponding to E1 or E1cb reaction. The effect of changing the reactant, solvent, or base in a particular elimination is expected to depend on whether the transition structure for the reaction is more like a carbanion (E1cb-like) or a carbocation (E1-like). Among the trends suggested by Bunnett are the following:[28]

1. Introduction of an α-aryl substituent or (to a smaller extent) an α-alkyl substituent should stabilize a developing carbocation and make the transition structure more E1-like.
2. Introduction of a β-aryl substituent should enhance the tendency toward E1cb reaction by stabilizing a developing carbanion at the β-carbon atom.
3. Introduction of a β-alkyl substituent should make the reaction more E1-like (or less E1cb-like). That is, an E1cb-like reaction should become more synchronous because the alkyl substituent would be expected to destabilize a developing carbanion and stabilize the incipient carbon-carbon double bond. An E1-like reaction should become even more E1-like.

4. Change to a better leaving group should make the reaction more nearly E1-like; conversely, change to a poorer leaving group should make the reaction more nearly E1cb-like.
5. Change to a more electronegative leaving group should lead to a transition structure with less carbocation character on the α-carbon atom and greater carbanion character on the β-carbon atom.

The last two trends are particularly relevant to the elimination reactions of the 2-hexyl halides. As indicated by the data in Table 10.7, leaving group ability in methanol solution follows the trend I > Br > Cl > F. In comparison with the other 2-hexyl halides, therefore, the transition structure for 2-hexyl fluoride should have relatively more carbanion character, so the stabilities of the developing carbon-carbon double bonds are less important in determining values of $\Delta G^{\ddagger}$ than is the case for the other 2-hexyl halides.[102,103]

Using the variable transition state model, the tendency for the 'onium compounds to give Hofmann orientation can be explained simply as being the result of the R_3N and R_2S moieties being relatively poor leaving groups. This model can also account for the failure to observe Hofmann orientation in the bimolecular elimination of trimethylamine from **16**. Steric repulsion (1,3-diaxial interactions) between $(CH_3)_3N^+$ and the other axial substituents, particularly the isopropyl group, would raise the energy of the reactive conformation of the starting material. This *steric compression* should therefore make trimethylamine a better leaving group in this molecule, so the elimination should be more nearly E2-like. As a result, the stability of the developing double bond lowers the energy of the pathway leading to 3-menthene, and a greater yield of the Saytzeff product is observed.[28]

A detailed experimental and theoretical study of the effects of isotopic substitution on the Hofmann elimination reaction reported by Eubanks and co-workers is consistent with the variable transition state model.[104] Data for ethoxide-promoted elimination of (2-phenylethyl)trimethylammonium-1-^{14}C and (2-phenylethyl)trimethylammonium-2-^{14}C bromide, as well as a review of previous kinetic isotope studies, led the authors to conclude that the mechanism occurs through an E2 mechanism with considerable E1cb character and with the β-proton more than half transferred to base in the transition structure (Figure 10.32).

Kinetic isotope effect data for derivatives with substituents on the benzene ring suggested that there is a progression of transition structures as the aryl substituent varies from electron withdrawing to electron donating.

---

[102]The enhanced E1cb character does not mean that the reaction is fully E1cb, however. Saunders, Jr., W. H.; Schreiber, M. R. *Chem. Commun.* **1966**, 145, found that there was no hydrogen isotope exchange in the elimination of 2-pentyl fluoride with ethoxide in CH_3CH_2OD solution.

[103]This argument can also rationalize the observation of more *cis*-2-hexene as the atomic number of the halogen decreases.

[104]Eubanks, J. R. I.; Sims, L. B.; Fry, A. *J. Am. Chem. Soc.* **1991**, *113*, 8821.

Figure 10.32 Transition structure proposed for Hofmann elimination of (2-phenylethyl)trimethylammonium ion. (Reproduced from reference 104.)

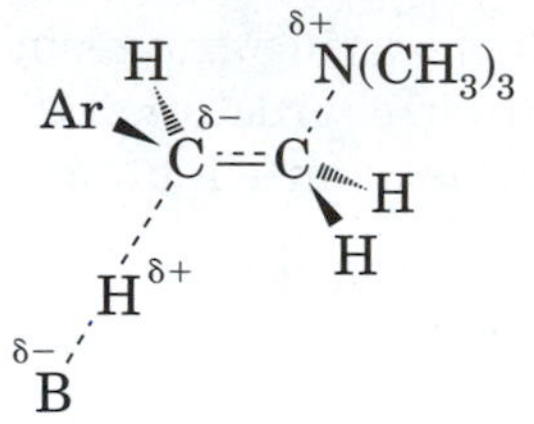

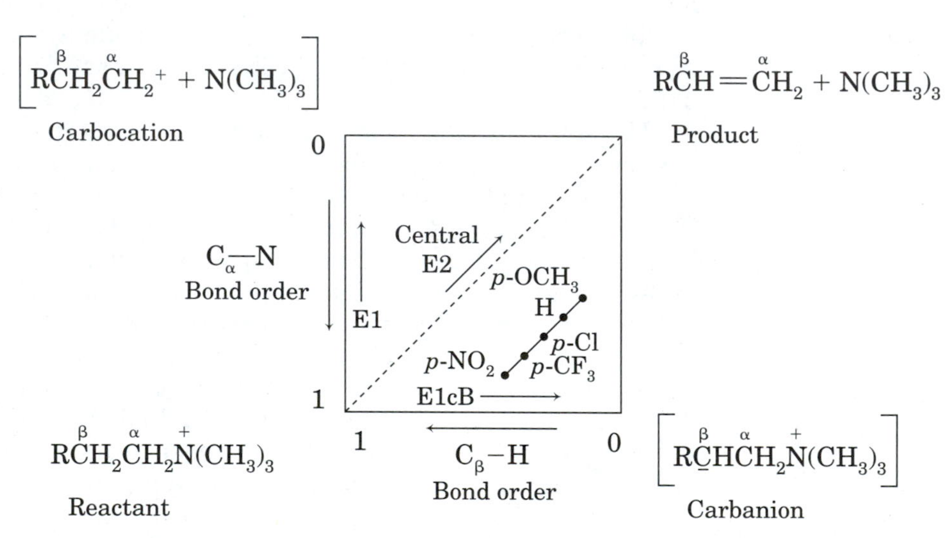

Figure 10.33 Transition state spectrum for elimination from substituted (2-phenylethyl)trimethylammonium ions. (Adapted from reference 104.)

This effect was attributed to stabilization of the carbanion character of the transition structure by electron withdrawing aryl substituents. As indicated by the points representing *p*-substituents in Figure 10.33, the investigators concluded that the transition structures become both more reactant-like and more E1cb-like as the substituents on the benzene ring become more electron withdrawing.[104]

10.3 Other 1,2-Elimination Reactions

Dehalogenation of Vicinal Dihalides

In each of the examples of 1,2-elimination reactions discussed above, one of the groups eliminated was a proton. However, there are many other types of elimination reactions. In particular, dehalogenation of vicinal dihalides has long been known.[105,106] Iodide ion can be used as the dehalogenating agent in both protic and aprotic solvents (equation 10.31). Chloride and bromide ions are effective in aprotic solvents such as DMF, but not in protic solvents, and the reaction of 1,2-dihalides with iodide is faster in DMF than in protic solvents. The overall process may be viewed either as the attack of a nucleophilic halide ion on a halogen atom or as a net two-electron reduction of the organic dibromide. In addition to halide ions, the reduction can also be accomplished with other nucleophiles or two-electron reducing agents, such as zinc or other metals.[107] Reactions with one-electron reducing agents have also been reported, so radical pathways for the elimination are possible.[108] While the substrate is most commonly a 1,2-dihalide, either or both of the halogens may instead be another leaving group, including hydroxy and alkoxy groups.[105,109]

$$\text{RCHBrCHBrR} \xrightarrow{\text{NaI}} \text{RHC}{=}\text{CHR} \qquad \textbf{(10.31)}$$

The stoichiometry of the elimination in equation 10.31 is[110]

$$\text{RCHXCHXR} + 3\,\text{I}^- \longrightarrow \text{RCH}{=}\text{CHR} + 2\,\text{X}^- + \text{I}_3^- \qquad \textbf{(10.32)}$$

Kinetic and stereochemical studies provided early clues to the mechanism of the reaction. The dehalogenation of a vicinal dibromide with iodide shows second order kinetics.[114]

$$\text{Rate} = k_2\,[\text{BrCH}_2\text{CH}_2\text{Br}][\text{I}^-] \qquad \textbf{(10.33)}$$

Substituent effects on the rates of elimination of aryl-substituted 1,2-diphenyl-1,2-dihaloethanes are small, suggesting that double bond formation is advanced in the transition structure and that most of the charge is localized on the departing halide ion.[107] Winstein and co-workers found the debromination of 2,3-dibromobutane with iodide ion to be highly stereospecific, with the meso diastereomer giving almost exclusively *trans*-2-butene, and the racemic diastereomer producing mostly *cis*-2-butene (Figure 10.34).[53] The results were taken as evidence for a concerted, anti elimination analogous to the E2 dehydrohalogenation reaction, as suggested in Figure 10.35.[111]

---

[105]At least one report in the literature dates to 1871. See the discussion in Mathai, I. M.; Schug, K.; Miller, S. I. *J. Org. Chem.* **1970**, *35*, 1733 and references therein.

[106]For an introduction, see reference 21, pp. 332–376.

[107]Baciocchi, E.; Schiroli, A. *J. Chem. Soc. B* **1969**, 554.

[108]Strunk, R. J.; DiGiacomo, P. M.; Aso, K.; Kuivila, H. G. *J. Am. Chem. Soc.* **1970**, *92*, 2849, have reported that tri-*n*-butyltin hydride effects the predominantly anti elimination of Br_2 from both (±)- and *meso*-2,3-dibromobutanes by a free radical chain reaction.

[109]Kochi, J. K.; Singleton, D. M. *J. Am. Chem. Soc.* **1968**, *90*, 1582.

[110]Goering, H. L.; Espy, H. H. *J. Am. Chem. Soc.* **1955**, *77*, 5023.

[111]Failure to observe complete stereospecificity may be the result of some S_N2 displacement of bromide or iodide prior to the elimination.

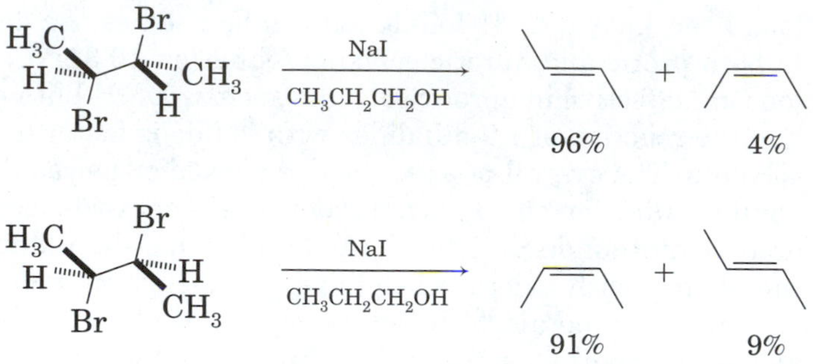

Figure 10.34 Stereospecific anti elimination in the debromination of *meso*- (top) and *racemic*-2,3-dibromobutane (bottom).

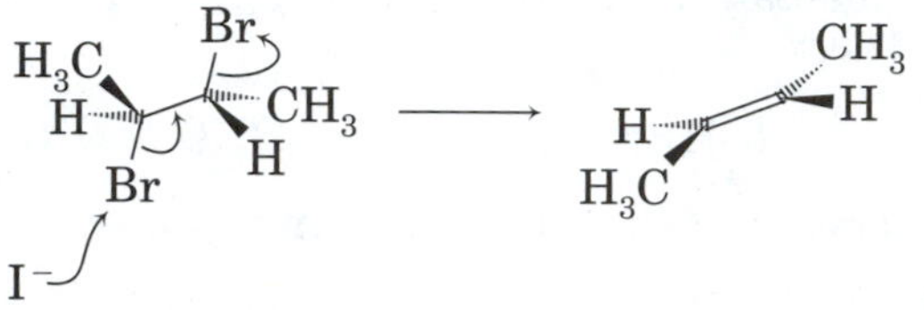

Figure 10.35 Concerted mechanism proposed for iodide-promoted dehalogenation reaction.

Barton and Rosenfelder found evidence for an *anti*-periplanar orientation of the two carbon-bromine bonds in iodine-promoted dehalogenation by determining the relative reactivity of two diastereomeric vicinal dibromide derivatives of cholanic acid. In the 11α,12β-dibromo compound **17**, each of the two bromine atoms is held in an axial conformation by the rigid steroid skeleton. On the other hand, each of the two bromine atoms is held in an equatorial conformation in the 11β,12α-dibromo isomer **18**. As a result, the conformation of the Br — C — C — Br grouping is anti in **17** but is gauche in **18**. When treated with sodium iodide in acetone, **17** underwent debromination readily, while **18** was unreactive under the same conditions.[112]

The conformational relationship of the bromine atoms is clear in **17** and **18** because of the rigidity of the steroid structure. However, individual cy-

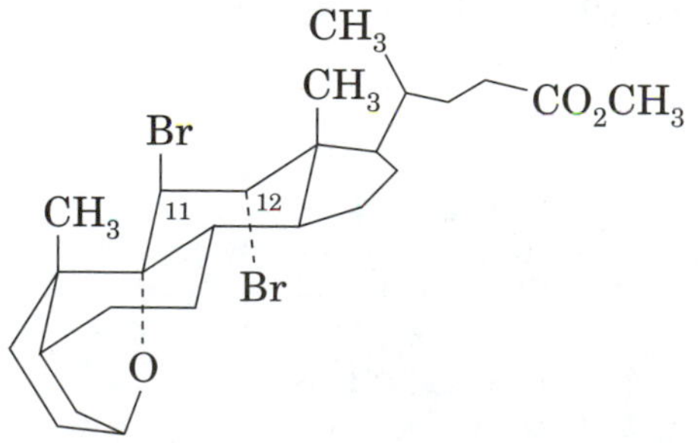

17 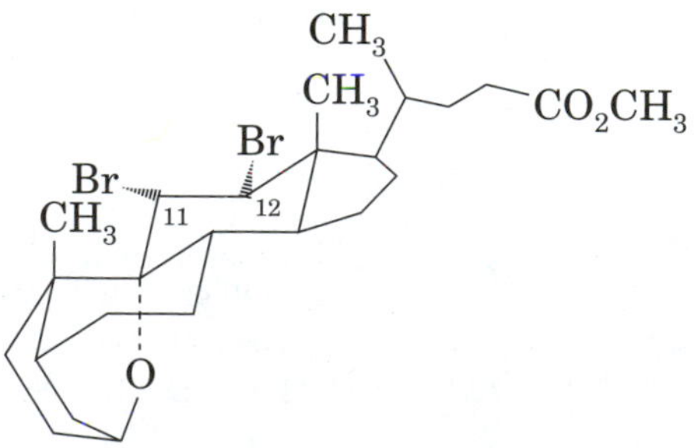 **18**

[112]Barton, D. H. R.; Rosenfelder, W. J. *J. Chem. Soc.* **1951**, 1048.

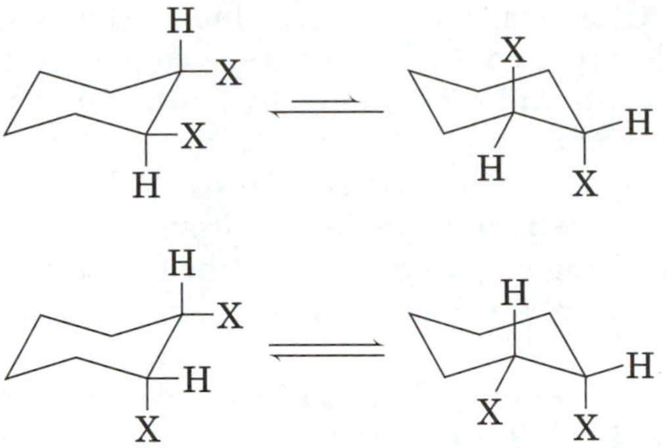

Figure 10.36 Conformational equilibria in *trans*- (top) and *cis*-1,2-dihalocyclohexane (bottom).

clohexane rings usually have considerable conformational mobility, and *trans*-1,2-dihalocyclohexanes exist as a mixture of diaxial and diequatorial conformers (Figure 10.36).

While the diequatorial conformer is expected to be the major conformer of the trans isomer, the *anti*-periplanar elimination suggested in Figure 10.35 can occur from the minor diaxial conformer. In the case of *cis*-1,2-dihalocyclohexanes, however, each of the two possible chair conformers has one axial and one equatorial halogen substituent. Therefore, it would appear that *cis*-1,2-dihalocyclohexanes should be unreactive. Nevertheless, iodide-promoted elimination does occur with cis isomers. For example, *cis*-1,2-dibromocyclohexane is debrominated by potassium iodide in methanol at 80°, although the trans isomer is debrominated more than eleven times faster under the same conditions. Furthermore, *cis*-1-bromo-2-chlorocyclohexane is dehalogenated at nearly the same rate as in the trans isomer. Goering and Espy explained the reactivity of the cis isomers by proposing that they undergo debromination by a two-step mechanism. In the first step (equation 10.34), the cis isomer undergoes a rate-limiting S_N2 reaction that produces a *trans*-1-iodo-2-cyclohexyl arenesulfonate, which then rapidly undergoes the concerted elimination (equation 10.35).[110]

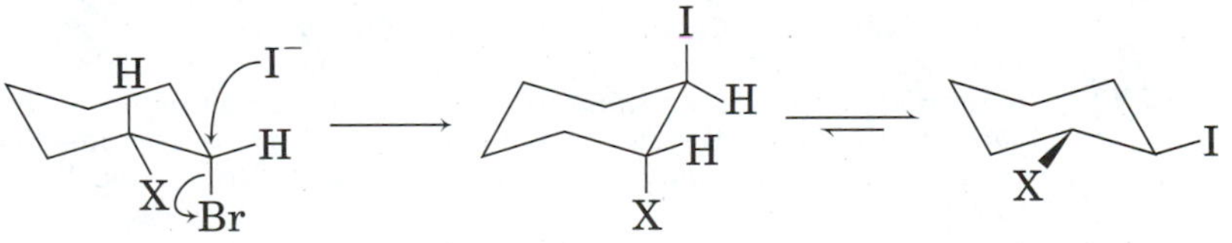

(10.34)

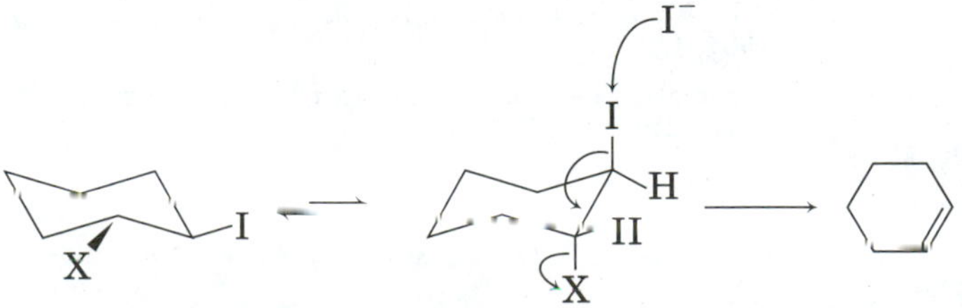

(10.35)

Similarly, Cristol and co-workers found that both *cis*- and *trans*-2-bromo-1-cyclohexyl arenesulfonates were converted to cyclohexene upon treatment with sodium iodide in propanol, although the trans isomers were more reactive. For example, *trans*-2-bromo-1-cyclohexyl tosylate was 62 times as reactive as the cis isomer, and *trans*-2-bromo-1-cyclohexyl brosylate was 48 times as reactive as its cis isomer. Again, the data suggested that the rate-limiting step in the case of the cis isomers is an S_N2 substitution of iodide for bromide, followed by anti elimination of the *trans*-1-halo-2-iodocyclohexane.[113,114]

Although eliminations promoted by iodide ion have been studied most extensively, a wide variety of other dehalogenating reagents have been used as well. The stereospecificity of the reaction depends on the reducing agent. Miller and co-workers found that all reducing agents convert *meso*-1,2-dibromo-1,2-diphenylethane to *trans*-stilbene but that the reduction of *racemic*-1,2-dibromo-1,2-diphenylethane gives yields of *cis*-stilbene ranging from 96% (with sodium iodide in acetone) to 0% (with, for example, $FeCl_2$ in DMF). The authors suggested that the failure to observe complete stereospecificity could arise from one or more alternative pathways for the elimination reaction, including the intermediacy of halonium ions such as those proposed for the addition of halogen to alkenes.[105,115] The intermediacy of bromonium ions in the debromination of vicinal dibromides is consistent with the formation of stilbene from stilbene dibromide in dichloromethane solution.[116,117]

Stereospecific debromination by a nearly syn pathway has also been observed with bis(trimethylsilyl)mercury and bis(trimethylgermyl)mercury.[118] A mechanistic scheme proposed for the debromination of 1,2-dibromoadamantane to form the highly reactive compound adamantene is illustrated in Figure 10.37.[119]

---

[113]Cristol, S. J.; Weber, J. Q.; Brindell, M. C. *J. Am. Chem. Soc.* **1956**, *78*, 598.

[114]Because relative rates of substitution of alkyl halides decrease but relative rates of elimination increase as the alkyl halide varies from 1° to 2° to 3°, some researchers have suggested that substitution-elimination may be a general pattern of reactivity for acyclic dihalides in which one halogen is attached to a 1° carbon. The dehalogenation of ethylene bromoiodide (ICH_2CH_2Br) is ten times as fast as is the debromination of ethylene bromide under the same conditions, so the substitution-elimination pathway seems likely to occur on other compounds with one Br on a primary carbon. Hine, J.; Brader, Jr., W. H. *J. Am. Chem. Soc.* **1955**, *77*, 361.

[115]Lee, C. S. T.; Mathai, I. M.; Miller, S. I. *J. Am. Chem. Soc.* **1970**, *92*, 4602.

[116]Bellucci, G.; Bianchini, R.; Chiappe, C.; Brown, R. S.; Slebocka-Tilk, H. *J. Am. Chem. Soc.* **1991**, *113*, 8012.

[117]See also Zheng, C. Y.; Slebocka-Tilk, H.; Nagorski, R. W.; Alvarado, L.; Brown, R. S. *J. Org. Chem.* **1993**, *58*, 2122.

[118]Bennett, S. W.; Eaborn, C.; Jackson, R. A.; Walsingham, R. W. *J. Organometal. Chem.* **1971**, *27*, 195.

[119]Cadogan, J. I. G.; Leardini, R. *J. Chem. Soc., Chem. Commun.* **1979**, 783.

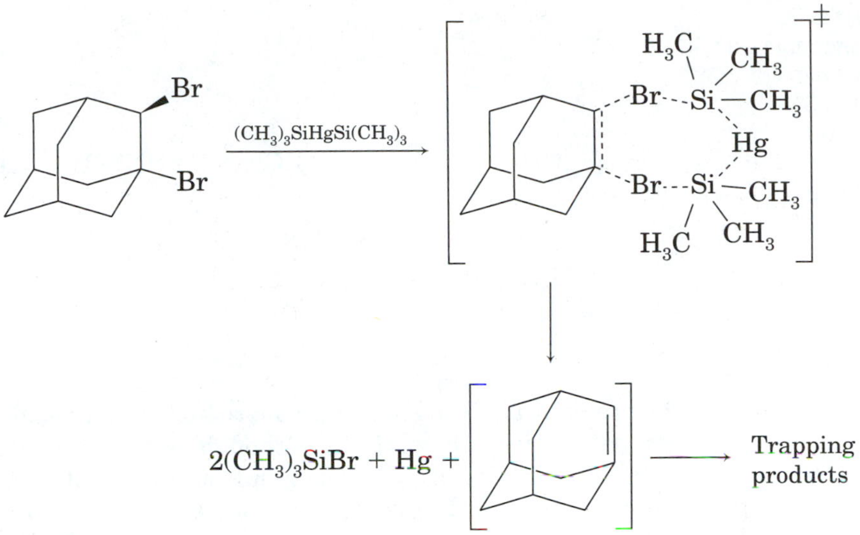

Figure 10.37 Syn elimination in the debromination of 1,2-dibromo-adamantane with bis(trimethylsilyl)-mercury. (Adapted from reference 119.)

Br, Br; $(CH_3)_3SiHgSi(CH_3)_3$ → $[\ \cdots Br \cdots Si(CH_3)_3 \cdots Hg \cdots Si(CH_3)_3 \cdots Br \cdots\ ]^{\ddagger}$

→ $2(CH_3)_3SiBr + Hg +$ [adamantene] → Trapping products

Dehydration of Alcohols

The dehydration of alcohols is one of the fundamental reactions of organic chemistry.[120] Loss of a proton from the carbon atom β to the OH group can lead to an alkene (equation 10.36).

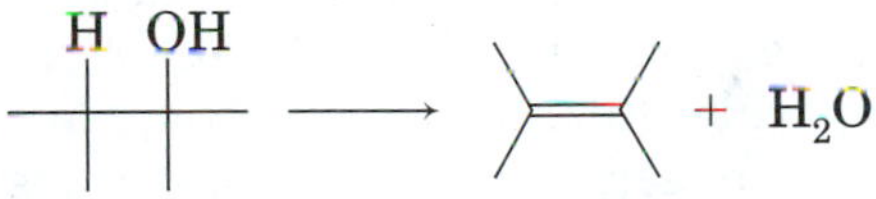

H OH → alkene $+\ H_2O$ **(10.36)**

The most common way to carry out the reaction is by heating the alcohol in the presence of a catalytic amount of a strong mineral acid such as sulfuric acid or phosphoric acid. The dehydration of 2-methylcyclohexanol is illustrated in equation 10.37.[121]

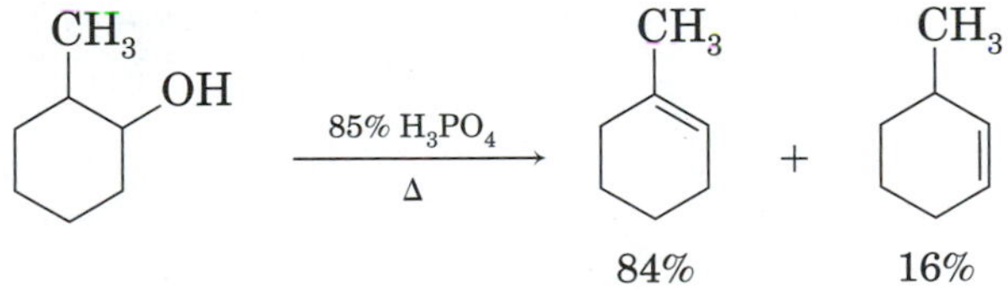

CH_3, OH; $\xrightarrow[\Delta]{85\%\ H_3PO_4}$ CH_3 (84%) + CH_3 (16%) **(10.37)**

[120]For a more detailed discussion, see (a) reference 21, pp. 221–274; (b) reference 20, pp. 145–156; (c) Knözinger, H. in *The Chemistry of the Hydroxyl Group*, Part 2; Patai, S., ed.; Wiley-Interscience. London, 1971; pp. 641–718.

[121]Taber, R. L.; Champion, W. C. *J. Chem. Educ.* **1967**, *44*, 620.

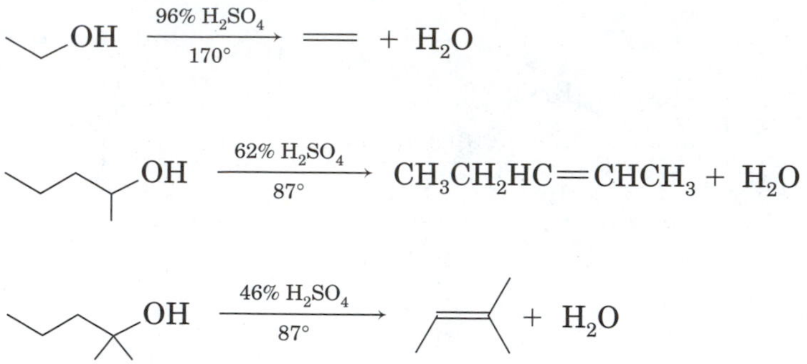

Figure 10.38 Conditions required for dehydration of 1°, 2°, and 3° alcohols with sulfuric acid. (Adapted from reference 122.)

$$CH_3CH_2OH \xrightarrow[170^\circ]{96\%\ H_2SO_4} CH_2{=}CH_2 + H_2O$$

$$CH_3CH_2CH_2CH(OH)CH_3 \xrightarrow[87^\circ]{62\%\ H_2SO_4} CH_3CH_2HC{=}CHCH_3 + H_2O$$

$$CH_3CH_2C(CH_3)_2OH \xrightarrow[87^\circ]{46\%\ H_2SO_4} CH_3CH{=}C(CH_3)_2 + H_2O$$

The preferential formation of the more highly substituted alkene is the common pattern in such reactions, although it must be considered possible that such a distribution may be the result of an equilibrium between the products as a result of the acidic conditions. That is, there may be an equilibrium-controlled and not a rate-controlled product distribution in acid-catalyzed reactions.

Many experimental observations are consistent with an E1 mechanism for the acid-catalyzed dehydration of alcohols. The relative rates of the reactions of 1°, 2°, and 3° alcohols parallel the stabilities of the corresponding carbocations. Consequently, higher temperatures or more concentrated acid solutions are required for dehydration of 1° than 2° than 3° alcohols, as indicated in Figure 10.38.[122] Furthermore, rearranged products are observed in reactions in which the intermediate carbocations can undergo alkyl or hydride shifts to form more stable carbocations.[123,124]

The simplest mechanism for a specific acid-catalyzed E1 dehydration of an alcohol is shown in equation 10.38. The rate-limiting step is the formation of the carbocation intermediate, which is followed by abstraction of a β-proton by base. In aqueous solutions of sulfuric acid, water is the most likely base for the proton abstraction.[125]

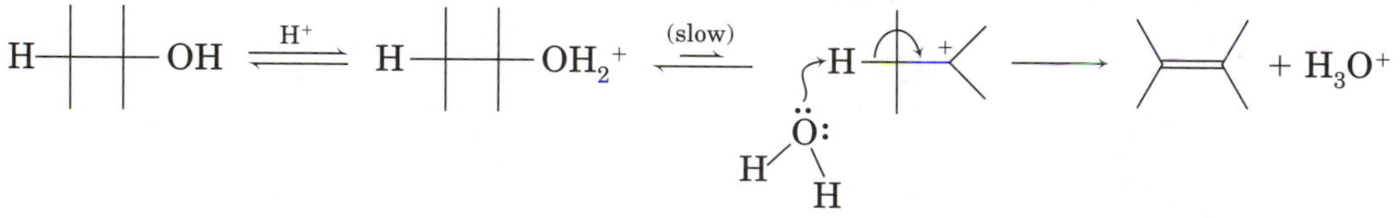

$$H{-}C{-}C{-}OH \xrightleftharpoons{H^+} H{-}C{-}C{-}OH_2^+ \xrightleftharpoons{(slow)} H{-}C{-}C^+ \;(\text{with } H_2\ddot{O}\text{:}) \longrightarrow C{=}C + H_3O^+ \quad (10.38)$$

[122] Fieser, L. F.; Fieser, M. *Advanced Organic Chemistry*; Reinhold Publishing Company: New York, 1961; pp. 138 *ff.*

[123] See, for example, Dostrovsky, I.; Klein, F. S. *J. Chem. Soc.* **1955**, 4401.

[124] Collins, C. J. *Quart. Rev. Chem. Soc.* **1960**, *14*, 357.

[125] The elimination can be subject to general base catalysis. Loudon and Noyce found general base catalysis in the dehydration of substituted 1-aryl-2-phenylethanols in aqueous dioxane solution with 1:1 dichloroacetic acid–sodium dichloroacetate buffers. Loudon, G. M.; Noyce, D. S. *J. Am. Chem. Soc.* **1969**, *91*, 1433.

Figure 10.39 S_N2 mechanism for acid-catalyzed exchange of OH in 1-propanol.

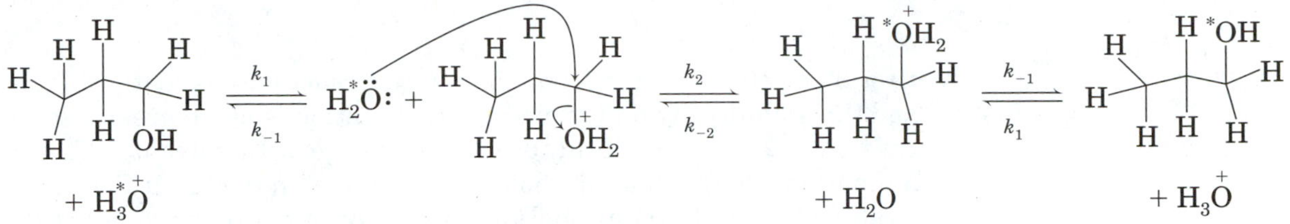

However, there is evidence that the dehydration reaction is much more complex than is suggested by equation 10.38 because for most alcohols exchange of OH with solvent is considerably faster than is elimination. Dostrovsky and Klein found that 1-butanol undergoes oxygen exchange with ^{18}O-labeled water three times faster than it undergoes elimination at 125°. Neopentyl alcohol, on the other hand, was found to undergo carbocation rearrangement and subsequent elimination of water 37 times faster than it exchanges oxygen with solvent.[123] Because 1° carbocations are unstable and because neopentyl systems react slowly in concerted substitution reactions, the results suggested an S_N2 mechanism for oxygen exchange with 1° alcohols, as is illustrated for 1-propanol in Figure 10.39.

There is evidence that the S_N2 mechanism for oxygen exchange also operates in 2° alcohols. For example, the rate of oxygen exchange of optically active 2-butanol in $H_2{}^{18}O$ was found to be exactly half the rate of racemization, as is expected for a Walden inversion (equation 10.39).[126]

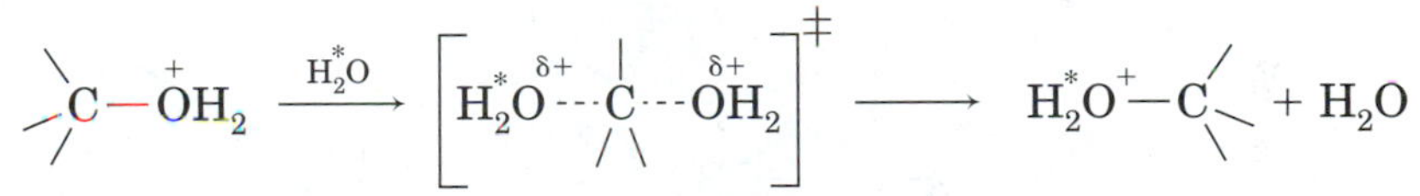

$$\text{C}-\overset{+}{\text{O}}\text{H}_2 \xrightarrow{\text{H}_2\overset{*}{\text{O}}} \left[\text{H}_2\overset{*}{\text{O}}\overset{\delta+}{\cdots}\text{C}\overset{\delta+}{\cdots}\text{OH}_2\right]^{\ddagger} \longrightarrow \text{H}_2\overset{*}{\text{O}}{}^{+}-\text{C} + \text{H}_2\text{O} \qquad \textbf{(10.39)}$$

An alternative mechanism for oxygen exchange could be dehydration of the alcohol to an alkene, followed by rehydration of the alkene with labeled water. However, that process would lead to racemization, not inversion, of the chiral center, and the rate of oxygen exchange would be more than half the rate of racemization. There is also another possibility that is consistent with the experimental data. A rate of oxygen exchange that is half the rate of inversion could also result from a mechanism involving a tightly solvated symmetric carbocation intermediate that collapses to reactant (protonated alcohol) or product (protonated alcohol with labeled oxygen) faster than one of the solvating water molecules can diffuse away from the carbocation center and be replaced by another water molecule from the solution (equation 10.40).[127]

[126] Bunton, C. A.; Konasiewicz, A.; Llewellyn, D. R. *J. Chem. Soc.* **1955**, 604; Bunton, C. A.; Llewellyn, D. R. *J. Chem. Soc.* **1957**, 3402.

[127] Dietze, P. E.; Jencks, W. P. *J. Am. Chem. Soc.* **1987**, *109*, 2057.

$$\underset{/}{\overset{\backslash}{}}\!C-\overset{+}{O}H_2 \xrightarrow{H_2\overset{*}{O}} \left[H_2\overset{*}{O}\cdots\overset{|+}{\underset{/\backslash}{C}}\cdots OH_2 \right] \rightleftharpoons H_2\overset{*}{O}-C\!\underset{\backslash}{\overset{/}{}} + H_2O \quad (10.40)$$

With a 3° alcohol, the tendency for an S_N2 reaction is greatly diminished, and the stability of an intermediate carbocation is much greater than is the case with 2° alcohols. Still, oxygen exchange with solvent is faster than elimination even with 3° alcohols. For example, in 0.09 N H_2SO_4 solution in water at 55°, *t*-butyl alcohol underwent oxygen exchange about 30 times faster than it underwent elimination.[128] At 75°, the ratio of the rate constants for exchange and elimination was about 21.[129] Because the S_N2 pathway is unlikely, it is probable that oxygen exchange occurs by an S_N1 mechanism with 3° alcohols.

Exchange of oxygen with solvent is an important process even with benzylic alcohols.[130] Noyce and co-workers found racemization of optically active 1,2-diphenylethanol in 50–60% aqueous sulfuric acid to be 58 times faster than the dehydration to *trans*-stilbene.[131] The results suggest that under these conditions there is reversible formation of a 1,2-diphenylethyl cation (equation 10.41), followed by rate-limiting detachment of a proton from the carbocation (equation 10.42).

$$X\text{-}C_6H_4\text{-}\overset{OH}{\overset{|}{C}}H\text{-}CH_2\text{-}C_6H_4\text{-}Y \;(\mathbf{19}) \underset{}{\overset{H^+}{\rightleftharpoons}} X\text{-}C_6H_4\text{-}\overset{+}{C}H\text{-}CH_2\text{-}C_6H_4\text{-}Y \;(\mathbf{20}) + H_2O \quad (10.41)$$

$$X\text{-}C_6H_4\text{-}\overset{+}{C}H-CH_2\text{-}C_6H_4\text{-}Y \xrightarrow{-H^+} (X\text{-}C_6H_4)HC{=}CH(C_6H_4\text{-}Y) \quad (10.42)$$

Furthermore, a study of the dehydration of a series of substituted 1,2-diarylethanols (**19**) showed very different effects of the substituents X and Y.[132] The Hammett correlation for the rate constant for dehydration was found to be

$$k_{X,Y} = -3.78(\sigma_X^+ + 0.23\sigma_Y) - 3.19 \quad (10.43)$$

[128]Dostrovsky, I.; Klein, F. S. *J. Chem. Soc.* **1955**, 791.

[129]That trend is consistent with the general observation that higher temperatures favor elimination over substitution. Higher acid concentrations also favor elimination. For discussions, see references 126 and 128.

[130]Grunwald, E.; Heller, A.; Klein, F. S. *J. Chem. Soc.* **1957**, 2604.

[131]Noyce, D. S.; Hartter, D. R.; Pollack, R. M. *J. Am. Chem. Soc.* **1968**, *90*, 3791.

[132]Noyce, D. S.; Hartter, D. R.; Miles, F. B. *J. Am. Chem. Soc.* **1968**, *90*, 3794.

This result suggests that the equilibrium population of the benzylic carbocation (**20**) is enhanced by a substituent X that can stabilize the carbocation through a resonance interaction, while the substituent Y has a much weaker effect. This result is incompatible with an alternative mechanism involving rate-limiting formation of a bridged cationic intermediate such as **21**, since in that case the X and Y substituents would be expected to have the same effect on cation stability.[133]

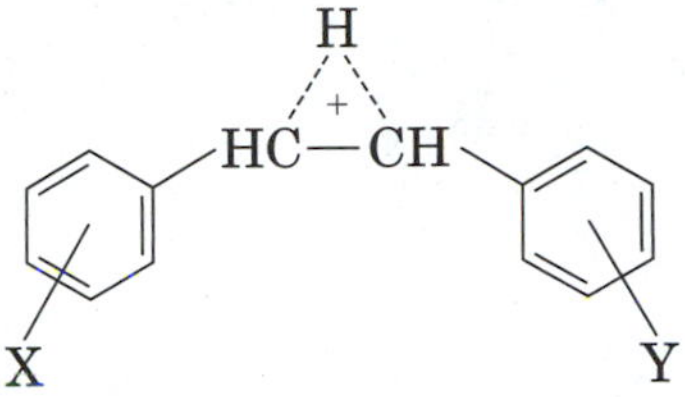

21

Because of the instability of 1° carbocations and the conclusion that 1° alcohols undergo acid-catalyzed oxygen exchange by an S_N2 process, there may be some question whether 1° carbocations are true intermediates in elimination reactions. The observation that neopentyl alcohol undergoes elimination more rapidly than it undergoes oxygen exchange with solvent suggests that a 1° carbocation is indeed formed, since S_N2 reactions are known to be quite slow in neopentyl systems. However, an alternative mechanistic possibility for dehydration of 1° alcohols is an acid-catalyzed E2 mechanism. Narayan and Antal proposed such a mechanism for the specific acid-catalyzed dehydration of 1-propanol with sulfuric acid in supercritical water (Figure 10.40).[134]

The concerted E2 process in Figure 10.40 seems not to apply to benzylic or 3° (alkyl) alcohols, which can react by an E1 pathway, but the situation for 2° alcohols in aqueous solution is less certain. Lomas studied the acid-catalyzed dehydration of 1,1′-diadamantylethanol to 1,1-bis(1-adamantyl)

Figure 10.40 Concerted E2 mechanism for dehydration of 1-propanol. (Adapted from reference 134.)

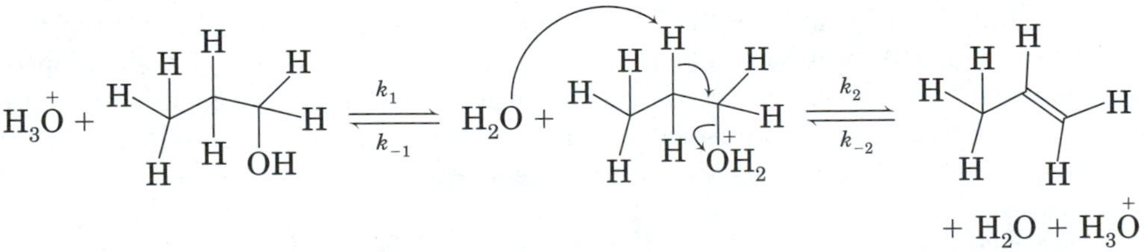

$H_3\overset{+}{O}$ + $CH_3CH_2CH_2OH$ $\underset{k_{-1}}{\overset{k_1}{\rightleftharpoons}}$ H_2O + $CH_3CH_2CH_2\overset{+}{O}H_2$ $\underset{k_{-2}}{\overset{k_2}{\rightleftharpoons}}$ $CH_3CH{=}CH_2$ + H_2O + $H_3\overset{+}{O}$

[133]For a discussion and leading references to proposals of a proton-alkene π complex as an intermediate in the dehydration of alcohols and the protonation of alkenes, see reference 125.

[134]Narayan, R.; Antal, Jr., M. J. *J. Am. Chem. Soc.* **1990**, *112*, 1927.

ethene in anhydrous acetic acid solution.[135] Dehydration of the parent compound and the trideuteriomethyl analogue produced a deuterium kinetic isotope effect, with k_H/k_D equal to 1.32. This result is not consistent with a mechanism in which the β-proton is lost in the rate-limiting step, but it is consistent with rate-limiting formation of the carbocation intermediate (with the k_H/k_D ratio reflecting a secondary kinetic isotope effect), as shown in equation 10.45.

$$H_3O^+ + Ad_2C(CH_3)OH \rightleftharpoons Ad_2C(CH_3)OH_2^+ + H_2O \quad (10.44)$$

$$Ad_2C(CH_3)OH_2^+ \xrightleftharpoons{\text{slow}} Ad_2C^+{-}CH_3 + H_2O \quad (10.45)$$

$$Ad_2C^+{-}CH_3 + H_2O \xrightleftharpoons{\text{fast}} Ad_2C{=}CH_2 + H_3O^+ \quad (10.46)$$

(Ad— = 1-adamantyl)

On the other hand, Dietze and Jencks concluded that the hydration of 1-butene and the oxygen exchange between solvent water and 2-butanol do not involve a common carbocation intermediate.[127] A common intermediate would be expected to partition to the same products under identical conditions, but the investigators found that different ratios of *cis*- to *trans*-2-butene are formed when 1-butene and 2-butanol are subjected to the same reaction conditions. Furthermore, the two reactants gave different ratios of products derived from deprotonation and water addition. Dietze and Jencks concluded that isomerization of 1-butene and oxygen exchange of 2-butanol may occur through parallel concerted mechanisms in which there is significant carbocation character in the transition structures. A possible explanation for the failure to observe a common carbocation intermediate is that ordinary 2° carbocations do not have a significant lifetime in the presence of water.[127]

Even though the dehydration reactions of many alcohols may occur by processes that do not produce solvent-equilibrated carbocations, it is clear that considerable carbocationic character does develop during the course of the reaction. This carbocationic character can lead to hydride, alkyl, and aryl shifts. For example, Schaeffer and Collins found that the dehydration of *cis*-2-phenylcyclohexanol (**22**) with phosphoric acid produced an 88% yield of 1-phenylcyclohexene (**23**) and a 2% yield of 3-phenylcyclohexene (**24**).[136]

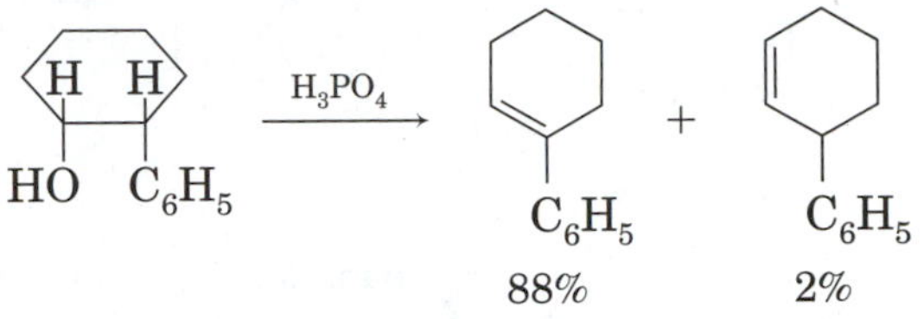

H_3PO_4 — C_6H_5 88% (**23**) + C_6H_5 2% (**24**) from **22** (HO, C_6H_5) **(10.47)**

However, the trans isomer (**25**) was found to give 21% of **23**, 9% of **24**, 6% of 4-phenylcyclohexene (**26**), 32% of 1-benzylcyclopentene (**27**), and about 20% of 1-benzalcyclopentane (**28**).

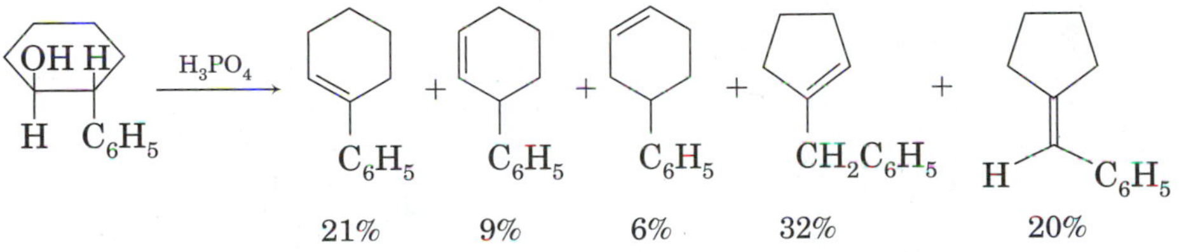

25 (H_3PO_4) → 21% **23** + 9% **24** + 6% **26** + 32% **27** ($CH_2C_6H_5$) + 20% **28** **(10.48)**

To explain these results, Schaeffer and Collins suggested that the two chair conformers of **25** could give rise to different reaction pathways. The diaxial conformer **25aa** has a phenyl group that is properly oriented for anchimeric assistance to the departure of water from the protonated alcohol, with the resulting formation of a symmetrical phenonium ion.

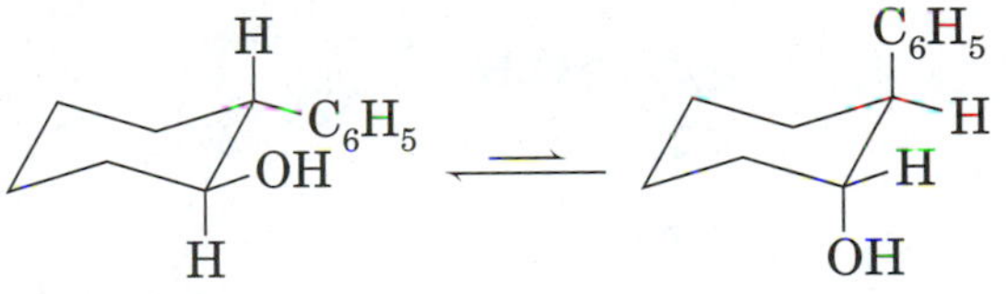

25ee ⇌ **25aa**

Loss of a proton from the phenonium ion produces **23** (equation 10.49). Confirmation of the phenyl migration pathway in the dehydration reaction was found in the dehydration of 2-phenylcyclohexanol-2-^{14}C, which produced 1-phenylcyclohexene labeled with ^{14}C at both C1 and C2. The mechanism shown in equation 10.49 cannot be the only pathway to **23**, however, because the symmetrical phenonium ion should produce product with the ^{14}C label distributed equally on C1 and C2. Instead, only about 25% of **23** was found to have the ^{14}C label on C2.

[135]Lomas, J. S. *J. Org. Chem.* **1981**, *46*, 412.

[136]Schaeffer, H. J.; Collins, C. J. *J. Am. Chem. Soc.* **1956**, *78*, 124.

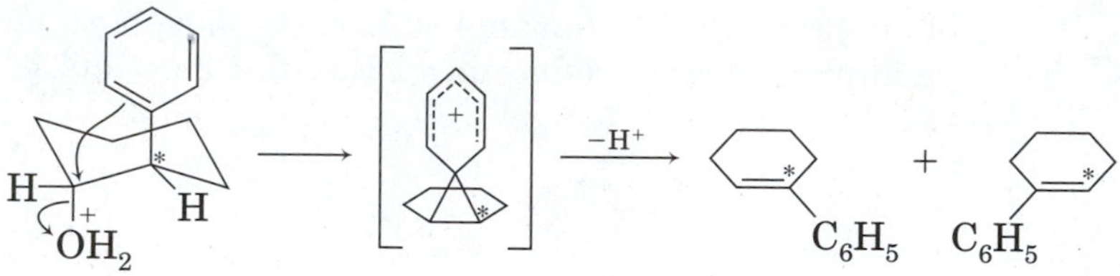

23 23 (10.49)

To explain the formation of some of the **23** without phenyl rearrangement and to explain the high yields of the other products, Schaeffer and Collins proposed that loss of water from the protonated diequatorial conformer **25ee** (the major conformer) leads to a 2-phenylcarbocation, which can lose a proton to form either unrearranged **23** or **24**. Reprotonation of **24** and subsequent deprotonation of the 3-phenylcyclohexyl carbocation could lead to **26**.

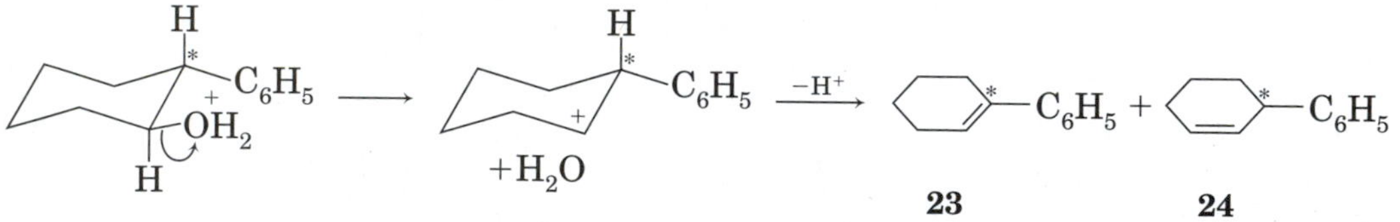

23 **24** (10.50)

The ring contraction products are explained by an alkyl shift involving C6. The C1—C6 bond is anti to the departing water molecule, which facilitates the alkyl shift shown in equation 10.51. The alkyl shift forms a benzylic carbocation which can lose a proton to form **28**. Reprotonation of **28** and subsequent deprotonation of the benzylcyclopentyl carbocation would lead to **27**.

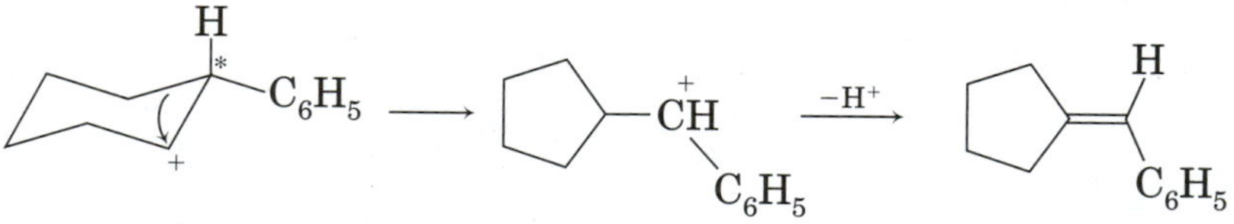

28 (10.51)

The products of dehydration of the cis isomer are quite different from those produced from the trans isomer because each of the two chair conformations of the cis isomer has one axial and one equatorial substituent (Figure 10.41). In neither conformation is the phenyl group or an adjacent carbon-carbon bond anti to the departing water molecule, so phenyl and

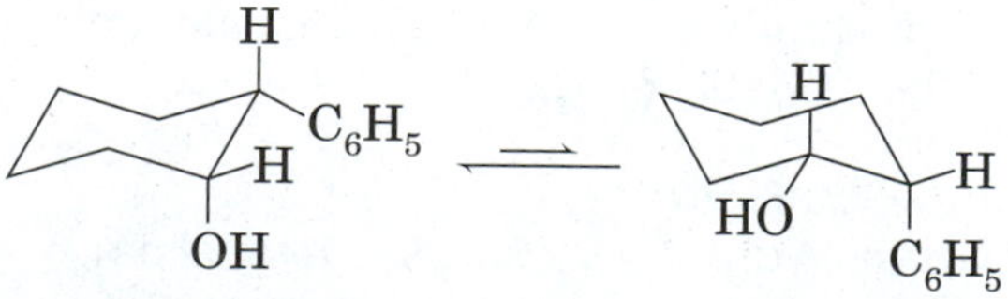

Figure 10.41 Chair conformations of *cis*-2-phenylcyclohexanol.

alkyl group shifts are not observed. Instead, the 2-phenylcarbocation leads to formation of the observed phenylcyclohexenes.

Although the dehydration of most alcohols is acid-catalyzed, base-catalyzed eliminations have also been observed when the elimination of water leads to a conjugated diene. For example, a concerted, E2 mechanism for the synthesis of myrcene (**29**) is shown in equation 10.52.[120c,137] A carbanion (E1cb) mechanism has been proposed for the dehydration of the ketol intermediate in the base-catalyzed aldol reaction of benzaldehyde and acetone (equations 10.53–10.55).[138,139]

$$\text{(H, OH)} \xrightarrow{HO^-} \left[HO^- \cdots H \cdots OH \right]^{\ddagger} \longrightarrow \mathbf{29} + HO^- + H_2O \qquad (10.52)$$

$$C_6H_5-\overset{\overset{O}{\|}}{C}-H + H_3C-\overset{\overset{O}{\|}}{C}-H \overset{^-OH}{\rightleftharpoons} C_6H_5-\underset{H}{\overset{OH}{C}}-CH_2-\underset{H}{C}{=}O \qquad (10.53)$$

$$C_6H_5-\underset{H}{\overset{OH}{C}}-CH_2-\underset{H}{C}{=}O + OH^- \rightleftharpoons C_6H_5-\underset{H}{\overset{OH}{C}}-\bar{C}H-\underset{H}{C}{=}O + H_2O \qquad (10.54)$$

$$C_6H_5-\underset{H}{\overset{OH}{C}}-\bar{C}H-\underset{H}{C}{=}O \rightleftharpoons \overset{C_6H_5}{\underset{H}{C}}{=}\overset{H}{\underset{CHO}{C}} + OH^- \qquad (10.55)$$

Heating alcohols in the gas phase to temperatures above 500° can lead to dehydration, but dehydrogenation to carbonyl compounds also occurs.[120(c)] Better yields of dehydration products are obtained if catalysts are added to the reaction mixture. Maccoll and Stimson found that heating *t*-butyl alco-

---

[137] Ohloff, G. *Chem. Ber.* **1957**, *90*, 1554.

[138] Noyce, D. S.; Reed, W. L. *J. Am. Chem. Soc.* **1959**, *81*, 624.

[139] See also Crowell, T. I. in *The Chemistry of Alkenes*; Patai, S., ed.; Wiley-Interscience: London, 1964; pp. 241–270.

hol in the gas phase at 315–422° with a catalytic amount of HBr produced isobutene and water.[140] The stoichiometry of the reaction is

$$(CH_3)_3COH + HBr \longrightarrow (CH_3)_2C{=}CH_2 + H_2O + HBr \qquad \textbf{(10.56)}$$

Similarly, heating 2-butanol in the gas phase with a catalytic amount of HBr at temperatures of 387–510° was found to produce a mixture of water, 1-butene, *cis*-2-butene, and *trans*-2-butene. The overall kinetic order of the reaction was second order, first order in alcohol and first order in HBr. The reaction was unaffected by additives expected to retard the rates of radical reactions, so a molecular reaction pathway seemed most likely.[141] Although several bimolecular mechanisms were considered, a mechanism involving the transition structure shown in equation 10.57 is particularly interesting because of the resemblance of this transition structure to those proposed for pyrolytic eliminations (page 695).[142]

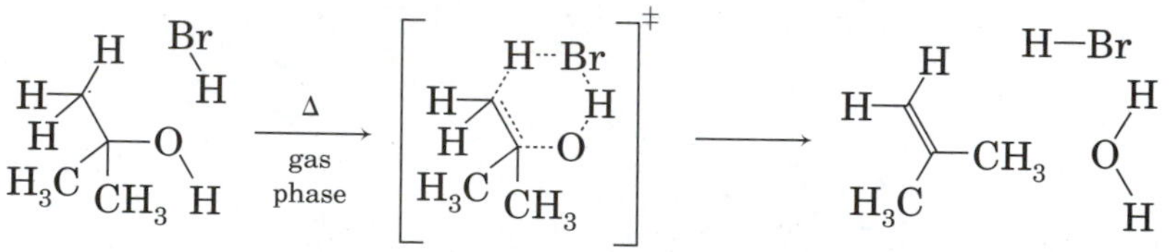

(10.57)

One of the most active areas of research in dehydration reactions is the use of heterogeneous catalysts such as metal oxides, alumina, zeolites, and other species to catalyze the elimination of water from alcohols. Not only do the catalysts lower the temperatures required for dehydration, but they can also alter product distributions. For example, dehydration of 2-butanol produced 45% of 1-butene when the dehydration was catalyzed by alumina, and 90% of 1-butene when zirconia was the catalyst.[143] A wide variety of mechanisms have been considered for these reactions.[120] The effect of the catalyst is a function of the nature of the acidic and basic sites on the catalyst surface, the size of openings into which organic molecules may fit, molecular shape, and reaction temperature.[144] In the case of zeolites, carbocations and alkyl silyl ethers have been detected as reactive intermediates.[145]

[140] Maccoll, A.; Stimson, V. R. *J. Chem. Soc.* **1960**, 2836.

[141] Failes, R. L.; Stimson, V. R. *J. Chem. Soc.* **1962**, 653.

[142] Reference 140. Another possibility for the transition structure was an ion pair composed of the protonated alcohol and bromide ion.

[143] Davis, B. H. *J. Org. Chem.* **1982**, *47*, 900.

[144] For example, see Yue, P. L.; Olaofe, O. *Chem. Eng. Res. Des.* **1984**, *62*, 167; Paukstis, E.; Jiratova, K.; Soltanov, R. I.; Yurchenko, E. N.; Beranek, L. *Collect. Czech. Chem. Commun.* **1985**, *50*, 643.

[145] See, for example, Stepanov, A. G.; Zamaraev, K. I.; Thomas, J. M. *Catal. Lett.* **1992**, *13*, 407.

Deamination of Amines

The reaction of 1° amines with nitrous acid produces diazonium ions (equation 10.58).[146,147]

$$R{-}NH_2 \xrightarrow[H_2O]{HONO} R{-}N(H){-}NO \rightleftharpoons R{-}N{=}N{-}OH \xrightarrow[-H_2O]{H^+} R{-}N_2^+ \quad (10.58)$$

The aryldiazonium ions derived from 1° aromatic amines are useful intermediates in the synthesis of a wide variety of substituted aromatic compounds. However, the alkyldiazonium ions derived from 1° aliphatic amines are regarded as not being synthetically useful because they generally lead to a mixture of products. For example, treatment of *n*-butylamine with sodium nitrite in aqueous HCl produced 1-butanol (25%), 2-butanol (13.2%), 1-chlorobutane (5.2%), 2-chlorobutane (2.8%), a mixture of butenes (36.5%), and unidentified higher boiling organic compounds (7.6%).[148] An explanation for the variety of products from 1° alkyldiazonium ions is shown in equation 10.59, in which loss of N_2 produces a carbocation that can then give a mixture of substitution, elimination, and rearrangement products.

$$R{-}N_2^+\ Cl^- \longrightarrow N_2 + R^+ \longrightarrow \begin{cases}\text{substitution}\\ \text{elimination}\\ \text{rearrangement}\end{cases} \quad (10.59)$$

There is evidence that carbocations do play a role in the reaction of some diazonium ions.[149] For example, the products of nucleophilic addition to *t*-butyl carbocations are essentially the same whether the carbocations are produced from aliphatic amines or from solvolysis of the corresponding alkyl halide.[150] However, the decomposition of the diazonium ion to form a carbocation should become less favorable as the stability of the carbocation decreases, and the formation of an ethyl cation from ethyldiazonium ion in the gas phase has been calculated to be endothermic by 13.9 kcal/mol.[151] The decomposition of a 1° alkyldiazonium ion to N_2 and a 1° carbocation in solution may be energetically more favorable, perhaps even exothermic, but it is questionable whether a free carbocation is formed from a 1° alkyldiazonium ion.[152] Furthermore, Streitwieser and Schaeffer reported that the dis-

---

[146]Ridd, J. H. *Quart. Rev. Chem. Soc.* **1961**, *15*, 418.

[147]White, E. H.; Woodcock, D. J. in *The Chemistry of the Amino Group*; Patai, S., ed.; Wiley-Interscience: London, 1968; 407–497; see especially pp. 461–480.

[148]Whitmore, F. C.; Langlois, D. P. *J. Am. Chem. Soc.* **1932**, *54*, 3441.

[149]For a review, see Friedman, L. in *Carbonium Ions*, Vol. II; Olah, G. A.; Schleyer, P. v. R., eds.; Wiley-Interscience: New York, 1970; pp. 655–713.

[150]Streitwieser, Jr., A. *J. Org. Chem.* **1957**, *22*, 861 and references therein.

[151]Ford, G. P.; Scribner, J. D. *J. Am. Chem. Soc.* **1983**, *105*, 349.

[152]The reaction is known to have an activation energy on the order of 3–5 kcal/mol (reference 153)

Table 10.8 Distribution of butenes from deamination and solvolysis reactions.

Substrate	Conditions			
NH_2	$NaNO_2$, aq. HCl	25%	19%	56%
OTs	HOAc	10.3%	43.2%	46.5%

(Data from reference 153.)

tribution of olefinic products from deamination of *sec*-butylamine was different from that obtained by solvolysis of the corresponding tosylate (Table 10.8).[153] The authors concluded that the product distribution from the deamination reaction must arise, at least in part, by some pathway other than a 2° carbocation intermediate.[154]

As an alternative to mechanistic schemes requiring carbocation intermediates, Streitwieser suggested that loss of N_2 from an alkyldiazonium ion may occur during, not before, bonding changes that lead to the reaction products.[150,153] As shown in Figure 10.42, Streitwieser proposed that N_2 loss may be concurrent with displacement by solvent, with loss of a proton to form an alkene, with a hydride shift to form a rearranged carbocation, or with an alkyl or aryl shift to form a different rearranged carbocation. The possibility that a free carbocation might also be formed was included in the reaction scheme.

One of the strongest arguments against the formation of 1° carbocations in the deamination of 1° amines comes from stereochemical studies. Brosch and Kirmse found that the nitrous acid deamination of optically active 1-[^{2}H]-1-butanamine (**30**) produced 1-butanol (**32**) with essentially complete inversion of configuration (equation 10.60).[155,156,157] This result suggests

---

[153]Streitwieser, Jr., A.; Schaeffer, W. D. *J. Am. Chem. Soc.* **1957**, *79*, 2888.

[154]It has been proposed in some cases that differences in product distributions from deamination reactions and those expected from solvolysis reactions expected to produce the same carbocations can be explained if the deamination produces a "hot" (distorted) carbocation. See, for example, Semenow, D.; Shih, C.-H.; Young, W. G. *J. Am. Chem. Soc.* **1958**, *80*, 5472. In other cases ion pairing effects have been suggested as important in determining the products in deamination reactions (reference 161).

[155]Brosch, D.; Kirmse, W. *J. Org. Chem.* **1991**, *56*, 907.

[156]Deamination of **30** was accompanied by the formation of 2-butanol (30%) as well as 1-butanol (51%).

[157]This study corrected a widely cited report (reference 153) that the reaction of optically active **30** produced **32** with 31% racemization. That result indicated that a 1° carbocation was a likely intermediate in the formation of at least some of the 1-butanol. As suggested in reference 155, this conclusion appears to have been based on a study that (unknown to the original investigators) was conducted with partially racemized **30**. The optical purity of the starting material could not have been detected with the experimental methods available at the time of the original investigation.

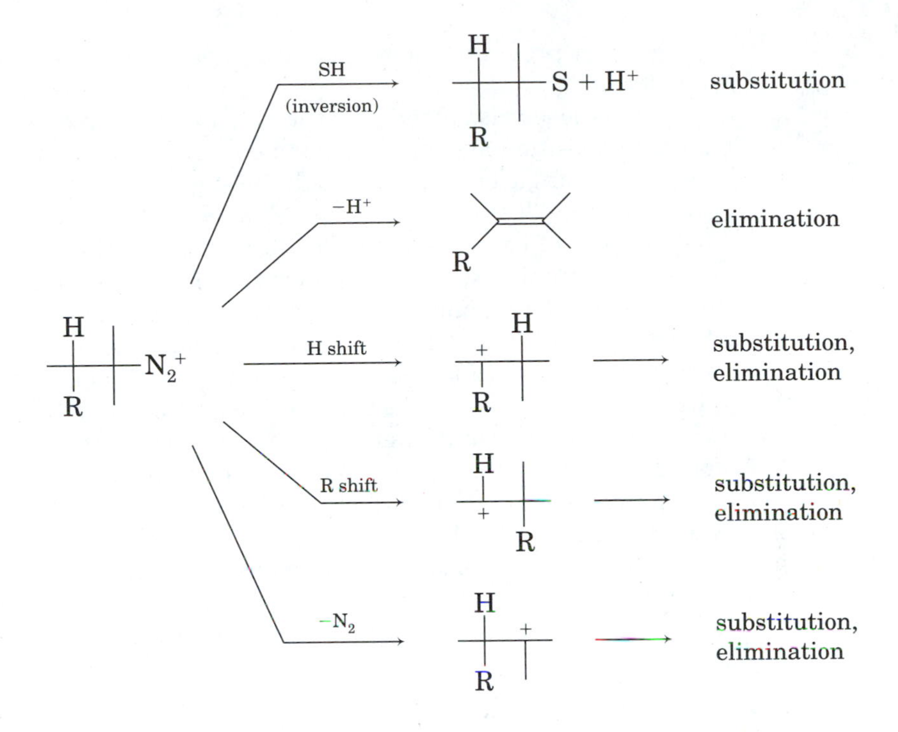

Figure 10.42 Reaction pathways in diazonium ion decomposition. (Adapted from reference 153.)

that the formation of **32** does not occur by capture of the 1° butyl cation by water but is, instead, the product of an S_N2 reaction on the diazonium ion **31**. Similar results were obtained in the study of an optically active secondary amine.[158]

$$\mathbf{30}\ (\text{NH}_2) \xrightarrow{\text{HNO}_2} \mathbf{31}\ (\text{N}_2^+) \xrightarrow{\text{H}_2\text{O}} \mathbf{32}\ (\text{OH}) \qquad (10.60)$$

Streitwieser rationalized the propensity for deamination reactions to give a wide variety of products with an argument based on the Hammond postulate. Because the transition states in the reaction shown in Figure 10.42 are likely to be *early* or *reactant-like*, then there should be only small differences in activation energies, and a wide variety of reactions should occur competitively. In contrast, solvolysis and other reactions have much higher activation energies (ca. 30 kcal/mol), so that the transition states are *later*. As a result, there should be much larger differences in activation energies for the various pathways available to the reactant, and therefore

[158]Radical pathways have also been implicated in the deamination of 1-octanamine in micellar media. Brosch, D.; Kirmse, H. *J. Org. Chem.* **1993**, *58*, 1118.

much larger preferences for one reaction pathway over another in solvolysis reactions.[153]

Streitwieser also noted that the small activation energies for diazonium ion decomposition are comparable to energy barriers for rotation about carbon-carbon single bonds. As a result, he suggested that the product distributions from the deamination of amines might be largely a consequence of the distribution of conformations in the diazonium ions.[153] Evidence for this view came from a study of the deamination of 3-phenyl-2-butanamine in acetic acid by Cram and McCarty.[159] The mixture of acetate products was reduced with $LiAlH_4$ to produce 1-methyl-1-phenylpropanol (the product of a phenyl migration to the site of the departing N_2 moiety) and 2-methyl-1-phenylpropanol (the product of a corresponding methyl migration), along with diastereomeric 3-phenyl-2-butanols. The product distribution was found to be a function of the stereochemistry of the starting amine. Product ratios from deamination of the threo diastereomer (**33**) were consistent with a 1.5 to 1 ratio of methyl to phenyl migration. However, product ratios from deamination of the erythro diastereomer (**34**) suggested an 8:1 preference for phenyl migration over methyl migration. (Solvolysis of the corresponding tosylates in acetic acid did not produce any product of methyl migration.)

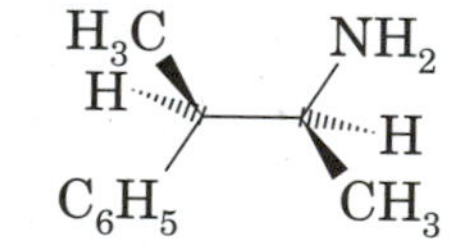

(+)-*threo*-3-phenyl-2-butanamine

33

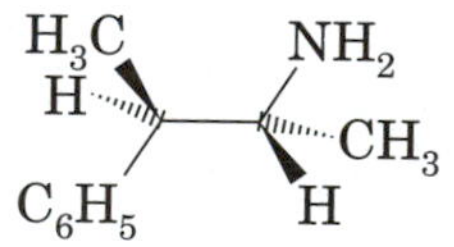

(+)-*erythro*-3-phenyl-2-butanamine

34

To explain these results, Cram and McCarty suggested that the rate constant for product formation from the carbocation intermediate is greater than the rate constant for rotation about the carbon-carbon single bond. Therefore, the product distribution is determined by the relative stability of the conformers of each enantiomer. As shown in Figure 10.43, the relative stability of the conformers of the (+)-*threo* diastereomer (top) suggests that migration of a methyl group may occur more often than migration of a phenyl group because of the greater population of conformers in which a methyl group is *anti*-periplanar to the departing N_2. Similarly, the greater population of conformers of the (+)-*erythro* diastereomer (bottom) in which a phenyl group is *anti*-periplanar to the departing N_2 is entirely consistent with the observation that phenyl migration is the dominant process in reactions of this diastereomer.[160,161,162]

---

[159] Cram, D. J.; McCarty, J. E. *J. Am. Chem. Soc.* **1957**, *79*, 2866.

[160] A similar conclusion was reached by Monera, O. D.; Chang, M.-K., Means, G. E. *J. Org. Chem.* **1989**, *54*, 5424, who investigated the deamination of 1-octanamine in a series of solutions ranging from pH 2 to pH 10. The relative distribution of products, 1-octanol, 2-octanol, 1-octene, and 2-octenes (80% trans, 20% cis) did not vary significantly as a function of pH, suggesting that the concentration of H^+ and HO^- do not strongly affect the product distribution. The authors interpreted the results in terms of conformationally determined reactions of the alkyldiazonium ion.

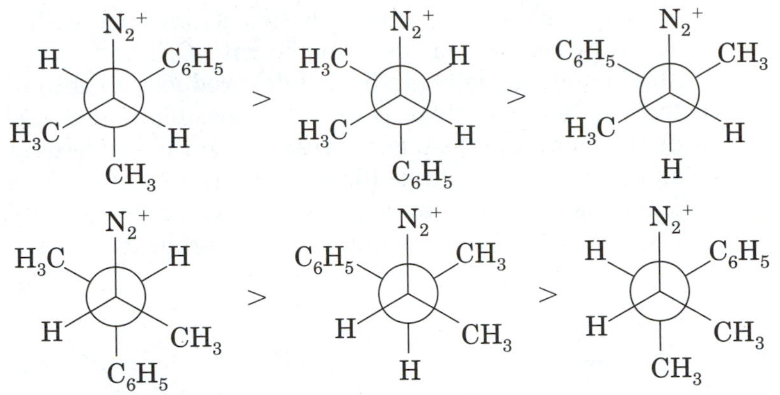

Figure 10.43
Relative stabilities of *threo-* (top) and *erythro-*3-phenyl-2-butanamine (bottom) conformers. (Adapted from reference 159.)

Pyrolytic Eliminations

Some intramolecular eliminations that occur thermally are known as ***pyrolytic eliminations***,[163] and many of these reactions result in syn elimination.[164,165] Often these reactions are carried out in the gas phase, where they are not affected by solvent, ions, or other species as they are in solution.[166] One of the most-studied pyrolytic eliminations is the Chugaev reaction (equation 10.61).[167,168] Reaction of an alcohol having a β-hydrogen atom

---

[161]Although the discussion in reference 159 was presented in terms of methyl or phenyl migration to a carbocation, after the loss of N_2, the arguments can also apply to concurrent migration and loss of N_2. The deamination of optically active 2-octanamine was found to produce 2-octanol with 24% net inversion, suggesting that some S_N2 reaction with solvent occurs in this 2° system also. Moss, R. A.; Talkowski, C. J.; Reger, D. W.; Powell, C. E. *J. Am. Chem. Soc.* **1973**, *95*, 5215. The reaction was sensitive to medium effects, and the stereochemistry changed to 6% retention of configuration in micellar media. However, Collins, C. J. *Acc. Chem. Res.* **1971**, *4*, 315, has argued that free carbocations are formed, at least with some 2° alkyl and benzylic amines.

[162]Central to this argument is the assumption that the rates of reaction of the incipient carbocations are significantly greater than the rates of rotation about the carbon-carbon single bonds. In situations in which conformational interconversion is faster than the rates of reaction of conformers, then the Curtin-Hammett principle applies. For a discussion of reaction rate constants and conformational equilibria, see Winstein, S.; Holness, N. J. *J. Am. Chem. Soc.* **1955**, *77*, 5562; Eliel, E. L. *J. Chem. Educ.* **1960**, *37*, 126 and references therein; Seeman, J. I.; Farone, W. A. *J. Org. Chem.* **1978**, *43*, 1854.

[163]Reference 21, pp. 377–482; reference 20, pp. 167–195;

[164]For a review of pyrolytic syn eliminations, see DePuy, C. H.; King, R. W. *Chem. Rev.* **1960**, *60*, 431.

[165]The term E_i (elimination, intramolecular, syn) was proposed for such reactions by Sahyun, M. R. V.; Cram, D. J. *J. Am. Chem. Soc.* **1963**, *85*, 1263.

[166]Nevertheless, there may be the complication of surface effects with the reaction vessel. See the discussion on page 702.

[167](a) O'Connor, G. L., Nace, H. R. *J. Am. Chem. Soc.* **1952**, *74*, 5454; (b) O'Connor, G. L.; Nace, H. R. *J. Am. Chem. Soc.* **1953**, *75*, 2118 and references therein; (c) Nace, H. R. Org. React. **1962**, *12*, 57.

[168]Chugaev, L. A. *Chem. Ber.* **1899**, *32*, 3332. An alternative spelling is "Tschugaeff" (reference 169), although "Chugaev" is used more commonly now.

with sodium or potassium (or with a strong base such as sodium amide), followed by reaction of the alkoxide with CS_2 and then S_N2 reaction with an alkyl halide (usually methyl iodide) produces a xanthate ester (**35**). The xanthate ester is purified as much as possible and is then heated, usually by distillation at atmospheric pressure, to induce decomposition to an alkene, along with carbon oxysulfide and a mercaptan.[167] This alkene synthesis is particularly useful because it provides a means of dehydration of an alcohol without the possible complications of carbocation rearrangements.[169,170,171]

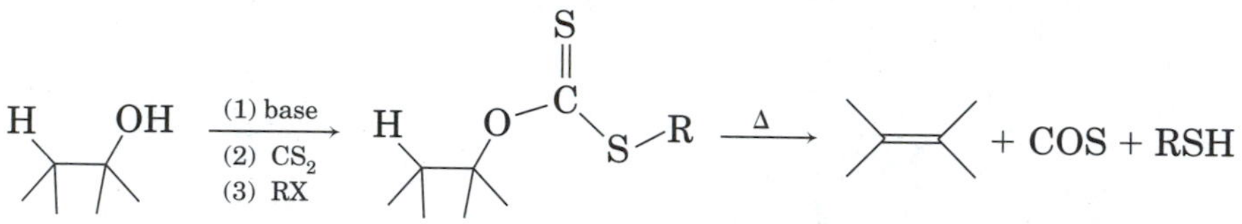

H OH —(1) base; (2) CS_2; (3) RX→ H O–C(=S)–S–R (**35**) —Δ→ alkene + COS + RSH **(10.61)**

Cram demonstrated that the Chugaev reaction is a syn elimination by showing that the formation of 2-phenyl-2-butenes by pyrolysis of the *S*-methyl xanthate of diastereomeric 3-phenyl-2-butanols is stereospecific, as shown in equations 10.62 and 10.63.[172,173,174]

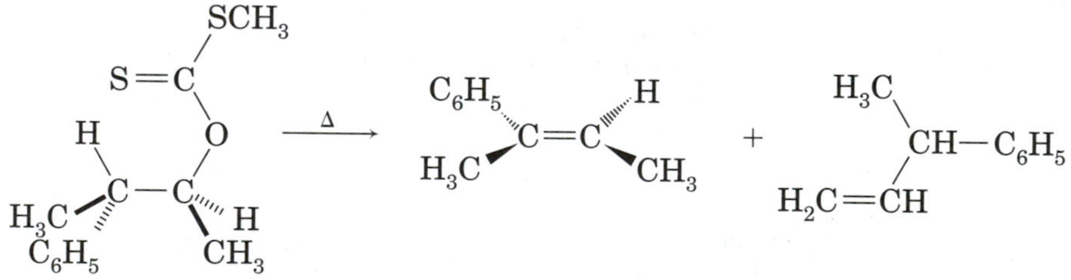

S=C(SCH₃)–O xanthate —Δ→ (C_6H_5)(H_3C)C=C(H)(CH_3) + H_3C–CH(C_6H_5)–CH=CH_2 **(10.62)**

S=C(SCH₃)–O xanthate —Δ→ (C_6H_5)(H_3C)C=C(CH_3)(H) + H_3C–CH(C_6H_5)–CH=CH_2 **(10.63)**

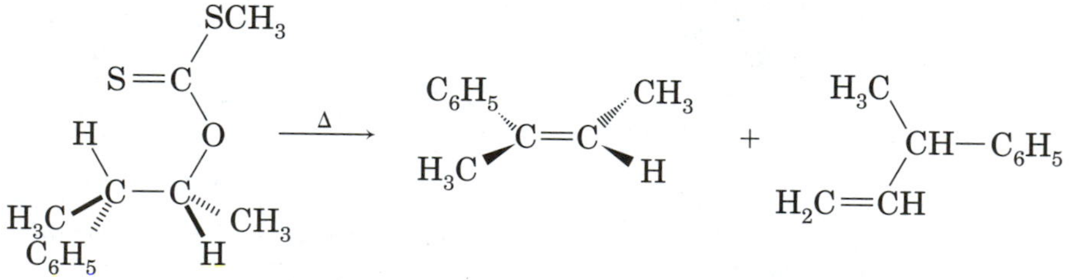

---

[169]See, for example, Stevens, P. G. *J. Am. Chem. Soc.* **1932**, *54*, 3732 and references therein.

[170]Schurman, I.; Boord, C. E. *J. Am. Chem. Soc.* **1933**, *55*, 4930.

[171]If more than one C_β-H bond is present, a mixture of alkenes can be produced. As discussed in reference 164, the product distribution depends on the number of C_β-H bonds that can lead to each product, steric factors in conformations in which each C_β-H bond is syn coplanar to the C_α-O bond, and the relative thermodynamic stabilities of the products.

[172]Cram, D. J. *J. Am. Chem. Soc.* **1949**, *71*, 3883. As noted in the equations, the elimination also produced 3-phenyl-1-butene (in 30–40% yield.)

[173]Syn elimination provides a means of determining stereochemical relationships in natural product structure elucidation. Barton, D. H. R. *J. Chem. Soc.* **1949**, 2174.

[174]In some cases product distributions suggest that some anti elimination may occur, although uncertainties about reactant purity can make it difficult to quantify the amount of non-syn elimination. For a discussion, see reference 164, pp. 444 *ff.*

Figure 10.44 Concerted mechanism proposed for Chugaev reaction.

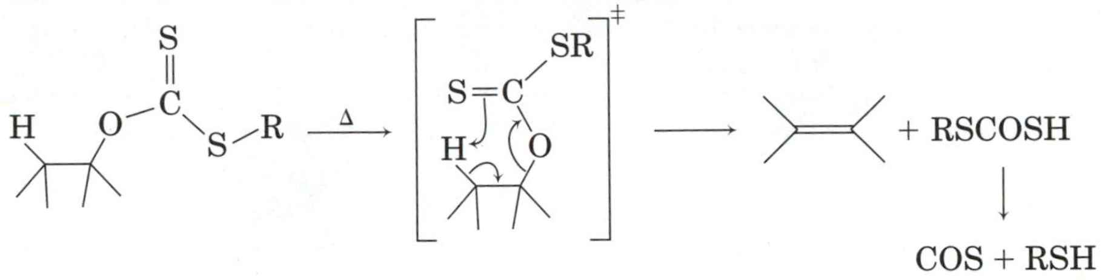

O'Connor and Nace determined that the E_a for pyrolysis of cholesteryl methyl xanthate is 32.9 kcal/mol, with a $\Delta S^{\ddagger}$ of − 4.7 eu.[167(a)] The low activation energy suggests a concerted reaction (in which bond forming and bond breaking occur simultaneously), and the negative activation entropy suggests a cyclic transition structure. Two slightly different mechanisms are consistent with the kinetic and stereochemical results. The mechanism currently accepted is a two-step process. The first, rate-limiting step is a concerted unimolecular elimination (E_i) involving hydrogen abstraction by the thione sulfur atom in a cyclic transition structure, while the second step is decomposition of the RSCOSH to COS and RSH (Figure 10.44).[175,176]

There is an alternative cyclic mechanism (Figure 10.45) involving hydrogen abstraction by the thioether sulfur atom that leads to the observed products in one step. However, Bader and Bourns reported a combination of carbon and sulfur isotope effects (Table 10.9) that are consistent with the mechanism in Figure 10.44 but are not consistent with the mechanism in Figure 10.45.[177,178]

Figure 10.45 Alternative mechanism for Chugaev reaction.

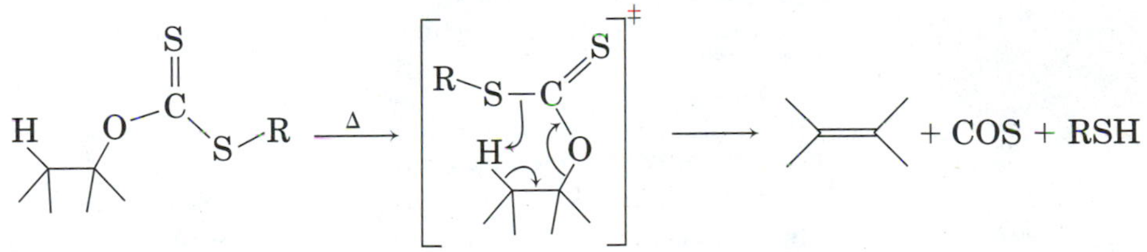

---

[175]The cyclic transition structure was proposed by Hurd and Blunck for pyrolysis of esters having a β-hydrogen atom (reference 180).

[176]In the IUPAC nomenclature system (reference 7), the E_i mechanism is known as a cyclo-$D_E D_N A_n$ mechanism.

[177]Bader, R. F. W.; Bourns, A. N. *Can. J. Chem.* **1961**, *39*, 348.

[178]In some cases, the presence of peroxide impurities may introduce an alternate radical mechanism for xanthate decomposition. Nace, H. R.; Manly, D. G.; Fusco, S. *J. Org. Chem.* **1958**, *23*, 687. Radical processes were also proposed for the formation of minor products in the pyrolysis of alkyl acetates. Shi, B.; Ji, Y.; Dabbagh, H. A.; Davis, B. H. *J. Org. Chem.* **1994**, *59*, 845.

Table 10.9 Kinetic isotope effects in pyrolysis of *S*-methyl-*trans*-2-methyl-1-indanyl xanthate.

	Percent isotope effect for the indicated label		
	$H_3C—S^*—C$ $^{32}S/^{34}S$	$C{=}S^*$ $^{32}S/^{34}S$	$C^*{=}S$ $^{12}C/^{13}C$
Predicted for the mechanism in			
Figure 10.44	≈0.0	≈0.7-1.0	≈0.0
Figure 10.45	≈1.2	≈0.0	3.0-4.0
Experimental	0.21 ± 0.07	0.86 ± 0.16	0.04 ± 0.06

(Table adapted from reference 167; data from reference 177.)

Generalizing the Chugaev reaction as shown in equation 10.64 suggests that many other organic compounds should pyrolyze to alkenes,[179]

$$\text{cyclic } b{=}a,\ c,\ H,\ d{-}e \xrightarrow{\Delta} [\ \text{cyclic transition structure}\ ]^{\ddagger} \longrightarrow c{=}b{-}a{-}H + d{=}e \qquad (10.64)$$

and unimolecular syn elimination has also been reported for alkyl esters,[180,181] carbonates and carbamates,[182] amides, vinyl ethers, and carboxylic acid anhydrides.[183] The mechanistic resemblance of several of these eliminations to the Chugaev reaction was noted by O'Connor and Nace, who found that activation entropies for the pyrolyses of cholesteryl carbamate, acetate, and methyl xanthate were very similar (Table 10.10).[184] Kinetic isotope effect studies indicate that the $C_\beta—H$ bond is being broken in the transition structure in these reactions also.[185] Furthermore, Curtin and

[179]If b is CH and a, c, d and e are all CH_2, then the reaction shown in equation 10.64 is a retro-ene reaction, which is discussed in more detail in Chapter 11.

[180]Hurd, C. D.; Blunck, F. H. *J. Am. Chem. Soc.* **1938**, *60*, 2419.

[181]In one case pyrolysis of an ester led to anti elimination. The results were explained on the basis of a mechanism incorporating anchimeric assistance: Smissman, E. E.; Li, J. P.; Creese, M. W. *J. Org. Chem.* **1970**, *35*, 1352. Anchimeric assistance was also proposed for a gas phase elimination of hydrogen chloride: Hernandez, J. A.; Chuchani, G. *Int. J. Chem. Kinet.* **1978**, *10*, 923; Chuchani, G.; Martin, I. *J. Phys. Chem.* **1986**, *90*, 431.

[182]Reference 167(b) and references therein.

[183]Reference 164 and references therein.

[184]However, the lower activation energy for the xanthate pyrolysis means that it occurs at lower temperatures than do the other pyrolyses. As a result, xanthate pyrolysis avoids the necessity of vapor phase pyrolysis that the other eliminations require, and the products are less likely to undergo further reaction at the lower temperature of xanthate pyrolysis. For a discussion, see reference 164.

[185]Data summarized in reference 164, p. 442.

Table 10.10 Activation parameters for pyrolysis of cholesteryl derivatives.

Compound	E_a (kcal/mol)	$\Delta S^{\ddagger}$ (eu)
Cholesteryl ethyl carbamate	41.0	−4.3
Cholesteryl acetate	44.1	−3.6
Cholesteryl methyl xanthate	32.9	−4.7

(Data from reference 182.)

Kellom demonstrated that ester pyrolysis occurs primarily through syn elimination by studying the pyrolysis of *erythro*- and *threo*-2-deuterio-1,2-diphenylethyl acetate (**36** and **39**, respectively).[186] Although the product of both reactions is *trans*-stilbene, the product of reaction of **36** retains almost all of the deuterium label (**37**), while the product of pyrolysis of **39** has lost most of the deuterium label (**40**).[187,188]

erythro-2-Deuterio-1,2-diphenylethyl acetate **36** → *trans*-α-deuteriostilbene **37** + **38** **(10.65)**

threo-2-Deuterio-1,2-diphenylethyl acetate **39** → *trans*-stilbene **40** + **41** **(10.66)**

Similarly, β-hydroxyalkenes undergo pyrolytic elimination (500°) in the gas phase to form an alkene and an aldehyde or ketone (equation 10.67).

[186]Curtin, D. Y.; Kellom, D. B. *J. Am. Chem. Soc.* **1953**, *75*, 6011.

[187]Quantitative interpretation of the percent of deuterium retention is subject to uncertainty because the stilbenes are not completely stable at the temperature of the pyrolysis reaction. See the footnote to Table I in reference 186.

[188]For a theoretical study of the gas phase pyrolysis of esters, see Lee, I.; Cha, O. J.; Lee, B.-S. *J. Phys. Chem.* **1990**, *94*, 3926.

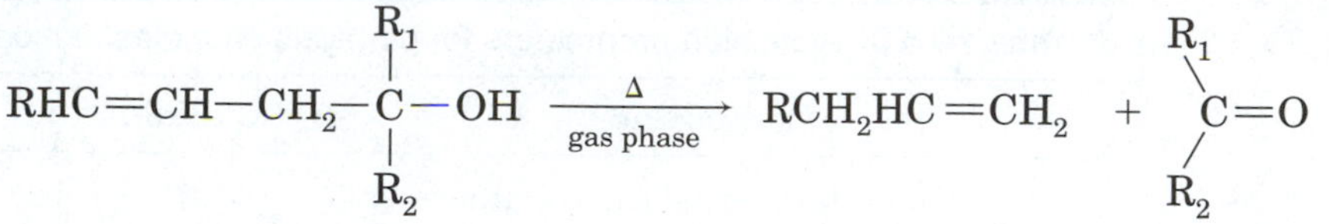

$$\mathrm{RHC{=}CH{-}CH_2{-}\underset{R_2}{\overset{R_1}{\overset{|}{\underset{|}{C}}}}{-}OH} \xrightarrow[\text{gas phase}]{\Delta} \mathrm{RCH_2HC{=}CH_2} + \mathrm{(R_1)(R_2)C{=}O} \quad (10.67)$$

Arnold and Smolinsky determined that the pyrolysis of 3-deuterio-3-ethyl-6-phenyl-5-hexen-3-ol (**42**) produced 3-deuterio-3-phenylpropene (**43**) and 2-butanone (**44**), which confirmed a syn elimination and which suggested a cyclic transition structure for the reaction (equation 10.68).[189]

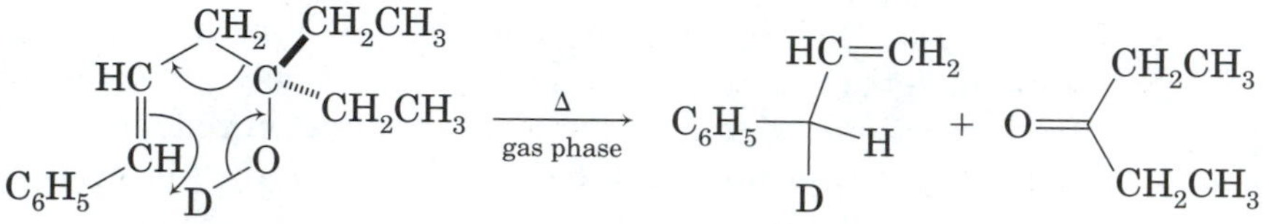

$$\mathbf{42} \xrightarrow[\text{gas phase}]{\Delta} \mathbf{43}\ (\mathrm{C_6H_5CHD{-}HC{=}CH_2}) + \mathbf{44}\ (\mathrm{O{=}C(CH_2CH_3)_2}) \quad (10.68)$$

This conclusion was supported by the observation that pyrolysis of 3-butenol has a $\Delta S^\ddagger$ of -8.8 eu, which is similar to the activation entropy values reported for pyrolysis of ethyl formate and for 3-butenoic acid, and the activation energies for all three pyrolyses are also similar, about 40 kcal/mol.[190]

Another well known concerted syn elimination is the ***Cope elimination***, which involves the thermal elimination of an alkene from an amine oxide (Figure 10.46).[191,192] Unlike the reactions discussed above, all of which have a six-membered ring in the transition structure, the Cope elimination has a five-membered ring in the transition structure. Cram and McCarty demonstrated that the stereochemistry of the Cope elimination is syn by showing that diastereomeric 2-amino-3-phenylbutane oxides produce different product distributions (Table 10.11).[193]

Figure 10.46 Concerted mechanism for the Cope elimination.

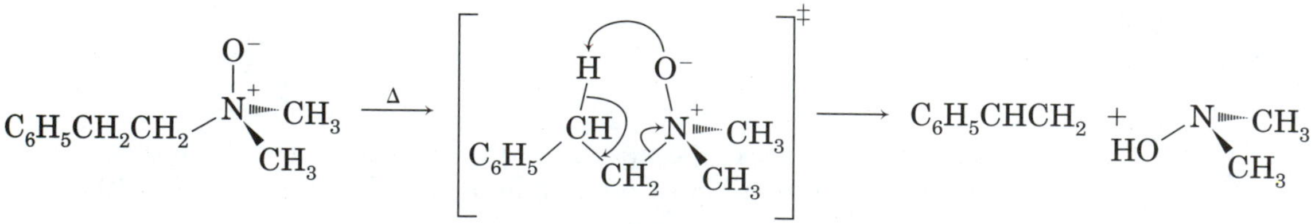

$$\mathrm{C_6H_5CH_2CH_2{-}\overset{+}{N}(O^-)(CH_3)_2} \xrightarrow{\Delta} [\text{cyclic transition structure}]^\ddagger \longrightarrow \mathrm{C_6H_5CHCH_2} + \mathrm{HO{-}N(CH_3)_2}$$

[189]Arnold, R. T.; Smolinsky, G. *J. Org. Chem.* **1960**, *25*, 129.

[190]Smith, G. G.; Yates, B. L. *J. Chem. Soc.* **1965**, 7242.

[191]Cope, A. C.; Foster, T. T.; Towle, P. H. *J. Am. Chem. Soc.* **1949**, *71*, 3929.

[192]Cope, A. C.; Trumbull, E. R. *Org. React.* **1960**, *11*, 317.

[193]Cram, D. J.; McCarty, J. E. *J. Am. Chem. Soc.* **1954**, *76*, 5740.

Table 10.11 Products from pyrolysis of diastereomers of 2-amino-3-phenylbutane oxide.

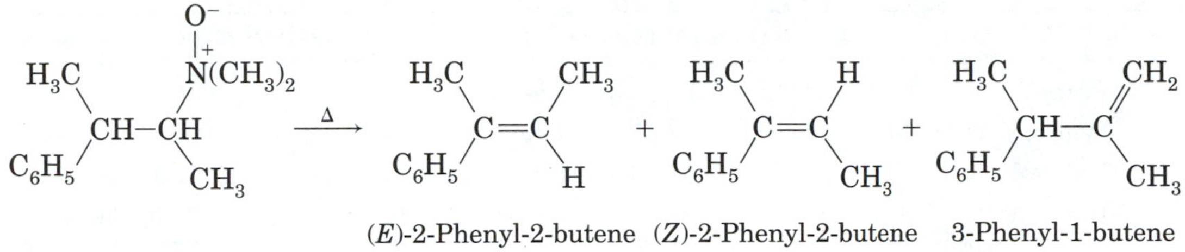

$$H_3C(C_6H_5)CH{-}CH(CH_3)\overset{+}{N}(CH_3)_2O^- \xrightarrow{\Delta} (C_6H_5)(H_3C)C{=}C(CH_3)H + (C_6H_5)(H_3C)C{=}C(H)CH_3 + (C_6H_5)(H_3C)CH{-}C({=}CH_2)CH_3$$

Diastereomer (racemic)	(*E*)-2-Phenyl-2-butene	(*Z*)-2-Phenyl-2-butene	3-Phenyl-1-butene
	Percent yield		
Erythro	4	89	7
Threo	93	0.2	7

Interestingly, the threo diastereomer undergoes pyrolysis at a lower temperature than does the erythro diastereomer. As shown in Figure 10.47, the authors suggested that Cope elimination of the threo diastereomer can occur through a transition structure in which a carbon-phenyl bond is eclipsed with a carbon-hydrogen bond. However, elimination of the erythro diastereomer requires a more strained transition structure in which a carbon-phenyl bond eclipses a carbon-methyl bond.

The regiochemistry of the Cope elimination is different from that of the Hofmann elimination. For many dialkylmethylamine oxides, the distribution of alkenes seems to be determined primarily by the number of β-hydrogen atoms available for abstraction on each of the alkyl groups.[194] As shown by selected data reproduced in Table 10.12, there seems to be only a slight

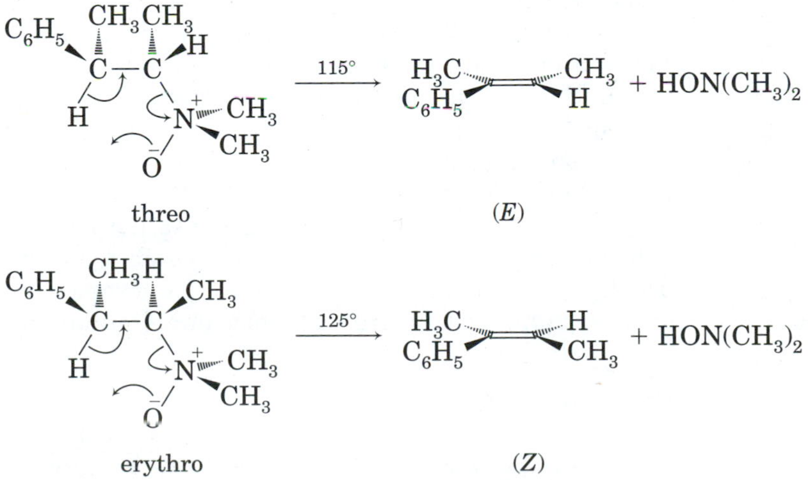

threo $\xrightarrow{115^\circ}$ (*E*)-$H_3C(C_6H_5)C{=}C(CH_3)H$ + $HON(CH_3)_2$

erythro $\xrightarrow{125^\circ}$ (*Z*)-$H_3C(C_6H_5)C{=}C(H)CH_3$ + $HON(CH_3)_2$

Figure 10.47 Stereospecificity of Cope elimination. (Adapted from reference 193.)

[194]Reference 192, particularly the discussion beginning on page 363.

Table 10.12 Product distributions in Cope eliminations from $R_1R_2CH_3N^+—O^-$.

Alkyl Groups		Olefinic Products			
R_1	R_2	Experimental results		Statistical prediction	
Ethyl	Propyl	62.5% ethene	37.5% propene	60% ethene	40% propene
Ethyl	Isopropyl	27.5% ethene	72.5% propene	33.3% ethene	67.7% propene
Ethyl	Isobutyl	67.6% ethene	32.4% isobutene	75% ethene	25% isobutene
n-Butyl	Isobutyl	64.8% 1-butene	35.2% isobutene	67.7% 1-butene	33.3% isobutene

(Data from a tabulation in reference 78.)

tendency for a more highly substituted alkene to be formed in preference to a less highly substituted alkene. Instead, the product distributions are closely predicted on the basis of the number of β-hydrogen atoms that could be abstracted on each alkyl group. If the β-carbon atom is substituted with a phenyl group (as is the case in the data reported in Table 10.11), then the elimination strongly favors the product formed by abstraction of that β-hydrogen atom. This regiochemical preference may result from the greater acidity of that hydrogen atom.[192]

One complication in all gas phase pyrolyses is uncertainty that the reaction does occur in the gas phase and not on the surface of the reaction vessel. In particular, Wertz and Allinger concluded from an analysis of experimental and calculated product yields that most of the experimental data previously reported for gas phase pyrolyses of esters, and perhaps other classes of compounds, might actually have resulted from surface-mediated reactions.[195] They argued that it may be impossible to inactivate glass surfaces, so that only when pyrolyses are conducted in stainless steel vessels that have carbonized interior surfaces and from which air is continuously excluded can true gas-phase results be obtained. They suggested that ionization of the ester at the glass surface leads to a glass-stabilized carbocation, which then loses a β-proton in the rate-limiting step of the reaction. Subsequent experimental work led Dabbagh and Davis to conclude that surface effects might be important in ester pyrolyses carried out in reactors containing glass beads if the beads have a high surface area, but not if they have a low surface area.[196]

Another complication with gas phase pyrolyses is that many possible nonconcerted reaction pathways are possible. For example, alkyl halides undergo elimination in the gas phase,[197] and some compounds, such as ethyl chloride, appear to undergo unimolecular elimination.[198] Maccoll cited stud-

[195]Wertz, D. H.; Allinger, N. L. *J. Org. Chem.* **1977**, *42*, 698.

[196]Dabbagh, H. A.; Davis, B. H. *J. Org. Chem.* **1990**, *55*, 2011.

[197]For a review, see Maccoll, A. *Chem. Rev.* **1969**, *69*, 33.

[198]Barton, D. H. R.; Howlett, K. E. *J. Chem. Soc.* **1949**, 165.

ies suggesting that the unimolecular decompositions involve transition structures with significant carbocation character.[197,199] Pyrolysis of (+)-2-chlorooctane in the gas phase at 325–385° was found to produce racemization of the starting material as well as elimination of HCl.[200] However, other compounds appear to react by radical chain mechanisms, and heterogeneous radical reactions often complicate studies that are not carried out in well seasoned (coated with a layer of organic material) vessels. Furthermore, there appears to be a significant radical (but not radical chain) component to the pyrolysis of sulfoxides.[201]

Problems

1. Ethoxide-promoted E2 elimination from 2-iodo-3-methylbutane produces 18% of the 1-alkene and 82% of the 2-alkene, while elimination from 2-bromo-2-methylbutane produces 29% of the 1-alkene and 71% of the 2-alkene. Similarly, E2 elimination of $(CH_3)_2S$ from dimethyl-*sec*-butylsulfonium ion produces 74% of the 1-alkene and 26% of the 2-alkenes, while the bimolecular elimination of $(CH_3)_2S$ from dimethyl-*t*-amylsulfonium produces 86% of the 1-alkene and 14% of the 2-alkene. In each case, an additional methyl substituent on the α-carbon atom leads to a greater proportion of the less highly substituted alkene. Explain the product distributions from all four reactions in terms of the factors that determine the distribution of products from E2 reactions.

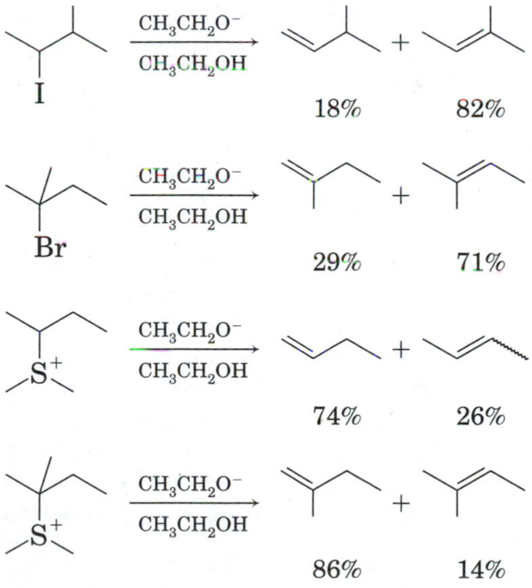

I; $CH_3CH_2O^-$ / CH_3CH_2OH → 18% + 82%

Br; $CH_3CH_2O^-$ / CH_3CH_2OH → 29% + 71%

S^+; $CH_3CH_2O^-$ / CH_3CH_2OH → 74% + 26%

S^+; $CH_3CH_2O^-$ / CH_3CH_2OH → 86% + 14%

[199] Maccoll, A. *Adv. Phys. Org. Chem.* **1965**, *3*, 91.

[200] Harding, C. J.; Maccoll, A.; Ross, R. A. *Chem. Commun.* **1967**, 289.

[201] However, at lower temperatures a concerted mechanism appears to be the major pathway for reaction. Kingsbury, C. A.; Cram, D. J. *J. Am. Chem. Soc.* **1960**, *82*, 1810.

2. Construct appropriate Newman projections for **8** and **9** (page 659) to rationalize the greater reactivity of **9** than **8** in dehydrohalogenation reactions.
3. A study of the stereochemistry of the dehydration of alcohols to alkenes over metal oxide catalysts utilized the compounds *erythro*-2-butanol-3-d_1 (**45**) and *threo*-2-butanol-3-d_1 (**46**). Predict the products expected from each of these reactants for both syn and anti elimination pathways.

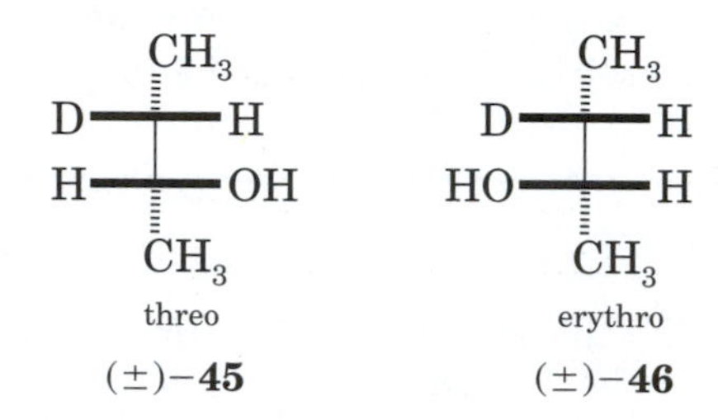

(±)-**45** (±)-**46**

4. A 20% yield of 2,2-dimethyltrimethylene oxide (**48**) was obtained as a result of an intramolecular substitution following removal of a proton from the OH group of 3-bromo-2,2-dimethyl-1-propanol (**47**) by KOH in aqueous solution. However, the major products of the reaction were isobutylene and formaldehyde. Propose a mechanism to account for the formation of these products.

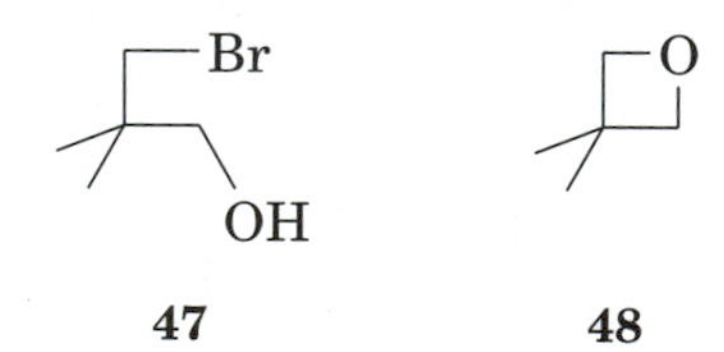

47 **48**

5. The product of addition of Br_2 to *trans*-cinnamic acid ((±)-*erythro*-2,3-dibromo-3-phenylpropanoic acid, **49**) can be decarboxylated to β-bromostyrene (1-bromo-2-phenylethene) by refluxing in water or by heating in the presence of sodium acetate in a variety of solvents. The product obtained by heating **49** in acetone in the presence of sodium acetate is the (*Z*) isomer of β-bromostyrene. In ethanol solution a mixture of (*Z*) and (*E*) isomers is formed, while in water the product is predominantly the (*E*) isomer. Propose a mechanism for the exclusive formation of the (*Z*) isomer of β-bromostyrene in acetone solution, and explain the effect of solvent on the stereoselectivity of the reaction.

(±)-**49**

6. Base-promoted 1,4-dehydrochlorination of 3-(1-chloroethyl)indene leads to a pair of diastereomeric dienes (equation 10.71). How could the compounds (1R*)-(1-$^2H_1$)-3-((1R*)-1-chloroethyl)indene (**50**) and (1R*)-(1-$^2H_1$)-3-((1S*)-1-chloroethyl)indene (**51**) be used to determine whether the reaction occurs by a syn or an anti pathway?

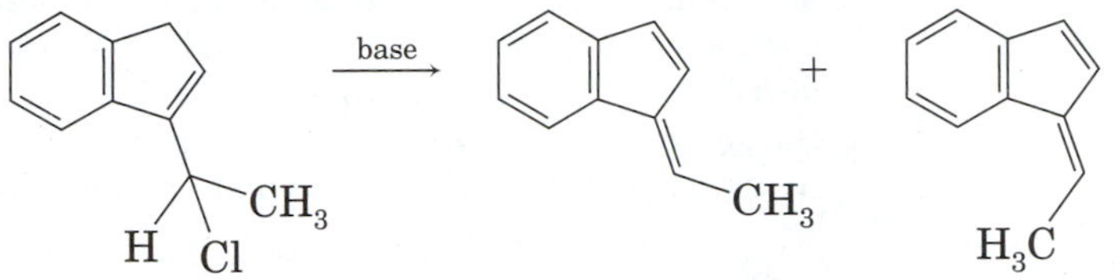

(10.69)

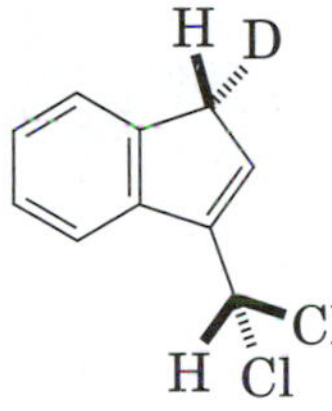

50 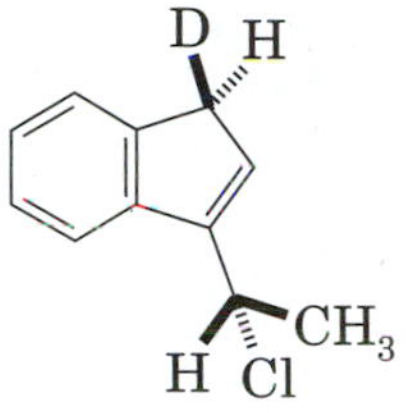 **51**

7. 2-[*p*-(2-Methyl-2-hydroxypropyl)phenyl]propionic acid (**52**) has been proposed to be a human metabolite of the analgesic ibuprofen (**53**). Show how you could synthesize **52** from **53** through a pathway involving (a) benzylic bromination with NBS, (b) dehydrobromination; (c) epoxidation with *m*-chloroperbenzoic acid, (d) 1,6-eliminative cleavage of the resulting 2-[*p*-(2-methyl-1,2-epoxypropyl) phenyl]propionic acid (**54**), and (e) catalytic hydrogenation.

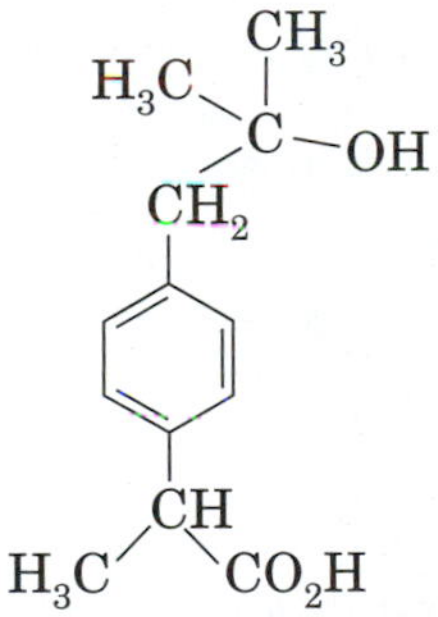

52 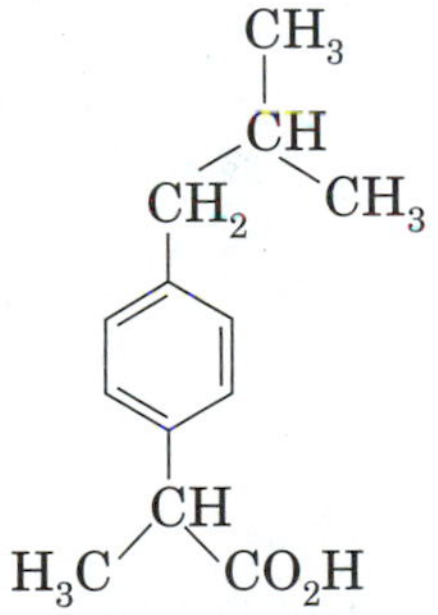 **53** 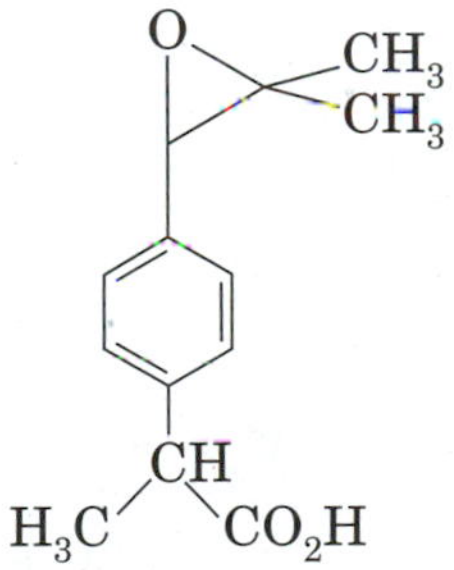 **54**

8. Reaction of 1,1-dichlorocyclohexane and of *cis*-1,2-dichlorocyclohexane with 0.1 M KOH in 4 : 1 (v : v) ethanol : water at 100° results in the formation of a compound with the molecular formula C_6H_9Cl. Under the same conditions, however, the product of reaction of *trans*-1,2-dichlorocyclohexane has the molecular formula C_6H_8.
   a. Propose a detailed mechanism to account for the formation of the products in each case, and explain why the product of the reaction of the trans isomer differs from the product formed by the other two isomers.

b. The relative reactivity of the three dichlorocyclohexanes under these reaction conditions is *cis*-1,2-dichlorocyclohexane (1.0), 1,1-dichlorocyclohexane (3×10^{-2}) and *trans*-1,2-dichlorocyclohexane (4.1×10^{-3}). Interpret these reactivities in terms of the mechanism proposed for each compound.

9. Concerted 1,2-eliminations occur with anti stereochemistry, and concerted 1,4-eliminations occur with predominantly syn stereochemistry. Predict the stereochemistry of concerted 1,6-, 1,8-, and 1,10-eliminations, and use a curved arrow description to justify each of your predictions.

10. Rationalize the difference in regioselectivity between the Hofmann elimination (equation 10.70) and Cope elimination (equation 10.71).

$$\text{Et(Me)}_2\text{N}^+\text{-Pr } ^-\text{OH} \xrightarrow[\text{neat}]{85^\circ - 150^\circ} \underset{98\%}{H_2C{=}CH_2} + \underset{2\%}{H_2C{=}CH{-}CH_3} \qquad \textbf{(10.70)}$$

$$\text{Et(Me)N}^+(\text{O}^-)\text{-Pr} \xrightarrow[\text{neat}]{85^\circ - 150^\circ} \underset{63\%}{H_2C{=}CH_2} + \underset{37\%}{H_2C{=}CH{-}CH_3} \qquad \textbf{(10.71)}$$

11. It has been proposed that the dehydration of alcohols catalyzed by perchloric, sulfuric, and acetic acids in methylene chloride solution could occur by a concerted elimination following formation of the alkyl ester of the inorganic acid. Show how the perchlorate ester shown in equation 10.72 could be used to effect concerted dehydration of the alcohol in a mechanism analogous to the mechanism of the pyrolytic eliminations discussed in the section beginning on page 695, and predict whether the reaction would exhibit syn or anti stereochemistry.

$$\text{H-C-C-OH} + HClO_4 \longrightarrow \text{H-C-C-O-Cl(=O)}_3 + H_2O \qquad \textbf{(10.72)}$$

12. Ketene can be formed by pyrolysis of acetic anhydride (equation 10.73).[202] Show how a cyclic transition structure similar to those drawn for other pyrolytic eliminations can explain this reaction.

$$H_3C{-}C(=O){-}O{-}C(=O){-}CH_3 \xrightarrow{\Delta} H_2C{=}C{=}O + H_3C{-}C(=O){-}OH \qquad \textbf{(10.73)}$$

13. As noted on page 689, β,γ-unsaturated alcohols can be dehydrated to conjugated dienes with hydroxide ion. α,β-Unsaturated alcohols can also be dehydrated to

[202]Wilsmore, N. T. M. *J. Chem. Soc.* **1907**, *91*, 1938.

conjugated dienes, as shown for the synthesis of 1,2-bis(methylene)cyclohexane in equation 10.74. Propose a mechanism for the reaction in equation 10.74 in which the second step is a concerted reaction analogous to a pyrolytic elimination reaction.

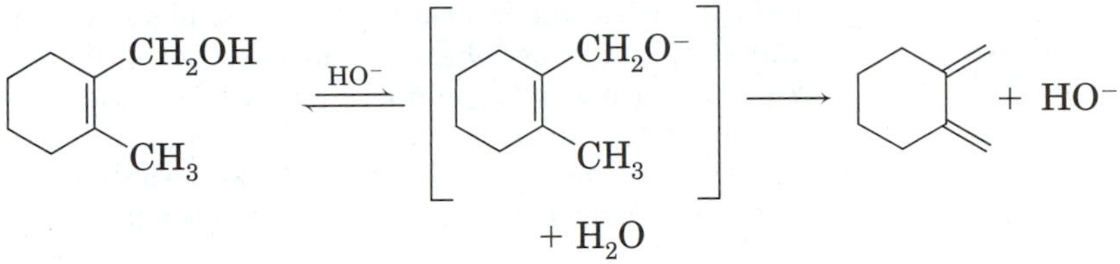

CH_2OH / CH_3 $\xrightleftharpoons{HO^-}$ [CH_2O^- / CH_3 + H_2O] ⟶ + HO^- **(10.74)**

14. The hydrocarbon α-cedrene (**55**) has been converted to its isomer β-cedrene (**56**) in 52% yield by a two step process involving first addition of HCl and then elimination of HCl. Propose conditions that would optimize the formation of **56** from the intermediate alkyl halide.

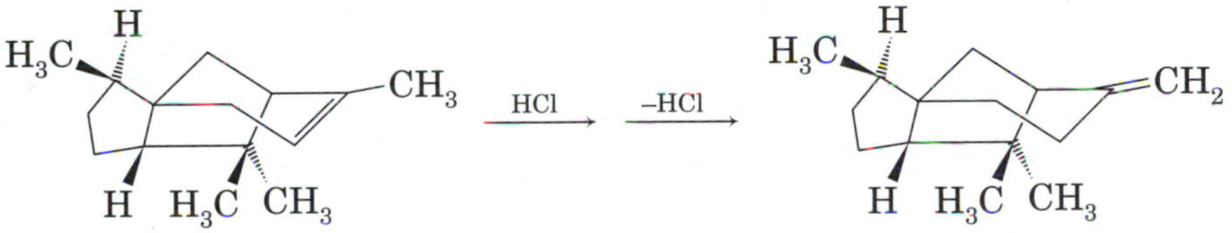

H, H_3C, CH_3, H, H_3C, CH_3 $\xrightarrow{HCl}$ $\xrightarrow{-HCl}$ H, H_3C, CH_2, H, H_3C, CH_3

α-Cedrene **55** β-Cedrene **56**

15. Rationalize the observation that 2β,3α-dibromocholestane readily undergoes debromination with iodide ion, but 3β,4α-dibromocholestane is unreactive under the same conditions.
16. Refluxing a solution of 1-(*p*-nitrobenzoyl)-2-benzylaziridine (**57**) in toluene for 24 hours produced a 91% yield of *N*-(*trans*-cinnamyl)-*p*-nitrobenzamide (**58**). Propose a mechanism for the rearrangement that is analogous to the mechanism of one of the elimination reactions discussed in this chapter.

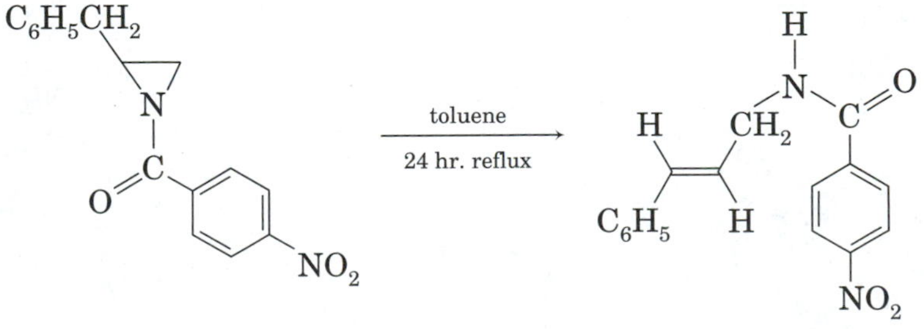

$C_6H_5CH_2$, N, C, O, NO_2 $\xrightarrow[\text{24 hr. reflux}]{\text{toluene}}$ H, N, C, O, H, CH_2, C_6H_5, H, NO_2

57 **58**

17. Treatment of *trans*-3-methoxycyclohexanecarboxylic acid with thionyl chloride produces the corresponding acid chloride. However, treatment of *cis*-3-methoxycyclohexanecarboxylic acid with thionyl chloride leads to the formation of

methyl *trans*-3-chlorocyclohexanecarboxylate and methyl 3-cyclohexene-1-carboxylate in a ratio of 3 : 5. Propose a mechanism to account for the formation of all of the products, and explain the difference in products observed for the cis and trans isomers.

18. Propose an explanation for each of the following observations.
    a. Pyrolysis of the methyl xanthate of *cis*-2-phenylcyclohexanol produces nearly 100% of 3-phenylcyclohexene, but pyrolysis of the methyl xanthate of *trans*-2-phenylcyclohexanol produces about 88% of 1-phenylcyclohexene and 12% of 3-phenylcyclohexene.
    b. The thermal decomposition of *trans*-2-methyl-1-tetralyl acetate (**59**) occurs at a lower temperature than does the decomposition of the cis isomer (**60**).

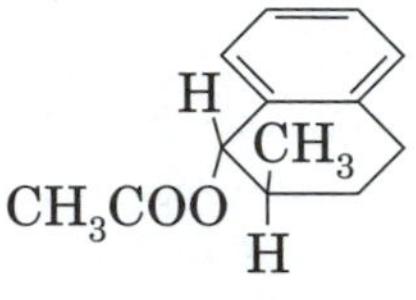

H, CH_3, CH_3COO, H — **59**

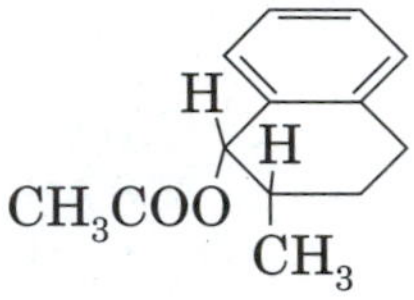

H, H, CH_3COO, CH_3 — **60**

    c. Heating the *N*-oxides of a series of *N*-methylazacycloalkanes (**61**) does not result in reaction when *n* is 3, but under the same conditions the intramolecular Hofmann reaction occurs when *n* is 4, and it occurs more readily when *n* is 5. Explain the dependence of the rate of the reaction on the size of the ring.

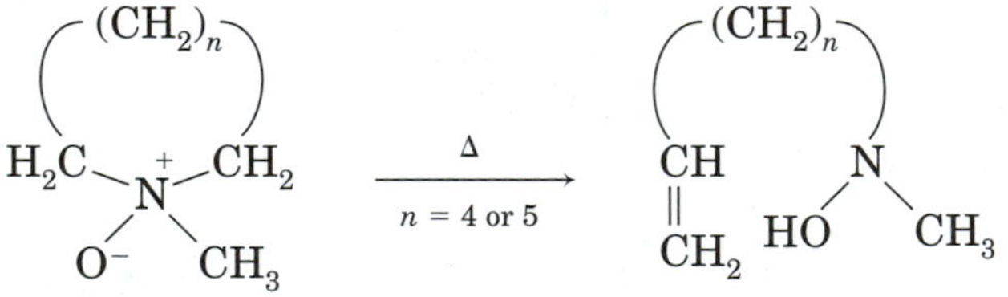

$(CH_2)_n$, H_2C, $\overset{+}{N}$, CH_2, O^-, CH_3 — **61** $\xrightarrow[n\,=\,4\text{ or }5]{\Delta}$ $(CH_2)_n$, CH=CH_2, N, HO, CH_3 — **62**

19. Treatment of 2-chlorocyclohexanone (**63**) with aqueous sodium hydroxide leads to the sodium salt of cyclopentanecarboxylic acid (**65**) in a reaction known as the **Favorskii rearrangement.**[203] The ketone **64** has been proposed to be an intermediate in the reaction. Propose a mechanism for the formation of **64** and categorize the mechanism of its formation as a type of elimination reaction.

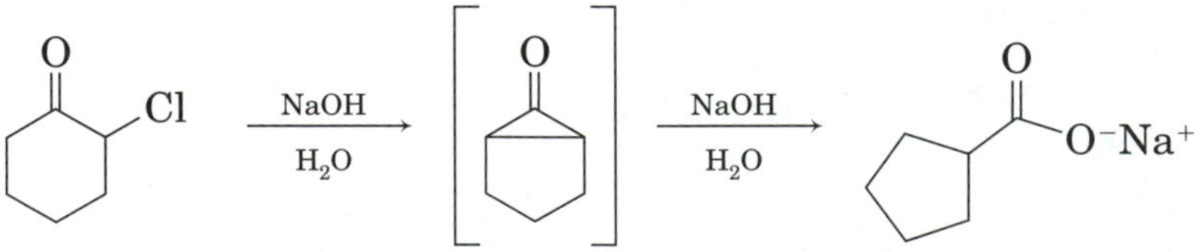

63 (O, Cl) $\xrightarrow[H_2O]{NaOH}$ [**64** (O)] $\xrightarrow[H_2O]{NaOH}$ **65** (O, O^-Na^+)

[203](a) Favorskii, A.; Bozhovskii, V. *J. Russ. Phys. Chem. Soc.* **1914**, *46*, 1097 [*Chem. Abstr.* **1915**, *9*, 1900]; (b) Loftfield, R. B. *J. Am. Chem. Soc.* **1951**, *73*, 4707.

20. Reaction of *trans*-4-*t*-butylcyclohexyltrimethylammonium chloride (**66**) with *t*-BuOK in *t*-BuOH at 75° produces *trans*-4-*t*-butylcyclohexyl-*N,N*-dimethylamine and *t*-butyl methyl ether (equation 10.75). Reaction of *cis*-4-*t*-butylcyclohexyltrimethylammonium chloride under the same conditions produces 90% of 4-*t*-butylcyclohexene and 10% of *cis*-4-*t*-butylcyclohexyl-*N,N*-dimethylamine (along with trimethylamine and methyl *t*-butyl ether, which are not shown in equation 10.76). Considering the chair conformation(s) accessible to each isomer, propose mechanisms to account for the differing products produced by the isomeric reactants.

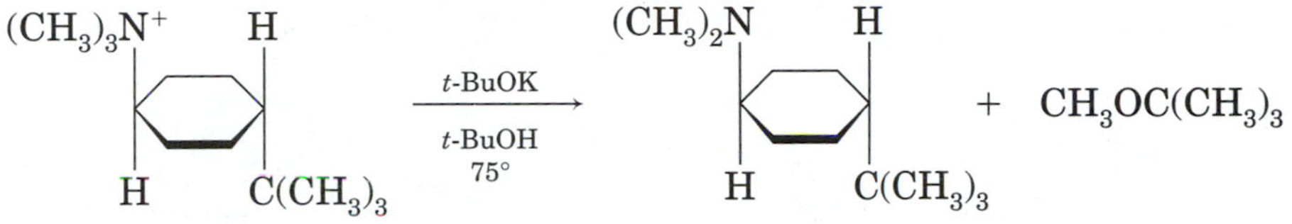

$(CH_3)_3N^+$ H / H $C(CH_3)_3$ (**66**) —*t*-BuOK, *t*-BuOH, 75°→ $(CH_3)_2N$ H / H $C(CH_3)_3$ (**67**) + $CH_3OC(CH_3)_3$ **(10.75)**

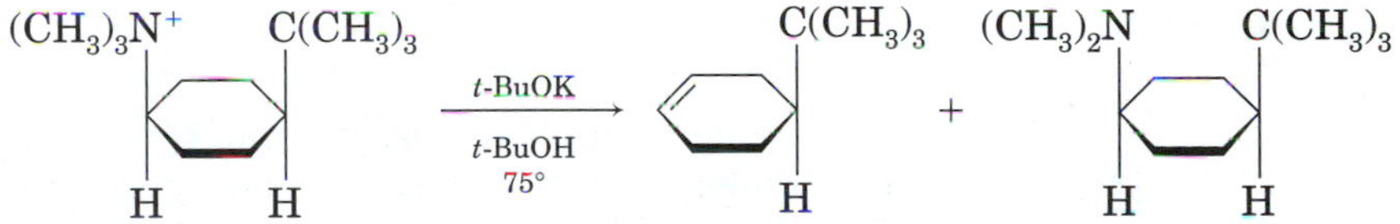

$(CH_3)_3N^+$ $C(CH_3)_3$ / H H (**68**) —*t*-BuOK, *t*-BuOH, 75°→ $C(CH_3)_3$ / H (**69**, 90%) + $(CH_3)_2N$ $C(CH_3)_3$ / H H (**70**, 10%) **(10.76)**

21. Dehydrobromination of 2-bromo-2,4,4-trimethylpentane with 4.0 M KOH in ethanol at 70° produces 86% of 2,4,4-trimethyl-1-pentene (**72**) and 14% of 2,4,4-trimethyl-2-pentene (**73**).[87] The heat of hydrogenation of **72** is −25.52 kcal/mol, while that of **73** is −26.79 kcal/mol.[204] Is this dehydrobromination an example of Saytzeff or Hofmann orientation? Justify your answer by citing one or more definitions of these terms.

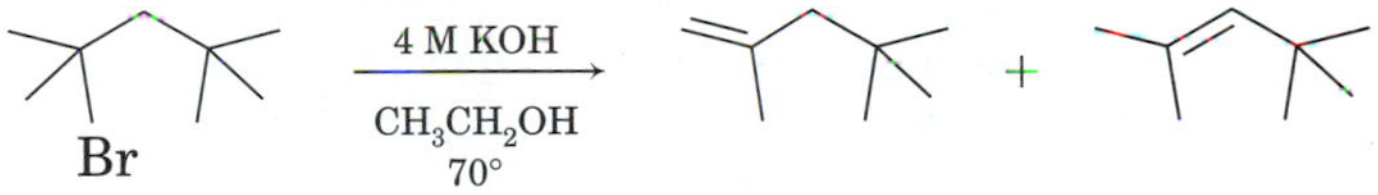

Br (**71**) —4 M KOH, CH_3CH_2OH, 70°→ **72** (86%) + **73** (14%)

[204]Turner, R. B.; Nettleton, Jr., D. E.; Perelman, M. *J. Am. Chem. Soc.* **1958**, *80*, 1430.

CHAPTER 11

Concerted Reactions

11.1 Introduction

The conversion of cyclobutene to 1,3-butadiene at 150° in the gas phase is notable for two reasons. First, the activation energy for the isomerization is only about 32.5 kcal/mol.[1] This is much lower than the activation energy expected for a reaction proceeding through initial homolysis of a carbon-carbon single bond as shown in Figure 11.1.[2] The ring strain present in the cyclobutene ring and the semi-allylic nature of the species shown as a diradical intermediate in Figure 11.1 should lower the activation energy for breaking the carbon-carbon single bond. However, Brauman and Archie used thermochemical data to estimate that such a diradical would be 47 kcal/mol higher in energy than the cyclobutene. Therefore, a reaction proceeding by the mechanism in Figure 11.1 should have an activation energy at least 15 kcal/mol higher than that which is observed for the reaction.[3]

Second, the reaction is highly stereospecific. Thermolysis of *trans*-1,2,3,4-tetramethylcyclobutene (**3**) gave only (*E,E*)-3,4-dimethyl-2,4-hexadi-

Figure 11.1 Hypothetical bond homolysis mechanism for conversion of cyclobutene to butadiene.

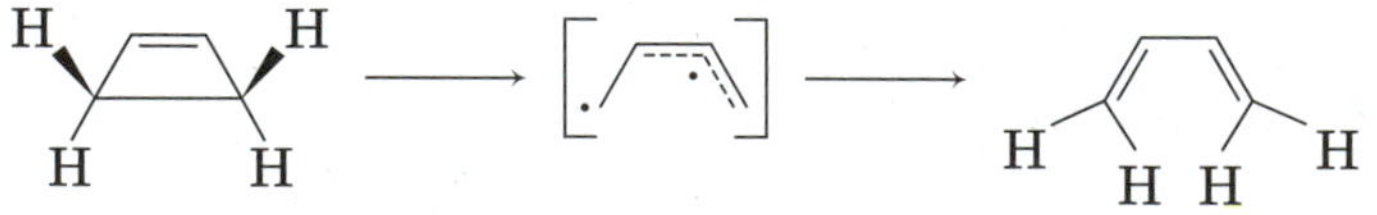

H H H H **1** → [ ] → H H H H **2**

[1]Cooper, W.; Walters, W. D. *J. Am. Chem. Soc.* **1958**, *80*, 4220.

[2]The corresponding activation energy for the pyrolysis of cyclobutane to ethene at temperatures near 400° is 62.5 kcal/mol. Benaux, C. T.; Kern, F.; Walters, W. D. *J. Am. Chem. Soc.* **1953**, *75*, 6196.

[3]Brauman, J. I.; Archie, Jr., W. C. *J. Am. Chem. Soc.* **1972**, *94*, 4262.

ene (**4**), and neither the (*Z,Z*) diastereomer (**5**), nor the (*E,Z*) diastereomer (**6**) were detected (equation 11.1).[4] Similarly, thermolysis of *cis*-1,2,3,4-tetramethylcyclobutene (**7**) was found to give only the (*E,Z*) product **6** (equation 11.2).[5,6]

Δ but neither nor (11.1)

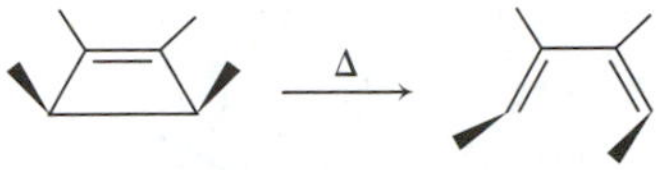

3 **4** **5** **6**

Δ (11.2)

7 **6**

In the reactions of **3** and **7**, the stereospecificity of the opening of the cyclobutene to the butadiene can be correlated with steric differences in the transition structures leading to the possible products.[7] The two methyl substituents are both above the plane of the cyclobutene ring in **7** and are both in the plane in the possible products, but the four possible transition structures differ in the location of these two methyl groups (Figure 11.2). In path (a) both methyl groups move in a clockwise fashion as the reaction proceeds from reactant to transition structure to product **6**. In path (b) both methyl groups move in a counterclockwise fashion, but the product is also the cis,trans isomer, **6**. Path (c) involves movement of one methyl group counterclockwise and the other one clockwise, which suggests steric hindrance of the two hydrogen atoms in the transition structure. Path (d) also involves movement of the two methyl groups in different directions, but this pathway would produce steric hindrance of two methyl groups in the transition structure. Thus, pathways (a) and (b) should be favored, so the steric factors correctly predict that **6** should be the major product of the reaction. Similar arguments apply to the reaction of **3**.

The same steric considerations should apply to the reverse process, the thermal closure of butadienes to cyclobutenes. Although the greater stability of butadiene makes conversion to a cyclobutene endothermic, evidence that it can occur is seen in the equilibration of *trans,trans*-2,3,4,5-tetraphenyl-2,4-hexadiene (**8**) and the cis,cis diastereomer (**10**) through *trans*-3,4-dimethyltetraphenylcyclobutene (**9**) in pyridine solution at 110° (equation 11.3). Similarly, the deuterium-labeled *cis,trans*-2,3,4,5-tetraphenyl-2,4-hexadiene **11** equilibrates with the isomeric **13**, presumably

---

[4]Criegee, R.; Noll, K. *Liebigs Ann. Chem.* **1959**, *627*, 1.

[5]Similar results for *cis*- and *trans*-3,4-dimethylcyclobutene were reported by Winter, R. E. K. *Tetrahedron Lett.* **1965**, 1207.

[6]Brauman and Archie (reference 3) were able to detect 0.005% of *trans,trans*-2,4-hexadiene from the pyrolysis of *cis*-3,4-dimethylcyclobutene at 280°.

[7]See the discussion in reference 5 and references therein.

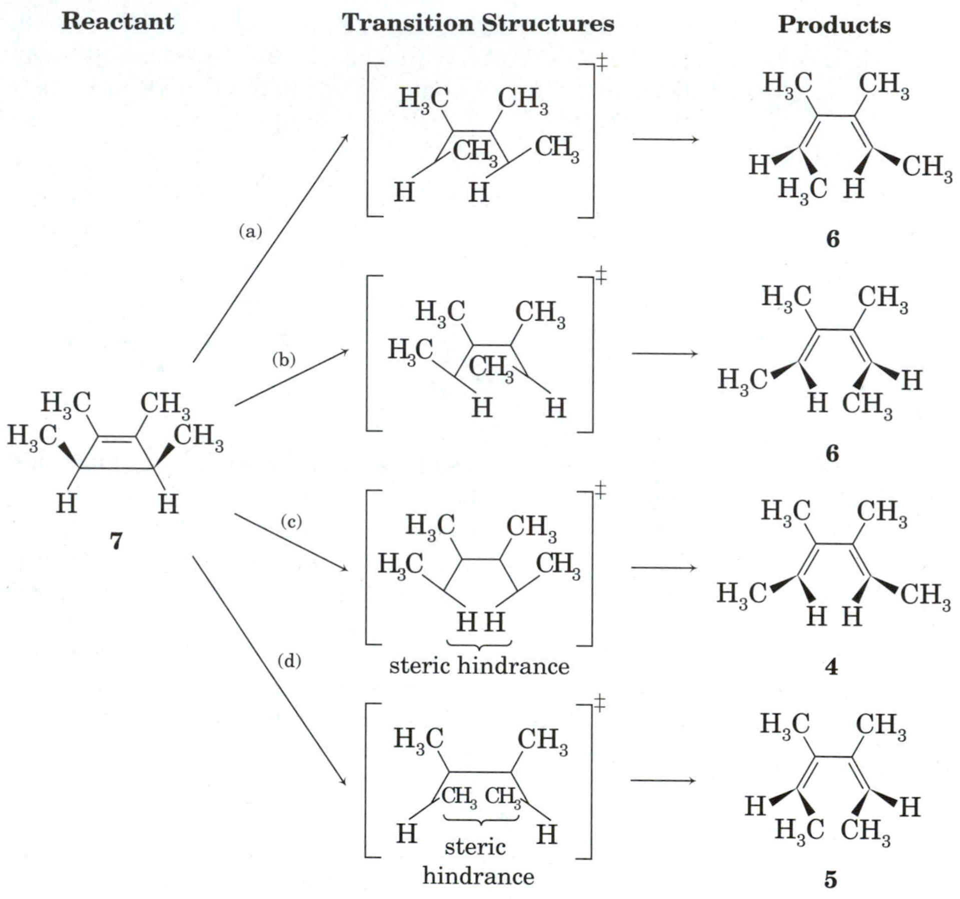

Figure 11.2 Possible steric effects in transition structures for thermal ring opening of *cis*-1,2,3,4-tetramethylcyclobutene (**7**).

through the cis cyclobutene **12** (equation 11.4), at temperatures greater than 110°. A long-term reaction of **11** revealed no evidence for isomerization to the cis,cis isomer, suggesting that any violations of the stereochemical preference for this reaction must occur no more than once in every 2.6×10^8 ring openings.[8]

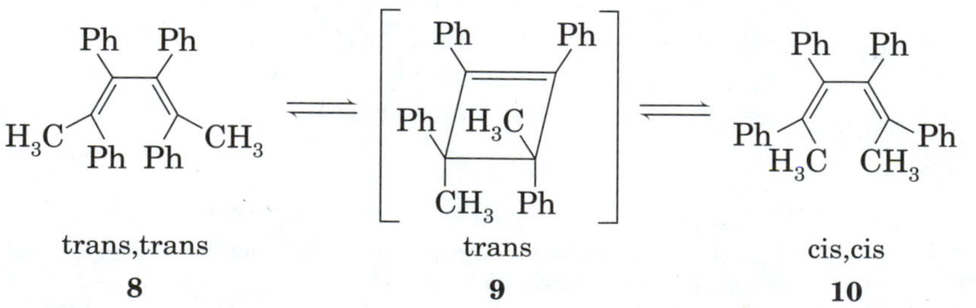

(11.3)

[8]Doorakian, G. A.; Freedman, H. H. *J. Am. Chem. Soc.* **1968**, *90*, 5310.

cis,trans **11** ⇌ [cis **12**] ⇌ trans,cis **13** **(11.4)**

A mechanistic model based on minimization of steric repulsion in transition structures may be able to rationalize results of these butadiene-cyclobutene interconversions, but that explanation cannot explain the results of analogous cyclohexadiene-hexatriene reactions. As shown in equation 11.5, it is observed that only products formed through apparent rotation of the two methyl groups in different directions (analogous to paths (c) or (d) in Figure 11.2) are formed thermally.[9] Furthermore, products with the opposite stereochemistry are formed photochemically upon irradiation of the cyclohexadiene (equation 11.6).[10,11,12]

14 —130°→ **15** **(11.5)**

16 —*hν*→ **17** **(11.6)**

---

[9]Vogel, E.; Grimme, W.; Dinné, E. *Tetrahedron Lett.* **1965**, 391. Also see Glass, D. S.; Watthey, J. W. H.; Winstein, S. *Tetrahedron Lett.* **1965**, 377; Marvell, E. N.; Caple, G.; Schatz, B. *Tetrahedron Lett.* **1965**, 385.

[10]Photochemical reactions (discussed in Chapter 12) are much more complicated than will be apparent from the discussions of orbital symmetry concepts here. See, for example, van der Lugt, W. T. A. M.; Oosterhoff, L. J. *J. Am. Chem. Soc.* **1969**, *91*, 6042. For further discussion and leading references, see Bernardi, F.; Olivucci, M.; Ragazos, I. N.; Robb, M. A. *J. Am. Chem. Soc.* **1992**, *114*, 2752.

[11]Focken, G. J. *Tetrahedron Lett.* **1962**, 549.

[12]A time-resolved UV resonance Raman study indicated excitation of 1,3-cyclohexadiene leads to vibrationally excited *di-s-cis*-1,3,5-hexatriene in 6 ps, which is less than the time required for vibrational cooling to the lowest vibrational level of that conformation or for relaxation to the *mono-s-cis* conformer: Reid, P. J.; Doig, S. J.; Wickham, S. D.; Mathies, R. A. *J. Am. Chem. Soc.* **1993**, *115*, 4754.

Investigators were unable to trap or detect intermediates in any of the reactions discussed here, and radical inhibitors were ineffective.[1] Therefore, the reactions appeared to be ***concerted***, meaning that bond breaking and bond forming processes occur simultaneously, rather than in step-wise fashion. A base of empirically derived knowledge enabled the products of many of these reactions to be predicted successfully, but for many years there was no satisfying theoretical explanation of the forces that determine which reactions proceed by what pathways. Thus some concerted reactions came to be known as no mechanism reactions.[13] The idea that a reaction might not have a mechanism is incompatible with the paradigms of contemporary organic chemistry, but the term does suggest what was for a long time the enigmatic nature of these transformations.

The period of mystery about these reactions came to an end—and a whole new era of theory and practice in organic chemistry began—with the publication of the first of a series of papers by Woodward and Hoffmann on the relationship between the stereochemistry of concerted reactions and the symmetry properties of molecular orbitals.[14,15,16,17] The interconversion of cyclobutene and butadiene is now one of many reactions Woodward and Hoffmann termed a ***pericyclic reaction***,[18] which is a reaction "in which concerted reorganization of bonding occurs throughout a cyclic array of continuously bonded atoms."[19]

Pericyclic reactions include a wide variety of transformations, including such well-known reactions as the Cope rearrangement and the Diels-Alder reaction (Figure 11.3). In the sections that follow we will discuss each of these as pericyclic reactions, but our focus will not be the application of these reactions in synthesis. Instead, the approach here will be to develop an explanation for each type of pericyclic reaction in terms of an intuitive model. Then we will consider the common features of these seemingly dis-

---

[13]For example, see the review "Rearrangements Proceeding Through 'No Mechanism' Pathways: The Claisen, Cope, and Related Rearrangements" by Rhoads, S. J. in *Molecular Rearrangements*, Vol. 1; de Mayo, P., ed.; Wiley-Interscience: New York, 1963; pp. 655–706.

[14]Woodward, R. B.; Hoffmann, R. *J. Am. Chem. Soc.* **1965**, *87*, 395.

[15]See also Woodward, R. B.; Hoffmann, R. *The Conservation of Orbital Symmetry*; Verlag Chemie/Academic Press: Weinheim, 1971.

[16]Hoffmann, R.; Woodward, R. B. *Acc. Chem. Res.* **1968**, *1*, 17.

[17]The 1981 Nobel Prize in chemistry was awarded jointly to Roald Hoffmann and to Kenichi Fukui (see page 772) for their studies of concerted reactions. It seems likely that R. B. Woodward would also have shared the prize (which would have been his second Nobel Prize in chemistry) if he had been living at the time of the award.

[18]Reference 15, p. 169.

[19]Commission on Physical Organic Chemistry, IUPAC, *Pure Appl. Chem.* **1994**, *66*, 1077. Constitutional isomers that are interconverted by pericyclic reactions—such as **6** and **7**—are known as **valence isomers**.

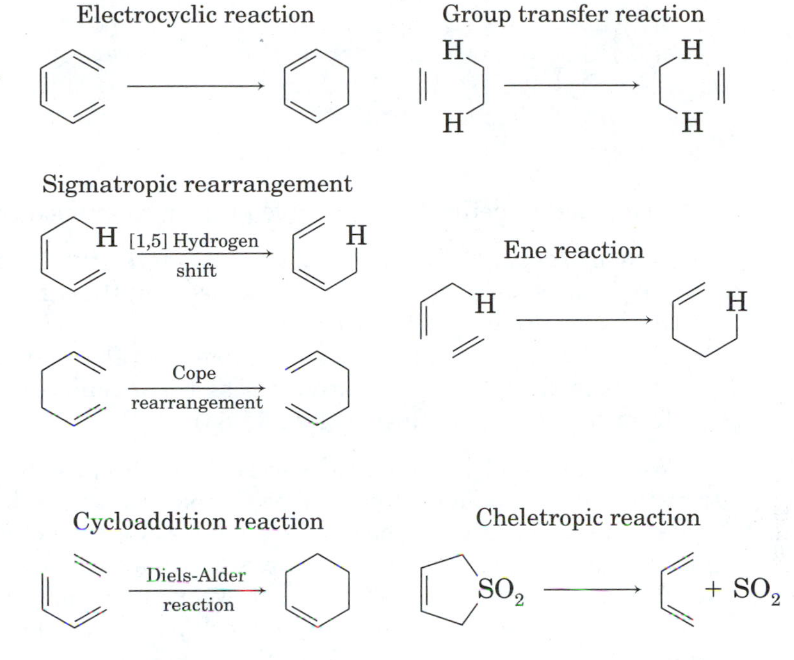

Figure 11.3 Examples of pericyclic reactions.

parate reactions in terms of an inclusive theory.[20,21] Finally, we will explore the use of a variety of complementary models to provide a basis for a further understanding of concerted reactions.

11.2 Electrocyclic Transformations

Definitions and Selection Rules

Woodward and Hoffmann defined the cyclobutene-butadiene interconversion as an ***electrocyclic transformation***—the formation of a single bond between the termini of an acyclic conjugated pi system containing k π-electrons or the reversal of such a reaction (Figure 11.4).[14,15,22]

---

[20]For a perspective on the development of theoretical models for pericyclic reactions, see Houk, K. N.; González, J.; Li, Y. *Acc. Chem. Res.* **1995**, *28*, 81.

[21]The emphasis here will be on the use of qualitative models to understand the basis for pericyclic reactions. For a survey of *ab initio* calculations for pericyclic reactions, see Houk, K. N.; Li, Y.; Evanseck, J. D. *Angew. Chem., Int. Ed. Engl.* **1992**, *31*, 682.

[22]For an extensive discussion of electrocyclic reactions, see Marvell, E. N. *Thermal Electrocyclic Reactions*; Academic Press: New York, 1980.

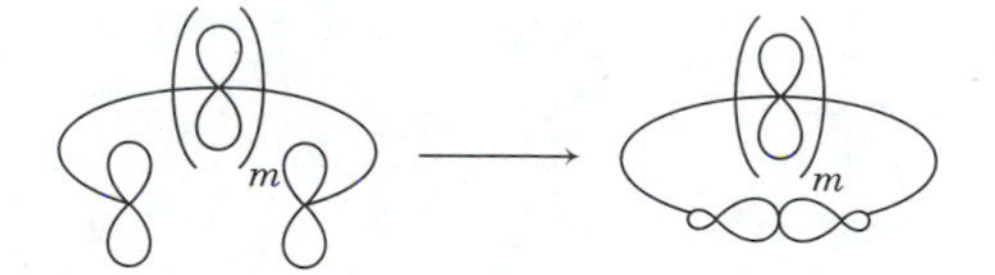

Figure 11.4 A generalized electrocyclic transformation; m is an integer.

They further defined two possible pathways for such reactions:

1. ***Disrotatory***, in which the substituents on the two termini rotate in different directions, one turning clockwise and the other turning counterclockwise, Figure 11.5 (left), and
2. ***Conrotatory***, in which the substituents on the π system termini both rotate in the same sense, either clockwise or counterclockwise, on going from reactant to product (Figure 11.5 (right)).

Woodward and Hoffmann observed that the products of electrocyclic transformations could be predicted on the basis of the highest occupied molecular orbital (HOMO) of the *acyclic* member of the reactant-product pair.[14] Specifically, they suggested that the reaction pathway is associated with the formation of a bonding relationship between the termini of the π chain, so that there is either $+ \cdots +$ or $- \cdots -$ overlap between the lobes of the two p orbitals that come together to form the new σ bond. An example is the interconversion of cyclobutene and butadiene. The Hückel MOs of butadiene are shown on the energy level diagram in Figure 11.6.[23] It is not necessary to know the actual wave function for each MO. Instead, only the *sign* of the coefficient at each carbon atom in each MO is significant here. Coefficients that are positive are denoted arbitrarily by a + sign on the top lobe of the p orbital and a − sign on the bottom lobe of the p orbital at that position. Coefficients that are negative are indicated by a − sign on the top lobe and a + sign on the bottom lobe of a certain p orbital.

The HOMO of 1,3-butadiene in the ground state is ψ_2. Simultaneous rotation about both the C1—C2 bond and about the C4—C3 bond in a *clock-*

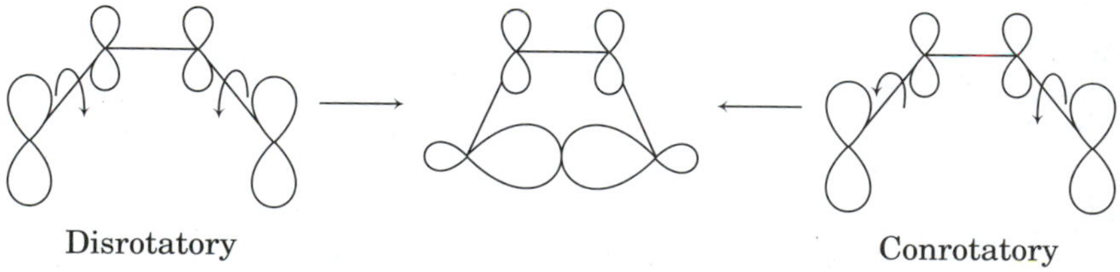

Disrotatory

Conrotatory

Figure 11.5 Disrotatory (left) and conrotatory (right) electrocyclic conversion of butadiene to cyclobutadiene.

[23]The σ framework in the acyclic structure is assumed to be invariant throughout the reaction. This MO analysis is only concerned with the Hückel MOs derived from the set of p orbitals in the acyclic species. The new σ bond formed in the cyclic structure is described in terms of a localized bonding σ orbital and a localized σ^* antibonding orbital formed by overlap of a pair of sp^3-hybrid orbitals formed by rotation of the two terminal p orbitals in the π system.

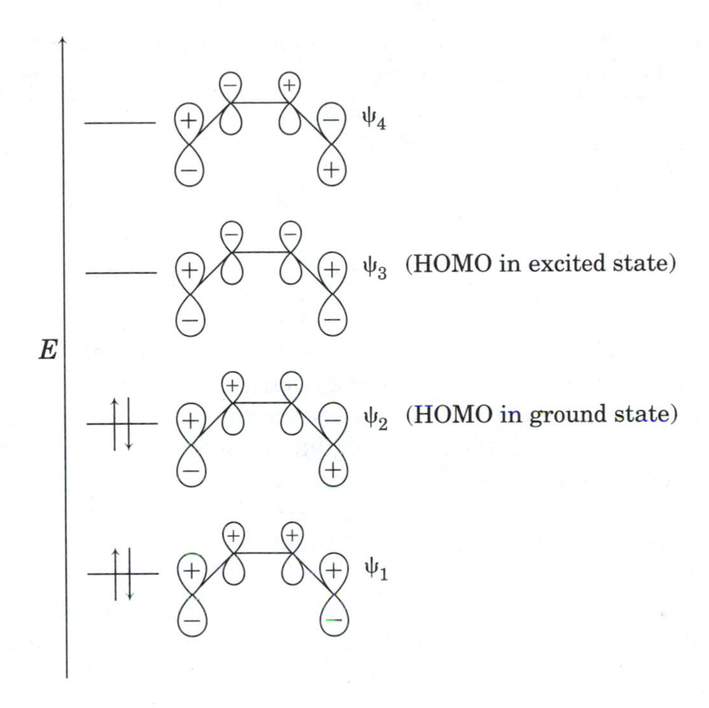

Figure 11.6 Orbital symmetries of butadiene orbitals. In the ground state, HOMO is ψ_2; in the excited state, HOMO is ψ_3.

wise fashion (as shown in Figure 11.7) would allow the two + lobes on the termini of ψ_2 to overlap in a bonding mode, so the reaction is predicted to take place by a conrotatory pathway.[24,25] Excitation of butadiene to its lowest excited state would involve promoting one electron from ψ_2 to ψ_3, so that ψ_3 becomes the HOMO. Now the symmetry properties of the termini of the HOMO are reversed, and the photochemical reaction should proceed by a disrotatory pathway (Figure 11.8).

Figure 11.7 Formation of σ bond by + ··· + overlap of *p* orbital lobes at termini of ψ_2 of butadiene through conrotatory reaction pathway (thermal reaction).

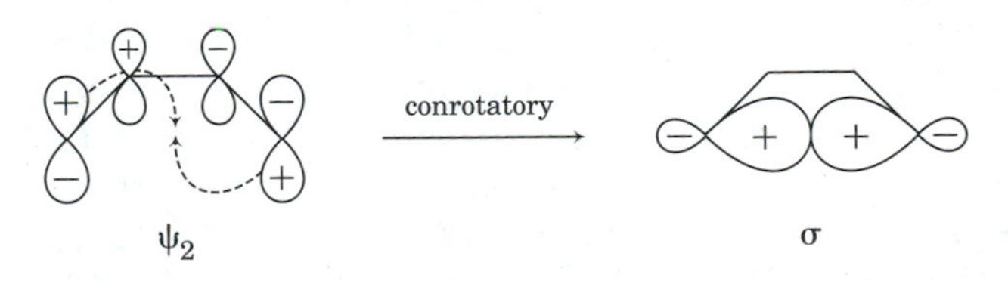

[24]The product of ring opening of cyclobutenes with a substituent at C3 only depends on the electronic properties of that substituent. For a discussion of the "torquoselectivity" of these reactions, see Niwayama, S.; Houk, K. N. *Tetrahedron Lett.* **1992**, *33*, 883; **1993**, *34*, 1251; **1994**, *35*, 527 and references therein.

[25]If the cyclobutene has an alkyl substituent at C3, both (*Z*) and (*E*) isomers are possible from the conrotatory pathway. Both steric effects and electronic effects of substituents elsewhere can influence the product distribution. For leading references, see Kallel, E. A., Wang, Y.; Spellmeyer, D. C.; and Houk, K. N. *J. Am. Chem. Soc.* **1990**, *112*, 6759.

Figure 11.8 + ··· + overlap of the terminal *p* orbital lobes of ψ_3 of photoexcited butadiene through disrotatory closure.

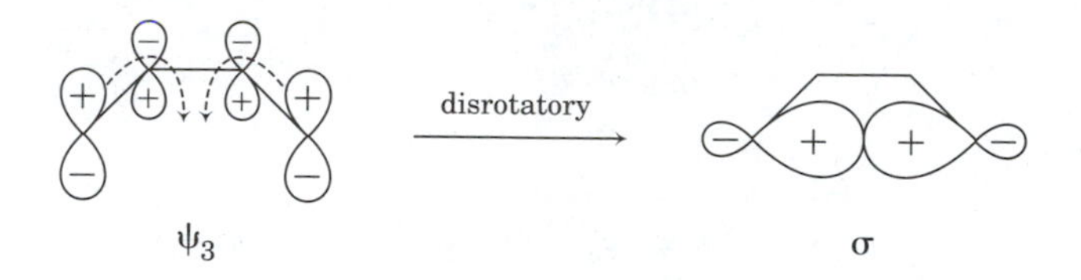

Unlike the steric explanation, this model can also rationalize the interconversion of hexatrienes and cyclohexadienes.[26] As shown in Figure 11.9, the ground state HOMO of hexatriene is ψ_3. Therefore the closure of hexatriene to cyclohexadiene should be disrotatory, so the opening of cyclohexadiene to hexatriene should also be disrotatory. In the excited state the HOMO of 1,3,5-hexatriene is ψ_4, so both the photochemical closure of hexatriene to cyclohexadiene and the reverse reaction, photochemical opening of cyclohexadiene to hexatriene, should be conrotatory.

Because of the alternating symmetry properties of the HOMOs of linear polyene systems (see Chapter 4), Woodward and Hoffmann were able to deduce the following **selection rule for electrocyclic transformations**:[27]

> If there are $4n + 2$ π electrons in a system of adjacent, parallel *p* orbitals undergoing an electrocyclic closure, then the reaction will be allowed thermally in the disrotatory fashion and allowed photochemically in the conrotatory fashion. If there are $4n$ π electrons, it will be allowed thermally conrotatory and photochemically disrotatory.

In the terminology of selection rules, in the ground state the disrotatory reaction involving $4n + 2$ π electrons (that is, 2, 6, 10, . . . , electrons) is *allowed*, while the conrotatory reaction involving $4n + 2$ π electrons is *forbidden*. Conversely, the ground state conrotatory reaction involving $4n$ electrons (that is, 4, 8, 12, . . . , electrons) is allowed, while the photochemical conrotatory reaction involving $4n$ electrons is forbidden.

Woodward and Hoffmann noted that the symmetry argument only indicates that one stereochemical pathway or the other is electronically favored, and that the forbidden pathway might be observed if molecular constraints inhibit the allowed mechanism. As an example, they cited the disrotatory opening of the dimethylbicyclo[3.2.0]heptene (**18**) to *cis,cis*-1,3-cycloheptadiene (**19**, Figure 11.10).[14] Here, the symmetry-allowed conrotatory pathway leading to the cis,trans isomer **20** is precluded by the angle strain of the

[26]Steric effects might influence product distributions when different products can be formed by allowed pathways. For example, both **4** and **5** could be formed by conrotatory electrocyclic opening of **3**. The fact that only **4** is produced suggests a steric contribution to the barrier for the formation of **5**.

[27]Adapted from reference 16.

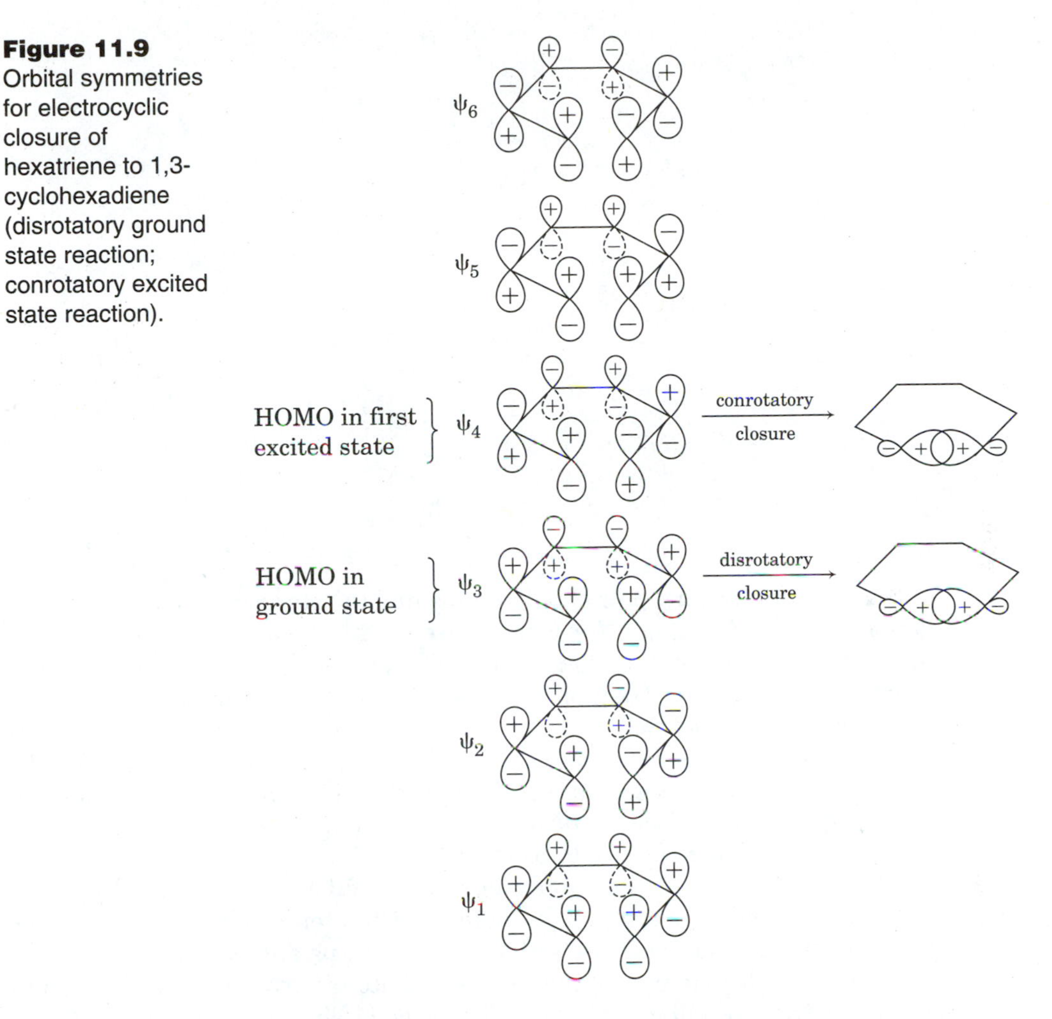

Figure 11.9 Orbital symmetries for electrocyclic closure of hexatriene to 1,3-cyclohexadiene (disrotatory ground state reaction; conrotatory excited state reaction).

electronically allowed product, which would have a trans double bond in a seven-membered ring. Apparently the opening of **18** takes place by a non-concerted process because the reaction occurs only at 400°,[28] much higher than the 200° needed for the opening of the *cis*-1,2,3,4-tetramethylcyclobut-1-ene, **7** (equation 11.2).

MO Correlation Diagrams

The HOMO *"simple symmetry argument"*[14] for the stereochemistry of electrocyclic reactions was supported by extended Hückel calculations indicating that the energy of the HOMO is dominant in thermal reactions. However, it is not intuitively obvious why the HOMO of the acyclic member

[28]Criegee, R.; Furrer, H. *Chem. Ber.* **1964**, *97*, 2949.

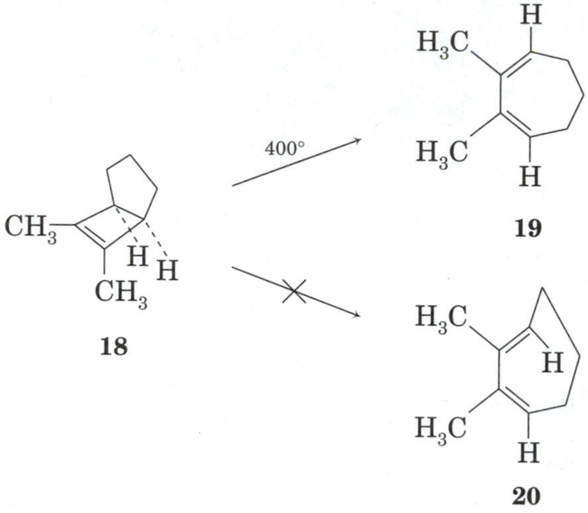

Figure 11.10
Effect of bridging on electrocyclic reactivity.

of the reactant/product pair should determine the stereochemistry of the reaction in either direction. Nor is it apparent why the analysis of only one of many MOs should lead to the same result as a detailed extended Hückel calculation.[29] Longuet-Higgins and Abrahamson supplemented the Woodward-Hoffmann approach by suggesting that the electrocyclic transformations should be considered as conversions of the *entire set of MOs of the reactant to the entire set of MOs of the product* in such a manner that the *symmetry of the molecular orbitals* is retained on going from reactant through transition structure to product.[29]

As a prelude to further analysis, it is useful to review one important property of molecular orbitals. As noted in Chapter 1, **symmetry-correct molecular orbitals** must be either symmetric or antisymmetric with respect to the full symmetry of the basis set of atomic orbitals that are used to construct the molecular orbitals.[30] Figure 11.11 shows the basis set atomic orbitals (i.e., the shapes of the *p* orbitals without designation of the mathe-

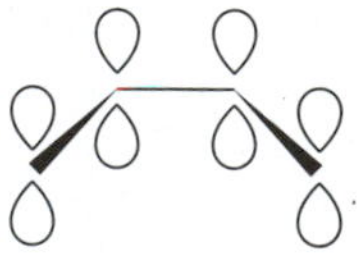

Figure 11.11
Basis set of atomic *p* orbitals of 1,3-butadiene.

[29]Longuet-Higgins, H. C.; Abrahamson, E. W. *J. Am. Chem. Soc.* **1965**, *87*, 2045.

[30]This analysis will need to consider only a sufficient number of molecular orbitals (and, thus, atomic orbitals in the basis set) and symmetry elements to distinguish between allowed and forbidden pathways. Moreover, it is not necessary here to consider the minor perturbation of molecular orbital symmetry that results from isotopic or alkyl substitution. In other words, to a first approximation the basis set orbitals of any conjugated diene are considered to be the same as those for 1,3-butadiene.

matical sign of the lobes) of the π system of 1,3-butadiene superimposed on the s-*cis* conformation of its σ skeleton. This basis set has a number of symmetry elements, but a plane of symmetry bisecting the C2—C3 bond and a C_2 rotation axis perpendicular to that bond will be most important here. Each molecular orbital can be designated as either symmetric or antisymmetric (but not *asymmetric*) with respect to the symmetry elements present in the basis set of orbitals.

Figure 11.12 shows the effect of a σ reflection (top) and of a C_2 rotation (bottom) on ψ_1 of butadiene. (The four carbon atoms of butadiene have been labeled A, B, C, and D so that the symmetry operations will be evident.) The sign of the coefficient of each atomic orbital having a positive coefficient in that wave function is indicated by labeling the top lobe of each of the *p* orbitals as + and the bottom lobe of each as −. The effect of the σ operation is to produce a set of orbitals that is identical in all respects to the beginning set, so ψ_1 is said to be symmetric (**S**) with respect to σ. The effect of a C_2 rotation on ψ_1 is also to put a *p* orbital in the same position where there was a *p* orbital before the operation. Now, however, the sign of the lobe in each position is the negative of the sign of the lobe in that position before the symmetry operation. Thus ψ_1 is antisymmetric (**A**) with respect to the C_2 operation.[31] Figure 11.13 shows the effect of the same two operations on ψ_2 of butadiene. This wave function is antisymmetric (A) with respect to the σ reflection, but it is symmetric (S) with respect to C_2.

Not only must we consider the symmetry properties of butadiene orbitals, but we must also consider the symmetry properties of both cyclobutene and the transition structure expected for the conversion of the reactant to product. The two pathways for the closure of butadiene to cyclobutene are illustrated in Figure 11.14. The C1—C2, C2—C3 and C3—C4 σ bonds are shown as solid lines. The *p* orbitals of butadiene and

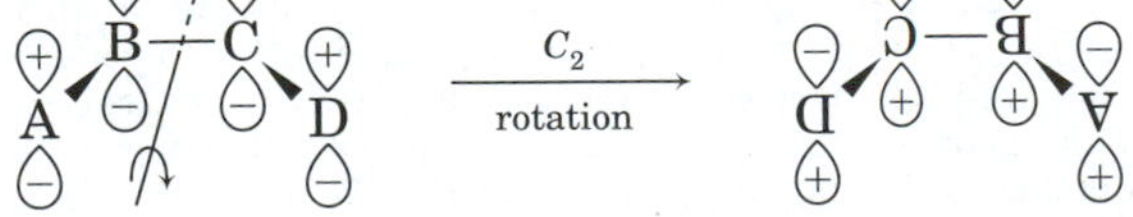

Figure 11.12
ψ_1 of butadiene is symmetric with respect to σ (top) but antisymmetric with respect to C_2 (bottom).

[31]The symmetry operation must either change the sign of *all* of the lobes or change the sign of *none* of them. It cannot change the sign of some but not others. Otherwise, either it is not a proper symmetry operation for that basis set or the MO under consideration is an improper MO, being asymmetric with respect to a proper symmetry operation.

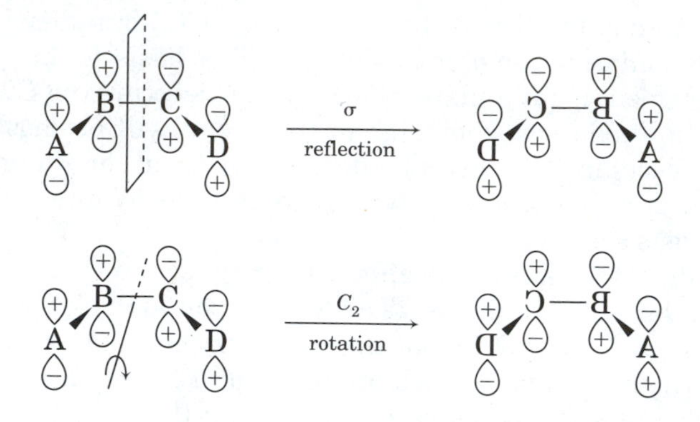

Figure 11.13 ψ_2 of butadiene is antisymmetric with respect to σ (top) but symmetric with respect to C_2 (bottom).

cyclobutene, as well as the sp^3 orbitals of the C3—C4 σ bond of cyclobutene, are represented by the shapes of the atomic p or sp^3 orbitals. Therefore this is only a basis set representation. Although there are many symmetry elements present in the representations of both butadiene and cyclobutene, in the conrotatory reaction the only symmetry element that is present continuously from reactant through transition structure to product is the C_2 rotation.[32,33] Similarly, only the σ reflection is present from reactant through transition structure to product for the disrotatory pathway.[34]

[32]C_2 symmetry is evident in the transition structure[33] calculated at the RHF/6-31G* level:

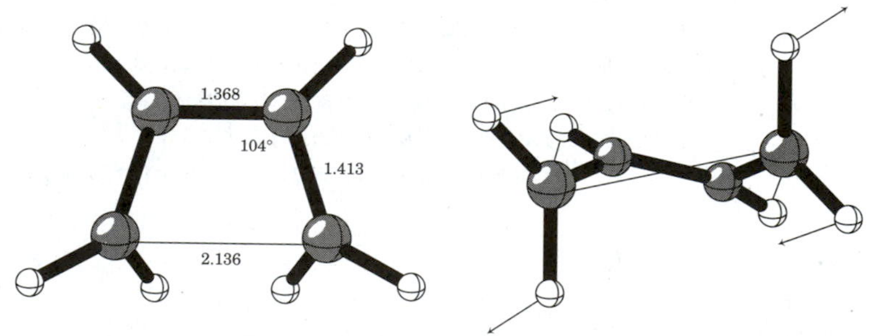

The drawings are adapted from Thomas IV, B. E.; Evanseck, J. D.; Houk, K. N. *J. Am. Chem. Soc.* **1993**, *115*, 4165. The geometry of the transition structure (angle in degrees, lengths in Angstroms) is given on the left. The drawing on the right shows a different view of the transition structure with arrows representing the transition vector (indicating the relative movement of atoms along the lowest energy path from the transition structure to product) for the conversion of butadiene to cyclobutene.

[33]A transition ***structure*** is a saddle point on a potential energy (enthalpy) surface for a reaction. A transition ***state*** corresponds to a free energy maximum on the path between reactant and product. The geometry of the transition structure corresponds closely to the geometry of the transition state if the barrier is relatively high and the entropy of the system does not vary rapidly near the geometry of the transition structure. If either of these conditions is not met, then the geometry of the transition structure may be very different from that of the transition state. For discussions, see references 19 and 21. Also see Bauer, S. H.; Wilcox, Jr., C. F. *J. Chem. Educ.* **1995**, *72*, 13.

[34]Lemal, D. M.; McGregor, S. D. *J. Am. Chem. Soc.* **1966**, *88*, 1335, suggested the terms *axisymmetric* and *sigmasymmetric* to describe transition structures of pericyclic reactions having a twofold (or higher) axis or a mirror plane, respectively. However, the terms *conrotatory* and *disrotatory*, which refer to a process rather than to a transition structure, have been more widely used by chemists.

Figure 11.14
Symmetry elements in basis sets for electrocyclic transition structures. (a) C_2 axis in conrotatory transition structure; (b) σ in disrotatory transition structure.

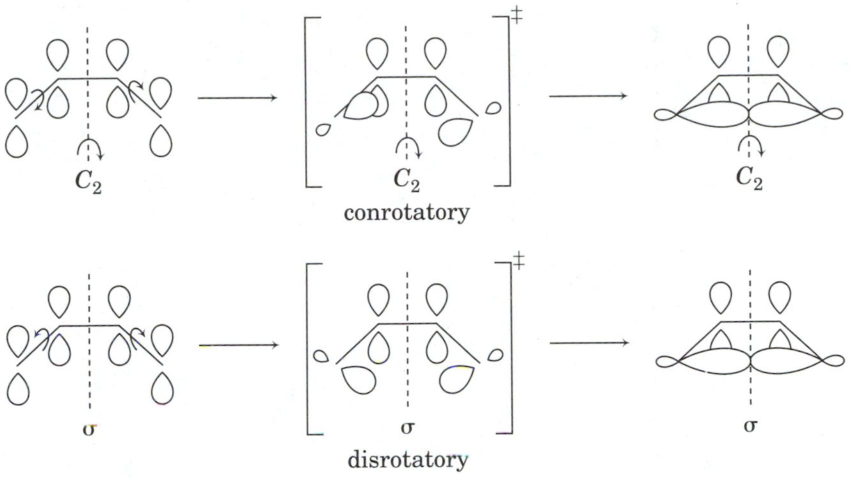

The symmetry with respect to the C_2 axis is indicated for each MO of butadiene and for each MO of cyclobutene in a **molecular orbital correlation diagram** (Figure 11.15). **Correlation lines**, such as the dashed line connecting ψ_1 of butadiene with π of cyclobutene, connect orbitals of reactant and product that have the same symmetry. (The correlation lines are drawn by connecting the lowest energy orbital on one side of the drawing with the lowest energy orbital of the same symmetry, S or A, on the other side of the drawing. Then the next-lowest energy pair of orbitals with the same symmetry are connected, and so on.) Therefore, ψ_1 of butadiene correlates with π of cyclobutene in the C_2 reaction pathway. Similarly, ψ_2 correlates with σ. Thus the conrotatory conversion of butadiene to cyclobutene causes each orbital of the butadiene to be transformed into an orbital of cyclobutene. Furthermore, in the electronic ground state of butadiene there are two electrons in ψ_1 and two electrons in ψ_2. In the electronic ground state of cyclobutene there are two electrons in σ and two electrons in π. Therefore, the correlation of ψ_1 with π and the correlation of ψ_2 with σ means that the entire set of bonding orbitals of the reactant correlates with the entire set of bonding orbitals of the product. As a result, there should not be a large electronic barrier for the reaction.

The MO correlation diagram applies to both the electrocyclic opening and closing reactions. For example, the opening of cyclobutene to butadiene is described by viewing Figure 11.15 from right to left. One aspect of the diagram that may not be intuitively obvious is the process by which the lowest energy orbital of cyclobutene, σ, which has density only on C1 and C4, is transformed into a butadiene orbital, ψ_2, with density on all four carbon atoms. Although initially the localized σ orbital is orthogonal to the *p* orbitals that comprise the double bond of cyclobutene, the conrotatory deformation removes the orthogonality and causes an interaction to occur. Thus, the first movement along the reaction coordinate mixes the π and σ orbitals

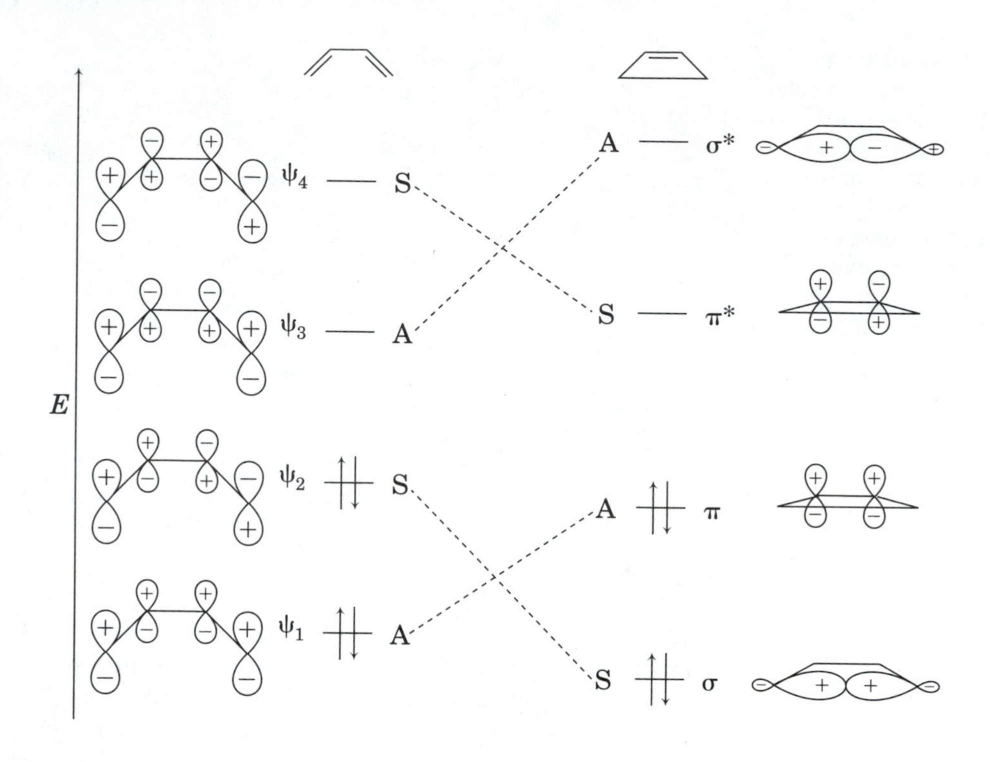

Figure 11.15 MO correlation diagram for conrotatory interconversion of butadiene and cyclobutene (C_2 symmetry). (Adapted from reference 15.)

of cyclobutene, and the orbitals that could be calculated for all geometries intermediate between reactant and product are also mixed. When the reaction is nearly complete, the sp^3 orbitals have been almost totally converted to *p* orbitals and are almost completely parallel to the other *p* orbitals.

The MO correlation diagram for the disrotatory pathway is shown in Figure 11.16. Now ψ_1 correlates with σ, but ψ_2 correlates with π*. Again, the ground electronic state of butadiene has two electrons in ψ_1 and two electrons in ψ_2. Thus the forced conversion of the ground state of butadiene into cyclobutene by a *disrotatory pathway* would lead to a cyclobutene molecule having two electrons in σ and two electrons in π*. The electronic configuration $\sigma^2\pi^{*2}$ represents a *doubly excited* electronic state of cyclobutene because it is a higher energy state than that which would be produced by promoting just one electron from π to π*. The forced disrotatory reaction should have a much higher activation energy than an alternative (conrotatory) pathway that produces ground state product.

State Correlation Diagrams

MO correlation diagrams such as those in Figure 11.15 and Figure 11.16 are useful tools, but they are not the final step in the analysis. What we really should be analyzing is the interconversion of molecular *states*, not orbitals,

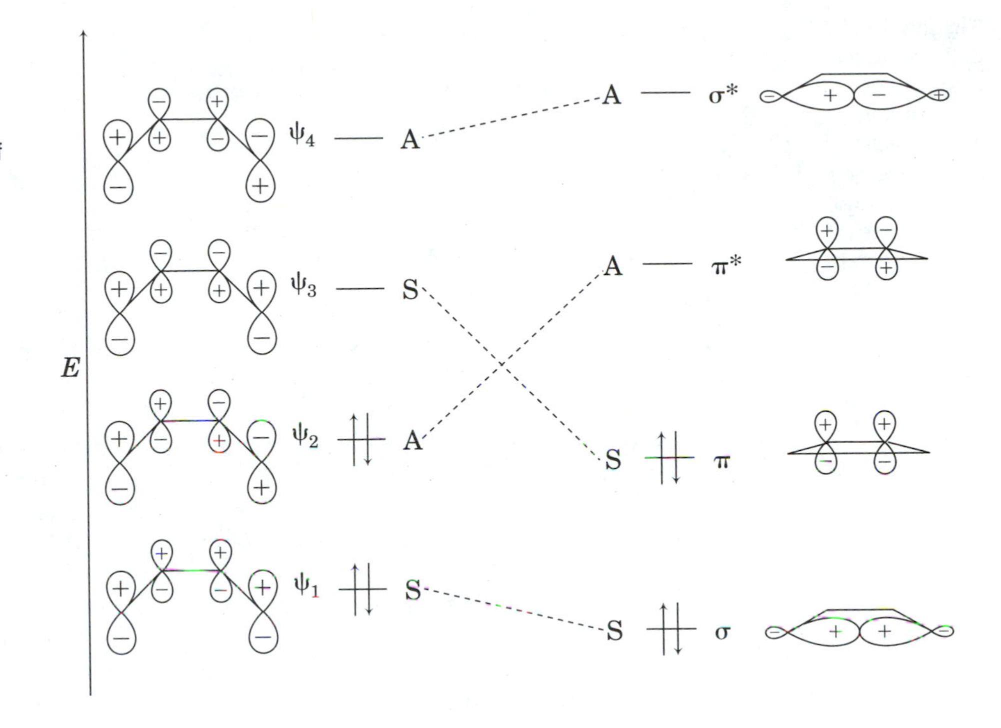

Figure 11.16 MO correlation diagram for disrotatory interconversion of butadiene and cyclobutene (σ symmetry). (Adapted from reference 15.)

since *we carry out chemical reactions on molecules* in their various electronic states, *not on individual MOs within molecules*. The state of a molecule is determined by its configuration, which is the product of the occupied MOs that comprise it.[35] Thus the ground state of cyclobutene can be described as $\sigma^2\pi^2$, while that of butadiene is $\psi_1{}^2\psi_2{}^2$. We can also distinguish different excited states, each formed by promotion of an electron from a bonding orbital to an antibonding orbital. Then we can represent the ground electronic state and various excited electronic states in an energy level diagram. Correlation of states on such a drawing then develops a **state correlation diagram**, which can be viewed as a reaction coordinate diagram for the electrocyclic transformation.

The correlation of states is determined by the total symmetry of each state, which is a function of the symmetry of each populated MO that characterizes that configuration. MOs that are symmetric with respect to a symmetry element are again designated as S, while those that are antisymmetric

[35]More precisely, *states* can be described by combinations of configurations. In the discussion here, however, each state is associated with only one configuration. For a discussion of configurations, see Turro, N. J. *Modern Molecular Photochemistry*; The Benjamin/Cummings Publishing Co., Inc.: Menlo Park, CA, 1978; pp. 17–24.

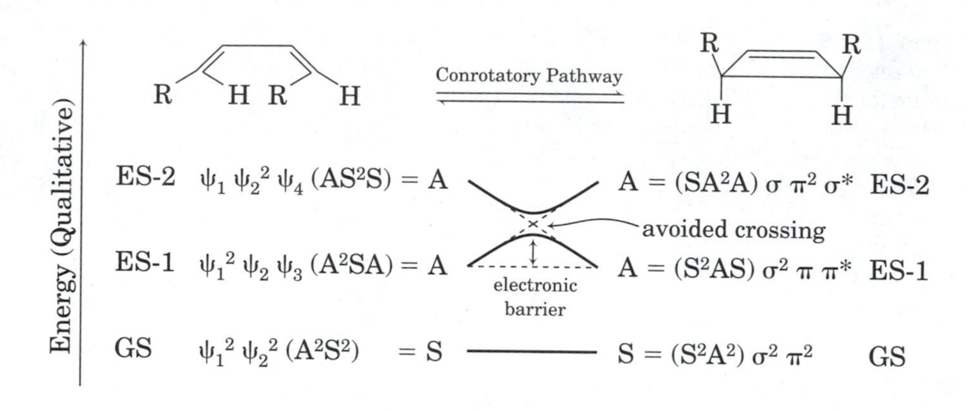

Figure 11.17 State correlation diagram for conrotatory electrocyclic interconversion of butadiene and cyclobutene. (Adapted from reference 29.)

are again represented by A.[36] Consider the state correlation diagram for the conrotatory pathway in Figure 11.17. The ground state (GS) of butadiene is the product of two S terms and two A terms, which mathematically is S, i.e., symmetric. Similarly, the GS of cyclobutene is $\sigma^2\pi^2$, which is also the product of two S terms and two A terms and is therefore also S. Assigning the total state symmetries for the ground and two excited states (chosen as described below) produces a state correlation diagram that represents the reactant states ranked according to energy on the left and the product states ranked according to energy on the right (Figure 11.17). Since the MOs of the ground state of butadiene are converted into the MOs of the ground state of cyclobutene, and since the GS of butadiene has the same total symmetry as the GS of cyclobutene, the ground state of butadiene correlates with the ground state of cyclobutene. Therefore, there should be no large activation energy arising from electronic interactions, and the reaction is allowed.

Two electronically excited states are shown for both reactant and product in Figure 11.17. In each case, the state labeled ES-1 is the excited state produced by moving one electron from the HOMO to the LUMO. For butadiene ES-1 has the configuration $\psi_1{}^2\psi_2\psi_3$. Its state symmetry, therefore, is the product of A^2SA, which is A. For cyclobutene ES-1 has the configuration $\sigma^2\pi\pi^*$, so its state symmetry is S^2AS, which is again A. However, ES-1 of butadiene does *not* correlate with ES-1 of cyclobutene, even though the state symmetries are the same. To understand why this is so, it is necessary to follow the occupied molecular orbitals from reactant to product in Figure 11.15. Following the correlation line from σ leads to ψ_2 of butadiene; following the correlation line for π leads to ψ_1; following the correlation line from

[36]The symbols **S** and **A** were used in reference 90 and are employed in the discussion here. However, there are alternatives for the representation of symmetric and antisymmetric wave functions and states; + or +1 can be used for S, and − or −1 can be used for A. The +1 and −1 designations may facilitate determination of the symmetry of a state, since the mathematics of multiplying signed numbers is more intuitive than the mathematics of multiplying bare signs. Thus the symmetry of the ground state can be represented as $(+1) \times (+1) \times (-1) \times (-1) = +1$.

π^* leads to ψ_4. The resulting configuration, $\psi_1\psi_2^2\psi_4$, is not ES-1 of butadiene but is, instead, the configuration labeled ES-2 in Figure 11.17. Similarly, following the orbitals populated in ES-1 of butadiene leads to the cyclobutene configuration labeled ES-2 in Figure 11.17. It is essential to note that the state designated ES-2 is *not* necessarily the next highest energy excited state after ES-1. Rather, it is a state that is chosen to complete the diagram because its orbital population would be produced from one of the states on the other side of the diagram.

Based on the state correlation diagram, we say that ES-1 of butadiene attempts to correlate with ES-2 of cyclobutene. The correct term is *attempts to correlate* because a principle known as the ***avoided crossing rule*** says that configurations of the same symmetry cannot cross on an energy level diagram.[37] The reason is that mixing of the configurations occurs when configurations of the same symmetry come close together in energy, just as MO mixing occurs when molecular orbitals interact in perturbational MO theory. In Figure 11.17 the crossing dashed lines indicate the correlations that would result if there were no configuration interaction, while the curved lines indicate the actual correlations that result from the avoided crossing.[29] Even though ES-1 of butadiene would actually be converted into ES-1 of cyclobutene (and vice versa) if the excited state of either could be forced to deform in a conrotatory fashion past the avoided crossing, the state correlation diagram indicates that there would be a high electronic barrier associated with the process. Since there is a lower energy (disrotatory) alternative, the photochemical reaction is said to be forbidden by the conrotatory pathway.

A state correlation diagram for the disrotatory pathway is shown in Figure 11.18. Now ES-1 of butadiene correlates with ES-1 of cyclobutene, so

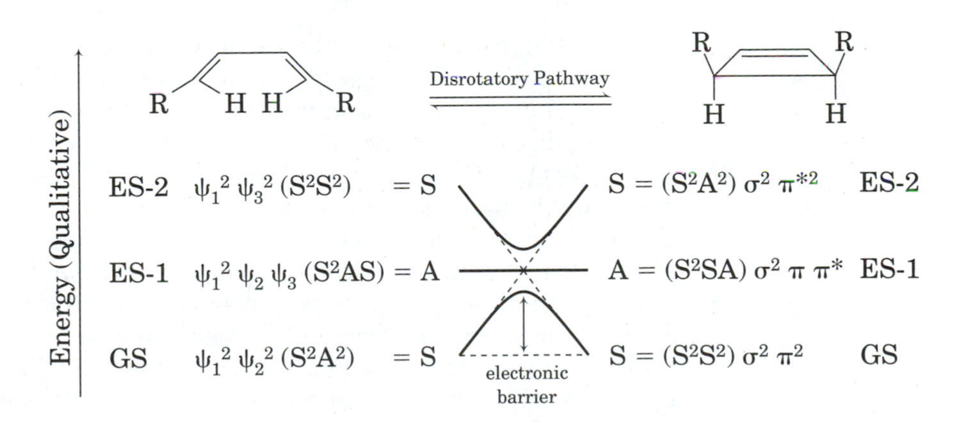

Figure 11.18 State correlation diagram for disrotatory interconversion of butadiene and cyclobutene. (Adapted from reference 29.)

[37]This rule is also known as the *no-crossing rule*. For a discussion, see Albright, T. A.; Burdett, J. K.; Whangbo, M.-H. *Orbital Interactions in Chemistry*; John Wiley & Sons, Inc.: New York, 1985; pp. 52–53.

there should be no high energy of activation resulting from orbital symmetry, and the photochemical reaction is allowed. On the other hand, GS of butadiene correlates with ES-2 of cyclobutene (and GS of cyclobutene with ES-2 of butadiene). There is an avoided crossing between the states of S symmetry, so there should be a high electronic barrier for a process that converts GS of butadiene to GS of cyclobutene by a pathway involving the avoided crossing. Therefore, the thermal reaction is forbidden by the disrotatory pathway.

The separation of the conrotatory and disrotatory processes into two separate diagrams here is artificial because the two reactions represent alternative pathways on the same potential energy surface. A molecule undergoing thermal activation is bent and twisted in many different directions through collisions with surfaces and with other molecules. In a sense, it can be said to "explore" the potential energy surface on which it resides. Because there is a much lower energy of activation for the conrotatory pathway than for the disrotatory pathway, a cyclobutene molecule in the electronic ground state is more likely to acquire sufficient energy of activation to complete the conrotatory reaction than to complete the disrotatory reaction. Therefore, the difference between allowed and forbidden is ultimately a difference in activation energies. Similar MO and state correlation diagrams can be drawn for other electrocyclic reactions, and the results are entirely consistent with the selection rule for electrocyclic reactions on page 718.

It is gratifying that the pathway predicted to be allowed for the opening of cyclobutenes to butadienes is experimentally observed. However, the fact that the allowed concerted pathway predicts the observed product means that we really do not know for certain whether the forbidden concerted pathway is higher or lower in energy than alternative, *nonconcerted* pathways (in which some bonds break or form before others). As Stephenson and Brauman pointed out,[38]

> The ring-chain tautomerism between cyclobutene and butadiene is perhaps the most familiar example of an allowed Woodward-Hoffmann process. This transformation invariably is discussed in every attempt to rationalize or teach the Woodward-Hoffmann orbital symmetry concepts. This popularity is due in large part to the existence of a geometrically well defined (and easily visualized) alternate, forbidden, electrocyclic pathway. Thus it is exceedingly simple to set up a nonallowed strawman, the disrotatory ring opening, and show that the allowed, conrotatory path is to be preferred. . . . It has been possible to evaluate an energy difference of ~15 kcal/mol between an allowed conrotatory process and *some* nonallowed pathway . . . [but] it is not possible to say with any degree of certainty whether the nonallowed path is diradical-like or a forbidden, concerted transformation (if indeed such a distinction can be made).

---

[38]Stephenson, Jr., L. M.; Brauman, J. I. *Acc. Chem. Res.* **1974**, *7*, 65.

11.3 Sigmatropic Reactions

Definitions and Examples

Woodward and Hoffmann termed the pericyclic rearrangement illustrated in Figure 11.19 a ***sigmatropic rearrangement*: the migration of a σ bond within a molecule**. Then a sigmatropic change of order [i,j] is defined as:[39,40]

> the migration of a σ bond, flanked by one or more π electron systems, to a new position, whose termini are atoms numbered *i* and *j* with regard to the original bonded loci (atoms 1 and 1′), in an uncatalyzed intramolecular process.

Thus the reaction in Figure 11.19 is a [1,5] sigmatropic rearrangement, specifically a [1,5] hydrogen shift, since the bond between atoms 1 and 1′ in the reactant moves to form a new bond between atoms 1′ and 5 in the product.[41]

Figure 11.19
A [1,5] sigmatropic rearrangement in 1,3-pentadiene-*1,1*-d_2.

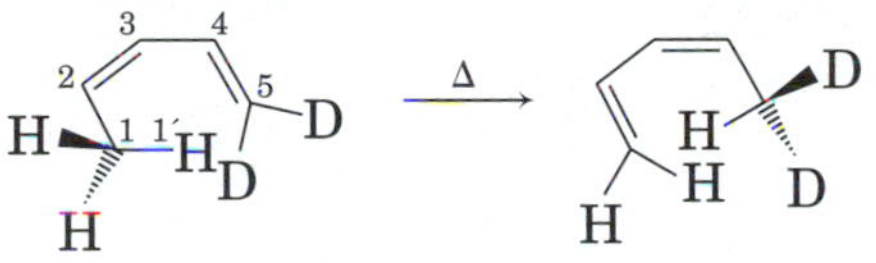

If it were not for the deuterium labeling in Figure 11.19, it would not be possible to tell that a reaction had occurred. (With only hydrogen atoms present, the reaction is termed a ***degenerate rearrangement***, meaning that the product is chemically identical to the reactant.[19]) However, the Cope rearrangement (Figure 11.20) and the Claisen rearrangement (Figure 11.21) are [3,3] sigmatropic reactions that have long been important synthetic tools.[42,43]

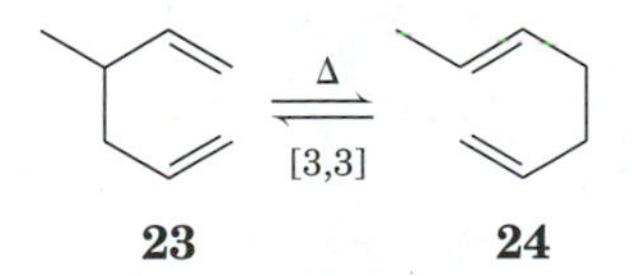

Figure 11.20
Cope rearrangement.

---

[39]This is a restatement of the definition given by Woodward, R. B.; Hoffmann, R. *J. Am. Chem. Soc.* **1965**, *87*, 2511.

[40]By convention, $i \leq j$. Also by convention, the notation for sigmatropic reactions uses square brackets for the number of atoms. In other pericyclic reactions, the numbers in square brackets refer to the number of electrons in each fragment. For a discussion, see reference 19.

[41]Roth, W. R.; König, J. *Liebigs Ann. Chem.* **1966**, *699*, 24.

[42]Rhoads, S. J.; Raulins, N. R. *Org. React.* **1975**, *22*, 1.

[43]Ziegler, F. E. *Chem. Rev.* **1988**, *88*, 1423.

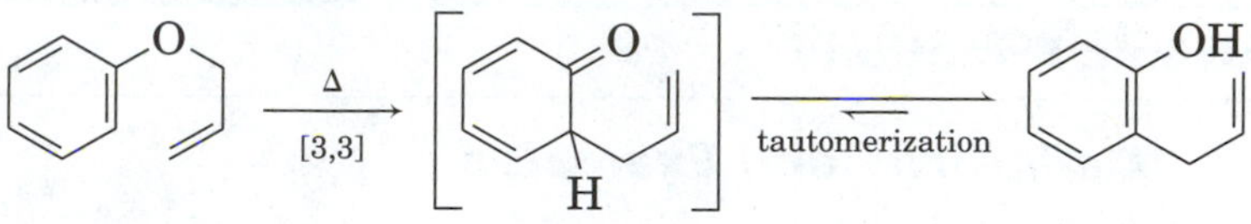

Figure 11.21 Claisen rearrangement.

In theory there are two different stereochemical pathways for $[1,j]$ sigmatropic reactions, as illustrated for the [1,5] hydrogen shift in Figure 11.22. In the first, both bond breaking and bond forming take place above the plane defined by the five carbon atoms. This is known as the **suprafacial** (above the face) **pathway**. Alternatively, the hydrogen atom might leave from above the plane and rebond from below it. This process is known as the **antarafacial** (opposite face) pathway.[44] Both processes are drawn to emphasize the distinction between antarafacial and suprafacial pathways, so the dashed lines in Figure 11.22 represent overall processes, not mechanisms. The representations in Figure 11.23 indicate that both processes require proper orientation of the atoms in the π system so that the hydrogen atom can be transferred from one atom to another without having to follow an apparently sinusoidal pathway.

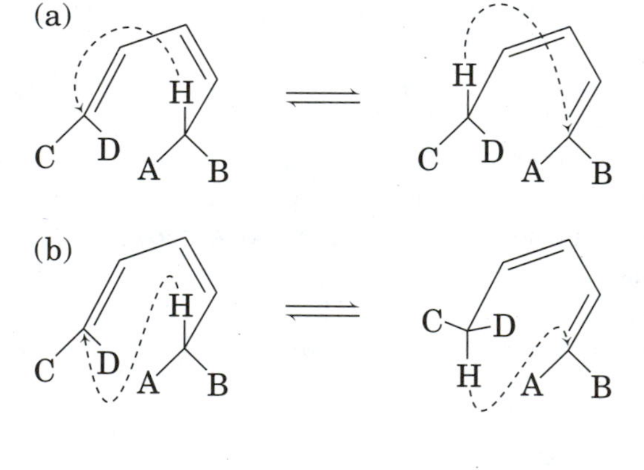

Figure 11.22 Suprafacial (a) and antarafacial (b) [1,5] hydrogen shifts. (Reproduced from figure 28 of reference 15.)

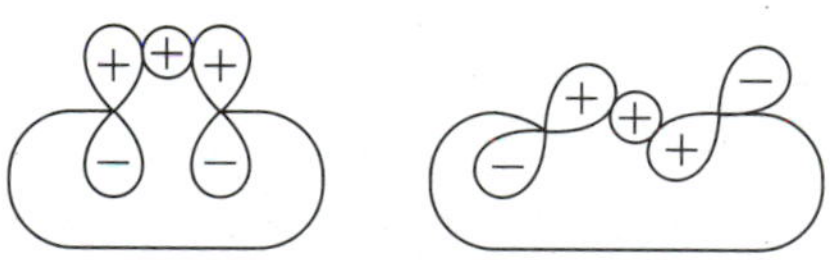

Figure 11.23 Molecular conformations suggested for suprafacial (left) and antarafacial (right) hydrogen shifts. (Adapted from reference 51(a).)

[44]Nickon, A.; Silversmith, E. F. *Organic Chemistry, the Name Game: Modern Coined Terms and Their Origins*; Pergamon Press: New York, 1987; p. 252, indicate that *antara* is a Sanskrit word meaning "the other."

Selection Rules for Sigmatropic Reactions

Sigmatropic reactions are considered to be concerted reactions, in which the bond breaking and bond forming changes take place at the same time. Nevertheless, it is useful to consider the orbital overlaps that would be possible in a mechanism that involves stretching the breaking bond *almost* to the point of radical formation. Consider first the hypothetical thermal (ground state) [1,3] hydrogen shift of a generalized propene derivative, as illustrated in Figure 11.24. We envision stretching of the C3—H bond to the point that the chemical system resembles a hydrogen atom and an allyl radical. Now the HOMO of H· is the hydrogen 1*s* orbital, and the HOMO of the allyl fragment is ψ_2. If the transition structure represents the dissociation of a stable bond, then the sign of the overlapping lobes that were responsible for that bond must have been the same, either both + or both −. It is customary to represent the sign of both lobes as +, but that choice is not essential.

As shown in Figure 11.25, ψ_2 (HOMO) of the allyl radical in the transition structure has a different sign on the top lobe of the *p* orbitals of the C1

Figure 11.24 Suprafacial (forbidden) and antarafacial (allowed) [1,3] hydrogen shifts in propene.

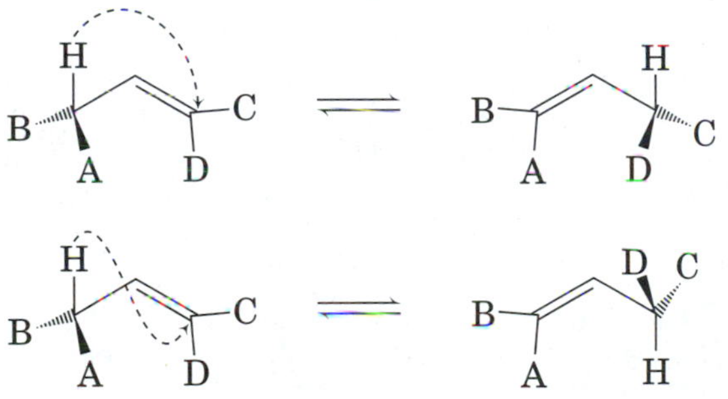

Figure 11.25 Suprafacial (forbidden) and antarafacial (allowed) 1,3-hydrogen shift in ground state propene.

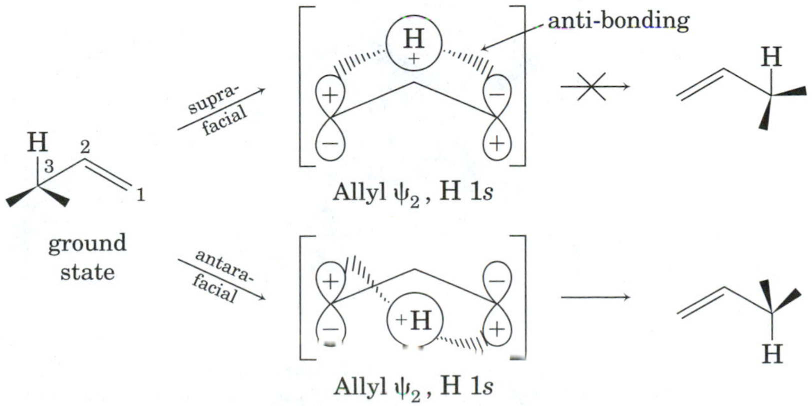

and C3 atoms. Thus a stable bond cannot begin to form by overlap of the orbital on hydrogen with the top lobe of the orbital of C1 at the same time there is any remaining bonding interaction between hydrogen and C3. In the antarafacial pathway on the other hand, there can be bonding between the bottom lobe of the *p* orbital on C1 and the 1*s* orbital of hydrogen at the same time there is some residual bonding between hydrogen and the top lobe of the *p* orbital on C3. Therefore bonding can be maintained as hydrogen is transferred from C1 to C3, so that pathway is said to be allowed by the principles of orbital symmetry. The concerted suprafacial pathway, which would lead to an antibonding relationship, is forbidden by the principles of orbital symmetry.

It must be emphasized that the antarafacial [1,3] hydrogen shift is allowed only by the principles of orbital symmetry. It is not necessarily compatible with the realities of bond angles and bond lengths. Indeed, the transition structure calculated for the antarafacial [1,3] hydrogen shift (Figure 11.26) is a highly contorted species calculated to have an energy similar to an allyl radical and a hydrogen atom, and the suprafacial [1,3] hydrogen shift is calculated to be a high energy species that resembles a trimethylene diradical. Therefore neither concerted pathway is feasible, and the thermal [1,3] hydrogen shift has not been found to occur by a concerted pathway.[21]

As was the case with electrocyclic reactions, the selection rules for sigmatropic reactions are reversed on going from ground states to excited states. If propene is photoexcited, the excitation energy is expected to reside on the allyl fragment,[45] so the HOMO of the allyl fragment is now ψ_3. As shown in Figure 11.27, the interaction of a hydrogen 1*s* orbital with ψ_3 of

Figure 11.26 Two views of the CASSCF/6-31G* transition structure for the antarafacial [1,3] hydrogen shift. (Reproduced from reference 21.)

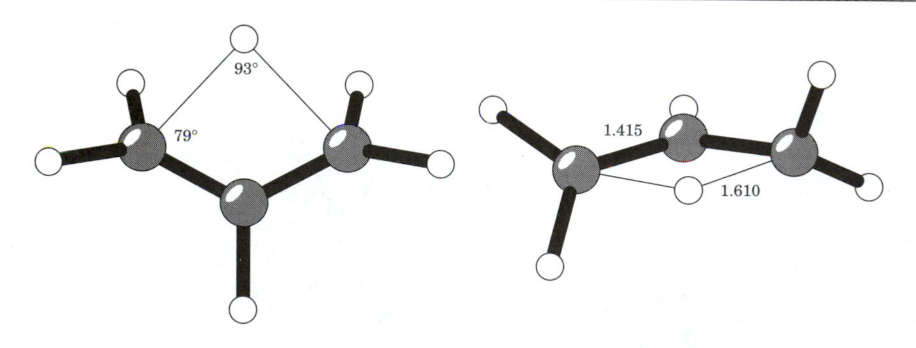

[45]The excitation energy of an allyl radical should be lower than that of a hydrogen atom, and the lowest energy excited state is expected to be the one from which reaction occurs (see Chapter 12).

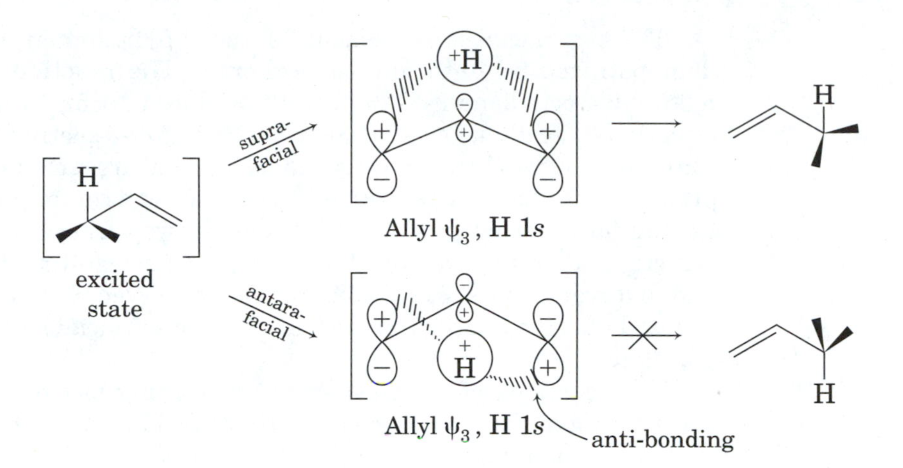

Figure 11.27 Suprafacial (allowed) and antarafacial (forbidden) [1,3] hydrogen shifts in electronically excited propene.

allyl allows bond breaking and bond forming to occur with a maximum of bonding interactions, so the *photochemical* [1,3] suprafacial rearrangement is allowed. Similarly, the photochemical antarafacial rearrangement is forbidden.

The selection rules for sigmatropic closures are also reversed if the π system is extended by one carbon-carbon double bond. As Figure 11.28 illustrates, the thermal [1,5] hydrogen shift is predicted to be allowed by the suprafacial pathway but forbidden by the antarafacial pathway. On the other hand, the photochemical reaction is predicted to be allowed by the antarafacial pathway but forbidden by the suprafacial pathway. Applying the same analysis to larger π systems shows that the selection rules for the [1,7] hydrogen shift are the same as those for the [1,3] hydrogen shift, and the rules for the [1,9] hydrogen shift are the same as for the [1,5] hydrogen shift.

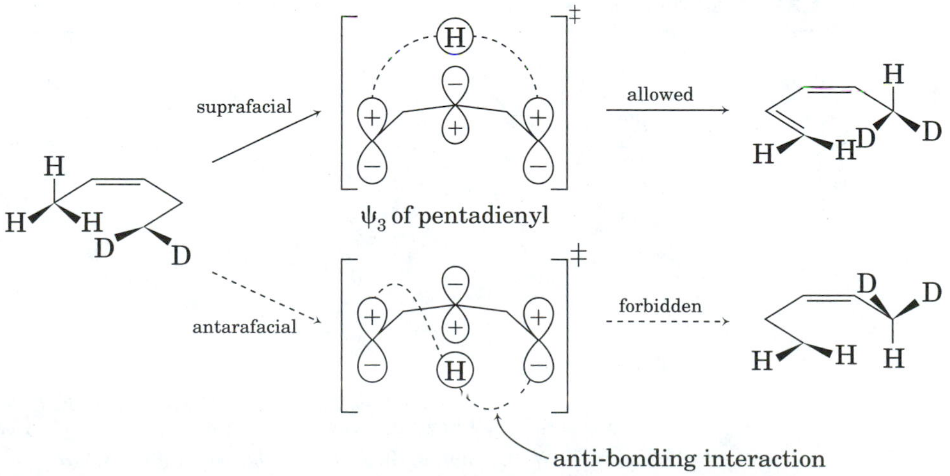

Figure 11.28 Suprafacial and antarafacial [1,5] hydrogen shifts in ground state 1,3-pentadiene-1,1-d_2.

suprafacial

antarafacial

ψ_3 of pentadienyl

allowed

forbidden

anti-bonding interaction

The suprafacial sterochemistry of the [1,5] hydrogen shift was elegantly demonstrated by Roth and co-workers.[46] The reaction of (*S*)-(2*E*,4*Z*)-6-methyl-2,4-octadiene-2-*d* (**29**) at 250° produced a 1.5 : 1 ratio of (*R*)-(3*E*,5*Z*)-3-methyl-3,5-octadiene-7-*d* (**30**) and (*S*)-(3*Z*,5*Z*)-3-methyl-3,5-octadiene-7-*d* (**28**). Both are products of a suprafacial [1,5] hydrogen shift. Apparently the preference for formation of **30** arises because of a conformational preference for the larger ethyl group to be oriented away from the other end of the structure. It is noteworthy that the products resulting from antarafacial [1,5] hydrogen shifts, **31** and **32**, were not observed. Roth and co-workers estimated that the energy preference for the suprafacial shift must be at least 8 kcal/mol.[47]

The reactions of 1,4-*bis*-(7-cycloheptatrienyl)benzene in Figure 11.31 can be rationalized as a series of thermal [1,5] and photochemical [1,7] suprafacial hydrogen shifts.[15,48] A [1,7] hydrogen shift has also been seen in the conversion of previtamin D_3 (**37**) to vitamin D_3 (**38**). The stereochemical nature of these reactions is not apparent in the absence of labeling, but the [1,7] hydrogen shift has been shown to be antarafacial in previtamin D_3

Figure 11.29 Stereochemistry of products formed by suprafacial [1,5] hydrogen shift.

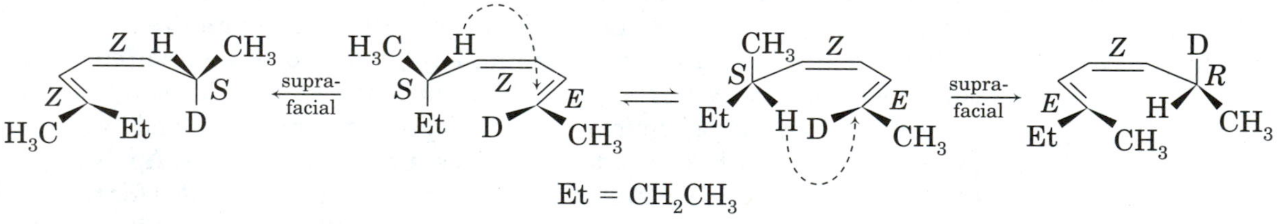

Figure 11.30 Predictions of stereochemistry of hypothetical products of antarafacial [1,5] hydrogen shift.

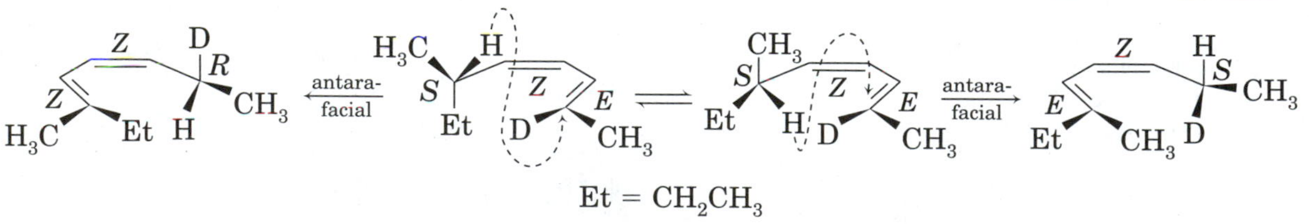

[46]Roth, W. R.; König, J.; Stein, K. *Chem. Ber.* **1970**, *103*, 426.

[47]For a detailed discussion of these results, see Carpenter, B. K. *Determination of Organic Reaction Mechanisms*; Wiley-Interscience: New York, 1984; pp. 180–184.

[48]Murray, R. W.; Kaplan, M. L. *J. Am. Chem. Soc.* **1966**, *88*, 3527.

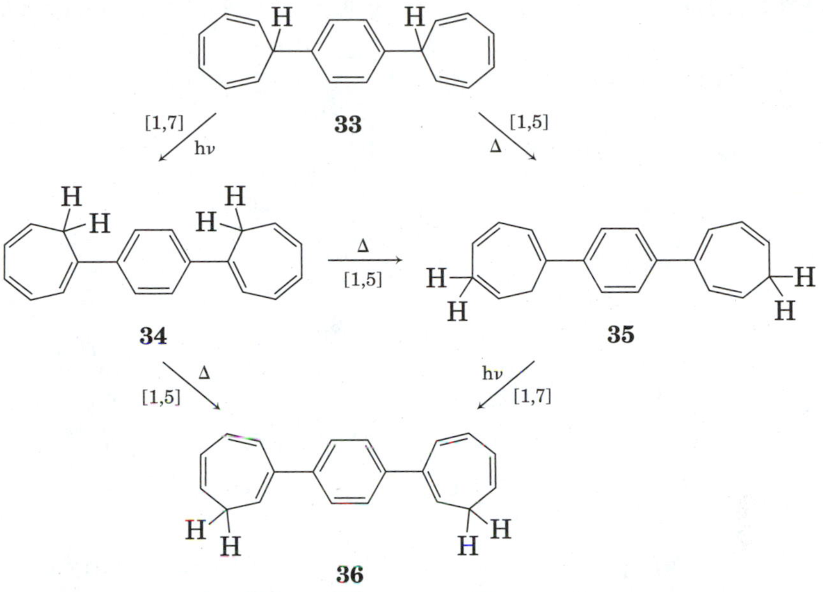

Figure 11.31 [1,5] Hydrogen shifts in 1,4-*bis*(7-cycloheptatrienyl) benzene. (Adapted from reference 15.)

model compounds, as in the reaction of the deuterium-labeled *cis*-isotachysterol **39** in Figure 11.33 (page 736).[49,50,51]

In sigmatropic [1,*j*] hydrogen shifts, it is necessary to consider only the stereochemistry of the reaction with regard to the conjugated π system. Because the hydrogen radical has one electron in a 1*s* orbital, it is spherically symmetric. Therefore the hydrogen atom has only one face, and all transformations must be suprafacial with respect to it. For migration of a group other than hydrogen, however, it is necessary to consider the stereochemistry of the reaction with respect to both components. For example the Cope rearrangement, which is a [3,3] sigmatropic reaction, can be either suprafacial or antarafacial with respect to each component.

The chair transition structure shown for the Cope rearrangement in Figure 11.34 (page 737) is suprafacial with respect to both components (the two allyl radicals drawn in the transition structure) because both bond breaking and bond forming take place below the plane defined by atoms C1, C2, and C3, and both bond breaking and bond forming take place above

---

[49]Hoeger, C. A.; Okamura, W. H. *J. Am. Chem. Soc.* **1985**, *107*, 268.

[50]Hoeger, C. A.; Johnston, A. D.; Okamura, W. H. *J. Am. Chem. Soc.* **1987**, *109*, 4690.

[51](a) Spangler, C. W. *Chem. Rev.* **1976**, *76*, 187, has summarized the activation parameters for many thermal [1,*j*] hydrogen shifts. For the suprafacial [1,5] hydrogen shift, typical values of $\Delta S^{\ddagger}$ are around -5 to -12 eu, with activation energies around 30–35 kcal/mol. For the antarafacial [1,7] hydrogen shift, the corresponding values are around -15 to -25 eu and ca. 15–25 kcal/mol. Therefore, it appears that the [1,7] hydrogen shifts have lower activation energies but more negative activation entropies than do the [1,5] hydrogen shifts. (b) For a discussion, see Gurskii, M. E.; Gridnev, I. D.; Il'ichev, Y. V.; Ignatenko, A. V.; Bubnov, Y. N. *Angew. Chem. Int. Ed. Engl.* **1992**, *31*, 781. Also see the discussion in Burnier, J. S.; Jorgensen, W. L. *J. Org. Chem.* **1984**, *49*, 3001.

Figure 11.32 [1,7] Hydrogen shift in vitamin D_3 synthesis. (Adapted from reference 50.)

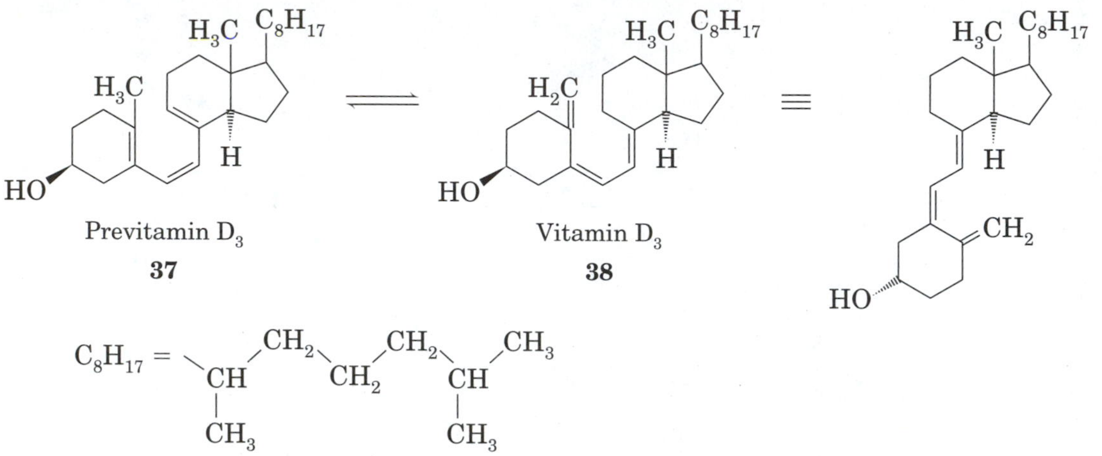

Previtamin D_3
37

Vitamin D_3
38

atoms C4, C5, and C6. Because of the symmetry of ψ_2 of each of the allyl fragments, bonding can begin between C1 and C6 as dissociation of the C3—C4 bond occurs, and the process shown is allowed by the principles of orbital symmetry. It may be sterically infeasible, but a reaction pathway that is antarafacial to both components (such as the one shown in Figure 11.35) will also produce a sigmatropic reaction that is allowed by the principles of orbital symmetry. A transition structure that is suprafacial with respect to both components is designated as the suprafacial-suprafacial pathway, and the one that is antarafacial with respect to both components would be the antarafacial-antarafacial pathway.[52]

Except for the restrictions due to orbital symmetry, it might be possible for a Cope rearrangement to occur by a pathway that is suprafacial with re-

Figure 11.33 Experimental evidence for an antarafacial [1,7] hydrogen shift in a vitamin D_3 analogue. (Adapted from reference 49.)

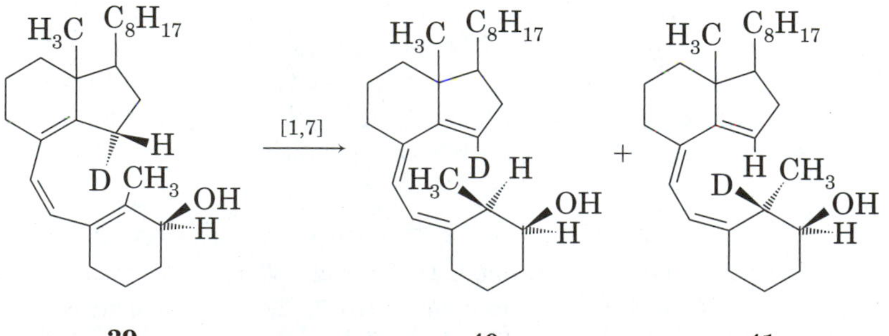

39 **40** **41**

[52]An antarafacial [3,3] sigmatropic rearrangement was proposed for one reaction by Miyashi, T.; Nitta, M.; Mukai, T. *J. Am. Chem. Soc.* **1971, 93**, 3441, but an alternative explanation for the experimental results was advanced by Baldwin, J. E.; Kaplan, M. S. *J. Am. Chem. Soc.* **1971**, *93*, 3969.

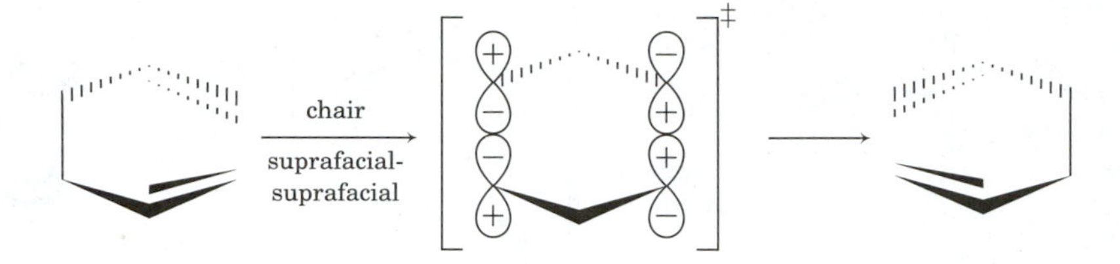

Figure 11.34 Suprafacial-suprafacial chair transition structure for the Cope rearrangement.

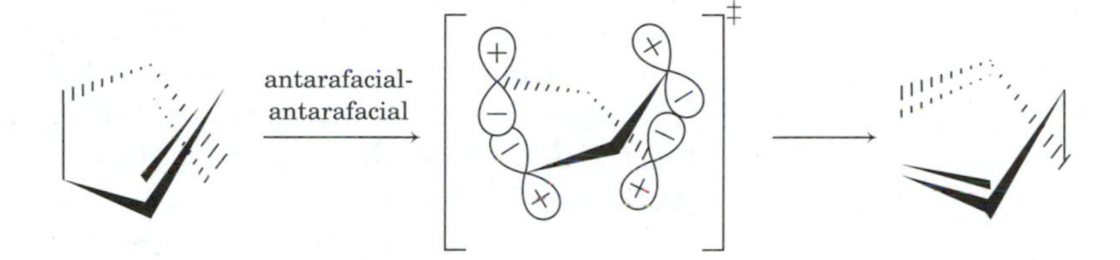

Figure 11.35 Antarafacial-antarafacial transition structure for Cope rearrangement.

spect to one component and antarafacial with respect to the other. Such pathways would be designated suprafacial-antarafacial or antarafacial-suprafacial. It is difficult to visualize, but such a process would require a rotation, say in the C5—C6 bond, as the C3—C4 bond dissociates. As shown in the example in Figure 11.36, however, such a mechanism would lead to an antibonding relationship between C6 and C1, so it is forbidden by the principle of conservation of orbital symmetry.[53]

The boat transition structure shown in Figure 11.37 is also a suprafacial-suprafacial pathway for the Cope rearrangement. Doering and Roth used stereochemical labels to distinguish between the suprafacial-suprafacial boat and chair transition structures for the Cope rearrangement of acyclic 1,5-dienes.[54] As shown in Figure 11.38, a chairlike transition structure for the Cope rearrangement of *rac*-3,4-dimethyl-1,5-hexadiene should

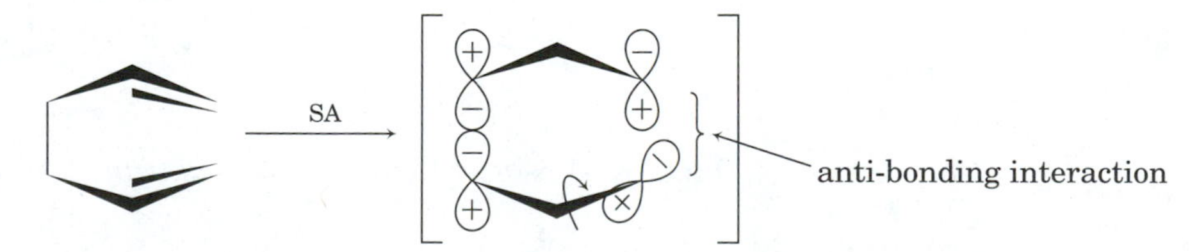

Figure 11.36 Forbidden suprafacial-antarafacial transition structure for Cope rearrangement.

---

[53]Even more complicated pathways can be imagined. Hansen, H.-J.; Schmid, H. *Tetrahedron* **1974**, 1959, considered seven possible transition structures for the Cope rearrangement, including boat and chair suprafacial-suprafacial and twist, cross, and plane antarafacial-antarafacial transition structures, and one anchor antarafacial-suprafacial transition structure.

[54]Doering, W. v. E.; Roth, W. R. *Tetrahedron* **1962**, *18*, 67.

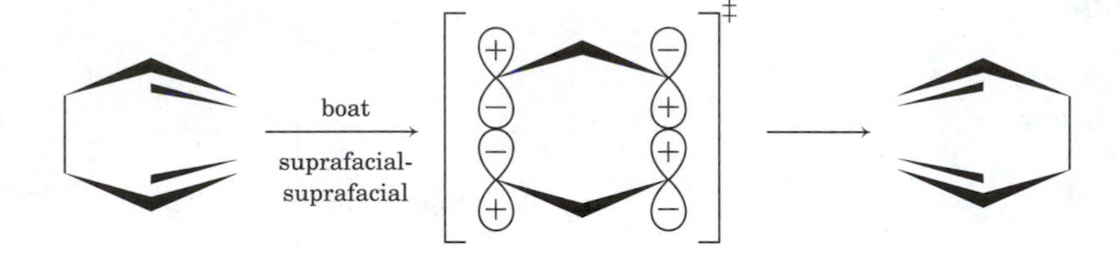

Figure 11.37 Suprafacial-suprafacial boat transition structure for the Cope rearrangement.

produce the trans,trans isomer of 2,6-octadiene, but the meso diastereomer of the reactant would produce the cis,trans isomer of the product. On the other hand, a boatlike transition structure would result in the formation of the trans,trans product from the meso starting material, but the cis,trans product would be produced from the racemic reactant (Figure 11.39). Doering and Roth found that 90% of the product obtained from the Cope rearrangement of the racemic starting material was the trans,trans isomer, confirming the chair transition structure for the reaction.[54] The boat transition structure is about 6 kcal/mol higher in energy, but reaction can occur by that pathway in molecules for which the chair pathway is destabilized sufficiently relative to the boat pathway. Other possible pathways appear to be much higher in energy and are not common routes for reaction.[55,56,57,58]

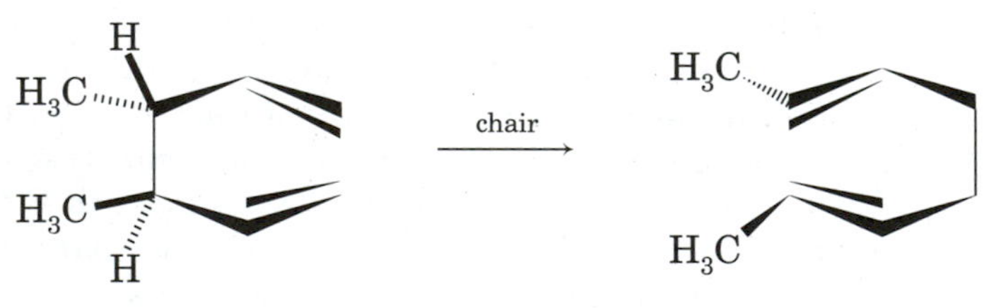

racemic-3,4-Dimethyl-1,5-hexadiene   *trans,trans*-2,6-Octadiene

H
H_3C
H
H_3C
chair
H_3C
H_3C

meso-3,4-Dimethyl-1,5-hexadiene   *cis,trans*-2,6-Octadiene

Figure 11.38 Products of Cope rearrangement of racemic (top) and meso (bottom) diastereomers of 3,4-dimethyl-1,5-hexadiene through chair transition structures.

---

[55]Shea, K. J.; Phillips, R. B. *J. Am. Chem. Soc.* **1980**, *102*, 3156.

[56]Gajewski, J. J.; Benner, C. W.; Hawkins, C. M. *J. Org. Chem.* **1987**, *52*, 5198.

[57]Similarly, a chair transition structure is 6.6 kcal/mol lower in energy than a boat transition structure in the Claisen rearrangement. Vance, R. L.; Rondan, N. G.; Houk, K. N.; Jensen, F.; Borden, W. T.; Komornicki, A.; Wimmer, E. *J. Am. Chem. Soc.* **1988**, *110*, 2314.

[58]See also Shea, K. J.; Stoddard, G. J.; England, W. P.; Haffner, C. D. *J. Am. Chem. Soc.* **1992**, *114*, 2635 and references therein.

Figure 11.39 Products of Cope rearrangement of racemic (top) and meso (bottom) diastereomers of 3,4-dimethyl-1,5-hexadiene through boat transition structures.

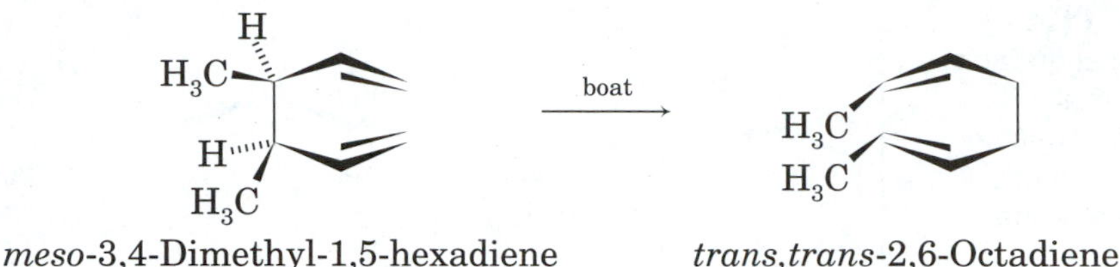

meso-3,4-Dimethyl-1,5-hexadiene *trans,trans*-2,6-Octadiene

racemic-3,4-Dimethyl-1,5-hexadiene *cis,trans*-2,6-Octadiene

As is the case for other pericyclic reactions, the selection rules for a thermal $[1,j]$ sigmatropic reaction are reversed for the photochemical reaction. If irradiation of a 1,5-hexadiene produces the excited state of one and only one of the two allyl components, then the [3,3] suprafacial-suprafacial reaction (Figure 11.40) is forbidden (as is the antarafacial-antarafacial pathway), while the antarafacial-suprafacial and suprafacial-antarafacial pathways are allowed (Figure 11.41). Analysis of higher sigmatropic reactions shows that the selection rules also reverse with the addition of a carbon-carbon double bond to either of the π systems. Thus the [3,5] sigmatropic reaction is thermally allowed to be suprafacial-antarafacial or antarafacial-suprafacial, and photochemically allowed to be suprafacial-suprafacial or antarafacial-antarafacial. Two of these reaction modes are illustrated in Figure 11.42. Because the symmetries of the linear polyene systems alternate in a regular way as each additional carbon-carbon double

Figure 11.40 Forbidden SS transition structure for photochemical Cope rearrangement.

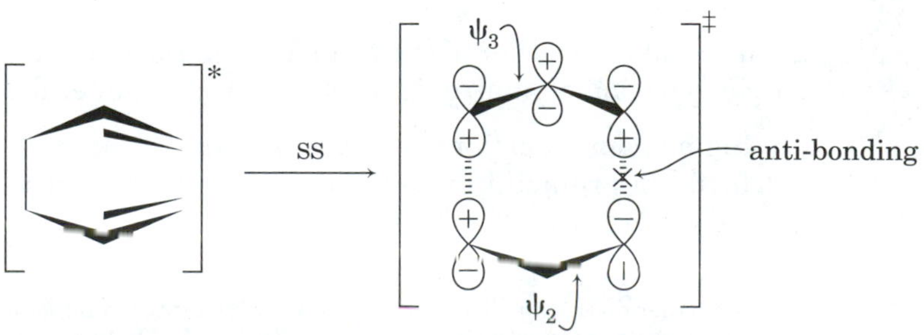

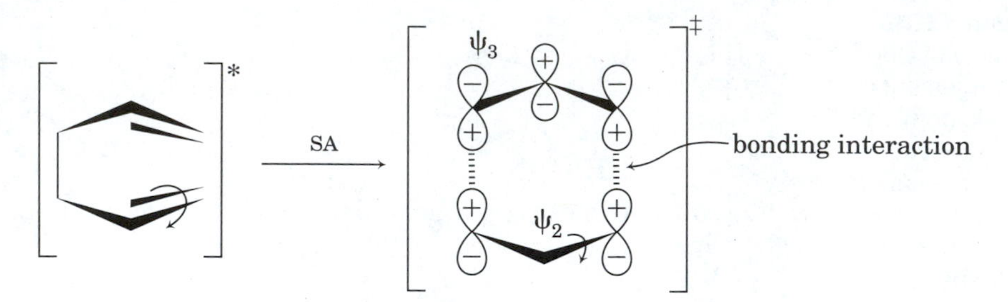

Figure 11.41 Allowed suprafacial-antarafacial transition structure for photochemical Cope rearrangement.

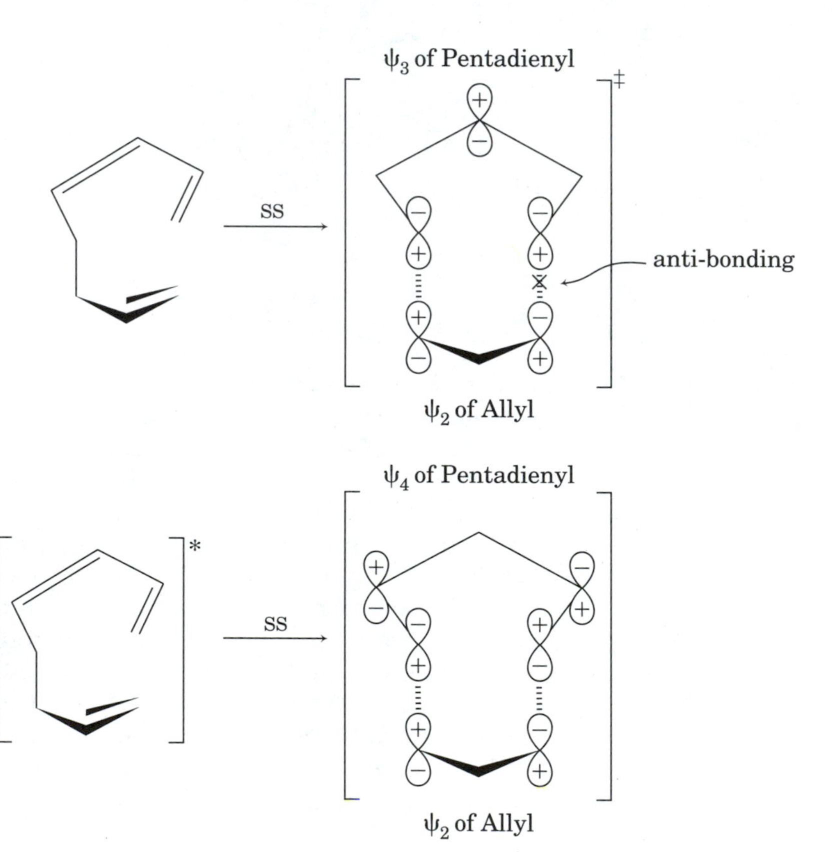

Figure 11.42 Suprafacial-suprafacial transition structures for the thermal (top) and photochemical (bottom) [3,5] sigmatropic reaction.

bond is added to either of the fragments in the transition structures, the selection rules for sigmatropic reactions may be generalized as follows:[59,60]

> Sigmatropic reactions of order $[i,j]$ are thermally allowed to be suprafacial-suprafacial or antarafacial-antarafacial and photochemically

[59]Reference 15, pp. 117 *ff.*

[60]For a further discussion of sigmatropic shifts and a method for generating qualitative potential energy surfaces to rationalize them, see Epiotis, N. D.; Shaik, S. *J. Am. Chem. Soc.* **1977**, *99*, 4936.

allowed to be antarafacial-suprafacial or suprafacial-antarafacial if $i + j = 4n + 2$. Conversely, they are thermally allowed to be antarafacial-suprafacial or suprafacial-antarafacial and photochemically allowed to be suprafacial-suprafacial or antarafacial-antarafacial if $i + j = 4n$.

Further Examples of Sigmatropic Reactions

The selection rules for sigmatropic reactions are powerful predictors of thermal rearrangements.[61] They explain clearly how some reactions can occur stereospecifically and with surprisingly low activation energies, while others do not. For example, *cis*-1,2-divinylcyclopropane (**42**) rearranges to *cis,cis*-1,4-cycloheptatriene (**43**) in a [3,3] sigmatropic reaction with a half-life of 25 minutes at 11° (Figure 11.43).[62,63] However, *trans*-1,2-divinylcyclopropane (**44**) reacts to give the same product only upon heating to well over 100°.[64] Apparently the reaction of **42** is so fast because the cis geometry places the two double bonds in the proper relationship for the Cope rearrangement. In the trans isomer, however, the termini of the two double bonds cannot overlap. Therefore, the reaction of **44** may proceed by a nonconcerted process involving diradicals formed by breaking the cyclopropane ring.[65]

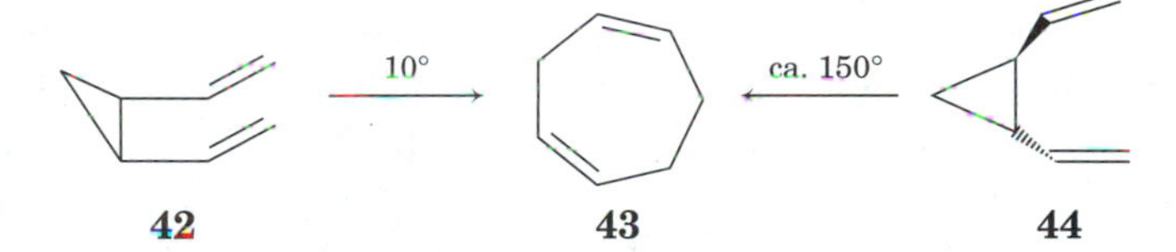

Figure 11.43 Thermal rearrangements of *cis*- and *trans*-1,2-divinylcyclopropane.

Both **42** and **44** are higher in energy than **43**. Sigmatropic rearrangements can also result in degenerate rearrangements of structures such as homotropilidene (**45**, Figure 11.44).[66] The classic example, however, is the molecule tricyclo[3.3.2.0^{4,6}]deca-2,7,9-triene (**46**, Figure 11.45), more com-

[61]Photochemical reactions are often more complex than would be predicted by the selection rules presented here. For a discussion, see Chapter 12.

[62]Schneider, M. P.; Rebell, J. *J. Chem. Soc., Chem. Commun.* **1975**, 283. Also see (a) Brown, J. M.; Golding, B. T.; Stofko, Jr., J. J. *J. Chem. Soc., Perkin Trans. 2* **1978**, 436; (b) Gajewski, J. J.; Hawkins, C. M.; Jimenez, J. L. *J. Org. Chem.* **1990**, *55*, 674.

[63]Initial studies were reported by Vogel, E. *Angew. Chem.* **1960**, *72*, 4; Vogel, E.; Ott, K.-H.; Gajek, K. *Liebigs Ann. Chem.* **1961**, *644*, 172.

[64]For leading references, see Gajewski, J. J.; Olson, L. P.; Tupper, K. J. *J. Am. Chem. Soc.* **1993**, *115*, 4548.

[65]Arai, M.; Crawford, R. J. *Can. J. Chem.* **1972**, *50*, 2158. Reclosure of the diradical intermediate to **42** would then allow rapid Cope rearrangement to **43**. See also the discussion in reference 64 concerning the rate-limiting step for this pathway.

[66]Doering, W. v. E.; Roth, W. R. *Tetrahedron* **1963**, *19*, 715.

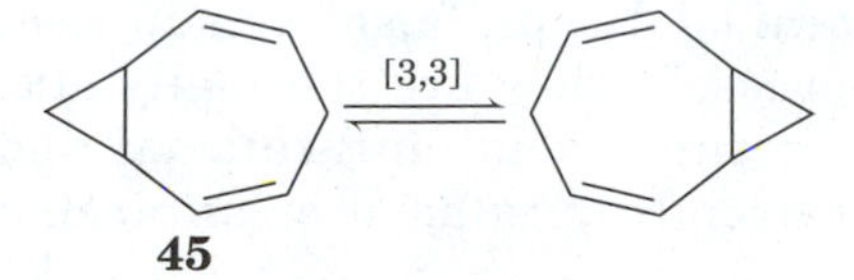

Figure 11.44 Degenerate sigmatropic rearrangement of homotropilidene.

Figure 11.45 Degenerate sigmatropic rearrangement in bullvalene. (Atom labels indicate the permutation arising from one [3,3] sigmatropic reaction.)

monly known by the trivial name bullvalene.[67] Although there are only ten carbon atoms in the molecule, the various permutations possible through [3,3] sigmatropic rearrangements produce about 1.2 million valence isomers.[68,69] The isomerization is so fast that the NMR spectrum is a broad peak at room temperature, and the spectrum sharpens to a single peak at 120°. However, at −85° the isomerization slows to the point that separate peaks for vinyl, cyclopropyl, and methine protons can be seen (Figure 11.46). Analysis of the spectrum indicates that the valence isomerization occurs at a rate of 540 sec^{-1} at 0°, with an activation energy of 12.8 kcal/mol.[68,70,71]

An interesting variation on the Cope rearrangement is the oxy-Cope reaction reported by Berson and Jones.[72] As shown in the general reaction in Figure 11.47, a Cope rearrangement on a 3-hydroxy-1,5-hexadiene produces an enol, which can tautomerize to the corresponding ketone or aldehyde. An

---

[67]See reference **44**, p. 131, for an account of the origin of this name.

[68]Schröder, G.; Oth, J. F. M.; Merényi, R. *Angew. Chem., Int. Ed. Engl.* **1965**, *4*, 752.

[69]Bullvalene is said to be **fluxional**, meaning that it undergoes rapid degenerate rearrangements (reference 19).

[70]Analysis by deuterium NMR on a sample dissolved in a liquid crystalline solvent gave a linear Arrhenius plot with a calculated rate constant at 300° of 4600 sec^{-1}, $\Delta H^{\ddagger}$ of 13.3 kcal/mol, and $\Delta S^{\ddagger}$ of + 2.5 eu: Poupko, R.; Zimmermann, H.; Luz, Z. *J. Am. Chem. Soc.* **1984**, *106*, 5391. The value of $\Delta S^{\ddagger}$ for a reaction measured in a liquid crystalline solvent may include a contribution from disorder introduced into the solvent as the reaction occurs.

[71]Valence tautomerization of bullvalene has also been studied in both the gas phase by Moreno, P. O.; Suarez, C.; Tafazzoli, M.; True, N. S.; LeMaster, C. B. *J. Phys. Chem.* **1992**, *96*, 10206, and in crystalline form by Schlick, S.; Luz, Z.; Poupko, R.; Zimmermann, H. *J. Am. Chem. Soc.* **1992**, *114*, 4315.

[72]Berson, J.; Jones, Jr., M. *J. Am. Chem. Soc.* **1964**, *86*, 5019.

Figure 11.46 Proton NMR spectrum of bullvalene as a function of temperature. (Reproduced from reference 68.)

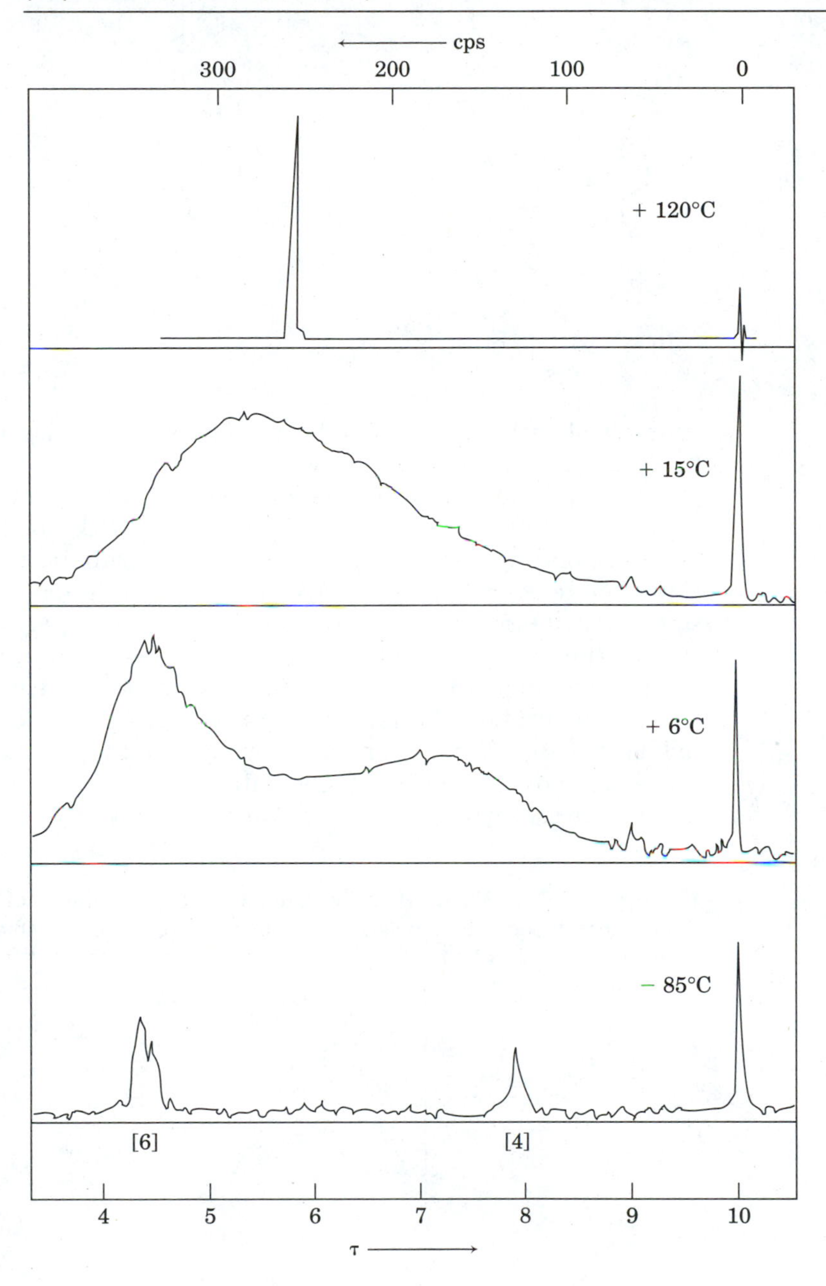

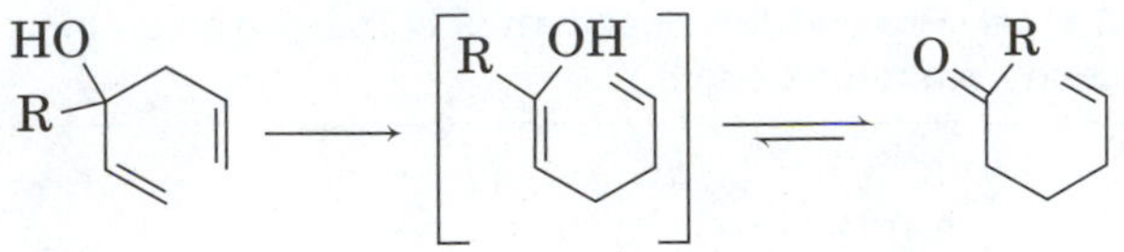

Figure 11.47 A general oxy-Cope rearrangement.

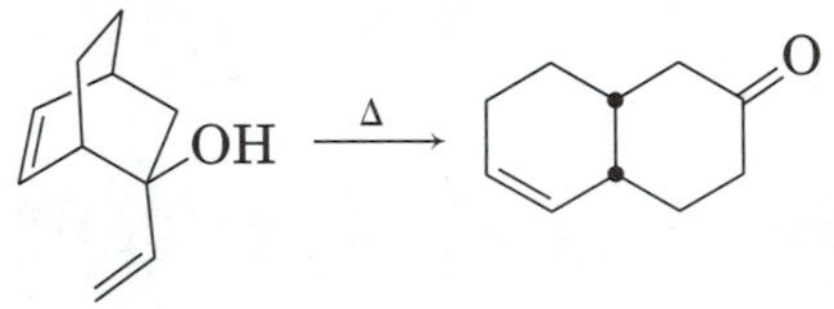

Figure 11.48 An example of the oxy-Cope rearrangement.

example of such a reaction is the conversion of the bicyclo-[2.2.2]octenol **47** to *cis*-$\Delta^{5,6}$-2-octalone (**48**) as shown in Figure 11.48.[73]

In a pericyclic reaction, the allowed pathway is lower in *electronic* energy because, as a result of the symmetries of the molecular orbitals involved, bonding can be maximized in the transition structure for the concerted reaction. However, MO and state correlation diagrams ignore any *steric* contribution to the total energies of the two pathways. The publication of the papers by Woodward and Hoffmann stimulated a number of chemists to probe the extent of orbital symmetry control by studying molecules for which the allowed pathway is sterically more difficult than the forbidden reaction. Berson considered whether a thermal [1,3] rearrangement could occur by a pathway that is antarafacial to the one-atom component. Unlike the case when a hydrogen atom migrates, a sigmatropic reaction in-

Figure 11.49 [1,3] methyl shift with antarafacial participation of methyl group. (Methyl substituents are labeled to emphasize inversion of the chiral center.)

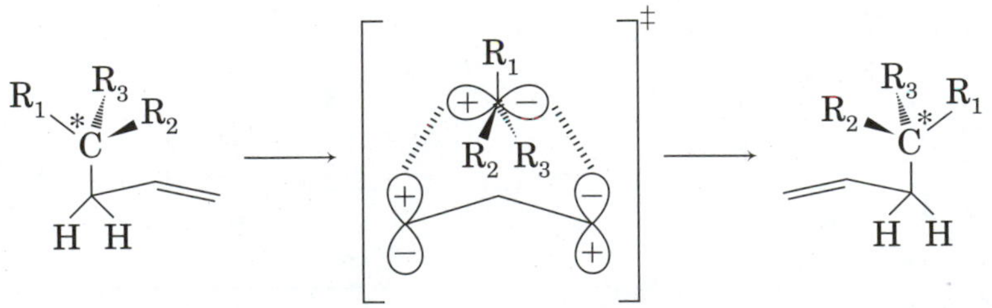

[73]For a discussion of the rate enhancement observed when an OH group bonded to a carbon atom located on a bond that is broken in a sigmatropic reaction is deprotonated, see (a) Wilson, S. R. *Org. React.* **1993**, *43*, 93; (b) Harris, N. J.; Gajewski, J. J. *J. Am. Chem. Soc.* **1994**, *116*, 6121.

volving migration of a one-carbon π system can be antarafacial with respect to the migrating atom. If a *p* orbital is visualized as a one-atom π system, with each lobe representing a different face of the π system, then antarafacial migration requires that bond breaking and bond forming occur on different lobes of the *p* orbital. Thus the [1,3] methyl shift shown in Figure 11.49 should be thermally allowed.

Berson and Nelson investigated the [1,3] rearrangement of *endo*-bicyclo-[3.2.0]-2-hepten-6-yl acetate-*exo*-7-*d* (**49**, Figure 11.50). The top pathway is suprafacial with respect to both the allyl component located in the five-membered ring and the migrating carbon atom above it. The bottom pathway is suprafacial with respect to the allyl fragment but is antarafacial with respect to the carbon atom, and it is the electronically allowed pathway. The two products can be distinguished by NMR because the two protium atoms in the —(CHOAc—CHD)— bridge are trans to each other in **50**, but they are cis to each other in **51**.[74] The product of the reaction was found to be **51**, indicating that orbital symmetry conservation is a greater factor than the need for a sterically unstrained transition structure in this case.[75,76,77]

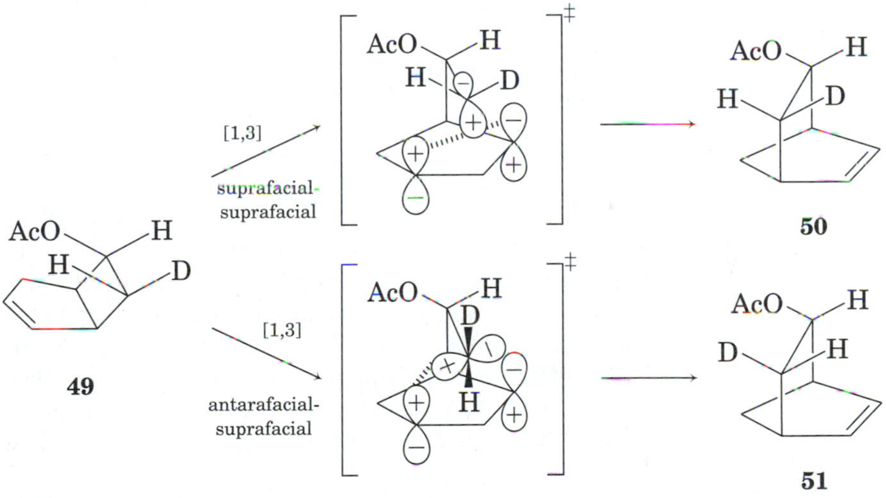

Figure 11.50 Stereochemical evidence for an antarafacial methyl shift.

[74]The term protium refers to the specific ^1H isotope of the hydrogen atom. Commission on Physical Organic Chemistry, IUPAC, *Pure Appl. Chem.* **1988**, *60*, 1115.

[75]Berson, J. A.; Nelson, G. L. *J. Am. Chem. Soc.* **1967**, *89*, 5504; Berson, J. A. *Acc. Chem. Res.* **1968**, *1*, 152.

[76]The NMR spectrum indicated that a very small amount (perhaps as much as 5%) of **50** may also have been present in the product, although it could have been produced by reaction of an isomer of the starting material.

[77]This result was confirmed by Klärner, F.-G.; Drewes, R.; Hasselmann, D. *J. Am. Chem. Soc.* **1988**, *110*, 297, who found that in a related system the [1,3] shift of a CH_2 group took place with predominant inversion, but that some product was formed by a nonconcerted diradical pathway.

11.4 Cycloaddition Reactions

Introduction

The pericyclic reactions discussed to this point have all been unimolecular rearrangements. The principles of conservation of orbital symmetry also apply to bimolecular cycloaddition reactions such as the Diels-Alder reaction.[78] As shown in Figure 11.51, the Diels-Alder reaction of 1,3-butadiene with ethene produces cyclohexene. In spite of numerous attempts, however, the analogous concerted cycloaddition of two ethene molecules to form cyclobutane does not occur thermally, even though that reaction is calculated to be exothermic by 18 kcal/mol.[79]

In the nomenclature of pericyclic reactions, the Diels-Alder reaction is defined as a **4 + 2 cycloaddition**, since there are four π electrons in one molecule and two π electrons in the other. The hypothetical dimerization of two ethene molecules would be a **2 + 2 cycloaddition**. Because σ bonds may also be considered to participate in cycloaddition reactions, we indicate that these cycloaddition reactions involve π systems by denoting them as $[_{\pi}4 + {}_{\pi}2]$ and $[_{\pi}2 + {}_{\pi}2]$ reactions, respectively.

Ethene Dimerization

Let us analyze the $[_{\pi}2 + {}_{\pi}2]$ cycloaddition first. Figure 11.52 shows an initial arrangement of two ethene molecules in which both the two σ bonds are parallel and the two sets of *p* orbitals are parallel. We view the reaction as the result of conversion of the four *p* orbitals into sp^3-hybrid orbitals. Just as is the case with electrocyclic reactions, cycloaddition reactions may be characterized as antarafacial (a) or suprafacial (s) with respect to each of the π systems. The cycloaddition shown in Figure 11.52 is suprafacial with respect to each component because bonding occurs on both of the bottom lobes of the *p* orbitals of the upper ethene, and bonding also occurs on both of the top lobes of the *p* orbitals of the lower ethene. This pathway for the ethene dimerization, which is termed a $[_{\pi}2_s + {}_{\pi}2_s]$ cycloaddition, leads to retention

(a) + $\xrightarrow{\Delta}$

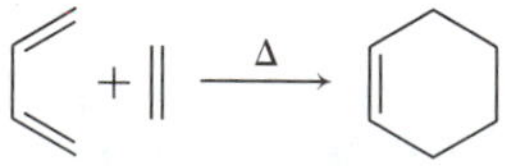

(b) + $\xrightarrow{\Delta}$ (crossed)

Figure 11.51 (a) Allowed [4 + 2] and (b) forbidden [2 + 2] concerted thermal cycloaddition reactions.

[78]Diels, O.; Alder, K. *Liebigs Ann. Chem.* **1928**, *460*, 98.

[79]Streitwieser, Jr., A. *Science* **1981**, *214*, 627. The Diels-Alder reaction of ethene and 1,3-butadiene is exothermic by 40 kcal/mol.

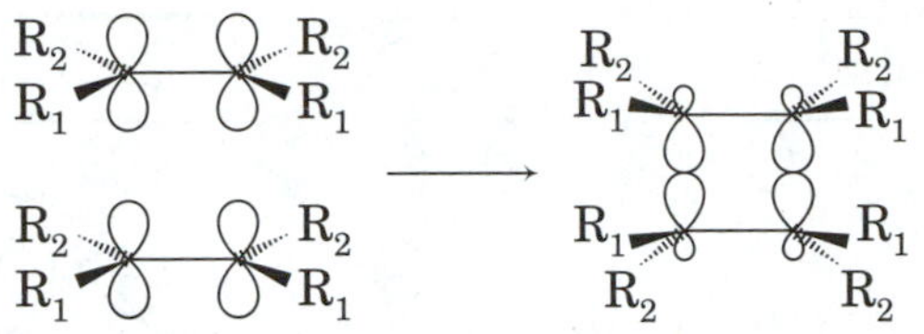

Figure 11.52 Retention of stereochemical relationships in $[_{\pi}2_s + {}_{\pi}2_s]$ cycloaddition.

of the stereochemical relationships of the alkene substituents. Later we will examine the $[_{\pi}2_a + {}_{\pi}2_a]$, the $[_{\pi}2_s + {}_{\pi}2_a]$, and the $[_{\pi}2_a + {}_{\pi}2_s]$ pathways for this same reaction.

The combined basis set of *p* orbitals (all four *p* orbitals) of the two ethene molecules, shown in more detail in Figure 11.53, has a number of symmetry elements. In addition to a horizontal plane of symmetry (σ_1) and a vertical plane of symmetry (σ_2), there is also a plane of symmetry bisecting the H—C—H bond angles, and there are a number of rotation axes. It is not necessary to consider all symmetry elements, however, just the minimum number of symmetry elements that both

1. are present in reactant, transition structure, and product, and
2. are needed to provide a distinction between allowed and forbidden pathways.

In constructing an MO correlation diagram for the $[_{\pi}2_s + {}_{\pi}2_s]$ reaction, it is not sufficient to consider just the MOs of the two ethene molecules individually. Either one of the two π bonds is, by itself, *asymmetric* with respect to σ_1. That is, it does not have the full symmetry of the basis set of atomic orbitals. Since the *p* orbitals that comprise the π bonds are degenerate in energy, they will split each other strongly. Thus, an adequate MO description of the reactants must be symmetry-correct for the *pair* of ethene units, not just for each ethene alone. Interaction of the set of four atomic *p* orbitals produces a set of four MOs for the extended system. These reactant MOs are designated both by their bonding or antibonding characteristics and by their symmetry with respect to both of the two planes of symmetry considered in this analysis. (The symmetry designation is given first with respect to σ_1, then with respect to σ_2.) Thus in Figure 11.54, π_{AS} is antisymmetric with respect to σ_1 but symmetric with respect to σ_2. The symmetry

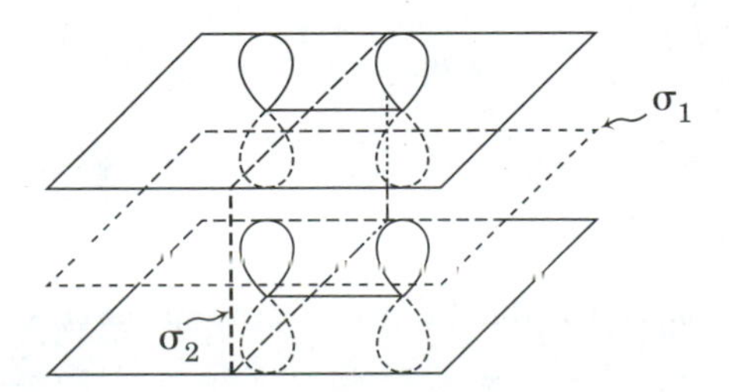

Figure 11.53 Basis set of *p* orbitals of two ethene molecules for [2 + 2] cycloaddition.

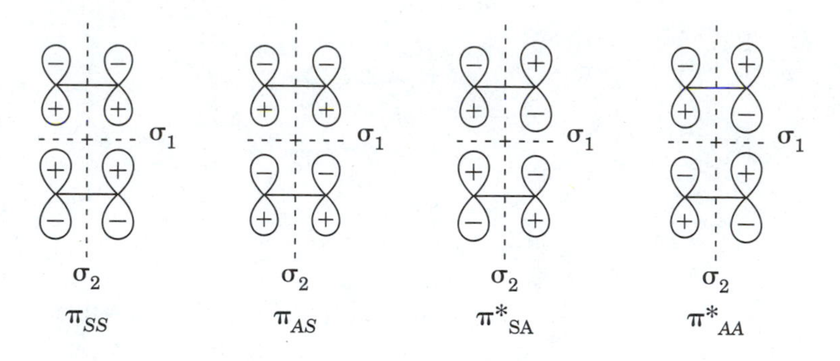

Figure 11.54 Symmetry-correct MOs for [2 + 2] cycloaddition of two ethenes. (Adapted from reference 90.)

designations of the other three orbitals of the reactant system are also shown in Figure 11.54.

Similarly, it is necessary to consider a set of cyclobutane σ MOs that are symmetry correct with respect to the same symmetry elements. Thus the combination of the four sp^3 atomic orbitals shown in Figure 11.55 produces two sets of σ (bonding) MOs and two sets of σ* (antibonding) MOs. Each of these σ orbitals is characterized with respect to the two mirror planes of symmetry that are preserved throughout the reaction, σ_1 and σ_2, in Figure 11.56.[80]

The MO correlation diagram for the cycloaddition reaction is shown in Figure 11.57. Here the artificial separation between the two π orbitals is introduced solely to indicate the two orbitals and their individual symmetry properties. The same is true of the arbitrary separation between the two π* orbitals, between the two σ orbitals, and between the two σ* orbitals. This MO correlation diagram shows that the thermal $[_{\pi}2_s + _{\pi}2_s]$ cycloaddition is forbidden, since the ground state of the reactant correlates with a high energy excited state of the product. That is, a process that could force the two

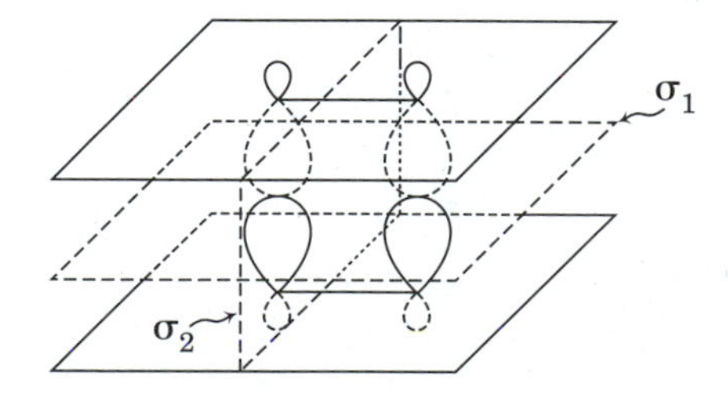

Figure 11.55 Basis set orbitals for cyclobutane produced by [2 + 2] cycloaddition.

[80]The symbol σ is used both for a σ orbital and for a mirror plane of symmetry.

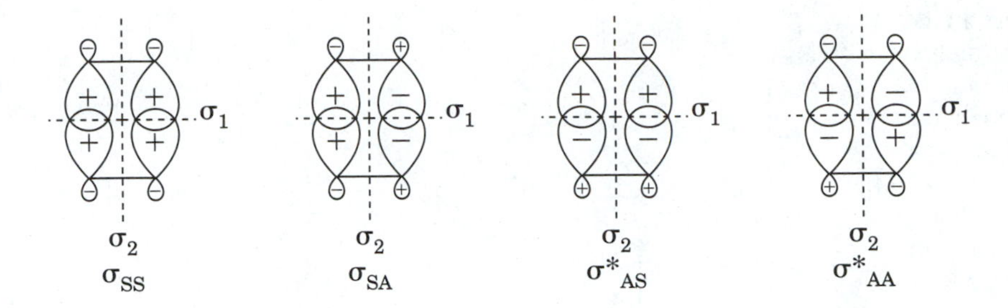

Figure 11.56 Symmetry correct orbitals for σ bonds in cyclobutene. (Adapted from reference 15.)

ethene molecules together in a suprafacial-suprafacial pathway would produce a cyclobutane molecule having the electronic configuration $\sigma^2\sigma^{*2}$, a doubly excited electronic state. Similarly, the reverse process (**cycloreversion** of cyclobutene to two ethenes) is also forbidden. However, promotion of an electron from one of the π orbitals to a π* orbital in a photochemical reaction allows the excited state of the reactant system to correlate with the excited state of the product system (Figure 11.58), so the photochemical cycloaddition should be allowed.[81]

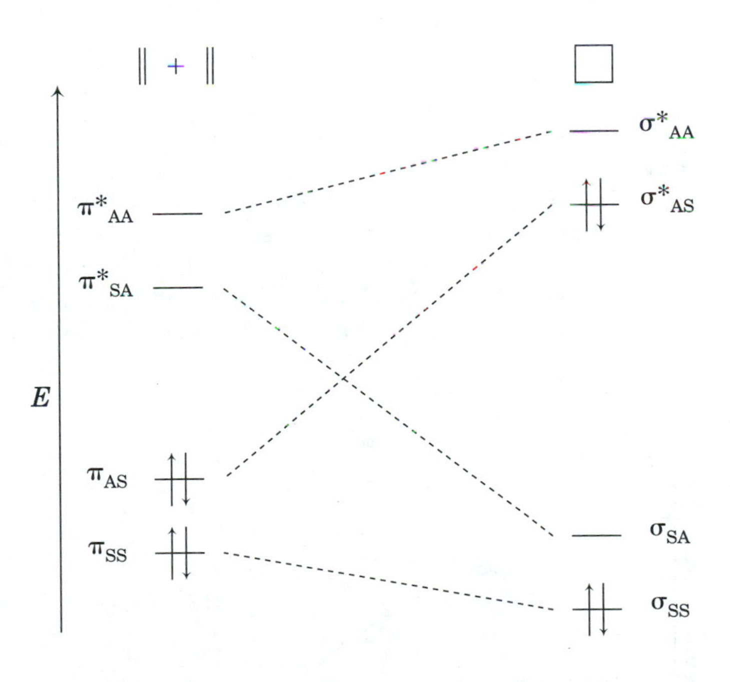

Figure 11.57 MO correlation diagram for thermal suprafacial-suprafacial [2 + 2] cycloaddition. (Adapted from reference 90.)

[81] The first excited state of the cyclobutane, having an electron in a σ* orbital, is much higher in energy than is the first excited state of the pair of ethenes, which has an electron in one of the π* orbitals.

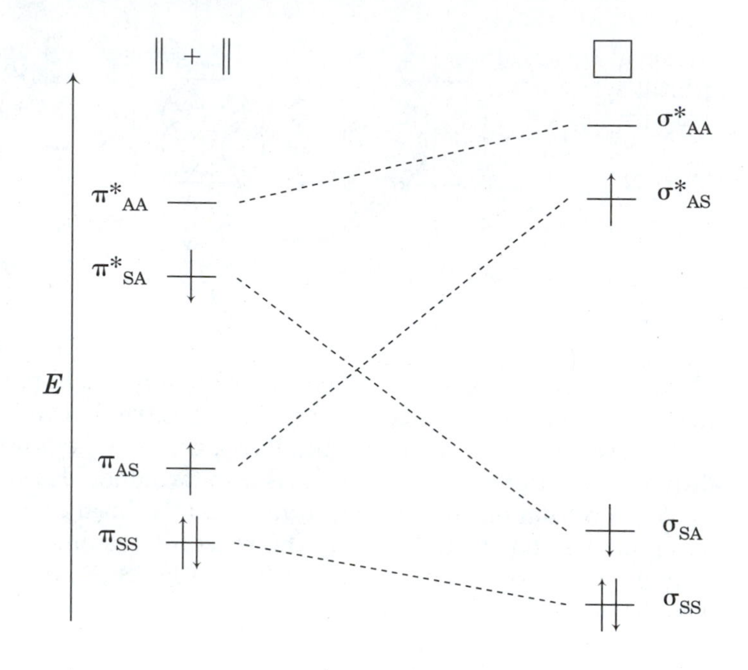

Figure 11.58 MO correlation diagram for photochemical suprafacial-suprafacial [2 + 2] cycloaddition. (Adapted from reference 90.)

The MO correlation diagrams in Figures 11.57 and 11.58 can be used to generate the state correlation diagram shown in Figure 11.59. It reemphasizes the conclusion that the ground state $[_{\pi}2_s + _{\pi}2_s]$ cycloaddition should have a high energy barrier because of the activation energy needed to raise the ground state molecule to the level of the avoided crossing. Therefore, the thermal reaction is forbidden. However, there is a correlation between the lowest-lying excited state of the reactant and the lowest-lying excited state of the product. There should not be a large activation energy for the photo-

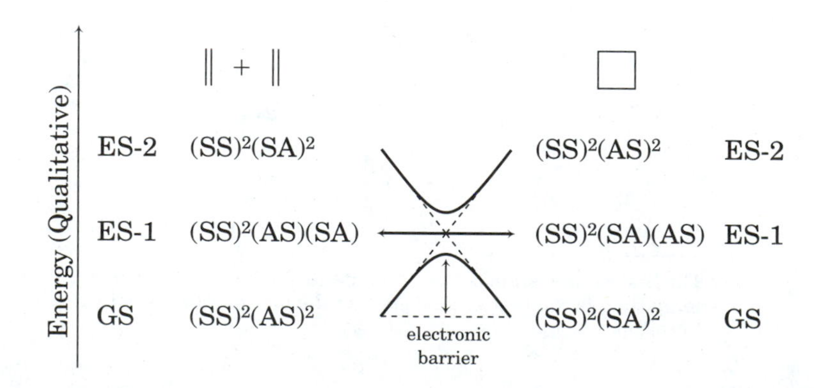

Figure 11.59 State correlation diagram for suprafacial-suprafacial [2 + 2] cycloaddition.

chemical cycloaddition, so it is allowed by the principles of conservation of orbital symmetry.

The Diels-Alder Reaction

The Diels-Alder reaction is a $[_{\pi}2_s + _{\pi}4_s]$ cycloaddition.[82,83] With a substituted diene and a substituted olefin (the dienophile), the reaction is highly stereospecific.[84] Configuration is retained in both the olefin and diene, making the reaction very useful in organic synthesis.[85,86] The *ab initio* transition structure calculated for the reaction (Figure 11.60) suggests an initial orientation of the reactants in which the diene and the double bond of the dienophile lie in parallel planes.[87] The minimum basis set of AOs needed to describe the reactant is a set of four *p* orbitals on the butadiene component plus a set of two *p* orbitals on the ethene component (Figure 11.61). For the product the minimum basis set is a set of four sp^3-hybrid orbitals (for the two σ bonds) and two *p* orbitals. The only symmetry element that is maintained from reactant through the transition structure to product is a plane of symmetry.

---

[82]The $_2\pi$ component need not be a carbon-carbon double bond. Other multiple-bonded groups, including alkynes and allenes, may also serve as dienophiles. For example, the reaction of cyclopentadiene with acetylenedicarboxylic acid was reported by Diels, O.; Alder, K.; Nienburg, H. *Liebigs Ann. Chem.* **1931**, *490*, 236:

CO_2H CO_2H CO_2H CO_2H

Ketenes undergo cycloadditions with alkenes in which the stereochemistry of the alkene is retained. For a review, see Hyatt, J. A.; Raynolds, P. W. *Org. React.* **1994**, *45*, 159. Both theoretical and experimental evidence support a reaction pathway in which the alkene and the ketene are perpendicular in the transition structure: Lovas, F. J.; Suenram, R. D.; Gillies, C. W.; Gillies, J. Z.; Fowler, P. W.; Kisiel, Z. *J. Am. Chem. Soc.* **1994**, *116*, 5285. For a discussion of cycloadditions involving ketenes, alkynes and allenes, see (a) Huntsman, W. D. *The Chemistry of Ketenes, Allenes and Related Compounds*, Part 2; Patai, S., ed.; Wiley-Interscience: Chichester, England, 1980; pp. 521–667; (b) Bastide, J.; Henri-Rousseau, O. *The Chemistry of the Carbon-Carbon Triple Bond*, Part 1; Patai, S., ed.; Wiley-Interscience: Chichester, England, 1978; pp. 447–522.

[83]For a critical discussion of the mechanism of the Diels-Alder reaction, see Sauer, J.; Sustmann, R. *Angew. Chem., Int. Ed. Engl.* **1980**, *19*, 779.

[84]For a discussion of the stereochemistry of the Diels-Alder reaction, see Martin, J. G.; Hill, R. K. *Chem. Rev.* **1961**, *61*, 537.

[85]For a discussion of synthetic aspects of the Diels-Alder and related reactions, see (a) Kloetzel, M. C. *Org. React.* **1948**, *4*, 1; (b) Holmes, H. L. *Org. React.* **1948**, *4*, 60; (c) Fringuelli, F.; Taticchi, A. *Dienes in the Diels-Alder Reaction*; Wiley-Interscience: New York, 1990; (d) Carruthers, W. *Cycloaddition Reactions in Organic Synthesis*; Pergamon Press: Oxford, England, 1990.

[86]Intramolecular Diels-Alder reactions are useful in synthesis. See, for example, Brieger, G.; Bennett, J. N. *Chem. Rev.* **1980**, *80*, 63.

[87]Li, Y.; Houk, K. N. *J. Am. Chem. Soc.* **1993**, *115*, 7478.

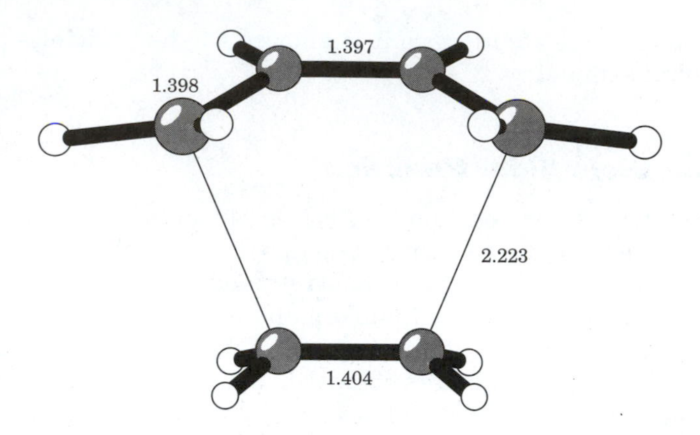

Figure 11.60 *Ab initio* transition structure for the Diels-Alder reaction of 1,3-butadiene and ethene. (Adapted from reference 87; distances are in Angstroms.)

Categorizing the diene and the alkene orbitals with respect to the plane of symmetry produces the symmetry designations shown on the left side of Figure 11.62. As in the case of ethene dimerization, the orbitals associated with the new σ bonds in cyclohexene must be described as symmetry correct combinations of sp^3-hybrid orbitals. The MO correlation diagram indicates that each of the bonding orbitals in the reactant correlates with a bonding orbital in the product. Since the entire set of bonding MOs of the reactant correlates with the entire set of bonding MOs of the product, there should not be a large electronic barrier for the cycloaddition, and the thermal Diels-Alder reaction is allowed.

The MO correlation diagram can be used to construct a state correlation diagram for the Diels-Alder reaction (Figure 11.63). Because the MOs populated in the ground state of the reactants are converted to MOs of the product in its ground state, and because the ground states of the reactants and product have the same symmetry, the ground state of the reactants correlates with the ground state of the product. Therefore, the thermal Diels-Alder reaction is allowed. The same diagram (viewed from right to left) applies to the reverse of a cycloaddition reaction, such as the cracking of dicyclopentadiene to cyclopentadiene, which is a retro-Diels-Alder reaction.[88,89]

---

[88]For leading references and a study of the effect of substituents on one retro-Diels-Alder reaction, see Chung, Y.-S.; Duerr, B. F.; Nanjappan, P.; Czarnik, A. W. *J. Org. Chem.* **1988**, *53*, 1334.

[89]By synthesizing a stable compound with similar structural features to the transition structure for the Diels-Alder reaction, researchers were able to produce antibodies to that analogue, and these antibodies were able to catalyze the Diels-Alder reaction by serving as entropy traps to bind the two reactants in the proper orientation for reaction, thus lowering their translational and rotational entropy: Braisted, A. C.; Schultz, P. G. *J. Am. Chem. Soc.* **1990**, *112*, 7430; Hilvert, D.; Hill, K. W.; Nared, K. D.; Auditor, M.-T. M. *J. Am. Chem. Soc.* **1989**, *111*, 9261. Catalytic antibodies have also been used to catalyze both the favored endo and the disfavored exo Diels-Alder adduct: Gouverneur, V. E.; Houk, K. N.; de Pascual-Teresa, B.; Beno, B.; Kanda, D.; Lerner, R. A. *Science* **1993**, *262*, 204. These papers provide references to catalytic antibodies that have been used to facilitate other concerted reactions.

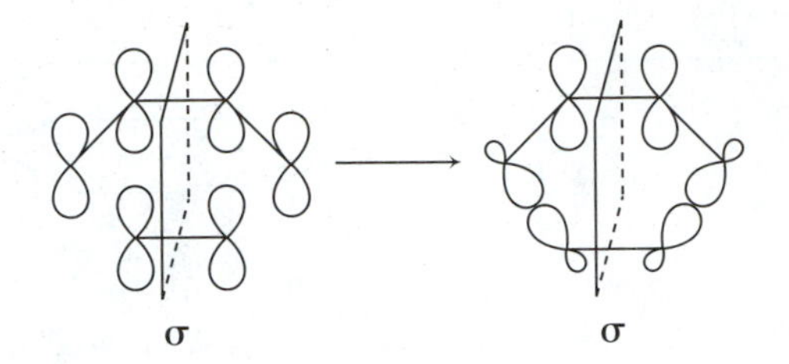

Figure 11.61
Basis set orbitals for reactants (left) and product (right) in Diels-Alder reaction.

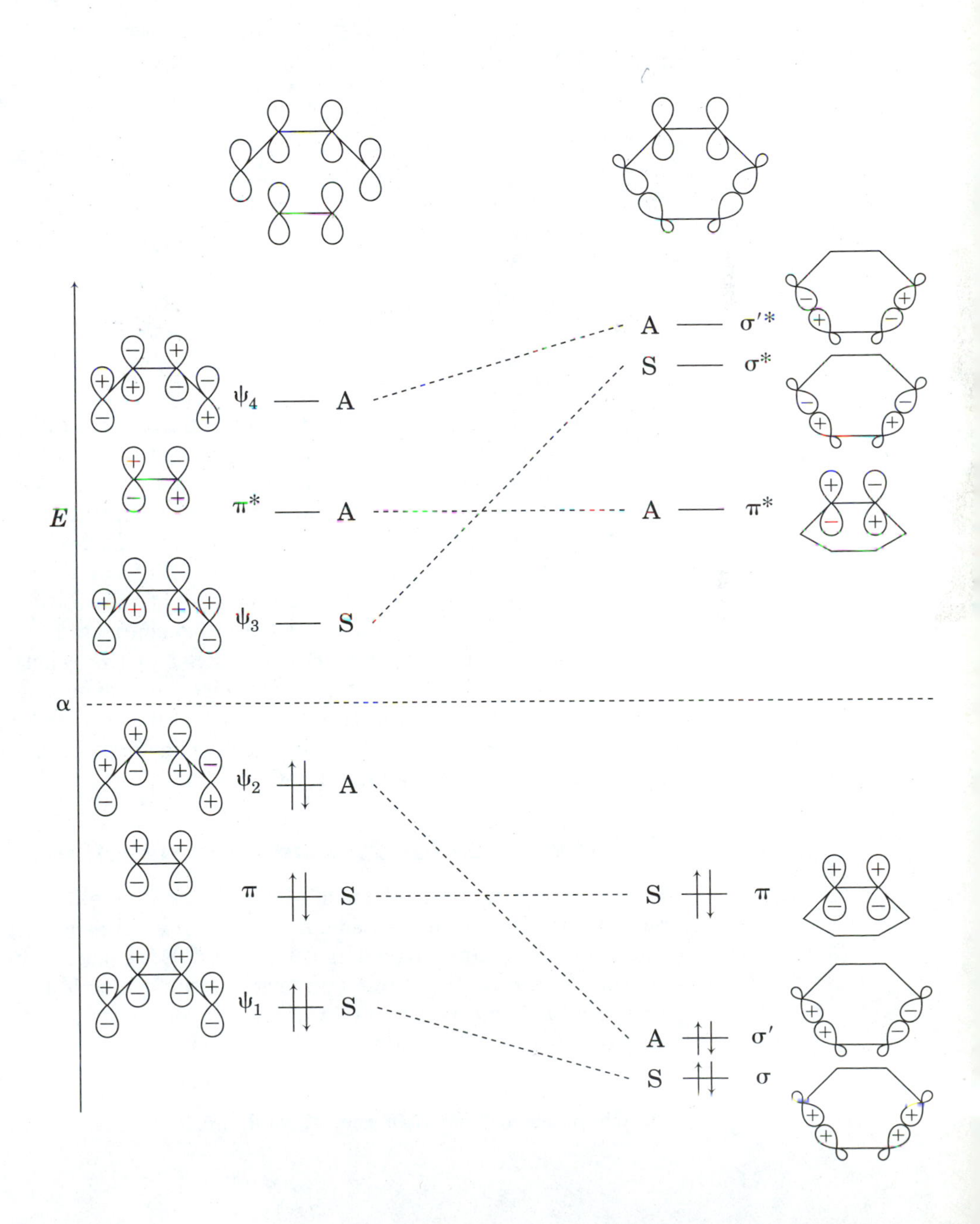

Figure 11.62
MO correlation diagram for a $[_{\pi}4_s + {}_{\pi}2_s]$ cycloaddition. (Adapted from reference 90.)

Figure 11.63 State correlation diagram for [$_{\pi}4_s$ + $_{\pi}2_s$] cycloaddition.

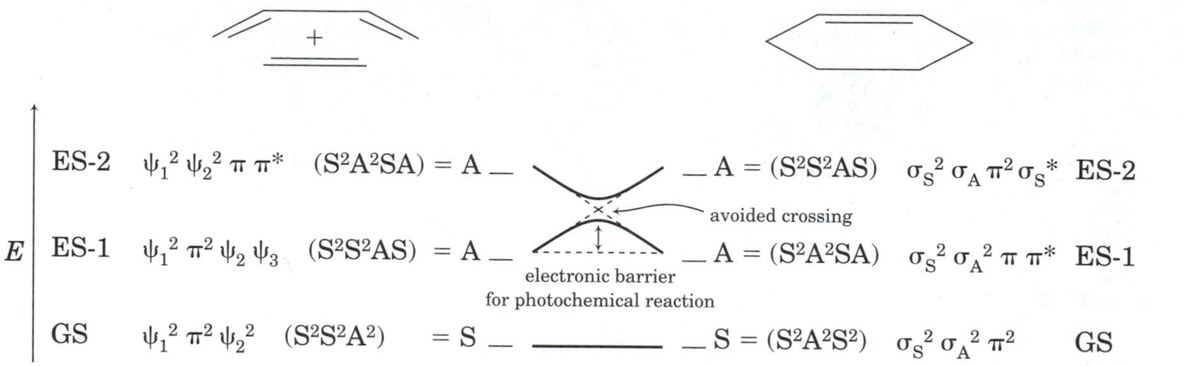

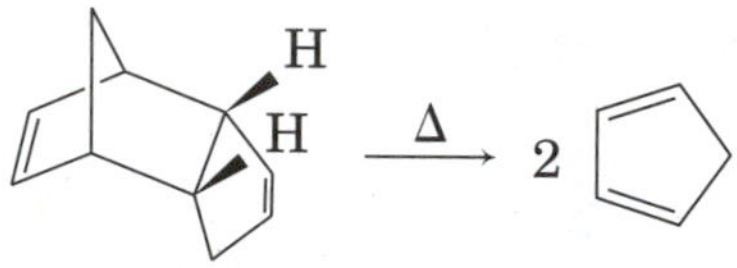

+

E

ES-2 $\psi_1^2\,\psi_2^2\,\pi\,\pi^*$ (S^2A^2SA) = A — — A = (S^2S^2AS) $\sigma_S^2\,\sigma_A\,\pi^2\,\sigma_S^*$ ES-2

avoided crossing

ES-1 $\psi_1^2\,\pi^2\,\psi_2\,\psi_3$ (S^2S^2AS) = A — — A = (S^2A^2SA) $\sigma_S^2\,\sigma_A^2\,\pi\,\pi^*$ ES-1

electronic barrier
for photochemical reaction

GS $\psi_1^2\,\pi^2\,\psi_2^2$ ($S^2S^2A^2$) = S — — S = ($S^2A^2S^2$) $\sigma_S^2\,\sigma_A^2\,\pi^2$ GS

H H Δ 2 **(11.7)**

52 **53**

The situation is very different for the photochemical [$_{\pi}2_S$ + $_{\pi}4_S$] cycloaddition, however. The MO correlation diagram indicates that ES-1 of the reactants (having the configuration $\psi_1{}^2\pi^2\psi_2\psi_3$) is converted to ES-2 of the product (having the configuration $\sigma_S{}^2\sigma_A\pi^2\sigma_S^*$). Therefore, ES-1 of the reactants is connected to ES-2 of the products by a dashed line. Similarly, the MO correlation diagram indicates that ES-2 of the reactants (having the configuration $\psi_1{}^2\pi\psi_2{}^2\pi^*$) is converted to ES-1 of the product (having the configuration $\sigma_S^2\sigma_A^2\pi\pi^*$), so these states are also connected by a dashed line on the state correlation diagram. Because ES-1 of the reactants and ES-2 of the product have the same total symmetry, however, there is an avoided crossing connecting ES-1 of the reactants with ES-1 of the product. Therefore, there should be a large electronic barrier for the photochemical Diels-Alder reaction, and it is symmetry-forbidden.

Selection Rules for Cycloaddition Reactions

From a more detailed analysis of MO and state correlation diagrams, Woodward and Hoffmann presented a set of selection rules for cycloaddition reactions, which are summarized in Table 11.1.[90] Here p and q are the number of electrons in the two π systems undergoing the cycloaddition reaction. When the sum of p and q is a member of the $4n$ series, then the reaction is thermally allowed to be suprafacial with respect to one of the π components and

[90]The table is adapted from Hoffmann, R.; Woodward, R. B. *J. Am. Chem. Soc.* **1965**, *87*, 2046.

Table 11.1 Selection rules for cycloaddition reactions.

$p + q$	Thermally Allowed	Photochemically Allowed
$4n$	$p_s + q_a$ or $p_a + q_s$	$p_s + q_s$ or $p_a + q_a$
$4n + 2$	$p_s + q_s$ or $p_a + q_a$	$p_s + q_a$ or $p_a + q_s$

antarafacial with respect to the other one. When the sum of p and q is a member of the $4n + 2$ series, then the reaction is thermally allowed when it is either suprafacial with respect to both components or antarafacial with respect to both. As usual, the selection rules are reversed for photochemical reactions.

The Diels-Alder reaction may be the best known cycloaddition, but other types of cycloaddition reactions are also synthetically important.[91] Consistent with the predictions of the selection rules, the $[_{\pi}6_s + _{\pi}4_s]$ cycloaddition has been seen in the reaction of tropone (**54**) with cyclopentadiene:[92]

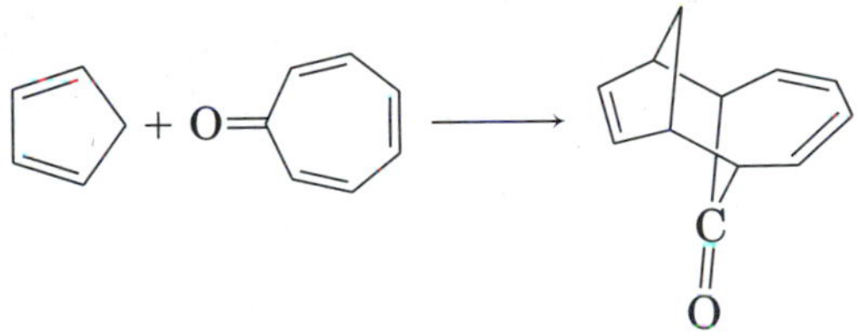

53 **54** **55** (11.8)

The $[_{\pi}8_s + _{\pi}2_s]$ cycloaddition has been observed in the addition of dimethyl fumarate (**57**) to 8,8-dimethylisobenzofulvene (**56**):[93,94]

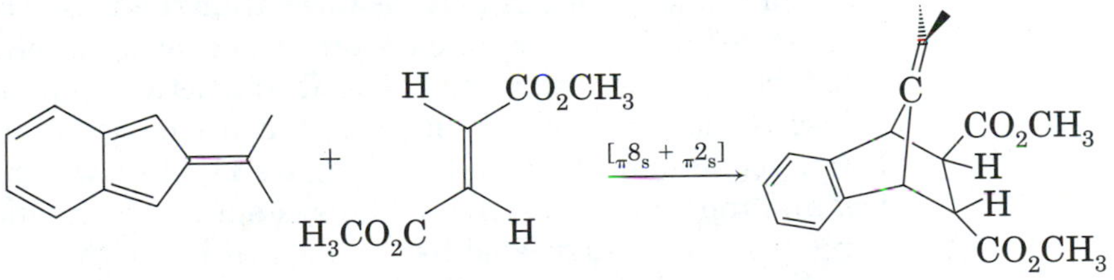

56 **57** **58** (11.9)

[91]Cycloaddition reactions involving more than six electrons in two π systems are known as *higher order* cycloadditions. Transition metals have been found to be effective in promoting higher order cycloadditions. For a discussion, see Rigby, J. R. *Acc. Chem. Res.* **1993**, *26*, 579.

[92]Cookson, R. C.; Drake, B. V.; Hudec, J.; Morrison, A. *Chem. Commun.* **1966**, 15.

[93]Tanida, H.; Irie, T.; Tori, K. *Bull. Chem. Soc. Jpn.* **1972**, *45*, 1999.

[94]Also see Russell, R. A.; Longmore, R. W.; Warrener, R. N. *J. Chem. Educ.* **1992**, *69*, 164.

The $[_\pi 8_s + _\pi 6_s]$ cycloaddition has been seen in the addition of 8,8-dimethylisobenzofulvene to tropone:[95]

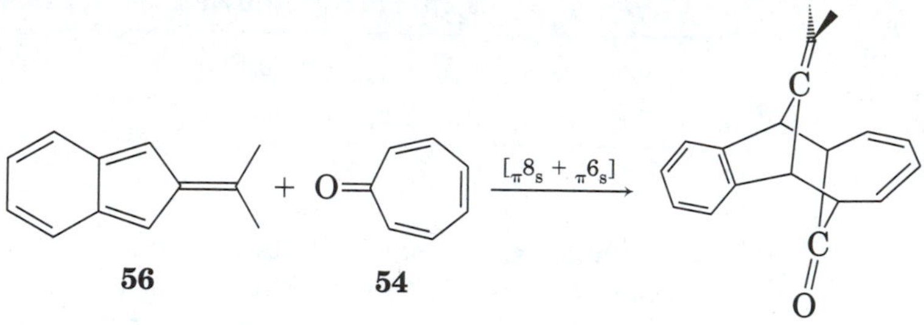

56 + O= **54** $\xrightarrow{[_\pi 8_s + _\pi 6_s]}$ **59** **(11.10)**

The dimerization of 8,8-dimethylisobenzofulvene is an example of a $[_\pi 8_s + _\pi 10_s]$ cycloaddition:[96]

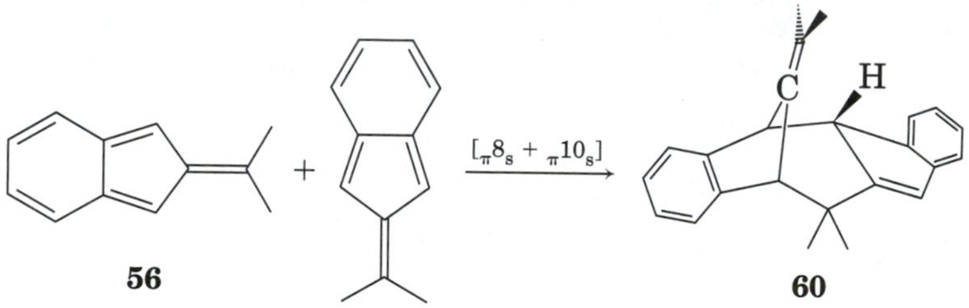

56 + **56** $\xrightarrow{[_\pi 8_s + _\pi 10_s]}$ **60** **(11.11)**

All of the cycloaddition reactions shown here have been examples of $[_\pi p_s + _\pi q_s]$ reactions. As with sigmatropic reactions, it is reasonable to ask whether cycloaddition reactions requiring antarafacial addition to π systems are sterically feasible. Clearly the $[p_a + q_a]$ process is possible for some values of p and q. One simply requires that the two π components be oriented at 90° with respect to each other so that a C_2 rotation axis is maintained from reactant through transition structure to product. The $[p_s + q_a]$ process is more difficult to envision, but a simple example illustrates the idea. Consider Figure 11.64(a) and Figure 11.64(b), which represent the side view and top view, respectively, of a system of two ethene units. If bonding changes occur as suggested by the dashed lines, the cycloaddition will be suprafacial with respect to the lower component and antarafacial with respect to the top component. A twisting about the carbon-carbon bond in the upper ethene must accompany the bonding change, so that substituents on the upper component that are initially cis to each other would become trans to each other in the cyclobutane product. This resulting stereochemistry is illustrated by the hypothetical $[_\pi 2_s + _\pi 2_a]$ cycloaddition of two *cis*-2-butene molecules (Figure 11.65).

---

[95]Paddon-Row, M. N.; Warrener, R. N. *Tetrahedron Lett.* **1974**, 3797.

[96]Warrener, R. N.; Paddon-Row, M. N.; Russell, R. A.; Watson, P. L. *Aust. J. Chem.* **1981**, *34*, 397.

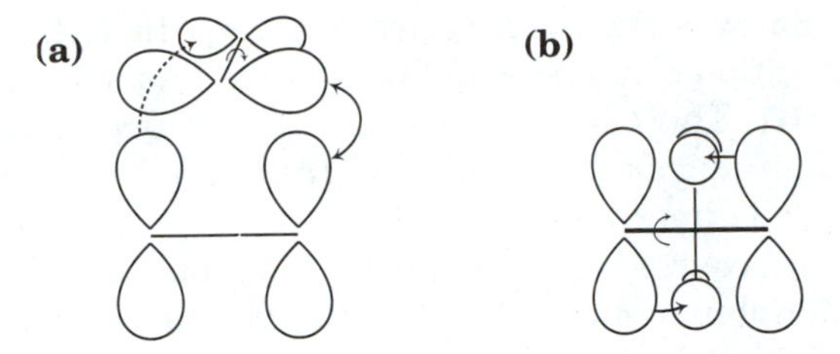

Figure 11.64 (a) Side and (b) top views of $[_{\pi}2_s + _{\pi}2_a]$ cycloaddition. (Adapted from reference 15.)

The dimerization of unstrained alkenes is not known to occur by the pathway shown in Figure 11.65, but there was a report that the dimerization of bicyclo[4.2.2]deca-*trans*-3,*cis*-7,9-triene (**61**, equation 11.12) produces as the major product **62**, in which the stereochemistry of the product is congruous with that predicted in Figure 11.65.[97] Two other (unidentified adducts were also formed in the reaction. However, Padwa and co-workers found the thermal dimerization of *cis,trans*-1,3-cyclooctadiene (**63**) to produce three dimers plus other products (equation 11.13).[98] Although the

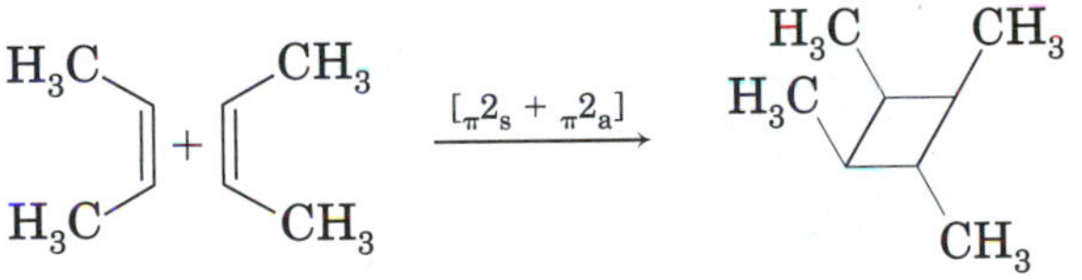

Figure 11.65 Hypothetical $[_{\pi}2_s + _{\pi}2_a]$ cycloaddition of *cis*-2-butene.

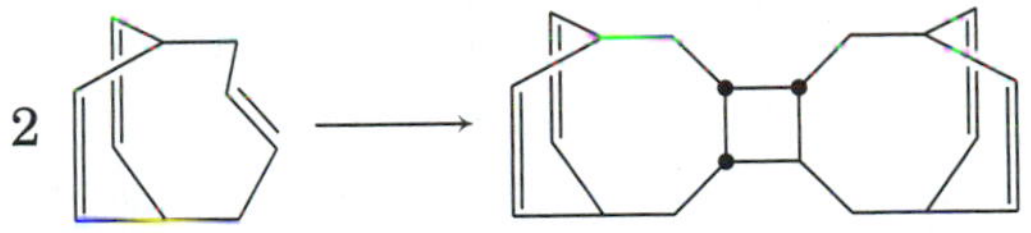

H_3C, CH_3, H_3C + CH_3 $\xrightarrow{[_{\pi}2_s + _{\pi}2_a]}$ H_3C, CH_3, H_3C, CH_3

2 **61** ⟶ **62** **(11.12)**

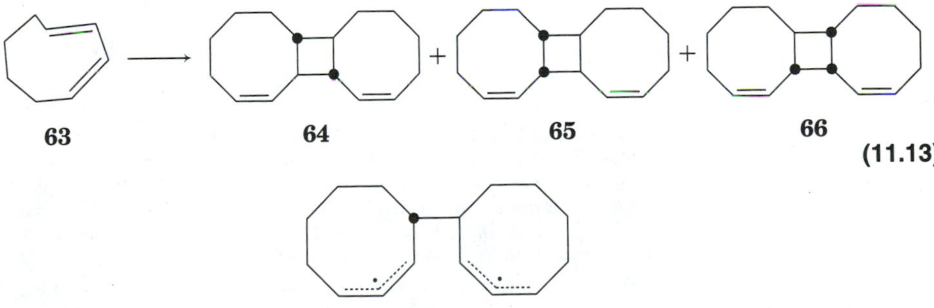

63 ⟶ **64** + **65** + **66** **(11.13)**

67

---

[97]Kraft, K.; Koltzenburg, G. *Tetrahedron Lett.* **1967**, 4357.

[98]Padwa, A.; Koehn, W.; Masaracchia, J.; Osborn, C. L.; Trecker, D. J. *J. Am. Chem. Soc.* **1971**, *93*, 3633.

major dimer, **66**, was the product predicted by the $[_{\pi}2_s + _{\pi}2_a]$ pathway, the product distribution could be explained on the basis of closure of a diradical intermediate, **67**. Therefore it appears that a concerted pathway is not required to explain formation of **62**, and there are no clear examples of concerted $[_{\pi}2_s + _{\pi}2_a]$ dimerization.[98,99,100]

Antarafacial concerted cycloaddition may be possible if one of the reactants has a nonplanar π system. Woodward and Hoffmann suggested that the addition of tetracyanoethylene to heptafulvalene, reported by Doering and co-workers, could occur by a $[_{\pi}14_a + _{\pi}2_s]$ pathway. The structure of the adduct, the stereochemistry of which was established by X-ray crystallography, confirms the antarafacial addition to the π system of heptafulvalene.[101]

$[_{\pi}14_a + _{\pi}2_s]$

68 **69** **70** **(11.14)**

---

[99]The formation of cyclobutanes from alkenes by radical pathways was established by the work of Bartlett and co-workers. For example, 1,1-dichloro-2,2-difluoroethene undergoes nonstereospecific cycloaddition with *trans,trans*-2,4-hexadiene at 100° to give an 82 : 18 mixture of diastereomeric cyclobutanes. The results were explained in terms of rotation about a carbon-carbon single bond in a diradical intermediate.

$F_2C{=}CCl_2$, 100°

82% 18%

Bartlett, P. D.; Montgomery, L. K.; Seidel, B. *J. Am. Chem. Soc.* **1964**, *86*, 617; Montgomery, L. K.; Schueller, K.; Bartlett, P. D. *J. Am. Chem. Soc.* **1964**, *86*, 622; Bartlett, P. D.; Montgomery, L. K. *J. Am. Chem. Soc.* **1964**, *86*, 628. Evidence for a very small amount (0.02%) of radical addition of butadiene to ethene to form vinylcyclobutane at 175° and 6000 psi was reported by Bartlett, P. D.; Schueller, K. E. *J. Am. Chem. Soc.* **1968**, *90*, 6071. For a review of the formation of cyclobutanes from thermal cycloaddition reactions, see Roberts, J. D.; Sharts, C. M. *Org. React.* **1962**, *12*, 1.

[100]There is also evidence that the tetramethylene diradical can be an intermediate in the thermal dimerization of two ethene molecules to cyclobutane, as well as in the reverse reaction. Pedersen, S.; Herek, J. L.; Zewail, A. H. *Science* **1994**, *266*, 1359, used femtosecond spectroscopy to measure the lifetime (700 fs) of the tetramethylene diradical formed by photochemical decarbonylation of cyclopentanone. For a discussion of this work, see Berson, J. A. *Science* **1994**, *266*, 1338.

[101]Doering, W. v. E. personal communication to Woodward, R. B.; Hoffmann, R. in reference 15, p. 85. However, the stereochemistry of addition to tetracyanoethylene is not distinguishable.

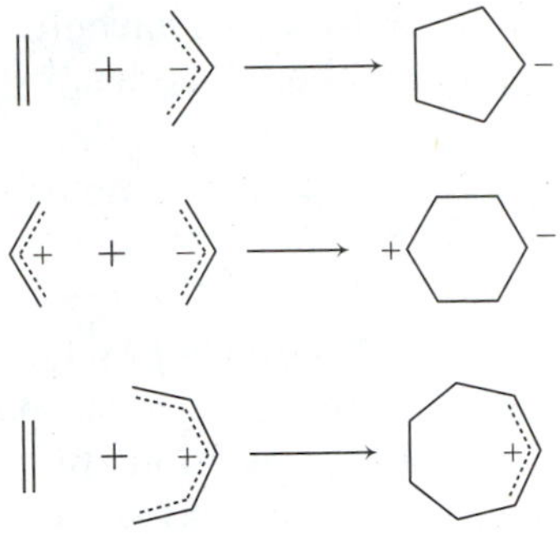

Figure 11.66 [4 + 2] cycloadditions. (Adapted from reference 15.)

The selection rule in Table 11.1 is developed for the number of *electrons* in the systems undergoing pericyclic change, not for the number of orbitals. Thus the addition of an allyl anion to an ethene, the addition of a pentadienyl cation to an ethene, and the addition of an allyl anion and an allyl cation are all [4 + 2] cycloadditions (Figure 11.66).[15,102]

We may further extend the analysis of concerted reactions as pericyclic processes by considering that a single *p* orbital, denoted by the symbol ω (omega), can be a participant in a pericyclic reaction. In this analysis, one lobe of the *p* orbital makes up the top face of a one-atom π system, while the other lobe makes up the bottom face. Thus the participation of a single *p* orbital is suprafacial if both cycloaddition processes involve only one of the two *p* orbital lobes, and it is antarafacial if the cycloaddition involves both. We may thus predict that the disrotatory opening of the cyclopropyl anion to an allyl anion (Figure 11.67) should take place via an $[_{\omega}2_a + _{\sigma}2_s]$ pathway. Conversely, the opening of the cation would be an $[_{\omega}0_s + _{\sigma}2_s]$ process, giving the opposite stereochemistry in the product. (Cycloadditions to σ bonds are described in more detail on page 769.)

An important type of cycloaddition reaction involving charges is known as a 1,3-dipolar cycloaddition. As illustrated by the schematic reaction in equation 11.15, a 1,3-dipole is a resonance-stabilized zwitterion that can

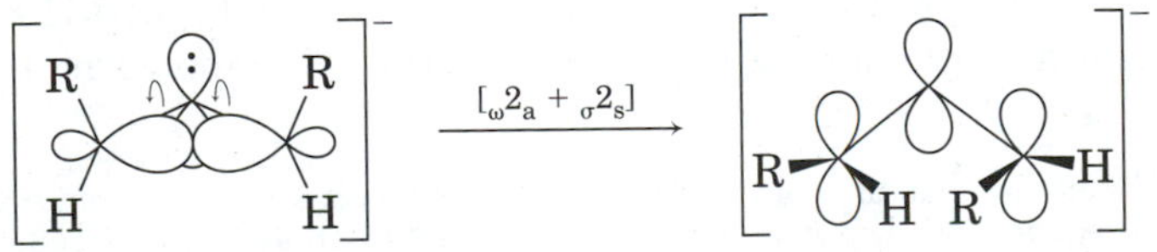

Figure 11.67 Pericyclic analysis of opening of cyclopropyl anion to allyl anion.

[102]For a discussion of cycloaddition reactions involving anions, see Staley, S. W. in *Pericyclic Reactions*, Vol. I; Marchand, A. P.; Lehr, R. E., eds.; Academic Press: New York, 1977; pp. 199–264; for a discussion of cycloaddition reactions of cations, see Sorensen, T. S.; Rauk, A. in *Pericyclic Reactions*, Vol. II; Marchand, A. P.; Lehr, R. E., eds.; Academic Press: New York, 1977; pp. 1–78.

add in a concerted fashion to an olefinic dipolarophile. Examples of 1,3-dipoles include ozone, nitrones, and carbonyl oxides.[85(d),103,104,105]

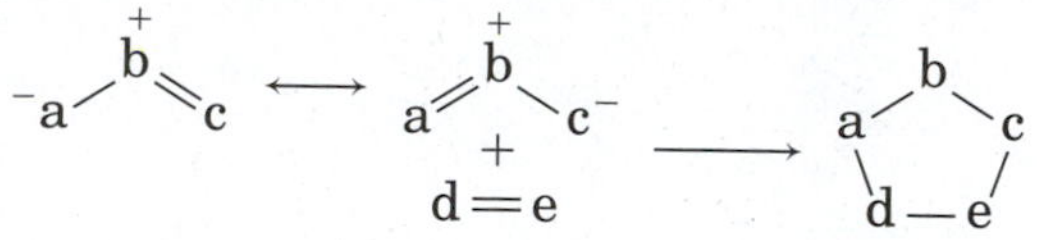

(11.15)

In a 1,3-dipolar cycloaddition, the 1,3-dipole provides four π electrons, so the allowed pathway is a $[_{\pi}4_s + {}_{\pi}2_s]$ process. The stereochemistry expected for this process has been observed in the reaction shown in equation 11.16. Electrocyclic opening of dimethyl 1-(4-methoxyphenyl)aziridine-2,3-dicarboxylate (**71**) produces the azomethine ylide **72**. Dimethyl acetylenedicarboxylate traps **72** by undergoing a concerted suprafacial-suprafacial 1,3-dipolar cycloaddition to form the *trans*-tetramethyl 1-(4-methoxyphenyl)-3-pyrroline-2,3,4,5-tetracarboxylate (**73**). The stereospecificity of the reaction was confirmed by the observation that the trans isomer of **71** produces the cis isomer of **73** under the same conditions.[106]

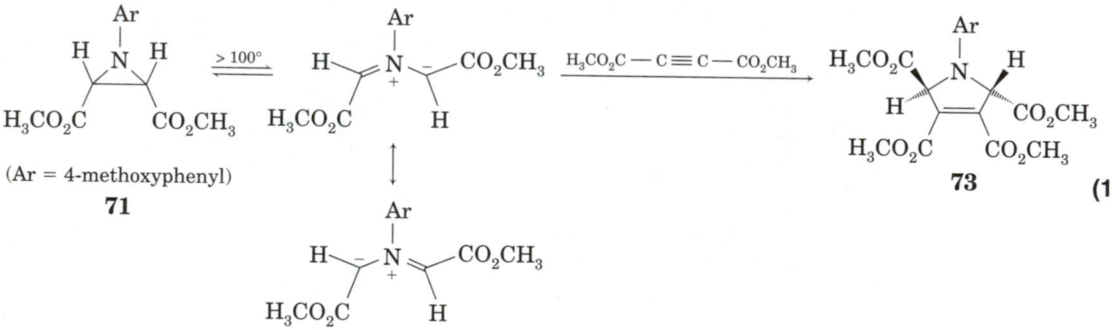

(Ar = 4-methoxyphenyl)
71

> 100°

$H_3CO_2C-C\equiv C-CO_2CH_3$

72

73

(11.16)

The Criegee mechanism for the ozonolysis of alkenes (illustrated in Figure 11.68) can be analyzed in terms of a series of three concerted reactions.[107] The addition of ozone to an alkene leads first to a 1,2,3-trioxacyclopentane structure known variously as an *initial ozonide, primary ozonide*, or *molozonide*. Ozone is a 4π electron system, so the 1,3-dipolar cycloaddition can occur through a $[_{\pi}4_s + {}_{\pi}2_s]$ cycloaddition. Figure 11.69 shows a molecular orbital correlation diagram for reaction of ozone with ethene.[108,109,110]

---

[103]For background information and a detailed discussion of the concerted pathway for this type of reaction, see Huisgen, R. *J. Org. Chem.* **1976**, *41*, 403 and references therein.

[104]See also Padwa, A., ed. *1,3-Dipolar Cycloaddition Chemistry*, Vols. 1 and 2; Wiley-Interscience: New York, 1984.

[105]For a discussion of 1,3-dipolar cycloadditions to alkynes, see Rutledge, T. F. *Acetylenes and Allenes. Addition, Cyclization, and Polymerization Reactions*; Reinhold Book Corporation: New York, 1969; pp. 253–260.

[106]Huisgen, R.; Scheer, W.; Huber, H. *J. Am. Chem. Soc.* **1967**, *89*, 1753.

[107]Criegee, R. *Rec. Chem. Prog.* **1957**, *18*, 111.

[108]Eckell, A.; Huisgen, R.; Sustmann, R.; Wallbillich, G.; Grashey, D.; Spindler, E. *Chem. Ber.* **1967**, *100*, 2192.

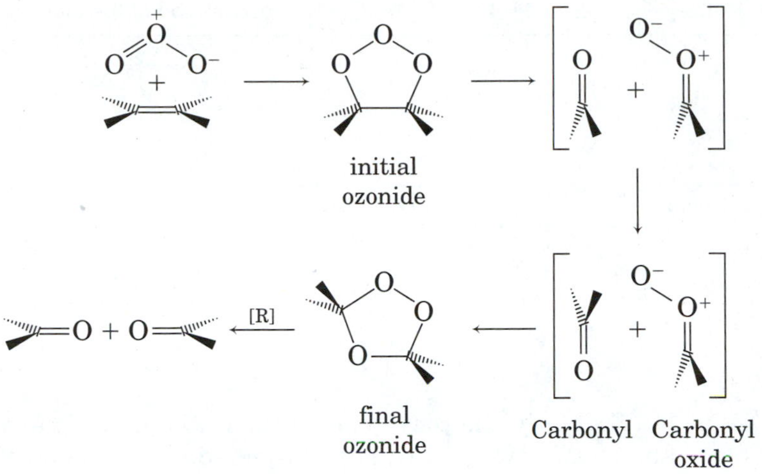

Figure 11.68
Criegee mechanism for ozonolysis of alkenes.

initial ozonide

final ozonide

Carbonyl

Carbonyl oxide

[R]

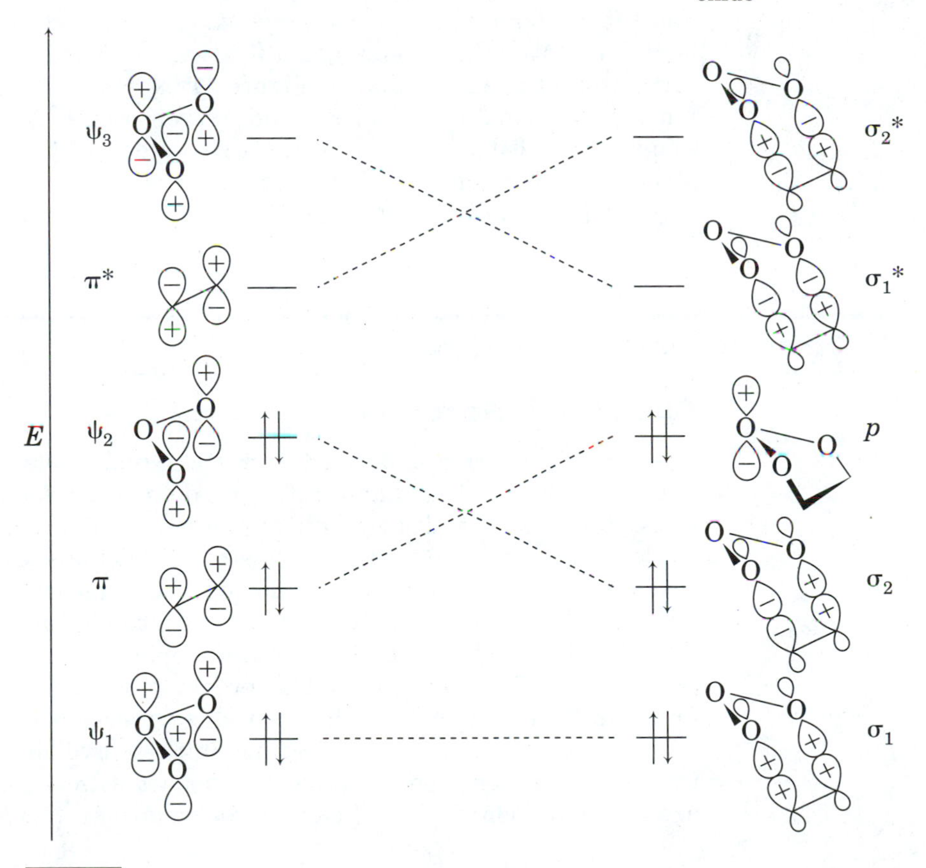

Figure 11.69
Molecular orbital correlation diagram for $[_{\pi}4_{s} + _{\pi}2_{s}]$ cycloaddition of ozone and ethene. (Adapted from reference 108.)

[109] Lattimer, R. P.; Kuczkowski, R. L.; Gillies, C. W. *J. Am. Chem. Soc.* **1974**, *96*, 348. This paper also discusses some alternate mechanisms proposed for the ozonolysis of alkenes.

[110] The cycloaddition can also be analyzed in terms of HOMO LUMO interactions (see page 772), with the interaction of the LUMO of ozone and the HOMO of the alkene being dominant. For a discussion, see Kuczkowski, R. L. in *1,3-Dipolar Cycloaddition Chemistry*, Vol. 2; Padwa, A., ed.; Wiley-Interscience: New York, 1984; pp. 197–276.

Figure 11.70 Retro-[$_{\pi}4_s$ + $_{\pi}2_s$] and [$_{\pi}4_s$ + $_{\pi}2_s$] reactions in conversion of the initial ozonide to final ozonide.

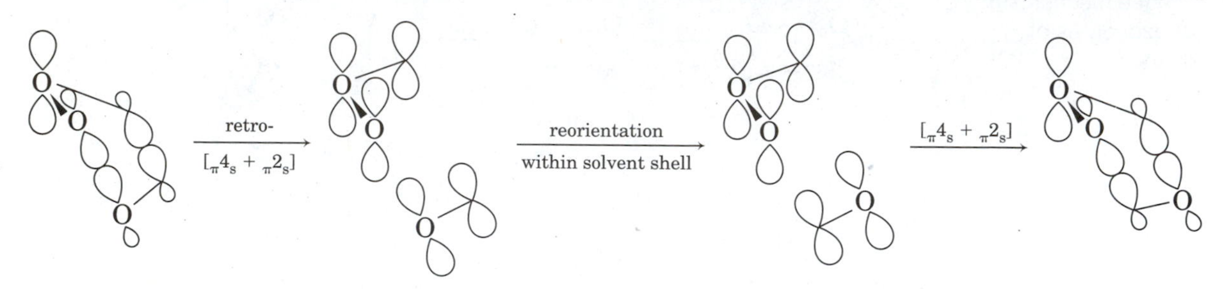

Similarly, the conversion of the initial ozonide to a carbonyl compound and a carbonyl oxide can be considered the reversion of a 1,3-cycloaddition, and the combination of the carbonyl oxide and the carbonyl compound can be viewed as another 1,3-cycloaddition (Figure 11.70). In both cases, an MO correlation diagram similar to Figure 11.69 can be constructed simply by moving positions of the carbon and oxygen atoms. Nonconcerted mechanisms are possible for each of the three steps in the Criegee ozonolysis mechanism, but Kuczkowski has summarized evidence suggesting that the reactions are concerted processes.[111]

11.5 Other Concerted Reactions

Cheletropic Reactions

The term *cheletropic* is derived from the Greek *chēlē* ("claw") and *tropos* ("turning").[112] Woodward and Hoffmann defined ***cheletropic reactions*** as those processes in which two σ bonds which terminate at a single atom are made, or broken, in concert.[113] Cheletropic reactions are also called *extrusion processes*, since a fragment appears to be squeezed out from a molecular system. Figure 11.71 illustrates a cheletropic reaction. The fragment C1—C*m* is a conjugated π system that separates from another molecular fragment, X. The reaction involves breaking two bonds, one from C1 to X and one from C*m* to X. Unless a redox process occurs simultaneously with the reaction, two of the electrons in these two bonds remain with the C1—C*m* π system and the other two depart with the group X. Examples are the extrusion of SO_2 illustrated in equations 11.17 and 11.18, which

[111]Kuczkowski, R. L. *Chem. Soc. Rev.* **1992**, *21*, 79 and references therein.

[112]Reference 44, p. 251.

[113]Reference 15, page 152.

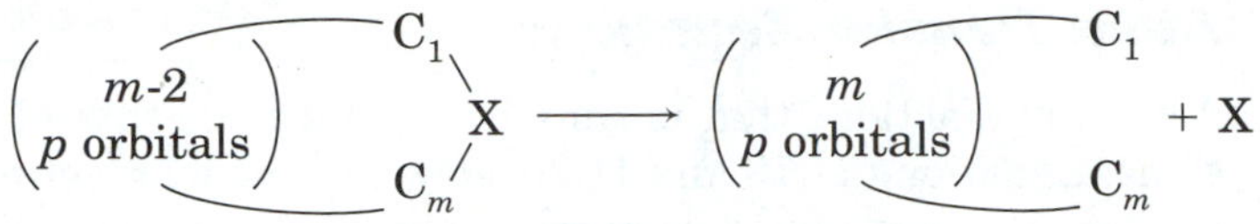

Figure 11.71 A generalized cheletropic reaction.

show the familiar reversal of stereochemistry as the π system is increased by two electrons.[114] The allowed pathway for the cheletropic reaction depends on the symmetry of the X orbitals with respect to the symmetry of the orbitals in the original molecule. Woodward and Hoffmann defined two types of pathways for such reactions—linear and nonlinear, depending upon whether rotation of one or the other fragment occurs during the reaction.[115] Selection rules for cheletropic reactions are given in Table 11.2, and a more detailed discussion can be found in reference 21.

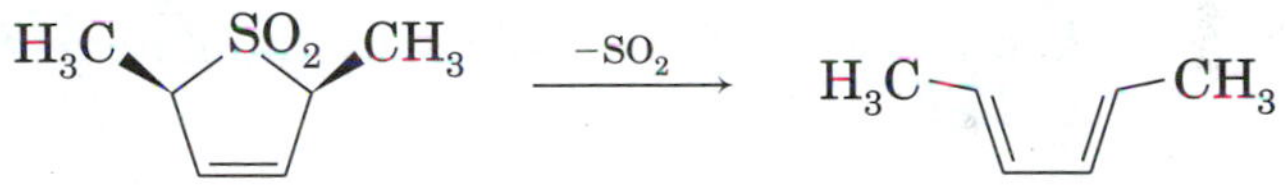

(11.17)

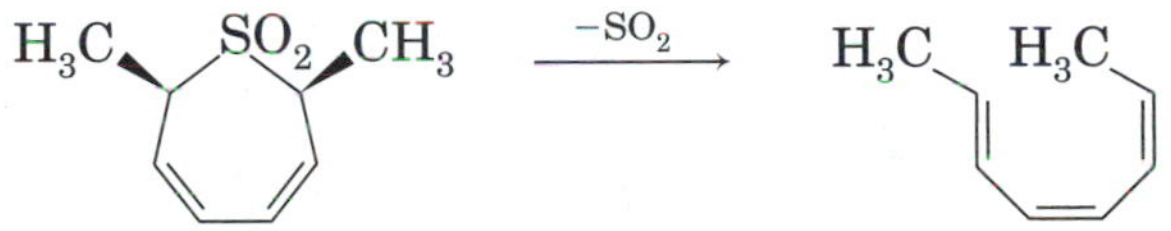

(11.18)

Table 11.2 Selection rules for cheletropic reactions.

	Allowed Thermal Reactions	
m	**Linear**	**Nonlinear**
$4n$	disrotatory	conrotatory
$4n + 2$	conrotatory	disrotatory

[114]Mock, W. L. in *Pericyclic Reactions*, Vol. II; Marchand, A. P.; Lehr, R. E., eds.; Academic Press: New York, 1977: pp. 141–179 and references therein.

[115]This simple statement greatly understates the detailed analysis necessary to understand the linear and nonlinear pathways. For a complete analysis, see reference 15, pp. 152 *ff*.

Atom Transfer Reactions

A class of reactions that is simpler to analyze is that of atom transfer reactions. Equations 11.19 and 11.20 show two model reactions, the transfer of a pair of hydrogen atoms from ethane to perdeuterioethene

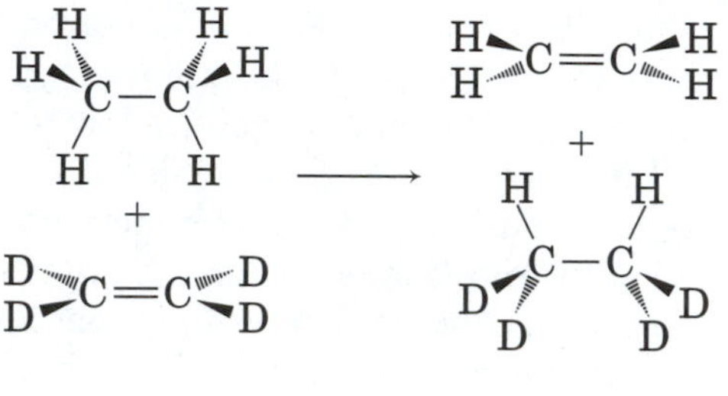

(11.19)

and the transfer of a pair of protium atoms from ethane to 1,1,4,4-tetradeuterio-1,3-butadiene.[74]

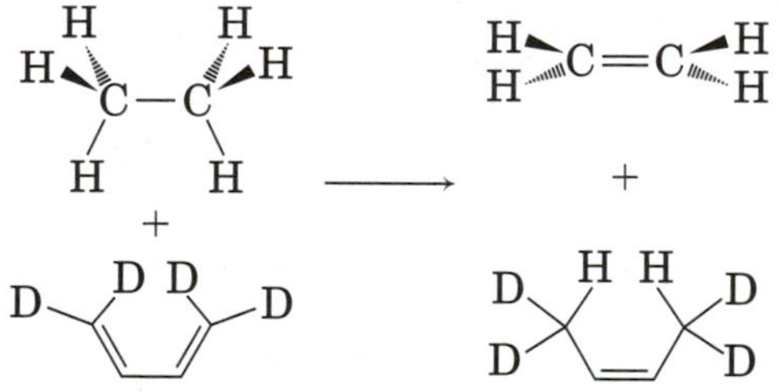

(11.20)

An MO correlation diagram for the reaction in equation 11.19 is shown in Figure 11.72, in which the MOs are categorized with respect to a plane of symmetry. As in the case of the Diels-Alder reaction, it is necessary to take combinations of the localized σ bond MOs to produce a set of MOs for the ethene component and then to classify these MOs as symmetric or antisymmetric with regard to the plane of symmetry. Therefore, σ_S represents a symmetric combination of two σ_{C-H} localized orbitals, while σ_A is the antisymmetric bonding combination. Since the entire set of bonding orbitals of the reactant correlates with the entire set of bonding orbitals of the product, the thermal atom transfer reaction is allowed by the principles of orbital symmetry. For the ethane plus butadiene case, however, the MO correlation diagram in Figure 11.73 (page 766) indicates that the ground state of the reactants correlates with a highly excited state of the products, so the thermal reaction should be forbidden.

Figure 11.74 (page 766) represents the general reaction for a synchronous transfer of two groups, one from each end of a system with m π centers

Figure 11.72
Orbital correlation diagram for ethane + ethene atom transfer reaction. (Adapted from reference 15.)

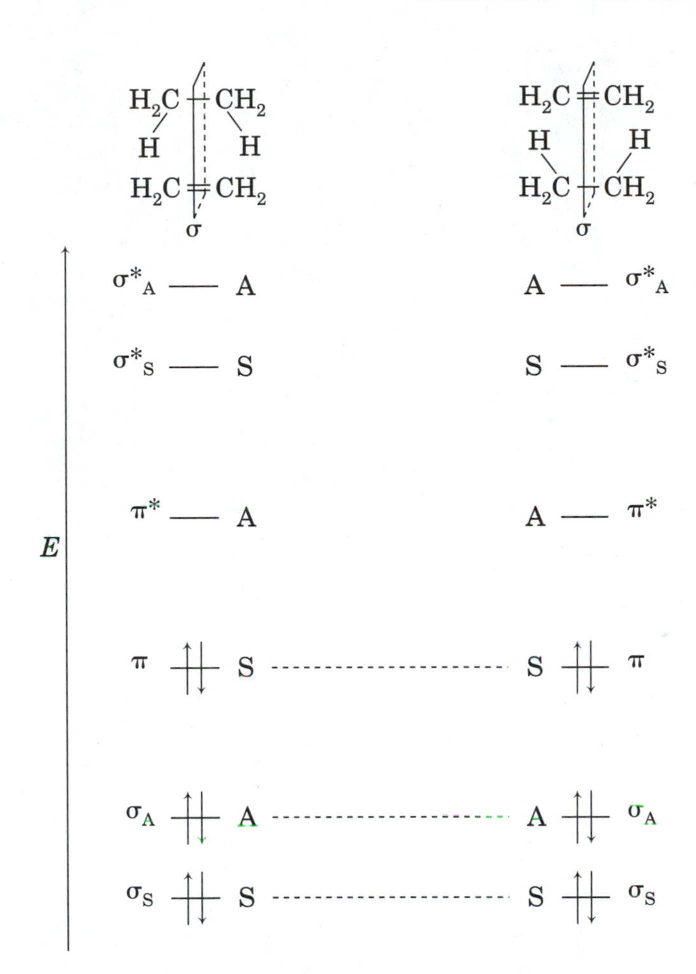

to a group with n π centers in a suprafacial process. In such a case the selection rules are as follows:[116]

- For suprafacial-suprafacial (or antarafacial-antarafacial) double group transfers, the transfer is thermally allowed when $p + q = 4n + 2$ and photochemically allowed when $p + q = 4n$.
- For suprafacial-antarafacial double group transfers, the selection rules are reversed.

An example is the transfer of hydrogen atoms from *cis*-9,10-dihydronaphthalene (**78**) to 1,2-dimethylcyclohexene (**79**) to give naphthalene and *cis*-1,2-dimethylcyclohexane, where p is 0 and q is 2 (Figure 11.75, page 766).[117] The analogous intramolecular rearrangement is known as a **dyotropic reaction.**[118]

[116]Adapted from reference 15.

[117]Doering, W. v. E.; Rosenthal, J. W. *J. Am. Chem. Soc.* **1967**, *89*, 4534.

[118]For leading references and a theoretical study, see Frontera, A.; Suñer, G. A.; Deyà, P. M. *J. Org. Chem.* **1992**, *57*, 6731.

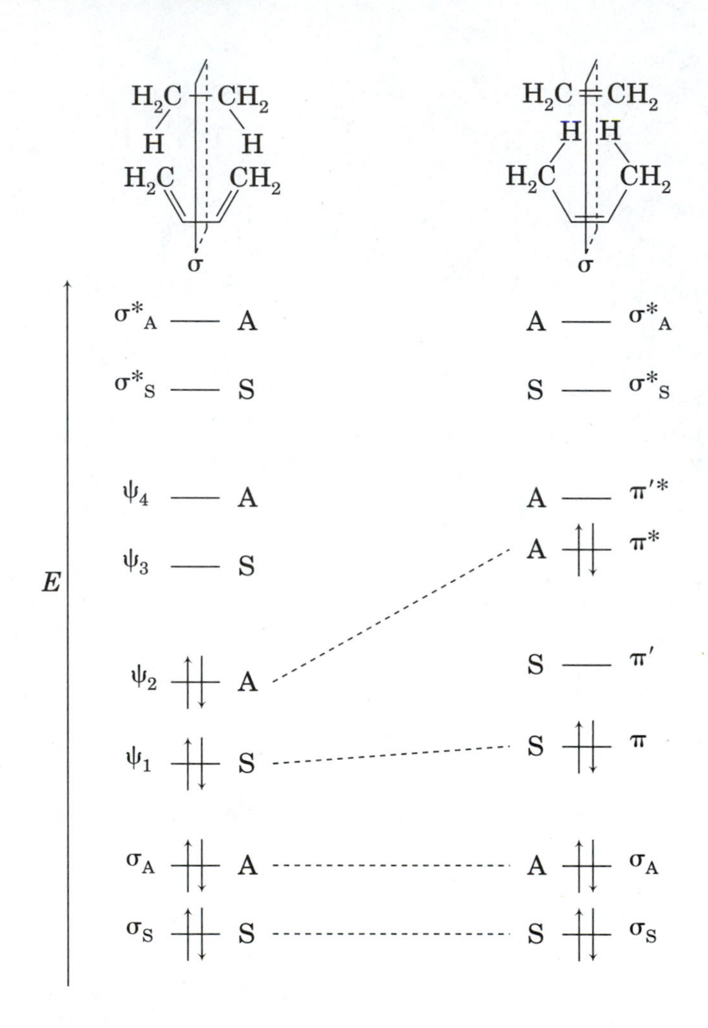

Figure 11.73 Orbital correlation diagram for ethane + butadiene atom transfer reaction. (Adapted from reference 15.)

Figure 11.74 Generalized reaction for synchronous suprafacial-suprafacial double atom transfer. (Adapted from reference 15.)

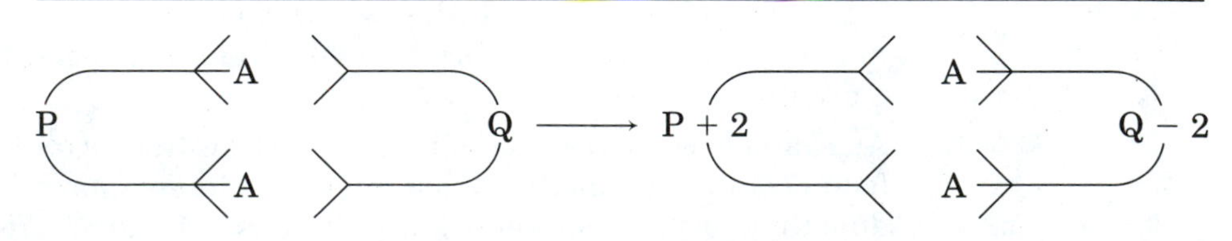

Figure 11.75 A synchronous H_2 transfer reaction.

H H H_3C H_3C CH_3 H H CH_3

78 **79** **80** **81**

Figure 11.76
Generalized ene reaction.

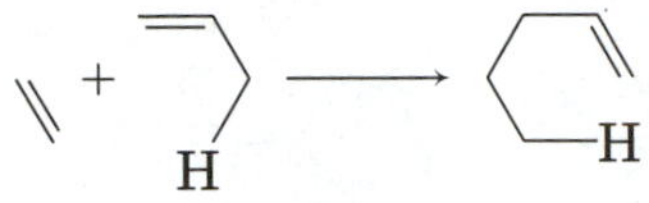

Ene Reactions

The **ene reaction**[119] involves the substituting addition of a compound with a double bond and an allylic hydrogen to a compound with a multiple bond in such a way that the allylic hydrogen is transferred from one atom to another, as shown by the general reaction in Figure 11.76.[120,121,122] A transition structure calculated at the RHF/3-21G level for the ene reaction between ethene and propene is shown in Figure 11.77.[123]

Typically the ene reaction involves an alkene substituted with electron withdrawing groups.[124] A specific example is the reaction between propene

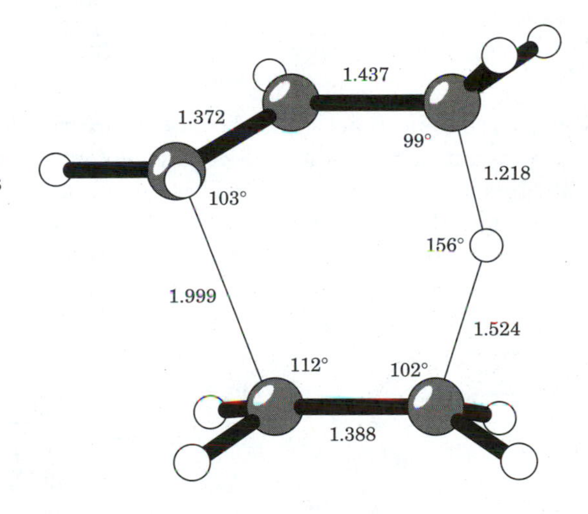

Figure 11.77
RHF/3-21G transition structure for the ene reaction of ethene and propene. (Distances are in Angstroms, angles in degrees. reproduced from Reference 21.)

---

[119]Alder, K.; Pascher, F.; Schmitz, A. *Chem. Ber.* **1943**, *76B*, 27.

[120]The reaction illustrated in Figure 11.76 suggests a transition structure similar to those proposed for the Cope rearrangement and other $4n + 2$ reactions. The possibility that this reaction might proceed through either a concerted mechanism or a step-wise mechanism has been considered. See, for example, Berson, J. A.; Wall, R. G.; Perlmutter, H. D. *J. Am. Chem. Soc.* **1966**, *88*, 187 and references therein. Also see (a) Keung, E. C.; Alper, H. *J. Chem. Educ.* **1972**, *49*, 97; (b) Nahm, S. H.; Cheng, H. N. *J. Org. Chem.* **1986**, *51*, 5093.

[121]For applications of the ene reaction in organic synthesis, see Hoffmann, H. M. R. *Angew. Chem., Int. Ed. Engl.* **1969**, *8*, 556; Oppolzer, W.; Snieckus, V. *Angew. Chem.* **1978**, *90*, 506. For a discussion of the ene reaction with organometallic reagents, see Dubac, J.; Laporterie, A. *Chem. Rev.* **1987**, *87*, 319.

[122]The application of computer-assisted mechanistic evaluation of the ene and retroene reactions to organic synthesis has been reported by Paderes, G. D.; Jorgensen, W. L. *J. Org. Chem.* **1992**, *57*, 1904.

[123]Loncharich, R. J.; Houk, K. N. *J. Am. Chem. Soc.* **1987**, *109*, 6947.

[124]For a review of applications of the asymmetric ene reaction in organic synthesis, see Mikami, K.; Shimizu, M. *Chem. Rev.* **1992**, *92*, 1021.

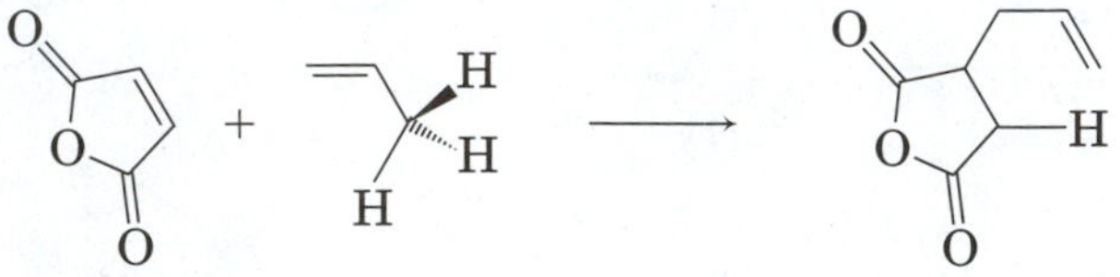

Figure 11.78
Ene reaction of propene and maleic anhydride.

82 83 84

and maleic anhydride to give allylsuccinic anhydride (**84**) in Figure 11.78.[119] Electron withdrawing groups are not always required, however, as shown by the dimerization of cyclopropene to give 3-cyclopropylpropene (equation 11.21).[125]

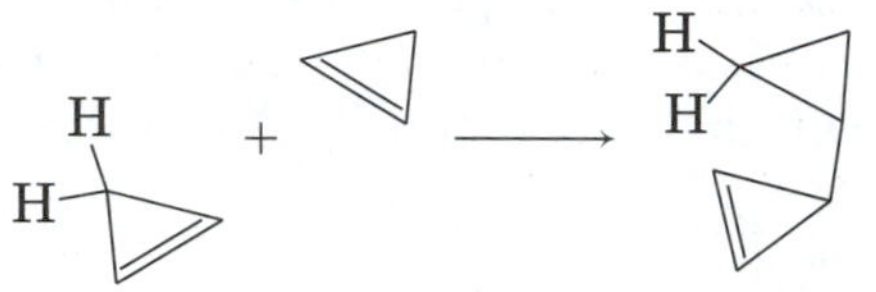

(11.21)

85 86

The reverse of the ene reaction is the **retroene reaction**. Figure 11.79 shows the conversion of *cis*-1-methyl-2-vinylcyclopropane (**87**) to *cis*-1,4-hexadiene (**88**) by such a process.[126,127] This reaction can also be described as a homo-[1,5] sigmatropic shift,[128] so the distinction between a sigmatropic reaction and an intramolecular ene reaction is a matter of definition.

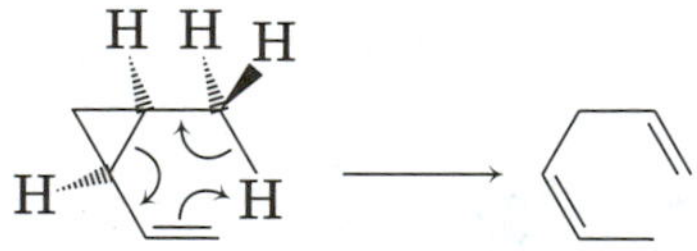

Figure 11.79
A retroene reaction.

87 88

11.6 A Generalized Selection Rule for Pericyclic Reactions

We have developed a different set of selection rules for electrocyclic, sigmatropic, cycloaddition, and other concerted reactions. However, the fundamental principle—the conservation of orbital symmetry—is the same in all cases. Now we will see that all of these reaction types can be considered to

[125]Dowd, P.; Gold, A. *Tetrahedron Lett.* **1969**, 85. Also see Baird, M. S.; Hussain, H. H.; Clegg, W. *J. Chem. Res. (S)* **1988**, 110.

[126]Roth, W. R.; König, J. *Liebigs Ann. Chem.* **1965**, *688*, 28.

[127]This reaction has been characterized as being one of the few thermal reactions that satisfy both kinetic and stereochemical criteria for concerted reactions: Berson, J. A. *Acc. Chem. Res.* **1972**, *5*, 406.

[128]Reference 15, p. 132.

be variants of cycloaddition reactions. To do so, we must note that a σ bond can participate in a cycloaddition process, just as can a π bond, with the following provision:[129]

> The participation of a σ bond is suprafacial if the cycloaddition occurs on the inside (overlapping) lobes of the two sp^3-hybrid orbitals or on their outside lobes.
> The σ bond cycloadds antarafacially if one cycloaddition occurs on an inside lobe and one occurs on an outside lobe.

To understand this definition, it is useful to depict cycloaddition reactions with dashed lines that represent bonding interactions that will be present in the ground state of the *product* of the reaction.[130] The effect of the cycloaddition shown for the $[{}_{\sigma}2_s + {}_{\pi}2_a]$ process in Figure 11.80 is to pull up the inner lobe of the sp^3-hybrid orbital on C3 so that it overlaps with the upper lobe of the p orbital on C2 and to pull down the inner lobe of the sp^3-hybrid orbital on C4 so that it overlaps with the lower lobe of the p orbital on C1. This results in the conrotatory process that was predicted earlier for the same reaction. The same reaction can also be described as a $[{}_{\sigma}2_a + {}_{\pi}2_s]$ cycloaddition, as shown on the bottom of Figure 11.80. Both processes are thermally allowed cycloaddition reactions.

It is not necessary to actually draw the p and sp^3-orbital lobes to describe electrocyclic reactions as cycloadditions. Instead, we may use lines to represent bonds and then mentally picture the (localized) orbitals used to construct those bonds. In these cases, arrowheads on the ends of the dashed lines clarify the orbital interactions.[130] Thus, Figure 11.81 is the equivalent of Figure 11.80.

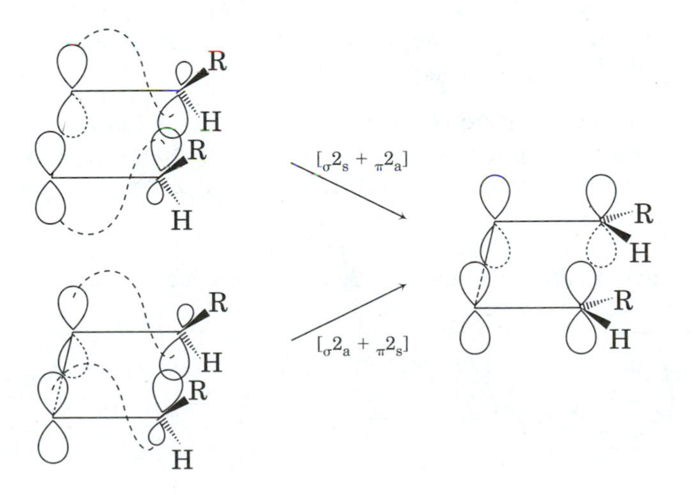

Figure 11.80 (Top) $[{}_{\sigma}2_s + {}_{\pi}2_a]$ and (bottom) $[{}_{\sigma}2_a + {}_{\pi}2_s]$ descriptions of cyclobutene ring opening.

---

[129]Reference 15, p. 70.

[130]Reference 15, p. 170.

Figure 11.81 Alternative representations for electrocyclic reactions considered as cycloadditions.

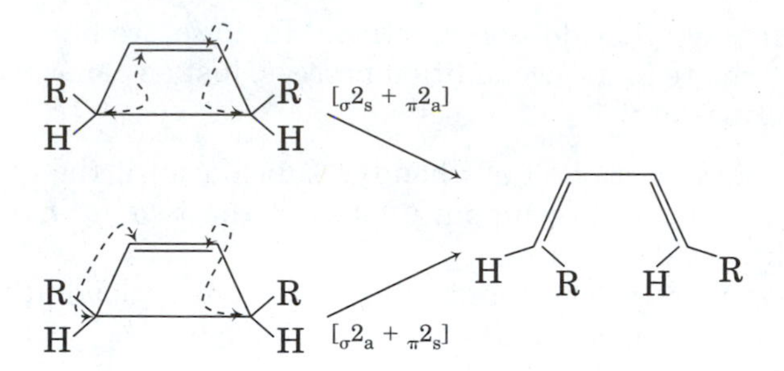

Similarly, a sigmatropic reaction can be described as a cycloaddition. For example, the antarafacial [1,3] hydrogen shift of propene in Figure 11.82 is seen to be equivalent to a $[_{\sigma}2_s + {_\pi}2_a]$ cycloaddition.

Figures 11.80 through 11.82 reemphasize the definition of pericyclic reactions (page 714), which refers to a cyclic array of continuously bonded atoms. The term *cyclic array* means that we may draw a closed curve through a reacting system of π systems (including isolated p orbitals) and σ bonds. Figure 11.83(a) is a redrawing of Figure 11.82 in which the closed curve is indicated by a continuous dashed line. Figure 11.83(b) and Figure 11.83(c) illustrate the electrocyclic opening of cyclobutene and the sigmatropic Cope rearrangement in similar fashion.

The selection rules for cycloadditions (Table 11.1, page 755) apply to pericyclic reactions described as cycloadditions. However, the rule may be generalized in concise form as follows:[131]

> A ground state pericyclic change is symmetry allowed when the total number of $(4n + 2)_s$ and $(4r)_a$ components is odd, where n and r are integers. If the total is even, the thermal reaction is forbidden, and the photochemical reaction is allowed. Conversely, if the total is odd, the photochemical reaction is forbidden.

Let us examine the application of this generalized selection rule to the various types of concerted reactions. The opening of cyclobutene to butadiene by the conrotatory pathway (Figure 11.80) was defined to be a $[_{\pi}2_s +$

Figure 11.82 Antarafacial [1,3] hydrogen shift of propene classified as a $[_{\sigma}2_s + {_\pi}2_a]$ cycloaddition reaction.

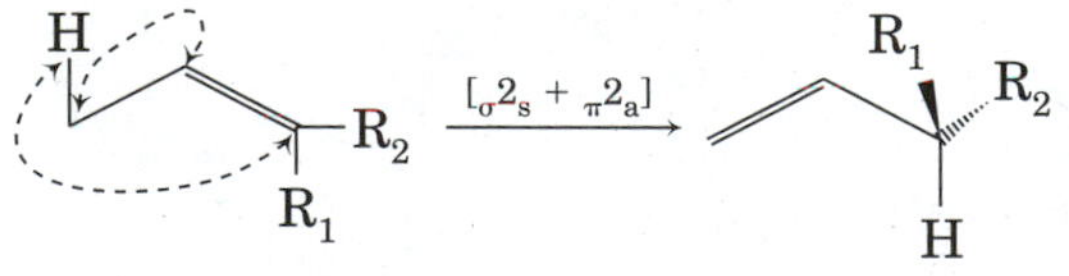

H R₂ R₁ $[_{\sigma}2_s + {_\pi}2_a]$ R₁ R₂ H

[131]Adapted from reference 15, pp. 170 *ff.*

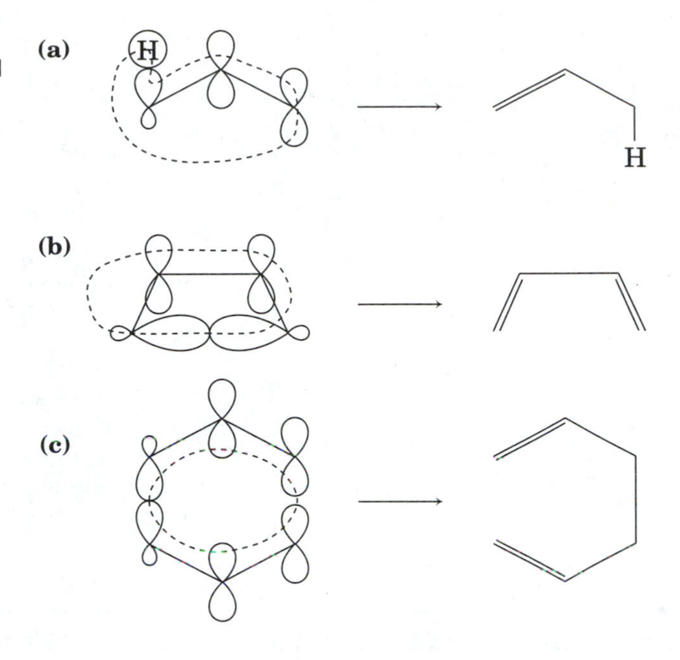

Figure 11.83 Illustration of closed curves in pericyclic reactions: (a) [1,3] hydrogen shift, (b) electrocyclic reaction, and (c) sigmatropic reaction.

${}_{\sigma}2_{a}$] cycloaddition. Therefore the number of $(4n + 2)_s$ components is 1 (because of the one ${}_{\pi}2_s$ component), and the number of $(4r)_a$ components is 0. Then the total of $(4n + 2)_s + (4r)_a$ components is $1 + 0 = 1$, which is odd, so the reaction is allowed. Similarly, the hypothetical $[{}_{\pi}2_s + {}_{\pi}2_a]$ dimerization of ethene (Figure 11.53) is allowed because the number of $(4n + 2)_s$ components is 1, the number of $(4r)_a$ components is 0, and the total is 1.

It should be emphasized that the application of the generalized selection rule does not require analyzing a concerted reaction in any particular way. All descriptions of a pericyclic reaction that predict the same stereochemistry in the product should lead to the same conclusion. For example, three descriptions of the Diels-Alder reaction are given in Figure 11.84. In Figure 11.84(a), the reaction is described as a $[{}_{\pi}2_s + {}_{\pi}4_s]$ cycloaddition. In this process there is one $(4n + 2)_s$ component and no $(4r)_a$ component, so the total (1) is an odd number, and the reaction is allowed. In Figure 11.84(b), each double bond of the diene is considered to be a separate unit, so the reaction is described as a $[{}_{\pi}2_s + {}_{\pi}2_s + {}_{\pi}2_s]$ process. That pathway has three $(4n + 2)_s$ components and no $(4r)_a$ component, so the total (3) is odd, and the reaction is allowed. Similarly, the $[{}_{\pi}2_s + {}_{\pi}2_a + {}_{\pi}2_a]$ cycloaddition in Figure 11.84(c) is also allowed.[132]

[132]The $[{}_{\pi}4_a + {}_{\pi}2_a]$ addition is allowed as a pericyclic reaction, but that would give a result stereochemically different from the Diels-Alder reaction.

Figure 11.84 Different analyses of the Diels-Alder reaction as a pericyclic reaction.

(a)

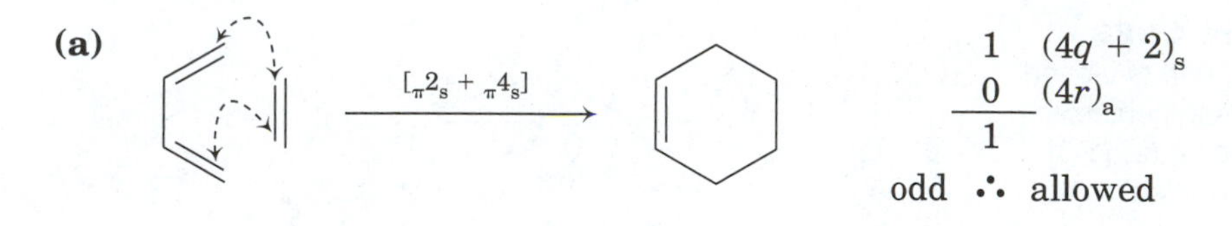

(b)

$[_{\pi}2_s + _{\pi}2_s + _{\pi}2_s]$

3 $(4q + 2)_s$
0 $(4r)_a$
3
odd ∴ allowed

(c)

$[_{\pi}2_s + _{\pi}2_a + _{\pi}2_a]$

1 $(4q + 2)_s$
0 $(4r)_a$
1
odd ∴ allowed

11.7 Alternative Conceptual Models for Concerted Reactions

In discussing concerted reactions, we first developed for each type of reaction a set of selection rules based on minimal molecular orbital models. We then proceeded to unify all of the reactions under one heading, pericyclic reactions, and to write one selection rule that includes all of the cases we have studied. As a result, we have combined several simple models, each of which was comprehensible by itself, into a general model that has lost much of its conceptual simplicity. It is easy to visualize the rotation of orbitals of the HOMO of a system undergoing electrocyclic closure (Figure 11.7). It is more difficult to visualize the theoretical significance of the total number of $(4n + 2)_s$ and $(4r)_a$ components. The generalized selection rule can be shown to be correct, but it does not lend itself to an intuitive sense of the physical phenomena involved. Additional insight into the nature of concerted reactions can be obtained from an examination of some of the complementary models that have been advanced to account for the same phenomena. Although all offer the same predictions, each model provides a different perspective for the basis of those predictions.

Frontier Molecular Orbital Theory

Fukui described pericyclic reactions in terms of frontier molecular orbital (FMO) theory.[133,134] If the interaction of the HOMO of one reactant with the

[133]Fukui, K. *Acc. Chem. Res.* **1971**, *4*, 57 and references therein. Fukui shared the 1981 Nobel Prize in chemistry with Roald Hoffmann. For a summary of the work that lead to the Nobel Prize, see reference 79.

[134]Salem, L. *J. Am. Chem. Soc.* **1968**, *90*, 543, 553, has presented a detailed PMO analysis of the interactions between two conjugated molecules that reaffirms the FMO predictions.

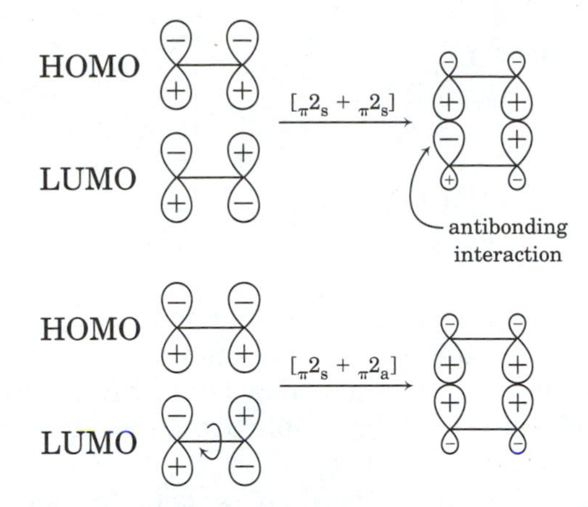

Figure 11.85
FMO analysis of the dimerization of two ethenes.

LUMO of the other reactant leads to bonding interactions, then the pericyclic reaction is said to be allowed (i.e., there should be a low energy pathway for the reaction). Conversely, if the HOMO-LUMO interaction leads to an antibonding interaction between atoms that must be bonded in the product, then the reaction is forbidden.[135]

Figure 11.85 shows an FMO analysis for two pathways for the dimerization of ethene. The HOMOs of the two molecules are degenerate, as are the LUMOs, so it does not matter which ψ_1 is used for HOMO and which ψ_2 is used as the LUMO. In Figure 11.85(a) the $[_\pi 2_s + _\pi 2_s]$ cycloaddition leads to an antibonding relationship between two atoms in the product. In Figure 11.85(b) the $[_\pi 2_s + _\pi 2_a]$ cycloaddition generates only bonding interactions in the product, so that process is allowed.

The FMO approach can be used to evaluate the two possible pathways for the opening of cyclobutene to butadiene (Figure 11.86). Using the LUMO of the σ bond and the HOMO of the π bond, it is apparent that the conrotatory opening leads to a bonding interaction between C1 and C2 as well as to a bonding relationship between C3 and C4. However, the disrotatory opening leads to an antibonding relationship between two of the atoms, so the

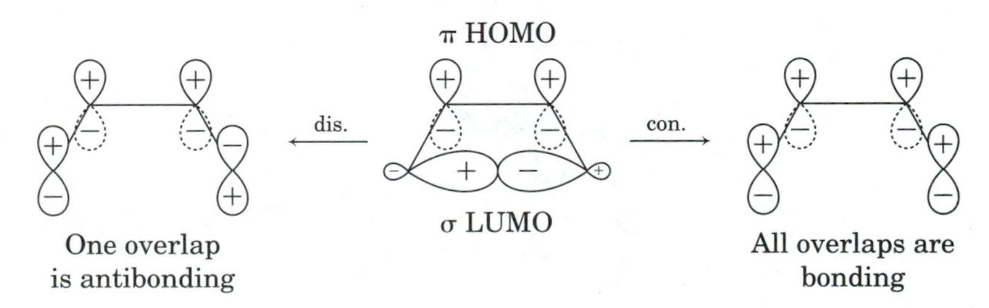

Figure 11.86
FMO analysis of the opening of cyclobutene to butadiene by disrotatory (left) and conrotatory (right) pathways.

---

[135]For a discussion of pericyclic reactions in terms of frontier MO theory, see Houk, K. N. in *Pericyclic Reactions*, Vol. II; Marchand, A. P.; Lehr, R. E., eds.; Academic Press. New York, 1977; pp. 181–271.

Figure 11.87 FMO analysis of the ene reaction. (Adapted from reference 122.)

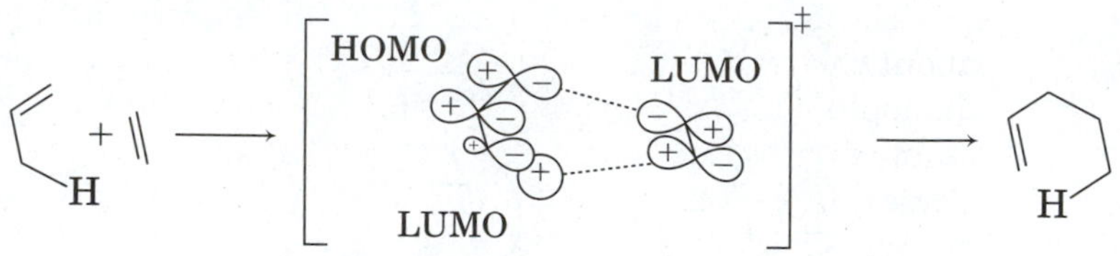

reaction is forbidden. Analysis of the reaction using the HOMO of the σ bond and the LUMO of the π bond leads to the same conclusion.[136] Figure 11.87 shows that the ene reaction can be described as the result of an interaction of the HOMO of the allylic double bond, the LUMO of the allylic C—H bond, and the LUMO of the eneophile.[137]

An FMO analysis of the Diels-Alder reaction is shown in Figure 11.88. Here there is a choice as to which reactant's HOMO and which reactant's LUMO are used because the energy levels in the two reactants are different. However, the result is the same no matter which combination is taken. Figure 11.88(a) shows that the combination of the HOMO of butadiene (ψ_2) with the LUMO of ethene (π^*) leads to bonding interactions in the new σ bonds, as does the interaction of the LUMO of butadiene (ψ_3) with the HOMO of ethene (π).

Figure 11.88 Two FMO analyses of the Diels-Alder reaction.

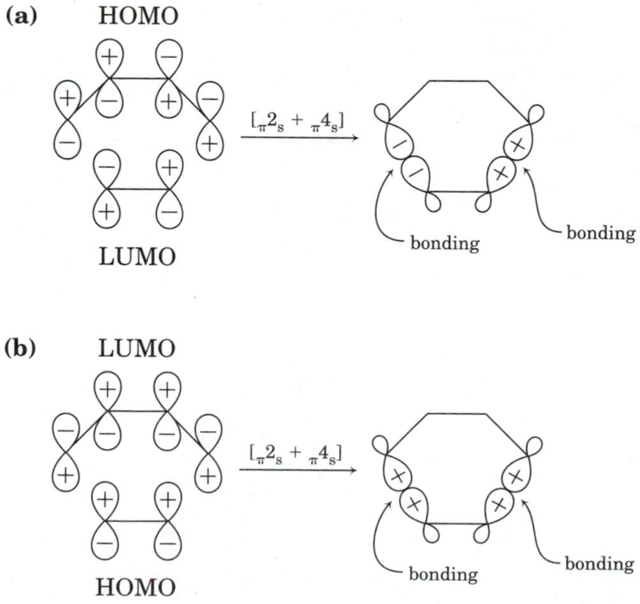

[136]For this analysis the σ bond and the π bond are considered to be two independent units, each with its HOMO and LUMO, even though they are in the same molecule.

[137]Inagaki, S.; Fujimoto, H.; Fukui, K. *J. Am. Chem. Soc.* **1976**, *98*, 4693.

The FMO method is also valuable in rationalizing two observations about the Diels-Alder reaction. The first is that the product of addition of a dienophile to a cyclic diene is usually the endo product rather than the more stable exo product, although varying yields of exo product can be formed.[138] For example, the cycloaddition of cyclopentadiene and cyclopentene gives both the endo product, **90**, and the exo product, **91** (equation 11.22).[139] As shown by the data in Table 11.3, formation of the endo product is favored under conditions of lower temperature and shorter reaction time, while the exo product is formed in higher percent yield at higher temperature and longer reaction time. Thus, **90** is the product of rate (kinetic) control, meaning that $\Delta G^{\ddagger}$ is lower for formation of **90** than for formation of **91**. However, **91** is the product of equilibrium control, meaning that it is lower in free energy than is **90**.[84]

Δ H H H H **(11.22)**

53 **89** **90** **91**

Woodward and Hoffmann proposed that the lower activation energies often observed for endo cycloaddition result from additional stabilization due to attractive interaction between *secondary orbitals* (i.e., lobes of *p* orbitals that are not converted to sp^3 orbitals by the reaction).[140] Figure 11.89 shows such an interaction for both types of HOMO-LUMO interactions in the endo dimerization of 1,3-butadiene. HOMO-LUMO interactions cannot be the only factor favoring endo product formation in the Diels-Alder reaction, however, because cyclopentene has no *p* orbitals other than those in the reacting alkene. Fox and co-workers proposed that steric destabilization of the exo transition structure can account for the preferential formation of

Table 11.3 Product distribution in Diels-Alder reaction of cyclopentadiene with cyclopentene. (Data from reference 139.)

Reaction Temperature	Reaction Time	% 90	% 91
198°	5.3 hours	97	3
200°	11 hours	81	9
300°	5 hours	46	50
300°	40 hours	29	72

[138] Alder, K.; Stein, G. *Angew. Chem.* **1937**, *50*, 510.

[139] Cristol, S. J.; Seifert, W. K.; Soloway, S. B. *J. Am. Chem. Soc.* **1960**, *82*, 2351

[140] Hoffmann, R.; Woodward, R. B. *J. Am. Chem. Soc.* **1965**, *87*, 4388.

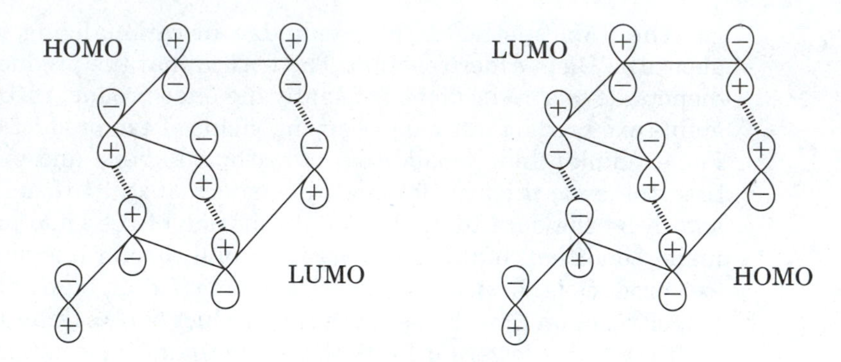

Figure 11.89 Secondary orbital overlap in the endo transition structure for the Diels-Alder dimerization of 1,3-butadiene.

90 in this case, and they have suggested that similar effects could be important for other dienes having sterically interactive substituents.[141]

The two HOMO-LUMO interactions shown in Figure 11.89 are equal in energy and are expected to be comparable in significance. More commonly the ${}_{\pi}2_s$ component of the Diels-Alder reaction is substituted with electron withdrawing groups (such as cyano or carbonyl), and the rate of the reaction is much faster with such dienophiles. In these cases it is expected that the dominant FMO interaction is between the LUMO of the olefin and the HOMO of the diene (Figure 11.90), since the energy difference between these two orbitals is usually much less than the energy difference between the diene LUMO and the dienophile HOMO. That expectation is consistent

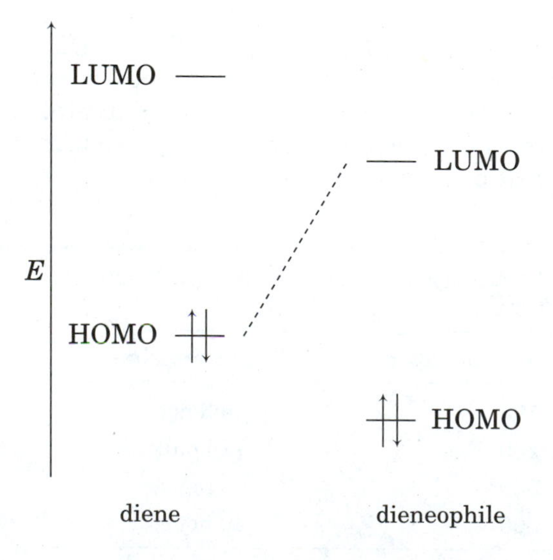

Figure 11.90 FMO interaction of diene HOMO and dienophile LUMO.

[141](a) Fox, M. A.; Cardona, R.; Kiwiet, N. J. *J. Org. Chem.* **1987**, *52*, 1469; (b) However, theoretical evidence for a dominant role of secondary orbital interactions in the stereochemistry of the Diels-Alder reaction of cyclopropene with a series of substituted 1,3-butadienes was reported by Apeloig, Y.; Matzner, E. *J. Am. Chem. Soc.* **1995**, *117*, 5375.

with the observation that, for many systems, the log of the rate of the Diels-Alder reaction correlates well with the inverse of the energy difference between the ionization potential of the diene and the electron affinity of the olefin (parameters that are expected to be related to the diene HOMO and olefin LUMO, respectively).[135] In addition, the rate of the Diels-Alder reaction is dramatically enhanced by the complexation of the dienophile with Lewis acid catalysts, which lowers the LUMO energy of the dienophile.[142,143] Diels-Alder reactions also occur readily in chain reactions involving dienophile radical cations.[144]

Another aspect of the Diels-Alder reaction that can be rationalized with FMO theory is the regiochemistry observed with unsymmetrical dienes or alkenes. Equation 11.23 shows the product distribution from the cycloaddition of 2-phenyl-1,3-butadiene and methyl acrylate. The "para" product (named by analogy with aromatic compounds) is the major product.[145] Both products can be formed by overlap of the HOMO of the diene with the LUMO of the olefin. However, the product distribution is predicted by the

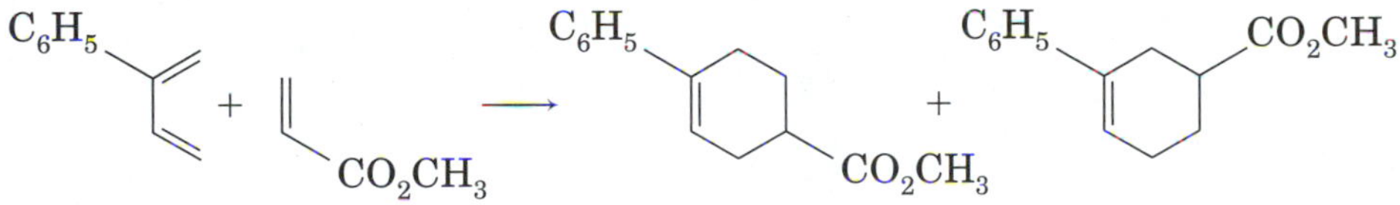

C_6H_5 + CO_2CH_3 ⟶ C_6H_5 CO_2CH_3 + C_6H_5 CO_2CH_3 **(11.23)**

para 80% meta 20%

---

[142](a) Yates, P.; Eaton, P. *J. Am. Chem. Soc.* **1960**, *82*, 4436; (b) For a discussion, see Guner, O. F.; Ottenbrite, R. M.; Shillady, D. D.; Alston, P. V. *J. Org. Chem.* **1987**, *52*, 391.

[143]The rates of Diels-Alder reactions are also influenced by solvent and pressure effects. The enhancement of reaction rates in water over those in hydrocarbon solvent are particularly significant. For leading references and a computer simulation study of the origin of the solvent effect, see Blake, J. F.; Jorgensen, W. L. *J. Am. Chem. Soc.* **1991**, *113*, 7430. Pressure effects are most important in intramolecular reactions. Diedrich, M. K.; Hochstrate, D.; Klärner, F.-G.; Zimny, B. *Angew. Chem., Int. Ed. Engl.* **1994**, *33*, 1079, reported that (*Z*)-1,3,8-nonatriene undergoes the competing intramolecular Diels-Alder reaction and a sigmatropic [1,5] hydrogen shift shown below. At 150° and 1 bar, the major product results from the hydrogen shift, but at 7.7 bar, the major product is the intramolecular Diels-Alder adduct.

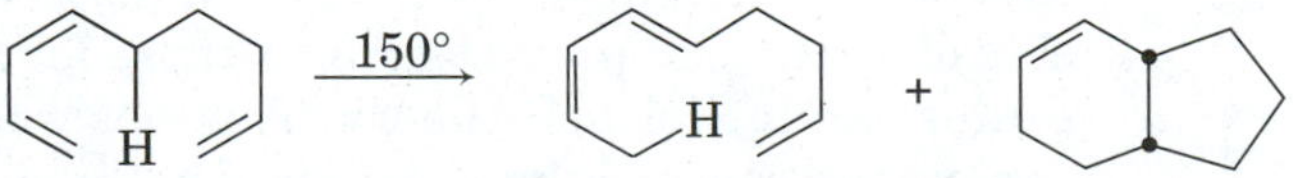

H —150°⟶ H +

[144]Bellville, D. J.; Wirth, D. D.; Bauld, N. L. *J. Am. Chem. Soc.* **1981**, *103*, 718; Bauld, N. L. *J. Am. Chem. Soc.* **1992**, *114*, 5800. Other pericyclic reactions also occur through the intermediacy of organic radical cations. For a discussion, see Bauld, N. L.; Bellville, D. J.; Harirchian, B.; Lorenz, K. T.; Pabon, Jr., R. A.; Reynolds, D. W.; Wirth, D. D.; Chiou, H.-S.; Marsh, B. K. *Acc. Chem. Res.* **1987**, *20*, 371.

[145]Houk, K. N. *Acc. Chem. Res.* **1975**, *8*, 361 and references therein.

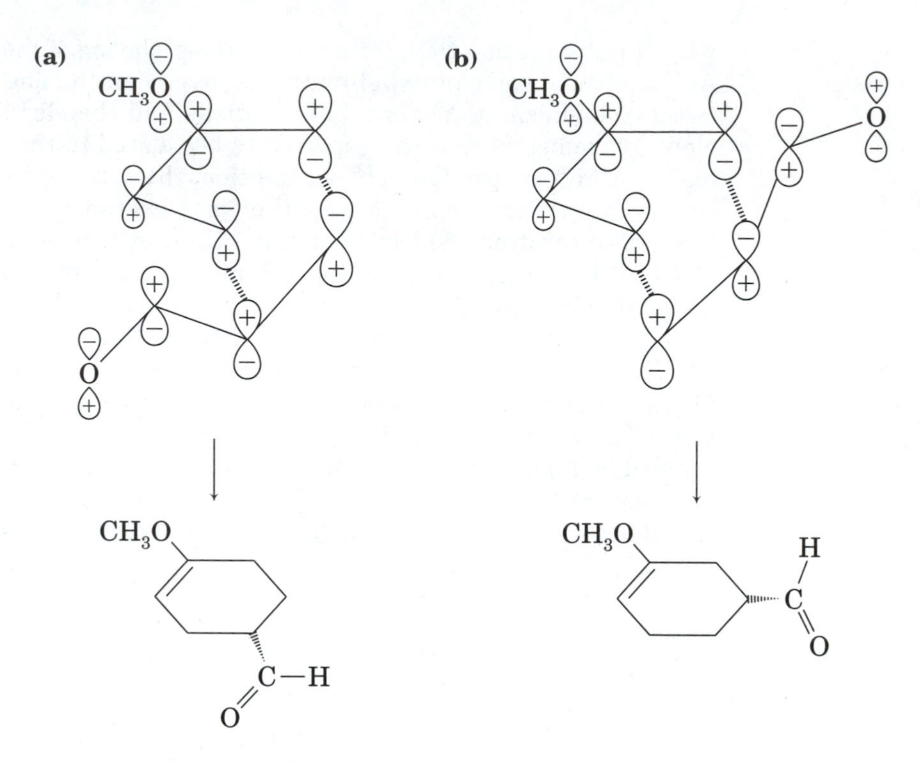

Figure 11.91 (a) Favorable interaction of larger lobe of HOMO of 2-methoxy-1,3-butadiene with larger lobe of LUMO of acrolein; (b) less favorable interaction of larger lobe of HOMO of 2-methoxy-1,3-butadiene with the smaller lobe of LUMO of acrolein.

rule that the major product is the one in which the larger lobe of the HOMO overlaps with the larger lobe of the LUMO in the transition structure.[146] Thus, in Figure 11.91(a) there is bonding overlap of the large lobe of the HOMO on C1 of the diene with the large lobe of the LUMO on C3 of acrolein. However, the orientation of reactants that leads to the minor product, shown in Figure 11.91(b), involves the less favorable overlap of the larger lobe of the HOMO on C1 of the diene with a very small lobe of the LUMO on C2 of acrolein.[147]

Hückel and Möbius Aromaticity of Transition Structures

Another approach to understanding concerted reactions is based on a parallel between the molecular orbitals of transition structures and the MOs of ground state species. We have seen numerous examples of selection rules that were different for $(4n + 2)$ and $(4n)$ systems, and these are the same sets of arithmetic progressions as those for aromatic and antiaromatic molecules in Hückel MO theory. This observation suggests that transition structures can also be considered to be either aromatic or antiaromatic

[146]Fleming, I. *Frontier Orbitals and Organic Chemical Reactions*; Wiley-Interscience: London, 1976; pp. 121 *ff*.

[147]The atomic orbital sizes shown in Figure 11.91 approximate the magnitude of the HOMO and LUMO coefficients determined for each position with a MOPAC (PM3) calculation.

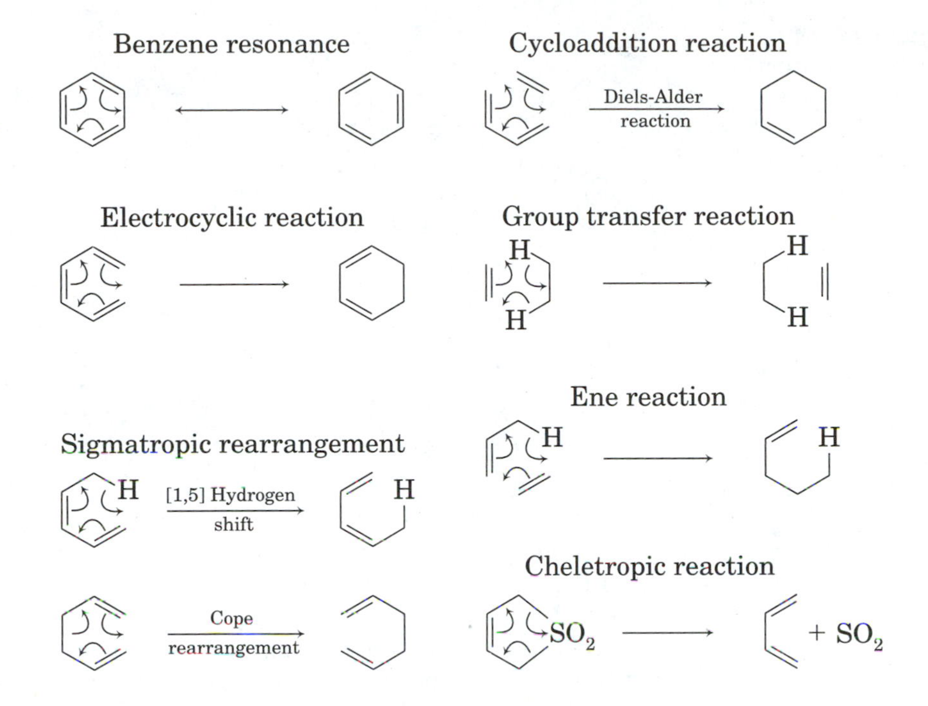

Figure 11.92 Curved arrow notation for benzene resonance and for aromatic transition structures in pericyclic reactions involving six electrons.

species.[148] In fact, the curved arrow notation for electron movement leading to the observed bonding changes for all of the allowed pericyclic reactions in Figure 11.3 resembles the curved arrow notation for conversion of one Kekulé resonance structure of benzene to the other (Figure 11.92).[149]

The view that the transition structures of allowed pericyclic reactions are aromatic is supported by a more detailed analysis. For example, both the suprafacial [1,5] hydrogen shift of 1,3-pentadiene and the disrotatory opening of 1,3-cyclohexadiene to 1,3,5-hexatriene (Figure 11.93) have transition structures with six electrons in a system composed of six basis set orbitals arranged in a cyclic array, just as does benzene. Thus these are **aromatic transition structures**. However, a very different situation is found for forbidden reactions, such as the disrotatory opening of cyclobutene to butadiene and the suprafacial [1,3] hydrogen shift in propene (Figure 11.94). In both cases the transition structures are shown with a set of four atomic orbitals arranged in a cyclic array, and there are a total of four electrons in the resulting MOs. Each system is like cyclobutadiene and is therefore an **antiaromatic transition structure**. Because reactions such as those in Figure 11.93 are allowed, while reactions such as those in Figure 11.94 are forbidden, Dewar proposed that systems containing $4n + 2$

[148]Evans, M. G.; Warhurst, E. *Trans. Faraday Soc.* **1938**, *34*, 614; Evans, M. G. *Trans. Faraday Soc.* **1939**, *35*, 824.

[149]Figure 11.92 is modeled after a figure in reference 21.

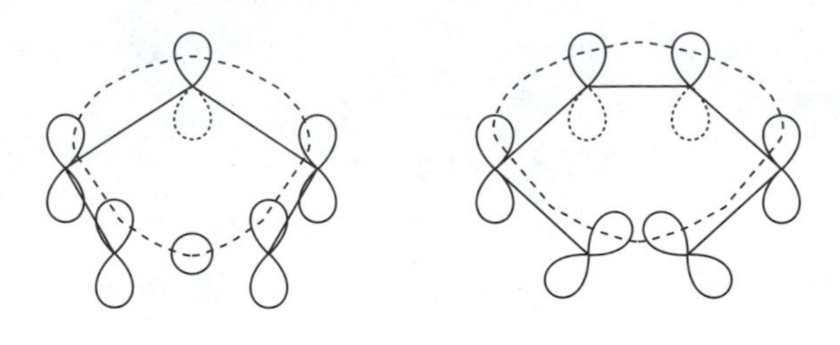

Figure 11.93 Aromatic transition structures.

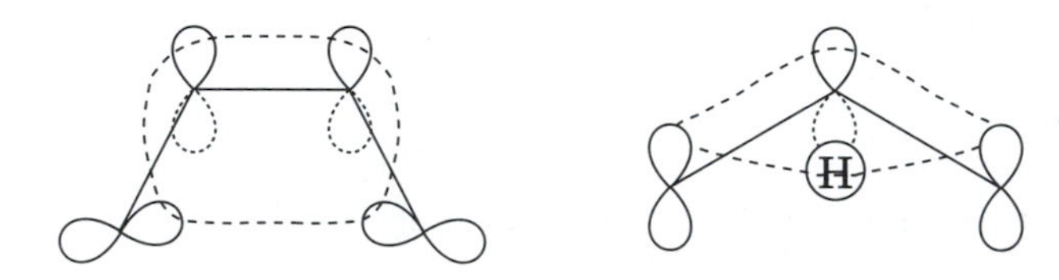

Figure 11.94 Antiaromatic transition structures.

electrons undergoing a pericyclic change prefer an aromatic transition state, while systems with $4n$ electrons avoid an antiaromatic transition state.[150]

The relationship between aromaticity and pericyclic reactions can be developed further by considering the type of orbital interactions that would be present in a novel structure proposed by Heilbronner.[151] Imagine a large cyclic array of n p orbitals arranged as shown in Figure 11.95. If each p orbital is twisted by $180°/n$ with respect to the orbital adjacent to it, then after

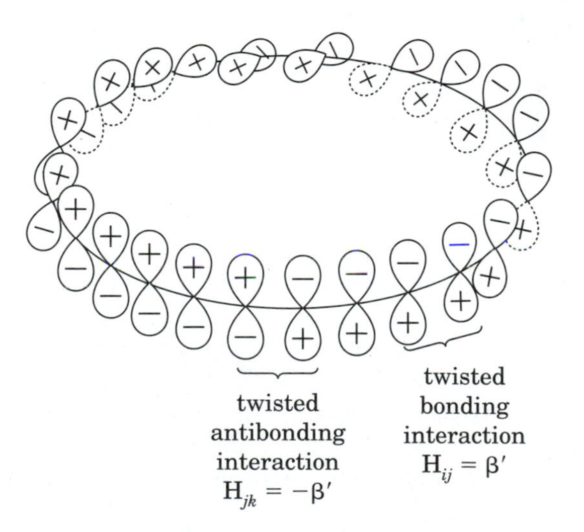

Figure 11.95 A hypothetical array of twisted p orbitals leading to a Möbius π system.

[150] See Dewar, M. J. S. *Tetrahedron Suppl.* **1966**, *8*, 75.

[151] Heilbronner, E. *Tetrahedron Lett.* **1964**, 1923.

Figure 11.96
A Möbius strip.

n p orbitals the ring would be complete, but with the first p orbital in an antibonding relationship with respect to the last. In other words, there will be one and only one $(+ \cdots -)$ interaction in the ring. Heilbronner noted that such a molecular system would be like a Möbius strip (Figure 11.96) in that it would have only one side.[152]

The molecular orbitals of a Möbius system can be calculated in almost exactly the same way as are Hückel MOs. If the resonance integral for a p orbital with another p orbital slightly twisted with respect to it is β (or β', since the orbitals are not completely parallel), then the resonance integral for the first and last p orbitals is $-\beta$. Solving the secular determinant gives the energy levels of such a system, as is illustrated in Figure 11.97. The top part of that figure shows the basis set orbitals, the secular determinant from HMO theory, and the resulting energy levels for Hückel cyclobutadiene. The bottom of Figure 11.97 shows the corresponding calculation for Möbius cyclobutadiene. (Note the antibonding overlap between the p orbitals on C1 and C2.) The secular determinant for this Möbius system now has two -1 elements, for the two resonance integrals for C1 and C2. Solving this determinant in the usual way now leads to two energy levels of $\alpha + 2\beta$ and two energy levels of $\alpha - 2\beta$.

As is often the case with simple HMO theory, it is the energy levels and not the MOs themselves that are of primary interest. Fortunately, Zimmerman has developed a circle mnemonic device (analogous to the circle mnemonic used with HMO theory) that provides a shortcut to finding the energy levels.[153] For Möbius MOs, the polygon corresponding to the cyclic molecule is inscribed in the circle of radius 2β with one *side* down (not with a corner down as was done for Hückel systems).[154] This procedure is illustrated in Figure 11.98 for both (a) Hückel MOs and (b) Möbius MOs for cyclopropenyl, cyclobutadiene, and benzene.

---

[152]Möbius strip molecules have been synthesized, although not with the kind of atomic orbital interactions described here. See the discussion in *Chem. Eng. News* **1982** (July 12), 21 *ff*. Barrelene has unique Möbiuslike overlap of its atomic orbitals: Zimmerman, H. E.; Grunewald, G. L.; Paufler, R. M.; Sherwin, M. A. *J. Am. Chem. Soc.* **1969**, *91*, 2330.

[153](a) Zimmerman, H. E. *J. Am. Chem. Soc.* **1966**, *88*, 1564; (b) Zimmerman, H. E. *Acc. Chem. Res.* **1971**, *4*, 272; (c) Zimmerman, H. E. in *Pericyclic Reactions*, Vol. 1; Marchand, A. P.; Lehr, R. E., eds.; Academic Press: New York, 1977, pp. 53–107. See also Shen, K. *J. Chem. Educ.* **1973**, *50*, 238.

[154]Here we are taking the Hückel and Möbius systems to have equal overlap and thus to have the same β value. Equal twisting seems especially reasonable when we are applying this reasoning to transition structures.

Figure 11.97
(a) Basis set orbitals, (b) secular determinant, and (c) energy levels for Hückel cyclobutadiene (top) and Möbius cyclobutadiene (bottom).

Hückel Cyclobutadiene

(a)

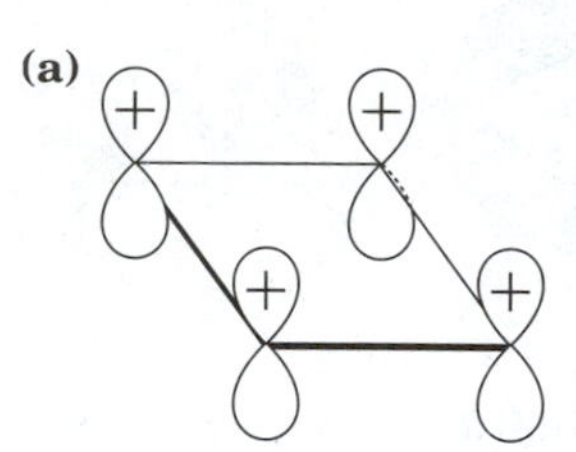

(b)

$$\begin{vmatrix} X & 1 & 0 & 1 \\ 1 & X & 1 & 0 \\ 0 & 1 & X & 1 \\ 1 & 0 & 1 & X \end{vmatrix} = 0$$

(c)

E

— $\alpha - 2\,\beta$

α

$\alpha + 2\,\beta$

Möbius Cyclobutadiene

(a)

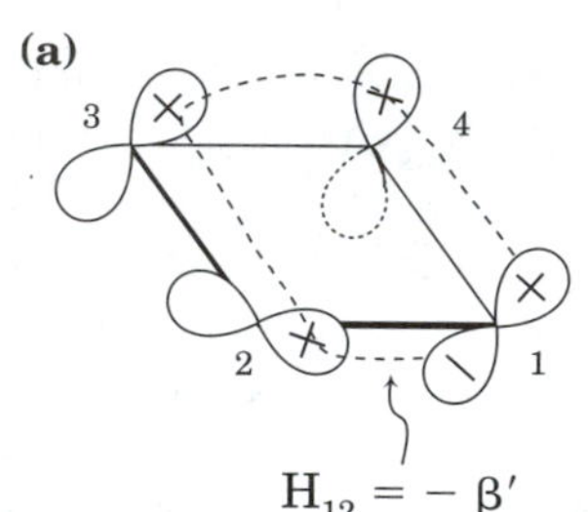

(b)

$$\begin{vmatrix} X & -1 & 0 & 1 \\ -1 & X & 1 & 0 \\ 0 & 1 & X & 1 \\ 1 & 0 & 1 & X \end{vmatrix} = 0$$

(c)

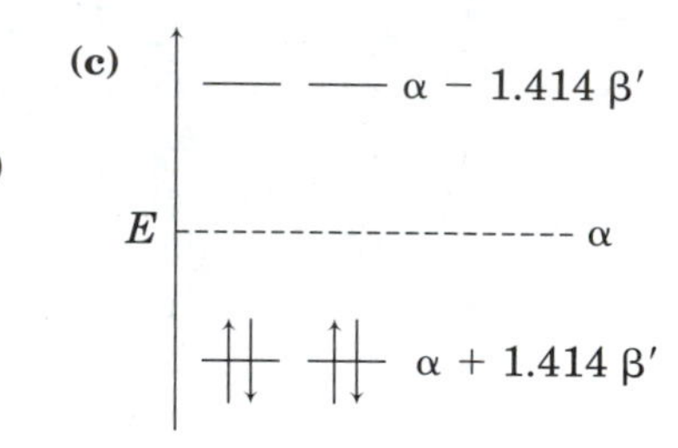

The three structures in Figure 11.98 illustrate the contrasts between Hückel and Möbius systems.

1. In HMO theory, the cyclopropenyl cation is aromatic (that is, it is a closed shell system with large delocalization energy), since both electrons are in the orbital with $E = \alpha + 2\,\beta$. The Hückel cyclopropenyl anion, however, is antiaromatic because it is an open shell system (having one electron in each of the $E = \alpha - \beta$ orbitals) with zero delocalization energy. In contrast, the Möbius cyclopropenyl anion is aromatic, since it is a closed shell system with all four electrons in bonding orbitals.
2. Möbius cyclobutadiene has four electrons in the two lowest energy levels ($E = \alpha + 1.414\,\beta$), as shown in Figure 11.97. Therefore Möbius cyclobutadiene should be a closed shell system with some delocalization stabilization; i.e., it should be aromatic (at least in comparison with a nondelocalized reference structure having a system of twisted p orbitals in a ring). Hückel cyclobutadiene is an antiaromatic, open shell species.
3. Hückel benzene is aromatic. In Möbius benzene, however, the first four electrons go into orbitals with energy levels at $E = \alpha + 1.732\,\beta$, and the last two go separately into orbitals at $E = \alpha$. Therefore, Möbius benzene is an open shell, unstable system.

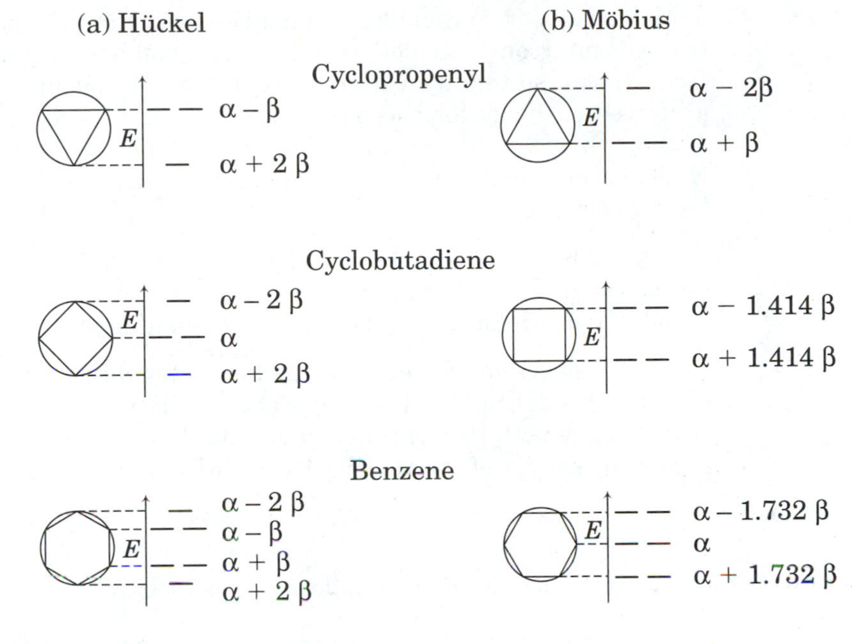

Figure 11.98 (a) Hückel and (b) Möbius energy levels developed for cyclopropenyl, cyclobutadiene, and benzene with the circle mnemonic. (Adapted from reference 153(b).)

Generalizing the results from these three examples leads to the conclusion that an array of $4n$ electrons having one positive-negative overlap is Möbius aromatic, while a system with $4n + 2$ electrons is Möbius antiaromatic.

The concept of Möbius aromaticity provides additional insight into the electronic nature of the transition structures in pericyclic reactions. The four pericyclic transition structures in Figures 11.94 and 11.93 were categorized according to their Hückel aromaticity. Figure 11.99 shows an analysis of the same four transition structures as Möbius systems. The conrotatory electrocyclic interconversion of butadiene and cyclobutene is a 4 electron system, so it is Möbius aromatic. Similarly the antarafacial [1,3] hydrogen shift is also a 4 electron system, so it too is Möbius aromatic. The antarafacial [1,5] hydrogen shift is a six electron system, so this transition structure

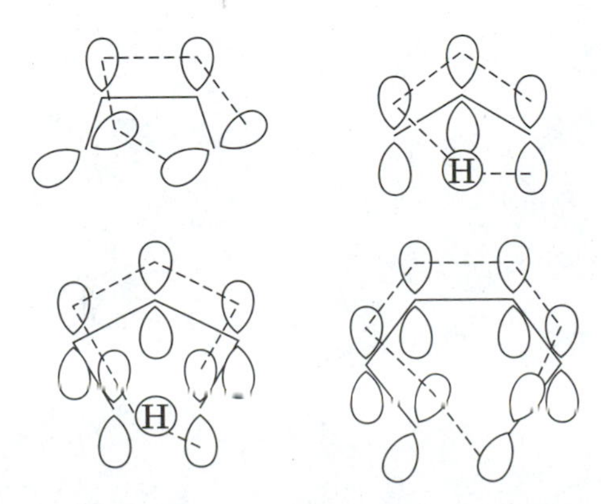

Figure 11.99 Aromatic (top) and antiaromatic (bottom) Möbius transition structures.

is Möbius antiaromatic, like Möbius benzene. Finally, the disrotatory electrocyclic interconversion of cyclohexadiene and hexatriene is also a six electron system, so this transition structure is also Möbius antiaromatic. The preferred pathway for the first two reactions is Möbius aromatic (Figure 11.99), while for the third and fourth reactions the preferred pathway is Hückel aromatic (Figure 11.93). Generalizing these results leads to the following selection rule.:[153]

> Systems containing $4n + 2$ electrons undergoing a pericyclic change prefer a Hückel aromatic transition structure, while systems with $4n$ electrons prefer a Möbius aromatic transition structure.

The distinction between the Hückel and Möbius pathways can be clarified by considering the thermal electrocyclic reaction of butadiene to cyclobutene by both the conrotatory and disrotatory pathways. Figure 11.100 shows the energy of the various MOs, with the disrotatory transition struc-

Figure 11.100 Hückel-Möbius MO reaction diagram for butadiene-cyclobutene. (Adapted from reference 153(b).)

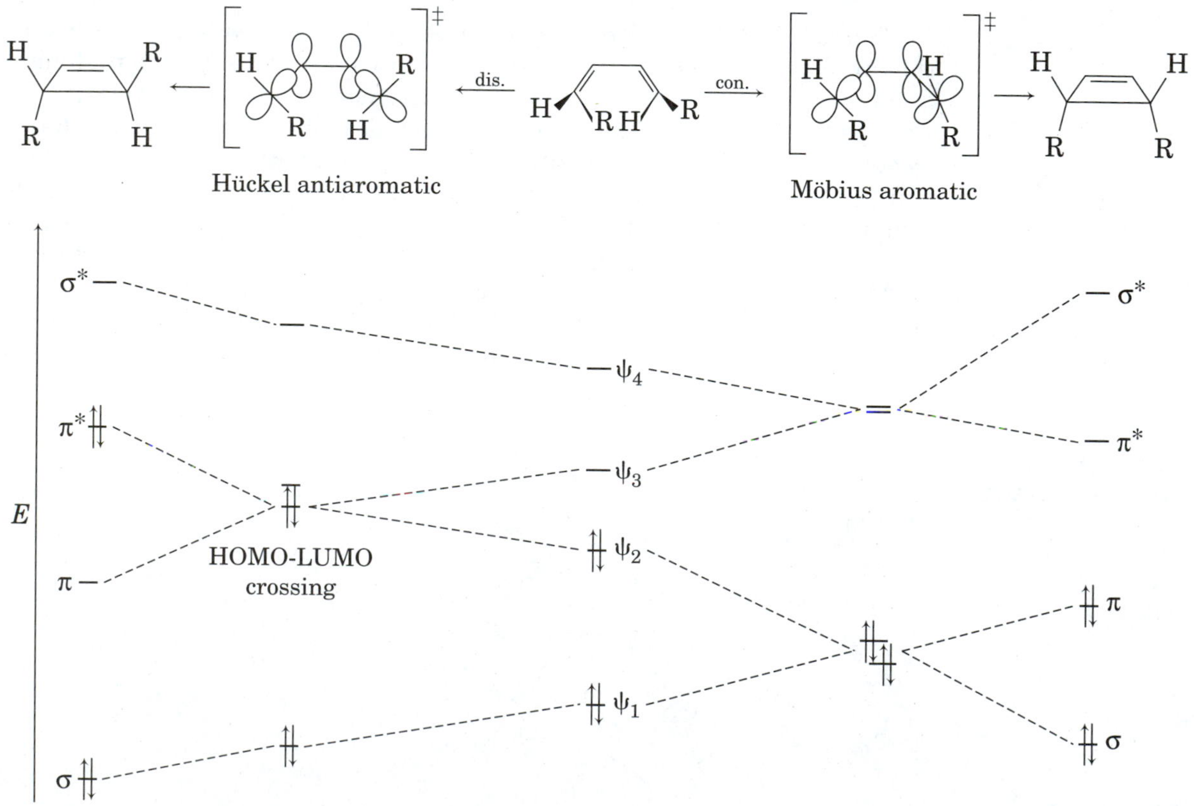

ture and its Hückel π system drawn on the left and with the conrotatory transition structure with its Möbius π system drawn on the right.[155] As noted by Zimmerman, degeneracies in either the Hückel or Möbius MOs indicate MO crossings,[153] so it is easy to follow MOs from reactant to product. Being able to assign energies to the MOs present in the transition structure for each pathway allows us to compare the change in energy level of each butadiene MO as the reaction proceeds along either of the concerted pathways and to estimate the transition state energy for each of the two pathways for the reaction.[156]

Dewar and co-workers noted that in a pericyclic reaction there is a HOMO and a LUMO for the reactant, transition structure, and product.[157] For an allowed reaction, the HOMO and LUMO do not cross in energy and, in fact, the difference in their energies remains fairly constant. For a forbidden reaction, on the other hand, there is a HOMO-LUMO crossing. Therefore,[157]

> The distinction between "allowed" and "forbidden" reactions . . . seems to be one of topology rather than symmetry, there being a qualitative distinction between pairs of isomers that can be interconverted by a pericyclic reaction without a HOMO-LUMO crossing and pairs that cannot be so interconverted without a HOMO-LUMO crossing.

Pairs of isomers that can be interconverted without a HOMO-LUMO crossing were termed *HOMOMERS*, while those whose interconversion requires a HOMO-LUMO crossing were termed *LUMOMERS*. In contrast to the other selection rules developed above, this classification system is independent of the path by which the interconversion takes place.

Synchronous and Nonsynchronous Concerted Reactions

Dewar noted that almost all analyses of concerted mechanisms implicitly assume symmetric transition structures. For example, the forbidden disrotatory interconversion of cyclobutene and butadiene assumed a pathway in which a C_2 rotation axis is maintained throughout the reaction. Dewar suggested that for forbidden processes, a lower energy pathway might be one in which no symmetry element was maintained.[158,159] For example, in the disrotatory electrocyclic closure of butadiene, the rotation about the C1—C2

---

[155]Although the drawing is not quantitative, the σ orbitals are drawn lower in energy than π orbitals, and σ^* orbitals are drawn higher than π^* orbitals. The resonance integral (β) is assumed to be the same for both Hückel and Möbius systems.

[156]Zimmerman has advocated this approach as a general tool in organic chemistry and has called it "MO following," noting that it is the molecular orbital theory counterpart to "electron pushing" in valence bond theory. For a more detailed discussion, see reference 153(c).

[157]Dewar, M. J. S.; Kirschner, S.; Kollmar, H. W. *J. Am. Chem. Soc.* **1974**, *96*, 5240.

[158]Dewar, M. J. S.; Kirschner, S. *J. Am. Chem. Soc.* **1974**, *96*, 5244.

[159]The question of symmetric transition structures in concerted reactions was also raised by McIver, Jr., J. W. *Acc. Chem. Res.* **1974**, *7*, 72.

bond might occur faster than the rotation about the C3—C4 bond. Although the reaction would still be ***concerted*** (meaning that bond breaking and bond forming processes occur simultaneously), the reaction might not be ***synchronous*** (meaning that all of the bond changes progress to the same extent in the transition structure).[160] Furthermore, it was also considered possible that nonsynchronous pathways might be lower in energy than synchronous pathways for allowed pericyclic reactions. Dewar and co-workers used MNDO methods to calculate the transition structures for several pericyclic reactions, including the Diels-Alder reaction and the Claisen and Cope rearrangements.[161] The calculations suggested that each of these reactions takes place in a nonsynchronous way.[162]

Dewar's conclusions stimulated considerable discussion among theoretical chemists. Borden and co-workers reported results of *ab initio* calculations indicating that the Diels-Alder reaction is, in fact, a synchronous reaction. However, these authors indicated that this result was obtained "only when a flexible basis set is used and when electron correlation is properly treated." Otherwise, a nonsynchronous process was observed.[163] Bernardi et al. also reported *ab initio* evidence for a synchronous Diels-Alder reaction.[164] Houk and co-workers reported experimental evidence that the Diels-Alder reaction is concerted and is most likely synchronous as well.[165,166] These authors pointed out that different computational methods have built-in biases toward different transition structures. This controversy reminds us that, as we have noted before, computational methods should be considered as useful tools that are developed from particular conceptual models and not as windows into reality.[167,168]

---

[160]For an account of work in this area, see the summary by Maugh II, T. H. *Science* **1984**, *223*, 1162.

[161]Dewar, M. J. S.; Griffin, A. C.; Kirschner, S. *J. Am. Chem. Soc.* **1974**, *96*, 6225; Dewar, M. J. S.; Wade, Jr., L. E. *J. Am. Chem. Soc.* **1977**, *99*, 4417; Dewar, M. J. S.; Pierini, A. B. *J. Am. Chem. Soc.* **1984**, *106*, 203; Dewar, M. J. S.; Healy, E. F. *J. Am. Chem. Soc.* **1984**, *106*, 7127.

[162]Dewar, J. J. S. *J. Am. Chem. Soc.* **1984**, *106*, 209.

[163]Osamura, Y.; Kato, S.; Morokuma, K.; Feller, D.; Davidson, E. R.; Borden, W. T. *J. Am. Chem. Soc.* **1984**, *106*, 3362.

[164]Bernardi, F.; Bottoni, A.; Robb, M. A.; Field, M. J.; Hillier, I. H.; Guest, M. F. *J. Chem. Soc., Chem. Commun.* **1985**, 1051.

[165]Houk, K. N.; Lin, Y.-T.; Brown, F. K. *J. Am. Chem. Soc.* **1986**, *108*, 554. A combination of theoretical and experimental isotope effects supports the view that the cycloaddition of ethene and 1,3-butadiene is both concerted and synchronous: Storer, J. W.; Raimondi, L.; Houk, K. N. *J. Am. Chem. Soc.* **1994**, *116*, 9675.

[166]For a discussion of experimental methods to evaluate concertedness of potential pericyclic reactions, see Lehr, R. E.; Marchand, A. P. in *Pericyclic Reactions*, Vol. I; Marchand, A. P., Lehr, R. E., eds.; Academic Press: New York, 1977; pp. 1–51.

[167]In these discussions we have considered explanations for concerted reactions primarily in terms of the principle of conservation of orbital symmetry as advanced by Woodward and Hoffmann, but there are other molecular orbital explanations that lead to the same conclusions. See, for example, (a) Weltin, E. E. *J. Am. Chem. Soc.* **1973**, *95*, 7650; **1974**, *96*, 3049; (b) Halevi, E. A. *Helvetica Chim. Acta* **1975**, *58*, 2136; (c) Day, A. C. *J. Am. Chem. Soc.* **1975**, *97*, 2431; (d) Trindle, C. *J. Am. Chem. Soc.* **1970**, *92*, 3251; (e) Langlet, J.; Malrieu, J.-P. *J. Am. Chem. Soc.* **1972**, *94*, 7254; (f) Mathieu, J.; Rassat, A. *Tetrahedron* **1974**, *30*, 1753; (g) Rassat, A. *Tetrahedron Lett.* **1975**, 4081; (h) He, F.-C.; Pfeiffer, G. V. *J. Chem. Educ.* **1984**, *61*, 948.

In contrast to the Diels-Alder reaction of ethene and 1,3-butadiene, there is evidence that other pathways may be competitive with the concerted [4 + 2] cycloaddition of two 1,3-butadiene molecules to form 4-vinylcyclohexene. Stephenson and co-workers found about a 10% loss of stereochemistry in the dimerization of *cis,cis*-1,4-dideuterio-1,3-butadiene.[169] *Ab initio* calculations by Li and Houk revealed that the concerted reaction may be only slightly lower in energy than step-wise processes involving diradical intermediates for this reaction.[87] Klärner and co-workers have reported that high pressure can suppress the extent of product formation by nonconcerted reaction.[170] Because the allowed and forbidden pathways generalized from state correlation diagrams reveal only the energetics of concerted reactions, the possibility that alternative, nonconcerted pathways may have lower or only slightly higher energies must always be considered.

Many pericyclic reactions are clearly nonsynchronous. Studies of kinetic isotope effects on the Claisen rearrangement of allyl phenyl ether have suggested that in the transition structure, the $C_{\alpha}-O$ bond is 50–60% broken, while the $C_{\gamma}-C_{ortho}$ bond is only 10–20% formed.[171] The relationship between bond forming and bond breaking in the analogous Cope rearrangement is expressed graphically by the reaction surface diagram shown in Figure 11.101. A fully synchronous reaction would follow the diagonal from reactant (lower left corner) to product (upper right corner). A reaction involving complete bond breaking before any bond forming would result in a path passing through the upper left corner and the intermediacy of two allylic radicals. A reaction involving complete bond formation before any bond breaking would involve a reaction pathway passing through the diradical intermediate in the lower right corner. *Ab initio* calculations suggested that the synchronous pathway applies for the Cope rearrangement of 1,5-hexadienes that do not bear radical- or ion-stabilizing substituents.[172] Based on α secondary kinetic isotope effects, however, Gajewski and Conrad concluded

---

[168]These discussions of concerted reactions have emphasized molecular orbital theory, but valence bond theory can produce the same result. For leading references to descriptions of concerted reactions in terms of valence bond theory, see (a) Goddard III, W. A. *J. Am. Chem. Soc.* **1972**, *94*, 793; (b) Mulder, J. J. C.; Oosterhoff, L. J. *J. Chem. Soc. D, Chem. Commun.* **1970**, 305, 307; (c) Silver, D. M.; Karplus, M. *J. Am. Chem. Soc.* **1975** *97*, 2645; (d) Herndon, W. C. *J. Chem. Educ.* **1981**, *58*, 371; (e) Bernardi, F.; Olivucci, M.; Robb, M. A. *Acc. Chem. Res.* **1990**, *23*, 405. For discussions of the relationship of valence bond theory and orbital symmetry concepts, see le Noble, W. J. *Highlights of Organic Chemistry: An Advanced Textbook*; Marcel Dekker, Inc.: New York, 1974; pp. 460–463. See also Berson, J. A. *Tetrahedron* **1992**, *48*, 3 (especially the discussion on pp. 14–17).

[169]Stephenson, L. M.; Gemmer, R. V.; Current, S. *J. Am. Chem. Soc.* **1975**, *97*, 5909.

[170]Klärner, F.-G.; Krawczyk, B.; Ruster, V.; Deiters, U. K. *J. Am. Chem. Soc.* **1994**, *116*, 7646.

[171]Kupczyk-Subotkowska, L.; Saunders, Jr., W. H.; Shine, H. J. *J. Am. Chem. Soc.* **1988**, *110*, 7153; Kupczyk-Subotkowska, L.; Subotkowska, W.; Saunders, Jr., W. H.; Shine, H. J. *J. Am. Chem. Soc.* **1992**, *114*, 3441. Similar results were obtained for the [3,3] rearrangement of allyl vinyl ether: Kupczyk-Subotkowska, L.; Saunders, Jr., W. H.; Shine, H. J.; Subotkowska, W. *J. Am. Chem. Soc.* **1993**, *115*, 5957.

[172]For an *ab initio* study and a summary of previous theoretical investigations, see Hrovat, D. A.; Morokuma, K.; Borden, W. T. *J. Am. Chem. Soc.* **1994**, *116*, 1072.

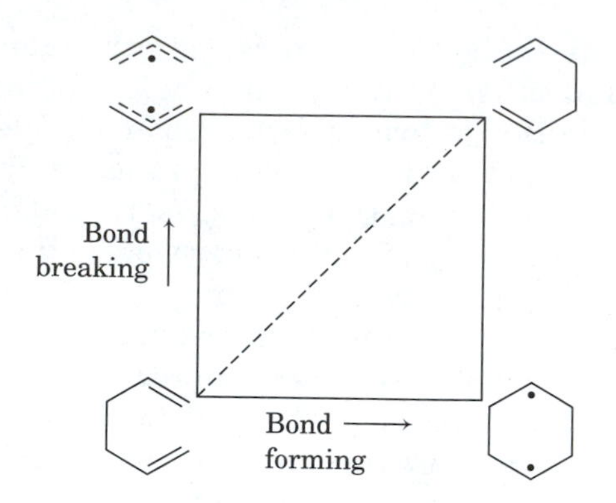

Figure 11.101 Reaction surface diagram for the Cope rearrangement. (Adapted from reference 173.)

that with radical stabilizing groups (such as phenyl) at C2 and C5, the transition structure is more like cyclohexane-1,4-diyl (the lower right corner), but that with radical stabilizing groups at C3, the transition structure lies nearer the upper left corner.[173,174,175,176]

The observation that an experimental result is consistent with the predictions of a pericyclic process does not indicate that the reaction is pericyclic. It must always be considered possible that other processes, perhaps unrecognized, could lead to the same product. This caution is especially relevant when a mechanism proposed for one molecular system is extended to another system with different structural features. For example, the thermal rearrangement of 1-phenylbicyclo[2.2.1]hexene-5-*d* (**92**) produces a 10 : 1 ratio of **93** : **94**. Since **93** is the product predicted for a symmetry-allowed [1,3] sigmatropic shift, this result might suggest that **93** is formed by a concerted, symmetry-allowed antarafacial [1,3] sigmatropic shift, while **94** arises from a competing radical or symmetry-forbidden concerted pathway. However, Newman-Evans and Carpenter determined that the product ratio of **93** to **94** is invariant over the temperature range from 80–165°, and a similar result was also observed for rearrangements of some other compounds. These investigators concluded that it is unlikely that two competing processes would have identical activation energies in several different reactants, which would be required in order for the product ratio to not change with temperature, so the results were interpreted to mean that there is one rate-limiting step for the reaction of **92** and that both **93** and **94** are formed via a singlet biradical intermediate.[177]

---

[173]Gajewski, J. J.; Conrad, N. D. *J. Am. Chem. Soc.* **1979**, *101*, 6693.

[174]Also see Houk, K. N.; Gustafson, S. M.; Black, K. A. *J. Am. Chem. Soc.* **1992**, *114*, 8565 and references therein.

[175]The intermediates are shown here as radicals, but ions might be involved instead—especially if there is a cation-stabilizing substituent on C2 and an anion-stabilizing substituent on C5.

[176]Evidence for a step-wise reaction resulting from stabilization of a 1,4-diyl intermediate has been reported by Wessel, T. E.; Berson, J. A. *J. Am. Chem. Soc.* **1994**, *116*, 495.

[177]Newman-Evans, R. H.; Carpenter, B. K. *J. Am. Chem. Soc.* **1984**, *106*, 7994.

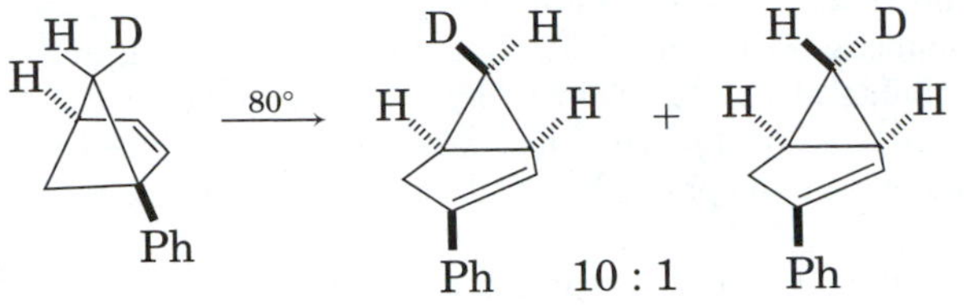

Figure 11.102 Rearrangement of **92** to a 10 : 1 ratio of **93** : **94**. (Adapted from reference 179.)

To explain the preferential formation of **93**, Carpenter and co-workers suggested that the reaction of **92** is not determined solely by the shape of the local potential energy surface at the intermediate. Rather, the authors proposed that dynamic effects—the conservation of linear and angular momentum in the motion of the nuclei—govern nuclear motion as the molecule passes through the transition structure and then proceeds to one or the other of the two products. According to this model, **93** is formed in higher yield because it results from a continuation of the nuclear motion that led to bond breaking and formation of the diradical.[178] The results of AM1 calculations of the potential energy surface for the rearrangement of **92** were consistent with this explanation.[179] Carpenter has concluded that a dynamic preference for inversion in ring-opening reactions of strained-ring compounds might be a result of dynamics and that in such a case a [1,*n*] sigmatropic shift need not be a pericyclic reaction at all.[179]

Problems

1. Using the HOMO method given by Woodward and Hoffmann in their first communication on orbital symmetry control of chemical transformations,[14] determine whether the thermal reaction shown in equation 11.24 should take place via a conrotatory or disrotatory pathway. Indicate the stereochemistry of the product formed by the allowed pathway. Do the same for the photochemical reaction.

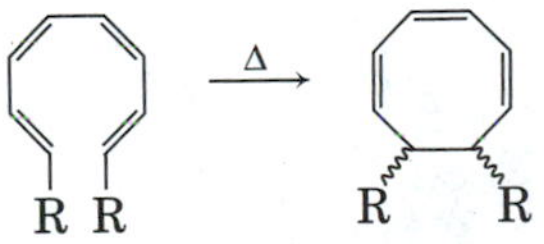

(11.24)

2. Use the symmetry and nodal properties of the appropriate molecular orbitals to determine whether the thermal [1,7] and [1,9] hydrogen shifts are allowed suprafacially or antarafacially.
3. Through the use of appropriate MO symmetry elements for reactants and products, determine whether the thermal $[_{\pi}2_s + {_{\pi}6_s}]$ and $[_{\pi}4_s + {_{\pi}4_s}]$ cycloaddition reactions are allowed or forbidden.

[178]Newman-Evans, R. H.; Simon, R. J.; Carpenter, B. K. *J. Org. Chem.* **1990**, *55*, 695.

[179]Carpenter, B. K. *J. Org. Chem.* **1992**, *57*, 4645 and references therein.

4. a. Construct a figure similar to Figure 11.35 (page 737) to demonstrate that the photochemical antarafacial-antarafacial [3,3] sigmatropic rearrangement is forbidden by the principles of orbital symmetry.
   b. Construct an MO correlation diagram similar to Figure 11.62 (page 753) to show that the photochemical Diels-Alder reaction is not allowed by the principles of orbital symmetry.
5. Use both MO and state correlation diagrams to predict the stereochemistry of the opening of the cyclopropyl anion to the allyl anion (Figure 11.67, page 759); do the same for the opening of the cyclopropyl cation to the allyl cation.
6. a. Using $-\beta$ instead of β for the type of *p-p* overlap illustrated in Figure 11.95 (page 780), calculate the energy levels of a Möbius cyclopropenyl system and verify that they are identical with those given by the circle mnemonic procedure (Figure 11.98, page 783).
   b. Having calculated the Möbius MOs in part a, consider again the opening of cyclopropyl systems to allyl systems. Draw an MO correlation diagram (as was done for butadiene-cyclobutene in Figure 11.100, page 784) and discuss the energetics and probable activation energies for the allowed and forbidden reactions in both cases. Discuss the diagram in terms of aromaticity of transition structures and in terms of the HOMO-LUMO crossing concept of Dewar.
7. For each of the following concerted reactions:

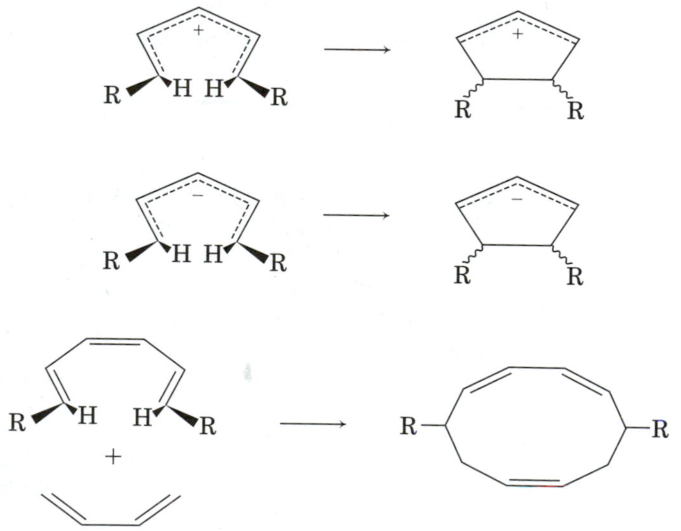

   a. Sketch one sterically feasible transition structure for a pathway that is allowed by the principles of orbital symmetry.
   b. Indicate the stereochemistry of substituents in the products.
   c. Classify the reaction using the usual formalism (e.g., ${}_{\pi}2_{a}$).
   d. Use correlation arrows to indicate the bonding change(s) at each center.
   e. Decide whether the reaction is thermally allowed or forbidden (stating the answer in terms of the generalized selection rule on page 770).
   f. Classify the transition structure as Hückel aromatic or Möbius aromatic.

8. Classify the following reaction as an allowed $[i,j]$ sigmatropic reaction, and state whether the pathway is suprafacial-suprafacial or suprafacial-antarafacial.

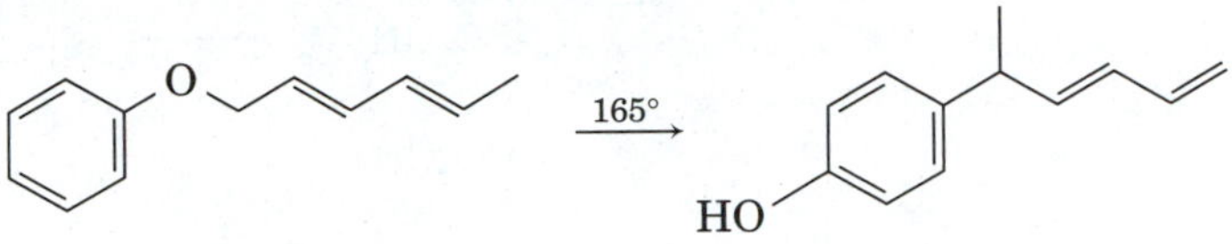

165°

HO

9. Explain the difference in reactivity between *cis*- and *trans*-1,2-divinylcyclobutane.

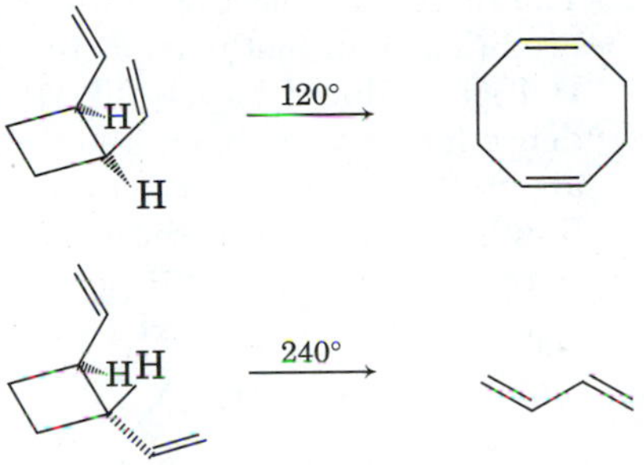

H H 120°

HH 240°

10. The reaction shown in equation 11.25 occurs readily, but the reaction shown in equation 11.26 does not. Explain this difference in reactivity.

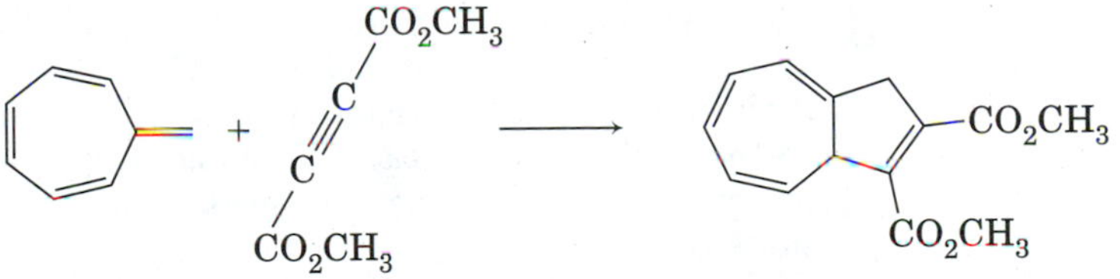

CO_2CH_3 C C CO_2CH_3 → CO_2CH_3 CO_2CH_3 **(11.25)**

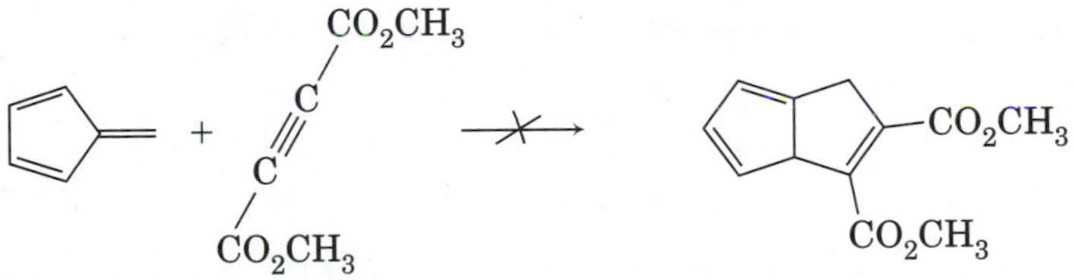

CO_2CH_3 C C CO_2CH_3 ↛ CO_2CH_3 CO_2CH_3 **(11.26)**

11. Describe each reaction below as the result of one or more allowed pericyclic reactions.

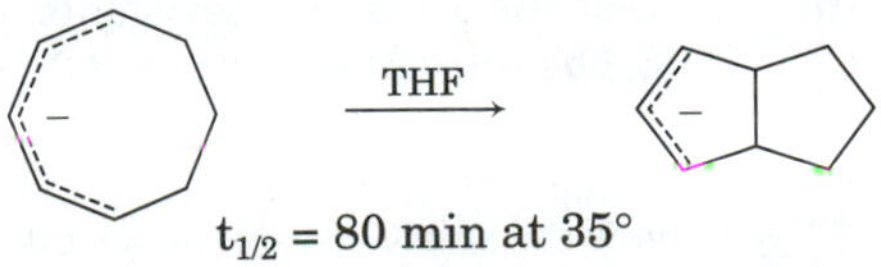

THF

$t_{1/2}$ = 80 min at 35°

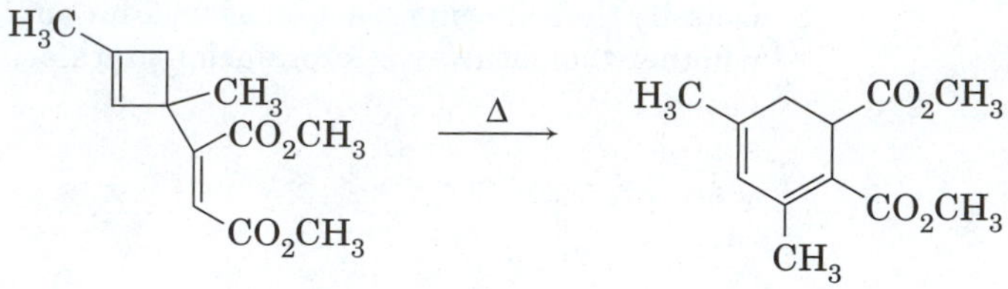

H_3C CH_3 CO_2CH_3 CO_2CH_3 $\xrightarrow{\Delta}$ H_3C CO_2CH_3 CO_2CH_3 CH_3

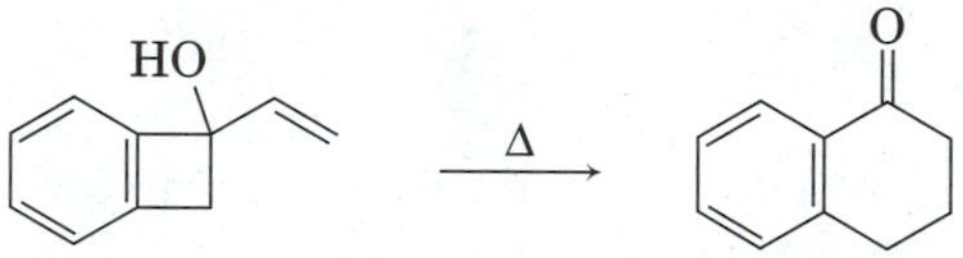

HO $\xrightarrow{\Delta}$ O

12. The reaction of bicyclo[3.1.0]hex-2-ene-*cis*-6-carboxaldehyde (**95**) with the ylide produced from methyltriphenylphosphonium salt and potassium *t*-butoxide in DMSO produced bicyclo[3.2.1]octadiene (**96**, equation 11.27). There was evidence for an unstable initial product that rearranged (with a rate constant of about 10^{-5} sec^{-1} in cyclohexane solution at room temperature) to produce the final product. What is the structure of the initial product, and why does it react so rapidly to produce **96**?

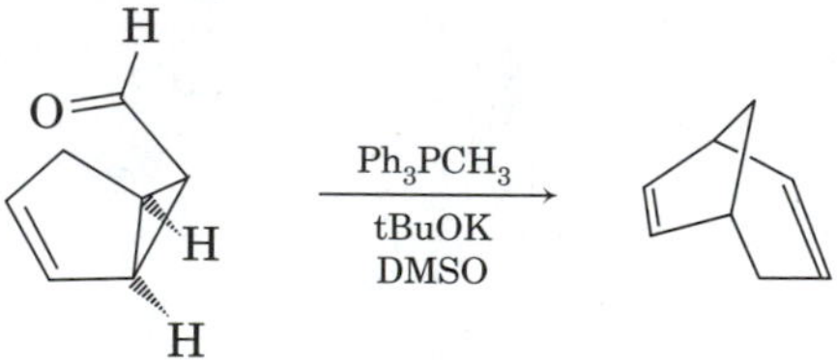

Ph_3PCH_3 / tBuOK / DMSO **(11.27)**

95 **96**

13. The Claisen rearrangement of allyl 2,6-dimethallylphenyl ether produces both 4-allyl-2,6-dimethallylphenol and 2-allyl-4,6-dimethallylphenol.[180] In addition, if the reaction is interrupted before completion, some methallyl 2-allyl-6-methallylphenyl ether can be isolated. Explain the formation of each of these products, and discuss the mechanism of the *p*-Claisen rearrangement in view of these results.
14. Is the suprafacial-suprafacial hydrogen atom transfer reaction illustrated in equation 11.28 an allowed or forbidden reaction?

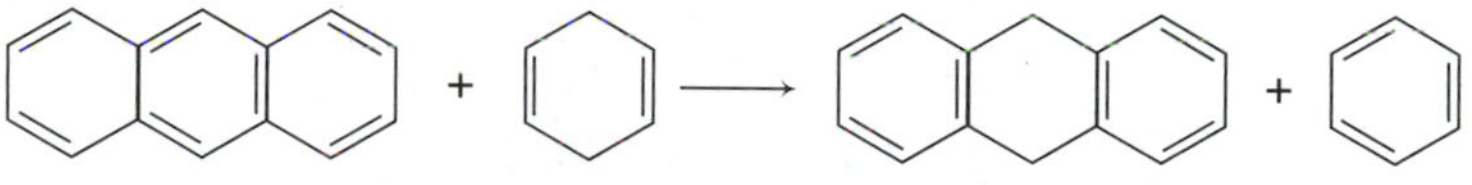

(11.28)

15. Benzyne can be produced by the reaction of magnesium with *o*-bromofluorobenzene in THF solution.[181] The reaction of benzyne so produced with 2,5-dimethyl-2,4-hexadiene did not lead to a Diels-Alder adduct. Instead, the product isolated from the reaction mixture was 2,5-dimethyl-3-phenyl-1,4-hexadiene. Propose a mechanism for the formation of this compound.
16. Consider the dianions chorismate and prephenate, which are shown in equation 11.29. The enzyme-catalyzed 3,3-sigmatropic shift of chorismate to prephenate

[180] Methallyl is the 2-methyl-2-propenyl group.

[181] Wittig, G. *Org. Synth., Coll. Vol. 4* **1963**, 964.

is an important step in a biosynthetic pathway leading to several aromatic compounds. The reaction also occurs, although considerably more slowly, at elevated temperatures in aqueous solution. To determine whether the [3,3] sigmatropic reaction involves a chairlike or boatlike transition structure, researchers prepared (*E*) and (*Z*) derivatives of chorismate by replacing one protium by tritium (represented by T or ^3H) at C9. They then followed the course of the reaction by monitoring the location of tritium in the product with the enzyme phenylpyruvate tautomerase, which preferentially releases the *pro-R* hydrogen atom of prephenate in the form of water.

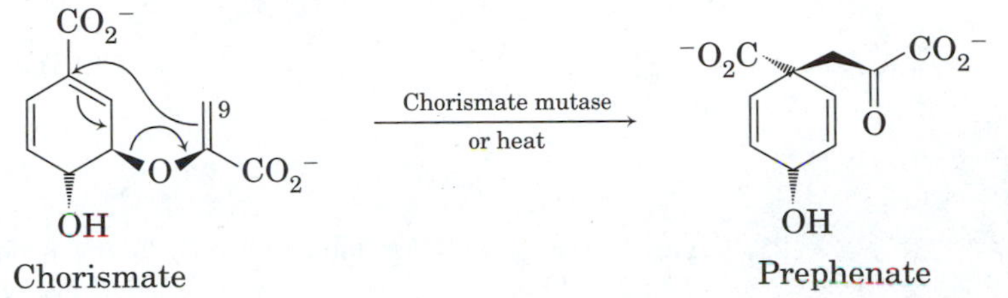

CO_2^- 9 O CO_2^- OH

Chorismate

Chorismate mutase
or heat

^-O_2C CO_2^- O OH

Prephenate

(11.29)

The investigators found that both the thermal and enzyme-catalyzed reactions of (*E*)-[9-^3H]-chorismate led to rapid release of tritiated water, whereas the reactions of the (*Z*) isomer did not. From this they concluded that both the enzyme-catalyzed and thermal reactions involve a chairlike transition structure.

a. Show that the chorismate to prephenate reaction is a [3,3] sigmatropic rearrangement.
b. Draw clearly the structures of the (*E*) and (*Z*) forms of [9-^3H] chorismate.
c. Show that the (*E*) isomer leads to the (*S*) configuration of the labeled carbon atom in prephenate if the transition structure for the reaction resembles a cyclohexane boat conformation, but the (*R*) configuration results if the transition structure is similar to cyclohexane chair conformation.
d. Show that removing the *pro-R* hydrogen (protium or tritium) atom from the labeled carbon atom of prephenate will produce tritiated water if the reaction of (*E*)-[9-^3H] chorismate does indeed involve the chair conformation in the transition structure.

17. The Claisen rearrangement of allyl 2,6-dimethylphenyl ether normally produces 4-allyl-2,6-dimethylphenol as the major product. However, when the reaction is carried out in molten maleic anhydride at 200°, two products having the molecular formula $C_{15}H_{16}O_4$ were isolated in addition to 4-allyl-2,6-dimethylphenol. What are the two compounds, and what does their isolation indicate about the mechanism of the reaction?
18. Explain the following reaction in terms of two consecutive pericyclic reactions, one an electrocyclic reaction and one a sigmatropic reaction.

CH_3 D D CH_3 —225°→ CH_3 CH_3 D D H

19. Both 2-acetoxy-*trans*-3-heptene and 4-acetoxy-*trans*-2-heptene were found to give a similar mixture of 1,3-heptadiene and 2,4-heptadiene upon gas phase pyrolysis at 350–360°. Furthermore, examination of recovered starting material revealed it

to consist of a mixture of 60% of the 2-acetoxy compound and 40% of the 4-acetoxy compound, no matter which had been the reactant. Rationalize these results.

20. The Diels-Alder reaction of (−)-pentadienedioic acid (HOOCCH=C=CHCOOH) with cyclopentadiene yields an adduct (**97**) having the absolute stereochemistry shown. What is the absolute configuration of (−)-pentadienedioic acid?

CO_2H H CO_2H H

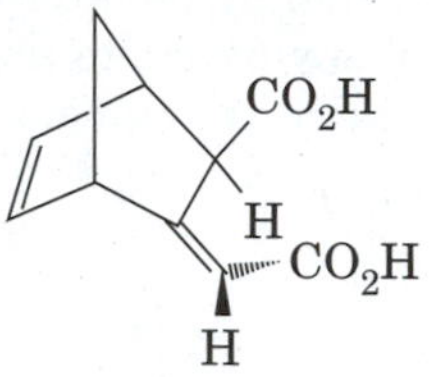

97

21. Upon heating, [3-²H]cyclobutene (**98**) forms products **99** and **100** in a ratio of 1.10 : 1.0. The rate of reaction of unlabeled cyclobutene under the reaction conditions is 8.93×10^{-5} sec^{-1}, while the rate of reaction of **98** is 8.17×10^{-5} sec^{-1}.

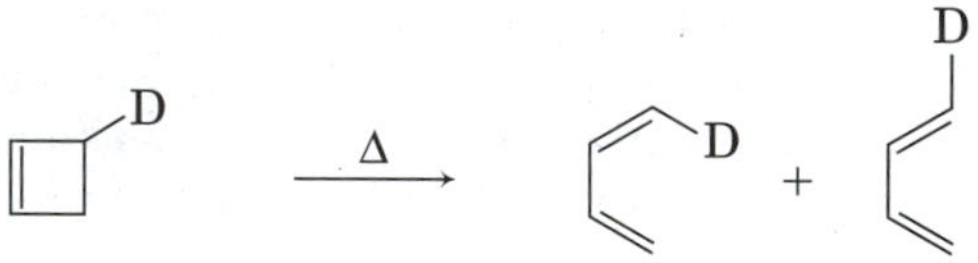

D Δ D + D

98 **99** **100**

a. Assume that the reaction takes place by a conrotatory pathway so that the deuterium atom can either go toward the inside to form **99** or go toward the outside to form **100**. What are the individual rate constants for the formation of **99** and **100**?
b. Using the results from part a, calculate the deuterium isotope effect (k_H/k_D) for the deuterium going inside, as well as k_H/k_D for the deuterium going outside. (Remember to divide the total rate of reaction of the nondeuterated compound by 2 to account for the two possible modes of reaction).
c. What kind of isotope effect is this (i.e., 1°, 2°; α, β, etc.)? Are the isotope effects normal or inverse? What does the result indicate about the degree to which a proton or deuterium experiences hybridization change in the transition structure if it is inside compared to when it is outside? Discuss the significance of this result with regard to the model usually assumed for the electrocyclic reaction.

CHAPTER 12

Photochemistry

12.1 Introduction

Energy and Electronic States

Photochemistry is broadly defined as the study of chemical reactions caused by light.[1] Typically light of interest to organic photochemists is in the ultraviolet-visible (Uv-vis) region of the spectrum, from 200 to 700 nm. Promotion of an electron from a lower energy to a higher energy molecular orbital as a result of absorption of Uv-vis light produces an electronically excited molecule. The high energy and the different electron distribution of the photoexcited molecule can lead to a variety of physical processes and chemical reactions. The concepts of structure and bonding, stereochemistry, reactive intermediates, and reaction type that were developed for thermal reactions also provide a foundation for the study of photochemical reactions. In particular, molecular orbital theory and valence bond theory are useful complementary models in organic photochemistry. MO theory is particularly effective in describing the spectroscopy of ground and excited states. Valence bond models provide useful depictions of electronically excited molecules in terms of ground state reactive intermediates.

Energy levels of molecules are quantized, as shown schematically in Figure 12.1, so only certain energy values are allowed.[2] Superimposed on each electronic state (E_0, E_1, . . .) is a set of vibrational energy levels (v_0, v_1, . . .) and, with smaller divisions still, a set of rotational energy levels (J_0,

[1]For a glossary of terms used in photochemistry, see (a) Pitts, Jr., J. N.; Wilkinson, F.; Hammond, G. S. *Adv. Photochem.* **1963**, *1*, 1; (b) Commission on Photochemistry, IUPAC, *Pure Appl. Chem.* **1988**, *60*, 1055.

[2]See, for example, Moore, W. J. *Physical Chemistry*, 3rd Ed.; Prentice-Hall, Inc.: Englewood Cliffs, NJ, 1962; p. 581 *ff.*

Figure 12.1 A generalized set of energy levels for an organic molecule. (Adapted from reference 2.)

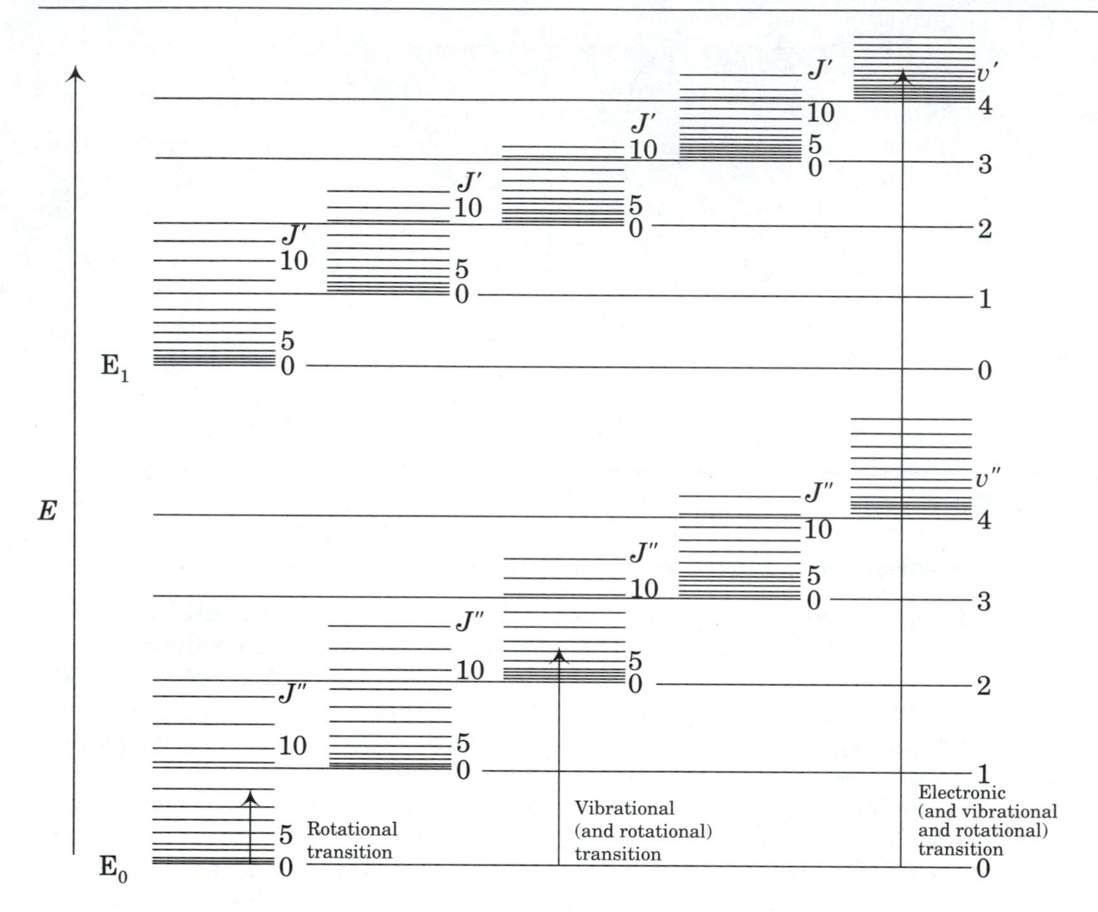

$J_1, \ldots$).[3] A molecule in its ground electronic state may be excited to a higher electronic, vibrational, or rotational energy level (as shown by the arrows in Figure 12.1) either by absorbing a photon of light corresponding exactly to the energy difference between the initial and final states or by accepting energy from another molecule, usually through a collision. A transition from one rotational energy level to another (within the same electronic and vibrational energy levels) requires only a small amount of energy, so the transition can be caused by microwave radiation. Changes from one vibrational level to another (within the same electronic state), require light in the infrared region of the spectrum. A transition from the ground electronic state

[3]By convention in physical chemistry, vibrational levels and rotational levels of the ground electronic state are denoted with double primes, as in v'' and J'', respectively. Vibrational levels and rotational levels of the excited state are denoted with single primes, as in v' and J'. Often, however, organic photochemists omit the double prime notation for the ground electronic state. Also, the vibrational quantum levels may be denoted as v or V instead of v.

to a higher electronic state generally requires light in the Uv-vis region, and such transitions give rise to photochemistry.

Photochemistry is inherently related to ***photophysics***, the study of those radiative and nonradiative processes that interconvert one electronic state with another electronic state without chemical change.[1(b),4] Central to both photochemistry and photophysics is the discussion of light in terms of its **energy**. Because light is quantized, it has properties like those of a particle, and an Avogadro's number of photons (or one mole of light quanta) defines a unit of measure termed the **Einstein**.[5,6] The relationship between the energy of light and its wavelength is given in equation 12.1.[7]

$$E\ (\text{kcal/mol}) = h\nu = hc/\lambda = 2.86 \times 10^4/\lambda\ (\text{nm}) \quad \textbf{(12.1)}$$

In addition to units of kcal/mol, energy is often given in electron volts (eV). It is also convenient to express the energy of spectroscopic transitions in wave numbers ($\bar{\nu}$), also known as reciprocal centimeters (because the units are cm^{-1}), since energy increases with increasing wave numbers.[8] It is important to distinguish between the *energy* of light and the *radiant intensity* of a particular light source.[9] A beam of low energy light with high radiant intensity (such as an infrared laser) is generally capable of causing only vibrational changes, although in a large number of molecules.[10] A weak beam of high energy light (as in a strongly filtered mercury vapor lamp) can cause electronic transitions, although in relatively fewer molecules.

A tabulation of light energies as a function of wavelength is given in Table 12.1. Light with a wavelength of 300 nm has an energy of ca. 96 kcal/mol, which is greater than the bond dissociation energy of many types

---

[4]Turro, N. J. *Modern Molecular Photochemistry*; The Benjamin/Cummings Publishing Co., Inc.: Menlo Park, CA, 1978; p. 4.

[5]Thus, for light of a single wavelength (monochromatic light), equation 12.1 allows calculation of the energy of light in kcal/mol of light quanta of that wavelength.

[6]For a more complete discussion of these relationships, see reference 4, pp. 8 *ff*.

[7]Organic photochemists often use units of nanometers (nm, 10^{-9} m) for wavelength, but another common unit is Angstroms (Å, 10^{-8} cm; 1 nm = 10 Å).

[8]These units may be interconverted by the relationship 1 eV = 23.06 kcal/mol = 8063 cm^{-1}. For a more complete discussion of units, see Gordon, A. J.; Ford, R. A. *The Chemist's Companion*; Wiley-Interscience: New York, 1972; p. 166.

[9]For further details, see reference 1. The unit of radiant power, the watt (W), is equal to 1 J sec^{-1}.

[10]In this discussion we are ignoring the possibility of **multiphoton excitation**, in which several photons of lower energy light can be absorbed within a short time to produce an electronically excited state. Multiphoton infrared excitation can lead to significant photochemical reactions. See, for example, (a) Lewis, F. D.; Buechele, J. L.; Teng, P. A.; Weitz, E. *Pure Appl. Chem*. **1982**, *54*, 1683; (b) reference 10; (c) Lupo, D. W.; Quack, M. *Chem. Rev*. **1987**, *87*, 181. For a review of high intensity laser photochemistry of organic compounds, see Wilson, R. M.; Schnapp, K. A. *Chem. Rev*. **1993**, *93*, 223. Absorption of intense UV laser radiation can lead to controlled explosive decomposition of materials, a process known as ***ablation***. For an introduction, see Srinivasan, R. *Science* **1986**, *234*, 559; for a specific application, see Srinivasan, R.; Ghosh, A. P. *Chem. Phys. Lett*. **1988**, *143*, 546.

Table 12.1 Energy content of different wavelengths of light

λ (nm)	$\bar{\nu}$ (cm^{-1})	ΔE (kcal/mol)	ΔE (eV)
200	50,000	143	6.20
250	40,000	114.4	4.96
300	33,333	95.5	4.14
350	28,571	81.7	3.54
400	25,000	71.5	3.10
450	22,222	63.6	2.76
500	20,000	57.2	2.48

of single bonds. Thus, light of this wavelength could be used to photodissociate organic molecules, provided that the excitation energy could be efficiently channeled into the dissociative pathway. Bond dissociation does occur in some cases, but most organic molecules can exhibit other physical and chemical processes as well.

Designation of Spectroscopic Transitions

In organic photochemistry, spectroscopic transitions are often designated by terms indicating the change in electron population of the molecular orbitals. Consider the model for the molecular orbitals of formaldehyde proposed by Mulliken, in which the carbon atom is considered to be sp^2-hybridized and the oxygen atom is considered to be unhybridized.[11,12] A simple pictorial representation of the atomic orbitals that are the basis set for the MOs is shown in Figure 12.2, and the energies of the calculated set of MOs are

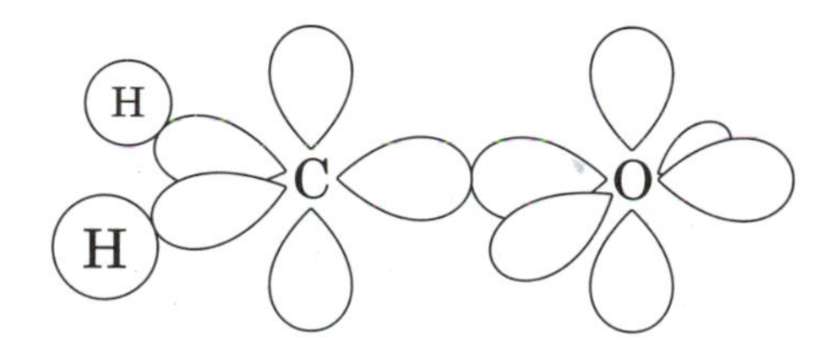

Figure 12.2 Basis set atomic orbitals for MOs of formaldehyde.

[11]Sidman, J. W. *Chem. Rev.* **1958**, *58*, 689. Also see the discussion in reference 4, pp. 21–22.

[12]This very simple representation of the molecular orbitals of formaldehyde provides a convenient model for describing the spectroscopy and photochemistry of carbonyl compounds, but more advanced representations may be more useful for other purposes. See, for example, Liang, M. *J. Chem. Educ.* **1987**, *64*, 124; Wiberg, K. B.; Marquez, M.; Castejon, H. *J. Org. Chem.* **1994**, *59*, 6817.

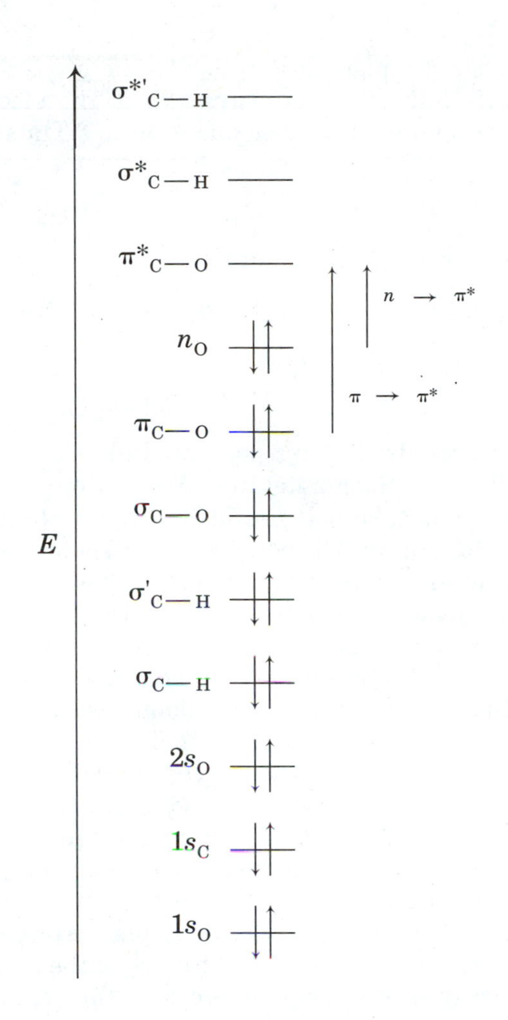

Figure 12.3 MO energy level diagram for formaldehyde showing $n \rightarrow \pi^*$ and $\pi \rightarrow \pi^*$ transitions.

given in Figure 12.3. The HOMO for formaldehyde is represented as a nonbonding orbital that is localized on oxygen and which lies in the plane of the molecule, while the LUMO is a π^*_{C-O} orbital. Thus, the lowest energy possible for an electronic transition is an $n \rightarrow \pi^*$ transition.[13] A higher energy transition is the $\pi \rightarrow \pi^*$ transition. In principle, many other electronic transitions may be seen by irradiating the compound with enough light to move an electron from a lower lying MO to the LUMO or from HOMO to a higher energy antibonding orbital. In practice, however, the two transitions described are the ones most typically responsible for the spectroscopic characteristics and photochemical reactions of carbonyl compounds.

---

[13]The notation used for these transitions is attributed to Michael Kasha (footnote 9, reference 1(a). Kasha's contributions to chemistry are summarized in *J. Phys. Chem.* **1991**, *95*, 10216.

In order to give a complete description to the process that occurs, it is necessary to specify fully the **configuration** of the electronic state—the population of electrons in each molecular orbital.[14] Thus the ground state (S_0) is[15]

$$(1s_O)^2(1s_C)^2(2s_O)^2(\sigma_{C-H})^2(\sigma'_{C-H})^2(\sigma_{C-O})^2(\pi_{C-O})^2(n_O)^2$$

Similarly, the lowest energy excited state is

$$(1s_O)^2(1s_C)^2(2s_O)^2(\sigma_{C-H})^2(\sigma'_{C-H})^2(\sigma_{C-O})^2(\pi_{C-O})^2\boldsymbol{(n_O)(\pi^*_{C-O})}$$

and the other excited state shown is

$$(1s_O)^2(1s_C)^2(2s_O)^2(\sigma_{C-H})^2(\sigma'_{C-H})^2(\sigma_{C-O})^2\boldsymbol{(\pi_{C-O})}(n_O)^2\boldsymbol{(\pi^*_{C-O})}$$

(In both cases the orbitals that are populated differently from the ground state are shown in bold.) Such a detailed description of the electronic configuration is cumbersome, so it is useful to adopt a shorthand notation to convey the same information. The notation specifies the configuration of the excited state by noting the type of transition that produced it. Thus the lower energy excited state is denoted an n,π^* state, and the higher energy excited state is called a π,π^* state.[16,17]

One additional bit of information must be added to the description of an electronic state. In S_0 all electrons exist in doubly occupied orbitals, and two electrons in the same orbital must have their spins paired. However, two electrons in different MOs may either have their spins paired or unpaired. If they are paired, the state is said to be a ***singlet state***, but if they are unpaired, it is said to be a ***triplet state***. States that have the same number of unpaired electrons are said to have the same ***multiplicity***.[18] The lowest energy excited singlet state is denoted S_1, and the next lowest energy excited singlet state is called S_2. The two lowest triplets are denoted T_1 and T_2. Thus, the general form of the notation is that S_0 is the ground singlet state, S_1 is the lowest energy excited singlet state, S_2 is the next lowest energy sin-

[14]More precisely, states can be described by combinations of configurations. In the discussion here, however, each state is associated with only one configuration. For a discussion of configurations, see reference 4, pp. 17–24.

[15]This term is pronounced S zero. In the configuration on the following line, the first term in parentheses is $1s_O$ (one s sub oh) meaning the 1*s* orbital on oxygen. Although they appear similar at first glance, the terms S_0 and $1s_O$ have very different meanings. The term S_0 refers to a designation of all occupied orbitals, while there term $1s_O$ indicates one particular orbital only.

[16]Alternatively, these states may be denoted n-π^* and π-π^*, respectively; see reference 1(b).

[17]The terms used in this sentence are usually pronounced as an n to π^* state and a π to π^* state. The electronic transitions leading to these states are usually written with arrows as $n \rightarrow \pi^*$ and $\pi \rightarrow \pi^*$ transitions, respectively.

[18]The multiplicity of a state is given by $m = 2\,S + 1$, where S is the total spin quantum number of the state (reference 1(b)). A species in which all the electrons are paired is called a singlet state ($m = 2 \times 0 + 1$), a radical with one unpaired electron is termed a doublet state ($m = 2 \times 1/2 + 1$), and a biradical or excited state with two unpaired electrons is a called triplet. For a discussion of the research of G. N. Lewis and the association of the triplet state with phosphorescence of organic compounds, see Kasha, M. *J. Chem. Educ.* **1984**, *61*, 204.

glet state, and so on. Similarly, T_1 is the lowest energy (excited) triplet state, T_2 is the next lowest energy triplet state, and so on.[19] Formaldehyde may have both singlet and triplet n,π^* states and both singlet and triplet π,π^* states.

Photophysical Processes

A useful way of representing the energies of the electronic states of a molecule is a **Jablonski diagram**, such as Figure 12.4.[20] The vertical scale represents potential energy. The horizontal scale has no particular significance; it allows us to separate the singlet and triplet state ***manifolds*** (sets of energy levels) so that the excited states are more clearly distinguished. Superimposed on each electronic state is a set of vibrational energy levels. For the sake of clarity, a set of rotational energy levels superimposed on each vibrational level is not shown.

There are two kinds of photophysical processes indicated in Figure 12.4. Those interconversions denoted by straight lines are **radiative processes**, which occur through the absorption or emission of light. Those indicated by

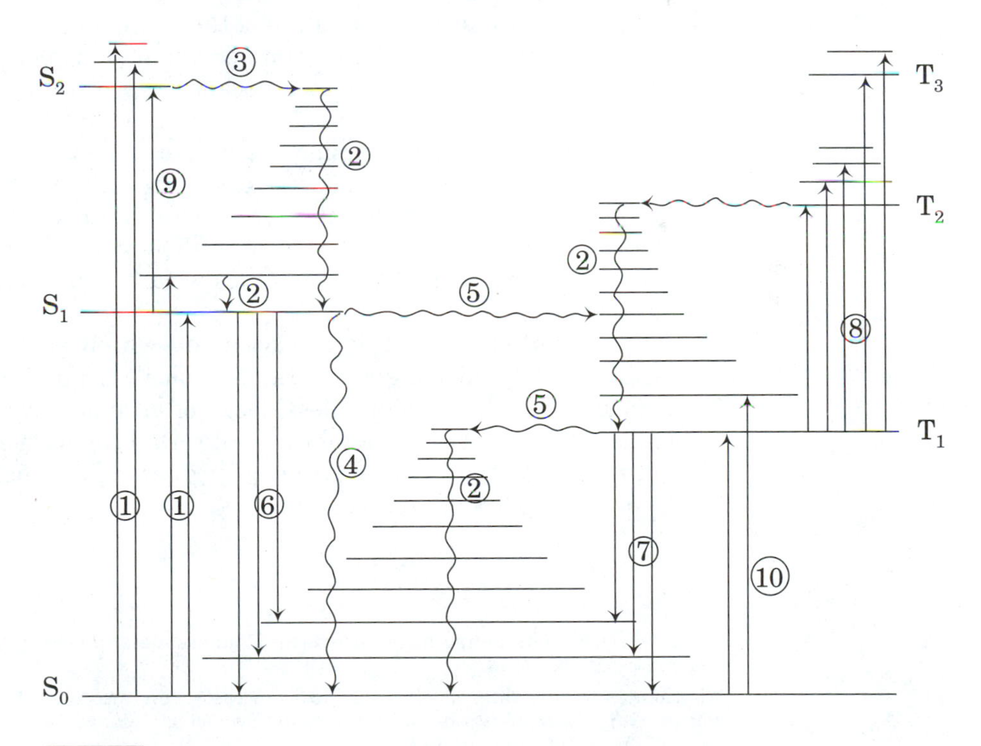

Figure 12.4 A generalized Jablonski diagram. (Adapted from reference 20.)

[19]It might be argued that the state called T_1 should be called T_0, since it is the lowest energy triplet state possible for a given molecule. However, the subscript 0 is reserved for the ground electronic state of a molecule. For those structures (such as diradicals) for which the ground state is a triplet state, then T_0 would be the ground state designation.

[20]For a discussion, see Wilkinson, F. *Quart. Rev. Chem. Soc.* **1966**, *20*, 403.

wavy lines are **nonradiative processes**, which occur without light being absorbed or emitted.[21] The numbers on the lines are keyed to the following definitions:

1. **Absorption of Light**. A ground state molecule (S_0) may absorb a photon of light, thus becoming converted to an excited state. The most likely transitions are $S_0 \rightarrow S_1$ or $S_0 \rightarrow S_2$, although S_0 to higher excited singlet state transitions are also possible.
2. **Vibrational Relaxation**. The absorption from S_0 to S_n involves an energy change from the 0th vibrational level of S_0 to any vibrational level of the excited state. However, the $v'' = 0$ vibrational level is the level most populated at room temperature for the ground electronic state of a molecule, and $v' = 0$ is also the vibrational level of the excited electronic state that is most likely to be populated at equilibrium. Unless the molecule dissociates before equilibrium can be obtained, there is a very rapid process (with a rate constant of about 10^{12} sec^{-1}) that relaxes the higher vibrational level of the excited state to its 0th vibrational level in condensed phases (i.e., solids or liquids).[22,23]
3. **Internal Conversion** is a nonradiative process that converts a higher electronic state into a lower state of the same multiplicity (a higher singlet state into a lower singlet state or a higher triplet state into a lower triplet state). The name arises because the process occurs internally, i.e., within the singlet manifold or within the triplet manifold. Note that the line for internal conversion is a horizontal one. This means that the 0th vibrational level of S_2 is converted into a vibrationally excited S_1 state having the same total energy. The vibrationally excited S_1 can then relax to its 0th vibrational level. The rate constants for internal conversion are fast ($>10^{10}$ sec^{-1}), especially when the two states are close in energy.[21]
4. **Radiationless Decay** is a process by which electronically excited states are returned to ground states (typically from S_1 to S_0) without the emission of light. Radiationless decay often has a slower rate constant (ca. $< 10^6$ sec^{-1}) than other forms of internal conversion because the energy gap between S_1 and S_0 is usually greater than that between S_2 and S_1 or other pairs of excited states.

---

[21]For a discussion of nonradiative (also termed radiationless) processes, see Freed, K. F. *Acc. Chem. Res.* **1978**, *11*, 74.

[22]Estimates for the rates of photochemical processes here are taken from the discussion by Porter, G. in *An Introduction to Photobiology*; Swanson, C. P., ed.; Prentice-Hall, Inc.: Englewood Cliffs, N. J.; 1969, pp. 1–22.

[23]In some gas phase studies it is thought that photochemistry can occur from the upper vibrational levels of the electronic excited state, but it is believed that vibrational relaxation occurs so rapidly in solution that most photochemical reactions in condensed phase take place from the 0th vibrational level of the electronically excited state. For an exception to this generalization, see Manring, L. E.; Peters, K. S. *J. Am. Chem. Soc.* **1984**, *106*, 8077.

5. **Intersystem Crossing**, the conversion of a singlet state into a triplet state (or vice versa), requires a spin flip of an electron. The probability of intersystem crossing depends, among other things, on the energy gap between the singlet and triplet states, so values of k_{isc} vary from 10^6 to 10^{10} sec^{-1}. The T_n state is lower in energy than the corresponding S_n state because of the lower electron repulsion for unpaired electrons.[24,25]

Processes 2–5 are nonradiative processes. The following are radiative processes.

6. **Fluorescence** is the emission of light from an excited state to a ground state with the same multiplicity. Usually the emission is $S_1 \rightarrow S_0$, and a generalization to that effect is known as Kasha's rule.[26] However, **anomalous fluorescence** ($S_2 \rightarrow S_0$) occurs in some compounds, among them azulene,[27] thiocarbonyl compounds,[28] and some gaseous polyenes.[29]
7. **Phosphorescence** is the emission of light from an excited state to a ground state with different multiplicity (usually from a triplet excited state to a singlet ground state). This process involves both an electronic state change and a spin flip so, like $S_0 \rightarrow T_n$ absorption, phosphorescence is a spin-forbidden process. (See the discussion on page 804.)
8. **Triplet-Triplet Absorption**. A molecule in a triplet excited state may absorb a photon to give a higher triplet state, so a Uv-vis spectrum may be obtained, and the time-dependence of the excited state decay may be monitored. (For further details, see the discussion of flash spectroscopy on page 817.) Triplet-triplet absorption spectroscopy can be an important technique for detecting triplet excited states.
9. **Singlet-Singlet Absorption**. Because triplet states may persist longer than singlet states, the only excitation of excited states possible during much of the development of photochemistry was triplet-triplet absorption. With the advent of picosecond and femtosecond spectroscopy,[30,31] it

---

[24]For a discussion of intersystem crossing, see (a) Turro, N. J. *J. Chem. Educ.* **1969**, *46*, 2; (b) McGlynn, S. P.; Azumi, T.; Kinoshita, M. *Molecular Spectroscopy of the Triplet State*; Prentice-Hall, Inc.; Englewood Cliffs, N. J., 1969.

[25]Although intersystem crossing is usually considered to involve singlet to triplet conversion, reverse intersystem crossing from T_2 to S_1 has been observed in anthracenes: Fukumura, K.; Kikuchi, K.; Koike, K.; Kokubun, H. *J. Photochem. Photobiol. A. Chem.* **1988**, *42*, 282.

[26]Kasha, M. *Discuss. Faraday Soc.* **1950**, *9*, 14.

[27]Beer, M.; Longuet-Higgins, H. C. *J. Chem. Phys.* **1955**, *23*, 1390.

[28]See, for example, Rao, V. P.; Ramamurthy, V. *J. Org. Chem.* **1988**, *53*, 332. For a review of thiocarbonyl photophysics, see Maciejewski, A.; Steer, R. P. *Chem. Rev.* **1993**, *93*, 67.

[29]Bouwman, W. G.; Jones, A. C.; Phillips, D.; Thibodeau, P., Friel, C.; Christensen, R. L. *J. Phys. Chem.* **1990**, *94*, 7429. Both $S_1 \rightarrow S_0$ and $S_2 \rightarrow S_0$ emission were observed from tetraenes and pentaenes in the gas phase.

[30]A **nsec** (nanosecond) is 10^{-9} sec. One **psec** (picosecond) is 10^{-12} sec. One **fsec** (femtosecond) is 10^{-15} sec.

[31]See, for example, Shank, C. V. in *Photochemistry and Photobiology*, Vol. 1; Zewail, A. H., ed.; Harwood Academic Publishers: New York, 1983; pp. 517–527; Repinec, S. T.; Sension, R. J.; Szarka, A. Z.; Hochstrasser, R. M. *J. Phys. Chem.* **1991**, *95*, 10380.

has become possible also to measure transitions from one excited singlet state to another, higher energy excited singlet state.

10. **Singlet-Triplet Absorption**, like phosphorescence, is a spin-forbidden process, so ordinarily $S_0 \rightarrow T_n$ transitions are not observed in Uv-vis spectroscopy. However, these transitions can be seen under certain conditions, as discussed further on page 809.

Selection Rules for Radiative Transitions

A detailed analysis of the factors that determine the probabilities of radiative transitions is beyond the scope of the present discussion. Briefly, however, three factors influence the probability of absorption or emission of light.[1(a)] One is based on symmetry considerations and the quantum mechanical formulation of transition moment integrals. If the initial and final electronic states differ in symmetry, then the transition is symmetry-allowed. If they are not, the transition is said to be **symmetry-forbidden**. The term *forbidden* does not necessarily mean that the transition cannot be observed, only that it is weak.

The second factor that determines the probability of a radiative electronic transition is the spin factor. Because an electron must both undergo an electronic transition and a spin flip during intersystem crossing, singlet-triplet absorption, or phosphorescence, these are said to be **spin-forbidden** processes. Singlet-triplet interconversions can be seen in many molecules, however, because pure singlet and triplet configurations are not complete descriptions for electronically excited states. Through a process known as **spin-orbit coupling**, some triplet character is mixed into pure singlet states, and some singlet character is mixed into triplet states. The greater the mixing of singlet and triplet character in either state, the greater will be the rate of the spin flipping process. The mixing of singlet and triplet configurations is thought to occur through the interaction of the magnetic moment of the spinning electron with the changing magnetic field (due to the nucleus) that the electron encounters as it moves in atomic orbitals.[32] Therefore, the heavier the nucleus, the greater the likelihood for a spin flip. Compounds incorporating high atomic number atoms can exhibit a **heavy atom effect** that results in higher rates of intersystem crossing. In addition, greater mixing occurs when the singlet and triplet states are closer in energy. For many organic compounds the two states S_1 and T_1 differ considerably in energy, so $S_1 \rightarrow T_1$ intersystem crossing is slow. For carbonyl compounds, however, the S_1–T_1 energy gap is often small, and $S_1 \rightarrow T_1$ intersystem crossing is much faster.

The third factor, which is often called the Franck-Condon term, is determined by the overlap between the nuclear coordinates in the initial and

[32]For an introduction to spin-orbit coupling, see reference 4.

final electronic states.[33] The greater the overlap (that is, the more nearly the nuclear coordinates in the initial state are the same as those in the final state), the more probable is the transition. This can more easily be visualized by considering the electronic and vibrational energy levels of a hypothetical molecule, as represented in Figure 12.5.[34,35] In the lower left portion of that figure, the energies of the states S_0 and S_1 are represented as functions of the internuclear distance $r_{A—B}$, where A and B are a pair of bonded

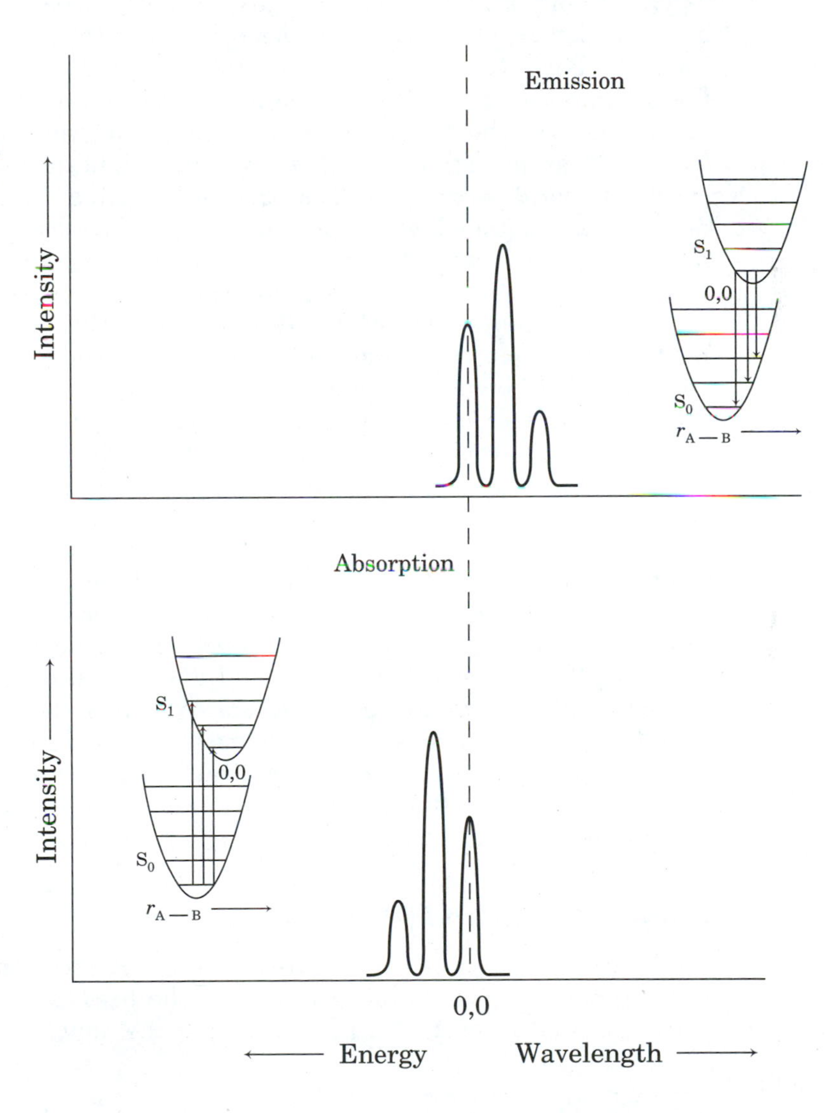

Figure 12.5 Schematic representations of the origins of Uv-vis absorption (bottom) and fluorescence (top) spectra.[34]

[33]Condon, E. U. *Am. J. Phys.* **1947**, *15*, 365 and references therein.

[34]A portion of Figure 12.5 was modified from a figure in Kohler, B. E. *Chem. Rev.* **1993**, *93*, 41

[35]See also Lee, S.-Y. *J. Phys. Chem.* **1990**, *94*, 4420.

atoms in the molecule. Because excitation promotes an electron from a nonbonding or bonding orbital to an antibonding orbital, bonding between any particular pair of atoms may be less in the excited state than in the ground state. Thus the equilibrium distance of the A—B bond may be longer in the excited state than in the ground state, so the curve corresponding to the electronic energy of the excited state is shown displaced to the right of the ground state curve.

Electronic transitions are said to be vertical, meaning that absorption and emission of light occurs with no movement of the nuclei (within the limits of the Born-Oppenheimer approximation). Since most molecules are in the 0th vibrational level of the ground electronic state, the Franck-Condon term measures the probability of the transition from the 0th vibrational level of the ground state to each vibrational level of the excited state, so the overlap integral varies with v'. Furthermore, each of these transitions is of different energy, so a high resolution absorption spectrum would appear as indicated in the lower portion of Figure 12.5.[36] Organic molecules in solution usually do not exhibit sharp absorption lines in Uv-vis spectra, however, because large molecules may have many closely-spaced vibrational bands and because rotational transitions and collisions (especially with solvent molecules) broaden the individual vibrational bands.

Tables of absorption spectra of organic compounds usually report Uv-vis spectral data in terms of the wavelength of maximum absorption (λ_{max}) and the molar absorptivity at maximum absorption (ϵ_{max}).[37] These values are characteristic for an organic compound in a particular solvent and may be used to help identify it.[38] For a photochemist, however, λ_{max} is only part of the information available from the spectrum. Often more important is the onset of absorption (the wavelength at which absorption begins), since that provides an upper limit for the 0,0 transition, the longest wavelength (lowest energy) absorption that can create the electronically excited state. Because vibrationally excited electronic states relax to the 0th vibrational level rapidly in condensed phases, photochemistry in solution usually occurs from the 0th vibrational level of an electronically excited state. Thus the energy associated with the 0,0 transition can be used to determine the energy available in the photoexcited molecule to drive a photochemical reaction.

Fluorescence and Phosphorescence

For most organic molecules, fluorescence is the spontaneous emission of light from the $v' = 0$ vibrational level of the first excited singlet state to some vibrational level ($v'' = 0, 1, 2, \ldots$) of the (singlet) ground electronic

[36]For a more detailed discussion, see Harris, D. C.; Bertolucci, M. D. *Symmetry and Spectroscopy*; Oxford University Press: New York, 1978; pp. 330 *ff.*

[37]The Beer-Lambert law states that $\epsilon = A/bc$, where A is the absorbance of the sample, b is the path length in cm, and c is the concentration in mol L^{-1}.

[38]*Cf.* Silverstein, R. M.; Bassler, G. C.; Morrill, T. C. *Spectrometric Identification of Organic Compounds*, 5th Ed.; John Wiley & Sons, Inc.: New York, 1991; pp. 289–314.

state.[39] As indicated in the top portion of Figure 12.5, the energy of the photon emitted in the $v' = 0$ to $v'' = 0$ fluorescence is the same as the energy of the photon absorbed in the $v'' = 0$ to $v' = 0$ transition if the geometry of the photoexcited molecule is nearly the same as that of the ground state molecule. However, all other fluorescence lines are at longer wavelengths (lower energy). Thus, the fluorescence and emission spectra should overlap at the 0,0 transition, providing confirmation of the 0,0 energy of the electronically excited state. In molecules such as anthracene, the σ bonding provides a molecular framework for the planar π system, and the π bonding results from population of many bonding MOs. Therefore promotion of one electron to an antibonding MO may not seriously distort the molecular geometry. In such cases there can be a nearly mirror image relationship between the absorption and fluorescence spectra of organic molecules, particularly when the spectra are plotted as intensity versus energy (cm^{-1}) instead of wavelength.[40] The similarity arises because the factors that make some $v'' = 0$ to $v' = x$ transitions more probable than others also make some $v' = 0$ to $v'' = x$ emissions more probable. Figure 12.6 shows the absorption and fluorescence spectra of anthracene plotted as intensity versus energy in cm^{-1}.[41]

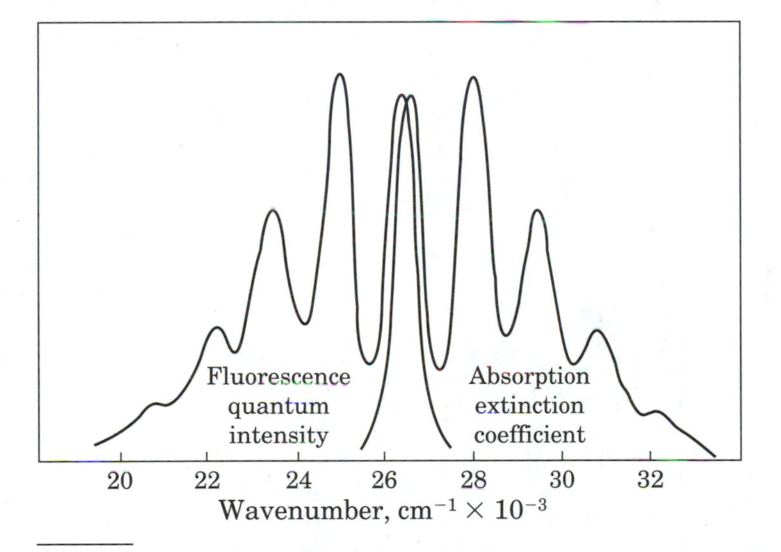

Figure 12.6 Mirror image relationship in absorption and fluorescence spectra of anthracene. (Reproduced from reference 41.)

[39]More precisely, fluorescence is the spontaneous emission of light from a higher energy excited state resulting in the formation of a lower energy state with the same multiplicity. Thus, a $T_2 \rightarrow T_1$ transition is also termed fluorescence. For example, Després, A.; Lejeune, V.; Migirdicyan, E.; Siebrand, W. *J. Phys. Chem.* **1988**, *92*, 6914, reported the fluorescence of photoexcited *m*-xylylene biradical,

which has a triplet ground state.

[40]For a discussion of the relationship between molecular geometry and absorption and fluorescence spectroscopy of π systems, see Berlman, I. B. *J. Phys. Chem.* **1970**, *74*, 3085.

[41]Bowen, E. J. *Adv. Photochem.* **1963**, *1*, 23. The spectral intensities are normalized to the same intensity for the 0,0 transition in order to facilitate the comparison.

There is generally a good mirror image relationship between the absorption and emission bands, although the slight difference in the 0,0 bands apparently arises from some relaxation of the initially formed excited state.[42]

For some molecules, however, the geometry of the ground and excited states may be very different. As a result, there may be a large difference between λ_{max} for absorption and λ_{max} for emission, and the 0,0 transition may be weak or not present. In the sterically hindered molecule fluorenylidene(9-anthryl)methane (**1**, Figure 12.7), for example, the geometry of the excited singlet state differs considerably from that of the ground state, and only a much lower energy, structureless emission is seen.[43] The reverse situation can also occur. The two benzene rings are not coplanar in the ground

1

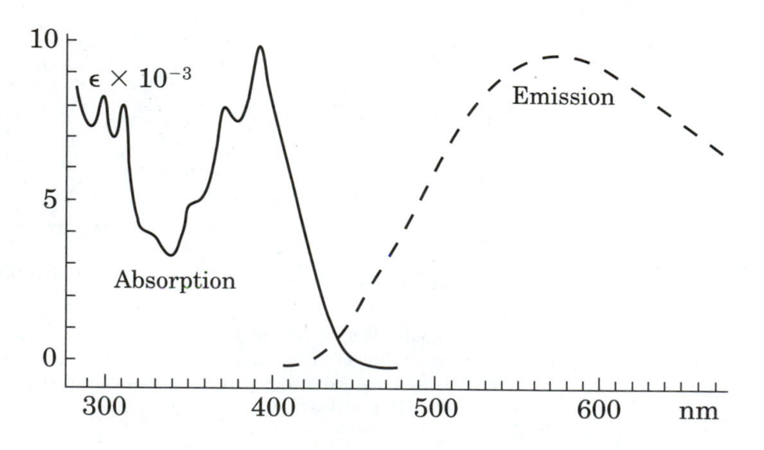

Figure 12.7 Effect of change in excited state geometry on fluorescence (- - -) spectrum of **1**. (Adapted from reference 43.)

[42]The energy difference between the λ_{max} values for absorption and emission is called the **Stokes shift**.

[43]Becker, H.-D.; Andersson, K. *J. Org. Chem.* **1983**, *48*, 4542.

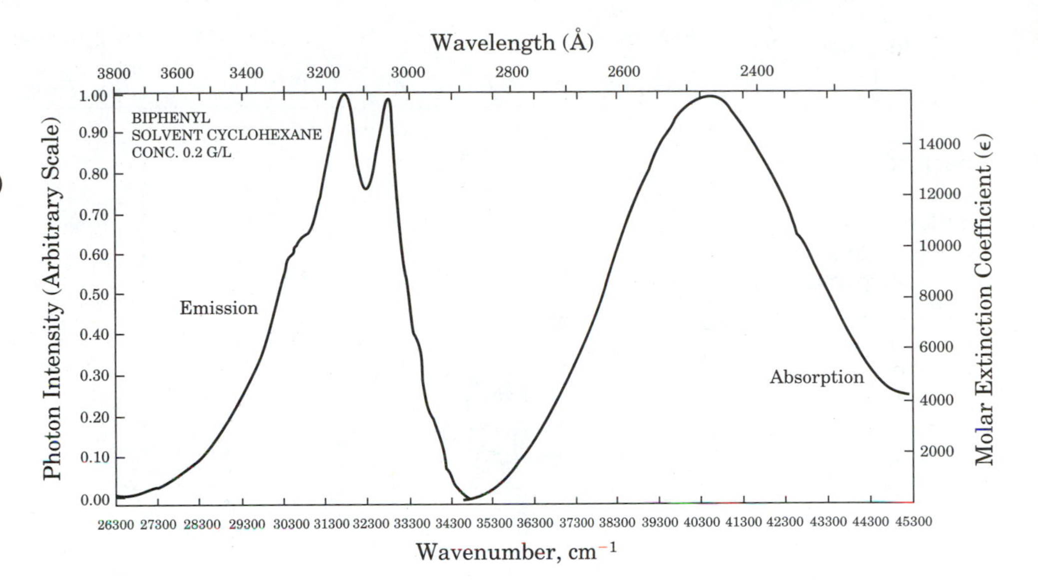

Figure 12.8 Absorption (right) and fluorescence (left) spectra of biphenyl. (Adapted from reference 44.)

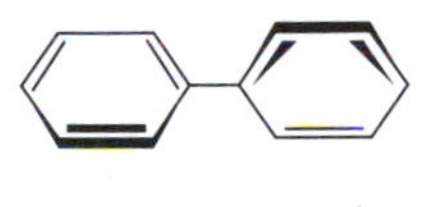

state of biphenyl, but intramolecular charge transfer produces a planar excited state with double bond character between the two rings. Thus, the absorbance of biphenyl (**2**) is structureless, while the fluorescence spectrum shows some vibrational structure (Figure 12.8).[44,45]

A similar relationship between geometry and a mirror image appearance of singlet-triplet absorption and phosphorescence is expected, but this is often difficult to determine experimentally. In fluid solution phosphorescence is usually reduced by diffusion-limited bimolecular interaction of the excited triplet compound and one or more ground state species. However, phosphorescence can often be observed by irradiating the compound in environments in which diffusion is quite slow, such as in an organic glass at liquid nitrogen temperature.[46] In addition, organic guests surrounded by host molecules can be protected from diffusional quenching. For example, Figure 12.9 shows a series of total emission spectra of 6-bromo-2-naphthol

---

[44]Berlman, I. B. *Handbook of Fluorescence Spectra of Aromatic Molecules*, 2nd Ed.; Academic Press: New York, 1971.

[45]Note that the horizontal scale in Figure 12.8 is linear in wavenumbers, so higher energy is to the right.

[46]A glass is a liquid medium that becomes extremely viscous but does not crystallize—and thus scatter light—at the temperature of the experiment. For example an EPA glass (made by mixing ether, isopentane and ethyl alcohol in a ratio of 5 : 5 : 2) remains transparent at −196° but is so viscous that diffusion is precluded. For details, see Murov, S. L. *Handbook of Photochemistry*; Marcel Dekker, Inc.: New York, 1973; pp. 90–92.

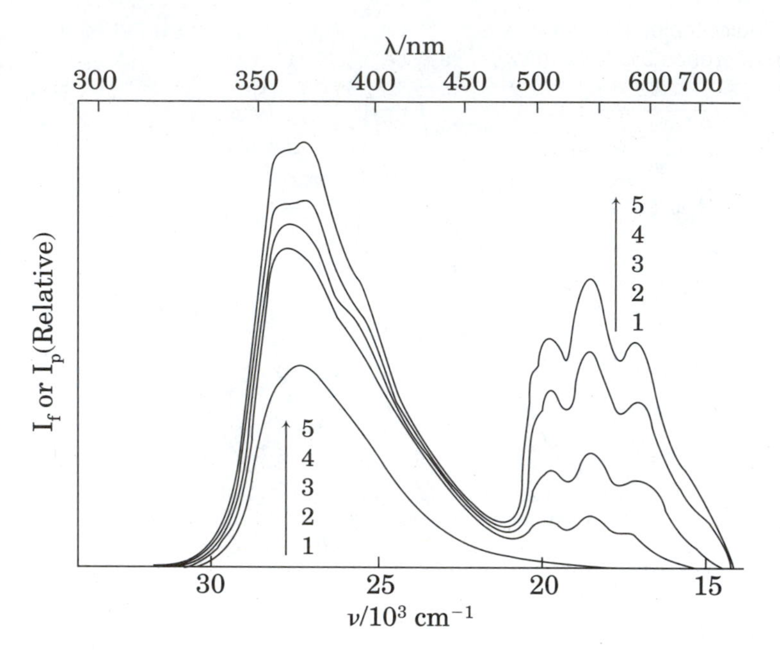

Figure 12.9 Observation of phosphorescence of 6-bromo-2-naphthol in a 1 : 2 complex with α-cyclodextrin in aqueous solution at room temperature. (Adapted from reference 48. Note that the *x*-axis is linear in cm^{-1}, so higher energies are to the left; λ values are shown above.)

at room temperature in aqueous solution containing varying concentrations of α-cyclodextrin.[47] In the absence of α-cyclodextrin (curve 1), there is no phosphorescence, so only fluorescence is seen. At a concentration of α-cyclodextrin of 1×10^{-2} M (curve 5), appreciable phosphorescence is detected. The phosphorescence is attributed to 6-bromo-2-naphthol molecules encapsulated in a 1 : 2 complex with two α-cyclodextrin molecules.[48]

Singlet-triplet absorption is ordinarily difficult to detect because the transition is spin-forbidden. Heavy atom solvents or oxygen perturbation have been used to induce singlet-triplet absorption, but absorptions observed in this manner are weak, and it is important to establish that the observed absorption is not due to artifacts resulting from the solvent or additive.[49,50]

---

[47]A total emission spectrum shows fluorescence and any phosphorescence that can be detected.

[48]Hamai, S. *J. Chem. Soc., Chem. Commun.* **1994**, 2243.

[49]For example, see (a) Evans, D. F. *J. Chem. Soc.* **1960**, 1735; (b) Rest, A. J.; Salisbury, K.; Sodeau, J. R. *J. Chem. Soc., Faraday Trans. 2* **1977**, *73*, 1396.

[50]Oxygen is a paramagnetic compound, so it enhances spin-orbit coupling and can enhance singlet-triplet conversions. Oxygen is a strong quencher of excited states, both singlet and triplet, but it is an especially effective quencher of triplet states due to their longer lifetimes. (See the discussion of quenching beginning on page 820.)

Figure 12.10 Singlet-triplet absorption spectrum and phosphorescence spectrum of naphthalene. (Reproduced from reference 51. Note that in these spectra longer wavelength (lower energy) is to the left.)

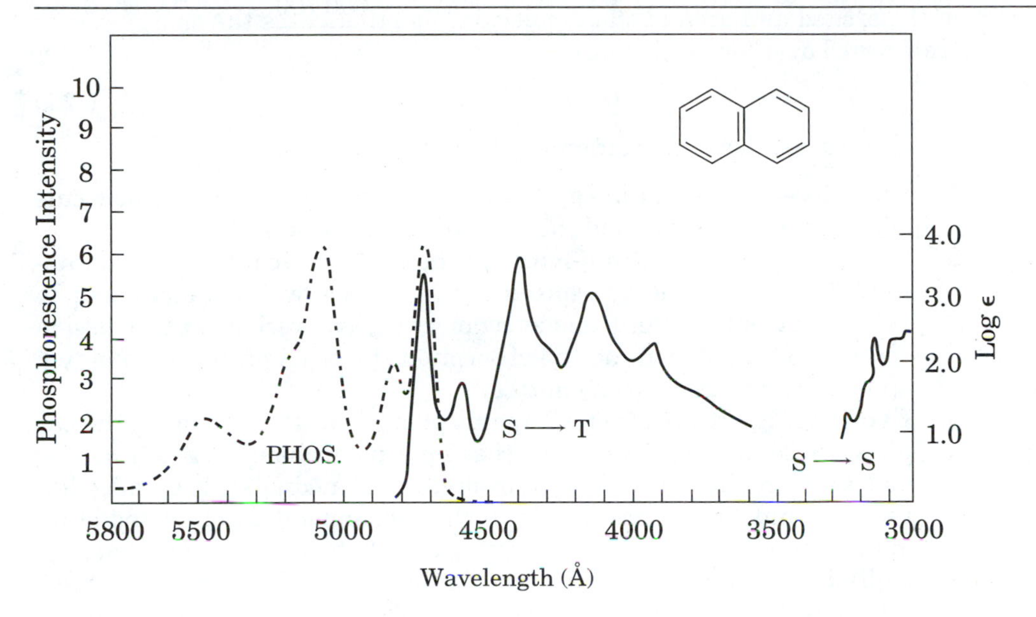

Marchetti and Kearns reported an alternative method in which the excitation spectrum of phosphorescence (that is, a spectrum indicating the relative efficiency of different wavelengths of excitation light in causing phosphorescence of organic compounds) is used to measure $S_0 \rightarrow T_1$ absorption in crystals or glasses.[51] The mirror image relationship between the singlet-triplet absorption so measured and the phosphorescence of naphthalene is shown in Figure 12.10.

Energy Transfer and Electron Transfer

Excited states are created by the absorption of a photon of light, but that is not the only way to produce them.[52] Electronically excited states can also be created by energy transfer (equation 12.2), just as vibrationally excited

[51]Marchetti, A. P.; Kearns, D. R. *J. Am. Chem. Soc.* **1967**, *89*, 768.

[52]In this section the phrase *excited state* will mean an electronically excited state unless there is a qualifying term, as in "vibrationally excited state."

states can be created by energy transfer from hot molecules or container walls.[53] In some cases the electronic energy is transferred by direct collision of the excited and ground state molecules; in other cases the energy can be transferred over longer distances.

$$\mathbf{D}^* + \mathbf{A} \longrightarrow \mathbf{D} + \mathbf{A}^* \tag{12.2}$$

There are at least four different kinds of energy transfer:[54]

1. **Radiative** or **trivial energy transfer**. In this case the donor molecule emits a photon of light and the acceptor molecule absorbs it.
2. **Förster transfer** or **long-range, single-step radiationless transfer**.[55] This type of energy transfer can take place over distances of up to 100 Å. Efficient Förster transfer requires a good overlap of the emission spectrum of the donor and the absorption spectrum of the acceptor, yet apparently a photon is not emitted.
3. **Exciton migration**. Much of organic photochemistry is done in the gas phase or solution. However, increasingly photochemists are studying molecules in the solid state or in organized media, such as micelles, vesicles, and liquid crystals. In such environments, an excited state molecule may transfer its energy very rapidly to a nearby molecule.
4. **Collisional energy transfer**. This is the process that is often of greatest interest to organic photochemists, since it allows the creation of excited triplet states that otherwise could not be produced by direct irradiation. In equation 12.3, for example, energy transfer from the triplet state of a donor molecule, $\mathbf{D}^{*(3)}$, to an acceptor molecule in its electronic ground state, **A**, produces the ground state of the donor and the triplet excited state of the acceptor. It is generally believed that collisional energy transfer occurs through exchange of electrons between the two species:

$$\mathbf{D}^{*(\uparrow\uparrow)} + \mathbf{A}_0^{(\uparrow\downarrow)} \longrightarrow \mathbf{D}_0^{(\uparrow\downarrow)} + \mathbf{A}^{*(\uparrow\uparrow)} \tag{12.3}$$

Collisional energy transfer is frequently called **sensitization**.[56] It is an important tool in photochemistry because it provides access to excited triplet states of compounds whose singlets do not intersystem cross efficiently (such as alkenes and conjugated polyenes). It also provides access directly to the triplet states of compounds for determination of the multiplicity of the state responsible for a photochemical reaction.

---

[53]Wilkinson, F. *Adv. Photochem.* **1964**, *3*, 241.

[54]For further details, see reference 4, pp. 296–311.

[55]Förster, T. *Discuss. Faraday Soc.* **1959**, *27*, 7.

[56]In common usage, *sensitization* means a process that occurs once, leading to later response. Thus an individual can become sensitized to an allergen, such as insect venom, through one exposure. Even though there may not be a severe allergic reaction upon initial exposure, the reaction can increase with each further contact. However, photochemists commonly use the term *sensitization* to refer simply to energy transfer.

Another important interaction between an electronically excited molecule and a ground state molecule is **electron transfer**. Because photoexcitation of a molecule promotes an electron from a lower energy bonding or nonbonding orbital to a higher energy antibonding orbital, the excited state molecule has very different redox properties from the ground state structure. It is a better electron donor (and thus has a less positive oxidation potential) because of the presence of the electron in a high energy orbital. It is also a better electron acceptor (and therefore has a less negative reduction potential) because of the vacancy in a lower energy orbital. As a result, photoexcited molecules can undergo electron transfer processes with ground state molecules that have a lower energy LUMO, as shown in Figure 12.11(a) as well as those that have a higher HOMO, as shown in Figure 12.11(b).

The radical ion pair $\mathbf{D}^{\cdot+}\ \mathbf{A}^{\cdot-}$ (or $\mathbf{D}^{\cdot-}\ \mathbf{A}^{\cdot+}$) is similar to the ion pairs formed through ground state reactions in that it can be a contact or intimate ion pair, a solvent-separated ion pair, or a pair of free radical ions.[57] However, its dynamics are also affected by the fact that it can exist as either a singlet or as a triplet, depending upon the spins of the two unpaired electrons. A radical ion pair can also be annihilated by *back electron transfer*

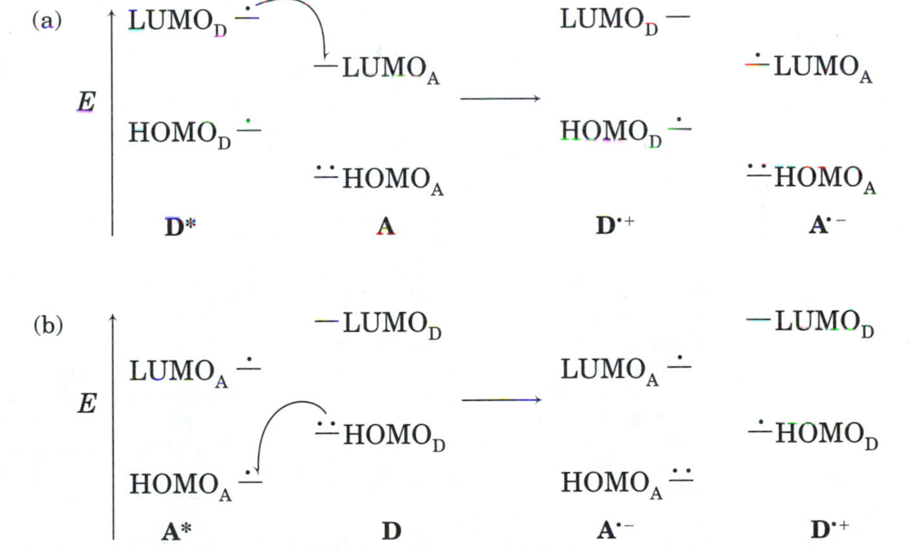

Figure 12.11 Creation of radical ion pairs by electron transfer: (a) from a photoexcited electron donor; (b) to a photoexcited electron acceptor. (Adapted from reference 58(e).)

[57]The initial charge transfer species is frequently termed an **exciplex** (**excited** com**plex**) if it is formed from two different species or an **excimer** (**exci**ted di**mer**) if it is formed from two molecules of the same kind. For discussions of photochemically generated radical ion pairs and for leading references, see Gould, I. R.; Young, R. H.; Mueller, L. J.; Farid, S. *J. Am. Chem. Soc.* **1994**, *116*, 8176; Gould, I. R.; Young, R. H.; Mueller, L. J.; Albrecht, A. C.; Farid, S. *J. Am. Chem. Soc.* **1994**, *116*, 8188; Vauthey, E.; Parker, A. W.; Nohova, B.; Phillips, D. *J. Am. Chem. Soc.* **1994**, *116*, 9182; Arnold, B. R.; Noukakis, D.; Farid, S.; Goodman, J. L.; Gould, I. R. *J. Amer. Chem. Soc.* **1995**, *117*, 4399.

from the radical anion to the radical cation, thus producing the electronic ground state of each species.[58,59]

Fundamentals of Photochemical Kinetics

Actinometry and Quantum Yield Determinations

The quantative study of photochemical reactions requires knowledge of the concentrations of reactants and products, the absorption spectra of the reactants, and the intensity of the irradiation at the wavelengths being absorbed. The determination of the intensity of light is known as ***actinometry***. Although some early work was by necessity done with direct measurement of light intensities, now photochemists routinely measure light intensities with a photochemical reference system with a known quantum efficiency by irradiating samples of the actinometry solution and of the compound to be studied under conditions of identical light intensity.[60,61] Among the most commonly used actinometers are the potassium ferrioxalate system:

$$2\,Fe^{+3} + C_2O_4^{\,2-} \xrightarrow[H_2O]{h\nu} 2\,Fe^{+2} + 2\,CO_2 \tag{12.4}$$

the benzophenone (**3**)–benzhydrol (**4**) system:

$$\underset{\mathbf{3}}{Ph_2C{=}O} + \underset{\mathbf{4}}{Ph_2CHOH} \xrightarrow{h\nu} \underset{\mathbf{5}}{Ph-\overset{\overset{\displaystyle HO}{|}}{\underset{\underset{\displaystyle Ph}{|}}{C}}-\overset{\overset{\displaystyle OH}{|}}{\underset{\underset{\displaystyle Ph}{|}}{C}}-Ph} \tag{12.5}$$

[58]For an introduction, see (a) Gust, D.; Moore, T. A. *Adv. Photochem.* **1991**, *16*, 1; *Top. Curr. Chem.* **1991**, *159*, 103–151; (b) Willner, I.; Willner, B. *Top. Curr. Chem.* **1991**, *159*, 153; (c) Lymar, S. V.; Parmon, V. N.; Zamarev, K. I. *Top. Curr. Chem.* **1991**, *159*, 1; (d) Fox, M. A. *Adv. Photochem.* **1986**, *13*, 237; *Top. Curr. Chem.* **1991**, *159*, 67; (e) Kavarnos, G. J. *Top Curr. Chem.* **1990**, *156*, 21. For a detailed discussion of photoinduced electron transfer, see Fox, M. A.; Chanon, M., eds. *Photoinduced Electron Transfer*, Parts A–D; Elsevier: Amsterdam, 1988.

[59]Although the energy transfer and electron transfer processes have been described in terms of bimolecular processes, the same processes can occur between two groups in the same molecule if one can be photoexcited independently of the other. In particular, photochemical electron transfer provides a means to investigate the mechanism of long-range electron transfer processes that may be important in the operation of many biological systems. For example, see Schmidt, J. A.; McIntosh, A. R.; Weedon, A. C.; Bolton, J. R.; Connolly, J. S.; Hurley, J. K.; Wasielewski, M. R. *J. Am. Chem. Soc.* **1988**, *110*, 1733.

[60]In some systems with only one irradiation chamber, samples of the actinometry solution and the compound under investigation are irradiated alternately. In such cases it is important to determine that the light intensity does not vary with time or, if it does, that the variation be known. Alternatively, an apparatus such as the "merry go round" allows for rotation of samples around the filtered output of a high intensity lamp. (Moses, F. G.; Liu, R. S. H.; Monroe, B. M. *Mol. Photochem.* **1969**, *1*, 245.) In other systems the samples rotate within a bank of lamps or they rotate at the output of a monochromator so that first one and then another sample is exposed to the nearly monochromatic light it passes. Alternatively, a beam splitter may be used so that lamp output can be monitored continuously during an irradiation: Zimmerman, H. E. *Mol. Photochem.* **1971**, *3*, 281.

[61]Reference 46, pp. 117–128.

the 2-hexanone (**6**) system:

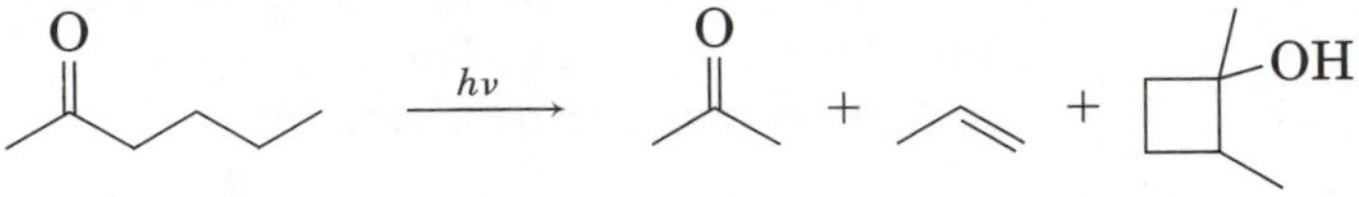

$$\text{6} \xrightarrow{h\nu} \text{7} + \text{8} + \text{9} \tag{12.6}$$

the sensitized diene isomerization system, as illustrated by the isomerization of 1,3-pentadiene (**10**):

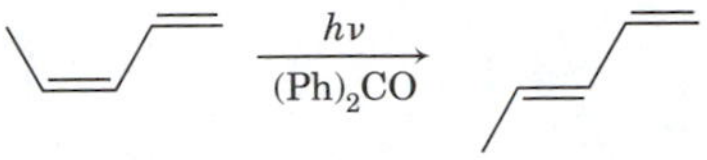

$$\text{10} \xrightarrow[(Ph)_2CO]{h\nu} \text{11} \tag{12.7}$$

and the benzophenone-sensitized photodimerization of 1,3-cyclohexadiene (**12**):[62]

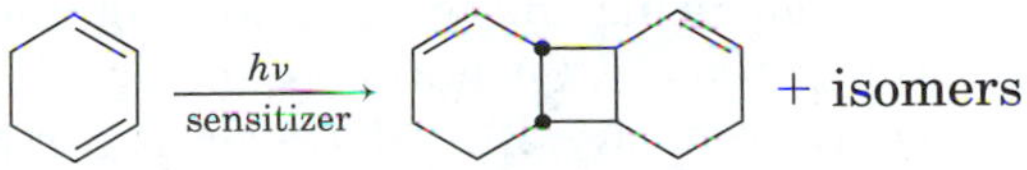

$$\text{12} \xrightarrow[\text{sensitizer}]{h\nu} \text{13} + \text{isomers} \tag{12.8}$$

Each actinometry system has certain advantages in terms of effective wavelength range, solubility, sensitivity, and ease of analysis.

A ***quantum yield***, also called a ***quantum efficiency***, of some photochemical process is defined as the number of photochemical events of that process that occur per photon of light absorbed. Consider the very simple photochemical reaction illustrated in equation 12.9:

$$\mathbf{A} \xrightarrow{h\nu} \mathbf{B} \tag{12.9}$$

The quantum yield of disappearance of reactant (**A**) is defined as

$$\Phi_{dis} = \frac{\text{molecules of } \mathbf{A} \text{ that disappear}}{\text{photons of light absorbed by } \mathbf{A}} \tag{12.10}$$

Similarly, the quantum yield of appearance of product (**B**) is given by:

$$\Phi_{app} = \frac{\text{molecules of } \mathbf{B} \text{ that are formed}}{\text{photons of light absorbed by } \mathbf{A}} \tag{12.11}$$

More commonly calculations are done on a molar basis, so

$$\Phi_{app} = \frac{\text{moles of } \mathbf{B} \text{ that are formed}}{\text{Einsteins of light that are absorbed}} \tag{12.12}$$

Instead of the simple hypothetical reaction in equation 12.9, organic molecules normally exhibit a number of photochemical and photophysical

[62] Vesley, G. F.; Hammond, G. S. *Mol. Photochem.* **1973**, *5*, 367.

processes that compete with each other. A minimal complication is illustrated in equation 12.13, in which the photoexcited molecule both fluoresceses and forms product. The rate constant for fluorescence is denoted k_f, while that for reaction is termed k_r.

$$\begin{array}{ccc} \mathbf{A} \xrightarrow{h\nu} & \mathbf{A^*} & \xrightarrow{k_r} \mathbf{B} \\ & \downarrow k_f & \\ & \mathbf{A} + h\nu' & \end{array} \tag{12.13}$$

Now

$$\Phi_f = \frac{\text{Einsteins emitted}}{\text{Einsteins absorbed}} \tag{12.14}$$

where Φ_f is the quantum yield of fluorescence. Φ_f can be determined directly by measuring the total number of photons absorbed per unit time (absorption intensity, I_a) and the number of photons emitted per unit time (fluorescence intensity, I_f).[63] However, it can also be calculated indirectly if certain rate constants are known.

Rate Constants for Unimolecular Processes

The spontaneous emission of light from an excited molecule is a unimolecular process that follows first order kinetics. Suppose a photoexcited molecule displays fluorescence as its only decay pathway. If at time t_0 there are N_0 photoexcited molecules, then at any subsequent time t the number of molecules still in the excited state is given by

$$N = N_0 e^{-k_f t} \tag{12.15}$$

where k_f is the rate constant for fluorescence. First order decay processes (such as nuclear fission) are often described in terms of the half-life. However photochemists usually discuss excited states in terms of the ***lifetime***, denoted as τ, which is defined as the time required for a population of excited states to decay to $1/e$ of its original value.[64,65] (The magnitude of $1/e$ is 0.368, so in a first order, unimolecular decay process the lifetime is longer than the half-life). In the example in equation 12.15, this condition is achieved when

---

[63]In many measurements, I_f is taken to be the magnitude of the detector (e.g., photomultiplier) response to fluorescence at a λ_{max}.

[64]The term *lifetime* is used by virtually all photochemists, although perhaps it is not as intuitive as another term, such as *characteristic time,* might be. To be specific, authors often use the symbol τ_s for singlet lifetime or τ_f for fluorescence lifetime instead of an unsubscripted τ in order to distinguish these terms from τ_p, which is the phosphorescence lifetime. When the symbol τ is used alone, it is important to ascertain which excited state is involved.

[65]The derivation of the relationship between τ and rate constants for decay has been given by Jaffé, H. H.; D'Agostino, J. T. *J. Chem. Educ.* **1970**, *47*, 14.

$t = 1/k_f$. Thus, $1/k_f$ is defined as the ***inherent fluorescence lifetime***, τ_s^0, which is the lifetime of the excited singlet state that would be observed if fluorescence were the only decay pathway. The inherent fluorescence lifetime can be calculated from the absorption spectrum of the compound.[66]

For most molecules fluorescence is only one of the processes that deactivate an electronically excited state. If a molecule can decay by both fluorescence (with rate constant k_f) or radiationless decay (with rate constant k_d), the time dependence of a population of excited states is given by

$$N = N_0 e^{-(k_f+k_d)t} \tag{12.16}$$

In this case,

$$\tau = 1/(k_f + k_d) \tag{12.17}$$

In general, if there are i first order, unimolecular decay processes that deactivate an excited singlet state, then

$$\tau_s = 1/(\sum_i k_i) \tag{12.18}$$

The quantum yield of fluorescence is a function of the rate constant for fluorescence divided by the rate constant for all decay processes.

$$\Phi_f = k_f/\sum_i k_i = k_f\tau_s \tag{12.19}$$

Here,

$$\Phi_f = k_f/(k_f + k_d) = \tau_s/\tau_s^0 \tag{12.20}$$

where τ_s^0 is determined from the absorption spectrum and τ_s is measured experimentally.

Flash Spectroscopy

Equation 12.18 describes not only the decay of excited singlet states but also excited triplet states or other photochemically generated reactive species that disappear by first order processes. If such species cannot be monitored by the emission of light, an alternative approach is to detect them through the absorption of light with a technique known as flash spectroscopy.[67] This technique uses a short duration pulse or flash of light to create a large population of transients that can then be monitored spectroscopically. In conventional flash spectroscopy, the lamp is filled with xenon or other gas that can be subjected to a several thousand volt charge between electrodes in the ends of the lamp. Discharging the lamp produces a bright flash of light last-

---

[66]For details, see Michl, J.; Bonačić-Koutecký, V. *Electronic Aspects of Organic Photochemistry*; Wiley-Interscience: New York, 1990; p. 74. In general, weak absorptions predict long inherent lifetimes, while strong absorptions are associated with high fluorescence rate constants and short inherent fluorescence lifetimes.

[67]Because the technique is used frequently to study photochemical reactions that lead to bond cleavage, it is often called **flash photolysis**. The term *photolysis* literally means breaking apart by light, however, so it technically would not apply to the measurement of triplet states that decay back to starting molecules without chemical change.

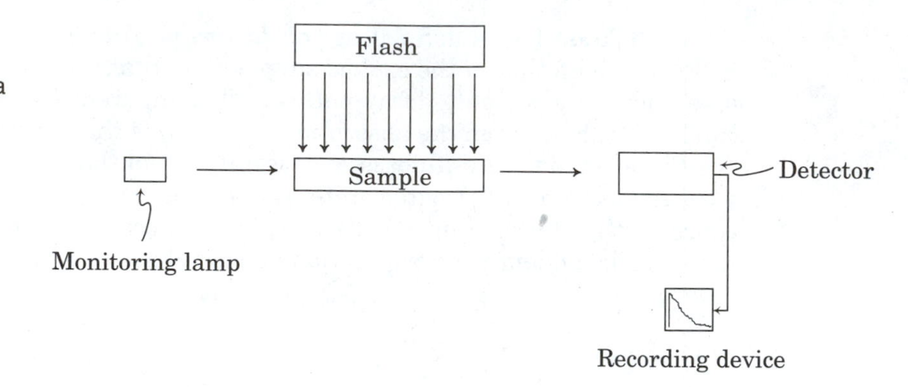

Figure 12.12
A schematic representation of a conventional flash spectrometer.

ing some microseconds. More recently lasers have been used to generate nanosecond to femtosecond flashes.

A schematic of a flash spectroscopy apparatus is shown in Figure 12.12. A solution of the sample is placed in a cell made of optical material that is transparent to the wavelengths of light to be used in the experiment. The cell is placed in an absorption spectrometer so that the absorption of monitoring light by the sample can be measured as a function of wavelength. To the extent that some singlet ground state molecules are converted to triplet states or other transients (such as radicals), the absorption of monitoring light by the sample after excitation is different from the absorption before the flash. Specifically, absorption by the ground state is less, while absorption by these transients appears for the first time.

For kinetic experiments, the monochromator in the detector is set for one wavelength, and the photomultiplier output voltage is monitored as a function of time after the excitation flash.[68] The schematic drawing in Figure 12.13 shows an experiment in which the monochromator is set for a wavelength at which the starting material does not absorb the monitoring light. At the time of the flash, scattered light from the flashlamp produces an apparent decrease in absorption, and the curve drops below the bottom of the display screen. In a few microseconds,[69] however, the flash is decaying and the photomultiplier voltage reflects absorption of monitoring light by the sample. If the absorption is due to a transient that decays with first

[68]In some cases it may be desirable to record the absorption spectrum of the transient. One method of recording such a spectrum involves measuring the intensity of absorption of monitoring light at a certain time after the flash as a function of wavelength passing through the emission monochromator. Another method involves using a second monitoring flash set off after the excitation flash and a photographic film to record the light passing through the monochromator. Such measurements can be important in establishing the identity of a transient that is an intermediate in a photochemical reaction.

[69]The ideal excitation source would provide a pulse of light that drops instantaneously from a high value to zero. Usually, however, the lamp has its own intensity profile, and the decay of very short-lived transients may be difficult to measure unless the effects of lamp decay are separated from the effects of transient absorption decay through a technique known as deconvolution.

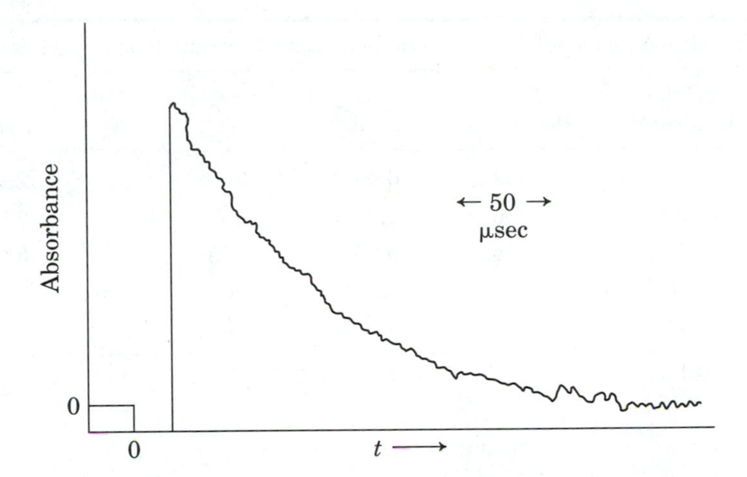

Figure 12.13 Transient absorption of light at one wavelength after flash excitation.

order kinetics, the detector records a first order decay curve. Plotting the natural logarithm of intensity (in arbitrary units) against the time ($t = 0$ arbitrarily taken as the maximum of the absorption) generates a plot such as that in Figure 12.14, which is based on the data in Figure 12.13. The value of τ is determined from the slope of a least squares analysis of the data points.

The kinetic methods described here, as well as others, make it possible to obtain detailed information about the unimolecular processes of electronically excited organic molecules. Table 12.2 lists some of the data available

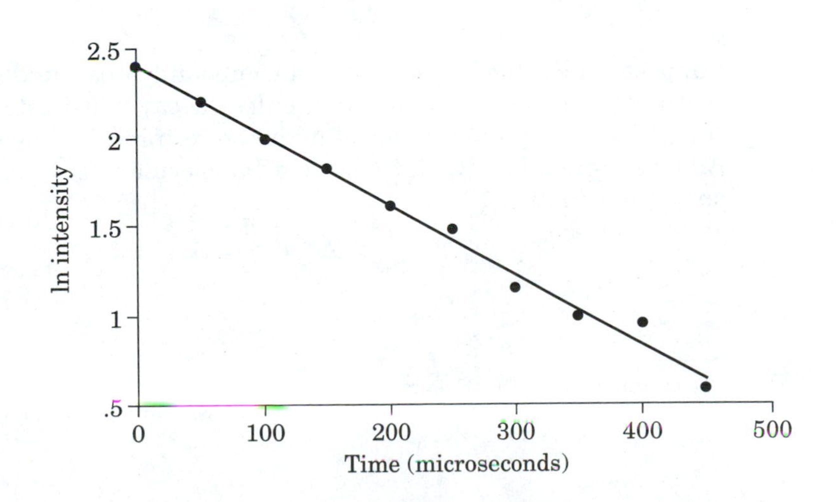

Figure 12.14 Plot of intensity versus time (microseconds) for decay of triplet-triplet absorption data in Figure 12.13.

Table 12.2 Photophysical data for selected compounds. (Except as noted, data are from reference 46.)

Compound	E (S_1) (kcal/mol)	E (T_1) (kcal/mol)	ϕ_f	ϕ_{isc}	ϕ_p	τ_f (nsec)	τ_p (sec)
Benzene	110	84	0.05	0.25[a]	0.23	29	6.3
Naphthalene	92	61	0.2	0.8	0.1	96	2.2
Anthracene	76	42	0.27	0.75		5	0.045
Acetone	88	ca. 80	ca. 10^{-3}	1.0	0.0001	2	0.04
Benzophenone	75	69	4×10^{-6}	1.0	0.74	5×10^{-3}	0.006

[a]Data from Carroll, F. A.; Quina, F. H. *J. Am. Chem. Soc.* **1976**, *98*,1.

for five compounds.[70] Note in particular that the differences between the energies of S_1 and T_1 are larger for benzene, naphthalene, and anthracene, but the differences are much smaller for acetone and benzophenone. The smaller energy gap leads to greater rate constants for intersystem crossing in the carbonyl compounds, and this expectation is supported by the near-unity quantum values of Φ_{isc}.

Bimolecular Decay of Excited States: Stern-Volmer Kinetics

In addition to unimolecular decay, photoexcited molecules may also exhibit bimolecular decay resulting from interactions with other (ground state) molecules. The interaction may take the form of a collisional energy transfer or sensitization process (equation 12.21) or as a quenching interaction, in which neither product is in the excited state (equation 12.22).

$$\mathbf{A}^* + \mathbf{Q} \longrightarrow \mathbf{A} + \mathbf{Q}^* \quad \textbf{(12.21)}$$

$$\mathbf{A}^* + \mathbf{Q} \longrightarrow \mathbf{A} + \mathbf{Q} \quad \textbf{(12.22)}$$

Suppose an excited singlet state of compound **A** can undergo fluorescence (with rate constant k_f) and radiationless decay (with rate constant k_d). In the absence of **Q**, the lifetime of **A*** is the reciprocal of the sum of k_f and k_d. Adding a quencher (**Q**) introduces a bimolecular decay pathway, which is shown in equation 12.23.

$$\mathbf{A}^* \xrightarrow{k_q[\mathrm{Q}]} \mathbf{A} \quad \textbf{(12.23)}$$

[70]Data from reference 46. Energy values for S_1 and T_1 are in kcal/mol. Lifetimes are in nsec for fluorescence in solution at room temperature and sec for phosphorescence at 77° K in an EPA glass. (See footnote 46).

Now the lifetime of **A*** in the presence of **Q**, τ_s', is given by[71]

$$\tau_s' = 1/(k_f + k_d + k_q[\mathbf{Q}]) \tag{12.24}$$

Dividing τ_s by τ_s' produces the following relationships:

$$\frac{\tau_s}{\tau_s'} = \frac{\Phi_f}{\Phi_f'} = \frac{I_f}{I_f'} = \frac{1/(k_f + k_d)}{1/(k_f + k_d + k_q[\mathbf{Q}])} \tag{12.25}$$

$$= \frac{k_f + k_d + k_q[\mathbf{Q}]}{k_f + k_d} = 1 + \frac{k_q[\mathbf{Q}]}{k_f + k_d} = 1 + k_q[\mathbf{Q}]\tau_s \tag{12.26}$$

Thus, plotting the ratio of lifetimes (τ_s/τ_s'), fluorescence quantum yields (Φ_s/Φ_s'), or fluorescence intensities (I_f/I_f') versus [**Q**] should yield a straight line with slope $k_q\tau_s$. If τ_s is known, then k_q can be determined. This correlation is known as the Stern-Volmer relationship, and is one of the most fundamental expressions in photochemistry.[72] Figure 12.15 shows a Stern-Volmer plot for the quenching of 1,4-dimethoxybenzene fluorescence by allyl chloride in acetonitrile solution.[73]

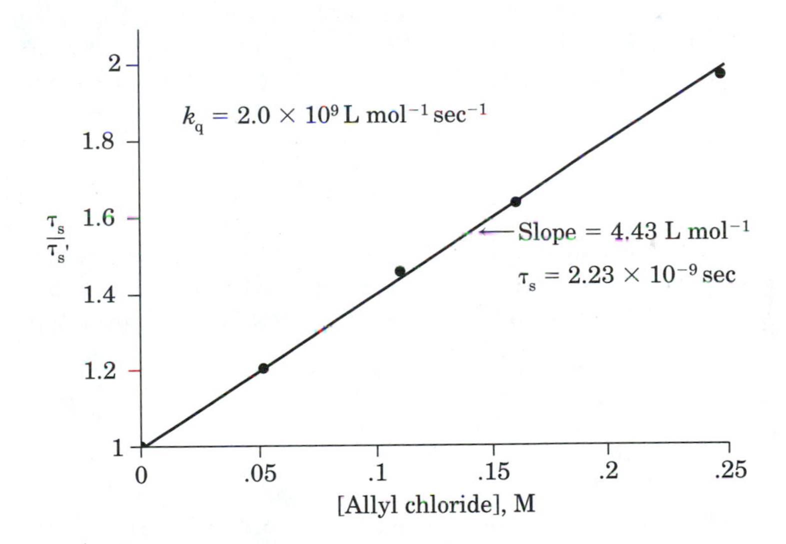

Figure 12.15 Stern-Volmer plot for quenching 1,4-dimethoxybenzene fluorescence by allyl chloride.

---

[71]In this formulation, the lifetime of the excited singlet state in the absence of **Q** is denoted τ_s, while that in the presence of **Q** is indicated as τ_s'. Often, however, τ^o is used to represent the lifetime of an excited state in the absence of quencher and τ to represent the lifetime of an excited state in the presence of quencher. In that case, the Stern-Volmer relationship becomes

$$\frac{\tau^\circ}{\tau} = \frac{\Phi^\circ}{\Phi} = \frac{I^\circ}{I} = 1 + k_q[\mathbf{Q}]\tau^\circ$$

[72]Shetlar, M. D. *Mol. Photochem.* **1974**, *6*, 191, has presented a generalized form of the Stern-Volmer equation for a system with multiple excited states. Green, N. J. B.; Pimblott, S. M.; Tachiya, M. *J. Phys. Chem.* **1993**, *97*, 196 have presented generalizations for cases in which the quenching requires description with a time-dependent rate constant.

12.2 Properties of Excited States

Electronically excited organic molecules are chemical entities, just as are ground state molecules. Even though they have short lifetimes, they are expected to have characteristic bond angles, dipole moments, bond strengths, and vibrational modes. Therefore they should have characteristic physical properties and chemical properties, just as ground state molecules do. Of course we cannot measure the boiling point of S_1 benzene or the heat of combustion of T_1 acetone, since these are bulk properties, and typically the concentration of excited molecules at any particular time is very low. Nonetheless, spectroscopic measurements can yield useful information about the excited states and about photochemical reactions.[74]

Acidity and Basicity in Excited States

Acidity and basicity are important properties of ground state molecules. If electronically excited states have lifetimes long enough for proton transfer reactions to take place, then the equilibria for such reactions can be determined. Consider the set of reaction coordinate diagrams shown in Figure 12.16. The lower curve represents a potential energy diagram for the dissociation of an acid, **A—H**.

$$\mathbf{A{-}H} \rightleftharpoons \mathbf{A^{-}} + \mathbf{H^{+}} \qquad \textbf{(12.27)}$$

The molecule **A—H** is presumed to contain an acidic proton and a π system that may absorb light; further, it is assumed that deprotonation of **A—H** leaves the π system intact. Thus, the ionization of a photoexcited **A—H**

[73]Unpublished data from Fitzgerald, M. C. In this experiment the solution was not degassed, so all values of τ reflect the effects of oxygen quenching.

[74]A particularly novel property of an electronically excited molecule is chirality that exists only because of photoexcitation. Miesen, F. W. A. M.; Wollersheim, A. P. P.; Meskers, S. C. J.; Dekkers, H. P. J. M.; Meijer, E. W. *J. Am. Chem. Soc.* **1994**, *116*, 5129, have reported the synthesis of optically pure 3-($^1n,\pi^*$)-(1*S*,6*R*)-bicyclo[4.4.0]decane-3,8-dione,

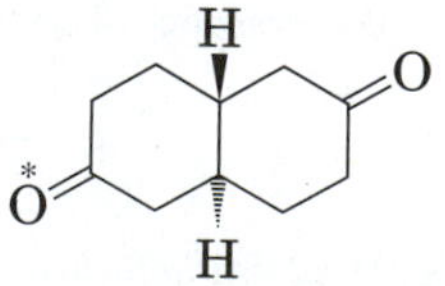

H
O
O*
H

which is chiral only in the excited state. The chirality was detected in the circular polarization of chemiluminescence associated with its synthesis from an optically active 1,2-dioxetane precursor.

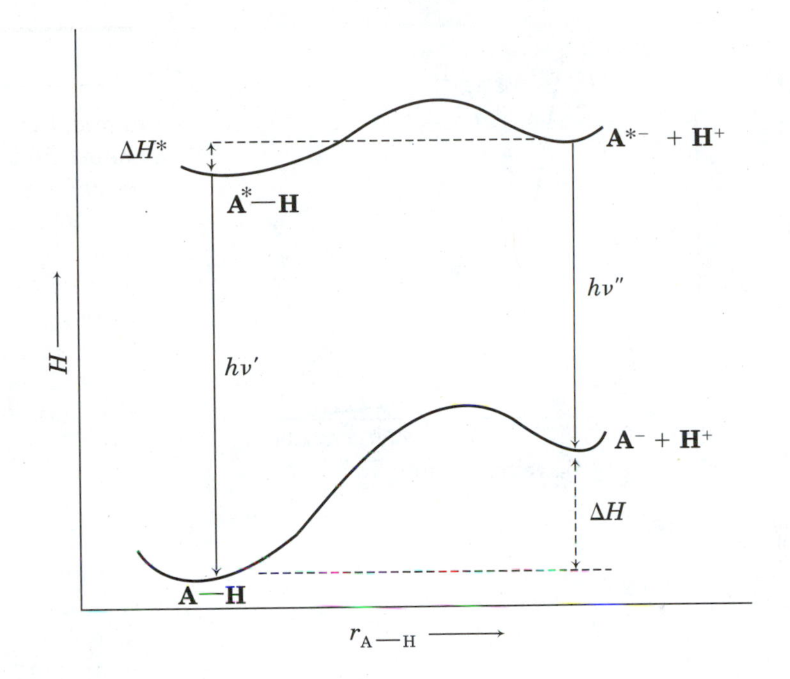

Figure 12.16 Potential energy surfaces for dissociation.

molecule can occur in such a way that the dissociation leaves the electronically excited state of the anion, **A**$^{*-}$,[75] which may then exhibit its characteristic fluorescence, with energy $h\nu''$. For example, curve A in Figure 12.17 shows the fluorescence spectra of 2-naphthol (**14**), which was recorded under acidic conditions, while curve B shows the fluorescence spectrum of the 2-naphthoate ion (**15**), which was recorded under basic conditions. At intermediate pH values, fluorescence spectra from both **14** and **15** may be seen. Figure 12.18 shows the fluorescence spectrum of β-naphthol in a series of solutions with pH values that increase from (A) to (G).[76,77]

$$[\mathbf{A}-\mathbf{H}]^* \rightleftharpoons [\mathbf{A}^-] + \mathbf{H}^+ \qquad \textbf{(12.28)}$$

[75]The reaction shown in equation 12.28 is said to be an **adiabatic reaction**, meaning a reaction that takes place on the excited state potential energy surface. A **nonadiabatic** process involves a change in electronic state. For a more detailed definition of these terms, see reference 1(a). See also (a) Becker, H.-D. *Pure Appl. Chem.* **1982**, *54*, 1589; (b) Förster, T. *Pure Appl. Chem.* **1970**, *24*, 443.

[76]Lawrence, M.; Marzzacco, C. J.; Morton, C.; Schwab, C.; Halpern, A. M. *J. Phys. Chem.* **1991**, *95*, 10294.

[77]See also (a) Kearwell, A.; Wilkinson, F. in *Transfer and Storage of Energy by Molecules*, Vol. 1; Burnett, G. M.; North, A. M., eds.; John Wiley & Sons, Inc.: New York, 1969; pp. 94–160; (b) Kelly, R. N.; Schulman, S. G. in *Molecular Luminescence Spectroscopy: Methods and Applications*, Part 2; Schulman, S. G., ed.; Wiley-Interscience: New York, 1988; 461–510; (c) Arnaut, L. G.; Formosinho, S. J. *J. Photochem. Photobiol., A: Chem.* **1993**, *75*, 1; Formosinho, S. J.; Arnaut, L. G. *J. Photochem. Photobiol., A: Chem.* **1993**, *75*, 21; (d) Marciniak, B.; Kozubek, H.; Paszyc, S. *J. Chem. Educ.* **1992**, *69*, 247.

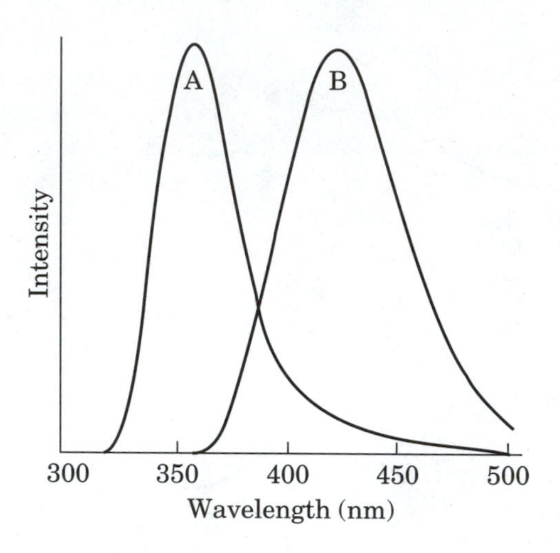

Figure 12.17 Fluorescence spectra of 2-naphthol in (A) 0.1 M $HClO_4$ and (B) 0.1 M NaOH solution. (Reproduced from reference 76.)

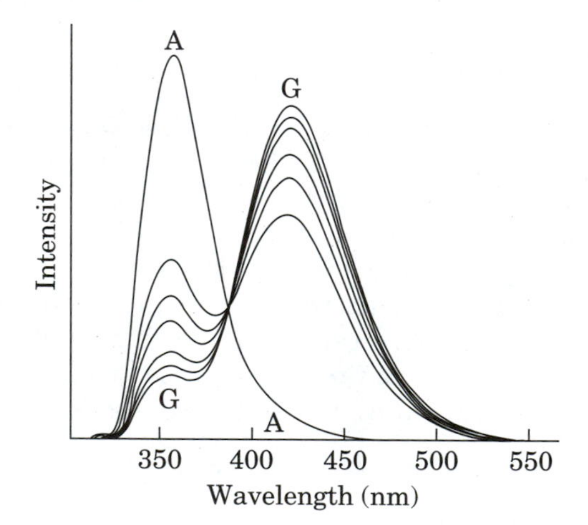

Figure 12.18 Fluorescence spectra of β-naphthol solutions as a function of pH. (Adapted from reference 76.)

OH

O$^-$

14 **15**

The energy difference between the ground state and excited state of **A—H** can be determined from the 0,0 band of the electronic transition. Similarly, the energy difference between the ground and excited state of $\mathbf{A}^-$ can be determined from its 0,0 band as well. The enthalpy of proton dissociation of ground state **A—H** is ΔH, while the enthalpy of proton dissociation of **A—H*** is ΔH^*.[78] Examination of Figure 12.16 leads to equation 12.29, which can be rewritten as equation 12.30.[79]

$$\Delta H^* + h\nu' = \Delta H + h\nu'' \quad \textbf{(12.29)}$$

$$\Delta H^* = \Delta H + (h\nu'' - h\nu') \quad \textbf{(12.30)}$$

If entropy changes are assumed to be approximately the same for dissociation in the ground and excited states, then equation 12.31 holds:

$$\Delta H^* - \Delta H \approx \Delta G^* - \Delta G \quad \textbf{(12.31)}$$

Now using the relationship between ΔG values and equilibria

$$\Delta G = 2.303\, RT\, \mathrm{p}K \quad \textbf{(12.32)}$$

produces the relationship

$$\mathrm{p}K^* - \mathrm{p}K = (h\nu'' - h\nu')/(2.303\, RT) \quad \textbf{(12.33)}$$

so pK^* can be determined from the known pK value.[80]

Table 12.3 shows some results for pK^* measurements for S_1 states of various molecules, as well as for pK^* values for T_1 states obtained by other methods.[81,82] The changes can be dramatic. For example, the S_1 state of 2-naphthol is almost one million times more acidic than its ground state,

---

[78]Note that ΔH^* is not related to $\Delta H^{\ddagger}$, the activation enthalpy for deprotonation.

[79]This procedure is known as the Förster cycle: Förster, T. *Naturwiss.* **1949**, *36*, 186; see also Weller, A. *Prog. Reaction Kinetics* **1961**, *1*, 189.

[80]There are also other methods used to determine this difference in pK values. For a more complete discussion of these relationships, see Ireland, J. F.; Wyatt, P. A. H. *Adv. Phys. Org. Chem.* **1976**, *12*, 131.

[81]Unless otherwise indicated, data are from Jackson, G.; Porter, G. *Proc. Roy. Soc. London* **1961**, *A260*, 13.

[82]See the discussion in (a) reference 80; (b) Porter, G. in *Reactivity of the Photoexcited Organic Molecule*, (Conference Proceedings); Wiley-Interscience: London, 1967; p. 79–117.

while the S_1 state of protonated 2-aminoanthracene is nearly 10^8 more acidic than its ground state![83,84] There are also changes in the acidity of the triplet states of these molecules, but they are generally much smaller.

These acidity changes can be rationalized with simple Lewis structure models for ground and excited state molecules.[85] The electronically excited state is formed by promoting an electron from HOMO to LUMO. If these two MOs have different atomic coefficients, then the effect will be to shift electron density (charge) from one part of the molecule to another. In other words, there will be a certain amount of intramolecular charge transfer that accompanies the excitation. In the case of 2-naphthol, the pK_a results and other data suggest that there is charge transfer away from the oxygen and toward the aromatic ring on going from S_0 to S_1. The 2-aminonaphthalene molecule is polarized in the same way. In the case of 2-naphthoic acid, the charge transfer is from the ring to the carboxylic acid group.

Figure 12.19 shows structures suggested by Jackson and Porter for the naphthalene derivatives described here.[81] The ground states are described as being the resonance hybrids of Type a resonance structures. The excited singlet states are said to have major contributions of Type b resonance structures in which charge distributions suggest acid-base properties very

Table 12.3 Acidity constants of ground and excited states. (Except as noted, data are from reference 80.)

Compound	Reaction	pK (S_0)	pK (S_1)	pK (T_1)
Naphthalene[a]	protonation	−4.0	11.7	−2.5
2-Naphthol	deprotonation	9.5	3.1	7.7 to 8.1
2-Naphthoic acid	deprotonation	4.2	8.2[b]	4.0[c]
2-Naphthylamine	protonation	4.1	−2.0	3.1 to 3.3

[a]Vander Donckt, E.; Lietaer, D.; Nasielski, J. *Bull. soc. chim. Belges* **1970**, *79*, 283.
[b]Kovi, P. J.; Schulman, S. G. *Anal. Chim. Acta* **1973**, *63*, 39.
[c]Reference 81.

[83]That is, the ground state amine is 10^8 more basic than is the excited state.

[84]Naphthols with electron withdrawing substituents such as cyano at C5 and C8 exhibit even more enhanced acidities in their excited singlet states and are capable of proton transfer to solvents such as alcohols in the absence of water. For example, the acidity of 5,8-dicyano-2-naphthol increases from 7.8 in the ground state to −4.5 in the excited singlet state. Tolbert, L. M.; Haubrich, J. E. *J. Am. Chem. Soc.* **1994**, *116*, 10593.

[85]An excited electronic state may be viewed as a resonance hybrid of ionic and radical resonance structures representing high energy ground state species. Craig, D. P. *Discuss. Faraday Soc.* **1950**, *9*, 5.

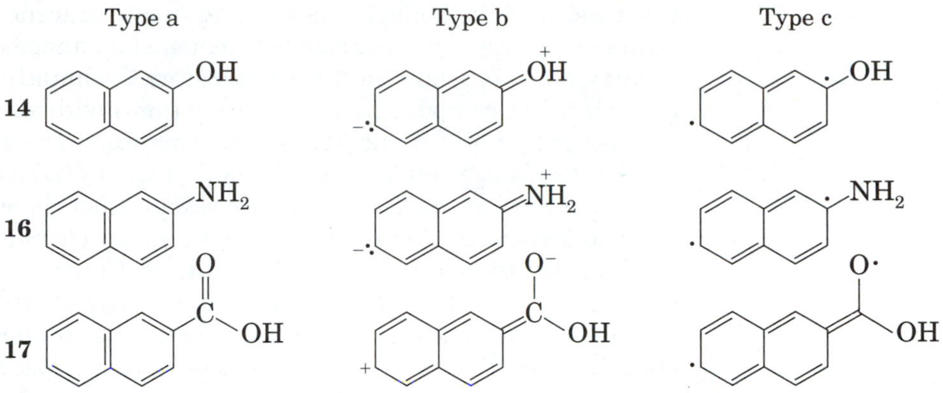

Figure 12.19 Important resonance contributors for excited states. (Reproduced from reference 81.)

different from those of the corresponding ground state molecules. The excited triplet states are described as having major contributions from Type c structures.[86]

Bond Angles and Dipole Moments of Excited State Molecules

For small molecules such as formaldehyde, careful vibrational analysis of the absorption spectrum, including rotational transitions, can yield information about bond lengths and bond angles in the electronically excited state.[87] The detailed methods of such analysis are beyond the scope of this discussion, but the results of such studies are valuable in understanding photochemical reactivity. As an example, let us consider the information available about the photoexcited carbonyl group. Based on the simple electronic energy diagram in Figure 12.3, the lowest energy transition in formaldehyde is predicted to be an $n \rightarrow \pi^*$ transition.[88] The spacing of the

---

[86]Because electronically excited singlet states are more likely to have their unpaired electrons on the same atoms than are triplet states (see reference 24), the triplet states are more likely to have biradicaloid character. For a more detailed discussion of the acid-base properties of photoexcited organic molecules, including carbon acids and carbon bases, see Wan, P.; Shukla, D. *Chem. Rev.* **1993**, *93*, 571.

[87]For a discussion of the dipole moments of larger molecules in the excited state, see Liptay, W. *Excited States* **1974**, *1*, 129.

[88]Experimentally, formaldehyde exhibits an $n \rightarrow \pi^*$ transition at 3.50 eV (80.7 kcal/mol), an $n \rightarrow \sigma^*$ transition at 7.09 eV (163 kcal/mol) and a $\pi \rightarrow \pi^*$ transition at 8.0 eV (184 kcal/mol): King, G. W. *Spectroscopy and Molecular Structure*; Holt, Rinehart and Winston, Inc.: New York, 1964; pp. 424 *ff.* and references therein. See the discussion in footnote 12.

progression of vibrational lines in the $n \rightarrow \pi^*$ transition of formaldehyde indicates that the C=O stretching frequency in the S_1 state is 1182 cm^{-1}, compared to 1746 cm^{-1} in the ground state.[89] Similarly, careful analysis of the rotational components of the transition indicates that the S_1 state is nonplanar by 20°.[90] Table 12.4 gives some experimental data for physical properties of the ground, S_1 (n,π^*) and T_1 (n,π^*) states of formaldehyde, and Figure 12.20 gives a valence bond representation of these states.[91,92]

The spectroscopy of other carbonyl compounds depends to some extent on the substituents attached to the carbonyl carbon atom. Benzophenone exhibits two broad bands: a very weak absorption with λ_{max} near 325 nm (ϵ in the range 20–200) and a more intense absorption (ϵ around 20,000) at about 250 nm. The longer wavelength peak is assigned to an $n \rightarrow \pi^*$ transition and the shorter wavelength peak to a $\pi \rightarrow \pi^*$ transition. The intensities support these assignments. For the $\pi \rightarrow \pi^*$ transition there are no symmetry, spin, or overlap prohibitions, so the transition is fully allowed and is intense. As was the case for formaldehyde, there is a very small overlap term for the $n \rightarrow \pi^*$ transition because to a first approximation the non-

Table 12.4 Physical properties of formaldehyde excited states. (Data from references 89, 90, and 91.)

Property/State	S_0	S_1	T_1
Geometry	planar	pyramidal	pyramidal
Δ (nonplanarity)	0°	20°	35°
C=O length	1.22 Å	1.32 Å	1.31 Å
ν C=O stretch	1746 cm^{-1}	1182 cm^{-1}	1251 cm^{-1}
<HCH	120°	122°	
Dipole moment	2.3 D	1.5 D	1.3 D

[89]Brand, J. C. D.; Williamson, D. G. *Adv. Phys. Org. Chem.* **1963**, *1*, 365.

[90]Robinson, G. W.; DiGiorgio, V. E. *Can. J. Chem.* **1954**, *36*, 31.

[91]Turro, N. J.; Dalton, J. C.; Dawes, K.; Farrington, G.; Hautala, R.; Morton, D.; Niemczyk, M.; Schore, N. *Acc. Chem. Res.* **1972**, *5*, 92.

[92]Conformational changes can also result from photoexcitation. For example, Tomer, J. L.; Spangler, L. H.; Pratt, D. W. *J. Am. Chem. Soc.* **1988**, *110*, 1615, reported different conformational preferences for the methyl group in S_1 and T_1 acetophenone.

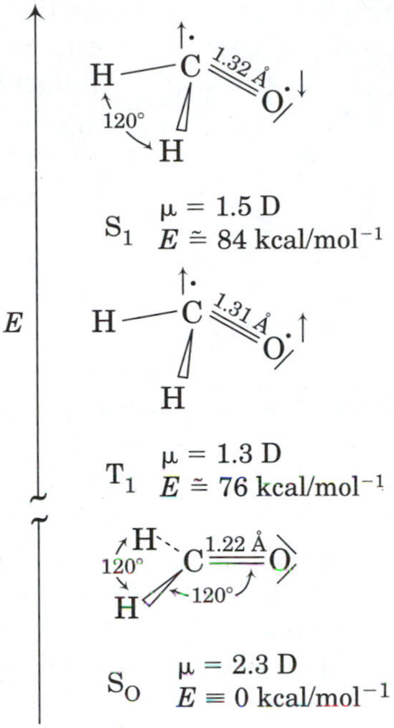

Figure 12.20 Representations of the electronic states of formaldehyde. (Adapted from reference 91.)

S_1 $\mu = 1.5$ D, $E \cong 84$ kcal/mol^{-1} (C–O 1.32 Å; 120°)

T_1 $\mu = 1.3$ D, $E \cong 76$ kcal/mol^{-1} (C–O 1.31 Å)

S_0 $\mu = 2.3$ D, $E \equiv 0$ kcal/mol^{-1} (C=O 1.22 Å; 120°)

bonding orbital and the π^* orbitals are orthogonal to each other. Some of the orthogonality is removed by vibrational deformation of the molecule, so the transition is observed but is weak.[93,94]

These assignments are reinforced by the effect of solvent polarity on the Uv-vis absorption spectrum. Since a polar ground state of a carbonyl compound is stabilized more by a polar solvent than is a less polar n,π^* excited state, the energy gap between the two states increases in a more polar solvent, causing the absorption to occur at a shorter wavelength. For the $\pi \rightarrow \pi^*$ transition, however, charge transfer takes place from the extended π system toward the oxygen atom, thus increasing the polarity of the molecule in the excited state. Therefore, the energy gap between the ground and excited states decreases, and the $S_0 \rightarrow S_2$ (π,π^*) absorption occurs at longer wave-

[93] Pople, J. A.; Sidman, J. W. *J. Chem. Phys.* **1957**, *27*, 1270.

[94] Even if the molecule is excited with light sufficiently high in energy to cause a $\pi \rightarrow \pi^*$ transition, the photochemistry of many aldehydes and ketones arises from the n,π^* state because the S_2 (π,π^*) state relaxes quickly to the S_1 (n,π^*) state. Furthermore, intersystem crossing is usually very rapid in carbonyl compounds, so much of the photochemistry arises from the $^3(n,\pi^*)$ state.

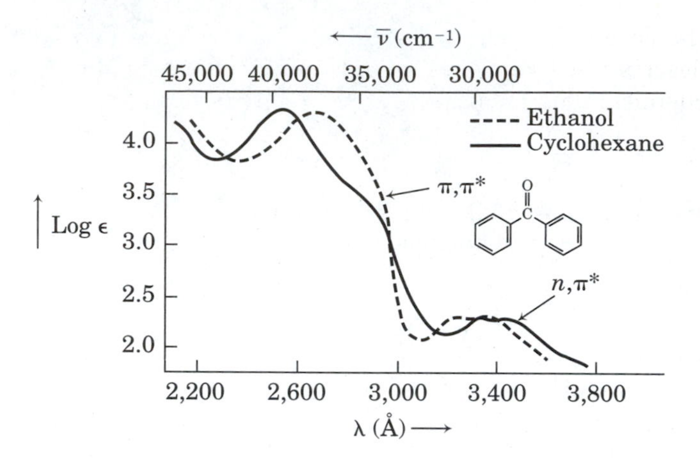

Figure 12.21 Solvent effect on benzophenone absorption. (Reproduced from reference 4.)

lengths. In general, a shift to the red with increasing solvent polarity is characteristic of $\pi \rightarrow \pi^*$ transitions, while a blue shift is characteristic of $n \rightarrow \pi^*$ bands.[95,96] These spectral shifts are illustrated by absorption spectra of benzophenone in Figure 12.21.

12.3 Representative Photochemical Reactions

Thermal reactions of organic molecules can be categorized according to functional groups in the reactants. In photochemical reactions the most important functional groups are the ***chromophores***—the functional groups that absorb light directly or accept energy by sensitization. An excited chromophore may either undergo photochemical reaction itself or may channel electronic or vibrational energy to another portion of the molecule. As was

---

[95]McConnell, H. *J. Chem. Phys.* **1952**, *20*, 700.

[96]In spectroscopic terms, the shift of an absorption maximum to shorter wavelength is called a **hypsochromic shift** or blue shift. The shift of an absorption maximum to longer wavelength is called a **bathochromic shift** or red shift. An increase in the intensity of an absorption band is called a **hyperchromic effect**, while a decrease in the intensity of a band is called a **hypochromic effect**. See, for example, Pavia, D. L.; Lampman, G. M.; Kriz, Jr., G. S. *Introduction to Spectroscopy: A Guide for Students of Organic Chemistry*; Saunders College Publishing: Philadelphia, 1979; pp. 191 *ff*. For a theoretical study of the $n \rightarrow \pi^*$ blue shift of acetone, see Gao, J. *J. Am. Chem. Soc.* **1994**, *116*, 9324.

the case with thermal reactions, both valence bond and molecular orbital descriptions are useful, complementary models to help us rationalize photochemical reactions.

Photochemical Reactions of Carbonyl Compounds

The n,π^* and π,π^* excited states of formaldehyde were discussed on page 799, so we consider first some photochemical reactions of carbonyl compounds.[97,98] It is useful to represent the n,π^* state of carbonyl compounds as shown in Figure 12.22, in which there is appreciable radical character on both the carbon atom and the oxygen atom and in which there is charge donation from oxygen to carbon.[99,91,100]

Figure 12.22 Resonance representation of carbonyl ground and excited states. (Adapted from Figure 2 of reference 91.)

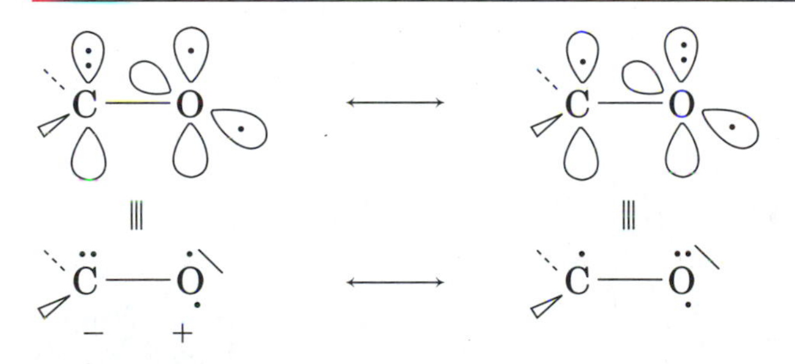

[97]The orbital model of formaldehyde on page 798 provides a simple basis for discussing the photochemical reactions of carbonyl compounds. For a more detailed discussion of the excited states of formaldehyde and acetaldehyde, see Hadad, C. M.; Foresman, J. B.; Wiberg, K. B. *J. Phys. Chem.* **1993**, *97*, 4293.

[98]Photochemical reactions of aldehydes and ketones are emphasized here, but other compounds containing carbon-oxygen double bonds also exhibit photochemical reactions. For a discussion of the photochemistry of esters and lactones, for example, see Suginome, H. in *The Chemistry of Acid Derivatives*, Supplement B, Vol. 2; Patai, S., ed.; Wiley-Interscience: Chichester, England, 1992; pp. 1107–1198.

[99]Zimmerman, H. E. *Adv. Photochem.* **1963**, *1*, 183. Figure 12.22 is a modified form of the representation suggested in this reference.

[100]For another analysis of the photochemistry of carbonyl compounds, see Formosinho, S. J.; Arnaut, L. G. *Adv. Photochem.* **1991**, *16*, 67.

The diradical Lewis structure on the right in Figure 12.22 provides a simple model that allows us to rationalize the major types of reactions that are characteristic of carbonyl compounds reacting from an n,π^* excited state:[101]

1. Norrish Type I Cleavage (α-Cleavage)
2. Hydrogen Abstraction, leading to
   a. Photoreduction
   b. Cyclobutanol Formation
   c. Norrish Type II Cleavage (β-Cleavage)
3. Oxetane Formation (Paterno-Buchi Reaction), and
4. Photochemical Electron Transfer

The α-cleavage reaction is the breaking of a bond α to the carbonyl group,[102] as shown schematically in equation 12.34, where the Lewis structure in brackets emphasizes the diradical character of the n,π^* state. Extended Hückel calculations indicate that the n orbital in formaldehyde has some density on the carbonyl hydrogen atoms.[103] By analogy, excitation of other aldehydes or of ketones should reduce electron density between the carbonyl carbon atom and a carbon atom bonded to it, thus weakening the bond and making dissociation more likely. With ketones, the resulting acyl radical can undergo loss of CO to produce another alkyl radical (equation 12.35).

$$R{-}C(=O){-}R \xrightarrow{h\nu} \left[R{-}\dot{C}(\dot{O}){-}R\right]^* \longrightarrow R{-}\dot{C}{=}O + R\cdot \qquad \textbf{(12.34)}$$

$$R{-}\dot{C}{=}O \longrightarrow R\cdot + CO \qquad \textbf{(12.35)}$$

Consider the photochemical reactions of cyclohexanone (**18**) in the gas phase (Figure 12.23).[104] Excitation produces the S_1 state, which intersystem crosses to the vibrationally excited T_1 state. In the gas phase the vibrational energy is not readily damped by collisions with other molecules, so α-cleavage produces a biradical with sufficient vibrational energy to lose CO. The resulting biradical can cyclize to cyclopentane, undergo intramolecular disproportionation to pentene, or fragment to ethene and propene. If intramol-

[101]These reactions are characteristic of nonconjugated carbonyl compounds; other reactions may be observed when conjugated carbonyl systems are photoexcited.

[102]Bamford, C. H.; Norrish, R. G. W. *J. Chem. Soc.* **1935**, 1504 and references therein.

[103]Jorgensen, W. L.; Salem, L. *The Organic Chemist's Book of Orbitals*; Academic Press: New York, 1973; p. 84.

[104]Srinivasan, R. *Adv. Photochem.* **1963**, *1*, 83.

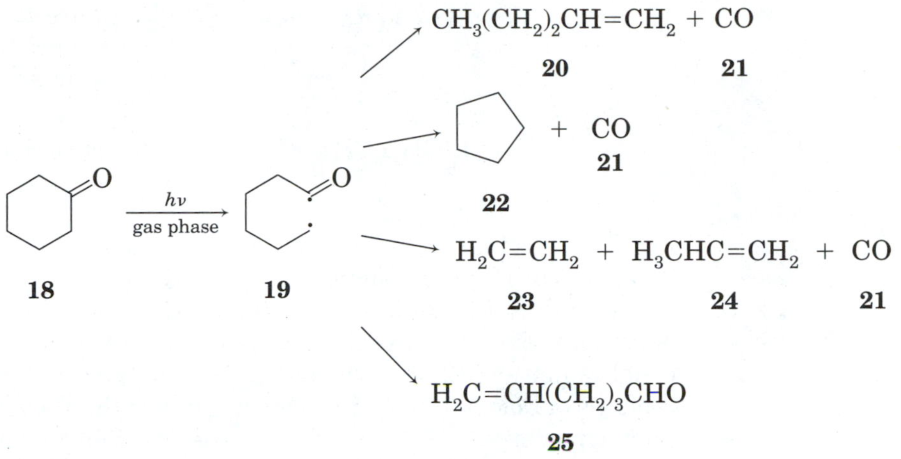

Figure 12.23 Photoreactions of cyclohexanone in the gas phase. (Adapted from reference 104.)

$\mathbf{18} \xrightarrow[\text{gas phase}]{h\nu} \mathbf{19}$

$CH_3(CH_2)_2CH{=}CH_2$ (**20**) + CO (**21**)

cyclopentane (**22**) + CO (**21**)

$H_2C{=}CH_2$ (**23**) + $H_3CHC{=}CH_2$ (**24**) + CO (**21**)

$H_2C{=}CH(CH_2)_3CHO$ (**25**)

ecular disproportionation occurs before loss of CO, then 5-hexenal can also be formed.

Photochemical reactions of carbonyl compounds in solution are usually simpler because the excess vibrational energy is rapidly quenched by collisions with solvent molecules. Therefore, only the energy of the 0th vibrational level of T_1 is available for reaction. When **18** is irradiated in solution, for example, the biradical produced by α-cleavage (**19**) does not eliminate CO. Instead, it undergoes intramolecular disproportionation by two different pathways. Abstraction of a hydrogen atom from the δ carbon atom by the radical on the acyl carbon atom produces the enal (**25**). If the radical centered on the ϵ carbon atom abstracts a hydrogen atom from the α carbon atom, then the reaction product is a ketene (**26**) (Figure 12.24). The ketene can react with hydroxylic solvents such as an alcohol to produce an ester (equation 12.36) or with water to produce an acid (equation 12.37). The latter process can produce an appreciable pH change in the reaction medium.[105]

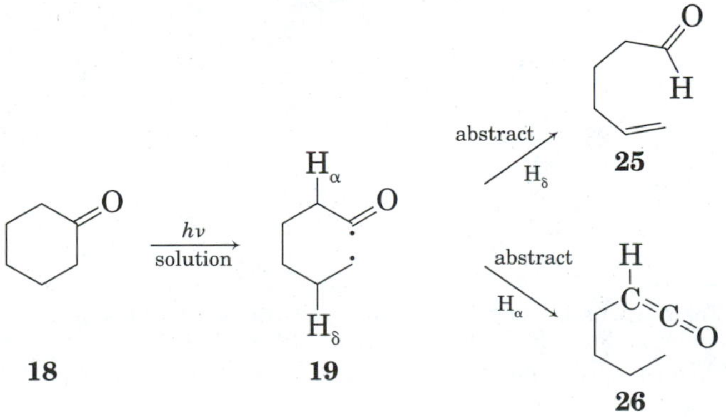

Figure 12.24 Photochemical reaction of cyclohexanone in solution.

$\mathbf{18} \xrightarrow[\text{solution}]{h\nu} \mathbf{19}$ (H_α, H_δ); abstract H_δ → **25**; abstract H_α → **26**

[105] Carroll, F. A.; Strouse, G. F.; Hain, J. M. *J. Chem. Educ.* **1987**, *64*, 84.

$$CH_3(CH_2)_3CH{=}C{=}O \xrightarrow{CH_3OH} CH_3(CH_2)_4CO_2CH_3 \quad (12.36)$$

26 **27**

$$CH_3(CH_2)_3CH{=}C{=}O \xrightarrow{H_2O} CH_3(CH_2)_4CO_2H \quad (12.37)$$

26 **28**

The depiction of an n,π^* excited state as a biradical suggests that photoexcited carbonyl compounds might also exhibit hydrogen abstraction reactions as do other radical species, as shown in equation 12.38.[106] This behavior is exhibited in the photoreduction of benzophenone in isopropyl alcohol (equation 12.39), which was first carried out by Ciamician on the rooftops of Bologna, Italy, and which is one of the classic reactions of organic photochemistry.[107] The products of the reaction are benzpinacol (**30**) and acetone, generated by the reactions in equations 12.40 through 12.42.[108,109]

$$R_2C{=}O \xrightarrow[(n,\pi^*)]{h\nu} [R_2\dot{C}{-}O\cdot]^* \xrightarrow{H{-}A} R_2\dot{C}{-}OH + A\cdot \quad (12.38)$$

$$2\,(C_6H_5)_2C{=}O + (CH_3)_2CH{-}OH \xrightarrow{h\nu} HO{-}C(C_6H_5)_2{-}C(C_6H_5)_2{-}OH + (CH_3)_2C{=}O \quad (12.39)$$

3 **29** **30** **7**

$$C_6H_5COC_6H_5 \xrightarrow{h\nu} [C_6H_5\dot{C}(O\cdot)C_6H_5]^* \xrightarrow{(CH_3)_2CHOH} C_6H_5\dot{C}(OH)C_6H_5 + H_3C\dot{C}(OH)CH_3 \quad (12.40)$$

3 **29** **31** **32**

---

[106]For a discussion of hydrogen atom abstraction by photoexcited carbonyl compounds, see Wagner, P.; Park, B.-S. *Org. Photochem.* **1991**, *11*, 227.

[107]Ciamician, G.; Silber, P. *Ber. dtsch. chem. Ges.* **1900**, *33*, 2911.

[108]Schönberg, A. *Preparative Organic Photochemistry*; Springer-Verlag: New York, 1968; p. 213.

[109]Ketones, aldehydes, and some other classes of compounds also undergo photoreduction by amines. For a review, see Cohen, S. G.; Parola, A.; Parsons, Jr., G. H. *Chem. Rev.* **1973**, *73*, 141.

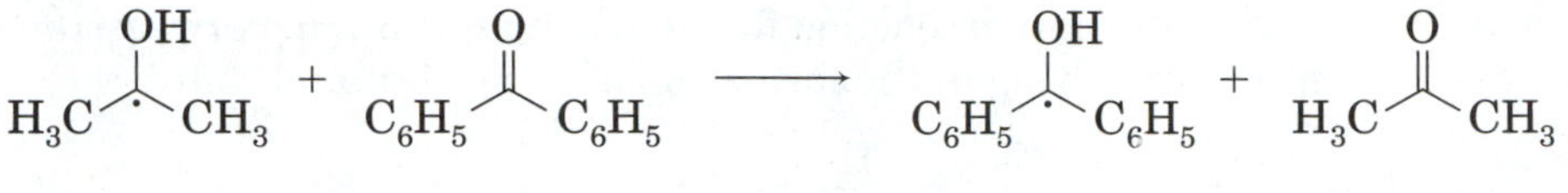

$$\mathbf{32} + \mathbf{3} \longrightarrow \mathbf{31} + \mathbf{7} \qquad (12.41)$$

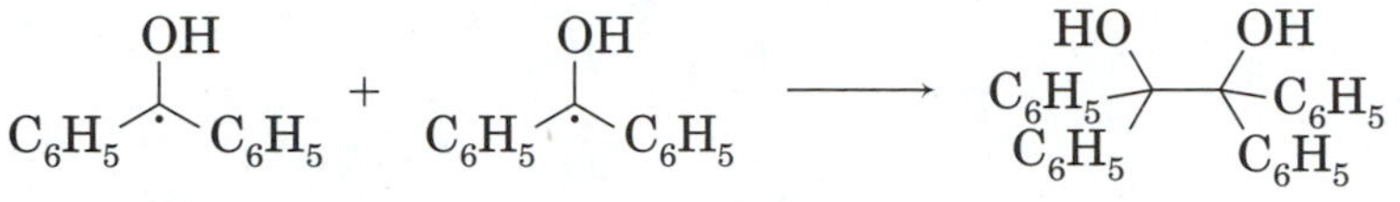

$$\mathbf{31} + \mathbf{31} \longrightarrow \mathbf{30} \qquad (12.42)$$

The hydrogen atom abstraction reaction can be intramolecular as well as intermolecular, provided both that the molecule can achieve the conformation required for the reaction to occur and that the excited state is sufficiently energetic to complete the reaction. A general example of the resulting process is given by the alkyl propyl ketone **33**. Abstraction of the γ hydrogen atom can take place via a facile six-membered transition state (equation 12.43), producing a biradical intermediate (**34**).

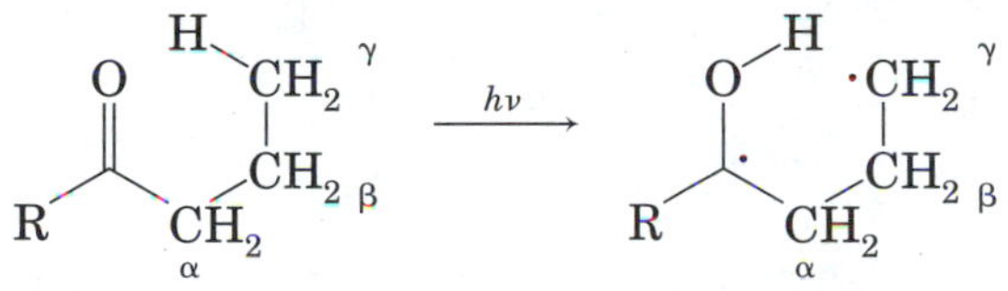

$$\mathbf{33} \xrightarrow{h\nu} \mathbf{34} \qquad (12.43)$$

The biradical is itself a high energy species, relative to stable ground state molecules, and it can further react by several competing pathways. One decay pathway is bond cleavage (equation 12.44) to produce an alkene (**35**) and an enol (**36**). The latter tautomerizes to another ketone (**37**). The overall process is known as **β-cleavage (Norrish Type II cleavage)**.[102]

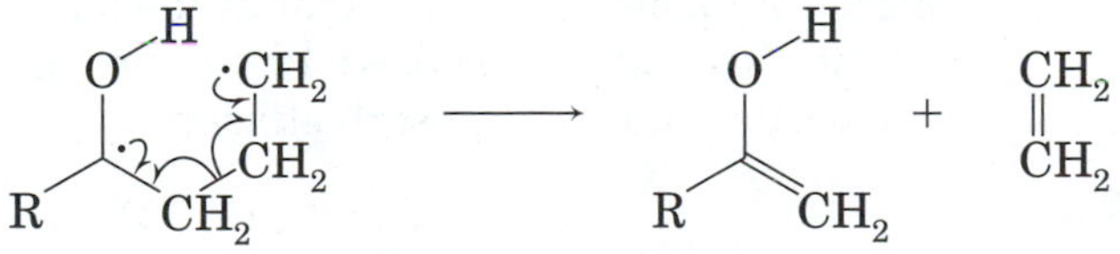

$$\mathbf{34} \longrightarrow \mathbf{35} + \mathbf{36} \qquad (12.44)$$

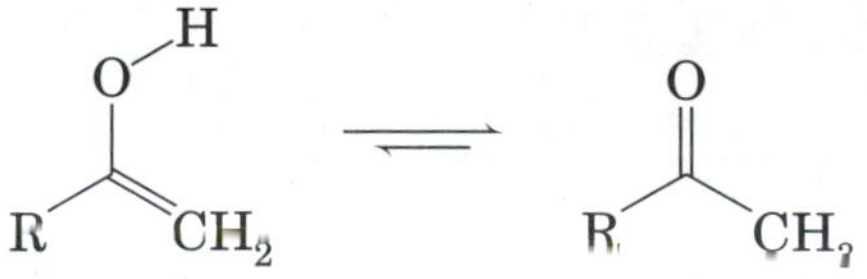

$$\mathbf{35} \rightleftharpoons \mathbf{37} \qquad (12.45)$$

An alternative reaction for **34** is bond formation between the two radical centers, resulting in the formation of a cyclobutanol (**38**).

34 → **38** (12.46)

Finally, the γ carbon atom can abstract the hydrogen atom from the oxygen atom, again via a six-membered ring transition structure, to reform the starting ketone (equation 12.47). This process accomplishes no net photochemistry, so the net process is to convert the energy of electronic excitation into vibrational energy of the ketone and solvent. However, if the starting ketone is optically active due to chirality at the γ carbon atom, then racemization can be observed.

34 → **33** (12.47)

Figure 12.25 shows the scheme described for the photochemical reaction of optically active γ-methylvalerophenone (**39**) by Wagner.[110] The rate constants for the individual processes can be determined from measurements of rate constants for product formation and for photophysical processes.

Intramolecular hydrogen abstraction can lead to useful photochemical syntheses. For example, irradiation of the α,β-unsaturated ketone **40** produces the dienol **41**, which tautomerizes to the β,γ-unsaturated ketone **42**.[111,112] The notable aspect of this reaction is that it converts a more stable conjugated enone to a less stable nonconjugated enone. The reactant, being more conjugated, absorbs light at longer wavelength than does the product, so the reaction can be driven by light that the reactant absorbs but which the product does not absorb efficiently.

40 $\xrightarrow{h\nu}$ **41** ⇌ **42** (12.48)

---

[110]Wagner, P. J. *Acc. Chem. Res.* **1971**, *4*, 168.

[111]Yang, N. C.; Jorgenson, M. J. *Tetrahedron Lett.* **1964**, 1203.

[112]Observation of dienol intermediates produced by the photochemical enolization of α,β-unsaturated ketones was reported by Duhaime, R. M.; Weedon, A. C. *Can. J. Chem.* **1987**, *65*, 1867.

Figure 12.25 Reaction pathways in the photochemistry of γ-methylvalerophenone. (Reproduced from scheme I, reference 110.)

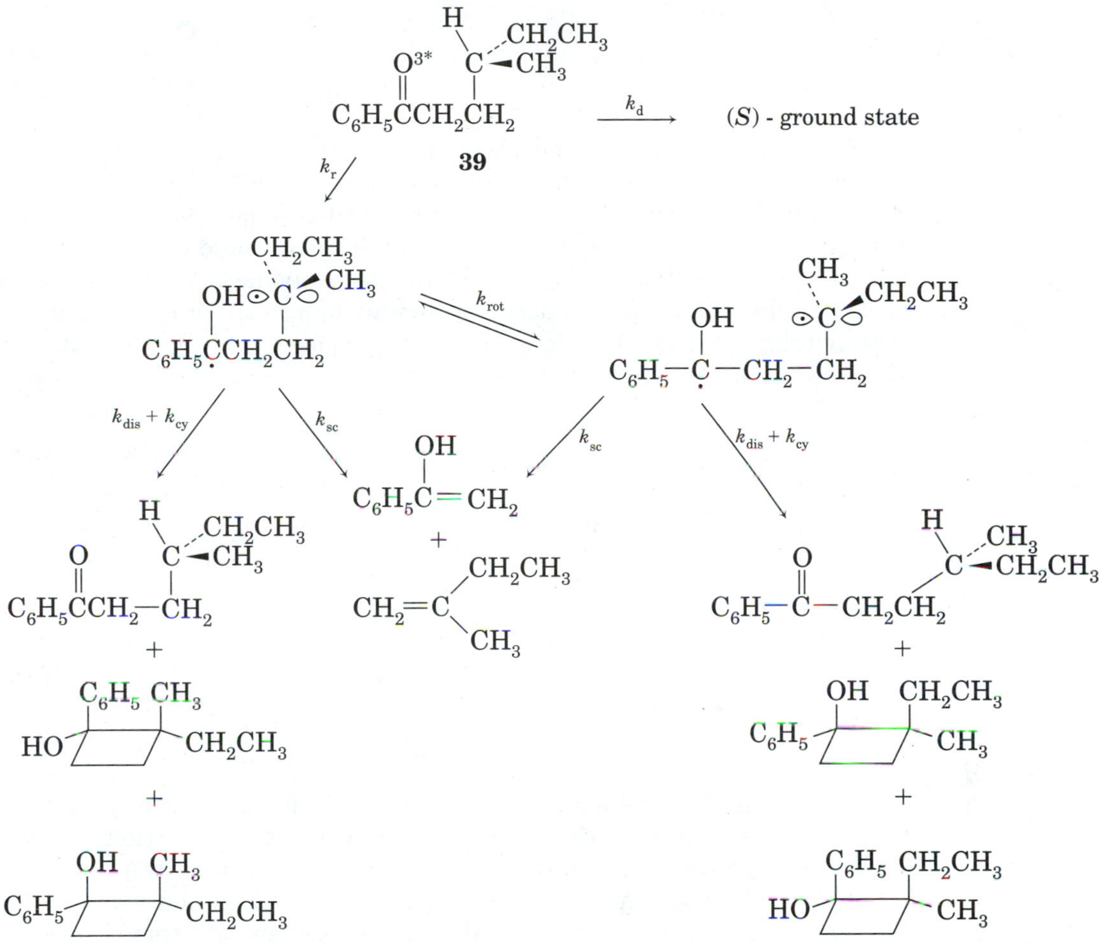

Photochemical oxetane formation is often called the Paterno-Buchi reaction.[113,114] The simple model of the carbonyl n,π^* excited state (Figure 12.22) suggests that addition of one end of the photoexcited carbonyl group to one end of a carbon-carbon double bond could produce a biradical intermediate. Support for this mechanism was obtained by Freilich and Peters,

[113]The reaction was reported by Paternó, E.; Chiefi, G. *Gazz. Chim. Ital.* **1909**, *39*, 431; Buchi, G.; Inman, C. G.; Lipinsky, E. S. *J. Am. Chem. Soc.* **1954**, *76*, 4327.

[114]For a review, see Arnold, D. R. *Adv. Photochem.* **1968**, *6*, 301.

who were able to spectroscopically detect the intermediate in the reaction of photoexcited benzophenone and dioxene.[115]

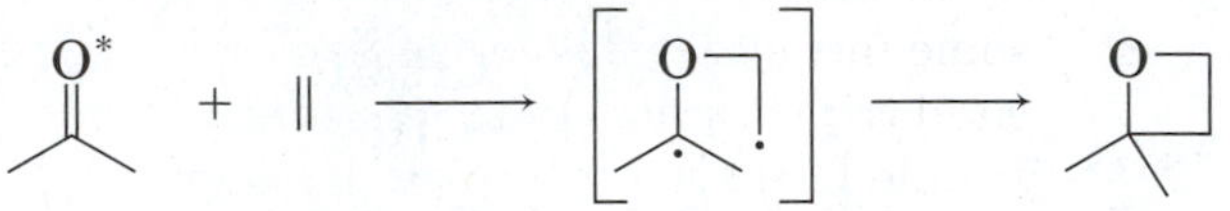

(12.49)

The regio- and stereochemistry of the Paterno-Buchi reaction depends on the structures of the reactants and on the electronic energy of the excited state carbonyl compound. With unsymmetrical alkenes, the products suggest preferential formation of the more highly substituted radical center in the biradical intermediate, but steric and electronic factors are also important. The Paterno-Buchi reaction provides synthetic entry into highly strained ring systems containing oxygen atoms, as exemplified by equations 12.50 and 12.51 (where Ar is 2-naphthyl).[116,117]

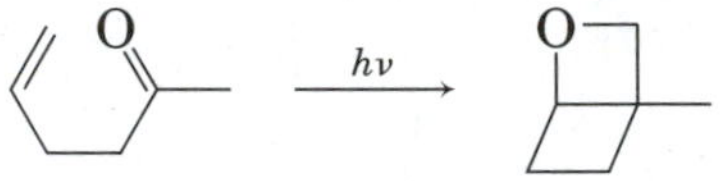

hν

43 **44** (12.50)

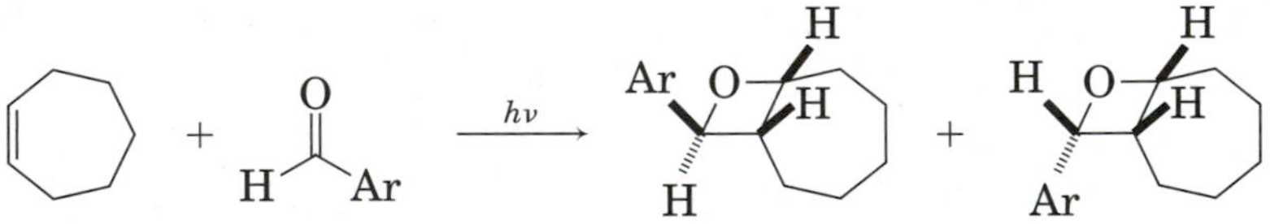

hν

45 **46** **47** **48** (12.51)

The biradical model for carbonyl photoreactivity is valid only for molecules which react from n,π^* excited states, in which the excitation energy is essentially localized on the carbonyl chromophore. That is certainly the case for aliphatic ketones and is true for some, but not all, aromatic ketones. If the carbonyl group is conjugated with a large π system or if there are conju-

[115]It was suggested that the biradical intermediate could be formed by way of an intimate radical ion pair:

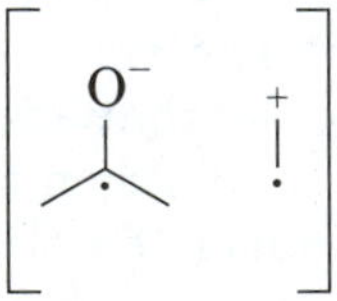

It was later considered more likely, however, that the biradical is formed directly from photoexcited ketone and olefin. The biradical may subsequently undergo heterolysis to form the intimate ion pair, which then dissociates to ketone and olefin. Freilich, S. F.; Peters, K. S. *J. Am. Chem. Soc.* **1981**, *103*, 6255; **1985**, *107*, 3819.

[116]Yang, N. C.; Chiang, W. *J. Am. Chem. Soc.* **1977**, *99*, 3163.

[117]Srinivasan, R. *J. Am. Chem. Soc.* **1960**, *82*, 775.

gating substituents on an aromatic ketone, then the lowest energy state of a given multiplicity could be the π,π^* state.[118] Furthermore, the lower polarity of the n,π^* triplet state relative to the π,π^* triplet state means that for some molecules a polar solvent can sometimes make the π,π^* the lowest excited triplet state. In such situations, the radical reactivity suggested by the simple Lewis structure model is greatly diminished.[119]

Photochemical Reactions of Alkenes and Dienes

The onset of absorption for a nonconjugated olefin is about 230 nm but is so gradual that it is difficult to determine an exact value. Furthermore, the λ_{max} is usually less than 200 nm, in a region of the spectrum known as the far Uv or the vacuum Uv that is beyond the wavelength limit of spectrophotometers that operate in air.[120] Therefore, only **end absorption** (that is, the onset or tail of the peak but not its maximum) is usually observed. According to HMO theory, the HOMO of a nonconjugated alkene is a π orbital, while the LUMO is a π^* orbital (Figure 12.26). Therefore the only Uv-vis transition should be a $\pi \rightarrow \pi^*$ transition.[121]

The simple orbital model in Figure 12.26 is not entirely consistent with the available experimental data, however. Indeed, Kropp has called the isolated carbon-carbon double bond "one of the most deceptively simple chromophores available to the organic chemist."[122] Even though the end absorption apparently includes the π,π^* transition, there is evidence that there may also be weak absorption due to $\pi \rightarrow \sigma^*$ and Rydberg transitions. In alkenes, the Rydberg state is envisioned as resulting from promotion of an electron from the π orbital to a carbon orbital that has the characteristics of a 3*s* orbital in helium. The radial distribution of this orbital is large

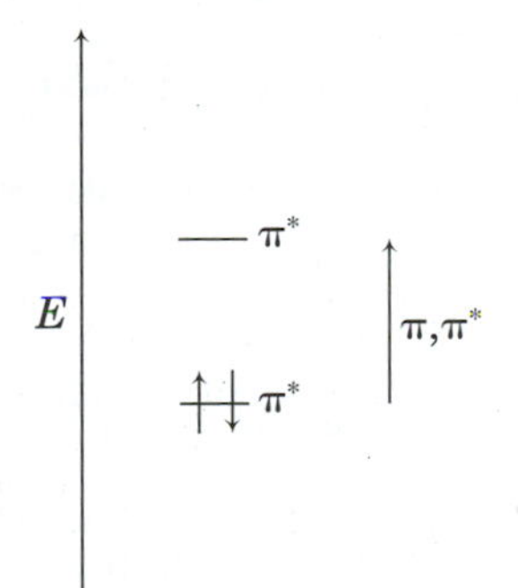

Figure 12.26 Origin of a π,π^* transition.

[118] Hammond, G. S.; Leermakers, P. A. *J. Am. Chem. Soc.* **1962**, *84*, 207.

[119] See the discussion in reference 4, pp. 380 *ff.*

[120] For a discussion of techniques and applications of far-UV photochemistry, see Leigh, W. J. *Chem. Rev.* **1993**, *93*, 487.

[121] For a theoretical study of the excited states of ethene, see Wiberg, K. B.; Hadad, C. M.; Foresman, J. B.; Chupka, W. A. *J. Phys. Chem.* **1992**, *96*, 10756.

[122] Kropp, P. J. *Mol. Photochem.* **1978–79**, *9*, 39.

in comparison with the $2p$ orbitals that are the basis set for the π orbital. In essence, this is an orbital around the entire molecule.[123,124,125]

The π,π^* singlet states of nonconjugated acyclic olefins do not fluoresce because rotation about the carbon-carbon bond occurs faster than the rate constant for fluorescence, and the relaxed (perpendicular) excited singlet has a negligible Franck-Condon overlap with the ground state.[126] Figure 12.27 shows the potential energy curves for the S_0, $^3\pi,\pi^*$ and $^1\pi,\pi^*$ states of an alkene as a function of rotation about the carbon-carbon (double) bond.[127,128,129] The $^1\pi,\pi^*$ excited state created by direct excitation of an alkene is a vertical state with the same geometry as the ground state molecule.[130] In the $^1\pi,\pi^*$ state the electrons in the singly occupied MOs are still spin paired, but to a first approximation the π^* orbital is antibonding to the same extent that the π orbital is bonding. Thus excitation removes the barrier to rotation about the former double bond, and rotation toward a 90° orientation of the two p orbitals lowers the energy of the excited state.[131]

---

[123]Merer, A. J.; Mulliken, R. S. *Chem. Rev.* **1969**, *69*, 639.

[124]The discussion of the Jablonski diagram (Figure 12.4) implied that internal conversion of upper excited singlet states is fast and that photochemical reactions arise only from the lowest lying excited state of a given multiplicity. That statement is generally true for organic molecules with extended π systems and large Franck-Condon factors for conversion of S_n into S_{n-x}. For simple alkenes, however, the overlap of the π,σ^*, Rydberg and π,π^* states is small, so internal conversion is slow. These states may exhibit their own characteristic photochemical reactions, and there may be significant wavelength effects on the distribution of photochemical products. An additional complication in the photochemical reactions of alkenes in the gas phase is the possible involvement of vibrationally excited (hot) electronic ground states produced from an excited state. For a review, see Collin, G. J. *Adv. Photochem.* **1988**, *14*, 135.

[125]Evidence supporting the existence of the Rydberg state has been reported by Hirayama, F.; Lipsky, S. *J. Chem. Phys.* **1975**, *62*, 576, who found weak fluorescence from a series of substituted ethenes.

[126]Fluorescence has been observed from conjugated polyenes. See the discussion in reference 29.

[127]Saltiel, J.; D'Agostino, J.; Megarity, E. D.; Metts, L.; Neuberger, K. R.; Wrighton, M.; Zafiriou, O. C. *Org. Photochem.* **1973**, *3*, 1.

[128]The $^1\pi,\pi^*$ designation refers to a singlet π,π^* state. Similarly, $^3\pi,\pi^*$ refers to a triplet π,π^* state. These states may also be designated as $^1(\pi,\pi^*)$ and $^3(\pi,\pi^*)$, respectively.

[129]The discussion here is a very simplified treatment of the excited states of simple alkenes. Theoretical calculations suggest that the twisted (90°) $^1\pi,\pi^*$ state can have considerable contribution from zwitterionic structures, particularly if the double bond has polar substituents or if there is pyramidalization of one of the π centers. See, for example, Wulfman, C. E.; Kumei, S. *Science* **1971**, *172*, 1061; Salem, L. *Science* **1976**, *191*, 822; Salem, L. *Acc. Chem. Res.* **1979**, *12*, 87; Buenker, R. J.; Bonačić-Koutecký, V.; Pogliani, L. *J. Chem. Phys.* **1980**, *73*, 1836; Brooks, B. R.; Schaefer III, H. F. *J. Am. Chem. Soc.* **1979**, *101*, 307; Michl, J.; Bonačić-Koutecký, V. (reference 66), pp. 206–216.

[130]An excited state with geometry identical to the ground state is sometimes called a *Franck-Condon state*.

[131]Although the simple HMO theory does not take such factors into account, there is considerable electron-electron repulsion present in the vertical excited state (that is, the electronically excited state that has nearly the same nuclear coordinates as the ground state had before excitation). Rotation makes the two p orbitals orthogonal, relieving the repulsion, and minimizing the energy of the system.

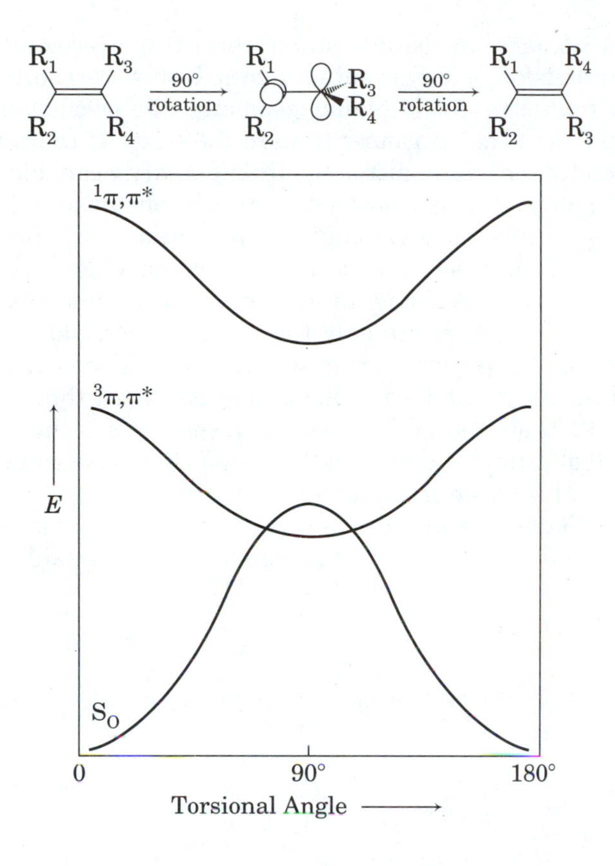

Figure 12.27 Torsional dependence of the electronic energies of the ground state and singlet and triplet π,π^* states of ethene. (Adapted from reference 127.)

Radiationless decay of the twisted excited singlet to either the cis or trans ground state can lead to isomerization of the photoexcited alkene. Furthermore, the rapid radiationless decay of the excited singlet state means that intersystem crossing from a π,π^* singlet to a triplet is extremely unlikely. However, olefinic triplets can be formed by sensitization. The potential energy curve for the $^3\pi,\pi^*$ state is lower than that for the $^1\pi,\pi^*$ state at all geometries, but it is also expected to rotate toward a 90° orientation and to resemble the biradical species represented in Figure 12.28.[132,133] Intersystem crossing from the twisted triplet to the singlet ground state also leads to either the cis or trans isomer of the alkene in the S_0 state. Therefore, sensitized excitation of alkenes can also lead to cis-trans isomerization.

---

[132]Note also that at 90° the energy of the π,π^* triplet is lower than that of the twisted ground state. Other calculations place the energy of the twisted triplet slightly higher than that of the twisted ground state (*cf.* Yamaguchi, Y.; Osamura, Y.; Schaefer III, H. F. *J. Am. Chem. Soc.* **1983**, *105*, 7506). In any case, the two states are expected to be close in energy at that geometry.

[133]For discussions of the 1,2-biradical nature of alkene triplets, see Caldwell, R. A.; Zhou, L. *J. Am. Chem. Soc.* **1994**, *116*, 2271; Caldwell, R. A.; Díaz, J. F.; Hrncir, D. C.; Unett, D. J. *J. Am. Chem. Soc.* **1994**, *116*, 8138.

Cis isomers can be less stable than trans isomers because of van der Waals repulsion of the cis substituents. If this steric interaction distorts the alkene from a perfectly planar geometry, the overlap of the two *p* orbitals may be less in the cis isomer than in the trans. If the double bond is part of an extended π system, distortion from planarity can lead to a larger HOMO-LUMO gap in the cis isomer. As a result, photoexcitation of the cis isomer may require shorter wavelength light than does excitation of the trans isomer. In an ideal case, an absorption spectrum of cis and trans isomers of an alkene would be as shown in Figure 12.29, in which excitation at a proper wavelength would excite only the trans isomer. More commonly, however, the absorption spectra overlap so that the best one can do is choose a wavelength at which one isomer has a higher ϵ than the other. As an example, Figure 12.30 shows the Uv absorption spectrum of the *cis-* (broken line) and *trans-* (solid line) isomers of *N,N*-dimethyl-*p*-methoxycinnamamide as a 5×10^{-5} M solution in methylene chloride.[134]

Irradiation of a mixture of cis and trans alkenes at a wavelength that is strongly absorbed by the trans isomer but only weakly absorbed by the cis

Figure 12.28 Orbital relationship in twisted olefinic π,π^* triplet.

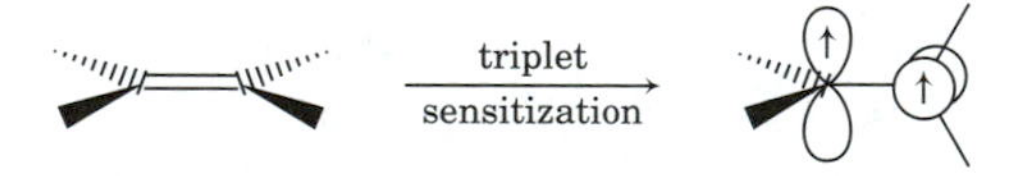

[134]Lewis, F. D.; Elbert, J. E.; Upthagrove, A. L.; Hale, P. D. *J. Org. Chem.* **1991**, *56*, 553.

[135]This generalization of the photoisomerization of nonconjugated alkenes does not apply to one of the best-known photochemical isomerizations, the conversion of 11-cis rhodopsin to all-trans product. The speed of the reaction (200 fs) has been related to distortion of the 11-cis double bond, which causes the rotation required for isomerization to be much less than 180°, and to steric interactions in the 11-cis isomer, which provide a force for rotation of the photoexcited cis isomer toward the geometry of the trans isomer. Wang, Q.; Schoenlein, R. W.; Peteanu, L. A.; Mathies, R. A.; Shank, C. V. *Science* **1994**, *266*, 422.

[136]The distribution of the cis and trans isomers formed from the twisted singlet is not entirely random, although a detailed explanation of the factors that determine the product ratios is beyond the scope of this discussion. In some cases the rate constant for formation of the cis isomer (k_c) is greater than for formation of the trans isomer (k_t). Because the cis isomer is higher in energy than the trans, there may be better overlap of the vibrational component of the wave function of the cis ground state with the vibrational component of the excited state than is the case with the trans isomer. Thus the Franck-Condon term for the twisted singlet to cis reaction could be more favorable than for the trans. There are numerous examples of photochemical reactions that produce the less stable of two products, a feature of photochemistry that makes it quite useful in organic synthesis.

[137]Some compounds exhibit one-way photoisomerization. For a review, see Arai, T.; Tokumaru, K. *Chem. Rev.* **1993**, *93*, 23.

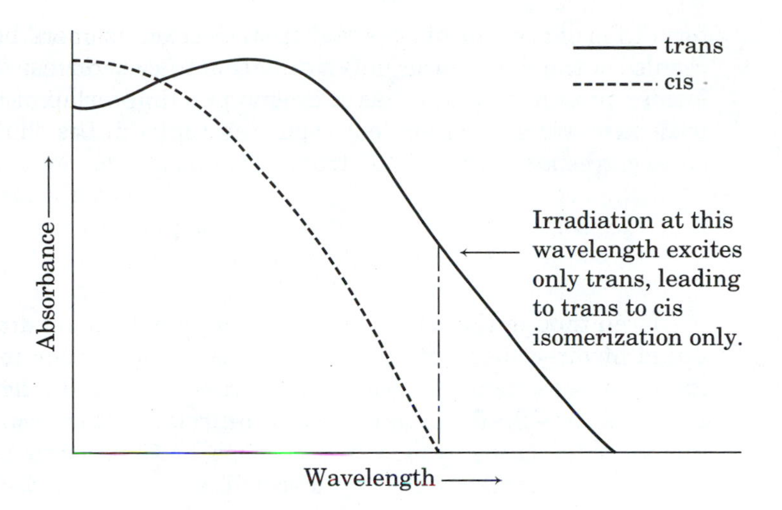

Figure 12.29 Idealized alkene absorption spectra.

isomer will allow photochemical conversion of trans alkene to cis alkene. A photoexcited trans molecule may relax to a twisted conformation and then decay either to the cis (**C**) or trans (**T**) isomer.[135,136,137] If the excited state decays to the cis isomer, the low absorbance by that isomer means that further reaction is not very probable. If it decays to the trans isomer, then it may be excited again later and thus have another chance to be converted to cis. Eventually, the system reaches a **photostationary state (pss)**, which is a function of wavelength of irradiation and molecular extinction coefficients as well as the rate constants for formation of the cis isomer (k_c) or for

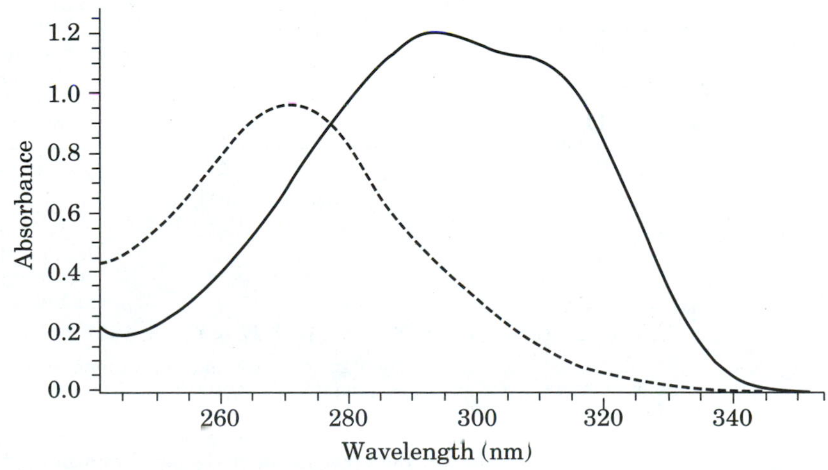

Figure 12.30 Uv spectra of *trans*- (solid line) and *cis*- (broken line) *N,N*-dimethyl-*p*-methoxycinnamamide. (Reproduced from reference 134.)

formation of the trans isomer (k_t) from the excited state. Figure 12.31 shows results of the direct (185 nm) irradiation of cyclooctene. The percent trans isomer present is plotted as a function of time for experiments beginning with pure trans (filled circles) or pure cis (open circles). The photostationary state corresponds to a 47 : 53 trans : cis mixture.[138]

$$\frac{[\mathbf{C}]_{\text{pss}}}{[\mathbf{T}]_{\text{pss}}} = \frac{\epsilon_{\text{trans}}}{\epsilon_{\text{cis}}} \cdot \frac{k_c}{k_t} \tag{12.52}$$

Even though the triplet states of simple alkenes ordinarily are not produced by direct irradiation, the $^3\pi,\pi^*$ states of alkenes can be formed indirectly by sensitization.[139] If one electron of an olefin triplet excited state undergoes a spin flip, the molecule can relax to the ground state of either the cis or the trans isomer. In equations 12.53 through 12.58, **S** is a sensitizer which can be excited by direct irradiation and which, among other

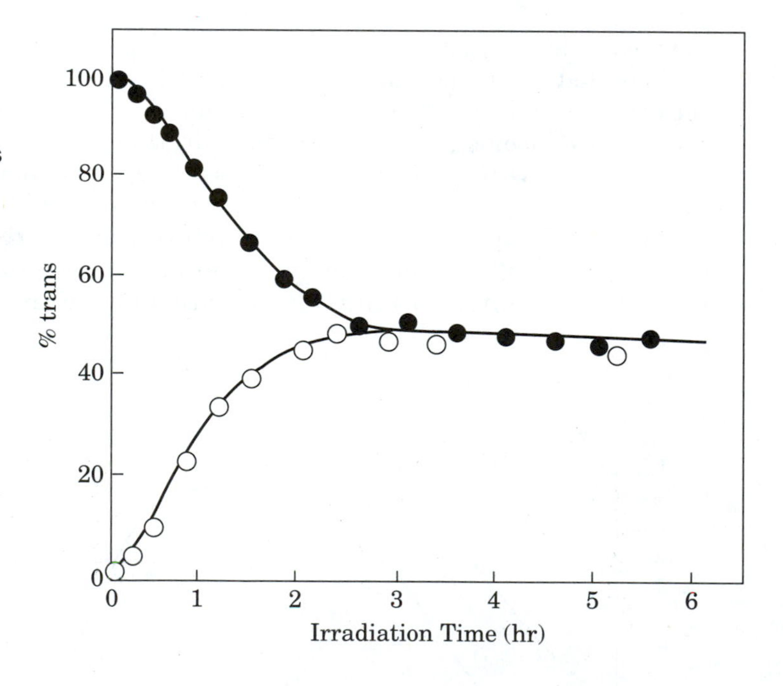

Figure 12.31 Photochemical production of photostationary state concentrations of *cis*- and *trans*-cyclooctene. (Adapted from reference 138.)

[138]Inoue, Y.; Takamuku, S.; Sakurai, H. *J. Phys. Chem.* **1977**, *81*, 7.

[139]Substituted alkenes can undergo intersystem crossing and subsequent isomerization from the triplet state. For example, Lewis, F. D.; Bassani, D. M.; Caldwell, R. A.; Unett, D. J. *J. Am. Chem. Soc.* **1994**, *116*, 10477, found that *trans*-1-phenylpropene does not isomerize from the singlet state, but at low temperatures it undergoes intersystem crossing and then isomerizes from the triplet state. On the other hand, *cis*-1-phenylpropene undergoes isomerization primarily from the singlet state at room temperature.

processes, undergoes intersystem crossing to its triplet state. If $\mathbf{S}^{*(3)}$ is higher in energy than $\mathbf{C}^{*(3)}$, then energy transfer can occur.[140,141]

$$\mathbf{S} \xrightarrow{h\nu} \mathbf{S}^{*(1)} \xrightarrow{k_{\mathrm{isc}}} \mathbf{S}^{*(3)} \quad \textbf{(12.53)}$$

$$\mathbf{S}^{*(3)} + \mathbf{C} \xrightarrow{k_{\text{et-c}}} \mathbf{C}^{*(3)} \equiv \mathbf{I}^{*(3)} \quad \textbf{(12.54)}$$

$$\mathbf{S}^{*(3)} + \mathbf{T} \xrightarrow{k_{\text{et-t}}} \mathbf{T}^{*(3)} \equiv \mathbf{I}^{*(3)} \quad \textbf{(12.55)}$$

$$\mathbf{I}^{*(3)} \xrightarrow{k_{\mathrm{c}}} \mathbf{C} \quad \textbf{(12.56)}$$

$$\mathbf{I}^{*(3)} \xrightarrow{k_{\mathrm{t}}} \mathbf{T} \quad \textbf{(12.57)}$$

Note that the intermediate triplet ($\mathbf{I}^{*(3)}$) is assumed to be the same species, no matter whether the cis isomer or the trans isomer was sensitized.[142] Now the photostationary state is a function of triplet decay ratios, k_{dt} and k_{dc}, and of the rate constants for energy transfer to the two isomers, $k_{\text{et-t}}$ and $k_{\text{et-c}}$.[143]

$$\frac{[\mathbf{C}]_{\mathrm{pss}}}{[\mathbf{T}]_{\mathrm{pss}}} = \frac{k_{\text{et-t}}}{k_{\text{et-c}}} \cdot \frac{k_{\mathrm{c}}}{k_{\mathrm{t}}} \quad \textbf{(12.58)}$$

Photochemical reactions can also occur by way of reactive ground state intermediates produced by a photochemical reaction. For example, irradiation of a solution of methylcyclohexene and an aromatic sensitizer such as *p*-xylene in methanol produces methyl 1-methylcyclohexyl ether (**52**) plus a small quantity of methylenecyclohexane (**53**, Figure 12.32). A plausible mechanism involves sensitized isomerization of the *cis*-1-methylcyclohexene to an unstable *trans*-cyclohexene intermediate (**50**). The highly strained **50** can rapidly add a proton from the solvent to form a 3° carbocation (**51**), which can then undergo nucleophilic addition (major product) or elimination (minor product). This mechanism is supported by the observation of trans cycloalkenes in trapping reactions[144] and by the detection of the

---

[140]For a kinetic analysis, see Hammond, G. S. *Kagaku to Kôgyô (Tokyo)* **1965**, *18*, 1464.

[141]The rate constants are given designations that suggest the processes involved. For example, $k_{\text{et-c}}$ is the rate constant for energy transfer to the cis isomer.

[142]In some cases there may be more than one minimum on an excited state potential energy surface, so this assumption may not be warranted.

[143]For a study of sensitized isomerizations of olefins, see Snyder, J. J.; Tise, F. P.; Davis, R. D.; Kropp, P. J. *J. Org. Chem.* **1981**, *46*, 3609.

[144]Goodman, J. L.; Peters, K. S.; Misawa, H.; Caldwell, R. A. *J. Am. Chem. Soc.* **1986**, *108*, 6803.

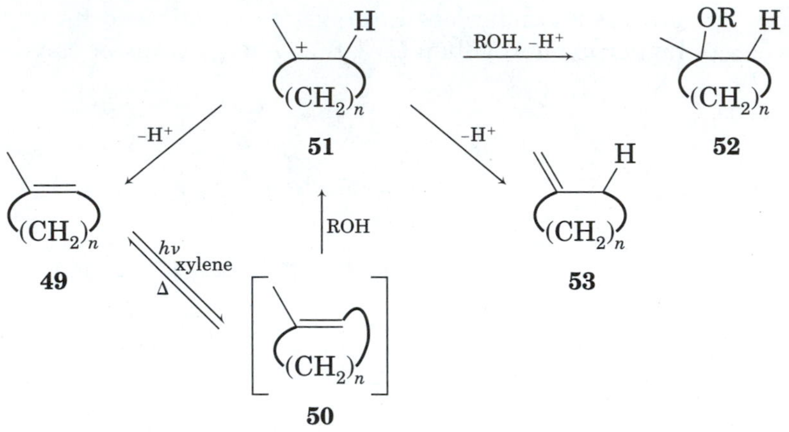

Figure 12.32 Photochemical isomerization and solvent incorporation with methylcyclohexene. (Adapted from reference 122.)

1-phenylcyclohexyl cation in the flash spectroscopy of 1-phenylcyclohexene in 1,1,1,3,3,3-hexafluoroisopropyl alcohol.[145] The role of the strained cycloalkene is further supported by the reaction of limonene. As shown in Figure 12.33, photohydration of limonene (**54**) produces an 89% yield of the alcohol **55**, whereas acid-catalyzed hydration produces only the diol **57**, in which both double bonds have reacted.[146]

Photochemical product formation can also arise from other reactive intermediates. Figure 12.34 shows the products from direct irradiation of

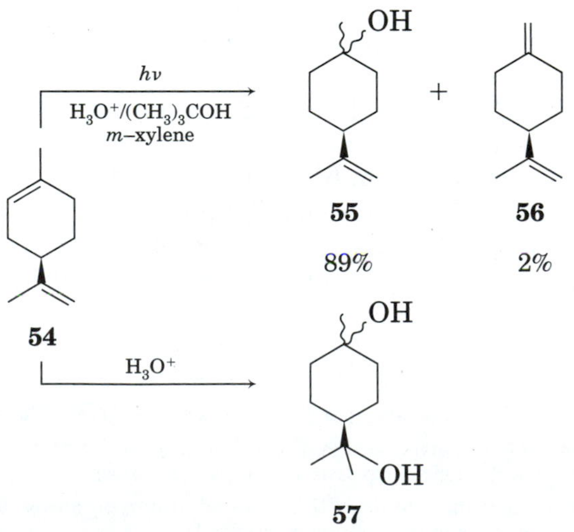

Figure 12.33 Photohydration of limonene. (Reproduced from reference 122.)

[145] Cozens, F. L.; McClelland, R. A.; Steenken, S. *J. Am. Chem. Soc.* **1993**, *115*, 5050.

[146] For a discussion of such reactions, see reference 122.

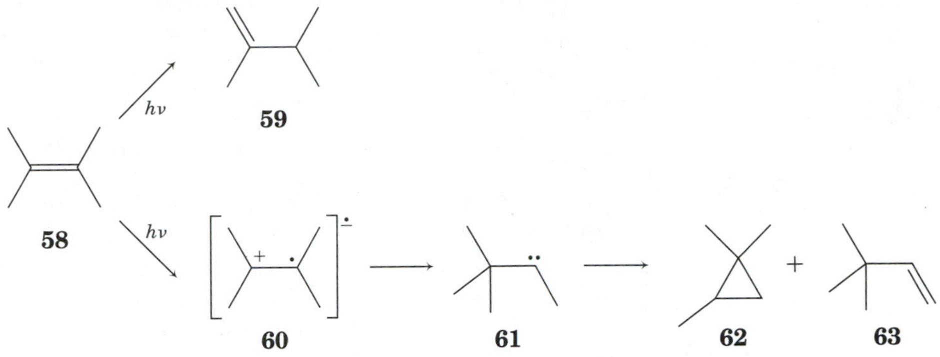

Figure 12.34
Products from direct irradiation of tetramethylethylene. (Adapted from reference 147(b).)

tetramethylethylene (**58**). Products **62** and **63** are thought to arise from the carbene **61**, which is produced by radiationless decay of the Rydberg state, **60**. There is apparently a different mechanism responsible for the formation of the isomerization product **59**, and a pathway involving a π,σ^* state has been suggested.[122,147]

The discussion in Chapter 11 considered photochemical reactions of alkenes in terms of orbital symmetry control of pericyclic reactions, but the possible intermediacy of multiple, independent alkene excited states complicates the analysis of photochemical reactions in terms of the Woodward-Hoffmann rules. For example, photochemical ring opening of cyclobutenes appears to be nonspecific.[148] Dauben and Haubrich found that irradiation of the cyclobutene **64** with 193 nm light produces not only the allowed *cis,cis,cis*-1,3,5-cyclodecatriene **67**, but also the cis,trans,cis isomer **65** and the cis,cis,trans isomer **66**.[149] Among the possible explanations for the forbidden products are

1. thermal formation from the highly strained, thermally allowed trans,trans,cis isomer;
2. thermal formation from a vibrationally excited ground state formed by radiationless decay of the excited state;
3. adiabatic opening of the cyclobutene to an electronically excited triene, which could then exhibit photochemical isomerization; or
4. reaction from some state other than a π,π^* state.

---

[147]Kropp, P. J.; Fravel, Jr., H. G.; Fields, T. R. *J. Am. Chem. Soc.* **1976**, *98*, 840. See also the discussion in (a) reference 122; (b) Cherry, W.; Chow, M.-F.; Mirbach, M. J.; Mirbach, M. F.; Ramamurthy, V.; Turro, N. J. *Mol. Photochem.* **1977**, *8*, 175; (c) Inoue, Y.; Mukai, T.; Hakushi, T. *Chem. Lett.* **1983**, 1665.

[148]Clark, B.; Leigh, W. J. *J. Am. Chem. Soc.* **1987**, *109*, 6086.

[149]Dauben, W. G.; Haubrich, J. E. *J. Org. Chem.* **1988**, *53*, 600.

64 → ($h\nu$, 193 nm) 65 + 66 + 67 **(12.59)**

Adam and co-workers have suggested that irradiation of cyclobutene with 185 nm light produces both π,π^* and Rydberg ($\pi,3s$) excited states. The π,π^* state is thought to lead to electrocyclic opening to butadiene, while the Rydberg state leads to carbene intermediates (**69** and **70**) that can fragment to give ethene and acetylene, methylenecyclopropane (**71**) and butadiene (Figure 12.35).[150,151,152]

The preceding analysis assumed that the two double bonds in **64** are independent, and the reaction was analyzed in terms of the double bond in the cyclobutene ring. However, nonconjugated double bonds may interact with each other, especially if they are physically near. For example, the di-π-methane rearrangement (equation 12.60), first reported by Zimmerman, has been the subject of extensive investigation.[153,154,155] The exact course of the reaction and its multiplicity depends on the nature of the substituent groups attached to the double bonds. A complete treatment of the reaction is beyond the scope of this discussion, but one point should be emphasized. Although the process may be formally represented as a $[_{\sigma}2 + _{\pi}2]$ cycloaddition, the observation of equal yields of the two products **77** and **78** from the irradiation of the deuterium-labeled compound **76** (Figure 12.36), suggests a mechanism involving a biradical intermediate.

74 → ($h\nu$, acetone sensitized) 75 **(12.60)**

1,5-, 1,6-, and 1,7-Dienes undergo sensitized intramolecular cycloaddition. The reactions can be interpreted in terms of stepwise addition of the

[150]Adam, W.; Oppenländer, T.; Zang, G. *J. Am. Chem. Soc.* **1985**, *107*, 3921.

[151]See also Leigh, W. J.; Zheng, K.; Clark, K. B. *J. Org. Chem.* **1991**, *56*, 1574; Prathapan, S.; Agosta, W. C. *Chemtracts-Org. Chem.* **1991**, *4*, 460.

[152]Leigh, W. J.; Zheng, K. *J. Am. Chem. Soc.* **1991**, *113*, 4019, reported evidence for photochemical disrotatory ring opening in one bicyclic system incorporating a cyclobutene ring, but the stereospecificity in such cases seems to depend on the structural features incorporated into the reactants. For a discussion of the relationship of orbital symmetry to the photochemistry of cyclobutene, see Leigh, W. J. *Can. J. Chem.* **1993**, *71*, 147.

[153]Zimmerman, H. E.; Grunewald, G. L. *J. Am. Chem. Soc.* **1966**, *88*, 183.

[154]Hixson, S. S.; Mariano, P. S.; Zimmerman, H. E. *Chem. Rev.* **1973**, *73*, 531.

[155]Zimmerman, H. E. *Org. Photochem.* **1991**, *11*, 1.

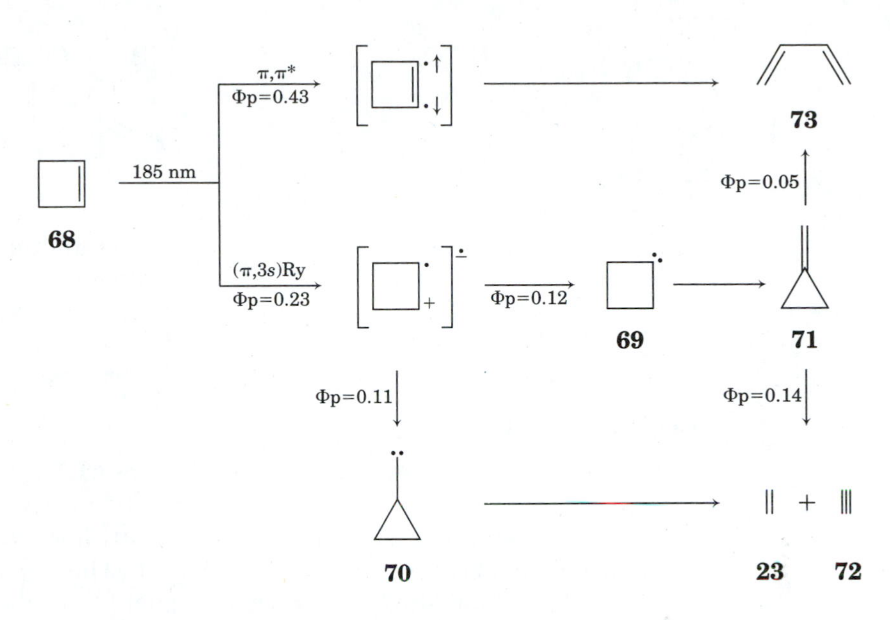

Figure 12.35 Photochemical reaction pathways in cyclobutene. (Adapted from reference 150.)

triplet of one olefin to the second olefin to form a 1,4-biradical intermediate that can then close to a cyclobutane ring through a kinetically controlled process.[156,157] Examples are shown in equations 12.61 and 12.62.

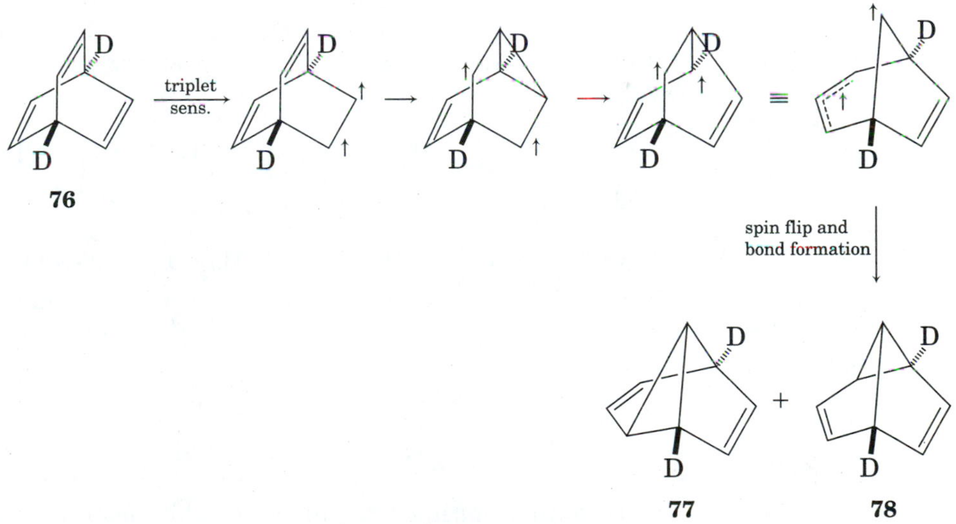

Figure 12.36 Mechanism proposed for the di-π-methane reaction. (Adapted from reference 154.)

[156]Scheffer, J. R.; Wostradowski, R. A. *J. Chem. Soc. D, Chem. Commun.* **1971**, 144; Scheffer, J. R.; Wostradowski, R. A.; Dooley, K. C. *J. Chem. Soc. D, Chem. Commun.* **1971**, 1217. Compounds with two functional groups capable of photochemical reaction are said to be bichromophoric. For a review, see De Schryver, F. C.; Boens, N.; Put, J. *Adv. Photochem.* **1977**, *10*, 359.

[157]For a theoretical study of photocycloaddition processes, see Bentzien, J.; Klessinger, M. *J. Org. Chem.* **1994**, *59*, 4887.

CO_2CH_3 … CO_2CH_3 $\xrightarrow[\text{sens.}]{h\nu}$ H_3CO_2C, H, H, CO_2CH_3 + H, CO_2CH_3, H, CO_2CH_3 (12.61)

79 **80** **81**

CO_2CH_3 … CO_2CH_3 $\xrightarrow[\text{sens.}]{h\nu}$ CO_2CH_3, CO_2CH_3 + CO_2CH_3, CO_2CH_3 (12.62)

82 **83c** **83t**

The excited states of conjugated dienes are simpler than those of non-conjugated olefins because the smaller HOMO-LUMO gap means that the lowest energy excited singlets and triplets should be π,π^* states (and not Rydberg states). However, the photochemistry of acyclic dienes is complicated by the possibility of s-trans and s-cis conformations of the ground state molecule and the excited states. In the vapor phase direct irradiation of butadiene leads to rearrangement to methylallene (**84**) and butyne (**85**) as well as fragmentation and polymerization, perhaps through thermal reaction of vibrationally excited products.[158] In solution, direct irradiation of 1,3-butadiene results in the formation of cyclobutene and bicyclobutane (**86**).[158,159,160]

$$\xrightarrow[\text{gas phase}]{h\nu} H_3C-CH=C=CH_2 + H_3C-CH_2-C\equiv CH + HC\equiv CH + H_2C=CH_2 + H_2 + \text{polymeric material} \quad (12.63)$$

73 **84** **85** **72** **23**

$$\xrightarrow[\text{solution}]{h\nu} \quad (12.64)$$

73 **68** **86**

Irradiation of conjugated dienes can also lead to cis,trans isomerization. Irradiation of cycloheptadiene (**87**) in methanol leads to the bicycloheptene **89** and to the methoxy ethers **90** and **91**. The products are explained by iso-

---

[158]Haller, I.; Srinivasan, R. *J. Chem. Phys.* **1964**, *40*, 1992.

[159]For a review and leading references, see Fonken, G. J. *Org. Photochem.* **1967**, *1*, 197.

[160]For a review of photochemical cycloaddition reactions of conjugated dienes and polyenes, see Dilling, W. A. *Chem. Rev.* **1969**, *69*, 845.

merization of the cis,cis reactant **87** to the cis,trans isomer **88**, which can undergo electrocyclic closure to **89** or add a proton from the solvent. Furthermore, addition of a proton from solvent can occur at either terminus of the strained (trans) alkene, leading to the carbocation intermediates and their ether trapping products (Figure 12.37).[161]

Photochemical cis,trans isomerization of conjugated dienes is usually carried out through triplet sensitization, and quantitative determination of cis,trans isomerization of conjugated dienes has been used to determine the quantum yield of intersystem crossing of aromatics.[162] At high diene concentrations, one diene triplet can undergo cycloaddition with a ground state diene before decay to the ground state occurs. Figure 12.38 shows the

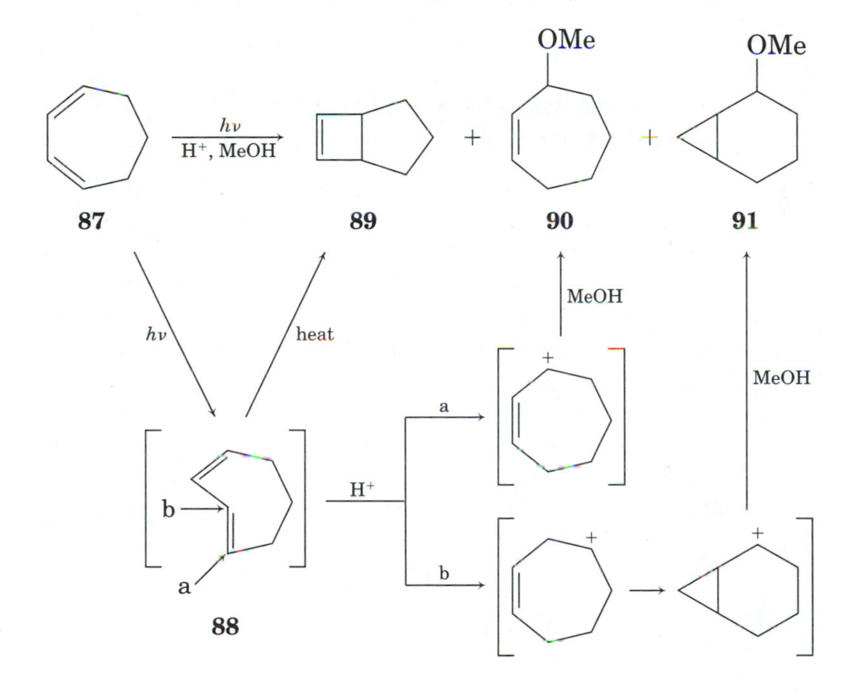

Figure 12.37 Photochemical reaction of 1,3-cycloheptadiene in methanol. (Reproduced from reference 161.)

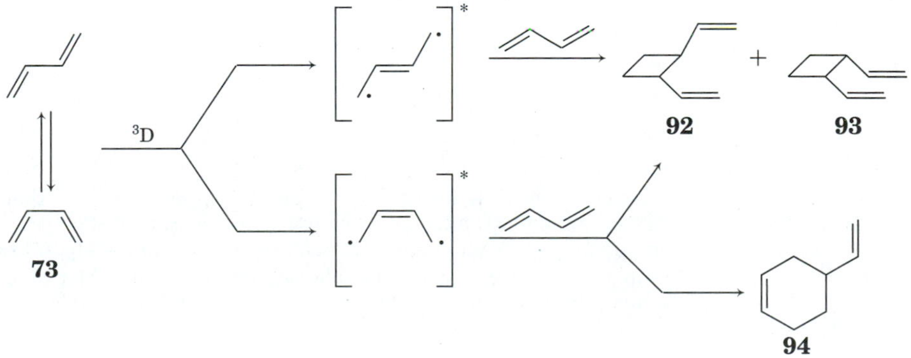

Figure 12.38 Pathways for triplet-sensitized reactions of 1,3-butadiene. (^{3}D is a triplet sensitizer; adapted from reference 127.)

[161] Inoue, Y.; Hagiwara, S.; Daino, Y.; Hakushi, T. *J. Chem. Soc., Chem. Commun.* **1985**, 1307.

[162] Lamola, A. A.; Hammond, G. S. *J. Chem. Phys.* **1965**, *43*, 2129.

pathways proposed for formation of *trans*- (**92**) and *cis*-1,2-divinylcyclobutane (**93**) and 4-vinylcyclohexene (**94**).[163] Similarly, sensitized dimerization of 1,3-cyclohexadiene produces the isomeric 1,2-addition products **13** and **95** and the 1,4-addition product **96**.[164]

$h\nu$ sens.

12 **13** + **95** + **96** (12.65)

Photochemical Reactions of α,β-Unsaturated Carbonyl Compounds

Structures with a carbon-carbon double bond conjugated with a carbonyl group exhibit characteristic reactions that can be rationalized with Lewis structure representations of the excited states that are combinations of those developed for the isolated carbonyl group and for the isolated carbon-carbon double bond. In most cases the lowest excited states, both singlet and triplet, are n,π^* states, but the π^* orbital now has density on the olefinic carbons, primarily on the β carbon atom, as well as on the carbonyl carbon atom and the oxygen atom. Thus the photochemistry of n,π^* excited states of enones may be rationalized with the resonance structures in equation 12.66.[165] This model makes it easy to understand the cycloaddition of an α,β-unsaturated ketone with an olefin by the mechanism shown in equation 12.67.[166]

$h\nu$ n,π^*

97 (12.66)

---

[163]See reference 127, p. 67.

[164]Valentine, D.; Turro, Jr., N. J.; Hammond, G. S. *J. Am. Chem. Soc.* **1964**, *86*, 5202.

[165]It is not necessarily the case in representations such as these that all resonance structures make an equal contribution to the reactivity of the excited state.

[166]Experimental evidence has been reported for the intermediacy of 1,4-biradicals in both intermolecular and intramolecular photocycloaddition of triplet alkenones with alkenes. For leading references, see Maradyn, D. J.; Weedon, A. C. *J. Am. Chem. Soc.* **1995**, *117*, 5359. For a theoretical study of the photochemical cycloaddition of acrolein and ethene, see Erickson, J. A.; Kahn, S. D. *Tetrahedron* **1993**, *49*, 9699. For a theoretical study of the regioselectivity of the photocycloaddition of triplet cyclohexenones to alkenes, see Broeker, J. L.; Eksterowicz, J. E.; Belk, A. J.; Houk, K. N. *J. Am. Chem. Soc.* **1995**, *117*, 1847.

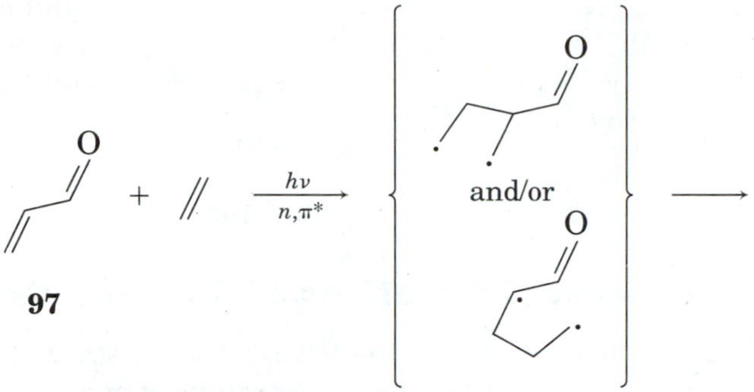

O + // —$h\nu$ / n,π^*→ {O, and/or, O} → O

97 97 (12.67)

An example of this type is the cycloaddition of cyclopentenone with cyclopentene (equation 12.68).[167]

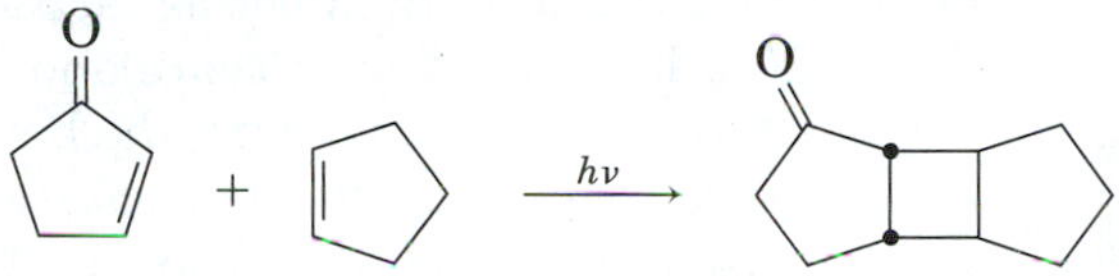

O + —$h\nu$→ O (12.68)

99 100 101

The intermediacy of species with diradical character in photocycloaddition of enones with alkenes is supported by the observation that the reaction of cyclohexenone with isobutylene produces the mixture of products shown in equation 12.69.[168] α,β-Unsaturated ketones can also dimerize through the same reaction pathway. For example, cyclopentenone gives approximately equal amounts of the head-to-tail dimer **109** and the head-to-head dimer **110** (equation 12.70).[169,170]

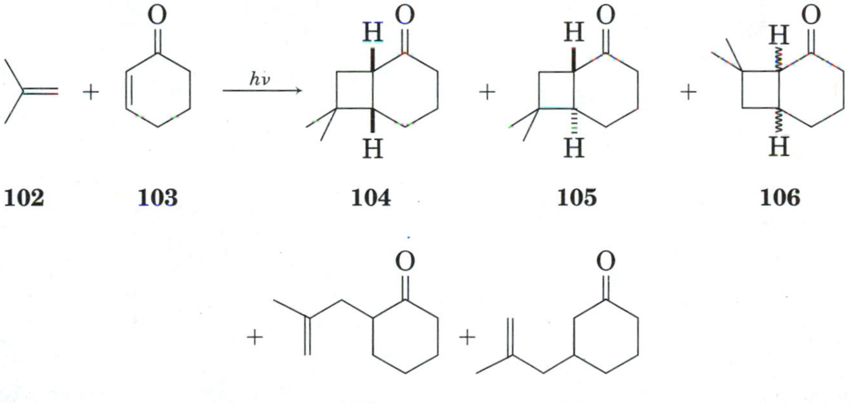

+ O —$h\nu$→ H O H + H O H + H O H

102 103 104 105 106 (12.69)

+ O + O

107 108

[167]Eaton, P. E. *J. Am. Chem. Soc.* **1962**, *84*, 2454.

[168]Corey, E. J.; Bass, J. D.; LeMahieu, R.; Mitra, R. B. *J. Am. Chem. Soc.* **1964**, *86*, 5570.

[169]Eaton, P. E. *J. Am. Chem. Soc.* **1962**, *84*, 2344.

[170]For a review of the photocycloadditions of enones and alkenes, see Schuster, D. I.; Lem, G.; Kaprinidis, N. A. *Chem. Rev.* **1993**, *93*, 3.

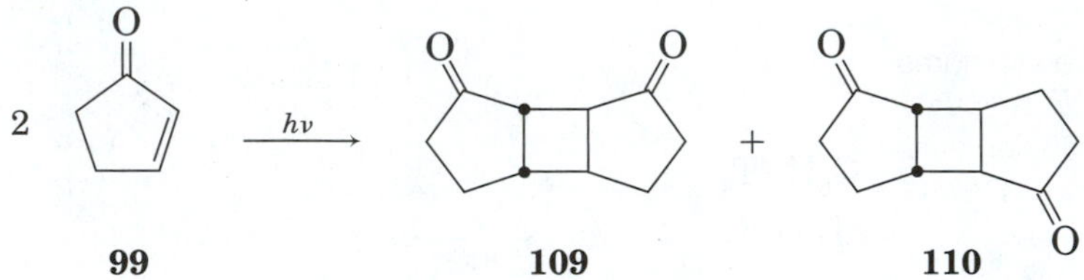

2 **99** $\xrightarrow{h\nu}$ **109** + **110** **(12.70)**

Photochemical Reactions of Aromatic Compounds

The HMO model predicts the lowest excited state of benzene to be a π,π^* state formed by promoting a π electron from one of the degenerate HOMOs to one of the degenerate LUMOs (Figure 12.39).[171,172] The geometry of the benzene ring is reinforced by the σ-bonded ring structure, so radiationless decay to a twisted state is not as fast as with alkenes. Therefore, benzene derivatives do exhibit fluorescence and intersystem crossing.

The larger the chromophore, the harder it is to visualize the electron redistribution accompanying the excitation in terms of simple valence bond pictures. However, it is still useful to use resonance structures of high energy species to suggest the chemical characteristics of the excited states of benzene. Bryce-Smith and Longuet-Higgins have suggested the radical structures in Figure 12.40 to explain the formation of fulvene (**112**) and

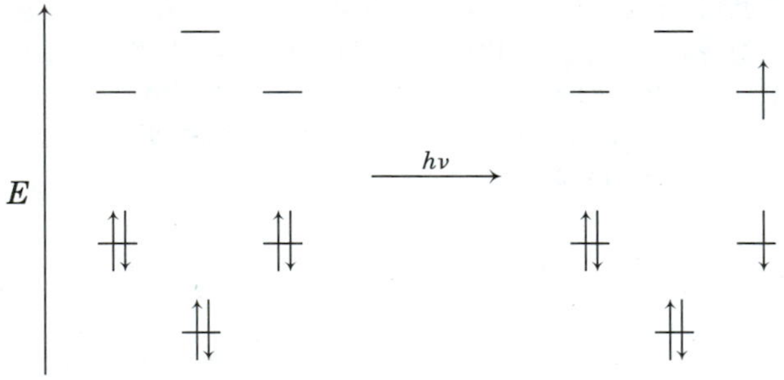

Figure 12.39 Schematic representation of a $\pi \rightarrow \pi^*$ transition in benzene.

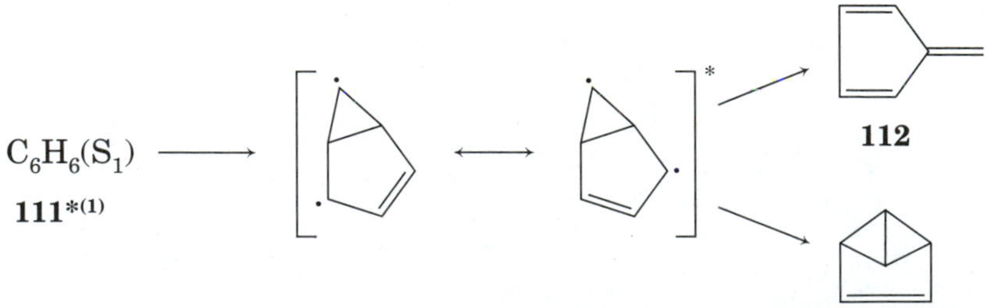

E $h\nu$

Figure 12.40 Radical resonance structures proposed for the intermediate in the formation of fulvene and benzvalene from benzene excited singlets. (Adapted from reference 174.)

$C_6H_6(S_1)$ **111**$^{*(1)}$ → **112**, **113**

[171]For a more detailed discussion of spectroscopic transitions in benzene, see Gilbert, A.; Baggott, J. *Essentials of Molecular Photochemistry*; CRC Press: Boca Raton, FL, 1991; pp. 355–357.

[172]Rydberg states can be observed in benzene and its derivatives as well. The $3R_g$ state of benzene was reported to have a lifetime of 70 ± 20 fsec and to decay to a vibrationally excited ground state; Wiesenfeld, J. M.; Greene, B. I. *Phys. Rev. Lett.* **1983**, *51*, 1745.

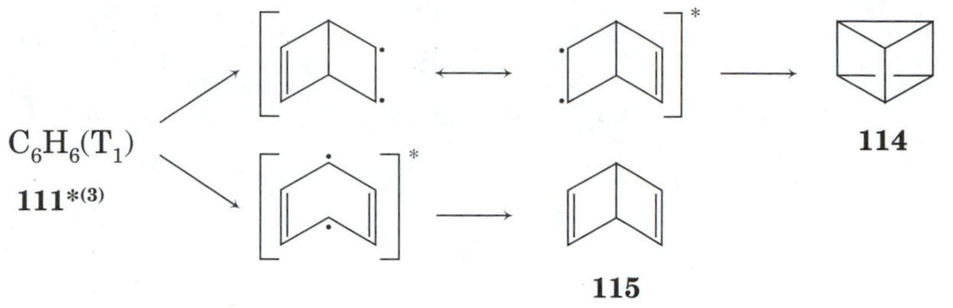

Figure 12.41 Resonance structures proposed to account for the formation of prismane and Dewar benzene from benzene triplets. (Adapted from reference 174.)

benzvalene (**113**) from the $^1\pi,\pi^*$ state of benzene and the intermediates in Figure 12.41 to explain the formation of prismane (**114**) and Dewar benzene (**115**) from the triplet state of benzene.[173,174]

Aromatic compounds also undergo photocycloaddition reactions with alkenes, leading to 1,2-, 1,3-, and (less often), 1,4-adducts, as generalized in equation 12.71.[175] Olefins having strongly electron withdrawing or donating substituents tend to give 1,2-photoaddition products, while olefins with alkyl substituents tend to give mostly 1,3-photoaddition.[176]

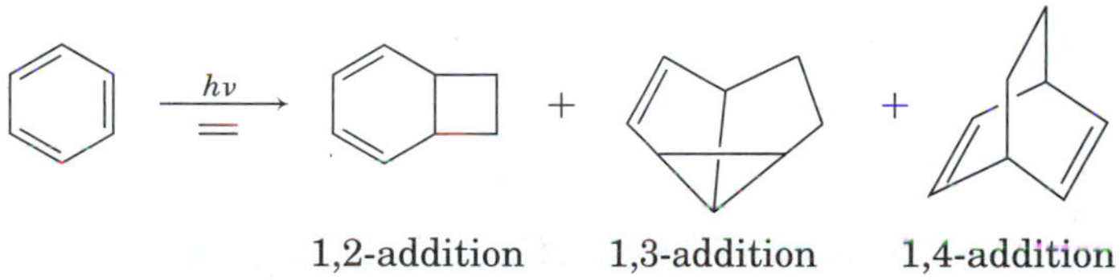

(12.71)

The reactions of polynuclear aromatics can also be rationalized by simple models analogous to those shown for benzene in Figure 12.40 and Figure 12.41. Two notable reactions are the dimerization of anthracenes (equation 12.72)[177] and the isomerization of *trans*-stilbene (**118**) to the cis isomer (**119**), followed by cycloaddition to a dihydrophenanthrene (**120**) and then oxidation to phenanthrene (**121**).[178]

---

[173]Bryce-Smith, D.; Longuet-Higgins, H. C. *Chem. Commun*. **1966**, 593.

[174]See also the discussion by Cundall, R. B. in *Transfer and Storage of Energy by Molecules*, Vol. 1; Burnett, G. M.; North, A. M., eds.; Wiley-Interscience: London, 1969; pp. 1–66, especially p. 42.

[175]Gilbert, A. in *Synthetic Organic Photochemistry*; Horspool, W. M., ed.; Plenum Press: New York, 1984; p. 1 and references therein.

[176]Bryce-Smith, D.; Gilbert, A.; Orger, B.; Tyrrell, H. *J. Chem. Soc., Chem. Commun*. **1974**, 334.

[177]For a summary of reactions of this type, see Greene, F. D.; Misrock, S. L.; Wolfe, Jr., J. R. *J. Am. Chem. Soc*. **1955**, *77*, 3852.

[178]Mallory, F. B.; Wood, C. S.; Gordon, J. T.; Lindquist, L. C.; Savitz, M. L. *J. Am. Chem. Soc*. **1962**, *84*, 4361; Moore, W. M.; Morgan, D. D.; Stermitz, F. R. *J. Am. Chem. Soc*. **1963**, *85*, 829. For a review, see Mallory, F. B.; Mallory, C. W. *Org. React*. **1984**, *30*, 1.

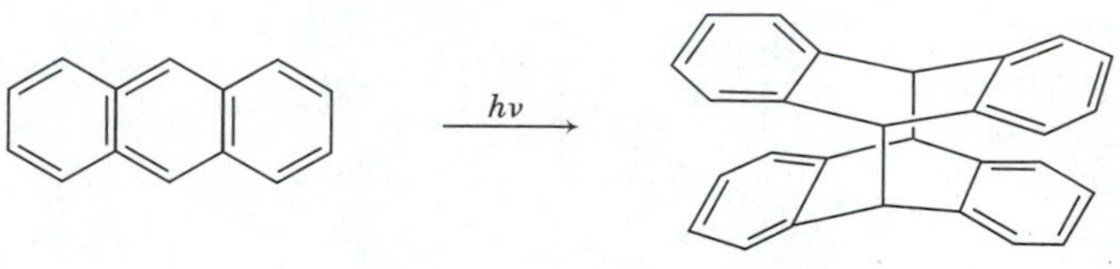

hν

116 **117** (12.72)

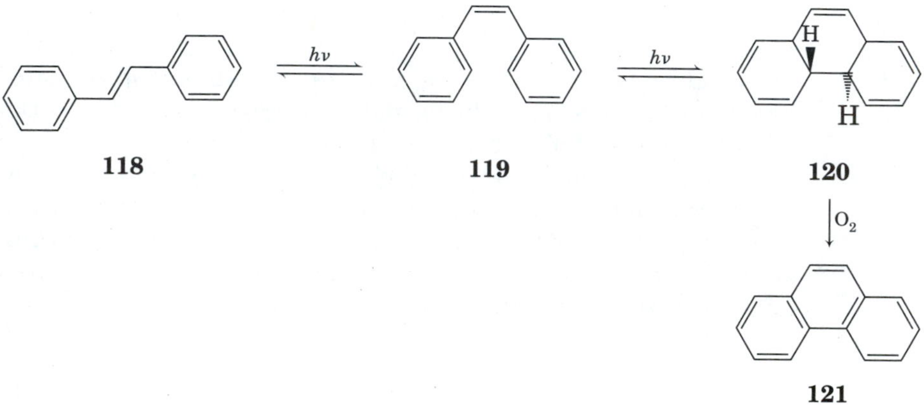

hν *hν*

H

H

118 **119** **120** (12.73)

O_2

121

Photosubstitution Reactions

Lewis structure models incorporating radical character are useful in the exposition of photochemical reactions of carbonyl compounds, alkenes and dienes, and aromatic compounds not substituted with polar groups. When heteroatoms are substituted onto the aromatic ring, however, considerable charge transfer character can be introduced into both the ground state and the excited state, and ionic resonance structures become more suitable as models for reactivity.[99,179] In nitrobenzene, for example, the nitro substituent group effectively serves as an empty *p* orbital attached to the ring, so the benzyl cation provides a rough model for its MOs. Thus the HOMO of nitrobenzene is similar to ψ_3 of the benzyl cation, and the LUMO is similar to ψ_4 (Figure 12.42). Promotion of an electron from HOMO to LUMO thus has the net effect of moving electron density from the positions meta to the nitro group to the ortho and para positions. In other words, the photoexcited nitrobenzene has even less electron density in the meta positions than does ground state nitrobenzene. This effect can be indicated with positive and negative charges in a resonance representation for the excited state (Figure 12.43).

---

[179]Zimmerman, H. E.; Sandel, V. R. *J. Am. Chem. Soc.* **1963**, *85*, 915.

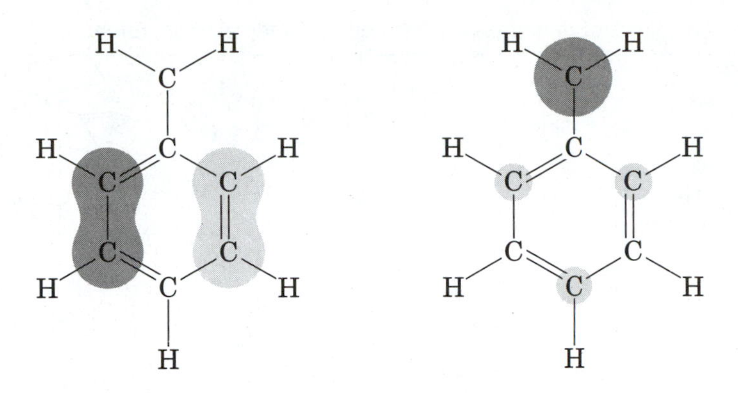

Figure 12.42 HOMO (left) and LUMO (right) of benzyl cation (PM3 calculation).

Figure 12.43 Resonance representation of S_1 of nitrobenzene (Adapted from reference 179.)

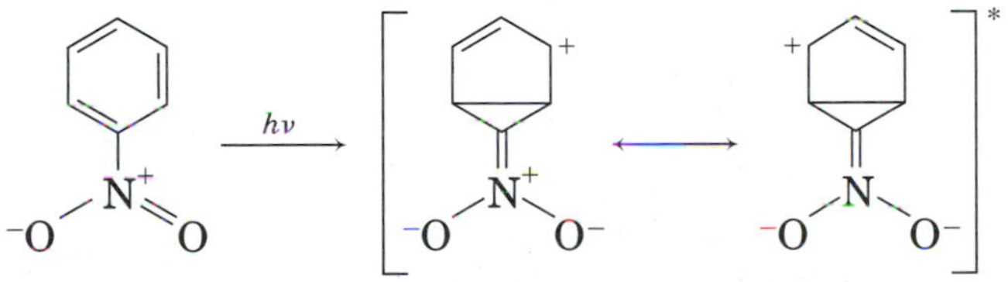

A methoxy substituent on a benzene ring has an effect that is opposite that of a nitro substituent, so the benzyl anion is a model for the MOs of anisole. Excitation removes an electron from ψ_4 and places it in ψ_5, which produces a charge difference suggested by the resonance representation shown in Figure 12.44. These simple models help us understand the photochemically induced solvolysis shown in Figure 12.45 in which *m*-nitrophenyl trityl ether (**122**) does not react in the dark but does readily undergo photosolvolysis.[180,181]

Figure 12.44 Valence bond representation of S_1 of anisole. (Adapted from reference 179.)

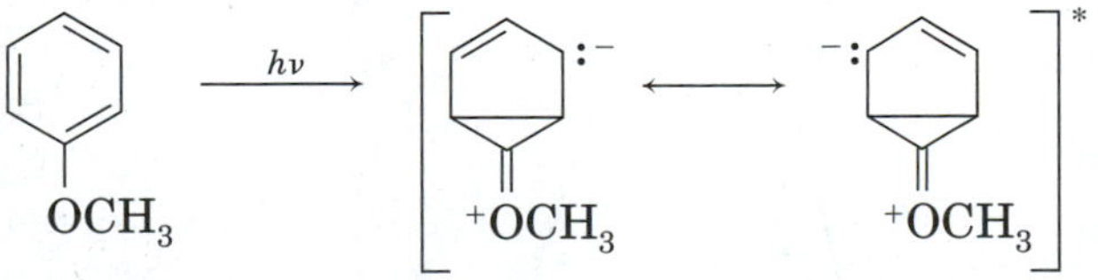

[180] Zimmerman, H. E.; Somasekhara, S. *J. Am. Chem. Soc.* **1963**, *85*, 922.

[181] Radical pathways may also be involved in photosubstitution reactions. See, for example, Bunce, N. J.; Cater, S. R.; Scaiano, J. C.; Johnston, L. J. *J. Org. Chem.* **1987**, *52*, 4214.

Figure 12.45 Photosolvolysis of *m*-nitrophenyl trityl ether. (Adapted from reference 180.)

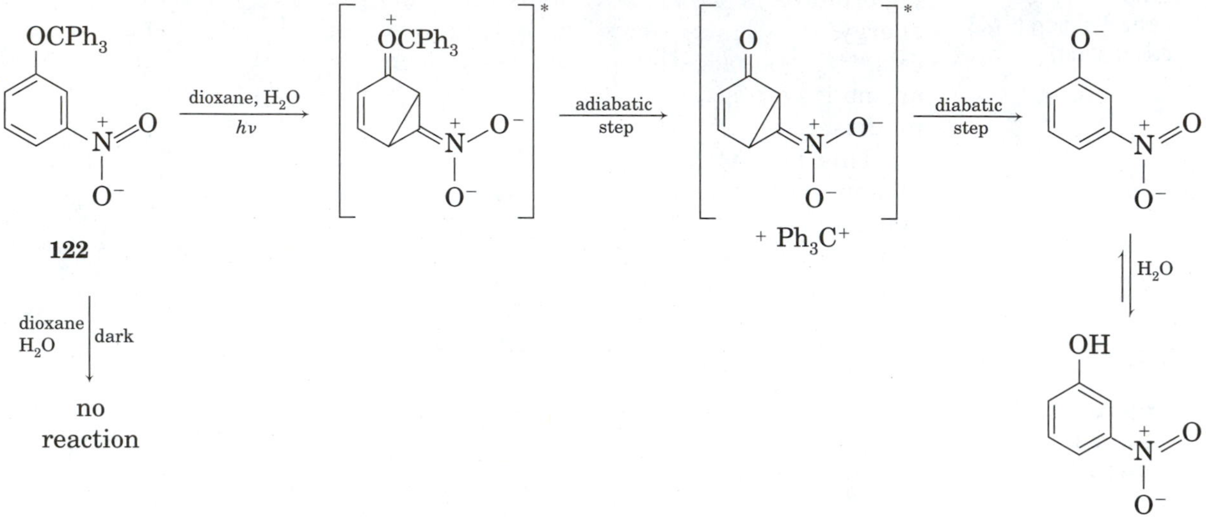

OCPh3
dioxane, H2O
hν
OCPh3
O−
O−
*
adiabatic
step
O
O−
O−
*
+ Ph3C+
diabatic
step
O−
O
O−
122
dioxane
H2O
dark
no
reaction
H2O
OH
O
O−

σ Bond Photodissociation Reactions

One of the conceptually simplest photochemical reactions is the photochemical dissociation of molecular chlorine.

$$Cl_2 \xrightarrow{h\nu} 2\ Cl\cdot \tag{12.74}$$

The absorption spectrum of molecular chlorine is shown in Figure 12.46. The λ_{max} is 330 nm with $\epsilon_{max} = 66$, but the onset of absorption is at least 478 nm. Thus the excitation energy (60 kcal/mol) is greater than the Cl—Cl dis-

80
70
60
50
40
30
20
10
ε
250
300
350
400
450
500
Wavelength (nm)

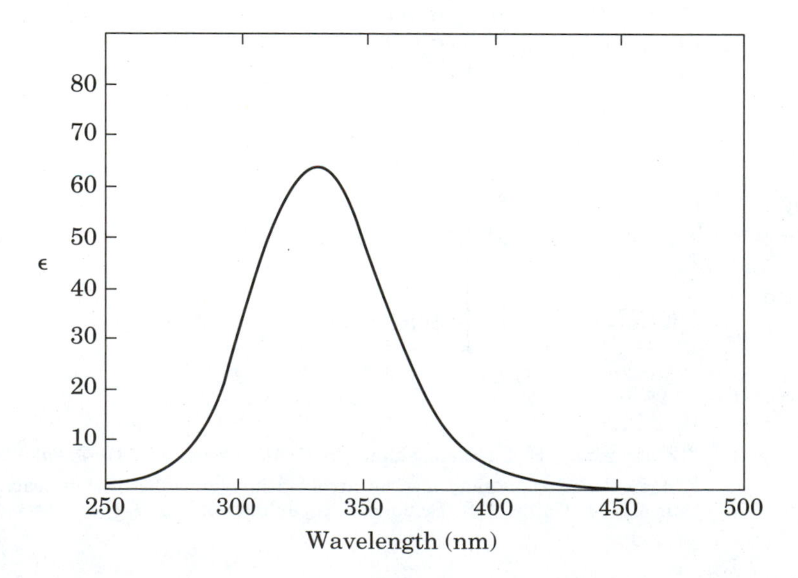

Figure 12.46 Absorption spectrum of Cl_2 in gas phase. (Adapted from reference 183.)

sociation energy (58 kcal/mol). In MO terms the excitation of an electron from a σ to a σ^* orbital creates an excited state with one electron in a bonding orbital and one electron in an antibonding orbital. With no net bonding energy, vibrational energy causes the two atoms to move apart as radicals.[182,183] The absorption band is essentially structureless because there are no longer discrete vibrational energy levels above the energy required for dissociation.

This chlorine photodissociation model does not transfer directly to the photolysis of organic molecules, such as the photodissociation reaction of benzene derivatives represented by the general reaction in equation 12.75. Here, the chromophore is the aromatic ring, which is assumed in the HMO treatment to be orthogonal to the σ bond that undergoes dissociation.

$$C_6H_5\text{-A—B} \xrightarrow{h\nu} C_6H_5\text{-A}\cdot + \text{B}\cdot \longrightarrow \text{products} \qquad \textbf{(12.75)}$$

Nevertheless, several examples of such reactions are known. One is the dissociation of toluene to benzyl and hydrogen radicals[184] (equation 12.76), which has been confirmed by the detection of benzyl radicals.

$$\underset{\textbf{123}}{C_6H_5\text{-}CH_2\text{—H}} \xrightarrow{h\nu} \underset{\textbf{124}}{C_6H_5\text{-}\dot{C}H_2} + \text{H}\cdot \qquad \textbf{(12.76)}$$

Two other reactions that appear to proceed through σ bond homolysis are the photo-Fries reaction (equation 12.77) and the photo-Claisen reaction (equation 12.78).[185] In the case of the photo-Fries reaction, both concerted and photodissociative mechanisms were proposed for the formation of the acylphenols. The photodissociative pathway was confirmed by Meyer and Hammond, who found that essentially no *o*- or *p*-hydroxyacetophenone was formed in the gas phase photolysis of phenyl acetate.[186] Instead, all products could be rationalized by recombination of phenoxy and methyl radicals (formed from decarbonylation of acyl radicals).

---

[182] To a first approximation there is no net bonding energy. A more detailed analysis would reveal some repulsion.

[183] More precisely, the excitation is believed to cause a vertical transition from the ground state ($^1\Sigma^+{}_g$) to a $^3\pi_{1u}$ dissociative excited state (a state with a negligible energy minimum). Calvert, J. G.; Pitts, J. N. *Photochemistry*; John Wiley & Sons, Inc.: New York, 1966; pp. 184, 226.

[184] Porter, G.; Strachan, E. *Trans. Faraday Soc.* **1958**, *54*, 1595.

[185] Kelly, D. P.; Pinhey, J. T.; Rigby, R. D. G. *Aust. J. Chem.* **1969**, *22*, 977.

[186] Meyer, J. W.; Hammond, G. S. *J. Am. Chem. Soc.* **1970**, *92*, 2187; *J. Am. Chem. Soc.* **1972**, *94*, 2219.

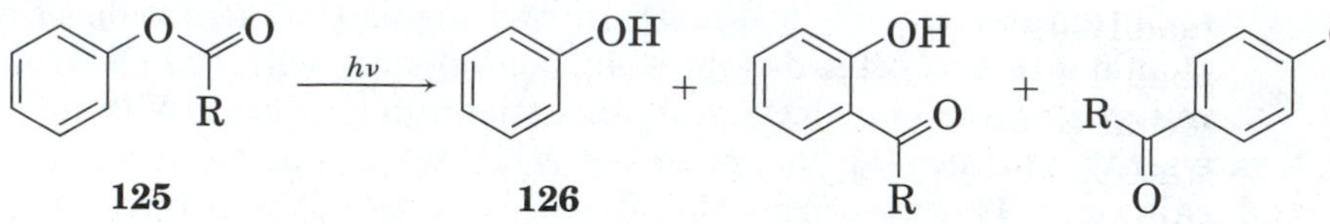

125 → 126 + 127 + 128 (12.77)

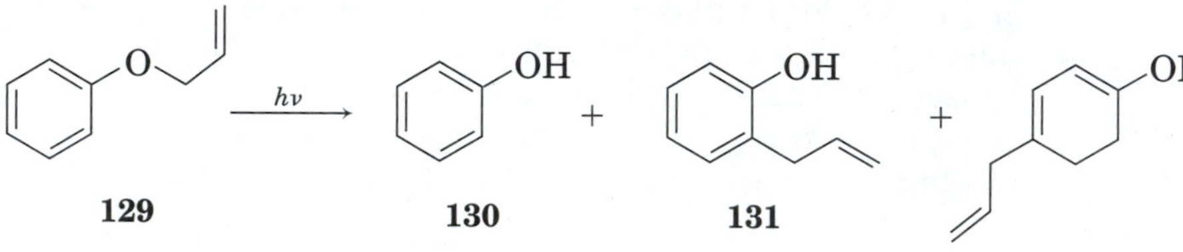

129 → 130 + 131 + 132 (12.78)

Similarly, a photodissociative mechanism for the photo-Claisen reaction was supported by observation of products expected from the recombination of radicals produced by photodissociation of 3-methyl-1-phenoxybut-2-ene (**133**, Table 12.5). In addition to phenol, products of the reaction are the rearranged ether **135**, the two γ,γ-dimethylallyl phenols **136** and **137**, and the two rearranged allyl phenols **138** and **139**, in ratios consistent with a reaction pathway involving radical recombination.[187]

Unlike the photodissociation of Cl_2, direct excitation of a phenyl ether or ester to a dissociative state does not seem likely because the absorption spectrum of toluene shows some vibrational structure, and fluorescence is

Table 12.5 Quantum yields for photoreactions of 2-methyl-4-phenoxybut-2-ene. (Adapted from reference 187.)

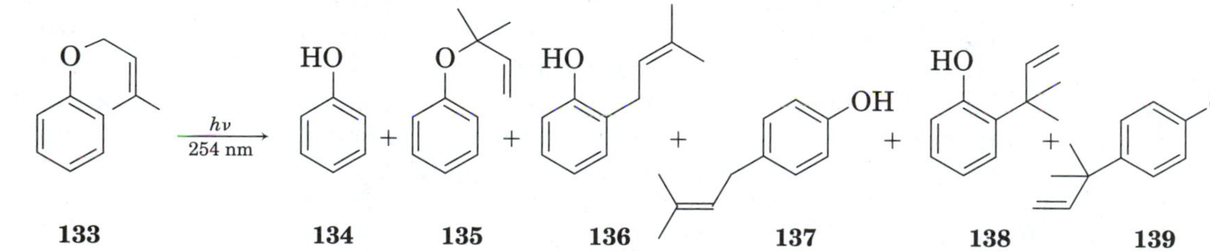

133 ($h\nu$, 254 nm) → Solvent	134	135	136	137	138	139
Cyclohexane	Φ_{app} = 0.10	0.05	0.06	0.05	0.02	0.03
2-Propanol	Φ_{app} = 0.11	0.03	0.18	0.15	0.05	0.07

[187]Carroll, F. A.; Hammond, G. S. *Israel J. Chem.* **1972**, *10*, 613.

readily observed.[188] Furthermore , the absorption and emission spectra of phenyl allyl ether are qualitatively similar to those of anisole. However, internal conversion of a bound (metastable) excited state to a dissociative state can lead to the observed reaction through a process known as **predissociation.**[189] Figure 12.47 shows a representation of the excited states of toluene as a function of the C—H bond distance.[190] There is an avoided crossing (represented by the dotted lines) between the aromatic π,π^* triplet and a σ,σ^* triplet, so that the lowest triplet has the energy shown by the solid line. In other words, T_1 for toluene has π,π^* character at short C—H distances and σ,σ^* character at longer C—H distances. Thus it is a predissociative excited state, and thermal activation provides enough energy to cross the small energy barrier.[191,192]

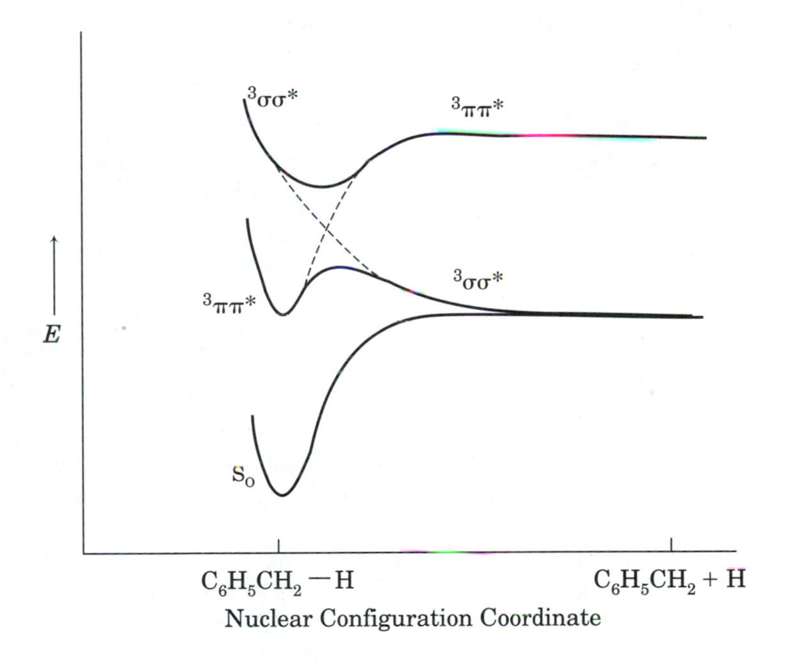

Figure 12.47 Predissociative model for toluene photodissociation. (Reproduced from reference 190.)

---

[188]Reference 183, pp. 310, 499.

[189]For a discussion, see Dixon, R. N. *Acc. Chem. Res.* **1991**, *24*, 16.

[190]Michl, J. *Top. Curr. Chem.* **1974**, *46*, 1.

[191]For a detailed discussion of predissociative states, see reference 88, pp. 245–247.

[192]The predissociative model is commonly used to explain photodissociation reactions. For example, Craig, B. B.; Weiss, R. G.; Atherton, S. J. *J. Phys. Chem.* **1987**, *91*, 5906, found evidence that the photochemical decomposition of *p*-nitrophenylacetate ion in water proceeds through the T_1 state. Also, Hilinski, E. F.; Huppert, D.; Kelley, D. F.; Milton, S. V.; Rentzepis, P. M. *J. Am. Chem. Soc.* **1984**, *106*, 1951, reported a study of the photodissociation of 1-chloromethylnaphthalene in hexane in which the 1-naphthylmethyl radical was detected by picosecond absorption and emission spectroscopy. The data were consistent with a mechanism in which the predissociative $T_n(\sigma,\sigma^*)$ state arises either directly from S_2 or from an upper triplet state, while the S_1 and T_1 states most likely do not lead to bond dissociation.

In polar solvents, α-halomethylaromatics give rise to photochemical reactions that can be described by the operation of both radical and ionic mechanisms. Equation 12.79 shows the results for irradiation of 1-chloromethylnaphthalene (1-$NpCH_2Cl$, **140**) in methanol.[193] The most obvious pathway for formation of the methyl ether **141** is heterolytic dissociation of the C — Cl bond to give a chloride ion and a 1-naphthylmethyl carbocation, the latter undergoing nucleophilic addition by the solvent.[194] However, the other three products appear to be formed via the 1-naphthylmethyl radical. Interestingly, sensitization and quenching studies, while not totally unambiguous, suggest that all of these products arise from the excited singlet state of **140**.

CH_2Cl (**140**) $\xrightarrow[CH_3OH]{h\nu}$ CH_2OCH_3 (**141**) + **142** + CH_2OH (**143**) + CH_3 (**144**)

(12.79)

The discussion here has emphasized photochemical homolytic dissociations, but photochemical reactions can also lead to the formation of ions, including carbocations and carbanions.[195,196] For example, irradiation of 4,4′-bis(dimethylamino)triphenylmethane leucohydroxide (**145**) in aqueous micellar solutions leads to the ejection of hydroxide ion and formation of 4,4′-bis(dimethylamino)triphenylmethyl cation (**146**, equation 12.80), which raises the pH of the medium from 5.4 to 10.0. The pH returns to 5.4 over a period of 15 minutes as the hydroxide ion recombines with the carbocation.[197]

---

[193] Arnold, B.; Donald, L.; Jurgens, A.; Pincock, J. A. *Can. J. Chem.* **1985**, *63*, 3140.

[194] In the photolysis of arylmethyl esters, an alternate route for the production of ions is the homolytic dissociation of the arylmethyl-oxygen bond, followed by electron transfer from the arylmethyl radical to the acetyl radical. For a discussion, see Pincock, J. A.; Wedge, P. J. *J. Org. Chem.* **1994**, *59*, 5587.

[195] For a review, see Das, P. K. *Chem. Rev.* **1993**, *93*, 119.

[196] For a discussion of the photochemical production of radical, carbocation and carbene intermediates from irradiation of alkyl halides in solution, see Kropp, P. J. *Acc. Chem. Res.* **1984**, *17*, 131.

[197] Irie, M. *J. Am. Chem. Soc.* **1983**, *105*, 2078.

$(CH_3)_2N$–C_6H_4–C(OH)(Ph)–C_6H_4–$N(CH_3)_2$ $\underset{\text{(dark, slow)}}{\overset{h\nu}{\rightleftharpoons}}$ $[(CH_3)_2NC_6H_4]_2C^+$–Ph + ^-OH **(12.80)**

145 **146**

Similarly, a carbocation intermediate has been proposed for the photochemical conversion of 9-phenylxanthen-9-ol (**147**) into the methyl ether **148** upon irradiation in methanol-water solution (equation 12.81).[198]

147 (Ph, OH) $\xrightarrow[\text{CH}_3\text{OH, H}_2\text{O}]{h\nu}$ [9-phenylxanthenyl cation]* + HO^- $\xrightarrow{\text{CH}_3\text{OH, H}_2\text{O}}$ **148** (Ph, OCH_3) **(12.81)**

Singlet Oxygen and Organic Photochemistry

The discussion of photochemistry here has necessarily focused on the excited states of organic compounds. However it is appropriate to discuss also some aspects of the photochemistry of molecular oxygen. Knowledge of organic photochemistry provides a basis for understanding oxygen photochemistry, and oxygen photochemistry leads to new and interesting reactions with organic compounds.

The detailed MO representation of the electronic structure of ground state O_2 and its excited singlet and triplet states is complex.[199] In essence, the ground state of O_2 is a triplet state designated as $^3\Sigma_g^-$. The first electronically excited state, 22 kcal/mol above the ground state, is a singlet state denoted as $^1\Delta_g$ (pronounced singlet delta gee) and commonly called *singlet oxygen*. There is also a higher energy electronic state, 38 kcal/mol above the ground state, which is termed $^1\Sigma_g^+$.[200] The latter species has a

---

[198]Wan, P.; Yates, K.; Boyd, M. K. *J. Org. Chem.* **1985**, *50*, 2881. The authors proposed that the carbocation shown in equation 12.81 is electronically excited, meaning that the ejection of hydroxide occurs on the excited state potential energy surface. Thus, the photoejection is proposed to be an adiabatic (footnote 75) photochemical reaction.

[199]For a detailed discussion of the structure of singlet oxygen, see Kasha, M. in *Singlet O_2*, Vol. I; Frimer, A. A., ed.; CRC Press: Boca Raton, FL, 1985; pp. 1–11.

[200]Kearns, D. R. *Chem. Rev.* **1971**, *71*, 395.

short lifetime (on the order of nsec in solution) and is of little chemical interest. The lifetime of singlet oxygen in solution varies considerably with solvent, from less than 10^{-6} sec in acetone or benzene to more than 10^{-4} sec in carbon tetrachloride or chlorofluorocarbon solvents.[201] The $^1\Delta_g$ state can be formed by thermal reactions or by sensitization with low energy organic sensitizers ($E_T > 22$ kcal/mol), as shown in equations 12.82 and 12.83.

$$S \xrightarrow{h\nu} S^{*(1)} \xrightarrow{k_{isc}} S^{*(3)} \qquad (12.82)$$

$$S^{*(3)} + O_2^{(3)} \longrightarrow S + O_2^{*(1)} \qquad (12.83)$$

There are two main chemical reactions of singlet oxygen that are of interest here.[202] The first is cycloaddition in a process that resembles a Diels-Alder reaction:

$$\mathbf{73} + O_2^{*(1)} \longrightarrow \mathbf{149} \qquad (12.84)$$

The second is allylic oxidation, which can also be visualized as a concerted ene process;[203]

$$O_2 + \mathbf{24} \longrightarrow \mathbf{150} \xrightarrow{[R]} \mathbf{151} \qquad (12.85)$$

A practical example of the use of singlet oxygen is provided by the synthesis of rose oxide (**154**) shown in equation 12.86.[204] The sensitizer used is rose bengal, a dye that can be used for singlet oxygen reactions induced by sunlight. The product of the ene reaction is reduced with Na_2SO_3 to a diol (**153**), which then undergoes spontaneous cyclization to the ether in cold acid solution. The second reaction is the cycloaddition used in one synthesis of cantharidin (**157**, equation 12.87). Addition of $^1O_2^*$ to **155** forms **156**, which is reduced and converted to **157**.[205]

---

[201]Monroe, B. M. in *Singlet O_2*, Vol. I; Frimer, A. A., ed.; CRC Press: Boca Raton, FL, 1985; pp. 177–224.

[202]For a discussion, see Frimer, A. A.; Stephenson, L. M. in *Singlet O_2*, Vol. II; Frimer, A. A., ed.; CRC Press: Boca Raton, FL, 1985; pp. 67–91.

[203]For a review, see Gollnick, K.; Kuhn, H. J. in *Singlet Oxygen*; Wasserman, H. H.; Murray, R. W., eds.; Academic Press: New York, 1979; pp. 287–427.

[204]Ohloff, G.; Klein, E.; Schenck, G. O. *Angew. Chem.* **1961**, *73*, 578. See also reference 108, p. 377.

[205]For a discussion and leading references, see reference 108, pp. 383 *ff*.

152 OH $\xrightarrow[\text{sens.}]{h\nu,\ O_2}$ $\xrightarrow{[R]}$ HO **153** OH $\xrightarrow{-H_2O}$ O **154** **(12.86)**

CH_3 O, O, CH_3 O **155** $\xrightarrow[\text{sens.}]{h\nu,\ O_2}$ CH_3 O, O, O, O, CH_3 O **156** ⟶ ⟶ H_3C O, O, O, H_3C O **157** **(12.87)**

The use of organic compounds to sensitize formation of singlet oxygen is becoming more important in organic synthesis, and may even have medical applications as well. Medical researchers are studying cancer treatments involving administration of dyes that are selectively taken up by tumor cells. Percutaneous (through the skin) irradiation of the dye with a bright light can produce cytotoxic singlet oxygen at the tumor site, minimizing the need to circulate toxic drugs throughout the patient's body.[206,207]

12.4 Some Applications of Organic Photochemistry

The last two reactions illustrate the use of organic photochemistry in synthesis. Photochemical reactions provide particularly valuable synthetic methodologies because frequently they are carried out under conditions that are neither acidic nor basic, enabling highly sensitive functional groups to be produced.[208] Moreover, photochemical reactions allow selective activation of one portion of a molecule (the chromophore). Often they involve high energy intermediates, so product distributions may differ from those observed in the synthesis of the same compounds by thermal reactions. Even though the quantum efficiencies of some photochemical reactions may not be high, the chemical yields are often higher than what might be achieved in thermal reactions from easily obtained starting materials.[209]

---

[206]Dougherty, T. J. *Adv. Photochem.* **1992**, *17*, 275.

[207]For a discussion of photomedicine and photodynamic therapy, see Dolphin, D. *Can. J. Chem.* **1994**, *72*, 1005.

[208]For a discussion of this point, see Jones, S. W.; Scheinmann, F.; Wakefield, B. J.; Middlemiss, D.; Newton, R. F. *J. Chem. Soc., Chem. Commun.* **1986**, 1260.

[209]Furthermore, the use of chiral sensitizers or of circularly polarized light can lead to asymmetric photochemical synthesis. For a review, see Inoue, Y. *Chem. Rev.* **1992**, *92*, 741.

The following additional examples offer a hint of the utility of organic photochemistry in synthesis.[210] Cycloaddition reactions provide a facile entry into highly strained ring systems and complex polycyclic molecules.[211] The [2 + 2] photochemical cycloaddition is a key step in the synthesis of cubane (**161**). An intramolecular photochemical [2 + 2] cycloaddition in **158** leads to **159**, which is converted to **161** by ground state processes.[212]

$h\nu$; base ; COOH ; COOH ; Br ; O

158 **159** **160** **161** **(12.88)**

Intramolecular cycloaddition also plays a central role in the synthesis of Dewar benzene (**164**),[213] the bicycloheptenol **166**,[214] and the tricyclic keto ethers **168** and **169**.[215] The photochemical retro Diels-Alder reaction has been used for a synthesis of bullvalene (**171**) from cyclooctatetraene dimer.[216]

$h\nu$; $Pb(OAc)_4$

162 **163** **164** **(12.89)**

---

[210]For detailed discussions of synthetic organic photochemistry, see Horspool, W. M., ed. *Synthetic Organic Photochemistry*; Plenum Press: New York, 1984; Srinivasan, R.; Roberts, T. D., eds.; *Organic Photochemical Syntheses*, Vol. 1; Wiley-Interscience: New York, 1971; Srinivasan, R., ed. *Organic Photochemical Syntheses*, Vol. 2; Wiley-Interscience: New York, 1976; reference 108; Ninomiya, I.; Naito, T. *Photochemical Synthesis*; Academic Press: London, 1989. See also Turro, N. J.; Schuster, G. *Science*, **1975**, *18*, 303.

[211]For a review of the photochemical synthesis of polycyclic structures, see (a) De Keukeleire, D.; He, S.-L. *Chem. Rev.* **1993**, *93*, 359; De Keukeleire, D. *Aldrichimica Acta* **1994**, *27*, 59.

[212]Eaton, P. E.; Cole, Jr., T. W. *J. Am. Chem. Soc.* **1964**, *86*, 962, 3157.

[213]van Tamlen, E. E.; Pappas, S. P. *J. Am. Chem. Soc.* **1963**, *85*, 3297.

[214]Chapman, O. L.; Pasto, D. J.; Borden, G. W.; Griswold, A. A. *J. Am. Chem. Soc.* **1962**, *84*, 1220.

[215]Ikeda, M.; Takahashi, M.; Uchino, T.; Ohno, K.; Tamura, Y.; Kido, M. *J. Org. Chem.* **1983**, *48*, 4241.

[216]Schröder, G. *Chem. Ber.* **1964**, *97*, 3140.

CH_3O —OH **165** $\xrightarrow{h\nu}$ CH_3O —OH **166** **(12.90)**

O O CH_2 $\xrightarrow{h\nu}$ O O H + O O H **(12.91)**

170 $\xrightarrow{h\nu}$ **171** + **111** **(12.92)**

There are many other applications of organic photochemistry. One is the development of a photographic system based on organic compounds rather than on silver salts. Such organic systems offer a grain size of molecular dimensions, not of inorganic crystal dimensions as is the case with silver halide film technology. Thus the organic films have, in theory at least, a much higher resolution than does silver halide film, although at the expense of decreased spectral sensitivity and film speed.[217] A general term for the process by which photochemical reactions lead to the formation of colored products (or, more generally, products of different color) is ***photochromism***.[218,219] Photography itself is only one aspect of the more general field of imaging systems, which also includes holography[220] and high density data storage. An advantage of some organic photochromic systems is the potential for photochemically reversible reactions so that light could be used to both store and erase information.[221] The development of organic photoconductive materials for use in xerography and photopolymerization is also an active area of research.[222]

---

[217]However, photochemically induced physical changes (such as a change in phase resulting from a photochemical reaction) may amplify the response of a photochemical information storage system. See, for example, (a) Tazuke, S.; Kurihara, S.; Yamaguchi, H.; Ikeda, T. *J. Phys. Chem.* **1987**, *91*, 249; (b) Zhang, M.; Schuster, G. B. *J. Am. Chem. Soc.* **1994**, *116*, 4852.

[218]For an introduction, see Brown, G. H., ed. *Photochromism* (*Techniques of Chemistry*, Vol. III); Wiley-Interscience: New York, 1971.

[219]For a review of organic photochemical systems for imaging, see Delzenne, G. A. *Adv. Photochem.* **1979**, *11*, 1.

[221]See, for example, Rieke, R. D.; Page, G. O.; Hudnall, P. M.; Arhart, R. W.; Bouldin, T. W. *J. Chem. Soc., Chem. Commun.* **1990**, 30.

[222]See, for example, (a) Monroe, B. M.; Weed, G. C. *Chem. Rev.* **1993**, *93*, 435; (b) Law, K. Y. *Chem. Rev.* **1993**, *93*, 449.

Chemiluminescence is the emission of light from electronically excited molecules produced from reactants in their ground electronic states.[223] Luminol (**172**) is a synthetic compound that has been studied extensively. The reaction is thought to generate the aminophthalate derivative **173** in an electronically excited state that can emit light.[224]

$$\mathbf{172} \xrightarrow[\text{base, } H_2O]{O_2} N_2 + [\mathbf{173}]^* \longrightarrow \text{fluorescence or energy transfer} \quad (12.93)$$

1,2-Dioxetanes are also chemiluminescent, and tetramethyl-1,2-dioxetane (**174**) is reported to glow at room temperature. The data suggest that decomposition of **174** produces two molecules of acetone, one in the ground electronic state and one in an electronically excited state (equation 12.94). Most of the excited acetone molecules are triplets, so the chemiluminescence observed in the absence of oxygen is acetone phosphorescence (equation 12.95). In the presence of oxygen the acetone triplets are quenched, however, so the emission is acetone fluorescence that is much weaker because only 1% as many excited singlet molecules as excited triplet molecules are produced by the decomposition of the 1,2-dioxetane.[225,226]

$$\mathbf{174} \longrightarrow \mathbf{7} + \mathbf{7^*} \quad (12.94)$$

$$\mathbf{7^*} \longrightarrow \mathbf{7} + h\nu \quad (12.95)$$

Dioxetanediones such as **176** may be intermediates in the chemiluminescent reaction of diaryl esters of oxalic acid, which decompose to 2 molecules of CO_2 and, in the presence of a fluorescent molecule (**F**), to the excited

[223] Reference 1(b). For an introduction, see Gundermann, K.-D.; McCapra, F. *Chemiluminescence in Organic Chemistry*; Springer-Verlag: Berlin, 1987.

[224] Merényi, G.; Lind, J.; Eriksen, T. E. *J. Biolumin. Chemilumin.* **1990**, *5*, 53.

[225] Turro, N. J.; Lechtken, P.; Schore, N. E.; Schuster, G.; Steinmetzer, H.-C.; Yekta, A. *Acc. Chem. Res.* **1974**, *7*, 97 and references therein.

[226] See also Turro, N. J. in *The Exciplex*; Gordon, M.; Ware, W. R., eds.; Academic Press: New York, 1975; pp. 165–186.

state of the fluorescer. In turn, the fluorescer may emit light, resulting in chemiluminescence.[227,228]

$$\underset{\textbf{175}}{\text{bis(2,4-dinitrophenyl) oxalate}} \xrightarrow{H_2O_2} \underset{\textbf{176}}{[\text{1,2-dioxetanedione}]} \xrightarrow{F} F^* + 2\,CO_2 \qquad \textbf{(12.96)}$$

$$\mathbf{F}^* \longrightarrow \mathbf{F} + h\nu' \qquad \textbf{(12.97)}$$

Other applications of fluorescence include the development of sensitive chemosensors for analysis of dilute solutions of inorganic cations and anions[229] and the study of the diffusion of individual molecules in solution at room temperature.[230] Fluorescent compounds have also been used as replacements of radioisotopes in the analysis of biological compounds.[231] In addition, photochemical reactions of organic compounds bound to electrodes have been used to improve amperometric immunosensors.[232] Photochemical reactions also offer alternative probes for the characterization of the microenvironments in diverse solid and liquid media, including crystals[233], zeolites, alumina, silica and clay surfaces, semiconductor surfaces, liquid crystals and host-guest inclusion complexes,[234] polymer films,[235] monolayers and supported multilayers of surfactant molecules,[236] and micelles.[237,238] Novel applications of photochemistry include the development of molecular tweezers for the photoreversible binding of metal ions[239] and the photoencapsulation of organic molecules in zeolite as an example of a molecular "ship in a bottle."[240]

---

[227] Rahut, M. M.; Bollyky, L. J.; Roberts, B. G.; Loy, M.; Whitman, L. R.; Iannotta, A. V.; Semsel, A. M.; Clarke, R. A. *J. Am. Chem. Soc.* **1967**, *89*, 6515.

[228] For a discussion, see Thrush, B. A. *J. Photochem.* **1984**, *25*, 9.

[229] Czarnik, A. W. *Acc. Chem. Res.* **1994**, *27*, 302.

[230] Nie, S.; Chiu, D. T.; Zare, R. N. *Science* **1994**, *266*, 1018.

[231] Mayer, A.; Neuenhofer, S. *Angew. Chem., Int. Ed. Engl.* **1994**, *33*, 1044.

[232] Willner, I.; Blonder, R.; Dagan, A. *J. Am. Chem. Soc.* **1994**, *116*, 9365. Also see the discussion in Freemantle, M. *Chem. Eng. News* **1994** (October 31), 16.

[233] Zimmerman, H. E.; Zhu, Z. *J. Am. Chem. Soc.* **1994**, *116*, 9757 and references therein.

[234] Ramamurthy, V., ed. *Photochemistry in Organized and Constrained Media*; VCH Publishers, Inc.: New York, 1991.

[235] Cui, C.; Weiss, R. G. *J. Am. Chem. Soc.* **1993**, *115*, 9820.

[236] Whitten, D. G. *Acc. Chem. Res.* **1993**, *26*, 502.

[237] Han, N.; Lei, X.; Turro, N. J. *J. Org. Chem.* **1991**, *56*, 2927.

[238] A general model of photochemistry in environments categorized as organized and confining media has been given by Weiss, R. G.; Ramamurthy, V.; Hammond, G. S. *Acc. Chem. Res.* **1993**, *26*, 530.

[239] Irie, M.; Kato, M. *J. Am. Chem. Soc.* **1985**, *107*, 1024.

[240] Lei, X.; Doubleday, Jr., C. E.; Zimmt, M. B.; Turro, N. J. *J. Am. Chem. Soc.* **1986**, *108*, 2444.

One of the ultimate goals of organic photochemistry is the storage of solar energy in chemical bonds, perhaps even to develop artificial photosynthesis.[58,241] The photochemical synthesis of strained structures that can revert to starting materials and release thermal energy upon demand has been studied extensively.[242,243] Even though photochemical electron-transport mechanisms have been developed, true artificial photosynthesis will require progress in the development of organized molecular assemblies (artificial chloroplasts) that can minimize recombination of photochemically generated charge separation.

Problems

1. An organic compound **A** absorbs light in the near Uv region of the spectrum. The onset of absorption is 375 nm. A dilute solution of **A** in an organic solvent exhibits fluorescence (onset = 25,900 cm^{-1}) and phosphorescence (onset 6800 Å). What are the energies of the S_1 and T_1 states of **A**?
2. The lowest-lying excited state of compound **B** exhibits photochemical reaction, fluorescence, and intersystem crossing. Measurements show that $\Phi_f = 0.3$ and $\Phi_{isc} = 0.5$. Irradiation of 100 mL of a 10^{-2} M solution of **B** with 10^{-3} Einsteins of light (all absorbed by **B**) gives a 5.0% conversion of **B** into product **C**.
   a. What is the quantum yield of product formation in this reaction?
   b. Assuming that all the product is formed from the S_1 state of **B**, what fraction of the S_1 population is *not* accounted for by fluorescence, intersystem crossing, or photochemical reaction?

   Show your reasoning for both parts.
3. Compound **D** exhibits triplet-triplet absorption in fluid solution at room temperature. Compound **Q** is found to quench the triplets of **D**. Table 12.6 shows intensity vs. time data for the triplet-triplet absorption of **D** as measured by flash spectroscopy with solutions varying in concentration of **Q**.

Table 12.6 Data for quenching of triplets of compound **O** by quencher **Q**.

[Q] = 0		[Q] = 0.1 M		[Q] = 0.16 M	
t	I	*t*	I	*t*	I
200	150	100	200	200	80
600	65	300	80	300	40
1000	30	400	50	400	25

I = intensity of absorption in arbitrary units; *t* = time after flash in microseconds (μsec).

[241] For example, see Tabushi, I. *Pure Appl. Chem.* **1982**, *54*, 1733.

[242] For an introduction, see Hautala, R. R.; King, R. B.; Kutal, C. *Solar Energy: Chemical Conversion and Storage*; Humana Press: Clifton, N.J., 1979.

[243] Yoshida, Z.-I. *J. Photochem.* **1985**, *29*, 27.

First analyze each of the three sets of data separately to show that for each different concentration of the quencher the decrease in intensity of T-T absorption follows first order kinetics. Next calculate the lifetime (τ) of the triplet state of **D** for each solution of quencher. Then use the results from each individual solution to construct a Stern-Volmer plot (by plotting τ/τ' versus [Q]) to determine k_q, the rate constant for quenching of **D** triplets by **Q**.

4. Compound **E** fluoresces with a λ_{max} of 399 nm at 30° in a 50% (v/v) dioxane-water solution. Under the same conditions, the anion of **E** fluoresces with a λ_{max} of 481 nm. The pK_a of **E** in the ground state is 11.53. What is its pK_a in the excited state responsible for fluorescence?

5. Irradiation of neat (−)-2-methylbutyl benzoate produces benzoic acid and 2-methyl-1-butene. In addition, the recovered starting material is found to be partially racemized. Propose mechanisms to account for the formation of the photoproducts.

6. Irradiation of 3-methylcyclopentanone in the gas phase produces carbon monoxide, ethene, propene, methylcyclobutane, 3-methyl-4-pentenal and 4-methyl-4-pentenal. Propose a mechanism for the formation of each of these products.

7. Both the S_1 and the T_1 states of 1-chloromethylnaphthalene (**140**, page 863) are quenched at the rate of diffusion (ca. 10^{10} M^{-1} sec^{-1}) by the conjugated diene 2,5-dimethyl-2,4-hexadiene. The lifetime of the S_1 state of **140** is reported to be 490 ps, while the lifetime of T_1 is reported to be 180 nsec. Calculate the percentage of singlet states and the percentage of triplet states that would be quenched at each of the following concentrations of quencher: 1 M, 0.1 M, 0.01 M, 0.001 M.

8. Both (*E*)-1-methylcyclooctene and (*Z*)-1-methylcyclooctene undergo photoisomerization upon direct irradiation (214 nm), and irradiation of either isomer leads to a photostationary state consisting of 23% of the (*E*) diasteromer and 77% of the (*Z*) diastereomer. The ratio of the extinction coefficients (ϵ_Z/ϵ_E) at 214 nm is 0.305.

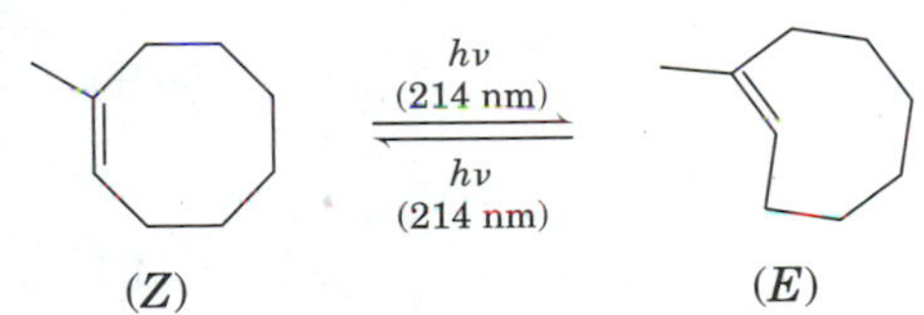

What is the decay ratio (k_{dE}/k_{dZ}) for the decay of the common twisted excited singlet state formed from either isomer to the (*E*) and (*Z*) isomers?

9. Irradiation of *trans*-2,6-dimethylcyclohexanone in the gas phase produces seven products, including three with the molecular formula $C_8H_{14}O$ and four with molecular formula C_7H_{14}. Give the structure of each of the products, and propose a mechanism to account for the formation of each.

10. The synthesis of rose oxide discussed on page 865 involves the singlet oxidation of citronellol to form 2,6-dimethyl-3-octene-2,8-diol (**153**). However, another diol is also formed, with a percent yield about half that of **153**. What is the other product, and what is the mechanism of its formation.?

11. a. Irradiation of **177** in toluene with added methanol leads to the formation of **178** and **179**. When the photochemical reaction is carried out with CH_3OD, each of the products has one deuterium atom incorporated into its structure. Where is the deuterium label found in each compound?

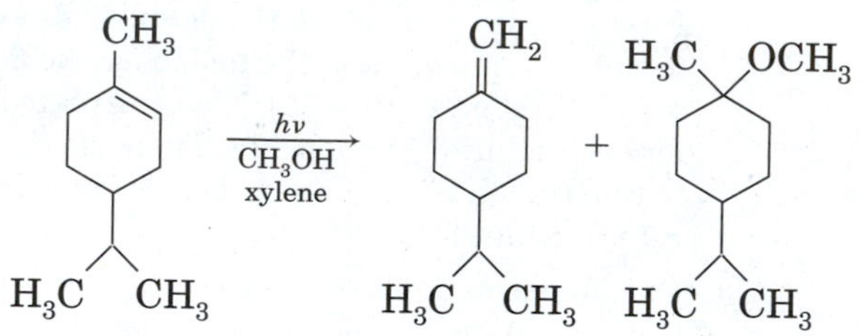

CH_3 / H_3C CH_3 **177** $\xrightarrow[\text{CH}_3\text{OH, xylene}]{h\nu}$ CH_2 / H_3C CH_3 **178** + H_3C OCH_3 / H_3C CH_3 **179**

b. Irradiation of *cis*-5-methyl-3-hexen-2-one (**40**) in CH_3OD produces 5-methyl-4-hexen-2-one (**42**, equation 12.48, page 836) in which one deuterium atom is incorporated into the structure. Where is the deuterium label located?

12. Irradiation of acetone in the presence of 2,3-dimethyl-2-butene leads to the formation of seven photoproducts. Two products are unsaturated ethers; two are unsaturated alcohols; one is an oxetane; two are hydrocarbons. Propose mechanisms and structures to account for the formation of all of the photoproducts.

13. Predict the major product(s) of each of the following reactions.

a. CH_2OAc / OCH_3 $\xrightarrow[\text{dioxane, H}_2\text{O}]{h\nu}$

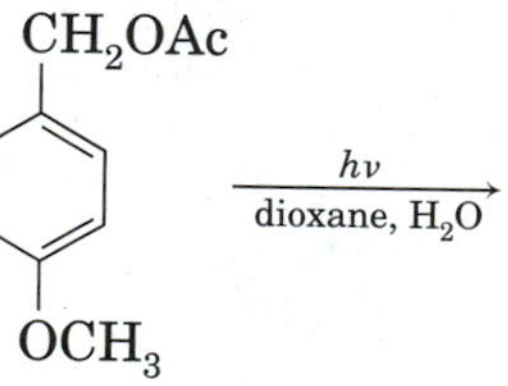

b. CH_2OAc / OCH_3 $\xrightarrow[\text{dioxane, H}_2\text{O}]{h\nu}$

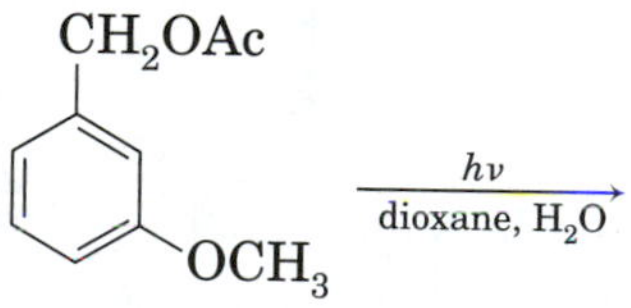

14. For each of the following reactions, propose a mechanism in which the first step is a Norrish Type I reaction (α cleavage):

a. O / HO $\xrightarrow{h\nu}$ O / HO

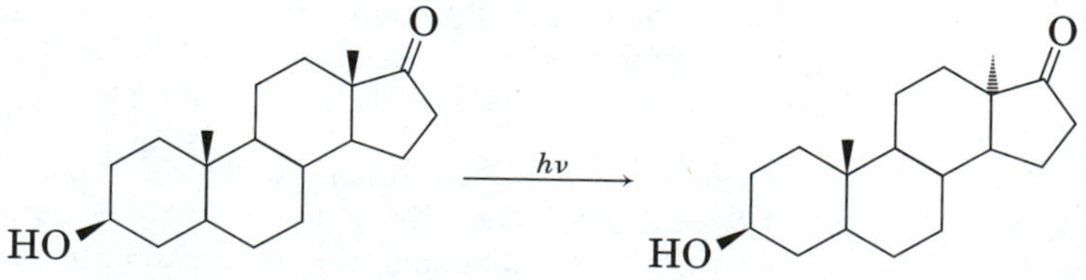

b.

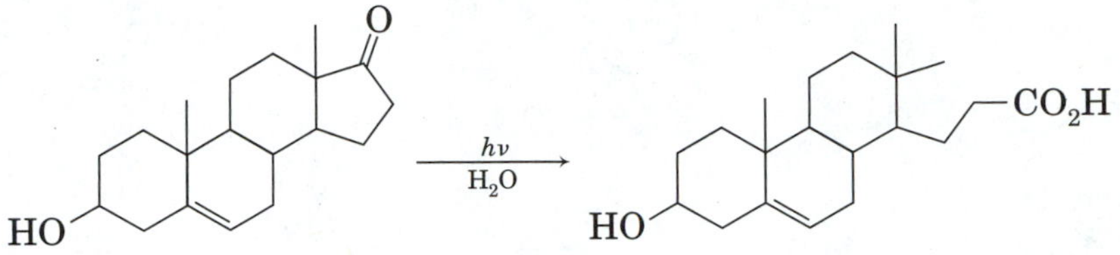

HO $\xrightarrow[H_2O]{h\nu}$ HO CO_2H

c.

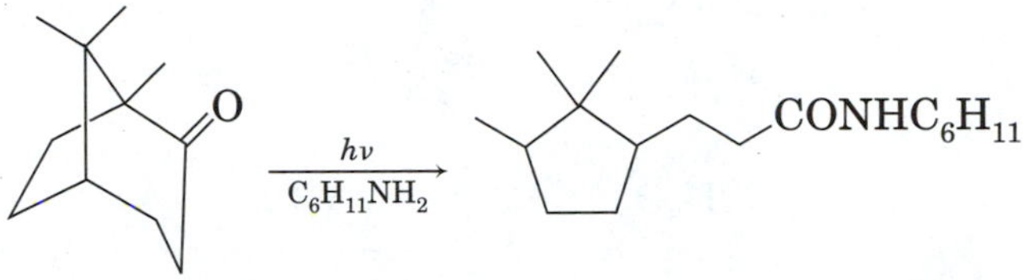

O $\xrightarrow[C_6H_{11}NH_2]{h\nu}$ $CONHC_6H_{11}$

d.

O $\xrightarrow[\text{hexane}]{h\nu}$ O H + O H

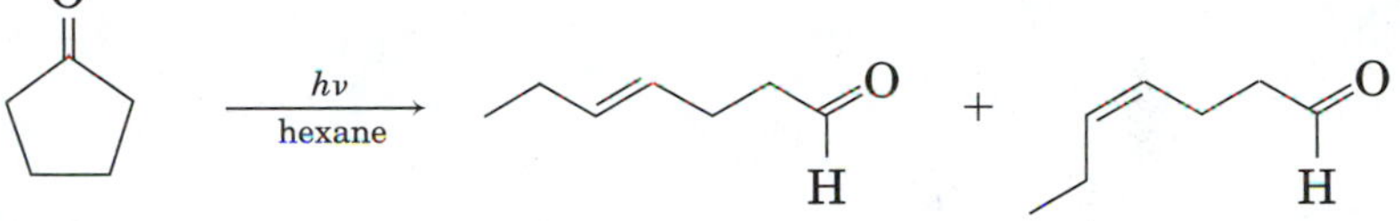

15. For each of the following reactions, propose a mechanism in which the first step is an intramolecular hydrogen abstraction:

a.

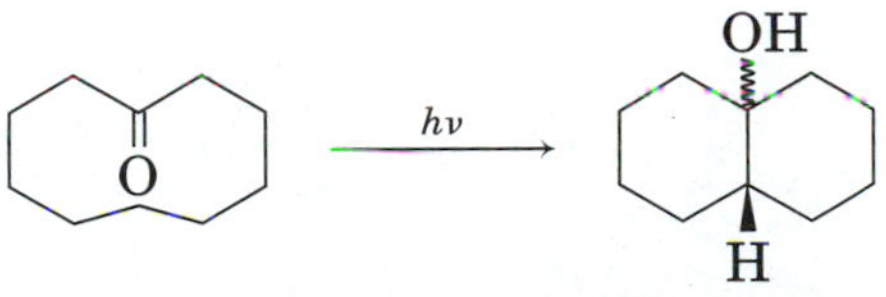

O $\xrightarrow{h\nu}$ OH H

b.

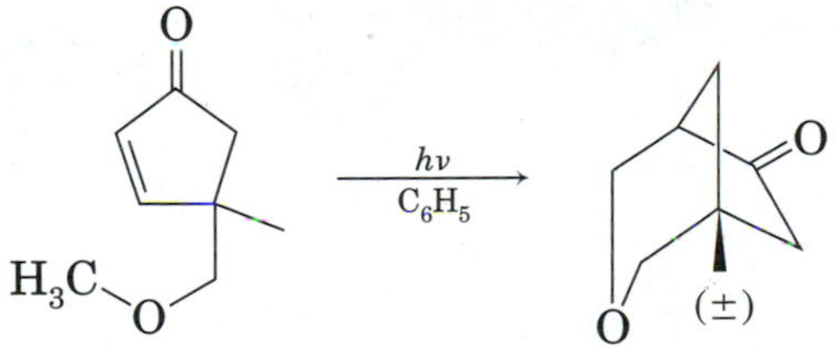

O H_3C O $\xrightarrow[C_6H_5]{h\nu}$ O O (±)

16. For each of the following reactions, propose a mechanism in which the first step is an intermolecular hydrogen abstraction:

a.

O $\xrightarrow[(CH_3)_3COH]{h\nu}$ $(CH_3)_3C$—O O

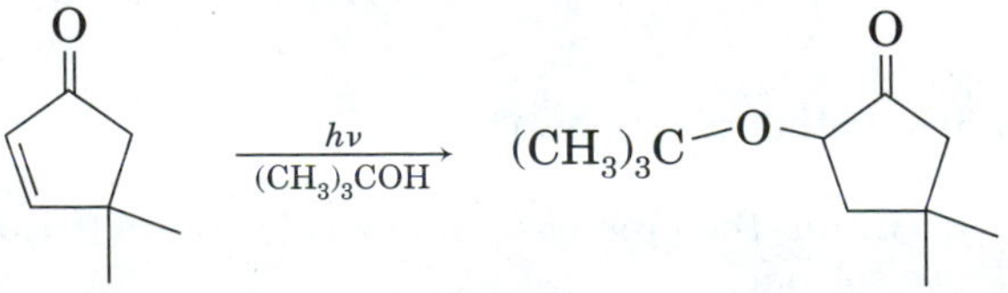

b. CH_3O ... $C{=}O$ $\xrightarrow[CH_3OH]{h\nu}$ CH_3O ... CH_2OH, C, OH ... CH_3O

CH_3O OCH_3

+ HO—C—C—OH

CH_3O OCH_3

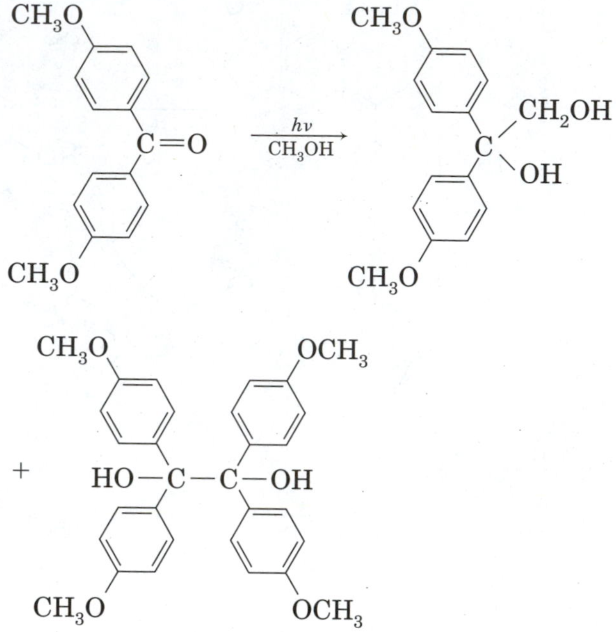

17. For each of the following reactions, propose a mechanism in which the first step is a σ bond photodissociation:

a. $\xrightarrow[CH_3CH_2OH]{h\nu}$ $CO_2CH_2CH_3$ OH

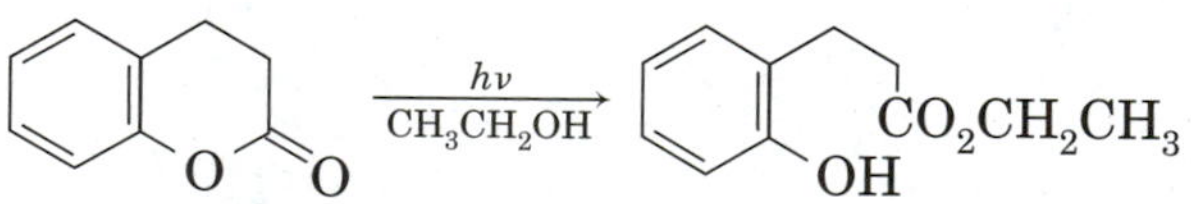

b. O, O, C_6H_5 $\xrightarrow[\text{ethanol}]{h\nu}$ O, HO

28%

+ OH O + OH

20% 14%

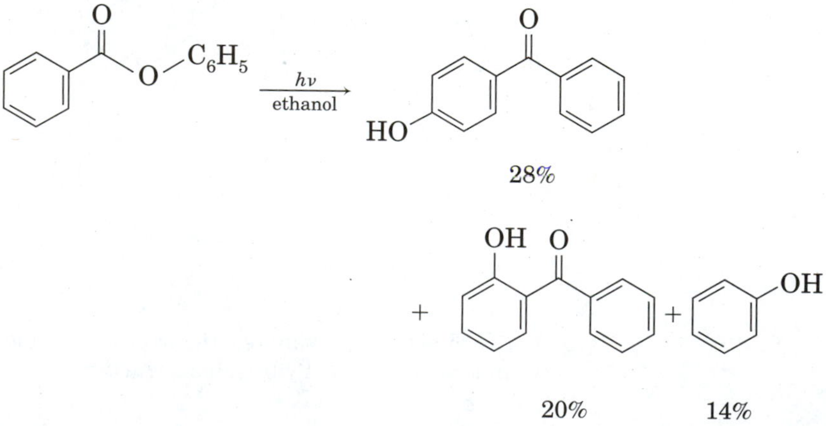

18. Irradiation of cyclobutyl phenyl ketone (**180**) in isopropyl alcohol leads to **181**, while irradiation in benzene leads to **182** and **183**. Propose mechanisms for formation of the products, and explain why different products are formed in the two solvents.

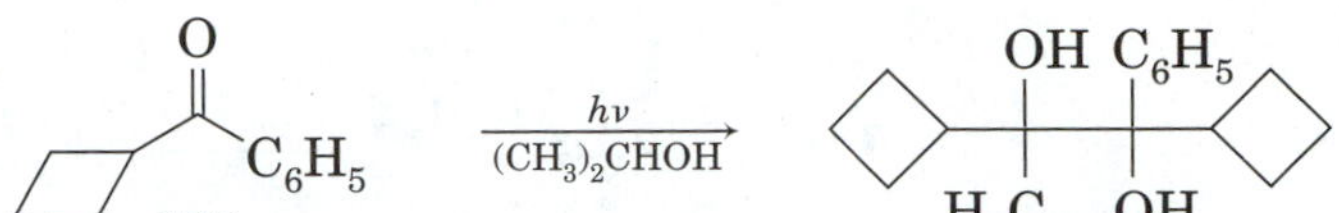

180 $\xrightarrow[(CH_3)_2CHOH]{h\nu}$ **181**

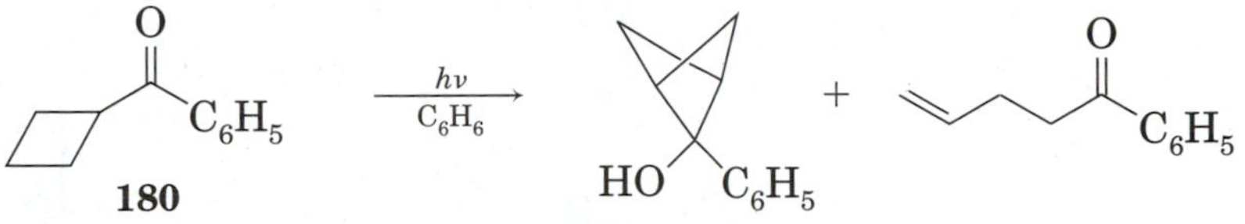

180 $\xrightarrow[C_6H_6]{h\nu}$ **182** + **183**

19. Propose a mechanism to account for the formation of the indicated product in each of the following photochemical reactions:

    a.

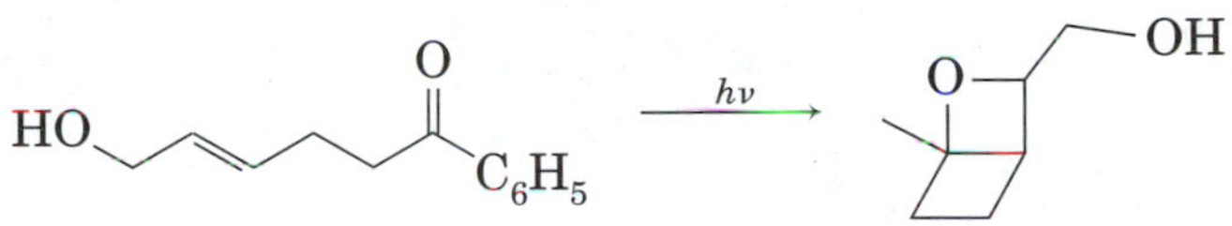

    b.

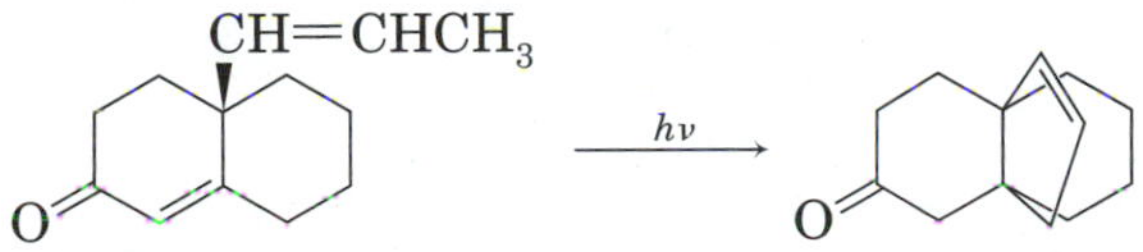

20. Irradiation of *cis*-2-propyl-4-*tert*-butylcyclohexanone in hexane produces 4-*tert*-butylcyclohexanone. However, irradiation of the trans isomer results in photoisomerization to the cis isomer. Propose a mechanism for the reaction observed for each compound, and explain why the two diastereomers react differently.

21. Propose a mechanism for the formation of the products shown in each of the following reactions:

    a.

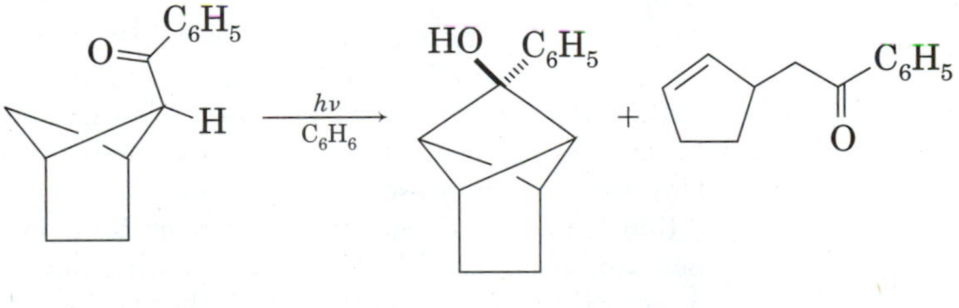

    b.

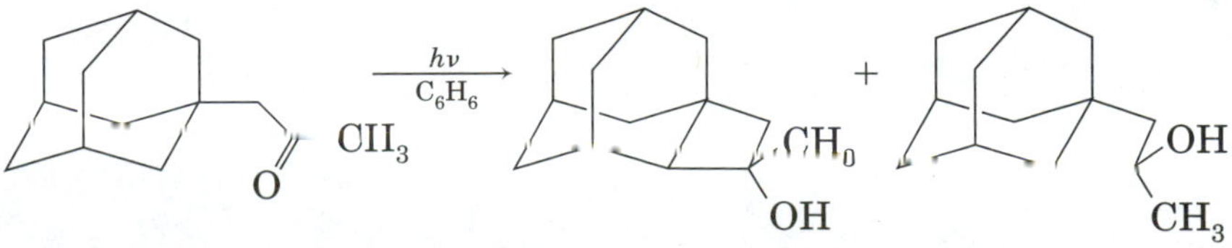

c.

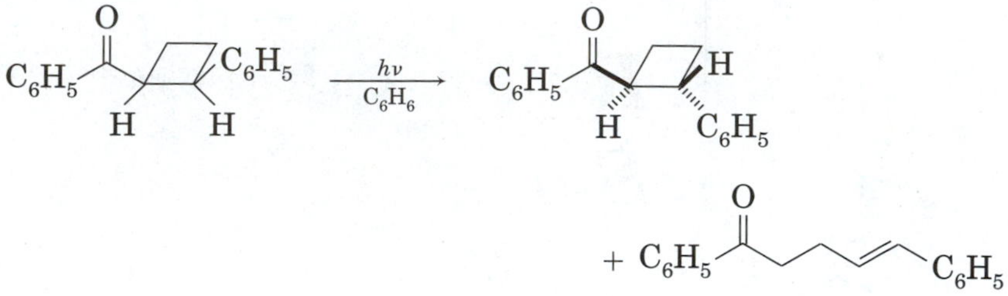

C_6H_5 C_6H_5 H H $\xrightarrow[C_6H_6]{h\nu}$ C_6H_5 H H C_6H_5 + C_6H_5 C_6H_5

d.

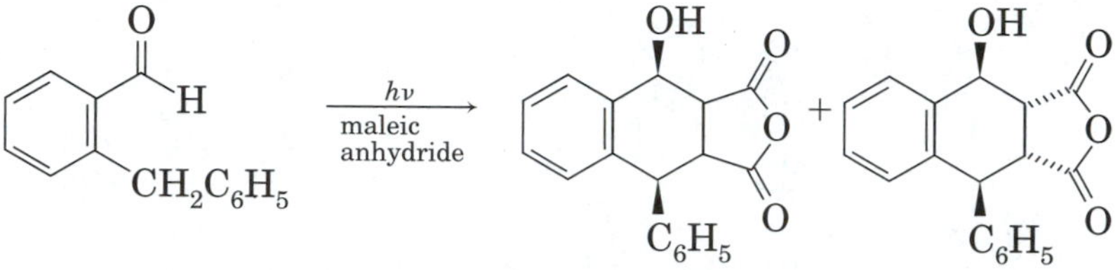

H $CH_2C_6H_5$ $\xrightarrow[\text{maleic anhydride}]{h\nu}$ OH O O O C_6H_5 + OH O O O C_6H_5

e.

O $\xrightarrow[C_6H_6]{h\nu}$ O + O + O (±)

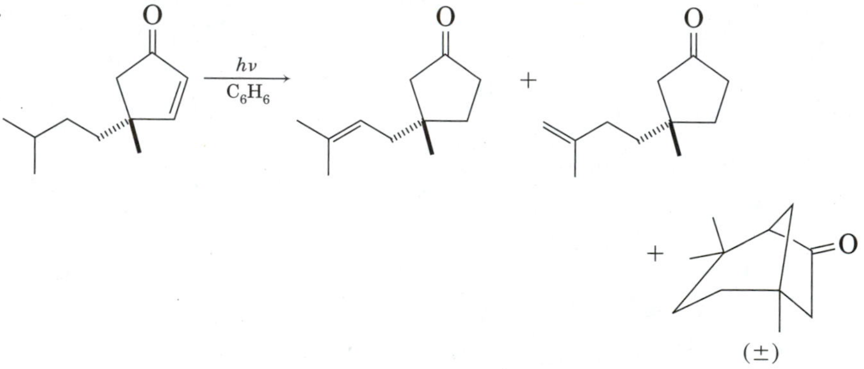

22. Irradiation of an equimolar mixture of benzene and 1,1-dimethoxyethene produces a 1 : 1 photoadduct. When the adduct is dissolved in methanol containing trace amounts of acid, 2,4,6-cyclooctatrienone is formed. Propose a structure for the 1 : 1 adduct and a mechanism for conversion of the adduct to the cyclooctatrienone.
23. The T_1 state (but not the S_1 state) of toluene sensitizes the photoisomerization of both (*E*)- and (*Z*)-2-heptene. Direct irradiation of either isomer of 1-phenyl-2-butene leads to photoisomerization. The fluorescence spectrum of 1-phenyl-2-butene is similar in shape to that of toluene but is slightly reduced in intensity. However, phosphorescence from 1-phenyl-2-butene was not detected under conditions in which phosphorescence from toluene could be readily observed. Propose an explanation for the photoisomerization of 1-phenyl-2-butene upon direct irradiation that is consistent with these observations.

References for Selected Problems

Chapter 1
Concepts and Models in Organic Chemistry

1. Asimov, I. *A Short History of Chemistry*; Anchor Books: Garden City, NY, 1965; Ihde, A. J. *The Development of Modern Chemistry*; Harper & Row: New York, 1964; Butterfield, H. *The Origins of Modern Science, 1300–1800*, Revised Edition; The Free Press: New York, 1965.
2. *Chem. Eng. News* **1986** (Sept. 1), pp. 4–5.
3. Reinecke, M. G. *J. Chem. Educ.* **1992**, *69*, 859 and references therein.
4. Bondi, J. *J. Phys. Chem.* **1964**, *68*, 441.
5. Kiyobayashi, T.; Nagano, Y.; Sakiyama, M.; Yamamoto, K.; Cheng, P.-C.; Scott, L. T. *J. Am. Chem. Soc.* **1995**, *117*, 3270.
6. Turner, R. B.; Goebel, P.; Mallon, B. J.; Doering, W. v. E.; Coburn, Jr., J. F.; Pomerantz, M. *J. Am. Chem. Soc.* **1968**, *90*, 4315. Also see Hautala, R. R.; King, R. B.; Kutal, C. in *Solar Energy: Chemical Conversion and Storage*; Hautala, R. R.; King, R. B.; Kutal, C., eds.; Humana Press: Clifton, NJ, 1979; p. 333.
7. Pilcher, G.; Parchment, O. G.; Hillier, I. H.; Heatley, F.; Fletcher, D.; Ribeiro da Silva, M. A. V.; Ferrão, M. L. C. C. H.; Monte, M. J. S.; Jiye, F. *J. Phys. Chem.* **1993**, *97*, 243.
8. Davis, H. E.; Allinger, N. L.; Rogers, D. W. *J. Org. Chem.* **1985**, *50*, 3601.
9. Wiberg, K. B.; Hao, S. *J. Org. Chem.* **1991**, *56*, 5108.
10. Fang, W.; Rogers, D. W. *J. Org. Chem.* **1992**, *57*, 2294.
11. Smyth, C. P. in *Physical Methods of Chemistry*, Vol. 1, Part 4; Weissberger, A.; Rossiter, B. W., eds.; Wiley-Interscience: New York, 1972; pp. 397–429.
12. Meek, T. L. *J. Chem. Educ.* **1995**, *72*, 17.
13. Owen, N. L.; Sheppard, N. *Trans. Faraday Soc.* **1964**, *60*, 634; Walters, E. A. *J. Chem. Educ.* **1966**, *43*, 134; Liberles, A. *J. Chem. Educ.* **1977**, *54*, 479; Bond, D.; Schleyer, P. v. R. *J. Org. Chem.* **1990**, *55*, 1003.
15. Mastryukov, V. S.; Schaefer III, H. F.; Boggs, J. E. *Acc. Chem. Res.* **1994**, *27*, 242; Gilardi, R.; Maggini, M.; Eaton, P. E. *J. Am. Chem. Soc.* **1988**, *110*, 7232.
18. Robinson, E. A.; Gillespie, R. J. *J. Chem. Educ.* **1980**, *57*, 329 (appendix, p. 333).
19. Newton, M. D.; Schulman, J. M.; Manus, M. M. *J. Am. Chem. Soc.* **1974**, *96*, 17.
21. Kass S. R.; Chou, P. K. *J. Am. Chem. Soc.* **1988**, *110*, 7899.

Chapter 2
Stereochemistry

3. **j.** Moore, W. R.; Anderson, H. W.; Clark, S. D.; Ozretich, T. M. *J. Am. Chem. Soc.* **1971**, *93*, 4932.
   k. Gerlach, H. *Helv. Chim. Acta* **1966**, *49*, 1291.
   l. Newman, P.; Rutkin, P.; Mislow, K. *J. Am. Chem. Soc.* **1958**, *80*, 465.
5. Seebach, D.; Lapierre, J.-M.; Skobridis, K.; Greiveldinger, G. *Angew. Chem., Int. Ed. Engl.* **1994**, *33*, 440.
6. **c.** Reynolds, K. A.; Fox, K. M.; Yuan, Z.; Lam, Y. *J. Am. Chem. Soc.* **1991**, *113*, 4339.
7. **d.** Reynolds, K. A.; Fox, K. M.; Yuan, Z.; Lam, Y. *J. Am. Chem. Soc.* **1991**, *113*, 4339.
8. **a.** Eaton, P. E.; Leipzig, B. *J. Org. Chem.* **1978**, *43*, 2483.
   b. and **c.** Halterman, R. L.; Jan, S.-T. *J. Org. Chem.* **1991**, *56*, 5253.

d. Deprés, J.-P.; Morat, C. *J. Chem. Educ.* **1992**, *69*, A232.

9. **a.** Cywin, C. L.; Webster, F. X.; Kallmerten, J. *J. Org. Chem.* **1991**, *56*, 2953.
b. Ingold, K. U. *Aldrichimica Acta* **1989**, *22*, 69.
c. Rychnovsky, S. D.; Griesgraber, G.; Zeller, S.; Skalitzky, D. J. *J. Org. Chem.* **1991**, *56*, 5161.
d. Cianciosi, S. J.; Ragunathan, N.; Freedman, T. B.; Nafie, L. A.; Baldwin, J. E. *J. Am. Chem. Soc.* **1990**, *112*, 8204.
e. and **f.** See the Eastman Fine Chemicals advertisement in *J. Org. Chem.* **1992**, *57* (20), on the page preceding the table of contents.
g. Bharucha, K. N.; Marsh, R. M.; Minto, R. E.; Bergman, R. G. *J. Am. Chem. Soc.* **1992**, *114*, 3120.
h. Naoshima, Y.; Munakata, Y.; Yoshida, S.; Funai, A. *J. Chem. Soc. Perkin Trans. 1* **1991**, 549.
i. Walborsky, H. M.; Goedken, V. L.; Gawronski, J. K. *J. Org. Chem.* **1992**, *57*, 410.
j. Rawson, D.; Meyers, A. I. *J. Chem. Soc., Chem. Commun.* **1992**, 494.
k. Kitching, W.; Lewis, J. A.; Perkins, M. V.; Drew, R.; Moore, C. J.; Schurig, V.; König, W. A.; Francke, W. *J. Org. Chem.* **1989**, *54*, 3893.
l. Freedman, T. B.; Cianciosi, S. J.; Ragunathan, N.; Baldwin, J. E.; Nafie, L. A. *J. Am. Chem. Soc.* **1991**, *113*, 8298.
m. Rao, A. V. R.; Gurjar, M. K.; Bose, D. S.; Devi, R. R. *J. Org. Chem.* **1991**, *56*, 1320.
n. Liu, C.; Coward, J. K. *J. Org. Chem.* **1991**, *56*, 2262.
o. Glattfeld, J. W. E.; Chittum, J. W. *J. Am. Chem. Soc.* **1933**, *55*, 3663.
p. Chattopadhyay, S.; Mamdapur, V. R.; Chadha, M. S. *J. Chem. Res. (M)* **1990**, 1818.
q. Hammarström, L.-G.; Berg, U.; Liljefors, T. *J. Chem. Res. (S)* **1990**, 152.
r. King, S. B.; Ganem, B. *J. Am. Chem. Soc.* **1994**, *116*, 562.
s. Moorthy, J. N.; Venkatesan, K. *J. Org. Chem.* **1991**, *56*, 6957.
t. Andersen, K. K.; Colonna, S.; Stirling, C. J. M. *J. Chem. Soc., Chem. Commun.* **1973**, 645.

10. **a.** Coke, J. L.; Shue, R. S. *J. Org. Chem.* **1973**, *38*, 2210.
b. Katsura, T.; Minamii, M. *Jpn. Kokai Tokkyo Hoho JP* 61,176, 557; see *Chem. Abstr.* **1986**, *106*, 66799f; for a discussion, see Salaün, J. *Chem. Rev.* **1989**, *89*, 1247.
c. and **d.** Floss, H. G.; Lee, S. *Acc. Chem. Res.* **1993**, *26*, 116.
e. Walborsky, H. M.; Impastato, F. J.; Young, A. E. *J. Am. Chem. Soc.* **1964**, *86*, 3283.
f. Skell, P. S.; Pavlis, R. R.; Lewis, D. C.; Shea, K. J. *J. Am. Chem. Soc.* **1973**, *95*, 6735.
g. Wiberg, K. B. *J. Am. Chem. Soc.* **1952**, *74*, 3891.

11. Jennings, W. B. *Chem. Rev.* **1975**, *75*, 307.

14. LeGoff, E.; Ulrich, S. E.; Denney, D. B. *J. Am. Chem. Soc.* **1958**, *80*, 622.

15. **b.** Miyamoto, K.; Tsuchiya, S.; Ohta, H. *J. Am. Chem. Soc.* **1992**, *114*, 6256.

16. Cinquini, M.; Cozzi, F.; Sannicolò, F.; Sironi, A. *J. Am. Chem. Soc.* **1988**, *110*, 4363.

17. Hoye, T. R.; Hanson, P. R.; Kovelesky, A. C.; Ocain, T. D.; Zhuang, Z. *J. Am. Chem. Soc.* **1991**, *113*, 9369

19. Polniaszek, R. P.; Dillard, L. W.; *Abstracts of the 203rd National Meeting of the American Chemical Society*, San Francisco, CA, April 5–10, 1992, Abstract ORGN 494.

20. Whitesides, G. M.; Kaplan, F.; Roberts, J. D. *J. Am. Chem. Soc.* **1963**, *85*, 2166.

21. Streitwieser, Jr., A.; Granger, M. R. *J. Org. Chem.* **1967**, *32*, 1528.

22. Hilvert, D.; Nared, K. D. *J. Am. Chem. Soc.* **1988**, *110*, 5593.

Chapter 3 Conformational Analysis and Molecular Mechanics

1. Allinger, N. L.; Miller, M. A. *J. Am. Chem. Soc.* **1961**, *83*, 2145; Beckett, C. W.; Pitzer, K. S.; Spitzer, R. *J. Am. Chem. Soc.* **1947**, *69*, 2488.

2. Eliel, E. L.; Allinger, N. L.; Angyal, S. J.; Morrison, G. A. *Conformational Analysis*, Wiley-Interscience: New York, 1965; p. 52 and references therein.

3. Booth, H.; Everett, J. R. *J. Chem. Soc., Perkin Trans. 2* **1980**, 255.

4. Juaristi, E.; Labastida, V.; Antúnez, S. *J. Org. Chem.* **1991**, *56*, 4802.

5. **b.** See Perrin, C. L.; Fabian, M. A. *Abstracts of the 209th National Meeting of the American Chemical Society*, Anaheim, CA, April 2-6, 1995, Abstract ORGN 6.
c. Jensen, F. R.; Bushweller, C. H.; Beck, B. H. *J. Am. Chem. Soc.* **1969**, *91*, 344.

6. Chupp, J. P.; Olin, J. F. *J. Org. Chem.* **1967**, *32*, 2297.

7. Huang, J.; Hedberg, K. *J. Am. Chem. Soc.* **1989**, *111*, 6909.

8. For literature values, see Stull, D. R.; Westrum, Jr., E. F.; Sinke, G. C. *The Chemical Thermodynamics of Organic Compounds*; John Wiley & Sons: New York, 1969; pp. 249–252; also see Cox, J. D.; Pilcher, G. *Thermochemistry of*

Organic and Organometallic Compounds; Academic Press: New York, 1970; p. 157 and references therein.

9. Whitlock, Jr., H. W. *J. Am. Chem. Soc.* **1962**, *84*, 3412; Gassman, P. G.; Yamaguchi, R. *J. Org. Chem.* **1978**, *43*, 4654; Eaton, P. E.; Cole, Jr., T. W. *J. Am. Chem. Soc.* **1964**, *86*, 3157; Hoeve, W. T.; Wynberg, H. *J. Org. Chem.* **1980**, *45*, 2925; Maier, G. *Angew. Chem., Int. Ed. Engl.* **1988**, *27*, 309 and references therein; Wiberg, K. B.; McMurdie, N.; McClusky, J. V.; Hadad, C. M. *J. Am. Chem. Soc.* **1993**, *115*, 10653.

10. Maier, G. *Angew. Chem., Int. Ed. Engl.* **1988**, *27*, 309.

11. Turner, R. B.; Nettleton, Jr., D. E.; Perelman, M. *J. Am. Chem. Soc.* **1958**, *80*, 1430; Brown, H. C.; Berneis, H. L. *J. Am. Chem. Soc.* **1953**, *75*, 10; Saunders, Jr., W. H.; Cockerill, A. F. *Mechanisms of Elimination Reactions*; Wiley-Interscience: New York, 1973; p. 173.

12. Golan, O.; Goren, Z.; Biali, S. E. *J. Am. Chem. Soc.* **1990**, *112*, 9300; Juaristi, E.; Labastida, V.; Antúnez, S. *J. Org. Chem.* **1991**, *56*, 4802.

13. and **14.** Barton, D. H. R.; Cookson, R. C. *Quart. Rev. Chem. Soc.* **1956**, *10*, 44 (especially p. 58); Johnson, W. S. *J. Am. Chem. Soc.* **1953**, *75*, 1498.

15. and **16.** Jensen, F. R.; Bushweller, C. H.; Beck, B. H. *J. Am. Chem. Soc.* **1969**, *91*, 344.

17. Booth, H.; Everett, J. R. *J. Chem. Soc., Perkin Trans. 2* **1980**, 255.

18. Eliel, E. L.; Manoharan, M. *J. Org. Chem.* **1981**, *46*, 1959.

21. Kuhn, L. P. *J. Am. Chem. Soc.* **1958**, *80*, 5950.

Chapter 4
Applications of Molecular Orbital Theory and Valence Bond Theory

3. Hückel molecular orbital calculations for many compounds of interest are reported in Heilbronner und Straub, *Hückel Molecular Orbitals: HMO*; Springer-Verlag: New York, 1966. An HMO program is provided on a disk included with Rauk, A. *Orbital Interaction Theory of Organic Chemistry*; Wiley-Interscience: New York, 1994.

9. Billups, W. E.; Lin, L.-J.; Casserly, E. W. *J. Am. Chem. Soc.* **1984**, *106*, 3698; Staley, S. W.; Norden, T. D. *J. Am. Chem. Soc.* **1984**, *106*, 3699; Norden, T. D.; Staley, S. W.; Taylor, W. H.; Harmony, M. D. *J. Am. Chem. Soc.* **1986**, *108*, 7912; Bachrach, S. M. *J. Org. Chem.* **1990**, *55*, 4961.

10. Staley, S. W.; Norden, T. D.; Taylor, W. H.; Harmony, M. D. *J. Am. Chem. Soc.* **1987**, *109*, 7641; Bachrach, S. M. *J. Org. Chem.* **1990**, *55*, 4961.

12. Schleyer, P. v. R. *J. Am. Chem. Soc.* **1985**, *107*, 4793; Vajda, E.; Tremmel, J.; Rozsondai, B.; Hargittai, I.; Maltsev, A. K.; Kagramanov, N. D.; Nefedov, O. M. *J. Am. Chem. Soc.* **1986**, *108*, 4352; Dorigo, A. E.; Li, Y.; Houk, K. N. *J. Am. Chem. Soc.* **1989**, *111*, 6942; Korth, H.-G.; Trill, H.; Sustmann, R. *J. Am. Chem. Soc.* **1981**, *103*, 4483; Thompson, T. B.; Ford, W. T. *J. Am. Chem. Soc.* **1979**, *101*, 5459; Mayr, H.; Förner, W.; Schleyer, P. v. R. *J. Am. Chem. Soc.* **1979**, *101*, 6032; Gobbi, A.; Frenking, G. *J. Am. Chem. Soc.* **1994**, *116*, 9275; Krusic, P. J.; Meakin, P.; Smart, B. E. *J. Am. Chem. Soc.* **1974**, *96*, 6211; Shaik, S. S.; Hiberty, P. C.; Lefour, J.-M.; Ohanessian, G. *J. Am. Chem. Soc.* **1987**, *109*, 363; Foresman, J. B.; Wong, M. W.; Wiberg, K. B.; Frisch, M. J. *J. Am. Chem. Soc.* **1993**, *115*, 2220; Wiberg, K. B.; Cheeseman, J. R.; Ochterski, J. W.; Frisch, M. J. *J. Am. Chem. Soc.* **1995**, *117*, 6535.

14. Herndon, W. C. *J. Am. Chem. Soc.* **1976**, *98*, 887 and references therein; Herndon, W. C. *J. Am. Chem. Soc.* **1974**, *96*, 7605.

15. Doering, W. v. E.; Knox, L. H. *J. Am. Chem. Soc.* **1954**, *76*, 3203; Breslow, R.; Hoffman, Jr., J. M. *J. Am. Chem. Soc.* **1972**, *94*, 2110; Katz, T. J. *J. Am. Chem. Soc.* **1960**, *82*, 3784, 3785.

17. Cox, J. D.; Pilcher, G. *Thermochemistry of Organic and Organometallic Compounds*; Academic Press: New York, 1970; p. 223.

18. Baird, N. C. *J. Chem. Educ.* **1971**, *48*, 509.

19. Iyoda, M.; Kurata, H.; Oda, M.; Okubo, C.; Nishimoto, K. *Angew. Chem., Int. Ed. Engl.* **1993**, *32*, 89.

20. Binsch, G. *Top. Stereochem.* **1968**, *3*, 97 (see especially pp. 134–146).

21. Sekiguchi, A.; Ebata, K.; Kabuto, C.; Sakurai, H. *J. Am. Chem. Soc.* **1991**, *113*, 7081.

22. Baird, N. C. *J. Chem. Educ.* **1971**, *48*, 509. (See especially p. 511.)

Chapter 5
Reactive Intermediates

1. Wentrup, C. *Reactive Molecules*; John Wiley & Sons: New York, 1984; p. 41.

2. Egger, K. W.; Cocks, A. T. *Helv. Chim. Acta* **1973**, *56*, 1537, particularly p. 1539.

3. Wentrup, C. *Reactive Molecules*; John Wiley & Sons: New York, 1984; pp. 33–34 and references therein.

4. Ayscough, P. B. *Electron Spin Resonance in Chemistry*; Methuen & Co.: London, 1967; p. 298; Bunce, N. J. *J. Chem. Educ.* **1987**, *64*, 907 (especially p. 910); Fleming, I. *Frontier Orbitals and Organic Chemical Reactions*; Wiley-Interscience: London, 1976; p. 60.

5. Walling, C.; Cooley, J. H.; Ponaras, A. A.; Racah, E. J. *J. Am. Chem. Soc.* **1966**, *88*, 5361.
6. Olah, G. A. *Carbocations and Electrophilic Reactions*; John Wiley & Sons: New York, 1974; p. 20.
7. Olah, G. A. *Carbocations and Electrophilic Reactions*; John Wiley & Sons: New York, 1974; p. 63; Olah, G. A.; White, A. M. *J. Am. Chem. Soc.* **1967**, *89*, 3591.
8. Olah, G. A.; Mateescu, G. D.; Wilson, L. A.; Gross, M. H. *J. Am. Chem. Soc.* **1970**, *92*, 7231.
9. Olah, G. A.; Prakash, G. K. S.; Williams, R. E.; Field, L. D.; Wade, K. *Hypercarbon Chemistry*; John Wiley & Sons: New York, 1987; p. 153 and reference therein.
10. Olah, G. A.; Prakash, G. K. S.; Sommer, J. *Superacids*; Wiley-Interscience: New York, 1985; p. 84; Saunders, M.; Vogel, P.; Hagen, E. L.; Rosenfeld, J. *Acc. Chem. Res.* **1973**, *6*, 53.
11. Walborsky, H. M.; Periasamy, M. P. *J. Am. Chem. Soc.* **1974**, *96*, 3711.
12. Zimmerman, H. E.; Zweig, A. *J. Am. Chem. Soc.* **1961**, *83*, 1196.
13. Crawford, R. J.; Erman, W. F.; Broaddus, C. D. *J. Am. Chem. Soc.* **1972**, *94*, 4298; Bates, R. B.; Ogle, C. A. *Carbanion Chemistry*; Springer-Verlag: Berlin, 1983; p. 24.
14. Fitjer, L.; Quabeck, U. *Angew. Chem., Int. Ed. Engl.* **1987**, *26*, 1023.
15. Oldroyd, D. M.; Fisher, G. S.; Goldblatt, L. A. *J. Am. Chem. Soc.* **1950**, *72*, 2407; Walling, C. in *Molecular Rearrangements*, Part I; de Mayo, P., ed.; Wiley-Interscience: New York, 1963; pp. 407–455 (particularly p. 440).
16. Moss, R. A.; Jones, Jr., M. in *Reactive Intermediates*, Vol. 1; Jones, Jr., M.; Moss, R. A., eds.; Wiley-Interscience: New York, 1978; pp. 67–116 (especially p. 97 and references therein); Wentrup, C. *Reactive Molecules*; John Wiley & Sons: New York, 1984; p. 238 and references therein; Hoffmann, R. W.; Reiffen, M. *Chem. Ber.* **1976**, *109*, 2565.
17. Perkins, M. J. *Adv. Phys. Org. Chem.* **1980**, *17*, 1.
18. Friedman, L.; Shechter, H. *J. Am. Chem. Soc.* **1960**, *82*, 1002.
19. Friedman, L.; Shechter, H. *J. Am. Chem. Soc.* **1961**, *83*, 3159.
20. Jones, R. R.; Bergman, R. G. *J. Am. Chem. Soc.* **1972**, *94*, 660.

Chapter 6
Methods of Studying Organic Reactions

1. Murray, R. W. *Acc. Chem. Res.*, **1968**, *1*, 313 and references therein; Murray, R. W.; Story, P. R.; Loan, L. D. *J. Am. Chem. Soc.* **1965**, *87*, 3025.
2. Long, F. A.; Pritchard, J. G. *J. Am. Chem. Soc.* **1956**, *78*, 2663.
4. Bartlett, P. D.; Trachtenberg, E. N. *J. Am. Chem. Soc.* **1958**, *80*, 5808.
5. Frost, A. A.; Pearson, R. G. *Kinetics and Mechanism*, 2nd ed.; John Wiley & Sons, Inc.: New York, 1961; pp. 101 and 105; Chung, Y.-S.; Duerr, B. F.; Nanjappan, P.; Czarnik, A. W. *J. Org. Chem.* **1988**, *53*, 1334.
6. LeFevre, G. N.; Crawford, R. J. *J. Org. Chem.* **1986**, *51*, 747.
7. Kluger, R.; Brandl, M. *J. Org. Chem.* **1986**, *51*, 3964.
8. Gano, J. E.; Lenoir, D.; Park, B.-S.; Roesner, R. A. *J. Org. Chem.* **1987**, *52*, 5636.
9. Wiberg, K. B.; Caringi, J. J.; Matturro, M. G. *J. Am. Chem. Soc.* **1990**, *112*, 5854.
10. Bartlett, P. D.; Wu, C. *J. Org. Chem.* **1985**, *50*, 4087.
11. **c.** Wiberg, K. B.; Slaugh, L. H. *J. Am. Chem. Soc.* **1958**, *80*, 3033.
12. Jones, J. M.; Bender, M. L. *J. Am. Chem. Soc.* **1960**, *82*, 6322.
13. Pocker, Y. *Proc. Chem. Soc.* **1960**, 17.
16. Overman, L. E.; Petty, S. T. *J. Org. Chem.* **1975**, *40*, 2779.
17. Dietze, P. E.; Underwood, G. R. *J. Org. Chem.* **1984**, *49*, 2492.
18. Whitworth, A. J.; Ayoub, R.; Rousseau, Y.; Fliszár, S. *J. Am. Chem. Soc.* **1969**, *91*, 7128.
19. Choe, J.-I.; Srinivasan, M.; Kuczkowski, R. L. *J. Am. Chem. Soc.* **1983**, *105*, 4703.
20. Hill, R. K.; Conley, R. T.; Chortyk, O. T. *J. Am. Chem. Soc.* **1965**, *87*, 5646.
21. Berliner, E.; Altschul, L. H. *J. Am. Chem. Soc.* **1952**, *74*, 4110; also see Leffler, J. E.; Grunwald, E. *Rates and Equilibria of Organic Reactions*; John Wiley and Sons, Inc.: New York, 1963; pp. 324–342.

Chapter 7
Acid-Base Catalyzed Reactions

1. Barlin, G. B.; Perrin, D. D. *Quart. Rev. Chem. Soc.* **1966**, *20*, 75.
2. Taft, R. W.; Bordwell, F. G. *Acc. Chem. Res.* **1988**, *21*, 463.
4. Wheeler, O. H. *J. Am. Chem. Soc.* **1957**, *79*, 4191.
6. Wiseman, J. S.; Abeles, R. H. *Biochem.* **1979**, *18*, 427.
7. Cardwell, H. M. E.; Kilner, A. E. H. *J. Chem. Soc.* **1951**, 2430 and references therein.
8. Levine, R.; Hauser, C. R. *J. Am. Chem. Soc.* **1944**, *66*, 1768.

9. Gutsche, C. D.; Redmore, D.; Buriks, R. S.; Nowotny, K.; Grassner, H.; Armbruster, C. W. *J. Am. Chem. Soc.* **1967**, *89*, 1235.

10. Smith, W. T.; McLeod, G. L. *Org. Syn. Coll. Vol. IV* **1963** 345. Also see Walker, J.; Wood, J. K. *J. Chem. Soc.* **1906**, *89*, 598.

11. Stefanidis, D.; Cho, S.; Dhe-Paganon, S.; Jencks, W. P. *J. Am. Chem. Soc.* **1993**, *115*, 1650.

12. Sørensen, P. E.; Jencks, W. P. *J. Am. Chem. Soc.* **1987**, *109*, 4675.

13. Bell, R. P.; Baughan, E. C. *J. Chem. Soc.* **1937**, 1947.

14. Hurd, C. D.; Saunders, Jr., W. H. *J. Am. Chem. Soc.* **1952**, *74*, 5324.

15. Drumheller, J. D.; Andrews, L. J. *J. Am. Chem. Soc.* **1955**, *77*, 3290.

16. Bender, M. L.; Chen, M. C. *J. Am. Chem. Soc.* **1963**, *85*, 30 (esp. p. 36); *ibid.*, 37 and references therein.

17. Reimann, J. E.; Jencks, W. P. *J. Am. Chem. Soc.* **1966**, *88*, 3973.

18. Olson, A. R.; Youle, P. V. *J. Am. Chem. Soc.* **1951**, *73*, 2468.

19. Hoz, S.; Livneh, M.; Cohen, D. *J. Org. Chem.* **1986**, *51*, 4537.

20. Zaugg, H. E.; Papendick, V.; Michaels, R. J. *J. Am. Chem. Soc.* **1964**, *86*, 1399; see also Johnson, S. L. *Adv. Phys. Org. Chem.* **1967**, *5*, 237.

21. Hauser, C. R.; Adams, J. T. *J. Am. Chem. Soc.* **1944**, *66*, 345.

22. Long, F. A.; Pritchard, J. G. *J. Am. Chem. Soc.* **1956**, *78*, 2663; Pritchard, J. G.; Long, F. A. *J. Am. Chem. Soc.* **1956**, *78*, 2667.

Chapter 8
Substitution Reactions

1. Ingold, C. K. *Structure and Mechanism in Organic Chemistry*, 2nd ed.; Cornell University Press: Ithaca, 1969; pp. 457 ff.

3. Magid, R. M.; Welch, J. G. *J. Am. Chem. Soc.* **1968**, *90*, 5211.

4. Smith, M. B.; Hrubiec, R. T.; Zezza, C. A. *J. Org. Chem.* **1985**, *50*, 4815.

5. Cowdrey, W. A.; Hughes, E. D.; Ingold, C. K. *J. Chem. Soc.* **1937**, 1208; Winstein, S.; Lucas, H. J. *J. Am. Chem. Soc.* **1939**, *61*, 1576. See also Hine, J. *Physical Organic Chemistry*, 2nd ed.; McGraw-Hill: New York, 1962; p. 143; Klyne, W.; Buckingham, J. *Atlas of Stereochemistry*, 2nd ed., Vol. 1; Oxford University Press: New York, 1978, p. 5.

6. McKenzie, A.; Clough, G. W. *J. Chem. Soc.* **1910**, *97*, 2564.

7. Brown, R. F.; van Gulick, N. M. *J. Org. Chem.* **1956**, *21*, 1046.

8. **a.** Cram, D. J. *J. Am. Chem. Soc.* **1949**, *71*, 3863. Also see Gould, E. S. *Mechanism and Structure in Organic Chemistry*; Holt, Rinehart and Winston: New York, 1959; p. 576 and references therein. **b.** Cram, D. J. *J. Am. Chem. Soc.* **1949**, *71*, 3875.

10. Winstein, S.; Lucas, H. J. *J. Am. Chem. Soc.* **1939**, *61*, 1581.

11. Chapman, J. W.; Strachan, A. N. *Chem. Commun.* **1974**, 293; Stock, L. M. *Prog. Phys. Org. Chem.* **1976**, *12*, 21.

12. Masci, B. *J. Org. Chem.* **1985**, *50*, 4081.

13. Brown, H. C.; Jensen, F. R. *J. Am. Chem. Soc.* **1958**, *80*, 2296.

14. Himeshima, Y.; Kobayashi, H.; Sonoda, T. *J. Am. Chem. Soc.* **1985**, *107*, 5286.

15. Grovenstein, Jr., E.; Kilby, D. C. *J. Am. Chem. Soc.* **1957**, *79*, 2972.

16. Truce, W. E.; Kreider, E. M.; Brand, W. W. *Org. React.* **1970**, *18*, 99.

17. Bunnett, J. F.; Rauhut, M. M.; Knutson, D.; Bussell, G. E. *J. Am. Chem. Soc.* **1954**, *76*, 5755; Samuel, D. *J. Chem. Soc.* **1960**, 1318; Rosenblum, M. *J. Am. Chem. Soc.* **1960**, *82*, 3796.

18. Roberts, J. D.; Semenow, D. A.; Simmons, Jr., H. E.; Carlsmith, L. A. *J. Am. Chem. Soc.* **1956**, *78*, 601.

20. Katritzky, A. R.; Laurenzo, K. S. *J. Org. Chem.* **1986**, *51*, 5039.

21. Bunnett, J. F.; Garbisch, Jr., E. W.; Pruitt, K. M. *J. Am. Chem. Soc.* **1957**, *79*, 385.

22. Roberts, J. D.; Vaughan, C. W.; Carlsmith, L. A.; Semenow, D. A. *J. Am. Chem. Soc.* **1956**, *78*, 611.

23. Roberts, J. D.; Semenow, D. A.; Simmons, Jr., H. E.; Carlsmith, L. A. *J. Am. Chem. Soc.* **1956**, *78*, 601.

Chapter 9
Addition Reactions

1. Buckles, R. E.; Bader, J. M.; Thurmaier, R. J. *J. Org. Chem.* **1962**, *27*, 4523.

2. Buckles, R. E.; Forrester, J. L.; Burham, R. L.; McGee, T. W. *J. Org. Chem.* **1960**, *25*, 24.

3. Bellucci, G.; Bianchini, R.; Chiappe, C. *J. Org. Chem.* **1991**, *56*, 3067.

4. Bedoukian, P. Z. *J. Am. Chem. Soc.* **1944**, *66*, 1325.

5. Tarbell, D. S.; Bartlett, P. D. *J. Am. Chem. Soc.* **1937**, *59*, 407.

6. Rolston, J. H.; Yates, K. *J. Am. Chem. Soc.* **1969**, *91*, 1477

7. van Tamelen, E. E.; Shamma, M. *J. Am. Chem. Soc.* **1954**, *76*, 2315.

8. Peterson, P. E.; Tao, E. V. P. *J. Am. Chem. Soc.* **1964**, *86*, 4503.
9. Fraenkel, G.; Bartlett, P. D. *J. Am. Chem. Soc.* **1959**, *81*, 5582.
10. Okuyama, T.; Sakagami, T.; Fueno, T. *Tetrahedron* **1973**, *29*, 1503. See also the discussion in Chwang, W. K.; Knittel, P.; Koshy, K. M.; Tidwell, T. T. *J. Am. Chem. Soc.* **1977**, *99*, 3395.
11. Bellucci, G.; Bianchini, R.; Vecchiani, S. *J. Org. Chem.* **1987**, *52*, 3355.
12. Fahey, R. C.; Schneider, H.-J. *J. Am. Chem. Soc.* **1968**, *90*, 4429.
13. Rolston, J. H.; Yates, K. *J. Am. Chem. Soc.* **1969**, *91*, 1469.
14. Cabaleiro, M. C.; Johnson, M. D. *J. Chem. Soc. B* **1967**, 565.
15. Rozen, S.; Brand, M. *J. Org. Chem.* **1986**, *51*, 3607.
16. Jacobs, T. L.; Searles, Jr., S. *J. Am. Chem. Soc.* **1944**, *66*, 686.
17. Halpern, J.; Tinker, H. B. *J. Am. Chem. Soc.* **1967**, *89*, 6427.
18. Kabalka, G. W.; Newton, Jr., R. J.; Jacobus, J. *J. Org. Chem.* **1978**, *43*, 1567.
19. **a.** Brown, H. C.; Geoghegan, Jr., P. *J. Am. Chem. Soc.* **1967**, *89*, 1522;
    b. Brown, H. C.; Hammar, W. J. *J. Am. Chem. Soc.* **1967**, *89*, 1524.
20. Brown, H. C.; Kurek, J. T. *J. Am. Chem. Soc.* **1969**, *91*, 5647.
21. Swern, D. *J. Am. Chem. Soc.* **1947**, *69*, 1692.
22. Modro, A.; Schmid, G. H.; Yates, K. *J. Org. Chem.* **1977**, *42*, 3673.
23. Ruasse, M.-F.; Motallebi, S.; Galland, B. *J. Am. Chem. Soc.* **1991**, *113*, 3440.

Chapter 10
Elimination Reactions

1. Dhar, M. L.; Hughes, E. D.; Ingold, C. K.; Mandour, A. M. M.; Maw, G. A.; Woolf, L. I. *J. Chem. Soc.* **1948**, 2093.
2. Cristol, S. J.; Hause, N. L. *J. Am. Chem. Soc.* **1952**, *74*, 2193.
3. Kibby, C. L.; Lande, S. S.; Hall, W. K. *J. Am. Chem. Soc.* **1972**, *94*, 214.
4. Searles, S.; Gortatowski, M. J. *J. Am. Chem. Soc.* **1953**, *75*, 3030.
5. Grovenstein, Jr., E.; Lee, D. E. *J. Am. Chem. Soc.* **1953**, *75*, 2639; Cristol, S. J.; Norris, W. P. *J. Am. Chem. Soc.* **1953**, *75*, 2645.
6. Ölwegård, M.; Ahlberg, P. *J. Chem. Soc., Chem. Commun.* **1989**, 1279; **1990**, 788.
7. Kurtz, R. R.; Houser, D. J. *J. Org. Chem.* **1981**, *46*, 202.
8. Goering, H. L.; Espy, H. H. *J. Am. Chem. Soc.* **1956**, *78*, 1454.
9. Nguyen Trong Anh, *Chem. Commun.* **1968**, 1089.
10. Cope, A. C.; LeBel, N. A.; Lee, H.-H.; Moore, W. R. *J. Am. Chem. Soc.* **1957**, *79*, 4720.
11. Gandini, A.; Plesch, P. H. *J. Chem. Soc.* **1965**, 6019.
12. O'Connor, G. L.; Nace, H. R. *J. Am. Chem. Soc.* **1953**, *75*, 2118.
13. Knözinger, H. in *The Chemistry of the Hydroxyl Group*, Part 2; Patai, S., ed.; Wiley-Interscience: London, 1971; pp. 660–661 and references therein.
14. Acharya, S. P.; Brown, H. C. *Chem. Commun.* **1968**, 305.
15. Barton, D. H. R.; Rosenfelder, W. J. *J. Chem. Soc.* **1951**, 1048.
16. Kashelikar, D. V.; Fanta, P. E. *J. Am. Chem. Soc.* **1960**, *82*, 4930.
17. Noyce, D. S.; Weingarten, H. I. *J. Am. Chem. Soc.* **1957**, *79*, 3093.
18. **a.** Alexander, E. R.; Mudrak, A. *J. Am. Chem. Soc.* **1950**, *72*, 1810.
    b. Alexander, E. R.; Mudrak, A. *J. Am. Chem. Soc.* **1950**, *72*, 3194.
    c. Cope, A. C.; LeBel, N. A. *J. Am. Chem. Soc.* **1960**, *82*, 4656.
19. Kende, A. S. *Org. React.* **1960**, *11*, 261.
20. Curtin, D. Y.; Stolow, R. D.; Maya, W. *J. Am. Chem. Soc.* **1959**, *81*, 3330.
21. Saunders, Jr., W. H.; Cockerill, A. F. *Mechanisms of Elimination Reactions*; Wiley-Interscience: New York, 1973; p. 173.

Chapter 11
Concerted Reactions

5. Longuet-Higgins, H. C.; Abrahamson, E. W. *J. Am. Chem. Soc.* **1965**, *87*, 2045. A discussion of the construction of orbital correlation diagrams using HMO theory was given by Dalton, J. C.; Friedrich, L. E. *J. Chem. Educ.* **1975**, *52*, 721.
8. Fráter, G.; Schmid, H. *Helv. Chim. Acta* **1968**, *51*, 190.
9. Vogel, E. *Liebigs Ann. Chem.* **1958**, *615*, 1; Hammond, G. S.; DeBoer, C. D. *J. Am. Chem. Soc.* **1964**, *86*, 899.
10. Doering, W. v. E.; Wiley, D. W. *Tetrahedron* **1960**, *11*, 183.
11. **a.** Bates, R. B.; McCombs, D. A. *Tetrahedron. Lett.* **1969**, 977.
    b. Pomerantz, M.; Wilke, R. N.; Gruber, G. W.; Roy, U. *J. Am. Chem. Soc.* **1972**, *94*, 2752.
    c. Arnold, B. J.; Sammes, P. G. *J. Chem. Soc., Chem. Commun.* **1972**, 1034; Arnold, B. J.; Sammes, P. G.; Wallace, T. W. *J. Chem. Soc., Perkin Trans. 1* **1974**, 415.
12. Brown, J. M. *Chem. Commun.* **1965**, 226.

13. Curtin, D. Y.; Johnson, Jr., H. W. *J. Am. Chem. Soc.* **1956**, *78*, 2611.

14. Dowd, P.; personal communication to Woodward and Hoffmann (reference 15 in Chapter 11), p. 143; Fleming, I.; personal communication to Woodward and Hoffmann, *ibid*.

15. Arnett, E. M. *J. Org. Chem.* **1960**, *25*, 324.

16. Copley, S. D.; Knowles, J. R. *J. Am. Chem. Soc.* **1985**, *107*, 5306; Dagani, R. *Chem. Eng. News* **1984** (May 28), 26.

17. Conroy, H.; Firestone, R. A. *J. Am. Chem. Soc.* **1956**, *78*, 2290.

18. Heimgartner, H.; Hansen, H.-J.; Schmid, H. *Helv. Chim. Acta* **1972**, *55*, 1385.

19. Greenwood, F. L. *J. Org. Chem.* **1959**, *24*, 1735; DePuy, C. H.; King, R. W. *Chem. Rev.* **1960**, *60*, 431.

20. Agosta, W. C. *J. Am. Chem. Soc.* **1964**, *86*, 2638.

21. Baldwin, J. E.; Reddy, V. P.; Schaad, L. J.; Hess, Jr., B. A. *J. Am. Chem. Soc.* **1988**, *110*, 8555.

Chapter 12
Photochemistry

4. Babu, M. K.; Rajasekaran, K.; Kannan, N.; Gnanasekaran, C. *J. Chem. Soc., Perkin Trans. 2* **1986**, 1721.

5. Pacifici, J. G.; Hyatt, J. A. *Mol. Photochem.* **1971**, *3*, 271.

6. Frey, H. M.; Lister, D. H. *Mol. Photochem.* **1972**, *3*, 323.

7. Arnold, B.; Donald, L.; Jurgens, A.; Pincock, J. A. *Can. J. Chem.* **1985**, *63*, 3140 and references therein.

8. Tsuneishi, H.; Inoue, Y.; Hakushi, T.; Tai, A. *J. Chem. Soc. Perkin Trans. 2* **1993**, 457.

9. Alumbaugh, R. L.; Pritchard, G. O.; Rickborn, B. *J. Phys. Chem.* **1965**, *69*, 3225.

10. Ohloff, G.; Klein, E.; Schenck, G. O. *Angew. Chem.* **1961**, *73*, 578; Schönberg, A. *Preparative Organic Photochemistry*; Springer-Verlag: New York, 1968; p. 377.

11. **a.** Kropp, P. J. *Mol. Photochem.* **1978-79**, *9*, 39.
b. Yang, N. C.; Jorgenson, M. J. *Tetrahedron Lett.* **1964**, 1203.

12. Carless, H. A. J. *J. Chem. Soc., Perkin Trans. 2* **1974**, 834.

13. Zimmerman, H. E.; Sandel, V. R. *J. Am. Chem. Soc.* **1963**, *85*, 915. Also see DeCosta, D. P.; Pincock, J. A. *J. Am. Chem. Soc.* **1993**, *115*, 2180.

14. **a.** Butenandt, A.; Poschmann, L. *Chem. Ber.* **1944**, *77B*, 394.
b. and **c.** Quinkert, G. *Angew. Chem. Int. Ed., Engl.* **1962**, *1*, 166.
d. Medary, R. T.; Gano, J. E.; Griffin, C. E. *Mol. Photochem.* **1974**, *6*, 107.

15. **a.** Barnard, M.; Yang, N. C. *Proc. Chem. Soc.* **1958**, 302.
b. Wolff, S.; Schreiber, W. L.; Smith, III, A. B.; Agosta, W. C. *J. Am. Chem. Soc.* **1972**, *94*, 7797.

16. **a.** Agosta, W. C.; Smith, III, A. B. *J. Am. Chem. Soc.* **1971**, *93*, 5513. See also Wolff, S.; Schreiber, W. L.; Smith, III, A. B.; Agosta, W. C. *J. Am. Chem. Soc.* **1972**, *94*, 7797.
b. Göth, H.; Cerutti, P.; Schmid, H. *Helv. Chem. Acta* **1965**, *48*, 1395.

17. Anderson, J. C.; Reese, C. B. *J. Chem. Soc.* **1963**, 1781.

18. Padwa, A.; Alexander, E.; Niemcyzk, M. *J. Am. Chem. Soc.* **1969**, *91*, 456; Padwa, A. *Acc. Chem. Res.* **1971**, *4*, 48.

19. **a.** Bahurel, Y.; Descotes, G.; Pautet, F. *Compt. Rend.* **1970**, *270*, 1528; Bahurel, Y.; Pautet, F.; Descotes, G. *Bull. Soc. Chim. France*, **1971**, *6*, 2222.
b. Nobs, F.; Burger, U.; Schaffner, K. *Helv. Chim. Acta* **1977**, *60*, 1607.

20. Turro, N. J.; Weiss, D. S. *J. Am. Chem. Soc.* **1968**, *90*, 2185; Dawes, K.; Dalton, J. C.; Turro, N. J. *Mol. Photochem.* **1971**, *3*, 71.

21. **a.** Padwa, A.; Eisenberg, W. *J. Am. Chem. Soc.* **1970**, *92*, 2590.
b. Gagosian, R. B.; Dalton, J. C.; Turro, N. J. *J. Am. Chem. Soc.* **1970**, *92*, 4752; Turro, N. J.; Dalton, J. C.; Dawes, K.; Farrington, G.; Hautala, R.; Morton, D.; Niemczyk, M.; Schore, N. *Acc. Chem. Res.* **1972**, *5*, 92.
c. Padwa, A.; Alexander, E.; Niemcyzk, M. *J. Am. Chem. Soc.* **1969**, *91*, 456; Padwa, A. *Acc. Chem. Res.* **1971**, *4*, 48.
d. Arnold, B. J.; Mellows, S. M.; Sammes, P. G.; Wallace, T. W. *J. Chem. Soc., Perkin Trans. 1* **1974**, 401. See also Durst, T.; Kozma, E. C.; Charlton, J. L. *J. Org. Chem.* **1985**, *50*, 4829.
e. Wolff, S.; Schreiber, W. L.; Smith, III, A. B.; Agosta, W. C. *J. Am. Chem. Soc.* **1972**, *94*, 7797.

22. Gilbert, A.; Taylor, G. N.; bin Samsudin, M. W. *J. Chem. Soc., Perkin Trans. 1* **1980**, 869.

23. Morrison, H.; Pajak, J.; Peiffer, R. *J. Am. Chem. Soc.* **1971**, *93*, 3978.

Credits

Chapter 1. Page 9, Figure 1.2 Reprinted with permission from Bader, R. F. W.; Carroll, M. T.; Cheeseman, J. R.; Chang, C. *J. Am. Chem. Soc.* **1987,** *109,* 7968. Copyright © 1987 American Chemical Society. **15, Table 1.7** Reprinted with permission from Benson, S. W.; Cruickshank, F. R.; Golden, D. M.; Haugen, G. R.; O'Neal, H. E.; Rodgers, A. S.; Shaw, R.; Walsh, R. *Chem. Rev.* **1969,** *69,* 279. Copyright © 1969 American Chemical Society. **17, Figure 1.5** Reprinted with permission from Arnett, E. M.; Amarnath, K.; Harvey, N. G.; Cheng, J. P. *Science,* **1990,** *247,* 423. Copyright © 1990 American Association for the Advancement of Science. **32, Figure 1.13** Reprinted with permission from "A Novel Pictorial Approach to Teaching MO Concepts in Polyatomic Molecules," by Hoffman, D. K.; Ruedenberg, K.; Verkade, J. G. *J. Chem. Educ.* **1977,** *54,* 590–595. Copyright © 1977 American Chemical Society. **34, Figure 1.16** Reprinted from *Chemical Physics Letters.* **1968** *1,* 613, Hamrin, K.; Johansson, G.; Gelius, U.; Fahlman, A.; Nordling, C.; Siegbahn, K., Ionization Energies in Methane and Ethane Measured by Means of ESCA, © 1968 with kind permission from Elsevier Science—NL, Sara Burgerhartstraat 25, 1055 KV Amsterdam, The Netherlands. **42, Figure 1.23** Reprinted with permission from Wiberg, K. B.; Hadad, C. M.; Breneman, C. M.; Laidig, K. E.; Murcko, M. A.; LePage, T. J. *Science,* **1992,** *252,* 1266. Copyright © 1992 American Association for the Advancement of Science. **43, Figure 1.24** From *Organic Reaction Mechanisms,* 2/e by Breslow, R. Copyright © 1969 by W. A. Benjamin, Inc. Reprinted by permission. **43, Footnote 167** Reprinted with permission from Patel, D. J.; Howden, M. E. H.; Roberts, J. D. *J. Am. Chem Soc.* **1963,** *85,* 3218. Copyright © 1963 American Chemical Society. **51, Figure 1.34** and **Figure 1.35** Reprinted with permission from "Models for the Double Bond," by Walters, E. A. *J. Chem. Educ.* **1966,** *43,* 134–137. Copyright © 1966 American Chemical Society. **52, Figure 1.36** Reprinted with permission from Hamilton, J. G.; Palke, W. E. *J. Am. Chem. Soc.* **1993,** *115,* 4159. Copyright © 1993 American Chemical Society.

Chapter 2. 67, Figure 2.9 Reprinted with permission from "A Flow-Chart Approach to Point Group Classification," by Carter, R. L. *J. Chem. Educ.* **1968,** *45,* 44. Copyright © 1968 American Chemical Society. **71, Figure 2.10** and **76, Figure 2.12** Reprinted with permission from "An Introduction to the Sequence Rule," by Cahn, R. S. *J. Chem. Educ.* **1964,** *41,* 116–125. Copyright © 1964 American Chemical Society. **97, Figure 2.22** Reprinted with permission from Djerassi, C.; Wolf, H.; Bunnenberg, E. *J. Am. Chem. Soc.* **1963,** *85,* 324. Copyright © 1963 American Chemical Society. **109, Figure 2.31** and **110, Figure 2.32, Table 2.3** Reprinted with permission from Mislow, K.; Siegel, J. *J. Am. Chem. Soc.* **1984,** *106,* 3319. Copyright © 1984 American Chemical Society.

Chapter 3. 129, Figure 3.6 Reprinted with permission from Jorgensen, W. L.; Buckner, J. K. *J. Phys. Chem.* **1987,** *91,* 6083. Copyright © 1987 American Chemical Society. **133, Figure 3.10** Reprinted with permission from "The Story Behind the Story: Why Did Adolf Baeyer Propose a Planar, Strained Cyclohexane Ring?" by Ramsay, O. B. *J. Chem. Educ.* **1977,** *54,* 563–564. Copyright © 1977 American Chemical Society. **134, Figure 3.13** and **Figure 3.14** Reprinted with permission from Stein, A.; Lehmann, C. W.; Luger, P. *J. Am. Chem. Soc.* **1992,** *114,* 7684. Copyright © 1992 American Chemical Society. **161, Figure 3.26** Reprinted with permission from Wiberg, K. B.; Murcko, M. A.; Laidig, K. E.; MacDougall, P. J. *J. Phys. Chem.* **1990,** *94,* 6956. Copyright © 1990 American Chemical Society.

Chapter 4. 190, Figure 4.7 From *Molecular Orbital Calculations,* by J. D. Roberts. Copyright © 1961 by W. A. Benjamin, Inc. Reprinted by permission. **193, Figure 4.11** and **206, Figure 4.24** *From Quantum Mechanics for Organic Chemists*, by Zimmerman, H. E. Copyright © 1975 by Academic Press, Inc. Reprinted by permission. **209, Figure 4.27** Reprinted with permission from Cox, E. G.; Cruickshank, D. W. J.; Smith, J. A. S. *Proc. Roy Soc A* **1958,** *247,* 1. Copyright © 1958 Royal Society of London.

Chapter 5. 250, Figure 5.6; 251, Figure 5.8; 252, Figure 5.9; and **253, Figure 5.10** Reprinted with permission from Dauben, W. G.; Funhaff, D. J. H. *J. Org. Chem.* **1988,** *53,* 5070. Copyright © 1988 American Chemical Society. **254, Figure 5.12** Reprinted with permission from "Textbook Errors, 131: Free Energy Surfaces and Transition State Theory," by Cruickshank, F. R.; Hyde, A. J.; Pugh, D. *J. Chem. Educ.* **1977,** *54,* 288–291. Copyright © 1977 American Chemical Society. **260, Figure 5.15** Reprinted with permission from "Introduction to the Interpretation of Electron Spin Resonance Spectra of Organic Radicals," by Bunce, N. J. *J. Chem. Educ.* **1987,** *64,* 907–914. Copyright © 1987 American Chemical Society. **261, Figure 5.16** Reprinted with permission from McConnell, H. M.; Heller, C.; Cole, T.; Fessenden, R. W. *J. Am. Chem. Soc.* **1960,** *82,* 766. Copyright © 1960 American Chemical Society. **261, Figure 5.17** and **262, Figure 5.18** Reprinted with permission from Fessenden, R. W.; Schuler, R. H. *J. Chem. Phys.,* **1963,** *39,* 2147. Copyright © 1963 American Institute of Physics. **264, Figure 5.24** Reprinted with permission from Beveridge, D. L.; Dobosh, P. A.; Pople, J. A. *J. Chem. Phys.,* **1968,** *48,* 4802. Copyright © 1968 American Institute of Physics. **276, Figure 5.31** Reprinted with permission from Schaefer III, H. F. *Science,* **1986,** *231,* 1100. Copyright © 1986 American Association for the Advancement of Science. **284, Figure 5.35** and **Figure 5.36** From *Hypercarbon Chemistry,* by G. A. Olah, G. K. S. Prakash, R. E. Williams, L. D. Field, and K. Wade. Copyright © 1987 by John Wiley & Sons, Inc. Reprinted by permission. **294, Figure 5.46** Reprinted with permission from Olah, G. A.; Donovan, D. J. *J. Am. Chem. Soc.* **1977,** *99,* 5026. Copyright © 1977 American Chemical Society. **298, Figure 5.51** Reprinted with permission from Ma, N. L.; Smith, B. J.; Pople, J. A.; Radom, L. *J. Am. Chem. Soc.* **1991,** *113,* 7903. Copyright © 1991 American Chemical Society.

Chapter 6. 321, Figure 6.3 Reprinted with permission from Porter, N.; Zuraw, P. *J. Chem. Soc., Chem. Commun.* **1985,** *1472.* Copyright © 1985 Royal Society of Chemistry. **322, Figure 6.4** Reprinted with permission from Porter, N.; Zuraw, P. *J. Chem. Soc., Chem. Commun.* **1985,** *1472.* Copyright © 1985 Royal Society of Chemistry. **326, Figure 6.11** and **327, Figure 6.12** Reprinted with permission from Chapman, O. L.; Tsou, U.-P.E.; Johnson, J. W. *J. Am. Chem. Soc.* **1987,** *109,* 553. Copyright © 1987 American Chemical Society. **347, Figure 6.27** Reprinted with permission from "A New Perspective on Kinetic and Thermodynamic Control of Reactions," by Snadden, R. B. *J. Chem. Educ.* **1985,** *62,* 653–655. Copyright © 1985 American Chemical Society. **351, Figure 6.32** Figure from *Spectroscopy and Molecular Structure,* by Gerald W. King, copyright © 1964 by Holt, Rinehart and Winston, Inc. and renewed 1992 by Gerald W. King, reproduced by permission of the publisher. **352, Figure 6.33; 355, Figure 6.35** and **Figure 6.36; 357, Figure 6.38;** and **360, Figure 6.41** Reprinted with permission from Wiberg, K. B. *Chem. Rev.* **1955,** *55,* 713. Copyright © 1955 American Chemical Society. **370, Figure 6.45** Reprinted with permission from Marriott, S.; Topsom, R. D. *J. Am. Chem. Soc.* **1985,** *107,* 2253. Copyright © 1985 American Chemical Society. **374, Figure 6.48** and **378, Figure 6.57** Reprinted with permission from Anderson, B. M.; Jencks, W. P. *J. Am. Chem. Soc.* **1960,** *82,* 1773. Copyright © 1960 American Chemical Society. **379, Figure 6.58** Reprinted with permission from Santerre, G. M.; Hansrote, Jr., C. J.; Crowell, T. I. *J. Am. Chem. Soc.* **1958,** *80,* 1254. Copyright © 1958 American Chemical Society.

Chapter 7. 413, Figure 7.1 Reprinted with permission from Jorgenson, M. J.; Hartter, D. R. *J. Am. Chem. Soc.* **1963,** *85,* 878. Copyright © 1963 American Chemical Society. **414, Figure 7.2** From *Physical Organic Chemistry: Reaction Rates, Equilibria and Mechanisms,* 2/e, by Hammett, L. P., p. 272. Copyright © 1970 by McGraw-Hill Book Company. Reprinted by permission. **425, Figure 7.5** Reprinted with permission from Jencks, W. P. *J. Am. Chem. Soc.* **1959,** *81,* 475. Copyright © 1959 American Chemical Society. **432, Figure 7.11** Reprinted with permission from Cordes, E. H. *Prog. Phys. Org. Chem.,* **1967,** *4,* 1. Copyright © 1967 John Wiley & Sons, Inc. **433, Figure 7.12** Reprinted with permission from Anderson, E.; Capon, B. *J. Chem. Soc. (B),* **1969,** *1033.* Copyright © 1969 Royal Society of Chemistry. **434, Figure 7.13** Reprinted with permission from Swain, C. G.; Brown, Jr., J. F. *J. Am. Chem. Soc.* **1952,** *74,* 2538. Copyright © 1952 American Chemical Society. **441, Figure 7.19** Reprinted with permission from Bender, M. L.; Thomas, R. J. *J. Am. Chem. Soc.* **1961,** *83,* 4189. Copyright © 1961 American Chemical Society. **448, Figure 7.26** Reprinted with permission from Weiner, S. J.; Singh, U. C.; Kollman, P. A. *J. Am. Chem. Soc.* **1985,** *107,* 2219. Copyright © 1985 American Chemical Society.

Chapter 8. 458, Figure 8.2; 459, Figure 8.3 and **Figure 8.4; 460, Figure 8.5** and **462, Figure 8.9** Reprinted with permission from Harris, J. M.; Shafer, S. G.; Moffatt, J. R.; Becker, A. R. *J. Am. Chem. Soc.* **1979,** *101,* 3295. Copyright © 1979 American Chemical Society. **466, Figure 8.12** Reprinted with permission from Bingham R. C.; Schleyer, P. v. R. *J. Am. Chem. Soc.* **1971,** *93,* 3189. Copyright © 1971 American Chemical Society. **470, Table 8.4** Reprinted with permission from Ritchie, C. D. *Acc. Chem. Res.* **1972,** *5,* 348. Copyright © 1972 American Chemical Society. **475, Figure 8.17** Reprinted with permission from Winstein, S.; Klinedinst, Jr., P. E.; Robinson, G. *J. Am. Chem. Soc.* **1961,** *83,* 885. Copyright © 1961 American Chemical Society. **476, Figure 8.19** and **477, Figure 8.20** Reprinted with permission from Jorgensen, W. L.; Buckner, J. K.; Huston, S. E.; Rossky, P. J. *J. Am. Chem. Soc.* **1987,** *109,* 1891. Copyright © 1987 American Chemical Society. **493, Figure 8.37** Reprinted with permission from Chandrasekkhar, J.; Jorgensen, W. L. *J. Am. Chem. Soc.* **1985,** *107,* 2974. Copyright © 1985 American Chemical Society. **494, Figure 8.38** Reprinted with permission from Gao, J. *J. Am. Chem. Soc.* **1991,** *113,* 7796. Copyright © 1991 American Chemical Society. **497, Figure 8.30** Reprinted with permission from King, J. F.; Tsang,

G. T. Y.; Abdel-Malik, M. M.; Payne, N. C. *J. Am. Chem. Soc.* **1985,** *107,* 2745. Copyright © 1985 American Chemical Society. **498, Figure 8.41** Reprinted with permission from Bach, R. D.; Coddens, B. A.; Wolber, G. J. *J. Org. Chem.* **1986,** *51,* 1030. Copyright © 1986 American Chemical Society. **501, Table 8.12** and **502, Table 8.13** Reprinted with permission from Wells, R. *Chem. Rev.* **1963,** *63,* 171. Copyright © 1963 American Chemical Society. **507, Figure 8.43** and **508, Figure 8.44** Reprinted with permission from Pross, A.; Shaik, S. S. *Acc. Chem. Res.* **1983,** *16,* 363. Copyright © 1983 American Chemical Society. **509, Figure 8.45** Reprinted with permission from Ashby, E. C.; Pham, T. N. *J. Org. Chem.* **1986,** *51,* 3598. Copyright © 1986 American Chemical Society. **512, Figure 8.47** Reprinted with permission from Pross, A. *Acc. Chem. Res.* **1985,** *18,* 212. Copyright © 1985 American Chemical Society. **512, Figure 8.48** Reprinted with permission from Ashby, E. C.; Park, B.; Patil, G. S.; Gadru, K.; Gurumurthy, R. *J. Org. Chem.* **1993,** *58,* 424. Copyright © 1993 American Chemical Society. **516, Figure 8.51** Reprinted by permission from Stock, L. M. *Prog. Phys. Org. Chem.* **1976,** *12,* 21. Copyright © 1976 John Wiley & Sons, Inc. **525, Figure 8.58** Reprinted with permission from Rosenblum, M. *J. Am. Chem. Soc.* **1960,** *82,* 3796. Copyright © 1960 American Chemical Society. **527, Figure 8.59** Reprinted with permission from Bacaloglu, R.; Bunton, C. A.; Cerichelli, G. *J. Am. Chem. Soc.* **1987,** *109,* 621. Copyright © 1987 American Chemical Society. **539, Figure 8.68** Reprinted with permission from Kim, J. K.; Bunnett, J. F. *J. Am. Chem. Soc.* **1970,** *92,* 7463. Copyright © 1970 American Chemical Society.

Chapter 9. 555, Figure 9.5 Reprinted with permission from Slebocka-Tilk, H.; Zheng, C. Y.; Brown, R. S. *J. Am. Chem. Soc.* **1993,** *115,* 1347. Copyright © 1993 American Chemical Society. **561, Figure 9.10** and **562, Figure 9.11** Reprinted with permission from Bellucci, G.; Chiappe, C.; Marioni, F. *J. Am. Chem. Soc.* **1987,** *109,* 515. Copyright © 1987 American Chemical Society. **563, Figure 9.12** Reprinted with permission from Bellucci, G.; Bianchini, R.; Chiappe, C.; Marioni, F.; Ambrosetti, R.; Brown, R. S.; Slebocka-Tilk, H. *J. Am. Chem. Soc.* **1989,** *111,* 2640. Copyright © 1989 American Chemical Society. **567, Figure 9.16** Reprinted with permission from Dubois, J.-E.; Chretien, J. R. *J. Am. Chem. Soc.* **1978,** *100,* 3560. Copyright © 1978 American Chemical Society. **570, Figure 9.17** Reprinted with permission from Yates, K.; McDonald, R. S.; Shapiro, S. A. *J. Org. Chem.* **1973,** *38,* 2460. Copyright © 1973 American Chemical Society. **574, Figure 9.19** Reprinted with permission from Ruasse, M.-F. *Acc. Chem. Res.* **1990,** *23,* 87. Copyright © 1990 American Chemical Society. **578, Table 9.3** Reprinted with permission from Poutsma, M. L. *J. Am. Chem, Soc.* **1965,** *87,* 4285. Copyright © 1965 American Chemical Society. **583, Figure 9.25** Reprinted with permission from Yamabe, S.; Minato, T.; Inagaki, S. *J. Chem. Soc., Chem. Commun.* **1988,** *532.* Copyright © 1988 Royal Society of Chemistry. **585, Figure 9.26** Reprinted with permission from Stavber, S.; Sotler, T.; Zupan, M.; Popovic, A. *J. Org. Chem.* **1994,** *59,* 5891. Copyright © 1994 American Chemical Society. **605, Figure 9.37** Reprinted with permission from Pasto, D. J.; Gontarz, J. A. *J. Am. Chem. Soc.* **1969,** *91,* 719. Copyright © 1969 American Chemical Society. **632, Figure 9.50** and **633, Figure 9.51** Reprinted with permission from Cram, D. J.; Kopecky, K. R. *J. Am. Chem. Soc.* **1959,** *81,* 2748. Copyright © 1959 American Chemical Society. **634, Figure 9.52** Reprinted with permission from Karabatsos, G. J. *J. Am. Chem. Soc.* **1967,** *89,* 1367. Copyright © 1967 American Chemical Society.

Chapter 10. 649, Figure 10.13 and **652, Figure 10.14** Reprinted with permission from Kwok, W. K.; Lee, W. G.; Miller, S. I. *J. Am. Chem. Soc.* **1969,** *91,* 468. Copyright © 1969 American Chemical Society. **656, Figure 10.17** Reprinted with permission from Bartsch, R. A.; Zavada, *J. Chem. Rev.,* **1980,** *80,* 453. Copyright © 1980 American Chemical Society. **656, Figure 10.18** Reprinted with permission from Dohner, B. R.; Saunders, Jr., W. H. *J. Am. Chem. Soc.* **1986,** *108,* 245. Copyright © 1986 American Chemical Society. **660, Figure 10.23** Reprinted with permission from Croft, A. P.; Bartsch, R. A. *J. Org. Chem.* **1994,** *59,* 1930. Copyright © 1994 American Chemical Society. **676, Figure 10.33** Reprinted with permission from Eubanks, J. R. I.; Sims, L. B.; Fry, A. *J. Am. Chem. Soc.* **1991,** *113,* 8821. Copyright © 1991 American Chemical Society. **681, Figure 10.37** Reprinted with permission from Cadogan J. I. G.; Leardini, R. *J. Chem. Soc., Chem. Commun.* **1979,** *783.* Copyright © 1979 Royal Society of Chemistry. **682, Figure 10.38** From *Advanced Organic Chemistry* by L. R. Fieser and M. Fieser. Copyright © 1961 by Reinhold Publishing Company. Adapted by permission. **685, Figure 10.40** Reprinted with permission from Narayan, R; Antal, Jr., M. M. *J. Am. Chem. Soc.* **1990,** *112,* 1927. Copyright © 1990 American Chemical Society. **693, Figure 10.42** Reprinted with permission from Streitwieser, Jr., A.; Schaeffer, W. D. *J. Am. Chem. Soc.* **1957,** *79,* 2888. Copyright © 1957 American Chemical Society. **695, Figure 10.43** Reprinted with permission from Cram, D. J.; McCarty, J. E. *J. Am. Chem. Soc.* **1957,** *79,* 2866. Copyright © 1957 American Chemical Society. **701, Figure 10.47** Reprinted with permission from Cram, D. J.; McCarty, J. E. *J. Am. Chem. Soc.* **1954,** *76,* 5740. Copyright © 1954 American Chemical Society.

Chapter 11. 724, Figure 11.15 and **725, Figure 11.16** From *The Conservation of Orbital Symmetry,* by Woodward, R. B. and Hoffmann, R. Copyright © 1970 by Verlag Chemie, GmbH, Weinhein/Bergstr. Reprinted by permission of VCH Publishers Inc. **726, Figure 11.17** and **727, Figure 11.18** Reprinted with permission from Longuet-Higgins, H. C.; Abrahamson, E. W. *J. Am. Chem. Soc.* **1965,** *87,* 2045. Copyright © 1965 American Chemical Society. **730, Figure 11.22** From *The Conservation of Orbital Symmetry,* by Woodward, R. B. and Hoffmann, R. Copyright © 1970 by Verlag Chemie, GmbH, Weinhein/Bergstr. Reprinted by permission of VCH Publishers Inc. **732, Figure 11.26** Reprinted with permission from Houk, K. N.; Li, Y.; Evanseck, J. D.

Angew. Chem., Int. Ed. Eng. **1992,** *31,* 682. Copyright © 1992 VCH Publishers Inc. **735, Figure 11.31** From *The Conservation of Orbital Symmetry,* by Woodward, R. B. and Hoffmann, R. Copyright © 1970 by Verlag Chemie, GmbH, Weinhein/Bergstr. Reprinted by permission of VCH Publishers Inc. **735, Figure 11.32** Reprinted with permission from Hoeger, C. A.; Johnston, A. D. *J. Am. Chem. Soc.* **1987,** *109,* 4690. Copyright © 1987 American Chemical Society. **736, Figure 11.33** Reprinted with permission from Hoeger, C. A.; Okamura, W. H. *J. Am. Chem. Soc.* **1985,** *107,* 268. Copyright © 1985 American Chemical Society. **743, Figure 11.46** Reprinted with permission from Schroder, G.; Oth, J. F. M.; Merenyi, R. *Angew. Chem., Int. Ed. Engl.* **1965,** *4,* 752. Copyright © 1965 VCH Publishers Inc. **749, Figure 11.56** From *The Conservation of Orbital Symmetry,* by Woodward, R. B. and Hoffmann, R. Copyright © 1970 by Verlag Chemie, GmbH, Weinhein/Bergstr. Reprinted by permission of VCH Publishers Inc. **749, Figure 11.57** and **750, Figure 11.58** Reprinted with permission from Hoffman, R.; Woodward, R. B. *J Am. Chem. Soc.* **1965,** *87,* 2046. Copyright © 1965 American Chemical Society. **752, Figure 11.60** Reprinted with permission from Li, Y.; Houk, K. N. *J. Am. Chem. Soc.* **1993,** *115,* 7478. Copyright © 1993 American Chemical Society. **753, Figure 11.62** Reprinted with permission from Hoffman, R.; Woodward, R. B. *J Am. Chem. Soc.* **1965,** *87,* 2046. Copyright © 1965 American Chemical Society. **757, Figure 11.64** and **759, Figure 11.66** From *The Conservation of Orbital Symmetry,* by Woodward, R. B. and Hoffmann, R. Copyright © 1970 by Verlag Chemie, GmbH, Weinhein/Bergstr. Reprinted by permission of VCH Publishers Inc. **761, Figure 11.69** Eckell, A.; Huisgen, R.; Sustmann, R.; Wallbillich, G.; Grashey, D.; Spindler, E. *Chem. Ber.* **1967,** *100,* 2192. **765, Figure 11.72** and **766, Figure 11.73** From *The Conservation of Orbital Symmetry,* by Woodward, R. B. and Hoffmann, R. Copyright © 1970 by Verlag Chemie, GmbH, Weinhein/Bergstr. Reprinted by permission of VCH Publishers Inc. **767, Figure 11.77** Reprinted with permission from Houk, K. N.; Li, Y.; Evanseck, J. D. *Angew. Chem., Int. Ed. Eng.* **1992,** *31,* 682. Copyright © 1992 VCH Publishers, Inc. **774, Figure 11.87** Reprinted with permission from Paderes, G. D.; Jorgensen, W. L. *J. Org. Chem.* **1992,** *57,* 1904. Copyright © 1992 American Chemical Society. **783, Figure 11.98** and **784, Figure 11.100** Reprinted with permission from Zimmerman, H. E. *Acc. Chem. Res.* **1971,** *4,* 272. Copyright © 1971 American Chemical Society. **788, Figure 11.101** Reprinted with permission from Gajewski, J. J.; Conrad, N. D. *J. Am Chem. Soc.* **1979,** *101,* 6693. Copyright © 1979 American Chemical Society. **789, Figure 11.102** Reprinted with permission from Carpenter, B. K. *J. Org. Chem.* **1992,** *57,* 4645. Copyright © 1992 American Chemical Society.

Chapter 12. 796, Figure 12.1 From *Physical Chemistry,* 3/e by Moore, W. J. Copyright © 1962 by Prentice-Hall, Inc. **801, Figure 12.4** Reprinted with permission from Wilkinson, F. *Quart. Rev. Chem. Soc.* **1966,** *20,* 403. Copyright © 1966 Royal Society of Chemistry. **807, Figure 12.6** Reprinted with permission from Bowen, E. J. *Adv. Photochem.* **1963,** *1,* 23. Copyright © 1963 John Wiley & Sons, Inc. **808, Figure 12.7** Reprinted with permission from Becker H.-D.; Andersson, K. *J. Org. Chem.* **1983,** *48,* 4542. Copyright © 1983 American Chemical Society. **809, Figure 12.8** From *Handbook of Fluorescence Spectra of Aromatic Molecules,* 2/e by Berlman, I. B. Copyright © 1971 Academic Press. Reprinted by permission. **811, Figure 12.10** Reprinted with permission from Marchetti, A.P.; Kearns, D. R. *J. Am. Chem. Soc.* **1967,** *89,* 768. Copyright © 1967 American Chemical Society. **813, Figure 12.11** Reprinted from Kavarnos, G. J. *Top. Curr. Chem.* **1990,** *156,* 21. Copyright © 1990 Springer-Verlag. **824, Figure 12.17** and **Figure 12.18** Reprinted with permission from Lawrence, M.; Marzzacco, C. J.; Morton, C.; Schwab, C.; Halpern, A. M. *J. Phys. Chem.* **1991,** *95,* 10294. Copyright © 1991 American Chemical Society. **827, Figure 12.19** Reprinted with permission from Jackson, G.; Porter, G. *Proc. Roy. Soc. London* **1961,** *A260,* 13. Copyright © 1961 Royal Society of London. **829, Figure 12.20** Reprinted with permission from Turro, N. J.; Dalton, J. C.; Dawes, K.; Farrington, G.; Hautala, R.; Morton, D.; Niemczyk, M.; Schore, N. *Acc Chem Res.* **1972,** *5,* 92. Copyright © 1972 American Chemical Society. **830, Figure 12.21** From *Modern Molecular Photochemistry* by Turro, N.J. Copyright © 1978 by Benjamin/Cummings Publishing Co., Inc. Reprinted by permission. **831, Figure 12.22** Reprinted with permission from Turro, N. J.; Dalton, J. C.; Dawes, K.; Farrington, G.; Hautala, R.; Morton, D.; Niemczyk, M.; Schore, N. *Acc Chem Res.* **1972,** *5,* 92. Copyright © 1972 American Chemical Society. **833, Figure 12.23** Reprinted with permission from Srinivasan, R. *Adv. Photochem.* **1963,** *1,* 83. Copyright © 1963 John Wiley & Sons, Inc. **837, Figure 12.25** Reprinted with permission from Wagner, P. J. *Acc. Chem. Res.* **1971,** *4,* 168. Copyright © 1971 American Chemical Society. **841, Figure 12.27** Reprinted with permission from Saltiel, J.; D'Agostino, J.; Megarity, E. D.; Metts, L.; Neuberger, K. R.; Wrighton, M.; Zafiriou, O. C. *Org. Photochem.* **1973,** *3,* 1. Copyright © 1973 Marcel Dekker, Inc. **843, Figure 12.30** Reprinted with permission from Lewis, F. D.; Elbert, J. E.; Upthagrove, A. L.; Hale, P. D. *J. Org. Chem.* **1991,** *56,* 553. Copyright © 1991 American Chemical Society. **844, Figure 12.31** Reprinted with permission from Inoue, Y.; Takamuku, S.; Sakurai, H. *J. Phys. Chem.* **1977,** *81,* 7. Copyright © 1977 American Chemical Society. **846, Figure 12.32** Reprinted with permission from Kropp, P. J. *Mol. Photochem.* **1978–79,** *9,* 39. Copyright © 1978–79 Marcel Dekker, Inc. **847, Figure 12.34** Reprinted with permission from Cherry, W.; Chow, M.-F.; Mirbach, J. J.; Mirbach, M. F.; Ramamurthy, V.; Turro, N. J. *Mol. Photochem.* **1977,** *8,* 175. Copyright © 1977 Marcel Dekker, Inc. **849, Figure 12.35** Reprinted with permission from Adam, W.; Oppenlander, T.; Zang, G. *J. Am. Chem. Soc.* **1985,** *107,* 3921. Copyright © 1985 American Chemical Society. **849, Figure 12.36** Reprinted with permission from Hixson, S. S.; Mariano, P. S.; Zimmerman, H. E. *Chem. Rev.* **1973,** *73,* 531. Copyright © 1973 American Chemical Society. **851, Figure 12.37** Reprinted with permission from Inoue, Y.; Hagiwara, S.; Daino, Y.; Hakushi, T. *J. Chem. Soc., Chem. Commun.* **1985,** *1307.* Copyright © 1985 Royal Society of Chemistry. **851, Figure 12.38** Reprinted with permission from Saltiel, J.; D'Agostino, J.; Megarity, E. D.; Metts, L,; Neuberger, K. R.; Wrighton, M.; Zafiriou, O. C. *Org. Photochem.* **1973,** *3,* 1. Copyright © 1973 Marcel Dekker, Inc. **854, Figure 12.40** and **855, Figure 12.41** From *Transfer and Storage of Energy by Molecules,* Vol 1, edited by Burnett, G. M.; North, A. M. Copyright © 1968 Wiley-Interscience. Reprinted by permission. **857, Figure 12.43** and **Figure 12.44** Reprinted with permission from Zimmerman, H. E.; Sandel, V. R. *J. Am. Chem. Soc.* **1963,** *85,* 915. Copyright © 1963 American Chemical Society. **858, Figure 12.45** Reprinted with permission from Zimmerman, H. E.; Somasekhara, S. *J. Am. Chem. Soc.* **1963,** *85,* 922. Copyright © 1963 American Chemical Society. **858, Figure 12.46** From *Photochemistry* by Calvert, J. G.; Pitts, J. N. Copyright © 1966 by John Wiley & Sons, Inc. Reprinted by permission. **860, Table 12.5** Reprinted with permission from Carroll, F. A.; Hammond, G. S. *Israel J. Chem.* **1972,** *10,* 613. Copyright © 1972 Laser Pages Publishing Ltd. **861, Figure 12.47** Reprinted from Michl, J. *Top. Curr. Chem.* **1974,** *46,* 1. Copyright © 1974 Springer-Verlag.

Name Index

Subject Index